Lecture Notes in Mechanical Engineering

Series Editors

Fakher Chaari, *National School of Engineers, University of Sfax, Sfax, Tunisia*
Francesco Gherardini $\circledD$, *Dipartimento di Ingegneria "Enzo Ferrari", Università di Modena e Reggio Emilia, Modena, Italy*
Vitalii Ivanov $\circledD$, *Department of Manufacturing Engineering, Machines and Tools, Sumy State University, Sumy, Ukraine*
Mohamed Haddar, *National School of Engineers of Sfax (ENIS), Sfax, Tunisia*

Editorial Board Members

Francisco Cavas-Martínez $\circledD$, *Departamento de Estructuras, Construcción y Expresión Gráfica Universidad Politécnica de Cartagena, Cartagena, Spain*
Francesca di Mare $\circledD$, *Institute of Energy Technology, Ruhr-Universität Bochum, Bochum, Germany*
Young W. Kwon, *Department of Manufacturing Engineering and Aerospace Engineering, Graduate School of Engineering and Applied Science, Monterey, USA*
Tullio A. M. Tolio, *Department of Mechanical Engineering, Politecnico di Milano, Milano, Italy*
Justyna Trojanowska, *Poznan University of Technology, Poznan, Poland*
Robert Schmitt, *RWTH Aachen University, Aachen, Germany*
Jinyang Xu, *School of Mechanical Engineering, Shanghai Jiao Tong University, Shanghai, China*

Lecture Notes in Mechanical Engineering (LNME) publishes the latest developments in Mechanical Engineering—quickly, informally and with high quality. Original research or contributions reported in proceedings and post-proceedings represents the core of LNME. Volumes published in LNME embrace all aspects, subfields and new challenges of mechanical engineering.

To submit a proposal or request further information, please contact the Springer Editor of your location:

Europe, USA, Africa: Leontina Di Cecco at Leontina.dicecco@springer.com
China: Ella Zhang at ella.zhang@cn.springernature.com
India, Rest of Asia, Australia, New Zealand: Swati Meherishi at swati.meherishi@springer.com

Topics in the series include:

- Engineering Design
- Machinery and Machine Elements
- Mechanical Structures and Stress Analysis
- Automotive Engineering
- Engine Technology
- Aerospace Technology and Astronautics
- Nanotechnology and Microengineering
- Control, Robotics, Mechatronics
- MEMS
- Theoretical and Applied Mechanics
- Dynamical Systems, Control
- Fluid Mechanics
- Engineering Thermodynamics, Heat and Mass Transfer
- Manufacturing Engineering and Smart Manufacturing
- Precision Engineering, Instrumentation, Measurement
- Materials Engineering
- Tribology and Surface Technology

Indexed by SCOPUS, EI Compendex, and INSPEC.

All books published in the series are evaluated by Web of Science for the Conference Proceedings Citation Index (CPCI).

To submit a proposal for a monograph, please check our Springer Tracts in Mechanical Engineering at https://link.springer.com/bookseries/11693.

Wenwei Wu · Jun Ding
Editors

Proceedings of the 22nd International Ship and Offshore Structures Congress (Volume 1)

Technical Committee Reports

Editors
Wenwei Wu
China Ship Scientific Research Center
Wuxi, Jiangsu, China

Jun Ding
China Ship Scientific Research Center
Wuxi, Jiangsu, China

ISSN 2195-4356   ISSN 2195-4364 (electronic)
Lecture Notes in Mechanical Engineering
ISBN 978-981-95-2667-3   ISBN 978-981-95-2668-0 (eBook)
https://doi.org/10.1007/978-981-95-2668-0

© The Editor(s) (if applicable) and The Author(s) 2026. This book is an open access publication.

Open Access This book is licensed under the terms of the Creative Commons Attribution-NonCommercial-NoDerivatives 4.0 International License (http://creativecommons.org/licenses/by-nc-nd/4.0/), which permits any noncommercial use, sharing, distribution and reproduction in any medium or format, as long as you give appropriate credit to the original author(s) and the source, provide a link to the Creative Commons license and indicate if you modified the licensed material. You do not have permission under this license to share adapted material derived from this book or parts of it.

The images or other third party material in this book are included in the book's Creative Commons license, unless indicated otherwise in a credit line to the material. If material is not included in the book's Creative Commons license and your intended use is not permitted by statutory regulation or exceeds the permitted use, you will need to obtain permission directly from the copyright holder.

This work is subject to copyright. All commercial rights are reserved by the author(s), whether the whole or part of the material is concerned, specifically the rights of translation, reprinting, reuse of illustrations, recitation, broadcasting, reproduction on microfilms or in any other physical way, and transmission or information storage and retrieval, electronic adaptation, computer software, or by similar or dissimilar methodology now known or hereafter developed. Regarding these commercial rights a non-exclusive license has been granted to the publisher.

The use of general descriptive names, registered names, trademarks, service marks, etc. in this publication does not imply, even in the absence of a specific statement, that such names are exempt from the relevant protective laws and regulations and therefore free for general use.

The publisher, the authors and the editors are safe to assume that the advice and information in this book are believed to be true and accurate at the date of publication. Neither the publisher nor the authors or the editors give a warranty, expressed or implied, with respect to the material contained herein or for any errors or omissions that may have been made. The publisher remains neutral with regard to jurisdictional claims in published maps and institutional affiliations.

This Springer imprint is published by the registered company Springer Nature Singapore Pte Ltd.
The registered company address is: 152 Beach Road, #21-01/04 Gateway East, Singapore 189721, Singapore

If disposing of this product, please recycle the paper.

Preface

The first volume contains the eight Technical Committee reports, and the second volume contains the reports of the eight Specialist Committees, presented and discussed at the 22nd International Ship and Offshore Structures Congress (ISSC 2025) in Wuxi (China), on September 22–26, 2025. The Official Discussers' reports and all floor discussions, including the replies by the committees, will be published after the Congress in electronic form.

The Standing Committee of the 22nd International Ship and Offshore Structures Congress comprises:

Chairman

Wenwei Wu — China

Co-chairman

Josko Parunov	Croatia
Yordan Garbatov	Portugal
Alex Babanin	Australia
Kim Branner	Denmark
Kazuhiro Iijima	Japan
Henk den Besten	Netherlands
Xiaozhi Wang	US
Bernt Leira	Norway
Myung-Hyun Kim	Republic of Korea
Patrick Kaeding	Germany
Sime Malenica	France
Andrea Ivaldi	Italy
Feargal Brennan	UK
Jan Czaban	Canada
Murilo Augusto Vaz	Brazil

On behalf of the Standing Committee, we would like to thank the sponsors of ISSC 2025.

vi Preface

Thanks to our Sponsors for their Financial Contribution.

September 2025

Wenwei Wu
Chairman

Jun Ding
Secretary

Contents

Committee I.1: Environment 1
    Sanne van Essen, Mariana Bernardino,
    Rüdiger Ulrich Franz von Bock und Polach, Ricardo Martins Campos,
    Guillaume de Hauteclocque, Sheng Dong, Thomas Berge Johannessen,
    Evert Lataire, Joong Soo Moon, Arttu Polojärvi, Jasna Prpic-Orsic,
    Erik Vanem, and Tingyao Zhu

Committee I.2: Loads 133
    Jungyong Wang, Arash Abbasnia, Louis Diebold, Andre Fujarra,
    Tomaso Gaggero, Spyros Hirdaris, Sangyeob Kim,
    Dimitrios Konispoliatis, Masayoshi Oka, Jose Miguel Rodrigues,
    Mahmud Sazidy, Florian Sprenger, Peter Wellens, and Guiyong Zhang

Committee II.1: Quasi-static Response 253
    Zhen Gao, Yooil Kim, Erick Alley, Krzysztof Wołoszyk, Zhiyu Jiang,
    Aaron Stanley, Hanbing Luo, Jerolim Andric, Marcelo Caire,
    Alessandro Sacchet, Eko Charnius Ilman, and Daisuke Yanagihara

Committee II.2: Dynamic Response 337
    D. Dessi, V. De Diego, S. Dhavalikar, M. Holtmann, S. J. Kim,
    L. Kaydihan, L. Moro, T. Pais, A. Paiva, G. Storhaug, H. Takahashi,
    S. Tavakoli, S. Wang, and B. Zao

Committee III.1: Ultimate Strength 452
    J. W. Ringsberg, L. Brubak, B.-Q. Chen, X. Chen, M. Chun, I. Darie,
    M. I. L. de Souza, M. Gaiotti, D. Georgiadis, M. Kõrgesaar,
    T. Magoga, A. M. Mohammad Zubar, K. Nahshon, T. Okafuji,
    M. Paredes, J. Romanoff, I. Schipperen, Y. Wang, A. Zamarin, and Z. Zhan

Committee III.2: Fatigue and Fracture 578
    H. Remes, G. An, M. Deul, P. Dong, P. Haselbach, S. Heggelund,
    P. Jurisic, A. Kahl, T. Kawabata, J. Liu, F. Prasetyo, M. Sicchiero,
    M. Soliman, M. Vicente del Amo, B. Yeter, J. Maljaars, S. Nakayama,
    A. Niraula, M. Ozdemir, J. Rodenburg, J. Roh, X. Song, F. Yanagimoto,
    and S. Wu

Committee IV.1: Design Methods, Principles and Criteria 728
    Y. Kawamura, J. Sales, P. Georgiev, J. Jelovica, W. Tang, A. Kolios,
    S. Vhanmane, M. Sidari, H. Amlashi, J. M. Kwon, A. Sobey, Y. Yan,
    and J. Sirkar

Committee IV.2: Material and Fabrication Technology 851
*Agnes Marie Horn, Jean-David Caprace, Matthias Krause,
Dora Tsiourva, Alessandro Caleo, Kunihiro Hamada,
Mojtaba Mokhtari, Iraklis Lazakis, Stéphane Paboeuf, and Bernt Leira*

Author Index 965

Committee I.1: Environment

Sanne van Essen[1(✉)], Mariana Bernardino[2],
Rüdiger Ulrich Franz von Bock und Polach[3], Ricardo Martins Campos[4],
Guillaume de Hauteclocque[5], Sheng Dong[6], Thomas Berge Johannessen[7],
Evert Lataire[8], Joong Soo Moon[9], Arttu Polojärvi[10], Jasna Prpic-Orsic[11],
Erik Vanem[12], and Tingyao Zhu[13]

[1] Delft, Netherlands
s.v.essen@marin.nl
[2] Instituto Superior Técnico (University of Lisbon), Lisbon, Portugal
[3] Hamburg, Germany
[4] University of São Paulo/Petrobras, Rio de Janeiro, Brazil
[5] Bureau Veritas, Paris, France
[6] Ocean University of China, Qingdao, China
[7] DNV, Oslo, Norway
[8] Ghent University, Ghent, Belgium
[9] Ulsan, Republic of Korea
[10] Aalto University, Espoo, Finland
[11] University of Rijeka, Rijeka, Croatia
[12] DNV, Oslo, Norway
[13] ClassNK, Tokyo, Japan

Committee Mandate. Concern for descriptions of the ocean environment, especially with respect to wave, current and wind, in deep and shallow waters, and ice, as a basis for the determination of environmental loads for structural design. Attention shall be given to statistical description of these and other related phenomena relevant to the safe design and operation of ships and offshore structures. The committee is encouraged to cooperate with the corresponding ITTC committee.

Keywords: Environment · waves · wind · ice · current · metocean conditions · effect on marine structures · climate · extreme conditions · statistics · metocean models · numerical models · data-driven models · laboratory observations · field observations · cyclones · rogue waves · scatter diagrams · wave-coupled phenomena

1 Introduction

The Environment committee of ISSC 2025 deals with advancements in the description of the ocean environment. We consider the environmental variables waves, currents, wind and ice and their interactions; other variables such as sea water temperature and salinity, soil properties, morphology etc. are not covered. Waves receive the most attention, as they have the largest influence on the performance of ships and floating offshore structures in open waters. The present report presents a selection of the relevant literature, building upon the previous report (Babanin *et al.*, 2022). As the mandate is broad, we decided on some focus areas.

1.1 Areas of Focus

The first focus is on applying environmental descriptions to the design of ships and marine structures. Since the ISSC specialises in these areas, its reports should be relevant to industry practitioners, both new and experienced. They will not always have a deep academic background in environmental theory but want to apply new insights or methods. We aimed to answer: how do advances in environmental science impact design loads and operations? While motion and load analysis fall under Committee I.2 Loads, we highlight where environmental advances may affect these areas.

We compiled examples of how environmental data is used in ship and marine structure design (see Table 1). This non-exhaustive overview, created by our committee and reviewed by other committee chairs, highlights the types of data, required detail, and the need to translate academic findings for industry use. It shows wide variation in both applications and input requirements. We also note a 'gap' between the simplicity desired by designers and the complexity of research outcomes. For instance, the industry's use of scatter diagrams for wave statistics oversimplifies the more complex relationships identified by research. Throughout the report, we highlight such gaps and emphasise that while naval architects may not be interested in model details, they are interested in the resulting improvement in wave spectral shape/wind speed/current profile/ice thickness predicted by that model. However, it is still important for engineers to understand the underlying physics and literature of new models or observation techniques to assess the quality of the data. We think that providing this information in a concise way is an important contribution of the present report. Translating scientific advancements into engineering-relevant properties is challenging but emphasises the relevance of the research. For example, understanding the details of breaking waves helps to predict the resulting loads on marine structures (e.g., Rognebakke et al., 2024 or Scharnke and Helder, 2023), and improved ocean wave statistics can reduce uncertainties, leading to lower safety factors and construction costs (e.g., Bitner-Gregersen et al., 2022). These examples highlight the practical importance of academic research in environmental modelling.

Our second focus is statistical modelling. This is very relevant for marine structure design and many committee members are experts in statistics, extreme values, and uncertainty estimation, particularly in relation to waves and ice. This also led us to conduct a benchmark on extreme value prediction methods.

The third focus is the impact of climate change on environmental descriptions, a rapidly evolving and critical topic. The changing environment also changes the statistics of environmental conditions, which may influence design loads on offshore structures (e.g., Lobeto et al., 2021, Lobeto et al., 2023 and Rosen et al., 2022). We expect this topic to further grow in importance.

Finally, we focus on innovative technologies in the world of the ocean environment: specifically, data-driven technologies and new observation methods in the laboratory or the ocean.

Table 1. (Non-exhaustive) examples of the required environmental information in the maritime industry. The examples for 'all marine structures except merchant ships' are the basis, any examples for other structures are additions to this.

Purpose	Examples of required environmental input
All marine structures	
Design for motion, structure response and fatigue	• Long-term wave statistics (scatter diagram or more advanced joint distributions, separation wind seas and swells) • Short-term wave statistics (frequency/directional spectral shapes) • Numerical/experimental irregular spectral wave generation, long and short-crested • Ability to mitigate unwanted spurious waves in numerical tool and wave basin
Design for extreme 'structure-related' events*	• Second- and higher-order wave modelling and wave breaking models, short- and long-term statistics of higher-order waves • Numerical and experimental deterministic wave generation of higher-order waves • Study on consistency between large-scale phase-averaged wave models and phase-resolved deterministic wave models
Design for extreme 'environment-related' events and conditions	• Separate statistics of tropical cyclones, rogue wave occurrences, ice statistics, icing risk etc
Monitoring/digital twin	• Hindcasts of environmental data, local field measurements
Operational decision-making	• Long term decisions on ship's route, speed and timing: statistical environmental forecasts including uncertainty assessment (horizon of days) - mainly wind and waves, extreme events such as cyclones, if possible, risk of rogue waves • Short-term decisions on offloading/set-down/speed reduction: deterministic real-time forecasting (horizon of minutes, e.g. based on ship radar input) - mainly waves
Future marine structure design, maintenance	• Expected effect of climate change on environmental statistics and hind-/forecasts
Merchant ships	
For strength and fatigue assessments	• Long- and short-term statistics encountered by merchant ships
Other structures in deep water	

(continued)

Table 1. (*continued*)

Purpose	Examples of required environmental input
Design of mooring or DP system	• Local current and wind statistics (mean directional speeds, depth profile, maybe gusts) • Numerical/experimental modelling of current and wind together with waves
Structures in shallow water (fixed/floating platforms, wave energy converters, ...)	
Design for motion, structure response and fatigue	• Waves, current and wind statistics including local bathymetry effects • Shallow-water wind, wave and current generation • Proper consideration of numerical and basin effects (e.g. basin resonance, reflections, measurement resolution)
Design of mooring or DP system	• Local long bound and free infragravity wave information (avoid resonance of mooring system with these waves)
Special case: offshore wind turbines, fixed or floating	
Design for motion, structure response and fatigue	• Long- and short-term wind statistics (mean wind speed, wind profile 100–300 m above the surface, wind turbulence and coherence) • Joint distributions of wind and waves • Numerical and experimental modelling of these properties
Monitoring/digital twin	• Measurement (e.g. lidar) of incoming wind field and wind turbine wakes, including accuracy assessment
Design for extreme 'environment-related' events and conditions	• Numerical model for wind field conditions in tropical cyclones
Special case: very large stationary structures (islands, airports, breakwaters, ...)	
Design for motion, structure response and fatigue	• Difference between point and spatial environmental statistics, including bathymetry effects and wave short-crestedness
Special case: structures in polar regions	
Design for ice loads	• Ice statistics • Numerical and experimental ice modelling

* *For example wave impacts, parametric roll, propeller ventilation*

1.2 A Note on the Influence of Marine Structures on the Environment

A topic that receives increased attention in the industry is the effect of marine structures on the environment (biodiversity, marine life etc.). We recognise that this is an important aspect to consider; examples include the effect of large structures on marine life, effect of wind turbines on birds, effect of sonar activity on the communication of marine mammals, ship propellers colliding with marine life, effect of pollution on biodiversity, etc. Although deemed relevant, we consider our present mandate to concern the effect of the ocean environment on ships and offshore structures, not vice versa. This topic is therefore out of scope of the present report.

1.3 Cooperation with ITTC

With regard to the cooperation with the ITTC on the topic of the environment, the ITTC special committee on modelling of environmental conditions presented their findings at the 29th ITTC congress (Iafrati et al., 2021). There was no follow-up of this special committee. One of our committee members is also part of the ITTC specialist committee on ice. The knowledge transfer on this topic was therefore natural. During the present ISSC term, the ITTC did not have a dedicated committee on waves, wind or currents (only on wind-assisted shipping, which does concern itself to some extent with numerical and experimental wind-wave modelling).

1.4 Lay-Out of the Report

The report is structured as follows: Sects. 2 to 4 cover environmental models, starting with metocean climate models (Sect. 2), phase-resolved and phase-averaged numerical models (Sect. 3) and data-driven models (Sect. 4). Laboratory and ocean observations are presented in Sects. 5 and 6. Sections 7 to 9 address special topics: environmental statistics (Sect. 7), extreme conditions like rogue waves and tropical cyclones (Sect. 8) and the new wave scatter diagram from the International Association of Classification Societies (IACS, Sect. 9). Conclusions are in Sect. 10. Appendix A contains the extreme value prediction benchmark, and Appendix B lists environmental data sources. Important topics such as hindcast models and wave breaking are covered across multiple chapters.

1.5 Symbols

Throughout the report, a few symbols are consistently used: significant wave height (H_s), mean wave period (T_m), peak wave period (T_p), zero up-crossing wave period (T_z) and mean wave direction (θ_m).

2 Large-Scale Metocean Climate Models

2.1 Introduction

Metocean climate information is useful for many activities related to the marine and coastal environments, such as planning and managing offshore platforms, ship structures, and offshore wind and wave energy resources (Casas-Prat et al., 2024). Understanding how the wind and wave regimes changed in the past and how they are expected to change in the future is of uttermost importance. In recent years, intensive research has been done into global and regional wave climate with the focus on trends, extremes and variability.

The availability of new by Coupled Model Intercomparison Project phase 6 (CMIP6) simulations leads to new and improved wave climate data. Up to the end of the 21st century, the most common approach to perform wind wave climate projections is by using global spectral wave models forced by the CMIP General Circulation Models (GCM) 10-m surface wind speeds. The purpose of CMIP simulations is to better understand past, present and future climate changes arising from natural, unforced variability or in response to changes in radiative forcing in a multi-model context and its output is the basis for the regular Assessment Report of the Intergovernmental Panel on Climate Change (IPCC, 2023).

However, the process of estimating the effect of a changing climate on the severity of future metocean conditions is plagued by large uncertainties. Quantifying these uncertainties and their sources, and the way they propagate through all the stages involved in assessing the effects is a difficult task, only possible due to the public availability of wave climate simulations ensembles.

2.2 Present Climate

The present-day climate is usually studied using simulated (reanalysis and hindcast) and observational data (in-situ observations and satellite altimeters), in both global and regional studies that focus on coastal impacts. Both types of data have advantages and disadvantages.

In-situ observations (including acoustic wave meters, optical wave meters, and wave buoys) are widely regarded as reliable sources of wind and wave data (see e.g., Miao et al., 2024). However, such measurements may suffer from inhomogeneities due to changes in equipment and location, sparsity of measurement locations and other accuracy issues (as discussed in Sect. 6). Data from satellites also has limitations, especially in the estimation of extreme values due to the impact of data corruption nearshore, the influence of coastline morphology and the spatio-temporal sampling achieved by altimeters. In general, available datasets are usually also shorter than the typical time scales that play a role in the design of ships and offshore structures. This introduces challenges related to extrapolation and extreme value estimation (as discussed in Sect. 7). Nevertheless, observations have a crucial importance, as they can be used to validate model simulations and be assimilated into hindcasts and reanalysis data sets.

Regarding the dynamical simulation of sea state conditions, hindcast and reanalysis datasets are generally used after local validation with in-situ or remote sensing data. Wave hindcasts usually refer to data that was simulated using a wave model, forced by wind provided by atmospheric hindcast data (numerical integration) or a reanalysis of atmospheric data. Wave reanalysis data results from the assimilation of wave data into the numerical wave model, which gives more complete and more coherent information about the sea state than can be obtained from observation data alone.

Statistical analysis performed over long time series can provide information on long-term trends, variability, and extreme events of metocean parameters. Also, spectral analysis can help understand the frequency content and energy distribution of waves and currents, aiding in the design and assessment of offshore structures (Mathewson, 2023).

Advancements in metocean data measurements and analysis have significantly improved the understanding and utilisation of metocean data in the offshore industry. In recent years, numerical models have become more sophisticated, incorporating higher resolutions and improved physics to accurately simulate metocean conditions and data assimilation techniques have been developed to integrate measured data into models, improving the accuracy of forecasts, hindcasts and reanalysis.

The sea surface wind field in the Yellow Sea and Bo Hai region was studied using the ERA5 wind field reanalysis data provided by the European Center for Medium-Range Weather Forecasts (ECMWF) in Bu et al. (2023) and Yu et al. (2023). The main analysis includes the spatial and temporal distribution characteristics of monthly mean wind direction and annual mean wind speed, and the spatial and temporal distribution characteristics of monthly mean and annual mean wind frequency.

Long-term trends of sea surface wind were studied in the northern South China Sea, using the same reanalysis data set by Hong and Zhang (2021) between 1979 and 2019. A decreasing trend in the annual mean wind speed in the coastal area and an increasing trend in the open sea was obtained. Significant correlations were also found between the variation of the wind field and El Niño–Southern Oscillation by wave coherence analysis.

Morim et al. (2022) present and describe a coordinated multi-product ensemble of present-day global wave fields. The presented coordinated global ensemble was derived from fourteen state-of-the-science global wave products obtained from different atmospheric reanalysis forcing and downscaling methods. The dataset, produced through the Coordinated Ocean Wave Climate Project (COWCLIP) phase 2, includes general and extreme statistics of H_s, T_m and θ_m computed across 1980–2014, at different frequency resolutions (monthly, seasonally, and annually).

Patra et al. (2023) present a dataset that represents historical ocean wave climate during the period 1960–2020, simulated using the numerical model WAVEWATCH III (WW3) forced by CMIP6 simulations corresponding to natural-only (NAT), greenhouse gas-only (GHG), aerosol-only (AER) forcings, combined forcing (natural and anthropogenic; ALL), and pre-industrial control conditions. This dataset is particularly useful to study for instance the relative contributions of natural and anthropogenic forcings to historical changes in the ocean wave climate.

More recently, Meucci et al. (2024) present a global wind wave climate model ensemble composed of eight spectral wave model simulations forced by three-hourly surface wind speed and daily sea ice concentration from eight different CMIP6 Global Climate Models (GCMs). The ensemble performance is evaluated against a reference global multi-mission satellite altimeter database and the recent ECMWF IFS Cy46r1 ERA5 wave hindcast, ERA5. For each ensemble member, three 30-year slices, one historical, and two future emission scenarios (SSP1-2.6 and SSP5-8.5) are available, and cover two distinct periods: 1985–2014 and 2071–2100. The authors conclude that this ensemble outperforms a previous CMIP5-forced wind wave climate ensemble, showing improved performance across all ocean regions.

At a more regional scale, Adell et al. (2023) studied the wind wave climate on the south coast of Sweden, using the Simulating WAves Nearshore (SWAN) model forced by ERA5 reanalysis data. Alonso and Solari (2021a) studied the wave climate of the Uruguayan coast, using the WW3 model forced by the CFSR reanalysis data set. Sartini and Antonini (2024) produced a high resolution spatial spectral wave climatology using hindcast data from the HOMERE database and the WW3 model for the French Atlantic Ocean. Trends in wave climate of the Mediterranean Sea, were analysed in the review paper of De Leo et al. (2024).

2.3 Climate Change

With the increasing interest in offshore wind energy in the last years, special interest has grown in the assessment of wind energy potential and offshore wind climate. Most studies evaluate possible changes in surface wind under climate change conditions using data from CMIP5 simulations, but with the new set of GHG emission and land used scenarios, CMIP6 simulations were produced.

The evolution of wind resources studied in the context of CMIP5 following the Representative Concentration Pathways (RCPs) climate-change simulations showed at global scale a southward shift in wind energy resources, under high carbon emissions. Based on these projections, changes in mean wind resources at the end of the 21st century would be small – generally, changes in mean energy density were expected to be less than 5% of the current values. The multi-model ensemble developed by Martinez and Iglesias (2024), based on GCMs from the CMIP6 project, predicts a widespread decline in future wind resources due to climate change with results pointing to changes of greater magnitude and more pronounced spatial variations at regional scales, than previous studies. These new projections reflect a higher climate sensitivity than did GCMs based on its predecessor project. With the IPCC newer and complex scenarios of climate change, the Shared Socioeconomic Pathways (SSPs), CMIP6 wind projections reveal significantly greater changes and more notable discrepancies across scenarios (both in the magnitude and spatial pattern of the changes). It has been surmised that the higher climate sensitivity anticipated in the models of the CMIP6 may be due to differences in the representation of clouds relative to the previous generation of models.

Sea surface winds over Southeast Asia are assessed by Herrmann et al. (2022) using an ensemble of the CMIP5 downscaled simulations performed over the 20th and 21st centuries under RCPs 4.5 and 8.5 scenarios within the Coordinated Regional Climate Downscaling EXperiment (CORDEX)-SEA project. Comparing with QuikSCAT satellite data results show that dynamical downscaling improves the sea surface wind speed representation, mainly by reducing its underestimation but the level of improvement depends on the RCM choice, GCM performance, and wind strength. Results also suggest that uncertainties related to the RCM choice are larger than those related to the GCM choice.

The wind energy resource over Europe was assessed by Carvalho et al. (2021), using CMIP6 simulations and changes from previous CMIP5 simulations were discussed. CMIP6 does not project an increase in wind resource for Northern Europe, showing a strong decline for practically all of Europe by the end of the century (SSP5-8.5). Unlike CMIP5, in CMIP6 stronger radiative forcing scenarios not only enhance the differences when compared to milder scenarios, but also change the spatial patterns of changes in the wind resource.

The same type of analysis was performed for the Northwest Passage by Qian and Zhang (2021), the Caribbean Sea by Bustos Usta and Torres Parra (2023), and the North Sea by Hahmann et al. (2022). Results show a decrease of wind resources in the region south of the Northwest Passage, with a significantly increase in the region north of 72°N (specifically in the Beaufort Sea). In the Caribbean Sea, wind intensification has different spatial patterns in the dry and wet seasons when compared to the annual mean. In the North Sea, while results indicate that annual mean wind speed and wind resources in northern Europe are not significantly affected by climate change between 2031–2050 compared to 1995–2014, the seasonal distribution of these resources is significantly altered.

Fernández-Alvarez et al. (2023) analysed the future changes in wind speed at 10 m above sea level (V10) in the North Atlantic Ocean and how these variations could affect offshore wind energy resources for three potential subregions. They dynamically downscaled three different future scenarios from a CMIP6 simulation, using the WRF-ARW atmospheric model. In the North Atlantic basin, surface wind speed is expected to decrease in the winter and spring seasons but increase in summer and autumn, mainly in tropical regions up to 30°N. Annually, the maximum increase is expected to be in the tropical region.

Lobeto et al. (2021) studied the behaviour of wind wave extremes in the end of the 21st century under RCP4.5 and RCP8.5, AR5 scenarios using a seven-member ensemble of wave climate simulations. Changes were estimated in H_s for return periods from 5 to 100 years by the end of the century after bias correction was applied to the simulated wave data. The authors obtained robust changes in extreme H_s over more than approximately 25% of the ocean surface and were able to conclude that increases in H_s extremes cover wider areas and are larger in magnitude than decreases for higher return periods.

Bernardino et al. (2021) simulate the global wave climate, forcing the WW3 model with global wind and ice-cover data from a RCP8.5 EC-Earth CMIP5 simulation, over a total of 120 years. This data was later used to assess the wind and wave climate in the North Atlantic Ocean, until the end of the 21st century including storm conditions, using a Lagrangean approach (D'Agostini et al., 2022). Using the total data set, a deeper study was performed by Bernardino et al. (2023) analysing mean conditions, extremes and variability of sea state conditions until the end of the 21st century. In these papers, the 120 years of simulation were divided into 4 periods representing the recent historical, the present and two future periods. The results show an increase in mean H_s, wave energy and cumulative wave energy in the South Atlantic, and an increase in variability and a decrease of mean H_s of total sea in the North Atlantic. Other regions also present changes but are less marked and less consistent through time. Regarding storm conditions in the North Atlantic, a decrease in the number of storms by the end of the 21st century is observed, although the observed events are more intense than in the historical period. The differences between the two periods are found to be more significant in the winter season.

A review paper was present by Casas-Prat et al. (2024), where historical and projected changes in the wind-wave climate over the world's oceans, and their impacts, are outlined. The main conclusions of this study are that there is a general agreement on a consistent historical increase in mean wave height of 1–3 cm/year in the Southern and Arctic Oceans, with extremes increasing by > 10 cm/year for the latter. Also, by 2100, mean wave height is projected to rise by 5–10% in the Southern Ocean and eastern tropical South Pacific, and by > 100% in the Arctic Ocean. The projected high increase in the Arctic Ocean is to be viewed in the context that the presence of ice significantly reduces wave formation. Projections based on simulations indicate a seasonally ice-free Arctic in the near future (Kim et al., 2023). By contrast, reductions in mean wave height up to 10% are expected in the North Atlantic and North Pacific, with regional variability and uncertainty for changes in extremes. Differences between 1.5 °C and warmer worlds reveal the potential benefit of limiting anthropogenic warming. Casas-Prat et al. (2024) also indicate that there are still challenges to be addressed in future work, as resolving global-scale climate change impacts on coastal processes and atmospheric–ocean–wave interactions. These challenges require a step-up in observational and modelling capabilities, including enhanced spatiotemporal resolution and coverage of observations, more homogeneous data products, multidisciplinary model improvement, and better sampling of uncertainty with larger ensembles.

2.4 Uncertainty Assessment

The uncertainty associated with the projected changes in wind waves is essential for offshore/coastal risk and adaptation estimates. The uncertainty is usually based on wave climate projection ensembles and assessed through the agreement in the sign and magnitude of the changes projected for the different ensemble members without the ability to reveal its sources. Research in analysing different sources of uncertainty and how it affects extreme values was developed in recent years.

Lobeto et al. (2023) investigated the epistemic uncertainty associated with wave propagation modelling in wave climate projections. Their study used a single-forcing, single-scenario, seven-member global wave climate projection ensemble. The three models used were third-generation spectral wave models that share a similar theoretical background (SWAN, WAM and WW3). The inter-model and intra-model uncertainties are independently quantified by assessing the differences between the projected changes from different wave models and different model parameterizations, respectively. Discrepancies are measured through the relative mean difference metric. The uncertainty was assessed to identify areas where it may have a greater impact, evaluating its magnitude, its projected changes, and the discrepancies among members. The study reveals divergences in projected changes from runs of different models and runs of the same model with different parameterizations over 75% of the ensemble mean change in several ocean regions, with the coasts of western North America, the Maritime Continent and the Arabian Sea showing the most significant wave modelling uncertainties.

Morim et al. (2023) quantify, at global scale, the uncertainties in contemporary extreme wave estimates across an ensemble of widely used global wave reanalyses/hindcasts supported by observations obtained from 64 wave buoys around the world. A comparison of the present-day uncertainties in extreme H_s estimates with projected future changes in extreme H_s, and associated uncertainties, was also obtained from a comprehensive ensemble of global wave projections. The authors find that contemporary uncertainties in 50-year return period wave heights ($H_{s,50}$) reach on average ~ 2.5 m in regions adjacent to coastlines and are primarily driven by atmospheric forcing. They also show that these uncertainty estimates dominate projected 21st-century changes in $H_{s,50}$ across ~ 80% of global ocean and coastlines.

The variability in estimates of 100-year return value of H_s, and changes in estimates over a period of time, was explored by Ewans and Jonathan (2023) for output of WW3 models from 7 representative CMIP5 models and one CMIP6 model. Focus was given to neighbourhoods of locations east of Madagascar and south of Australia, where a previous study of CMIP5-derived output reported significant decrease and increase in H_s respectively, in RCP scenarios RCP4.5 and RCP8.5. A large variation between return value estimates was found from different GCMs, and from variation of longitude and latitude within each neighbourhood, for estimates based on samples corresponding to $\leq$ 165 years of model output. It was concluded that these sources of uncertainty tend to be larger than those due to typical modelling choices, such as choice of threshold for peak-over-threshold (POT) or block length for block maxima (BM). Careful specification of the extreme value model and choice of thresholds (for POT analysis) or block length (for BM analysis) is critical.

3 Numerical Models at Engineering Scale

3.1 Waves

In the last three years, both phase-resolving wave models and spectral wave models were developed further. The advances in these research fields are summarized in Fig. 1, and discussed below.

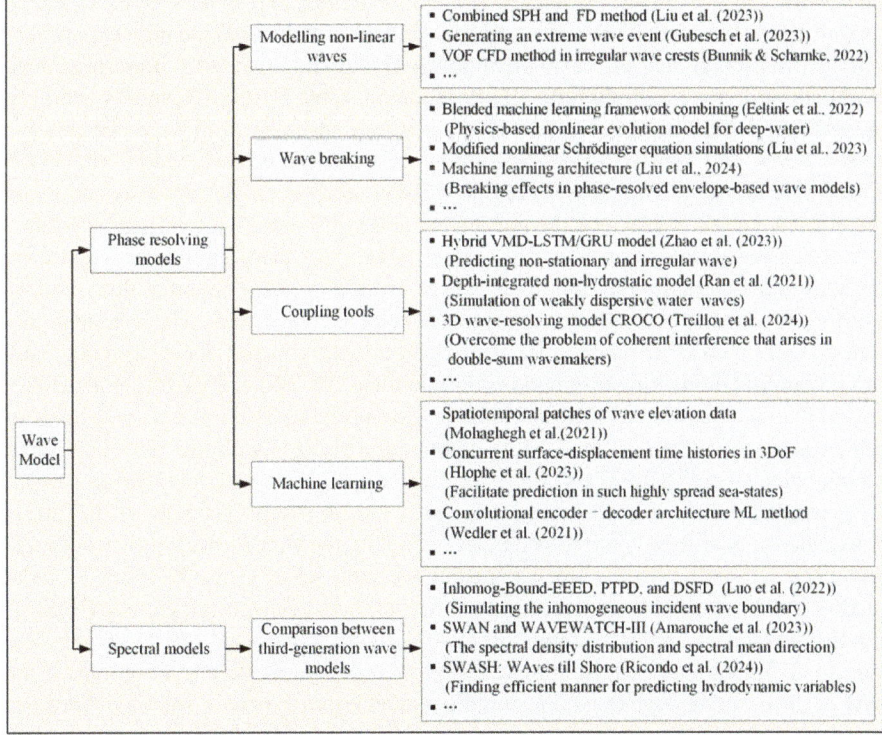

Fig. 1. Overview of advances in wave model research

3.1.1 Phase-Resolving Models

Exercises such as the comparative study of the interaction of focused waves with floating structures in Ransley et al (2021) illustrated that the accuracy of the predicted response to waves correlates strongly with the accuracy of the incident wave generation. Combining that with the knowledge that phase-resolved wave modelling still faces challenges (related to e.g., modelling of breaking waves, reflections, shallow-water effects, higher-order effects in steep waves and high computational costs), this explains the active field of study of the phase-resolved wave models.

There have been significant advances in fully non-linear computational fluid dynamics (CFD) modelling of non-linear waves during recent years. These advances have been mostly related to particle-based Smoothed Particle Hydrodynamics (SPH) and grid-based Eulerian Navier-Stokes (NS) methods. Fouques et al. (2021) presented proposed qualification criteria and on-going evaluation efforts for the wave solutions generated from either potential-flow based or CFD-based codes. Pákozdi et al. (2022) proposed guidelines for the use of a numerical wave tank (NWT) to generate waves as input to a CFD simulation in the context of wave impact on a gravity-based structure. K. Liu *et al.* (2023) presented a coupling algorithm for the simulation of incompressible free-surface flows over the near and far fields by solving the Navier-Stokes equations. Gubesch et al. (2023) presented the results of an experimental and numerical investigation into the generation of realistic model scale extreme waves for survivability testing of offshore structures. Bunnik and Scharnke (2022) showed that a volume-of-fluid CFD method can achieve a good agreement with experiments in (extreme) irregular wave crests for a duration of three hours, within a reasonable computational time. In (van der Plas et al., 2022), it is shown that the same CFD code can be used to accurately replicate high experimental wave events, and their effect on the motions of a floating fish farm. Bandringa et al. (2021) showed that multi-phase unsteady incompressible RANS (Reynolds-averaged Navier-Stokes) CDF is also capable of replicating non-linear shallow-water extreme regular waves, and their effect on the motions of a moored buoy. Jaouën et al. (2022) showed that the same code can accurately predict steep deep-water waves combined with current, and their effect on the motions of and drift loads on a semi-submersible.

Besides the developments in CFD wave modelling, there have been significant advances in the modelling of non-linear waves using non-linear potential flow models. Such models can also be used to generate phase-resolved wave fields. They are generally more efficient than CFD models, but they cannot directly model wave breaking (only through empirical breaking dissipation formulations). W. Wang, Pákozdi, et al. (2022) showed that it is possible to generate three-hour long- or short-crested wave sequences with good accuracy against reasonable computational costs with the fully non-linear potential flow model REEF3D:NLPF (Pákozdi et al., 2023). (Numerical) waves are usually generated either on deep or shallow water, but W. Wang, Kamath, *et al.* (2022) showed that it is also possible to accurately simulate the transformation of waves and their spectra from deep to shallow water using the same REEF3D:NLPF model. Landet and Bloch Helmers (2024) combined a non-linear higher-order spectral method (HOSM) with linear response amplitude operators (RAO), to identify critical sea states and wave trains to be further analysed using more advanced tools ('screening'). A hot topic is the study of the onset of wave breaking, and the related improvement of the wave breaking formulations in non-linear potential flow models. Y. Liu *et al.* (2023) compared different breaking models to experimental results, showing that most existing models are able to model breaking in irregular spectral waves, but none of them produce the correct dissipation in focused wave groups. They therefore propose an updated breaking model focused wave groups. A new HOSM wave breaking model, based on the ratio of the water particle velocity in the wave crest and the local crest velocity combined with an eddy viscosity model, is validated against CFD calculations and experiments by Gramstad et al. (2023). This validation shows that the new model is able to produce steep sea state crest statistics in good agreement with the model tests, and in reasonable agreement with the CFD calculations. Ducrozet et al. (2023) also proposed an improved eddy viscosity-based wave breaking model for the open-source higher-order spectral model HOSM-NWT. An alternative to such physics-based breaking models is to use machine learning (ML) to capture breaking effects, for inclusion in non-linear wave models. Such a model was developed by Eeltink et al. (2022) and Y. Liu *et al.* (2024). Their model seems to have broken the problem into two in a similar way as the conventional physics-based models do: one part which detects whether the wave is breaking, and another which captures the subsequent behaviour of the waves. The presence of current may change the breaking inception, which was studied by Touboul and Banner (2021). They investigated breaking of two-dimensional deep water non-linear wave packets propagating in the presence of a background current that varies linearly with depth using a boundary integral element method, finding that previous observations about the crest slowdown still hold in the presence of current. However, the shape of the breaking crest appears to be significantly affected by the current.

While the techniques based on CFD or non-linear potential flow have advanced in modelling detailed wave and loading events, they remain computationally expensive. Non-linear potential flow methods are faster but cannot be reliably applied to modelling of wave breaking and stability. Therefore, coupling tools with different levels of fidelity seems a promising approach for long-term wave impact assessment. Cao *et al.* (2021) investigated how different slope limiters influence nearshore wave simulation with an emphasis on depth-limited wave breaking simulation in the open source NHWAVE model. L. Zhao *et al.* (2023) proposed a VMD-LSTM/GRU model by combining the advantages of the long short-term memory (LSTM) model, the gate recurrent units (GRU), and the variational mode decomposition (VMD) technique. Ran et al. (2021) proposed a convergent two-dimensional depth-integrated non-hydrostatic shallow water model discretized using the discontinuous Galerkin (DG) method, which can account for the dispersive effects by including a non-hydrostatic pressure component and can be used for the simulation of weakly dispersive water waves. Treillou et al. (2024) presented the implementation of a single-sum wavemaker in the three-dimensional wave-resolving model CROCO, to overcome the problem of the coherent interference (also called 'checkerboard') that arises in double-sum wavemakers of wave-resolving models. To be able to assess the life-time statistics of extreme waves (and wave-induced responses on structures), high-fidelity CFD methods can also be combined with lower-order models in a multi-fidelity approach. Such 'screening' or 'adaptive sampling' methods are discussed in Sect. 7.2.

The full phase-resolving wave prediction can also be carried out using ML. Mohaghegh et al. (2021) overcame some challenges of traditional phase-resolved prediction of ocean waves - related to computational times and reconstruction of the wave field based on measurements through an advanced ML technique based on spatio-temporal patches of the time history of surface wave elevation data in the domain. Hlophe et al. (2023) investigated the benefit of exploiting concurrent surface-displacement time histories measured by a buoy in three degrees of freedom (DoF) to facilitate prediction in such highly spread sea-states. Wedler et al. (2023) explored the potential of an ML approach based on a fully convolutional encoder–decoder architecture for the efficient and accurate prediction of water waves. The high-order spectral (HOS) method forms the foundation for the generation of the training data. The HOS method is applied to generate long-crested irregular wave data with consecutive orders of non-linearity starting from first order up to fourth order. A follow-up study by Klein et al. (2024) extended this study using laboratory data (see Sect. 5).

3.1.2 Spectral Models

The scientific community has developed a series of phase-averaged spectral models called Third Generation wave models, including the WAve Model (WAM), WW3, SWAN and the non-hydrostatic Simulating WAves till Shore (SWASH) model for coastal waters. Some comparison has been carried out to compare the application range of different models and to help improve the computation performance. Luo et al. (2022) established the 'Inhomog-Bound-EEED', 'Inhomog-Bound-PTPD' and 'Inhomog-Bound-DSFD' approaches to simulate the inhomogeneous incident wave boundary in coastal areas and verified them using a coupling of a Boussinesq wave model with a theoretical diffraction wave model and with the SWAN wave model. The proposed three methods have good applicability to simulate the spatially inhomogeneous multidirectional irregular wave incident boundary. Amarouche, Akpınar, Rybalko, *et al.* (2023) assessed the SWAN and WW3 models regarding the directional wave spectra estimates based on measurements conducted in the eastern Black Sea. Ricondo et al. (2024) presented a site-specific metamodel based on numerical simulations from SWASH, that aims to simplify the prediction of hydrodynamic variables along cross-shore profiles. To accomplish this, a large synthetic database of offshore wave and sea level conditions is created and downscaled using numerical modelling together with sampling, selection, and interpolation techniques. Vasarmidis et al. (2024) analysed the governing equations of SWASH and derived the linear and non-linear solutions up to third order of all dependent variables, considering one to four vertical layers.

Generally, the present wave models can predict the waves reasonably well, but their accuracy can be improved with regional characteristics for specific local seas and higher temporal and spatial resolutions. Fernández, Calvino and Dias (2021) investigated the performance of WW3 (version 6.07) using the ST6 source term package in the context of wave growth and dissipation in the west coast of Ireland. The model parameters are validated using measurements from four local buoys. To assess the SWAN and WW3 models regarding the directional wave spectra estimates, Amarouche et al. (2021) presented a spatial calibration of SWAN model parameters based on satellite observations for the Black and Azov Seas. Lira-Loarca et al. (2022) presented a wave model based on an unstructured grid in the Mediterranean Sea using WW3 v6.07. A methodology for coastal risk assessment is developed, leveraging the detailed information provided by the model in shallow waters and relying on a 'Storm Power Index', 'Coastal Vulnerability Index' and 'Risk Index'. Soran et al. (2022) conducted spatial calibration of hindcast WW3 by investigating its H_s quality using different source term parameterizations (ST1-ST4 and ST6) in the Black Sea against the 2020-year satellite observations covering the entire Black Sea. They concluded that the calibrated ST1, ST3 and ST4 forced with ERA5 wind could be recommended for long-term wave simulation, and ST6 for extreme wave simulation in the Black Sea. Aydoğan and Ayat (2021) conducted a similar study in which the performance of the SWAN model forced by different wind resources ERA5, CFSv2, and ERA-I was tested to quantify the improvement in H_s and T_p prediction induced by ERA5 wind fields regarding the Black Sea. They concluded that the ST6 and the winds forced by ERA5 provided an increased accuracy of H_s prediction in the Black Sea. Moreover, Feng and Chen (2021) established a new wave hindcast system based on the SWAN model by using the ERA5 wind reanalysis data for the western inner shelf of the Yellow Sea. Their results showed that the ERA5 reanalysis wind outperformed its early generation for the inner shelf and could better reproduce the extreme wave simulation. Silva et al. (2022) used scatterometer wind fields to force the SWAN model for the Portuguese Atlantic islands of the Azores and Madeira with high spatial resolutions of 0.05° × 0.05° and 0.02° × 0.02°. Their results demonstrated that H_s was well reproduced by the SWAN model. Kazeminezhad and Ghavanini (2023) proposed and evaluated an operational wave forecasting system based on WW3 forced with Global Forecast System winds by using different parameterizations ST2-ST4 and ST6 for Arabian Sea. By comparing the forecasted H_s with the altimeter over Arabian Sea and buoy along the northwest of the Indian Ocean, it was shown that the ST3 and ST4 packages had better performance than the others for forecasting H_s in the entire Arabian Sea. Lucero et al. (2023) used ML models to improve a wave hindcast database for H_s, T_m and θ_m based on WW3 for the Chilean coast (where the WW3 hindcast was considered to have a poor performance). Their study showed that ML techniques are a fast and effective way to improve existing wave hindcast databases at relatively low cost.

Jiang et al. (2022) implemented a high-resolution unstructured SWAN model for the Changjiang River Estuary (CRE). Five different winds are adopted to assess their quality over the CRE, including ERA5, CFSv2, GFS, CCMP-NRT and a newly available high resolution (9 km) wind product from APRCP (Asia-Pacific Regional Coupled Prediction System). This study provides a good start to build a hindcast and forecast system for the CRE region. W. Zhang et al. (2023a) comparatively evaluated two reanalysis wind fields, simulate waves driven by CFSv2 and ERA5, and test the simulation performance of the ST6 source term package of the SWAN model in the Bay of Bengal (BOB). Siadatmousavi et al. (2024) integrated data assimilation with bulk and partitioned spectral parameters into the WW3 wave model. The dependency of future forecasts on the current state in numerical models leads to an error flow in the results. Different wind input and wave dissipation packages of WW3 were assessed for the Persian Gulf. To characterise long-term statistics of wave energy resources (mainly wave height) in coastal and offshore waters for the use of marine renewable energy in detail, high-resolution models in regional areas are desirable. These were therefore studied by many researchers. Webb et al. (2020) conducted a wave and current assessment based on a large-scale WW3 simulation in which several key components to improve model skill were included for the sea areas around Japan. The nearshore resolution is 1 km with a 0.0100° × 0.0125° latitude-longitude grid cell. The well tested long-term wave statistics around the coast of Japan covered a 30-year period (1994–2022). Allahdadi *et al.* (2021) developed an unstructured SWAN model with a high resolution of 0.2 km within 20 km off the coast of the Gulf of Mexico, Puerto Rico, and the U.S. Virgin Islands (GPRVI) region, to provide a reliable setting for a long-term wave energy characterisation for this region. Albuquerque et al. (2021) presented a validated high-resolution partitioned wave hindcast of New Zealand waters, to improve the understanding of the local wave climate. J. Li *et al.* (2022a; 2022b) reanalysed a 40-year (1979–2018) hindcast of H_s with the SWAN model for the Bohai Sea and northern Yellow Sea, with a spatial resolution of 0.1° in both the longitude and latitude as the initial wave height database. García Medina et al. (2023) developed a wave model based on WW3 and an unstructured SWAN model with a 0.2 km resolution to generate a 42-year (1979–2020) hindcast dataset suitable for the wave climate and the wave energy resource characterization of the American Samoa coast. Moreover, Neary and Ahn (2023) developed a 31-year global (0.5° resolution) and coastal US (1/15° resolution) hindcast model based on WW3 combined with the US NOAA buoy network to generate and map the global/coastal US distribution of the mean and extreme H_s. Furthermore, Christakos et al. (2024) developed a 30-year period (1990–2019) wave dataset covering a big part of the North Atlantic and the Arctic Ocean with a high spatial (3 × 3 km) resolution using the advanced NORA3 model, to identify long-term trends of H_s and wave energy-rich areas in the Arctic Ocean (where developing marine renewable energy is getting more and more attention).

Wave models are widely used in wave hindcast studies, for the purpose of long-term study of wave climate and wave energy distribution. Bieman et al. (2023) proposed an innovative hybrid modelling approach to improve the accuracy of operational wave forecasts. An operational wave model is combined with a ML model, which is trained using wave measurements within the wave model domain. F. Wang *et al.* (2024a; 2024b; 2024c; 2024d) incorporated the wave-generated turbulence dissipation term together with the improved post-breaking spectrum term in the MASNUM wave model as new dissipation terms. The report of the 30th ITTC (International Towing Tank Conference) Specialist Committee on Wind Powered and Wind Assisted Ships (Werner et al., 2024) included a review of hindcast models for the long-term statistics of wind and waves from the point of view of applicability for the evaluation of wind assisted ships at design stage.

Recently, wave models were also combined with neural networks and other ML methods to further reduce the computation error. Huang et al. (2022) proposed a regional wind wave prediction surrogate model based on a convolutional neural network (CNN), which takes historical wind and wave data as input to realise the prediction of current waves. Luo et al. (2023) proposed a novel data-driven model for efficient spatial–temporal forecasting of H_s, T_m, and θ_m in the Pearl River Estuary in Southern China. N. Wang *et al.* (2022a; 2022b; 2022c) developed a novel hybrid approach by integrating a physics-based SWAN model with ML algorithms to predict wind waves in a shallow estuary.

Some concern exists in the community about the accuracy of the wave spectral models: they may provide spectra that are too broad compared to those derived from wave measurements (e.g. Ardag and Resio, 2019; Cavaleri et al., 2020). Advances have been made (e.g., machine learning-enhanced spectral tuning) recently. For example, Jonathan et al. (2024) deals with an assessment of trimodal wave spectral parameters using three machine learning (ML) methods for an offshore site in the Norwegian Sea, and found that the Extra trees method has the best performance and is very accurate for predicting spectral parameters of three wave systems; Martzikos et al. (2025) investigates the prediction of sea surface elevation spectra from pressure measurements using frequency-specific artificial neural networks (ANNs) and found that incorporating spectral features of pressure signals into ANN architectures provides a robust and efficient framework for sea surface spectra prediction.

3.2 Ice

Recent advances in ice modelling, both on geophysical scale and on engineering scale, are summarised in Fig. 2 and discussed below.

3.2.1 Geophysical Scale

Geophysical scale modelling, or sea ice dynamics modelling, is typically performed at spatial scales of hundreds of kilometres or more and over temporal scales of months. The primary goal has been to integrate the ice models into earth system models, where thermodynamics and ocean-ice-climate interactions have a significant role. In recent studies, many new geophysical scale sea ice models have been proposed. Viscous-plastic rheology or elastic-viscous-plastic rheology is used in most geophysical-scale sea ice models currently in use. Ólason et al. (2022) presented a new brittle rheology and an accompanying numerical framework for large-scale sea-ice modelling, utilising the Bingham-Maxwell constitutive model and Maxwell-Elasto-Brittle rheology. It addresses the problem of excessive thickening of ice in model runs longer than one winter and a high computational cost. At present, many geophysical scale sea ice models do not resolve floes and represent sea ice as a continuum. Manucharyan and Montemuro (2022) introduced SubZero, a conceptually new sea ice model geared to explicitly simulate the life cycles of individual floes by using complex discrete elements with time-evolving shapes. The SubZero model was studied in idealised experiments, which provided a valuable alternative sea ice model for simulating floe interactions. In addition to the representation of physical processes, the numerical discretization of sea ice dynamics on the model grid is particularly challenging. Mehlmann and Gutjahr (2022) proposed a new sea ice dynamics discretisation for describing sea ice flow in a global three-dimensional Cartesian system. The equations are approximated on a computational grid, that consists of triangles covering the surface of the sphere. The new sea ice dynamics show typical drift patterns.

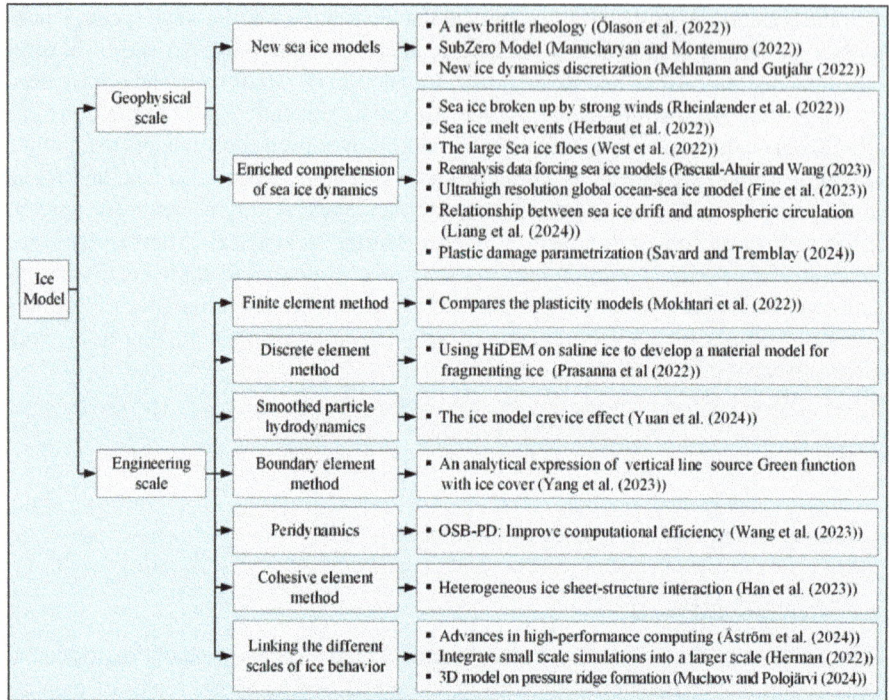

Fig. 2. Overview of advances in ice model research

Recent advancements in sea ice research have significantly enriched our comprehension of sea ice dynamics and modelling. The loss of thick multiyear sea ice in the Arctic leads to weaker sea ice that is more easily broken up by strong winds. Rheinlænder *et al.* (2022) investigated the driving forces behind sea ice breakup events during winter by using neXtSIM sea ice model. Their simulations successfully reproduce the timing, location, and propagation of sea ice leads associated with wind-driven breakup and highlight the importance of accuracy of the atmospheric forcing, sea ice rheology, and changes in sea ice thickness. The ocean is suggested to play a major role in the ongoing winter decay of the sea ice cover in the western Eurasian Basin. Herbaut et al. (2022) provided insights into ice-ocean interactions in winter north of Svalbard, enhancing understanding of significant sea ice melt events by using a high-resolution sea ice-ocean model. Their research shows that despite the prominent role of the wind in the winter sea ice melt events, a stronger Atlantic Water current would lift the warm water reservoir closer to the surface, facilitating the heat transfer to the sea ice. The discrete element method (DEM) can provide detailed descriptions of sea ice dynamics that explicitly model floes and discontinuities in the ice. West et al. (2022) developed a DEM using the ParticLS software library, capturing dynamic sea ice patterns and deformation nuances. The large ice floes are modelled as continuous objects made up of many bonded particles that can interact with each other, deform, and fracture. The model is tested on ice advecting through an idealised channel and through Nares Strait. Reanalysis data combine an atmospheric model and data assimilation algorithms to create continuous and gridded spatial–temporal datasets, which are often used for forcing sea ice models. Pascual-Ahuir and Wang (2023) compared MITgcm-ECCO2 results using JRA25 and ERA5, noting overestimations of Arctic summer ice and Antarctic winter ice underestimations. Optimizing with Green functions improved results, while ERA5 better represented Arctic Sea ice. Precise sea ice dynamics are vital in coupled ice-ocean models. Fine et al. (2023) used an ultrahigh resolution global ocean-sea ice model to investigate the potential influence of the upper Arctic Ocean on Arctic Sea ice. The model's horizontal resolution varies from 8 km at the equator to 2 km at the poles. The simulation reproduces observed distributions of seasonal sea-ice thickness and concentration realistically, and volume, fresh water, and heat transports through key passages are realistic, lying within observationally determined ranges. Liang et al. (2024) analysed an Arctic ice-ocean coupled simulation to investigate the relationship between wintertime sea ice drift and atmospheric circulation, and examined the driving force terms in the sea ice dynamic equation. Their research suggested that developing more sophisticated internal ice stress expression in ice models is crucial for accurately projecting future sea ice changes. Savard and Tremblay (2024) implemented a plastic damage parametrization, separate from the elastic damage within the elasto-brittle framework, within the standard viscous–plastic (VP) sea ice model to research the sensitivity of sea ice deformation statistics to plastic damage.

3.2.2 Engineering Scale

One of the most significant challenges in studying the ice load and vibration response of structures during the interaction between sea ice and marine structures is the lack of a comprehensive and accurate description of the mechanical properties of sea ice and the overall dynamic behaviour of structures. In engineering scale ice mechanics, the most employed simulation techniques include the DEM, the finite element method (FEM), smoothed particle hydrodynamics (SPH), peridynamics, the boundary element method (BEM), the cohesive element method (CEM), and so on. There is still no universal or widely accepted material model for crushing ice because of its exceptionally complex behaviour. Mokhtari et al. (2022) systematically compared the plasticity models both theoretically (in terms of their constitutive laws) and numerically in eight single-element tests and 28 different simulations of a physical cone crushing test in a laboratory scale through ABAQUS. Prasanna et al. (2022) used the particle-based model Helsinki DEM (HiDEM) on saline ice to develop a material model for fragmenting ice. They found that traditional ice modelling schemes are incomplete: local crushing of the ice is not valid as a generic failure mode for fragmented ice under compression. Rather, shear failure as described by Mohr-Coulomb theory is shown to be the dominant failure mode.

The FEM-SPH adaptive method is a new numerical simulation method developed in recent years. Yuan et al. (2024) investigated the ice model crevice effect on the vertical water-entry of a sphere through SPH simulations. The results illustrate the cavity evolution process and determine the sphere's trajectory, effectively constructing the three-dimensional shape of the cavity under different values of the crevice spacing in the ice model. The numerical calculation of panel integrals associated with the Green's function, satisfying the water surface condition with ice cover, is the basis for BEM to solve hydrodynamic problems of ocean structures in ice regions. To improve the efficiency and accuracy of BEM hydrodynamic calculations of structures in water with ice cover, Yang *et al.* (2023a) derived an analytical expression of the vertical line source Green's function with ice cover. Peridynamics is an emerging theory of solid mechanics, which shows high solution accuracy for computing fracture, fragmentation, and large-scale deformation of ice. As a non-local meshless method, peridynamics can accurately solve ice-structure interaction, but requires longer computational times to solve the equations of motion than the original meshed methods. To further improve the computational power of the ship-ice interaction model, C. Wang *et al.* (2023a; 2023b; 2023c; 2023d; 2023e) proposed algorithm optimisations in the ordinary state-based peridynamics (OSB-PD) to improve computational efficiency. Coupling the CEM with FEM is currently an important research topic in the field of ice-structure interactions, due to its capability of fracture simulation, which merged continuum mechanics with non-linear fracture mechanics. S. Han et al. (2024) established a model to simulate the ice-structure interactions based on the CEM combined with random field. According to the research, the heterogeneity of ice sheet led to significant uncertainties in the numerical simulations.

Linking the different scales of ice behaviour is becoming one of the key goals for sea ice simulations. Achieving this requires simulation tools that can model large sea areas with high resolution or provide effective methods for analysing simulation results from the aspect of scaling. This is supported by advances in high-performance computing, enabling explicit three-dimensional modelling of ice dynamics at meter-scale resolution over extensive sea areas. Åström et al. (2024) applied the HiDEM originally developed for glacier calving, to sea-ice breakup and dynamics. The code is highly optimised to utilise high-end supercomputers and a few tens of graphics processing units (GPUs), to achieve an extreme time and space resolution. Another option used is to integrate mechanical phenomena, or results from small scale simulations, into a larger scale setting meaningful way. This can be done by, for example, including various small scale ice failure modes into a simulation tool describing detailed interactions of larger ice features. Idealised discrete-element simulations were used by Herman (2022) to research granular effects in sea ice rheology in the marginal ice zone. Further, engineering scale simulations can be used to produce results which can then be used for informing large-scale modelling. For example, Muchow and Polojärvi (2024) presented DEM simulations on pressure ridge formation, then demonstrate that Cauchy-Froude scaling applies for translating laboratory-scale results on ridging to full-scale scenarios.

3.3 Current

This chapter reviews the key developments in ocean current (circulation) modelling. These advances are summarised in Fig. 3. The interaction between waves and currents is of great importance to coastal engineers who study the hydrodynamics of nearshore areas. Hydrodynamic loads on coastal structures, scour around offshore structures, flow fields near pipelines, and the dispersal of pollutants are typical examples that require an enhanced understanding of wave-current interactions.

3.3.1 Global Ocean Models

Global ocean models are mathematical simulations used to study ocean behaviour such as currents and temperature. Terrazas Silva et al. (2024) explored the interconnections between the West Mexican Current (WMC) and the Costa Rica Coastal Current (CRCC) in the Eastern Tropical Pacific. Utilising the Hybrid Coordinated Ocean Model (HYCOM) over a 19-year period, the authors discovered that the interconnection between these currents is influenced by the seasonal migration of the Tehuantepec Bowl (TB). Y. Zhang *et al.* (2023a; 2023b; 2023c) introduced a novel method for comparing Argo float observations with simulated trajectories from HYCOM. By analysing the surface and subsurface flow fields, the study found that HYCOM generally captured the basic characteristics of ocean currents but exhibited regional differences, particularly in the Gulf Stream and North Atlantic Warm Current. The research demonstrated the importance of using real-world observational data to improve model accuracy. Hidayah et al. (2021) applied a hydro-oceanographic model to analyse the current dynamics around Bawean Island in the Java Sea. This study highlighted the weak current velocities in the area, with significant seasonal variations influenced by monsoon patterns. The validation of the model using ADCP measurements revealed a moderate accuracy, emphasizing the need for further refinement in small-scale coastal current modelling.

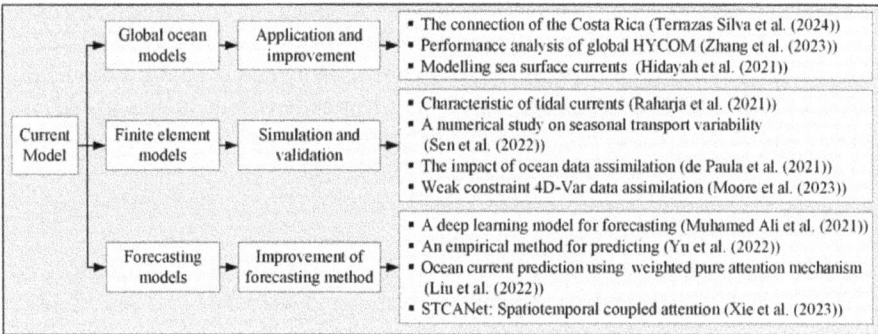

Fig. 3. Overview of advances in current model research

3.3.2 Finite Element Models

Finite element models (FEM) are used to simulate ocean currents by dividing ocean areas into small grids, or cells, and applying physical equations to calculate the speed and direction of the current within each cell. Raharja et al. (2021) used the Finite Volume Coastal Ocean Model (FVCOM) to simulate tidal current circulation in the Lombok Strait, Indonesia. The model, driven by tidal elevations, showed strong agreement with observations, revealing dominant influences from M2 and K1 tidal components, with distinct current patterns at different depths. Sen et al. (2022) applied the Regional Ocean Modelling System (ROMS) to study the seasonal variability of the North Indian Ocean's boundary currents. The study highlighted the peak transport of the East India Coastal Current (EICC) and Somali Current (SC) and the role of wind stress in their semi-annual reversals. de Paula et al. (2021) utilised ROMS with four-dimensional variational data assimilation (4D-Var) to simulate ocean circulation in the southwestern South Atlantic. Assimilation of SSH, SST, and in situ data significantly improved model accuracy, particularly in representing eddy dipole events. Moore et al. (2023) introduced a saddle-point formulation of weak constraint 4D-Var data assimilation in ROMS, improving computational efficiency through time-parallelisation. This method reduced run-time while maintaining accurate simulations of the California Current System.

3.3.3 Forecasting Models

Accurate forecasting of ocean surface currents is crucial for the planning of marine activities, including fisheries, shipping, and pollution control. Muhamed Ali et al. (2021) developed a deep learning model using Long Short-Term Memory (LSTM) to predict ocean currents in the Gulf of Mexico. Their method extended the useful forecast period to more than 4 days with improved accuracy and correlation (0.6). Yu et al. (2022) developed an empirical method for predicting the South China Sea Warm Current (SCSWC) using wind stress data. By leveraging the most dominant Empirical Orthogonal Function (EOF) mode of sea surface height (SSH), the study achieved prediction accuracies of up to 75% during the winter seasons. This research provided a promising approach for forecasting complex ocean currents based on wind-driven dynamics. J. Liu et al. (2022) introduced a deep learning model based purely on an optimised attention mechanism, enhancing the prediction performance of ocean currents by focusing on key elements, making it more reliable over large temporal and spatial ranges. C. Xie *et al.* (2023a; 2023b) proposed STCANet, a deep network model with spatiotemporal coupling and attention mechanisms for ocean current prediction. It outperformed existing models like LSTM, ARIMA, and CNN_GRU in both temporal and spatial accuracy.

3.4 Wind

Wind models were developed quickly in recent years, especially for wind energy engineering. Recent advances in these research fields are summarized in Fig. 4 and discussed below.

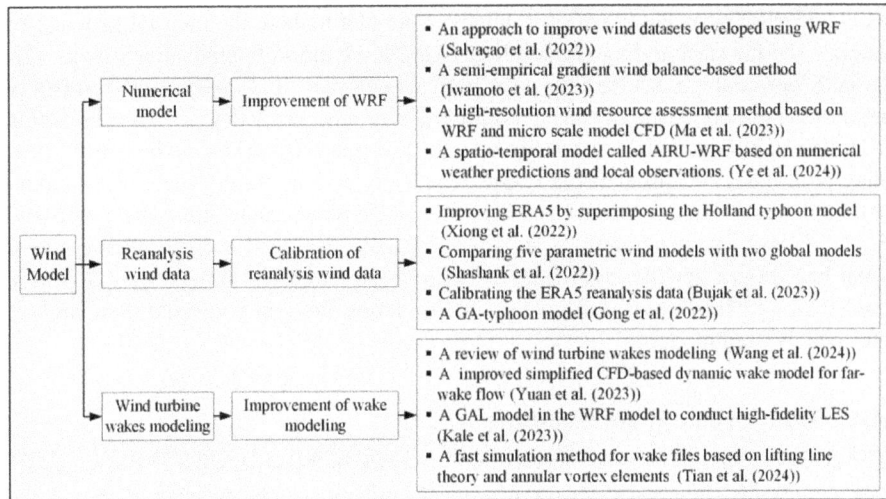

Fig. 4: Overview of advances in wind model research

3.4.1 Weather Research Forecasting Model

The numerical weather prediction (NWP) models are one of the most widely used tools for wind assessment (see also Sects. 2 and 3). Various studies have further pointed out the Weather Research Forecasting (WRF) model as an effective wind assessment tool. Salvação et al. (2022) presented an approach to improve wind datasets developed using WRF by combining its predictions with remotely sensed wind observations in enhanced wind speed analyses that leads to blended winds. Based on the WRF model, Iwamoto et al. (2023) developed a semi-empirical gradient wind balance-based method to modify both the wind and pressure fields. Ye et al. (2024) propose a spatio-temporal model called AIRU-WRF, which combines NWPs with local observations in order to make wind speed forecasts that are short-term (minutes to hours ahead), and of high resolution, both spatially (site-specific) and temporally (minute-level). Recently, the WRF model was utilised in the wind energy resources field. Ma et al. (2023) studied a high-resolution wind resource assessment method based on mesoscale atmospheric model WRF and micro scale model CFD.

3.4.2 Reanalysis of Wind Data

In the field of wind numerical simulation, the 'reanalysis data' are widely used. Examples of models in this category are provided by NCEP and ECMWF. Usually, parametric and global model wind fields are the two primary sources to acquire wind and pressure fields to estimate storm surges. The parametric models usually depict a better wind field near the cyclone centre. Unlike the parametric models, the wind fields from global models have better accuracy in areas far away from the eye of the cyclone. Shashank et al. (2022) compared five parametric wind models to assess their performance in the Bay of Bengal. The study highlights that a blended model using the generalised asymmetric Holland model and WRF provides superior accuracy in simulating both the internal structure of cyclones and the surrounding wind field, making it the most effective approach for this region. Xiong et al. (2022) found that the ERA5 reanalysis wind speeds are significantly smaller than the observations during typhoons, and improved ERA5 by superimposing the Holland typhoon model wind field onto ERA5 to create a hybrid wind field. Similarly, Bujak et al. (2023) calibrated the ERA5 reanalysis data by tuning the white-capping coefficient in the Janssen white-capping formulation when establishing the ERA5 and ALADIN forced wave models. Gong *et al.* (2022a) developed a data-driven typhoon model based on a genetic algorithm, to improve the empirical typhoon wind model. Results showed improved performance in simulating the typhoon wind compared to other widely used models.

3.4.3 Wind Turbine Wake Modelling

Wind energy technology has advanced rapidly over the past few decades. However, large-scale wind farms often generate significant wake effects, which can severely impact the performance of wind turbines (WTs). Consequently, the development of accurate WT wake models is crucial for the optimal design and control of wind farms. A number of methods are used for numerical simulations and analyses of aerodynamics and wakes: CFD, blade element momentum (BEM) theory, and the free vortex wake (FVW) method. The main software tools used for wake simulation include OpenFOAM, QBlade, OpenFAST, FLORIS, WindSE, ANSYS, WindSim, and STAR-CCM + (L. Wang *et al.*, 2024a; 2024b; 2024c; 2024d). Numerous scholars have improved various aspects of wake modelling. For instance, Qian et al. (2022) proposed a new hybrid method in which the near-wake flow is simulated using a CFD model based on the Navier–Stokes equations, with the turbine represented by actuator lines, while the far-wake flow is modelled using an improved simplified CFD-based dynamic wake model. The two models are coupled at a section downwind of the turbine, and the method is validated by the results of full CFD simulations across the entire domain. Kale et al. (2023) implemented a generalised actuator line model in the WRF model to conduct high-fidelity large-eddy simulations of turbulent wind fields in stratified atmospheric boundary layer flows, examining the effects of atmospheric stability on wake characteristics and aerodynamic responses of wind turbines. Tian et al. (2024) proposed a fast simulation method for wind turbine wake fields by combining lifting line theory with annular vortex elements to simulate bound circulation on the blades and the evolution of the wake field.

3.5 Wave-Coupled Phenomena

Interactions of the environmental phenomena discussed above also need to be considered in modelling, when relevant for the specific structure or area of operation. We focus on the interaction of waves with wind and current (also called wave-coupled phenomena). Some interesting work on these topics was presented at the 5th workshop on waves and wave-coupled processes (https://events.ecmwf.int/event/364/).

A review of existing wind–wave coupling models and parameterisations used for large-eddy simulation of the marine atmospheric boundary layer is presented by Deskos et al. (2021). This also presents a discussion about the modelling knowledge gaps and computational challenges ahead. Janssen and Bidlot (2023) analysed the problem of wind–wave interaction with emphasis on strong winds. In these events it is assumed that non-linearity is so large that the slope of the wind waves has reached a limiting steepness. Results show that in a young, steep wind sea, the background roughness length almost vanishes, giving a reduced drag. In addition, it is shown that for steep waves, the slowing down of the wind by waves is a non-linear process; hence, the growth rate of the waves by wind depends in a non-linear fashion on the wave spectrum. For strong winds, it is found that this non-linear effect gives a further reduction of the wind input (as waves are typically steep in such conditions). Therefore, in these extreme circumstances, the drag coefficient decreases with increasing wind speed. Aiming to find a method that would better capture the intricacies of wind-wave interactions, especially in swell-dominated conditions, Ning and Paskyabi (2024) integrated a wave-induced stress parameterisation method into the open-source LES code, Parallel Large-eddy simulation Model (PALM), and assessed its performance under low wind conditions with varying wave ages and directions. Results show that the swell-induced stress is the dominant force in the wave boundary layer (WBL) at low wind speed, with a magnitude that varies significantly with wave direction relative to the wind. This variation also influences the dynamic equilibrium within the WBL, modifying the wind shear, and these effects persist up to the top of the boundary layer flow. Downscaled wind-wave forecasting in enclosed basins is challenging, especially due to the necessity of accurate wind forcing. Barbariol et al. (2022) therefore present an improved wave forecasting system for the Adriatic Sea based on WW3, forced by the global IFS-ECMWF forecast. They use wind and wave observations from different sources to calibrate and validate the wave forecasts. Comparison with other forecasting systems in the area highlights the importance of the wind forcing accuracy and the wave model calibration. On a smaller scale, Cimarelli et al. (2023) studied wind-wave interaction at low Reynold's numbers, using the OpenFOAM CFD model. They aimed to obtain a fully coupled solution of the wind–wave problem, based on first principles and realistic air and water properties. By reducing the complex wind-wave interaction problem to a two-phase open channel flow driven by a constant pressure gradient (where the wind is turbulent and the water is almost quiescent), it becomes governed by a single parameter, the wind friction Reynolds number. This very simply approach provides some insight that may explain field observations, such as the large scatter of the drag coefficient data in field measurements on offshore structures.

The interaction of long-crested waves with currents in deep and finite water depth is studied by Kumar and Hayatdavoodi (2023) using CFD. Various wave conditions are considered by systematically changing the wave height and the wavelength. Several current profiles are studied as polynomial functions of water depth following the profiles and magnitudes of available ocean current data. Both following and opposing currents are considered, and in total, 26 wave-current configurations are investigated. Results of the numerical wave-current tank are compared to existing laboratory measurements, with an overall good agreement. Elobeid et al. (2024) investigated the implications of wave-current interaction on the dynamic responses of a semi-submersible platform for floating offshore wind turbines under operational and extreme conditions. Firstly, two analytical models based on Airy wave theory are developed to analyse the effects of current interaction with regular and irregular waves. Then, these models are integrated with the engineering tool OrcaFlex for the coupled aero-hydro-servo-elastic analysis. It is shown that translational motions (surge, sway and heave) are affected by wave-current interaction, with mean and maximum values decreasing under a following current and increasing under an opposing current. Rotational motions (roll, pitch, yaw) are not significantly affected. Xu *et al.* (2023a) established an inhomogeneous wave and current environment along the Qiongzhou strait close to China, through a numerical hydrodynamic model which is validated using measured data from literature. Ye and Hu (2023) developed a new analytical model to incorporate vertical velocity distribution and full energy dissipation terms in energy flux balance equations. This model is applied to interpret the data from a new flume experiment of vegetation in combined wave-current flows. New CD-Re relationships were derived by applying the new model. X. Zhang *et al.* (2024a, 2024b; 2024c; 2024d; 2024e) presented a fourth-order-accurate finite volume solver for dealing with the problem of wave-current interaction with a horizontal cylinder near free surface. It solves the filtered Navier-Stokes equations which are on a collocated grid. They used the method to simulate the interaction between solitary waves, uniform currents, and a horizontal cylinder near the free surface. The results reveal a clear correlation between the vertical forces on the cylinder and the current velocity in cases of wave-following current interactions. Halsne et al. (2024) use the WAM spectral wave model to evaluate the influence of a strong tidal current (> 3 m/s) in the Moskstraumen in Norway on the occurrence of extreme waves. The study shows that for a bimodal swell and wind sea state, the most intense interactions occur when the wind sea opposes the tidal current. The H_s, spectral steepness and extreme wave crests can significantly increase due to this tidal current. On a more fundamental level, Ellingsen and Y. Li (2017) and Y. Li and Ellingsen (2019) propose a new dispersion relation for waves that propagate in a shear current. This relation can be used to improve wave propagation modelling in numerical wave-current interaction models.

4 Data-Driven Models

The rapid expansion of artificial intelligence (AI) in recent years has significantly impacted environmental sciences. Machine learning (ML) models have become integral to virtually all areas of oceanography, meteorology, and ocean engineering, and this trend shows no signs of slowing down. AI's contributions have been profound in the development of supervised regression models for forecasting. The advent of extensive GPU-based computational infrastructures, combined with the availability of large datasets and accurate reanalysis data, has enabled deep learning models to achieve rapid improvements in accuracy.

Data-driven models are reshaping how research centres and private companies operate, increasing the demand for highly skilled professionals in statistics, software engineering, big data management, and high-performance computing. This shift requires more efficient interdisciplinary collaboration, as oceanographers and engineers now work closely with data scientists, mathematicians, statisticians, and software engineers. This reorganisation of work, combined with the growing commercial relevance of meteorology and oceanography applications, has spurred the private sector's expansion, especially among 'big tech' companies. Examples of sectors experiencing rapid growth while relying on metocean forecasts include offshore renewable energy, maritime transportation, oil and gas exploration, the fishing industry, and insurance and risk management related to extreme events.

Despite the popularization of AI and the development of new algorithms with excellent performance, a high number of research projects and scientific papers are accompanied by common methodological mistakes. Some examples include: (i) developing deep networks for problems that could be better solved with simpler models; (ii) incorrect selection of input variables and architectures, reflecting a lack of knowledge about the physical processes involved; (iii) ignoring the need to work with vector components of circular or cyclic variables, such as direction and time (hour of the day or day of the year); (iv) building regression models to predict reanalysis data and validating them against reanalysis data, instead of using archives of forecast data; (v) improperly handling the impact of noise; (vi) building forecast models of environmental variables based solely on single-point time series, neglecting advection; and (vii) not considering the problem of data imbalance when targeting extreme events.

Several interesting and successful studies in different areas of environmental sciences are presented below, divided into atmospheric and oceanic applications.

4.1 Atmospheric Modelling

The data-driven models have been developed with diverse architectures and dimensions, conceived to work aside physics-based (dynamic) models, combining the physics with statistical learning, or designed to completely replace traditional numerical models. The inclusion of AI components into numerical weather prediction (NWP) models is not new. Krasnopolsky et al. (2005, 2006) introduced a fast neural network emulation of atmospheric longwave radiation in a climate model, demonstrating 20 years ago the efficient synergetic combination of dynamic models with machine components. In a similar study, Krasnopolsky et al. (2010) argue the longwave and shortwave atmospheric radiation are the most time-consuming components of model physics, which can be significantly improved using neural networks – with impact on weather prediction. Krasnopolsky et al. (2013) built an ensemble of neural networks to emulate stochastic convection parameterization in NWP models. These are examples of hybrid models, combining dynamic models (physics) with data-driven models. In this case, the ML can be included as a component of the dynamic models or designed as a post-processing tool (e.g. Campos et al., 2019), commonly used for bias correction (Rasp and Lerch, 2018; Campos et al., 2020).

Moving to entire solutions using AI, applied to weather forecasting, there are impressive models recently developed, namely GraphCast, FourCastNet, Pangu-Weather, FuXi, and ECMWF Artificial Intelligence/Integrated Forecasting System (AIFS). These models are all discussed below. Schultz *et al.* (2021) discusses whether it is possible to completely replace the current numerical weather models and data assimilation systems with deep learning approaches. Lam et al. (2023) explain the main motivation and advantages of full AI models. The authors argue that traditional NWP models used to predict weather are large, complex, and computationally demanding and do not learn from past weather patterns.

Lam et al. (2023) presented an alternative weather forecast system, GraphCast, that harnesses machine learning and graph neural networks (GNNs) to process spatially structured historical data. Lam et al. (2023) and Conti (2024) describe GraphCast learns from four decades (1979–2017) of curated atmospheric data provided by the European Centre for Medium-Range Weather Forecast (ECMWF), finding patterns in the weather that it can then exploit to make forecasts. It predicts hundreds of weather variables, over 10 days at 0.25 degree resolution globally, in under one minute. Lam et al. (2023) show that GraphCast outperforms the most accurate operational deterministic systems on 90% of 1380 verification targets, and its forecasts support better severe event prediction, including tropical cyclones, atmospheric rivers, and extreme temperatures.

Kurth et al. (2022) highlights physics-based NWP limits accuracy due to high computational cost and strict time-to-solution limits. Pathak et al. (2022) and Kurth et al. (2022) introduce a new AI model, FourCastNet, which can predict global weather and generate medium-range forecasts five orders-of-magnitude faster than traditional NWP. FourCastNet (Fourier Forecasting Neural Network) is a global data-driven weather forecasting model, trained using ERA5 reanalysis data, which provides accurate short to medium-range global predictions at 0.25° resolution. It accurately forecasts high-resolution, fast-timescale variables such as the surface wind speed, precipitation, and atmospheric water vapor. Pathak et al. (2022) and Kurth et al. (2022) showed that FourCastNet matches the forecasting accuracy of the widely-used ECMWF Integrated Forecasting System (IFS) at short lead times for large-scale variables, while outperforming IFS for variables with complex fine-scale structure, including precipitation. A key feature is that FourCastNet produces a week-long forecast in less than 2 s, which enables rapid and inexpensive large-ensemble forecasts with thousands of ensemble-members for improving probabilistic forecasting.

Bi et al. (2022, 2023) describe Pangu-Weather, an artificial-intelligence-based method for accurate, medium-range global weather forecasting. They discuss that three-dimensional deep networks equipped with earth-specific priors (assumptions or constraints embedded in the model) are effective at dealing with complex patterns in weather data, and that a hierarchical temporal aggregation strategy reduces accumulation errors in medium-range forecasting. Pangu-Weather was built using 43 years of hourly global weather data from the 5th generation of ECMWF reanalysis (ERA5) data, used to train a few deep neural networks with about 256 million parameters in total. It has spatial resolution of 0.25° × 0.25° and, according to Bi et al. (2022), it is the first AI-based model that outperforms state-of-the-art NWP models in terms of accuracy of all factors (e.g., geopotential, specific humidity, wind speed, temperature, etc.) and in all time ranges (from one hour to one week). Most importantly, the methodology developed also works well with extreme weather forecasts and ensemble forecasts. Furthermore, when initialized with reanalysis data, the accuracy of tracking tropical cyclones is also higher than that of ECMWF high-resolution forecast (HRES) according to Bi et al. (2023).

L. Chen *et al.* (2023a; 2023b; 2023c; 2023d) recognizes the new AI models with superior performance than ECMWF-HRES; however, they emphasize the challenge remains in mitigating the accumulation of forecast errors for longer effective forecasts, such as achieving comparable performance to the ECMWF ensemble in 15-day forecasts. L. Chen *et al.* (2023a; 2023b; 2023c; 2023d) present FuXi, a cascaded ML weather forecasting system that provides 15-day global forecasts at a temporal resolution of 6 h and a spatial resolution of 0.25° × 0.25°. FuXi is developed using 39 years of the ECMWF ERA5 reanalysis dataset and the performance evaluation demonstrates that it is comparable to ECMWF ensemble mean (EM) in 15-day forecasts. It is especially relevant for longer forecast ranges, when compared to the other AI models.

Besides AI models developed by technology companies, the large forecast centres, such as the NOAA and ECMWF, have also been intensively developing AI technologies. The new ECMWF Artificial Intelligence Forecasting System (AIFS) is introduced by Lang et al. (2024). AIFS is based on a GNN encoder and decoder, and a sliding window transformer processor, and is trained on the ERA5 reanalysis and ECMWF's operational NWP analyses. It has a flexible and modular design and supports several levels of parallelism to enable training on high-resolution input data. The current version has 1° x 1° of spatial resolution which is now under improvement. AIFS is run four times daily alongside ECMWF's physics-based NWP model and forecasts are available to the public under ECMWF's open data policy.

One derived application of deep learning weather forecast models that has received significant attention is the prediction of hurricanes and typhoons, due to their obvious life-threatening and economic impacts. Rüttgers et al. (2022) and Gao et al. (2018) provide examples of predicting hurricane track and intensity. Gao et al. (2018) employed a long short-term memory (LSTM) neural network, trained with typhoon observations from 1949 to 2011, to predict typhoon tracks. The results show that the new algorithm provides desirable 6–24-h nowcasting of typhoon tracks with improved precision. The references cited above regarding AI weather prediction models also highlight strong performance in predicting extreme events and hurricane tracks.

In summary, GraphCast, FourCastNet, Pangu-Weather, FuXi, and ECMWF-AIFS have demonstrated that AI-based methods can surpass conventional NWP methods, while also revealing new directions for improving deep learning weather forecast systems in the future. Schultz et al. (2021) believe that it is not inconceivable that numerical weather models may one day become obsolete, but several fundamental breakthroughs are needed before this goal comes into reach. It is important to note that these AI models begin their integration with analyses derived from physics-based numerical models. Thus, strictly speaking, they are not entirely machine learning frameworks, as their initial conditions are based on dynamical models with complex data assimilation.

As we are in the early stages of these new technologies, it is important to validate and compare operational forecasts from different independent groups worldwide, free from any commercial bias. Ren et al. (2021) evaluated and compared several deep learning-based weather prediction (DLWP) systems, analysing their advantages and disadvantages by comparing them with conventional NWP methods. Charlton-Perez et al. (2024) investigated how well current machine learning models can simulate high-impact weather events by comparing short- to medium-range forecasts of Storm Ciarán. They demonstrated that the four machine learning models considered (FourCastNet, Pangu-Weather, GraphCast, and FourCastNet-v2) produce forecasts that accurately capture the synoptic-scale structure of the cyclone. However, Charlton-Perez et al. (2024) argued that the ability of these models to resolve the finer details crucial for issuing weather warnings is mixed, with all the machine learning models underestimating the peak wind amplitudes associated with the storm. Only some of the models resolve the warm-core seclusion, and none capture the sharp bent-back warm frontal gradient. Rasp et al. (2020) and Garg et al. (2022) presented benchmark datasets for data-driven weather forecasting, covering both deterministic and mid-range probabilistic forecasts. They proposed simple and clear evaluation metrics that enable direct comparison between different methods.

4.1.1 Subseasonal Forecasts and Climate Prediction

Subseasonal forecasts refer to weather predictions made for periods ranging from 2 weeks to 2 months (typically 2 to 6 weeks ahead). They are crucial for planning operations that require information about weather conditions beyond the short-term (days) but not as long as seasonal predictions (months). Skillful subseasonal forecasts are crucial for various sectors of society but pose a significant scientific challenge (Chen et al., 2024a; 2024b). Zhu et al. (2018) discuss the need for post-processing bias correction of subseasonal operational products, as bias is inherent in physics-based forecast models at this time scale. Among several strategies adopted, a simple and effective one, using a lagged ensemble, is presented by Vitart and Takaya (2021). Scheuerer et al. (2020) developed a statistical post-processing framework that uses an artificial neural network (ANN) to establish relationships between NWP ensemble forecasts and gridded observed 7-day precipitation accumulations. Their goal is to model the increase or decrease in the probabilities for different precipitation categories relative to their climatological frequencies. Guan et al. (2015) highlighted the need for a long training period (>5 years) to avoid a large impact on bias correction from an extreme year case and to maintain a broader diversity of weather scenarios.

L. Chen et al. (2024a) conducted a survey to consolidate the current understanding of ML applications in weather and climate prediction through an extensive review of more than 20 methods highlighted in the existing literature. While ML demonstrates significant capabilities in short-term weather prediction, its application in medium-to-long-term climate forecasting remains limited, constrained by factors such as complex climate variables and data limitations. L. Chen et al. (2024b) discussed that, recently, machine learning-based weather forecasting models have outperformed the most successful numerical weather predictions generated by the ECMWF at short- to mid-term range, but they have not yet surpassed conventional models at subseasonal timescales. They introduced a new AI model, FuXi, previously described in this chapter, adapted for the Subseasonal-to-Seasonal timescale (FuXi-S2S). The model was trained on 72 years of daily statistics from the ERA5 reanalysis data, and their results show that it outperforms ECMWF's state-of-the-art model.

For climate prediction, Jeon and Kim (2024) introduced technologies that utilize artificial intelligence to predict long-term changes and short- to medium-term extreme weather events. These include the prediction of the El Niño-Southern Oscillation (ENSO) using convolutional neural networks, sea level rise using LSTM networks and support vector machines (SVMs), and the prediction of typhoons using various AI techniques. Reichstein et al. (2019) highlighted the importance of spatio-temporal features for gaining a better understanding of earth system science problems, thereby improving the predictive ability of seasonal forecasting and the modelling of long-range spatial connections across multiple timescales. Ham et al. (2019) employed a deep-learning approach to produce skilful ENSO forecasts for lead times of up to one and a half years. They used transfer learning to train a convolutional neural network (CNN) first on historical simulations and then on reanalysis data from 1871 to 1973. The authors showed that, during the validation period from 1984 to 2017, the all-season correlation skill of the Nino3.4 index from the CNN model was much higher than that of current state-of-the-art dynamical forecast systems.

To address the issue of coarse climate models suffering from inherent bias due to ignored sub-grid scales, Sorensen et al. (2024) proposed a framework to non-intrusively debias coarse-resolution climate predictions using neural networks. This study is particularly interesting because they developed a learning method that enables the correction of dynamics and the quantification of extreme events with return periods longer than those present in the training data.

4.1.2 Wind Speed and Wind Power Forecasting

In addition to the large deep learning (DL) weather models described so far, there are many other specific developments and applications, resulting in numerous publications. One prominent topic is the development of machine learning for wind speed and wind power forecasting. The use of renewable wind energy to generate electricity has attracted increasing attention; however, the intermittent nature of wind speed poses challenges to the safety and stability of electric power grids when wind power is integrated on a large scale. Y. Wang *et al.* (2021a; 2021b) reviewed and compared AI models in terms of computational cost and accuracy, and summarized challenges and future research directions in wind energy forecasting.

Campos et al. (2022) proposed a hybrid model that combines a traditional NWP model with neural networks (NN) for mid- and long-range ensemble forecasting. The main NN output was the residual of wind speed, defined as the difference between the arithmetic ensemble mean, derived from the ECMWF predictions, and the observations. This approach corrected the original ECMWF bias from -0.3 to -1.4 m/s to values between -0.1 and 0.1 m/s, while also reducing the RMS by 10 to 30%. AlShafeey and Csaki (2024) also addressed the formulation and validation of forecasts focusing on long-term predictive accuracy.

Jamii et al. (2022) proposed a machine learning-based paradigm to predict wind power (WP) generation and load demand, using wind speed, temperature, and atmospheric pressure as input variables. They compared four models: LASSO, decision tree (DT), relevance vector machines (RVM), and kernel ridge regression (KRR). Moreno et al. (2024) considered two decomposition techniques for wind speed forecasting: singular spectral analysis in the time domain and variational mode decomposition in the frequency domain, which were combined in a hybrid structure incorporating multi-stage decomposition and time series prediction. Sulaiman and Mustaffa (2024) presented an innovative approach that combines deep learning (DL) with Teaching-Learning-Based Optimization (TLBO) to accurately predict wind power output. Mollick et al. (2024) identified the most accurate machine learning algorithm for short-term wind speed forecasting, evaluating 14 different regression-based machine learning models. Their results highlight the exceptional predictive performance of a boosting-based ensemble method known as categorical boosting.

For a coastal wind farm, G. Gu et al. (2024a) used a Random Forest (RF) model to predict wind power generation. This method is simple, efficient, and delivers excellent practical results, as well as the Extreme Gradient Boosting (XGBoost) model. The authors concluded that the RF model outperforms simple linear statistical models. They also advocate for the use of the traditional numerical atmospheric model WRF (Weather Research and Forecasting), which can effectively capture and forecast peak generating power, thereby meeting daily management requirements. Benti et al. (2023) presented a review of current advances and prospects in forecasting renewable energy generation using machine learning and deep learning techniques. The authors highlight challenges and future research directions in the field, such as addressing uncertainty and variability in renewable energy generation, data availability, and model interpretability. Alves et al. (2023) reviewed 23 articles from 1983 to 2023 on machine learning for wind speed and direction nowcasting. Their results underscore the novel effectiveness of machine learning, demonstrating that deep learning models surpassed traditional methods in improving the accuracy of wind speed and direction forecasts. Lagomarsino-Oneto et al. (2023) also conducted several comparisons and showed that, when accurately designed, shallow algorithms can be competitive with deep architectures.

Finally, an interesting study by Dorsay et al. (2023) developed proxy observations of surface wind using a globally distributed network of wave buoys. They investigated three potential methods for estimating wind speed from wave spectra obtained from small, low-cost spherical wave buoys: physics-motivated parameterizations, inverse modelling (estimating wind speed from spectral energy balance), and a data-driven approach using artificial neural networks.

4.1.3 Wind Profile

Accurate estimation of the wind profile, especially in the lowest few hundred meters of the atmosphere, is of great significance for weather forecasting - another area that has benefited from AI research. This also affects wind forecasts at various altitudes, which are essential for wind turbine operations. B. Liu et al. (2024) proposed a novel method called the PLM-RF method, which combines the power-law method (PLM) with the random forest (RF) algorithm to extend wind profiles beyond the surface layer. They applied the new method to three atmospheric radiation measurement sites for independent validation, and the wind profiles estimated by the PLM-RF model were found to be consistent with Doppler wind lidar observations, confirming that the PLM-RF model has good applicability. Beu and Landulfo (2024) used the LSTM recurrent neural network and Lidar datasets over complex terrain to estimate the wind profile up to 230 m. The results showed that the LSTM outperformed the power-law method as the distance from the surface increased. García-Gutiérrez et al. (2023) presented a novel methodology that estimates the wind profile within the atmospheric boundary layer by using a neural network along with predictions from a mesoscale model in conjunction with a single near-surface measurement. The authors explain that a major advantage of this solution, compared to others available in the literature, is that it requires only near-surface measurements for prediction once the neural network has been trained.

4.1.4 Solar Energy Forecasting

Solar energy is now more popular and important economically than ever been. Accurate forecasting of energy generation from (floating) solar power plants is crucial in terms of economics due to this uncertainty in the output in different seasons and the change in meteorological conditions. Pattanaik et al. (2024) analysed the ability of machine learning algorithms to accurately anticipate solar power for the upcoming hour and hourly days in advance, by investigating the following machine learning models: Naïve Bayes Algorithm, Multilayer Perceptron Theorem (MLP), and LSTM. Abumohsen et al. (2024) selected the LSTM, Gated Recurrent Units (GRU), Recurrent Neural Networks (RNN), Random Forest (RF), Support Vector Regression (SVR), Bi-directional LSTM (Bi-LSTM), and Convolutional Neural Network (CNN) for the forecast of solar power generation. The authors concluded that, the hybrid model combining CNN-LSTM-RF demonstrated superior accuracy with R2 of 92%, a RMSE of 0.07 kW, and a Mean Absolute Error (MAE) of 0.05 kW.

4.2 Oceanic Modelling

Like the applications in the atmosphere, data-driven methods are also used in various oceanic contexts, including ocean waves (including rogue waves), currents, downscaling coastal models, and storm surges. Forecasting is the component that benefits most from AI. Even tsunami prediction has been developed using neural networks (Y. Wang et al., 2023) with a quality index of approximately 90%.

4.2.1 Wave Modelling and Forecast

While physics-based numerical wave models such as WW3 represent the state of the art for hindcasting and forecasting, they require intensive computational resources. Numerous studies have explored wave forecasting using AI, mostly based on single-point time-series predictions of H_s. Hu et al. (2021) applied novel machine learning (ML) methods based on XGBoost and a LSTM recurrent neural network to predict wave height and period under near-idealized wave growth conditions on Lake Erie. For short-range forecasting, Zhan, Li and Zhu (2023) introduced a novel, lightweight machine learning forecasting architecture that leverages frequency domain information. Their methodology demonstrated superior performance compared to other models, such as the autoregressive moving average (ARMA), LSTM, and XGBoost models. Ali and Prasad (2019) proposed a slightly more complex model using an extreme learning machine (ELM) coupled with an improved complete ensemble empirical mode decomposition method with adaptive noise (ICEEMDAN). This method incorporates the historical lagged series of H_s as the model's predictor to forecast future H_s.

Surrogate models have the potential to completely replace numerical models with significantly lower computational costs. Minuzzi and Farina (2023) presented a new deep learning training framework for forecasting H_s by applying an LSTM model trained with reanalysis data and observations. They considered four different lead times: 6, 12, 18, and 24 h. Their results showed that the LSTM forecast could serve as a surrogate for the computationally expensive physical models and as an alternative to reanalysis data. O'Donncha et al. (2019), building on the approach of James et al. (2018), aimed to improve the accuracy of computationally lightweight surrogate models by updating forecasts based on their historical accuracy relative to sparse observation data. The authors demonstrated that machine learning models could replicate SWAN-simulated wave conditions in less than 1/1000th of the computational time.

Combining dynamical models and ML models has proven to provide excellent results in wave forecasting. Costa et al. (2023) and Pinto *et al.* (2023), in North Atlantic and South Atlantic respectively, successfully developed post-processing neural networks to predict the residual component (model – observation) of H_s and improve the accuracy of wave estimates, following the methodology proposed by Campos et al. (2019, 2020). The hybrid model proposed by Bieman et al. (2023) resulted in a significant average error decrease compared to dynamic models, with approximately 30% smaller error in spectral wave height and period. The problem of choosing between multiple operational forecasts was discussed by Campos *et al.* (2021a), who developed a random forest model for selecting the best wave prediction - a very simple and lightweight approach.

The benefit of ensemble and probabilistic forecasts, compared to deterministic forecasts, is also evident in wave prediction. These should also be accompanied with appropriate probabilistic error metrics (Astfalck et al., 2023). Campos et al. (2020) developed a new method to post-process an ensemble forecast, providing a non-linear average of the ensemble members using multilayer perceptron neural networks. The model was trained with more than 2 million observations from altimeters, significantly reducing both systematic and scatter errors. Figure 5 shows the benefit of the neural network post-processing applied to NOAA's global wave ensemble forecast. In the plot, the error for the NN forecast at day 10 is the same as for day 8 associated with ensemble mean and day 6 for the deterministic run, illustrating the importance of ensemble forecasts combined with machine learning algorithms.

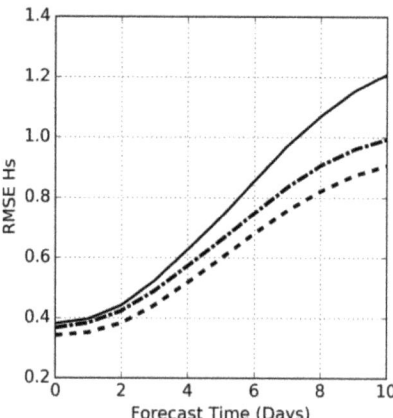

Fig. 5. Root Mean Square Error (RMSE) of H_s (in meters) as a function of forecast lead time (days) up to 10 days. This compares the results from the deterministic forecast (solid line), the ensemble forecast (ensemble mean, dash-dotted line), and the neural network post-processing (dashed line).

Campos et al. (2019) and Minuzzi and Farina (2024) demonstrated that using an ensemble of neural networks provides better results than single ML models. The latter presented an interesting methodology that leverages multiple different artificial neural network architectures, namely, MLP, RNN, LSTM, CNN, and a hybrid CNN-LSTM. In summary, a hybrid approach using an ensemble of machine learning models combined with an ensemble of dynamic models ensures optimal wave forecast accuracy. A good explanation of how to generate an ensemble of neural networks can be found in Krasnopolsky and Lin (2012).

Applied to extreme events, S. Chen (2019) developed ML methods to perform probabilistic forecasting of coastal wave height. These probabilistic forecasts include both a deterministic forecast and the probability distribution of the forecast error. A support vector machine was used to develop a real-time forecasting model for generating deterministic wave height predictions. The forecast errors from the deterministic model were then used to create a database for generating probabilistic forecasts using a modified fuzzy inference model. Gong *et al.* (2022b) developed a hybrid multilayer perceptron neural network and a hybrid genetic expression programming model with a switch layer to forecast typhoon waves. Also focused on probabilistic forecast of extreme waves, Campos et al. (2024) developed and compared machine learning methods applied in the probabilistic domain, demonstrating the capability of improving forecasts of hurricane-force winds and waves. Hybrid modelling, where an ML algorithm predicts the residual component, has been shown to improve the forecast of extreme events, as previously described.

Most studies presented so far have emphasised the accuracy of approaches based on in situ wave measurements, but such methods are site-specific and lack the spatial and temporal fidelity required for wave forecasts. Recent studies have worked towards overcoming these limitations. J. Wang *et al.* (2023a; 2023b; 2023c; 2023d; 2023e) developed a machine learning model to emulate wave fields using a spatio-temporal approach trained with ECMWF data. W. Zhang *et al.* (2024a, 2024b; 2024c; 2024d; 2024e) proposed a spatiotemporal deep-learning method to correct gridded H_s forecasts from ECMWF. This method is based on a trajectory gated recurrent unit deep neural network and performs real-time rolling corrections for the 0–240 h H_s forecasts from ECMWF-IFS. Patanè et al. (2024) introduced a deep learning architecture to create a digital twin for monitoring buoys, aimed at estimating H_s using spatial and temporal information about the wind field in the area of interest. Their results demonstrate that a multi-block hybrid deep neural network, consisting of convolutional layers for spatial feature extraction and short-term memory layers for modelling the involved dynamics, outperforms other empirical, numerical, machine learning, and deep learning methods in the literature. Kordatos et al. (2024) developed a machine learning-statistics framework combining radial basis function neural networks (RBFNNs) with 2D geospatial interpolation to predict H_s in nested fine-resolution domains, avoiding intensive computations. They successfully predicted H_s at high resolution using inputs from low-resolution WAM (the physics-based numerical wave model) simulations.

Jiang et al. (2024) provide an interesting discussion on the application of ML to wave forecasts. They argue that the extent to which complex AI models can outperform traditional methods has largely been overlooked. They compared five different models:

AutoRegressive (AR), XGBoost, ANN, LSTM, and WaveNet. The results suggest that the performance differences among these models are negligible, indicating that all the AI models have only 'learned' the linear auto-regression from the data. It indicates, as do previously cited papers, that simple models and shallow networks can, in many cases, perform as well as deep networks and complex systems. Additionally, the authors warned that many recent studies used signal decomposition methods for such time series prediction, and most of them decomposed the test sets, which is incorrect.

In addition to several different methodologies around ML applied to wave forecasts described so far, statistical models can also provide an automatic calibration and uncertainty quantification in waves dynamical downscaling (Alonso and Solari, 2021b).

A significant effect of ocean waves in coastal areas, wave overtopping, impacts both the safety of individuals and the integrity of port facilities. Coastal structures are often designed with a maximum allowable wave overtopping discharge, making accurate prediction crucial. Carro et al. (2024) developed a predictive tool based on artificial intelligence to forecast wave overtopping over a breakwater, aiming to balance safety and port efficiency. Bieman et al. (2021) introduced a new model for predicting mean wave overtopping discharge using XGBoost. Their results indicated that the XGBoost model generally outperforms other methods for test datasets with normally incident waves.

For phase-resolved models, Y. Liu *et al.* (2024) developed a machine learning architecture to model wave breaking, enabling more detailed breaking physics to be captured. They showed that it can be trained on focused wave groups but can also capture breaking in random waves and modulated plane waves – being a potential contribution to phase-resolved envelope-based wave models such as the non-linear Schrödinger.

4.2.2 Rogue Waves

Rogue waves present a considerable risk to marine and offshore structures, making the ability to identify and predict them extremely important. We focus here on data-driven methods for rogue waves; dedicated Sect. 8.1 emphasises on their physics. Breunung and Balachandran (2024) state that while the occurrence of oceanic rogue waves at sea is generally acknowledged, reliable rogue wave forecasts are still unavailable. They developed neural networks trained with an extensive dataset designed to distinguish between waves that precede an extreme wave and those that do not. With this approach, three out of four rogue waves could be correctly predicted 1 min ahead of time. When the advance warning time is extended to 5 min, the ratio of accurate predictions drops to seven out of ten rogue waves. Bitner-Gregersen et al. (2024) worked on improving warning criteria for extreme and rogue waves. They investigated the use of integrated wave parameters, the coupling of a phase-averaged wave spectral model with a phase-resolving wave model, and the application of machine learning methodology.

Häfner et al. (2023) investigated symbolic models for oceanic rogue waves using data-driven approaches including causal analysis, deep learning, parsimony-guided model selection, and symbolic regression. Regarding important variables and future selection for such machine learning models, Cicon et al. (2023) made a significant contribution. Unlike other studies in the field, they argued that the Benjamin Feir Index (BFI) has limited predictive power in the real ocean and that rogue waves are largely generated by bandwidth-controlled linear superposition (for a detailed discussion of these two generation phenomena, see Sect. 8.1). They found that the bandwidth parameter crest-trough correlation shows the highest univariate correlation with rogue wave probability, which is crucial information for developing future forecasting systems.

Physics-informed neural networks (PINNs) have emerged in several different areas and have shown to be a powerful method for rogue waves. Peng et al. (2022) developed multi-layer PINNs to explore data-driven rogue periodic waves, breather waves, soliton waves, and periodic wave solutions of the well-known Chen–Lee–Liu equation. This study marked the first time that a data-driven rogue periodic wave was learned to solve the partial differential equation. The numerical results indicate that PINNs effectively generate solutions for the rogue periodic wave, breather wave, soliton wave, and periodic wave within the Chen–Lee–Liu equation. Sun et al. (2024) proposed data-driven solutions for rogue waves in the focusing and variable coefficient non-linear Schrödinger equations via deep learning. Their numerical experiments demonstrated that PINNs capture the non-linear features of rogue wave solutions very well, a finding also investigated by R. Wang et al. (2021a; 2021b). Zhong et al. (2022) examined several critical factors, such as neural network depth and the number of training points, affecting the performance of the PINNs algorithm applied to rogue waves. The results from Zhong et al. (2022) provide valuable insights into the application of neural networks for understanding rogue wave structures in non-linear wave systems.

4.2.3 Storm Surge

Storm surge, the abnormal rise in sea level caused by intense atmospheric disturbances, poses significant risks to coastal areas and has become a critical topic in forecasting. Qin et al. (2024) proposed a multi-input and multi-output (MIMO) neural network to forecast storm surge time series along the southeast coast of China. The authors noted that, despite a slight underestimation of peak values and some temporal shifts in certain typhoon cases, the results demonstrate the accuracy of ANN in short-term forecasting for mild to moderate storm surges. Lockwood et al. (2022) examined ANN models to predict storm surge levels based on hurricane characteristics along the U.S. Gulf and East Coasts. Their findings indicated that ANN models can accurately predict storm surges, achieving a RMSE below 0.2 m and correlation coefficients above 0.85. They also used ANN models to assess the sensitivity of storm surge levels to variations in hurricane characteristics and local geophysical features. In another similar study, W. Xie et al. (2023a; 2023b) applied convolutional neural networks to forecast storm surges, incorporating two-dimensional wind field information and merging it with local water level features to produce more time-efficient and intelligent forecasts.

4.2.4 Ocean Currents

Despite the significant efforts by the ocean modelling community, predicting ocean currents remains a challenge to this day. Forecasts often quickly diverge from observed conditions throughout the water column, becoming unreliable. Berlinghieri et al. (2023) employed a machine-learning technique known as a Gaussian process to estimate currents. This method can make predictions even when data are sparse, providing more accurate forecasts of ocean currents, which can be beneficial for search and rescue operations and monitoring plastic and oil spills. Muhamed *et al.* (2021) proposed a new approach to ocean current prediction based on deep learning, evaluated on the energetic currents of the Gulf of Mexico's Loop Current (LC) at multiple spatial and temporal scales. They used a LSTM neural network to predict the evolution of the velocity field in each plane along all three directions. Their results indicated that the useful forecast period extended beyond four days, with a RMSE of less than 0.05 cm/s and a correlation coefficient of 0.6.

Zeng et al. (2015) developed artificial neural networks to forecast sea surface height (SSH) in the Gulf of Mexico (GoM) to predict Loop Current variations and its eddy shedding process. The methodology involved applying empirical orthogonal function (EOF) analysis to decompose long-term satellite-observed SSH into spatial patterns (EOFs) and time-dependent principal components (PCs). A non-linear autoregressive network was then developed to predict the major PCs of the GoM SSH in the future. Sinha and Abernathe (2021) explored machine learning algorithms as an alternative approach to infer surface currents from satellite-observable quantities. They trained ML models with SSH, sea surface temperature (SST), and wind stress data derived from available primitive equation ocean general circulation model (GCM) simulation outputs as inputs to predict surface currents (u, v). Sinha and Abernathe (2021) described various training strategies using convolutional filters (2D/3D) to understand the effect of each input feature on the neural network's ability to accurately represent surface flow. The study's model sensitivity analysis revealed that, in addition to SSH, some form of geographic information is essential for making accurate predictions of surface currents with deep learning models.

Immas et al. (2021) proposed two predictive tools using deep learning techniques, LSTM neural network and a Transformer, to perform real-time in-situ predictions of ocean currents at any location. Comparisons with Harmonic Method predictions at various locations in the territorial sea of the United States show that both models provide state-of-the-art accuracy without having been trained with data from these sites, making them suitable for practical applications in different locations. Choi et al. (2022) developed a novel method to enhance the accuracy of a real-time ocean forecasting system. This method consists of a real-time restoration system for satellite ocean temperature using a deep generative inpainting network (GIN) and the assimilation of satellite data with the initial fields of the numerical ocean model. Choi et al. (2022) concluded that the proposed approach can provide more accurate forecasts with efficient operation times.

Finally, Pourzangbar et al. (2023) provide an extensive review of over 200 journal papers focusing on the use of machine learning algorithms for promoting sustainable management of marine and coastal environments. Areas of application include data collection and analysis, pollutant and sediment transport, image processing and deep learning, and the identification of potential regions for aquaculture and wave energy activities.

5 Laboratory Measurements and Observations

This chapter focuses on the modelling of environmental conditions in laboratory facilities. We start with a review of some new facilities that were opened during the mandate period, after which we describe advances in experimental modelling of waves, ice, current and wind. Modelling of these environmental conditions is often assessed in conjunction with an application to a marine structure, as details of the experimental goal in many cases determine the important aspects of the environmental modelling. Some of the discussed publications therefore also include the hydrodynamic (or aerodynamic) responses of the structures to the modelled waves, wind etc., but we only selected publications with an emphasis on the details, generation mechanisms or underlying principles of the environmental modelling. For publications focussing on the hydrodynamic response of the tested structures, we refer to the report of Committee I.2 Loads.

5.1 New Environmental Laboratory Facilities

New laboratory facilities for experiments in waves, ice, current and/or wind were opened or started construction in several institutes within the mandate period. Not all of them are linked to publications; we list the ones that are. In Belgium, a new Coastal and Ocean Basin was constructed and opened in 2023, with dimensions 30×30 m, a variable water depth up to 1.4 m and a central pit with a depth of 4.5 m. In this basin, waves and currents can be generated in any relative direction (ITTC, 2023, Streicher et al., 2024). In Singapore, a new Ocean Basin was opened in 2022, with dimensions 60×48 m, a variable water depth up to 12 m and a central 10 m diameter pit with a depth of 50 m. Waves and current can be generated in arbitrary directions (ITTC, 2022). Bayle *et al.* (2024) describe a new Delta Transport Process Laboratory, which is an experimental facility for surface and internal wave-induced currents under rotation.

5.2 Waves

When waves are generated in a basin, there are two types of uncertainty to consider: aleatoric and epistemic uncertainties. Statistical uncertainties are related to the natural variability of the measured quantities, in this case the stochasticity of the waves. This is discussed for laboratory waves in Sect. 5.2.1 (and more general in Sects. 7 and 8). Epistemic uncertainties can for instance be related to measurement accuracy, design of the set-up and basin effects. These uncertainties have been receiving increased attention for laboratory wave modelling in recent years. They are discussed in Sects. 5.2.2 to 5.2.4. Next, Sect. 5.2.5 discusses advances in deterministic generation of wave events, Sect. 5.2.6 treats wave breaking experiments, and Sect. 5.2.7 considers shallow-water wave generation.

5.2.1 Statistical Variability

These will always be present when you do experiments, as waves and wave-induced responses are stochastic. When you are interested in mean values or standard deviations of the wave or response properties, it means that a longer test duration decreases the statistical uncertainty. When you are interested in extremes, a longer test duration both increases probability of occurrence of the extremes themselves and decreases uncertainty. These effects are discussed and quantified experimentally in van Essen, Scharnke and Seyffert (2023) and van Essen et al. (2024) for waves around a sailing ship, and in Bunnik and Scharnke (2022) and Scharnke et al. (2023) for waves impacting on a stationary deck box.

5.2.2 Basin Effects

Then we move on to the contributions to epistemic uncertainty in basin wave modelling. One important issue with the modelling of waves in a wave basin is the difference between waves generated by wind (as in the ocean), and waves generated by a wavemaker (as in a basin). Obviously, the aim is to be as close as possible to the ocean waves, to perform realistic tests with marine structures. However, a wave generator is only able to input wave energy at the edge of the basin, whereas wind can input energy continuously. Wind influences the onset of wave breaking (Buckley and Veron, 2019). This, combined with non-linear propagation effects, means that the statistics of the waves in a basin vary as a function of distance from the wavemaker, as described by e.g., Canard et al. (2022). A way to 'move' the desired statistics to a given location in a basin is proposed in Canard et al. (2024). It also means that the breaking characteristics of the waves in a basin are different than at sea, which influences the steepest waves that can be made as well as the shape of the spectral tail (see Scharnke et al., 2022). Related to this, waves generated in a basin generally have a 'cut-off frequency' that depends on the wavemaker geometry and wave direction. This means that no waves can be generated above this frequency. Tang *et al.* (2022b) use fully non-linear potential flow simulations to show that this (unavoidable) practice can in some cases generate a wave field that is 'out of equilibrium', where the non-linear physics produce significantly more extreme or rogue waves than are observed in the case where the full spectral tail was included in the initial conditions.

Even when this is the intention, waves generated in a basin are generally not perfectly long-crested. Unwanted basin effects such as oblique spurious waves can be caused by the difference in shape of the wave generator flaps and the orbital wave motions, by shoaling on side beaches or reflections (see e.g., Laflèche et al., 2023). When oblique long-crested waves are made, unwanted wave effects may be generated in the corner of two wave maker segments. Such unwanted effects can be mitigated using wave generators with multiple flaps or higher-order correction of the wavemaker flap motions (see e.g., Laflèche et al., 2024). Windt et al. (2021) assess the ability of second-order Stokes' wave theory to describe the kinematics of regular waves in a short wave flume, by measuring wave kinematics with Particle Image Velocimetry (PIV), and comparing them to Stokes theory. The results show error values of the order of 10–20%, depending on the amount of non-linearity in the waves and the reflection coefficient of the wave flume. This indicates the potential inaccuracies in any wave tank if second-order wave theory is used to derive the kinematics from the free surface elevation, without having detailed knowledge of the reflection characteristics and other basin effects. The wavemaker shape has a direct impact on the wave heights and periods that can be generated, but also on the quality of the waves obtained in the basin. Fouques et al. (2022) relate the quality of the wave field in a basin to the wavemaker geometry, where quality is defined as the predictability of the wave field and the minimization of spurious wave energy. They also provide some recommendations on wavemaker design, to minimize unwanted wave effects. The effect of reflections and excitation of basin modes on the wave field in a basin can be reduced using 'active reflection compensation' (ARC). Ellwood et al. (2021) experimentally investigate the effect of second-order wave generation and ARC on waves around coastal structures.

5.2.3 Scale Effects

Scale effects are another contributing factor to epistemic uncertainties in waves and wave-induced responses. We know that breaking waves (and waves impacting a marine structure) may entrap air. As it is impossible to scale both air and water correctly in a model test, this introduces scale effects. This is studied by Scharnke and Helder (2023). This publication shows that the variability in the shape of breaking wave crests is larger at scale 50 than at scale 25 that the crest surface is much smoother at scale 50. There were also differences in the breaking process between waves in atmospheric and depressurised air conditions. Such differences are most likely related to changes in surface instability at different scales and different air pressures (see also e.g., van Meerkerk (2021) and Remmerswaal (2023)). The shape of the wave crest has an important effect on the resulting impact load on marine structures. In the study of Scharnke and Helder (2023), it is concluded that depressurisation of the facility for wave impact experiments should only be considered if loading due to entrapped air dominates the global loading (as opposed to direct water impact loading). It is also concluded for the tested case (a deck box in steep impacting waves) that the measured loads at scale 50 are conservative compared to those at scale 25. A larger model is better, but this is obviously often limited by other considerations such as the maximum wave generated by the wavemaker or the avoidance of blockage effects by side walls.

5.2.4 Instrumentation and Measurement Accuracy

Yet another contribution to epistemic uncertainties is related to the way the waves are measured in a basin. Various experimental studies in ocean engineering and related fields require wave measurements, and measurement of the free surface shape and the wave kinematic data is of great interest in many engineering/research applications. The used instrumentation influences the accuracy and resolution of the results. Gomit, Chatelier and David (2022) present an overview of the state-of-the-art of non-intrusive measurement techniques for free surface flow and describe the principles and implementations of the different categories of methods. They conclude amongst other things that data assimilation techniques that combine experimental data and computational methods are promising to obtain accurate free surface measurements. A comparison of different types of equipment to measure the surface of complex three-dimensional free-surface flows including air entrainment (such as breaking waves or sloshing waves) is also provided by Rak et al. (2023). The review includes manometers, wave probes and point gauges with different working principles, electromagnetic sensors, ultrasonic sensors, laser ranging and triangulation and high-speed imaging. This includes point- and area-measurements, as well as invasive and non-invasive techniques. It was concluded that accurate spatial and temporal measurements in such conditions are only possible using high-resolution non-contact methods, such as LIDAR-based laser scanning, reconstruction from cameras with overlapping fields of view or laser triangulation. The accuracy of the free-surface and air cavity reconstruction of aerated sloshing fluids using two different tomographic reconstruction techniques (combined with optical measurements) is discussed by Tödter et al. (2023). It was concluded that sloshing-induced ventilated gas pockets could be captured, which shows that the considered measurement and reconstruction techniques are suitable for the quantification of aerated multiphase flows. Tukker et al. (2023) provide a measurement uncertainty model for ultrasonic wave measurement sensors, which can be used to estimate the uncertainties of waves measured around marine structures during experiments. G. Zhao *et al.* (2023a; 2023b) compares the measurement of waves generated by a passing ship using optical stereo-imaging with the measurement using conventional resistance-type wave probes. They show that the accuracy of these systems in measuring wave height can be similar, while the optical system provides an area- rather than a point-measurement. D. Li *et al.* (2022a; 2022b) propose a stereo-imaging method for the spatial-temporal measurement of waves in the laboratory, based on binocular stereo vision and digital image processing. They generated a series of regular and focused waves in a wave flume to examine the feasibility and reliability of the method by comparison with the measurements by wave probes. Combining wave height measurements from instruments based on different principles might lead to less uncertain results. A similar study was done by Le Page et al. (2024), who use a novel approach with short waves and an adapted lighting system to 'roughen' the water surface and make it suitable for stereo-imaging. Sener et al. (2023) use this idea and compare measurements taken with capacitance-type and ultrasonic-type wave height gauges for the same wave models, and analyse the uncertainty in wave modelling. Their findings highlight that identifying differences in measurement principles is significant for wave modelling.

Not only the location of the free surface is interesting to measure in experiments with marine structures; the need to validate CFD calculations has created an increasing demand to (also) measure the wave/fluid kinematics. PIV is the established non-invasive optical technique to provide accurate velocity estimates within a fluid. Maceas et al. (2021) describes a method aimed both at speeding up the PIV computations, and minimizing the divergences (noise) in the output velocity fields. When a continuous laser is used to illuminate the flow, light deficiency and particle blurring may increase the loss of correlation between consecutive PIV snapshots. Lopez-Gavilan and Barrero-Gil (2023) developed mathematical expressions for the added loss of correlation, which allows to estimate a relationship between the main variables of the PIV set up and the achievable spatial resolution.

5.2.5 Deterministic Generation of Wave Events

Historically, experimental waves have mainly been generated in a 'spectral' way, meaning that the wave spectrum was calibrated. In the last decades, this has shifted more and more towards deterministic wave generation, meaning that the exact wave sequences at a specific location in the basin are calibrated. The incentives for this include the need to validate CFD results in detail, the use of efficient screening methods (see Sect. 7), the need to reduce the influence of unwanted basin effects, or the 'replay' of offshore wave measurements (such as the Draupner wave, or conditions at the time of marine accidents). The deterministically generated waves can be irregular time traces, or focused wave events. Sprenger et al. (2022) provide an overview of the work of one of the leading experts in this field, including the basic principles of deterministic wave generation. Lakshman et al. (2023) provides a review of experimental generation methods for three-dimensional focused wave events. It is known that non-linear physics modify the shape of the largest events in random wave realisations. When a screening approach is applied to define targeted deterministic wave groups that capture the most important physics of extreme events, it is therefore important to know whether these non-linear physics are similar in the groups and the random wave fields. Tang and Adcock (2022c) compare the non-linear changes from the extreme events in random wave fields with the non-linear changes during the evolution of deterministic wave groups. They conclude that carefully designed wave groups are likely capturing very similar non-linear physics, which is promising for screening approaches. Recent developments in data-driven methods also found their way to experimental wave generation. The calibrated generation of waves in a basin remains a challenging procedure, that often requires multiple iterations before the waves meet quality criteria on spectral shape, wave height and/or period. Klein et al. (2024) applied a convolutional neural network to relate target experimental wave sequences at a specific location in a basin to the corresponding wavemaker control signals. Such a technique may be used to reduce the number of required wave calibration iterations for new projects.

5.2.6 Wave Breaking

The breaking of waves can be depth-induced, or steepness-induced. Especially steepness-induced breaking remains a complex and not well-understood topic in wave modelling (as explained in Scct. 3). This is closely related to the uncertainties in the spectral tail discussed above. As also mentioned in Sect. 3, wave breaking inception models and their validation are extensively studied. Spectral bandwidth influences the inception of wave breaking. This was studied experimentally by Cao et al. (2023), concluding that wave groups of larger bandwidth break at a lower 'global' wave steepness. Consequently, such waves lose relatively more energy. McAllister et al. (2023) argues that the relation between spectral bandwidth and the global wave steepness at which breaking starts is too complex to parameterise in a general way, but shows numerically that the local surface slope of maximally steep non-breaking waves, of all simulated spectral bandwidths, approaches a limit of $1/\tan(\pi/3) \approx 0.5774$. Another parameter that strongly influences the inception of wave breaking (and consequently the maximum wave height that can be obtained) is wave directionality. McAllister *et al.* (2024) derive an empirical relation between wave directional spreading or the crossing angle of two wave systems, and the maximum wave steepness that can be measured before breaking. They show that three-dimensional wave breaking for larger directionality or crossing angles can lead to significantly higher waves than two-dimensional wave breaking. The publication includes and empirical formulation for this relation. Depending on the occurrence of such high-spreading sea states on the ocean, and the effects of wave irregularity and water depth, this may have important consequences for design codes of maritime and offshore structures (e.g. for wave-in-deck loads or green water), and for instance rogue wave warning systems (see Sect. 8). Govender et al. (2023) present experimental measurements of mean and turbulent velocities in a strongly plunging wave and used PIV and bubble image velocimetry to obtain the associated fluid velocities. The mean velocities were used to estimate the forward and reverse mass fluxes, from which an estimate of the average fluid density in crest of the waves was obtained. This publication provides observations of both time-dependent and phase-averaged features of velocities, turbulence kinetic energy (TKE), and the wave and turbulence shear stresses over the wave cycle of plunging breakers. Deng et al. (2023) measured the wave kinematics and inline forces induced by Draupner-type breaking freak waves experimentally, and recommended the preferable wave stretching model for predictions of wave kinematics and forces of freak waves.

5.2.7 Shallow-Water Wave Generation

Experimental wave generation on shallow water can be associated with additional difficulties. Low-frequency infragravity (free and bound) waves increase in energy on shallow water, and beaches are typically not able to dampen such long waves properly. On top of that the resonant modes of wave basins may coincide with both the typical low-frequency bound wave frequencies, and with the natural frequencies of mooring systems at the typically observed model scales. This may lead to large resonant structure motions, which are not representative for the ocean. Finally, some basins use a 'ramp' as transition from the deep-water wave generator to a shallower floor. This ramp may introduce additional unwanted free waves. Düz et al. (2022) discusses a potential way to mitigate these unwanted effects. The established second-order wave generation theory becomes more important on shallow water, as the second-order wave contributions are larger. Mortimer et al. (2022) validates this theory explicitly for isolated wave groups, which provide a demanding test on the correct generation of second-order bound waves and the stroke length of the wavemaker. Pierella et al. (2021) show that, contrary to what is sometimes believed, second-order wave correction is also important to obtain a more uniform wave field over the basin length in highly non-linear irregular waves (not only in regular waves). Watanabe et al. (2020) combine the study of shallow water, focusing and breaking waves, to formulate general wave focusing conditions for arbitrary water depth and bathymetry. The technique can be used to control the focus location and time in wave experiments with an arbitrary bottom structure, which can be useful to study the effect of extreme waves on structures in shallow water.

5.3 Ice

Classic model ice is developed for ships interacting and especially breaking ice. For this purpose, the model scale ice properties must be scaled, that is, not only the ice thickness is scaled according to the geometrical scaling of the ship, but also the ice strength and other material properties, and the velocities used in the experiments are scaled. For scaling the ice properties, it has become customary to use the so-called Cauchy-Froude scaling (Timco, 1984). Other interaction scenarios might require different scaling approaches or another model ice type following the idea of case-based scaling (von Bock und Polach et al., 2020).

Some of the most recent work on ice tank testing has moved away from Cauchy-Froude scaling. Ham Puolakka and Hendrikse (2024) describe a model ice for experimentation related to ice-induced vibrations of narrow and compliant vertical-walled offshore structures, such as offshore wind turbines with monopile foundations. Crucial feature of the developed model ice was the indentation-rate-dependent behaviour. In the analysis it was found that neither the aspect ratio nor shape appeared to influence the development of ice-induced vibrations. A qualitative verification is done by numerical reproductions of full-scale ice-induced vibrations of the Molikpaq platform and Norströmsgrund lighthouse. Owen, Hammer and Hendrikse (2023b) investigated peak ice load magnitudes on offshore structures in an ice basin. They found that the peak loads on compliant structures cannot surpass those on rigid structures under identical ice conditions, contradicting earlier assumptions. In a related laboratory study on the failure of freshwater ice under haversine loading, Owen, Hammer and Hendrikse (2023a) found two distinct mechanical behaviours of damaged and confined ice samples during each loading cycle: brittle at high velocity and non-brittle at low velocity. At low velocity, ice fracture was interrupted, leading to stress relaxation until velocity increased and a peak load was recorded. It was suggested that this behaviour may help to explain the frequency lock-in regime in ice-induced vibration.

Lemström, Polojärvi, Tuhkuri, (2022a) and Lemström, Polojärvi, Puolakka, et al. (2022b) conducted laboratory-scale experiments on ice-structure interaction processes in shallow water. The ice properties in the experiments were not scaled, but varied, yet the magnitude of ice loads was not found to be proportional to ice strength. The interaction process was divided into two phases. The analysis indicated that the weight of the incoming ice and the ice rubble pile-up had a dominant role during the first phase showing an increasing ice load, with ice failure against the rubble pile occurring during the second phase, where the ice load remained constant. The ice properties appeared to have an impact on the ice failure process development, while the normalization introduced provided information on the development of this process.

Recent work on model-scale testing on ships has also studied scaling. Matala (2021) studied the properties of brash ice, a rarely explored topic critical for ice-going commercial vessels, which often operate in brash ice channels. While existing guidelines define only the channel thickness and width, the study focused on how the other brash ice parameters influence the ice behaviour around the ship hull. The study found that vessel resistance in brash ice is significantly affected by these properties, suggesting the need for updating the standardisation of model brash ice to ensure experiment results. Matala and Suominen (2022, 2023) continued the study on brash ice further and introduced a Channel number. The Channel number relates to the scaling of the forces contributing to vessel resistance in brash ice channels, particularly by improving the modelling of resistance from displaced ice fragments. This Channel number, designed to supplement the Cauchy-Froude scaling, was found to be more accurate for brash ice conditions than the Cauchy-Froude scaling by itself.

Further, recent experiments on fracture of columnar freshwater ice by Gharamti et al. (2024) indicate that the ice thickness has no impact on the fracture behaviour of ice. The study used large, warm ice samples ranging from 10 to 40 cm in thickness. While the load at fracture increased linearly with the ice thickness, the thickness did not affect the most important fracture properties, such as the apparent fracture toughness, fracture energy, crack opening displacement, notch sensitivity, or process zone size. Further, experiments with granular laboratory ice, that is neither model ice nor scaled, were compiled in Böhm, Herrnring and von Bock und Polach (2022) and can serve as a reference for material modelling of ice.

5.4 Current

Some offshore structures are sensitive for currents, such as moored structures or structures that use dynamic positioning. The traditional way to exert current on such structures in basin experiments is to use very large current pumps. Otter et al. (2022) proposes an alternative method to simulate current loading and wave-current interactions during scale model tests, by using a dynamic winch which is controlled using a software-in-the-loop approach. The idea of this method is that it is cheaper and more versatile than traditional physical current generation in a basin, as it allows for a wider range of test conditions and can be applied in any wave basin. They applied this approach to a floating offshore wind turbine, showing that the winch actuator is capable of reliably emulating the drag force exerted by a current on the platform over a range of test conditions.

5.5 Wind

Scale model testing of wind for the marine and offshore industry historically focused on the effect of wind loads on the performance of dynamic positioning systems, or on added resistance of ships sailing in wind and waves. However, recent developments in the industry have changed this to assessing the performance of wind-assisted shipping and (floating) offshore wind turbines. This also changed the requirements and boundary conditions for experimental wind modelling in wave basins. We exclude pure wind modelling in wind tunnels from the present report, as experimental aerodynamics are a field of study on their own. The report of the 30th ITTC Specialist Committee on Wind Powered and Wind Assisted Ships (Werner et al., 2024) includes a review of methods and facilities for hybrid (experimental-numerical) testing of the hydrodynamic and aerodynamic performance of wind-powered or wind-assisted ships. The ITTC also provides "recommended procedures and guidelines" for model tests of offshore wind turbines (ITTC, 2021), including recommendations for the calibration of the environmental conditions. Gueydon et al. (2020) provides a review of the state of the art of testing floating offshore wind turbines in basins, with a focus on the experimental wind modelling. Modelling wind loads on a wind turbine in a Froude-scaled laboratory set-up can be done in two ways: using a physical rotor in a wind field, or using hybrid methods where the wind turbine loads are computed and applied by actuators to the model. Gueydon et al. (2023) provides some examples of experimental test set-ups for the second option, with or without wind fans. It also emphasises why monitoring of the loads transferred to the floater adds up to the quality of the test results and confidence in the experiments. Another application of such a hybrid set-up is discussed by Fontanella et al. (2020), who use a hardware-in-the-loop methodology that allows for the recreation of the effect of a realistic three-dimensional wind field on an offshore wind turbine. Guo et al. (2024) presents a 1:70 model test study in a wave basin, exploring the coupling effect of wind, waves and currents on the performance of a 12 MW semi-submersible floating wind turbine. They show remarkable interaction effects of the different environmental conditions on the wind turbine responses.

5.6 Wave-Coupled Phenomena

Similar as in Sect. 3.5, here we discuss studies on the interaction of waves with currents, wind or ice. However, now we focus on laboratory studies rather than numerical modelling. We start with wave-current interactions. Miranda et al. (2023) present a simple passive wave absorption system, designed specifically for flumes with a combined wave-current generation facility. Draycott et al. (2018) explore the impact of current on the wave field in a laboratory, and they demonstrate a simple methodology for re-creating combined wave-current scenarios that were measured at sea. Draycott et al. (2022) use scale model experiments to assess the effect of current on the measurements of a wave buoy, which is useful information for evaluation of the quality of wave buoy measurements (see also Sect. 6). Besides evaluation of some 'conventional' marine structures that are sensitive to currents, the experimental modelling of currents is also essential for tidal turbines. Z. Zhang *et al.* (2023a; 2023b; 2023c) evaluate the effect of current and wave interactions on the power fluctuation and wake characteristics of a tidal stream turbine, showing that the presence of waves significantly affects the operating flow and resulting turbine performance. X. Zhang *et al.* (2024c) present PIV observation results of the turbulence in waves interacting with currents. These observations provide a better insight into the details of the vortices generated in such interacting flows, which in turn may be used to improve wave-current interaction modelling. In the context of ocean cleaning operations, one of the questions is how the movement of floating marine litter interacts with ocean currents and wave-induced Stokes drift forces. Calvert et al. (2024) present laboratory experiments on the drift of floating marine litter in different wave conditions, reporting very different drift speeds for different object shapes in the same waves.

Continuing on the topic of wave-wind interactions, Bruch et al. (2021) discuss experiments in a wave-wind laboratory, to investigate the role of wind-wave interaction on the formation of spray generation on the sea surface. They tested four types of wave forcing and five wind speeds, concluding that spray generation increases with increasing wind-induced wave breaking, and is highest for steep and heavily-breaking waves. Larger droplet spray generation is best correlated with the wave-slope variance; smaller droplet generation is best correlated with a non-dimensional number combining the wave-slope variance with the friction velocity cubed. Townsend et al. (2023) provide input for the inlet wind velocity used in numerical models for offshore wind engineering in combined wind and wave conditions, using experiments that study the interaction of the atmospheric boundary layer (specifically the wind velocity profile) with a fixed wave geometry. Application of this improved inlet velocity profile shows that wind forces on an offshore wind turbine are significantly greater than with the conventional model, especially in large waves. Z. Zhang *et al.* (2024a) aimed to understand the effect of wind forcing on steep unidirectional waves better. For this purpose, they performed experiments with waves in a large basin, interacting with turbulent wind generated by wind fans. They found that the effect of wind forcing on wave H_s varies with the initial wave steepness. Wind forcing increases the growth of waves with an initially small steepness, but reduces the growth of large, steep waves. As expected, it was found that the wave energy input increases the high-frequency tail of the wave spectra (an effect that further increases with fetch).

To study the interaction of waves with ice, Park et al. (2022) performed experiments with models of freely drifting ice floes, made of low-density polyethylene, in a wave basin. They measured the wave transmission, as well as the mean ice floe drift velocity. The results showed that the wave transmission increased as the wave period increased, decreased as the ice concentration and ice-covered length increased (as expected) and decreased as the wave steepness increased. The mean ice drift velocity normalized by wave celerity increased as the wave steepness increased. A set of experiments with model ice by Passerotti et al. (2022) investigated the change of wave spectra with their progression through the ice and derived an empirical attenuation coefficient with a power law dependence on the wave frequency comparable to field measurements. This finding must account that not all relevant mechanical properties of model ice scale as intended with sea ice (von Bock und Polach et al., 2021).

6 Ocean Measurements and Observations

6.1 Waves

As discussed in the previous chapter, laboratory measurement of waves has important advantages. It is for example possible to control the wave conditions, measure at many locations, target extreme conditions, and reduce sampling uncertainties by generating long ergodic/stationary conditions. All of these things are hard to attain in ocean measurements (see e.g., Bitner-Gregersen et al. (2021)). However, laboratory wave measurements also have downsides: basin effects, limited fetch, wave generation by wave generators instead of wind etc. Wave measurements at the ocean are therefore still an important and necessary source of information.

In design, for long-term lifetime assessment of the performance of marine structures at sea, it has been generally accepted that metocean climate variability can be discretised in stationary periods of 3 to 6 h. The use of the 3 to 6 h assumption of stationarity is seldom justified, especially for ships with forward speed (which may also change their heading or speed within this duration). This assumption should therefore be considered carefully when assessing marine structures' performance at sea (see e.g., Nielsen and Ikonomakis, 2021).

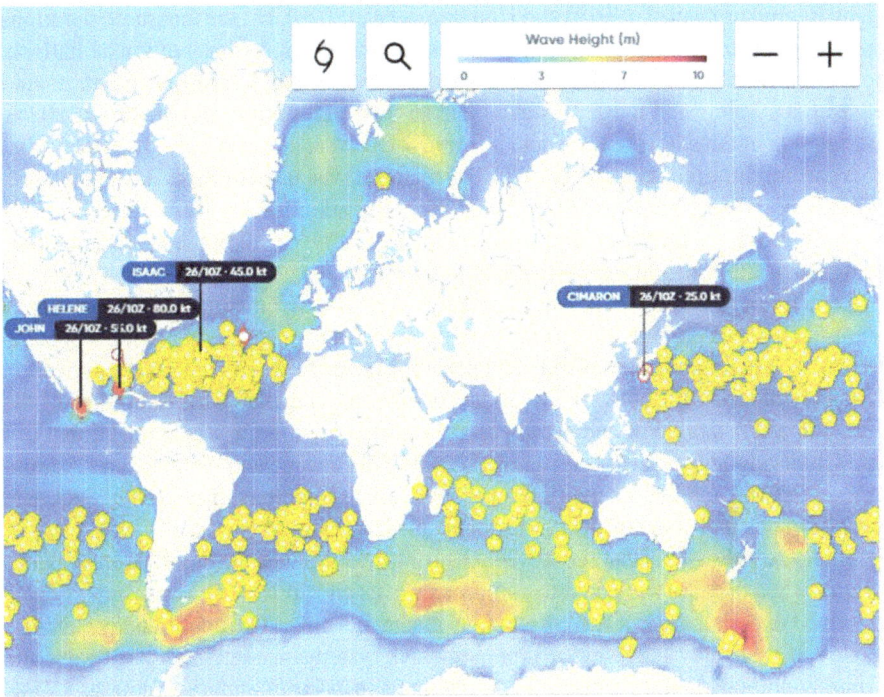

Fig. 6. SOFAR spotter buoy network (source: https://www.sofarocean.com/products/spotter#s-dashboard, reproduced with permission)

6.1.1 Buoy Measurements

The traditional method of conducting wave measurements mostly relied on the use of buoys. However, deploying buoys can be both costly and cumbersome, as they are often expensive and difficult to retrieve at sea. The introduction of cheaper, smaller buoys that communicate using GPS has simplified this process. They can function as Lagrangian drifters as demonstrated in Rainville et al. (2023) by using 'microSWIFT' buoys, or they can be moored at a given location by using a range of different mini buoys as demonstrated by Collins et al. (2024). The smaller buoys also present some drawbacks. Their smaller diameter and lower weight change the hydrodynamic characteristics of the buoys compared to the traditional larger buoys. This makes them suitable for a different range of wave periods and, unfortunately, more sensitive for biases introduced by a mooring system. Several studies have compared the performance of the lightweight SOFAR Spotter buoys to existing buoys such as the Datawell waverider. Overall, the studies by Lancaster et al. (2021), Beckman and Long (2022), Collins et al. (2024), Ewans and Collins (2024) and Yue Ding et al. (2024) show that the Spotter buoys measure comparable wave height, period, spectrum and main direction with the accuracy of conventional buoys, but that their communication was sometimes interrupted by tilting or submersion—typically manifested as low-frequency noise or drop-outs—and that the moored small buoys have some issues with measurement of the directional spreading. Wave data from the globally distributed Spotter buoys are now available sequentially, e.g. by Raghukumar et al. (2019) and Houghton et al. (2023), but the measured durations are still quite short. A worldwide network of such Spotter buoys was built in a relatively short time. Data from this global network of 400–500 buoys (see Fig. 6), measured between 2019 and March 2022, is available for research purposes at (SOFAR, 2024). The dataset contains hourly wave and inferred wind data, including H_s, peak direction, wave spectra, inferred wind speed, etc. This data set was used by e.g., Houghton et al. (2023), who present a method to combine buoy measurements with satellite altimeter data to improve H_s forecasts, and by Davis et al. (2023) to study the wave properties during hurricanes Ian and Fiona in 2022.

As mentioned above, buoys are generally used for two distinctly different applications. Dedicated in-situ wave (and sometimes current) measurements for a specific project are usually done with moored buoys at the project location. Another application of instrumented drifting buoys, wave gliders or other moving platforms is monitoring the ocean climate. A large program in this context is the Argo program within the Global Climate Observing System (GCOS) run by the National Oceanic and Atmospheric Administration (NOAA) and its international partners. This program is discussed in Sect. 6.5, which explains that some of these buoys also measure waves.

There are indications that the wave measuring performance of (moored) wave buoys changes when they are subjected to combined waves and current. Draycott *et al.* (2021) performed experiments in a wave basin to evaluate this effect for moored buoys, using a simple spherical buoy and mooring arrangement in systematic combinations of wave and current conditions and directions. They observed that vortex-induced motions (VIM) of the buoys can be significant, and sensitive to both the current speed and mooring arrangement. Wave amplitude results still generally matched independent wave gauge measurements, with better agreement in co-linear wave-current conditions than in opposing conditions. However, as discussed above such VIM will influence the reported wave directional spreading by such buoys (Ewans and Collins, 2024). Pillai et al. (2021) argue that wave buoys in current do not measure the sea state power and wave steepness accurately and present a method to correct measured wave data from a buoy for these effects, thus reducing its uncertainty.

6.1.2 Ship-as-Wave-Buoy Measurements

Contrary to the encountered sea states, which were obtained by combining AIS and hindcast wave datasets, many recent studies have used measured ship motions or structural stresses combined with the known ship response characteristics to estimate encountered short-term wave statistics or sea states (X. Chen et al., 2021; Bisinotto et al., 2023; Nelli et al., 2023; Zago et al., 2023). This approach is commonly referred to as 'ship as a wave buoy'. An outlook of what could be gained when many of the existing ships are instrumented to measure variables relevant for atmospheric, oceanic and biogeochemical studies is provided by Rosa et al. (2021). One of the most common methods in these studies was the wave spectrum method, previously studies by Nielsen and Dietz (2020) and X. Chen et al. (2020). This method was expanded to estimate the multi-peaked directional wave spectrum, by which not only H_s and relative wave direction, but also T_m could be well assessed by X. Chen et al. (2021), and was further expanded to estimate the stochastic wave spectra in which uncertainty in transfer functions of a ship could be involved by (X. Chen et al., 2023b).

To improve the accuracy of predicting the encountered sea states from vessel motions using the directional wave spectrum method, Nelli et al. (2023) proposed a technique that uses freely available satellite altimeter data. This technique calibrates the response amplitude operator (RAO) of the ship and reduces uncertainties when key RAO data is missing. They used the method to estimate the sea state conditions encountered by the icebreaker Akademik Tryoshnikov during the Antarctic Circumnavigation Expedition. The estimated conditions were compared with the measurements from a marine radar, showing a good agreement across various parameters, including H_s, wave periods, and θ_m. Moreover, to increase the accuracy and to reduce the computational cost when predicting the encountered sea states by ships, various machine learning-based approaches have been recently developed (P. Han et al., 2022; Mittendorf et al., 2022; Nielsen et al., 2023). Since these methods only provide short-term wave statistics, currently they are used in digital twins of ship structures and for hull monitoring to ensure the safe and efficient operation of ships and offshore structures (Esmailian et al., 2022; Lee and Kim, 2022; Fujikubo et al., 2024). Further, Komoriyama et al. (2023) discussed a Kalman filter technique for identifying wave profiles encountered by a ship with no forward speed from measured ship responses. This technique was then applied to a series of tank test results for long-crested irregular heading and oblique waves. The idea to use existing ships to provide wave data is very attractive and efficient. However, it should be kept in mind that ship-as-a-wave-buoy measurements also have some major drawbacks for certain applications. Due to the relatively large size of the average sea-going ship, it hardly moves in short and/or steep waves (response amplitude operator approaching zero). In such conditions, hardly any ship motions can be measured, and the false impression could be given that the waves are small. Wave buoys have similar drawbacks, but due to their smaller dimensions this is only problematic for very short waves (typically below 1 s wave period). This threshold wave period is much longer for most ships.

6.1.3 In-Situ Measurements from Platforms

The number of relevant publications related to in-situ wave measurements from platforms was limited over the mandate period. Miao et al. (2024) use observational data from eight locations in the Bohai Sea close to China between 1998 and 2022 to characterise the H_s conditions in this area. Six of these locations were equipped with wave buoys, and two with acoustic wave height meters. However, this publication does not elaborate on measurement accuracy of the different equipment types. Magnusson et al. (2021) discusses a comparison between three wave sensors: a radar, a laser, and a Waverider buoy, all three measuring at the Ekofisk platform in the North Sea. The comparison shows that, in general, all three sensors provide similar measurements of the integral wave properties and frequency spectra. However, there are some significant differences between the details of the measurements, which could impact design and operations, forecast verification and climate monitoring.

6.1.4 Radar Measurements from Ships and Short-Term Wave Prediction

Wave radars can be mounted on a platform or the coastline, but they can also be ship-mounted. Traditionally this was done using downward looking radars, measuring the relative wave elevation close to the hull accurately. However, obtaining undisturbed wave elevation from relative wave elevation requires deterministic knowledge about the ship motions, the radiated and diffracted waves and dynamic swell-up. An alternative method is to use a wave radar to resolve the wave field at some distance around the ship. This opens the door to new applications, such as short-term wave prediction and operational guidance (see e.g., Naaijen et al., 2018). This application requires special types of radar, targeted at high resolutions at a short range. One option is using an X-band pulse radar, which can be (a modified version of) the normal navigation radar. It can adequately measure wave lengths and propagation velocities but performs badly at predicting wave height. Methods to improve this wave height estimation are proposed by Huang et al. (2021) (using machine learning methods) and by Lee et al. (2024) (leveraging the shadowing characteristics in the radar images). Both methods seem to improve the observed H_s from the radar images, validated against numerical references and buoy observations. Alternatively, an FMCW radar uses continuous signals with a changing frequency instead of pulses. This allows for a very low radar power. By using both the radar reflection intensity and Doppler analysis, it can adequately provide the surface velocities (combination of vessel speed, current and wave orbital velocities), see e.g., Rathod et al. (2018). However, its use is in an early experimental state. The ship motion forecasting challenge is not only related to the processing of radar data to get usable wave measurements, but also to the subsequent propagation of this wave data to the ship location. Chen *et al.* (2023a; 2023b; 2023c; 2023d) present a machine-learning based method to perform this propagation. Simpson et al. (2020) present a method that does both the radar data processing and the forecasting, by coupling observations from an X-band marine radar with a phase-resolving wave model. It deterministically fits the wave model parameters to the measurements, and then propagates the wave field in space and time within the model.

6.1.5 Remote Sensing Measurements

Advancements in metocean data measurements and analysis have significantly improved the understanding and utilisation of metocean data in the offshore industry. In the past years, numerical models have become more sophisticated, incorporating higher resolutions and improved physics to accurately simulate metocean conditions and data assimilation techniques have been developed to integrate measured data into models, improving the accuracy of forecasts, hindcasts and reanalysis. Regarding metocean data (surface wind and waves), developments in recent years are related to improvements in remote sensing data due to new satellite missions (Aouf et al., 2021; Asharaf et al., 2021; Lin and Portabella, 2022; Cagigal et al., 2023; Zuo et al., 2024) and to algorithm development to reconstruct missing data in large gaps, in space and time, due to satellite sensors blocked by clouds (Barth et al., 2022; Bu et al., 2023).

6.1.6 Shallow-Water Wave Measurements

Together, wave buoys, satellites and hindcast models provide world-wide coverage of deep-water wave conditions. However, shallow-water wave observations remain sparse and often inaccessible. Such wave conditions vary significantly, due to effects of the coastline, bathymetry, presence of structures etc. Appendix B therefore also includes a few local wave databases in coastal areas with significance for marine structures. A well-documented example is described by Kinsela et al. (2024). This publication presents a growing dataset of measurements from moored wave buoys in shallow waters (<35 m) in southeast Australia. It includes over 7000 days of measurements at 20 locations. The dataset includes time series of spectral wave parameters, as well as buoy displacement data. Another example is the dataset described by Ludka et al. (2019), who provide beach profiles and waves measured at three Californian beaches between 2001 and 2016. The waves in this database are provided by a 'buoy-driven regional wave model'. The concurrent measurement of the beach profiles and the wave parameters enables analysis of the effect of bathymetry changes on the waves.

In coastal areas, long (infragravity) waves with periods > 20–30 s are generally much more energetic than in deep water. This is due to the increase in bound higher-order low-frequency wave energy in shallow water, interaction of these waves with the bathymetry and coastline, or free long waves originating from distant sources (Rijnsdorp, Reniers and Zijlema, 2021). Such long waves are not accurately measured using conventional wave buoys (see e.g., van Essen, Ewans and McConochie, 2018), and biofouling effects may worsen the general measurement accuracy of buoys in coastal area (Campos et al., 2021b). Alternatively, instruments such as acoustic doppler profilers (ADPs) and an acoustic doppler velocimeters (ADVs) can be used to measure long waves in coastal areas. Rutten et al. (2024a, 2024b) present a dataset of continuous measurements of synchronised surface elevation, velocity and pressure were recorded at 2–4 Hz by such ADPs and ADVs for a five-month duration in the Dutch North Sea. Nine storms, with H_s above 2.5 m for a duration of at least six hours, were recorded during this time. This dataset can be used to investigate wave transformation, wave non-linearity and wave directionality for wind waves, swell waves and infragravity waves, in 'normal' and storm conditions.

6.2 Ice

One of the most significant full-scale experimental campaigns on sea ice in recent years was the year-long Multidisciplinary Drifting Observatory for the Study of the Arctic Climate (MOSAiC) expedition, which was conducted from 2019 to 2020, with publications based on the findings of the expedition having begun to emerge in the past few years (Fig. 7). The primary objective of the expedition was to improve global climate models, but it also focused on on gathering detailed data on sea ice dynamics (Nicolaus et al., 2022); one central question was how sea ice moves and deforms. The equipment used allowed for the observations of phenomena such as ice fractures, compression and compressive stresses within the ice field, and pressure ridging. Overall, MOSAiC studied sea ice dynamics across spatial scales, ranging from centimeters to tens of kilometers, considering variability across these scales. Similarly to all other data from the expedition, the sea ice data are publicly available (Macfarlane et al., 2023).

The ice observations made during MOSAiC focused on several topics. Examples include Salganik et al. (2023), who studied ice ridges and especially the thermodynamics of ridge keels. They were able to observe three different mechanisms of ridge consolidation, related to either congelation, snow-slush, or meltwater. These mechanisms led to faster than expected consolidation of the ridge keel. Further work on thermodynamics was done by Zampieri et al. (2024), who used MOSAiC observations to improved insight of surface energy balance over thin ice for providing more reliable predictions on sea ice thickness in modelling. Ice thickness predictions were also in the focus of F. Gu *et al.* (2024b), who used the measurements from the expedition to develop a modified parameterisation of freezing temperature for the bottom of the ice sheet, which significantly improved the agreement between modelling results and observations.

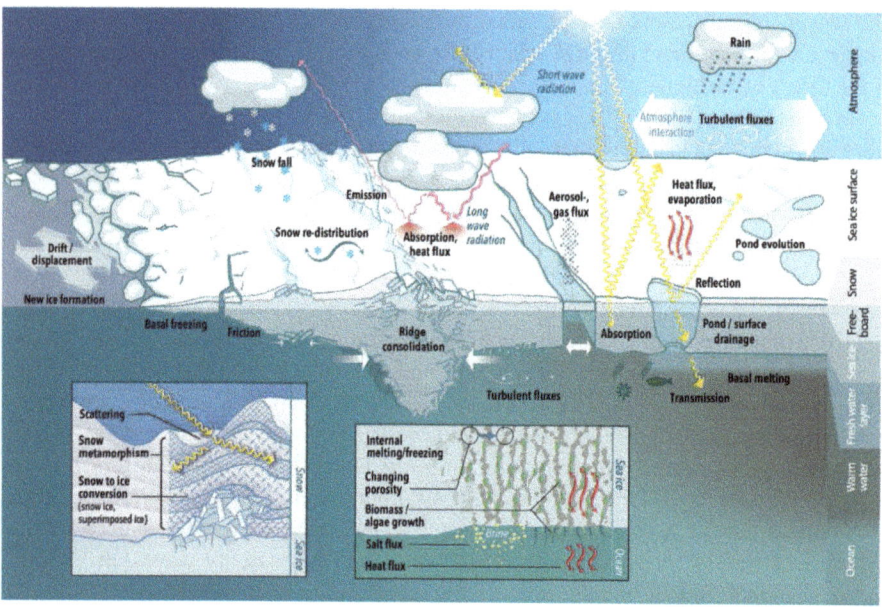

Fig. 7. Sea ice and snow related processes studied during the MOSAiC expedition (source: (Nicolaus et al., 2022), reproduced with permission under CC-BY licence)

The sea ice dynamics studies by using buoy arrays and satellite remote sensing remain important. Whereas the measurements have often focused on large-scale phenomena, there is a shift towards capturing small-scale events as well, including individual failure events and ice floe-to-ice floe interactions. This trend complements recent advances in high-resolution numerical modelling and aims for material model development. Example of this is the study by Parno et al. (2022), who performed first contemporaneous strain and stress measurements on an approximately 1 km^2 sea ice area in the Beaufort Sea. Related to similar themes, Hutchings et al. (2024) performed measurements on deformation scaling by using buoy array that allowed to analyse sea ice deformation across scales over four orders of magnitude with minimum scale being only hundreds of meters. They found that similar relationship does not hold for regions of kilometer scale or less than for larger areas. This result is of importance for modelling of sea ice, since deformation scaling is a metric commonly used in model development and validation. Recent observational analysis related to model development of models also includes the use of satellite borne data, as Ringeisen, Hutter and Von Albedyll (2023) analysed crack angles based on satellite images and above-described MOSAiC data.

6.3 Current

Aijaz et al. (2023) compared ocean currents from eight global models against observations from drifting buoys to verify them and study their performance. The verification could be considered independent of the observations, since the currents were not assimilated into the models. They assessed the effects of incorporating Stokes drift and tidal currents from separate global wave and tide models on the verification of model currents and found that incorporating Stokes drift significantly enhanced the performance of the models. By adding wave-induced Coriolis-Stokes forcing to the classical Ekman layer, Hui et al. (2022) estimated the wave-modified ocean surface currents in the northwestern Pacific. These currents combined classical Ekman currents (from Cross-Calibrated Multi-Platform (CCMP) wind data), geostrophic currents (from MADT), and wave-induced currents (from ERA-Interim wave data). When compared to Lagrangian drifter data and observations, the model matched both zonal and meridional currents with high correlation. Although the model underestimates current speeds in strong currents and coastal areas, it was shown to accurately capture surface currents over large areas in the deep ocean. Brolly (2023) used probabilistic neural networks to form estimates for full probability distributions behind the statistics related to ocean dynamics. This allowed them to efficiently estimate multiple statistics and offer a more comprehensive description for these statistics. They could then show that their approach allows to accurately simulate drifter trajectories, replicating the clustering seen in ocean gyres and which linked to garbage patch formation. Sainte-Rose, Pham and Pavalko (2023) also focused on debris transport by studying data from two drifting buoys released in May–June 2019 in the North-East Pacific to explore surface convergence and the persistence of debris within the Great Pacific Garbage Patch (GPGP). By comparing the trajectories of the buoys with local circulation patterns, a plastic dispersal model, and sea-level anomalies, they were able to identify a clear dependency between converging drifter movements and sea areas with high plastic debris density with one of the drifter trajectories also highlighting the persistent accumulation of debris within the GPGP.

6.4 Wind

Remote sensing missions often include both wave and wind measurements; most references in Sect. 6.1.5 are therefore also relevant for wind. The Cross-Calibrated Multi-Platform (CCMP) data are combination of ocean surface vector wind retrievals at 10 m altitude, from multiple types of satellite microwave sensors and a background field from reanalysis over the world's oceans. The resulting dataset is available every six hours. It is closely tied to the satellite retrievals where they are available. Where satellite retrievals are not available, CCMP is statistically consistent with satellite winds (Mears et al., 2022). Scatterometer satellites provide very stable ocean vector wind data records. However, sensors may still drift or experience step changes. A method to improve scatterometer wind measurements is presented by Ricciardulli and Manaster (2021). By combining measurements from multiple Advanced Scatterometers (ASCAT) and radiometers for comparison, drifting sensors can be identified. This way, it is shown that post-processed wind fields measured by ASCAT scatterometers become more stable over time, and well cross-calibrated with each other.

Buoys with anemometers can also provide observations of wind speed over the ocean. They are routinely used as a source of validation data for satellite wind products. However, the movement of buoys in high waves and the airflow over waves might cause inaccurate readings, raising concern when buoys are used as a source of wind speed validation data. Wright et al. (2021) evaluate the relative accuracy of buoy winds through comparing winds measured with a buoy to winds from ASCAT satellite measurements and the ERA5 hindcast model. This comparison shows there is a small, but statistically significant difference between height-adjusted winds from buoys with 4- and 5-m-high anemometers compared to the same ASCAT wind speed ranges in high seas. However, this result does not follow conventional arguments for wave sheltering of buoy winds, whereby the lower anemometer height winds are distorted more than the higher anemometer height winds in high winds and high seas. It was concluded that wave sheltering is not significantly affecting the winds from buoys with anemometers at 4 and 5 m for winds under 18 m/s. Further differences between buoy and ASCAT winds are observed in high swell conditions, motivating the need to consider the possible effects of sea state on ASCAT winds.

Ricciardulli et al. (2022) describe wind measurements made in 2021 by the National Oceanic and Atmospheric Administration (NOAA), using a novel uncrewed surface vehicle 'Saildrone'. Such instruments were deployed to monitor regions of the Atlantic Ocean and Caribbean Sea frequented by tropical cyclones (Foltz et al., 2022), and one of them crossed the path of hurricane Sam (Category 4) on 30 September 2021. This provided surface-ocean videos of conditions in the core of a major hurricane, reporting near-surface winds as high as 40 m/s. Ricciardulli et al. (2022) presents an analysis and interpretation of the Saildrone ocean surface wind measurements in hurricane Sam. The measurements were compared to different other datasets, including a buoy, satellite measurements (SMAP and AMSR2), radar measurements (ASCAT and RadarSat2) and wind models. The Saildrone winds show good consistency with the satellite observations. Especially strength of the winds matches the reference measurements well. The paper also reviews the collective consistency among all measurement sources, by describing the uncertainty of each wind dataset and discussing potential sources of systematic errors.

A comprehensive dataset of wind and surface wave conditions within tropical cyclones is provided by Tamizi and Young (2024). The dataset combines data measured by buoys and remote sensing (satellite altimeters, scatterometers, and radiometers) with estimates of tropical cyclone tracks and wind field parameters from best track archives. It provides data for each tropical cyclone from each of the data sources, for 2927 global tropical cyclones over the period from 1985 to 2017. Global statistics of the observations are provided, along with data on the geographic distribution of tropical cyclones within the database.

6.5 Argo Fleet of Drifting Buoys

The review by Johnson et al. (2022) describes the status, main uses and future applications of Argo, the largest in-situ oceanic observation platform in the world. Argo is a network of drifters, initially launched in year 2000. The network now includes about 4000 floats, with the number constantly increasing and the collected data is publicly available in real time. Within the Argo program, about 1000 drifters are deployed each year, to maintain a global $5° \times 5°$ gridded array of ~ 1300 drifters. In September 2024, around 70% of the $5° \times 5°$ bins in the world's oceans was covered by Argo floats. Most of these buoys measure sea surface temperature (SST) or sea level pressure (SLP). 4 of them measure wind and 37 measure waves, the latter of which are mostly located in the North Atlantic and close to Japan. The importance of the Argo program cannot be overstated, as the generated data have become widely used to improve forecasts for the ocean, weather, and climate. Argo has supported a variety of ocean climatology studies over large range of temporal resolutions, confirmed existing theories and led to new discoveries about ocean circulation and water mass formation. It is now becoming increasingly important for studies on ocean variability over time, including the redistribution of heat and salt, changes in ocean currents, and the evolution of marine heatwaves. The data produced by Argo has also recently been analysed and used to derive improved data on eddies (Rykova, 2023), also through machine-learning supported approaches (Lyman and Johnson, 2023). Machine-learning based approaches together with biogeochemical-Argo have also been used in combination with data from Argo for new insights into the regional distribution of oxygen levels in the ocean (Sharp et al., 2023). The study by Su et al. (2023) highlighted significant discrepancies between modelled and Argo-observed estimates on ocean circulation, with high-latitude ocean regions showing major underestimations in current velocities and directions. The findings emphasise the need for more accurate studies on global circulation models, which are crucial for predicting climate-related factors including sea-level rise. A recent development is the 'deep Argo' that measures from the sea surface to the seafloor, for monitoring deep-ocean warming and circulation. Recently it supported a study, which confirmed that warming in depths below 2000 m accounted for about 10% of total ocean heat uptake (Johnson and Purkey, 2024). Overall, in combination with other data sources, Argo has clarified key climate phenomena such as the contributions of ocean warming to sea-level rise and the strengthening of the water cycle due to atmospheric warming. With climate and weather models advancing Argo observations can be expected to remain critical. Further, the future important role of the Argo network appears to be ensured by the work on the advanced technologies including machine learning integrations used as part of the instrumentation within the network (Johnson and Fassbender, 2023). It was noted that it has been a challenge to maintain and expand such drifter buoy networks during the Covid pandemic (Boyer et al., 2023; Sprintall et al., 2024).

6.6 Wave-Coupled Phenomena

Here we discuss the interactions of waves with either currents, wind or ice. This is similar as in Sects. 3.5 and 5.6, but now focussing on ocean observations. We start with the interaction of wind and waves. Using wind and wave data collected from a buoy off the coast of Taiwan, Tsai and Hsieh (2023) examine the effect of sea states on wind velocity fluctuations in the longitudinal direction near the sea surface under near-neutral stratification, characterized by inverse wave age and observed that that wind waves and swells modify the overlying turbulence structure differently. Wind-swell coupling results in greater variations and values of the scaled velocity standard deviation and turbulence intensity with increasing wind speed, as well as a more pronounced peak in the scaled wind velocity spectrum compared to pure wind sea conditions. The authors conclude that near-surface turbulence is driven by eddy impingement on the sea surface, associated with wave induced fluctuations in the lower part of the eddy surface layer. Cavaleri et al. (2024) explore the earliest stages of wind wave generation in the open sea, based on measurements in the Adriatic Sea for a very low range of wind speeds. They suggest that the minimal wind speed for the appearance of the first wavelets is close to 1.8 m/s. The wavelets quickly disappear as soon as the developing wind waves take a leading role. It is suggested that this process is due to the strong spatial gradients in the surface orbital velocity, which impede the instability mechanism at the base of the wavelet formation. In later stages of the wave development, these gradients decrease, and wavelets reappear. On the topic of wave-ice interaction, Rabault et al. (2023) present a database of in-situ observations of sea ice drift and waves in ice. A total of 15 deployments were performed over a period of 5 years in both the Arctic and Antarctic, involving 72 instruments. These provide GPS drift tracks and measurements of waves in ice. The dataset is suitable to be used for instance to tune sea ice drift models, investigate wave damping by sea ice, and help calibrate other sea ice measurement techniques.

7 Statistical Modelling

Statistical modelling and probabilistic descriptions of metocean conditions are important for the design and structural reliability assessment of ships and other marine structures. They provide necessary input to identify what conditions the structure is expected to experience and hence what environmental loads they should be designed to withstand. More accurate statistical modelling may allow for more optimised design, where unnecessary conservatism can be avoided by a reduction in safety factors and associated building costs (Bitner-Gregersen et al., 2022). This chapter of the report reviews recent developments in the theory and application of statistical modelling of relevant environmental variables describing the ocean conditions. The focus is mostly on statistical modelling applied to such data, but also important theoretical and methodological developments from other applications may be included, if deemed relevant. The structure of the section follows a similar structure as the previous ISSC report on the environment and addresses different relevant aspects of statistical modelling: long- and short-term statistics, extreme value analysis, multivariate analysis, spatio-temporal analysis and non-stationary analysis. This distinction into topics is arguably somewhat arbitrary, and some papers contribute to several of these aspects, but this division is still believed to be useful. The review is limited to recent developments since the previous ISSC report (Babanin *et al.*, 2022), see also Vanem et al. (2022), and focuses on the environmental description; statistical models for loads and responses are out of scope of this section.

7.1 Wave Scatter Diagrams and Statistics for Design and Operation of Ships

Long-term sea state statistics and short-term wave data from measurements or numerical modelling are crucial for designing and operating merchant ships and offshore structures. As discussed in Sects. 2 and 3, several global wave hindcast datasets have become available and significantly improved over the last decade, including those by Bidlot et al. (2019), Hersbach et al. (2020), Stefanakos (2021) for ERA5/ECMWF, Alday et al. (2021) for IOWAGA/IFREMER and Q. Liu et al. (2021) for WW3-ST6v/U. These datasets offer extensive global coverage and high spatial and temporal resolutions, validated against in-situ observations and satellite altimeters. They provide comprehensive long-term wave statistics, including H_s, θ_m and various wave periods. However, challenges remain in their practical application due to regional characteristics, resolution limitations as discussed in Sect. 3, and uncertainties as noted by Bitner-Gregersen et al. (2022) and Kodaira et al. (2023).

Wave data can introduce significant uncertainty in long-term environmental assessments, so selecting datasets specific to a location is crucial. Extreme waves are rare, making it challenging to validate significant wave heights H_s beyond 8–10 m. Different uncertainties arise from wave distribution models and fitting techniques, which can affect ship and offshore platform design. For instance, Kodaira et al. (2023) found that discrepancies in maximum H_s values from various models reached nearly 5 m in the Northeast Atlantic.

Recent advancements in information technology have enabled the archiving of wave frequency-directional spectra by some meteorological offices, enhancing insights into extreme wave directional variability (Bitner-Gregersen et al., 2022); (DNV, 2024). While validated long-term wave statistics from hindcast datasets can typically inform the design of offshore and coastal structures, merchant ships operate globally, often relying on IMO requirements (IMO, 2010) for North Atlantic conditions and long-term sea state scatter diagrams with a 25-year return period.

At the start of the mandate period, ship designers used the IACS standard wave scatter diagram Rec.34 Rev.1 (IACS, 2001) as basis for design. This recommendation has been criticised for inaccuracies based on the underlying decades-old visual observations. In contrast, global numerical hindcast datasets have significantly improved, with various studies (Sasmal et al. (2021); de Hauteclocque et al. (2020); de Hauteclocque et al. (2023); Kodaira et al. (2023); Fujimoto et al. (2024)) validating long-term wave statistics for practical applications. Comparisons of buoy and altimeter data by Ribal and Young (2019) and Dodet et al. (2020) confirm that many hindcast datasets align well with measurements. Long-term wave hindcast statistics over an area may differ from those actually encountered by merchant ships, as noted in Sect. 6. These 'encountered wave statistics' may be more relevant for the safe and practical design and operation of vessels. Since the late 2010s, various studies have estimated these statistics by matching Automatic Identification System (AIS) data from merchant ships in the North Atlantic with hindcast datasets (see e.g., Austefjord (2019), Miratsu et al. (2019) and Oka et al. (2019)). Austefjord et al. (2023) combined AIS data from over 21,000 cargo ships above 90 m length in the North Atlantic (2013–2020) with IOWAGA wave data to update the standard scatter diagram in IACS Rec.34 using contemporary long-term statistics. This update is discussed in detail in Sect. 9. Nielsen and Ikonomakis (2021) studied the encountered wave statistics (H_s, T_z) along the routes of about 200 Maersk Line container ships over a 3-year period worldwide in combination with ERA5 data. They compared the observed wave heights with those from the Global Wave Statistics of (Hogben, 1986) and found that the weather routing and seamanship affected the results. Ruth and Thompson (2022) combined almost 100 million hours of worldwide AIS data over a 7-year period from about 2600 ships together with ERA5 to get the encountered wave statistics for structural fatigue analysis. They concluded that the probabilities of ship speed and relative wave heading profiles encountered by ships were also very important for structural fatigue analysis except the encountered H_s and T_m. Furthermore, Chirosca and Rusu (2022) analysed mean and extreme sea states over a 20-year period using ERA5 data for six major European shipping routes, which were determined using AIS and observations from ships. By calculating the probabilities of encountering these sea states along the routes, ships can avoid the most hazardous areas, ensuring the safety of passengers and cargo. In a related study, Petranović et al. (2021) conducted an operability analysis for an individual passenger ship operating in the Adriatic Sea. Using a hindcast database, they identified sea states where the ship could not perform its normal operations.

Due to the limited availability of AIS data, typically no more than 10 years, researchers have developed statistical storm avoidance models by combining AIS and hindcast wave datasets for a 25-year design life. Sasmal et al. (2021) used three years of AIS data from tankers, bulkers, and container ships in the North Atlantic to propose a model that rederived 25-year long-term wave statistics (H_s and T_z) based on ERA5/IOWAGA/TodaiWW3-NK. Fujimoto et al. (2024) expanded this model to include seasonal variations, utilising eight years of AIS data from approximately 7,600 ships and ERA5 wave data to update encountered wave statistics globally. To mitigate the influence of the short AIS data period, de Hauteclocque et al. (2023) combined 20 million ship positions collected over 30 years by volunteer observation ships with IOWAGA data, providing practical wave inputs for designing ship structures. This integration of AIS and hindcast data has become a standard approach in the maritime industry. Miratsu et al. (2022) compared various responses (motion, vertical bending moment and pressure) from 75 ships using encountered scatter diagrams from Sasmal et al. (2021) with area-based diagrams from hindcast datasets. They found an average factor of about 0.85 in the 25-year long-term predictions, with lower motions and loads based on the encountered wave statistics. Oka and Ma (2023) conducted a similar analysis on vertical bending moments, noting that the effects are more pronounced globally than in the North Atlantic.

The wave scatter diagrams derived from long-term wave statistics encountered by merchant ships are generally discrete. For practical applications, Bitner-Gregersen *et al.* (1995) proposed a joint model using a 3-parameter Weibull distribution for H_s and a conditional log-normal distribution for wave period. However, because the Weibull distribution may not adequately capture the bulk and tail of encountered H_s from hindcast datasets, several alternative models have been introduced. Austefjord et al. (2023) proposed a mixture of Weibull distributions, while de Hauteclocque et al. (2023) combined log-normal and Weibull distributions, and Fujimoto et al. (2024) used a generalised extreme value (GEV) distribution. Both publications employed a generalised split normal distribution for the conditional wave period.

Interest in shallow water and coastal wave statistics has grown due to increased navigation of autonomous ships, renewable energy wind farm designs, and harbour operability. For instance, Lucas et al. (2023) reanalysed hindcast data to characterize 12-year long-term sea states (H_s, T_m and θ_m) near Portugal. James and Panchang (2022) compared short-term wave statistics in UK coastal waters from 40 Datawell buoys in water depths around 10 m with different short-term statistical models (Rayleigh/Forristal/Longuet-Higgins/Naess/Boccotti/van Vledder/Klopman distributions), identifying the van Vledder model as the most accurate. Umesh and Behera (2021) improved coastal predictions on India's east coast using SWAN and SWAN-SWASH, showing that their results better matched buoy measurements. van Eeden et al. (2022) conducted a sensitivity analysis on grid sizes using SWAN and the "Dutch Offshore Wind Atlas" for long-term extreme H_s statistics along the Belgian North Sea coast. It was found that grid refinement in the wind database only benefits the wave predictions when considering complex bathymetric effects in the shallower areas of the coast. Climate change is also prompting significant shifts in long-term wave statistics in shallow and coastal areas. Research by Alsaaq and Shamji (2022) and Gramcianinov et al. (2023) investigated extreme wave trends in the Red Sea and South Atlantic, respectively, while Simonetti and Cappietti (2023) assessed changes in wave energy distribution along the Mediterranean coastline. Wei (2021) applied AI to forecast wind-wave time series along the US Atlantic Coast. The model was first trained using measured NOAA buoy data, then used to forecast H_s, T_m and θ_m. This yielded accurate predictions of H_s and T_m associated with storm events at short forecast lead times.

7.2 Long- and Short-Term Statistics

Environmental processes are often assumed piecewise stationary for modelling purposes. That is, a long-term probabilistic description of environmental variables often involves the combination of a model for the long-term variation of environmental conditions (i.e. integrated variables describing e.g. sea states or wind conditions) and a conditional model for the short-term variability of variables such as individual wave heights and instantaneous wind speeds. In this section of the report, a review of statistical modelling of both long-term environmental conditions and the short-term environmental variables is presented. A recent review of probabilistic models for environmental conditions, including wind and wave parameters, is presented in Ramezani et al. (2023).

Qin (2022) presents challenges in the probabilistic modelling for long-term significant wave height. He formulated a novel framework of a probabilistic model for long-term significant wave height to facilitate accurate probabilistic analysis and further decision making in engineering practice. The proposed novel framework identified homogeneous clusters of wave records, including the definition of the boundary and the probabilistic description of different clusters. Vanem and Fazeres-Ferradosa (2022) proposes a truncated, translated Weibull distribution for the long-time distribution of H_s in shallow waters and demonstrates that the model fits shallow water data quite well.

The statistical distribution of wave crest elevation is considered as a key input for the design and operations of offshore structures. For example, in offshore engineering, the estimation of a design crest height regarding a specified probability of exceedance for the defined duration or the design return-period is used to define the deck elevation, maintain an effective air gap and avoid wave-in-deck loading. Therefore, further studies on the long- and short-term statistics of wave crests have been conducted in recent years. Van Essen, Scharnke and Seyffert (2023) and Scharnke et al. (2023) investigated how many short-term sea states, measurement points and seed repeats for a deterministic short-term sea state are needed by tank tests to derive a reliable and repeatable probability distribution of the design wave, extreme design loads (such as green water loading, slamming, or air gap impacts which are typically strongly non-linear). Based on the investigated results, they provided guidelines for the convergence of most probable maximum (MPM) wave crest heights, MPM green water wave impact forces and MPM wave-in-deck loads on a stationary deck box for a ferry. Moreover, Zve et al. (2023) explored the competing non-linear processes that define the largest crest heights in uni-directional random seas. They first explored how the near-resonant interactions affect the crest heights arising in broad-banded, non-breaking, uni-directional seas in a wide range of effective water depths. They also quantified the role of the bound-wave interactions. Their numerical calculations concluded that $k_p d = 1.363$ (k_p being the wavenumber of the spectral peak frequency and d the water depth) indeed defines the boundary between energy focusing and defocusing for realistic Jonswap sea-states, irrespective of the spectral bandwidth and steepness. However, it was also concluded that the bound-wave contributions increased the largest crest heights, while the near-resonant interactions reduce them for $k_p d < 1.363$. The obtained results have important implications for describing crest-height distributions and appropriateness of second-order models for practical engineering calculations. Furthermore, Petrova et al. (2022) provide results on the probabilistic structure of experimental irregular uni-directional sea states that are subjected to the effects of non-linear focusing. In their study, they also investigated the evolution of the statistical distributions in space and time focusing on the crest maxima of waves exceeding a prescribed threshold, to add to the attempts for reliable modelling and prediction of extremely large waves at sea. Note that the findings on wave crest distributions in these experimental studies may be subject to some differences with conditions at sea, as waves can only be generated on the sides of the basin. This was discussed in detail in Sect. 5.

Wave parameters such as the skewness and kurtosis are often used in the evaluation of non-linear surface waves such as freak/rogue waves. Gramstad and Lian (2024) therefore calculated the skewness and kurtosis for short-term wave statistics for many sea states, covering a wide range of covariates (steepness, water depth, directional spreading and frequency bandwidth). Based on their results, they developed an efficient and convenient numerical method for calculation of the sea surface skewness and kurtosis for arbitrary wave spectra, to include these covariates into higher-order distributions for crest heights, wave heights and surface elevation. For free surface elevation of a non-linear irregular water wave field, Fuhrman et al. (2023) re-visited the derivation of the probability density function (pdf) in Longuet-Higgins (1963) utilising both moment and cumulant generating functions. They found that the second-order pdf can be represented exactly in terms of the Airy function through a change of variables coupled with complex analysis. The second-order pdf modified by them predicted increased probability of extreme positive surface elevations typical of e.g. rogue waves with good accuracy for directionally spread irregular seas in both finite and infinite water depths.

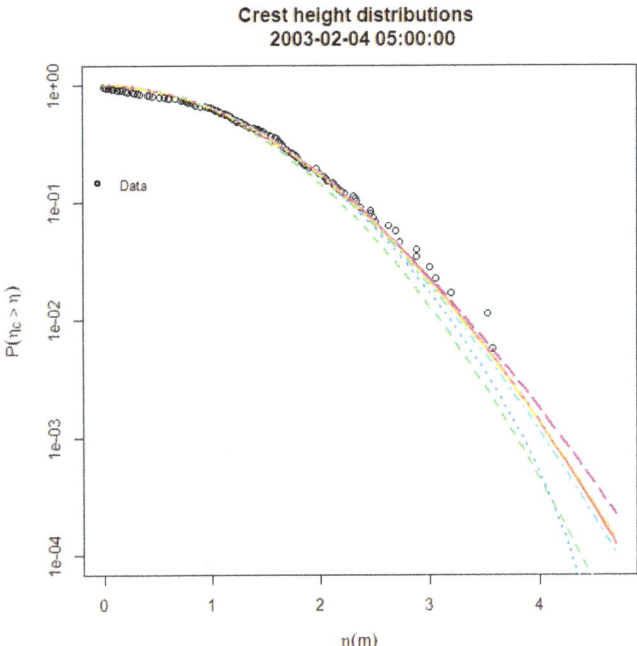

Fig. 8. Comparing various state-of-the-art crest height distributions to data

There is a variety of the state-of-the-art short-term wave height distributions models, see e.g. Figure 8. Each of these models is successful in at least one water depth regime, but not always in intermediate and shallow water depths or in steep sea-states because of wave breaking. Karmpadakis et al. (2022) addressed the short-term distribution of zero-crossing wave heights in intermediate and shallow water depths. They provided new physical insights on the effects of non-linearity, directionality, reduced effective water depth and finite spectral bandwidth. They proposed a new short-term wave height distribution through analysing a large database of experimental simulations of short-crested sea-states on flat bed bathymetries.

The proposed distribution is validated against field data recorded by wave radars in intermediate and shallow water depth locations in the southern North Sea. A new crest height distribution for non-linear and breaking waves in different water depths is also proposed in Karmpadakis and Swan (2022).

Freak/rogue waves have received relatively high attention in recent years, as unexpected occurrences of such waves can be dangerous for ships and offshore structures, navigation and operational activities in the ocean. Some publications on data-driven modelling of rogue waves were discussed in Sect. 4.2.2, and publications on their physics and statistics are discussed in detail in the dedicated Sect. 8.

7.3 Extreme Value Analysis and Extreme Wave Statistics

Extreme weather events and their uncertainties are receiving increasing attention because of their impact on safety of ships and other structures at sea. An important step that often occurs in design is the estimation of an extreme design wave based on recorded or hindcast data. This usually involves selecting and fitting a suitable probability distribution to the wave height data and extrapolating this distribution to determine a suitable design wave, e.g. the so-called '50- or 100-year wave'. This is a characteristic large wave height that can be expected with a certain low probability during the lifetime of the structure. Two important statistical problems can be distinguished in the prediction of wave extremes. The first problem is that the predicted extreme values for, say, 100 years must be based on data collected over a relatively short period, for example 10 years. The second issue is extrapolation of the observed data into its extreme region, typically lying well beyond even the most extreme of the available observation.

Improved knowledge of how to reliably describe extreme weather conditions at sea has a direct impact on reducing the safety factor and consequently the cost of marine structures. Bitner-Gregersen et al. (2022), developed by the ISSC-ITTC joint group "Uncertainties in Wave Modelling", discusses uncertainties associated with the long-term description of integrated wave parameters such as H_s and the zero-crossing/ spectral wave period, which can have a significant impact on the return values used in design and marine operations. A review of probabilistic methods of extreme value assessment in hydro-climatological applications is presented in Nerantzaki and Papalexiou (2022).

de Hauteclocque et al. (2023) proposed a new fitting technique for the probability distribution of H_s. The model is a combination of three different distributions, fitted separately to the core and extreme values of the data. For the core, the authors recommend a mixed distribution with two components: log-normal and Weibull, while the tail is fitted with the Generalized Pareto distribution. For the conditional distribution of the wave periods, the model is based on Generalised split normal distribution with the corresponding dependency functions.

Amarouche, Akpınar, Kamranzad, *et al.* (2023a; 2023b) investigated the sensitivity of the estimated extreme H_s to the analysed data sources and periods. In the study, the analysis of extreme waves was performed with two different models, i.e.,: Annual Maximum fitting to the Generalized Extreme Value Distribution (AM-GEV) and peak-over-threshold fitted to the Generalized Pareto Distribution (POT-GPD), and with four different wave data sources; (i) a 60-year wave hindcast dataset developed with the Simulating Wave Nearshore (SWAN) model forced by Japanese 55-year Reanalysis (JRA-55) wind reanalysis, (ii) the ECMWF fifth generation wave reanalysis (ERA5) for 40 years, (iii) the satellite observations calibrated and provided by the Institut Français de Recherche pour l'Exploitation de la Mer (IFREMER) for 20 years and (iv) the wave buoy measurements provided by Copernicus for 40 and 30 years. The authors stated that the extreme wave estimates can change by more than 20% over most of the global ocean depending on the used wave sources, and therefore the choice of suitable sources must be carefully considered. A new way to de-cluster wave data for POT modelling based on the extremal index is proposed in Oikonomou et al. (2020).

Non-stationary Extreme Value Analysis (NEVA) allows to determine the probability of exceedance of extreme sea states considering trends in the time series of data at hand, see also the discussions in Sect. 7.6. De Leo et al. (2021) analysed the reliability of NEVA of H_s and T_p under the assumption of linear trend for time series of annual maxima (AM) H_s in the Mediterranean Sea. A methodology to assess the significance of the results of the non-stationary model employed is proposed. Görmüş, Ayat and Aydoğan (2022) evaluated the extreme waves in the Mediterranean and the Black Sea. Annual

Maximum Series (AMS), and Partial Duration Series (PDS) of the H_s are used from the ERA5 dataset. Generalized Extreme Value (GEV), Gumbel, Weibull, Lognormal, and Generalized Pareto Distribution (GPD) models are used to predict H_s for 50, 100, and 500-years of return periods.

Non-linear wave-induced response of structures is often very sensitive for details of the wave excitation (height, period, steepness, breaking etc.). To resolve such response including all flow details, it is usually required to apply high-fidelity modelling. Such models cannot be run for long enough to assess lifetime loads on marine structures. Most extreme value prediction methods for highly non-linear structure response therefore rely on multi-fidelity wave modelling, where only critical sea states or critical wave events are run with high-fidelity methods. A review of extreme value prediction methods for non-linear waves and wave impact loads is provided by van Essen and Seyffert (2023). These methods include response-conditioning (or design wave) methods, screening methods and adaptive sampling methods. In the first category, response-conditioning methods, Seyffert (2022) applied the 'NL-DLG' design wave method to generate an ensemble of wave profiles, conditioned to contain rare wave groups. It can also be applied to other non-linear responses to efficiently generate critical wave profiles for these responses. Most available design wave methods generate linear Gaussian wave profiles, which is not always sufficient for highly non-linear structure response. Kim et al. (2022) defined a new design wave method, which can generate higher-order wave events by incorporating a higher-order spectral (HOS) model.

The interest in applying adaptive sampling methods to wave-induced response increased after the publications of Mohamad and Sapsis (2018) and Gramstad et al. (2020). Guth and Sapsis (2022) applied adaptive sampling combined with Gaussian Process surrogate regression to generate design wave groups (consisting of a set of Karhunen–Loève (KL) wave components) for non-linear ship response. This can be seen as a combined response-conditioning and adaptive sampling method. The method was also applied to obtain wave loads on a monopile in Guth, Katsidoniotaki and Sapsis (2024) showing that high-fidelity simulation of the resulting critical wave groups leads to reasonable response distributions (but tends to underestimate the tails). One possible reason is that the number of considered wave components in the design waves has to be maximised for the method to be efficient. One way to overcome this is to combine screening with adaptive sampling instead. This way the critical wave groups are not synthesized based on input conditions, but selected ('screened') from a long lower-fidelity simulation. Van Essen, Scholcz and Seyffert (2023) defines a new method that combines screening with (multi-fidelity) Gaussian Process Regression and adaptive sampling. This pilot study also showed that the method could work for a simplified problem that selects critical linear wave events to predict the full second-order wave crest height distribution. One drawback with most multi-fidelity extreme value prediction methods is the necessity to define input for high-fidelity wave modelling based on selected or generated wave events with a lower fidelity. Some ideas to reduce this problem are to use event matching procedures such as described in Gramstad et al. (2023), to use coarse mesh CFD as a low-fidelity screening tool as demonstrated in e.g., van Essen et al. (2021) or to use direct coupling methods as done in e.g., Kamath et al. (2023).

H. Liu et al. (2024) investigate the climatology, variability and trends of extreme wind and wave events based on ERA5 reanalysis data spanning from 1940 to 2021. This study also examines the relationship between large-scale climate models and extreme wind and wave events and discusses the possible underlying mechanisms, thus providing insight into the main differences in extreme weather conditions in different regions.

7.4 Multivariate Analysis and Joint Distributions

Reliability-based design of offshore structures usually requires joint probabilistic models to describe the long-term environmental scenario. Traditionally, wave scatter diagrams have been used for this purpose, but more information can be obtained by more careful joint models of the relevant environmental variables. The full probabilistic modelling of wind and wave parameters for a site in the South China Sea usually attacked by typhoons is studied by Song et al. (2022), see also Song et al. (2024). The full probabilistic model of the environmental variables is developed using the C-vine copula method. An alternative joint model based on generative adversarial networks was proposed in Song et al. (2023). The application of the full probabilistic model to the fatigue analysis of a floating offshore wind turbine (FOWT) is illustrated through an example. Simão, Sudati Sagrilo and Videiro (2022) presented a multi-dimensional joint model alternative for the probabilistic long-term environmental data description, including directional variables. The joint model aims to statistically describe the main environmental parameters for the probabilistic design of marine structures, such as floating production, storage, and offloading vessels (FPSOs) systems. Liao et al. (2022) focuses on the statistical characteristics of directional wave climate in the seasonal ice zone of the Barents Sea. The joint distributions of H_s, T_m and θ_m were constructed using a mixture trivariate distribution model. The ocean data were classified into four groups based on the relative weights of the energy content of wind wave and swell fields. Yang et al. (2023b) focuses on the bivariate distribution of wind speed and air density with a mixture copula model, each component of which was constructed using a Weibull distribution for the wind speed, a log-normal distribution for the air density, and a Gaussian copula function for the description of the dependency structure. The optimum component number and maximum likelihood estimate of the mixture model were determined using the Bayesian information criterion and expectation-maximisation algorithm, respectively. Copulas and Bayesian inference were applied to model H_s and T_p in Duan et al. (2024). H. Wang et al. (2023a; 2023b; 2023c; 2023d; 2023e) proposed joint distribution of wind speed and direction over complex terrains based on non-parametric copula models.

Non-parametric approaches to joint distribution modelling have become popular in recent years (Q. Han et al., 2018). Non-parametric kernel density estimators were combined by copulas to form multivariate model of wave and wind conditions in Wen et al. (2024). Non-parametric marginals combined with non-parametric copulas for the dependence structure is proposed in Latif and Simonovic (2022). However, as shown in Vanem, Lande and Fekhari (2024a; 2024b), non-parametric models may fail to capture the tails of the distributions well, so some caution is advised if such models are to be used for extreme value analysis. A similar conclusion was reached by Meng and Li (2024).

The uncertainty of design parameters for marine structures based on multivariate models were analysed in G. Liu et al. (2023), which suggest that such uncertainty analysis should be considered during model selection. A joint probabilistic approach is presented in García-Rojo (2004) for establishing joint distributions of environmental variables from combined information from different measurement locations. The approach has been found to compare well with the commonly used Measure-Correlate-Predict method.

7.4.1 Multivariate Extreme Value Analysis and Environmental Contours

The extremes of multiple variables are difficult to assess and even the definition of a multivariate extreme value is ambiguous. Recent reviews of multivariate extreme value modelling are reported in e.g. Engelke and Ivanovs (2021) and Nolan (2024).

Environmental contours are a pragmatic and widespread method to estimate the long-term extreme response of marine structures. It is essentially a way to describe multi-variable extreme environmental conditions. Over the years, a range of approaches have been proposed. A benchmarking study was recently conducted by Haselsteiner et al. (2021) to compare the various methods using a common set of data. Nine teams of researchers contributed to the benchmark. The analysis of the submitted contours highlighted significant differences between contours derived via different methods. de Hauteclocque et al. (2022) extended this benchmark study by providing a quantitative assessment of the contours submitted to the exercise. Three main reasons for the discrepancies were identified: first, the statistical model used to model the joint probability; second, the effect of serial correlation, which is generally neglected; and last, the varying assumptions made by the different environmental contour methods. In the recent years, progress has been made on each of those items.

On the modelling of the extreme joint distribution, a promising approach, coined "SPAR" (Semi-Parametric Angular-Radial) is introduced in Mackay and Jonathan (2024), and with implementation details and practical examples in Murphy-Barltrop et al. (2024) and Mackay, Murphy-Barltrop and Jonathan (2025). The approach relies on a transformation of variables to polar coordinates; the tail of the radial variable is then modelled using General Pareto distribution, whose parameters are conditional on the angle. The inference can then be seen as a set of univariate, non-stationary problems.

On the serial correlation issue, discussed specifically in Mackay et al. (2021), an approach for partially accounting for serial dependence in the construction of environmental contours is proposed by Vanem (2023a), (2023b) based on simulating a time series of a primary variable that preserves both its marginal distribution and auto-correlation structure. The approach discussed in Mackay and de Hauteclocque (2023), based on a collection of univariate fits, can also address this serial correlation issue.

Finally, about the different assumptions of the environmental contour, Mackay and Haselsteiner (2021) clarifies the distinction between different kinds of contours available in the literature. Approaches like "highest density contours" (Haselsteiner et al., 2017) and ISORM, Inverse Second Order Reliability Method, (Chai and Leira, 2018) have an overestimation bias. Furthermore, the overestimation bias increases dramatically with the number of dimensions. This explains why the original IFORM, Inverse First Order Reliability Method, and its derivatives (direct sampling and direct-IFORM) are preferred in practical applications.

Other aspects of environmental contours have also been investigated in recent years. We can cite Huseby et al. (2021) which improves on the direct sampling approach of

(Huseby, Vanem and Natvig (2013), (2015)) by introducing a more robust approach to extract the contour, with a more rigorous mathematical foundation. Closely related, Hafver, Agrell and Vanem (2022) ingeniously observed that direct sampling contours can be extracted using Voronoi cells. Further to this finding, Mackay and de Hauteclocque (2023) combines this with the direct-IFORM method (Derbanne and de Hauteclocque, 2019) and propose a robust way to derive contours in high numbers of dimensions, with an application in 4D. Environmental contours for multivariate extremes are also discussed in Simpson and Tawn (2024), involving a transformation to Laplace space and a definition of contours in radial-angular coordinates. Three-dimensional environmental contours accounting for sampling methods and season were presented in Meng and Li (2024).

For analysing the uncertainty related to the construction of environmental contours, Y. Zhao and Dong (2023b) assess the uncertainty on the extreme mooring loads of floating system considering short-term variability including significant differences due to marginal distribution fitting, parameter estimation methods and joint models. The metocean conditions conditioned on extreme structural responses were studied in Speers et al. (2024), which report that environmental contours might not be conservative in every situation and suggested that contours may need to be calibrated for the structural response being analysed in order to reduce bias. Similar results were reported in Seyffert and Kana (2020). The environmental contour approach to assess extreme structural response was compared to a sequential sampling approach (Gramstad et al., 2020) in Wang et al. (2024a; 2024b; 2024c; 2024d), suggesting that the latter might be preferred in certain situations where the assumptions implicit in the environmental contour method is violated. An alternative approach based on scenario optimization is presented in Crespo, Agrell and Vanem (2024). An improved IFORM for long-term extreme response analysis was proposed in Wang et al. (2024a; 2024b; 2024c; 2024d) and compared to environmental contours. It showed that extreme responses tend to occur inside the environmental contours, indicating that design sea conditions may not lie on the contours.

Huseby et al. (2015), Hafver et al. (2022) and Zhao and Dong (2021) investigated the effect of statistical models for constructing the bivariate distribution of metocean data on design loads and reliability assessment of offshore structures. The variabilities in short-term extreme response and the contributions of all sea states have been ignored, and this has caused unreliable results. Considering concerns over the sensitivity of the hydrodynamic performance of floating structures to wave data, Zhao and Dong (2022) evaluated the effect of environmental conditions on the short-term extreme response parameters and present a full long-term analysis for the failure probability evaluation of floating structures. Vanem (2023b), Zhao and Dong (2023a) and Vanem, Fekhari, et al. (2024) provided realistic multivariate models of environmental parameters to accurately describe the statistical characteristics of metoceanic conditions based on copula models and construct environmental contours for the reliability-based design of marine structures. An extension of the conditional extremes models (Heffernan and Tawn, 2004) for bivariate mixture models is presented in Tendijck et al. (2023), where the extremal dependence between the variables might be a mixture of simpler bivariate distributions.

7.5 Spatial and Temporal Statistics

Since both individual waves and sea states evolve in space and time, the simultaneous statistics of spatial and temporal behaviour of individual waves and sea states are important in many practical applications. For example, in the conclusions from the ExWaMar project concerned with developing warning criteria for rogue waves (Bitner-Gregersen et al., 2024), it is emphasised that space-time data should be employed in the development of rogue wave warning criteria, both in order to reduce the sampling variability of measurements and because point crest statistics underestimates the statistics of crests over an area small enough to be relevant for ships and offshore structures. With reduced cost and the availability of dedicated open-source software, stereo video imagery is increasingly used for wave measurements and it is necessary to employ spatio-temporal techniques to analyse the data. Likewise, with the availability of inexpensive free-drifting wave buoys (see Sect. 6), the estimation of sea state parameters based on a distributed set of buoy measurements require analysis in time and space. A description of sea state parameters in space and time can also be important in quantifying fatigue and extreme loads on ships in transit. In addition, there is an increased interest in large floating structures such as energy storage hubs, floating solar panels or even islands and airports (see also the Ocean Space Utilisation Committee). Such installations span larger areas than traditional marine structures, requiring more emphasis on space-time wave statistics.

7.5.1 Spatial and Temporal Statistics of Sea State Parameters

Smit et al. (2021) used hourly H_s data provided by approximately 60 free-drifting directional wave buoys distributed in the Northern Pacific to investigate the effect on wave forecast quality of using data assimilation techniques. Using the WW3 wave model and a sequential optimal data assimilation technique, the accuracy of the wave forecast was investigated and compared with the accuracy without using data assimilation. It was demonstrated that the use of a simple assimilation technique gave a 27% reduction in the root mean square error of H_s forecast compared with the case where no data assimilation was used.

Building on the methodology for conditional extremes developed by Heffernan and Tawn (2004), Shooter et al. (2021), (2022) have developed spatial conditional extreme models for storms. Shooter et al. (2021) consider storm peak H_s in the North Sea characterized by direction and distance and finds reasonable agreement with hindcast results. Shooter et al. (2022) develops a multivariate spatial conditional extreme (MSCE) model suitable for investigating the characteristics of joint sea state parameters given the occurrence of an extreme value of one of the parameters at one location. A joint model is conditioned on extreme satellite wind speed and developed for satellite wind speed, hindcast wind speed and hindcast H_s over a satellite trajectory in the North Atlantic. It is concluded that the spatial dependence of all three quantities decays over 600–800 km.

Nielsen (2022) used the ERA5 reanalysis database to investigate the variation in sea state parameters for ships sailing on four typical ocean crossing routes. H_s, T_z and θ_m were considered for different vessel speeds and different methods of interpolation of sea state parameters. It is highlighted that the sea state parameters are varying rapidly along the ship route and that analysis of sea state parameters should be based on bilinear interpolation between the nearest grid points rather than the nearest neighbour approach.

Hildeman, Bolin and Rychlik (2021) use the stochastic partial differential equation (SPDE) approach in combination with a spatial deformation method to incorporate non-stationarity and anisotropy to model H_s. The model was fitted to the ERA-Interim hindcast data set for the North Atlantic and it was demonstrated how the model could be employed to estimate fatigue damage and wave height encountered for a ship in transit. Although the present model does not consider the temporal evolution of H_s, this could be achieved by introducing a space-time separable covariance function.

Çelik (2022) and Altunkaynak, Çelik and Mandev (2023) applied fuzzy logic techniques to the problem of short time forecasting of H_s and compared with several measurement stations in the North Pacific and North Atlantic. To improve the forecasting quality of machine learning techniques, it is useful to preprocess the data and attempt to sort the time series into stochastic and deterministic parts before developing the fuzzy logic algorithm. The authors have investigated preprocessing strategies based on the wavelet transform, singular value decomposition (Çelik, 2022) and singular spectrum analysis (SSA) (Altunkaynak, Çelik and Mandev, 2023) in addition to unprocessed data. They found that the SSA preprocessing algorithm was superior with the ability to provide good estimates of H_s with a lead time of up to 12 h over a range of water depths.

Whereas long-term analysis of extreme wave responses has traditionally been carried out by integrating all sea states consisting of blocks of piece-wise constant sea state parameters, long-term analysis of sea states is increasingly carried out for discrete storms where the evolution of the sea state parameters during the storm is of interest. Tendijck et al. (2024) use Markov processes to model extreme excursions in multivariate time series around the storm peak introducing distinct peak, pre-peak and post-peak periods. Using a simple response parameter for a northern North Sea location, it is concluded that the approach improves the description of extreme excursions in a storm compared with a historical storm matching approach.

A bivariate regional frequency analysis of H_s and wave periods is reported in Vanem (2020), (2021) extending previous applications of univariate regional frequency analyses, in order to exploit spatial data in extreme value analysis, see also Bai, Ruan and Wang (2023). A spatial non-stationary extreme value analysis over the Mediterranean Sea is presented in De Leo et al. (2021).

A time-series model for H_s that preserves both the marginal distribution and the autocorrelation structure is proposed in Vanem (2023b), (2023a), inspired by the work of Papalexiou (2018), Papalexiou and Serinaldi (2020). This can account for the effect of serial correlation in the estimation of extreme values and can also be important in applications where the sequence of wave loadings is important. Time series of H_s is separated into deterministic and probabilistic components in W. Huang et al. (2024) to establish non-stationary time series models accounting for climate variability.

7.5.2 Spatial and Temporal Description of Individual Waves

Bitner-Gregersen and Gramstad (2021) studied the spatio-temporal sampling variability of the sea state recorded in the North Sea in 2018 where the 'Justine Three Sisters' rogue waves were observed. Using a high order spectral (HOS) model over a 3.5 × 3.5 km domain to third order in wave steepness, they simulated 500 30 min simulations and extracted data from 256 evenly spaced points throughout the domain. It is concluded that sampling variability over 30 min records in the statistics of individual waves are significant and should be considered carefully in design. The difference between statistics of crest height over an area and a point is large. They obtain a good fit to the distribution of surface elevation using Gram-Charlier series and recommend further investigations of the limitations and applications of the Gram-Charlier series.

Benetazzo et al. (2021) used stereo-imaging observations from a tropical storm in the Northwest Pacific to analyse the spatio-temporal extreme value statistics of maximum crests and wave heights. Using a consistent formulation for spatio-temporal extreme values based on the Tayfun and Bocotti (second-order/autocovariance) distributions for crest and wave height respectively, they compared extreme values over several 20 min records over a 140 × 120 m area, with predictions based on spectral estimates from ECMWF. The spectra provided by ECMWF were compared with the spectra found from the stereo measurements. The authors found that the stereo-imaging results provided wave spectral parameters in good agreement with the ECMWF predictions. The largest wave heights and crests were located to the north-east of the eye of the storm whereas the potential for rogue waves is found to the south/south-west. The agreement between measurements and predictions of extreme crests and wave heights are reasonable although it is acknowledged that the distributions used in the comparison are correct up to second order only.

In a subsequent analysis, Davison et al. (2022) used a HOS model to third order in wave steepness to simulate the same storm and compare with the stereo-imaging results. This study used spectral input from ECMWF at different stages in the storm generating a large number of 20 min records of surface elevation, and studied the spatio-temporal extreme crest focusing on the effect of crossing swell and wind seas. They found good agreement between the numerical results and measurements both for point statistics and area statistics. They find an increased space-time crest height probability for wind sea and swell systems crossing at 160 deg and propose a consistent formulation for wave steepness in crossing seas, suggesting that it may be useful for predicting rogue wave prone sea states.

In an analysis of large storm wave measurements in the Eastern Mediterranean, Knobler et al. (2022) present a novel analysis of waves in space and time focusing on the potential risks posed by large waves. Using the Euler characteristics of Gaussian fields as a starting point and incorporating and modifying the third-order Tayfun-Fedele model for crest height at a fixed point, a formulation for crest height over an area is proposed, which takes into account vessel area. In a striking example where also forward speed is taken into account, it is demonstrated that the probability of encountering a critical crest height with probability 10^{-5} at a fixed point is increased to 1/4000 for a small vessel and 1/1260 for a large vessel.

Malila et al. (2023) present an analysis of 18 years of high-quality laser altimeter measurements from the central North Sea. This dataset is complemented by stereo-video observations at the same location containing five individual storms collected in one winter season. The stereo-video data provides an opportunity to investigate the spatio-temporal evolution of the largest waves and compare with the laser point measurements. They observed a clear slowdown of the largest waves close to the position of maximum crest elevation. This is not found in linear or second-order theory, but consistent with higher-order non-linear wave predictions and laboratory observations. By investigating crest-area statistics with areas varying from zero to 60x60 m, they found differences between point and area crest height statistics up to a factor of 1.6, consistent with the linear and second-order crest-area predictions.

In order to obtain stable tail statistics of non-linear wave properties in space and time, Tang and Adcock (2022a, 2022b) have investigated methods for approximating the wave statistics using a limited number of deterministic wave groups calculated numerically using the Modified Non-linear Schrodinger equation (MNLS). By conditioning the initial conditions of the MNLS simulations, they conclude that this approach can provide accurate space-time statistics of wave properties at a fraction of the computational cost of direct non-linear Monte Carlo simulations.

Tang and Adcock (2021) investigated the usefulness of data driven methods for estimating crest distributions over space and time. Two simple machine learning approaches were employed to link sea state parameters to crest distributions: a simple fit to Gumbel parameters and a Random Forest approach. Data were generated using second-order theory and the results are compared with established crest distribution models. It was concluded that the data driven approach could be useful for establishing crest distribution models, particularly the Random Forest method performed well compared with the other methods.

7.6 Non-stationary Analysis and Covariate Effects

Environmental conditions will often be dependent on many factors such as season of the year, spatial location, long-term trends (e.g. due to climate change) and prevailing wind or wave directions. Thus, the *iid* assumption (i.e., that the data are independent and identically distributed) will generally not be fulfilled, and stationary models to describe the environment may strictly speaking not be correct if the conditions are non-stationary. These non-stationarities could be important and statistical models incorporating them may represent a notable improvement compared to stationary models. The effect of non-stationarities would be important for both univariate and joint models, and for extreme value models and distribution models for all the data. The importance of accounting for non-stationarity in risk and reliability assessments are discussed in Radfar and Galiatsatou (2023), which finds that considering non-stationarity in extreme coastal events is important, and even more so when the dependence structure of the wave parameters is accurately modelled. Failure probabilities of coastal structures may be underestimated by up to 33% with stationary models that are not accounting for climatic trends. Similar results were found by Baldan et al. (2022), who report that estimates of return levels of extreme sea levels are more conservative if non-stationary models are assumed. Obviously, the best model will be case-specific, but at least for some applications, non-stationarities will be important and may influence inference about environmental conditions. A review of methods to detect, attribute and manage non-stationarities in weather extremes is presented in Slater et al. (2021), which distinguishes between two types of non-stationarities; trends and abrupt changes and between different symptoms, such as non-stationarities in mean levels, variability or frequency. It is stressed that departure from stationarity should be detected and tested before engineering design is adjusted for such effects.

One way of accounting for non-stationarities is to include covariates (explanatory variables) in the statistical models. Another is to pre-process the data to remove the non-stationary effects. Yet another approach could be to use time-series or spatial fields to account for autocorrelations and dependencies in space and time. One example of the latter is the statistical model for H_s presented in Vanem (2023b), where non-stationarities due to temporal dependencies at different timescales are modelled by first pre-processing the data to account for seasonal dependencies and then a time-series model is established that preserves the short-term serial correlation. This is demonstrated to correct for the positive bias know to occur when ignoring serial correlation (Mackay et al., 2021). Different interpolation schemes to account for spatio-temporal variation of sea state parameters along ship routes are presented in Nielsen (2022). However, in this subsection, the main focus will be on statistical models accounting for non-stationary effects by adding covariates to the models.

A spatio-temporal model for H_s is proposed in Obakrim et al. (2023), where wind fields are used as predictors. Both wind-sea and swell are accounted for by combining a local predictor for the wind sea and a global predictor for the swell in a linear regression model. Hence, non-stationarities in space and time are modelled using wind field data as covariates. A vector autoregressive model with exogeneous input is used in Raudiya, Rohmawati and Adytia (2021) for modelling and forecasting ocean waves based on wind information.

Seasonality is another source of non-stationarity that may be accounted for in the statistical models. This can be accounted for by pre-processing the data by estimating the seasonal effects and then fit a stationary model on the residuals (see e.g. Athanassoulis and Stefanakos (1995), Vanem (2018)). Alternatively, non-stationary models accounting for the seasonality in different ways may be established. A rather cumbersome process for modelling seasonality in metocean data is proposed in Ma and Zhang (2024), where the data are segmented into short periods where the conditions are assumed to be stationary. Seasonality is modelled in extreme value models by D'Arcy et al. (2023), where skew-surges are modelled by a non-stationary generalized Pareto distribution with daily covariates for the tails and peak-tides are allowed to vary over months and years. Dependence is accounted for by a tidal covariate and temporal dependencies are adjusted for by the sub-asymptotic extremal index. Time series of H_s are decomposed into stationary and non-stationary components in Nasir et al. (2023), where the stationary part is modelled as an autoregressive model and non-stationary components describing seasonality and trend is modelled by finite Fourier series expansion. Wind speed time series are decomposed into deterministic and stochastic components in Yang and Dong (2023), by combining signal decomposition methods and recurrence quantification analysis. This is reported to be able to capture complicated non-stationary patterns without relying on covariate information.

Metocean conditions are often associated with directions, and directional models are often useful in design. Hence, storm direction (θ_m) is included as a covariate in the non-stationary models proposed in Malliouri et al. (2023) for modelling of coastal storms. They establish a directional extreme value model where the parameters of the generalized Pareto distributions are allowed to vary smoothly by direction, using a finite order Fourier expansion. This can be used to obtain return values for H_s as a function of storm direction. An approach to jointly estimate the parameters of frequency-direction wave spectra from three-dimensional buoy time series are outlined in Grainger et al. (2023), based on a multivariate extension of the debiased Whittle likelihood. This is reported to yield a significant improvement in performance compared to least-squares or moment-matching techniques.

Climate change and large-scale climate variability is a driver for non-stationarities in metocean data, and several papers are dealing with how to incorporate this into probabilistic models. An investigation into the uncertainties in estimating the effect of climate change on the 100-year return values of H_s using non-stationary extreme value models is presented in Ewans and Jonathan (2023). They report that there is large variability in return value estimates from different climate models (general circulation models) and that these sources of uncertainty seem to be larger than typical modelling choices for extreme value models. Variation in return value estimate of $\pm 15\%$ are reported. The influence of climate variability on ocean wave energy was investigated by Sardana, Kumar and Rajni (2024). A non-stationary model with different climate variability indices as covariates was assumed, indicating that both the mean and extreme wind-sea and swell wave energy are influenced by the different climate modes (ENSO, NAO, IOD, SAM). These studies highlight the importance of accounting for climate change and climate variability in statistical models of the ocean environment.

Four different approaches to model the influence of climate variability on global ocean waves were explored in Liu *et al.* (2023), i.e., linear regression, composite analysis, empirical orthogonal functions (EOF) and wavelet analysis. The linear regression model uses six common climate oscillations as regressors for ocean waves. The composite analysis determines the mean values of the wave parameters for years when the climate index anomalies are outside ± 1 standard deviation. EOF analysis decompose the data into spatial and temporal patterns and calculates the correlation between the principal components and the climate indices. Wavelet analysis decomposes the data into the time-frequency domain to identify the dominant periods and how they evolve over time, and to analyse the correlation between the wave climate and the climate index anomalies. Other climate variability indices were used as covariates in non-stationary GEV distributions for extreme H_s and wind speeds over the Indo-Pacific Ocean in Kumar et al. (2024). Non-stationary models combining a Poisson distribution for the frequency of typhoons and Gumbel, Weibull and generalized Pareto for H_s are presented in Li et al. (2024). Varying typhoon frequencies are modelled by different Poisson rates for different time periods, and non-stationarities are modelled by assuming time-dependent linear functions for the location parameter in the models for H_s. A notable difference between return value estimates obtained from stationary models are reported. Non-stationary GPD models are also presented in D'Arcy, Tawn and Sifnioti (2022) for skew-surges, with covariates accounting for both climate change and seasonality.

Non-stationarities may have a particular impact on estimates of extreme events, and several recent papers have reported applications of non-stationary extreme value analysis for metocean data. A recent literature review of applications of non-stationary extreme value analysis in coastal regions are presented in Radfar, Galiatsatou and Wahl (2023). Non-stationary GEV models are introduced to model extreme wind loads in Arif et al. (2022), where the GEV model parameters are modelled by different functions of time, similar to the models in Vanem (2015); see also De Leo et al. (2021). A non-linear peak-over-threshold model is proposed in Barlow et al. (2023) as a piecewise-linear model in one- or two-dimensional covariates related to season and direction. Another seasonal-directional extreme value model for individual wave heights is proposed in Bohlinger et al. (2023), where cyclic cubic splines are used to model the effect of varying season and direction. A multivariate (wind speed and H_s) spatial conditional extremes model is presented in Shooter et al. (2022), involving non-stationary directional-seasonal marginal extreme value analyses at specific locations and extremal spatial dependence between the locations.

As this brief literature survey suggests, several applications of non-stationary modelling of metocean variables have been described. Typically, non-stationarities are related to spatio-temporal variabilities at different scales; seasonality, directionality, climate variability and trends due to climate change. A particular focus of several studies is on the non-stationarities of extremes. Several methods exist for taking such effects into account, and the preferred approach will be highly case-specific. These should be considered in cases where departure from stationarity is deemed to be important, and the *iid* assumption cannot easily be defended.

8 Extreme Events and Conditions

8.1 Rogue Waves

Rogue waves, also known as freak waves, are waves which have unusually large crest or wave heights in the sea states in which they appear. A practical and much used definition of a rogue wave is a wave with a crest larger than 1.25 H_s or a temporal wave height larger than 2.0 H_s (DNV, 2024, based on Haver, 2000). They have been observed in all water depths, with or without the presence of current or variable bathymetry. If rogue waves occur with a sufficiently high frequency in sea states which are sufficiently severe and frequent enough to be relevant for design, they can be governing for the integrity of ships and offshore structures. As a result, the presence of rogue waves and the prediction of the sea states in which they may occur continues to receive considerable attention. The physics underlying the presence of rogue waves in some sea states have been the subject of discussion. Are rogue waves a separate population driven by non-linear modulation processes or merely rare occurrences of the established linear superposition principle accounting for space-time affects and local non-linearities? In the latter case, rogue waves are largely accounted for in established design practices. Section 4.2.2 is also dedicated to rogue waves, but that discussion focuses on the data-driven technologies used to predict them rather than the physics of the waves themselves.

The presence of non-linear modulational instability effects to explain the presence of rogue waves in the ocean is an active topic of study. Akhmediev (2023) and Tikan et al. (2022) provide recent reviews of the mathematics underlying the presence of rogue waves generated by non-linear modulational instabilities. Whereas long-crested sea states with narrow frequency bandwidth are prone to these non-linear modulational instabilities, investigations of sea states with realistic directional and frequency bandwidths have yielded less evidence of the presence of these effects to explain rogue waves. Nevertheless, there are published reports of rogue wave events which are attributed to non-linear modulation effects. Onorato et al. (2021) report a rogue wave event measured in the Bay of Biscay where a buoy measured a 27.8 m wave height in an H_s of 11 m. Analysing the buoy measurements using the non-linear Schrödinger equation, it was concluded that the wave was part of a highly non-linear wave packet extending over 1 km with individual crests propagating faster than 100 km/hr and which may have persisted for 20 min or more. Cavaleri et al. (2021) report a rogue wave event at the CNR-ISMAR oceanographic tower at 16 m depth offshore Venice in the Northern Adriatic Sea. After a critical review of the available data, it was concluded that the wave crest was at least 5.1 m in an H_s of 2.4 m with a T_p of 5.8 s. The authors note the remarkable persistence of this wave observed on the cameras with little change in crest height after 20 s and 200 m propagation. Toffoli *et al.* (2024) reported direct observations of surface waves from a stereo camera system across the Southern Ocean in the Australian winter aboard the South African icebreaker S.A. Agulhas II. Using the collected data the authors documented that modulation instability exists in the ocean.

In contrast, Gemmrich and Cicon (2022), in an analysis of a rogue wave observed at 45 m water depth offshore Vancouver Island, found no evidence that modulational instability played a significant role in the formation of the rogue wave. Buoy measurements of a crest height of 11.96 m in an H_s of 6.05 m were analysed and it was found using simulations and wavelet analysis of the buoy record that linear superposition together with the bound wave structure in intermediate water depth could explain the occurrence of the extreme wave at reasonable probability levels. Knobler et al. (2022) present analysis of deep-water buoy measurements during two large storms in the Eastern Mediterranean. The largest waves were found to display similar characteristics to the Draupner, Andrea and (estimated) El Faro rogue waves. The authors attribute the formation of the largest waves to focusing by linear dispersion enhanced by second-order non-linearities. Teutsch and Weisse (2023) investigated a continuous dataset which consists of the buoy and radar wave elevation data of different frequency resolutions, from eight measurement stations in the southern North Sea. They also concluded that the modulational instability is not the most probable mechanism to generate rogue waves in their dataset in the real seas from the southern North Sea.

Ji et al. (2022) investigated the relation between the occurrence of freak waves at real sea states and the statistical characteristics of wave trains using the observed wave data of multiple sea areas. They also explored the probable reasons for freak wave generation at real sea conditions. They found that the relation between occurrence probability of freak wave and kurtosis could be described by the MER (Modified Edgeworth Rayleigh) model proposed by Mori and Janssen (2006). Furthermore, they analysed the generation types of freak waves based on the wavelet energy spectrum and time process of wave surface for typical freak waves in the observation data and confirmed that the freak waves at real sea states can be attributed both to modulation instability and linear superposition.

Malila et al. (2023) investigate extreme wave crests in an 18-year record of high-frequency laser altimeter wave measurements at a central North Sea location. The observed rogue waves in this dataset are visualised in Fig. 9. It is concluded that significant deviations from second-order crest observations are observed for large and steep sea states with small directional spreading. The data set is complemented by stereo-video observations at the same location during five winter storms providing useful information about the space-time evolution of the largest events. From the stereo-video observations, they observe a slow-down near the time and location of maximum crest elevation for both breaking and nonbreaking extreme crests. As a result of this slow-down, it is concluded that an estimate of steepness based on time series and linear dispersion can significantly overestimate the crest steepness of an individual wave.

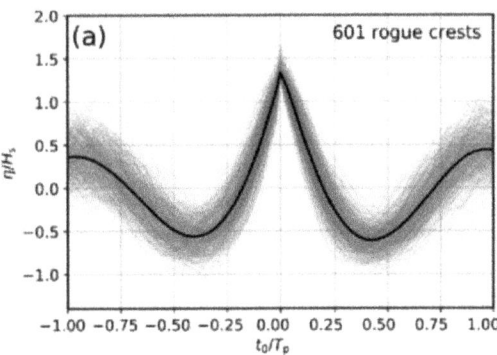

Fig. 9. Temporal profiles of all verified rogue wave crests ($C/H_s \geq 1.25$) in an 18-year North Sea dataset of Malila et al. (2023) - individual crest profiles in gray, and ensemble average profile in black. © American Meteorological Society, used with permission.

As also mentioned in Sect. 5.2.6, McAllister *et al.* (2024) experimentally evaluated the influence of wave directionality and crossing angles of mixed wave conditions on wave breaking and resulting crest heights. They concluded that directional spreading or a crossing angle can significantly delay breaking, and this increase maximum crest heights (which may be one of the causes of freak waves). These conclusions were based on short-crested and crossing focused wave events; whether similar conclusions are valid for irregular realistic wave conditions needs to be evaluated. Majumder et al. (2022) also investigated the effect on freak wave occurrences due to the cross-sea conditions based on observed buoy data in the north Indian Ocean. They confirmed that there were 55 freak wave events between 2009–2017. All of the freak wave events occurred in a combined sea state of swell and wind sea.

A decrease in water depth can also give rise to non-linear amplification of the crest height, yielding an increased probability of rogue wave formation. Lyu et al. (2023) develop an evolution model for a directional wave field based on the depth-modified non-linear Schrödinger equation and investigate numerically the distributions of wave heights over a slow transition from deep to shallow waves. It is concluded that whereas the directionality of the waves reduces the non-linear crest and wave heights, the presence of a changing bathymetry retains an important role in non-linear crest amplification.

Tang et al. (2023) carried out an experimental and numerical study of random, directional waves interacting with an abrupt depth transition. A JONSWAP spectrum was investigated with a relatively narrow directional spectrum in the experiments. In the fully non-linear potential solver, the directional spreading was extended to realistic directional spreading. It was concluded that whereas directional spreading significantly reduces the non-linear amplification of the crest and the presence of rogue waves, it does not remove the formation of rogue waves over an abrupt depth transition.

Significant efforts have been put into investigating sea states prone to rogue wave formation and developing warning criteria for rogue waves. As also mentioned in Sect. 4.2.2, Bitner-Gregersen et al. (2024) report the findings from the ExWaMar (2016–2020) project dedicated to the development of warning criteria for increased risk of rogue waves based on input from existing weather forecast models. Field data, laboratory data and numerical simulations were analysed for uni- and bi-modal sea states in deep and intermediate water depths. It was found that the sampling variability makes it very challenging to identify reliable warning criteria for rogue waves, particularly when temporal point data are employed. The authors emphasise the need to employ information in both space and time and consider the difference between point statistics and area statistics of crests since even a relatively small area will have crest statistics which is significantly underestimated by the crest distribution at a point. It is concluded that efforts to predict the occurrence of rogue waves based on wave forecast input should be based on: identifying sea state parameters which indicate an increased probability of rogue wave occurrence, coupling a phase averaged forecast wave model to a phase resolved numerical model, and/or by the use of machine learning. L. Wang et al. (2023) also investigated the quantitative relation between the occurrence probability of freak waves and sea-state parameters under bi-modal spectral sea-states. They developed a quantitative relation between the occurrence probability of freak waves and kurtosis in co-propagating mixed waves. This relation may enable a fast prediction of the occurrence probability of freak waves for a given bimodal sea state, which may be very useful for practical applications.

Cicon et al. (2023) investigated the suitability of several sea state parameters to predict the occurrence of rogue waves. Analysing data from several open ocean and coastal buoys in the Northeast Pacific, they show that whereas the BFI had limited ability to predict the occurrence of rogue waves, a frequency bandwidth parameter expressed as the crest-trough correlation had a high correlation with rogue wave probability. In line with the previous paper by Gemmrich and Cicon (2022), Cicon et al. (2023) propose a rogue wave risk assessment model based on this parameter and demonstrate that a typical regional WW3 wave model give reasonable predictions for the parameter.

As also mentioned in Sect. 4.2.2, Häfner et al. (2023) employed machine learning to the challenge of identifying sea states prone to rogue wave formation. The authors applied a neural network to the Free Ocean Wave Dataset consisting of measurements from 158 wave buoys in deep and shallow US waters (Häfner et al., 2021). They focused on wave height and $H_s > 1$ m. The aim of the study was to obtain an mathematical equation for rogue wave probability while retaining the neural network's predictive capabilities, so that the results can be interpreted in the context of existing wave theory. They concluded that a predictive model must include crest-trough correlation (frequency bandwidth), sea state steepness, directional bandwidth and depth. It was further concluded that the Rayleigh distribution for wave height (trough-crest) was an upper bound to rogue wave height and that there are distinct differences between rogue wave formation in deep and shallow water.

In design of ships and offshore structures, it is of interest to have statistical distributions of the surface elevation, crest and wave height which include non-linearities and the presence of rogue waves. Nederkoorn and Seyffert (2022) investigated the likelihood of rogue wave occurrence by integrating well established short-term distributions of crest and wave height in a point and over an area integrated over a scatter diagram of H_s and T_m. The method was applied to a floating spar and it was concluded that the risk of encountering rogue waves increases significantly if even the relatively modest horizontal extent of the structure is included in the statistical analysis.

Fuhrman et al. (2023) and Klahn et al. (2024) have revisited the classical approximate Gram-Charlier series solution to the probability density function (pdf) for the free surface elevation of waves with arbitrary frequency and directional bandwidth. Fuhrman et al. (2023) finds an exact solution to the second order pdf whereas Klahn et al. (2024) extends this to higher orders by deriving an ordinary differential equation which defines the pdf and which may be solved numerically. The results are worked out to fifth order and good agreement with experiments and numerical simulations are found for realistic directional spectra at different water depths.

As also mentioned in Sect. 7.2, Gramstad and Lian (2024) have developed an efficient numerical method for solving the integral expressions for skewness and kurtosis including bound waves up to third order. The method has been compared with experimental and numerical results and applied to a large number of sea states within a broad range of water depths, directional and frequency bandwidths, and sea state steepness. These parameters are used in proposed expressions for skewness and kurtosis which are shown to provide a marked improvement over existing parametrisations.

It is conceivable that sea states prone to rogue waves are sea states where the non-linear amplification of the crest and wave height is larger than the limiting effects of wave breaking. Analytical models for non-linear surface wave elevation neglect the effect of breaking waves, non-linear spectral models rely on a simplified breaking model to dissipate energy approximately over a breaking event. An accurate representation of both non-linear amplification and wave breaking should be included in accurate distributions of crest and wave heights. Karmpadakis and Swan (2022) and Karmpadakis et al. (2022) have developed empirical distributions for, respectively, crest heights and wave heights in JONSWAP spectra considering higher order non-linear effects and the dissipative effect of wave breaking. The models are validated against laboratory and field data for a wide range of water depths, sea state steepnesses and directional bandwidths. It is demonstrated that the new models provide a significant improvement over existing models, particularly for the steepest sea states.

For problems which are governed by crest or wave height alone, reliable short-term distributions which include the possibility of rogue waves may be integrated over all sea states to give a good estimate of the long-term crest or wave height with a given return period. For problems such as wave-in-deck loads (impacts on offshore platform decks), other properties of the wave may be important. A breaking wave with a slightly lower crest elevation than the largest non-breaking wave may be governing for wave-in-deck loads, since the velocities in the crest are significantly larger in a breaking wave. Such cases pose a formidable analysis challenge: if the governing properties for design cannot be determined it may be necessary to run Monte Carlo simulations of individual waves in a long-term analysis of all severe sea states in order to identify the governing wave for each failure mode.

Gramstad et al. (2023) developed a novel approach to long term analysis of undisturbed wave properties considering wave breaking and the irregular, directional and non-linear nature of ocean waves. By sampling large linear events in space and using this as input to a high order spectral potential solver (HOSM), it was possible to carry out a long-term Monte Carlo simulation of 10^6 years of wave data. The resulting crest distribution was in good agreement with model test results (for individual sea states) and established empirical crest distributions which consider both non-linear crest amplifications and the dissipative effects of wave breaking. A simple load model was included in the analysis in order to demonstrate the suitability of this approach applied to long term analysis of wave in deck loads. In this study, the HOSM model was enhanced with a wave-breaking model and the kinematics of the largest waves were compared with CFD results. It was concluded that whereas the HOSM model provide a marked improvement on weakly non-linear methods, it should be extended to include CFD simulations for problems which are sensitive to breaking wave kinematics. Other (multi-fidelity) methods aimed at the prediction of long-term extreme values of highly non-linear marine structure responses, governed by wave details, can be found in Sect. 7.

Zeeberg et al. (2024) report the development in the AWARE project of an empirical model for local evolution of the wave surface and underlying kinematics in time and space, which considers both the non-linear crest amplification and wave breaking. The model is computationally effective and can be applied to all large crests encountered in a long-term analysis of linear waves. The model has been extensively validated against model test results, both of the surface elevation and the underlying kinematics, and also of the resulting wave load on a large number of structures. This model, in combination with linear Monte Carlo simulations of the long-term wave environment is currently being used to assess the structural reliability of structures on the Danish continental shelf.

8.2 Extreme Ice Events and Climate Change in Polar Seas

Prabhu et al. (2021) presented results of diagnostic analysis of the relationship between Indian Summer Monsoon Rainfall (ISMR) and sea ice area (SIA) for the period 1983–2015, which demonstrate a robust relationship over the two sectors of Antarctica: Western Pacific Ocean (WPO) and Bellingshausen and Amundsen Seas (BAS). Large-scale atmospheric circulation modulated by equatorial Pacific Sea Surface Temperature (SST) signature is suggested as a possible link between Antarctic Sea ice and ISMR. Chen et al. (2022) investigated the accessibility of the Northern Sea Route and Northwest Passage under global warming of 2 °C and 3 °C. The results indicate that the changes under global warming in sea ice improve the navigability of the Arctic passages. Ships in Polar Class 6 may be unimpeded along two Arctic passages in November from 2 °C warming onward, whereas ordinary ships may be capable of passing the Northern Sea Route with global warming of 3 °C, with maximum potential in September. To evaluate the potential of sea ice to drive or amplify ocean variability, Steinsland et al. (2023) reconstruct the sea surface hydrography and sea ice variability in the Labrador Sea, a region influenced by the subpolar gyre (SPG) and where deep-water formation occurs. They infer that sea ice variability throughout Marine Isotope Stage 5e (MIS 5e) was coupled with the variability of the SPG. Especially the location of a proximal Marginal Ice Zone (MIZ) to the Labrador Sea convection region could have been important for SPG dynamics. Pascual-Ahuir and Wang (2023) compare the sea ice properties simulated by the Estimating the Circulation and Climate of the Ocean, Phase II (ECCO2) model forced by Japanese 25-year Reanalysis (JRA25) and forced by the recently released the 5th major atmospheric reanalysis produced by European Centre for Medium-Range Weather Forecasts (ERA5). The optimised simulation reveals that ERA5 predicts a closer representation of the Arctic Sea ice extent and thickness than JRA25. Despite the improvements, further work is necessary for the Antarctic, where the lack of reliable observational data and limitations of the ocean model hinders the optimization. Sterlin et al. (2023) implemented in the ocean–sea ice model NEMO-LIM3 a variable formulation of the atmospheric and oceanic neutral drag coefficients over sea ice that includes the form drag effect and examined the impact of this formulation on the ice cover and ocean surface properties over the past decades in the Arctic and Antarctic. All these impacts on the ocean surface properties demonstrate the relevance of form drag for polar ocean modelling.

8.3 Tropical Cyclones

Extreme waves forced by extreme winds, such as tropical cyclones (TCs), pose a great threat to ocean and coastal structures. The study on the influence of TCs on ocean waves is therefore crucial in the engineering community. TCs are a generic term for rotating storm systems that form over warm ocean waters, while hurricanes and typhoons refer to these storms in specific regions: hurricanes in the Atlantic and northeastern Pacific, and typhoons in the northwestern Pacific. The primary difference lies in their location and regional naming conventions. We use the term TC here; other areas of this report may use the synonyms.

To date, a variety of literatures associated with TCs have emerged. Chen *et al.* (2023a) summarized the significant progress related to internal TC intensity change processes over 2018–2022 from the World Meteorological Organization's 10th International Workshop on Tropical Cyclones (IWTC-10). Although there have been well-earned gains in the understanding of intensity change processes intrinsic to the TC system, the journey is not complete. Schenkel et al. (2023) summarised the current understanding and recent updates to TC outer size and structure forecasting and research primarily since 2018 as part of the IWTC-10.

For detecting and estimating cyclone track, Mohit *et al.* (2023) developed a small regionally based probabilistic method. The random number of cyclone eye point data was taken from the output of the dynamic climate simulations model AGCM (Atmospheric General Circulation Model) and DPDF (Database for Policy Decision-Making for Future Climate Change) cyclone track data. This virtual cyclone path shows that the direction of the present and the future cyclone path can provide more information for the cyclone disaster prevention. In addition, a new method based on mathematical morphology for detecting and tracking TC in gridded reanalysis data is proposed and evaluated by Wu and Duan (2023). It is demonstrated that the proposed method can reproduce the TC tracks for both over ocean and after landfall with good accuracy and efficiency. Mukherjee and Ramakrishnan (2023) attempt to understand cyclone Shaheen's track, intensity, and synoptic features with the GPU-based Advanced Research Weather Research and Forecasting Model v-3.8.1 (ARW v-3.8.1) on the Google cloud platform (GCP). Results indicated that Shaheen had a strong low-level (850 hPa) relative vorticity of approximately 1.93×10^{-3}/s on October 2, 2021, during the matured severe cyclonic stage, which got reduced to 7.79×10^{-4}/s on October 4, 2021, as the system weakened.

As for other TC activities, Xu et al. (2022) calculated all TC wind fields affecting the eastern sea area of China (ESAC) during 1949–2019 using the Fujita-Takahashi formula and thus obtained a new TC dataset with a spatial resolution of $0.1° \times 0.1°$. Based on this dataset, the spatial-temporal variations of TC activity were analysed, which can be used to provide a scientific basis for disaster prevention and mitigation planning and adjustment of key coastal protection zones, as well as to evaluate the value of marine sedimentary records better. Hong, Hu and Kareem (2023) proposed a model for predicting TC intensity evolution by utilising the conditional Wasserstein generative adversarial network with gradient penalty (CWGAN-GP) for TC-related risk assessment. Xu et al. (2023b) improve the modelling of TC intensification by incorporating non-breaking wave-induced turbulence and sea spray from breaking waves into an atmosphere-ocean-wave coupled model. Observed improvements suggest that an accurate characterization of ocean wave-coupled processes is crucial for advancing our understanding of severe weather events in both, the atmosphere and ocean.

To assess regional changes in TCs and their future impacts, Bruyère et al. (2022) used large climate model ensembles to increase sample size and explore historical variability and future changes in regional TC behaviour. Results point to rapidly increasing risks of damaging TC winds and major TC flooding, as well as a heightened risk of water ingress through wind-driven rain. Li et al. (2023) utilised hourly reanalysis data from the ECMWF ERA5 dataset covering the period from 2001–2020, alongside the wave risk assessment method to analyse the 20-year variability of hazardous waves in the Northwest Pacific (NWP). The extreme wave height shows an increasing trend in high- and low-latitude areas of the NWP. The maximum annual increase rate of 0.013 m/year for extreme wave height exists in the region of the Japanese Archipelago. Grossmann-Matheson et al. (2024) conducted a global study of extreme value (1 in 100-year return period) waves generated by tropical cyclones across all TC basins. Results demonstrate that values in the North Atlantic and Western Pacific basins are the largest globally, which is partly due to the relative intensities and frequencies of occurrence of storms in these basins, but also because the typical velocities of forward movement of storms are larger and hence can sustain the generation of larger waves.

For the protection of ocean and coastal structures, understanding the statistical characteristics of wave parameters during tropical storm is necessary. For this purpose, Benetazzo et al. (2021) performed extreme value analysis of individual maximum sea surface elevations (crest heights) and maximum crest-to-trough wave heights in the Northwestern Pacific during tropical storm Kong-Rey (2018). They highlight the sea areas around the storm centre where the spatio-temporal highest waves are more likely, and, via scale analysis, the principal mechanisms responsible for the occurrence of extreme conditions in bi-modal (composed of wind-sea and swell) and short-crested storm seas. Van

Vloten et al. (2022) described a methodology that allows to obtain deterministic estimations of the tail probability distribution of maximum H_s using long collections of high-fidelity tracks that reproduce similar historical diversity and frequency trends. A hybrid approach significantly reduces computational resources by enabling to narrow the number of non-stationary numerically simulated cases forced with vortex-type wind fields parameterized using the Holland Dynamic Model.

Many scholars have made great efforts to improve the accuracy of wave fields simulation during TCs. Davison et al. (2022) investigated realistic sea states during typhoon Kong-Rey (2018) using an ensemble of numerical simulations. Results illustrated that in specific conditions, space-time extreme crest heights in crossing seas can be larger than in unimodal seas due to second-order bound wave interactions between the wind sea and the swell. Yang, Liang and Shao (2022) explored the comparative advantages of different wind input and dissipation schemes available in the third-generation wave model, WW3 in most of the China Seas, where TCs frequently occur. Results demonstrated that the parameterisation Source Term 6–4.18 (ST6–4.18) performs well, especially during strong wind. By studying the input-dissipation source term packages in the spectral wave models under TCs in the Bay of Bengal domain, Uma and Sannasiraj (2023) found that the influence of the drag coefficient on the input source term is pronounced. Grossmann-Matheson et al. (2023) developed a parametric model for computationally efficient wave height estimation within TC conditions. The resulting model is ideal for engineering applications which require simulation of numerous TC cases.

9 Updated IACS Scatter Diagram

This final chapter covers the regulatory evolution in environmental conditions. The evolution of rules can be seen as the final, practical step of the work reported in the previous chapters. By design, rule changes lag behind the latest theoretical developments. In the last three years, one major recommendation for wave data to be considered when assessing a merchant ship design has been updated: IACS released Revision 2 of Recommendation 34 (IACS, 2022). This recommendation provides the joint probability of significant wave height and wave period to be considered, along with spectral shape and operational profile (relative heading and ship speed). The theoretical background is published by IACS (https://iacs.org.uk/) and in (Austefjord et al., 2023).

The data underlying the first revision of Rec.34 were visual observations on board ships from the second half of the 20th century, gathered and post-processed (Hogben, 1986). While this provided a valuable resource at the time, it is now considered outdated. See for instance (Bitner-Gregersen *et al.*, 1995), that documents inaccuracies in visual estimates of wave period. The research done over the past decades, as documented in the ISSC reports, now allows for a more accurate description of the statistics of the wave environment. In the present report, this is mainly reflected in Sect. 3, 6 and 7. In 2022, IACS benefited from accurate hindcast data (Rascle and Ardhuin, 2013), from altimeter and buoy measurements, from statistical models, and from a vast AIS (Automatic Identification System) database, which were not available 20 years ago.

When revising Rec.34, IACS' fundamental assumptions were kept identical: the wave statistics must describe the North Atlantic (as required by IMO GBS), as actually encountered by vessels. The fact that the North Atlantic represents the worst area for ship structure has been verified (for both fatigue and extreme conditions). To obtain wave data as actually encountered by vessels, the AIS position database was synchronised with hindcast data.

Since the sources of wave data used for this second revision of Rec.34 are very different from the first one, differences in the final results of the scatter diagram are to be expected. It turns out that Rec.34 Rev.2 is, on its own, more favourable than Rec.34 Rev.1. This does not mean that waves are more benign in 2020 than before, but rather that the visual observations and their subsequent post-processing significantly overestimated the waves. From Fig. 10 and Fig. 11, a striking difference is the shape of the contours; the new revision present a sharp steepness limit (lacking in Rev.1), which is consistent with the physical limit due to wave breaking.

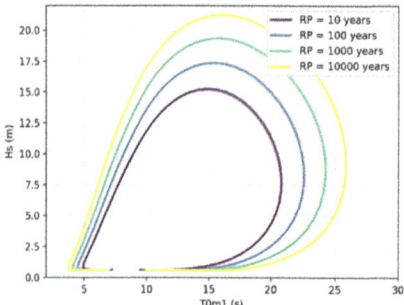 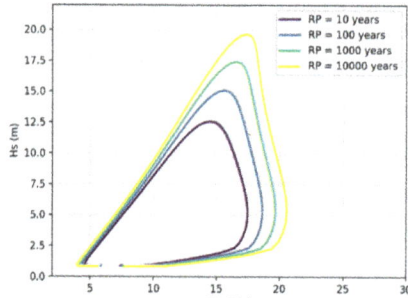

Fig. 10. Contours from Rec.34 Rev.1   **Fig. 11.** Contours from Rec.34 Rev.2

Within the IACS and classification context, Rec.34 is the first element considered in the road to the final hull scantling. On the wave-load side (seakeeping), significant progress has also been made; using state-of-the-art tools will significantly modify the loads (an increase can be expected according to (Derbanne et al., 2016)); see also the report of Committee I.2 Loads. Besides, over time, IACS benefits from extensive feedback from its fleet. Calibrations are, implicitly or preferably explicitly, included in the rules, and can be expected to be revised based on new inputs. Overall, everything else equal, the new Rec.34 is less severe than the previous one; however, the progress in downstream calculations is such that the final scantling required by future IACS rules might be more reasonable and balanced for all structural members of ships.

Outside the classification context, some caution should be exercised: users directly applying Rec.34 (i.e., not using it implicitly through IACS rules) are recommended to carefully check the calibration and safety factors when updating the wave statistics from Rev.1 to Rev.2. In general, when using a scatter diagram (or a more complex joint distribution metric) as the basis for designing a marine structure, it is essential to understand the underlying assumptions of the data and their implications for the entire design process. This is also valid for the new IACS scatter diagram. Important to note is that safety factors, models, design and operational experiences were all adapted to the previous, more severe scatter diagram. The difference between the wave data encountered by ships and stationary wave data is probably partly accounted for already in the present safety factors in design of marine structures. It is therefore important to adapt the full design process, and to further study and monitor the effect of the new IACS recommendation on the design process and safety of future marine structures.

10 Conclusions

Metocean research, which applies environmental science to the maritime industry, has been an active field for decades. This continued during the mandate period, resulting in several advancements in the research on waves, ice, wind and current. This review highlights these developments and their implications for structural and ocean engineering, aiming to make academic findings more accessible to industry practitioners. Throughout the report, we focus on applying new research, the effects of climate change, statistics and uncertainty, and emerging technologies. 'The environment' encompasses a broader range of factors than just the waves, ice, wind, and currents in the mandate. While other factors are relevant for some structures, topics such as temperature and salinity effects, sediment transport, pure meteorological studies, and the impact of marine structures on the environment fall outside the scope of this report. Below is a summary of each chapter.

In Sect. 2, we review metocean climate studies, focusing on both present and future conditions. Present-day climate is assessed using global wind and wave predictions from reanalysis datasets like ERA5 and CFSR, as well as observations. New datasets have been made available through the COWCLIP project. For regional analysis, downscaled data is used to provide more local insights. Future wind and wave climate is evaluated using data from the CMIP5 and CMIP6 climate simulations, which help assess trends, extremes, and variability. Ensemble analysis is employed to explore the uncertainties associated with these climate simulations.

Section 3 covers advances in numerical models for waves, ice, wind, and currents. Analytical, potential flow, and CFD models remain the key methods for phase-resolved sea wave predictions. Wave breaking, and particularly the modelling of breaking inception and dissipation, is a challenging aspect in non-linear potential flow models. There is growing interest in speeding up computations using GPUs, machine learning, and multi-fidelity methods. In ice modelling, viscous-plastic and elastic-viscous-plastic rheologies are now coupled with global ocean models to better simulate sea ice. At the engineering scale, research focuses on accurately describing sea ice mechanics and the dynamic behaviour of structures. Research on reconstructing wind fields based on reanalysis data is gaining importance. Wind field reconstruction from reanalysis data is gaining importance, and wave-current interaction remains a key challenge in current modelling due to its effects on marine structures.

Section 4 focuses on data-driven methodologies, particularly the growing impact of Artificial Intelligence (AI) in environmental sciences. Machine learning (ML) models, powered by advanced GPUs and large datasets, have significantly improved forecasting accuracy, with deep learning models showing notable progress. AI is now being used for weather forecasting and hybrid models that combine traditional physics-based approaches with ML are gaining traction. AI applications span all time scales, from nowcasting and short- to mid-range forecasts (including tropical cyclones) to climate predictions. Such applications include wind forecasts for renewable energy, surface wind profiles, extreme events, rogue waves, ocean currents, and storm surges. Given the early stage of these technologies, independent validation remains essential.

Section 5 reviews advance in laboratory measurements and observations, focusing on both techniques and the data collected. The need to validate complex numerical models, along with progress in hybrid numerical-experimental methods, has led to more deterministic wave experiments replicating specific events. This trend drives the development of improved techniques like stereo imaging and advanced Particle Image Velocimetry (PIV). While past research centred on statistical uncertainties, recent studies are shifting toward systematic uncertainties, exploring basin effects, spurious waves, measurement errors, and repeatability and the effect of the basin and test set-up design. Key studies have also aimed to better understand wave breaking, including the impact of wave directionality. In ice modelling, recent attention has focused on case-based scaling rather than general rules.

Section 6 reviews advance in measurements and observations at sea. The use of affordable satellite-connected wave buoys has increased for environmental research and ship trials, as they are now more accessible and easier to deploy. A key recent focus is the impact of bad weather avoidance on long-term environmental statistics for merchant ships. Traditional methods, relying on scatter diagrams from the most severe locations, can lead to overly conservative ship designs. There's growing interest in improving this by using actual ship trajectories, AIS data with wave hindcasts, or ship-as-wave-buoy techniques. Ship radar measurements are also being explored for short-term motion prediction, which can be used to assist on-board decision-making. Ice observations during the mandate period were heavily influenced by data from the year-long MOSAiC Arctic expedition (2019–2020).

Section 7 focuses on statistical modelling in environmental studies. Several new approaches to statistical modelling of the environment have been proposed, driven by increased data availability (observations, hindcasts, experiments) and enhanced computing power. Climate models are increasingly used to generate joint distributions of environmental variables, shifting the focus from limited measurement durations to the accuracy of these models. There is a trend towards more complex models that incorporate non-stationary and space-time effects, increased dimensionality, and storm statistics. For instance, when establishing wave height distributions, there's a growing need to separate data for directional or seasonal effects. Additionally, the maximum crest elevation under a platform can be significantly higher than at a central point, necessitating adjustments in design load estimations. New methodologies for assessing multivariate extremes and specific models for shallow water and coastal areas have also been developed, enhancing the applicability of statistical modelling for specific marine structures at particular locations. However, the variety of models available can complicate the selection process. The extreme value benchmark in Appendix A and the data sources listed in Appendix B aim to simplify some steps of this process.

Section 8 addresses extreme events such as rogue waves, polar conditions, and tropical cyclones. Rogue waves, defined as significantly higher than surrounding waves, are attributed to non-linear modulational effects and rare wave-focusing instances. There have been observations during the reporting period of extreme waves in relatively moderate sea states, which is difficult to explain by linear dispersion. However, a consensus is emerging that the largest waves, crucial for ship and offshore structure design, fit within conventional distributions when considering space-time effects and local non-linearities. Research has identified a frequency bandwidth parameter that correlates with rogue wave occurrences, and new short-term distributions for crest and wave heights have been proposed to improve predictions of these events. Accurate estimation of maximum ice loads on ships is also vital, leading to various probabilistic approaches for long-term ice load predictions based on field data. In light of global warming, understanding changes in sea ice properties is essential. Additionally, extreme waves from tropical cyclones pose significant risks to marine structures. To assess their impact, various statistical wave properties and extreme value analysis methods have been explored, comparing models and parameterisation approaches to enhance cyclone modelling efforts.

Section 9 covers the new wave scatter diagram recommended by IACS in Rec.34 Rev.2. Unlike Rev.1, which relied on visual observations, Rev.2 uses a blend of well-validated hindcast data, altimeter and buoy measurements, statistical models, and a comprehensive AIS ship tracking database. The updated scatter diagram is considered more benign and aligns with wave breaking limits. However, outside the classification context, users applying Rec.34 directly (rather than through IACS rules) should exercise caution and carefully verify their calibration and safety factors when transitioning from Rev.1 to Rev.2. Further research is needed to understand the implications of this update for future marine structure design and safety.

We conducted a benchmark of open-source extreme value extrapolation methods, detailed in Appendix A. This benchmark serves as an introductory resource for newcomers in the maritime industry. We compared various tools for extrapolating extreme values over longer return periods using two datasets: hindcast H_s and experimental slamming loads. Most analysis settings were fixed to assess the implementation differences across packages. The results from all tested packages were generally consistent, aside from some variations in confidence intervals. However, their ease-of-use varied significantly. We know that choices regarding settings like block size, peak threshold, fitting method, and minimiser can greatly influence outcomes. Future benchmarks could explore similar test cases with more flexibility in analysis settings.

We compiled a list of open environmental data sources in Appendix B. While many datasets are available from various observations across different regions and timeframes, they remain largely unknown to the maritime industry. These sources can be valuable in the early design stage when local metocean conditions are not well understood. Although not exhaustive, our list offers readers access to some potentially hard-to-find datasets.

Finally, several key themes emerge from our chapters. One of them is the impact of wave breaking on load distributions. The industry's recognition that the largest ocean waves can break at any water depth raises significant concerns. It is no longer sufficient to consider only expected crest height at a prescribed return period and rely on the wave properties associated with that height; we must also evaluate smaller waves that may break in critical positions. This is particularly important for the offshore sector, where inadequate air gaps in jacket structures (either by design or due to excessive seabed subsidence) can pose risks. A number of empirical crest height distributions and methods for incorporating wave breaking into long-term analysis have been proposed to address this issue. Additionally, the effect of climate change on the design inputs for marine structures is increasingly recognised. Improved models that incorporate these impacts are becoming more common, aided by data-driven techniques. There is a growing awareness in the industry, reflected in dedicated conference symposia and a rise in related publications. Monitoring this evolution is vital, and we recommend that the next committee prioritises this issue.

Appendix A: Extreme Value Analysis Benchmark

Although floating structures are engineered to endure extreme conditions, expertise in "extreme value theory" and broader statistical methods is not a core focus within the maritime industry. As a result, methods and tools that are widely used by statisticians are not as well-utilised as they could be.

This benchmark illustrates the availability of open-source tools that implement advanced extreme value theory. Many of these tools are reliable and reasonably user-friendly, allowing practitioners to adopt cutting-edge techniques in extreme value analysis without substantial development effort. The primary audience for this benchmark is naval architects interested in textbook methods for extreme value analysis, rather than experienced statisticians. For a more thorough, theoretical review of tools for extreme value modeling, see the recent work by Belzile et al. (2023). The objective here is to guide newcomers to extreme value statistics in practical applications, leveraging the contributions of the open-source community.

The benchmark focuses on univariate, stationary extreme value modelling. The two most classical approaches for such work are Peak-Over-Threshold (POT), and Block-Maxima (BM) (Coles, 2001). The implementation of those two methods, in their more standard flavour, by different open-source packages is compared. Two very different datasets are used to evaluate the different software packages:

- Wave height time series (26 years), obtained from the IOWAGA model (see Fig. 12), for latitude $= 57.5°$ and longitude $= -23.25°$;
- Slamming loads on the bow of a ferry vessel (24 h), obtained from experiments with a ferry done within the Cooperative Research Ships (CRS) project SCREAM (see Fig. 13).

The datasets, as well as the detailed instruction for the benchmark can be found at: https://github.com/ISSC-ENV/EVA_BENCHMARK.

For the two cases, the objective is to estimate the return level (of slamming loads and significate wave height) with the two approaches (BM and POT), with parameters that are specified (Table 2).

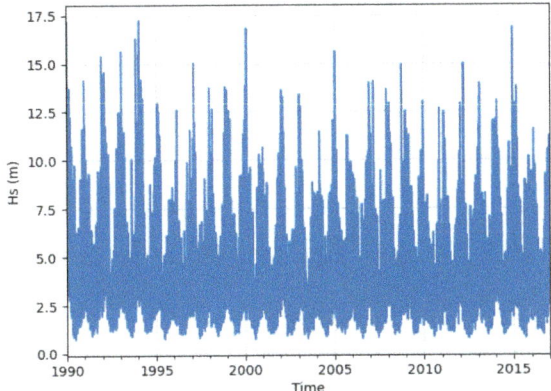

Fig. 12. Time series of H_S

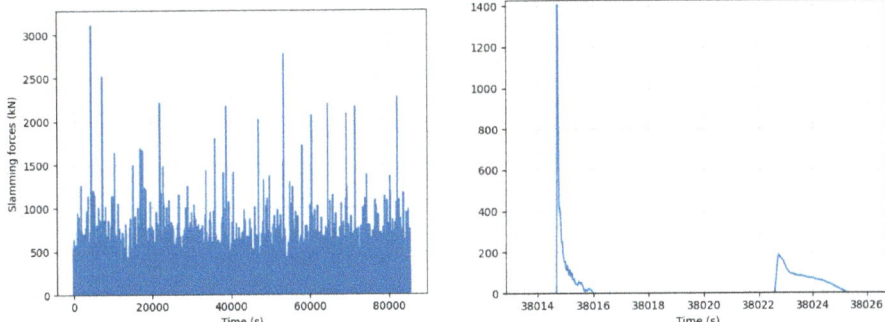

Fig. 13. Full slamming times series (left), and one of the impacts (right)

Table 2. Parameter settings

	Slamming	Wave height
Threshold (POT)	1200 kN	12.5 m
De-clustering windows	100 s	2 days
Block size (BM)	3500 s	1 year

Table 3. List of investigated packages

Packages	Language	Licence	Maintenance/last check or commit
pyExtremes	Python	MIT	V (2024)
Scikit-Extreme	Python	MIT	X (2019)

(continued)

Table 3. (*continued*)

Packages	Language	Licence	Maintenance/last check or commit
Wafo	Matlab	GPL	X (2021)
Snoopy	Python	GPL	V (2024)
Ismev	R	GPL	V (2024)
extRemes	R	GPL	V (2024)

A.1 Packages Considered

When looking for packages to include in the benchmark, it appeared quite clearly that most of the available resources are available in "R" language; (Belzile et al., 2023) and (Gilleland and Katz, 2016) provide extensive list of packages in R for extreme value analysis. On the other hand, the committee members skills in terms of practical implementation lean much more towards Python, Matlab or Fortran, in which resources exist, but are less comprehensive, with a limited user base (same conclusion reached in (El Khazen et al. 2020)). The tested packages are listed in Table 3. This list is not exhaustive, but provides some packages in each of the languages. Most of those packages include more advanced features that will not be investigated in this benchmark. For instance, extRemes and pyExtreme includes some Bayesian inference, and extReme can deal with non-stationary analysis.

A.2 Results

A.2.1 Block Maxima

Figure 14 presents the return level predicted by the different packages for the slamming and wave height case. The agreement between all packages is as expected very good. The reasons for those slight discrepancies were nevertheless identified:

- Duration of a "year": using 365.24 days or 365/366 days leads to differences in block maxima. In the dataset, there is one storm occurring at 31st of December at 21h, which is then counted differently depending on the way to define a year;
- The negative log-likelihood is not always perfectly minimised with default setting;
- The differences are however considered negligible for the two tested cases;
- Slight difference in the default way to define the return period in Snoopy.

Looking at intermediate results like the coefficient from the GEV distribution shown in Table 4, it can be checked the different packages constantly estimate the same parameters.

Committee I.1: Environment 91

Table 4. Estimated parameters with the BM approach for the slamming case

Package	Shape	Location	Scale
Snoopy	0.111045	1618.645614	459.608864
Pyextremes	0.120076	1578.746287	477.529930
extRemes	−0.111099	1618.555000	459.614600

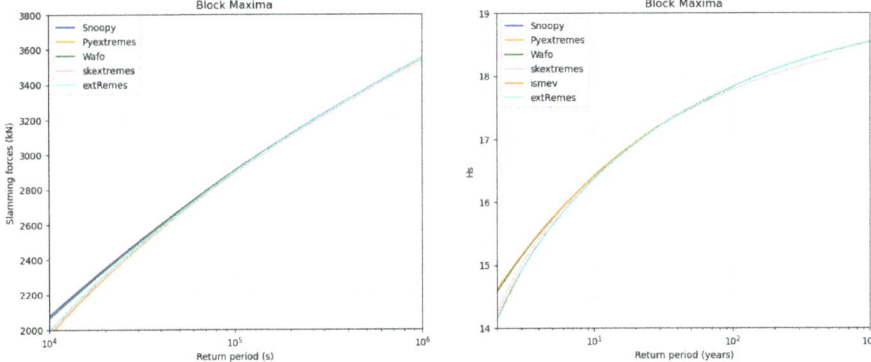

Fig. 14. Block maxima mean results for slamming forces (left) and H_S (right)

A.2.2 Peak-Over-Threshold

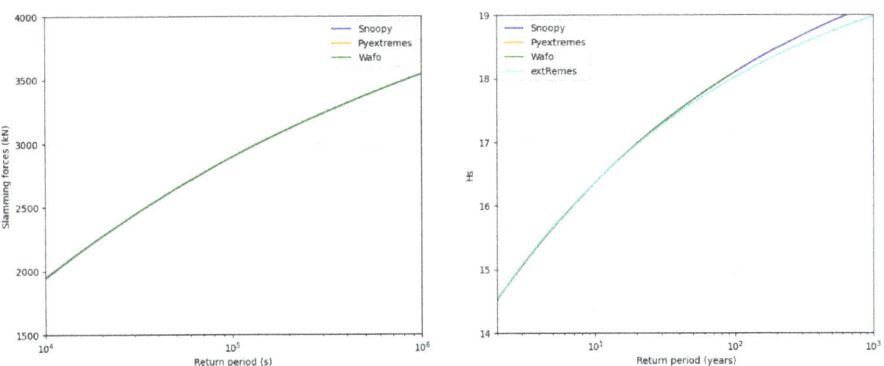

Fig. 15. Peak-over-threshold mean results for slamming forces (left) and H_S (right)

Figure 15 shows the results using the peak-over-threshold. As for block maxima, the agreement is very good, and the slight differences can be explained:

The way to de-cluster the data can differ slightly from one package to the other;

Same as for block-maxima, the negative log-likelihood is not always perfectly minimised with default setting. The differences are however considered negligible for the two tested cases.

A.3 Confidence Intervals

All the package investigated have a confidence interval estimation feature. Several methods are available (e.g. (Coles, 2001)). Three methods will be compared: delta-approach, bootstrapping and profile likelihood.

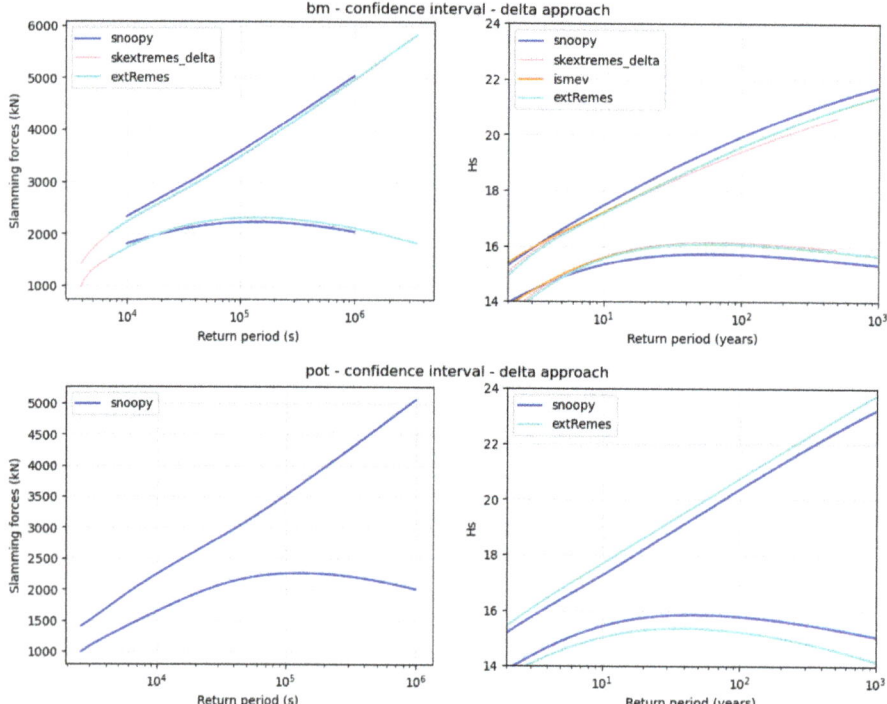

Fig. 16. Confidence intervals estimated with BM (top) and POT (bottom), both combined with the delta method

A.3.1 Delta Approach

Among the packages investigated, four of them include estimates of the confidence interval with the delta method. The agreement between the packages is very good (see Fig. 16), although we can observe that Snoopy is slightly off; the reason for this is not yet identified.

A.3.2 Bootstrap Approach

Four contributions included confidence intervals estimated using bootstrapping. As opposed to the delta approach, where all package provided consistent results, the scatter is here much larger (see Fig. 17). The main reason is the sensitivity of the bootstrapping approach to the minimiser settings. Indeed, among the bootstrapped sample, it can happen that the maximum likelihood estimator (MLE) has difficulty to converge. This may lead to a heavy shape factor, in which case the tolerance or bounds of the minimiser can lead to significant differences. The scatter here observed is thus considered as coming from the sensitivity intrinsic to the method, rather than a problem with the implementation.

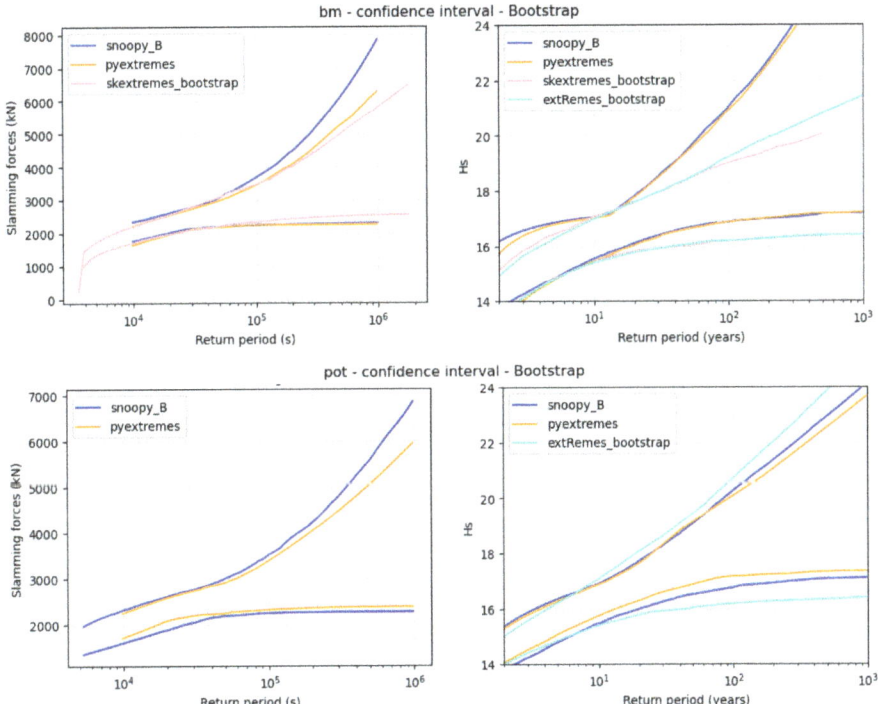

Fig. 17. Confidence intervals estimated with BM (top) and POT (bottom), both combined with bootstrapping

A.3.3 Likelihood Profile

Wafo and extRemes can also be used to provide confidence intervals estimated by the "likelihood profile" approach, with very comparable results for the block maximum data (Fig. 18).

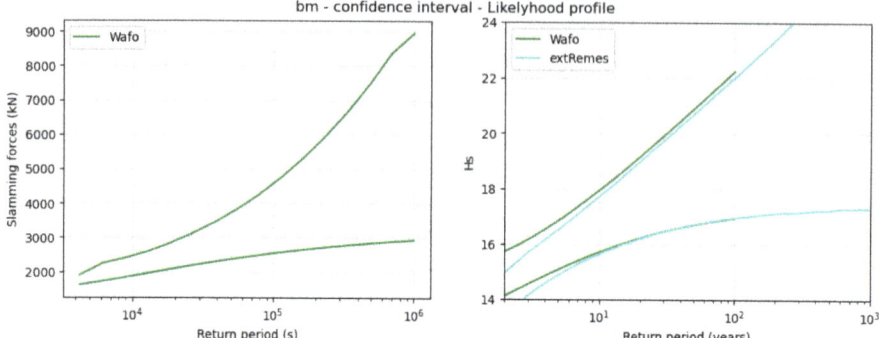

Fig. 18. Confidence intervals estimated with BM combined with likelihood profile

A.4 Discussion

Extreme values extrapolated using different packages were compared for a given method. The results from the different implementations were consistent, with only slight discrepancies. In this section, the results of the different methods are compared. For conciseness, this will be performed only for the H_s case, using the extRemes package. Figure 19 compares the BM to the POT approach for the wave height data, and Fig. 20 compares different methods for estimating the confidence interval. While the results from the different approaches are comparable, the scatter is larger than the difference observed between the different implementations of a given approach.

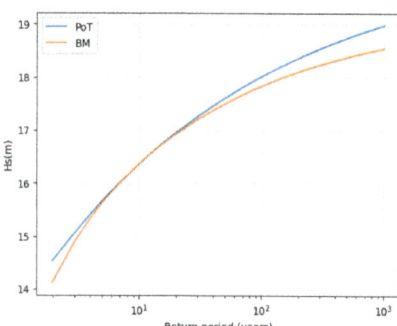
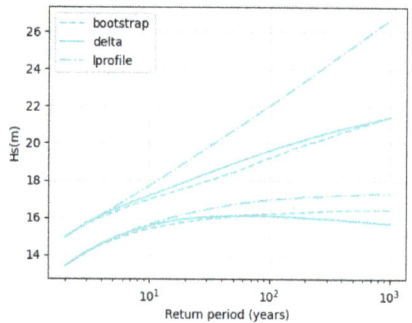

Fig. 19. comparing POT to BM for the H_s case

Fig. 20. Different estimate of confidence interval, with extReme and block-maxima approach for the H_s case

Furthermore, the different approaches can be used with different parameters. Choice of threshold or block-size is well known to be crucial to an accurate extreme value prediction. In practice, most of the packages offer some tools to assess this sensitivity. Figure 21 shows an example of such sensitivity analysis using the Snoopy package.

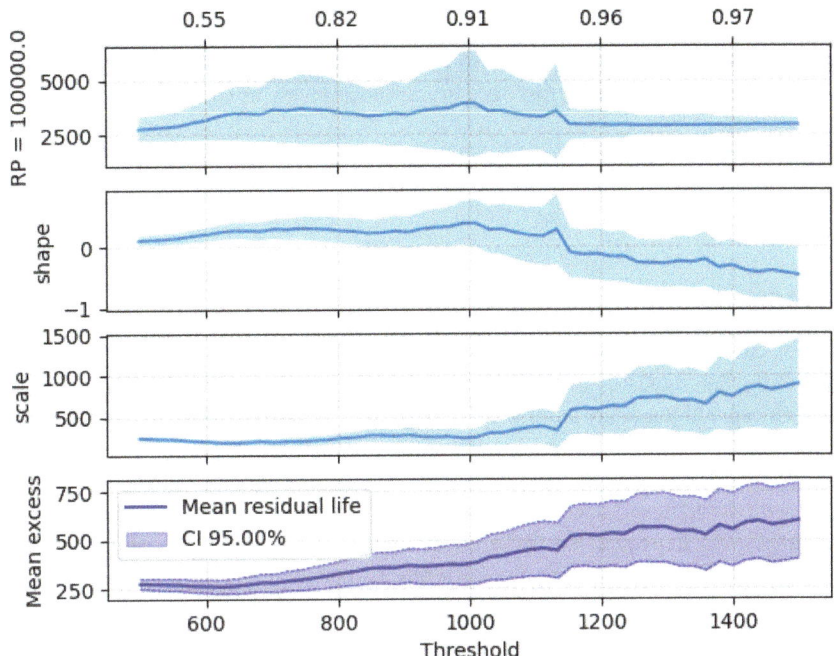

Fig. 21. Threshold sensitivity in POT approach

A.5 Practical Considerations

While results from all tested packages are consistent, the ease-of-use can vary significantly. For each of the packages investigated, some practical considerations are listed below. Those comments are admittedly somewhat subjective and can depend on user habits with each programming language.

Scikit-Extreme: The package provides reliable result; however, the API is not flexible enough to handle various time series. Time scale in something else than years is surprisingly difficult to manage. To provide results to the benchmark on the slamming case, some hack and customisation where necessary. The package is not maintained anymore, and the former developer recommend a switch to pyExtremes.

pyExtremes: The installation is easy, and the documentation clear. Getting results for the benchmark took less than an hour (with good python knowledge, but not prior experience with pyExtremes). The only trick necessary to obtain the above presented results was to scale down the slamming forces.

Snoopy: The package has been used by its main developer; the ease of use thus not be objectively evaluated. The package is public, with sources on gitlab.com, but the user base is limited.

exTremes: The package was easy to use, at least for the wave height case. It was tested by two persons, including one with absolutely no prior experience with "R". For the POT on the slamming case, a difficulty was encountered to handle the small-time scale.

ismev: The package was easy to use for the H_s case (tested by someone with absolutely no prior experience with "R"). To provide results to the benchmark, a little customisation was however necessary to extract return levels in csv files rather than in a plot.

Wafo: Easy installation and usage. Documentation Ok. The package is however not maintained since 2015. Note that its Python translation "pyWafo" has been investigated, but the installation process failed.

A.6 Conclusions

Text-book extreme value analysis is readily available in several packages, and several programming languages. Admittedly, most of the resources are in "R" language, which is not a widespread language among naval architects. Some packages do exist in Python, with, however, a weaker user base.

All packages investigated are considered reliable enough for the basic case investigated, with sometime a need for classical precaution (scaling of very large variable for instance). As expected, the differences coming from the implementation details were found much weaker than the differences between the different approach or parametrisation (threshold or block size for instance).

Appendix B: Publicly Available Environmental DATA Sets

There is nowadays a large number of environmental data available that is highly valuable for designing and analysing marine structures. However, many of these resources are not widely advertised, and relevant datasets may go unnoticed - especially by naval architects who are not deeply involved in wave modelling or other specialised fields. This appendix aims to address this gap by listing datasets known to the committee members. While not comprehensive, this resource may provide a helpful starting point for accessing critical environmental data.

B.1 Hindcast Datasets

B.1.1 ERA5

ERA5 is probably the most used dataset used for work related to marine structures. It is well maintained, with a professional API. The data can be either retrieved manually (i.e. with a web-browser) or with well-documented scripts. Note that extreme wave heights are probably slightly underestimated (de Hauteclocque et al., 2020). Reference publications: Hersbach et al. (2020) and Bidlot et al. (2019). Data available at https://cds.climate.copernicus.eu/

B.1.2 Ifremer Dataset (IOWAGA)

Ifremer provides several datasets generated using WaveWatch III. Three main variants are available, the original one (refers as IOWAGA) is using CFSR wind forcing, but ERA-INTERIM forcing is also available. Recently, a version using ERA5 forcing and other improvements (like wave-current) has been added. The variant using CFSR forcing has been used by IACS to update its wave standard. Reference publications: Rascle and Ardhuin (2013) and Alday et al. (2021). Data available at https://www.umr-lops.fr/Donnees/Vagues/sextant#/. A direct FTP link is also available at ftp://ftp.ifremer.fr/ifremer/ww3/HINDCAST.

B.1.3 Copernicus

The "Copernicus" wave dataset, different from ERA5 and referred to as "Waverys". Reference publication: Law-Chune et al. (2021). Data available at: https://resources.marine.copernicus.eu.

B.1.4 NOAA

NOAA provides a global wave hindcast for the period (1990–2009). Note that the extreme waves are significantly underestimated (de Hauteclocque et al., 2020). Data available at https://polar.ncep.noaa.gov/waves/hindcasts/nopp-phase2.php.

B.1.5 Other Sources Using Hindcast Data

Morim et al. (2022) provide a dataset of global wave fields, obtained from 14 state-of-the-art hindcasts. The dataset includes general and extreme statistics of H_s, T_m and θ_m between 1980–2014, at different frequency resolutions (monthly, seasonally, and annually). See also Sect. 2. The full global dataset can be accessed via the data repository Australian Ocean Data Network (AODN) at https://doi.org/https://doi.org/10.26198/3kkc-2g71.

Patra et al. (2023) provide a dataset of the historical ocean wave climate between 1960–2020, simulated using the numerical model WW3 forced by CMIP6 simulations. See also Sect. 2. The full global wave climate dataset can be accessed at https://doi.org/https://doi.org/10.12770/0983962b-4acc-4f8f-9484-e2195029b87b.

B.2 Measurements

B.2.1 Wave Buoys

There exists a significant amount of national or regional website hosting buoy data. Hopefully, there is nowadays an effort to make those data more visible and easier to recover. Most of north-America's buoy are referenced by the NDBC (National Data Buoy Center) at https://www.ndbc.noaa.gov/. Historical data are well organized, and can easily be downloaded (manually, or with a script) at https://www.ndbc.noaa.gov/historical_data.shtml.

Along this line, the "European Marine Observation and Data Network" also reference wave buoys world-wide. Live data can be viewed directly, and links to hosting organization can also be found at https://emodnet.ec.europa.eu/geoviewer/.

Kinsela et al. (2024) provide a dataset of wave buoy measurements in shallow waters (<35 m) in southeast Australia. It includes over 7000 days of measurements at 20 locations. The dataset includes time series of spectral wave parameters, as well as buoy displacement data. See also Sect. 6. The publication includes a number of links to the data files via the Australian Ocean Data Network.

Data from the Dutch part of the North Sea is available from Rijkswaterstaat (RWS). Results from different hindcast databases were compared to buoy measurements from the RWS buoys in the Southern North Sea (close to The Netherlands) by van Essen and Peters (2017). Data available at at https://waterberichtgeving.rws.nl and https://waterinfo.rws.nl/#!/kaart/golfhoogte/.

Ludka et al. (2019) provide a dataset of beach profiles and measured waves at three Californian beaches between 2001–2016. See also Sect. 6. The publication includes a number of links to the data files.

B.2.2 Wave Altimeters

Altimeters are considered as a reliable reference for wave height. Data can be found at several places. One easy place to get started is the CCI-Seastate project website, which gather the raw data together with processed and calibrated dataset. Reference publication: Dodet et al. (2020). Data available at https://climate.esa.int/en/projects/sea-state/data/.

B.2.3 Infragravity Waves

Rutten *et al.* (2024a) describe a dataset of infragravity wave measurements, consisting of synchronised surface elevation, velocity and pressure at 2–4 Hz for a five-month duration in the Dutch North Sea. See also Sect. 6. The full dataset can be accessed via the data repository of Delft University of Technology at https://doi.org/10.4121/233f11ff-7804-4777-8b32-92c4606e56d8.

B.2.4 Ice and Wave-Ice Interactions

Macfarlane *et al.* (2023) provides sea ice data from the MOSAiC Arctic expedition of 2019–2020. See also Sect. 6. The publication includes a number of links to the data files.

Rabault et al. (2023) provides a database of sea ice drift and waves in ice measurements. A total of 15 deployments were performed over a period of 5 years in both the Arctic and Antarctic, involving 72 instruments. These provide GPS drift tracks and measurements of waves in ice. See also Sect. 6. Processed data is available via the server of the Norwegian Meteorological Institute in the Arctic Data Center (ADC) database at https://adc.met.no/datasets/https://doi.org/10.21343/azky-0X44.

B.2.5 Tropical Cyclones

Tamizi and Young (2024) provide a dataset of wind and surface wave conditions within tropical cyclones, combining data measured by buoys and remote sensing (satellite altimeters, scatterometers, and radiometers) with estimates of tropical cyclone tracks and wind field parameters. It contains 2927 global cyclones between 1985–2017. See also Sect. 6. The dataset can be downloaded at https://doi.org/10.26188/24471688.

B.2.6 Vessel Position and Sensors

The ICOADS database contains measured and observed weather data on board of ships. This dataset contains vessel positions (latitude/longitude), and can thus be used to retrieve statistics on weather encountered by vessels (as opposed to weather at fixed positions). Note that the vessels in the ICOADS dataset are not a fully representative sample of the world-wide fleet; the share of container ships is for instance larger that the world-wide one. Reference publication: Freeman et al. (2017). Data available at https://rda.ucar.edu/datasets/ds548.0/#!description.

For vessel positions, the AIS data are much more comprehensive (from 2014 at least), but this data is not freely available.

B.3 Raw List of Additional Links

For completeness, below are links to sources of environmental data that the committee members have not personally used.

- https://manati.star.nesdis.noaa.gov/datasets/ASCATData.php
- https://www.remss.com/measurements/ccmp/
- https://www.hycom.org/
- Buoys (mostly included in https://emodnet.ec.europa.eu/geoviewer/):

  - https://cdip.ucsd.edu/m/stn_table/
  - https://meetnetvlaamsebanken.be/
  - https://candhis.cerema.fr/
  - https://www.SOFARocean.com/mx/SOFAR-spotter-archive

References

Abumohsen, M., Owda, A.Y., Owda, M., Abumihsan, A.: Hybrid machine learning model combining of CNN-LSTM-RF for time series forecasting of Solar Power Generation. e-Prime – Adv. Electr. Eng. Electron. Energy **9**, 100636 (2024). https://doi.org/10.1016/j.prime.2024.100636

Adell, A., Almström, B., Kroon, A., Larson, M., Uvo, C.B., Hallin, C.: Spatial and temporal wave climate variability along the south coast of Sweden during 1959–2021. Reg. Stud. Mar. Sci. **63**, 103011 (2023). https://doi.org/10.1016/j.rsma.2023.103011

Aijaz, S., et al.: Verification and intercomparison of global ocean Eulerian near-surface currents. Ocean Model. **186**, 102241 (2023). https://doi.org/10.1016/J.OCEMOD.2023.102241

Akhmediev, N.: Waves that appear from nowhere. Proc. R. Soc. Victoria **135**(2), 64–68 (2023). https://doi.org/10.1071/RS23011

Albuquerque, J., Antolínez, J.A.A., Gorman, R.M., Méndez, F.J., Coco, G.: Seas and swells throughout New Zealand: a new partitioned hindcast. Ocean Model. **168**, 101897 (2021). https://doi.org/10.1016/j.ocemod.2021.101897

Alday, M., Accensi, M., Ardhuin, F., Dodet, G.: A global wave parameter database for geophysical applications. Part 3: Improved forcing and spectral resolution. Ocean Model. **166**, 101848 (2021). https://doi.org/10.1016/j.ocemod.2021.101848

Ali, M., Prasad, R.: Significant wave height forecasting via an extreme learning machine model integrated with improved complete ensemble empirical mode decomposition. Renew. Sustain. Energy Rev. **104**, 281–295 (2019). https://doi.org/10.1016/j.rser.2019.01.014

Allahdadi, M.N., He, R., Ahn, S., Chartrand, C., Neary, V.S.: Development and calibration of a high-resolution model for the Gulf of Mexico, Puerto Rico, and the U.S. Virgin Islands: implication for wave energy resource characterization. Ocean Eng. **235**, 109304 (2021). https://doi.org/10.1016/j.oceaneng.2021.109304

Alonso, R., Solari, S.: Comprehensive wave climate analysis of the Uruguayan coast. Ocean Dyn. **71**(8), 823–850 (2021). https://doi.org/10.1007/s10236-021-01469-6

Alonso, R., Solari, S.: Automatic calibration and uncertainty quantification in waves dynamical downscaling. Coast. Eng. **169**, 103944 (2021). https://doi.org/10.1016/j.coastaleng.2021.103944

Alsaaq, F., Shamji, V.R.: Extreme wind wave climate off Jeddah Coast, the Red Sea. J. Marine Sci. Eng. **10**(6), 748 (2022). https://doi.org/10.3390/jmse10060748

AlShafeey, M., Csaki, C.: Adaptive machine learning for forecasting in wind energy: a dynamic, multi-algorithmic approach for short and long-term predictions. Heliyon **10**(15), e34807 (2024). https://doi.org/10.1016/j.heliyon.2024.e34807

Altunkaynak, A., Çelik, A., Mandev, M.B.: Hourly significant wave height prediction via singular spectrum analysis and wavelet transform based models. Ocean Eng. **281**, 114771 (2023). https://doi.org/10.1016/j.oceaneng.2023.114771

Alves, D., Mendonça, F., Mostafa, S.S., Morgado-Dias, F.: The potential of machine learning for wind speed and direction short-term forecasting: a systematic review. Computers **12**(10), 206 (2023). https://doi.org/10.3390/computers12100206

Amarouche, K., et al.: Spatial calibration of an unstructured SWAN model forced with CFSR and ERA5 winds for the Black and Azov Seas. Appl. Ocean Res. **117**, 102962 (2021). https://doi.org/10.1016/j.apor.2021.102962

Amarouche, K., Akpınar, A., Rybalko, A., Myslenkov, S.: Assessment of SWAN and WAVEWATCH-III models regarding the directional wave spectra estimates based on Eastern Black Sea measurements. Ocean Eng. **272**, 113944 (2023). https://doi.org/10.1016/j.oceaneng.2023.113944

Amarouche, K., Akpınar, A., Kamranzad, B., Khames, G.-E.-Y.: Global extreme wave estimates and their sensitivity to the analysed data period and data sources. Mar. Struct. **92**, 103494 (2023). https://doi.org/10.1016/j.marstruc.2023.103494

Aouf, L., Hauser, D., Chapron, B., Toffoli, A., Tourain, C., Peureux, C.: New directional wave satellite observations: towards improved wave forecasts and climate description in Southern Ocean. Geophys. Res. Lett. **48**(5), 1–10 (2021). https://doi.org/10.1029/2020GL091187

Ardag, D., Resio, D.T.: Inconsistent spectral evolution in operational wave models due to inaccurate specification of nonlinear interactions. J. Phys. Oceaonogr **2019**(49), 705–722 (2019)

Arif, M., Khan, F., Ahmed, S., Imtiaz, S.: Extreme wind load analysis using non-stationary risk-based approach. Saf. Extreme Environ. **4**(3), 247–255 (2022). https://doi.org/10.1007/s42797-022-00064-2

Asharaf, S., Waliser, D.E., Posselt, D.J., Ruf, C.S., Zhang, C., Putra, A.W.: CYGNSS ocean surface wind validation in the tropics. J. Atmos. Oceanic Technol. **38**(4), 711–724 (2021). https://doi.org/10.1175/JTECH-D-20-0079.1

Astfalck, L., Bertolacci, M., Cripps, E.: Evaluating probabilistic forecasts for maritime engineering operations. Data-Centric Eng. **4**(15), e15 (2023). https://doi.org/10.1017/dce.2023.11

Åström, J., Robertsen, F., Haapala, J., Polojärvi, A., Uiboupin, R., Maljutenko, I.: A large-scale high-resolution numerical model for sea-ice fragmentation dynamics. Cryosphere **18**(5), 2429–2442 (2024). https://doi.org/10.5194/tc-18-2429-2024

Athanassoulis, G.A., Stefanakos, Ch.N.: A nonstationary stochastic model for long-term time series of significant wave height. J. Geophys. Res. Oceans **100**(C8), 16149–16162 (1995). https://doi.org/10.1029/94JC01022

Austefjord, H.N.: IACS Evaluation of Recommendation No. 34. 1st Int. Symposium on Extreme Maritime Weather: Towards Safety of Life at Sea and a Sustainable Blue Economy (2019)

Austefjord, H.N., de Hauteclocque, G., Johnson, M.C., Zhu, T.: Update of wave statistics standards for classification rules. Advances in the analysis and design of marine structures. In: 9th MARSTRUCT Conference, pp. 43–52 (2023). https://doi.org/10.1201/9781003399759-5

Aydoğan, B., Ayat, B.: Performance evaluation of SWAN ST6 physics forced by ERA5 wind fields for wave prediction in an enclosed basin. Ocean Eng. **240**, 109936 (2021). https://doi.org/10.1016/j.oceaneng.2021.109936

Babanin, A.V., et al.: Committee I.1: environment. In: International Ship and Offshore Structure Conference (ISSC) (2022). https://doi.org/10.5957/ISSC-2022-COMMITTEE-I-1

Bai, G., Ruan, Z., Wang, J.: The bivariate region frequency extreme value analysis of significant wave height and mean wave period in the South China Sea. Ocean Eng. **276**, 114151 (2023). https://doi.org/10.1016/j.oceaneng.2023.114151

Baldan, D., Coraci, E., Crosato, F., Ferla, M., Bonometto, A., Morucci, S.: Importance of non-stationary analysis for assessing extreme sea levels under sea level rise. Nat. Hazard. **22**(11), 3663–3677 (2022). https://doi.org/10.5194/nhess-22-3663-2022

Bandringa, H., Jaouën, F., Helder, J., Bunnik, T.: On the validity of CFD for simulating a shallow water CALM buoy in extreme waves. In: OMAE Conference (2021). https://doi.org/10.1115/OMAE2021-62738

Barbariol, F., et al.: Wind-wave forecasting in enclosed basins using statistically downscaled global wind forcing. Front. Mar. Sci. **9**, 1002786 (2022). https://doi.org/10.3389/fmars.2022.1002786

Barlow, A.M., Mackay, E., Eastoe, E., Jonathan, P.: A penalised piecewise-linear model for non-stationary extreme value analysis of peaks over threshold. Ocean Eng. **267**, 113265 (2023). https://doi.org/10.1016/j.oceaneng.2022.113265

Barth, A., Alvera-Azcárate, A., Troupin, C., Beckers, J.-M.: DINCAE 2.0: multivariate convolutional neural network with error estimates to reconstruct sea surface temperature satellite and altimetry observations. Geosci. Model Dev. **15**(5), 2183–2196 (2022). https://doi.org/10.5194/gmd-15-2183-2022

Bayle, P.M., et al.: Delta transport processes laboratory. CoastLab 2024: Physical Modelling in Coastal Engineering and Science (2024). https://doi.org/10.59490/coastlab.2024.757

Beckman, J.N., Long, J.W.: Quantifying errors in wind and wave measurements from a compact, low-cost wave buoy. Front. Mar. Sci. **9**, 966855 (2022). https://doi.org/10.3389/fmars.2022.966855

Belzile, L.R., Dutang, C., Northrop, P.J., Opitz, T.: A modeler's guide to extreme value software. Extremes **26**(4), 595–638 (2023). https://doi.org/10.1007/s10687-023-00475-9

Benetazzo, A., et al.: On the extreme value statistics of spatio-temporal maximum sea waves under cyclone winds. Prog. Oceanogr. **197**, 102642 (2021). https://doi.org/10.1016/j.pocean.2021.102642

Benti, N.E., Chaka, M.D., Semie, A.G.: Forecasting renewable energy generation with machine learning and deep learning: current advances and future prospects. Sustainability **15**(9), 7087 (2023). https://doi.org/10.3390/su15097087

Berlinghieri, R., et al.:. Gaussian processes at the Helm(holtz): a more fluid model for ocean currents (2023). arXiv:2302.10364. https://doi.org/10.48550/arXiv.2302.10364

Bernardino, M., Gonçalves, M., Guedes Soares, C.: Marine climate projections toward the end of the twenty-first century in the North Atlantic. J. Offshore Mech. Arctic Eng. **143**(6), 1–10 (2021). https://doi.org/10.1115/1.4050698

Bernardino, M., Gonçalves, M., Campos, R.M., Guedes Soares, C.: Extremes and variability of wind and waves across the oceans until the end of the 21st century. Ocean Eng. **275**, 114081 (2023). https://doi.org/10.1016/j.oceaneng.2023.114081

Beu, C.M.L., Landulfo, E.: Machine-learning-based estimate of the wind speed over complex terrain using the long short-term memory (LSTM) recurrent neural network. Wind Energ. Sci. **9**, 1431–1450 (2024). https://doi.org/10.5194/wes-9-1431-2024

Bi, K., Xie, L., Zhang, H., Chen, X., Gu, X., Tian, Q.: Accurate medium-range global weather forecasting with 3D neural networks. Nature **619**, 533–538 (2023). https://doi.org/10.1038/s41586-023-06185-3

Bi, K., Xie, L., Zhang, H., Chen, X., Gu, X., Tian, Q.: Pangu-Weather: a 3D high-resolution model for fast and accurate global weather forecast (2022). arXiv:2211.02556. https://doi.org/10.48550/arXiv.2211.02556

Bidlot, J.R., Lemos, G., Semedo, A.: ERA5 reanalysis & ERA5 based ocean wave hindcast. In: 2nd International Workshop on Waves, Storm Surges and Coastal Hazards (2019)

Bieman, J.P., van Gent, M.R.A., van den Boogaard, H.F.P.: Wave overtopping predictions using an advanced machine learning technique. Coast. Eng. **166**, 103830 (2021). https://doi.org/10.1016/j.coastaleng.2020.103830

Bieman, J.P., Ridder, M.P., Mata, M.I., Nieuwkoop, J.C.C.: Hybrid modelling to improve operational wave forecasts by combining process-based and machine learning models. Appl. Ocean Res. **136**, 103583 (2023). https://doi.org/10.1016/j.apor.2023.103583

Bisinotto, G.A., Sparano, J.V., Simos, A.N., Cozman, F.G., Ferreira, M.D., Tannuri, E.A.: Sea state estimation based on the motion data of a moored FPSO using neural networks: an evaluation with multiple draft conditions. Ocean Eng. **276**, 114235 (2023). https://doi.org/10.1016/j.oceaneng.2023.114235

Bitner-Gregersen, E.M., Cramer, E.H., Korbijn, F.: Environmental description for long-term load response of ship structures (ISOPE-I-95–335). In: 5th ISOPE Conference (1995)

Bitner-Gregersen, E.M., Gramstad, O.: Statistical description of nonlinear waves. In: OMAE Conference (2021). https://doi.org/10.1115/OMAE2021-62867

Bitner-Gregersen, E.M., Gramstad, O., Magnusson, A.K., Malila, M.: Extreme wave events and sampling variability. Ocean Dyn. **71**, 81–95 (2021). https://doi.org/10.1007/s10236-020-01422-z

Bitner-Gregersen, E.M., et al.: Uncertainties in long-term wave modelling. Mar. Struct. **84**, 103217 (2022). https://doi.org/10.1016/j.marstruc.2022.103217

Bitner-Gregersen, E.M., et al.: Rogue waves: results of the ExWaMar project. Ocean Eng. **292**, 116543 (2024). https://doi.org/10.1016/j.oceaneng.2023.116543

Bohlinger, P., Economou, T., Aarnes, O.J., Malila, M., Breivik, Ø.: A general framework to obtain seamless seasonal–directional extreme individual wave heights—showcase ekofisk. Ocean Eng. **270**, 113535 (2023). https://doi.org/10.1016/j.oceaneng.2022.113535

Böhm, A.M., Herrnring, H., von Bock und Polach, R.U.F.: Data from uniaxial compressive testing of laboratory-made granular ice. Data Brief **42**, 108236 (2022). https://doi.org/10.1016/j.dib.2022.108236

Boyer, T., et al.: Effects of the pandemic on observing the global ocean. Bull. Am. Meteor. Soc. **104**(2), E389–E410 (2023). https://doi.org/10.1175/BAMS-D-21-0210.1

Breunung, T., Balachandran, B.: Prediction of freak waves from buoy measurements. Sci. Rep. **14**, 16048 (2024). https://doi.org/10.1038/s41598-024-66315-3

Brolly, M.T.: Inferring ocean transport statistics with probabilistic neural networks. J. Adv. Model. Earth Syst. **15**(6) (2023). https://doi.org/10.1029/2023MS003718

Bruch, W., et al.: Sea-spray-generation dependence on wind and wave combinations: a laboratory study. Bound.-Layer Meteorol. **180**, 477–505 (2021). https://doi.org/10.1007/s10546-021-00636-y

Bruyère, C.L., et al.: Using large climate model ensembles to assess historical and future tropical cyclone activity along the Australian east coast. Weather Clim. Extremes **38**, 100507 (2022). https://doi.org/10.1016/j.wace.2022.100507

Bu, J., Yu, K., Ni, J., Huang, W.: Combining ERA5 data and CYGNSS observations for the joint retrieval of global significant wave height of ocean swell and wind wave: a deep convolutional neural network approach. J. Geodesy. **97**(8), 1–22 (2023). https://doi.org/10.1007/s00190-023-01768-4

Buckley, M.P., Veron, F.: The turbulent airflow over wind generated surface waves. Eur. J. Mech. B/Fluids **73**, 132–143 (2019). https://doi.org/10.1016/j.euromechflu.2018.04.003

Bujak, D., Lončar, G., Carević, D., Kulić, T.: The Feasibility of the ERA5 forced numerical wave model in fetch-limited basins. J. Mar. Sci. Eng. **11**(1), 59 (2023). https://doi.org/10.3390/jmse11010059

Bunnik, T., Scharnke, J.: Statistical variation of the 3-h maximum crest height in a survival sea state. In: OMAE Conference (2022). https://doi.org/10.1115/OMAE2022-79045

Bustos Usta, D.F., Torres Parra, R.R.: Projected wind changes in the Caribbean Sea based on CMIP6 models. Clim. Dyn. **60**(11–12), 3713–3727 (2023). https://doi.org/10.1007/s00382-022-06535-3

Cagigal, L., Méndez, F.J., van Vloten, S.O., Rueda, A., Coco, G.: Wind wave footprint of tropical cyclones from satellite data. Int. J. Climatology **43**(1), 372–381 (2023). https://doi.org/10.1002/joc.7764

Calvert, R., et al.: A laboratory study of the effects of size, density, and shape on the wave-induced transport of floating marine litter. J. Geophys. Res. Oceans **129**, e2023JC020661 (2024). https://doi.org/10.1029/2023JC02066

Campos, R.M., Costa, M.O., Almeida, F., Guedes Soares, C.: Operational wave forecast selection in the Atlantic Ocean using random forests. J. Mar. Sci. Eng. **9**, 298 (2021). https://doi.org/10.3390/jmse9030298

Campos, R.M., Figurskey, D., Mehra, A.: Experiments on machine learning post-processing models applied to probabilistic wave forecasts. In: 3rd Symposium on Community Modeling and Innovation; 104th AMS Annual Meeting, Baltimore, MD, 28 January–1 February 2024 (2024)

Campos, R.M., Islam, H., Ferreira, T.R.S., Guedes Soares, C.: Impact of heavy biofouling on a nearshore heave-pitch-roll wave buoy performance. Appl. Ocean Res. **107**, 102500 (2021). https://doi.org/10.1016/j.apor.2020.102500

Campos, R.M., Krasnopolsky, V., Alves, J.H.G.M., Penny, S.G.: Improving NCEP's global-scale wave ensemble averages using neural networks. Ocean Model. **149**, 101617 (2020). https://doi.org/10.1016/j.ocemod.2020.101617

Campos, R.M., Krasnopolsky, V., Alves, J.H.G.M., Penny, S.G.: Nonlinear wave ensemble averaging in the Gulf of Mexico using neural networks. J. Atmos. Ocean. Technol. **2019**(36), 113–127 (2019). https://doi.org/10.1175/JTECH-D-18-0099.1

Campos, R.M., Palmeira, R.M.J., Pereira, H.P.P., Azevedo, L.C.: Mid-to-long range wind forecast in Brazil using numerical modeling and neural networks. Wind **2**(2), 221–245 (2022). https://doi.org/10.3390/wind2020013

Canard, M., Ducrozet, G., Bouscasse, B.: Varying ocean wave statistics emerging from a single energy spectrum in an experimental wave tank. Ocean Eng. **246**, 110375 (2022). https://doi.org/10.1016/j.oceaneng.2021.110375

Canard, M., Ducrozet, G., Bouscasse, B.: Generation of controlled irregular wave crest statistics in experimental and numerical wave tanks. Ocean Eng. **310**, 118676 (2024). https://doi.org/10.1016/j.oceaneng.2024.118676

Cao, R., Padilla, E.M., Callaghan, A.H.: The influence of bandwidth on the energetics of intermediate to deep water laboratory breaking waves. J. Fluid Mech. **971**, A11 (2023). https://doi.org/10.1017/jfm.2023.645

Carro, H., et al.: Machine learning tool for wave overtopping prediction based on the safety-operability ratio. Ocean Eng. **312**(Part 1), 119006 (2024). https://doi.org/10.1016/j.oceaneng.2024.119006

Carvalho, D., Rocha, A., Costoya, X., deCastro, M., Gómez-Gesteira, M.: Wind energy resource over Europe under CMIP6 future climate projections: what changes from CMIP5 to CMIP6. Renew. Sustain. Energy Rev. **151**, 111594 (2021). https://doi.org/10.1016/j.rser.2021.111594

Casas-Prat, M., et al.: Wind-wave climate changes and their impacts. Nat. Rev. Earth Environ. **5**(1), 23–42 (2024). https://doi.org/10.1038/s43017-023-00502-0

Cavaleri, L., Barbariol, F., Benetazzo, A.: Wind-wave modelling: where we are, Where we go. J. Mar. Sci. Eng. **2020**(8), 260 (2020). https://doi.org/10.3390/jmse8040260

Cavaleri, L., Barbariol, F., Bastianini, M., Benetazzo, A., Bertotti, L., Pomaro, A.: An exceptionally high wave at the CNR-ISMAR oceanographic tower in the Northern Adriatic Sea. Sci. Data **8**(1) (2021). https://doi.org/10.1038/s41597-021-00825-x

Cavaleri, L., Langodan, S., Pezzutto, P., Benetazzo, A.: The earliest stages of wind wave generation in the open sea. J. Phys. Oceanography **54**, 755–766 (2024). https://doi.org/10.1175/JPO-D-23-0217.1

Çelik, A.: Improving prediction performance of significant wave height via hybrid SVD-Fuzzy model. Ocean Eng. **266**, 113173 (2022). https://doi.org/10.1016/j.oceaneng.2022.113173

Chai, W., Leira, B.J.: Environmental contours based on inverse SORM. Mar. Struct. **60**, 34–51 (2018). https://doi.org/10.1016/j.marstruc.2018.03.007

Charlton-Perez, A.J., et al.: Do AI models produce better weather forecasts than physics-based models? A quantitative evaluation case study of Storm Ciarán. npj Clim. Atmos. Sci. **7**, 93 (2024). https://doi.org/10.1038/s41612-024-00638-w

Chen, S.-T.: Probabilistic forecasting of coastal wave height during typhoon warning period using machine learning methods. J. Hydroinf. **21**(2), 343–358 (2019). https://doi.org/10.2166/hydro.2019.115

Chen, X., Okada, T., Kawamura, Y., Mitsuyuki, T.: Estimation of on-site directional wave spectra using measured hull stresses on 14,000 TEU large container ships. J. Mar. Sci. Technol. (Japan) **25**(3), 690–706 (2020). https://doi.org/10.1007/s00773-019-00673-w

Chen, X., Okada, T., Kawamura, Y., Mitsuyuki, T.: Estimation of directional wave spectra and hull structural responses based on measured hull data on 14,000 TEU large container ships. Mar. Struct. **80**, 103087 (2021). https://doi.org/10.1016/j.marstruc.2021.103087

Chen, J., Kang, S., You, Q., Zhang, Y., Du, W.: Projected changes in sea ice and the navigability of the Arctic Passages under global warming of 2 °C and 3 °C. Anthropocene **40**, 100349 (2022). https://doi.org/10.1016/j.ancene.2022.100349

Chen, J., Milne, I., Taylor, P.H., Gunawan, D., Zhao, W.: Forward prediction of surface wave elevations and motions of offshore floating structures using a data-driven model. Ocean Eng. **281**, 114680 (2023). https://doi.org/10.1016/j.oceaneng.2023.114680

Chen, L., et al.: FuXi: a cascade machine learning forecasting system for 15-day global weather forecast. npj Clim. Atmos. Sci. **6**, 190 (2023). https://doi.org/10.1038/s41612-023-00512-1

Chen, X., et al.: Research advances on internal processes affecting tropical cyclone intensity change from 2018–2022. Tropical Cyclone Res. Rev. **12**(1), 10–29 (2023). https://doi.org/10.1016/j.tcrr.2023.05.001

Chen, X., Takami, T., Oka, M., Kawamura, Y., Okada, T.: Stochastic wave spectra estimation (SWSE) based on response surface methodology considering uncertainty in transfer functions of a ship. Mar. Struct. **90**, 103423 (2023). https://doi.org/10.1016/j.marstruc.2023.103423

Chen, L., Han, B., Wang, X., Zhao, J., Yang, W., Yang, Z.: Machine learning methods in weather and climate applications: a survey. Appl. Sci. **13**(21), 12019 (2024a). https://doi.org/10.3390/app132112019

Chen, L., et al.: A machine learning model that outperforms conventional global subseasonal forecast models. Nat. Commun. **15**, 6425 (2024b). https://doi.org/10.1038/s41467-024-50714-1

Chirosca, A.M., Rusu, L.: Characteristics of the wind and wave climate along the European seas focusing on the main maritime routes. J. Mar. Sci. Eng. **10**(1), 75 (2022). https://doi.org/10.3390/jmse10010075

Choi, Y., Park, Y., Hwang, J., Jeong, K., Kim, E.: Improving ocean forecasting using deep learning and numerical model integration. J. Marine Science and Eng. **10**(4), 450 (2022). https://doi.org/10.3390/jmse10040450

Christakos, K., Lavidas, G., Gao, Z., Björkqvist, J.V.: Long-term assessment of wave conditions and wave energy resource in the Arctic Ocean. Renew. Energy **220**, 119678 (2024). https://doi.org/10.1016/j.renene.2023.119678

Cicon, L., Gemmrich, J., Pouliot, B., Bernier, N.: A probabilistic prediction of rogue waves from a WAVEWATCH III model for the Northeast Pacific. Weather Forecast. **38**(11), 2363–2377 (2023). https://doi.org/10.1175/WAF-D-23-0074.1

Cimarelli, A., Romoli, F., Stalio, E.: On wind-wave interaction phenomena at low Reynolds numbers. J. Fluid Mechanics **956**, A13 (2023). https://doi.org/10.1017/jfm.2023.4

Coles, S.: An Introduction to Statistical Modeling of Extreme Values. Springer, London (2001). https://doi.org/10.1007/978-1-4471-3675-0

Collins, C.O., et al.: Performance of moored GPS wave buoys. Coast. Eng. J. **66**(1), 17–43 (2024). https://doi.org/10.1080/21664250.2023.2295105

Conti, S.: Artificial intelligence for weather forecasting. Nat. Rev. Electr. Eng. **1**, 8 (2024). https://doi.org/10.1038/s44287-023-00009-2

Costa, M.O., Campos, R.M., Guedes Soares, C.: Enhancing the accuracy of metocean hindcasts with machine learning models. Ocean Eng. **287**, 115724 (2023). https://doi.org/10.1016/j.oceaneng.2023.115724

Crespo, L.G., Agrell, C., Vanem, E.: A scenario optimization approach to the prediction of extreme structural responses to environmental loading. In: OMAE 2024, Volume 2: Structures, Safety, and Reliability (2024). https://doi.org/10.1115/OMAE2024-131595

D'Agostini, A., Bernardino, M., Guedes Soares, C.: Projected wave storm conditions under the RCP8.5 climate change scenario in the North Atlantic Ocean. Ocean Eng. **266**, 112874 (2022). https://doi.org/10.1016/j.oceaneng.2022.112874

D'Arcy, E., Tawn, J.A., Joly, A., Sifnioti, D.E.: Accounting for seasonality in extreme sea-level estimation. Ann. Appl. Stat. **17**(4), 3500–3525 (2023). https://doi.org/10.1214/23-AOAS1773

D'Arcy, E., Tawn, J.A., Sifnioti, D.E.: Accounting for climate change in extreme sea level estimation. Water **14**(19), 2956 (2022). https://doi.org/10.3390/w14192956

Davis, J.R., et al.: Saturation of ocean surface wave slopes observed during hurricanes. Geophys. Res. Lett. **50**(16) (2023). https://doi.org/10.1029/2023GL104139

Davison, S., Benetazzo, A., Barbariol, F., Ducrozet, G., Yoo, J., Marani, M.: Space-time statistics of extreme ocean waves in crossing sea states. Front. Mar. Sci. **9**, 1002806 (2022). https://doi.org/10.3389/fmars.2022.1002806

de Hauteclocque, G., Zhu, T., Johnson, M., Austefjord, H., Bitner-Gregersen, E.: Assessment of global wave dataset for long term response of ships. In: OMAE Conference (2020). https://doi.org/10.1115/OMAE2020/18874

de Hauteclocque, G., Mackay, E., Vanem, E.: Quantitative comparison of environmental contour approaches. Ocean Eng. **245**, 110374 (2022). https://doi.org/10.1016/j.oceaneng.2021.110374

de Hauteclocque, G., Maretic, N.V., Derbanne, Q.: Hindcast based global wave statistics. Appl. Ocean Res. **130**, 103438 (2023). https://doi.org/10.1016/j.apor.2022.103438

De Leo, F., Besio, G., Briganti, R., Vanem, E.: Non-stationary extreme value analysis of sea states based on linear trends: analysis of annual maxima series of significant wave height and peak period in the Mediterannean Sea. Coastal Eng. **167**, 103896 (2021). https://doi.org/10.1016/j.coastaleng.2021.103896

De Leo, F., Briganti, R., Besio, G.: Trends in ocean waves climate within the Mediterranean Sea: a review. Clim. Dyn. **62**(2), 1555–1566 (2024). https://doi.org/10.1007/s00382-023-06984-4

de Paula, T.P., Lima, J.A.M., Tanajura, C.A.S., Andrioni, M., Martins, R.P., Arruda, W.Z.: The impact of ocean data assimilation on the simulation of mesoscale eddies at São Paulo plateau (Brazil) using the regional ocean modeling system. Ocean Model. **167**, 101889 (2021). https://doi.org/10.1016/j.ocemod.2021.101889

Deng, Y., Zhu, C., Wang, Z.: A comparative study of wave kinematics and inline forces on vertical cylinders under Draupner-type freak waves. Ocean Eng. **288**, 115959 (2023). https://doi.org/10.1016/j.oceaneng.2023.115959

Derbanne, Q., Storhaug, G., Shigunov, V., Xie, G., Zheng, G.: Rule formulation of vertical hull girder wave loads based on direct computation. In: 13th PRADS Conference (2016)

Derbanne, Q., de Hauteclocque, G.: A new approach for environmental contour and multivariate de-clustering. In: OMAE Conference (2019). https://doi.org/10.1115/OMAE2019-95993

Deskos, G., Lee, J.C.Y., Draxl, C., Sprague, M.A.: Review of wind-wave coupling models for large-eddy simulation of the marine atmospheric boundary layer. J. Atmos. Sci. **78**(10), 3025–3045 (2021). https://doi.org/10.1175/JAS-D-21-0003.1

DNV. DNV RP-C205 - Environmental conditions and environmental loads (2024)

Dodet, G., et al.: The sea state CCI dataset v1: towards a sea state climate data record based on satellite observations. Earth Syst. Sci. Data **12**(3), 1929–1951 (2020). https://doi.org/10.5194/essd-12-1929-2020

Dorsay, C., Egan, G., Houghton, I.A., Hegermiller, C., Smit, P.B.: Proxy observations of surface wind from a globally distributed network of wave buoys. J. Atmos. Ocean. Technol. **40**, 1591–1603 (2023). https://doi.org/10.1175/JTECH-D-23-0044.1

Draycott, S., et al.: Re-creation of site-specific multi-directional waves with non-collinear current. Ocean Eng. **152**, 391–403 (2018). https://doi.org/10.1016/j.oceaneng.2017.10.047

Draycott, S., Pillai, A.C., Gabl, R., Stansby, P.K., Davey, T.: An experimental assessment of the effect of current on wave buoy measurements. Coast. Eng. **174**, 104114 (2022). https://doi.org/10.1016/j.coastaleng.2022.104114

Duan, X., Wang, S., Liu, D., Shi, J., Wu, Y., Zhou, X.: A statistical analysis method for significant wave height and spectral peak frequency considering the random and time-varying effects based on copula function and Bayesian inference. Ocean Model. **190**, 102390 (2024). https://doi.org/10.1016/j.ocemod.2024.102390

Ducrozet, G., Wang, Y., Derakhti, M.: Enhanced wave breaking modelling in a High-Order Spectral model. In: Breaking Waves Workshop B'Waves, Bordeaux, France, 30 May–1 June 2023 (2023)

Düz, B., Scharnke, J., de Wilde, J.: Mitigating spurious low-frequency waves in a model basin with a ramped floor. In: OMAE Conference (2022). https://doi.org/10.1115/OMAE2022-78434

El Khazen, M. W., Gogonel, A., Cucu-Grosjean, L.: Work-in-progress: lessons learnt from creating an extreme value library in Python. In: Proceedings - Real-Time Systems Symposium, 2020-December (2020). https://doi.org/10.1109/RTSS49844.2020.00052

Eeltink, D., et al.: Nonlinear wave evolution with data-driven breaking. Nat. Commun. **13**(1), 2343 (2022). https://doi.org/10.1038/s41467-022-30025-z

Ellingsen, S.Å., Li, Y.: Approximate dispersion relations for waves on arbitrary shear flows. J. Geophys. Res. Oceans **122**, 9889–9905 (2017). https://doi.org/10.1002/2017JC012994

Ellwood, G., et al.: Experimental study on the role of second-order wave generation and active reflection compensation on overtopping and damage of rubble mound breakwaters. In: Australasian Coasts & Ports Conference (2021)

Engelke, S., Ivanovs, J.: Sparse structures for multivariate extremes. Ann. Rev. Stat. Appl. **8**(1), 241–270 (2021). https://doi.org/10.1146/annurev-statistics-040620-041554

Esmailian, E., Steen, S., Koushan, K.: Ship design for real sea states under uncertainty. Ocean Eng. **266**, 113127 (2022). https://doi.org/10.1016/j.oceaneng.2022.113127

Ewans, K., Collins, C.: A comparison of wave directional spreading measurements made with a Spotter buoy and a directional Waverider buoy in parallel. In: OMAE Conference (2024). https://doi.org/10.1115/OMAE2024-126998

Ewans, K., Jonathan, P.: Uncertainties in estimating the effect of climate change on 100-year return value for significant wave height. Ocean Eng. **272**, 113840 (2023). https://doi.org/10.1016/j.oceaneng.2023.113840

Feng, X., Chen, X.: Feasibility of ERA5 reanalysis wind dataset on wave simulation for the western inner-shelf of Yellow Sea. Ocean Eng. **236**, 109413 (2021). https://doi.org/10.1016/j.oceaneng.2021.109413

Fernández-Alvarez, J.C., Costoya, X., Pérez-Alarcón, A., Rahimi, S., Nieto, R., Gimeno, L.: Dynamic downscaling of wind speed over the North Atlantic Ocean using CMIP6 projections: implications for offshore wind power density. Energy Rep. **9**, 873–885 (2023). https://doi.org/10.1016/j.egyr.2022.12.036

Fernández, L., Calvino, C., Dias, F.: Sensitivity analysis of wind input parametrizations in the WAVEWATCH III spectral wave model using the ST6 source term package for Ireland. Appl. Ocean Res. **115**, 102826 (2021). https://doi.org/10.1016/j.apor.2021.102826

Fine, E.C., et al.: Arctic ice-ocean interactions in an 8-to-2 kilometer resolution global model. Ocean Model. **184**, 102228 (2023). https://doi.org/10.6075/J0XK8F

Elobeid, M., Pillai, A.C., Tao, L., Ingram, D., Hanssen, J.E., Mayorga, P.: Implications of wave–current interaction on the dynamic responses of a floating offshore wind turbine. Ocean Eng. **292**, 116571 (2024). https://doi.org/10.1016/j.oceaneng.2023.116571

Foltz, G.R., Zhang, C., Meinig, C., Zhang, J. A., Zhang, D.: An unprecedented view inside a hurricane. Eos **103** (2022). https://doi.org/10.1029/2022EO220228

Fontanella, A., et al.: A hardware-in-the-loop wave-basin scale-model experiment for the validation of control strategies for floating offshore wind turbines. J. Physics: Conf. Series **1618**(3), 032038 (2020). https://doi.org/10.1088/1742-6596/1618/3/032038

Fouques, S., et al.: Qualification criteria for the verification of numerical waves - Part 1: potential-based numerical wave tank (PNWT). OMAE Conf (2021). https://doi.org/10.1115/OMAE2021-63884

Fouques, S., Akselsen, A., Harris, T., Brett, K.: On the quality of laboratory water waves generated mechanically by flap wavemakers. In: OMAE Conference (2022). https://doi.org/10.1115/OMAE2022-81324.

Freeman, E., et al.: ICOADS Release 3.0: a major update to the historical marine climate record. Int. J. Climatol. **37**, 2211–2232 (2017). https://doi.org/10.1002/joc.4775

Fuhrman, D.R., Klahn, M., Zhai, Y.: A new probability density function for the surface elevation in irregular seas. J. Fluid Mechanics **970**, 669 (2023). https://doi.org/10.1017/jfm.2023.669

Fujikubo, M., et al.: A digital twin for ship structures - R&D project in Japan. Data-Centric Eng. **5**, 3 (2024). https://doi.org/10.1017/dce.2024.3

Fujimoto, W., Ishibashi, K., Zhu, T.: Analyzing AIS and wave hindcast data for global wave scatter diagrams with seasonality. Ocean Eng. (2024)

Gao, S., et al.: A nowcasting model for the prediction of typhoon tracks based on a long short term memory neural network. Acta Oceanologica Sinica **37**, 8–12 (2018). https://doi.org/10.1007/s13131-018-1219-z

García Medina, G., Yang, Z., Li, N., Cheung, K.F., Lutu-McMoore, E.: Wave climate and energy resources in American Samoa from a 42-year high-resolution hindcast. Renew. Energy **210**, 604–617 (2023). https://doi.org/10.1016/j.renene.2023.03.031

García-Gutiérrez, A., López, D., Domínguez, D., Gonzalo, J.: Atmospheric boundary layer wind profile estimation using neural networks, mesoscale models, and LiDAR measurements. Sensors **23**(7), 3715 (2023). https://doi.org/10.3390/s23073715

García-Rojo, R.: Algorithm for the estimation of the long-term wind climate at a meteorological mast using a joint probabilistic approach. Wind Eng. **28**(2), 213–223 (2004). https://doi.org/10.1260/0309524041211378

Garg, S., Rasp, S., Thuerey, N.: Weatherbench probability: a benchmark dataset for probabilistic medium-range weather forecasting along with deep learning baseline models (2022). arXiv: 2205.00865. https://doi.org/10.48550/arXiv.2205.00865

Gemmrich, J., Cicon, L.: Generation mechanism and prediction of an observed extreme rogue wave. Sci. Rep. **12**(1), 1718 (2022). https://doi.org/10.1038/s41598-022-05671-4

Gharamti, I.E., Ahmad, W., Puolakka, O., Tuhkuri, J.: Thickness-independent fracture in columnar freshwater ice: an experimental study. Eng. Fract. Mech. **298**, 109906 (2024). https://doi.org/10.1016/j.engfracmech.2024.109906

Gilleland, E., Katz, R.W.: ExtRemes 2.0: an extreme value analysis package in R. J Stat. Softw. **72**, 1–39 (2016). https://doi.org/10.18637/jss.v072.i08

Gomit, G., Chatellier, L., David, L.: Free-surface flow measurements by non-intrusive methods: a survey. Exp. Fluids **63**(6), 94 (2022). https://doi.org/10.1007/s00348-022-03450-5

Gong, Y., Dong, S., Wang, Z.: Development of a coupled genetic algorithm and empirical typhoon wind model and its application. Ocean Eng. **248**, 110723 (2022a). https://doi.org/10.1016/j.oceaneng.2022.110723

Gong, Y., Dong, S., Wang, Z.: Forecasting of typhoon wave based on hybrid machine learning models. Ocean Engineering **266**(Part 3), 11293 (2022b). https://doi.org/10.1016/j.oceaneng.2022.112934

Görmüş, T., Ayat, B., Aydoğan, B.: Statistical models for extreme waves: comparison of distributions and Monte Carlo simulation of uncertainty. Ocean Eng. **248**, 110820 (2022). https://doi.org/10.1016/j.oceaneng.2022.110820

Govender, K., Mukaro, R., Mocke, G.: Laboratory measurements of mean and turbulence velocities and shear stresses through the wave roller in strong plunging waves. Coast. Eng. **180**, 104254 (2023). https://doi.org/10.1016/j.coastaleng.2022.104254

Grainger, J.P., Sykulski, A.M., Ewans, K., Hansen, H.F., Jonathan, P.: A multivariate pseudo-likelihood approach to estimating directional ocean wave models. J. R. Stat. Soc. Ser. C: Appl. Stat. **72**(3), 544–565 (2023). https://doi.org/10.1093/jrsssc/qlad006

Gramcianinov, C.B., Staneva, J., de Camargo, R., da Silva Dias, P.L.: Changes in extreme wave events in the southwestern South Atlantic Ocean. Ocean Dyn. **73**(11), 663–678 (2023). https://doi.org/10.1007/s10236-023-01575-7

Gramstad, O., Agrell, C., Bitner-Gregersen, E., Guo, B., Ruth, E., Vanem, E.: Sequential sampling method using Gaussian process regression for estimating extreme structural response. Mar. Struct. **72**, 102780 (2020). https://doi.org/10.1016/j.marstruc.2020.102780

Gramstad, O., Johannessen, T.B., Lian, G.: Long-term analysis of wave-induced loads using High Order Spectral Method and direct sampling of extreme wave events. Mar. Struct. **91**, 103473 (2023). https://doi.org/10.1016/j.marstruc.2023.103473

Gramstad, O., Lian, G.: Parametrization of sea surface skewness and kurtosis with application to crest distributions. J. Fluid Mech. **979**, 1047 (2024). https://doi.org/10.1017/jfm.2023.1047

Grossmann-Matheson, G., Young, I.R., Alves, J.H., Meucci, A.: Development and validation of a parametric tropical cyclone wave height prediction model. Ocean Eng. **283**, 115353 (2023). https://doi.org/10.1016/j.oceaneng.2023.115353

Grossmann-Matheson, G., Young, I.R., Meucci, A., Alves, J.H.: Global tropical cyclone extreme wave height climatology. Sci. Rep. **14**(1), 4167 (2024). https://doi.org/10.1038/s41598-024-54691-9

Gu, G., et al.: Wind power forecasting based on a machine learning model: considering a coastal wind farm in Zhejiang as an example. Int. J. Green Energy **21**(11), 2551–2558 (2024a). https://doi.org/10.1080/15435075.2024.2319228

Gu, F., Kauker, F., Yang, Q., Han, B., Fang, Y., Liu, C.: Effects of Freezing Temperature Parameterization on Simulated Sea-Ice Thickness Validated by MOSAiC Observations (2024b). https://doi.org/10.1029/2024GL108281

Guan, H., Cui, B., Zhu, Y.: Improvement of statistical postprocessing using GEFS reforecast information. Weather Forecast. **30**(4), 841–854 (2015). https://doi.org/10.1175/WAF-D-14-00126.1

Gubesch, E., Abdussamie, N., Penesis, I., Chin, C.: Physical and numerical modelling of extreme wave conditions. Ocean Eng. **283**, 115055 (2023). https://doi.org/10.1016/j.oceaneng.2023.115055

Gueydon, S., Bayati, I., Bosman, R., van Kampen, W., de Ridder, E.-J.: Wind turbine load monitoring during model-test in a wave basin. In: OMAE Conference (2023). https://doi.org/10.1115/OMAE2023-103266

Gueydon, S., Bayati, I., de Ridder, E.J.: Discussion of solutions for basin model tests of FOWTs in combined waves and wind. Ocean Eng. **209**, 107288 (2020). https://doi.org/10.1016/j.oceaneng.2020.107288

Guo, J., Liu, M., Fang, Z., Chen, W., Pan, X., Yang, L.: An experimental study on the influence of wind-wave-current coupling effect on the global performance of a 12 MW semi-submersible floating wind turbine. Ocean Eng. **304**, 117795 (2024). https://doi.org/10.1016/j.oceaneng.2024.117795

Guth, S., Katsidoniotaki, E., Sapsis, T.P.: Statistical modeling of fully nonlinear hydrodynamic loads on offshore wind turbine monopile foundations using wave episodes and targeted CFD simulations through active sampling. Wind Energy **27**(1), 75–100 (2024). https://doi.org/10.1002/we.2880

Guth, S., Sapsis, T.P.: Wave episode based Gaussian process regression for extreme event statistics in ship dynamics: between the Scylla of Karhunen-Loève convergence and the Charybdis of transient features. Ocean Eng. **266**, 112633 (2022). https://doi.org/10.1016/j.oceaneng.2022.112633

Häfner, D., Gemmrich, J., Jochum, M.: FOWD: a free ocean wave dataset for data mining and machine learning. J. Atmos. Oceanic Technol. **38**(7), 1305–1322 (2021). https://doi.org/10.1175/JTECH-D-20-0185.1

Häfner, D., Gemmrich, J., Jochum, M.: Machine-guided discovery of a real-world rogue wave model. Proc. Natl. Acad. Sci. USA **120**(48) (2023). https://doi.org/10.1073/pnas.2306275120

Hafver, A., Agrell, C., Vanem, E.: Environmental contours as Voronoi cells. Extremes **25**(3), 451–486 (2022). https://doi.org/10.1007/s10687-022-00437-7

Hahmann, A.N., García-Santiago, O., Peña, A.: Current and future wind energy resources in the North Sea according to CMIP6. Wind Energy Sci. **7**(6), 2373–2391 (2022). https://doi.org/10.5194/wes-7-2373-2022

Halsne, T., Benetazzo, A., Barbariol, F., Christensen, K.H., Carrasco, A., Breivik, Ø.: Wave modulation in a strong tidal current and its impact on extreme waves. J. Phys. Oceanography **54**(1), 131–151 (2024). https://doi.org/10.1175/JPO-D-23-0051.1

Ham, Y.-G., Kim, J.-H., Luo, J.-J.: Deep learning for multi-year ENSO forecasts. Nature **573**, 568–572 (2019). https://doi.org/10.1038/s41586-019-1559-7

Hammer, T.C., Puolakka, O., Hendrikse, H.: Scaling ice-induced vibrations by combining replica modeling and preservation of kinematics. Cold Reg. Sci. Technol. **220**, 104127 (2024). https://doi.org/10.1016/j.coldregions.2024.104127

Han, Q., Hao, Z., Hu, T., Chu, F.: Non-parametric models for joint probabilistic distributions of wind speed and direction data. Renew. Energy **126**, 1032–1042 (2018). https://doi.org/10.1016/j.renene.2018.04.026

Han, P., Li, G., Skjong, S., Zhang, H.: Directional wave spectrum estimation with ship motion responses using adversarial networks. Mar. Struct. **83**, 103159 (2022). https://doi.org/10.1016/j.marstruc.2022.103159

Han, S., Yang, B., Yang, B., Zhang, G.: Numerical simulation of heterogeneous ice sheet-structure interaction based on cohesive element method. Appl. Ocean Res. **145**, 103942 (2024). https://doi.org/10.1016/j.apor.2024.103942

Haver, S.: Evidences of the existence of freak waves. In: Olagnon, M., Athanassoulis, G. (eds.) Proceedings of Rogue Waves 2000, Ifremer, 29–30 November 2000, Brest, pp. 129–140 (2000)

Haselsteiner, A.F., et al.: A benchmarking exercise for environmental contours. Ocean Eng. **236**, 109504 (2021). https://doi.org/10.1016/j.oceaneng.2021.109504

Haselsteiner, A.F., Ohlendorf, J.-H., Wosniok, W., Thoben, K.-D.: Deriving environmental contours from highest density regions. Coast. Eng. **123**, 42–51 (2017). https://doi.org/10.1016/j.coastaleng.2017.03.002

Heffernan, J.E., Tawn, J.A.: A conditional approach for multivariate extreme values (with discussion). J. R. Stat. Soc. Ser. B: Stat. Methodol. **66**(3), 497–546 (2004). https://doi.org/10.1111/j.1467-9868.2004.02050.x

Herbaut, C., Houssais, M.N., Blaizot, A.C., Molines, J.M.: A role for the ocean in the winter sea ice distribution north of Svalbard. J. Geophys. Res. Oceans **127**(6) (2022). https://doi.org/10.1029/2021JC017852

Herman, A.: Granular effects in sea ice rheology in the marginal ice zone. Phil. Trans. R. Soc. A: Math. Phys. Eng. Sci. **380**(2235) (2022). https://doi.org/10.1098/rsta.2021.0260

Herrmann, M., Nguyen-Duy, T., Ngo-Duc, T., Tangang, F.: Climate change impact on sea surface winds in Southeast Asia. Int. J. Climatology **42**(7), 3571–3595 (2022). https://doi.org/10.1002/joc.7433

Hersbach, H., et al.: The ERA5 global reanalysis. Q. J. R. Meteorol. Soc. **146**(730), 1999–2049 (2020). https://doi.org/10.1002/qj.3803

Hidayah, Z., Wirayuhanto, H., Norma Sari, Z.R., Wardhani, M.K.: Modelling sea surface currents in the eastern coast of Bawean Island, East Java. IOP Conf. Series: Earth Environ. Sci. **925**(1), 012006 (2021). https://doi.org/10.1088/1755-1315/925/1/012006

Hildeman, A., Bolin, D., Rychlik, I.: Deformed SPDE models with an application to spatial modeling of significant wave height. Spatial Stat. **42**, 100449 (2021). https://doi.org/10.1016/j.spasta.2020.100449

Hlophe, T., Taylor, P.H., Kurniawan, A., Orszaghova, J., Wolgamot, H.: Phase-resolved wave prediction in highly spread seas using optimised arrays of buoys. Appl. Ocean Res. **130**, 103435 (2023). https://doi.org/10.1016/j.apor.2022.103435

Hogben, N.: Global wave statistics. In: British Maritime Technology (BMT) (1986)

Hong, B., Zhang, J.: Long-term trends of sea surface wind in the Northern South China sea under the background of climate change. J. Marine Sci. Eng. **9**(7) (2021). https://doi.org/10.3390/jmse9070752

Hong, X., Hu, L., Kareem, A.: A tropical cyclone intensity prediction model using conditional generative adversarial network. J. Wind Eng. Ind. Aerodyn. **240**, 105515 (2023). https://doi.org/10.1016/j.jweia.2023.105515

Houghton, I.A., Penny, S.G., Hegermiller, C., Cesaretti, M., Teicheira, C., Smit, P.B.: Ensemble-based data assimilation of significant wave height from SOFAR Spotters and satellite altimeters with a global operational wave model. Ocean Model. **183**, 102200 (2023). https://doi.org/10.1016/j.ocemod.2023.102200

Hu, H., van der Westhuysen, A.J., Chu, P., Fujisaki-Manome, A.: Predicting Lake Erie wave heights and periods using XGBoost and LSTM. Ocean Model. **164**, 101832 (2021). https://doi.org/10.1016/j.ocemod.2021.101832

Huang, W., Yang, Z., Chen, X.: Wave height estimation from X-Band nautical radar images using temporal convolutional network. IEEE. J. Sel. Topics Appl. Earth Observat. Remote Sens. **14**, 11395–11405 (2021). https://doi.org/10.1109/JSTARS.2021.3124969

Huang, L., Jing, Y., Chen, H., Zhang, L., Liu, Y.: A regional wind wave prediction surrogate model based on CNN deep learning network. Appl. Ocean Res. **126**, 103287 (2022). https://doi.org/10.1016/j.apor.2022.103287

Huang, W., Zhu, X., Jin, Y., Shen, X.: Nonstationary modelling of significant wave height using time series decomposition method. Ocean Eng. **310**, 118731 (2024). https://doi.org/10.1016/j.oceaneng.2024.118731

Hui, Z., Li, Y., Sun, J., Yu, L., Ju, X., Xiong, X.: Validation and error analysis of wave-modified ocean surface currents in the northwestern Pacific Ocean. J. Oceanol. Limnol. **40**(4), 1289–1303 (2022). https://doi.org/10.1007/s00343-021-1182-y

Huseby, A.B., Vanem, E., Natvig, B.: Alternative environmental contours for structural reliability analysis. Struct. Saf. **54**, 32–45 (2015). https://doi.org/10.1016/j.strusafe.2014.12.003

Huseby, A.B., Vanem, E., Agrell, C., Hafver, A.: Convex environmental contours. Ocean Eng. **235**, 109366 (2021). https://doi.org/10.1016/j.oceaneng.2021.109366

Huseby, A.B., Vanem, E., Natvig, B.: A new approach to environmental contours for ocean engineering applications based on direct Monte Carlo simulations. Ocean Eng. **60**, 124–135 (2013). https://doi.org/10.1016/j.oceaneng.2012.12.034

Hutchings, J.K., Bliss, A.C., Mondal, D., Elosegui, P.: Sea ice deformation is not scale invariant over length scales greater than a kilometer. Geophys. Res. Lett. **51**, 12 (2024). https://doi.org/10.1029/2024GL108582

IACS. Standard Wave Data (2001)

IACS. Standard Wave Data. Rec. No. 34 1992/Rev.2 (2022)

Iafrati, A., et al.: Specialist committee on modelling of environmental conditions. In: 29th International Towing Tank Conference (ITTC), vol. II, pp. 747–765 (2021). https://ittc.info/media/10940/volume-ii.pdf

Immas, A., Do, N., Alam, M.-R.: Real-time in situ prediction of ocean currents. Ocean Eng. **228**, 108922 (2021). https://doi.org/10.1016/j.oceaneng.2021.108922

IMO. Annex 12 resolution MSC.296(87) (adopted on 20 May 2010): adoption of the guidelines for verification of conformity with goal-based ship construction standards for bulk carriers and oil tankers (2010)

IPCC. Summary for Policymakers: Synthesis Report. Climate Change 2023: Synthesis Report. Contribution of Working Groups I, II and III to the Sixth Assessment Report of the Intergovernmental Panel on Climate Change, pp. 1–34 (2023)

ITTC. RP 7.5-02-07-03.8 - Model tests for offshore wind turbines (2021). https://www.ittc.info/media/9747/75-02-07-038.pdf

ITTC. TCOMS Ocean Basin facility description (2022). https://ittc.info/media/10242/tcoms-ocean-basin-facility-description-002.pdf

ITTC. Coastal and Ocean Basin (COB) Oostende factsheet (2023). https://ittc.info/media/10409/cob-factsheet.pdf

Iwamoto, T., Takagawa, T., Shibayama, T., Esteban, M., Mäll, M.: A proposal of a semi-empirical method for modifying the atmospheric pressure and wind fields of tropical cyclones. Coast. Eng. J. **65**(3), 418–432 (2023). https://doi.org/10.1080/21664250.2023.2228005

James, J. P., Panchang, V.: Investigation of wave height and period distributions in coastal environments. J. Geophys. Res. Oceans **127**(5) (2022). https://doi.org/10.1002/essoar.10508475.1

James, S.C., Zhang, Y., O'Donncha, F.: A machine learning framework to forecast wave conditions. Coast. Eng. **137**, 1–10 (2018). https://doi.org/10.1016/j.coastaleng.2018.03.004

Jamii, J., Mansouri, M., Trabelsi, M., Mimouni, M.F., Shatanawi, W.: Effective artificial neural network-based wind power generation and load demand forecasting for optimum energy management. Front. Energy Res. Sec. Wind Energy **10**, 89841 (2022). https://doi.org/10.3389/fenrg.2022.898413

Janssen, P.A.E.M., Bidlot, J.-R.: Wind–wave interaction for strong winds. J. Phys. Oceanogr. **53**(3), 779–804 (2023). https://doi.org/10.1175/JPO-D-21-0293.1

Jaouën, F., Koop, A., Bunnik, T.: URANS predictions of drift loads on a semi submersible in steep waves. In: OMAE Conference (2022). https://doi.org/10.1115/OMAE2022-79001

Jeon, S., Kim, J.: Artificial intelligence to predict climate and weather change. JMST Adv. **6**, 67–73 (2024). https://doi.org/10.1007/s42791-024-00068-y

Ji, X., Li, A., Li, J., Wang, L., Wang, D.: Research on the statistical characteristic of freak waves based on observed wave data. Ocean Eng. **243**, 110323 (2022). https://doi.org/10.1016/j.oceaneng.2021.110323

Jiang, H., Zhang, Y., Qian, C., Wan, X.: Comment on papers using machine learning for significant wave height time series prediction: Complex models do not outperform auto-regression. Ocean Model. **189**, 102364 (2024). https://doi.org/10.1016/j.ocemod.2024.102364

Jiang, Y., et al.: Modeling waves over the Changjiang River Estuary using a high-resolution unstructured SWAN model. Ocean Model. **173**, 102007 (2022). https://doi.org/10.1016/j.ocemod.2022.102007

Johnson, G.C., et al.: Argo - Two decades: global iceanography, revolutionized. Ann. Rev. Mar. Sci. **14**, 379–403 (2022). https://doi.org/10.1146/annurev-marine-022521-102008

Johnson, G.C., Fassbender, A.J.: After two decades, Argo at PMEL, looks to the future. Oceanography **36**(2–3), 54–59 (2023). https://doi.org/10.5670/oceanog.2023.223

Johnson, G.C., Purkey, S.G.: Refined estimates of global ocean deep and abyssal decadal warming trends. Geophys. Res. Lett. **51**(18) (2024). https://doi.org/10.1029/2024GL111229

Jonathan, P., Wilson, G.A., Jesús, P.Y., Toapanta-Ramos, F.: Assessment of trimodal wave spectral parameters using machine learning methods and vessel response statistics to enhance safety of marine operations. Ocean Eng. **311**, 118921 (2024). https://doi.org/10.1016/j.oceaneng.2024.118921

Kale, B., Buckingham, S., van Beeck, J., Cuerva-Tejero, A.: Comparison of the wake characteristics and aerodynamic response of a wind turbine under varying atmospheric conditions using WRF-LES-GAD and WRF-LES-GAL wind turbine models. Renew. Energy **216**, 119051 (2023). https://doi.org/10.1016/j.renene.2023.119051

Kamath, A., Wang, W., Pakozdi, C., Bihs, H.: Identification and investigation of extreme events using an arbitrary lagrangian–eulerian approach with a laplace equation solver and coupling to a Navier–Stokes Solver. J. Offshore Mech. Arctic Eng. **145**(6) (2023). https://doi.org/10.1115/1.4057014

Karmpadakis, I., Swan, C.: A new crest height distribution for nonlinear and breaking waves in varying water depths. Ocean Eng. **266**, 112972 (2022). https://doi.org/10.1016/j.oceaneng.2022.112972

Karmpadakis, I., Swan, C., Christou, M.: A new wave height distribution for intermediate and shallow water depths. Coast. Eng. **175**, 104130 (2022). https://doi.org/10.1016/j.coastaleng.2022.104130

Kazeminezhad, M.H., Ghavanini, F.A.: Operational wave forecasting for extreme conditions in the Arabian Sea – a comparison with buoy and satellite data. Ocean Eng. **275**, 114152 (2023). https://doi.org/10.1016/j.oceaneng.2023.114152

Kim, S., et al.: Numerical and experimental study of a FORM-based design wave applying the HOS-NWT nonlinear wave solver. Ocean Eng. **263**, 112287 (2022). https://doi.org/10.1016/j.oceaneng.2022.112287

Kim, Y.-H., Min, S.-K., Gillett, N.P., Notz, D., Malinina, E.: Observationally-constrained projections of an ice-free Arctic even under a low emission scenario. Nat. Commun. **14**, 3139 (2023). https://doi.org/10.1038/s41467-023-38511-8

Kinsela, M.A., et al.: Nearshore wave buoy data from southeastern Australia for coastal research and management. Scientific Data **11**(1), 190 (2024). https://doi.org/10.1038/s41597-023-02865-x

Klahn, M., Zhai, Y., Fuhrman, D.R.: Heavy tails and probability density functions to any nonlinear order for the surface elevation in irregular seas. J. Fluid Mechanics **985**, 304 (2024). https://doi.org/10.1017/jfm.2024.304

Klein, M., et al.: Data-driven generation of tailored wave sequences. In: OMAE Conference (2024). https://doi.org/10.1115/OMAE2024-129690

Knobler, S., Liberzon, D., Fedele, F.: Large waves and navigation hazards of the Eastern Mediterranean Sea. Sci. Rep. **12**(1), 16511 (2022). https://doi.org/10.1038/s41598-022-20355-9

Kodaira, T., Sasmal, K., Miratsu, R., Fukui, T., Zhu, T., Waseda, T.: Uncertainty in wave hindcasts in the North Atlantic Ocean. Mar. Struct. **89**, 103370 (2023). https://doi.org/10.1016/j.marstruc.2023.103370

Komoriyama, Y., Iijima, K., Tatsumi, A., Fujikubo, M.: Identification of wave profiles encountered by a ship with no forward speed using Kalman filter technique and validation by tank tests - long-crested irregular wave case. Ocean Eng. **271**, 113627 (2023). https://doi.org/10.1016/j.oceaneng.2023.113627

Kordatos, I., Donas, A., Galanis, G., Famelis, I., Alexandridis, A.: Significant wave height prediction in nested domains using radial basis function neural networks. Ocean Eng. **305**, 117865 (2024). https://doi.org/10.1016/j.oceaneng.2024.117865

Krasnopolsky, V.M., Fox-Rabinovitz, M.S., Chalikov, D.V.: New approach to calculation of atmospheric model physics: accurate and fast neural network emulation of longwave radiation in a climate model. Mon. Weather Rev. **133**(5), 1370–1383 (2005). https://doi.org/10.1175/MWR2923.1

Krasnopolsky, V.M., Fox-Rabinovitz, M.S.: Complex hybrid models combining deterministic and machine learning components for numerical climate modeling and weather prediction. Neural Netw. **19**(2), 122–134 (2006). https://doi.org/10.1016/j.neunet.2006.01.002

Krasnopolsky, V.M., Fox-Rabinovitz, M.S., Hou, Y.T., Lord, S.J., Belochitski, A.A.: Accurate and fast neural network emulations of model radiation for the ncep coupled climate forecast system: climate simulations and seasonal predictions. Mon. Weather Rev. **138**, 1822–1842 (2010). https://doi.org/10.1175/2009MWR3149.1

Krasnopolsky, V.M., Lin, Y.: A neural network nonlinear multimodel ensemble to improve precipitation forecasts over continental US. Adv. Meteor. **2012**, 649450 (2012). https://doi.org/10.1155/2012/649450

Krasnopolsky, V.M., Fox-Rabinovitz, M.S., Belochitski, A.A.: Using ensemble of neural networks to learn stochastic convection parameterizations for climate and numerical weather prediction models from data simulated by a cloud resolving model. Adv. Artif. Neural Syst. **2013**(1), 485913 (2013). https://doi.org/10.1155/2013/485913

Kumar, A., Hayatdavoodi, M.: On wave–current interaction in deep and finite water depths. J. Ocean Engineering and Marine Energy **9**(3), 455–475 (2023). https://doi.org/10.1007/s40722-023-00278-x

Kumar, P., Yadav, A., Sardana, D., Prasad, R.: Extreme wave height response to climate modes and its association with tropical cyclones over the Indo-Pacific Ocean. Ocean Eng. **296**, 116789 (2024). https://doi.org/10.1016/j.oceaneng.2024.116789

Kurth, T., et al.: FourCastNet: Accelerating Global High-Resolution Weather Forecasting using Adaptive Fourier Neural Operators (2022). arXiv:2208.05419v1. https://doi.org/10.48550/arXiv.2208.05419

Laflèche, S., Christakos, K., Ommani, B., Fouques, S., Kristiansen, T.: Experimental reproduction of inhomogeneous fjord waves. Coast. Eng. **189**, 104492 (2024). https://doi.org/10.1016/j.coastaleng.2024.104492

Laflèche, S., Ommani, B., Kristiansen, T., Fouques, S.: Wave inhomogeneities in a wave basin of constant water depth. Ocean Eng. **283**, 115007 (2023). https://doi.org/10.1016/j.oceaneng.2023.115007

Lagomarsino-Oneto, D., et al.: Physics informed machine learning for wind speed prediction. Energy **268**, 126628 (2023). https://doi.org/10.1016/j.energy.2023.126628

Lakshman, R., Sriram, V., Sundar, V.: A review on directional focusing waves: generation methods toward 3D idealization of rogue or extreme waves in laboratory, pp. 21–32 (2023). https://doi.org/10.1007/978-981-19-9913-0_3

Lam, R., et al.: Learning skillful medium-range global weather forecasting. Science **382**, 1416–1421 (2023). https://doi.org/10.1126/science.adi2336

Lancaster, O., Cussu, R., Boulay, S., Hunter, S., Baldock, T.E.: Comparative wave measurements at a wave energy site with a recently developed low-cost wave buoy (Spotter), ADCP and pressure loggers. J. Atmos. Ocean. Technol. (2021). https://doi.org/10.1175/JTECH-D-20-0168.1

Landet, T., Bloch Helmers, J.: QSR: fast estimation of semi-linear vessel responses in realistic non-linear irregular waves. In: 34th ISOPE Conference (2024)

Lang, S., et al.: AIFS - ECMWF's data-driven forecasting system (2024). arXiv:2406.01465. https://doi.org/10.48550/arXiv.2406.01465

Latif, S., Simonovic, S.P.: Nonparametric approach to Copula estimation in compounding the joint impact of storm surge and rainfall events in coastal flood analysis. Water Res. Manag. **36**(14), 5599–5632 (2022). https://doi.org/10.1007/s11269-022-03321-y

Law-Chune, S., Aouf, L., Dalphinet, A., Levier, B., Drillet, Y., Drevillon, M.: WAVERYS: a CMEMS global wave reanalysis during the altimetry period. Ocean Dyn. **71**, 357–378 (2021). https://doi.org/10.1007/s10236-020-01433-w

Le Page, S., Tassin, A., Caverne, J., Ducrozet, G.: A particle-free stereo-video free-surface reconstruction method for wave-tank experiments. Exp. Fluids **65**, 157 (2024). https://doi.org/10.1007/s00348-024-03887-w

Lee, C., Kim, Y.: Local response estimation of a seagoing vessel using onboard measurement data. Mar. Struct. **86**, 103298 (2022). https://doi.org/10.1016/j.marstruc.2022.103298.4

Lee, J.-H., Nam, Y.-S., Lee, J., Liu, Y., Kim, Y.: Estimation of significant wave height using wave-radar images. J. Mar. Sci. Eng. **12**(7), 1134 (2024). https://doi.org/10.3390/jmse12071133

Lemström, I., Polojärvi, A., Puolakka, O., Tuhkuri, J.: Load distributions in the ice-structure interaction process in shallow water. Ocean Eng. **258**, 111730 (2022). https://doi.org/10.1016/j.oceaneng.2022.111730

Lemström, I., Polojärvi, A., Tuhkuri, J.: Model-scale tests on ice-structure interaction in shallow water: global ice loads and the ice loading process. Mar. Struct. **81**, 103106 (2022). https://doi.org/10.1016/j.marstruc.2021.103106

Li, Y., Ellingsen, S.Å.: A framework for modelling linear surface waves on shear currents in slowly varying waters. J. Geophys. Res. Oceans **124**, 2527–2545 (2019). https://doi.org/10.1029/2018JC014390

Li, D., Xiao, L., Wei, H., Li, J., Liu, M.: Spatial-temporal measurement of waves in laboratory based on binocular stereo vision and image processing. Coast. Eng. **177**, 104200 (2022). https://doi.org/10.1016/j.coastaleng.2022.104200

Li, J., Shao, Z., Liang, B., Lee, D.: Regional assessment of extreme significant wave heights in the Bohai Sea and northern Yellow Sea. Appl. Ocean Res. **123**, 103182 (2022). https://doi.org/10.1016/j.apor.2022.103182

Li, R., et al.: Analysis of the 20-year variability of ocean wave hazards in the Northwest Pacific. Remote Sens. **15**(11), 2768 (2023). https://doi.org/10.3390/rs15112768

Li, J., et al.: Evaluations of extreme wave heights around Hainan Island and their uncertainty induced by decadal variations of input variables. Ocean Eng. **294**, 116705 (2024). https://doi.org/10.1016/j.oceaneng.2024.116705

Liang, X., et al.: The linkage between wintertime sea ice drift and atmospheric circulation in an Arctic ice-ocean coupled simulation. Ocean Model. **189**, 102362 (2024). https://doi.org/10.1016/j.ocemod.2024.102362

Liao, Z., Huang, W., Dong, S., Li, H.: Modelling trivariate distribution of directional ocean data in the Barents Sea seasonal ice zone. Ocean Eng. **260**, 111745 (2022). https://doi.org/10.1016/j.oceaneng.2022.111745

Lin, W., Portabella, M.: Characterizing global sea surface local wind variability from ASCAT data. IEEE Trans. Geosci. Remote Sens. **60**, 1–10 (2022). https://doi.org/10.1109/TGRS.2022.3228317

Lira-Loarca, A., Cáceres-Euse, A., De-Leo, F., Besio, G.: Wave modeling with unstructured mesh for hindcast, forecast and wave hazard applications in the Mediterranean Sea. Appl. Ocean Res. **122**, 103118 (2022). https://doi.org/10.1016/j.apor.2022.103118

Liu, Q., et al.: Global wave hindcasts using the observation-based source terms: description and validation. J. Adv. Model. Earth Syst. **13**(8) (2021). https://doi.org/10.1029/2021MS002493

Liu, J., Yang, J., Liu, K., Xu, L.: Ocean current prediction using the weighted pure attention mechanism. J. Mar. Sci. Eng. **10**(5), 592 (2022). https://doi.org/10.3390/jmse10050592

Liu, K., et al.: Coupling SPH with a mesh-based Eulerian approach for simulation of incompressible free-surface flows. Appl. Ocean Res. **138**, 103673 (2023). https://doi.org/10.1016/j.apor.2023.103673

Liu, G., Zhou, X., Kou, Y., Wu, F., Zhao, D., Xu, Y.: Uncertainty analysis for the calculation of marine environmental design parameters in the South China Sea. J. Oceanology and Limnology **41**(2), 427–443 (2023). https://doi.org/10.1007/s00343-022-2052-y

Liu, J., Meucci, A., Young, I.R.: A comparison of multiple approaches to study the modulation of ocean waves due to climate variability. J. Geophys. Res. Oceans **128**(9) (2023). https://doi.org/10.1029/2023JC019843

Liu, Y., et al.: Comparison of breaking models in envelope-based surface gravity wave evolution equations. Phys. Rev. Fluids **8**(5), 054803 (2023). https://doi.org/10.1103/PhysRevFluids.8.054803

Liu, B., et al.: Extending the wind profile beyond the surface layer by combining physical and machine learning approaches. Atmos. Chem. Phys. **24**, 4047–4063 (2024). https://doi.org/10.5194/acp-24-4047-2024

Liu, H., Li, D., Chen, Q., Feng, J., Qi, J., Yin, B.: The multiscale variability of global extreme wind and wave events and their relationships with climate modes. Ocean Eng. **307**, 118239 (2024). https://doi.org/10.1016/j.oceaneng.2024.118239

Liu, Y., Eeltink, D., van den Bremer, T.S., Adcock, T.A.A.: A machine learning architecture for including wave breaking in envelope-type wave models. Ocean Eng. **305**, 118009 (2024). https://doi.org/10.1016/j.oceaneng.2024.118009

Lobeto, H., Menendez, M., Losada, I.J.: Future behavior of wind wave extremes due to climate change. Sci. Rep. **11**(1), 1–12 (2021). https://doi.org/10.1038/s41598-021-86524-4

Lobeto, H., et al.: On the assessment of the wave modeling uncertainty in wave climate projections. Environ. Res. Lett. **18**(12) (2023). https://doi.org/10.1088/1748-9326/ad0137

Lockwood, J.W., Lin, N., Oppenheimer, M., Lai, C.-Y.: Using neural networks to predict hurricane storm surge and to assess the sensitivity of surge to storm characteristics. J. Geophys. Res. Atmos. **127**, e2022JD037617 (2022). https://doi.org/10.1029/2022JD037617

Longuet-Higgins, M.S.: The generation of capillary waves by steep gravity waves. J. Fluid Mech. **16**, 138–159 (1963)

Lopez-Gavilan, P., Barrero-Gil, A.: On the limits of Particle Image Velocimetry with continuous wave lasers. Exp. Thermal Fluid Sci. **144**, 110873 (2023). https://doi.org/10.1016/j.expthermflusci.2023.110873

Lucas, C., Silva, D., Guedes Soares, C.: Climatic directional wave spectra in coastal sites. Coast. Eng. **180**, 104255 (2023). https://doi.org/10.1016/j.coastaleng.2022.104255

Lucero, F., Stringari, C.E., Filipot, J.F.: Improving WAVEWATCH III hindcasts with machine learning. Coast. Eng. **185**, 104381 (2023). https://doi.org/10.1016/j.coastaleng.2023.104381

Ludka, B.C., et al.: Sixteen years of bathymetry and waves at San Diego beaches. Scientific Data **6**(1), 161 (2019). https://doi.org/10.1038/s41597-019-0167-6

Luo, L., Liu, S., Li, J.: Simulation of wave incident boundary for spatially inhomogeneous multidirectional irregular waves. Ocean Eng. **244**, 110366 (2022). https://doi.org/10.1016/j.oceaneng.2021.110366

Luo, Y., et al.: Wave field predictions using a multi-layer perceptron and decision tree model based on physical principles: a case study at the Pearl River Estuary. Ocean Eng. **277**, 114246 (2023). https://doi.org/10.1016/j.oceaneng.2023.114246

Lyman, J.M., Johnson, G.C.: Global high-resolution random forest regression maps of ocean heat content anomalies using in situ and satellite data. J. Atmos. Ocean. Technol. **40**(5), 575–586 (2023). https://doi.org/10.1175/JTECH-D-22-0058.1

Lyu, Z., Mori, N., Kashima, H.: Freak wave in a two-dimensional directional wavefield with bottom topography change. Part 1. Normal incidence wave. J. Fluid Mech. **959**, A19 (2023). https://doi.org/10.1017/jfm.2023.73

Ma, J., Liu, F., Xiao, C., Wang, K., Liu, Z.: High resolution wind resource assessment method based on mesoscale atmospheric model and CFD technology. Wind Eng. **47**(3), 627–638 (2023). https://doi.org/10.1177/0309524X221149476

Ma, P., Zhang, Y.: A time-varying copula approach for describing seasonality in multivariate ocean data. Mar. Struct. **94**, 103567 (2024). https://doi.org/10.1016/j.marstruc.2023.103567

Maceas, M., Osorio, A.F., Bolanos, F.: A methodology for improving both performance and measurement errors in PIV. Flow Meas. Instrum. **77**, 101846 (2021). https://doi.org/10.1016/j.flowmeasinst.2020.101846

Macfarlane, A.R., et al.: A database of snow on sea ice in the central Arctic collected during the MOSAiC expedition. Sci. Data **10**(1) (2023). https://doi.org/10.1038/s41597-023-02273-1

Mackay, E.B.L., de Hauteclocque, G., Vanem, E., Jonathan, P.: The effect of serial correlation in environmental conditions on estimates of extreme events. Ocean Eng. **242**, 110092 (2021). https://doi.org/10.1016/j.oceaneng.2021.110092

Mackay, E.B.L., Haselsteiner, A.F.: Marginal and total exceedance probabilities of environmental contours. Mar. Struct. **75**, 102863 (2021). https://doi.org/10.1016/j.marstruc.2020.102863

Mackay, E.B.L., de Hauteclocque, G.: Model-free environmental contours in higher dimensions. Ocean Eng. **273**, 113959 (2023). https://doi.org/10.1016/j.oceaneng.2023.113959

Mackay, E.B.L., Jonathan, P.: Modelling multivariate extremes through angular-radial decomposition of the density function. Arxiv, pre-print (2024)

Mackay, E.B.L., Murphy-Barltrop, C.J.R., Jonathan, P.: The SPAR model: a new paradigm for multivariate extremes: application to joint distributions of metocean variables. J. Offshore Mech. Arctic Eng. **147**(1) (2025). https://doi.org/10.1115/1.4065968

Magnusson, A.K., Jensen, R.E., Swail, V.R.: Spectral shapes and parameters from three different wave sensors. Ocean Dyn. **2021**(71), 893–909 (2021). https://doi.org/10.1007/s10236-021-01468-7

Majumder, S., Remya, P.G., Nair, T.M.B., Sirisha, P.: Analysis of meteorological and oceanic conditions during freak wave events in the Indian Ocean. Ocean Eng. **259**, 111920 (2022). https://doi.org/10.1016/j.oceaneng.2022.111920

Malila, M.P., et al.: Statistical and dynamical characteristics of extreme wave crests assessed with field measurements from the North Sea. J. Phys. Oceanogr. **53**(2), 509–531 (2023). https://doi.org/10.1175/JPO-D-22-0125.1

Malliouri, D.I., Moraitis, V., Petrakis, S., Vandarakis, D., Hatiris, G.-A., Kapsimalis, V.: A non-stationary and directional probabilistic analysis of coastal storms in the Greek seas. Water **15**(13), 2455 (2023). https://doi.org/10.3390/w15132455

Manucharyan, G.E., Montemuro, B.P.: SubZero: a sea ice model with an explicit representation of the floe life cycle. J. Adv. Model. Earth Syst. **14**(12) (2022). https://doi.org/10.1029/2022MS003247

Martinez, A., Iglesias, G.: Global wind energy resources decline under climate change. Energy **288**, 129765 (2024). https://doi.org/10.1016/j.energy.2023.129765

Martzikos, N., Malara, G., Arena, F.: A neural network framework for extrapolating sea surface spectra from wave pressure signals. Ocean Eng. **334**, 121619 (2025). https://doi.org/10.1016/j.oceaneng.2025.121619

Matala, R.: Investigation of model-scale brash ice properties. Ocean Eng. **225**, 108539 (2021). https://doi.org/10.1016/j.oceaneng.2020.108539

Matala, R., Suominen, M.: Investigation of vessel resistance in model scale brash ice channels and comparison to full scale tests. Cold Reg. Sci. Technol. **201**, 103617 (2022). https://doi.org/10.1016/j.coldregions.2022.103617

Matala, R., Suominen, M.: Scaling principles for model testing in old brash ice channel. Cold Reg. Sci. Technol. **210**, 103857 (2023). https://doi.org/10.1016/j.coldregions.2023.103857

Mathewson, J.H.: Met-ocean data measurements and analysis for offshore structures and operation. J. Marine Science **05**(02), 2–5 (2023)

McAllister, M.L., Draycott, S., Calvert, R., Davey, T., Dias, F., van den Bremer, T.S.: Three-dimensional wave breaking. Nature (2024)

McAllister, M.L., Pizzo, N., Draycott, S., van den Bremer, T.S.: The influence of spectral bandwidth and shape on deep-water wave breaking onset. J. Fluid Mech. **974**, A14 (2023). https://doi.org/10.1017/jfm.2023.766

Mears, C., Lee, T., Ricciardulli, L., Wang, X., Wentz, F.: Improving the accuracy of the Cross-Calibrated Multi-Platform (CCMP) ocean vector winds. Remote Sens. **14**(17), 4230 (2022). https://doi.org/10.3390/rs14174230

Mehlmann, C., Gutjahr, O.: Discretization of Sea ice dynamics in the tangent plane to the sphere by a CD-Grid-type finite element. J. Adv. Model. Earth Syst. **14**(12) (2022). https://doi.org/10.1029/2022MS003010

Meng, X., Li, Z.-X.: 3-Dimensional environmental contours of winds and waves accounting for different sampling methods and seasonal effects. Ocean Eng. **304**, 117724 (2024). https://doi.org/10.1016/j.oceaneng.2024.117724

Meucci, A., Young, I.R., Trenham, C., Hemer, M.: An 8-model ensemble of CMIP6-derived ocean surface wave climate. Scientific Data **11**(1), 1–12 (2024). https://doi.org/10.1038/s41597-024-02932-x

Miao, Q., et al.: A study on wave climate variability along the nearshore regions of Bohai Sea based on long term observation data. Ocean Eng. **304**, 117947 (2024). https://doi.org/10.1016/j.oceaneng.2024.117947

Minuzzi, F.C., Farina, L.: A deep learning approach to predict significant wave height using long short-term memory. Ocean Model. **181**, 102151 (2023). https://doi.org/10.1016/j.ocemod.2022.102151

Minuzzi, F.C., Farina, L.: Artificial neural networks ensemble methodology to predict significant wave height. Ocean Eng. **300**, 117479 (2024). https://doi.org/10.1016/j.oceaneng.2024.117479

Miranda, F., et al.: A novel and simple passive absorption system for wave-current flumes. Alex. Eng. J. **71**, 463–477 (2023). https://doi.org/10.1016/j.aej.2023.03.067

Miratsu, R., Fukui, T., Matsumoto, T., Zhu T.: Quantitative evaluation of ship operational effect in actually encountered sea states (2019)

Miratsu, R., Sasmal, K., Kodaira, T., Fukui, T., Zhu, T., Waseda, T.: Evaluation of ship operational effect based on long-term encountered sea states using wave hindcast combined with storm avoidance model. Mar. Struct. **86**, 103293 (2022). https://doi.org/10.1016/j.marstruc.2022.103293

Mittendorf, M., Nielsen, U.D., Bingham, H.B., Storhaug, G.: Sea state identification using machine learning—a comparative study based on in-service data from a container vessel. Mar. Struct. **85**, 103274 (2022). https://doi.org/10.1016/j.marstruc.2022.103274

Mohaghegh, F., Murthy, J., Alam, M.R.: Rapid phase-resolved prediction of nonlinear dispersive waves using machine learning. Appl. Ocean Res. **117**, 102920 (2021). https://doi.org/10.1016/j.apor.2021.102920

Mohamad, M.A., Sapsis, T.P.: Sequential sampling strategy for extreme event statistics in nonlinear dynamical systems. Proc. Natl. Acad. Sci. **115**(44), 11138–11143 (2018). https://doi.org/10.1073/pnas.1813263115

Mohit, M.A., Al Towhiduzzaman, M., Khatun, M.R., Nasrin, M.S.: Tropical cyclone data calibration for the estimation of a path of a cyclone along the coast of Bangladesh. Geosyst. Geoenviron. **2**(2) (2023). https://doi.org/10.1016/j.geogeo.2022.100159

Mokhtari, M., Kim, E., Amdahl, J.: Pressure-dependent plasticity models with convex yield loci for explicit ice crushing simulations. Mar. Struct. **84**, 103233 (2022). https://doi.org/10.1016/j.marstruc.2022.103233

Mollick, T., Hashmi, G., Sabuj, S.R.: Wind speed prediction for site selection and reliable operation of wind power plants in coastal regions using machine learning algorithm variants. Sustain. Energy Res **11**, 5 (2024). https://doi.org/10.1186/s40807-024-00098-z

Moore, A.M., Arango, H.G., Wilkin, J., Edwards, C.A.: Weak constraint 4D-Var data assimilation in the Regional Ocean Modeling System (ROMS) using a saddle-point algorithm: application to the California Current Circulation. Ocean Model. **186**, 102262 (2023). https://doi.org/10.1016/j.ocemod.2023.102262

Moreno, S.R., Seman, L.O., Stefenon, S.F., Coelho, L.S., Mariani, V.C.: Enhancing wind speed forecasting through synergy of machine learning, singular spectral analysis, and variational mode decomposition. Energy **292**, 130493 (2024). https://doi.org/10.1016/j.energy.2024.130493

Mori, N., Janssen, P.A.E.M.: On kurtosis and occurrence probability of freak waves. J. Phys. Oceanogr. **36**(7), 1471–1483 (2006). https://doi.org/10.1175/JPO2922.1

Morim, J., et al.: A global ensemble of ocean wave climate statistics from contemporary wave reanalysis and hindcasts. Scientific Data **9**(1), 1–8 (2022). https://doi.org/10.1038/s41597-022-01459-3

Morim, J., Wahl, T., Vitousek, S., Santamaria-Aguilar, S., Young, I.R., Hemer, M.: Understanding uncertainties in contemporary and future extreme wave events for broad-scale impact and adaptation planning. Sci. Adv. **9**(2), 1–13 (2023). https://doi.org/10.1126/sciadv.ade3170

Mortimer, W., Raby, A., Antonini, A., Greaves, D., van den Bremer, T.S.: Correct generation of the bound set-down for surface gravity wave groups in laboratory experiments of intermediate to shallow depth. Coast. Eng. **174**, 104121 (2022). https://doi.org/10.1016/j.coastaleng.2022.104121

Muchow, M., Polojärvi, A.: Three-dimensional discrete element simulations on pressure ridge formation. EGUsphere **2024**, 1–15 (2024). https://doi.org/10.5194/egusphere-2024-831

Muhamed Ali, A., Zhuang, H., VanZwieten, J., Ibrahim, A.K., Chérubin, L.: A deep learning model for forecasting velocity structures of the loop current system in the Gulf of Mexico. Forecasting **3**(4), 934–953 (2021). https://doi.org/10.3390/forecast3040056

Mukherjee, P., Ramakrishnan, B.: Investigation of unique Arabian Sea tropical cyclone with GPU-based WRF model: a case study of Shaheen. J. Atmos. Solar-Terrestrial Phys. **246**, 106052 (2023). https://doi.org/10.1016/j.jastp.2023.106052

Murphy-Barltrop, C.J.R., Mackay, E., Jonathan, P.: Inference for bivariate extremes via a semi-parametric angular-radial model. Arxiv, pre-print (2024)

Naaijen, P., van Oosten, K., Roozen, K., van 't Veer, R.: Validation of a deterministic wave and ship motion prediction system. In: OMAE Conference (2018). https://doi.org/10.1115/OMAE2018-78037

Nasir, F., et al.: Significant wave height modelling and simulation of the monsoon-influenced South China Sea coast. Ocean Eng. **277**, 114142 (2023). https://doi.org/10.1016/j.oceaneng.2023.114142

Neary, V.S., Ahn, S.: Global atlas of extreme significant wave heights and relative risk ratios. Renew. Energy **208**, 130–140 (2023). https://doi.org/10.1016/j.renene.2023.03.079

Nederkoorn, T.P., Seyffert, H.C.: Long-term rogue wave occurrence probability from historical wave data on a spatial scale relevant for spar-type floating wind turbines. Ocean Eng. **251**, 110955 (2022). https://doi.org/10.1016/j.oceaneng.2022.110955

Nelli, F., Derkani, M.H., Alberello, A., Toffoli, A.: A satellite altimetry data assimilation approach to optimise sea state estimates from vessel motion. Appl. Ocean Res. **132**, 103479 (2023). https://doi.org/10.1016/j.apor.2023.103479

Nerantzaki, S.D., Papalexiou, S.M.: Assessing extremes in hydroclimatology: a review on probabilistic methods. J. Hydrology **605**, 127302 (2022). https://doi.org/10.1016/j.jhydrol.2021.127302

Nicolaus, M., et al.: Overview of the MOSAiC expedition: snow and sea ice. Elementa: Sci. Anthropocene **10**(1), 46 (2022). https://doi.org/10.1525/elementa.2021.000046

Nielsen, U.D., Mittendorf, M., Shao, Y., Storhaug, G.: Wave spectrum estimation conditioned on machine learning-based output using the wave buoy analogy. Mar. Struct. **91**, 103470 (2023). https://doi.org/10.1016/j.marstruc.2023.103470

Nielsen, U.D., Dietz, J.: Ocean wave spectrum estimation using measured vessel motions from an in-service container ship. Mar. Struct. **69**, 102682 (2020). https://doi.org/10.1016/j.marstruc.2019.102682

Nielsen, U.D., Ikonomakis, A.: Wave conditions encountered by ships—a report from a larger shipping company based on ERA5. Ocean Eng. **237**, 109584 (2021). https://doi.org/10.1016/j.oceaneng.2021.109584

Nielsen, U.D.: Spatio-temporal variation in sea state parameters along virtual ship route paths. J. Oper. Oceanogr. **15**(3), 169–186 (2022). https://doi.org/10.1080/1755876X.2021.1872894

Ning, X. and Paskyabi, M.B.: Parameterization of wave-induced stress in large-eddy simulations of the marine atmospheric boundary layer. J. Geophys. Res. Oceans **129**(9) (2024). https://doi.org/10.1029/2023JC020722

Nolan, J.P.: Modeling multivariate extremes. WIREs Comput. Stat. **16**(2), 1652 (2024). https://doi.org/10.1002/wics.1652

Ólason, E., et al.: A new brittle rheology and numerical framework for large-scale sea-ice models. J. Adv. Model. Earth Syst. **14**(8) (2022). https://doi.org/10.1029/2021MS002685

Obakrim, S., Ailliot, P., Monbet, V., Raillard, N.: Statistical modeling of the space–time relation between wind and significant wave height. Adv. Stat. Climatol. Meteorol. Oceanography **9**(1), 67–81 (2023). https://doi.org/10.5194/ascmo-9-67-2023

O'Donncha, F., Zhang, Y., Chen, B., James, S.C.: Ensemble model aggregation using a computationally lightweight machine-learning model to forecast ocean waves. J. Mar. Syst. **199**, 103206 (2019). https://doi.org/10.1016/j.jmarsys.2019.103206

Oikonomou, C.L.G., Gradowski, M., Kalogeri, C., Sarmento, A.J.N.A.: On defining storm intervals: Extreme wave analysis using extremal index inferencing of the run length parameter. Ocean Eng. **217**, 107988 (2020). https://doi.org/10.1016/j.oceaneng.2020.107988

Oka, M., Takami, T., Ma, C.: Evaluation method for the maximum wave load based on AIS and hindcast wave data (2019)

Oka, M., Ma, C.: Long-term prediction for vertical bending moment utilising the AIS data and global wave data. J. Mar. Sci. Technol. (Japan) **28**(3), 719–731 (2023). https://doi.org/10.1007/s00773-023-00949-2

Onorato, M., et al.: Observation of a giant nonlinear wave-packet on the surface of the ocean. Sci. Rep. **11**(1), 23606 (2021). https://doi.org/10.1038/s41598-021-02875-y

Otter, A., Flannery, B., Murphy, J., Desmond, C.: Current simulation with software in the loop for floating offshore wind turbines. J. Physics: Conf. Series **2265**(4), 042028 (2022). https://doi.org/10.1088/1742-6596/2265/4/042028

Owen, C.C., Hammer, T.C., Hendrikse, H.: Hysteresis and dichotomous mechanics in cyclic crushing failure of confined freshwater columnar ice. Cold Reg. Sci. Technol. **209**, 103816 (2023). https://doi.org/10.1016/j.coldregions.2023.103816

Owen, C.C., Hammer, T.C., Hendrikse, H.: Peak loads during dynamic ice-structure interaction caused by rapid ice strengthening at near-zero relative velocity. Cold Reg. Sci. Technol. **211**, 103864 (2023). https://doi.org/10.1016/j.coldregions.2023.103864

Parno, J., Polashenski, C., Parno, M., Nelsen, P., Mahoney, A., Song, A.: Observations of stress-strain in drifting sea ice at floe scale. J. Geophys. Res. Oceans **127**(5) (2022). https://doi.org/10.1029/2021JC017761

Passerotti, G., et al.: Interactions between irregular wave fields and sea ice: a physical model for wave attenuation and ice breakup in an ice tank. J Phys. Oceanogr. **52**(7), 1431–1446 (2022). https://doi.org/10.1175/JPO-D-21-0238.1

Patra, A., Dodet, G., Accensi, M.: Historical global ocean wave data simulated with CMIP6 anthropogenic and natural forcings. Sci. Data **10**(1), 1–10 (2023). https://doi.org/10.1038/s41597-023-02228-6

Pákozdi, C., et al.: Joint-industry effort to develop and verify CFD modeling practice for predicting wave impact. In: OMAE Conference (2022). https://doi.org/10.1115/OMAE2022-79152

Pákozdi, C., Kamath, A., Wang, W., Martin, T., Bihs, H.: Efficient calculation of hydrodynamic loads on offshore wind substructures including slamming forces. J. Offshore Mech. Arctic Eng. **145**(2), 021901 (2023). https://doi.org/10.1115/1.4055701

Papalexiou, S.M.: Unified theory for stochastic modelling of hydroclimatic processes: preserving marginal distributions, correlation structures, and intermittency. Adv. Water Resour. **115**, 234–252 (2018). https://doi.org/10.1016/j.advwatres.2018.02.013

Papalexiou, S.M., Serinaldi, F.: Random fields simplified: preserving marginal distributions, correlations, and intermittency, with applications from rainfall to humidity. Water Res. Res. **56**(2) (2020). https://doi.org/10.1029/2019WR026331

Park, S.B., et al.: Experimental study of wave transmission and drift velocity using freely floating synthetic ice floes. Ocean Eng. **251**, 111058 (2022). https://doi.org/10.1016/j.oceaneng.2022.111058

Pascual-Ahuir, E.G., Wang, Z.: Optimized sea ice simulation in MITgcm-ECCO2 forced by ERA5. Ocean Model. **183**, 102183 (2023). https://doi.org/10.1016/j.ocemod.2023.102183

Patanè, L., Iuppa, C., Faraci, C., Xibilia, M.G.: A deep hybrid network for significant wave height estimation. Ocean Model. **189**, 102363 (2024). https://doi.org/10.1016/j.ocemod.2024.102363

Pathak, J., et al.: FourCastNet: a global data-driven high-resolution weather model using adaptive Fourier Neural operators (2022). arXiv:2202.11214. https://doi.org/10.48550/arXiv.2202.11214

Pattanaik, S.S., Sahoo, A.K., Panda, R., Behera, S.: Life cycle assessment and forecasting for 30 kW solar power plant using machine learning algorithms. e-Prime – Adv. Electr. Eng. Electron. Energy **7**, 100476 (2024). https://doi.org/10.1016/j.prime.2024.100476

Peng, W.-Q., Pu, J.-C., Chen, Y.: PINN deep learning method for the Chen–Lee–Liu equation: Rogue wave on the periodic background. Commun. Nonlinear Sci. Numer. Simul. **105**, 106067 (2022). https://doi.org/10.1016/j.cnsns.2021.106067

Petranović, T., Mikulić, A., Katalinić, M., Ćorak, M., Parunov, J.: Method for prediction of extreme wave loads based on ship operability analysis using hindcast wave database. J. Mar. Sci. Eng. **9**(9), 1002 (2021). https://doi.org/10.3390/jmse9091002

Petrova, P.G., Guedes Soares, C., Aguiar, T.C.G.R., Esperança, P.T.T.: Statistical distributions of nonlinear waves from random laboratory wave fields. Ocean Eng. **243**, 110170 (2022). https://doi.org/10.1016/j.oceaneng.2021.110170

Pierella, F., Bredmose, H., Dixen, M.: Generation of highly nonlinear irregular waves in a wave flume experiment: Spurious harmonics and their effect on the wave spectrum. Coast. Eng. **164**, 103816 (2021). https://doi.org/10.1016/j.coastaleng.2020.103816

Pillai, A.C., Davey, T., Draycott, S.: A framework for processing wave buoy measurements in the presence of current. Appl. Ocean Res. **106**, 102420 (2021). https://doi.org/10.1016/j.apor.2020.102420

Pinto, P.M.G.M., Campos, R.M., Gallo, M.N., Ribeiro, C.E.P.: Predicting significant wave height with artificial neural networks in the South Atlantic Ocean: a hybrid approach. Ocean Dyn. **73**, 303–315 (2023). https://doi.org/10.1007/s10236-023-01546-y

Pourzangbar, A., Jalali, M., Brocchini, M.: Machine learning application in modelling marine and coastal phenomena: a critical review. Front. Environ. Eng., Sec. Environ. Impact Assess. **2**, 1235557 (2023). https://doi.org/10.3389/fenve.2023.1235557

Prabhu, A., Mandke, S.K., Kripalani, R.H., Pandithurai, G.: Association between Antarctic Sea ice, Pacific SST and the Indian summer monsoon: an observational study. Polar Sci. **30**, 100746 (2021). https://doi.org/10.1016/j.polar.2021.100746

Prasanna, M., Polojärvi, A., Wei, M., Åström, J.: Modeling ice block failure within drift ice and ice rubble. Phys. Rev. E **105**(4), 045001 (2022). https://doi.org/10.1103/PhysRevE.105.045001

Qian, G.W., Song, Y.P., Ishihara, T.: A control-oriented large eddy simulation of wind turbine wake considering effects of Coriolis force and time-varying wind conditions. Energy **239**, 121876 (2022). https://doi.org/10.1016/j.energy.2021.121876

Qian, H., Zhang, R.: Future changes in wind energy resource over the Northwest Passage based on the CMIP6 climate projections. Int. J. Energy Res. **45**(1), 920–937 (2021). https://doi.org/10.1002/er.5997

Qin, J.: Evolving probabilistic modeling for long-term significant wave heights with a focus on extremes. Renew. Energy **187**, 362–370 (2022). https://doi.org/10.1016/j.renene.2022.01.069

Qin, Y., Wei, Z., Chu, D., Zhang, J., Du, Y., Che, Z.: Artificial neural network-based multi-input multi-output model for short-term storm surge prediction on the southeast coast of China. Ocean Eng. **300**, 116915 (2024). https://doi.org/10.1016/j.oceaneng.2024.116915

Rabault, J., et al.: A dataset of direct observations of sea ice drift and waves in ice. Sci. Data **10**, 251 (2023). https://doi.org/10.1038/s41597-023-02160-9

Raharja, I.M.D., Radjawane, I.M., Hendrawan, I.G.: Characteristic of tidal currents in the lombok strait using 3D FVCOM numerical model. IOP Conf. Ser. Earth Environ. Sci. **925**(1), 012002 (2021). https://doi.org/10.1088/1755-1315/925/1/012002

Radfar, S., Galiatsatou, P.: Influence of nonstationarity and dependence of extreme wave parameters on the reliability assessment of coastal structures - A case study. Ocean Eng. **273**, 113862 (2023). https://doi.org/10.1016/j.oceaneng.2023.113862

Radfar, S., Galiatsatou, P., Wahl, T.: Application of nonstationary extreme value analysis in the coastal environment – a systematic literature review. Weather Clim. Extremes **41**, 100575 (2023). https://doi.org/10.1016/j.wace.2023.100575

Raghukumar, K., Chang, G., Spada, F., Jones, C., Janssen, T., Gans, A.: Performance characteristics of "spotter," a newly developed real-time wave measurement buoy. J. Atmos. Oceanic Technol. **36**(6), 1127–1141 (2019). https://doi.org/10.1175/JTECH-D-18-0151.1

Rainville, E., Thomson, J., Moulton, M., Derakhti, M.: Measurements of nearshore ocean-surface kinematics through coherent arrays of free-drifting buoys. Earth Syst. Sci. Data **15**(11), 5135–5151 (2023). https://doi.org/10.5194/essd-15-5135-2023

Rak, G., Hočevar, M., Kolbl Repinc, S., Novak, L., Bizjan, B.: A review on methods for measurement of free water surface. Sensors **23**(4), 1842 (2023). https://doi.org/10.3390/s23041842

Ramezani, M., Choe, D.-E., Heydarpour, K., Koo, B.: Uncertainty models for the structural design of floating offshore wind turbines: a review. Renew. Sustain. Energy Rev. **185**, 113610 (2023). https://doi.org/10.1016/j.rser.2023.113610

Ran, G., Zhang, Q., Li, L.: A depth-integrated non-hydrostatic model for nearshore wave modelling based on the discontinuous Galerkin method. Ocean Eng. **232**, 108661 (2021). https://doi.org/10.1016/j.oceaneng.2021.108661

Ransley, E.J., et al.: Focused wave interactions with floating structures: a blind comparative study. Proc. Inst. Civil Eng. Eng. Comput. Mech. **174**(1), 46–61 (2021). https://doi.org/10.1680/jencm.20.00006

Rascle, N., Ardhuin, F.: A global wave parameter database for geophysical applications. Part 2: model validation with improved source term parameterization. Ocean Model. **70**, 174–188 (2013). https://doi.org/10.1016/j.ocemod.2012.12.001

Rasp, S., Dueben, P.D., Scher, S., Weyn, J.A., Mouatadid, S., Thuerey, N.: WeatherBench: a benchmark data set for data-driven weather forecasting. J. Adv. Model. Earth Syst. **12**, e2020ms002203 (2020). https://doi.org/10.1029/2020ms002203

Rasp, S., Lerch, S.: Neural networks for postprocessing ensemble weather forecasts. Mon. Weather Rev. **146**, 3885–3900 (2018). https://doi.org/10.1175/MWR-D-18-0187.1

Rathod, A.H., Mali, S., Gaikwad, C.J., Heddallikar, A.: Estimating radar backscatter from water wave surface using X band FMCW radar. In: 2nd International Conference on Electronics, Materials Engineering & Nano-Technology (IEMENTech), pp. 1–6 (2018). https://doi.org/10.1109/IEMENTECH.2018.8465256

Raudiya, F., Rohmawati, A.A., Adytia, D.: Non-stationary order of vector autoregression in significant ocean wave forecasting. In: 2021 9th International Conference on Information and Communication Technology (ICoICT), pp. 326–330 (2021). https://doi.org/10.1109/ICoICT52021.2021.9527502

Reichstein, M., et al.: Deep learning and process understanding for data-driven Earth system science. Nature **566**, 195–204 (2019). https://doi.org/10.1038/s41586-019-0912-1

Remmerswaal, R.: Numerical modelling of variability in liquid impacts. PhD thesis, University of Groningen (2023). https://doi.org/10.33612/diss.562133319

Ren, X., et al.: Deep learning-based weather prediction: a survey. Big Data Res. **23**, 100178 (2021). https://doi.org/10.1016/j.bdr.2020.100178

Rheinlænder, J.W., et al.: Driving mechanisms of an extreme winter sea ice breakup event in the Beaufort Sea. Geophys. Res. Lett. **49**(12) (2022). https://doi.org/10.1029/2022GL099024

Ribal, A., Young, I.R.: 33 years of globally calibrated wave height and wind speed data based on altimeter observations. Sci. Data **6**(1), 77 (2019). https://doi.org/10.1038/s41597-019-0083-9

Ricciardulli, L., Manaster, A.: Intercalibration of ASCAT scatterometer winds from MetOp-A, -B, and -C, for a stable climate data record. Remote Sens. **13**(18), 3678 (2021). https://doi.org/10.3390/rs13183678

Ricciardulli, L., Foltz, G.R., Manaster, A., Meissner, T.: Assessment of Saildrone extreme wind measurements in hurricane Sam using MW satellite sensors. Remote Sens. **14**(12), 2726 (2022). https://doi.org/10.3390/rs14122726

Ricondo, A., Cagigal, L., Pérez-Díaz, B., Méndez, F.J.: HySwash: a hybrid model for nearshore wave processes. Ocean Eng. **291**, 116419 (2024). https://doi.org/10.1016/j.oceaneng.2023.116419

Rijnsdorp, D.P., Reniers, A.J.H.M., Zijlema, M.: Free infragravity waves in the North Sea. J. Geophys. Res. Oceans **126**(8) (2021). https://doi.org/10.1029/2021JC017368

Ringeisen, D., Hutter, N., Von Albedyll, L.: Deformation lines in Arctic sea ice: intersection angle distribution and mechanical properties. Cryosphere **17**(9), 4047–4061 (2023). https://doi.org/10.5194/tc-17-4047-2023

Rognebakke, O., Ruth, E., Landet, T., Austefjord, H.: Impact loads on ships from breaking waves. In: OMAE Conference (2024). https://doi.org/10.1115/OMAE2024-126610

Rosa, T.L., Piecho-Santos, A.M., Vettor, R., Guedes Soares, C.: Review and prospects for autonomous observing systems in vessels of opportunity. J. Mar. Sci. Eng. **9**(4), 366 (2021). https://doi.org/10.3390/jmse9040366

Rosen, J., Kilner, A., Gumley, J., Jayasinghe, K., Thurumella, H.: The impact of climate change on offshore operations and design considerations for offshore vessels and installations. In: OMAE Conference (2022). https://doi.org/10.1115/OMAE2022-79274

Ruth, E., Thompson, I.: Comparing design assumptions with hindcast wave conditions and measured ship speed and heading. Ocean Eng. **247**, 110613 (2022). https://doi.org/10.1016/j.oceaneng.2022.110613

Rutten, J., et al.: Continuous wave measurements collected in intermediate depth throughout the North Sea storm season during the RealDune/REFLEX experiments. Data **9**, 70 (2024). https://doi.org/10.3390/data9050070

Rutten, J., et al.: Data collection underlying the publication: Continuous Wave Measurements Collected in Intermediate Depth throughout the North Sea Storm Season during the RealDune/REFLEX Experiments. Version 1. 4TU.ResearchData. collection (2024b). https://doi.org/10.4121/233f11ff-7804-4777-8b32-92c4606e56d8.v1

Rüttgers, M., Jeon, S., Lee, S., You, D.: Prediction of typhoon track and intensity using a generative adversarial network with observational and meteorological data. IEEE Access **10**, 48434–48446 (2022). https://doi.org/10.1109/ACCESS.2022.3172301

Rykova, T.: Improving forecasts of individual ocean eddies using feature mapping. Sci. Rep. **13**(1), 6216 (2023). https://doi.org/10.1038/s41598-023-33465-9

Salganik, E., et al.: Different mechanisms of Arctic first-year sea-ice ridge consolidation observed during the MOSAiC expedition. Elementa **11**(1), 1–12 (2023). https://doi.org/10.1525/elementa.2023.00008

Salvação, N., Bentamy, A., Guedes Soares, C.: Developing a new wind dataset by blending satellite data and WRF model wind predictions. Renew. Energy **198**, 283–295 (2022). https://doi.org/10.1016/j.renene.2022.07.049

Sainte-Rose, B., Pham, Y., Pavalko, W.: Persistency and surface convergence evidenced by two maker buoys in the Great Pacific Garbage Patch. J. Mar. Sci. Eng. **11**(1), 68 (2023). https://doi.org/10.3390/jmse11010068

Sardana, D., Kumar, P.: Influence of climate variability modes over wind-sea and swell generated wave energy. Ocean Eng. **291**, 116471 (2024). https://doi.org/10.1016/j.oceaneng.2023.116471

Sartini, L., Antonini, A.: On the spectral wave climate of the French Atlantic Ocean. Ocean Eng. **304**, 117900 (2024). https://doi.org/10.1016/j.oceaneng.2024.117900

Sasmal, K., Miratsu, R., Kodaira, T., Fukui, T., Zhu, T., Waseda, T.: Statistical model representing storm avoidance by merchant ships in the North Atlantic Ocean. Ocean Eng. **235**, 109163 (2021). https://doi.org/10.1016/j.oceaneng.2021.109163

Savard, A., Tremblay, B.: On the sensitivity of sea ice deformation statistics to plastic damage. Cryosphere **18**(4), 2017–2034 (2024). https://doi.org/10.5194/tc-18-2017-2024

Scharnke, J., Ewans, K., Kirezci, C., Babanin, A.: High frequency tail in measured wave spectra of steep to breaking sea states. In: OMAE Conference (2022). https://doi.org/10.1115/OMAE2022-79053

Scharnke, J., Helder, J.: Scale effects and variability in wave-in-deck type of impact loading, more insights into the results of the BreaKin JIP. In: OMAE Conference (2023). https://doi.org/10.1115/OMAE2023-104288

Scharnke, J., van Essen, S.M., Seyffert, H.C.: Required test durations for converged short-term wave and impact extreme value statistics - Part 2: Deck box dataset. Mar. Struct. **90**, 103411 (2023). https://doi.org/10.1016/j.marstruc.2023.103411

Schenkel, B.A., et al.: Recent progress in research and forecasting of tropical cyclone outer size. Tropical Cyclone Res. Rev. **12**(3), 151–164 (2023). https://doi.org/10.1016/j.tcrr.2023.09.002

Scheuerer, M., Switanek, M.B., Worsnop, R.P., Hamill, T.M.: Using artificial neural networks for generating probabilistic subseasonal precipitation forecasts over California. Mon. Weather Rev. **148**(8), 3489–3506 (2020). https://doi.org/10.1175/MWR-D-20-0096.1

Schultz, M.G., et al.: Can deep learning beat numerical weather prediction? Phil. Trans. R. Soc. A **379**, 20200097 (2021). https://doi.org/10.1098/rsta.2020.0097

Sen, R., Pandey, S., Dandapat, S., Francis, P.A., Chakraborty, A.: A numerical study on seasonal transport variability of the North Indian Ocean boundary currents using Regional Ocean Modeling System (ROMS). J. Phys. Oceanogr. **15**(1), 32–51 (2022). https://doi.org/10.1080/1755876X.2020.1846266

Sener, M.Z., Yoon, H.K., Nguyen, T.T.D., Park, J., Kose, E.: An experimental study on capacitive and ultrasonic measurement principles and uncertainty assessment in laboratory wave measurements. Ocean Eng. **285**, 115320 (2023). https://doi.org/10.1016/j.oceaneng.2023.115320

Seyffert, H.C.: Generating an ensemble of mutually exclusive and exhaustive waves targeted for extreme responses. Ocean Eng. **243**, 110172 (2022). https://doi.org/10.1016/j.oceaneng.2021.110172

Seyffert, H.C., Kana, A.A.: Response-based reliability contours for complex marine systems considering short and long-term variability. Appl. Ocean Res. **103**, 102332 (2020). https://doi.org/10.1016/j.apor.2020.102332

Sharp, J.D., Fassbender, A.J., Carter, B.R., Johnson, G.C., Schultz, C., Dunne, J.P.: GOBAI-O_2: temporally and spatially resolved fields of ocean interior dissolved oxygen over nearly 2 decades. Earth Syst. Sci. Data **15**(10), 4481–4518 (2023). https://doi.org/10.5194/essd-15-4481-2023

Shashank, V.G., Sriram, V., Sannasiraj, S.A.: Improvements in wind field hindcast for storm surge predictions in the Bay of Bengal: a case study for the tropical cyclone Varadah. Appl. Ocean Res. **127**, 103324 (2022). https://doi.org/10.1016/j.apor.2022.103324

Shooter, R., Ross, E., Ribal, A., Young, I.R., Jonathan, P.: Multivariate spatial conditional extremes for extreme ocean environments. Ocean Eng. **247**, 110647 (2022). https://doi.org/10.1016/j.oceaneng.2022.110647

Shooter, R., Tawn, J., Ross, E., Jonathan, P.: Basin-wide spatial conditional extremes for severe ocean storms. Extremes **24**(2), 241–265 (2021). https://doi.org/10.1007/s10687-020-00389-w

Siadatmousavi, S.M., Yaghoobi Kalourazi, M., Khosh Kholgh, A.: Improving the WAVEWATCH-III wave model results using data assimilation in the Persian Gulf. Ocean Eng. **300**, 117460 (2024). https://doi.org/10.1016/j.oceaneng.2024.117460

Silva, D., Gonçalves, M., Bentamy, A., Guedes Soares, C.: Assessment of the use of scatterometer wind data to force wave models in the North Atlantic Ocean. Ocean Eng. **266**, 112803 (2022). https://doi.org/10.1016/j.oceaneng.2022.112803

Simão, M.L., Sudati Sagrilo, L.V., Videiro, P.M.: A multi-dimensional long-term joint probability model for environmental parameters. Ocean Eng. **255**, 111470 (2022). https://doi.org/10.1016/j.oceaneng.2022.111470

Simonetti, I., Cappietti, L.: Mediterranean coastal wave-climate long-term trend in climate change scenarios and effects on the optimal sizing of OWC wave energy converters. Coast. Eng. **179**, 104247 (2023). https://doi.org/10.1016/j.coastaleng.2022.104247

Simpson, A., Haller, M., Walker, D., Lynett, P., Honegger, D.: Wave-by-wave forecasting via assimilation of marine radar data. J. Atmos. Oceanic Technol. **37**(7), 1269–1288 (2020). https://doi.org/10.1175/JTECH-D-19-0127.1

Simpson, E.S., Tawn, J.A.: Inference for new environmental contours using extreme value analysis: ES Simpson, JA Tawn. J. Agric. Biol. Environ. Stat. **30**(3), 638–662 (2025). https://doi.org/10.1007/s13253-024-00612-2

Sinha, A., Abernathe, R.: Estimating ocean surface currents with machine learning. Front. Mar. Sci., Sec. Ocean Observation **8** (2021). https://doi.org/10.3389/fmars.2021.672477

Slater, L.J., et al.: Nonstationary weather and water extremes: a review of methods for their detection, attribution, and management. Hydrol. Earth Syst. Sci. **25**(7), 3897–3935 (2021). https://doi.org/10.5194/hess-25-3897-2021

Smit, P.B., et al.: Assimilation of significant wave height from distributed ocean wave sensors. Ocean Model. **159**, 101738 (2021). https://doi.org/10.1016/j.ocemod.2020.101738

SOFAR. SOFAR Spotter Archive (2024).: https://www.SOFARocean.com/mx/SOFAR-spotter-archive. Accessed 26 Sept 2024

Song, Y., Chen, J., Sørensen, J.D., Li, J.: Multi-parameter full probabilistic modeling of long-term joint wind-wave actions using multi-source data and applications to fatigue analysis of floating offshore wind turbines. Ocean Eng. **247**, 110676 (2022). https://doi.org/10.1016/j.oceaneng.2022.110676

Song, Y., Hong, X., Sun, T., Zhang, Z.: Joint probabilistic modeling of extreme wind-wave conditions under typhoon impact and applications to extreme response analysis of floating offshore wind turbines. Eng. Struct. **318**, 118686 (2024). https://doi.org/10.1016/j.engstruct.2024.118686

Song, Y., Hong, X., Xiong, J., Shen, J., Xu, Z.: Probabilistic modeling of long-term joint wind and wave load conditions via generative adversarial network. Stoch. Env. Res. Risk Assess. **37**(7), 2829–2847 (2023). https://doi.org/10.1007/s00477-023-02421-4

Soran, M.B., Amarouche, K., Akpınar, A.: Spatial calibration of WAVEWATCH III model against satellite observations using different input and dissipation parameterizations in the Black Sea. Ocean Eng. **257**, 111627 (2022). https://doi.org/10.1016/j.oceaneng.2022.111627

Sorensen, B.B., Charalampopoulos, A., Zhang, S., Harrop, B.E., Leung, L.R., Sapsis, T.P.: A non-intrusive machine learning framework for debiasing long-time coarse resolution climate simulations and quantifying rare events statistics. J. Adv. Model. Earth Syst. (2024). https://doi.org/10.1029/2023MS004122

Speers, M., Randell, D., Tawn, J., Jonathan, P.: Estimating metocean environments associated with extreme structural response to demonstrate the dangers of environmental contour methods. Ocean Eng. **311**, 118754 (2024). https://doi.org/10.1016/j.oceaneng.2024.118754

Sprenger, F., Kosleck, S., Klein, M.: Wave riding through time – the contributions of Günther F. Clauss to the field of ocean wave research. In: OMAE Conference (2022). https://doi.org/10.1115/OMAE2022-79170

Sprintall, J., et al.: COVID impacts cause critical gaps in the Indian Ocean observing system (2024). https://doi.org/10.1175/BAMS-D-22-0270.1

Stefanakos, C.: Global wind and wave climate based on two reanalysis databases: ECMWF ERA5 and NCEP CFSR. J. Marine Sci. Eng. **9**(9), 990 (2021). https://doi.org/10.3390/jmse9090990

Steinsland, K., Grant, D.M., Ninnemann, U.S., Fahl, K., Stein, R., de Schepper, S.: Sea ice variability in the North Atlantic subpolar gyre throughout the Last Interglacial. Quatern. Sci. Rev. **313**, 108198 (2023). https://doi.org/10.1016/j.quascirev.2023.108198

Sterlin, J., Tsamados, M., Fichefet, T., Massonnet, F., Barbic, G.: Effects of sea ice form drag on the polar oceans in the NEMO-LIM3 global ocean–sea ice model. Ocean Model. **184**, 102227 (2023). https://doi.org/10.1016/j.ocemod.2023.102227

Streicher, M., et al.: Evaluation of the accuracy of the generated wave fields in the coastal & ocean basin (Cob). In: CoastLab 2024: Physical Modelling in Coastal Engineering and Science (2024). https://doi.org/10.59490/coastlab.2024.793

Su, F., et al.: Widespread global disparities between modelled and observed mid-depth ocean currents. Nat. Commun. **14**(1), 2089 (2023). https://doi.org/10.1038/s41467-023-37841-x

Sulaiman, M.H., Mustaffa, Z.: Enhancing wind power forecasting accuracy with hybrid deep learning and teaching-learning-based optimization. Cleaner Energy Syst. **9**, 100139 (2024). https://doi.org/10.1016/j.cles.2024.100139

Sun, J., Dong, H., Liu, M., Fan, Y.: Data-driven rogue waves solutions for the focusing and variable coefficient nonlinear Schrödinger equations via deep learning. Chaos **34**, 073134 (2024). https://doi.org/10.1063/5.0209068

Tamizi, A., Young, I.R.: A dataset of global tropical cyclone wind and surface wave measurements from buoy and satellite platforms. Sci. Data **11**(1), 106 (2024). https://doi.org/10.1038/s41597-024-02955-4

Tang, T., Adcock, T.A.A.: Data driven analysis on the extreme wave statistics over an area. Appl. Ocean Res. **115**, 102809 (2021). https://doi.org/10.1016/j.apor.2021.102809

Tang, T., Adcock, T.A.A.: A reduced order model for space–time wave statistics using probabilistic decomposition–synthesis method. Ocean Eng. **259**, 111860 (2022a). https://doi.org/10.1016/j.oceaneng.2022.111860

Tang, T., Adcock, T.A.A.: Estimating space–time wave statistics using a sequential sampling method and Gaussian process regression. Appl. Ocean Res. **122**, 103127 (2022b). https://doi.org/10.1016/j.apor.2022.103127

Tang, T., Adcock, T.A.A.: A reduced order parameterization of random wave fields with deterministic wave groups. In: OMAE Confeence (2022c). https://doi.org/10.1115/OMAE2022-80636

Tang, T., Barratt, D., Bingham, H.B., van den Bremer, T.S., Adcock, T.A.A.: The impact of removing the high-frequency spectral tail on rogue wave statistics. J. Fluid Mech. **953**, A9 (2022). https://doi.org/10.1017/jfm.2022.961

Tang, T., et al.: The influence of directional spreading on rogue waves triggered by abrupt depth transitions. J. Fluid Mech. **972**, 737 (2023). https://doi.org/10.1017/jfm.2023.737

Tendijck, S., Eastoe, E., Tawn, J., Randell, D., Jonathan, P.: Modeling the extremes of bivariate mixture distributions with application to oceanographic data. J. Am. Stat. Assoc. **118**(542), 1373–1384 (2023). https://doi.org/10.1080/01621459.2021.1996379

Tendijck, S., Jonathan, P., Randell, D., Tawn, J.: Temporal evolution of the extreme excursions of multivariate kth order Markov processes with application to oceanographic data. Environmetrics **35**(3), 2834 (2024). https://doi.org/10.1002/env.2834

Terrazas Silva, M.A., Salas de León, D.A., Machain Castillo, M.L., Monreal Gómez, M.A.: The connection of the Costa Rica coastal current with the West Mexican current in the Gulf of Tehuantepec. Cont. Shelf Res. **279**, 105294 (2024). https://doi.org/10.1016/j.csr.2024.105294

Teutsch, I., Weisse, R.: Rogue waves in the southern north sea—the role of modulational instability. J. Phys. Oceanogr. **53**(1), 269–286 (2023). https://doi.org/10.1175/JPO-D-22-0059.1

Tian, Y., Zhong, Y., Liu, H., Liu, W., Kong, F., Chen, H.: A new fast simulation method of wind turbine wake based on annular vortex element. Renew. Energy **229**, 120765 (2024). https://doi.org/10.1016/j.renene.2024.120765

Tikan, A., et al.: Prediction and manipulation of hydrodynamic rogue waves via nonlinear spectral engineering. Phys. Rev. Fluids **7**(5) (2022). https://doi.org/10.1103/PhysRevFluids.7.054401

Timco, G.W.: Ice forces on structures: physical modelling techniques. In: 7th IAHR International Symposium on Ice, pp. 117–150 (1984)

Tödter, S., et al.: On the applicability of the tomographic reconstruction technique for measurements of multiphase flows during sloshing model tests. In: 7th Conference on Advanced Model Measurement Technology for the Maritime Industry (AMT) (2023)

Toffoli, A., Alberello, A., Clarke, H., Nelli, F., et al.: Observations of rogue seas in the Southern Ocean. Phys. Rev. Lett. **132**(15), 154101 (2024). https://doi.org/10.1103/PhysRevLett.132.154101

Touboul, J., Banner, M.L.: On the breaking inception of unsteady water wave packets evolving in the presence of constant vorticity. J. Fluid Mech. **915**, A16 (2021). https://doi.org/10.1017/jfm.2021.65

Townsend, J.F., Xu, G., Jin, Y., Yu, E., Wei, H., Han, Y.: On the development of a generalized atmospheric boundary layer velocity profile for offshore engineering applications considering wind–wave interaction. Ocean Eng. **286**(part 2), 115621 (2023). https://doi.org/10.1016/j.oceaneng.2023.115621

Treillou, S., Marchesiello, P., Baker, C.M.: Correction of coherent interference in wave-resolving nearshore models and validation with experimental data. Ocean Model. **189**, 102369 (2024). https://doi.org/10.1016/j.ocemod.2024.102369

Tsai, Y.S., Hsieh, C.M.: Effect of sea states on the wind velocity fluctuations near the surface under near-neutral conditions. Appl. Ocean Res. **139**, 103691 (2023). https://doi.org/10.1016/j.apor.2023.103691

Tukker, J., Hensse, J., Schümann, M.: Measurement quality of airborne ultrasonic wave measurement technology. In: 7th Conference on Advanced Model Measurement Technology for the Maritime Industry (AMT) (2023)

Umesh, P.A., Behera, M.R.: On the improvements in nearshore wave height predictions using nested SWAN-SWASH modelling in the eastern coastal waters of India. Ocean Eng. **236**, 109550 (2021). https://doi.org/10.1016/j.oceaneng.2021.109550

Uma, G., Sannasiraj, S.A.: Assessment of input and dissipation source terms in the spectral wave model during tropical cyclones of varying intensity in Bay of Bengal. Ocean Eng. **285**, 115181 (2023). https://doi.org/10.1016/j.oceaneng.2023.115181

van der Plas, P., Serraris, J.W., Helder, J.A., Hanssen, F.-C.W., Bjar, L.F., Krogenes, K.O.: Reconstruction of fish farm model tests in CFD for detailed analysis of internal sloshing and wave loading in extreme sea states. In: OMAE Conference (2022). https://doi.org/10.1115/OMAE2022-80520

van Eeden, F., Klonaris, G., Verbeurgt, J., Troch, P., De Wulf, A.: Sensitivities in wind driven spectral wave modelling for the Belgian Coast. J. Mar. Sci. Eng. **10**(8) (2022). https://doi.org/10.3390/jmse10081138

van Essen, S.M., Peters, H.C.: Design wave and wind environment for minimum power requirements of vessels in the southern North Sea. In: RINA Conference - Influence of EEDI on Ship Design and Operation (2017)

van Essen, S.M., Ewans, K., McConochie, J.: Wave buoy performance in short and long waves, evaluated using tests on a hexapod. In: OMAE Conference (2018). https://doi.org/10.1115/OMAE2018-77092

van Essen, S.M., et al.: Screening wave conditions for the occurrence of green water events on sailing ships. Ocean Eng. **234**, 109218 (2021). https://doi.org/10.1016/j.oceaneng.2021.109218

van Essen, S.M., Scharnke, J., Seyffert, H.C.: Required test durations for converged short-term wave and impact extreme value statistics - Part 1: Ferry dataset. Mar. Struct. **90**, 103410 (2023). https://doi.org/10.1016/j.marstruc.2023.103410

van Essen, S. M., Scholcz, T., Seyffert, H.C.: Prediction of short-term non-linear response using screening combined with multi-fidelity Gaussian Process Regression. In: OMAE Conference (2023). https://doi.org/10.1115/OMAE2023-100954

van Essen, S.M., Seyffert, H.C.: Finding dangerous waves—review of methods to obtain wave impact design loads for marine structures. J. Offshore Mech. Arctic Eng. **145**(6) (2023). https://doi.org/10.1115/1.4056888

van Essen, S.M., Bunnik, T., Scharnke, J.: Statistical uncertainty of ship response to waves as a function of test duration. In: OMAE Conference (2024). https://doi.org/10.1115/OMAE2024-122486

van Meerkerk, M.: Variability in wave impacts: an experimental investigation. Delft University of Technology (2021)

van Vloten, S.O., Cagigal, L., Rueda, A., Ripoll, N., Méndez, F.J.: HyTCWaves: A hybrid model for downscaling tropical cyclone induced extreme waves climate. Ocean Model. **178**, 102100 (2022). https://doi.org/10.1016/j.ocemod.2022.102100

Vanem, E.: Non-stationary extreme value models to account for trends and shifts in the extreme wave climate due to climate change. Appl. Ocean Res. **52**, 010 (2015). https://doi.org/10.1016/j.apor.2015.06.010

Vanem, E.: A simple approach to account for seasonality in the description of extreme ocean environments. Marine Syst. Ocean Technol. **13**(2–4), 63–73 (2018). https://doi.org/10.1007/s40868-018-0046-6

Vanem, E.: Bivariate regional extreme value analysis for significant wave height and wave period. Appl. Ocean Res. **101**, 102266 (2020). https://doi.org/10.1016/j.apor.2020.102266

Vanem, E.: Bivariate regional frequency analysis of sea state conditions. In: OMAE Conference (2021). https://doi.org/10.1115/OMAE2021-61988

Vanem, E., Zhu, T., Babanin, A.: Statistical modelling of the ocean environment – a review of recent developments in theory and applications. Mar. Struct. **86**, 103297 (2022). https://doi.org/10.1016/j.marstruc.2022.103297

Vanem, E., Fazeres-Ferradosa, T.: A truncated, translated Weibull distribution for shallow water sea states. Coast. Eng. **172**, 104077 (2022). https://doi.org/10.1016/j.coastaleng.2021.104077

Vanem, E.: Analysing multivariate extreme conditions using environmental contours and accounting for serial dependence. Renew. Energy **202**, 470–482 (2023a). https://doi.org/10.1016/j.renene.2022.11.033

Vanem, E.: Analyzing Extreme sea state conditions by time-series simulation accounting for seasonality. J. Offshore Mech. Arctic Eng. **145**(5) (2023b). https://doi.org/10.1115/1.4056786

Vanem, E., Fekhari, E., Dimitrov, N., Kelly, M., Cousin, A., Guiton, M.: A joint probability distribution for multivariate wind-wave conditions and discussions on uncertainties. J. Offshore Mech. Arctic Eng. **146**(6), 061708 (2024a). https://doi.org/10.1115/1.4064498

Vanem, E., Lande, Ø., Fekhari, E.: A simulation study on the usefulness of the bernstein copula for statistical modeling of metocean variables. In: Structures, Safety, and Reliability, vol. 2 (2024b). https://doi.org/10.1115/OMAE2024-121159

Vasarmidis, P., Klonaris, G., Zijlema, M., Stratigaki, V., Troch, P.: A study of the non-linear properties and wave generation of the multi-layer non-hydrostatic wave model SWASH. Ocean Eng. **302**, 117633 (2024). https://doi.org/10.1016/j.oceaneng.2024.117633

Vitart, F., Takaya, Y.: Lagged ensembles in sub-seasonal predictions. Q. J. R. Meteorol. Soc. **147**(739) (2021). https://doi.org/10.1002/qj.4125

von Bock und Polach, R.U.F., Ziemer, G., Klein, M., Hartmann, M.C.N., Toffoli, A., Monty, J.: Case based scaling: recent developments in ice model testing technology. In: OMAE Conference (2020). https://doi.org/10.1115/OMAE2020-18320

von Bock und Polach, R.U.F., Klein, M., Hartmann, M.: A new model ice for wave-ice interaction. Water **13**(23), 3397 (2021). https://doi.org/10.3390/w13233397

Wang, R.-Q., Ling, L., Zeng, D., Feng, B.-F.: A deep learning improved numerical method for the simulation of rogue waves of nonlinear Schrödinger equation. Commun. Nonlinear Sci. Numer. Simul. **101**, 105896 (2021). https://doi.org/10.1016/j.cnsns.2021.105896

Wang, Y., Zou, R., Liu, F., Zhang, L., Liu, Q.: A review of wind speed and wind power forecasting with deep neural networks. Appl. Energy **304**, 117766 (2021). https://doi.org/10.1016/j.apenergy.2021.117766

Wang, N., Chen, Q., Zhu, L., Sun, H.: Integration of data-driven and physics-based modeling of wind waves in a shallow estuary. Ocean Model. **172**, 101978 (2022a). https://doi.org/10.1016/j.ocemod.2022.101978

Wang, W., Kamath, A., Pákozdi, C., Bihs, H.: A numerical investigation on wave spectrum identification and transformation from deep to shallow waters for the coastal and offshore installations. In: OMAE Conference (2022b). https://doi.org/10.1115/OMAE2022-79010

Wang, W., Pákozdi, C., Kamath, A., Bihs, H.: Representation of 3-h offshore short-crested wave field in the fully nonlinear potential flow model REEF3D::FNPF. J. Offshore Mech. Arctic Eng. **144**(4) (2022c). https://doi.org/10.1115/1.4053774

Wang, Y., Imai, K., Miyashita, T., Ariyoshi, K., Takahashi, N., Satake, K.: Coastal tsunami prediction in Tohoku region, Japan, based on S-net observations using artificial neural network. Earth Planets Space **75**, 154 (2023). https://doi.org/10.1186/s40623-023-01912-6

Wang, C., Cao, C., Ye, L., Wang, C., Guo, C.Y.: An efficient peridynamic method and its MPI parallelization for simulating the continuous icebreaking process. Ocean Eng. **279**, 114460 (2023). https://doi.org/10.1016/j.oceaneng.2023.114460

Wang, H., Xiao, T., Gou, H., Pu, Q., Bao, Y.: Joint distribution of wind speed and direction over complex terrains based on nonparametric copula models. J. Wind Eng. Ind. Aerodyn. **241**, 105509 (2023c). https://doi.org/10.1016/j.jweia.2023.105509

Wang, J., Aouf, L., Yin, H.: Wave data-driven forecasting model based on deep learning and its operational status. In: 3rd International Workshop on Waves, Storm Surges, and Coastal Hazards. Notre Dame, IN, USA. Oral presentation (2023c). https://waveworkshop.nd.edu/

Wang, L., Ding, K., Zhou, B., Li, J., Liu, S., Tang, T.: Quantitative prediction of the freak wave occurrence probability in co-propagating mixed waves. Ocean Eng. **271**, 113810 (2023e). https://doi.org/10.1016/j.oceaneng.2023.113810

Wang, F., Yang, Y., Yin, X., Jiang, X., Sun, M.: Improving wave modeling performance by incorporating wave-generated turbulence dissipation and improved post-breaking spectrum. Ocean Model. **188**, 102311 (2024a). https://doi.org/10.1016/j.ocemod.2023.102311

Wang, H., Gramstad, O., Schär, S., Marelli, S., Vanem, E.: Comparison of probabilistic structural reliability methods for ultimate limit state assessment of wind turbines. Struct. Saf. **111**, 102502 (2024b). https://doi.org/10.1016/j.strusafe.2024.102502

Wang, J., Bai, Z., Xie, B., Gui, J., Gong, H., Zhou, Y.: Improved inverse first-order reliability method for analyzing long-term response extremes of floating structures. J. Marine Sci. Appl. **24**(3), 552–566 (2025c). https://doi.org/10.1007/s11804-024-00459-6

Wang, L., et al.: Wind turbine wakes modeling and applications: past, present, and future. Ocean Eng. **309**, 118508 (2024d). https://doi.org/10.1016/j.oceaneng.2024.118508

Watanabe, Y., Tsuda, Y., Saruwatari, A.: Wave packet focusing in shallow water. Coast. Eng. J. **62**(2), 336–349 (2020). https://doi.org/10.1080/21664250.2020.1756033

Webb, A., Waseda, T., Kiyomatsu, K.: A high-resolution, long-term wave resource assessment of Japan with wave–current effects. Renew. Energy **161**, 1341–1358 (2020). https://doi.org/10.1016/j.renene.2020.05.030

Wedler, M., Stender, M., Klein, M., Hoffmann, N.: Machine learning simulation of one-dimensional deterministic water wave propagation. Ocean Eng. **284**, 115222 (2023). https://doi.org/10.1016/j.oceaneng.2023.115222

Wei, Z.: Forecasting wind waves in the US Atlantic Coast using an artificial neural network model: Towards an AI-based storm forecast system. Ocean Eng. **237**, 109646 (2021). https://doi.org/10.1016/j.oceaneng.2021.109646

Wen, Z., Wang, F., Wan, J., Wang, Y., Yang, F., Guo, C.: Copula-based joint tropical cyclone-induced wind and wave risk analysis: considering the effect of uncertainty using Bayesian inference. Nat. Haz. **120**(15), 14355–14380 (2024). https://doi.org/10.1007/s11069-024-06709-8

Werner, S., et al.: Specialist committee on wind powered and wind assisted ships. In: Proceedings of the 30th International Towing Tank Conference (ITTC), vol. II (2024)

West, B., O'Connor, D., Parno, M., Krackow, M., Polashenski, C. (2022). Bonded discrete element simulations of sea ice with non-local failure: applications to Nares Strait. J. Advances in Modeling Earth Systems, 14(6). https://doi.org/10.1029/2021MS002614

Windt, C., Untrau, A., Davidson, J., Ransley, E.J., Greaves, D.M., Ringwood, J.V.: Assessing the validity of regular wave theory in a short physical wave flume using particle image velocimetry. Exp. Thermal Fluid Sci. **121**, 110276 (2021). https://doi.org/10.1016/j.expthermflusci.2020.110276

Wright, E.E., Bourassa, M.A., Stoffelen, A., Bidlot, J.-R.: Characterizing buoy wind speed error in high winds and varying sea state with ASCAT and ERA5. Remote Sens. **13**(22), 4558 (2021). https://doi.org/10.3390/rs13224558

Wu, T., Duan, Z.: A new and efficient method for tropical cyclone detection and tracking in gridded datasets. Weather Clim. Extremes **42**, 100626 (2023). https://doi.org/10.1016/j.wace.2023.100626

Xie, C., Chen, P., Man, T., Dong, J.: STCANet: Spatiotemporal coupled attention network for ocean surface current prediction. J. Ocean Univ. China **22**(2), 441–451 (2023). https://doi.org/10.1007/s11802-023-5269-2

Xie, W., Xu, G., Zhang, H., Dong, C.: Developing a deep learning-based storm surge forecasting model. Ocean Model. **182**, 102179 (2023). https://doi.org/10.1016/j.ocemod.2023.102179

Xiong, J., Yu, F., Fu, C., Dong, J., Liu, Q.: Evaluation and improvement of the ERA5 wind field in typhoon storm surge simulations. Appl. Ocean Res. **118**, 103000 (2022). https://doi.org/10.1016/j.apor.2021.103000

Xu, C., et al.: Spatial-temporal distribution of tropical cyclone activity on the eastern sea area of China since the late 1940s. Estuar. Coast. Shelf Sci. **277**, 108067 (2022). https://doi.org/10.1016/j.ecss.2022.108067

Xu, G., Chen, X., Xue, S., Townsend, J.F., Chen, X., Tang, M.: Numerical assessment of non-uniform terrain and inhomogeneous wave–current loading effects on the dynamic response of a submerged floating tunnel. Ocean Eng. **288**, 115942 (2023a). https://doi.org/10.1016/j.oceaneng.2023.115942

Xu, X., et al.: Tropical cyclone modeling with the inclusion of wave-coupled processes: sea spray and wave turbulence. Geophys. Res. Lett. **50**(24) (2023b). https://doi.org/10.1029/2023GL106536

Yang, H., Liang, B., Shao, Z.: Study on the influence range of tropical cyclones on ocean waves. Ocean Eng. **266**, 112864 (2022). https://doi.org/10.1016/j.oceaneng.2022.112864

Yang, Y., Zhang, F., Zhu, R., Li, Y.: Study on vertical line source Green's function for hydrodynamic calculations of ocean structures in water with ice cover. Ocean Eng. **276**, 114193 (2023a). https://doi.org/10.1016/j.oceaneng.2023.114193

Yang, Z., Dong, S.: A novel decomposition-based approach for non-stationary hub-height wind speed modelling. Energy **283**, 129081 (2023). https://doi.org/10.1016/j.energy.2023.129081

Yang, Z., Huang, W., Dong, S., Li, H.: Mixture bivariate distribution of wind speed and air density for wind energy assessment. Energy Convers. Manage. **276**, 116540 (2023b). https://doi.org/10.1016/j.enconman.2022.116540

Ye, W., Hu, Z.: A new model for quantifying wave damping by vegetation in combined wave–current flow. Ocean Eng. **288**, 116119 (2023). https://doi.org/10.1016/j.oceaneng.2023.116119

Ye, F., Brodie, J., Miles, T., Aziz Ezzat, A.: AIRU-WRF: A physics-guided spatio-temporal wind forecasting model and its application to the U.S. Mid Atlantic offshore wind energy areas. Renew. Energy **233**, 119934 (2024). https://doi.org/10.1016/j.renene.2023.119934

Yu, Y., Wang, Y.-T., Yang, R.-P., Li, C.-H.: Wind climate analysis in the Yellow Sea and Bohai Sea based on ERA5 reanalysis data. In: Environmental Science and Engineering. Springer, Heidelberg (2023). https://doi.org/10.1007/978-3-031-25284-6_7

Yu, Z., Fan, Y., Metzger, E.J.: An empirical method for predicting the South China Sea Warm Current from wind stress using Ekman dynamics. Ocean Model. **174**, 102030 (2022). https://doi.org/10.1016/j.ocemod.2022.102030

Yuan, Q., Gong, Z., Zhao, Z., He, J.: Ice model crevice effect on vertical water-entry of a sphere. Ocean Eng. **300**, 117425 (2024). https://doi.org/10.1016/j.oceaneng.2024.117425

Yue, D., Taylor, P.H., Hlophe, T., Zhao, W.: Comparison of two types of wave buoys: linear and second-order motion. In: MAE Conference (2024). https://doi.org/10.1115/OMAE2024-127016

Zago, L., Simos, A.N., Kawano, A., Kogishi, A.M.: A new vessel motion based method for parametric estimation of the waves encountered by the ship in a seaway. Appl. Ocean Res. **134**, 103499 (2023). https://doi.org/10.1016/j.apor.2023.103499

Zampieri, L., Clemens-Sewall, D., Sledd, A., Hutter, N., Holland, M.: Modeling the winter heat conduction through the sea ice system during MOSAiC (2024). https://doi.org/10.1029/2023GL106760

Zeeberg, A.R., Tychsen, J., Nielsen, J.S., Cholley, J.-M.: A Monte Carlo based model for estimating the reliability of offshore structures exposed to environmental loading. In: OMAE Conference (2024). https://doi.org/10.1115/OMAE2024-130321

Zeng, X., Li, Y., He, R.: Predictability of the loop current variation and eddy shedding process in the Gulf of Mexico using an artificial neural network approach. J. Atmos. Oceanic Technol. **32**(5), 1098–1111 (2015). https://doi.org/10.1175/JTECH-D-14-00176.1

Zhan, K., Li, C., Zhu, R.: A frequency domain-based machine learning architecture for short-term wave height forecasting. Ocean Eng. **287**, 115844 (2023). https://doi.org/10.1016/j.oceaneng.2023.115844

Zhang, W., Zhao, H., Chen, G., Yang, J.: Assessing the performance of SWAN model for wave simulations in the Bay of Bengal. Ocean Eng. **285**, 115295 (2023). https://doi.org/10.1016/j.oceaneng.2023.115295

Zhang, Y., et al.: Performance analysis of global HYCOM flow field using Argo profiles. Int. J. Digital Earth **16**(1), 3537–3560 (2023). https://doi.org/10.1080/17538947.2023.2252407

Zhang, Z., et al.: Power fluctuation and wake characteristics of tidal stream turbine subjected to wave and current interaction. Energy **264**, 126185 (2023). https://doi.org/10.1016/j.energy.2022.126185

Zhang, W., et al.: A deep-learning real-time bias correction method for significant wave height forecasts in the Western North Pacific. Ocean Model. **187**, 102289 (2024). https://doi.org/10.1016/j.ocemod.2023.102289

Zhang, X., Huo, J., Zhang, M., Bai, J., Zou, L.: On wave-current interaction with a horizontal cylinder near free surface by a fourth-order-accurate finite volume compact solver. Ocean Eng. **302**, 117666 (2024). https://doi.org/10.1016/j.oceaneng.2024.117666

Zhang, X., Huo, J., Zhang, M., Xie, Z.: Solitary wave-current forces on a horizontal cylinder by a fourth-order-accurate finite volume compact solver. Ocean Eng. **294**, 116788 (2024). https://doi.org/10.1016/j.oceaneng.2024.116788

Zhang, X., Simons, R., Zheng, J., Zhang, C.: Investigation on the turbulent structures in combined wave-current boundary layers. Ocean Eng. **306**, 118073 (2024). https://doi.org/10.1016/j.oceaneng.2024.118073

Zhang, Z., et al.: Laboratory study of wind impact on steep unidirectional waves in a long tank. Phys. Rev. Fluids **9**, 104801 (2024). https://doi.org/10.1103/PhysRevFluids.9.104801

Zhao, Y., Dong, S.: Design loads and reliability assessment of marine structures considering statistical models of metocean data. Ocean Eng. **241**, 110099 (2021). https://doi.org/10.1016/j.oceaneng.2021.110099

Zhao, Y., Dong, S.: Comparison of environmental contour and response-based approaches for system reliability analysis of floating structures. Struct. Saf. **94**, 102150 (2022). https://doi.org/10.1016/j.strusafe.2021.102150

Zhao, G., Dai, S., Jia, L., Yuan, Z. (2023). Three-dimensional measurement of surface disturbances in Kelvin wake using stereo-vision principle. 7th Conf. on Advanced Model Measurement Technology for The Maritime Industry (AMT)

Zhao, L., Li, Z., Qu, L., Zhang, J., Teng, B.: A hybrid VMD-LSTM/GRU model to predict non-stationary and irregular waves on the east coast of China. Ocean Eng. **276**, 114136 (2023). https://doi.org/10.1016/j.oceaneng.2023.114136

Zhao, Y., Dong, S.: Multivariate probability analysis of wind-wave actions on offshore wind turbine via copula-based analysis. Ocean Eng. **288**, 116071 (2023). https://doi.org/10.1016/j.oceaneng.2023.116071

Zhao, Y., Dong, S.: Uncertainty analysis of extreme mooring loads associated with environmental contours and peak tension distributions. Mar. Struct. **89**, 103369 (2023). https://doi.org/10.1016/j.marstruc.2023.103369

Zhong, M., Gong, S., Tian, S.-F., Yan, Z.: Data-driven rogue waves and parameters discovery in nearly integrable P-T-symmetric Gross-Pitaevskii equations via PINNs deep learning. Physica D **439**, 133430 (2022). https://doi.org/10.1016/j.physd.2022.133430

Zhu, Y., et al.: Toward the improvement of subseasonal prediction in the national centers for environmental prediction global ensemble forecast system. J. Geophys. Res. Atmos. **123**, 6732–6745 (2018). https://doi.org/10.1029/2018JD028506

Zuo, H., Stoffelen, A., Rennie, M., Bay Hasagar, C.: The contribution of Aeolus wind observations to ECMWF sea surface wind forecasts. J. Geophys. Res. Atmos. **129**(6) (2024). https://doi.org/10.1029/2023JD039555

Zve, E.S., Swan, C., Hughes, G.O.: Crest-height statistics in finite water depth. Part 1: the role of the nonlinear interactions in uni-directional seas. Ocean Eng. **289**, 116369 (2023). https://doi.org/10.1016/j.oceaneng.2023.116369

Open Access This chapter is licensed under the terms of the Creative Commons Attribution-NonCommercial-NoDerivatives 4.0 International License (http://creativecommons.org/licenses/by-nc-nd/4.0/), which permits any noncommercial use, sharing, distribution and reproduction in any medium or format, as long as you give appropriate credit to the original author(s) and the source, provide a link to the Creative Commons license and indicate if you modified the licensed material. You do not have permission under this license to share adapted material derived from this chapter or parts of it.

The images or other third party material in this chapter are included in the chapter's Creative Commons license, unless indicated otherwise in a credit line to the material. If material is not included in the chapter's Creative Commons license and your intended use is not permitted by statutory regulation or exceeds the permitted use, you will need to obtain permission directly from the copyright holder.

Committee I.2: Loads

Jungyong Wang[1(✉)], Arash Abbasnia[2], Louis Diebold[3], Andre Fujarra[4], Tomaso Gaggero[5], Spyros Hirdaris[6], Sangyeob Kim[7], Dimitrios Konispoliatis[8], Masayoshi Oka[9], Jose Miguel Rodrigues[10], Mahmud Sazidy[11], Florian Sprenger[12], Peter Wellens[13], and Guiyong Zhang[14]

[1] St. John's, Canada
Jungyong.Wang@nrc-cnrc.gc.ca
[2] Instituto Superior Técnico (University of Lisbon), Lisbon, Portugal
[3] Bureau Veritas, Paris, France
[4] Federal University of Santa Catarina, Florianópolis, Brazil
[5] University of Genoa, Genova, Italy
[6] Athens, Greece
[7] Pusan, South Korea
[8] National Technical University of Athens, Athens, Greece
[9] National Maritime Research Institute, Tokyo, Japan
[10] SINTEF Ocean, Trondheim, Norway
[11] Defence Research and Development Canada, Ottawa, Canada
[12] University of Rostock, Rostock, Germany
[13] Delft University of Technology, Delft, the Netherlands
[14] Dalian University of Technology, Dalian, China

Committee Mandate. Concern for the environmental and operational loads from waves, wind, current, ice, slamming, sloshing, green water, weight distribution, and other operational factors. Consideration shall be given to deterministic and statistical load predictions based on model experiments, full-scale measurements and theoretical methods. Uncertainties in load estimations shall be highlighted. The committee is encouraged to cooperate with the corresponding ITTC committee.

Keywords: Wave loads · Current loads · Wind loads · Ice loads · Wave-current interaction · Airgap · Slamming · Sloshing · Wave-in-deck · Uncertainty · Potential theory · CFD · Field methods · Model tests · Full-scale measurements · Data-Driven methods · Ships · Offshore structures · Wind turbines · Mooring systems · Benchmark

1 Introduction

This report comprehensively reviews the types of loads acting on ships and offshore structures over the past three years (2021–2024), highlighting key research trends and emerging methodologies. By addressing hydrodynamic forces, wind and current loads, ice loads, and associated uncertainties, the report serves as a valuable resource for understanding the forces influencing the design, analysis, and operation of ships and offshore structures.

© The Author(s) 2026
W. Wu and J. Ding (Eds.): ISSC 2025, LNME, pp. 133–252, 2026.
https://doi.org/10.1007/978-981-95-2668-0_2

Wave loads, discussed in Sect. 2, represent one of this field's most extensively studied areas. Sections 2.1 and 2.2 focus on wave loads acting on ships and offshore structures, respectively. For both sections, common attention has been given to potential theory methods and field methods. For ships (Sect. 2.1), full-scale measurement, sloshing, and slamming research were addressed. For offshore structures (Sect. 2.2), wave-current interactions, airgap and wave-in-deck loads, and mooring systems were reviewed. These loads are also examined through model tests and data-driven methods, reflecting the growing use of machine learning and artificial intelligence in marine engineering.

Current and wind loads, covered in Sect. 3, are essential considerations in marine operations, as they often overlap with wave loads to create combined environmental forces. The section explores current loads in Sect. 3.1 and wind loads in Sect. 3.2, with dedicated subsections addressing the unique challenges posed to ships and offshore structures.

Ice loads, discussed in Sect. 4, are particularly relevant for Arctic and sub-Arctic operations. This section is divided into sections addressing ice loads on ships (Sect. 4.1) and offshore structures (Sect. 4.2). Most recent research in this domain involves numerical approaches, with studies introducing advanced methods to simulate ice failure behaviour, including ice fracture. Given the increasing demand for wind turbines in ice-covered waters, an ice-induced vibration section has been included, offering a detailed review. Generally, the scarcity of full-scale data remains challenging for validating these studies.

Section 5 addresses the discussion of uncertainty for both experimental and numerical approaches. An overview of uncertainty analysis methods is provided to illustrate the different approaches. Section 6 provides a summary as well as key findings for each section.

This committee also performed benchmark tests for numerical prediction of slamming experiments. Four different numerical approaches were used, and the results are discussed in Appendix A.

2 Wave Loads

2.1 Wave Loads on Ships

Wave loads are critical in ship design and operation, influencing structural integrity and performance. This section provides an overview of the recent developments and research results contributing to this field's progress. The survey covers numerical, experimental, and data-driven approaches and highlights phenomena such as slamming and sloshing. Linear and non-linear potential flow methods can model fluid-structure interaction in ideal fluids at low computational cost. Typical application examples encompass pressure-induced loads, wave-induced motions and added resistance. Field methods complement this approach by capturing viscous effects, turbulence, and flow separation, with continuous developments that allow for increasingly high-fidelity simulations of wave-structure interaction. Lately, data-driven methods are rapidly evolving, utilizing machine learning and large datasets to improve predictions of wave loads and structural responses, offering the potential for enhanced efficiency and accuracy.

Nevertheless, experimental methods, particularly model tests with segmented and elastic hulls, are crucial to validate theoretical predictions and investigate hydroelasticity, whipping, and springing. Advances in instrumentation have improved the accuracy of these tests in controlled laboratory environments. To verify both numerical and experimental observations and to gain knowledge with respect to fatigue assessment and operational safety, full-scale measurements using onboard monitoring are required. Techniques like indirect load estimation and machine learning extract valuable insights from this data. One of the particular phenomena that are addressed in this section is slamming. This highly localized and nonlinear load, caused by impulsive wave impacts, is explored using advanced simulations and experiments to address its effects on structural safety and fatigue. Another phenomenon that has been the focus of investigation over the recent years is sloshing, causing challenges, particularly for liquid cargo ships, due to internal fluid dynamics affecting stability and loads. Besides the effects caused by sloshing, mitigation strategies through numerical and experimental methods are discussed. This section integrates these phenomena and approaches, providing a comprehensive view of the latest research work on wave loads to provide guidance.

2.1.1 Potential Theory Method

Potential flow methods are commonly used to model physical processes where viscous effects, turbulence or flow separation are not essential or play only a minor role. Typical applications where such methods can be applied to determine wave loads on ships include added resistance, wave excitation forces, pressure distribution, and slamming loads. Solving the interaction between the waves and advancing floating bodies requires formulating the motion of fluid particles oscillating randomly in reality. Therefore, the superposition attribute of potential flow has been deployed to develop computational methodologies for implicitly solving the equations of motion for fluid particles and advancing floating bodies. Thus, potential flow solvers are faster and computationally cheaper than the viscous models, whilst the range of fidelity of the potential flow solutions extends over a vast domain of applications.

Potential flow theory can be used to calculate the wave interaction with advancing or stationary floating bodies (such as ships) using the boundary integral equation. Hence, the Laplace equation governs the flow field and the boundary conditions are required to yield a unique solution. The computational boundary encompasses the free surface boundary, body boundary, sea-bottom and lateral walls such as waterways or channels. Hence, various analytical and numerical approaches have been developed to treat the complexities of the moving boundaries (the free surface and the body). The frequency domain solution has been developed by taking advantage of the superposition of the velocity potential function to determine the exciting loads on ships advancing in irregular waves in arbitrary heading angles. The frequency domain solution has been classified as linear/weakly nonlinear potential flow solutions due to simplifying the boundary conditions, such as the linearised free surface boundary conditions. By using the Fast Fourier Transform (FFT) and the impulse function, the frequency-domain solution can be converted to the time domain. Moreover, fully nonlinear potential flow models have been developed since a couple of decades ago, wherein the time-domain solution is obtained from the exact boundary conditions. The computational cost for the fully nonlinear models is remarkably higher than the frequency-domain solution.

Linear and Weakly Nonlinear Potential Flow Theory

Strip theory has been known as the fastest potential solver and is widely used to determine the dynamic of slender bodies advancing in a range of wave frequencies. Therefore, the strip theory has been appreciated in comparative studies on ship manoeuvring (Waskito et al., 2022) and the added resistance of advancing ships in different heading angles (Lee et al., 2021b). To implement the far-field method to approximate the added resistance, the strip theory was modified by Amini-Afshar & Bingham (2021) to calculate high-frequency solutions. Moreover, a modification was made by Duan et al. (2022) to approximate the dynamic of ships in shallow water and beam sea encountering angle. (Nielsen et al., 2021) used the strip theory to tune the transfer function for wave interaction with ships in different heading angles. Since the longitudinal component of the body normal vector is neglected in the slender body assumption, the two-dimensional boundary integral equation based on the second Green's identity has been implemented to obtain the hydrodynamic coefficients of the ship section. To elevate the accuracy of the body boundary condition, the three-dimensional boundary integral equation has been employed to take the exact normal vector components into account. Thus, three-dimensional Green's functions, which stand for the velocity potential of a source pulsating underneath the free surface to satisfy the linearised free surface boundary conditions, and the image method has been employed to impose the constant water depth and lateral walls. Submergence of the singular points leads to neglecting the instantaneous body boundary condition in the linear potential theory. The quadrilateral flat panel method was developed by Abbasnia et al. (2024) in which Fredholm's boundary integral equation was employed to obtain the strength of pulsating source in uniform currents, and the calculated hydrodynamic coefficients were inserted into the equation motion of rigid bodies in six degrees of freedom over a range of frequencies. Analogously, Yang et al. (2023, 2022) used the curvilinear quadratic element to calculate the response of advancing ships in the head sea. A hybrid three-dimensional linear potential flow model was developed by Dong et al. (2024) to approximate the near-field and the far-field fluid flow induced by advancing ships in head seas. The near-field solution was used to compute the ship's response, and the far-field solution was used to approximate the added resistance, which is equivalent to the steady drift force. However, the developed model was verified for a monohull, and Yao et al. (2023) extended the application to a trimaran.

Fully Nonlinear Potential Flow Theory

Fully nonlinear free surface boundary conditions and the instantaneous body boundary condition can be imposed into the solution of the boundary value problem using the Mixed Eulerian-Lagrangian (MEL) method. Three-dimensional, fully nonlinear wave tanks have been developed based on the different approaches of the MEL method, and various implicit approaches have been outlined to calculate the time derivative of the velocity potential. A development was made by Tian et al. (2024) for the instantaneous wave interaction with a real ship hull over a range of frequencies. The cumulative chord cubic parameter spline method was deployed to enhance the mesh re-generation and the time marching. Sustainability and the convergence of the numerical solution drastically depend on the accuracy of the intersection between the exact free surface and the instant wetted surface. The response of an advancing real ship form in irregular waves using a fully nonlinear wave tank was determined by Tang et al. (2021). Further development was carried out by Tang et al. (2023) and Sun et al. (2021) to calculate the bending moment and shear force along advancing flexible ships in nonlinear waves. Additionally, Wang et al. (2024) performed further development, in which the torsion of the ship structure was investigated using the solution of a fully nonlinear potential model.

Since there is a massive cost of computation for the fully nonlinear potential model, only few attempts at deploying parallel computation to tackle bottlenecks such as those by Dombre et al. (2019) and Abbasnia & Guedes Soares (2020) have been made.

2.1.2 Field Methods

This section is about assessing wave loads on ships using field methods. Field methods can be used to determine ship slamming in waves, but that topic already has its own section and is therefore avoided here as much as possible. Another topic often encountered in the literature in the context of field methods is hydro-elasticity; however, hydro-elasticity also already has its own section and is therefore not discussed here. The emphasis is also on ships as offshore structures are discussed in the next section.

Sometimes, field methods are used without it being clear what the advantages of the method are compared to (nonlinear) potential flow methods. For instance, Chen et al. (2022e) verified field method results against a strip theory potential flow code and a comparison between two ships in waves that could have been performed with a weakly nonlinear panel method. Such applications, however, demonstrate the applicability of field methods for loads and motions in waves, even when their application is not strictly necessary. This is also exemplified by the results of Mandru et al. (2024), who compared experimental data to a field method and analyzed field method results at model-scale and full-scale. The results in terms of motions in waves and the wave-added resistance showed little difference. According to the authors, differences between model-scale and full-scale were observed in the velocity distribution of velocities in the propeller plane due to viscous effects.

The advantages of field methods for ships in waves over potential flow methods, that have a longer tradition, are twofold: 1. viscous effects are addressed, and 2. large deformations of the free surface, such as those involved in wave breaking, are modelled. Note that field methods typically specialize in one of the aforementioned advantages and not both. Viscous effects for ships in waves are highly relevant when roll (and the notorious roll-damping) is considered. This reporting period has seen a number of articles in this category, so they deserve to be discussed as a group.

Large deformations of the free surface are encountered by ships, for instance, when loaded by green water or when undergoing slamming events. During green water loading, a significant relative wave height at the ship's bow, or sometimes along the sides or at the stern, develops into a flow on deck that can interact with structures on deck and cause damage. During slamming in waves, the ship itself causes large free surface deformations along the hull when a large (but negative) relative wave height initially brings part of the ship out of the water. The way free surface development takes place along the ship hull upon slamming defines how the load on the ship develops.

In any categorization, there is the risk of omitting relevant accounts of field methods for wave loads on ships. For this reason, a number of articles will be presented before moving on to the final part of this section, in which new field methods with potential future applications for ships are discussed. 'New' indicates novel techniques for dealing with the free surface or with more physics than incompressible one-phase or two-phase flows commonly used in current ship studies.

Roll Damping of Ships in Waves

The challenge with roll damping is that it is dominated by flow features that are several orders of magnitude smaller than a typical measure of the size of the ship. The ship requires a fairly large computational domain, whereas the flow features that induce damping need a very fine grid size. These requirements combined lead to a number of unknowns that are too large to solve within a reasonable time span during the engineering of a ship. At present, simulations provide comparatively good results for roll amplitudes (and hence roll damping) when compared to model test results. Note that this does not necessarily imply that simulation results are also good predictors of full-scale roll motions because the Reynolds numbers differ significantly at model-scale and full-scale.

Ghamari et al. (2022) overcame the challenge of roll damping in field methods by proposing a hybrid strategy for a ship cross-section with a keel (which induces flow features) and rounded bilges (which are typically more challenging because the location of flow separation is part of the solution itself). A field method, validated effectively using experiments presented in the article, was applied to find the linear and quadratic damping coefficients from roll decay simulations with the ship. The coefficients were then incorporated in the ship motion solver of a potential flow method for wave interaction with structures. When exposed to regular waves, the roll motion amplitudes of the ship in the potential flow solver with the additional damping coefficients were within 10% of the roll motion amplitudes from an experiment.

For a ship with a moonpool, Duan et al. (2023) proposed a strategy to obtain improved damping coefficients that include the effect of the moonpool on roll motions from more highly resolved decay tests so that they could be incorporated in simulations of the ship in waves at lower resolutions. Response Amplitude Operators (RAO) of lateral accelerations for a range of wave frequencies were determined from the simulations and compared with measurements from experiments. Differences between the two were not larger than around 5%. Although the authors' main intention was to exclude waves from their evaluation of a ship's roll motion at forward speed, the procedure proposed by Sumislawski & Abdel-Maksoud (2023) can be used to find the roll damping coefficients that are necessary to assess the correct roll motion behaviour of a ship at forward speed in waves. Harmonic Excitation of Roll Motion (HERM) uses an actuator to induce roll motions of the ship. For their study, a system based on the gyroscopic effect of rotating disks was developed to make the HERM procedure more efficient. The system was used to evaluate the effect of the rudder on roll damping, to determine the added resistance of the ship due to the roll motion, and to investigate parametric rolling. The latter was performed through a field method so that the omission of waves allowed for shorter simulations and, therefore, a more efficient evaluation of the ship's performance.

Parametric roll was also investigated using a field method described by Liu et al. (2024). The final roll motion amplitude of the ship in head waves at several forward velocities of the ship from the simulations was compared to measurements from an experiment, and both were found to be within 10% of each other. The parametric roll was then investigated with different orientations of the ship's course with respect to the direction of the waves. The final investigation evaluated roll motion mitigation in the form of bilge keels, showing that bilge keels reduced the roll motions under the study's conditions by between 25 and 45%.

Green Water on Ships

Representing green water loading in field methods is challenging because the free surface near the ship undergoes abrupt changes, with the wave that exceeds the deck level developing in a bore-type flow that can cause impacts with deck structures. After impact, a vertical jet of water can form, and the free surface development is decisive for how the load on the deck structure develops. The aforementioned processes need to be resolved with high resolution in both space and time to lead to results that can be trusted. Green water loading can be deemed so challenging that authors go to lengths to try to avoid it. For instance, Islam & Guedes Soares (2022) artificially increased their deck level when assessing the sea margin and the added resistance in waves using a field method. The simulation results typically overpredict added resistance compared to the experiment, but only by a small margin so that they can be used to formulate the conclusion. The conclusion was that the rule of thumb for a sea margin of 25% is not sufficient in moderately steep waves, and the authors advise revisiting this rule.

Green water was the specific subject of study for Sun et al. (2023a). The authors used a field method to evaluate a Tublehome vessel with a wave baffle on deck in regular waves with four different steepness values. First, the suitability of the software for representing green water loading was established by comparing the simulation results to measurements of the water level on the deck and pressure from experiments. Then systematic simulations were performed, leading to the following findings: steeper waves did not lead to a longer duration of green water, but the steepness significantly affected the water level on deck. When plunging-type events occurred on deck, they led to more water being accumulated on deck behind the baffle. The conclusions state that the reported setup was suitable for investigating green water but that more work is necessary to investigate the effect of wave length, baffle shape and baffle inclination.

Notable Applications of Field Methods for Wave Loads

The following articles do not demonstrate the advantages of field methods mentioned above but show methods or applications that can be beneficial when using field methods to their advantage. Jeon et al. (2022) developed a wave maker for circular (or partial circular) domains that can be used to create directionally focused waves for emulating 'rogue' waves. The article shows ship motions due to the focused wave loads. However, it would be interesting to see if the directional wave focusing technique can be used to study green water loading in extreme waves.

In Gong et al. (2022), a field method was used to study broaching of a trimaran. Broaching in waves can lead to highly aperiodic motion of the ship and instability (more commonly for monohulls than for trimarans naturally). From their systematic set of simulations, the authors found that steep waves and a wave phase velocity close to the ship's forward velocity were necessary conditions broaching to occur.

New Field Methods for Wave Loads on Ships

Field methods typically solve for two-phase flow with the free surface as an interface between phases. In heavy seas, however, breaking and overturning waves entrain air in water, where it remains for several wave periods. The mixture of gas bubbles and water can be referred to as aerated water. With the typical percentages of 1 to 2% air by volume in water, the mixture is more compressible than air itself. The compressibility can significantly affect the pressure during wave impacts, green water impacts, or slamming events. van der Eijk & Wellens (2024) investigated a novel formulation incorporating aerated water as a third phase in a field method. The field method was compared to a purpose-built experiment involving wedges impacting the free surface of homogeneously aerated water. Their findings showed that increasing the percentage of aeration reduces the maximum impact pressure during slamming; 1% by volume of air in water reduces the pressure peak during slamming by 15% compared to the pressure in water without air. The pressure peaks of the field method simulations compared well with the pressure peaks in the experiment. Pressure variations after the impact could be attributed to density waves propagating through the aerated medium in both the experiments and simulations, and the simulations could be used to visualize the density waves. With aeration, simulations of slamming of ships in waves can be modelled more realistically in future.

2.1.3 Model Tests

Introduction

Model testing is a crucial component of the design process of ships. While the standard testing program usually addresses the determination of required power in calm water conditions, the standard structural ship design process follows class regulations and usually does not require experimental campaigns. However, model tests to measure the structural response of a ship in waves are essential for validating numerical models and ensuring the structural integrity and safety of ships, particularly for growing hull sizes and extreme environmental and operating conditions. Large ships with high elasticity and low natural frequencies can experience significant dynamic responses, such as whipping from impulsive wave loads and springing from short-period waves, potentially leading to severe ultimate loads and fatigue damage. Model tests with segmented hull models offer a more robust approach for analyzing ship hydroelasticity compared to traditional seakeeping tests. Yet, the impact of certain parameters on hydroelastic responses remains unclear. Several recent studies have looked into this field, and their findings are summarized in the following section: *Model and experiment design*. Several researchers have been investigating particular aspects of structural responses in waves by performing test campaigns. In recent notable efforts, the focus was mainly on large container ships. The work summarized in the section *Experimental studies* addressed nonlinearities related to wave steepness variation, the impact of forward speed and loading conditions and extreme wave heights.

Model and Experiment Design

Models used in hydroelastic experiments need to accurately reflect both the hull geometry and elastic characteristics of the prototype. Typically, two types of models are used: segmented (further distinguished in flexible joint and flexible backbone models) and fully elastic hulls. Grammatikopoulos (2023) reviewed the development of ship models used in hydroelastic experiments. The study highlights the importance of flexible models for investigating dynamic responses in slender ships and high-speed craft, discussing the advantages and limitations of the different model types in replicating ship stiffness and dynamic responses. The challenges of scaling and modelling in hydroelastic experiments are pointed out, including the difficulties in achieving full similarity between the model and full-scale ship due to physical constraints. The author also examined the application of these models in studying nonlinear responses such as springing, whipping, and antisymmetric vibrations, emphasizing the importance of correct scaling for accurate predictions. Looking ahead, further development of fully elastic models, improvements in measurement techniques like strain field measurement, and better calibration methods to enhance the accuracy of hydroelastic experiments are required.

Segmented models are the most used approach for estimating wave-induced loads, especially for new designs where established design load estimates are not yet available. Ibrahim et al. (2022) detailed their efforts to design, construct, and test a 6-foot segmented model of a medium-sized semi-displacement vessel. Their model was equipped with a backbone beam to replicate the structural stiffness characteristics of the full-scale ship. The instrumentation of the model includes strain gauges placed on the beam, local pressure sensors, and advanced Fiber Bragg Grating (FBG) strain and pressure sensors.

In order to investigate the effects of the number of segments, propulsion mode, and backbone configuration on ship hydroelastic behaviour, Chen et al. (2022b) developed a three-dimensional nonlinear time-domain hydroelastic method to account for the nonlinearity of the instantaneous wetted surface and slamming forces. From an analysis of natural frequencies and vertical bending moments (VBM), the authors conclude that self-propelled hull models with variable cross-section backbones representing the full-scale cross-sections' characteristics are more effective and reliable for studying nonlinear hydroelastic behaviour.

The design and fabrication of segmented ship models significantly impact wave-induced loads. Si et al. (2024) investigate three different segmented models of a 20,000 TEU ultra-large container ship, each with distinct support frames and backbones. Models A and B share similar fabrication methods but differ in backbone type and scale, while Model C employs steel framing validated by finite element methods and connected via welding. These models' dry and wet modes were measured and compared, revealing that design and fabrication methods notably influence ship model modes and wave-induced loads. Model A's lower natural frequencies were attributed to its lower frame stiffness (C_F), stiffness of the support crossbeams (C_{SCB}), and stiffness of the connections among the various members (C_C), impacting the transmission of wave-induced loads. Adequate values for CF, CSCB, and CC are crucial for effective load transmission and achieving similarity in vertical stiffness. FE analyses indicated that natural frequencies converge when segment numbers exceed ten, providing sufficient crossbeam stiffness. Additional uncertainty analysis highlighted significant variances in test uncertainties, stressing the need for precise control over repeated tests, moment of inertia, calibration coefficients, and wave height in future experiments.

While most studies focus on scaling for vertical bending modes, container ships also experience significant torsional moments due to large deck openings. Chen et al.

(2024) introduced a fully elastic ship-shaped model designed to replicate the vertical bending, horizontal bending, and torsional vibration modes of the S175 container ship. The model, equipped with strain gauges, enabled the measurement of critical hydroelastic responses, including vertical bending moments, vertical shear forces, and torsional moments. Comprehensive testing under different wave conditions and forward speeds showed that the model's structural responses align well with numerical simulations. The results show the benefits of this approach, noting that while the model produced reasonable bending and torsional modes, some fabrication limitations affected the final plate thickness. Future improvements in fabrication methods could enhance model accuracy. Overall, the authors demonstrated the model's effectiveness in capturing linear and non-linear responses, including slamming-induced whipping and springing effects in vertical bending and torsional modes, thus providing valuable insights into container ship behaviour under varied wave conditions.

Besides the hull and backbone properties, pressure sensors play a crucial role in the experiments. Commonly, strain gauges are used in ship model tests to measure elastic deformation, while the application of Fiber Bragg Grating sensors (FBG) is still relatively new in this field. FBG sense pressure based on the detection of changes in the wavelength of light reflected from a periodic variation in the refractive index of the fiber core, which shifts in response to external physical parameters such as strain or temperature. The advantages of FBG lie in their accuracy and sensitivity, but also in their small size which allows for distributed measurements with minimum intrusion. While their receptiveness to strain is beneficial in hydroelastic ship model testing, temperature sensitivity is somewhat problematic. Xu et al. (2022) developed a (FBG) pressure sensor to measure multi-point loads on ship hulls during wave experiments. The sensor detects changes in optical fiber responses due to hydraulic pressure and incorporates a steel diamond structure for effective temperature self-compensation using a single FBG. The authors showed that the sensor achieves suitable sensitivity and precision with a low-temperature sensitivity. Initial model tests in static and dynamic conditions with 15 sensor units installed on a ship model hull demonstrate the applicability to measure and monitor pressure distributions.

Given the stochastic nature of waves, it is crucial to determine how long and how many test realizations (seeds) are needed to obtain reliable extreme values for design purposes. Recent research by Scharnke et al. (2023) covered long-duration experiments with a ferry hull (see Fig. 1) to understand the convergence of most probable maximum (MPM) values and the number of seeds required for accurate statistics. Their work reveals that for convergence of wave crest MPM values, around 5–14 seeds with a 3-h duration or 8–22 seeds with a 1-h duration are needed. For example, the convergence requirements for green water impact forces on a ferry, which have a higher impact frequency, differ from those for wave-in-deck impacts on a stationary deck box with a lower impact frequency. The analysis shows that using only 1 or 2 seeds results in substantial inaccuracies, with root mean square errors (RMSE) of approximately 11% of significant wave height for wave crests and 5%–24% for green water forces. The study also evaluated the impact of fitting and extrapolation on convergence, finding that fitting introduces minor biases but does not significantly affect the number of seeds required.

The results provide practical guidelines for determining the number of seeds required to achieve reliable extreme value estimates in wave load experiments.

Fig. 1. Ferry model hull with relative wave elevation probes, force panels on the accommodation and force panels and local pressure sensors in the bow flare (Scharnke et al., 2023)

Experimental Studies

As the steepness of the waves that a ship encounters influences the motion and structural responses, experimental studies are required to investigate nonlinearities in both waves and responses systematically. A study by Kim et al. (2023b) investigated wave-induced motions and loads on a container ship model in 120° oblique regular waves, focusing on the influence of wave steepness. The experiments utilized a model at a scale of 1:65 with nine segments and a mooring system consisting of four horizontally arranged spring lines to maintain the model's heading angle. The authors aimed to understand how wave steepness affects nonlinear responses, including vertical bending moments (VBM) and horizontal bending moments (HBM) near amidships and six-degree-of-freedom motions at the centre of gravity. Various wave series with different steepness levels were tested, revealing that steeper waves increase higher-order harmonic components and slamming events in bending moments, deviating significantly from linear responses. The mooring system's impact on asymmetric HBM and average yaw motion was analyzed, showing that the system's restoring moments correlated with these responses. The findings indicate that nonlinear effects such as slamming significantly alter VBM and HBM responses in steep wave conditions, with differences in mooring line tensions influencing yaw motion. The results suggest that while the rigid model assumption is validated in head waves, further investigations are required for oblique wave conditions due to stronger slamming and structural vibrations. Further experiments were carried out by Kim et al. (2024) with a rigid container vessel in irregular waves. Long-duration tests, totalling approximately 25 h of data per sea state, were carried out across five different sea states (H_s from 6 m to 17 m) to calculate statistically significant extreme values of hogging and sagging. It was found that the ratio between linear and non-linear VBM, known as the non-linear factor, can be modelled as a function of the linear value alone, regardless of the sea state. This finding is significant as it facilitates the development of efficient design assessment methodologies using increased design sea states or design waves. Detailed investigations revealed that the non-linear factor depends mainly on the linear VBM value at a given return period, validating the approach of using increased design sea-states for experimental data. Although the conclusions are based on a single vessel, which limits their general applicability, they support previous studies and suggest that the non-linear factor remains consistent across different sea states.

An experimental campaign with a 10,000 TEU container ship has been performed by Zhang et al. (2022i). The authors designed a segmented ship model with adjustable cross-sections to ensure similarity in vertical, horizontal, and torsional stiffness along its length. The study included tests with both head and oblique waves. Frequency spectrum analysis revealed significant high-frequency wave loads affecting springing phenomena, with oblique waves exhibiting higher proportions of these loads. Non-linear wave-induced sag and hog ratios in vertical bending moments were attributed to high-frequency components. Bow flare slamming was predominant in a fully loaded condition, while stern and stem slamming became more pronounced as the draft decreased. Although stiffness has minimal effect on the frequency of slamming, it was found that reduced stiffness results in increased severity of mean slamming pressures and loads.

Ship forward speed relative to the incoming waves affects the impact of wave loads and the structural responses of ships. Zou et al. (2024) investigated bow flare slamming loads and whipping responses of a ship using a segmented model in regular waves. The model was designed to replicate the hull's vertical and horizontal bending stiffnesses. Temporal variations in harmonic components were analyzed using continuous wavelet transform. Key findings include that higher sailing speeds significantly impact the magnitude and distribution of bow flare slamming pressures, causing increased asymmetry in slamming loads under oblique waves. At higher speeds, the distribution of pressures on the bow flare became more uneven, with more significant impact on the waveward side. The whipping responses lead to noticeable periodic changes in harmonic strengths around the model's natural wet frequency, intensifying vertical bending moment asymmetry at midship. Repeated tests confirmed the experimental stability but highlighted significant variability in bow-slamming loads affecting higher-order harmonics.

Ship sizes have been increasing over the last years, with associated challenges for structural integrity under extreme environmental conditions. A study by Tang et al. (2022c) investigated the nonlinear bending moments of an ultra-large container ship in severe wave conditions. An experimental campaign with a 1:65 scaled model with a variable cross-section backbone beam was performed, focusing on the analysis of vertical bending moments (VBM, with strain gauges used for measurements) at various Froude numbers, wave headings, and wave heights. Key aspects investigated include resonance frequencies, high-order harmonic components, and the asymmetry of hogging and sagging moments. The backbone beam's design and calibration were crucial for accurate results. The model's natural frequencies were compared with theoretical predictions, showing satisfactory agreement. The authors also investigated the effects of wave height on VBM asymmetry, particularly in extreme waves (H_s up to 25 m) where high-frequency nonlinear loads become significant. Results indicate that sagging moments can be up to three times larger than hogging moments in severe conditions. Finally, the study highlights the variation in phase differences between wave-frequency and high-order harmonics, which differ significantly from moderate sea conditions.

By measuring ship responses in waves, it is also possible to draw conclusions on the wave parameters and subsequently derive further responses that may not have been measured. This approach might be of particular interest in real ship operation. However, establishing such a method under controlled laboratory conditions is favourable. Komoriyama et al. (2023) proposed a Kalman filter (KF) technique for identifying wave profiles encountered by a stationary ship from measured responses. The method was applied to tank test data with an acrylic resin ship model, measuring wave elevation, six degrees of freedom of ship motions, and hull strains using FBG sensors in long-crested irregular waves. The KF technique was validated by comparing identified wave profiles with reference measurements. Accuracy improved by selecting appropriate inputs and combining multiple signals. The method also estimated non-measured ship responses and predicted future wave profiles a few cycles ahead. Accuracy was affected by frequencies with low transfer function (TF) amplitudes. Combining multiple measurement signals enhanced accuracy, especially in low TF regions (see Fig. 2).

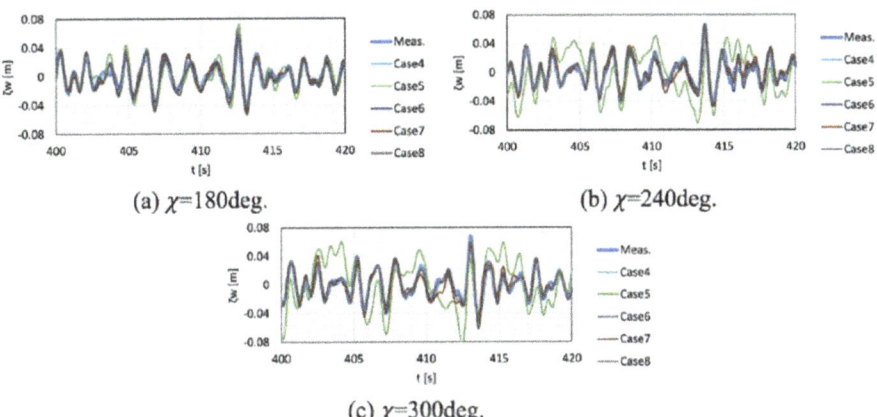

Fig. 2. Comparison of measured wave profiles with wave profiles that have been derived from different combinations of other measured signals (Komoriyama et al., 2023)

2.1.4 Full-Scale Measurement

Full-scale measurements, namely onboard monitoring, are an effective technology that contributes to a ship's health evaluation and design improvement, and research and development have progressed. As the technology has already reached the stage of practical application, the number of research papers has decreased compared to previous years. The main research topics in this field include the development of methods for estimating unmeasured responses and environmental conditions such as waves and ship conditions through data assimilation; the development of methods for evaluating designs and design criteria based on the data; and the formulation of standards and regulations based on the evaluation results.

As it is not always feasible to measure all parameters and responses that are relevant directly on board a ship, indirect estimation approaches can be used. Gebrezgabir et al. (2023) proposed a method to indirectly estimate slam and wave load responses using measurement data of strains, accelerations and angular velocities. It was applied to full-scale sea trial data of the 98 m high-speed catamaran HSV 2 Swift. The responses were reconstructed for different ship headings, speeds and sea conditions. Tested using only the transmissibility matrix without knowledge of sea conditions, excellent agreement was achieved, particularly for more significant loading events. This technique, in combination with finite element analysis (FEA), reduces the number of sensors, reduces data noise, improves data quality by recovering missing data, and improves data quality by reducing the number of sensors, reducing data noise, and recovering missing data. This makes it possible to estimate stress at any point. Chen et al. (2021b) proposed a method to estimate ocean wave spectra using stress and motion components measured by onboard monitoring. A method was proposed to estimate wave spectra of arbitrary shapes using nonlinear programming, one of the mathematical programming methods for solving optimization problems. The wave spectrum was estimated based on monitoring data of a 14000 TEU container ship. The results were compared for 1) using vertical bending stress and horizontal bending stress, 2) using vertical bending stress, horizontal bending stress, and double bottom bending stress. The selection of an appropriate combination of accuracy and response was discussed by comparing with Hindcast and radar data. Takami et al. (2024) proposed an approach to identify stability parameters based on onboard response measurements, namely linear and nonlinear roll damping coefficients, in conjunction with the natural roll frequency. Response measurements (heave, pitch, sway, etc.) from waves other than roll were used to estimate the encounter wave profile. Then, stability parameters that best reproduce the measured roll motion were identified through optimization. Good results were obtained for both long- and short-crested irregular waves.

Contributing to evaluating methods for fatigue design of hull structures, Warren et al. (2022) assessed the applicability of classification society rules for high-speed catamarans. A comparison was made of the long-term distribution of stresses measured on a 111-m-long wave-piercing catamaran ferry during operation in the Canary Islands with the load spectrum estimated using the method recognized by the Classification Society DNV GL (DNV-GL, 2015). In addition, an improved distribution fitting method for fatigue analysis was proposed based on the conversion of the stress history measured in the time domain into a suitable stress spectrum and a Weibull fitting method for cyclic stress in the low-stress and high-stress regions due to slamming. The results showed that the classification society method is very conservative regarding fatigue estimation compared to fatigue results based on measured data. Engelbrecht & Bekker (2023) investigated how to set the threshold for discomfort caused by vibrations due to slamming impacts based on onboard monitoring. Data were collected from two slamming-prone vessels. Vertical acceleration measurements were conducted near the ship's working and living areas. Human responses were collected from passengers using a daily survey. Observers on the bridge performed momentary slamming vibration comfort ratings during a series of intentionally induced slamming events. As a result, the limit value based on the RMS value converged to the same value (0.03 m/s^2) regardless of the cumulative time. This value is higher than the ISO and Classification Society standard value (0.01 m/s^2) and allows vibration.

2.1.5 Slamming

When a ship operates in rough seas, its bow may lift and then slap on waves because of the relative motions between the hull and the wave surface. Due to high vertical velocities during the re-entry of the ship hull in water, impulsive loads associated with the high hydrodynamic pressure peaks occur. This phenomenon is called 'slamming', and it is strongly nonlinear. When the global hull girder vibrations occur because of an impulsive wave loading, such as a wave slam at the bow (bow-slamming) or stern (stern-slamming), the phenomenon is denoted by the term whipping. Whipping is a transient phenomenon due to excessive impulsive loading in the bow or stern of the vessel. According to classical ship theory, the 2-node natural frequency dominates the dynamic response resonant mode during whipping, leading to hull girder stress variations (see Fig. 3).

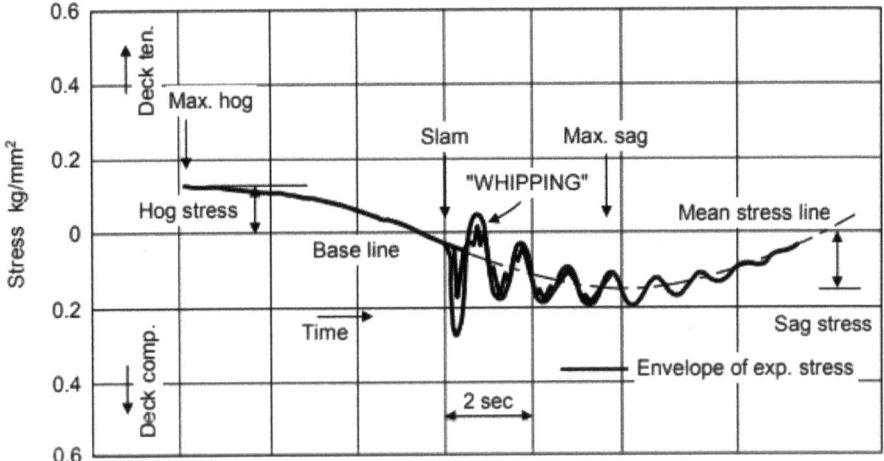

Fig. 3. Whipping stress due to slamming (ABS, 2024).

From a practical perspective, springing and whipping events manifest simultaneously in confused seas. Since associated wave-induced resonances are superimposed on wave frequency loading, they may contribute to fatigue damage accumulation (Hirdaris et al., 2023). Slamming tends to occur with waves in the ship's bow sections, in which case the shape of the bow is of paramount importance, although there are ship forms that may be equally prone to bow and stern slamming. Research challenges relate to (a) accurate use and validation of advanced numerical methods for the prediction of local hydrodynamic pressures, especially in both symmetric and asymmetric planes, (b) methods for the prediction of global whipping responses and their impact on fatigue strength, and (c) understanding the influence of symmetric and antisymmetric higher vibration modes on global response. As applicable, this section reviews lessons from accidents, key advances of relevance local (bow and stern induced), and global slamming loads. Results from experiments and full-scale measurements (FSM) are also included.

Learning from Accidents

Accident records and FSM present the pathways for research and design updates, especially for the case of novel ship designs for which empirical methods may not be valid and computational tools may have limitations. Despite genuine attempts from both ISSC (2018) and ITTC (2021) communities, the experience based on FSM is not systematically developed and remains relatively scarce, particularly for ships operating in extreme conditions. This is because FSM data processing requires a large number of human resources and know-how for the rapid and meaningful processing of big data streams. Within the reporting period of this committee, practical results of relevance to slamming and whipping are limited to experiments (wedge drop tests for local slamming and segmented model tests for whipping) as discussed in relevant parts of the report (Chen et al., 2019; Hosseinzadeh et al., 2023b, 2023a; Li et al., 2022a; Lu et al., 2023; Zou et al., 2024).

The confidential nature of the vast majority of available data records also prohibits the transparent use and implementation of lessons learned. In relatively recent years, increased stresses have been recognized as relevant for the loss of ultra-large container ships MSC Napoli and MOL Comfort (see Fig. 4). The former was a United Kingdom-flagged ship that developed a hull breach due to rough seas and slamming in the English Channel on January 18, 2007. An investigation conducted by the UK Marine Accident Investigation Branch (MAIB), Det Norske Veritas (DNV) and Bureau Veritas (BV) demonstrated that she broke due to inadequate frame strengthening in the engine room area. The investigation recognized that wave loads were amplified by 30% during the accident, possibly due to whipping (MAIB, 2008). The adverse effects of an extreme slamming case also influenced the M/S Estonia accident in 1994. During her last journey in the Baltic Sea, due to slamming, the ship lost her bow visor JAIC (Joint Accident Investigation Commission) (1997). The damage to her watertight front door led to her capsize and the loss of 852 lives.

Fig. 4. Recent Container ship accidents (a) MSc Napoli – © gcaptain.com (b) MOL Comfort - © ship engineer worldpress.com.

The MOL Comfort experienced a fracture of the midship part while in the Indian Ocean on June 17, 2013. The ship was split into two halves and sank. The accident report suggested that the ship's structural strength could be exceeded because of wave loads (ClassNK, 2014; MJB, 2015). The analysis accounted for uncertainties associated with lateral load effects on hull girder strength, yield stress deviations, welding-induced residual stress, weather, operational conditions, whipping effects and cargo loads, etc. It was recommended that future rule requirements should account for the effects of lateral and whipping loads to evaluate ship structural strength. These investigations led to the revision of IACS longitudinal strength requirements and the selective introduction of Classification Society notations. The new longitudinal strength standard for container ships URS 11 A (IACS, 2015) introduced a partial safety factor that accounts for the influence of increased stresses along the double bottom of container ships. This new standard may be considered important especially in the hogging condition when a double-bottom topology may be exposed to high compressive stresses that can exceed the buckling capacity of shell structures or double-bottom plating and consequently trigger the progressive collapse of the hull. The same standard introduced a nonlinear correction factor to update the vertical wave-induced bending moment.

Forensic analysis demonstrated that the influence of whipping-induced loads on hull girder strength is still a matter of research, and Classification Society procedures still differ. Although whipping is included in the functional requirements for large container vessels, because of the uncertainties influencing design factors, it is left up to the individual assurance body to consider its relevance. For example, DNV (2015) classification guidelines introduced a partial safety factor of 0.9, reducing the effectiveness of whipping during collapse. BV does not use that reduction, claiming that there is no evidence for a factor of 0.9 (Derbanne et al., 2016). This could be attributed to uncertainties associated with (1) existing computational tools and hydroelastic modelling approaches; (2) the definition of the influence of dynamic failure modes, which may be different from those understood within the context of rigid body ship dynamics; (3) the lack of consensus on the definition of representative design or extreme operational conditions; (4) limited understanding of methods that can be used to model the influence of extreme events on fatigue accumulation based on data from in-service experience.

Local Slamming Loads

Local slamming loads are severe transient and may cause local structural damage during extreme events (Hirdaris et al., 2014). The phenomenon was first recognized by Yamamoto et al. (1985), who reported damages due to bow flare slamming on an 819 TEU container ship in cyclone conditions. At the time, the authors concluded that an 840 kPa impact pressure in the way of a 13.7 m circular area might lead to structural damage.

In the early days, theoretical research was driven by von Karman (1929) and Wagner (1932). A comprehensive research summary on ship-slamming physics is presented by Kapsenberg (2011). Wagner's model is based on the potential flow theory. It applies within the context of ideal and incompressible fluid flow that is generally used to estimate the pressure distribution on ideally stiff wedges. Zhao & Faltinsen (1993) solved the similarity flow for wedges by a nonlinear boundary element method with a jet flow approximation. Their results agreed with the simple asymptotic solution by Wagner (1932) for small deadrise angles. Faltilsen et al. (1996) extended this approach to 2D arbitrary geometries. These papers provided important research output regarding the value of analytical formulations and potential flow-based simulations, and the ideas presented are implemented in classification guidelines (DNVGL 2021).

With the advent of super-computing power, advanced numerical methods have been employed to better understand the influence of fluid-induced nonlinearities and strong fluid-structure interaction. On the hydrodynamics front, numerical models are based on the Arbitrary Lagrangian-Eulerian (ALE) scheme (Stenius et al., 2007, 2011; Wang & Guedes Soares, 2014; Wang & Soares, 2012), the Finite Volume Method (FVM) (Southall et al., 2014, 2015), Smoothed Particle Hydrodynamics (SPH) (Alexandru et al., 2007; Farsi & Ghadimi, 2016; Veen & Gourlay, 2012), and the Volume of Fluid (VOF) method (Southall et al., 2014). Recently published work by Tavakoli & Hirdaris (2023) confirms that accurate numerical modelling of the time-dependent nature of hydroelasticity-induced slamming, along with violent fluid-structure interaction and free surface evolution, remains challenging. The authors suggest that plate deformations may significantly affect dynamic response, especially in oblique conditions. This is because rolling free surfaces and waves may influence the distribution of hydrodynamic loads.

Strong coupling FSI methods are used to study the influence of flexible fluid–structure interactions on wedge-shape structures. In their traditional format, they combine the Boundary Element Methods (BEM) or Reynolds Averaged Navier Stokes (RANS), Computational Fluid Dynamics (CFD) and FEM (Yan et al., 2022c), see Fig. 5. More advanced numerical schemes such as ALE (Stenius et al., 2011; Wang & Guedes Soares, 2014) and SPH (Panciroli & Porfiri, 2015) may also be coupled with FEA.

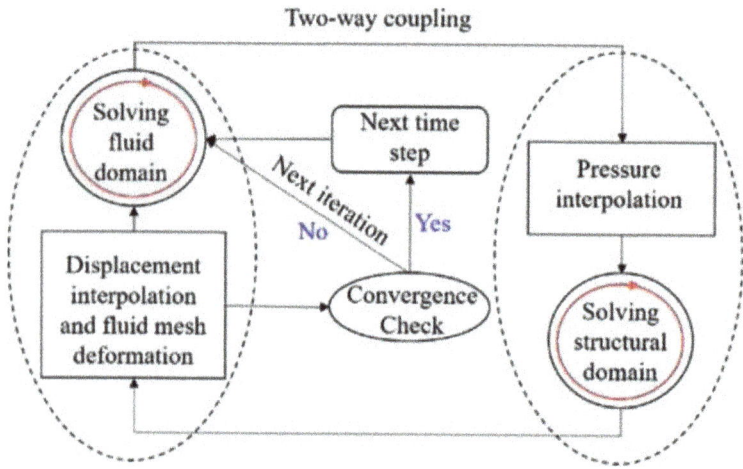

Fig. 5. Two-way coupled FSI modelling method as per (Yan et al., 2022b)

Recently, a benchmark study for flat stiffened plates by Truong et al. (2021) reviewed uncertainties associated with modelling and computation. Yan et al. (2022a) and Yan et al. (2022c) presented a detailed validation of two-way coupled Finite Volume (FVM) and FEA methods. The authors studied the complete history of the flat plate water entry problem. Experimental and numerical uncertainties were analyzed according to the ASME V&V method (ASME 2009) and confirmed the validity of the simulation-based approach. A sequel study compared the water entry fluid actions for the cases of a flat plate impinging on the water at different velocities and the case of a symmetric wedge idealizing slamming at different deadrise angles. In this study, numerical predictions have been compared against available experimental data by Tödter et al. (2020) for flat plates and the hydrodynamic pressure design curves implemented in classification guidelines for wedge-like structures (DNV-GL (2015)) as shown in Fig. 6. The results generally showed good agreement. This work confirms that a two-way coupled FVM-based CFD-FEA modelling procedure does not change the trend of pressure design curves already implemented in classification rules for stiff wedges. The smaller the deadrise angles, the larger the differences in fluid pressures and forces between stiff and flexible structures. The similarity in the pressure distributions of flexible wedges is not obvious.

Because of the complexity of three-dimensional flows, such as fluid separation, fluid breaking, and bubble flows, the accurate prediction of local slamming loads in three dimensions is still very limited. Recently, Xie et al. (2024) presented results from drop experiments idealizing the asymmetrical slamming load characteristics of a truncated ship bow under the combination of heel and pitch orientations. In this work, an angle device was used to independently adjust the heel and pitch orientation of the test model. The results suggest that air cavities and jet flows may cause multiple oscillatory pressure peaks. Abnormal pressure characteristics appear in some locations, where the pressure decreases with the increment of water entry velocity.

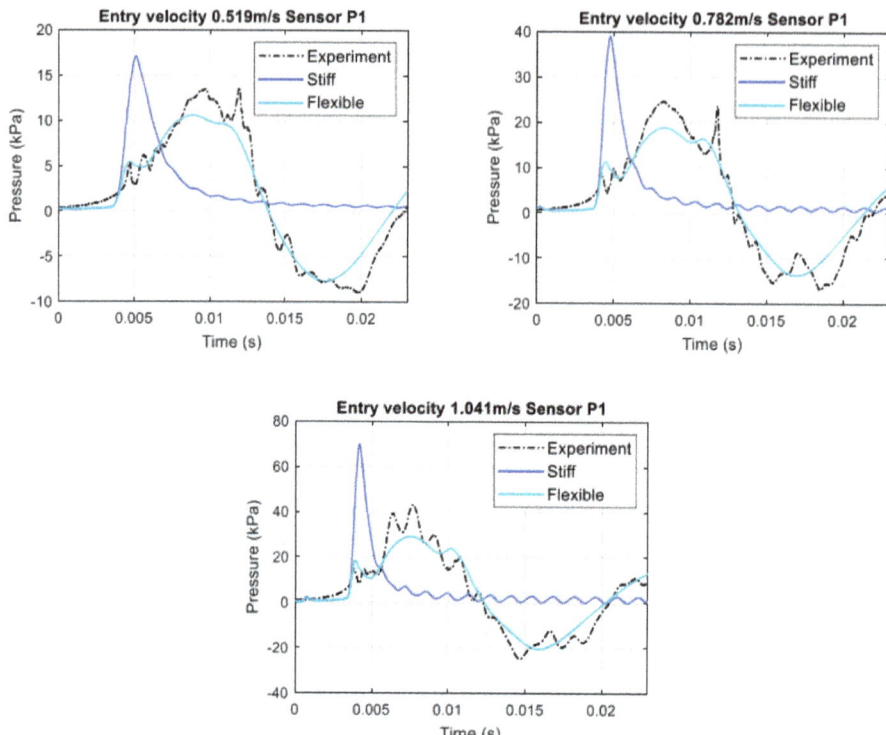

Fig. 6. Comparison of stiff and flexible plate simulations carried out by Yan et al. (2022c) with experimental (flexible plate) results presented by Tödter et al. (2020)

Severe green water effects coupled with slamming events in confused seas are rarely researched. This is because of the complexity associated with strong nonlinear interactions between waves and ships. Recently, Liu et al. (2022c) presented comparisons against theoretical and experimental data to verify the reliability and accuracy of their time domain method. Results indicate that the response of peak pressures in the bow and deck area was affected by high-frequency components, while peak pressures in the way of the superstructure were affected by low frequencies. Ha et al. (2024) presented a numerical model for predicting slamming impact loads on the bow flare of a ship-type floating production storage and offloading (FPSO) model under breaking waves. To investigate the relation between the slamming impact load and the free motion of the FPSO model, the breaking impact simulations were performed under fixed positions and the free motion of the FPSO model.

Since slamming is primarily a local loading event, it may affect the seakeeping qualities and strength of high-speed crafts. For example, high-speed ferries typically have operational limitations based on maximum motions, accelerations and slamming occurrence. Monohulls tend to suffer high slamming pressures near the lowermost chine from the bow up to 1/3 of the hull length and towards amidships (Kapsemberg & Brizzolara, 1999). Past analyses of the slamming loads on fast ferries operating at as-built conditions concluded that fluid-structure interactions were not affecting local pressure loads for hulls built in aluminum or steel. However, fluid-structure interactions may indeed play an important role for high-speed hulls built in reinforced plastic (e.g., see Honey et al. (2021)). Cross-deck bottom slamming is an important design constraint for high-speed catamarans, which can be equipped with wave-piercing bows. Shabani et al. (2018) mitigate slamming loads and to avoid large clearance from the waterline. (Hosseinzadeh et al., 2023b, 2023a) present two-part companion papers that deal with the experimental and numerical studies of the impact-induced loads and structural responses of a three-dimensional non-prismatic aluminum wedge with a stiffened panel during free-fall water entry. The effects of water impact velocity, deadrise angle, mass of the wedge, and bending stiffness on the slamming pressures and structural responses are discussed in detail. Two-way coupling methods are assessed and used to simulate the water entry problem. Initially, an explicit nonlinear finite element method with a Multi-Material Arbitrary Lagrangian-Eulerian (MMALE) solver is employed to evaluate the elastic response of the structure following a free-fall water impact. Subsequently, the hydroelastic slamming problem is modelled using a two-way coupled technique with a k-ε turbulence model and implicit unsteady solver for both the fluid and structural domains (Star-CCM + /ABAQUS). The importance of FSI simulation is assessed using a hydroelasticity factor (R_F), which is found to have a significant effect on the unstiffened bottom for all impact velocities studied. For a stiffened bottom panel, hydroelasticity is only significant at high-impact velocities.

In another recent study (Tavakoli et al., 2024) confirmed that local slamming loads can also be important to the seakeeping performance of planning craft. This is because, as the speed increases, the hydrodynamic lift forces that support the weight of such vessels magnify, and buoyancy gradually converges to zero while the displaced volume diminishes. In rough sea conditions, this can result in relatively large wave-induced motions and vertical accelerations that can be way greater than gravity. During what is known as the unsteady planning phase, large slamming forces may also arise, causing strains (see; (Gilbert et al., 2023; Hosseinzadeh et al., 2023b, 2023a; Tavakoli et al., 2023), and this may damage the vessel's bow region. Monohedral planning craft configurations (i.e., crafts with constant deadrise angle along their entire length) are subject to low deadrise angles in their rear part and larger toward the bow, where slamming loads can also be excessive. A solution can be given by developing optimal step hull configurations (see Bonci & de Jong (2023)). Recently, advanced numerical methods such as the 2D + t model introduced by Garme (2023) were used to account for the influence of one-way coupling in design. A two-way coupled method using unsteady Reynolds-averaged Navier-Stokes equation is presented by Diez et al. (2022), who also applied multidisciplinary design optimization (MDO) for the estimation of loads on a deep-V planning-hull grillage panel subject to slamming loads in regular waves. Notably, the effects of one-versus two-way coupling are negligible for both the original/traditional and optimized grillages (as per the hydroelasticity factor R). In contrast, the impact of FE boundary conditions on the analysis and optimization outcomes is significant, confirming the need for proper calibration of the FE model in FSI and MDO studies.

Slamming-Induced Whipping Loads

Zou et al. (2024) conducted an experimental investigation using a segmented model to determine bow flare slamming loads and whipping responses of a ship in regular waves. A continuous wavelet transform was employed to capture temporal variations in the harmonic components. Their study revealed that the whipping responses significantly increased the asymmetry of the vertical bending moment (VBM) at amidships at high sailing speeds.

Lu et al. (2023) introduced a 3-D nonlinear time-domain hydroelastic analysis method for ship wave loads considering asymmetric slamming. The hydrodynamic idealization is based on a 3-D Rankine panel method and structural dynamics are implemented by FEA. The interaction between the hull structure and the flow field was considered to predict the hydroelastic response of the hull girder. The motions and sectional loads of the ship are expressed by modal superposition. After calculating the asymmetric slamming loads using the Modified Logvinovich Model (MLM), the hydroelastic response of a 21,000-TEU container ship was analyzed. This incorporates the horizontal bending-torsional coupling vibration.

Li et al. (2022a) presented an experimental investigation of stern slamming and global whipping-induced responses using a segmented model test of a cruise ship with a scale ratio of 1:60. Experimental random uncertainty was analyzed by considering the model responses in a series of regular waves. It is found that different types of response have different levels of dispersion. In a sequel study, Li et al. (2024a) developed a time-domain hydroelastic numerical model for investigating the slamming and whipping loads imposed on a ship travelling at different forward speeds. The numerical model integrates a beam model, the 3-D Rankine panel model, and the 2-D modified Logvinovich model. The tilt angle of the 2-D profile in the slamming model is determined by the direction of relative velocity, which considers the forward sailing speed. It is shown that the first-order elastic mode mainly affects the dynamic response. The authors conclude that slamming forces may be sensitive to the longitudinal interval of the slamming profiles, especially in the way of the vessel's stern.

Parunov et al. (2024) presented a benchmark study on motion and global wave loads for the case of a Canadian Patrol Frigate in regular waves. Nine institutions participated in the benchmark with six codes, quantifying the hydroelastic responses. The seakeeping methods employed include non-linear strip theory, 3D boundary element method formulated in frequency and time domain, and computational fluid dynamics (CFD). Euler and Timoshenko beams are used to model the hull girder stiffness. Experimentally based methods, CFD and momentum theories, are employed to calculate slamming loads. It was found that fully coupled CFD and finite element method (FEM) provide results consistent with measurements, but such simulations are prohibitively computationally expensive.

2.1.6 Sloshing

Sloshing is a physical phenomenon that is to be considered for any moving vehicle or structure containing a liquid with a free surface. This is a vast topic covering many applications: propellant slosh in spacecraft/plane tanks and rockets, liquid slosh for trucks, offshore units and ships transporting liquids. For ships transporting liquids, sloshing is a large-scale phenomenon involving some very small-scale effects. Indeed, overall volume and geometries, fluid densities come into play as well as many local parameters such as local geometry of the tank's boundaries gas/liquid physical and thermodynamic properties (density ratio, surface tension, compressibility, phase change).

Following this physical description, one can identify two main topics: global (overall liquid motions within the tank) and local flows (impact pressures). The first topic, global flow, is of primary interest since the internal liquid motions inside the ship tank(s) may influence the ship's motions. Experimental and numerical calculation aspects will be covered. In addition, sloshing mitigation techniques will be presented. The second topic deals with local flows (free surface, kinematics) and impact pressures that may occur during liquid impacts. These impact pressures need to be quantified, especially for ships transporting cryogenic liquids. Indeed, loss of containment (due to extreme impact pressure) could have severe consequences for the crew and the ship. Also, numerical calculations and experimental results are covered.

Global Flow

The first topic deals with the evaluation of the global flow within the tank(s) in order to assess the possible influence of these internal liquid motions (and their forces) on the ship's stability and motions. The loss of stability due to liquid motions is usually calculated by the (analytical) evaluation of the free-surface correction of the metacentric height, and no dynamic calculation is required. One exception concerns the ore carrier. Indeed, due to recent major accidents involving ships carrying liquefied solid bulk cargoes, work has been carried out to better understand the fluidization of these solid cargoes, which can lead to capsizing. A transparent material was developed to be a substitute for iron ore (non-transparent) that allows to better visualize and understand this fluidization process through further sloshing model tests Chen & Zhang (2024).

For seakeeping applications, the influence of the internal liquid motions on the ship motions (or floating unit) depends on the tank's shape, dimensions, position, arrangement (internal structures or not), and liquid density. Going in order of increasing complexity, here are different configurations that have been studied in the literature.

First, for small tank's dimensions, for instance, a swimming pool (typically 20m) with respect to the ship ones (typically 200m), the influence of the internal liquid motions on seakeeping can be neglected (Wu et al., 2023b). These (very) different dimensions lead to a decoupling between the swimming pool natural periods and the ship motions ones resulting in quasi-static motions of the free surface. Thus, for small ship motions, a linear numerical model (neglecting nonlinear effects) was developed and has shown to be applicable by giving results close to CFD results (Qi et al., 2023). For such a linear model, the damping (due to bottom friction) can be approximated by a linear equivalent damping, which depends on the fluid velocity squared (Neugebauer et al., 2023).

The second configuration is related to large tanks such as spherical or prismatic LNG tanks (but without large internal structures such as baffles). These large tanks have a significant influence on the ship motions (Caputano & Goodwin, 2023; Wang et al., 2023e, 2024c). In Wang et al. (2023e), a typical FLNG equipped with three (prismatic) tanks was studied (usually, one or two tanks were considered). It was shown that the most conservative roll RAO (for a large range of frequencies) was associated with three tanks equally filled: changing one filling level in one of the three tanks significantly reduced the roll RAO. This result, which needed to be confirmed for other fillings distributions than the tested ones, was interesting since it confirmed that the iso-filling assumption made for the LNGC/FSRU/FLNG sloshing analysis was conservative. Regarding numerical methods that can adequately handle such type of coupling effects for large tanks (without internal structures), potential flow theory (frequency and time domain) and CFD (volume of fluids + VOF, Smooth Particle Hydrodynamics (SPH)) are widely used for both interior (liquid motions inside the tank(s)) and exterior problems (seakeeping), and are in excellent agreement with the model tests (Jiao et al., 2024; Wang et al., 2024c).

The last configuration for coupled seakeeping is related to tanks with internal structures such as anti-roll tanks. Kapsenberg & Carette (2023) showed that Free-Surface anti-roll tanks were more efficient than U-tanks (larger damping over a broader range of frequencies). The fluid flow characteristics inside such anti-roll tanks (traveling bore and the complex internal flow including flow separation due to the internal baffles) explained why CFD methods were required, and potential methods were unreliable. Finally, the Effective Gravity Angle (EGA, a combination of the local acceleration and the roll angle) was introduced and proved to be a governing parameter for such anti-roll tanks.

Local Flow

The second topic deals with the evaluation of the local flow (free surface, fluid velocities and impact pressures) within the tank(s). The final objective of such investigation is to assess the design loads to be considered for the design of the tank and ship's structures. To do this, sloshing model tests and numerical calculations considering water and air are reviewed. Indeed, for the sake of simplicity, many sloshing model tests (and so the corresponding CFD calculations) were carried out using the most straightforward available fluids, water and air. In Bardazzi et al. (2024), comparisons between sloshing model tests on a quasi-2D tank and one-phase SPH calculations were carried out. The studied filling level was close to the critical depth for which the resonant sloshing dynamics changed from "hard spring" to "soft spring" as filling increased. Very good agreement between CFD and experiments was observed for the free-surface elevation time series and the complex sloshing regimes (doubling frequency, tripling frequency bifurcations, asymmetric and chaotic regimes). Their study confirmed that even one-phase CFD calculations captured the global flow, the detailed local flow, and the complex sloshing regimes.

Other sloshing model tests focused on the free-surface profiles that can lead to significant impact pressures (Shen et al., 2024). It was confirmed that flip-through impacts (transition between liquid impact with an air pocket and wave running up along the wall) gave rise to the most significant pressures and stresses. However, pressure measurements showed poor repeatability, and impact pressures needed to be studied statistically. This stochastic nature of sloshing impact pressures and the complexity of the liquid impact physics explained why discrepancies were observed between experimental and numerical sloshing impact pressures (Pilloton et al., 2022). This led to original experimental sloshing campaigns using different fluids than air and water to study, among others, the effects of gas-to-liquid density ratio and phase transition on impact pressures. The first consisted of small-scale model tests using a non-flammable liquid with a boiling point of 34 °C at atmospheric pressure (Lee et al., 2021a). Condensation was observed, but the authors estimated that the influence of density ratio was more dominant than the effect of phase transition on the modification of impact pressures. In parallel, a larger-scale experimental campaign (named SLING) was launched. The objective was to better understand the source of variability and the main governing parameters (gas-to-liquid density ratio, compressibility, phase change) of the sloshing impact pressures (Ezeta et al., 2023). A flume tank (equipped with a wave maker) was installed in a high-end autoclave to study phase change during sloshing impacts. This multiphase facility also allowed the disentangling of the influence of the gas-to-liquid density ratio and the gas compressibility parameters. The first results confirmed the importance of the influence of the gas-to-liquid density ratio on sloshing impact pressures (the higher this density ratio is, the lower the pressures). The final aim of these studies was to improve the representativeness of the sloshing model and so the scaling of the sloshing model tests (Fan et al., 2024; Jain et al., 2021; van der Meer, 2022). However, in these model tests, raised edges (such as corrugations) were rarely represented due to small-scale issues. To circumvent this issue, dedicated numerical calculations were carried out to better understand the loads acting on these raised edges (Audiffren et al., 2024). In the studied configuration, the most significant loads occur on the first hit corrugation. In addition to these numerical calculations, machine learning approaches, based on the existing sloshing model tests database, were applied to predict sloshing activity that ship operators can use. Finally, the statistical post-processing of the sloshing impact pressures is the last step of a sloshing analysis. It was shown that discrepancies existed among the different mathematical formulations for the long-term pressures (Ahn, 2023; Ahn et al., 2024). Also, the importance of considering the hydro-elastic responses of the containment system and the supporting hull structure was highlighted (Malenica et al., 2022; Park et al., 2024).

2.1.7 Data-Driven Methods

Introduction

Recent technology developments facilitate the possibility of creating large datasets throughout a ship's lifetime. From the early design phase to the operation at sea, designers and operators have nowadays the possibility to utilize the information in this data to acquire new knowledge. The data can generally come from different sources, e.g., numerical simulations, model tests or full-scale measurements. Traditionally, ship designers have used experience and data from previous projects as well as regression-based empirical methods to create a preliminary design that fulfils the client's requirements and regulatory boundary conditions. Data-driven methods allow for more efficient, systematic utilization of large datasets for such tasks. Ship owners are under pressure to fulfil IMO emission regulations for their fleets. Due to rapidly advancing sensor and telemetry technology onboard ships, many ship owners possess large amounts of ship operational data. Data-driven methods, often based on Machine Learning (ML), offer significant potential for more efficient, safer, and environmentally friendly ship operations. Furthermore, operational data provides valuable input for the design of the next generation of ships. Several methods are available to efficiently and systematically exploit a given data resource, and new approaches are under development.

A general review of currently available ML approaches for application in the ship design process was given by Huang et al. (2022e). It shows the different ML application fields in ship design and operation. While the applications for routing and fuel consumption prediction were rated relatively mature, autonomous shipping was highlighted as a developing field with an increasing need for algorithm development. When it comes to ship design, it was found that most models so far focus on resistance reduction as the standard optimization goal. For example, Ao et al. (2023) proposed a deep-learning neural network model to predict the total resistance of KCS hull form variations. Using geometry modification parameters and Free-Form Deformation (FFD) techniques, the authors created a dataset to train and test the model, which is capable to predict hull resistance during preliminary design. The research concluded that deep learning enhances hull design efficiency and accuracy in the early design phase. However, there are not many application cases where data-driven methods have been applied for other important design aspects, such as stability and structural integrity so far. As highlighted in Fig. 7, approaches that can be classified as supervised and reinforced learning are among those that are best suited for wave load applications. In contrast, unsupervised learning algorithms are well suited for performance predictions such as fuel consumption.

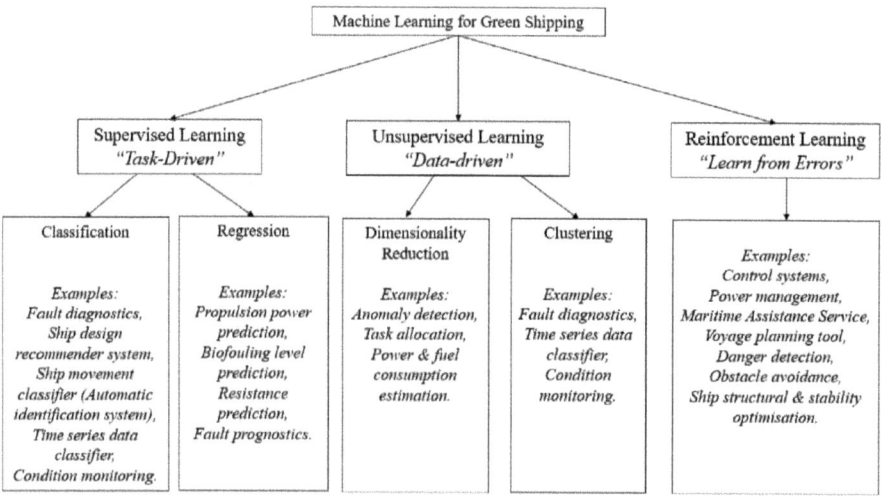

Fig. 7. ML strategies and application examples as a toolbox towards 'green shipping' from Huang et al. (2022e)

However, there have been early notable studies on ML applications in this field, such as Rogers et al. (1998), who developed an approach based on neural networks to determine dynamic bending moments without knowledge of the actual wave parameters. Their method is based on input from the ship's heave and pitch motion. Another early example is the optimization of a bulk carrier with two conflicting objectives (weight and fatigue) performed by Cui et al. (2012) where a simple reinforced learning approach called Q-learning was used. Based on a combination of multi-objective particle swarm optimization, reinforced learning, and CAE software, the authors were able to provide an improved design with respect to the original design for every chosen objective.

A review of recent research activities shows that the work on data driven methods for the estimation of the impact of wave loads on structural responses is still rather sparse. However, there is an increasing number of activities related to applying ML algorithms to predict the seakeeping performance of ships from simulations, model tests, or full-scale measurements. Besides offering possibilities and advantages in selected applications, data-driven approaches require special care regarding data pre-processing. This topic has become more critical as it has become easier to access large amounts of data via sensors and black box algorithms.

Structural Responses

In ship structural design, the prediction of the vertical bending moment (VBM) and the total longitudinal stress (TLS) is crucial. Current state-of-the-art methods offer approaches to determine these properties at different fidelity levels. Low-fidelity methods provide efficient and practical applications with low accuracy, and vice versa for high-fidelity methods. A study from Jiang et al. (2023a) introduced a multi-fidelity regression model using an artificial neural network (MF-ANN), which combines low-fidelity data for initial modelling and high-fidelity data for corrections, significantly enhancing prediction accuracy while offering efficiency and robustness. The proposed model was successfully validated through VBM predictions, while TLS predictions demonstrate that even minimal high-fidelity data can achieve high accuracy.

In another example, Wang et al. (2024b) investigated a real-time prediction method for wave-induced hull girder loads using ship motion data as input. Two Recurrent Neural Network (RNN) models were developed to predict VBM, horizontal bending moment (HBM), and torsional moment (TM) at the mid-ship of a large container ship. An improved model incorporating an error correction strategy was also explored to enhance prediction accuracy. The authors investigated the influence of RNN parameters such as the window length of the time series, optimizer, number of layers, and neurons on the prediction accuracy. Their study found that a 3-layer GRU network is optimal for TM and HBM, while a single-layer LSTM network is better for VBM, highlighting different data relevance. LSTM (Long Short-Term Memory) models were designed to effectively capture long-term dependencies in sequential data by using memory cells and gating mechanisms to regulate the flow of information. They are particularly useful for tasks involving time series data and other sequences where context from previous inputs is crucial for accurate predictions. GRU (Gated Recurrent Unit) models simplify the LSTM architecture by combining the forget gate (which determines what information to discard from the previous state) and the input gate (which decides what new information to incorporate) into a single update gate, making it computationally more efficient while still effectively capturing dependencies in sequential data. They can also be used for time series data, providing a balance between performance and complexity. Verification showed that prediction models perform best under the same irregular sea state conditions, with RMSE below 5.2%, and perform less accurately with varying sea conditions. The error correction strategy significantly improved the models, reducing RMSE for HBM from 10.1% to 6.8% and for TM from 7.5% to 5.8%. The improved prediction models show enhanced performance across different conditions.

A data-driven approach that does not use ML but can estimate long-term extreme values of wave loads on a ship along a defined route involving hindcast wave database and ship operability analysis was shown by Petranović et al. (2021). In their approach, the authors determined the transfer functions of wave-induced motions and loads through closed-form expressions. It identifies the most probable extreme wave loads for short-term sea states, calculates annual maximum values, and fits them to a Gumbel distribution. These distributions were then statistically combined along the route, considering sea states as either independent or fully correlated. This method aids in determining probable wave loads for long return periods, which are essential for ship structural design. The results closely align with other long-term distribution methods when assuming statistical independence among sea states. It also allows for considering wave zone correlations and the sensitivity of wave loads to operational criteria. Numerical results indicate that operability criteria are not met in 2–4% of sea states. Avoiding these reduces extreme bending moments by a factor of 2, reducing ship speed in these states cuts bending moments by 20%, and assuming complete correlation among sea states reduces bending moments by 8–21%.

Motion Responses

Over the last few years, there has been some noticeable development of data-driven methods to predict the impact of wave loads on the motion behaviour of ships. Three application scenarios can be distinguished: Purely data-driven models, physical-informed models and calibration models for low-fidelity software tools. The following selected examples highlight the progress in these fields.

A data-driven framework introduced by Ren et al. (2023) can identify 6-degree-of-freedom (DOF) ship models based on ship motion measurements and corresponding wave load estimations. The proposed model incorporates hydrodynamic coefficients, thruster inputs, mooring systems, and wind and wave excitation loads. Sparse regression was utilized to fit these measurements, ensuring that the reduced-order system captures the main features of the system's dynamics while maintaining the physical integrity of the candidate functions. The authors verified their method through experiments and comparative studies, demonstrating its potential for application to other floating structures. This work highlights the importance of balancing model accuracy and complexity and shows that the algorithm can achieve satisfactory short-term motion predictions through direct integration. However, the real-time application requires significant improvements in computational speed. Future research is suggested to focus on enhancing estimation accuracy, reducing computation time, and improving robustness against noisy measurements, especially in scenarios involving time-varying speeds and headings.

Another recent example is a study by Lee et al. (2023), where an integrated artificial neural network system was developed to predict ship motions in varying sea states. The system combines an LSTM encoder-decoder to capture motion-induced wave effects with a convolutional neural network (CNN) for analyzing spatiotemporal wave-field data. The LSTM encoder processes historical motion data to encode the memory effects of radiated waves. At the same time, the LSTM decoder utilizes this encoded information along with anticipated wave field data to predict future ship motions. Validation using a physics-based seakeeping program demonstrates the system's robustness in making deterministic predictions in new sea states.

While purely data-driven methods can provide satisfactory results, physics-informed neural networks (PINN) offer greater accuracy in many application cases. PINNs integrate principles from physics into the architecture and training of neural networks. They are designed to directly enforce physical laws, such as conservation principles or governing equations, into their learning process. By embedding prior knowledge of physical behaviour, PINNs can effectively learn from limited data and generalize to new scenarios more robustly. This approach not only enhances the interpretability of neural network predictions but also improves their accuracy by constraining solutions to adhere to known physical constraints. One recent application example is the study by Schirmann et al. (2022), who investigated the effectiveness of linear response regression (RR) and nonlinear neural network (NN) models in predicting vessel heave, pitch, and roll using multidirectional wave data. The inclusion of physics-based motion predictions in these models was highlighted, demonstrating improved performance in reducing the Mean Squared Error (MSE) compared to models without physics-based inputs. Their work revealed that NN models outperform RR models when physics-based information is included. This can be attributed to their ability to capture nonlinear relationships. However, physic-informed RR models can predict heave and pitch motions reasonably well.

Another field where research activities are observed is the application of ML algorithms to enhance the accuracy of low-fidelity software results by using training data from low- and high-fidelity codes. Kim et al. (2023a) developed a method to improve the accuracy of efficient ship motion models. In their study, they calibrated low-fidelity models, which use constant coefficient ordinary differential equations using data from high-fidelity simulations. Forced motion calculations determined radiation forces, and fixed-body calculations assessed diffraction forces, which were then regressed to obtain appropriate coefficients for the low-fidelity models. The proposed calibration method significantly improved the performance of low-fidelity models, ensuring their outputs align closely with high-fidelity models through penalized regression techniques to stabilize coefficients.

Data Processing

Large amounts of data offer new possibilities for ship owners and designers to improve both the operation of current ships and the design of the next generation of ships. However, especially on-board and weather data pose significant challenges regarding data quality and availability, directly impacting ML model results. The methodologies proposed by ISO19030 alone might not be sufficient to address these issues. While the significance of the field grows, not much focus has been put on this in terms of research so far. A study by Jabary et al. (2023) discussed the development and implementation of a data processing system for ship operational data. The system integrated raw operational data from multiple sources, such as onboard measurements and external data, into a unified database. To ensure data quality, the raw data underwent pre-processing steps, including data filtering and imputation, to address missing values and outliers. Data quality issues like missing data, outliers, and different data formats and frequencies presented significant challenges, which were mitigated through the pre-processing procedures. The processed data was then used for performance analyses, with the aim of optimizing ship operations for critical performance indicators. The general observations and conclusions regarding data quality and availability also apply to other ML applications, such as wave and structural loads.

2.2 Wave Loads on Offshore Structures

Significant advancements have been made in understanding and analysing wave loads on offshore structures, driven by the growing demand for accurate and reliable design tools to ensure the efficiency, robustness, and safety of these structures operating in complex marine environments. This section reviews state-of-the-art developments, organized into key thematic subsections as in Sect. 2.1. Unlike ships, offshore structures are designed to maintain a fixed position during operation. While the fundamental theoretical principles may overlap with those applied to ships, the specific requirements for offshore structures introduce unique challenges related to wave-induced phenomena and their modelling. For instance, first- and second-order wave and current loads are critical for the design of mooring systems and station-keeping. In contrast, ship design focuses more on calm water resistance and added resistance in waves, which are essential for seakeeping and manoeuvrability. These distinctions highlight the specialized approaches for analysing and simulating wave loads on offshore structures.

2.2.1 Potential Theory Method

Potential flow theory has been exploited to calculate the exciting wave loads on such structures and their response to the wave interaction with fixed and floating structures. Concerning ships, different boundary conditions are considered for offshore structures, such as the stiffness of the station-keeping system, to determine the dynamic response of offshore structures. Moreover, structures supporting offshore energy systems, such as floating solar panels, can be very large, including hinged modules tethered to the sea bottom (Zheng et al., 2024). Eventually, the time-domain response of offshore structures is required to approximate the coupled dynamics of offshore structures, and correspondingly, the nonlinearity of wave loads on the offshore structure must be considered for studying the resonance modes.

Linear and Weakly Nonlinear Potential Flow Theory.

Hao & Pan (2022) used Green's function of submerged pulsating singular points to implement a panel method solver for computing potential flow due to wave interaction with offshore structures with a thin shell geometry. Dipole elements were employed to model the body boundary condition on the thin shell geometry as a drop of the potential, which can be exploited to include porosity-related effects. Moreover, numerical studies on a two-layer panel model with small gaps suggest that the mesh size requirement in DNV-RP-C205 needs revision. Hence, the linearized free surface boundary conditions and the linear body boundary conditions are satisfied in the solution. The diffraction and radiation potential functions have been calculated in the frequency domain, which imposes the constant water depth condition. The image method has been deployed to address the sea bottom boundary condition and linearized free surface boundary conditions.

A comprehensive study was conducted by Cong et al. (2021) on the wave run-up on the bottom-mounted cylinder subject to bi-directional irregular waves using the second-order potential flow model. The second-order velocity potential induced by the sum and difference pairs of wave components was computed through Quadratic Transfer functions (QTF). Prior experimental and numerical studies had verified the solution for the single cylinder, and the model was here used to determine the second-order wave elevation near the cylinder near the wall. A comparative study was presented to demonstrate the second-order wave run-up on the cylinder, as shown in Fig. 8. The significance of representing the free surface to the second order is remarkable in the free surface's evolution and the surface elevation's magnitude. The impact of the second-order terms is amplified for oblique sea heading angles where the interference of directional frequency pairs is considered.

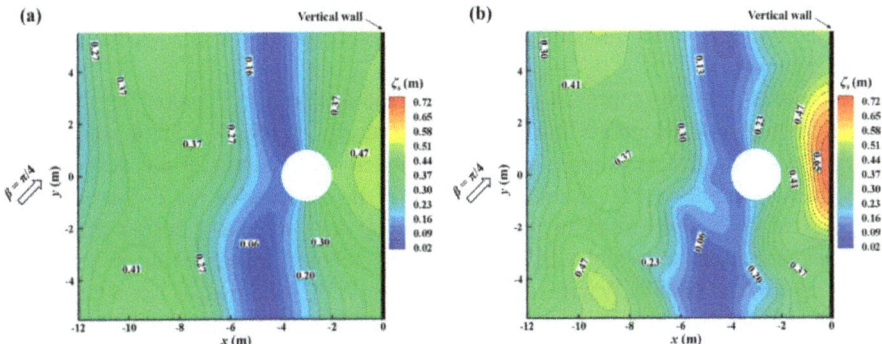

Fig. 8. Comparative study of the wave elevation amplitude distribution around the cylinder near the vertical wall for linear solution (a) and second-order solution (b) in a regular wave (Ka = 0.48)

For an array of truncated cylinders in a row distributed in equal space, the solution of the second-order velocity potential was computed by (Huang et al., 2021) to estimate the wave elevation within the cylinder and sum and difference wave forces due to frequency pairs of irregular waves. Indeed, Huang & Teng (2020) work was extended for bi-chromatic incident waves, and some improvements in the numerical algorithms were included where the wave propagating and evanescent modes were derived. Since the up- and down-wave vertical surfaces are allowed to be placed close to the body, the calculation of the free surface integral can be performed in a small region, and an analytical method was developed to directly evaluate the up- and down-wave vertical surface integrals. After solving the boundary integral equation, the sum- and difference-frequency radiation and diffraction potentials were computed explicitly. Numerical results for the quadratic transfer functions of the second-order free surface elevation, pressure and force were demonstrated for a series of geometries in channels.

The effect of incident waves on the piston-like behaviour of the free surface between two fixed barges was investigated by Gao et al. (2022)). A linear potential model using a higher-order boundary element method (BEM) was employed to determine the effect of the incident wave angle on the resonant fluid motion within the gap for two fixed square corner barges. An adaptive Gaussian method was used to improve the accuracy for nearly singular integration. The results show that gap resonant peaks at other incident wave angles would be more critical than the beam sea case, which has been widely deemed the most crucial case. Moreover, the second mode peak value can overcome the first mode peak value at specific conditions. The resonant frequency and peak value increase while the gap width decreases.

Fully coupled aero-hydrodynamic modelling of floating offshore wind turbines was simulated by Deng et al. (2023), in which the time-domain second-order potential model was developed to introduce the instantaneous hydrodynamic pressure and corresponding wave loads in the equation of motion. The free surface was included in the boundary integral equation, and the perturbation approach was implemented to calculate the second-order velocity potential. Furthermore, the time-domain second-order wave loads associated with the aerodynamic load were introduced into the equation of motion, with the stiffness matrix of the mooring line taken into account on the left-hand side of the equation of motion.

A coupled-dynamic approach of a modularized design of a Tension Leg Platform (TLP) structure which supports the power transmission plant for a wind farm was studied by Abbasnia et al. (2024). A three-dimensional linear potential model was developed to study the buoyant de-centralization of modules on the response of the TLP structure in long-crested waves, with slackline occurrence being considered as a design criterion.

The potential flow model has also been implemented to approximate resonance modes in fixed and floating Wave Energy Converters (WEC) of the type Oscillating Water Column (OWC). Since the free surface inside the OWC chamber may be influenced by the Power-Take Off (PTO) system through the pneumatic pressure of the trapped air, the efficiency of the OWC devices has become a point of interest to study the performance of OWCs for various geometrical properties of OWC chamber, sea-bottom variations and the PTO mechanism. Srinu et al. (2024) developed a two-dimensional linear potential flow to study the performance of a fixed OWC for a wavy sea bottom. The linear pneumatic model was used to solve the velocity potential inside the chamber, and the resonance mode of the piston-like free surface was investigated. Moreover, George et al. (2021) used the two-dimensional potential flow model to optimize the U-shaped oscillating water column (U-OWC) device based on an artificial neural network model.

Fully Nonlinear Potential Flow Theory

The resonance mode of the free surface within the trapped free surface among the double-body structure was investigated by developing a 2D fully nonlinear NWT in which a high-order finite element method (FEM) was employed to solve the potential flow (Huang et al., 2022b).

Kim et al. (2021) developed a 3D, fully nonlinear numerical wave tank (NWT) to approximate the response of a freely floating cylinder in nonlinear waves. The decomposition method was used to calculate the time derivative of the velocity potential implicitly, and the linear BEM formulation was used to calculate the kinematics of the water particles. The results were compared to prior studies to demonstrate the range of validity of the NWT. A FEM solver was developed to solve the fully nonlinear potential flow for twin-cylinder oscillating in currents to find out modes of resonance of the free surface. Similarly, a two-dimensional fully nonlinear model was developed by Abbasnia et al. (2022) based on a high-order BEM to determine the dynamic response of a twin-cylinder supporting a floating solar panel. The numerical model was verified by a set of experimental records.

Abbasnia et al. (2021) developed a two-dimensional, fully nonlinear model to simulate free surface elevation within a fixed Oscillating Water Column (OWC) device. The spring-like behaviour of the trapped air due to the PTO mechanism was taken into account for the chamber free surface. A series of investigations were made to study the dynamic response of a floating pontoon OWC near the breakwater in coastal zones using the three-dimensional, fully nonlinear potential model (Cheng et al., 2022b, 2022a). The results were compared with experiments to verify the numerical results.

Cong et al. (2024) conducted a feasibility study for enhancing the wave energy capture of OWC devices through the utilization of Multi-Chamber Multi Turbine (MCMT) technology. Unlike traditional single-chamber OWCs, MCMT OWCs comprise multiple chamber modules that can coordinate during different wave phases to optimize wave energy cultivation. By considering coupling effects in a linear reciprocal relationship between the air pressure and air-flow movement in different chamber modules, a numerical model was developed to assess the functional performance of three-dimensional MCMT OWCs of arbitrary geometric shapes using a fully nonlinear higher-order BEM. The study focuses on representative MCMT OWC designs with annular or rectangular cross-sections. Two specific scenarios were investigated: an annular MCMT OWC integrated into the monopile foundation of an offshore wind turbine and a rectangular MCMT OWC integrated into a barge-type breakwater. Numerical analyses revealed that dividing the chamber into multiple modules can convert sloshing-mode free-surface movement into separate piston-mode movements, thereby enhancing wave energy capture.

2.2.2 Field Methods

Field methods provide various techniques and tools for solving the Navier-Stokes equations, which are essential for simulating wave-structure interactions in ocean and coastal engineering. These methods are categorized into Lagrangian, Eulerian, and Lagrangian-Eulerian combined techniques. Modelling wave-structure interactions poses significant numerical challenges due to highly deformed, moving interfaces with complex physics such as wave breaking and fluid turbulence. Additionally, intricate interactions occur between waves and both rigid and flexible structures, as well as non-linear interactions between fluids and solids with different physical properties. Eulerian methods, which depend on a mesh grid, have been extensively developed and applied, with tools like *OpenFOAM*, *Star-CCM+*, *REEF3D*, *ANSYS/Fluent*, and *ANSYS/CFX* being prominent examples.

The use of *OpenFOAM* has significantly advanced the study of hydrodynamic loads on offshore structures through various complex interactions and modelling techniques. Researchers have effectively investigated wave-structure interactions, including the examination of nonlinear wave fields and amplification harmonics around fixed tandem cylinders, as well as the development of improved screen force models for hydrodynamic loads on net panels for fish cages and offshore fish farms (Mohseni & Guedes Soares, 2022a, 2022b; Wang et al., 2022b). Coupling methods between potential and viscous flow solvers have enhanced computational efficiency while accurately capturing hydrodynamic loads (Robaux & Benoit, 2022). Studies on floating structures have validated models for floating body motion with mooring dynamics, providing critical insights into the dynamic responses of barge-type platforms and the significance of nonlinear drag forces (Chen & Hall, 2022; Otori et al., 2023). Hydrodynamic coefficients and induced flows around heave plates have been determined with high accuracy (Pinguet et al., 2022). Empirical and numerical methodologies, such as the Morison approach and domain decomposition strategies, have been assessed for predicting hydrodynamic forces (Aliyar et al., 2022; Clément et al., 2022). High-fidelity Computational Fluid Dynamics (CFD) models have been created for WECs in extreme wave conditions and dual-chamber oscillating water column devices, demonstrating robustness (Gadelho & Guedes Soares, 2022; Katsidoniotaki et al., 2023). Additionally, research into wave loading on TLP-type floating wind turbines and wake interactions between spar-type floating offshore wind turbines has underscored the capability to predict dynamic responses and aerodynamic loads (Mohseni & Guedes Soares, 2024). Moreover, novel methodologies based on *OpenFOAM*, such as *qaleFOAM* for simulating focusing waves and their interactions with fixed/moving cylinders, and the *FOWT-UALM-SJTU* solver for coupled aero-hydrodynamic simulations of Floating Offshore Wind Turbines (FOWT), have been developed to address specific challenges in modelling complex offshore systems. (Gong et al., 2021; Huang et al., 2023b; Yu et al., 2023c).

Similarly, the application of *STAR-CCM+* has provided significant insights into the hydrodynamic loads on offshore structures. Researchers have extensively explored wave-structure interactions and the effects of platform motion on the aerodynamic performance and wake properties of FOWTs, employing advanced techniques such as Dynamic Fluid Body Interaction (DFBI) and Volume Of Fluid (VOF) methods(Alkhabbaz et al., 2024; Califano et al., 2023). The influence of hydroelastic effects on wave-induced forces and structural responses has been highlighted through flexible structure modelling and fluid-structure interaction (FSI) simulations, demonstrating the significant impact of hydroelasticity on the overall load and bending moments of monopile structures (Gu et al., 2023; Thome et al., 2023). Investigations into nonlinear difference-frequency wave loads on semi-submersible FOWTs have underscored the importance of accurate uncertainty quantification and the impact of higher harmonics on overall hydrodynamic loads (Wang et al., 2021a; Zeng et al., 2023). Advanced aero-hydrodynamic analyses have demonstrated the benefits of novel biomimetic and fractal designs in improving the stability and efficiency of semi-submersible platforms, as well as optimizing WECs integrated with floating wind turbines, showing significant improvements in power generation efficiency through heaving viscosity damping correction (Huang et al., 2023a; Zhang et al., 2022f). The hydrodynamic responses of floating platforms under irregular waves have been validated using coupling approaches between High-Order Spectral (HOS) and CFD models, enhancing the accuracy of low-frequency responses and improving the predictive capability for extreme wave conditions (Zhang et al., 2023f) (Fig. 9).

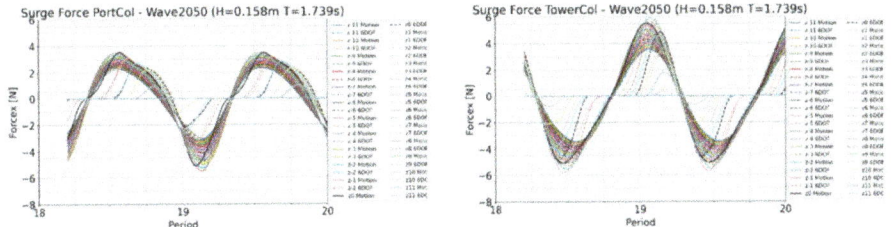

Fig. 9. Comparison of the sectional surge force on the port (left) and tower (right) column of a FOWT floater in Califano et al. (2023)

Building on the strengths of these established tools, *REEF3D* and *ANSYS/Fluent* have also been instrumental in advancing hydrodynamic studies of various offshore structures. *REEF3D* has been employed to investigate hydrodynamic characteristics and responses under oscillatory flow conditions, providing insights into parameters such as blockage ratios and Keulegan–Carpenter numbers on square cylinders (Dutta et al., 2022). It has also been used to analyze the dynamic responses and mooring forces of submersible steel-frame offshore fish farms under regular wave conditions, highlighting the sensitivity of surge motions to wave parameters and the effects of structural variations on motion responses and mooring forces (Wang et al., 2022c). *ANSYS/CFX* been also applied to assess the impact of wind and wave loads on floating photovoltaic systems, where numerical simulations and finite element analysis were conducted to evaluate stress distributions and ensure the structural integrity of these systems under various conditions (Choi et al., 2023).

In addition to commercial and open-source tools, in-house codes have also played a crucial role in advancing our understanding of hydrodynamic behaviour. A hybrid model combining fully nonlinear potential flow and Navier-Stokes equations has been developed to simulate complex wave-structure interactions in large domains, achieving good agreement with experimental data and showcasing the advantages of such hybrid models (Saincher & Sriram, 2022). Another study utilized a two-dimensional RANS CFD model with a VOF method to investigate hydrodynamic interactions between adjacent floating bodies, highlighting the sensitivity of hydrodynamic coefficients to the displacement of floaters (Jiang et al., 2024). Additionally, the hydrodynamics of a submersible steel-frame offshore fish farm in regular waves was analyzed using an open-source CFD code coupled with an in-house FEM code, demonstrating the accuracy of the coupled CFD-FEM approach in predicting the response of floating offshore wind turbines under regular wave conditions (Wang et al., 2022h).

Furthermore, particle-based methods have significantly advanced the study of wave-structure interactions and hydrodynamic loads on offshore structures. The particle-based method has been enhanced for 3D simulations, enabling accurate modelling of wave interactions with fixed structures through a hybrid approach combining potential and viscous flow models such as Meshless Local Petrov-Galerkin method with Rankine source function (MLPG_R) model (Agarwal et al., 2021). The SPH framework has been validated for various applications, including the DeepCwind semi-submersible floating platform, demonstrating high fidelity in predicting hydrodynamic loads and capturing nonlinear behaviours under different wave conditions (Tagliafierro et al., 2023). SPH simulations have also accurately modelled the interactions between steep and breaking waves with vertical cylinders, revealing complex relationships between wave phases and peak forces (Yang et al., 2023d). Further, SPH studies on sea access roads and floating boxes have provided detailed insights into horizontal and uplift wave forces, validating the method's accuracy against experimental data (Chen et al., 2022f; Vineesh & Sriram, 2023). The SPH method has shown potential for full-scale applications, accurately simulating the hydrodynamics of floating offshore wind turbines and capturing the dynamic responses and mooring line forces under various wave conditions (Tan et al., 2023). Additionally, SPH modelling of wave interactions with multi-float structures has offered valuable insights into the hydrodynamic performance and structural responses, demonstrating its capability in engineering practice (He et al., 2023).

Finally, the Boussinesq equation has been effectively applied to study hydrodynamic loads and wave interactions with offshore structures, leveraging its ability to handle nonlinear and dispersive wave effects efficiently over large domains. A mathematical model using higher-order Boussinesq equations was developed to analyze solitary wave interactions with a floating pontoon, validated against CFD simulations using *OpenFOAM* and *OceanWave3D*, demonstrating good agreement and highlighting the model's ability to capture complex wave profiles and pressure distributions (Mohapatra et al., 2022). A hybrid modelling approach that couples a finite-element Boussinesq model (FEBOUSS) with a Meshless Local Petrov-Galerkin model (MLPG_R) for Navier-Stokes equations efficiently simulates wave-structure interactions, especially for directional waves, by combining the strengths of both models (Agarwal et al., 2022). The near-trapping phenomenon in a four-cylinder array was investigated using a Boussinesq model with a cut-cell technique, accurately predicting high free surface elevations and wave loads, which are critical for the safety assessment of offshore structures (Ning et al., 2022).

2.2.3 Wave-Current Interaction

Floating structures are affected by various environmental factors such as waves, currents, and winds. In many cases, the hydrodynamic performance of floating structures has been focused only on waves. Nevertheless, in recent years, the current effects have been thoroughly studied in the literature.

The importance of considering wave-current interaction effects of wave loads on cylindrical bodies has been investigated (Ghadirian et al., 2021; Xin et al., 2023; Yang et al., 2021). It was shown that waves in the presence of opposing currents are causing a nonlinear effect of the free surface elevation and inline force component. In contrast, current colinear waves are the most linear, causing a wave height decrease with increasing current velocity. Furthermore, the coupled effects between surface waves and subsurface current play a more important role in the hydrodynamic loads than the surface elevation. A higher-order FEM has been applied by Huang et al. (2022b) to analyze the nonlinear wave resonance generated by oscillations of twin cylinders in a uniform current. They concluded that the resonance which is caused by wave-current effect occurs at all first-order resonant frequencies for both symmetric and antisymmetric motions of the cylinders.

Offshore Wind

Offshore wind energy harnessing is a major renewable energy technology. During the design process of an offshore Wind Turbine (WT), it is necessary to properly predict the energy yield and structural loads. Offshore WT foundations can be categorized as gravity-based foundations, suction caissons, monopiles, and floating structures with mooring systems. Of these foundations, monopiles have been the most widely used for shallow water, either as isolated bodies or as part of Jacket foundations. Regarding their wave-current interaction phenomena, numerical models taking into consideration different wave-current interaction angles and focusing on the stress inside the seabed and the liquefaction area around the pile have been considered in the literature (Lin & Guo, 2022; Wei et al., 2022). The Reynolds-averaged Navier-Stokes equations with the k-ε turbulence model have been used, and Biot's poro-elastic theory has been applied to simulate the transient seabed response. As far as floating structures with mooring systems are concerned, the nonlinear stationary response of a spar-type WT under current and wave interactions has also been examined (Silva et al., 2021). These interactions have been estimated using statistical quadratization theory, which describes the nonlinear system as an equivalent polynomial of quadratic order. The main advantage of the theory is the low computational cost, which is lower than corresponding time domain simulations while producing accurate estimations of the platform's responses. Additionally, a semi-submersible floating platform, encompassing either a WT or a WT and a fish cage, under the actions of wind, current and wave loads has been studied (Fan et al., 2023; Zou et al., 2023). The hydrodynamic responses of the platform, under the action of aero-hydrodynamic coupling, have been evaluated, highlighting their interaction with the aerodynamic loads of the blades. Concerning the effect of the fish nets on the platform's displacements, these can suppress the low-frequency motions but have almost no influence on the wave-frequency motion. A vertical cylindrical body can also be applied as a monopile for offshore wind turbine support. Buljac et al. (2022) determined the environmental load effects on a monopile concurrently subjected to wind, wave and currents. They showed that the wind predominantly causes surge direction structural loads at the natural frequency of the wind turbine, whereas currents enhance the dynamic loads at low frequencies due to nonlinear hydrodynamic interactions.

Aquaculture

Fish cages, which operate individually or as an array, are increasingly being considered to be deployed in open sea areas to have higher water quality and fewer conflicts with other maritime activities. Therefore, the development of a fish farm which can withstand the current and wave loads is a prerequisite for the industry. The wave-current interactions with the mooring system of a fish cage have been studied in Cheng et al. (2021) and Liu & Guedes Soares (2023). These analyses concluded that for a current velocity of less than 0.5 m/s, the mooring forces are still within the safe range even if a mooring line breaks, whereas in order to avoid cage volume reduction due to large current velocities, the weight of the bottom ring of the cage and/or the net solidity should be selected appropriately.

Nevertheless, an increase in the ring's weight results in larger mooring loads, whilst a decrease in the net solidity may lead to the fish escape. Regarding the type of netting materials and their connections, numerical studies have been conducted focusing on the axial forces on the connection points between the net and ropes/steel connectors (Føre et al., 2022; Hu et al., 2022b; Slagstad et al., 2023; Xie et al., 2023b). Findings from these studies include that for large stiffness of the connectors, the vibration displacement of the net becomes small. On the other hand, with small stiffness, the motion of the rope becomes significant which increases fatigue damage.

Tidal Energy

Tidal energy is a renewable and predictable energy source for which pre-commercial type models and full-scale devices have been developed. Recently, many studies have focused on various forms of tidal current turbines. Specifically, attention has been paid to tidal rotors undergoing prescribed uncoupled surge, heave, and pitch motions in the presence of regular waves concluding that the heave motion reduces the mean power coefficient by about 9%, whereas the pitch motion causes periodic fluctuations in the damping coefficient of the power coefficient (Huang et al., 2022a; Wang et al., 2022g). Also, several studies have reported that the frequency and amplitude of the surging motion can increase the power output of a horizontal-axis tidal turbine (Wang et al., 2021c; Zhang et al., 2023e). Furthermore, Zilic de Arcos et al. (2023) applied a coupled model for the force-motion interactions between a tidal rotor and a floating platform. Their analysis showed load fluctuations due to phase interactions between waves and platform motions, which, at certain ranges of wave periods, can cause a reduction in fatigue damage and improve the quality of power delivery. A parametric study on the dynamic loads of a tidal current turbine with and without considering wave-current interactions, using higher-order theories coupled with a modified BEM, has been developed by El-Shahat et al. (2021). From their analysis, it was concluded that maximum loading ranges were 170% of the average out-of-plane bending moment and 99% of the average in-plane bending moment without interactions, whereas with wave-current interactions, these values were reduced to 110% and 69% for out-of-plane and in-plane bending moments, respectively. In addition to the developed tidal turbines, a hybrid wave and current energy converter has been presented to harvest energy from both ocean wave and current simultaneously (Chen et al., 2022d). This device can improve the electric power output by 38% - 71% for regular- and 79% for irregular- waves compared to a WEC.

Submerged Tunnels

Submerged floating tunnels are innovative deep-water transportation infrastructures for crossing waterways. Compared to coastal bridges subjected to strong winds and waves, submerged tunnels have a more significant potential for development in a strait or bay area with high tropical cyclone activity (Chen et al., 2023b; Xu et al., 2023). However, like other coastal structures, floating tunnels are subjected to wave-current actions. In this regard, several studies involving segment tests and numerical simulations were conducted to examine the motion response of a submerged floating tunnel under wave-current actions (Jeong et al., 2022; Won et al., 2021; Yang et al., 2023e; Zhou et al., 2023a). The horizontal response of a submerged floating tunnel is significantly amplified when the wave force direction is the same as the direction of the current force, whereas non-uniform terrain may mitigate the displacement distortion at both ends of the tunnel (Xu et al., 2023). Furthermore, with respect to combined earthquake and wave-current actions, recent studies have shown that the response of the submerged tunnel is strongly affected by the direction of the earthquake action (i.e., horizontal earthquake action increases the responses of the system compared to the vertical). Additionally, waves can weaken the tension amplification of earthquakes by up to 10% and increase the maximum horizontal acceleration of the tunnel by 40% (Luo et al., 2021; Wu et al., 2021).

Floating Bridges

Floating bridges are often located in exposed areas and are subjected to combined environmental loads. Analyses have shown that the hydrodynamic loads of wave-current combined flow are much larger than those of the wave-only flow, while when waves and currents are travelling in the same direction, the responses of the bridge are significantly amplified (Huynh et al., 2023). Specifically, under 1-year load cases, the amplifications in the standard deviations of the acting loads and bending moments are 30%–52%. On the other hand, under 100-year load cases, these values are doubled (Dai et al., 2022). Huang et al. (2022g) concluded that the hydrodynamic pressure on the structure caused by the earthquake increases with increasing water depth, whereas the combined effect of seismic and waves differs from that of only seismic action. Hence, the influence of the travelling wave effect cannot be ignored. It should be noted, however, that the vertical displacements of the bridge pontoons caused by heavy moving loads (vehicle loads) are significantly larger than those caused by environmental loads (Miao et al., 2021). In addition, several studies have been presented recently which are focused on basin tests of a scaled floating bridge model under the combination of wave, current, and wind loads (Rodrigues et al., 2022; Viuff et al., 2023; Xiang et al., 2023). From the experimental analysis, it was concluded that, generally, the current reduces the bridge's pontoon motion responses related to the horizontal motion of the bridge, while the vertical pontoon motions are amplified. Figure 10 shows floating bridge experimental model tested by Rodrigues et al. (2022). The model was subjected to concurrent waves and current and prescribed wind loads time series. The wind pre-computed loads were obtained from numerical simulations and applied using cables connecting each pontoon to an individual actuator.

Mooring Buoys

Cylindrical and spherical buoys have also been used as catenary anchor leg mooring buoy structures which interact with waves, underwater currents, and wind. Amaechi et al. (2022) conducted a numerical study to evaluate the buoy's motion responses. Regarding the effect of the current speed on the body's hydrodynamics, it was concluded that the buoy's motions are influenced by the increase of the current, which also increases the effective tension in the mooring lines.

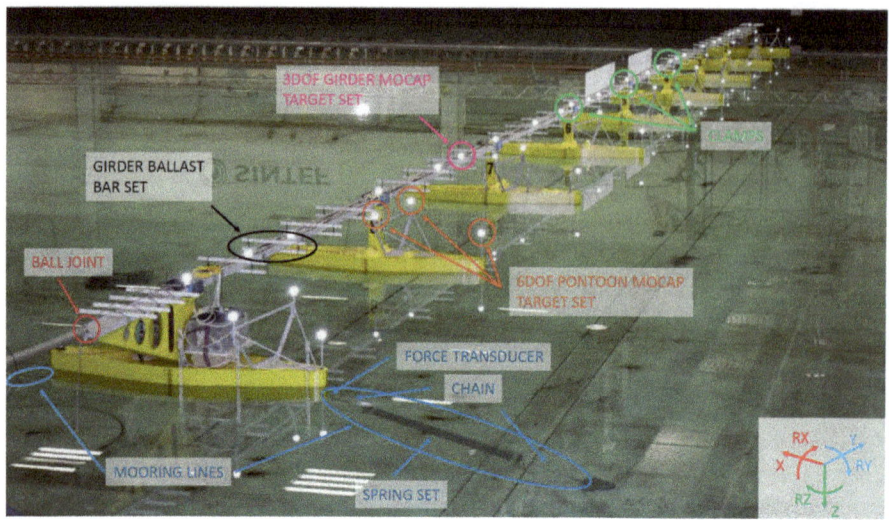

Fig. 10. Photo of the floating bridge experimental model from Rodrigues et al. (2022)

2.2.4 Airgap and Wave-in-Deck Loads

The structural safety of offshore platforms is critically dependent on understanding and managing of the airgap and Wave-In-Deck (WID) loads. Recent studies have mainly focused on quantifying these factors to enhance platform design and safety protocols. This review synthesizes the findings from multiple studies conducted between 2021 and 2024, highlighting advancements in modelling and risk assessment procedures related to WID in offshore structures.

Airgap refers to the vertical distance between the lowest part of an offshore platform's deck and the crest of the highest expected wave. This gap is critical for preventing WID occurrences, which can lead to severe structural damage. WID loads are the forces exerted on the platform's deck when waves exceed the airgap. These loads can be extremely high, especially during storms or in regions with harsh ocean conditions. The impact can be divided into three phases: the initial impulsive force phase, an oscillatory phase, and a subsequent negative pressure phase. These phases can cause both global and local structural damage, including denting, cracking, and, in severe cases, platform collapse.

A Complex Phenomenon

WID loads are inherently highly non-linear, not only due to the complex nature of the wave forces but also because of the intricate fluid-structure interaction caused by the irregular geometry of the structures subjected to these loads, such as the topsides of oil and gas rigs. As a result, simplified methods are often employed in practice, as outlined in standards like the silhouette method from the American Petroleum Institute recommended practice (API, 2014).

However, these procedures are extremely conservative, potentially leading to an overestimation of the loads on structures by as much as 30%–75%, as reported by Wang et al. (2023c). This emphasizes the value of applying high-fidelity numerical methods, such as Smoothed Particle Hydrodynamics (SPH), to capture better the complex nonlinearities involved. These advanced methods can effectively handle phenomena like air compressibility effects (de Oliveira Costa et al., 2021), the influence of platform deck appendages such as balconies (Fonseca de Carvalho e Silva et al., 2023), and the accurate modelling of freak waves (Huo et al., 2021; Luo et al., 2022). Interestingly, the application of data-driven methods remains largely absent in direct WID load predictions.

One ought to keep in mind that the ultimate goal of assessing WID loads and conducting airgap studies is to ensure the structural integrity of the offshore unit while protecting its crew and cargo in the most efficient way possible, typically by reducing costs. While this may seem obvious, it is crucial to recognize that the assessment must be robust enough to allow for generalization and quick design iterations based on its findings, yet not overly conservative, as this could lead to unnecessary waste of resources and materials. With this approach in mind, some researchers have revisited traditional procedures – "recipes" for WID loads assessment, which are found in recommended practices (API, 2014; DNV-GL, 2019), among others – to develop more efficient and robust methods.

Procedural Aspects of Assessment of WID Loads

The accuracy and effectiveness of the methods applied are crucial for the industry, particularly for operators, in decision-making processes and in outlining action items as part of their business risk management, as noted by Azman et al. (2021). These authors employed pushover analysis, also known as Ultimate Strength Analysis, to determine the Reserve Strength Ratio (RSR) of an offshore platform, incorporating WID loads into the analysis in accordance with ISO (2020). Similarly, Lihin et al. (2022) proposed a revised methodology for assessing fixed offshore structures under WID loading, using the silhouette method calibrated with field measurements from a structural health monitoring system installed on the structure. To improve the reliability assessment of offshore structures concerning WID loads, Ma & Swan (2023b) presented findings that enhance the physical understanding of waves occurring in realistic design sea-states, highlighting that many 'design wave events' will involve breaking waves, regardless of water depth. Their analysis of a large database of WID events revealed that recommended practices consistently under-predict the maximum WID loads. Additionally, these researchers argue that their recently developed Lagrangian Momentum Absorption (LMA) model provides highly accurate predictions by integrating fully nonlinear wave inputs with the structural openness/porosity (Ma & Swan, 2020).

Application of Nonlinear Potential Flow Methods

To support further advancements in assessment methods and procedures, various approaches have been employed in recent years to calculate air gap responses and WID loads. Higher-order potential field methods remain highly relevant for the numerical modelling of these problems. For example, Wang et al. (2022e) used the Green-Naghdi potential flow theory to study the airgap of a structure up to the second order. Similarly, Ren et al. (2021) conducted a study on wave surface elevation around a multi-column offshore structure, investigating wave run-up and peak surface elevation caused by wave interaction with rounded-corner square columns of different corner ratios. Additionally, Wei et al. (2023) compared the Newman approximation with the Pinkster approximation for computing second-order wave loads, which influence the air gap calculation for a semi-submersible platform.

In the specific case of FOWTs, Arya & Ranjan Behera (2023) studied freak wave impacts on a SPAR platform using a potential flow theory-based numerical model in the time domain. The motions of the SPAR FOWT and airgap effects were analyzed using the *FAST* software, although the study did not include the effects of non-linear wave loads, viscosity, or drag force. Despite these limitations, the study provides a valuable foundation for further development.

Field Methods

The application of field methods for calculating airgaps remains a prevalent research focus, particularly concerning WID flow and the corresponding loads in wave-structure interaction problems. Numerical formulations based on Smoothed Particle Hydrodynamics (SPH) are particularly popular due to their meshless nature, which is advantageous when dealing with discontinuous fluids. For instance, Fei et al. (2024) derived a relationship between wave forces and characteristic velocity during wave impacts on deck structures using SPH-based simulations. Similarly, Sasikala et al. (2021) employed Weakly Compressible Smoothed Particle Hydrodynamics (WCSPH) with two types of boundary conditions—Repulsive and Dynamic—to solve the slamming problem on a horizontal deck.

In general, Computational Fluid Dynamics (CFD) studies, including but not limited to SPH, are often validated against experimental data, with observed errors ranging from 3% to 15%, as noted by (Nizamani et al., 2022). These studies typically focus on detailed geometries of existing designs or case studies involving simple boxes or plates. They almost invariably conclude that existing standards are conservative regarding WID loads and offer alternative formulations for key variables.

Several examples illustrate the current state of the art. Deng et al. (2023b) investigated slamming on the wave-ward side of a deck box on a semi-submersible subjected to an extreme wave, finding the loads to be much lower than standard predictions. Yao et al. (2023b) found a correlation between the significant incident wave height on a mooring-stabilized semi-submersible and the peak pressures experienced by the platform, consistent with findings from Huo et al. (2021), who reported a dramatic increase in wave slamming pressure, particularly for freak waves.

In many cases, a dam break problem is used to approximate WID loads on topsides or in geometries suitable for calibrating numerical models, as demonstrated by Mu et al. (2023). However, this approach is not always appropriate for modelling WID (Hernández-Fontes et al., 2021). Zago et al. (2023) emphasized the need to distinguish between two main classes of flow, each requiring distinct modelling approaches. The first class involves an essentially two-dimensional flow with a single wave front traversing the entire deck, which can be modelled using a linear wave theory-based dam break approach to predict the type and propagation of the overtopping wave. The second class, involving multiple wave fronts originating from all sides of the platform, is better handled using a 3D SPH method.

Model Tests

The required test durations for achieving converged short-term wave and impact extreme value statistics are crucial factors in designing model test campaigns, as recently investigated by (Scharnke et al., 2023). Model tests are typically used to validate or calibrate numerical solvers for specific configurations, especially concerning extreme sea states. They also serve to explore novel phenomena or designs and to draw conclusions about the parameters governing these phenomena.

For the first type of test, several recent studies are noteworthy. Fonseca de Carvalho e Silva et al. (2023) analyzed loads on a balcony structure subjected to extreme wave impacts, investigating alternative configurations that could attenuate these impact loads (see Fig. 11). Studies on flare barriers and wave run-up (Liu et al., 2022b, 2023a), along with experimental validations (Nizamani et al., 2022; Yao et al., 2023b; Zago et al., 2023), also stand out in this context.

Regarding the exploration of novel phenomena, notable examples include the findings of Hernández-Fontes et al. (2021) on flow classification, Guo et al. (2021b) on the classification and comparison of wave impact modes on semi-submersibles, Mohajernasab et al. (2023) on the effects of air pockets and Ma & Swan (2023a) on the influence of topside porosity.

Additionally, it is important to highlight some studies that are significant for their comprehensive test campaigns or their focus on real structures, regardless of whether they fit into the aforementioned categories: AlMashan et al. (2021), Nam et al. (2022), Zhang et al. (2022), and (Luo et al., 2022).

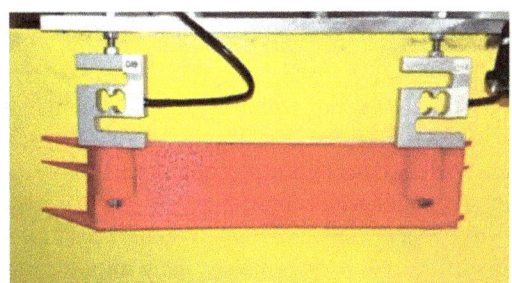

Fig. 11. 3D printed mooring balcony attached to the scaled model for wave tank experiments vie load cells to capture impact loads in Fonseca de Carvalho e Silva et al. (2023)

2.2.5 Mooring System

Station-keeping systems are required for floating structures to limit their excursion and orientation of the structures under the action of wave, current, and wind environmental loads. Several studies have been presented in the literature in recent years, focusing on the design, analysis, and simulation of different types of mooring systems.

Moored FOWTs form the leading innovation in the ocean energy sector. Shared mooring lines are a method to reduce the station-keeping cost of floating farms due to a reduced number of anchor deployments. Specifically, the mooring lines situated in the interior of the farm connect directly between adjacent floating platforms rather than extending to individual anchors (Wilson et al., 2021). Shared mooring systems can exhibit higher platform offsets compared to individual ones (Gözcü et al., 2022; Munir et al., 2021), whereas the responses of the system cannot be independent due to the hydrodynamic coupling between the floating structures. However, shared moorings can offer some benefits in fatigue and failure tolerance compared to a conventional mooring approach, which is captured under mooring coupled dynamic analysis since quasi-static analysis may lead to overestimation of the durability of shared moorings (Lozon & Hall, 2023; Zhang & Liu, 2023). As far as isolated floating FOWTs are concerned, recent studies have focused on the NREL 5MW OC4 semi-submersible FOWT (Huang & Yang, 2021; Liu et al., 2022a; Neisi et al., 2022; Xu et al., 2021b) with different mooring line materials, mooring components (clump weight and buoy) and anchors (drag embedment anchor and suction anchor) under various conditions of wind gust and average wind speeds. In addition, Zong et al. (2023) conducted a sensitivity analysis on the structural damage and deformation of the floating WT under bow and side impact from vessels. They concluded that although the FOWT may not be critically damaged by ship impact, the equipment inside may fail to work properly.

Moreover, the IEA 15MW WT mounted on the UMaine VolturnUS-S reference platform was also examined in the literature (Chueh et al., 2023; Wang et al., 2023b). Specifically, attention has been paid to effectively improving the survivability of the WT by proposing a novel constant tension mooring system, instead of a conventional catenary chain, for shallow waters and a power-control strategy which reduces the displacements of the platform. Regarding recent advances in the simulation and analysis of floating FOWTs, the developed models deal with the wind, wave, mooring dynamics, platform motions, and turbine structural dynamics. Coupled aerodynamic-hydrodynamic modelling has commonly been undertaken using fully coupled aero-hydro-servo-elastic-mooring tools *FAST* and *OpenFAST* developed by NREL which are applied either to numerical software, i.e., *ANSYS AQWA, WAMIT, OrcaWave, DNVGL WADAM* etc., or *STAR-CCM+* and *OpenFOAM* CFD simulation tools (Pillai et al., 2022; Ramzanpoor et al., 2023; Subbulakshmi & Verma, 2023; Wang et al., 2022d; Yan et al., 2023). Nonetheless, other hybrid potential-viscous flow models have been developed with excellent correlation with commercial software (Yu et al., 2023c). Also, Ferri et al. (2022) have developed a numerical model to optimize platform and mooring system characteristics towards the reduction of the dynamic response of a 10 MW FOWT.

The investigation of the technological and economic feasibility of multi-purpose floating structures, which combine offshore wind turbines with WECs, is a promising alternative for reducing the levelized cost of energy and enhancing the performance of renewable technologies. Although offshore multi-purpose floating systems have not yet achieved comprehensive commercial maturity, the feasibility of combining a floating wind turbine and a WEC has already been investigated. Specifically, a floating structure which encompasses an array of Oscillating Water Column (OWC) devices, moored with tension tethers, and supports a 10MW FOWT has been numerically simulated using an in-house developed servo-hydro-aero-elasto-dynamic formulation for installation in a North Sea or Mediterranean Sea location (Konispoliatis et al., 2021, 2022b). The same hybrid FOWT has been considered by Kardakaris et al. (2023) with a catenary chain line instead of a TLP mooring system to examine the effect of the mooring type on the system's power efficiency. Also, the chain mooring characteristics of a hybrid system which encompasses a 5MW WT and three-point absorbers WECs have been examined by Men et al. (2023). Their analysis showed that although the mooring tensions at the windward side chain increased, they did not seem very sensitive to the converter's size.

Apart from steel-wired mooring systems, hybrid lines composed of synthetic fibre and steel ropes (Yu et al., 2023a) as well as fully synthetic lines (da Cruz et al., 2023;

Nguyen & Thiagarajan, 2022; Pillai et al., 2022) have been considered for mooring applications in offshore structures. Polyester fibre ropes offer a reduced cost per breaking strength and better fatigue properties than steel lines. However, in damaged conditions, the residual strength of ropes decreases with the increase of the damage level (Lian et al., 2022). Regarding the mechanical characteristics of the fibre rope, the focus has been given to rope stiffness under empirical and neural network predictions (Zhang et al., 2023b) and different model practices (Sørum et al., 2023a). When compared to polyester moored systems, nylon ropes have the potential to reduce the mooring cost by increasing their fatigue life and reducing their cross-section area compared to fibre lines (Sørum et al., 2023b; Verde & Lages, 2023), whereas they (nylon moorings) also allow greater motion responses which interfere optimally with wave energy harvesting activities (Depalo et al., 2022; Xu et al., 2021a, 2021c; Xu & Guedes Soares, 2023).

Regarding wave energy absorption, recent studies have focused on the effect of mooring types (i.e., taut, catenary, TLP) on wave power efficiency. Concerning floating point absorber WECs, TLP moorings are proven to be more suitable for long-wavelength installation locations, whereas taut moorings seem suitable for locations dominated by short and medium wavelengths (Meyer et al., 2023). Regarding oscillating water column (OWC) devices, taut moorings have attained the best performance, followed by the TLP and the catenary moorings (Gubesch et al., 2022b). However, TLP moorings can be prone to a brief loss of tension, leading to large snapping forces of short duration on the tendons, a phenomenon presented by Miškov et al. (2023). A taut mooring line is also qualified to moor an attenuator-type WEC. Nevertheless, elastic mooring cables seem preferable to inelastic ones due to their ability to reduce the snap loads of the converter (Stansby et al., 2022; Zhao et al., 2023). Nonetheless, independently from the specific nature of the converter and the mooring type, a crucial aspect in wave energy extraction regards the implementation of optimal control strategies. In this context, model-based control techniques towards the maximization of power capabilities have been performed by Papini et al. (2023), Windt et al. (2021), and Carapellese et al. (2022).

Moored porous structures such as fish cage systems and permeable floating bodies have been dealt with in the realm of potential flow theory using a Boundary Element method (BEM) to simulate the wave effect and the discrete lumped-mass model for the determination of the mooring dynamic behaviour (Konispoliatis et al., 2022a; Ma et al., 2022b, 2023). Here, the mooring line is discretized into a number of finite elements, the forces on which are determined and assembled into a symmetric banded global system. Due to the increased interest in CFD modelling, which typically solves the fluid flow problem using Navier-Stokes equations, several authors have studied the importance of including viscosity in moored structures. Specifically, focus has been given to *OpenFOAM* and its coupling with *MoorDyn* (Aliyar et al., 2022a; Chen & Hall, 2022; Jiang & el Moctar, 2023) as well as with improved *MoorDyn* solvers for nonlinear effects simulation (Jeon et al., 2023). In addition, apart from commercial software, in-house FEMs have been incorporated with CFD (Huang et al., 2022c)), allowing accurate solutions for dynamic interactions between structures and their mooring systems. Nevertheless, although high-fidelity CFD models attain the best agreement with measured experimental data, low-fidelity linear wave theory models can predict values within the same order of magnitude of CFD and experiments with much lower computational time (Engström et al., 2023).

Furthermore, the research was also focused on moored floating breakwaters. These are built to protect water areas from wave attacks. Compared to the bottom-seated breakwater systems, floating breakwaters have an easier installation, are less expensive in deep water applications, and are less dependent on the seabed conditions. Regarding the applied methodologies for the analysis and simulation of moored breakwaters, mesh-based CFD models (Chen et al., 2022c; Wei & Yin, 2022), Boundary Elements Method (BEM) (Liang et al., 2022b), and Smoothed Particle Hydrodynamics method (SPH) have been applied for the simulation of the free surface, the flow and the wave breaking. The latter method was widely adopted due to the flexible free-surface boundary condition. In this approach, the fluid is discretized into a set of Lagrangian particles, which are entitled with physical quantities such as density, pressure, etc. (Cui et al., 2022; Han & Dong, 2023). However, the solid boundary implementation is adversely affected by the Lagrangian nature and kernel interpolation. Hence, various algorithms have been proposed for floating breakwaters and have been verified experimentally. The fictitious particles method has been developed to construct solid boundary by placing fictitious (or virtual or ghost) particles in the solid domain (Chen et al., 2023c). Each fixed ghost particle corresponds to an interpolation point in the fluid domain, the velocity and pressure of which are evaluated from the fluid particles around the interpolation point. Unlike the latter approach, the dynamic boundary particles (DBP) keep stationary relative to the solid boundary and participate in the same equations of state and continuity as fluid particles. A modified DBP was implemented in a floating breakwater with taut, slack and hybrid mooring systems (Guo et al., 2022).

Lastly, fatigue assessment of mooring systems involves considerable uncertainties with respect to both load and capacity. To better understand and quantify the fatigue capacity and degradation of mooring lines, several original fatigue tests on mooring chains of different material grades under artificial or natural corrosion conditions, with or without cathodic protection, have been performed in recent years to study the effect of the corrosion condition on fatigue resistance (Aursand et al., 2023; Lone et al., 2023; Qvale et al., 2021, 2022; Zhang et al., 2022h). The results proved that the corrosion has a detrimental impact on fatigue resistance, whilst fatigue capacity increases as the mean loads are reduced. Furthermore, extended S-N curve formulations have been proposed by Mendoza et al. (2022)), accounting for not only prior and future fatigue loads but also mean load and corrosion effects (Lone et al., 2022a, 2022b). Also, the mooring fatigue performance of a typical semi-submersible platform using chain and wire ropes has been numerically simulated by He et al. (2022) and Yang et al. (2023a)), who highlighted the decrease of fatigue life under the increase of corrosion rates. Apart from fatigue analysis, the failure of a mooring line is of utmost concern due to uncertainty in the line's behaviour and lack of structural robustness. Mooring line failure conditions have been recently examined for offshore fish cages (Hou et al., 2022; Tang et al., 2022a), wind turbine structures (Sun et al., 2023b; Zhang et al., 2022b), semi-submersible (Mao et al., 2022), and TLP platforms (Wu et al., 2022). Probabilistic assessment methodologies and deep neural network approaches have been developed, indicating that the failure of one mooring line significantly affects the probability of failure of the remaining mooring lines due to the increase of the structure's motions and tensions in the remaining lines. Aiming to reduce the number of short-term simulations needed to evaluate multi-dimensional fatigue damage integral, a univariate dimension-reduction method has been developed by Ibarra et al. (2022) that transforms the environmental conditions into standard Gaussian variables for fatigue damage determination. Also, Shao et al. (2023) studied the power performance of a wave park of two types of heaving WECs and the accumulated fatigue damage of their mooring lines, using relative tension-based fatigue analysis and rainflow counting methods. Their analysis has focused on the effect of the wave heading angle and the arrangement of the lines towards the incoming wave train in the amount of fatigue damage accumulated by mooring lines.

2.2.6 Model Tests

Advanced model testing remains an indispensable approach in studying wave loads on offshore structures. These empirical assessments provide crucial insights that directly inform the iterative design and validation of theoretical models and engineering practices. Incorporating these empirical results into data-driven methodologies underscores their critical role in the evolving technological landscape of marine engineering.

Research on classical offshore platforms has focused on the hydrodynamic impacts of wave-induced slamming and breaking waves, which is critical for understanding structural loads during extreme sea states. For instance, an investigation into the effects of freak waves on floating platforms through a series of flume experiments examined various air-gap conditions, identifying four distinct wave impact patterns that revealed how wave interactions with the platform could result in significant motions, tension forces in tethers, and impact pressures at various points on the structure (Luo et al., 2022). Further research explored the interaction between internal solitary waves (ISWs) and FPSOs, focusing on oblique wave impacts. Through experimental testing combined with numerical simulations, the primary forces exerted by ISWs were identified, leading to the development of a simplified prediction model. This study provided regression formulas for the friction coefficient and correction factors, crucial for accurately predicting wave-induced loads on FPSOs, thereby offering valuable data for improving FPSO design and operational strategies in ISW-prone environments (Zhang et al., 2022). Additionally, a comprehensive experimental study under the Norwegian SLADE KPN research project simulated steep, breaking waves to analyze their impact on offshore platforms and wind turbines, focusing on local structural responses by measuring pressure and strain to characterize hydroelastic effects. These findings are instrumental in refining safety standards and design guidelines for offshore structures subjected to such conditions (Ahani et al., 2022).

Research has highlighted the importance of mooring system integrity under varying environmental loads in FOWTs, with models for those needing to be properly validated. For example, flume experiments on a 5 MW barge-type FOWT under mooring line failure scenarios showed that traditional models often overestimate dynamic responses, underscoring the need to account for damping effects in design (Tang et al., 2023a). Similarly, the hydroelastic response of FOWTs has been a focal point of investigation, particularly concerning the nonlinear and higher-order hydrodynamic loads that can induce significant vibrations. A study on a spar-type 10 MW FOWT with a flexible tower and platform demonstrated that second-order hydrodynamic loads could lead to pronounced springing and ringing effects, which are critical considerations for the design of future, larger-scale systems (Leroy et al., 2022). Experiments also revealed that aerodynamic loading influences FOWT hydrodynamics by suppressing pitch resonance and amplifying surge frequency vibrations, offering insights for optimizing FWT designs (Wen et al., 2022). Research on TLP configurations, such as the CENTEC-TLP, has shown that these designs can effectively minimize system responses to wave loads, even under extreme conditions, thereby enhancing the overall structural resilience and reliability of FOWTs (Hmedi et al., 2022).

Additionally, experimental studies on concurrent wind, wave, and current loads on monopile-supported OWT models demonstrated significant discrepancies when these loads were analyzed in isolation, emphasizing the need for comprehensive assessments considering their combined effects (Buljac et al., 2022). The adaptability and performance of mooring systems have been further explored through model tests and numerical simulations, particularly under cyclic loading conditions typical in marine environments. These studies have provided critical insights into fatigue life and failure mechanisms, essential for developing more resilient mooring designs (Fonseca et al., 2022). Finally, research on the hydrodynamic effects of heave plates, commonly used in semisubmersible FOWTs, provided critical data on parameters like KC numbers, oscillation frequencies, and plate geometries, informing design improvements (Zhang et al., 2023d).

The exploration of WECs, particularly Oscillating Water Columns (OWC), has advanced the understanding of efficient wave energy capture, with extensive experimental work driving these advancements. Studies integrating OWC systems into multifunctional marine structures, such as floating breakwaters, have employed scale models to optimize geometric configurations for maximum energy extraction. These tests have demonstrated these systems' dual functionality, which can both extract energy efficiently and provide robust coastal protection under varying marine conditions (Cheng et al., 2022a). Research has also focused on the shape optimization of back-buoy oscillating water columns, using both numerical simulations and experimental validation to improve primary conversion efficiency by examining the effects of various bottom-corner geometries (Jalani et al., 2022). Furthermore, model tests have been crucial in assessing the operational efficiencies and energy transduction capabilities of OWC installations, offering detailed performance metrics that are critical for optimizing these systems (Gubesch et al., 2022a). Studies on the nonlinear wave loads on quad-OWC devices, validated through wave tank experiments, have provided critical insights into optimizing the spatial arrangement of OWC arrays to reduce wave forces and improve platform stability (Zhou et al., 2023b).

Studies on aquaculture structures have meticulously quantified the effects of wave-induced motions using physical model tests, offering insights to enhance structural integrity and stability. For instance, experiments on gravity fish cages under regular wave conditions yielded valuable data on hydrodynamic loads and motion responses, crucial for improving aquaculture design and safety (Liu et al., 2023d). In-depth analyses of hinged multi-body aquaculture platforms have revealed how different stiffness and wave periods affect their dynamic behaviour, offering guidance on optimizing mooring configurations and structural designs (Ma et al., 2022a). Further research has explored the design and implementation of robust mooring systems, which are essential for ensuring the survivability and operational efficiency of aquaculture platforms in harsh marine environments. Comparative investigations between closed and semi-closed fish cages under various wave conditions provided insights into their hydrodynamic performance, highlighting differences in global responses and survival under extreme conditions. These studies underscore the importance of experimental validation in understanding the complex interactions between aquaculture structures and their marine environments, providing a foundation for developing more resilient and efficient designs (Shen et al., 2022).

Investigations into the hydroelastic behaviour of Very Large Floating Structures (VLFS) under low-frequency loads have utilized scale models to assess performance and stability under various sea states, ensuring their deployment in maritime applications. For example, model tests of hydroelastic truncated floating bridges provided critical insights into the dynamic response of these large structures, highlighting the importance of careful experimental design and system identification for accurately replicating the behaviour of full-scale structures (Rodrigues et al., 2022). The dynamic responses of other floating platforms to marine conditions have also been critically examined through a combination of model tests and numerical simulations. For instance, research on semi-submerged cylinders under combined steady and oscillatory flow conditions shed light on the hydrodynamic forces acting on these structures, with empirical models developed to predict these forces accurately (Hu et al., 2023). Studies have explored the impact of marine growth on the hydrodynamic behaviour of cylinders, using experimental methods to quantify the effects of different types of roughness, such as mussels and corals, on drag and inertia coefficients under wave and current loading (Marty et al., 2022). Additionally, experimental and numerical investigations on wave impacts on box-girder bridges provided detailed analyses of wave-induced forces and moments, contributing to the development of robust design strategies for coastal bridge structures (Zhu et al., 2022). Moreover, long-term model testing has been pivotal in evaluating the durability and safety of VLFS under cyclic and extreme loads, offering guidance for their long-term deployment in marine environments (Scharnke et al., 2023). Figure 12 shows the convergence test results of non-dimensional wave crests and impact force peaks for a wave tank model test case study from Scharnke et al. (2023). N is the number of seeds, and the lines denote the different test cases denoting different severities and duration of the seastates.

Studies on kinematic behaviour using advanced techniques like Particle Image Velocimetry (PIV) have provided detailed insights into structure-fluid interactions. For instance, experiments on internal solitary waves interacting with horizontal cylinders in stratified fluids, validated through PIV, advanced the understanding of wave-flow coupling and led to accurate force models for these interactions (Wang et al., 2022f). Physical model tests combined with PIV have also been employed to investigate the effects of solitary waves on elastic submerged plates, successfully capturing detailed turbulence fields and wave loads. Measurements of surface elevation, plate motion, velocity, and turbulence fields have allowed for a comprehensive understanding of the hydrodynamic behaviour of different plate materials under solitary wave conditions (Hsiao & Hsiao, 2022).

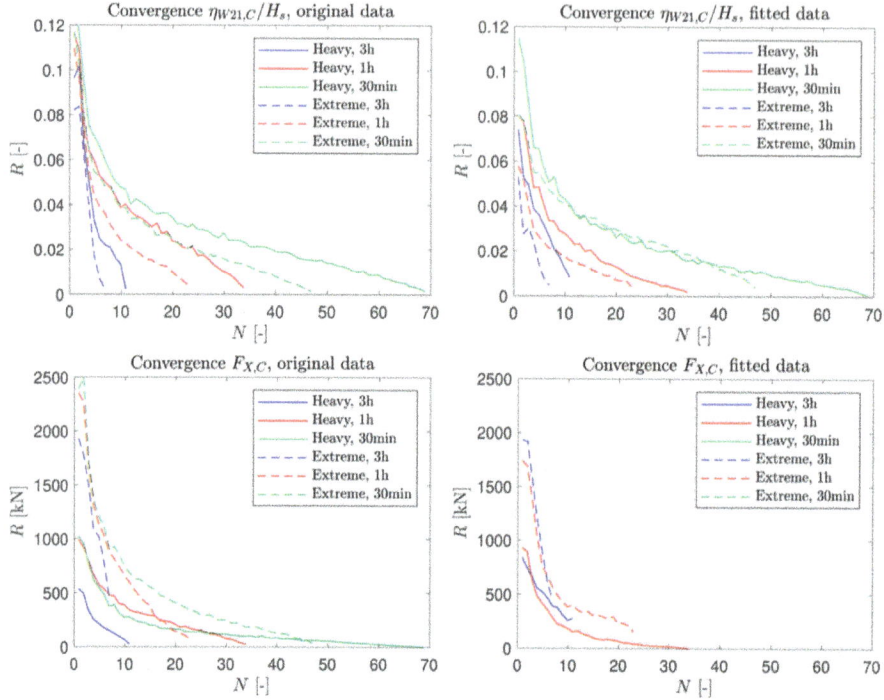

Fig. 12. Results for the convergence study from Scharnke et al. (2023)

2.2.7 Data-Driven Methods

The integration of data-driven methodologies in marine hydrodynamics and offshore engineering has significantly increased in the last three years. Previous efforts, such as those by Sclavounos & Ma (2018) and Pena-Sanchez et al. (2020), have already shown the value of applying machine learning techniques in predicting and modelling wave loads on offshore structures. These studies highlighted the potential of integrating artificial intelligence (AI) and machine learning with traditional marine engineering practices to enhance predictive accuracy and computational efficiency.

In 2021, Guo et al. (2021a) introduced a Long Short-Term Memory (LSTM) based model focused on predicting the heave and surge motions of semi-submersible vessels. The study emphasized the importance of real-time, accurate predictions for offshore operational safety and efficiency, demonstrating the model's capability to predict future motions with an average accuracy close to 90%. The use of LSTM, a recurrent neural network, was particularly notable for its ability to process time-series data, making it well-suited for predicting the dynamic responses of offshore structures to wave loads. Ma & Sclavounos (2021) studied the use of Support Vector Machines (SVM) in modelling nonlinear hydrodynamic loads on fixed cylinders in shallow water by analyzing experimental data from which (nonlinear) transfer functions were established. This work illustrated the potential of SVM regression methods in capturing complex marine hydrodynamic loads, offering insights into the field's understanding of fatigue and extreme statistics analysis for offshore structures. Pena & Huang (2021) presented the application of deep learning, specifically Generative Adversarial Networks (Wave-GAN), for predicting nonlinear regular wave loads and run-up on fixed cylinders. The study demonstrated Wave-GAN's ability to accurately predict wave-induced pressures for unseen wave conditions, by utilizing Computational Fluid Dynamics (CFD) simulations to generate training and testing datasets. A "grey-box modelling" approach by Pitchforth et al. (2023)), which aimed to enhance the accuracy of wave loading predictions on offshore structures, was published in the same year. By integrating Morison's equation with a data-driven Gaussian process NARX model, the study achieved significant improvements in predictive accuracy and model extrapolation capabilities, especially in scenarios with limited data coverage. The work underscores the value of merging physics-informed machine learning with traditional offshore engineering practices.

In 2022, the research on applications of data-driven methods continued to diversify. Katsidoniotaki et al. (2022) developed a digital twin to predict extreme loads on a wave energy conversion system's mooring system under harsh wave conditions, demonstrating the digital twin model's ability to predict mooring force with high accuracy and drastically reduced computation time, while Li et al. (2022b) introduced a Temporal Convolution Network (TCN) model for predicting wave runup on semi-submersible platforms, employing a data-driven approach to establish the nonlinear mapping of wave-structure interactions, optimizing the input tensor space considering both temporal and spatial dependencies. Liang et al. (2022a) combined Computational Fluid Dynamics (CFD) with Deep Reinforcement Learning (DRL) to investigate wave dissipation by actively controlling flat plate breakwaters against regular waves. Utilizing an in-house NWT to simulate fluid-structure interactions, the model applied an Artificial Neural Network (ANN) to learn wave dissipation strategies through continuous wave-plate interaction. The study demonstrated the model's adaptability and ability to learn control strategies automatically, presenting a significant advancement in the application of AI techniques for enhancing coastal protection measures.

In 2023, (Chen et al., 2023a) presented an Auto-Regressive model for forward predicting surface wave elevations and the motions of offshore floating structures (see Fig. 13). The model, validated with numerically synthesized wave data, indicated that band-pass filtering could improve prediction accuracy considerably, especially with narrower spectral bandwidths (as would be expected). Other studies further advanced the application of machine learning in predicting mooring line tensions and the motions of moored barges, respectively, using reservoir computing models (Yang et al., 2023b, 2023c).

In summary, the last three years have been characterized by a rapid expansion in the application of advanced data-driven methodologies and machine learning techniques in marine hydrodynamics and offshore engineering. These developments build upon the preliminary work established previously and showcase a notable shift towards more sophisticated, accurate, and efficient predictive models. This progression signifies a promising future for the integration of AI and machine learning in enhancing the design, analysis, and operation of offshore structures, ensuring their resilience against the dynamic challenges posed by marine environments.

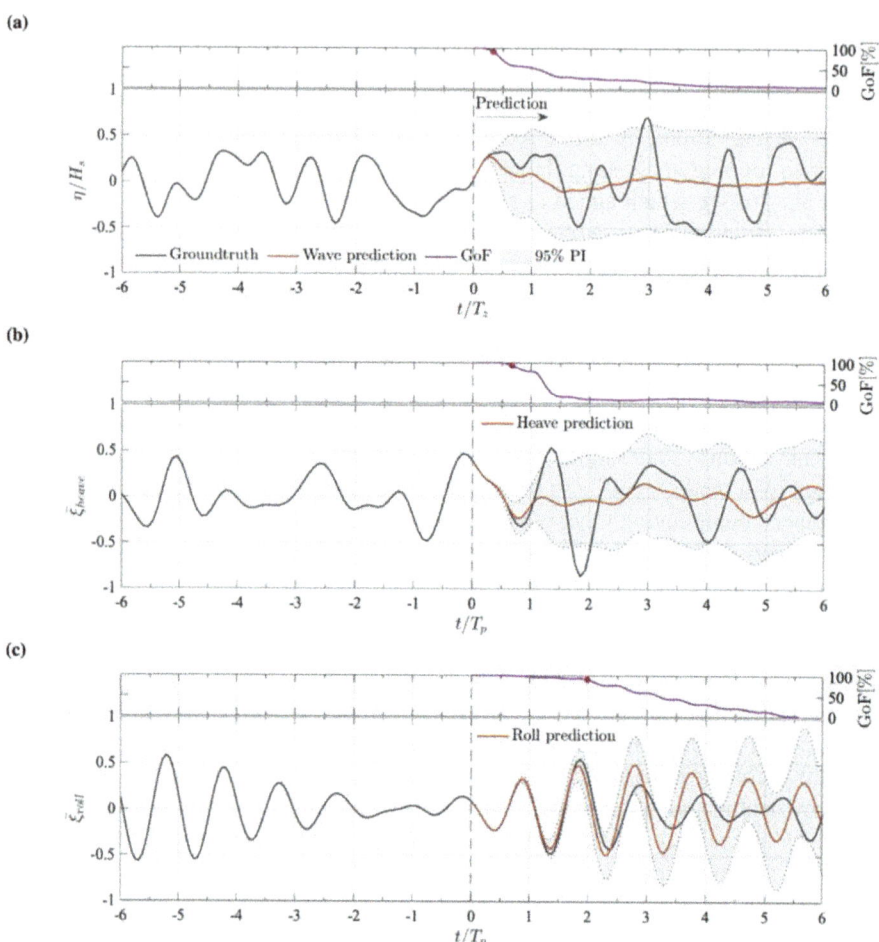

Fig. 13. Wave tank data prediction results on (a) surface wave elevation (b) heave motion, and (c) roll motion of a scaled model in a wave tank under the action of waves in Chen et al. (2023a).

3 Current Loads and Wind Loads

Historically, the main loads attributed to current and wind acting on ships and offshore structures are the drag forces (ahead or lateral/oblique) and the yawing moment in the free surface plane, first considered according to a static or quasistatic nature. From this perspective, the key impacts of these loads are the mean drift and their role in providing damping for the slow drift because waves coexist. To estimate these loads and assess their impact on the system (whether it is a ship or an offshore structure), methods derived from detailed experimental research, such as those proposed by API

(American Petroleum Institute), DNV (Det Norske Veritas), or OCIMF (Oil Companies International Marine Forum), are commonly employed. These methods determine the forces and moments experienced by the system because of its movement relative to the translational velocity induced by currents or wind speeds. In fact, an additional drag force may be considered because of the yaw rate, which is particularly significant for the components of systems operating in water.

Inaccuracies in estimating the current and wind loads on ships can lead to an imprecise calculation of their power requirements, difficulties in maintaining position during manoeuvres or cargo handling, increased reliance on tugboats for port operations or near other structures, and an increased risk of incidents and shortcomings during certain tasks. On the other hand, inaccuracy in estimating these loads on offshore systems, usually anchored, can lead to the impairment of the mooring system itself, the catastrophic failure of subsystems and lines of prospection and production (drills, risers, and umbilicals, mainly), as well as the imposition of additional difficulty in their operations assisted by PSV, for example.

Due to the growing dimensions of modern ships and offshore structures, there is an increased need to meticulously consider the force and momentum coefficients resulting from wind and/or current effects, particularly in conjunction with wave dynamics and varying environmental conditions at operational sites such as ports enclosed by tall structures, sheltered areas, and shallow-water channels. In the study by Torre et al. (2021), a large cruise ship anchored in a port was examined with its force and momentum coefficients in a wind tunnel to assess the potential impact of nearby civil engineering works on the reduction of these coefficients.

Furthermore, the growing scale of the systems has also impeded purely experimental studies in towing tanks or wind tunnels because of the requirement for progressively smaller models. On the other hand, this difficulty opens space for decisive support of investigations based on CFD simulations, recognized as increasingly complete and precise, as found in Yoo et al. (2022), where wind loads on the FPSO side by side with a shuttle tanker are evaluated by simulations using Reynolds-averaged Navier-Stokes (RANS) equations and the SST k-ω turbulence model; as in El Moctar et al. (2023), where the effects of wind are examined through experimental and numerical methods (URANS and two turbulence models), demonstrating that the configuration of deck structures significantly impacts aerodynamic forces and moments.

Another recent scenario that has received significant attention is that current and wind are considered in offshore wind power generation systems; see Edwards et al. (2024). In contrast to typical issues, the loads in this situation exhibit dynamic features, often cyclic, that may result in resonant and/or unstable dynamic effects. According to these aspects, the crucial problem currently refers to the aerohydrodynamic behaviour of floating offshore wind turbines (FOWTs), investigated numerically and experimentally from the point of view of the wind turbine, for example, in Chen et al. (2022a) and Wang et al. (2023d), and for the floating platform as found in Yin et al. (2022).

As pointed out in Bandizadeh Sharif et al. (2023), the large class of fluid-structure interaction (FSI) problems includes flow-induced vibrations (FIV), which come from the significant influence of marine currents and winds on a series of offshore structures; see Fig. 14.

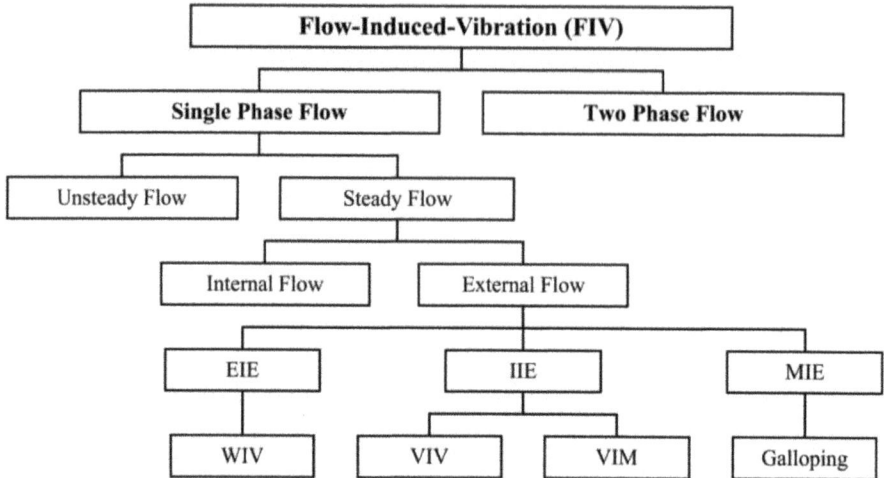

Fig. 14. A possible FIV classification, as proposed in Bandizadeh Sharif et al. (2023).

Focusing solely on single-phase flows in a steady external regime, FIV can be associated with extraneously induced excitations (EIE), for instance, the wake-induced vibration (WIV) phenomenon, with instability-induced excitations (IIE), mainly vortex-induced vibrations (VIV) and vortex-induced motions (VIM); and with motion-induced excitations (MIE), as those found in the galloping phenomenon. This presents a challenging situation since excitations do not occur in isolation, and the required design approach may vary from destructive, when the objective is to reduce the damaging effects caused by these phenomena, to useful, when FIV is used for renewable energy generation; see Fig. 15.

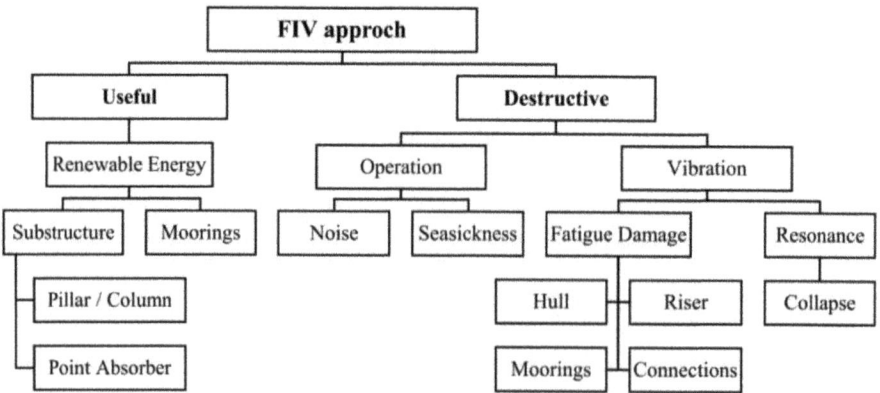

Fig. 15. The two main approaches concerning FIV in designs, according to Bandizadeh Sharif et al. (2023), are useful and destructive.

More information on the analysis of current and wind forces on ships and offshore structures is provided in the following subsections.

3.1 Current Loads

As previously stated, the need for more accurate predictions of current loads associated with the greater size of vessels has driven investigations where tests in towing tanks and/or wind tunnels are specifically conducted to validate increasingly accurate and more comprehensive CFD simulations. This method is exemplified in the work of Mauro et al. (2023), where the authors proposed an initial design phase of drill ships using high-fidelity CFD calculations. The results were comparable to those obtained using conventional methods, namely API-1984, DNV-GL2018, and OCIMF-1994/2010. The need for greater safety in the manoeuvring or dynamic positioning of ships in the current has required intense research to achieve an increasingly accurate determination of force coefficients, as presented numerically in Kim et al. (2022a), Aydın et al. (2022), Song et al. (2023), and analytically in Zhang et al. (2023a), where the results showed the important influence of this environmental agent on the manoeuvring performance of ships.

The unpredictable nature of currents poses a risk to structures in marine environments, shortening their lifespan by hastening existing failures or causing faster deterioration over fewer cycles due to material fatigue. This is why investigations of the effects of marine currents are so important for the design of offshore lines (mainly risers, umbilicals, and mooring lines).

In the study by de Carvalho et al. (2022), theoretical models were used to calculate the tension in mooring lines and the compression of the fenders on the dolphins within a port infrastructure. The authors suggested that utilizing a nonlinear model could be beneficial for engineers and analysts in estimating the forces and dynamic amplification coefficients needed in the simplified (linear static) approaches outlined in the standards. Also, in analytical terms, according to Amouzadrad et al. (2023), a moored structure offered enhanced stability compared to a free-edge structure, as it exhibits greater deflection than the free structure. Structural displacements increase as current speeds increase, which is caused by elevated hydrodynamic loads on the structure in the upstream region. As the current speed increases, the increase in transmission coefficients is due to the more significant current generating larger hydrodynamic forces, resulting in increased displacements that cause more wave energy to pass beneath the structure. Furthermore, the number of resonating patterns decreases as the wavenumber increases, which is expected as the wave reflection decreases in the upstream region. In addition, the floating structure became more stable for lower current speed values and higher mooring stiffness.

During the fatigue life of offshore lines, plastic deformation accumulates from the multiaxial stress-strain response of the material, subsequently resulting in cyclic accumulation of plastic damage. Marine currents, as well as concomitant waves and winds play an important role in analyzing accumulated damage. For example, using FEM-based commercial software, Gemilang et al. (2021) investigated the catenary lines' low cycle fatigue in a spread mooring configuration from an FPSO unit. The comparison between the SN method and the low cycle analysis showed that the first method was conservative only when the average load was 20% below the minimum breaking load.

It also showed that increasing the spreading radius of lines could exacerbate the fatigue problem and lead to faster failure. The significance of marine currents is also suggested in the study by Zhang et al. (2022g), in which the authors performed fatigue assessments on oil offloading lines (OOL) that connect an FPSO to a catenary anchor leg mooring (CALM) buoy.

As an alternative to commercial software, Sivaprasad et al. (2023) proposed an artificial neural network (ANN) to estimate fatigue damage in a top-tensioned riser (TTR) subjected to vortex-induced vibration (VIV). Although it is a 2D analysis that does not consider waves, wind and material variations, as well as considering a uniform current and a linear soil model, the validated ANN model showed acceptable performance, with a maximum difference of 9.6%. In the same sense as the alternatives in Wu et al. (2023a)), a cluster analysis based on the Gaussian mixture model (GMM) was applied to analyze the responses of the full-scale riser. The method can adequately separate 242 measurement events into 12 clusters, reducing the parameter dimensions so that the underlying physics can be better understood. Furthermore, Sun et al. (2022a) introduced a technique that relied on the average conditional exceedance rate (ACER) method for a precise estimation of extreme values of current loads.

When considering the production of sustainable energy, the potential to harness marine currents is also taken into account. The behaviour of a submerged turbine exposed to sea waves and currents was studied by Lloyd et al. (2021), using the numerical CFD method and experiments in a test tank to confirm the results. Five different flow conditions were examined, including two conditions that involved both current and waves, and three conditions involving only current. The variability in loads was considerably higher when looking at waves and currents, with the maximum loads aligning with the peaks and troughs of the wave surface. The results also showed that when waves were included, the amplitude of thrust/torque fluctuation was 35 times greater than in tests with currents alone.

Wind turbines are another source of clean energy generation and are usually installed in places where regular currents and tides can be significant with respect to flow-induced vibrations (FIV).

Fujarra et al. (2022) investigated the basic principles regarding how initial roll or pitch angles influence the flow-induced motion (FIM) amplitudes of a low aspect ratio floating circular cylinder. The 6 DOF amplitude response, coupling trajectories, and frequency spectra were analyzed. Experiments with four initial angles demonstrated that the initial pitch influences the yaw motion, leading to geometric asymmetry that enhances the yaw displacements. The yaw motion increased as the reduced velocity was increased, making it essential to correctly model and analyse VIM on floating offshore wind turbines (FOWT).

The complexity associated with the design of the FOWT was discussed in Edwards et al. (2024), where the authors emphasized the importance of employing a combined numerical and experimental methodology to develop such systems. The intricacy is inherent in the experimental method due to the significant interaction between the wind turbine's aerodynamics and the floating offshore structure's hydrodynamics, making it challenging to fully model at a smaller scale.

The study by Otter et al. (2022) offered a review of modelling approaches for FOWTs, with an emphasis on numerical techniques and software tools. The authors noted that the selection of a computational method was based on the balance between accuracy and computational efficiency. Low-fidelity methods were used during the preliminary design phase, mid-fidelity techniques were suitable for analyzing global dynamics, and high-fidelity approaches were necessary for detailed examinations. In addition, full-physical and hybrid test methods were used to validate and calibrate the numerical models, depending on factors such as testing facility, budget, and uncertainty; see Fig. 16. Full-physical tests can model all aspects of the wind turbine, including aerodynamic damping, but are expensive and require a dedicated model for each turbine design. Hybrid testing is cheaper, more versatile, and does not require bulky wind generation equipment, allowing for the simulation of turbines at various scales. Trends and gaps in current research are identified, such as comparative studies and improving the accuracy of numerical engineering tools. The paper also concludes that the traditional modelling approach remains essential for FOWT designs.

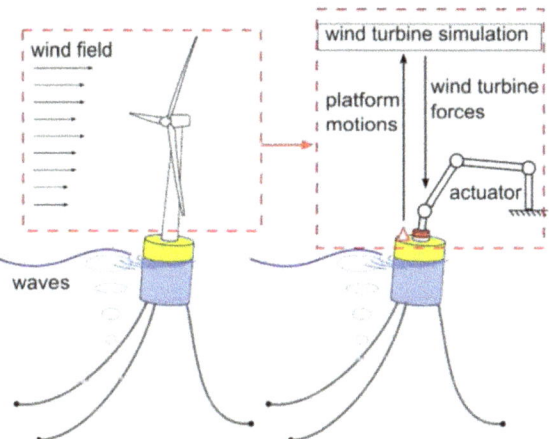

Fig. 16. The two approaches for testing FOWT according to Otter et al. (2022): (left) full-physical versus (right) hybrid testing.

Sergiienko et al. (2022) examined and evaluated scale-down methods within the same framework of the FOWT, considering data from 9 reference wind turbines and eight semi-submersible platforms (SSP) for 5–15 Megawatts to identify trends and scaling laws related to their physical properties. The mass, power and thrust forces of the wind turbine scale quadratically with the diameter of the rotor, and these characteristics do not depend on the design of the drive train. The main challenge for larger wind turbines was tower design, as the "soft-stiff" tower design was not applicable to floating offshore wind energy systems with power ratings higher than 10–12 Megawatts. Existing SSPs have a variety of design solutions, with a draught of less than 20 m and natural frequencies of around 20 and 30 s. However, the crucial geometric factor influencing the design is not associated with its waterplane area but with the product of the offset column diameter and the distance to the offset column.

Based on a practical approach to reduce FIV, Bandizadeh Sharif et al. (2023) provided an extensive review of both numerical and experimental studies on suppression systems in circular cylindrical marine structures, focusing on devices such as strake, spoiler, fairing, splitter plate, shroud, dimple, ribbon, bump, and travelling wave wall to reduce vortex shedding. According to the authors, the helical strake was the most widely used suppression device, demonstrating the most effective performance in reducing such vibrations. The optimal design could reduce FIV by 95–100%, therefore reducing the fatigue damage of these structures.

The mitigation aspects were also investigated in Gonçalves et al. (2023a), where based on small-scale experiments, the authors emphasized the critical role of the design of the heave plate (HP) in reducing flow-induced motion (FIM) and indicated that only the largest HP positively reduces dynamic behaviours due to wave and current incidences, making it essential for FOWT developments. In Gonçalves et al. (2023b), the impact of draught, pontoon height, and column face length on the FIM of an SSP with four square section columns was experimentally investigated. The findings indicated that a reduction in the column-pontoon ratio resulted in lower transverse motion amplitudes.

Fluid-structural interactions such as FIV can happen with wave excitation, a co-existence that is crucially important for offshore system design. This issue was experimentally investigated by Du et al. (2022), considering a large-scale model of a tension leg platform (TLP) towed at different velocities concomitant to irregular waves. The study found that 0.32D transverse amplitudes were highest at Vr = 7.0 (where Vr is a nondimensional parameter usually applied for FIV studies), while the inline and yaw motion experienced increasing amplitudes with reduced velocities. The most significant responses were at 0 and 22.5 degrees of heading, while no critical heading was recognized for surge motion, and the VIM amplitudes were significantly suppressed under nonlinear stiffness and greater damping. In a similar sense, Rueda-Bayona et al. (2023) took into account the VIV excitation concomitant to those coming from winds and irregular waves, proposing a methodology based on the design of experiment (DOE), the analysis of variance (ANOVA) and the near-hydrodynamic field analysis to correctly evaluate the damped natural periods and responses due to the nonlinear fluid-structure interactions.

Two FOWTs were numerically investigated in Li et al. (2023) (SSP-type and barge) using OpenFOAM and an in-house model to consider the simultaneous wave and current excitation. The primary findings indicated that waves could amplify VIM when there was a significant angle between the current and the wave and that the highest wave height during lock-in was not the most hazardous scenario. Rather, a smaller wave height with a longer period could provoke even greater motion. Three-dimensional CFD simulations with an SSP-type FOWT in model and full-scale were also studied by Jiang et al. (2023b), indicating a significant influence of scale and highlighting the improved effectiveness of the CFD method for VIM studies. The FOWT based on the OC4 platform (SSP of three columns without pontoons) was numerically investigated under different headings in Liu et al. (2023c), revealing that the VIM "lock-in" and transverse response persist up to the nondimensional reduced velocity of 30, with a maximum of 0.8D in amplitude, for incidences of 0 and 90 degrees, while smaller amplitudes were observed for 180 degrees due to the interaction between the upstream vortices and the downstream side column. A similar approach was used by Zhao et al. (2022), considering CFD simulations of a four-column SSP with pontoons and revealing the differences between VIM at 0 and 45 degrees of current heading. The aspects of strake mitigation were numerically investigated by Kharazmi & Ketabdari (2022) for a large-scale spar platform, mainly discussing the arc angle and the number of entrances of these helical strakes with respect to VIM attenuation.

Still, in the context of FIV (or FIM) experiments, an unconventional measurement technique was presented by Leal et al. (2023) and Fujarra et al. (2023) based on the usage of inertial measurement units (IMU). The authors stated that the frequencies and amplitudes of FIV measured with the cost-effective IMU method closely matched those obtained using a significantly pricier optical motion capture system. They reported an average relative error of less than 2.8%, which supports the validity of using inexpensive hardware, although it requires more advanced signal processing techniques.

3.2 Wind Loads

3.2.1 Wind Loads on Ships

Kobayashi et al. (2022) stated that numerical simulation was increasingly employed with wind tunnel testing and regression formulas in the ship design phase to predict the wind load acting on ships (vessel and superstructure). The numerical approaches, however, required more secure, standardized procedures that addressed the clarification of computational variances depending on simulation conditions, particularly regarding the blockage effect, the computational estimation capacity for stationary and non-stationary regimes, wall functions applied to the ship surfaces, the difference of wind profiles recovered by the normalization based on the momentum integration in a shear flow, and the effect of Reynolds number.

Considering this need for verification and validation (V&V), Deng et al. (2022) studied the wind loads on a large container ship both experimentally and numerically under various conditions to assess aerodynamic performance and find methods to decrease wind resistance. A similar approach based on wind tunnel tests and CFD simulations was conducted by Dao et al. (2023) to study wind loads on the topsides of a liquefied natural gas carrier (LNGC), a floating production storage and offloading ship (FPSO) and a jack-up platform (JUP). Overall, favourable validation outcomes were obtained, indicating that the reliability of the experiments and CFD models can be used as a benchmark for future research on offshore wind loading.

A fully numerical investigation of the aerodynamic drag of navy ships in different positions was presented by Islam et al. (2022). Considering the full and model scales, this study showed that the scale effect was negligible for the surge and sway forces but more pronounced for the yaw moment. Although this was possible, the authors concluded that further in-depth investigation remains necessary, including comparison with experimental results and sea trials. In Van He et al. (2023), the wind loads on an original passenger ship were also investigated only by considering CFD simulations. Based on the pressure and velocity distributions around the hull, several new geometries with different frontal accommodation shapes were evaluated to reduce wind drag and improve aerodynamic performance.

Therefore, recent studies allow one to conclude that CFD simulations can provide reliable data for aerodynamic loads acting on ships, which are usable for continuous improvement of aerodynamic performance and manoeuvres under the action of wind.

Research on wind loading on ships was also intensified in the last period due to the IMO resolution (International Maritime Organization (IMO) (2023)), which aimed to reduce ship emissions and achieve net zero greenhouse gas emissions from international shipping by around 2050. Consequently, the maritime industry and commercial transport by ships strove to quickly minimize overall emissions using all available means. In addition to alternative solutions such as cleaner fuels and enhanced ship designs, the resurgence of wind-assisted ship propulsion (WASP) was notable, including those based on rigid sails and towing kites, but mainly using Flettner rotors. As discussed in Khan et al. (2021), theoretical advances in performance prediction programs (PPPs) were notable in this context but still lacked parameters and validation from experiments. This represented one of the future challenges for WASP, along with the precise assessment of environmental and economic impacts and the often-overlooked structural consequences.

In this sense, a methodology based on the application of previously computed wind forces in CFD to a small-scale model using a cable-driven parallel robot (CDPR) was presented in Sauder & Alterskjær (2022) to test the Flettner rotors installed in a very large ore carrier (VLOC) under highly controlled wind loads, still due to a simplistic aerodynamic model. Changes in vessel speed, roll angle, drift angle, and rudder operation in response to the incoming wind angle and speed were recorded and found to qualitatively agree with earlier WASP studies. As expected by the authors, this methodology allowed for a comprehensive validation of the hydrodynamic performance of these new designs. In addition, it facilitated quick comparative evaluations of different configurations, setups, or sail reefing/control strategies. Other factors that could be investigated using this method included course stability when using sails, reduction in manoeuvrability with sails, the positive impact of sails on damping wave-induced motions, and the effect on added resistance in waves.

Kume et al. (2022) also evaluated the aerodynamic characteristics of a ship with Flettner rotors, comparing results from wind tunnel tests and CFD simulations. RANS-based results of the WASP scaled by Reynolds number (Re) agreed qualitatively and quantitatively with those of the wind tunnel tests. Moreover, as proposed by previous works, the CFD results conducted at full-scale Re indicated an excellent performance in numerically predicting wind loads on a full-scale rotor if the L_{pp}-based Re was more than one million.

An investigation based on systematic simulations of Flettner rotors applied as WASP was found in Monteiro (2024). A wide range of spin ratios α (relations of velocities in the cylinder wall and the far field was considered under bi- and three-dimensional CFD models. Regardless of the approach, three flow regimes were observed for different spin ratios. For a spin ratio of 2.5, the 2D lift coefficient (C_L) results aligned closely with those of other studies. However, at higher spin ratios, C_L values were significantly overestimated, whereas the drag coefficients (C_D) were for the 2D discrepancy (particularly focused on the amount of flow deflection), although there was no significant improvement in C_D. The torque coefficients were underestimated but followed the same pattern of experimental results, and changes in domain heights did not significantly alter the results.

3.2.2 Wind Loads on Offshore Structures

The energy available over the sea is highly suitable for power generation, as offshore winds are quite abundant with high speeds near the surface, enabling efficient harnessing of this source. As mentioned in Buljac et al. (2022), accurate estimation of energy production and loads in offshore wind turbines (OWTs) is crucial during the development of new offshore wind farms. Frequently, a beneficial increase in wind power and an advantageous increase in wind energy production, with respect to the size of the turbine and its support system leads to a decrease in structural loads. In this scenario, the researchers have examined small-scale tests conducted at the Wind-Wave-Current Tank (WWCT) laboratory at Newcastle University, UK. The combined effects of waves, currents, and wind on a monopile-supported OWT model were investigated more realistically, as this structure remains the most widely used in the industry. As a result, the current was the main contributor to the overall mean load, with a lower mean wind contribution, while waves dominated load fluctuations. Moreover, (Buljac et al., 2022) found how much influence the tower had on environmental loads, which was generally prominent for prototype OWTs in the cutoff wind speed range and beyond. At the natural frequency of the turbine, the wind primarily induced structural loads in the surge direction, whereas currents amplified dynamic loads at lower frequencies because of non-linear hydrodynamic interactions.

More discussion of significant advances in OWT technology was found in the review by Asim et al. (2022), where critical aspects were assessed to identify knowledge gaps and to facilitate future efforts to understand the intricate behaviours of these systems, mainly those regarding aerodynamic characteristics, dynamic response, structural integrity, mooring cable design, ground scouring, and cost modelling.

Another important concern for OWTs is the structural integrity of their support structures, mainly in terms of failure mechanisms and physics-based modelling techniques. Yeter & Garbatov (2022) reviewed the structural integrity of the OWT support structures, covering a historical background, fatigue and fracture phenomena, crack estimation, and models to predict the remaining life. The paper reviewed influential studies on structural integrity and reliability, which are essential for evaluating the remaining life and managing the lifecycle. It also discussed a risk-based framework for assessing structural integrity, considering the uncertainties of modelling, loads, and material properties.

As a conclusion point on wind and current loads, it is crucial to recognize that innovative methods to enhance the efficiency of sea-based energy production are continuously being developed and evaluated. An example of research in this direction can be found in the numerical investigation of El Beshbichi et al. (2022), which compared the dynamic performance of multirotor FOWT on a spar, SSP and TLP. Six different load scenarios were analyzed, revealing that yaw motion was crucial for these systems, especially for the spar-type FOWT. However, in the SSP and TLP cases, this motion was although the high standard deviations in tower base bending moments heightened the fatigue damage risk in the TLP configuration. In addition, significant tendon loads can raise fatigue-related issues and limit state performance. The sizeable mean pitch angle and yaw standard deviation can decrease the quality of the electric power output. Severe storm conditions lead to a substantial rise in the response standard deviation, particularly affecting SSP-type FOWT.

4 Ice Loads

4.1 Ice Loads on Ships

Reliable prediction of ice loads remains a critical issue for the design and safe operation of ships navigating in the Arctic and ice-infested waters. While navigating through ice, ships are expected to encounter a variety of sea and glacial ice features, requiring the estimation of ice loads. These ship ice loads can be categorized into global and local ice loads. The global ice loads, also known as ice resistance, are important for evaluating overall ship performance, whereas local ice loads are crucial for assessing structural integrity and safety. During this reporting period, researchers have continued to use various modelling techniques, model-scale laboratory experiments, and full-scale trials and measurements. Most of these efforts build on previously developed modelling approaches and are primarily based on existing and published experimental and trial data. The COVID-19 pandemic may have indirectly affected the research type during this reporting time.

4.1.1 Full-Scale Measurements and Trials

Full-scale measurements and trials provide realistic and reliable ice load data by capturing actual environmental and operating conditions experienced by ships. They also serve as benchmarks for validating and calibrating various analytical, numerical and semi-empirical ice load models and model-scale test results.

Sinsabvarodom et al. (2021) analyzed full-scale test data of a station-keeping vessel operating in drifting ice in the Bay of Bothnia in March 2017. The study employed an ice resistance method to estimate global ice loads and determine short-term extreme mooring loads for the vessel under various ice management scenarios.

Influence Coefficient Matrix (ICM) is an ice load prediction method based on a coefficient matrix and is widely employed to convert strain data into ice pressures. This method is often used to estimate the ice loads experienced by ships during full-scale trials. Böhm et al. (2021) investigated the effect of ICM on the accuracy of shear strain-based local ice load measurements onboard the SA Agulhas II. Using a numerical model of a ship grillage system, consisting of a plate and five frames, under various ice load conditions, the study revealed that the number and location of strain gauges significantly affect the ICM and, consequently, the accuracy of the ice load measurements. While the common instrumentation practice is suitable for measuring ice loads within the instrumented area, the study emphasized the importance of optimizing instrumentation setups to achieve improved ice load measurements across various loading scenarios.

The far-field identification method is another approach to predicting ice loads in full-scale measurements. Wang et al. (2023a) employed a radial basis function neural network (RBFNN) based far-field identification method to monitor the ice-induced strains onboard the RV Xue Long 2 during a research expedition. The effect of noise on the identification results of ice loads on the RV Xue Long 2 bow shoulder was investigated numerically and eliminated using a noise injection learning algorithm. The method's feasibility was verified through a model-scale test, where the testing bow shoulder panel was rigidly fixed on a pedestal and loaded with a hydraulic tensioning jack. The method requires further elaboration and full-scale validation. He et al. (2021) utilized Green's function to monitor the time history of far-field dynamic ice loads onboard a ship during the full-scale measurement of Tianen's 2019 Arctic voyage. The study indicated that the Green's function strongly depends on the strain signal intensity. They employed a regularization method to avoid ill-posed problems and to obtain a numerically stable solution.

4.1.2 Model-scale Experiments

Model-scale experiments are often preferred over full-scale measurements and trials as they provide a cost-effective, controlled, and detailed analysis of ice loads and the underlying physics of the ship-ice interaction process. Several model-scale experiments were conducted to explore ship ice loads during this reporting period.

Wan et al. (2023) conducted model-scale ice tests on an Arc7 ice-class LNG carrier to investigate ice resistance under various ice and operating conditions. The tests involved towing the model LNG carrier with a dynamometer through level ice, broken ice and brash ice fields to analyze ice failure modes and ice-hull interaction characteristics. Meanwhile, Xue et al. (2024) utilized a combined experimental and numerical approach to predict ice resistance in broken ice fields. The study employed synthetic polypropylene model ice, and tests were conducted in the conventional non-refrigerated towing tank. The test results were incorporated with CFD-DEM-based simulations to account for the fluid effect, such as the changes in fluid velocity and ship waves. Both studies compared their ice resistance results with existing empirical and semi-empirical formulas.

Li et al. (2021) performed model-scale tests to analyze the escort performance of ships navigating through narrow ice channels created by icebreakers. The test results were then utilized to benchmark numerical simulations of ten sample ships performing escort operations. The study explored the effect of channel width and ice thickness on the resistance and attainable speed.

Sun & Huang (2021) and Müller et al. (2024) investigated ice impact loads in model-scales. Sun & Huang (2021) assessed spatial and temporal variations of ice loads during ship-ice glancing impact tests in an ice tank. The study analyzed tactile sensor data from the test results to classify various local ice failure processes such as pure crushing, spalling and non-simultaneous crushing. It proposed an approach to generate migrating non-uniform ice loads based on Gaussian functions. Müller et al. (2024) investigated the influences of initial ice contact shape on the ice failure modes and resulting impact loads against a rigid steel plate through small-scale drop tower experiments conducted at the Hamburg University of Technology. The study considered 27 cylindrical ice specimens with varying end shapes of cone, dome, wedge, ellipse and other shapes. The experiments were conducted at a ship-ice interaction speed of 2000 mm/s.

Guo et al. (2023) evaluated the ramming performance of an icebreaking research vessel (IRV) model in thick artificial ice at different ship speeds. While these tests may not capture all aspects of the ship ramming process, they provided valuable data for future studies on ramming performance.

4.1.3 Modelling and Simulation Effort

Nowadays, modelling and simulations are the primary research tools for investigating ship-ice interaction processes and predicting ice loads for ships operating in ice. During this reporting period, these modelling and simulation efforts have been predominant in predicting ice loads compared to full-scale and model-scale approaches. These efforts included a range of analytical, semi-empirical, numerical and probabilistic methods.

Analytical and Semi-empirical Approach

Various analytical and semi-empirical approaches, including several machine learning algorithms for predicting ship resistance and global ice loads in different ice features, were employed by several studies (Cai et al., 2022; Ryan et al., 2021; Sun et al., 2022b; Uto et al., 2024; Zhou et al., 2022). Uto et al. (2024) proposed a ship resistance methodology that coupled the World Meteorological Organization (WMO) egg code to account for the natural variability of sea ice. The methods derived mean, maximum and minimum resistances assuming three independent sub-ice regimes in each ice regime. The thickness and size of the ice floes within the ice regimes were estimated based on the stage of development and form of the sea ice in the WMO egg code. The methodology was validated against the full-scale thrust measurement data from the ice trials onboard the icebreaker P/V SOYA.

Sun et al. (2022b) applied an Artificial Neural Network (ANN) model based on 140 data sets from 18 sea trials and model-scale tests to develop a ship resistance model. The study found that seven features related to ship design and ice mechanical properties (ship length, ship breadth, ship draft, stem angle, ship speed, ice thickness and ice flexural strength) along with a Radial Basis Function-Particle Swarm Optimization algorithm (RBF-PSO), could provide a generalized prediction model for ice resistance. Similarly, Ryan et al. (2021) utilized a set of existing ice resistance formulas along with open-water resistance, and fuel consumption formulas to develop an Arctic Ship Performance Model (ASPM). The ASPM was implemented in the MATLAB platform using a machine learning algorithm and validated against full-scale measurement data from the previous Arctic Cargo ship voyage across the Northern Sea Route (NSR). Sazonov & Dobrodeev (2021) proposed a modified semi-empirical method to predict ice resistance in level and broken ice channels for large vessels. The method was validated against model test data from the ice basin of the Krylov State Research Centre.

Cai et al. (2022) employed an image analysis technique to estimate the circumferential crack size of broken ice, a vital parameter to simulate different level ice-breaking processes. The study adopted a deep learning YOLACT model, an instance segmentation model, for improved detection of broken ice floes in the images. A new image processing algorithm was used to analyze the detected ice floes that quantifies the circumferential cracks. Zhou et al. (2022) also used a semi-empirical "circumferential crack model" to calculate the dynamic global ice load in level ice. The study statistically processed the measured ice properties data using the Nataf Transformation to derive the ice parameters required for numerical simulation.

Jiang et al. (2023c) investigated the effect of sea waves and hydrodynamic interaction on added mass and, consequently, on ice impact loads under various sea states. Added mass was estimated with and without accounting for hydrodynamic interactions using a multi-body hydrodynamic code and Potential flow theory, respectively. The Pierson-Moskowitz Spectrum (P-M Spectrum) was applied to evaluate added mass in various sea states. The added mass was then integrated into Popov's model for estimating ship ice impact loads. The results indicated that the sea waves affect the ice loads, while the influence of hydrodynamic interactions is minimal. The study also considered wave-induced motions on ice loads.

Nan et al. (2021) proposed a Long Short-Term Memory (LSTM) networks-based model as an alternative to traditional time and frequency domain inversion models to invert ice loads experienced by ships. The refined grey wolf optimizer algorithm was utilized to optimize the structures of the LSTM networks. The model was trained with ice load data generated by the finite element method.

Numerical Approach

Li & Huang (2022) comprehensively reviewed existing numerical approaches for simulating the ship-ice interaction process and predicting ship performance and local ice loads in broken ice fields. The study analyzed the efficiency and accuracy of these approaches and highlighted the need for further research in ship-ice interaction simulation. Han et al. (2021) presented a simulation technique combining FEM and CFD to predict icebreaking performance in pack ice conditions. This study modelled the pack ice as a rigid body with real-time control of hydrostatic and hydrodynamic forces acting on the pack ice. The hydrostatic force was expressed as a nonlinear restoring force, while the hydrodynamic force was determined using a drag coefficient validated through CFD. Chen et al. (2021a) investigated continuous-level icebreaking using traditional FEM and cohesive element methods (CEM). The FEM was employed to simulate bending fracture, while CEM was used to model the ship-ice interaction process. When compared to existing empirical formulas and model test results, the combined FEM and CEM method demonstrated better accuracy for thinner ice than for thicker ice.

Several studies employed the coupled CFD-DEM method in analyzing ship resistance in ice. DEM defines ship-ice and ice-ice contacts in this coupled method, while CFD captures the hydrodynamics. Zhang et al. (2022d) and Watanabe et al. (2023) applied this method to study ship resistance in brash ice. Zhang et al. (2022d) used the VOF method to simulate the free surface of the water, whereas the latter study applied the lattice Boltzmann method (LBM). Both studies compared their results against model tests and FSICR empirical formulations. Xie et al. (2023a) also used the CFD-DEM coupling approach to study the self-propulsion performance of an ice-strengthened bulk carrier model, which was towed by a carriage with a running propeller in brash ice. Tang et al. (2022b) and Xiong et al. (2023) implemented the coupled CFD-DEM method in STAR-CCM + to investigate ship resistance in pack ice with various ice concentrations and ship speeds for the KRISO Container Ship (KCS) model and the icebreaker Xue Long 2, respectively. Huang et al. (2022d) employed the CFD-DEM approach to study pre-sawn ice resistance, whereas Hu et al. (2022a) utilized an in-house sphere-based DEM combined with CFD to simulate the ship performance in the Arctic marginal ice zone (MIZ).

Ren & Park (2023) developed a numerical model based on the moving particle semi-implicit (MPS) method to predict the ice resistance of a model icebreaker during continuous level icebreaking. The model assumed the level ice to be a continuous isotropic elastic material with a maximum tensile stress failure criterion for the bending and part of the crushing fractures. Crushing fracture due to the shear stress was ignored.

Chen et al. (2023d) numerically investigated the effect of the ice flexural strength on the level ice resistance for icebreakers. The study considered an ice material model previously developed by Sazidy (2015). It changed the critical simulation parameters within the model to achieve different flexural strength values and to analyze their effect on the ice resistance. The study employed a value criterion of the influencing parameters such as the plastic failure strain, Young's modulus, and compressive yield strength within a three-dimensional data space and then applied this criterion to a numerical simulation of an icebreaker running in level ice. The simulation results were compared with semi-empirical models of ice resistance.

Xuan et al. (2021) and Xuan et al. (2023) numerically investigated the global ice load characteristics during manoeuvring captive motion for the icebreaker Xue Long in level ice and broken ice fields, respectively. The model in Xuan et al. (2023) combined DEM and a drag model, while the later-level ice load model utilized a traditional FEA approach with an element erosion technique to simulate ice failure. Both the models were validated against towing test data from the icebreaker Araon model. Yang et al. (2024) simulated the turning and Zig-Zag manoeuvring motions in broken ice fields by coupling the non-smooth discrete element method (NDEM) with the 3-DOF Maneuvering Modeling Group (MMG) model. They investigated the effect of ice concentration, size, thickness, ship speed and rudder angle on the ship's manoeuvring trajectory and steering flexibility.

Gu et al. (2022) applied a circumferential crack fracture-based numerical model to analyze the turning performance of icebreaker AHTS/IB Tor Viking II in level ice. The model was based on the previous research and enhanced accuracy by incorporating ice bending and crushing failure processes. Liu & Ji (2021) investigated the ice resistance of ships in escort operations in level ice using the dilated polyhedron-based DEM (DPDEM), which applied an improved bond-failure criterion to enhance the stability of the simulation for the sea ice-breaking process. Erceg et al. (2022) provided a numerical model to predict local ice loads resulting from ship-ice interaction in level ice, using a dedicated contact algorithm that included contact detection, force evaluation and cusp formation for arbitrary bow shapes at the waterline. The ice sheet was modelled as a homogeneous plate resting on an elastic foundation.

Ni et al. (2024) used the Cohesive Element Method (CEM) to simulate ship collisions with rafted ice. Considering the freezing strength and porosity of ice, the rafted ice was modelled as a multilayer isotropic elastoplastic material. The numerical results were compared with model-scale data from ice-freezing strength experiments conducted at the Svalbard University Laboratory. In a separate study, Ni et al. (2022) modified the length-to-thickness ratio in the original cohesive zone formulation based on several J-integral assumptions and existing research, providing an enhanced formula for simulating Mode I type crack propagation in ice. The accuracy of this formula was assessed using an FEA model of a double cantilever beam and experimental results. Liu et al. (2023b) also investigated the fracture behaviour in level ice using a couped bond-based peridynamic and traditional FEA (PD-FEA) through simulations of the cantilever beam and Mode I crack failure. Yu et al. (2023b) also employed a cohesive element formulation for modelling a wedge-shaped ice indenter within a user-defined subroutine VUMAT in ABAQUS/EXPLICIT to study the dynamic responses of the stiffened panel against the indenter impact. The tri-axial ice model was considered as an isotropic, rate-dependent, elastic-plastic material that fails in a quasi-brittle manner.

Jin et al. (2021) numerically simulated the low-speed icebreaking process for a hovercraft based on a simplified circumferential icebreaking pattern. The model was compared against existing model test data, and the influence of ice thickness and bending strength on ice loads was examined.

Statistical Approach

Zhang et al. (2023c) employed a bivariate reliability approach to analyze bow ice load data for an oil tanker operating in the Arctic environment. The data were derived from simulations of collisions between the oil tanker bow and 2D crushed ice fields with randomly distributed ice debris. The crushed ice fields were modeled using a cellular automata model. The results were then compared against Asymptotic and Gumbel logistics models.

Zhong et al. (2023) developed a methodology to monitor ice-induced hull pressure for ships navigating through ice floe fields. The methodology is based on statistical analysis of ice pressure data from ice model tests and coupled CFD-DEM numerical simulations. The model tests were carried out in the towing tank at Harbin Engineering University (HEU), where the ice model was made of non-refrigerated synthetic polypropylene, and the ice pressure was measured using tactile pressure sensors. The methodology was applied to a single-ship case study; however, further investigation is required to verify its robustness.

Li et al. (2024b) provided a combined statistical and numerical approach to predict extreme ice loads on various parts of the icebreaking hull. The study utilized a rejection sampling technique to generate stochastic ice fields that account for the variability of sea ice thickness. The interaction between the icebreaking hull and these stochastic ice fields was then numerically simulated to estimate extreme ice loads. The numerical results were validated against existing ice tank test data from the Xue Long 2 model.

4.2 Ice Loads on Offshore Structures

4.2.1 General

Ice load is a common force experienced by offshore structures located in the Arctic or ice-infested regions. Many factors, such as environmental conditions, ice properties, and structural characteristics, influence the ice load acting on an offshore structure. The mechanical properties of ice are complex, and the physical processes and failure modes of ice interaction with different offshore structures vary. The variable failure modes of ice and its complex mechanical properties make it challenging to estimate the ice load on offshore structures. Ice loads can cause significant damage to offshore structures. When a structure collides with ice, a dynamic impact load is generated, resulting in structural damage. Continuous pulsating loads can cause vibrations in the structure, which may lead to fatigue failure over time.

The comprehensive interaction between ice and structures is still challenging to predict. However, specific characteristics of ice forces have been observed in model-scale or full-scale tests. Crushing failure is the primary failure mode between ice and vertical structures. The ductile failure of ice typically occurs at a relatively low velocity, and ice loads in time histories exhibit saw-tooth patterns. At a high velocity, the brittle failure occurs, and spalling or flaking propagates on the periphery of the contact interface, inducing a "line-like" ice load in the centre of the interface. For the inclined structure, bending failures also occur during interaction.

Currently, numerical simulations and experiments remain two pivotal approaches for studying ice loads on offshore structures. In numerical simulations, two main methods are used: one simplifies the ice-offshore structure interaction process into a mechanical model, and the other simulates the detailed ice-structure interaction based on a specific numerical algorithm. Experimental research primarily focuses on model-scale tests, while full-scale observations have become relatively rare in recent years.

4.2.2 Modelling Effort (Analytical, Numerical and Statistical Approach)

Theoretical models are classical approaches for dealing with ice and structure interactions. They are derived from analytical models or data from full-scale or model-scale tests. These methods enable rapid prediction of ice resistance and are often used in the initial stages of engineering design and load prediction. Yu and Low (2022) developed a phenomenological model based on a stress-strain relationship equation that captures ductile-brittle transition phenomena under different strain rates. The model was validated against rigid indentation tests, and the predictions were found to match the experimental data well regarding effective pressure, peak pressure, and the time series of ice loads.

Because of the increasing computational power in recent years, numerical simulations have become the most popular approach to investigate ice loads on offshore structures. Truong & Jang (2022) developed a numerical model for level ice-structure interaction problems and validated it by comparing the estimated results with relevant ice collision test data as well as existing simulation results. Using this model, the effects of strain gauge arrangements and corresponding load cells on the estimation of local ice pressures acting on actual offshore structures were investigated. These efforts can improve the accuracy and reliability of the Influence Coefficient Method (ICM), which is an ice load prediction method based on a coefficient matrix, as described earlier. Makarov et al. (2022) reviewed various ice load calculation methods for simulating ice-structure interaction. By comparison, the authors concluded that numerical methods such as FEM and DEM were the most frequently used and provided recommendations on the use of SPG (Smooth Particle Galerkin Method) for the ice-structure interaction modelling. Song et al. (2023) used the nonlinear FEM to investigate the dynamic response of monopile-supported offshore wind turbines. The dynamic response of ice-structure interaction was analyzed, and the influence of ice drift velocity and average wind speed was examined using a statistical method. The simulation results were compared with the ISO 19906 and International Electrotechnical Commission (IEC) standards to gain insight into ice loads. The predictions obtained by the proposed method agreed well with these two international design standards, especially for the maximum ice load.

With the increasing demand for energy and the development and utilization of renewable energy, the advantages of offshore wind energy have become increasingly prominent, especially in polar areas, due to the abundant and high-quality wind. However, building offshore wind turbines in these regions also faces unprecedented challenges because extreme ice loads often lead to collapse or damage to structures. Therefore, in recent years, many researchers have conducted in-depth studies about the effect of ice loads on the structure of offshore wind turbines (OWTs).

Ji & Yang (2022) presented a coupling model of the discrete element method (DEM) and the finite element method (FEM) to study the impact of ice loads and ice-induced vibrations on the monopile foundation of OWT in the process of ice crushing. The results demonstrated that the ice thickness was positively correlated with the ice force on the structure, which was the intuitive representation of the ice load energy. The effect of speed on the ice force model appeared only for large ice thicknesses. Liu et al. (2022d) studied structures' dynamic response and damage during the ice-structure interaction using the fully coupled model with the fluid-structure interaction (FSI) method. The numerical results were consistent with the calculated ice force from the ice force spectrum fitting method. They also showed that the existence of fluid affected the collision process as well as the ice-crushing behaviour. Wang et al. (2022a) established a fully coupled model of a monopile OWT structure with an anti-ice cone. The dynamic response of the OWT under the combined wind and ice load was studied by using the cohesive element method (CEM). The angle of the anti-ice structure had an important effect on the dynamic ice load and motion response of the OWT structure. Changes in the cone angle could influence the pattern of ice destruction, and the optimal ice-breaking cone angle for the OWT in the operating state is 60°. Barooni et al. (2022) constructed a fully coupled numerical model of a floating OWT. Its dynamic response to ice-induced loads was investigated comprehensively using several simulation studies. The ice load had little effect on the output power when the wind turbine operated at rated wind speeds. Shi et al. (2023) used a numerical method based on the three-dimensional interaction model to predict the ice dynamics and structural responses. The area damage rate and fatigue damage rate of the monopile foundation were proposed for further study under wind and ice loads. The results showed that the ice thickness had a more significant effect on OWT damage and fatigue than ice speed. Wu et al. (2024) introduced an OWT dynamic analysis framework that included soil-structure interaction (SSI), random ice loads, and aerodynamic loads. The results showed that when considering SSI, the OWT with a flexible foundation demonstrated notably greater tower-top displacements in comparison to a fixed foundation when subjected to ice loads. The bending moment at the mudline of the OWT under combined ice-aerodynamic loads exhibited an increasing trend compared to the cases of individual ice or aerodynamic loads.

In addition to wind resources, the Arctic region is also rich in oil and gas resources, so the development and utilization of these resources have significant development prospects. Global warming has accelerated the melting of ice and snow in this region, creating favorable conditions for oil and gas exploration. However, the impact of ice loads on offshore platforms also poses challenges to the exploitation of these resources. At present, some progress has been made in the research of offshore platforms under ice loads, mainly aiming at the floating offshore platform adapted to the polar environment. Zhang et al. (2022a) adopted a fully coupled time-domain algorithm to study the dynamic performance of semi-submersible platforms under ice and wave conditions. The paper indicated that the optimized structure could significantly reduce the peak and average ice load, the surge motion and pitch motion of the platform, and the tension of the mooring system. Zhu & Ji (2022) presented the spherical discrete element method (DEM) to simulate and estimate the ice load and mooring force on the moored structure during the ice-structure interaction. The numerical results were compared with nonlinear finite element calculations as well as model-scale experiments. It was demonstrated that the maximum ice load and mooring force had a linear correlation with the ice thickness. The mean and maximum values of the structure's motion also increased as the ice thickness increased. Han et al. (2022) constructed a numerical simulation model of a large-volume gravity-based structure (GBS) towing process assisted by multiple tugs in polygon-shaped broken ice floes. The dynamics of the towing system demonstrated that ice characteristics such as ice thickness, ice concentration and floe size affected the ice load of the multi-body coupled towing system. The forward speed of tugs and GBS also decreased with the increase in ice thickness, ice concentration and floe size.

4.2.3 Model-Scale Experiments

Compared with full-scale measurements, the cost of model-scale tests can be significantly lower, and they can be repeated to explore the influence of different parameters. Herrnring et al. (2021) performed ice extrusion and double pendulum tests and compared the results with the maximum ice force and contact pressure obtained from numerical simulations. The finite element numerical model was based on the Mohr-Coulomb node splitting (MCNS) model to simulate the ice-structure interaction problem dominated by crushing and focused on the brittleness of the ice model. In the process of verification, the MCNS model agreed well with the experimental results. Lemström et al. (2022) conducted laboratory-scale experiments on ice-structure interaction processes in shallow water by pushing a ten-meter-wide ice sheet against an inclined structure of the same width. The results showed that the ice loading process on the inclined structure in shallow water consisted of two distinct phases: the initial phase with a linearly increasing ice load due to an increase in ice mass above the water line and the steady-state stage with an approximately constant load. Tian et al. (2022) carried out model tests to investigate the ice-loading mechanism on a typical structure. They analyzed the failure behaviour of level ice as well as the influence of ice loads on the structures. Numerical simulations based on the cohesive element model were adopted to evaluate the effects of the crack propagation and the crushing process of level ice. From the study, they found that the global ice load of a vertical structure was random, while the global ice load on an inclined plate structure presented a distinct pulsation characteristic and strong periodicity due to the relatively constant frequency of the bending failure. Hendrikse et al. (2022) conducted model tests at the Aalto University's ice tank. They collected a large amount of data on ice-structure interaction, mainly from ice failing against a vertically sided cylindrical pile. The dataset included various model test results from offshore wind turbine structures, structural models representing a series of single- and multi-degree-of-freedom oscillators, scaled dynamic models of the Norströmsgrund lighthouse, and the Molikpaq caisson structure. The results of the experiment provided a reference for the subsequent relevant research. Van den Berg et al. (2022) analyzed the structural responses during intermittent crushing through an experiment of ice-structure interaction. The structural model was represented by a hybrid test device that combines physical and numerical components to simulate the behaviour of a specific test structure in real time. The experimental results showed that the median ice forcing frequency was linearly related to the ice drift velocity. Owen et al. (2023) investigated the peak load-velocity dependence of the model ice used at Aalto University's ice tank. The development of frequency lock-in and intermittent crushing in the model ice was analyzed using a single-degree-of-freedom (DOF) structure. The study found that peak loads on compliant structures cannot exceed peak loads on rigid structures in the same ice conditions, with the only difference being that the peak loads on compliant structures occur at higher ice drift speeds in the far field due to the change in relative velocity.

4.2.4 Full-Scale Measurement and Trials

Due to the difficulty and high cost of full-scale measurements in harsh conditions, relevant studies are rare, making the data extremely valuable. These data can be utilized to inform new designs, analyze important phenomena, and validate new methods. Wang and Zhang (2023) designed and installed an isolation cone system on a mooring dolphin platform in the Bohai Sea. Field tests showed that the acceleration of the ice-induced vibration of the platform structure decreased significantly and that the isolation cone system significantly reduced the ice-induced vibration of the structure.

4.2.5 Ice-Induced Vibration

In cold regions, ice loads constitute a significant factor affecting structural safety. Ice-induced vibration can be harmful to offshore structures and their associated equipment. Model-scale experiments are an important method for studying these issues, as they provide valuable data and verify the effectiveness of new designs and numerical methods. Hammer et al. (2023) conducted an experimental study on ice-induced vibrations of an offshore wind turbine on a monopile foundation. A real-time hybrid test setup was used to incorporate the specific modal properties of an offshore wind turbine at the point of ice action. Experiments showed that the vibration caused by ice depended on the speed of ice drift. The interaction at low-speed results in intermittent crushing, and the multi-modal interaction regime gives rise to the largest bending moments of the support structure. Liu et al. (2023) studied two platform designs to reduce ice-induced vibration and improve the platforms' self-centering ability. Both platforms were tested in the laboratory, and the data were used to develop numerical simulations. The results showed that the ice-induced vibrations of the two novel jacket platforms with concrete-filled double-skin legs were reduced under ice load conditions, and the self-centring offshore platform with concrete-filled double-skin legs had a better control effect. Hammer and Hendrikse (2023) experimentally investigated the effect of wind and ice loading direction inconsistency on the development of ice-induced vibration in offshore wind turbines. The test results showed that misaligned scenarios result in the development of sustained ice-induced vibrations in the direction of the ice load. Furthermore, steady-state high-amplitude ice-induced vibrations could grow more efficiently in a misaligned scenario than in an aligned scenario.

Several researchers have studied the ice-induced vibration problem using various mathematical or numerical methods. Abramian et al. (2022) presented a mathematical analysis of an extended model describing a sea ice-induced frequency lock-in for vertically-sided offshore structures. The results showed that a frequency lock-in regime occurred during ice-induced vibrations when the dominant ice force frequency was close to the natural frequency of the structure. Gagnon (2022) used LS-Dyna to develop a full-scale three-dimensional numerical simulation model of ice-induced vibrations in structures and applied it to the Molikpaq facility for an ice encroachment event in the Beaufort Sea. The ice model was made of crushable foam material incorporating regular spallation events. The model manifested typical characteristics of ice-crushing in the brittle regime. A narrow horizontal hard zone (relatively intact high-interface-pressure ice) was present in the mid-height region of the ice-edge contact area. The model could be configured for various types of structures (offshore platforms, offshore wind turbines, moored structures, etc.) that interact with ice sheets or other ice masses. Long et al. (2023) used the discrete element method (DEM) and the parallel bond model to investigate the interaction between sea ice and offshore wind turbines (OWTs). They analyzed the ice-induced vibration of OWTs at various ice velocities. The results showed that brittle failure of sea ice occurred at higher ice drift speeds, leading to random structure vibration. At lower ice drift speeds, both brittle and ductile failure of sea ice resulted in self-excited vibrations. Zhang et al. (2024) used the cohesive element method (CEM) to analyze the ice-induced vibration of offshore structures. They studied the vibration response of the sea ice-jacket platform under both unprotected and protected conditions with ice-breaking cones. The results showed that the motion response of offshore platforms exhibits a positive correlation with the impact speed of the ice, while the sensitivity of this impact was found to be minimal. Furthermore, the influence of different ice directions on the vibration response of multi-leg offshore platforms was significant, and the shielding effect had a major impact on the platform's response.

With the development of interdisciplinary numerical models, machine learning methods have also been applied to simulate ice-induced vibration phenomena and predict the structural response. Zhang et al. (2022b) used gated recurrent neural networks (GRNN) to predict and suppress the response of offshore platforms under ice loads. The numerical results showed that the ice-induced vibration response prediction method based on the Gated Recurrent Unit Network (GRU) can predict the structural response with satisfactory accuracy. The ice-induced vibration response control method based on the Long-Short-Term Memory Network (LSTM) and GRU network effectively learns the Linear Quadratic Regulator (LQR) method and achieves a significant control effect. Chen et al. (2023) optimized a short-term prediction model for structural response using machine learning theory. A prediction model was established based on the field monitoring data from the Bohai JZ20-2MUQ platform. By utilizing the strong self-learning ability of neural networks for data prediction, the ice-induced vibration (IIV) response prediction model for a specific platform can be realized with high prediction accuracy, relying solely on theoretical analysis and engineering experience for sample feature extraction.

5 Uncertainty

Uncertainty quantification in the evaluation of loads on ships and offshore structures covers several different aspects. In the reviewed literature, uncertainty estimation is carried out for numerical methods, model-scale, and full-scale experiments. Various available standards for the quantification and expression of uncertainty have been used, such as ISO (ISO/IEC Guide 98-3:2008 - Uncertainty of Measurement—Part 3: Guide to the Expression of Uncertainty in Measurement (GUM:1995), n.d.) (ISO/IEC Guide 98-3:2008 - Uncertainty of Measurement—Part 3: Guide to the Expression of Uncertainty in Measurement (GUM:1995), n.d.) standards, ITTC standards, ISSC standards, and American Society of Mechanical Engineers standards.

5.1 Ships

5.1.1 Slamming Loads

The quantification of slamming loads with full-scale measurements is quite a challenging task. For such a reason, model-scale tests and numerical simulations are the most common way for slamming load evaluation. The complexity of fluid-structure interaction simulations and experiments requires a careful estimation of the uncertainties. In (Yan et al., 2022c), a systematic comparison of experiments with fluid-structure interaction (FSI) simulations for flat plate water entry was presented. Particular focus was attributed to hydroelasticity and air trapping effects, quantification of the experimental and numerical uncertainties, and the validity of modelling assumptions for the prediction of bottom slamming-induced loads. Verification and validation to estimate errors were carried out using the American Society of Mechanical Engineers standard. The experimental part was carried out using an electrically driven, vertically moving platform made of polyoxymethylene copolymers (POM-C).

The platform had three pressure sensors, one accelerometer inside the test body, and two strain gauges at the bottom. The numerical simulation followed the two-way fluid-flexible-structure interaction (FFSI) implicit coupling scheme introduced by Lakshmynarayanana & Hirdaris (2020). The FSI was modelled using the STAR CCM + /ABAQUS interface. Generally, good agreement between experiments and numerical simulations was found for both the rigid and hydroelastic cases. The influence of air entrapment was investigated by monitoring the presence of air during experiments with a high-speed camera. The uncertainty estimation for both the experimental and numerical parts was carried out. The uncertainty estimation comprised the accuracy of the sensors, the acquisition system, and the uncertainties due to the repetition of experiments (30 repetitions were performed). Discretization uncertainties were quantified using the Grid Convergence Index (GCI) method. The errors were lower for quantities not influenced by local effects (forces) in comparison with pressures and strains that were affected by larger errors. The authors used the American Society of Mechanical Engineers standard for "Verification and Validation" to estimate the errors in both the experimental and numerical parts. The main findings and suggestions were: for local slamming, a high number of experiments was recommended; the validation uncertainty depended on the parameters taken into account; the percentage validation uncertainty tended to be smaller for higher impact speeds in the case of pressure and was roughly independent of the velocity in the case of force.

Regarding a purely numerical simulation, Wang et al. (2021b) presented an analysis of numerical uncertainty due to discretization in the Arbitrary Lagrangian-Eulerian (ALE) Finite Element method for the prediction of water-slamming loads. The paper quantified uncertainty using two ITTC-recommended methods, which used the independent grid and time-based discretization as well as a constant Courant-Friedrichs-Lewy (CFL) number-based discretization approach. The study found that the uncertainty values obtained from the free-falling models were closer to those from experiments. The same conclusions regarding the better accuracy of the Courant-Friedrichs-Lewy (CFL) method in comparison with the ITTC method were drawn, but with a different application, as shown in Islam & Guedes Soares (2021).

5.1.2 Global Loads

Global loads on ships can be due to waves, grounding, or collision. The study of global load effects can be carried out by numerical simulations, considering the hull girder as either a rigid body or a flexible structure and taking into account both linear and non-linear effects. On the other hand, experimental model-scale tests can be carried out in dedicated facilities. The main uncertainties are related to the transfer function evaluation for linear effects, while the correct simulation of hydroelasticity, whipping, and springing effects are the challenges for non-linear effects estimation. Regarding experiments, the number of repetitions and the generation of waves are the main topics.

Regarding linear effects, the results of a benchmark study organized by the ISSC-ITTC Joint Committee on global linear wave loads on a container ship with forward speed were presented by Parunov et al. (2022). When linear effects were considered, the main uncertainties were linked to evaluating transfer functions. To address this, the paper analyzed different seakeeping codes and their impact on the long-term extreme vertical wave bending moments. Fifteen seakeeping codes were tested, belonging to one of the three methods: strip theory, 3D frequency-domain method, and 3D time-domain method. Quantification of the uncertainty of the transfer functions was carried out in the form of the Frequency Independent Model Error (FIME) given by the Eq. (1).

$$\varphi = \frac{\sum_i \widehat{H_i} H_i}{\sum_i H_i^2} \tag{1}$$

where, $\widehat{H_i}$ are the measured transfer functions $\widehat{H_i}$ is the calculated transfer function as Total Difference, as Kim & Kim (2016) determined. The uncertainty of the experimental results was expressed following the ISO Guide (ISO, 1995). Uncertainties in the seakeeping codes have been evaluated for the heave, pitch, and vertical wave bending moment at midship for three heading angles. The paper found that strip theory codes showed lower dispersion both compared to the average of the other codes and the experimental values.

The uncertainties in the transfer functions were assessed through the Frequency Independent Model Error (FIME) (see Eq. 1) and other more sophisticated methods.

The paper concluded that the FIME was the most promising method. Both strip theory and 3D methods gave reasonably good results for linear wave loads. More considerable uncertainties were found in the non-linear response, such as the hydro-elastic response.

The uncertainties in non-linear responses were investigated in Hirdaris et al. (2023) regarding the importance of uncertainties in hull girder loads influenced by flexible fluid-structure interactions. The paper mentioned several accidents in which hydroelasticity could have played a significant role. Nevertheless, the influence of springing and whipping effects on structural design was not taken into account in a coordinated way due to substantial uncertainties linked with the modelling of the phenomenon. The paper concluded that despite 40 years of research, a clear understanding of hydroelastic uncertainties in wave-induced loads remained elusive.

Challenges include theoretical assumptions, data processing, and limited industry-academia collaboration. Structural models offer detailed stress information, while linear frequency domain methods are mature and time-domain approaches are emerging. Computational Fluid Dynamics (CFD) methods are still evolving but highlight the importance of coupling in hull girder modelling. Uncertainty quantification methods are crucial for design but face challenges due to nonlinearities and data ambiguity.

A fully coupled CFD-FEM method to estimate ship wave loads and hydroelastic responses was presented by Huang et al. (2022f). The method was tested on a standard S175 containership, comparing results with available experimental data. Uncertainty analyses were conducted to determine the influence of fluid grid density, time step, fluid domain size, fluid viscosity, and hull structural element number. Uncertainties in the paper were studied using two methods based on Richardson Extrapolation (Correction Factor [(Stern et al., 2001)] and the Grid Convergence Index (GCI) using a factor of safety (FS)[(Celik et al., 2008)]). The main findings were that a CFD solver using the Euler Overlay Method (EOM)wave generation method can accurately and steadily produce 5th-order Stokes waves. It suggested that a shorter downstream region (1.5 times the ship length) did not significantly affect wave calculations and ship responses compared to the recommended 3L length (3 times the ship length), reducing computational workload. A smaller time step (recommended < 1/100 of 2-node natural vibration period) ensured accurate high-frequency whipping load reproduction. The study also highlighted significant differences in ship wave loads obtained by different methods, emphasizing the complexity of factors influencing whipping responses, including structural model variations and external factors such as wave type and steepness.

Non-linear effects in global response for collision and grounding were presented in Kim et al. (2022b). In the paper, a benchmark study comparing the structural dynamic response by explicit nonlinear FEA approaches and the semi-numerical super-element method for grounding and collision for passenger ship was presented. In the paper, four scenarios for ship-ship collision were considered, and five different scenarios for grounding. The modelling methods accounted for hydrodynamic actions associated with operational and environmental conditions. The results were similar regarding maximum response for collision and grounding, except for oblique collision and groundings involving sharp rock. Uncertainties were reduced using the same inputs to the MCOL program (an LS-DYNA® module) and the same fracture criteria. Another important source of uncertainties was represented by the extent of the FEM model (partial or full ship) and by the fracture strains, well as by the meshing size.

Regarding experimental approaches, van Essen et al. (2023) reported an analysis of the required test durations for converged short-term waves and impact extreme value statistics using experimental analysis. In part 1 of the paper, guidelines were given for the convergence of most probable maximum wave crest heights and most probable maximum green water wave impact forces on a ferry. In contrast, in part 2, similar results for wave-in-deck loads on a stationary deck box were given. The key parameters for a good estimation were the exposure duration (e.g., 3 h for offshore structures) and the number of seeds (3 h stationary periods). Convergence plots were given in the paper that could be used to find the number of seeds needed to achieve a certain accuracy. A second aim of the paper was to understand if fitting (3-parameter Weibull fit) could influence the number of required seeds for convergence, but it was found to have a negligible influence. A final remark was that a large number of seeds (significant time and cost-consuming) were required to have reasonable estimates of impact load distributions, suggesting the need to investigate different methods.

5.1.3 Mooring

In Abdelwahab & Soares (2023), the uncertainty modelling of experimental results for a physical model of a tanker moored to a terminal inside a port was presented. During the experiments, 6 DOF motions and loads on the mooring lines were measured. The uncertainty evaluation followed the ITTC guidelines (ITTC, 2011). The uncertainty analysis of the experiments taken into account is listed in Fig. 17.

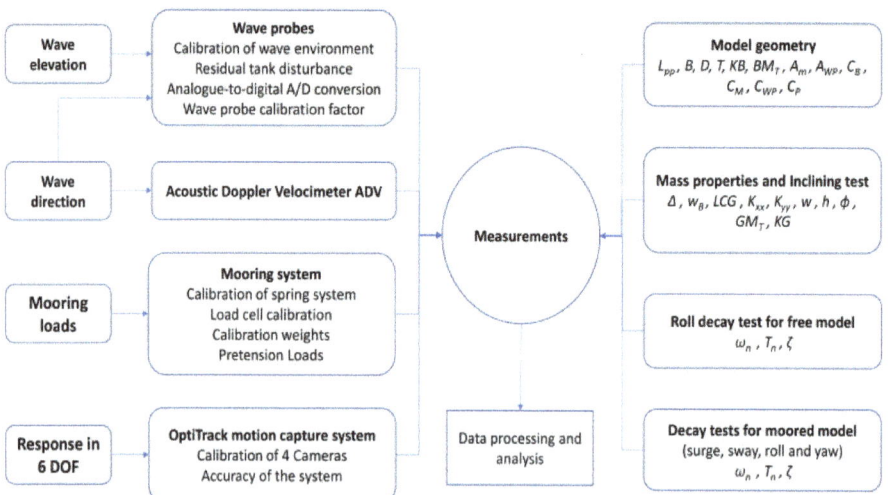

Fig. 17. Uncertainty sources for the model experiments from (Abdelwahab & Soares, 2023).

The authors found that type B uncertainties, related to wave measurements, model geometry, and mass distribution, were more significant than Type A uncertainties. Uncertainties in wave measurements in shallower waters involved non-linear effects and extremes. Non-linear influences in mooring systems, exciting forces, and ship response also impacted mooring load uncertainties. Performing systematic experimental uncertainty analysis was challenging but crucial for improving test quality despite its time and cost requirements. The results could benefit future research at the same facility by providing valuable insights into physical model uncertainties.

5.1.4 Others

In Gaidai et al. (2022), a novel structural reliability method was presented, particularly suitable for multidimensional structural responses. The method was applied to a container ship subjected to large deck panel stresses and extreme roll angles. The method was benchmarked with onboard measured time histories. The methodology presented could efficiently assess the reliability of high-dimensional dynamic systems, which is particularly useful when dealing with limited data sets. It was applied to real-time data from measured wave heights and vessel reliability during transatlantic voyages. The method produced reasonable confidence intervals and did not require restarting simulations after a system failure, making it a valuable tool for various nonlinear dynamic systems reliability studies across different engineering fields.

In Miratsu et al. (2022), the effect of ship operation on sea loads was studied using a statistical storm avoidance model of merchant ships to reconstruct the sea states encountered over 25 years in the North Atlantic. The uncertainties of the ship operation effect were investigated for different hindcasts (TodaiWW3-NK, IOWAGA, and ERA5). The sea states encountered were estimated by AIS data applying a storm avoidance model (Sasmal et al., 2021). The findings indicated that the ship operational effect averages between 0.82–0.88 with a standard deviation of 0.01–0.03, resulting in a 10–20% reduction in long-term motion and wave load predictions at a 10^{-8} probability level of exceedance corresponding to a minimum of a 25-year return period. However, this estimate carried uncertainty linked to various factors such as hindcast products, ship characteristics (ship length and type), and storm avoidance model settings.

5.2 Offshore Structures

In Islam & Guedes Soares (2021), an assessment of different uncertainty analysis methods commonly used for uncertainty quantification in Computational Fluid Dynamics (CFD) was carried out and a comparison constant Courant–Friedrichs–Lewy (CFL) number-based approach for uncertainty estimation to the ITTC recommended grid and time-independent procedures was given. Four different uncertainty estimation procedures were presented. CFD-related uncertainty quantification in wave load simulation was carried out using a fixed vertical cylinder as a case study. The study directly compared the uncertainty estimation of three methods: Factor of Safety, Correction Factor, and Least-squares root approach. Comparisons were made regarding forces and moments for the three methods, both for the Courant–Friedrichs–Lewy (CFL) number-based approach for uncertainty estimation and the ITTC-recommended grid and time-independent procedures. The paper concluded that the CFL number-based simulations were more stable and provided better convergence of results in all the cases. Regarding the verification study, the paper concluded that each of the four uncertainty analysis methods had pros and cons, and all were equally applicable for uncertainty assessment in CFD studies. A summary of the results is given in Fig. 18.

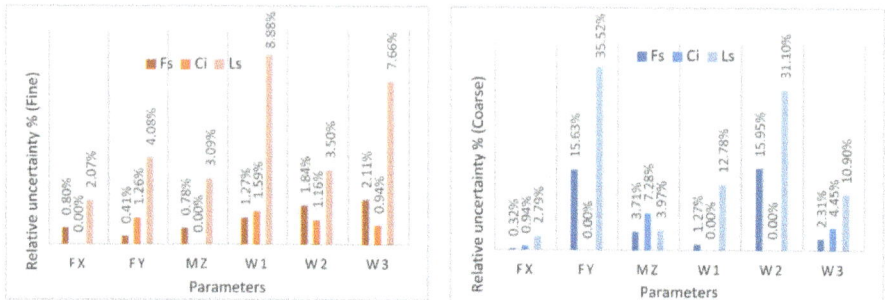

Fig. 18. Relative comparison among uncertainty at constant CFL number, using the Factor of safety, Correction factor, and Least-squares root methods by Islam & Guedes Soares (2021)

In Hallak et al. (2022), epistemic uncertainties to estimate the minimum necessary air gap for a semi-submersible accommodation unit in stormy conditions were assessed. The number of uncertainties in the air gap evaluation process of a semi-submersible platform were numerous, but the paper focused on model uncertainties. The uncertainty evaluation of the modelling was carried out by taking into account the uncertainties resulting from the binned values of scatter diagrams and from the assumption that the highest significant wave height sea state was the critical sea state. To investigate this, environmental contours were taken into account instead of a single value. Regarding the uncertainty implied in the finite set of control points, Kriging interpolation (Kleijnen, 2009) was used, allowing adaptation of the uncertainty propagation techniques to assess the variances of the estimates in points different from the control points. The requirements for operational and survival air gaps were defined in the paper as the return periods of 10 years (operational) and 100 years (survivability) plus a margin. The sea states were characterized with JONSWAP unidirectional wave spectra. For all scenarios considered, the actual critical sea states had significantly lower heights and main periods compared to the top points of the environmental contours. In the paper, it was found that FOSM methods always retained high levels of uncertainty due to the uncertainties in the estimated critical period when directly obtained from scatter diagrams. In conclusion, the paper suggested that the number of critical points had a strong effect on the uncertainties and that for long-term analysis, environmental contours had to be taken into account to reduce uncertainties.

5.3 Waves Description

In Bitner-Gregersen et al. (2022), a comparison of the uncertainty evaluation in wave data and models used for the design and laboratory testing of marine structures given by the two organizations ISSC (International Ship and Offshore Structures Congress) and ITTC (International Towing Tank Conference) was presented. In particular, the analysis focused on long-term wave descriptions. ISSC and ITTC use different definitions of uncertainties. ITTC uses the ISO definition (Type A cannot be reduced, Type B can be reduced) while ISSC uses the definitions applied by the Society for Risk Analysis community (aleatory and epistemic), and in the paper, the ISSC definition was used. One of the main sources of uncertainties was represented by the dataset, which had to be carefully selected for the specific area of interest. A finding of the paper was that many datasets were not well-calibrated for the extremes. Another important parameter to be carefully checked was the time extension of the database, which had to be not too short. The paper highlighted the importance of considering new databases such as ECMWF ERA5, Ifremer IOWAGA and Japanese TodaiWW3-NK to update the IACS scatter diagrams. In addition, frequency analysis could be carried out to find sea areas with similar statistical characteristics, improving the dataset subdivision of the scatter diagrams suggested by IACS.

In Zhang et al. (2022c), an investigation of the influence of initial uncertainties in predicting the geometrical properties of maximum waves in a short-term sea state was given. The uncertainty was investigated using Monte Carlo simulations. The Chalikov-Sheinin (CS) model was used to model the wave. Uncertainties induced by the initial conditions were evaluated by six wave profile parameters (the horizontal asymmetry parameter, the vertical asymmetry parameter, the length of maximum wave L, the maximum crest-to-trough height and the block coefficient). The maximum crest, trough, and crest-to-trough heights in a sea state were only sensitive to the amplitudes of a certain range of free waves (around the peak wavenumber). The randomness caused by the initial random-phase/amplitude model (natural uncertainty) was very large. The behaviour of the three asymmetry parameters was very high, while the maximum crest was less influenced.

Vanem et al. (2022) reviewed the latest findings regarding the Statistical modelling of the ocean environment. The review covered long-term and short-term statistics in deep and shallow water, extreme value and extreme wave statistics, multivariate analysis and joint distributions, machine learning applications and spatial and temporal statistics.

6 Conclusion

The question of determining and analysing relevant wave load scenarios that a ship might encounter in its operational lifetime and the respective consequences on ship design and operation is a topic of continuous development and research.

Within the reporting period of this committee, progress has been made in different fields related to wave loads on ships. The key insights from Sect. 2.1 are summarized below.

Potential flow theory remains a popular approach for modelling wave-induced loads on ships, offering efficient solutions for linear and weakly nonlinear problems. Recent advances in frequency-domain and time-domain methods provide approaches for improved modelling of complex boundary conditions, such as those encountered in shallow and confined waters. Recent work has focused on hybrid and fully nonlinear models to improve accuracy, particularly for added resistance and structural responses in irregular wave conditions. However, fully nonlinear approaches remain computationally demanding, prompting ongoing efforts to optimize their efficiency through parallel computation.

Field methods for wave loads on ships have been discussed, specifically in the context of roll damping and green water loading. When applied for estimating roll damping field methods compare well with model-scale experiments. Note that this does not guarantee predictive value at full scale. Field methods with advanced free surface transport formulation allow for the systematic study of green water in a range of wave steepness to obtain improved deck structure designs. Future field methods for slamming of ships in waves may include aeration compressibility due to entrained air in the water to get lower, more realistic pressure maxima during slamming.

Experimental methods continue to play a critical role in validating numerical models and improve our understanding of hydroelastic responses. Recent studies highlight the importance of segmented and fully elastic ship models, which enable detailed investigations into nonlinear phenomena such as whipping, springing, and slamming. Advances in instrumentation, such as Fiber Bragg Grating sensors, have the potential to enhance the accuracy of pressure and strain measurements. Efforts to reduce uncertainties in test setups, such as controlling wave steepness and improving model stiffness scaling, contribute to improving the reliability of experimental results.

Full-scale measurements, particularly through onboard monitoring systems, provide invaluable data for understanding operational loads and validating design criteria. Methods for indirect load estimation, using strain and motion data, have advanced significantly, allowing accurate reconstructions of unmeasured responses. These techniques enhance fatigue assessments and structural health monitoring, but challenges persist in processing and integrating large data streams into design and operational frameworks.

Data-driven approaches have the potential to significantly increase the efficiency of the prediction and analysis of wave loads and ship responses. Recent studies show that machine learning models, such as neural networks and recurrent models, offer promising tools for real-time load predictions and operational optimization. Multi-fidelity methods that combine low- and high-fidelity data have enhanced the accuracy of structural response predictions. Despite these advances, issues related to data quality, pre-processing, and integration into design workflows require further attention.

Slamming remains a critical phenomenon for structural integrity, particularly in rough seas. Recent research activities have improved the modelling and prediction of slamming loads through coupled fluid-structure interaction methods and advanced computational techniques. Experimental data from wedge-drop and segmented model tests provide valuable insights into local pressure distributions and global whipping responses. However, further work is needed to address uncertainties in asymmetric slamming and its impact on fatigue and failure modes.

For liquid cargo carriers, sloshing dynamics are critical in design and operation due to their impact on ship stability, motion behaviour and structural loads. Numerical and experimental studies have advanced understanding of global tank dynamics and local impact pressures. Innovative mitigation strategies, such as optimizing tank geometry and incorporating internal structures, appear to be promising. However, challenges remain in scaling sloshing tests and accounting for phase transitions in cryogenic fluids.

Similarly, below are the key findings from wave loads on offshore structures in Sect. 2.2.

Potential theory methods have seen refinements in linear and nonlinear formulations for efficient wave load predictions, mainly targeting wave run-up in structural elements, gap resonance, and prediction of free surface elevation in Oscillating Water Columns (OWC) type of Wave Energy Converters (WEC) resorting mainly to Boundary Element Methods, and the decomposition of the potential into first and second order problems while applying a perturbation based approach to its solution and coupling first order models with aerodynamic models for the OWC case studies.

Field methods for solving Navier-Stokes equations continue to be instrumental in understanding wave-structure interactions, addressing challenges such as highly deformed interfaces, wave breaking, and turbulence. Tools like OpenFOAM, STAR-CCM +, and REEF3D have facilitated innovative research into hydrodynamic loads, mooring dynamics, and wave-induced responses on offshore structures, complemented by hybrid and particle-based modelling approaches that enhance accuracy and efficiency in simulating complex offshore systems.

Recent studies have emphasized the critical role of wave-current interactions in determining hydrodynamic loads and structural responses in offshore structures. Key advancements include the investigation of wave-current effects on offshore wind turbines, aquaculture fish cages, tidal energy devices, submerged tunnels, floating bridges, and mooring buoys, showcasing enhanced predictive models and experimental approaches to optimize performance, reduce structural fatigue, and ensure stability under combined environmental loads.

The structural safety of offshore platforms relies heavily on accurately understanding and managing airgap and Wave-In-Deck (WID) loads, which have been the focus of extensive research in recent years. In high-fidelity numerical modelling, experimental validations and procedural improvements have enhanced the prediction of these complex, nonlinear phenomena. This leads to more efficient and robust platform designs while reducing conservatism in traditional assessment methods and better ensuring structural integrity under extreme wave conditions.

Regarding moorings, recent studies have advanced the design, analysis, and simulation of various mooring systems, including innovative solutions like shared moorings for floating wind farms, hybrid mooring materials combining synthetic fibres and steel ropes, and optimized configurations for renewable energy applications, while also addressing challenges related to fatigue resistance, corrosion effects, and dynamic responses under complex load conditions.

Advanced model testing remains a crucial activity in the study of wave loads on offshore structures, providing empirical insights that refine theoretical models and engineering practices. Recent research has explored wave impacts on offshore platforms, floating wind turbines (FOWTs), WECs, aquaculture structures, and very large floating structures (VLFS), using scale models, innovative techniques like Particle Image Velocimetry (PIV), and numerical validation to improve design, safety, and operational efficiency of the structures.

Over the past three years, the integration of data-driven methodologies and machine learning techniques into marine hydrodynamics and offshore engineering has led to significant advancements. Applications such as predicting wave-induced motions and hydrodynamic loads with models like LSTM and Wave-GAN, as well as the use of digital twins and reinforcement learning for mooring systems and coastal protection, have demonstrated notable improvements in predictive accuracy, computational efficiency, and adaptability, fostering the development of more resilient and optimized offshore structures. However, these advancements have not translated into a substantial enhancement in the fundamental understanding of wave-structure interaction. It suggests that while machine learning has proven valuable as a tool, it has not yet been instrumental in driving deeper theoretical insights in this domain.

In Sect. 3, the analysis of current and wind loads on ships and offshore structures was provided. The key findings emphasize the growing relevance of accurate load predictions, especially with modern vessels and offshore systems' increasing dimensions and operational complexity. Precise estimation of current and wind loads significantly influences and is critical to improve ship design and manoeuvrability, ensure stability of mooring systems, and increase energy generation efficiency and performance of offshore wind turbines. The integration of high-fidelity CFD simulations has proven essential for addressing the limitations of traditional experimental methods, particularly in the early phases of complex projects. Studies have shown that underestimating these aerohydrodynamic efforts can lead to premature structural fatigue and increased operational risks, among other problems. However, the combined effects of currents, wind and wave loads continue to pose a significant challenge to the design and operability of modern marine systems. In this context, future trends are related to the continuous evolution of hybrid approaches, numerical and experimental, promising enhanced accuracy in predicting and managing such environmental loads.

In Sect. 4, research on ice loads on ships and offshore structures has highlighted advancements across full-scale measurements, model-scale experiments, and modelling and simulation techniques. In Sect. 4.1, full-scale studies validated methods for estimating global and local ice loads under realistic conditions for ice loads on ships. They emphasized the importance of optimizing instrumentation setups for accurate measurements. Model-scale experiments provided cost-effective insights into ice resistance, ship performance in narrow ice channels, and impact loads, with findings showing that factors like ice thickness, channel width, and ship speed significantly affect ship-ice interactions. Simulations using advanced techniques such as coupled CFD-DEM, cohesive element methods, and machine learning algorithms enhanced the accuracy of predictions for ship resistance, ice impact loads, and manoeuvring in ice. Statistical approaches were applied to analyze extreme ice load scenarios and variability in sea ice conditions. These findings collectively contribute to improving the safety, efficiency, and design of ships operating in ice-covered waters.

Section 4.2 provided research on ice loads and their effects on offshore structures. Generally, ice-structure interactions are complex and influenced by environmental conditions, ice properties, and structural characteristics. Crushing failure is identified as a primary failure mode, with distinct ductile and brittle behaviours under varying velocities. Numerical simulations, including FEM, DEM, and coupled methods like FSI, dominate the field, offering insights into ice-induced vibrations, structural responses, and optimal designs for mitigating ice impacts. Studies on offshore wind turbines (OWTs) reveal the significance of ice thickness, drift velocity, and anti-ice cone angles in dynamic responses and fatigue resistance. Machine learning approaches have emerged as effective tools for predicting and controlling ice-induced vibration responses, enhancing accuracy and efficiency. Model-scale experiments remain crucial for exploring ice-loading mechanisms and validating numerical models, while full-scale observations, though rare, provide invaluable data for real-world applications. These efforts collectively improve the understanding of ice loads, advancing the design and safety of offshore platforms and wind turbines in polar regions.

In Sect. 5, Uncertainty quantification in the assessment of loads on ships and offshore structures was explained. Recent literature has examined slamming, global, and mooring loads on ships (Sect. 5.1), highlighting the substantial challenges posed by full-scale measurements of slamming loads. It has been demonstrated that a synergistic approach combining model-scale tests and advanced fluid-structure interaction simulations can effectively quantify slamming load uncertainties, underscoring the need for meticulous validation and favouring methods such as the CFL approach to enhance accuracy. Furthermore, an in-depth investigation of global loads on ships under various conditions has revealed that while linear wave load responses can be reliably evaluated using established numerical methods and transfer function assessments, significant uncertainties remain in non-linear dynamics. These uncertainties are influenced by hydroelasticity, fluid-structure interactions, and the complexities associated with experimental setups, indicating the necessity for improved methodologies and collaborations to refine load predictions and structural designs. Additionally, innovative reliability and statistical modelling techniques have been introduced to effectively address uncertainties in structural responses and the effects of ship operations, thereby providing valuable insights for enhancing the reliability and accuracy of marine engineering analyses.

In the context of offshore structures (Sect. 5.2), the studies underscored the critical importance of uncertainty quantification methods in CFD, highlighting the advantages of using the CFL number-based approach to achieve stability and convergence while emphasizing the necessity of incorporating environmental contours and accounting for multiple critical points to effectively manage uncertainties in assessing the air gap for semi-submersible platforms.

Moreover, in the Waves Description subsection (Sect. 5.3), the studies collectively highlighted the vital role of precise uncertainty evaluation methods in analyzing wave data and modelling marine structures, showcasing the differing definitions of uncertainties as articulated by ISSC and ITTC, the significance of dataset quality and calibration, the considerable influence of initial conditions on wave properties, and the substantial advancements in the statistical modelling of ocean environments.

This committee also performed a benchmark study (in Appendix A). Numerical simulations of flat plate impacts were conducted using CFD, SPH for rigid implications, and a two-way coupled FFSI approach with ABAQUS and STAR CCM+. A hybrid STAR CCM + method and an incompressible fluid solver in LS-DYNA were also used for hydroelastic water impacts. The simulations matched the dimensions and properties of experimental test bodies. Results showed similar peak loads between simulations and experiments, except the stiffer LS-DYNA FFSI—simulated void fractions beneath impacting plates aligned well with experimental high-speed imaging. Differences in peak pressures, forces, and strains were due to varying assumptions and high-frequency vibration modes. Future work should focus on uncertainty modelling and prediction.

Benchmark Study

Introduction

In rough seas, ocean-going ships are subject to impact-induced loads that may induce vibrations, and noise and affect structural integrity, especially in the ship's bow and stern areas. The same problem is practically relevant to the dynamics of high-speed vessels prone to bow or bottom-slamming-induced loads (Tavakoli et al., 2024).

Objectives

This section presents comparisons between experiments and simulations for the case of flat-water plate entry (Yan et al., 2022c). The focus of the benchmark is on rigid body and hydro-elastic dynamics using the following methods:

- A two-way coupled flexible fluid-structure interaction (FFSI) model based on ABAQUS and STAR CCM + simulations presented by Yan et al. (2022c).
- A hybrid hydroelasticity method was introduced by the National Research Council of Canada (NRC) (Sitek et al., 2021). Whereas this approach is similar to Yan et al. (2022c), it combines fluid and solid solvers inherently available in STARCCM +.
- Two different methods by the National Maritime Research Institute of Japan (NMRI) using the LSDYNA explicit fluid structure interactions scheme (Livermore Software Technology Corporation, 2014) and Smoothed Particle Hydrodynamics.

The outcomes of the above methodologies are compared with the results obtained from the experiments presented by Tödter et al. (2019).

Experimental Configurations

The test setup consisted of a vertically moving platform powered by an electric motor, allowing freely adjustable platform velocities. The topologies and sensors considered are shown in Fig. 19 and Fig. 20 and described in Table 1.

Several key points for easy reference are summarized in this section.

- The platform was positioned above the test section of a circulating water channel of 6.0 m length and 1.5 m width, with the water depth set at 0.55 m. A control loop kept the velocity of the test body constant during water impacts.
- The body's walls and flanges consisted of rigid aluminium plates. The quadratic flat bottom plate of 0.09 m^2 area had a thickness of 4.7 mm and was made of polyoxymethylene copolymers (POM-C) having an elasticity modulus of approximately 2.8 GPa.

- The rigid test body had an aluminium bottom plate of 12.0 mm thickness, which was welded to the walls of the test body. The bottom's shape and its area were the same for the elastic and the rigid test cases.
- Inside the test body, two stiffeners were fastened to the rigid bottom to minimize deformations during impact. Three pressure sensors and one accelerometer were placed inside both test bodies. Additionally, the elastic bottom was equipped with two strain gauges.
- Three absolute pressure sensors, Kulite XTM-190(M) depicted as P1, P2, and P3 in Fig. 1, were placed along the centre line of the bottom plates. Following hammer tests, the lowest natural plate frequency was estimated at 147.1 Hz in air, 53.0 Hz on the water surface, and 47.1 Hz when submerged.
- For the elastic bottom plate, small sleeves made of POM-C with a diameter of 10 mm and a total height of 10.92 mm were glued to the bottom plate at the sensor positions to allow the sensor mounting.
- The bottom plate thickness for the rigid test body was reduced to 10.92 mm over a circular area of 20 mm diameter at these sensor positions. The sensor's circular measurement area had a diameter of 3.8 mm.
- As the sensors recorded absolute pressures, the atmospheric pressure had to be subtracted from every measurement.
- A PCB 353B34 accelerometer was located a distance of 30 mm from the bodies' walls above the bottom plate to decouple accelerometer measurements from structural responses (depicted as Acc in Figure A1 and Figure A2). This accelerometer was able to measure acceleration peaks up to 50 × acceleration of gravity of 9.81 m/s^2.
- Two HBM 1-XY-93-3/350 strain gauges (depicted as S1, S2 in Fig. 19) were mounted along the center line of the elastic bottom plate at a distance of 100 and 150 mm between bottom plate edges and strain gauge centres, respectively.
- Two additional strain gauges (not shown in Fig. 19) were placed at the elastic test bodies' walls to monitor the temperature compensation of strain gauge measurements.

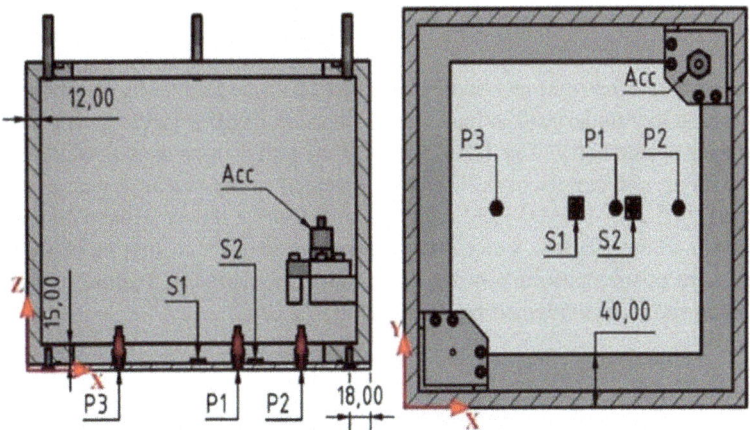

Fig. A1. Side view (left) and top view (right) of the test body with an elastic bottom and the arrangement of pressure sensors P1, P2 and P3, strain gauges S1 and S2 and accelerometer Acc (Displayed values are in mm).

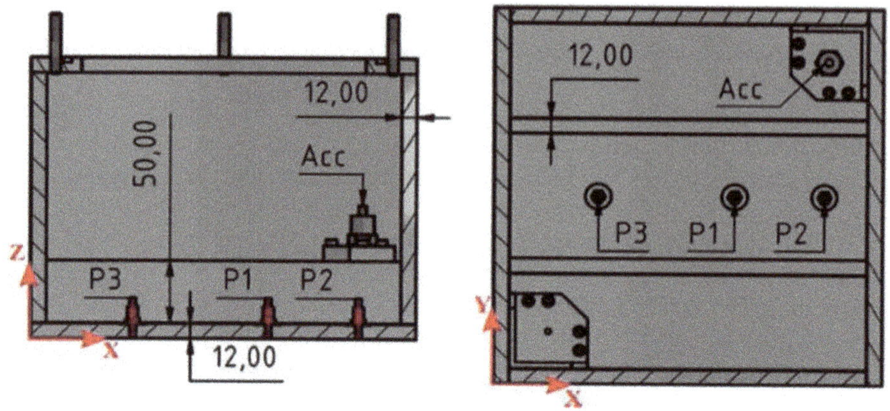

Fig. A2. Side view (left) and top view (right) of the test body with a rigid bottom and the arrangement of pressure sensors P1, P2 and P3 and accelerometer Acc (Displayed values are in mm).

Table 1. Coordinates specifying the positions of the three pressure sensors, the two strain gauges and the accelerometer ([a]Elastic, [b]Rigid).

Sensor	X [mm]	Y [mm]	Z [mm]
Pressure Sensor P1	185	150	0.0
Pressure Sensor P2	240[a]/255[b]	150	0.0
Pressure Sensor P3	80	150	0.0
Strain Gauge S1	150	150	4.7
Strain Gauge S2	200	150	4.7
Accelerometer	258	258	74[a] / 81.7[b]

The elastic and the rigid test bodies weighed 18.45 and 20.32 kg, respectively, and they were mounted underneath a load cell to measure integral forces acting on the test bodies during water entry. The HBM U3/10 load cell, which measured forces up to 10.0 kN with an uncertainty of 0.2%, was connected to the platform via an extension. A Ballupp BTL5-A11-M1000-P-32 type magnetostrictive sensor with an uncertainty of 0.02% measured the platform's motion to yield a reproducible resolution of 4.0 μm. The time derivative of the platform's motion determined its velocity. Further details on the experimental setup are explained by Tödter et al. (2019).

Benchmark Cases

The benchmark cases are summarized in Table 2 and focused on benchmarks of pressures (sensor location P1), plate entry impact forces, elastic strains (sensor locations S1, S2) and air entrapment effects in the way of the plate bottom under rigid body and elastic conditions. The modelling and simulation boundary conditions are highlighted in Table 3. A summary of the modelling assumptions in structural and fluid domains is given in Table 4. All methods except the NMRI SPH accounted for flexible dynamics by two-way coupling. The NMRI SPH method considered rigid body dynamics only. The ABAQUS FEA method presented in Yan et al. (2022c) assumed pinned (point A, Figure A3), fixed (point B, Figure A3) and restricted displacements in x- and y-directions (point C, Fig. 21), respectively. In NRC simulations, the pinned boundary conditions were unavailable due to the use of solid elements. They used fixed boundary conditions at the A and C locations.

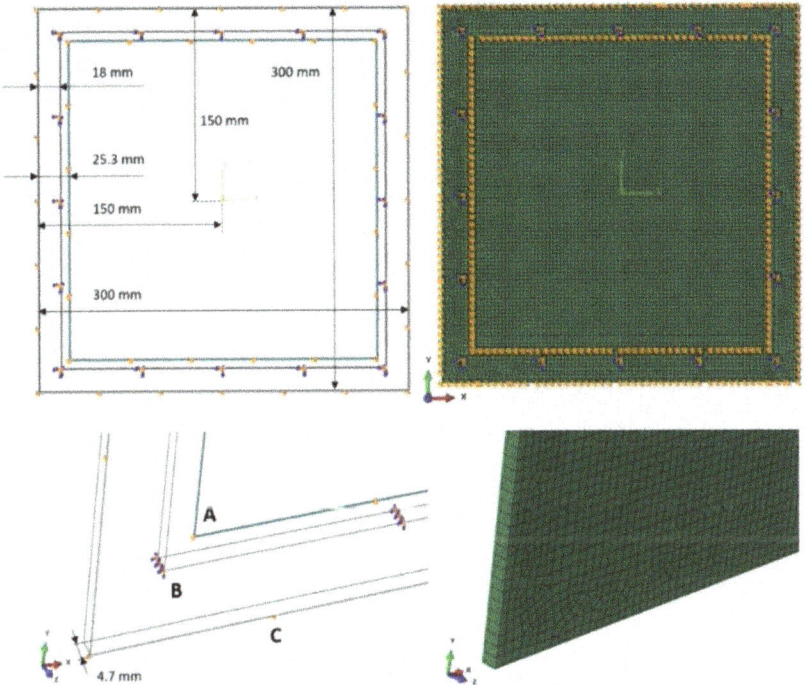

Fig. A3. Three-dimensional plate FE model assuming pinned (point A), fixed (point B) and restricted displacements in x- and y-directions (point C) (Yan et al., 2022c). In NRC simulations, the pinned boundary conditions were unavailable due to the use of solid elements. Hence, they used fixed boundary conditions at the A and C locations.

For NMRI-LSDYNA, the simulation is conducted using the ICFD function of LSDYNA, which is based on the Finite Element Method. The influence of air is not considered in the analysis. Water is treated as an incompressible fluid (Livermore Software Technology Corporation, 2014). For the FFSI analysis, implicit coupling is utilized. The boundary conditions used are precisely the same as those applied in the ABAQUS FEM method. For NMRI–SPH, the WCSPH model (Monaghan, 1994) is developed based on the open-source code DualSPHysics (Domínguez et al., 2022). In the WCSPH model, the influence of air is not considered, and water is modelled using the weakly compressible assumption governed by the equation of state. The Wendland kernel is employed for spatial interpolation. The boundary condition is implemented by directly applying the entry velocity to the rigid plate.

Table 2. Benchmark cases

Case	Benchmark			
	Focus	*Plate model*	*Location*	*Entry velocities*
1	Pressure	Rigid	P1	0.519 m/s,
2		Elastic		0.782 m/s,
				1.041 m/s
3	Impact force	Rigid	Entry point	0.782 m/s
4		Elastic		
5	Strain	Elastic	S1, S2	1.041 m/s
6	Air entrapment	Rigid	Plate bottom	1.041 m/s for
7		Elastic		impact moment
				2. 5 ms
				5.833ms
				4.833 ms
				30.83 ms

Table 3. Boundary conditions in fluid domain

Boundary surfaces	Boundary conditions			
	[1]	NRC	NMRI - LSDYNA	NMRI - SPH
Plate	wall	wall	moving body	moving body
Side (box/plate side)	wall	wall	moving body	moving body
Bottom	velocity inlet	velocity inlet	wall (free slip)	wall (free slip)
Top	pressure outlet	pressure outlet	air is not considered	air is not considered
Symmetry (or tank side)	symmetry	velocity inlet	tank side	tank side

Table 4. Summary of modelling assumptions in fluid and structural domains

Method	Material modelling	Structural modelling	Fluid Modelling	Coupling	BC
Yan et al. (2022c)	Isotropic	ABAQUS FEA using 50,660 8-node continuum shell elements (SC8R)	STAR CCM + RANS CFD see Fig. 22	Two-way coupling with morphing	see Fig. 21 and Table 3
NRC	Isotropic	STAR CCM + FEA (8-node) using 92,000 Hexahedral elements	RANS CFD	Two-way coupling with morphing	FEA used fixed boundary conditions at the A and C locations (see Fig. 21 and Table 3)
NMRI LSDYNA	Isotropic	FEA (8-node) using 7,500 Hexahedral elements	Incompressible CFD (Livermore Software Technology Corporation, 2014)	Two-way coupling with morphing	Moving boundary with the predefined boundary velocity (see Table 3)
NMRI SPH	Only Rigid	-	WCSPH (Domínguez et al., 2022)	No coupling	Moving boundary with the predefined boundary velocity (see Table 3)

Comparisons and Discussion

Figure 22 illustrates comparisons of simulations with experimental data in the way of pressure sensor P1 for test cases corresponding to three different impact velocities (0.519, 0.782, 1.041m/s). The left column graphs directly compare results from rigid body simulations, while the right column graphs demonstrate the comparisons when flexible dynamics are considered. In both rigid and flexible domains, the NMRI (LS DYNA) simulations overestimate the experimental and other simulation trends. This could be attributed to the stiffness of the structural dynamic model. Neglecting air influence and assuming incompressible water may make the NMRI (LS DYNA) simulation more sensitive to entry velocity, especially in cases where the body is regarded as rigid. However, for NMRI SPH, although air influence is also not considered, the weakly compressible water model can somewhat mitigate this sensitivity to entry velocity. The differences between simulation peaks and experiments increase with increasing entry velocity. In the rigid domain, the simulation peak values appear larger than experimental peak values at the onset of impact. Overall, the pressure results from the simulations seem to be relatively smoother than the experimental results, where small fluctuations are captured. This could be attributed to the influence of higher-order vibration modes, which were also observed in experiments and in the way of the first two crests of numerical and experimental pressure results depicted by sensor P1, differences of pressure amplitudes indicated that the plate vibrated upwards with larger amplitudes in the way of water entry. The differences between peak values become smaller when comparing hydroelastic CFD results from Yan et al. (2022c) and NRC to experimental results. Figure 23 demonstrates impact force comparisons between simulations and experiments for 0.782 m/s entry velocity when the body is rigid and hydroelastic. Forces rise and decay in the same fashion as pressures. When excluding the LSDYNA simulations, the maximum experimental impact force in the way of the band range seems higher than the STAR CCM + CFD simulations. This could be attributed to uncertainties associated with the entry velocities, e.g., the numerical records of the lowest entry velocity (0.519 m/s) deviated slightly from experimental data, where the vibration frequency was smaller than that measured in the test tank.

The larger pressure in rigid or flexible CFD simulations could be attributed to the boundary conditions, uncertainties in impact velocity and the structural rigidity of the structure. Regarding the latter, it is noted that in CFD simulations, the rigidity of the structure is considered either infinite (for rigid body simulations) or suitably adjusted for hydroelastic simulations while in experiments, the aluminium structure is not physically completely rigid, and the experimental aluminium flange on top of the plate has a broader influencing area (and, therefore, additional rigidity). For example, in the experiments, the additional rigidity from the aluminium flange on top of the plate limited the amplitude of upward vibrations. Therefore, the air trapped under the plate during experiments has been lower. When the entry velocity increases, the air trapped under the plate during pressure peak time increases. The differences observed between experimental and computational results decay as hydroelastic dynamics reach a quasi-steady state.

Air entrapment during impact was recorded by a high-speed imaging camera during experiments. To better understand the air entrapment process, experiments were compared against simulation results. Special focus was attributed to qualitative comparisons of relevance to the volume fraction of air and its distribution on the plate at different time instances. Air entrapment features idealized by experiments and simulations are directly comparable for cells where air occupies over 50%. Figure 25 and Fig. 26 show comparisons of results for the rigid and flexible cases with a water entry velocity of 1.041 m/s. In Fig. 25, experimental results are compared with numerical results from (Yan et al., 2022c) and NRC. Stage (a) represents impact at the start of the water entry when the space under the plate bottom is almost full of air. Stage (b) demonstrates an initial period when the plate enters the water, and most air is still trapped at the centre of the plate (bright white area in the experiments). Simulation results show similar features, namely: (1) the volume fraction of air near the plate central area is nearly 100%, so there is a large area of trapped air in the way of the plate centre; (2) near the plate edge, the volume fraction of air decreases to about 95% as compared to the stage (a). It may be concluded that some air is escaping from the central air pockets towards the edge of the plate. This phenomenon conveys the influence of fluctuating water near the plate edge, as shown in the experimental graph at stage (b). Stage (c) corresponds to the moment after the first pressure peak. The pressure decreases to around 20% of its peak value, and the air volume under the plate decreases. The simulated area of trapped air in the way of the plate central area decreases significantly, and the average volume fraction of air is around 93%. More air is escaping through the edges while the area of fluctuated water is getting larger. The latter is well shown in both experiments and simulations (dark, bubbly area in the experiments and the colour transition from orange to yellow in the simulations). A review of stages (d) and (e) demonstrates that in both experiments and simulations, the air underwater diminishes gradually as we move away from the centre of the plate. In simulations, the average air volume fraction near the plate centre decreases from about 80% at stage (d) to about 55% at stage (e).

Consequently, the percentages of air entrapped in central cells are lower than in stages (b) and (c). It is noted that the simulation method assumed infinite rigidity. In experiments, some local deformations in the way of the plate centre may exist, possibly leading to slightly different levels of air entrapment. However, based on overall comparisons, their influence is not significant.

In Fig. 26, columns display the volume of the fraction of air from experiments and simulations. Notably, the column Wet/Dry displays a version of structural deformations following simulations according to the methodology (Yan et al., 2022c). The structural deformation characteristics are not visible in the hybrid NRC simulation presented in the last column. Stage (a) represents impact at the start of the water entry when the space under the plate bottom is almost full of air. Stages (b) and (c) demonstrated a gradual process whereby the pressure reaches approximately 70% of its first peak. The maximum structural displacements are 3.4 mm during stage (b) and 3.9 mm during stage (c). The entrapped air is in the way of the centre area of the deformed plate (white and dark red areas in the experiment and simulations, respectively). Stage (d) demonstrates the moment the plate vibrates vertically from its crest to its neutral position. Stage (e) demonstrates the moment the plate vibrates from the neutral axis to the bottom. At stage (d), the maximum structural displacement is 2.6 mm. During deformation, air is pushed out from the central area, forming a ring-shaped distribution. The phenomenon is observed in both experiments and simulations (see the dark, bubbly ring in the experiment and the dark red ring in the simulations at stage (d)). The latter observation suggests that the air volume fraction with an average value of 0.9 is distributed partially in the centre and partially as a ring shape between the plate centre and edge. After one vibration cycle, air and water mix and lumpy air bubbles form. The simulated area is slightly wider than the observed in experiments (Fig. 24).

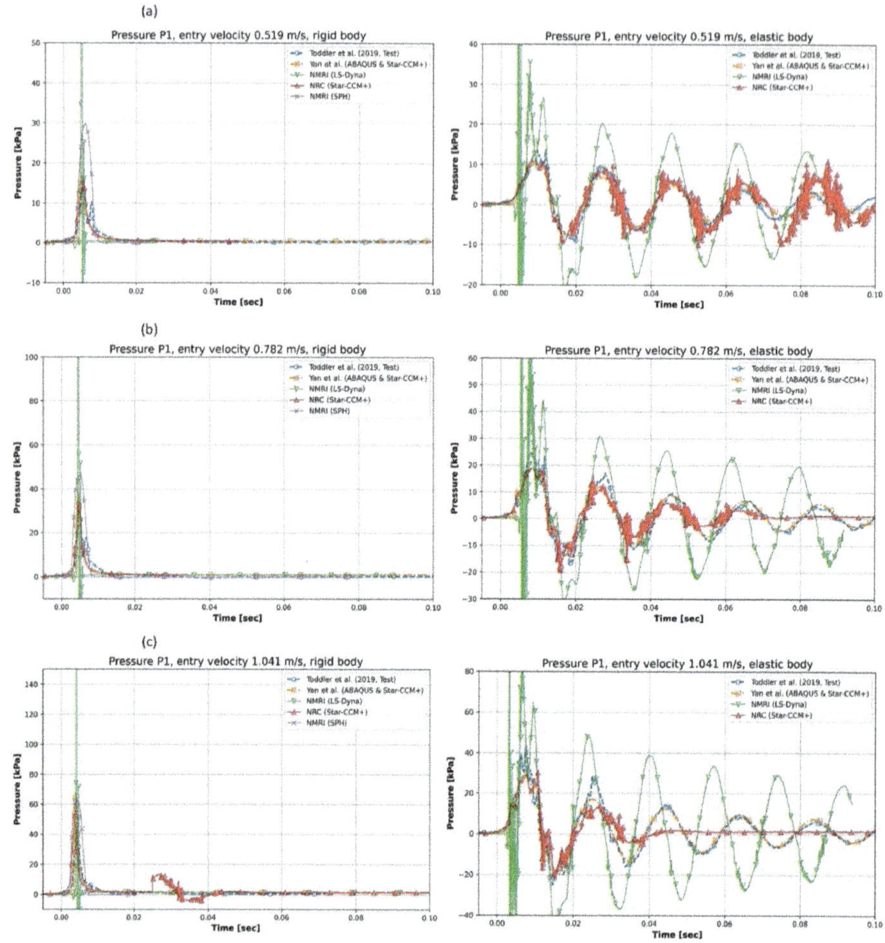

Fig. A4. Comparison of rigid versus hydroelastic responses in pressure sensor P1 at different entry velocities.

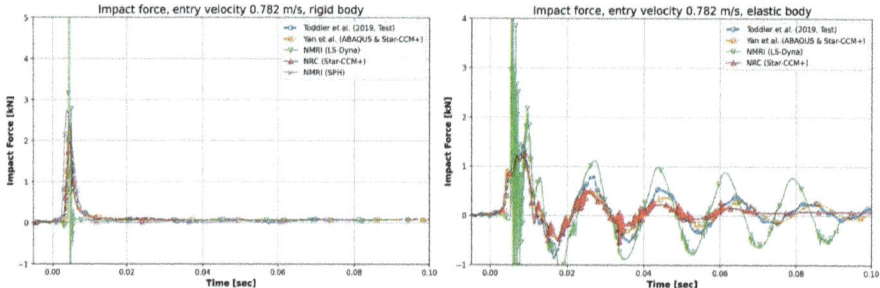

Fig. A5. Rigid versus hydroelastic impact forces at entry velocity 0.782 m/s

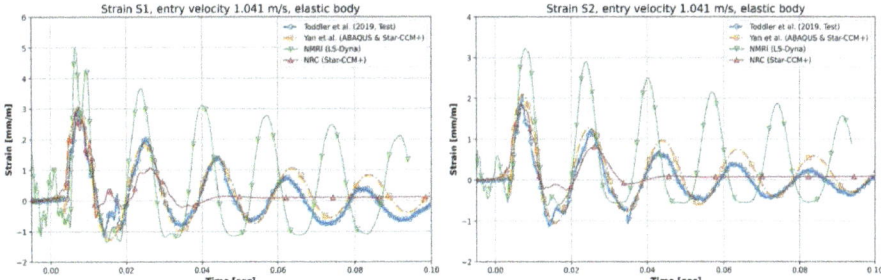

Fig. A6. Strains at sensors S1 and S2 when hydroelastic impact is considered at entry velocity 1.041 m/s.

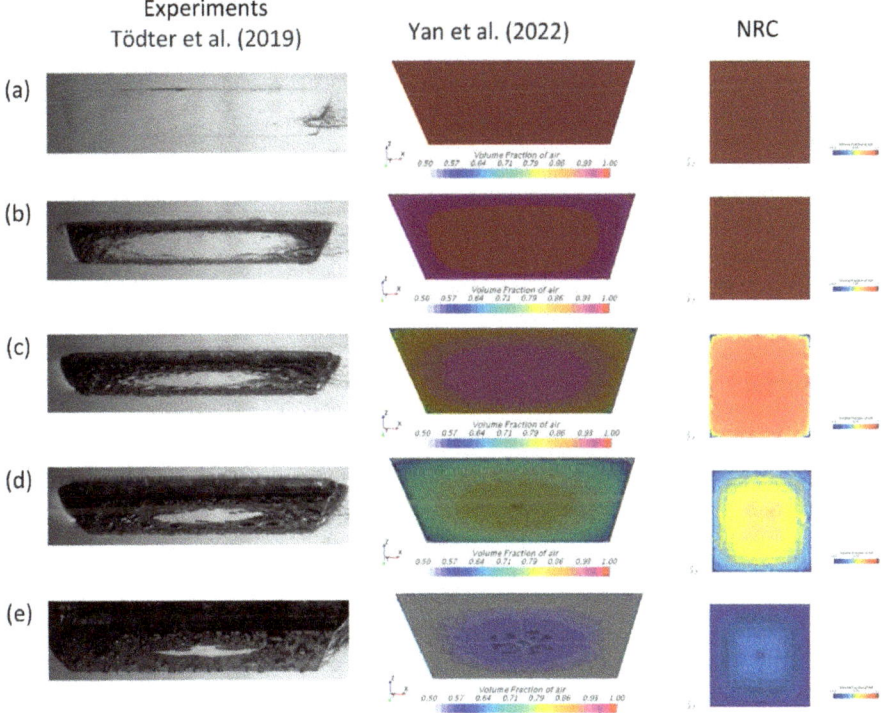

Fig. A7. Air entrapment underneath the rigid plate bottom at different time instants (*t*) of the water entry (1.041 m/s) for (a) impact, (b) $t = 2.500$ ms, (c) $t = 5.833$ ms, (d) $t = 10.832$ ms, (e) $t = 30.830$ ms. B: air entrapment in CFD simulation.

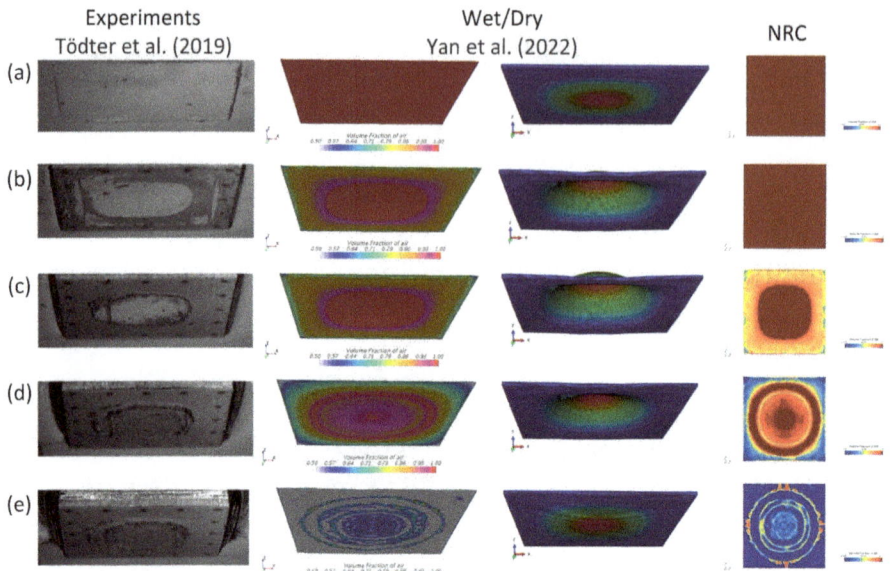

Fig. A8. Air entrapment underneath the elastic plate bottom at different time instants (t) of water entry (1.041 m/s) for (a) impact, (b) $t = 5.500$ ms, (c) $t = 5.833$ ms, (d) $t = 10.832$ ms, (e) $t = 30.830$ ms. Simulations in columns 2 and 4 display the air entrapment in FFSI simulation. The ABAQUS structural dynamic displacement depicted in column 3 assumes a deformation scale factor of 15.

Conclusions

Numerical simulations of flat plate impacts were carried out to analyze the capability of the applied methods. Computational Fluid Mechanics (CFD) and Smoothed Particle Hydrodynamics (SPH) methods were used to model rigid impacts. A two-way coupled Flexible Fluid-Structure Interaction (FFSI) approach was implemented, using ABAQUS for the structural domain and STAR CCM + for the fluid domain (Yan et al., 2022c). A hybrid STAR CCM + method for the fluid domain (Sitek et al., 2021) was employed, and an incompressible fluid solver implemented in LS DYNA (Livermore Software Technology Corporation, 2014) was applied for hydroelastic water impacts. The modelled test bodies had the exact dimensions and properties as those experimentally investigated by (Tödter et al., 2019). Time histories of simulations and experiments attained similar peak loads, excluding LS DYNA FFSI, that appeared significantly stiffer.

The simulated void fraction underneath the impacting plates compared favourably to cavities experimentally captured by high-speed imaging. The uncertainties associated with the peaks of pressures, forces and strains are considered to reflect the differences in assumptions between numerical simulations and experiments. Also, the simulated peak loads seem to be affected by high-frequency vibration modes and their interactions with boundary conditions, highlighting sensitivity to modelling details.

It is recommended that future work on uncertainty modelling prediction is assessed along the lines of the American Society of Mechanical Engineers (ASME) standard for Verification and Validation (V&V) (ASME, 2009. As part of this process, well-documented experiments and close cooperation between experimentalists and simulation engineers for validation are highly recommended.

ASME. (2009). Standard for verification and validation in computational fluid dynamics and heat transfer: ASME V&V 20, The American Society of Mechanical Engineers. Retrieved from https://www.asme.org/codes-standards/find-codes-standards

Domínguez, J. M., Fourtakas, G., Altomare, C., Canelas, R. B., Tafuni, A., García-Feal, O., et al. (2022). DualSPHysics: from fluid dynamics to multiphysics problems. Computational Particle Mechanics, 9(5), pp. 867–895. Retrieved from https://doi.org/10.1007/s40571-021-00404-2

Livermore Software Technology Corporation. (2014). ICFD theory manual: Incompressible fluid solver in LS-DYNA. Retrieved from https://ftp.lstc.com/anonymous/outgoing/inaki/docs/pdf_icfd/ICFD_theory.pdf

Monaghan, J. J. (1994). Simulating Free Surface Flows with SPH. Journal of Computational Physics, 110(2), pp. 399–406. Retrieved from https://www.sciencedirect.com/science/article/pii/S0021999184710345

Sitek, M., Bojanowski, C., Bergeron, A., Licht, J. (2021). Involute Working Group – FSI Analysis of Fuel Plates Using Finite Volume and Finite Element Methods. United States. Retrieved from https://www.osti.gov/biblio/1845461

Tavakoli, S., Zhang, M., Kondratenko, A. A., Hirdaris, S. (2024). A review on the hydrodynamics of planing hulls. Ocean Eng. 303.

Tödter, S., el Moctar, O., Neugebauer, J., Schellin, T. E. (2019). Experimentally Measured Hydroelastic Effects on Impact-Induced Loads During Flat Water Entry and Related Uncertainties. Journal of Offshore Mechanics and Arctic Engineering, 142(1). Retrieved from https://doi.org/10.1115/1.4044632

Yan, D., Mikkola, T., Lakshmynarayanana, A., Tödter, S., Schellin, T. E., Neugebauer, J., et al. (2022). A study into the FSI modelling of flat plate water entry and related uncertainties. Marine Struct. 86. Retrieved August 29, 2023

References

Abbasnia, A., Guedes Soares, C.: OpenMP parallelism in computations of three-dimensional potential numerical wave tank for fully nonlinear simulation of wave-body interaction using NURBS. Eng. Anal. Boundary Elem. **117**, 321–331 (2020)

Abbasnia, A., Karimirad, M., Friel, D., Whittaker, T.: Fully nonlinear dynamics of floating solar platform with twin hull by tubular floaters in ocean waves. Ocean Eng. **257** (2022)

Abbasnia, A., Kosleck, S., Buhr, E., Lutz, M., Adam, F.: Volumetric design for hydrodynamic response of modularized lightweight TLP structures under irregular waves (2024a)

Abbasnia, A., Rezanejad, K., Guedes Soares, C.: Adaptive fully nonlinear potential model for the free surface under compressible air pressure of oscillating water column devices. Eng. Anal. Boundary Elem. **133**, 153–164 (2021)

Abbasnia, A., Sutulo, S., Soares, C.G.: Frequency-domain 3D computer program for predicting motions and loads on a ship in regular waves. J. Mar. Sci. Appl. **23**(1), 64–73 (2024)

Abdelwahab, H.S., Soares, C.G.: Experimental uncertainty of a physical model of a tanker moored to a terminal in a port. Marine Struct. **87** (2023)

Agarwal, S., Sriram, V., Murali, K.: Three-dimensional coupling between Boussinesq (FEM) and Navier–Stokes (particle based) models for wave structure interaction. Ocean Eng. **263** (2022)

Agarwal, S., Sriram, V., Yan, S., Murali, K.: Improvements in MLPG formulation for 3D wave interaction with fixed structures. Comput. Fluids **218** (2021)

Ahani, A., Abrahamsen, B.C., Greco, M.: Experimental analysis of high and steep wave impacts and related hydroelastic effects relevant for offshore structures in steel (2022). https://ntnuopen.ntnu.no/ntnu-xmlui/handle/11250/3050174. Accessed 4 Sept 2023

Ahn, Y.: Sloshing loads estimation using a genetic programming. In: Proceedings of the Thirty-Third International Ocean and Polar Engineering Conference, Melbourne (2023)

Ahn, Y., et al.: Analyzing experimental data of 6-dof motions and sloshing impacts in lng carriers. In: Proceedings of the ASME 2024 43rd International Conference on Ocean, Offshore and Arctic Engineering, Singapore (2024)

Alexandru, I., et al.: Comparison of experimental and numerical impact loads on ship-like sections. Adv. Marine Struct. 339–349 (2007)

Aliyar, S., Ducrozet, G., Bouscasse, B., Venkatachalam, S., Ferrant, P.: Efficiency and accuracy of the domain and functional decomposition strategies for the wave-structure interaction problem. Ocean Eng. **266** (2022)

Alkhabbaz, A., et al.: The aero-hydrodynamic interference impact on the NREL 5-MW floating wind turbine experiencing surge motion. Ocean Eng. **295** (2024)

AlMashan, N., Neelamani, S., Al-Houti, D.: Experimental investigations on wave impact pressures under the deck and global wave forces and moments on offshore jacket platform for partial and full green water conditions. Ocean Eng. **234** (2021)

Amaechi, C.V., Wang, F., Ye, J.: Numerical studies on CALM buoy motion responses and the effect of buoy geometry cum skirt dimensions with its hydrodynamic waves-current interactions. Ocean Eng. **244** (2022)

Amini-Afshar, M., Bingham, H.B.: Added resistance using Salvesen–Tuck–Faltinsen strip theory and the Kochin function. Appl. Ocean Res. **106** (2021)

Amouzadrad, P., Mohapatra, S.C., Guedes Soares, C.: Hydroelastic response to the effect of current loads on floating flexible offshore platform. J. Marine Sci. Eng. **11**(2) (2023)

Ao, Y., Li, Y., Gong, J., Li, S.: An artificial intelligence-aided design (AIAD) of ship hull structures. J. Ocean Eng. Sci. **8**(1), 15–32 (2023)

API. API Recommended Practice 2A-WSD (2014)

Arya, T., Ranjan Behera, M.: Numerical Investigation Of A Spar Type Floating Offshore Wind Turbine Platform Under Extreme Waves (2023). http://asmedigitalcollection.asme.org/OMAE/proceedings-pdf/IOWTC2023/87578/V001T01A013/7234398/v001t01a013-iowtc2023-119363.pdf

Asim, T., Islam, S.Z., Hemmati, A., Khalid, M.S.U: A review of recent advancements in offshore wind turbine technology. Energies **15**(2) (2022)

ASME. Standard for verification and validation in computational fluid dynamics and heat transfer: ASME V&V 20. The American Society of Mechanical Engineers (2009). https://www.asme.org/codes-standards/find-codes-standards

Audiffren, C., Nugue, J., Marcer, R., Couty, N., Laurent, T.: Simulation of liquid hydrogen jet impacts on a corrugated membrane. In: Proceedings of the Thirty-Fourth International Ocean and Polar Engineering Conference. Rhodes (2024)

Aursand, M., Haagensen, P.J., Skallerud, B.H.: Remaining fatigue life assessment of corroded mooring chains using crack growth modelling. Marine Struct. **90** (2023)

Aydın, Ç., Ünal, U.O., Sarıöz, K.: Computation of environmental loads towards an accurate dynamic positioning capability analysis. Ocean Eng. **243**, 110201 (2022). Accessed 18 Oct 2024

Bandizadeh Sharif, M., Ghassemi, H., He, G., Karimirad, M.: A review of the flow-induced vibrations (FIV) in marine circular cylinder (MCC) fitted with various suppression devices. Ocean Eng. **289**(P1), 116261 (2023). https://doi.org/10.1016/j.oceaneng.2023.116261

Bardazzi, A., Lugni, C., Faltinsen, O.M., Durante, D., Colagrossi, A.: Different scenarios in sloshing flows near the critical filling depth. J. Fluid Mech. **984**, A73 (2024)

Barooni, M., Nezhad, S.K., Ali, N.A., Ashuri, T., Sogut, D.V.: Numerical study of ice-induced loads and dynamic response analysis for floating offshore wind turbines. Marine Struct. **86**(September), 103300 (2022). https://doi.org/10.1016/j.marstruc.2022.103300

El Beshbichi, O., Xing, Y., Ong, M.C.: Comparative dynamic analysis of two-rotor wind turbine on spar-type, semi-submersible, and tension-leg floating platforms. Ocean Eng. **266**, 112926 (2022). https://www.sciencedirect.com/science/article/pii/S0029801822022090?pes=vor. Accessed 22 Feb 2023

Bitner-Gregersen, EM., et al.: Uncertainties in long-term wave modelling. Marine Struct. **84** (2022)

Böhm, A.M., von Bock und Polach, R.U.F., Herrnring, H., Ehlers, S.: The measurement accuracy of instrumented ship structures under local ice loads using strain gauges. Marine Struct. **76** (2021)

Bonci, M., de Jong, P.: High-speed RHIB seakeeping analysis using non-linear time domain simulations and systematic hull parametrization. In: Proceedings of the 13th Symposium on High Speed Marine Vehicles (2023)

Buljac, A., Kozmar, H., Yang, W., Kareem, A.: Concurrent wind, wave and current loads on a monopile-supported offshore wind turbine. Eng. Struct. **255** (2022)

Cai, J., Ding, S., Zhang, Q., Liu, R., Zeng, D., Zhou, L.: Broken ice circumferential crack estimation via image techniques. Ocean Eng. **259** (2022)

Califano, A., Berthelsen, P.A., Magalhaes Duque Da Fonseca, N.M.: Effect of body motion on the wave loads computed with CFD on the INO-WINDMOOR floater. In: Journal of Physics: Conference Series. Institute of Physics (2023)

Caputano, E., Goodwin, P.: Sloshing and free surface correction in hydrodynamic analyses. In: Proceedings of the Thirty-third (2023) International Ocean and Polar Engineering Conference, Ottawa, pp. 2667–2673 (2023)

Carapellese, F., Pasta, E., Paduano, B., Faedo, N., Mattiazzo, G.: Intuitive LTI energy-maximising control for multi-degree of freedom wave energy converters: the PeWEC case. Ocean Eng. 256 (2022)

de Carvalho, A. van L.B., Campello, E.M.B., Franzini, G.R., Skaf, K.J.: An assessment of mooring systems' forces of ships berthed at dolphins. Ocean Eng. **253**(March), 111090 (2022). https://doi.org/10.1016/j.oceaneng.2022.111090

Celik, I.B., Ghia, U., Roache, P.J., Freitas, C.J., Coleman, H., Raad, P.E.: Procedure for estimation and reporting of uncertainty due to discretization in CFD applications. J. Fluids Eng. Trans. ASME **130**(7), 0780011–0780014 (2008)

Chen, C., Ma, Y., Fan, T.: Review of model experimental methods focusing on aerodynamic simulation of floating offshore wind turbines. Renew. Sustain. Energy Rev. **157**(May 2021), 112036 (2022a). https://doi.org/10.1016/j.rser.2021.112036

Chen, D., Feng, X., Hou, C., Chen, J.F.: A coupled frequency and time domain approach for hydroelastic analysis of very large floating structures under focused wave groups. Ocean Eng. **255**, 111393 (2022b)

Chen, H., Hall, M.: CFD simulation of floating body motion with mooring dynamics: coupling MoorDyn with OpenFOAM. Appl. Ocean Res. **124** (2022)

Chen, J., Milne, I., Taylor, P.H., Gunawan, D., Zhao, W.: Forward prediction of surface wave elevations and motions of offshore floating structures using a data-driven model. Ocean Eng. **281**, 114680 (2023a)

Chen, J.S., et al.: VHF radar observations of sea surface in the Northern Taiwan strait. J. Atmos. Oceanic Tech. **36**(2), 297–315 (2019)

Chen, J., Zhang, J., Wang, G., Zhang, Q., Guo, J., Sun, X.: Numerical simulation of the wave dissipation performance of floating box-type breakwaters under long-period waves. Ocean Eng. **266** (2022c)

Chen, P., Zhang, J.: Study on the liquefaction mechanism of shipping iron ore by adopting perspective materials. In: Proceedings of the Thirty-Fourth International Ocean and Polar Engineering Conference, Rhodes, pp. 2704–2710 (2024)

Chen, R., Huang, W., Chen, X., Kang, M.: Prediction of ice resistance of icebreaker during continuous icebreaking. Chin. J. Ship Res. **16**(5), pp. 101–108+120 (2021a)

Chen, S., et al.: Design, dynamic modeling and wave basin verification of a hybrid wave–current energy converter. Appl. Energy **321** (2022d)

Chen, S., Zou, B., Han, C., Yan, S.: Comparative study on added resistance and seakeeping performance of x-bow and wave-piercing monohull in regular head waves. J. Marine Sci. Eng. **10**(6) (2022e)

Chen, X., Xu, G., Chen, Z., Zhu, L., Cai, S.: A study of the interaction between depression internal solitary waves and submerged floating tunnels in stratified fluids. Appl. Ocean Res. **132** (2023b)

Chen, Y., Meringolo, D.D., Liu, Y.: SPH study of wave force on simplified superstructure of open-type sea access road. Ocean Eng. **249** (2022f)

Chen, Y., Liu, Y., Meringolo, D.D., Ming Hu, J.: Study on the hydrodynamics of a twin floating breakwater by using SPH method. Coast. Eng. **179** (2023c)

Chen, Y., Zhang, S., Chen, W.K., Magee, A.: Design, construction and testing of a fully elastic ship model for investigating hydro-elastic responses of container ships. Mar. Struct. **98**, 103663 (2024)

Chen, Z., He, Y., Ren, Y., Liu, Y.: Analysis and realization of the influence of sea ice flexural strength on ice resistance in numerical simulation of icebreaking by icebreaker. Ocean Eng. **273**(February), 113995 (2023d). https://doi.org/10.1016/j.oceaneng.2023.113995

Chen, Z., Zhang, C., Zhao, C., Chen, X., Liu, H.: Shipboard coherent microwave radar, pp. 1–5 (2021b)

Cheng, H., Li, L., Ong, M.C., Aarsæther, K.G., Sim, J.: Effects of mooring line breakage on dynamic responses of grid moored fish farms under pure current conditions. Ocean Eng. **237** (2021)

Cheng, Y., et al.: Experimental and numerical analysis of a hybrid WEC-breakwater system combining an oscillating water column and an oscillating buoy. Renew. Sustain. Energy Rev. **169** (2022a)

Cheng, Y., et al.: Experimental and numerical investigation of WEC-type floating breakwaters: a single-pontoon oscillating buoy and a dual-pontoon oscillating water column. Coast. Eng. **177** (2022b)

Choi, S.M., Park, C.D., Cho, S.H., Lim, B.J.: Effects of various inlet angle of wind and wave loads on floating photovoltaic system considering stress distributions. J. Clean. Prod. **387** (2023)

Chueh, C.J., Chien, C.H., Lin, C., Lin, T.Y., Chiang, M.H.: Dynamic co-simulation analysis and control of an IEA 15 MW offshore floating semi-submersible wind turbine under taiwan offshore-wind-farm conditions of wind and wave. J. Marine Sci. Eng. **11**(1) (2023)

Class, N.K.: Investigation report on structural safety of large container ships (2014)

Clément, C., Bozonnet, P., Vinay, G., Pagnier, P., Nadal, A.B., Réveillon, J.: Evaluation of Morison approach with CFD modelling on a surface-piercing cylinder towards the investigation of FOWT Hydrodynamics. Ocean Eng. **251** (2022)

Cong, P., Ning, D., Teng, B.: Enhancement of the energy capture performance of oscillating water column (OWC) devices using multi-chamber multi-turbine (MCMT) technology. Energy Convers. Manage. **322** (2024)

Cong, P., Teng, B., Bai, W.: Second-order wave run-up on a vertical cylinder adjacent to a plane wall based on the application of quadratic transfer function in bi-directional waves. Marine Struct. **76** (2021)

da Cruz, D.M., Penaquioni, A., Zangalli, L.B., Bastos, M.B., Bastos, I.N., da Silva, A.L.N.: Non-destructive testing of high-tenacity polyester sub-ropes for mooring systems. Appl. Ocean Res. **134** (2023)

Cui, H., Turan, O., Sayer, P.: Learning-based ship design optimization approach. CAD Comput. Aided Design **44**(3), 186–195 (2012)

Cui, J., Chen, X., Sun, P.N., Li, M.Y.: Numerical investigation on the hydrodynamic behavior of a floating breakwater with moon pool through a coupling SPH model. Ocean Eng. **248** (2022)

Dai, J., Abrahamsen, B.C., Viuff, T., Leira, B.J.: Effect of wave-current interaction on a long fjord-crossing floating pontoon bridge. Eng. Struct. **266**, 114549 (2022)

Dao, M.H., et al.: Wind tunnel and CFD studies of wind loadings on topsides of offshore structures. Ocean Eng. **285**(P1), 115310 (2023). https://doi.org/10.1016/j.oceaneng.2023.115310

Deng, R., Song, Z., Ren, H., Li, H., Wu, T.: Investigation on the effect of container configurations and forecastle fairings on wind resistance and aerodynamic performance of large container ships. Eng. Appl. Comput. Fluid Mech. **16**(1), 1279–1304 (2022). https://doi.org/10.1080/19942060.2022.2086177

Deng, S., Liu, Y., Ning, D.: Fully coupled aero-hydrodynamic modelling of floating offshore wind turbines in nonlinear waves using a direct time-domain approach. Renew. Energy **216** (2023a)

Deng, Y., Li, H., Wang, Z.:. A numerical study of wave impacts on a semi-submersible (2023b). http://asmedigitalcollection.asme.org/OMAE/proceedings-pdf/OMAE2023/86878/V005T06A052/7041067/v005t06a052-omae2023-106568.pdf

Depalo, F., Wang, S., Xu, S., Guedes Soares, C., Yang, S.H., Ringsberg, J.W.: Effects of dynamic axial stiffness of elastic moorings for a wave energy converter. Ocean Eng. **251** (2022)

Derbanne, Q., Storhaug, G., Shigunov, V., Xie, G., Zheng, G.: Rule formulation of vertical hull girder wave loads based on direct computation. In: PRADS 2016 - Proceedings of the 13th International Symposium on PRActical Design of Ships and Other Floating Structures (2016)

Diez, M., et al.: Experimental and computational fluid-structure interaction analysis and optimization of deep-V planing-hull grillage panels subject to slamming loads – part I: regular waves. Marine Struct. **85**, 103256 (2022). https://www.sciencedirect.com/science/article/pii/S0951833922000958

DNV. Fatigue and ultimate strength assessment of container ships including whipping and springing (2015)

DNV-GL. Fatigue Assessment of Ship Structures (DNVGL-CG-0129) (2015)

DNV-GL. Offshore Technical Guidance Prediction of air gap for column stabilised units (2019)

Dombre, E., Harris, J.C., Benoit, M., Violeau, D., Peyrard, C.: A 3D parallel boundary element method on unstructured triangular grids for fully nonlinear wave-body interactions. Ocean Eng. **171**, 505–518 (2019)

Domínguez, J.M., et al.: DualSPHysics: from fluid dynamics to multiphysics problems. Comput. Part. Mech. **9**(5), 867–895 (2022). https://doi.org/10.1007/s40571-021-00404-2

Dong, G., Yao, C., Yu, J., Sun, X., Feng, D.: A frequency domain hybrid Green function method for seakeeping and added resistance performance of ships advancing in waves. Eng. Anal. Bound. Elem. **168** (2024)

Du, Z., et al.: An experimental investigation on vortex-induced motion (VIM) of a tension leg platform in irregular waves combined with a uniform flow. Appl. Ocean Res. **123**(December 2021), 103185 (2022). https://doi.org/10.1016/j.apor.2022.103185

Duan, F., Ma, N., Gu, X.C., Zhou, Y.H., Wang, S.M.: A fast time domain method for predicting of motion and excessive acceleration of a shallow draft ship in beam waves. Ocean Eng. **262** (2022)

Duan, F., Ma, N., Gu, X., Zhou, Y.: An improved method for predicting roll damping and excessive acceleration for a ship with moonpool based on computational fluid dynamics method. J. Offshore mech. Arctic Eng.-Trans. ASME **145**(5) (2023)

Dutta, D., Bihs, H., Afzal, M.S.: Computational fluid dynamics modelling of hydrodynamic characteristics of oscillatory flow past a square cylinder using the level set method. Ocean Eng. **253** (2022)

Edwards, E.C., Holcombe, A., Brown, S., Ransley, E., Hann, M., Greaves, D.: Trends in floating offshore wind platforms: a review of early-stage devices. Renew. Sustain. Energy Rev. **193**, 114271 (2024). https://linkinghub.elsevier.com/retrieve/pii/S1364032123011292

van der Eijk, M., Wellens, P.: An efficient pressure-based multiphase finite volume method for interaction between compressible aerated water and moving bodies. J. Comput. Phys. **514**(November 2023), 113167 (2024). https://doi.org/10.1016/j.jcp.2024.113167

El-Shahat, S.A., Li, G., Fu, L.: Investigation of wave–current interaction for a tidal current turbine. Energy **227** (2021)

Engelbrecht, M., Bekker, A.: A discomfort threshold for impulsive whole-body vibration on a slamming-prone vessel. Appl. Ergon. **109** (2023)

Engström, J., Shahroozi, Z., Katsidoniotaki, E., Stavropoulou, C., Johannesson, P., Göteman, M.: Offshore measurements and numerical validation of the mooring forces on a 1:5 scale buoy. J. Marine Sci. Eng. **11**(1) (2023)

Erceg, S., Erceg, B., von Bock und Polach, F., Ehlers, S.: A simulation approach for local ice loads on ship structures in level ice. Marine Struct. **81**(February 2021), 1–15 (2022)

van Essen, S.M., Scharnke, J., Seyffert, H.C.: Required test durations for converged short-term wave and impact extreme value statistics — part 1: ferry dataset. Mar. Struct. **90**, 103410 (2023)

Ezeta, R., Kimmoun, L., Brosset, L.: Influence of ullage pressures on wave impacts induced by solitary waves in a flume tank, in: ISOPE (Ed.), Proceedings of the Thirty-Third International Ocean and Polar Engineering Conference (2023)

Faltilsen, R., Aarsnes, O.M., Zhao, J.V.: Water entry of arbitrary two dimensional sections with and without flow separation. In: Symposium on Naval Hydrodynamics, pp. 408–423 (1996)

Fan, Q., Xu, Y., Xie, Q., Zhang, M., Ren, H., Sun, T.: Investigation of the dynamic response of a floating wind-aquaculture platform under the combined actions of wind, waves and current. J. Ocean Eng. Sci. (2023)

Fan, Y.L., Jain, U., van der Meer, D.: Air-cushioning below an impacting wave-structured disk: free-surface deformation and slamming load. Phys. Rev. Fluids (2024)

Farsi, M., Ghadimi, P.: Effect of flat deck on catamaran water entry through smoothed particle hydrodynamics. Proc. Inst. Mech. Eng. Part M: J. Eng. Maritime Environ. **230**(2), 267–280 (2016). https://doi.org/10.1177/1475090214563960

Ferri, G., Marino, E., Bruschi, N., Borri, C.: Platform and mooring system optimization of a 10 MW semisubmersible offshore wind turbine. Renew. Energy **182**, 1152–1170 (2022)

Fonseca de Carvalho e Silva, D., Gouveia Telles de Menezes, F., Schmidt, D., Cardozo de Mello, P.: Slamming effects on fpso balconies: numerical and experimental analysis of alternative configurations for impact attenuation (2023). http://asmedigitalcollection.asme.org/OMAE/proceedings-pdf/OMAE2023/86878/V005T06A010/7041100/v005t06a010-omae2023-105066.pdf

Fonseca, N., Nybø, S., Rodrigues, J.M., Gallego, A., Garrido, C.: Identification of wave drift forces on a floating wind turbine sub-structure with heave plates and comparison with predictions. In: Proceedings of the International Conference on Offshore Mechanics and Arctic Engineering - OMAE, p. 8 (2022). /OMAE/proceedings-abstract/OMAE2022/85932/1147962. Accessed 19 Jan 2023

Føre, H.M., Endresen, P.C., Bjelland, H.V.: Load coefficients and dimensions of raschel knitted netting materials in fish farms. J. Offshore Mech. Arctic Eng. **144**(4) (2022)

Fujarra, A.L.C., Leal, A.P., Carnier, R.M., Gonçalves, R.T., Suzuki, H.: Validation of a low-cost IMU for flow-induced vibration tracking in offshore systems. J. Braz. Soc. Mech. Sci. Eng. **45**(7), 1–17 (2023). https://doi.org/10.1007/s40430-023-04275-x

Gadelho, J.F.M., Guedes Soares, C.: CFD study of a dual chamber floating oscillating water column device. Ocean Eng. **261** (2022)

Gaidai, O., Fu, S., Xing, Y.: Novel reliability method for multidimensional nonlinear dynamic systems. Marine Struct. **86** (2022)

Gao, S., Cong, P., Teng, B.: Effect of incident wave angle to gap resonance between two fixed barges. Ocean Eng. **257** (2022)

Gebrezgabir, S., Holloway, D.S., Ali-Lavroff, J.: Slam and wave load response reconstruction in high speed catamarans using transmissibility on full scale sea trials. Ocean Eng. **271** (2023)

Gemilang, G.M., Reed, P.A.S., Sobey, A.J.: Low-cycle fatigue assessment of offshore mooring chains under service loading. Marine Struct. **76**, 102892 (2021). https://linkinghub.elsevier.com/retrieve/pii/S0951833920301854

George, A., Cho, I.H., Kim, M.H.: Optimal design of a u-shaped oscillating water column device using an artificial neural network model. Processes **9**(8) (2021)

Ghadirian, A., Vested, M.H., Carstensen, S., Christiensen, E.D., Bredmose, H.: Wave-current interaction effects on waves and their loads on a vertical cylinder. Coastal Eng. **165** (2021)

Ghamari, I., Mahmoudi, H.R., Hajivand, A., Seif, M.S.: Ship roll analysis using CFD-derived roll damping: numerical and experimental study. J. Marine Sci. Appl. **21**(1), 67–79 (2022)

Gilbert, C., Gilber, J., Javaherian, J.: Water entry of a flexible wedge: how flexural rigidity influences spray root and pressure wave propagation. Phys. Rev. Fluids **8**(9) (2023)

Gonçalves, R.T., Malta, E.B., Simos, A.N., Hirabayashi, S., Suzuki, H.: Influence of heave plate on the flow-induced motions of a floating offshore wind turbine. J. Offshore Mech. Arctic Eng. **145**(3) (2023a). https://doi.org/10.1115/1.4056345. Accessed 18 Oct 2024

Gonçalves, R.T., Marques, M.A., Da Silva, L.S.P., Hirabayashi, S., Suzuki, H.: Experimental study of the draft on the flow-induced motions (FIM) of a semi-submersible platform with four square columns. In: Proceedings of the International Conference on Offshore Mechanics and Arctic Engineering - OMAE. American Society of Mechanical Engineers Digital Collection (2023b). https://doi.org/10.1115/OMAE2023-102704. Accessed 18 Oct 2024

Gong, J., Li, Y., Cui, M., Yan, S., Ma, Q.: Study on the surf-riding and broaching of trimaran in oblique stern waves. Ocean Eng. **266**(4) (2022)

Gong, J., Yan, S., Ma, Q., Li, Y.: Numerical simulation of fixed and moving cylinders in focusing wave by a hybrid method. Int. J. Offshore Polar Eng. **31**(1), 102–111 (2021)

Gözcü, O., Kontos, S., Bredmose, H.: Dynamics of two floating wind turbines with shared anchor and mooring lines. In: Journal of Physics: Conference Series. Institute of Physics (2022)

Grammatikopoulos, A · A review of physical flexible ship models used for hydroelastic experiments. Marine Struct. **90**(February) (2023)

Gu, N., Liang, D., Zhou, X., Ren, H.: A CFD-FEA method for hydroelastic analysis of floating structures. J. Marine Sci. Eng. **11**(4) (2023)

Gu, Y., Zhou, L., Ding, S., Tan, X., Gao, J., Zhang, M.: Numerical simulation of ship maneuverability in level ice considering ice crushing failure. Ocean Eng. **251**(March), 111110 (2022). https://doi.org/10.1016/j.oceaneng.2022.111110

Gubesch, E., Abdussamie, N., Penesis, I., Chin, C.: Maximising the hydrodynamic performance of offshore oscillating water column wave energy converters. Appl. Energy **308**, 118304 (2022a). Accessed 4 Sept 2023

Gubesch, E., Abdussamie, N., Penesis, I., Chin, C.: Effects of mooring configurations on the hydrodynamic performance of a floating offshore oscillating water column wave energy converter. Renew. Sustain. Energy Rev. **166** (2022b)

Guo, C., Zhang, C., Wang, C., Wang, C.: Experimental Study On IRV ramming artificial model ice. J. Marine Sci. Eng. **11**(10) (2023)

Guo, W., Zou, J., He, M., Mao, H., Liu, Y.: Comparison of hydrodynamic performance of floating breakwater with taut, slack, and hybrid mooring systems: an SPH-based preliminary investigation. Ocean Eng. **258** (2022)

Guo, X., Zhang, X., Tian, X., Li, X., Lu, W.: Predicting heave and surge motions of a semi-submersible with neural networks. Appl. Ocean Res. **112** (2021a)

Guo, Y., Xiao, L., Wei, H., Li, X., Li, L., Deng, Y.: Classification and comparison of wave impact modes on semi-submersibles. China Ocean Eng. **35**(2), 161–175 (2021b)

Ha, Y.J., Nam, B.W., Kim, K.H., Hong, S.Y.: Numerical study for slamming loads on the bow of a ship-type FPSO model under breaking waves. Ocean Eng. **299** (2024)

Hallak, T.S., Teixeira, Â.P., Guedes Soares, C.: Epistemic uncertainties on the estimation of minimum air gap for semi-submersible platforms. Marine Struct. **85** (2022)

Han, D., Paik, K.-J., Jeong, S.-Y., Choung, J.: Prediction of the ice resistance of icebreakers using explicit finite element analyses with a real-time load control technique. Ocean Eng. **240** (2021)

Han, X., Dong, S.: Interaction between regular waves and floating breakwater with protruding plates: Laboratory experiments and SPH simulations. Ocean Eng. **287** (2023)

Han, Y., Zhu, X., Zhou, L.: Numerical simulation of multi-tug towing of a gravity-based structure in broken sea ice. Ocean Eng. **261** (2022)

Hao, L., Pan, Z.: Translating and pulsating Green function with an ice cover. J. Eng. Math. **134**(1) (2022)

He, M., Liang, D., Ren, B., Li, J., Shao, S.: Wave interactions with multi-float structures: SPH model, experimental validation, and parametric study. Coast. Eng. **184** (2023)

Van He, N., Van Hien, N., Bui, N.T.: Analyis of aerodynamic performance of passenger ship with different frontal accommodations using CFD. Ocean Eng. **270**(January), 113622 (2023). https://doi.org/10.1016/j.oceaneng.2023.113622

He, S., Chen, X., Kong, S., Ji, S.: Measurement and identification of ice loads on hull structures in far field based on dynamic effects. Chin. J. Ship Res. **16**(5), 54–63 (2021)

He, W., Xie, L., Wang, S., Hu, Z., Xie, D., Wang, C.: Preliminary assessment of the mooring fatigue performance of a semi-submersible platform in time-domain utilizing fracture mechanics-based approach. Appl. Ocean Res. **129** (2022)

Hernández-Fontes, J.V., Mendoza, E., Hernández, I.D., Silva, R.: A detailed description of flow-deck interaction in consecutive green water events. J. Offshore Mech. Arctic Eng. **143**(4) (2021)

Herrnring, H., Kubiczek, J.M., Ehlers, S.: The ice extrusion test: a novel test setup for the investigation of ice-structure interaction–results and validation. Ships Offshore Struct. **15**(S1), S1–S9 (2021). https://doi.org/10.1080/17445302.2020.1713437

Hirdaris, S.E., et al.: Loads for use in the design of ships and offshore structures. Ocean Eng. **78**, pp. 131–174 (2014). https://linkinghub.elsevier.com/retrieve/pii/S0029801813003557

Hirdaris, S., et al.: Review of the uncertainties associated to hull girder hydroelastic response and wave load predictions. Marine Struct. **89**, 103383 (2023). Accessed 22 Feb 2023,

Hmedi, M., et al.: Experimental analysis of CENTEC-TLP self-stable platform with a 10 MW turbine. J. Marine Sci. Eng. **10**(12) (2022)

Honey, L., Judge, C.Q., Gilbert, C.M.: Slamming events of a planing hull and wedge water entry experiment: comparisons and insights on fluid-structure interaction. In: SNAME International Conference on Fast Sea Transportation, p. D021S003R003 (2021). https://doi.org/10.5957/FAST-2021-003

Hosseinzadeh, S., Tabri, K., Topa, A., Hirdaris, S.: Slamming loads and responses on a non-prismatic stiffened aluminium wedge: Part II. Numerical simulations. Ocean Eng. **279**, 114309 (2023a)

Hosseinzadeh, S., et al.: Slamming loads and responses on a non-prismatic stiffened aluminium wedge: part I. Experimental study. SSRN Electron. J. **279**, 114309 (2023b)

Hou, H.M., Liu, Y., Dong, G.H., Xu, T.J.: Reliability assessment of mooring system for fish cage considering one damaged mooring line. Ocean Eng. **257** (2022)

Hsiao, Y., Hsiao, S.C.: Experimental study on the interaction of solitary wave with elastic submerged plate. Ocean Eng. **261** (2022)

Hu, B., Liu, L., Wang, D.: Prediction of performance of a non-icebreaking ship in marginal ice zone. J. Hydrodynam. **34**(2), pp. 315–328 (2022a)

Hu, S., Zhang, S., Liu, M., Dong, S., Zheng, R.: Hydrodynamic response research of a cage system under waves and currents based on modal analysis. Ocean Eng. **243** (2022b)

Hu, T., et al.: Experimental investigation on hydrodynamic forces of semi-submerged cylinders in combined steady flow and oscillatory flow. Ocean Eng. **268** (2023)

Huang, B., et al.: The effects of heave motion on the performance of a floating counter-rotating type tidal turbine under wave-current interaction. Energy Convers. Manage. **252** (2022a)

Huang, H.C., Yang, Y.F., Zhu, R.H., Wang, C.Z.: Nonlinear wave resonance due to oscillations of twin cylinders in a uniform current. Appl. Ocean Res. **121** (2022b)

Huang, H., Gu, H., Chen, H.C.: A new method to couple FEM mooring program with CFD to simulate Six-DoF responses of a moored body. Ocean Eng. **250** (2022c)

Huang, H., Liu, Q., Yue, M., Miao, W., Wang, P., Li, C.: Fully coupled aero-hydrodynamic analysis of a biomimetic fractal semi-submersible floating offshore wind turbine under wind-wave excitation conditions. Renew. Energy **203**, 280–300 (2023a)

Huang, J., Teng, B.: The second-order interaction of monochromatic waves with three-dimensional bodies in a channel. Ocean Eng. **217** (2020)

Huang, J., Teng, B., Cong, P.W.: Second-order sum- and difference-frequency bichromatic wave forces on three-dimensional bodies in a channel. Ocean Eng. **224** (2021)

Huang, L., Li, F., Li, M., Khojasteh, D., Luo, Z., Kujala, P.: An investigation on the speed dependence of ice resistance using an advanced CFD+DEM approach based on pre-sawn ice tests. Ocean Eng. **264** (2022d)

Huang, L., Pena, B., Liu, Y., Anderlini, E.: Machine learning in sustainable ship design and operation: a review. Ocean Eng. **266** (2022e)

Huang, S., Jiao, J., Guedes Soares, C.: Uncertainty analyses on the CFD–FEA co-simulations of ship wave loads and whipping responses. Marine Struct. **82** (2022f)

Huang, W.H., Yang, R.Y.: Water depth variation influence on the mooring line design for fowt within shallow water region. J. Marine Sci. Eng. **9**(4) (2021)

Huang, Y., Wang, P., Zhao, M., Zhang, C., Du, X.: Dynamic response of sea-crossing bridge under combined seismic and wave-current action. Structures **40**, 317–327 (2022g)

Huang, Y., Zhao, W., Wan, D.: Wake interaction between two spar-type floating offshore wind turbines under different layouts. Phys. Fluids **35**(9) (2023b)

Huo, F., Yang, H., Yao, Z., An, K., Xu, S.: Study on slamming pressure characteristics of platform under freak wave. J. Marine Sci. Eng. **9**(11) (2021)

Huynh, L.E., Chu, C.R., Wu, T.R.: Hydrodynamic loads of the bridge decks in wave-current combined flows. Ocean Eng. **270** (2023)

IACS. The international association of classification Societies longitudinal strength standard S11 (2015)

Ibarra, M.A.C., Simão, M.L., Videiro, P.M., Sagrilo, L.V.S.: Long-term fatigue analysis of mooring lines considering wind-sea and swell waves using the univariate dimension-reduction method. Appl. Ocean Res. **118** (2022)

Ibrahim, A.M., Morabito, M.G., Beaver, W.: Development of a segmented model for a medium-sized semi-displacement vessel. In: SNAME Maritime Convention. SNAME, Houston (2022)

International Maritime Organization (IMO). Revised GHG reduction strategy for global shipping adopted, p. 1 (2023). https://www.imo.org/en/MediaCentre/PressBriefings/pages/Revised-GHG-reduction-strategy-for-global-shipping-adopted-.aspx, https://www.imo.org/en/MediaCentre/PressBriefings/pages/Revised-GHG-reduction-strategy-for-global-shipping-adopted-.aspx. Accessed 18 Oct 2024

Islam, H., Guedes Soares, C.: Assessment of uncertainty in the CFD simulation of the wave-induced loads on a vertical cylinder. Marine Struct. **80** (2021). Accessed 29 Aug 2023

Islam, H., Soares, C.G.: Head wave simulation of a KRISO container ship model using openFOAM for the assessment of sea margin. J. Offshore Mech. Arctic Eng.-Trans. ASME **144**(3) (2022)

Islam, H., Sutulo, S., Guedes Soares, C.: Aerodynamic load prediction on a patrol vessel using computational fluid dynamics. J. Marine Sci. Eng. **10**(7) (2022)

ISO.x ISO/IEC Guide 98-3:2008 - Uncertainty of measurement—Part 3: Guide to the expression of uncertainty in measurement (GUM:1995) (2023)

ISSC. ISSC Report. Proceedings of the 20th international ship and offshore structures congress, 1 (2018)

ITTC. ITTC – Recommended Procedures and Guidelines - Verification and validation of linear and weakly nonlinear seakeeping computer codes. 7.5-02-07-02.5 (Revision 01). International Towing Tank Conference. Lyngby, Denmark, p. 17 (2011)

ITTC. ITTC Report. Proceedings of the 21st international towing tank symposium (2021)

Jabary, W., et al.: Development of machine learning approaches to enhance ship operational performance evaluation based on an integrated data model. In: 33rd International Ocean and Polar Engineering Conference, Ottawa (2023)

JAIC (Joint Accident Investigation Commission). Final report on the capsizing on 28 September 1994 in the Baltic Sea of the Ro-ro passenger vessel MV Estonia. Edita Books, Helsinki (1997)

Jain, U., Vega-Martinez, P., van der Meer, D.: Air entrapment and its effect on pressure impulses in the slamming of a flat disc on water (2021)

Jalani, M.A., et al.: Experimental study on a bottom corner of the floating WEC. Ocean Eng. **243** (2022)

Jeon, W., Park, S., Cho, S.: Moored motion prediction of a semi-submersible offshore platform in waves using an OpenFOAM and MoorDyn coupled solver. Int. J. Naval Archit. Ocean Eng. **15** (2023)

Jeon, W., Park, S., Jeon, G.-M., Park, J.-C.: Computational study on rogue wave and its application to a floating body. Appl. Sci.-Basel **12**(6) (2022)

Jeong, K., Min, S., Jang, M., Won, D., Kim, S.: Feasibility study of submerged floating tunnels with vertical and inclined combined tethers. Ocean Eng. **265** (2022)

Ji, S., Yang, D.: Ice loads and ice-induced vibrations of offshore wind turbine based on coupled DEM-FEM simulations. Ocean Eng. **243**(July 2021), 110197 (2022). https://doi.org/10.1016/j.oceaneng.2021.110197

Jiang, C., et al.: A multi-fidelity prediction model for vertical bending moment and total longitudinal stress of a ship based on composite neural network. J. Hydrodyn. **35**(1), 27–35 (2023a)

Jiang, F., Yin, D., Califano, A., Berthelsen, P.A.: Application of CFD on VIM of semi-submersible FOWT: a case study. J. Phys. Conf. Ser. **2626**(1), 1–11 (2023)

Jiang, Z., Li, F., Suominen, M., Tavakoli, S., Kujala, P., Hirdaris, S.: An investigation of the influence of added mass on the ICE loads in various sea states. In: Proceedings of the International Conference on Offshore Mechanics and Arctic Engineering - OMAE, 6(August) (2023c)

Jiang, Z., Li, F., Tavakoli, S., Kujala, P., Suominen, M., Hirdaris, S.: A viscous investigation on the hydrodynamic coefficients and wave loads under the interaction of side-by-side cylinders in regular waves. Ocean Eng. **297** (2024)

Jiao, J., Zhao, M., Jia, G., Ding, S.: SPH simulation of two side-by-side LNG ships' motions coupled with tank sloshing in regular waves. Ocean Eng. **297** (2024)

Jin, J., Zhou, L., Ding, S., Gu, Y.: Numerical simulation of the ice breaking process for hovercraft. J. Marine Sci. Eng. **9**(9) (2021)

Kapsemberg, G., Brizzolara, S.: Hydro-elastic effects of bow flare slamming on a fast monohull. In: International Fast Ship Conference FAST 1999 (1999)

Kapsenberg, G., Carette, N.: A consistent method to design and evaluate the performance of anti-roll tanks for ships. Ship Technol. Res. **70**(2), 117–145 (2023)

Kapsenberg, G.K.: Slamming of ships: where are we now? Philos. Trans. Roy. Soc. A: Math. Phys. Eng. Sci. **369**(1947), 2892–2919 (2011)

Kardakaris, K., Konispoliatis, D.N., Soukissian, T.H.: Theoretical evaluation of the power efficiency of a moored hybrid floating platform for wind and wave energy production in the Greek seas. AIMS Geosci. **9**(1), 153–183 (2023) von Karman, T.H.: The impact on seaplane floats during landing. NACA Technical Note, NACA-TN-32 (1929)

Katsidoniotaki, E., Psarommatis, F., Göteman, M.: Digital twin for the prediction of extreme loads on a wave energy conversion system. Energies **15**(15), 5464 (2022). https://www.mdpi.com/1996-1073/15/15/5464/htm. Accessed 4 Sept 2023

Katsidoniotaki, E., Shahroozi, Z., Eskilsson, C., Palm, J., Engström, J., Göteman, M. : Validation of a CFD model for wave energy system dynamics in extreme waves. Ocean Eng. **268** (2023)

Khan, L.J., Macklin, J.J.R., Peck, B.C.D., Morton, O., Souppez, J.-B.R.G.: A review of wind-assisted ship propulsion for sustainable commercial shipping. In: Wind Propulsion Conference, 15th–16th September 2021, London, UK (2021)

Kharazmi, R., Ketabdari, M.J.: Numerical modeling to develop strake design of Spar platform for Vortex-Induced motions suppression. Ocean Eng. **250**(March), 111060 (2022). https://doi.org/10.1016/j.oceaneng.2022.111060

Kim, D., Tezdogan, T., Incecik, A.: A high-fidelity CFD-based model for the prediction of ship manoeuvrability in currents. Ocean Eng. **256**, 111492 (2022a). Accessed 18 Oct 2024

Kim, M., Pipiras, V., Reed, A. M., Weems, K.: Calibration of low-fidelity ship motion programs through regressions of high-fidelity forces. Ocean Eng. **290** (2023a)

Kim, S., Bouscasse, B., Ducrozet, G., Delacroix, S., De Hauteclocque, G., Ferrant, P.: Experimental investigation on wave-induced bending moments of a 6,750-TEU containership in oblique waves. Ocean Eng. **284** (2023b)

Kim, S., de Hauteclocque, G., Bouscasse, B., Lasbleis, M., Ducrozet, G.: Experimental analysis of extreme wave loads on a containership. Ocean Eng. **306** (2024)

Kim, S.J., Kim, M.H., Koo, W.: Nonlinear hydrodynamics of freely floating symmetric bodies in waves by three-dimensional fully nonlinear potential-flow numerical wave tank. Appl. Ocean Res. **113** (2021)

Kim, S.J., et al.: Comparison of numerical approaches for structural response analysis of passenger ships in collisions and groundings. Marine Struct. **81** (2022b)

Kim, Y., Kim, J.-H.: Benchmark study on motions and loads of a 6750-TEU containership (2016). https://doi.org/10.1016/j.oceaneng.2016.04.015

Kleijnen, J.P.C.: Kriging metamodeling in simulation: a review. Eur. J. Oper. Res. **192**(3), 707–716 (2009)

Kobayashi, H., et al.: CFD assessment of the wind forces and moments of superstructures through RANS. Appl. Ocean Res. **129**(September), 103364 (2022). https://doi.org/10.1016/j.apor.2022.103364

Komoriyama, Y., Iijima, K., Tatsumi, A., Fujikubo, M.: Identification of wave profiles encountered by a ship with no forward speed using Kalman filter technique and validation by tank tests - long-crested irregular wave case -. Ocean Eng. **271** (2023)

Konispoliatis, D.N., Chatjigeorgiou, I.K., Mavrakos, S.A.: Hydrodynamics of a moored permeable vertical cylindrical body. J. Marine Sci. Eng. **10**(3) (2022a)

Konispoliatis, D.N., et al.: Refos: a renewable energy multi-purpose floating offshore system. Energies **14**(11) (2021)

Konispoliatis, D.N., Manolas, D.I., Voutsinas, S.G., Mavrakos, S.A.: Coupled dynamic response of an offshore multi-purpose floating structure suitable for wind and wave energy exploitation. Front. Energy Res. **10** (2022b)

Kume, K., Hamada, T., Kobayashi, H., Yamanaka, S.: Evaluation of aerodynamic characteristics of a ship with flettner rotors by wind tunnel tests and RANS-based CFD. Ocean Eng. **254**(March), 111345 (2022). https://doi.org/10.1016/j.oceaneng.2022.111345

Lakshmynarayanana, P.A.K., Hirdaris, S.: Comparison of nonlinear one- and two-way FFSI methods for the prediction of the symmetric response of a containership in waves. Ocean Eng. **203**, 107179 (2020). Accessed 30 Aug 2023

Leal, A.P., Condino Fujarra, A.L., Carnier, R.M., Gonçalves, R.T., Suzuki, H.: Flow-induced vibration analysis using a low-cost inertial measurement unit. In: Proceedings of the International Conference on Offshore Mechanics and Arctic Engineering - OMAE. American Society of Mechanical Engineers Digital Collection (2023). https://doi.org/10.1115/OMAE2023-105138. Accessed 18 Oct 2024

Lee, J., Ahn, Y., Kim, Y.: Experimental study on effect of density ratio and phase transition during sloshing impact in rectangular tank. Ocean Eng. **242**, 110105 (2021a)

Lee, J.H., Kim, Y., Kim, B.S., Gerhardt, F.: Comparative study on analysis methods for added resistance of four ships in head and oblique waves. Ocean Eng. **236** (2021b)

Lee, J.-H., Lee, J., Kim, Y., Ahn, Y.: Prediction of wave-induced ship motions based on integrated neural network system and spatiotemporal wave-field data. Phys. Fluids **35**(9) (2023)

Lemström, I., Polojärvi, A., Tuhkuri, J.: Model-scale tests on ice-structure interaction in shallow water, Part I: global ice loads and the ice loading process. Marine Struct. **81**(October 2021), 103106 (2022). https://doi.org/10.1016/j.marstruc.2021.103106

Leroy, V., Delacroix, S., Merrien, A., Bachynski-Polić, E.E., Gilloteaux, J.C.: Experimental investigation of the hydro-elastic response of a spar-type floating offshore wind turbine. Ocean Eng. **255** (2022)

Li, F., Huang, L.: A review of computational simulation methods for a ship advancing in broken ice. J. Marine Sci. Eng. **10**(2) (2022)

Li, F., Suominen, M., Kujala, P.: Ship performance in ice channels narrower than ship beam: model test and numerical investigation. Ocean Eng. **240** (2021)

Li, H., Zou, J., Deng, B., Liu, R., Sun, S.: Experimental study of stern slamming and global response of a large cruise ship in regular waves. Marine Struct. **86**(April), 103294 (2022a). https://doi.org/10.1016/j.marstruc.2022.103294

Li, H., Zou, J., Peng, Y., Zhou, X., Lu, L., Sun, S.: Numerical study of slamming and whipping loads in moderate and large regular waves for different forward speeds. Marine Struct. **94**, 103563 (2024a). https://www.sciencedirect.com/science/article/pii/S095183392300196X

Li, L., Han, G., Ji, S.: Statistical analysis of ice load on icebreaker ship based on stochastic ice fields. J. Marine Sci. Eng. **12**(3) (2024b)

Li, X., Xiao, Q., Wang, E., Peyrard, C., Gonçalves, R.T.: The dynamic response of floating offshore wind turbine platform in wave-current condition. Phys. Fluids **35**(8) (2023)

Li, Y., Peng, T., Xiao, L., Wei, H., Li, X.: Wave runup prediction for a semi-submersible based on temporal convolutional neural network. J. Ocean Eng. Sci. (2022b). Accessed 4 Sept 2023

Lian, Y., et al.: An upper and lower bound method for evaluating residual strengths of polyester mooring ropes with artificial damage. Ocean Eng. **262** (2022)

Liang, H., Zhao, E., Qin, H., Mu, L., Su, H.: A model coupling CFD and DRL: investigation on wave dissipation by actively controlled flat plate. IEEE Access **10**, pp. 98290–98308 (2022a). Accessed 4 Sept 2023

Liang, J., Liu, Y., Chen, Y., Li, A.: Experimental study on hydrodynamic characteristics of the box-type floating breakwater with different mooring configurations. Ocean Eng. **254** (2022b)

Lihin, Z.H., Khan, R., Ishak, M.I.M., Rahman, A.A., Kar, S., Abdul Rahman, A.R.: Revised methodology for assessing fixed offshore structures with wave in deck loading. In: Offshore Technology Conference Asia, OTCA 2022. Offshore Technology Conference (2022)

Lin, J., kun Guo, Y.: Numerical investigation of non-cohesive seabed response around a mono-pile by crossing wave-current. Water Sci. Eng. **15**(1), 57–68 (2022)

Liu, H., Zhao, C., Ma, G., He, L., Sun, L., Li, H.: Reliability assessment of a floating offshore wind turbine mooring system based on the TLBO algorithm. Appl. Ocean Res. **124** (2022a)

Liu, J., Liu, M., Xiao, L., Wei, H., Li, X.: Experimental study on the effect of flare barriers on wave run-up and motion response of a semi-submersible platform. Ocean Eng. **281** (2023a)

Liu, J., Xiao, L., Yang, L., Liu, M.: Experimental study on the effect of flare barriers on wave run-up and wave loads of a rounded-square column. Ocean Eng. **260** (2022b)

Liu, L., Ji, S.: Dilated-polyhedron-based DEM analysis of the ice resistance on ship hulls in escort operations in level ice. Marine Struct. **80** (2021)

Liu, L., et al.: Study on numerical simulation and mitigation of parametric rolling in a container ship under head waves. BRODOGRADNJA **75**(3) (2024)

Liu, R., Xue, Y., Lu, X.: Coupling of finite element method and peridynamics to simulate ship-ice interaction. J. Marine Sci. Eng. **11**(3), 481 (2023b)

Liu, X., Liu, F., Ren, H., Chen, X., Xie, H.: Experimental investigation on the slamming loads of a truncated 3D stern model entering into water. Ocean Eng. **252**, 110873 (2022c). https://doi.org/10.1016/j.oceaneng.2022.110873

Liu, Y., Ge, D., Bai, X., Li, L.: A CFD study of vortex-induced motions of a semi-submersible floating offshore wind turbine. Energies **16**(2) (2023c)

Liu, Y., Shi, W., Wang, W., Li, X., Qi, S., Wang, B.: Dynamic analysis of monopile-type offshore wind turbine under sea ice coupling with fluid-structure interaction. Front. Marine Sci. **9** (2022d)

Liu, Z., Guedes Soares, C.: Sensitivity analysis of the cage volume and mooring forces for a gravity cage subjected to current and waves. Ocean Eng. **287** (2023)

Liu, Z., Xu, H., Guedes Soares, C.: Experimental study on the mooring forces and motions of a fish cage under regular waves. Ocean Eng. **280** (2023d)

Livermore Software Technology Corporation. ICFD theory manual: Incompressible fluid solver in LS-DYNA (2014). https://ftp.lstc.com/anonymous/outgoing/inaki/docs/pdf_icfd/ICFD_theory.pdf

Lloyd, C., et al.: Validation of the dynamic load characteristics on a tidal stream turbine when subjected to wave and current interaction. Ocean Eng. **222**, 108360 (2021). Accessed 18 Oct 2024

Lone, E.N., Mainçon, P., Gabrielsen, Ø., Sauder, T., Larsen, K., Leira, B.J.: Analysis of S-N data for new and corroded mooring chains at varying mean load levels using a hierarchical linear model. Mar. Struct. **91**, 103466 (2023)

Lone, E.N., Sauder, T., Larsen, K., Leira, B.J.: Probabilistic fatigue model for design and life extension of mooring chains, including mean load and corrosion effects. Ocean Eng. **245** (2022a)

Lone, E.N., Sauder, T., Larsen, K., Leira, B.J.: Fatigue reliability of mooring chains, including mean load and corrosion effects. Ocean Eng. **266** (2022b)

Lozon, E., Hall, M.: Coupled loads analysis of a novel shared-mooring floating wind farm. Appl. Energy **332** (2023)

Lu, L., Ren, H., Li, H., Zou, J., Chen, S., Liu, R.: Numerical method for whipping response of ultra large container ships under asymmetric slamming in regular waves. Ocean Eng. **287**(P2), 115830 (2023). https://doi.org/10.1016/j.oceaneng.2023.115830

Luo, G., Zhang, Y., Pan, S., Ren, Y.: Dynamic response analysis of submerged floating tunnels to coupled wave-seismic action. J. Eng. Mech. **38**(2), 211–220 (2021)

Luo, M., Rubinato, M., Wang, X., Zhao, X.: Experimental investigation of freak wave actions on a floating platform and effects of the air gap. Ocean Eng. **253**, 111192 (2022)

Ma, C., Bi, C.W., Xu, Z., Zhao, Y.P.: Dynamic behaviors of a hinged multi-body floating aquaculture platform under regular waves. Ocean Eng. **243** (2022a)

Ma, C., Zhao, Y.P., Bi, C.W.: Numerical study on hydrodynamic responses of a single-point moored vessel-shaped floating aquaculture platform in waves. Aquacult. Eng. **96** (2022b)

Ma, C., Zhao, Y.-P., Bi, C.-W., Xie, S.: Numerical study on dynamic analysis of a nine-module floating aquaculture platform under irregular waves. Ocean Eng. **285**, 115253 (2023). https://linkinghub.elsevier.com/retrieve/pii/S0029801823016372

Ma, L., Swan, C.: The effective prediction of wave-in-deck loads. J. Fluids Struct. **95** (2020)

Ma, L., Swan, C.: An experimental study of wave-in-deck loading and its dependence on the properties of the topside structure. Marine Struct. **88** (2023a)

Ma, L., Swan, C.: Wave-in-deck loads: an assessment of present design practice given recent improvements in the description of extreme waves and the nature of the applied loads. Ocean Eng. **285** (2023b)

Ma, Y., Sclavounos, P.D.: Support vector machines model of the nonlinear hydrodynamics of fixed cylinders. J. Offshore Mechan. Arctic Eng. **143**(5) (2021)

MAIB. Report on the investigation of the structural failure of MSC Napoli, marine accident investigation Branch - MAIB. Carlton House Carlton Place, Southampton (2008)

Makarov, O., Bekker, A., Li, L.: Comparative analysis of numerical methods for the modeling of ice–structure interaction problems. Continuum Mech. Thermodyn. **34**(6), 1621–1639 (2022)

Malenica, S., Seng, S., Diebold, L., Scolan, Y.-M., Korobkin, A.A., Khabakhpasheva, T.: On three dimensional hydroelastic impact. In: 9th International Conference on HYDROELASTICITY IN MARINE TECHNOLOGY. Roma (2022)

Mandru, A., Rusu, L., Bekhit, A., Pacuraru, F.: Numerical study of a model and full-scale container ship sailing in regular head waves. INVENTIONS **9**(1) (2024)

Mao, Y., Wang, T., Duan, M.: A DNN-based approach to predict dynamic mooring tensions for semi-submersible platform under a mooring line failure condition. Ocean Eng. **266** (2022)

Marty, A., Schoefs, F., Damblans, G., Facq, J. V., Gaurier, B., Germain, G.: Experimental study of two kinds of hard marine growth effects on the hydrodynamic behavior of a cylinder submitted to wave and current loading. Ocean Eng. **263** (2022)

Mauro, F., Valentina, E. Della, Ferrari, V., Begovic, E. A method for early-stage design current loads determination on drill-ships. Ocean Eng. **287**(September) (2023)

van der Meer, D.: Linear stability analysis of a time-divergent slamming flow. J. Fluid Mech. **934** (2022). https://doi.org/10.1017/jfm.2021.1064

Men, J., Yan, F., Wang, Y., Wang, L., Zheng, X., Wang, W.: An evaluation of mooring system in a wind-wave hybrid system under intact and accident states. Ocean Eng. **283** (2023)

Mendoza, J., Haagensen, P. J., Köhler, J.: Analysis of fatigue test data of retrieved mooring chain links subject to pitting corrosion. Marine Struct. **81** (2022)

Meyer, J., Windt, C., Sinn, P., Hildebrandt, A: On the mooring methodology of heaving point absorber arrays. Ocean Eng. **281** (2023)

Miao, Y., Chen, X., Ye, Y., Ding, J., Huang, H.: Numerical modeling and dynamic analysis of a floating bridge subjected to wave, current and moving loads. Ocean Eng. **225** (2021)

Miratsu, R., Sasmal, K., Kodaira, T., Fukui, T., Zhu, T., Waseda, T.: Evaluation of ship operational effect based on long-term encountered sea states using wave hindcast combined with storm avoidance model. Marine Struct. **86** (2022)

Miškov, V., Dragić, M., Tomin, V., Hofman, M.: Sea trials of Sigma wave energy converter – slacking and snapping of TLP mooring lines. Marine Struct. **89** (2023)

MJB. Maritime Bureau of Japan's Ministry of Land, Infrastructure, Transport and Tourism Final report of committee on large container ship safety (2015)

El Moctar, O., Lantermann, U., Shigunov, V., Schellin, T.E.: Experimental and numerical investigations of effects of ship superstructures on wind-induced loads for benchmarking. Phys. Fluids **35**(4) (2023)

Mohajernasab, S., Drobyshevski, Y., Abdussamie, N., Ojeda, R.: Experimental investigation into the effects of air pockets on wave-in-deck loads of offshore structures. Marine Struct. **90** (2023)

Mohapatra, S.C., Islam, H., Hallak, T.S., Soares, C.G.: Solitary wave interaction with a floating pontoon based on boussinesq model and CFD-based simulations. J. Marine Sci. Eng. **10**(9) (2022)

Mohseni, M., Guedes Soares, C.: Numerical investigation of inline wave force on a truncated vertical cylinder with different cross-sections in regular head waves. Ocean Eng. **251** (2022a)

Mohseni, M., Guedes Soares, C.: Numerical simulation of wave interaction with a pair of fixed large tandem cylinders subjected to regular, non-breaking waves. J. Offshore Mech. and Arctic Eng. **144**(3) (2022b)

Mohseni, M., Guedes Soares, C.: CFD analysis of wave loading on a 10 MW TLP-type offshore floating wind turbine in regular waves. Ocean Eng. **301** (2022)

Monaghan, J.J.: Simulating free surface flows with SPH. J. Comput. Phys. **110**(2), 399–406 (1994). https://www.sciencedirect.com/science/article/pii/S0021999184710345

Monteiro, L.F.G.: High-fidelity analysis of flettner rotors aerospace engineering examination committee (2024)

Müller, F., Böhm, A., Herrnring, H., von Bock und Polach, F., Ehlers, S.: Influence of the ice shape on ice-structure impact loads. Cold Regions Sci. Technol. **221**(January) (2024)

Munir, H., Lee, C.F., Ong, M.C.: Global analysis of floating offshore wind turbines with shared mooring system. IOP Conf. Ser.: Mater. Sci. Eng. **1201**(1), 12024 (2021)

Nam, B.W., Hong, S.Y., Kim, H.J.: An experimental study on characteristics of relative wave elevations around a tension leg platform in irregular waves. Appl. Ocean Res. **122** (2022)

Nan, M.Y., Hu, J.J., Wang, X.L., Vladimir, Y.: Study on Ice load inversion based on LSTM networks. Chuan Bo Li Xue/J. Ship Mech. **25**(12), 1675–1684 (2021)

Neisi, A., Ghassemi, H., Iranmanesh, M., He, G.: Effect of the multi-segment mooring system by buoy and clump weights on the dynamic motions of the floating platform. Ocean Eng. **260** (2022)

Neugebauer, J., et al.: Sloshing-Induced Impact Loads and Free Surface Elevation in a Pool in Waves. In: Proceedings of the Thirty-Third International Ocean and Polar Engineering Conference, Ottawa, pp. 2623–2630 (2023)

Nguyen, N., Thiagarajan, K: Nonlinear viscoelastic modeling of synthetic mooring lines. Marine Struct. **85** (2022)

Ni, B., Wang, Y., Xu, Y., Chen, W.: Numerical simulation of ship collision with rafted ice based on cohesive element method. J. Mar. Sci. Appl. **23**(1), 127–136 (2024)

Ni, B., Xu, Y., Huang, Q., You, J., Xue, Y.: Application of improved cohesive zone length formula in ice mode I crack propagation. Chin. J. Ship Res. **17**(3), 58–66 (2022)

Nielsen, U.D., Mounet, R.E.G., Brodtkorb, A.H.: Tuning of transfer functions for analysis of wave–ship interactions. Marine Struct. **79** (2021)

Ning, D., Xu, J., Chen, L., Cong, P., Zhao, M., Jiang, C.: Boussinesq modelling of near-trapping in a four-cylinder array. Ocean Eng. **248** (2022)

Nizamani, M., Nizamani, Z., Nakayama, A., Osman, M.: Analysis of loads caused by waves on the deck near the free surface of the offshore platform using computational fluid dynamics. Ships Offshore Struct. **17**(9), 1964–1974 (2022)

de Oliveira Costa, D., Araújo Perim, J., Guedes Camargo, B., Sena Sales Junior, J., Carlos Fernandes, A.: Wave impact loads on the bottom of flat decks (2021). http://asmedigitalcollection.asme.org/OMAE/proceedings-pdf/OMAE2021/85161/V006T06A012/6780540/v006t06a012-omae2021-62990.pdf

Otori, H., Kikuchi, Y., Rivera-Arreba, I., Viré, A.: Numerical study of hydrodynamic forces and dynamic response for barge type floating platform by computational fluid dynamics and engineering model. Ocean Eng. **284** (2023)

Otter, A., Murphy, J., Pakrashi, V., Robertson, A., Desmond, C.: A review of modelling techniques for floating offshore wind turbines. Wind Energy **25**(5), 831–857 (2022)

Panciroli, R., Porfiri, M.: Analysis of hydroelastic slamming through particle image velocimetry. J. Sound Vib. **347**, 63–78 (2015)

Papini, G., Paduano, B., Pasta, E., Carapellese, F., Mattiazzo, G., Faedo, N.: On the influence of mooring systems in optimal predictive control for wave energy converters. Renew. Energy 119242 (2023). https://linkinghub.elsevier.com/retrieve/pii/S0960148123011576

Park, C.-J., Lee, J.-K., Kim, Y.: Study on hydroelastic responses of membrane-type LNG cargo containment structure under impulsive sloshing loads of different media. J. Marine Sci. Eng. (2024)

Parunov, J., et al.: Benchmark on the prediction of whipping response of a warship model in regular waves. Marine Struct. **94**, 103549 (2024). https://www.sciencedirect.com/science/article/pii/S095183392300182X

Parunov, J., et al.: Benchmark study of global linear wave loads on a container ship with forward speed. Marine Struct. **84** (2022)

Pena, B., Huang, L.: Wave-GAN: A deep learning approach for the prediction of nonlinear regular wave loads and run-up on a fixed cylinder. Coast. Eng. **167**, 103902 (2021). Accessed 4 Sept 2023

Pena-Sanchez, Y., Windt, C., Davidson, J., Ringwood, J.V.: A critical comparison of excitation force estimators for wave-energy devices. IEEE Trans. Control Syst. Technol. **28**(6), 2263–2275 (2020)

Petranović, T., Mikulić, A., Katalinić, M., Ćorak, M., Parunov, J.: Method for prediction of extreme wave loads based on ship operability analysis using hindcast wave database. J. Marine Sci. Eng. **9**(9) (2021)

Pillai, A.C., et al.: Anchor loads for shallow water mooring of a 15 MW floating wind turbine — Part I: chain catenary moorings for single and shared anchor scenarios. Ocean Eng. **266** (2022)

Pilloton, C., Colagrossi, A., Marrone, S., Bardazzi, A.: Three hours real–time SPH simulation of sloshing flows inside a LNG ship with realistic severe sea-state forcing. In: SPHERIC 2022, Xi'an (2022)

Pinguet, R., Benoit, M., Molin, B., Rezende, F.: CFD analysis of added mass, damping and induced flow of isolated and cylinder-mounted heave plates at various submergence depths using an overset mesh method. J. Fluids Struct. **109** (2022)

Pitchforth, D.J., Mills, R.S., Rogers, T.J., Tygesen, U.T., Cross, E.J.: Physics-informed gaussian processes for wave loading prediction. In: Structural Health Monitoring 2023: Designing SHM for Sustainability, Maintainability, and Reliability - Proceedings of the 14th International Workshop on Structural Health Monitoring, pp. 2205–2214 (2023). https://www.scopus.com/inward/record.uri?eid=2-s2.0-85182281713&partnerID=40&md5=c8871a735c8f6aeb8ab6d664947396af

Qi, Y., Söring, H., El Moctar, O., Neugebauer, J.: An efficient numerical approach to predict waves in swimming pools on ships. In: Proceedings of the Thirty-Third International Ocean and Polar Engineering Conference, Ottawa, pp. 2603–2608 (2023)

Qvale, P., Nordhagen, H.O., Ås, S.K., Skallerud, B.H.: Effect of long periods of corrosion on the fatigue lifetime of offshore mooring chain steel. Marine Struct. **85** (2022)

Qvale, P., Zarandi, E. P., Ås, S.K., Skallerud, B.H.: Digital image correlation for continuous mapping of fatigue crack initiation sites on corroded surface from offshore mooring chain. Int. J. Fatigue **151** (2021)

Ramzanpoor, I., Nuernberg, M., Tao, L.: Coupled aero-hydro-servo-elastic analysis of 10MW TLB floating offshore wind turbine. J. Ocean Eng. Sci. (2023)

Ren, D., Park, J.C.: Particle-based numerical simulation of continuous ice-breaking process by an icebreaker. Ocean Eng. **270**(January), 113478 (2023). https://doi.org/10.1016/j.oceaneng.2022.113478

Ren, X., Tao, L., Liang, Y., Han, D.: Nonlinear wave surface elevation around a multi-column offshore structure. Ocean Eng. **238** (2021)

Robaux, F., Benoit, M.: Assessment of one-way coupling methods from a potential to a viscous flow solver based on domain- and functional-decomposition for fixed submerged bodies in nonlinear waves. Eur. J. Mech. B/Fluids **95**, 315–334 (2022)

Rodrigues, J.M., Viuff, T., Økland, O.D.: Model tests of a hydroelastic truncated floating bridge. Appl. Ocean Res. **125** (2022)

Rogers, F., Haddara, M., Molyneux, D.: Dynamic bending moment identification using neural networks. In: 25th American Towing Tank Conference. SNAME (1998)

Rueda-Bayona, J.G., Guzmán, A., Eras, J.J.C.: Vortex-induced vibration effect of extreme sea states over the structural dynamics of a scaled monopile offshore wind turbine. J. Ocean Eng. Marine Energy **9**(2), 359–376 (2023). https://doi.org/10.1007/s40722-022-00272-9

Ryan, C., Huang, L., Li, Z., Ringsberg, J.W., Thomas, G.: An Arctic ship performance model for sea routes in ice-infested waters. Appl. Ocean Res. **117** (2021)

Saincher, S., Sriram, V.: A three dimensional hybrid fully nonlinear potential flow and Navier Stokes model for wave structure interactions. Ocean Eng. **266**, 112770 (2022)

Sasikala, N., Sannasiraj, S. A., Manasseh, R.: 3D simulation of wave slamming on a horizontal deck using WCSPH. In: Lecture Notes in Civil Engineering (2021)

Sasmal, K., Miratsu, R., Kodaira, T., Fukui, T., Zhu, T., Waseda, T.: Statistical model representing storm avoidance by merchant ships in the North Atlantic Ocean. Ocean Eng. **235**, 109163 (2021)

Sauder, T., Alterskjær, S.A.: Hydrodynamic testing of wind-assisted cargo ships using a cyber–physical method. Ocean Eng. **243**(June 2021), 110206 (2022). https://doi.org/10.1016/j.oceaneng.2021.110206

Sazidy, M.: Development of velocity dependent ice flexural failure model and application to safe speed (2015)

Sazonov, K., Dobrodeev, A.: Ice Resistance assessment for a large size vessel running in a narrow ice channel behind an icebreaker. J. Mar. Sci. Appl. **20**(3), 446–455 (2021)

Scharnke, J., van Essen, S.M., Seyffert, H.C.: Required test durations for converged short-term wave and impact extreme value statistics–Part 2: deck box dataset. Marine Struct. **90** (2023). Accessed 29 Aug 2023

Schirmann, M.L., Collette, M.D., Gose, J.W.: Data-driven models for vessel motion prediction and the benefits of physics-based information. Appl. Ocean Res. **120** (2022)

Sclavounos, P.D., Ma, Y.: Artificial intelligence machine learning in marine hydrodynamics. In: Proceedings of the International Conference on Offshore Mechanics and Arctic Engineering - OMAE (2018)

Sergiienko, N.Y., da Silva, L.S.P., Bachynski-Polić, E.E., Cazzolato, B.S., Arjomandi, M., Ding, B.: Review of scaling laws applied to floating offshore wind turbines. Renew. Sustain. Energy Rev. **162**(March), 112477 (2022). https://doi.org/10.1016/j.rser.2022.112477

Shabani, B., Lavroff, J., Davis, M.R., Holloway, D.S., Thomas, G.A.: Slam loads and kinematics of wave-piercing catamarans during bow entry events in head seas. J. Ship Res. **62**(03), 134–155 (2018). https://doi.org/10.5957/JOSR.180001

Shao, X., Ringsberg, J.W., Yao, H.-D., Li, Z., Johnson, E., Fredriksson, G.: A comparison of two wave energy converters' power performance and mooring fatigue characteristics – one WEC vs many WECs in a wave park with interaction effects. J. Ocean Eng. Sci. (2023). https://linkinghub.elsevier.com/retrieve/pii/S2468013323000384

Shen, L., Wei, Z., Ji, S., Ivanov, D.: Experimental investigation of the flip-through impact: The generation mechanism and statistical characteristics. Ocean Eng. **294**, 116690 (2024)

Shen, Y., Firoozkoohi, R., Greco, M., Faltinsen, O.M.: Comparative investigation: closed versus semi-closed vertical cylinder-shaped fish cage in waves. Ocean Eng. **245** (2022)

Si, H., et al.: Effects of design and fabrication of segmented ship models on test results of wave-induced loads. Ocean Eng. **305** (2024)

Silva, L.S.P., Cazzolato, B., Sergiienko, N.Y., Ding, B.: Nonlinear dynamics of a floating offshore wind turbine platform via statistical quadratization—mooring, wave and current interaction. Ocean Eng. **236** (2021)

Sinsabvarodom, C., Leira, B.J., Chai, W., Næss, A.: Short-term extreme mooring loads prediction and fatigue damage evaluation for station-keeping trials in ice. Ocean Eng. **242**, 109930 (2021) (2021). https://linkinghub.elsevier.com/retrieve/pii/S0029801821012750

Sitek, M., Bojanowski, C., Bergeron, A., Licht, J.: Involute working group – FSI analysis of fuel plates using finite volume and finite element methods. United States (2021). https://www.osti.gov/biblio/1845461

Sivaprasad, H., et al.: Fatigue damage prediction of top tensioned riser subjected to vortex-induced vibrations using artificial neural networks. Ocean Eng. **268**(December 2022), 113393 (2023). https://doi.org/10.1016/j.oceaneng.2022.113393

Slagstad, M., Yu, Z., Amdahl, J.:A simplified approach for fatigue life prediction of aquaculture nets under waves and currents. Ocean Eng. **272** (2023)

Song, S., Jung, J.H., Jung, S.J., Jung, D.: Numerical investigation of current loads of side-by-side moored vessel. In: Proceedings of the International Offshore and Polar Engineering Conference, OnePetro, pp. 3677–3681 (2023). https://dx.doi.org/. Accessed 18 Oct 2024

Sørum, S.H., Fonseca, N., Kent, M., Faria, R.P.: Modelling of synthetic fibre rope mooring for floating offshore wind turbines. J. Marine Sci. Eng. **11**(1) (2023a)

Sørum, S.H., Fonseca, N., Kent, M., Faria, R.P.: Assessment of nylon versus polyester ropes for mooring of floating wind turbines. Ocean Eng. **278** (2023b)

Southall, N., Choi, S., Lee, Y., Hong, C., Hirdaris, S., White, N.: Impact analysis using CFD - a comparative study. In: Proceedings of the International Offshore and Polar Engineering Conference (2015). http://www.scopus.com/inward/record.url?eid=2-s2.0-84944675448&partnerID=MN8TOARS

Southall, N., Lee, Y., Johnson, M., Hirdaris, S., White, N.: Towards a pragmatic method for prediction of whipping: wedge impact simulations using OpenFOAM. In: Proceedings of the International Offshore and Polar Engineering Conference (2014)

Srinu, D., Venkateswarlu, V., Vijay, K.G., Atmanand, M.A.: Hydrodynamic analysis of oscillating water column in the presence of seabed undulations. J. Marine Sci. Technol. (Jpn.) **29**(2), 404–417 (2024)

Stansby, P., et al.: Experimental study of mooring forces on the multi-float WEC M4 in large waves with buoy and elastic cables. Ocean Eng. **266** (2022)

Stenius, I., Rosén, A., Kuttenkeuler, J.: Explicit FE-modelling of hydroelasticity in panel-water impacts. Int. Shipbuild. Prog. **54**, 111–127 (2007)

Stenius, I., Rosén, A., Kuttenkeuler, J.: Hydroelastic interaction in panel-water impacts of high-speed craft. Ocean Eng. **38**(2–3), 371–381 (2011)

Stern, F., Wilson, R.V., Coleman, H.W., Paterson, E.G.: Comprehensive approach to verification and validation of CFD simulations—part 1: methodology and procedures. J. Fluids Eng. Trans. ASME **123**(4), 793–802 (2001)

Subbulakshmi, A., Verma, M.: Transient response reduction of floating offshore wind turbine subjected to sudden mooring line failure. Ocean Eng. **271** (2023)

Sumislawski, P., Abdel-Maksoud, M.: Advances on numerical and experimental investigation of ship roll damping. J. Hydrodyn. **35**(3), 431–448 (2023)

Sun, B., Zhao, B., Xu, Y., Ma, S., Duan, W.: Numerical study of green water on a tumblehome vessel in strong nonlinear regular waves. J. Marine Sci. Appl. **22**(1, SI), 102–114 (2023a)

Sun, J., Gaidai, O., et al.: Extreme riser experimental loads caused by sea currents in the Gulf of Eilat. Probab. Eng. Mech. **68**(November 2021) (2022a)

Sun, J., Huang, Y.: Investigations on the ship-ice impact: Part 2. Spatial and temporal variations of ice load. Ocean Eng. **240**(February), 109686 (2021). https://doi.org/10.1016/j.oceaneng.2021.109686

Sun, K., et al.: Dynamic response analysis of floating wind turbine platform in local fatigue of mooring. Renew. Energy **204**, 733–749 (2023b)

Sun, Q., Zhang, M., Zhou, L., Garme, K., Burman, M.: A machine learning-based method for prediction of ship performance in ice: Part I. ICE resistance. Marine Struct. **83** (2022b)

Sun, S.L., Wang, J.L., Li, H., Chen, R.Q., Zhang, C.J.: Investigation on responses and wave climbing of a ship in large waves using a fully nonlinear boundary element method. Eng. Anal. Bound. Elem. **125**, 250–263 (2021)

Tagliafierro, B., et al.: Numerical validations and investigation of a semi-submersible floating offshore wind turbine platform interacting with ocean waves using an SPH framework. Appl. Ocean Res. **141** (2023)

Takami, T., Dam Nielsen, U., Juncher Jensen, J., Maki, A., Matsui, S., Komoriyama, Y.: Onboard identification of stability parameters including nonlinear roll damping via phase-resolved wave estimation using measured ship responses. Mech. Syst. Signal Process. **210**(December 2023) (2024)

Tan, Z., Sun, P.N., Liu, N.N., Li, Z., Lyu, H.G., Zhu, R.H.: SPH simulation and experimental validation of the dynamic response of floating offshore wind turbines in waves. Renew. Energy **205**, 393–409 (2023)

Tang, H.J., Yao, H.C., Yang, R.Y.: Experimental and numerical studies on successive failures of two mooring lines of a net cage subjected to currents. Ocean Eng. **266** (2022a)

Tang, H.J., Yao, H.C., Yang, R.Y.: Experimental and numerical study of a barge-type floating offshore wind turbine under a mooring line failure. Ocean Eng. **278** (2023a)

Tang, X., Zou, M., Zou, Z., Li, Z., Zou, L.: A parametric study on the ice resistance of a ship sailing in pack ice based on CFD-DEM method. Ocean Eng. **265** (2022b)

Tang, Y., Sun, S.L., Abbasnia, A., Guedes Soares, C., Ren, H.L.: A fully nonlinear BEM-beam coupled solver for fluid–structure interactions of flexible ships in waves. J. Fluids Struct. **121** (2023b)

Tang, Y., Sun, S.L., Ren, H.L.: Numerical investigation on a container ship navigating in irregular waves by a fully nonlinear time domain method. Ocean Eng. **223** (2021)

Tang, Y., Sun, S.L., Yang, R.S., Ren, H.L., Zhao, X., Jiao, J.L.: Nonlinear bending moments of an ultra large container ship in extreme waves based on a segmented model test. Ocean Eng. **243** (2022c)

Tavakoli, S., Hirdaris, S.: the hydroelastic slamming in oblique seas. In: Proceedings of the ASME 2023 42nd International Conference on Ocean, Offshore and Arctic Engineering. Ocean Engineering, p. 5 (2023)

Tavakoli, S., Mikkola, T., Hirdaris, S.: A fluid–solid momentum exchange method for the prediction of hydroelastic responses of flexible water entry problems. J. Fluid Mech. **965**, A19 (2023)

Tavakoli, S., Zhang, M., Kondratenko, A.A., Hirdaris, S.: A review on the hydrodynamics of planing hulls. Ocean Eng. **303** (2024)

Thome, M., el Moctar, O., Schellin, T.E.: Assessment of hydrodynamic loads on an offshore monopile structure considering hydroelasticity effects. J. Marine Sci. Eng. **11**(2) (2023)

Tian, J., Zhu, P.Q., Ding, J., Sun, S.L.: Fully nonlinear interaction between a ship-type floater and waves. Ocean Eng. **309** (2024)

Tian, Y., Yu, C., Gang, X., Kong, S., Ji, S., Wang, Y.: Experimental investigation on ice loading mechanism of typical offshore structures under the action of level ICE (2022). www.isope.org

Tödter, S., el Moctar, O., Neugebauer, J., Schellin, T.: Experimentally measured hydroelastic effects on impact-induced loads during flat waterentry and related uncertainties. J Offshore Mech Arct Eng. **142**, 011604 (2020)

Tödter, S., el Moctar, O., Neugebauer, J., Schellin, T.E.: Experimentally measured hydroelastic effects on impact-induced loads during flat water entry and related uncertainties. J. Offshore Mech. Arctic Eng. **142**(1) (2019). https://doi.org/10.1115/1.4044632

Torre, S., Burlando, M., Ruscelli, D., Repetto, M.P., Camauli, G.: Wind tunnel experimental investigation of the aerodynamic coefficients reduction due to sheltering surroundings on a cruise ship moored in port. J. Wind Eng. Industr. Aerodyn. **218**(June), 104731 (2021). https://doi.org/10.1016/j.jweia.2021.104731

Truong, D.D., Jang, B.S.: Estimation of ice loads on offshore structures using simulations of level ice-structure collisions with an influence coefficient method. Appl. Ocean Res. **125**(December 2021), 103235 (2022). https://doi.org/10.1016/j.apor.2022.103235

Truong, D.D., et al.: Benchmark study on slamming response of flat-stiffened plates considering fluid-structure interaction. Marine Struct. **79** (2021)

Uto, S., Matsuzawa, T., Shimoda, H., Wako, D., Toyota, T.: Formulation and validation of resistance prediction scheme for ships in ice regime described in WMO egg code. Cold Regions Sci. Technol. 221(February), 104159 (2024). https://doi.org/10.1016/j.coldregions.2024.104159

Vanem, E., Zhu, T., Babanin, A.:Statistical modelling of the ocean environment – A review of recent developments in theory and applications. Marine Struct. **86** (2022).

Veen, D., Gourlay, T.: A combined strip theory and smoothed particle hydrodynamics approach for estimating slamming loads on a ship in head seas. Ocean Eng. **43**, 64–71 (2012)

Verde, S., Lages, E N.: A comparison of anchor loads, planar displacement, and rotation for nylon and polyester moored systems for a 15 MW floating wind turbine in shallow water. Ocean Eng. **280** (2023)

Vineesh, P., Sriram, V.: Numerical investigation of wave interaction with two closely spaced floating boxes using particle method. Ocean Eng. **268** (2023)

Viuff, T., Ravinthrakumar, S., Økland, O.D., Grytå, O.A., Xiang, X.: Experimental study of floating bridge global response when subjected to waves and current. Appl. Ocean Res. **138** (2023)

Wagner, H.: Uber stoss und Gleitvorgange an der Oberache von Flussigkeiten. Z. Angew. Math. Mech. **12**, 192–215 (1932)

Wan, Z., Yuan, Y., Tang, W.: Experimental investigation on ice resistance of an arctic LNG carrier under multiple ice breaking conditions. Ocean Eng. **267** (2023)

Wang, B., Liu, Y., Zhang, J., Shi, W., Li, X., Li, Y.: Dynamic analysis of offshore wind turbines subjected to the combined wind and ice loads based on the cohesive element method. Front. Marine Sci. **9** (2022a)

Wang, G., Martin, T., Huang, L., Bihs, H.: An improved screen force model based on CFD simulations of the hydrodynamic loads on knotless net panels. Appl. Ocean Res. **118** (2022b)

Wang, G., Martin, T., Huang, L., Bihs, H.: Numerical investigation of the hydrodynamics of a submersible steel-frame offshore fish farm in regular waves using CFD. Ocean Eng. **256** (2022c)

Wang, H., Duan, W., Chen, J., Tian, C.: Numerical predictions of motions and coupled bending and torsional vibrations of containerships in regular head and oblique waves. Eng. Anal. Bound. Elem. **166** (2024a)

Wang, J., Chen, X., Sun, K., Ji, S.: Far-field identification of ice loads on ship structures by radial basis function neural network. Ocean Eng. **282**(April), 115072 (2023a). https://doi.org/10.1016/j.oceaneng.2023.115072

Wang, K., Chu, Y., Huang, S., Liu, Y.: Preliminary design and dynamic analysis of constant tension mooring system on a 15 MW semi-submersible wind turbine for extreme conditions in shallow water. Ocean Eng. **283** (2023b)

Wang, L., Robertson, A., Jonkman, J., Yu, Y. H.: Uncertainty assessment of CFD investigation of the nonlinear difference-frequency wave loads on a semisubmersible FOWT platform. Sustain. (Switz.) **13**(1), 1–25 (2021a). https://www.scopus.com/record/display.uri?eid=2-s2.0-85099183562&origin=resultslist&sort=plf-t&src=s&st1=%22+Uncertainty+assessment+of+CFD+investigation+of+the+nonlinear+difference-frequency+wave+loads+on+a+semisubmersible+FOWT+platform%22&sid=7d230c92a7

Wang, L., et al.: Validation of CFD simulations of the moored DeepCwind offshore wind semisubmersible in irregular waves. Ocean Eng. **260** (2022d)

Wang, Q., Liaot, K. P., Fan, N., Duan, W.Y., Ma, Q.W.: Numerical analysis of flow kinematics in green water on deck. Int. J. Offshore Polar Eng. **32**(1), pp. 58–65 (2022e)

Wang, Q., et al.: Real-time prediction of wave-induced hull girder loads for a large container ship based on the recurrent neural network model and error correction strategy. Int. J. Naval Archit. Ocean Eng. **16** (2024b)

Wang, S.D., Wei, G., Du, H., Wang, X.L., Xu, J N.: Experimental investigation on the three-dimensional oblique interaction of an internal solitary wave with a horizontal finite-length cylinder. J. Fluids Struct. **111** (2022f)

Wang, S., Guedes Soares, C.: Numerical study on the water impact of 3D bodies by explicit finite element method. Ocean Eng. (2014)

Wang, S., Islam, H., Guedes Soares, C.: Uncertainty due to discretization on the ALE algorithm for predicting water slamming loads. Marine Struct. **80**, 103086 (2021b)

Wang, S., Li, C., Zhang, Y., Jing, F., Chen, L.: Influence of pitching motion on the hydrodynamic performance of a horizontal axis tidal turbine considering the surface wave. Renew. Energy **189**, 1020–1032 (2022g)

Wang, S., Soares, C.G.: Article in The International Journal of Maritime Engineering (2012). https://www.researchgate.net/publication/260869000

Wang, S., Zhang, Y., Xie, Y., Xu, G., Liu, K., Zheng, Y.: The effects of surge motion on hydrodynamics characteristics of horizontal-axis tidal current turbine under free surface condition. Renew. Energy **170**, pp. 773–784 (2021c)

Wang, W. hua, Zhao, Z. han, Wang, L. jian, Huang, Y.: Improved analysis method on extreme upwell of irregular waves for the airgap of floating platforms based on OTG-13. Ocean Eng. **274** (2023c)

Wang, X., et al.: A review of aerodynamic and wake characteristics of floating offshore wind turbines. Renew. Sustain. Energy Rev. **175**(October 2022), p. 113144 (2023d). https://doi.org/10.1016/j.rser.2022.113144

Wang, Y., Chen, H.C., Koop, A., Vaz, G.: Hydrodynamic response of a FOWT semi-submersible under regular waves using CFD: verification and validation. Ocean Eng. **258** (2022h)

Wang, Y., Jiang, S., Zhou, T.: Experimental investigation on response of ship rolling motion coupled with liquid sloshing. In: Proceedings of the ASME 2023 42nd International Conference on Ocean, Offshore and Arctic Engineering, Melbourne (2023e)

Wang, Y., Jiang, S., Zhou, T.: Experimental and numerical investigation on ship rolling motion response coupled with liquid sloshing. In: Proceedings of the ASME 2024 43rd International Conference on Ocean, Offshore and Arctic Engineering OM, Singapore (2024c)

Warren, M.F., et al.: Fatigue estimation on a high-speed wave piercing catamaran during normal operations. Trans. Roy. Inst. Naval Archit. Part A: Int. J. Maritime Eng. **164**(1 A), A55–A67 (2022)

Waskito, K. T., Sasa, K., Chen, C., Kitagawa, Y., Lee, S. W.: Comparative study of realistic ship motion simulation for optimal ship routing of a bulk carrier in rough seas. Ocean Eng. **260** (2022)

Watanabe, S., Hu, C., Aoki, T.: Coupled lattice boltzmann and discrete element simulations of ship-ice interactions. IOP Conf. Ser.: Mater. Sci. Eng. **1288**(1), 012015 (2023)

Wei, K., Yin, X.: Numerical study into configuration of horizontal flanges on hydrodynamic performance of moored box-type floating breakwater. Ocean Eng. **266** (2022)

Wei, S., Liang, Z., Cui, L., Zhai, H., Jeng, D.: Numerical study of seabed response and liquefaction around a jacket support offshore wind turbine foundation under combined wave and current loading. Water Sci. Eng. **15**(1), 78–88 (2022)

Wei, W., Ling, A., Liang, Y.: Applicable analysis of the environment condition for the air gap of the semi-submersible platform (2023). www.isope.org

Wen, B., Jiang, Z., Li, Z., Peng, Z., Dong, X., Tian, X.: On the aerodynamic loading effect of a model Spar-type floating wind turbine: an experimental study. Renew. Energy **184**, 306–319 (2022)

Wilson, S., Hall, M., Housner, S., Sirnivas, S.: Linearized modeling and optimization of shared mooring systems. Ocean Eng. **241** (2021)

Windt, C., Faedo, N., Penalba, M., Dias, F., Ringwood, J.V.: Reactive control of wave energy devices – the modelling paradox. Appl. Ocean Res. **109** (2021)

Won, D., Seo, J., Kim, S. Dynamic response of submerged floating tunnels with dual sections under irregular waves. Ocean Eng. **241** (2021)

Wu, J., et al.: Analysis of full-scale riser responses in field conditions based on Gaussian mixture model. J. Fluids Struct. **116**, 103793 (2023a). https://doi.org/10.1016/j.jfluidstructs.2022.103793

Wu, J., Yu, Y., Cheng, S., Li, Z., Yu, J.: Probabilistic multilevel robustness assessment framework for a TLP under mooring failure considering uncertainties. Reliab. Eng. Syst. Saf. **223** (2022)

Wu, S., Huang, Y., Ding, Y., Wang, G., Li, J., Guo, Y.: Study on the influence of swimming pool shape of large cruise ship on its sloshing performance, in: Proceedings of the Thirty-Third International Ocean and Polar Engineering Conference, Ottawa, pp. 2609–2615 (2023b)

Wu, Z., Wang, D., Ke, W., Qin, Y., Lu, F., Jiang, M.: Experimental investigation for the dynamic behavior of submerged floating tunnel subjected to the combined action of earthquake, wave and current. Ocean Eng. **239** (2021)

Xiang, S., Cheng, B., Li, D., Tang, M., Zeng, Z.: Structural dynamic performance of floating continuous beam bridge under wave and current loadings: an experimental study. Appl. Ocean Res. **137** (2023)

Xie, C., Zhou, L., Ding, S., Liu, R., Zheng, S.: Experimental and numerical investigation on self-propulsion performance of polar merchant ship in brash ice channel. Ocean Eng. **269**(January), 113424 (2023a). https://doi.org/10.1016/j.oceaneng.2022.113424

Xie, H., Dai, X., Liu, F., Liu, X.: Experimental study on the slamming pressure distribution of a 3D stern model entering water with pitch angles. Ocean Eng. **291** (2024)

Xie, W., Liang, Z., Jiang, Z., Tu, S., Chen, W., Zhang, H.: Hydrodynamic behaviors of a spring-mounted fishing net in wave-current combined flows. Ocean Eng. **287** (2023b)

Xin, Z., Li, X., Li, Y.: Coupled effects of wave and depth-dependent current interaction on loads on a bottom-fixed vertical slender cylinder. Coastal Eng. **183** (2023)

Xiong, Z., Wu, X., Li, Y.: Research on ice resistance of ships sailing in floating ice floes. J. Marine Sci. Technol. (Taiwan) **31**(4), 460–470 (2023)

Xu, G., Chen, X., Xue, S., Townsend, J. F., Chen, X., Tang, M.: Numerical assessment of non-uniform terrain and inhomogeneous wave–current loading effects on the dynamic response of a submerged floating tunnel. Ocean Eng. **288** (2023)

Xu, G., He, B., Li, H., Gui, X., Li, Z.: FBG pressure sensor in pressure distribution monitoring of ship. Opt. Express **30**(12), 21396–21409 (2022)

Xu, K., Larsen, K., Shao, Y., Zhang, M., Gao, Z., Moan, T.: Design and comparative analysis of alternative mooring systems for floating wind turbines in shallow water with emphasis on ultimate limit state design. Ocean Eng. **219** (2021a)

Xu, S., Guedes Soares, C.:. Parametric study on the short-term extreme mooring tension of nylon rope for a point absorber. Ocean Eng. **267** (2023)

Xu, S., Wang, S., Guedes Soares, C.: Experimental study of the influence of the rope material on mooring fatigue damage and point absorber response. Ocean Eng. **232** (2021b)

Xu, S., Wang, S., Guedes Soares, C.: Experimental investigation on the influence of hybrid mooring system configuration and mooring material on the hydrodynamic performance of a point absorber. Ocean Eng. **233** (2021c)

Xuan, S., Zhan, C., Liu, Z., Feng, B., Chang, H., Wei, X.: Numerical investigation of global ice loads of maneuvering captive motion in ice floe fields. J. Marine Sci. Eng. **11**(9) (2023)

Xuan, S., Zhan, C., Liu, Z., Zhao, Q., Guo, W.: Numerical research on global ice loads of maneuvering captive motion in level ice. J. Marine Sci. Eng. **9**(12) (2021)

Xue, Y., et al.: A combined experimental and numerical approach to predict ship resistance and power demand in broken ice. Ocean Eng. **292**(August 2023) (2024)

Yamamoto, Y., Iida, K., Fukasawa, T., Murakami, T., Arai, M., Ando, A.: Structural damage analysis of a fast ship due to bow flare slamming. Int. Shipbuild. Prog. **32**, 124–136 (1985)

Yan, D., Garbatov, Y., Guedes Soares, C.: Review on uncertainties in fatigue loads and fatigue life of ships and offshore structures. Ocean Eng. **264** (2022a)

Yan, D., Mikkola, T., Kujala, P., Hirdaris, S.: Hydroelastic analysis of slamming induced impact on stiff and flexible structures by two-way CFD-FEA coupling. Ships Offshore Struct. 1–13 (2022b)

Yan, D., et al.: A study into the FSI modelling of flat plate water entry and related uncertainties. Marine Struct. **86** (2022c). Accessed 29 Aug 2023,

Yan, X., Chen, C., Yin, G., Ong, M.C., Ma, Y., Fan, T.: Numerical investigations on nonlinear effects of catenary mooring systems for a 10-MW FOWT in shallow water. Ocean Eng. **276** (2023)

Yang, B., Zhang, G., Rao, H., Wang, S., Yang, B., Sun, Z.: Numerical simulation of the maneuvering performance of ships in broken ice area. Ocean Eng. **294**(December 2023), 116783 (2024). https://doi.org/10.1016/j.oceaneng.2024.116783

Yang, Y., Li, Y., Wang, J., He, Z., Dong, X., Yuan, X.: Time-varying dynamic performance of a floating platform-mooring system based on the mechanical experimental data derivation method for corroded wire ropes. Ocean Eng. **280** (2023a)

Yang, Y., Peng, T., Liao, S.: Predicting future mooring line tension of floating structure by machine learning. Ocean Eng. **269**, 113470 (2023b). Accessed 31 Mar 2024

Yang, Y., Peng, T., Liao, S.: Predicting 3-DoF motions of a moored barge by machine learning. J. Ocean Eng. Sci. **8**(4), 336–343 (2023c). Accessed 31 Mar 2024

Yang, Y., Stansby, P.K., Rogers, B.D., Buldakov, E., Stagonas, D., Draycott, S.: The loading on a vertical cylinder in steep and breaking waves on sheared currents using smoothed particle hydrodynamics. Phys. Fluids **35**(8) (2023d)

Yang, Z., et al.: Experimental study on the solitary wave-current interaction and the combined forces on a vertical cylinder. Ocean Eng. **236** (2021)

Yang, Z., et al.: Experimental study on the wave-induced dynamic response and hydrodynamic characteristics of a submerged floating tunnel with elastically truncated boundary condition. Marine Struct. **88** (2023e)

Yao, C., Dong, G., Sun, X.S., Zheng, Y., Feng, D.: Numerical study on motion and added resistance of a trimaran advancing in waves based on hybrid Green function method. Appl. Ocean Res. **130** (2023a)

Yao, Z., Huo, F., Zhu, Y., Tang, C., Jia, K., Li, D., Ma, Y.: Experimental and numerical investigation on slamming mechanism of a mooring column-stabilised semi-submersible. Processes **11**(3) (2023b)

Yeter, B., Garbatov, Y.: Structural integrity assessment of fixed support structures for offshore wind turbines: a review. Ocean Eng. **244**(October 2021), 110271 (2022). https://doi.org/10.1016/j.oceaneng.2021.110271

Yin, D., et al.: State-of-the-art review of vortex-induced motions of floating offshore wind turbine structures. J. Marine Sci. Eng. **10**(8) (2022)

Yoo, J.H., Schrijvers, P., Koop, A., Park, J.C.: CFD prediction of wind loads on fpso and shuttle tankers during side-by-side offloading. J. Mar. Sci. Eng. **10**(5) (2022). https://www.scopus.com/inward/record.uri?eid=2-s2.0-85132836359&doi=10.3390%2Fjmse10050654&partnerID=40&md5=b152aee4d606ad575407fdc6d5b62a99

Yu, J., Cheng, X., Fan, Y., Ni, X., Chen, Y., Ye, Y.: Mooring design of offshore aquaculture platform and its dynamic performance. Ocean Eng. **275** (2023a)

Yu, T., Liu, K., Wang, G., Liu, J., Wang, Z.: A tri-axial ice model for simulating ice-stiffened panel impact: experiments and numerical modeling. Marine Struct. **88**(November 2021), 103358 (2023b). https://doi.org/10.1016/j.marstruc.2022.103358

Yu, Z., Ma, Q., Zheng, X., Liao, K., Sun, H., Khayyer, A.: A hybrid numerical model for simulating aero-elastic-hydro-mooring-wake dynamic responses of floating offshore wind turbine. Ocean Eng. **268** (2023c)

Zago, V., Dalrymple, R. A., Almashan, N., Bilotta, G., Al-Houti, D., Neelamani, S.: Characterization and modeling of greenwater overtopping of a sea-level deck. Ocean Eng. **275** (2023)

Zeng, X., Shi, W., Feng, X., Shao, Y., Li, X.: Investigation of higher-harmonic wave loads and low-frequency resonance response of floating offshore wind turbine under extreme wave groups. Marine Struct. **89** (2023)

Zhang, A., Chuang, Z., Liu, S., Zhou, L., Qu, Y., Lu, Y.: Dynamic performance optimization of an arctic semi-submersible production system. Ocean Eng. **244**(July 2021) (2022a)

Zhang, C., Wang, S., Xie, S., He, J., Gao, J., Tian, C.: Effects of mooring line failure on the dynamic responses of a semisubmersible floating offshore wind turbine including gearbox dynamics analysis. Ocean Eng. **245** (2022b)

Zhang, D., Chu, X., Wu, W., He, Z., Wang, Z., Liu, C.: Model identification of ship turning maneuver and extreme short-term trajectory prediction under the influence of sea currents. Ocean Eng. **278**, 114367 (2023a). Accessed 18 Oct 2024

Zhang, H., Liao, X., Shi, H., Babanin, A., Soares, C.G.: Effect of initial condition uncertainty on the profile of maximum wave. Marine Struct. **82** (2022)

Zhang, H., Zeng, J., Tong, S., Jin, B., Chou, C., Li, H., Dong, H.: Dynamic stiffness of polyester fiber mooring ropes: experimental investigation based on radial basis function neural networks. Ocean Eng. **280** (2023b)

Zhang, J., Gaidai, O., Ji, H., Xing, Y.: Operational reliability study of ice loads acting on oil tanker bow. Heliyon **9**(4), e15189 (2023c). https://doi.org/10.1016/j.heliyon.2023.e15189

Zhang, J., Zhang, Y., Shang, Y., Jin, Q., Zhang, L.: CFD-DEM based full-scale ship-ice interaction research under FSICR ice condition in restricted brash ice channel. Cold Regions Sci. Technol. **194**(September 2021), 103454 (2022d). https://doi.org/10.1016/j.coldregions.2021.103454

Zhang, L., Shi, W., Zeng, Y., Michailides, C., Zheng, S., Li, Y.: Experimental investigation on the hydrodynamic effects of heave plates used in floating offshore wind turbines. Ocean Eng. **267** (2023d)

Zhang, N., Xiao, L., Guo, Y., Yang, L., Chen, G.: Parametric study of wave impact pressure impulse and characteristic pressure on a square column with overhanging deck. Ocean Eng. **258** (2022e)

Zhang, R.-R., et al.: Experimental investigation and prediction model of the loads exerted by oblique internal solitary waves on FPSO (2022). https://doi.org/10.1007/s13344-022-

Zhang, X., Li, B., Hu, Z., Deng, J., Xiao, P., Chen, M.: Research on size optimization of wave energy converters based on a floating wind-wave combined power generation platform. Energies **15**(22) (2022f)

Zhang, X., Ni, W., Sun, L.: Fatigue analysis of the oil offloading lines in FPSO system under wave and current loads. J. Marine Sci. Eng. **10**(2) (2022g)

Zhang, X., Wu, W., Fu, H., Li, J.: The effect of corrosion evolution on the stress corrosion cracking behavior of mooring chain steel. Corros. Sci. **203** (2022h)

Zhang, Y., Liu, H.: Coupled dynamic analysis on floating wind farms with shared mooring under complex conditions. Ocean Eng. **267** (2023)

Zhang, Y., et al.: Quantifying the surge-induced response of a floating tidal stream turbine under wave-current flows. Energy **283** (2023e)

Zhang, Y., Xu, H., Law, Y., Santo, H., Magee, A.: Hydrodynamic analysis and validation of the floating DeepCwind semi-submersible under 3-h irregular wave with the HOS and CFD coupling method. Ocean Eng. **287** (2023f)

Zhang, Z., Ma, N., Lin, Y., Gu, X., Shi, Q.: Experimental study on non-linear wave loads of a ship model with variable cross-section backbone in irregular waves. In: 32nd International Ocean and Polar Engineering Conference, pp. 2315–2322. International Society of Offshore and Polar Engineers (2022i)

Zhao, C., Stansby, P., Johanning, L.: OrcaFlex predictions for a multi-float hinged WEC with nonlinear mooring systems: Elastic mooring force and dynamic motion. Ocean Eng. **286** (2023)

Zhao, R.-X., Faltinsen, O.M.: Water entry of two-dimensional bodies. J.. Fluid Mech **246**, 593–612 (1993). https://api.semanticscholar.org/CorpusID:121224301

Zhao, W., Wei, Z., Wan, D.: Numerical investigation on the effects of current headings on vortex induced motions of a semi-submersible. J. Hydrodyn. **34**(3), 382–394 (2022)

Zheng, Z., et al.: Motion response and energy harvesting of multi-module floating photovoltaics in seas. Ocean Eng. **310** (2024)

Zhong, K., et al.: Direct measurements and CFD simulations on ice-induced hull pressure of a ship in floe ice fields. Ocean Eng. **272** (2023)

Zhou, J., Liu, J., Guo, A.: Investigation of the dynamic behavior of the flow–structure system of submerged floating tunnels under wave and current actions. Coast. Eng. **183** (2023a)

Zhou, L., Gu, Y., Ding, S., Liu, R.: Prediction of ice-resistance distribution for R/V xuelong using measured sea-ice parameters. Water (Switzerland) **14**(4) (2022)

Zhou, Y., Ning, D., Chen, L., Mayon, R., Zhang, C.: Experimental investigation on an OWC wave energy converter integrated into a floating offshore wind turbine. Energy Convers. Manage. **276** (2023b)

Zhu, D., Dong, Y., Frangopol, D.M.: Experimental and numerical investigation on wave impacts on box-girder bridges. Struct. Infrastruct. Eng. **18**(10–11), 1379–1397 (2022)

Zhu, H., Ji, S.: Discrete element simulations of ice load and mooring force on moored structure in level ice. CMES – Comput. Model. Eng. Sci. **132**(1), 5–21 (2022)

Zilic de Arcos, F., Vogel, C.R., Willden, R.H.J.: A numerical study on the hydrodynamics of a floating tidal rotor under the combined effects of currents and waves. Ocean Eng. **286** (2023)

Zong, S., Liu, K., Zhang, Y., Yan, X., Wang, Y.: The dynamic response of a floating wind turbine under collision load considering the coupling of wind-wave-mooring loads. J. Marine Sci. Eng. **11**(9), 1741 (2023). https://www.mdpi.com/2077-1312/11/9/1741

Zou, J., Li, H., Sun, Z., Han, B., Wang, Z.: Experimental analysis of bow flare slamming and whipping responses in a ship at different sailing speeds. Ocean Eng. **305** (2024)

Zou, Q., Lu, Z., Shen, Y.: Short-term prediction of hydrodynamic response of a novel semi-submersible FOWT platform under wind, current and wave loads. Ocean Eng. **278** (2023)

ISO/IEC Guide 98-3:2008 - Uncertainty of measurement — Part 3: Guide to the expression of uncertainty in measurement (GUM:1995). https://www.iso.org/standard/50461.html. Accessed 20 Mar 2024

Open Access This chapter is licensed under the terms of the Creative Commons Attribution-NonCommercial-NoDerivatives 4.0 International License (http://creativecommons.org/licenses/by-nc-nd/4.0/), which permits any noncommercial use, sharing, distribution and reproduction in any medium or format, as long as you give appropriate credit to the original author(s) and the source, provide a link to the Creative Commons license and indicate if you modified the licensed material. You do not have permission under this license to share adapted material derived from this chapter or parts of it.

The images or other third party material in this chapter are included in the chapter's Creative Commons license, unless indicated otherwise in a credit line to the material. If material is not included in the chapter's Creative Commons license and your intended use is not permitted by statutory regulation or exceeds the permitted use, you will need to obtain permission directly from the copyright holder.

Committee II.1: Quasi-static Response

Zhen Gao[1](✉), Yooil Kim[2], Erick Alley[3], Krzysztof Wołoszyk[4], Zhiyu Jiang[5], Aaron Stanley[6], Hanbing Luo[7], Jerolim Andric[8], Marcelo Caire[9], Alessandro Sacchet[10], Eko Charnius Ilman[11], and Daisuke Yanagihara[12]

[1] Norwegian University of Science and Technology (NTNU), Trondheim, Norway
zhengaosjtu@sjtu.edu.cn
[2] Inha University, Incheon, Republic of Korea
[3] Naval Postgraduate School, Monterey, USA
[4] Gdansk University of Technology, Gdansk, Poland
[5] University of Agder, Grimstad, Norway
[6] London, UK
[7] Tianjin University, Tianjin, China
[8] University of Zagreb, Zagreb, Croatia
[9] Federal University of Rio de Janeiro, Rio de Janeiro, Brazil
[10] Genoa, Italy
[11] Institut Teknologi Bandung, Bandung, Indonesia
[12] Kyushu University, Fukuoka, Japan

Committee Mandate. Concern for quasi-static response of ships and offshore structures, as required for safety and serviceability assessments. Attention shall be given to uncertainty quantification of the quasi-static load and response analysis approaches, and their limitations, including the exact and approximate methods for derivation of different acceptance criteria.

Keywords: Corrosion · direct calculations · experiments and testing · fatigue assessment · finite element analysis · motion analysis · optimization · offshore structures · probabilistic approach · progressive collapse · quasi-static · reliability analysis · residual strength · ship structures · strength assessment · stress analysis · structural integrity · structural response · surrogate model · uncertainty analysis

1 Introduction

ISSC II.1 Quasi-Static Response is a traditional technical committee under ISSC that deals with the ships and offshore structures' responses of quasi-static nature under various environmental conditions and loads.

The mandate that was assigned to the committee remained unchanged from the previous term, which was "Concern for quasi-static response of ships and offshore structures, as required for safety and serviceability assessments. Attention shall be given to uncertainty quantification of the quasi-static load and response analysis approaches, and their limitations, including the exact and approximate methods for derivation of different acceptance criteria."

The definition of quasi-static response analysis method was discussed and concluded in the previous ISSC report (ISSC, 2022), which is cited here. "In structural response analysis, a method may be considered quasi-static where the effects of structural dynamics (structural inertia and damping) may be neglected. In this regards the time component, or time derivatives, may be neglected. To adopt a quasi-static method, the true time dependant loading must be sufficiently slow in relation to the structural response not to coincide with resonant response frequencies. Due to this 'slow' progression, during analysis the system may be considered in a static equilibrium at all time instances. These points are true for many quasi-static analyses, where loading may be through incremental application of force or displacement to a structure, and static equilibrium of the system is achieved before the next increment is applied. Therefore, time associated with the loading is only implied and not explicitly included in the assessment. In other words, the structural responses at any time instant will be only determined by the loads at that time instant, and the structural responses have no memory effect."

In general, quasi-static response analysis methods have been well established with respect to the excitation load frequencies and the resonant frequencies of the considered ship or marine structure's dynamics. Under the consideration that the excitation frequencies of the loads are smaller than the lowest natural frequency of the structure, the dynamic amplification of the responses for given loads is small and close to one. In such cases, the excitation loads are mainly balanced with the restoring or stiffness of the system, inducing very limited dynamic effects. Such methods are very common traditionally for structural design of ships, bottom-fixed structures and even floating offshore structures, and can be easily applied. On the other hand, when the loading frequencies are close to the resonant frequencies of the system, due to the cancellation of the inertial and restoring/stiffness effects, the dynamic effect in connection with damping will dominate the responses of the system. Moreover, if the loading frequencies are much higher than the resonant frequencies, the inertial effect will dominate, leading to a balance between the excitation loads and the inertial loads. Obviously under these two conditions, a quasi-static response analysis method is not applicable, and mostly underestimates the responses.

In principle, this ISSC committee should address only the structural response analysis methods under the quasi-static assumptions. However, the nature of quasi-static responses is closely related to the loading patterns and frequencies, a chapter on modelling of quasi-static loads or slowly-varying loads that lead to quasi-static response features were also considered. From the structural design point of view, it is inevitable to obtain the structural responses of deformations, stresses, and in some cases damages, so that a design check could be conducted by comparing the characteristic response or load effect with the characteristic strength or resistance. For floating structures, it is typical to separate the motion analysis (which in most cases relies on a dynamic analysis) and the structural stress analysis (for which in many cases a quasi-static analysis suffices unless there is a need to consider local vibrations of the structure components). In case of significant aero-elasticity or hydro-elasticity, coupled analysis of loads and motion/structural deformation responses become necessary. Moreover, it is also typical that a design check for engineering applications is often made with separate assessments of structural responses and strengths, unless an analysis considering the interaction of environmental loads and structural failure mechanisms needs to be made in refined assessments. Therefore, in this report, we shall discuss the quasi-static approaches following these traditions.

However, in general, in the last decades, there were very limited research on the development of new quasi-static response analysis methodologies, and the focus was the application of such methods to novel marine structures. In particular, in recent years, with the increased computation capacity and software maturity, most of the publications, due to the novelty requirements, dealt with either linear or nonlinear hydrodynamic and/or structural problems considering dynamic analysis methods, based on either a frequency-domain or a time-domain formulation. As a result, the issue of using a quasi-static response approach was often considered as a part of the research work in a scientific publication and as a comparative approach with respect to more advanced dynamic analysis methods. Nevertheless, this committee has reviewed the recent literature in which quasi-static analysis approach was either developed, applied or compared in various situations. However, due to the above-mentioned observations, it is the current committee's suggestion that this committee could be well combined with the committee II.2 Dynamic Response.

In recent years, quasi-static response analysis methods that were developed and applied to ships and offshore structures, have been reviewed extensively in the ISSC committee II.1 Quasi-Static Response. In the last two terms (ISSC 2018, ISSC 2022), a similar ISSC report structure was used, which consisted of the chapters on load modelling, structural modelling and response analysis, strength assessment with quasi-static approaches, response uncertainty assessment and structural reliability applications, as well as ship and offshore structure rules and software systems. In this report, we will also review the quasi-static approaches for loading, response analysis, strength assessment, reliability analysis as well as uncertainty assessment of the quasi-static approaches in different chapters. A similar report structure was used for this term except for the rules and standards chapter which has been removed. The development of quasi-static response analysis methods is relatively mature in such design rules and standards and therefore there were limited publications related to this topic published in the term. An overview of this report is presented below.

Section 2 discusses the load modelling for ships or marine structures, with focus on the operational/design loads, abnormal/accidental loads, the recently developed surrogate models, as well as the load estimation models based on test or field measurements.

Section 3 is the core of the report, dealing with quasi-static structural response analysis methods, from the most simplified rule-based assessments to the analyses based on the first principles and direct calculations, and to the optimization analyses. The issues of modelling structural responses under the degrading environments were also discussed.

Quasi-static approaches for structural strength assessment are discussed in Sect. 4, including buckling and ultimate strength of ships and offshore structures, residual and fatigue strength assessments.

Section 5 deals with the uncertainty assessment for quasi-static approaches or assumptions that are considered in the assessment of environmental conditions, loads, responses and structural strengths.

Main conclusions from this term's work and recommendations for future research as well as for future committee are made in Sect. 6.

During this term, outside the ISSC community, Artificial Intelligence technology is being readily used for literature reviews and associated reporting. However, due to the specific technical nature of this committee, the use of these tools is not considered effective. Therefore, this report is written by the committee members based on a manual review and summary of the selected literatures.

2 Load Modelling

Quasi-static load modelling forms a crucial framework for designing marine and offshore structures. Assessing the forces acting over a structure's lifespan is essential to ensure its integrity and safety. This is also the first-principle considerations that are often used for design of marine structures. This section discusses the evolution of methods used to define and simulate quasi-static loads. Environmental forces, such as those from wind, waves, and currents, are inherently variable and dynamic, and consist of multiple frequency components. Although these forces change over time, they can be approximated as "static" values when their dominating frequency is significantly lower than the system's fundamental natural frequency, thus avoiding any dynamic influence on the system's behaviour. In other words, the maximum response occurs at the same time that the maximum load occurs, which simplifies the procedure for design analysis.

Quasi-static loads that are covered in this section are categorized into two kinds, and they are normal operational loads and accidental loads. Normal operational loads include wave loads, wind loads, ice loads. Abnormal accidental loads cover sloshing and slamming loads together with collision and grounding. Data driven models as well as experimental approaches and full-scale measurement are also handled.

2.1 Operational/Design Loads

This section summarizes the various types of operational/design loads and recent research efforts conducted since the committee's last ISSC report (ISSC 2022). Operational/design loads can be defined as the loads that a structure is typically expected to encounter in routine operations during its lifetime, considering the statistical methodologies for factoring the occurrence of each load component and providing safety mechanisms via probabilities of non-exceedance and appropriate reserve factors.

2.1.1 Wave Loads and Extreme Wave Loads

Wave loads are a critical consideration factor in design and operation of ships and offshore structures. Extreme wave loads refer to the substantial forces exerted on offshore structures due to the action of unusually high ocean waves, often caused by extreme weather events such as storms or hurricanes. These loads directly affect the hydrostatic and hydrodynamic response performance of marine structures. Precise estimation of the wave loads onto marine structures is vital in understanding and assessing the structure's performance. This subsection summarizes recent research regarding wave loads and extreme wave loads. In most cases, wave loads are considered as continuous dynamic loads that interacts with marine structures' motion and structural responses, unless the loads have the impulsive nature so that a time-integration of the loads can be obtained first and followed by a dynamic response analysis for local or global hydroelastic problems.

Waves Loads

Matsui et al. (2023) proposed a formula for predicting long-term vertical bending moment (VBM) responses of ships based on response amplitude operation (RAO) characteristics. The research outlined the relationship between long-term response and the peak value of the response amplitude operator (RAO). It introduced a method based on peak RAO values, simplifying the VBM prediction by using hull form parameters such as length, breadth, and draft, similar as the rule-based design format or a surrogate model. The formula was validated using numerical data from 154 ship models, proving its accuracy and applicability to various ship types.

Kim et al. (2023) experimentally studied wave-induced motions and loads on a 1/65 scaled rigid 9-segmented containership model in oblique regular waves. The research analysed nonlinear impacts on vertical and horizontal bending moments near amidships and 6-degree-of-freedom motions at the centre of gravity. The study also assessed the nonlinear effects due to wave steepness, finding that steeper waves contribute significantly to higher-order harmonic components, including slamming events. A mooring system was used to maintain the model's heading angle, and experiments with and without the mooring system revealed its influence on horizontal bending moments and yaw motions.

Extreme Wave Loads

Kim et al. (2024) presented a detailed investigation into the vertical bending moments (VBM) experienced by a containership in extreme wave conditions. The research was conducted through model-scale experiments using a rigid 6750-TEU container ship. Five different sea states were studied, with significant wave heights (Hs) ranging from 6 m to 17 m. The objective was to understand the non-linear effects associated with wave loads and how they deviate from linear predictions. A key finding is that the non-linear factor, representing the deviation between linear and nonlinear VBM, can be predicted based on linear values alone. This finding simplifies the design process, offering potential improvements to current methodologies, such as "design sea state" approaches.

Jiang et al. (2023) investigated the behaviour of a modular floating structure (MFS) under wave-induced forces. The study employs a coupled mooring–joint–viscous flow solver to model the complex dynamics involving mooring lines, joint restrictions, and nonlinear wave interactions. Two types of joints—rigid and flexible—are analysed to assess their influence on the system's motion responses and load distributions. Key findings suggest that while surge motions between the two bodies remain nearly identical for both joint types, significant differences arise in pitch motions, particularly with the flexible joint, where dynamic pitch behaviour is more pronounced. Heave motions were minimally affected by the joint type. Additionally, forces acting on the mooring lines and joints were evaluated, showing that wave steepness plays a critical role in amplifying higher-order wave components, which in turn affects the load distribution on the structure. The research highlights the importance of nonlinear wave effects, particularly in steep waves, for accurately predicting the hydrodynamic behaviour of moored, multibody offshore systems.

Takami et al. (2023) proposed a method to predict extreme vertical bending moments of ships in nonlinear wave fields using a combination of Higher Order Spectrum Method (HOSM) and First-order Reliability Method (FORM). Nonlinear wave effects on MPWE profiles and reliability indices for extreme VBM were examined, emphasizing sensitivity of VBM transfer functions to high-frequency waves impacting reliability index distributions. The research focuses on the impact of nonlinear wave components on extreme wave-induced VBM distributions. By comparing the results with Monte Carlo simulations and linear wave solutions, the study demonstrates that the combination of HOSM and FORM provides efficient and accurate extreme value predictions.

H. Zhang et al. (2022a) studied ship responses to abnormal waves generated by different methods, comparing vertical reactions of an LNG carrier to waves targeting various hull sections at different speeds. Parameters related to wave spatial profiles are suggested as indicators of wave generation mechanisms, hydrodynamic characteristics, and potential extreme ship reactions. The study compares three types of abnormal waves—generated via linear and nonlinear focusing methods—with similar temporal characteristics to the New Year Wave, a known rogue wave. The research reveals that while the temporal profiles of these waves may be alike, their spatial profiles differ significantly, leading to variations in the hydrodynamic forces acting on the ship. The location of the wave and the ship's forward speed notably influence the vertical response, with nonlinear waves producing smaller vertical motions compared to their linear counterparts. The findings underscore the importance of understanding wave generation mechanisms when designing for extreme sea conditions.

2.1.2 Wind Loads

Assessment of wind loads on marine structures is crucial for their safe and efficient operation, especially as offshore engineering structures become more complex and larger in size. Wind loads can significantly affect the motions of ships and offshore platforms, posing stability and safety risks and impeding manoeuvrability. Because of the complex nature of wind, it is necessary to use experimental methods, e.g., wind tunnel tests, or computational methods, e.g., computational fluid dynamics (CFD), to accurately characterize wind loads and eventually to propose regression formulas. Such methods typically assume a quasi-static wind loading condition when the wind speed varies and a distributed or an overall drag force is applied. In some extreme cases, one may just consider the maximum wind speed for a reference period, followed by a static wind load and quasi-static response analysis for structural design. Recent research has focused on improving the assessment of wind loads, with studies highlighting the need to consider atmospheric turbulence and the impact of wind on offshore wind turbines.

Kozmar et al. (2022) reviews current approaches to wind load assessments in the design of offshore wind turbines. It notes that international standards often rely on simplified, rule-based methods, which may lead to significant underestimations of wind loads on larger turbines. These methods are derived for small height and do not account for atmospheric turbulence or other complex wind characteristics that affect structural behaviour. The authors compare standard approaches with more realistic wind data simulations to quantify discrepancies. They found that the current standards could significantly underestimate the key parameters, such as axial displacements and pitching moments and these discrepancies could lead to errors in predicting the structural integrity and fatigue life of offshore wind turbines, emphasizing the need for revised standards that incorporate more accurate wind load assessments.

Dao et al. (2023) explores wind loadings on various offshore platforms, including an LNG carrier (LNGC), a Floating Production Storage and Offloading (FPSO) vessel, and a Jack-Up Platform (JUP). The study combines wind tunnel experiments and CFD simulations to assess wind load coefficients and pressure distributions on the topsides of these structures. The authors conducted sensitivity analyses on mesh types, turbulence models, and wind profiles to ensure accuracy. The results show good agreement between experimental data and CFD predictions, offering insights into wind loading variations across different wind directions and platform geometries. The research highlights how complex topside structures influence wind loads and internal flow patterns, particularly on the FPSO with its porous blocks.

For complex superstructures, ITTC (2021) provides a guideline on CFD-based determination of wind resistance coefficients, without any specific inputs on uncertainty analysis or computational conditions. To address the gaps, Kobayashi et al. (2022) proposed a new approach for grid sensitivity of hulls and superstructures and validated the effect of computational conditions for several types of ships.

Thies and Fakiolas (2022) discussed the increasing popularity of wind-assisted propulsion systems for cargo ships and carriers, such as Flettner rotors and wind sails to save energy and reduce greenhouse gas emissions. Kume et al. (2022) evaluated the aerodynamic characteristics of Flettner rotors with the superstructures coexisting by wind tunnel tests and RANS-based numerical simulations. Kramer and Steen (2022) investigated the added resistance due to side forces from windsails for a cargo ship and developed a physical model combining CFD, manoeuvring theory, discrete lifting line, and empirical models. The resistance, primarily caused by the rudder, was found to be comparable to the resistance added by waves. The research examined the effectiveness of several appendages, such as bilge keels and dynamic keels, in reducing sail-induced resistance. A key finding is that dynamic keels significantly improved fuel savings, while fixed appendages increased friction. Additionally, controlling the sails to limit side forces reduced resistance and maintained high fuel savings.

For floating offshore wind turbines in a farm, wind loads on a wind turbine are related to the wake behaviour. Arabgolarcheh et al. (2023) studied the wake interaction of a tandem wind farm configuration using a Navier-Stokes actuator line by prescribing the motion of the upstream turbine and by fixing the downstream turbine. It was found that the peak-to-peak thrust and power variations depend on modelling the discrete nature of the blades. Carmo et al. (2024) presented an investigation on the mutual interaction between the motions of floating wind turbines and wakes using FAST.Farm, which is a mid-fidelity engineering tool based on the dynamic wake meandering model (Jonkman and Shaler, 2021). A small three-unit array of floating wind turbines was analysed across a wide range of environmental conditions and the impact of compliance of the floating substructure was assessed. Wind loads analysis for floating wind farms is still a growing research area.

2.1.3 Ice Loads

Ice load refers to the physical forces exerted on a ship or marine structure by ice during collisions or pressure exertion. It significantly impacts the structural integrity, safety, and navigation capabilities of vessels, as well as the stability and functionality of marine structures. Accurate estimating and management of ice loads are crucial for designing vessels and structures that can withstand such pressures, preventing damage and ensuring safety. The magnitude of ice loads is determined based on the ice breaking criteria, which typically involve a quasi-static and simplified assumption. That means the dynamics of ice are neglected. However, the ice loads may induce significant dynamics of the structures. Due to the harsh and inaccessible nature of polar regions, field data on ice load is limited, and most existing studies rely on theoretical modelling or simulations. To improve ice load prediction and management, extensive experimental research and data collection are necessary. This will establish safer and more efficient design standards for polar marine activities, balancing economic pursuits with environmental protection.

Erceg et al. (2022) developed a simulation approach for ship-ice interaction in level ice, addressing limitations in existing numerical methods. The study presents a contact algorithm for arbitrary bow shapes, enabling local ice load calculations across the bow. Truong and Jang (2022) investigated the influence coefficient method (ICM) for estimating ice loads on offshore structures. They developed a numerical model for level ice-structure interactions using the Drucker-Prager plasticity model and element erosion technique.

Jeon and Kim (2022) developed a methodology for calculating fatigue damage in arctic vessels colliding with level ice. They used FEM simulations with a modified Drucker-Prager model and element erosion to accurately predict local ice loads on the hull. They calculated cumulative fatigue damage at various speeds, providing insights into ice-structure interactions and vessel durability in arctic conditions. Lemström et al. (2022a) conducted model-scale experiments to study level ice interaction with a wide inclined structure in shallow water. Analysis of local peak load events revealed that the narrow local peak load events caused the highest local line loads, with magnitudes increasing with ice strength. Xu et al. (2022) used the extended finite element method to study level ice fracture mechanisms during ship-ice collisions. Analysis revealed that initial cracks in both modes relate to local tensile failure, with maximum tensile hydrostatic stress coinciding with crack initiation points.

Gu et al. (2022) investigated the manoeuvrability of polar ships operating in ice-covered areas through numerical simulations. They utilized a method that incorporates circumferential crack fracture of level ice to calculate ice loads and predict ship movements during turns. Ji and Yang (2022) examined the effects of drifting sea ice on offshore wind turbines (OWTs) using a coupled discrete element method (DEM) and finite element method (FEM). Simulations of a 3.3-MW OWT under various ice conditions revealed insights into ice-induced vibrations and structural responses. Lemström et al. (2022b) conducted laboratory-scale experiments to study ice-structure interactions in shallow water by pushing a ten-meter-wide ice sheet against an inclined structure. The results provided insights into the dominant role of incoming ice weight in the initial phase and buckling in the steady-state phase.

Liu and Ji (2022) compared the sphere-based discrete element method (SDEM) and the dilated-polyhedron-based discrete element method (DPDEM) for analysing ship–ice interactions. Results indicated that SDEM performs better in simulating ice crushing failure, although both methods produced similar mean ice resistances. Ren and Park (2023) developed a numerical model using the moving particle semi-implicit (MPS) method to predict ice resistance during continuous ice-breaking operations by an icebreaker. The algorithm effectively predicted ice resistance and fracture occurrences, with relative errors of approximately 12% and 30% under different boundary conditions, demonstrating its potential for accurate simulation of ice-breaking processes. Zhong et al. (2023) developed a procedure to predict ice-induced hull pressure for ships navigating in flow ice fields, anticipated to become more common due to global warming. Results showed comparable probability distributions of ice pressure from experiments and simulations.

Zhou et al. (2023) investigated the effects of propeller action on brash ice loads and pressure fluctuations for a reversing polar ship using a coupled CFD-DEM model. Higher reversing speeds diminish these effects, reducing propeller efficiency and altering pressure fluctuation patterns. Chen et al. (2023) investigated the effect of sea ice flexural strength on ice resistance for icebreakers using numerical simulations. By adjusting parameters in the sea ice material model, including plastic failure strain, Young's modulus, and compressive yield strength, they analyzed how changes in flexural strength impact ice resistance. X. Wang et al. (2024) used a dilated-polyhedron-based discrete element method (DPDEM) to study interactions between a self-propelled ice-resistant platform (IRSPP) and drifting sea ice. The study validated DEM parameters by comparing numerical ice resistance of previous icebreakers with the Lindqvist formula. S. Han et al. (2024) developed a numerical model using the cohesive element method combined with a random field to simulate interactions between sea ice and offshore structures. The study showed that the location of ice failure and the shape of ice fragments depend on the distribution of ice properties, with cracks forming in weaker regions.

2.2 Abnormal and Accidental Loads

2.2.1 Sloshing and Slamming Loads

Sloshing Loads

In recent years, the community of ships and offshore structures continued the extensive research work on sloshing and slamming load and load effect analyses. Assessing sloshing loads has been a significant issue in designing the cargo containment system (CCS) of liquefied natural gas carriers (LNGC). In order to assess the long-term sway loads on LNG carriers, Ahn et al. (2023) adopted a 6-DOF irregular sloshing model test with the aim to analyse short and long-term methods under different maritime classification guidelines, as shown in Fig. 1. The authors also compare the results, provide possible reference values and explore the experimental uncertainties of the sloshing phenomenon.

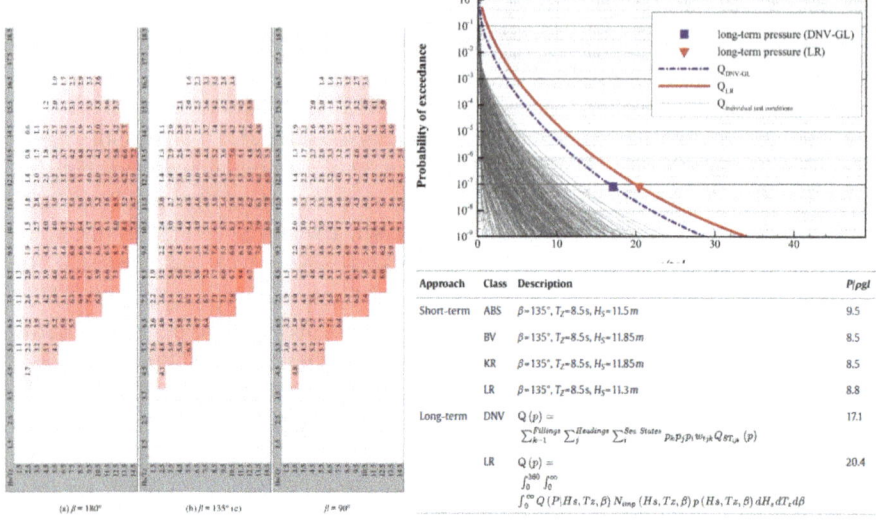

Fig. 1. Exceedance probabilities and long-term pressure of different guidance notes and comparison of the sloshing loads for maritime classification societies, from Ahn et al. (2023)

To validate a method for grouping sea states that reduces the experimental situation when assessing sloshing loads using a long-term method, Ahn et al. (2023) conducted extensive sloshing model experiments to classify sea states using 16 methods and compare long-term exceedance probabilities and pressures, as shown in Fig. 2.

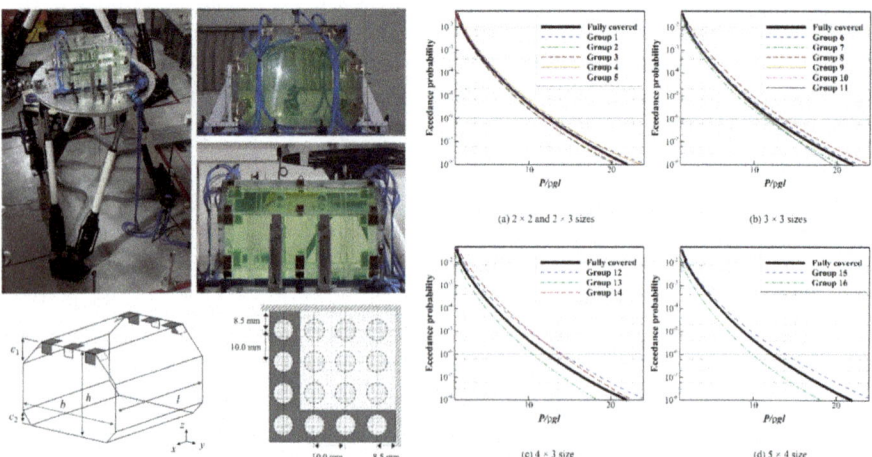

Fig. 2. Probability of exceedance of long-term approaches. Fully covered sea states (black thick and solid line) and grouped covered sea states, from Ahn et al. (2023)

To predict sloshing pressure for No. 1 and No. 2 tanks of an LNG carrier, Lee et al. (2022) performed sloshing model tests on the foremost two tanks simultaneously, which were installed on a motion platform. They analysed the differences in sloshing impact pressures by position in each tank under various loading conditions, wave heading angles, and sea states.

Tödter et al. (2023) experimentally investigate the impact of short-crested seas and swell on sloshing induced loads. The impact loads and counts were correlated with wave spreading functions, tank filling levels, and tank natural sloshing frequencies. These studies contribute to a better understanding of sloshing phenomena in LNG carriers and provide valuable data for the design and safety assessment of cargo containment systems.

Slamming Loads

Estimating the simultaneous structural responses caused by hydrodynamic loads during high impact water entry is a challenging task. Experiments are an important way to study the entry slamming pressure, structural response, and hydroelastic effects. In this ISSC review period, experimental work included assessment of high-speed ship-SWATH, a wedge-shape section with stiffened panels, a bow flare section and a 3D bow segment.

For investigating the dynamic ultimate load carrying capacity of reinforced plates with different stiffnesses under impact loads, Xia et al. (2024) performed free-drop tests and numerical simulations on two simplified models of stiffened plates to analyse the effects of different factors, such as stiffness, location, and entry velocity, on the plastic deformation of the structure. The distribution pattern of structural strain and plastic deformation trend of different model structures are shown in Fig. 3.

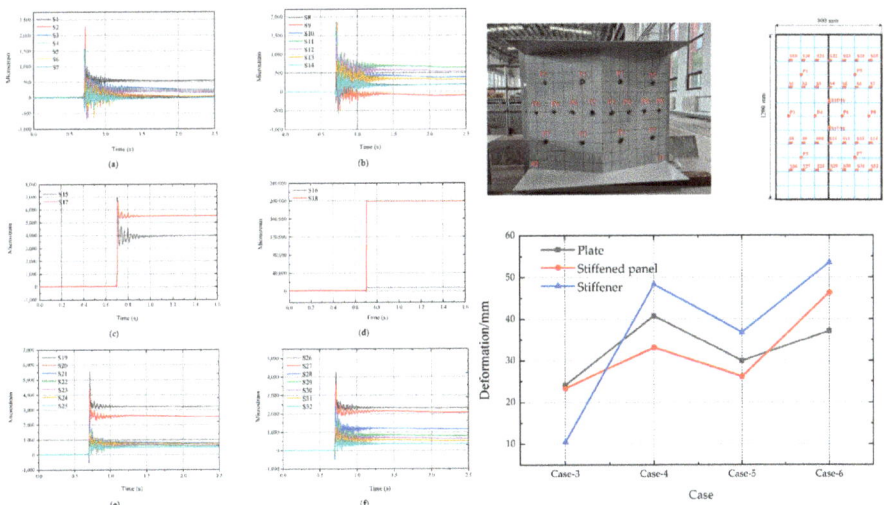

Fig. 3. Time series of structural microstrain and deformation trend of different members after each model test, from Xia et al. (2024)

Xie et al. (2024) investigated the impact load characteristics of a truncated ship bow under asymmetrical impact through drop tests. They analysed the load characteristics of the ship bow under asymmetric water impact and discussed the influence of air cavities and jet flows on pressure oscillation. They also investigated the influence of heel and pitch angles on impact pressure.

Shan et al. (2024) conducted a series of drop tests on a 3D bow flare model to study the dynamic characteristics of water impact. The slamming pressure characteristics of different heights and sections of the bow flare were compared, and the influence of velocity on slamming pressure were analysed. The suitability of the direct calculation approach was evaluated by comparing experimental findings with the theoretical calculations of Wagner and DNV guidelines. An experimental study on the wet deck slamming of a SWATH ship in regular waves was carried out by Ma et al. (2023). The characteristics of slamming pressure and the influence of wave height and wavelength parameters were analysed, as shown in Fig. 4. It was found that the slamming pressure on the wet deck exhibited non-stationary oscillatory features and was affected by various factors.

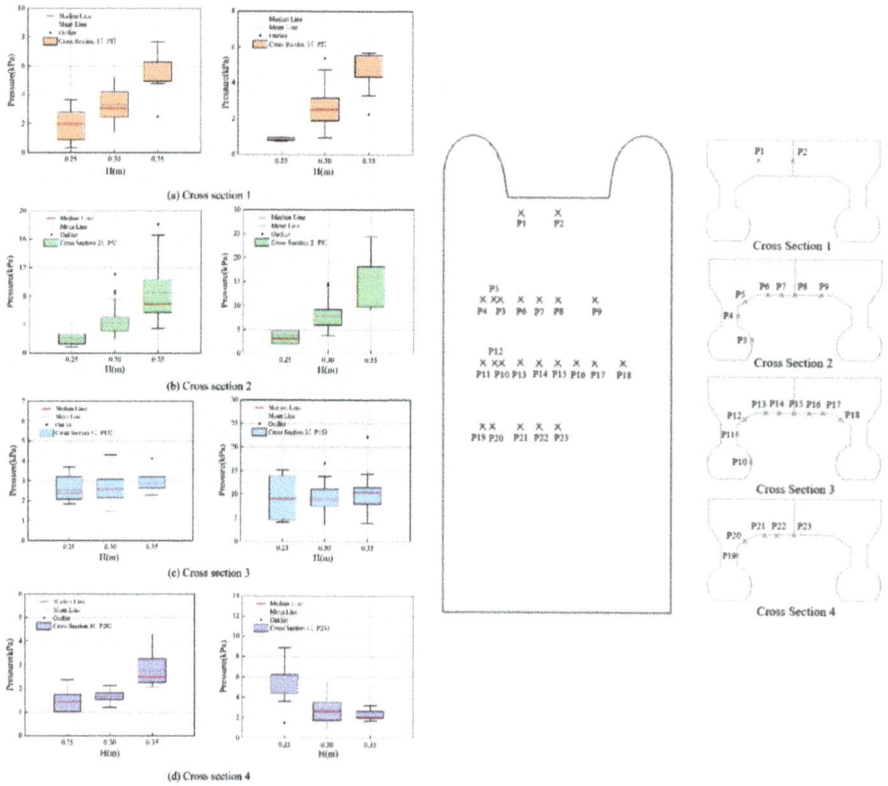

Fig. 4. Comparison of peak slamming pressure at different wave heights(left) and different wave length (right), from Ma et al. (2023)

Pan et al. (2024) designed a large scaled model of a trimaran and conducted freedrop water entry experiments to analyse the slamming pressure and strain in the wet deck areas of the connecting bridges. The ranges of maximum slamming pressure and the danger area for the design of the trimaran were determined.

Hydroelastic effect of the structure were investigated experimentally by Hosseinzadeh et al. (2023a) using a free-drop in-water of a three-dimensional non-prismatic aluminium wedge and by Han et al. (2024a) by using elastic wedges with different deadrise angles. These studies included investigation of the influence of impact velocity of the water flow, the deadrise angle, the mass of the wedge, and the bending stiffness on the slamming pressures and the structural response, as shown in Fig. 5.

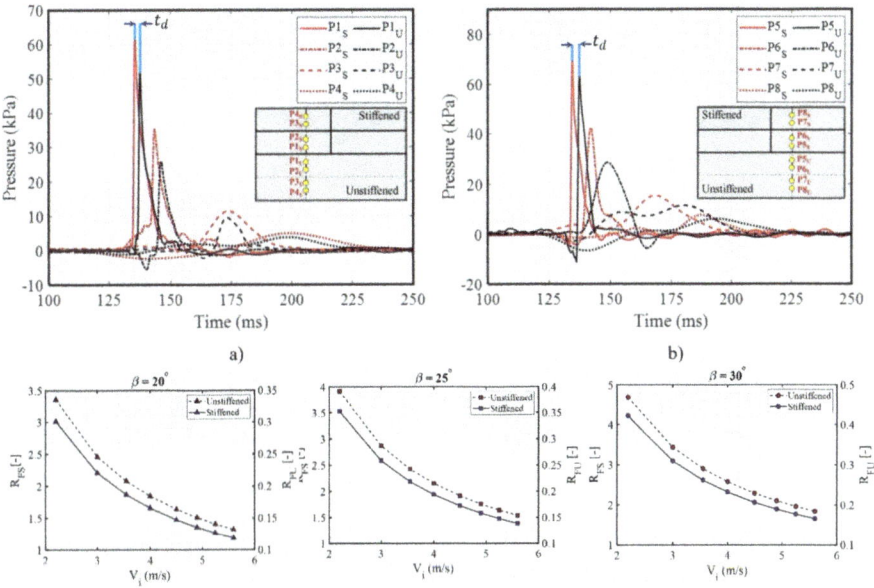

Fig. 5. Comparison of measured pressure of 100 [cm] initial drop height on different locations at a) β = 27deg and b) β = 23.5deg, and nondimensional hydroelasticity factor at different deadrise angles versus the impact velocity for both unstiffened (RFU) and stiffened (RFS) plates of the wedge, from Han et al. (2024a)

In addition to the above experimental studies, in recent years Xiao et al. (2024), Hosseinzadeh et al. (2023a), Jiao et al. (2024), Chen et al. (2023) have used numerical methods to solve for the slamming pressure and local structural response. A two-way coupled fluid-structure interaction method was proposed by Xiao et al. (2024) for the study of the slamming load and structural response of a non-prismatic reinforced steel wedge. The coupling library preCICE is introduced to realize the two-way coupling between computational fluid dynamics and finite element analysis simulation, which can accurately and efficiently calculate the hydroelastic slamming problem of elastic wedges. Two different two-way coupling methods are evaluated and compared to simulate the water ingress problem by Hosseinzadeh et al. (2023b). The explicit nonlinear finite element method with the MMALE solver as well as the k-ε turbulence model and Star- CCM + /ABAQUS were used to model the fluid and structural domains in a two-way coupling methods. It is concluded that the bottom plate deformation affects hydrodynamic loads during slamming, and the impact-induced loads depend on the water impact velocity and the flexibility of the bottom plates. The result of the effects of structural stiffness and impact velocity on the deflection of the bottom plate studied using the two-way coupling method are shown in Fig. 6.

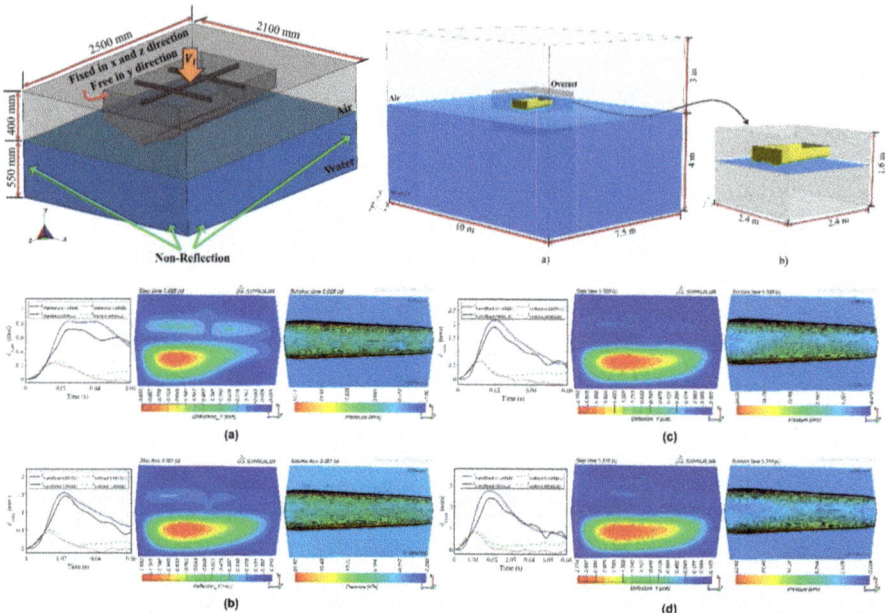

Fig. 6. Time history of the maximum deflection of the bottom unstiffened and stiffened plates computed by two different coupling method (left), bottom deflection distribution (middle), and pressure distribution on the bottom of the wedge (right) with: a) 2.20 [m/s] initial impact velocity; b) 3.00 [m/s] initial impact velocity; c) 3.55 [m/s] initial impact velocity; d) 4.00 [m/s] initial impact velocity, from Hosseinzadeh et al. (2023b)

Jiao et al. (2024) investigated the hydrodynamics of three-dimensional wedge asymmetry and oblique water impact by using a CFD-FEM two-way coupled numerical tool that couples the commercial software STAR-CCM + and Abaqus. The numerical simulation well predicts both the fluid and structural dynamic characteristics such as fluid cavity and structural stress concentration of asymmetric impact problems. Figure 7 presents the evolution of the volume fraction of water and cavity and the stresses at two typical velocity angle cases. It can be seen that the presence of the cavity on the low-pressure side leads to different local stresses.

The fluid-structure interaction problem and the slamming loads on a wedged grillage structure free-falling into 5th order Stokes plane progressive wave under gravity effect is investigated by a partitioned CFD-FEM two-way coupled numerical method by Z. W. Chen et al. (2023).

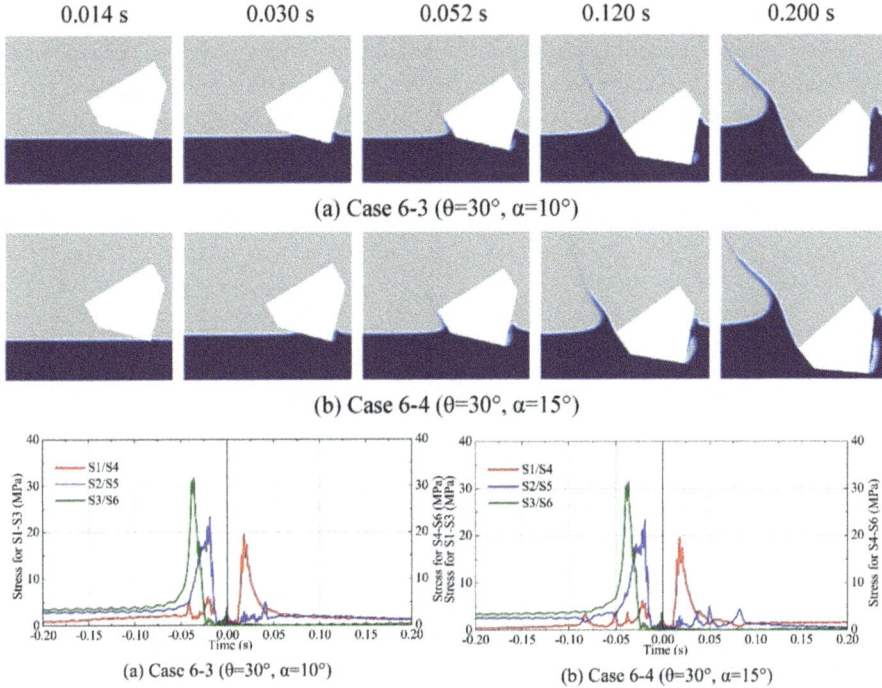

Fig. 7. Evolution of volume fraction of water at different velocity angles and stress at symmetric points, from Jiao et al. (2024)

2.2.2 Collision and Grounding

Accidental events like collisions and grounding may cause critical conditions to ships and offshore structures and the design of a structure against accidental loads aims at avoiding catastrophic outcomes including casualties, damages to environments, and economic losses (Ladeira et al., 2023).

As an increasing number of offshore wind farms are located close to ship traffic routes, the risk of collisions between ships and offshore wind turbines (OWTs) increases (Song et al., 2021). For such collisions, modelling and understanding of the physical processes is critical, and Figure illustrates a classification of the phenomenon where coupling effects including the wind load effects, hydrodynamic forces, soil-structure interaction, ice-structure interaction, and mooring line coupling can be important aspects for modelling the external dynamics and internal mechanics. In general, analytical, numerical, and experimental methods can be used to evaluate the structural responses of OWTs during collision events. Simulations for different ship collision directions in combination with different wind loads are also compared, as shown in Fig. 8.

Liu et al. (2022) considered jacket-supported wind turbines and conducted a parametric study consisting of 27 scenarios on the impact responses of ships with offshore wind turbines (OWTs). The study identified the most dangerous impact scenarios for the jacket legs. Su and Fang (2023) proposed the use of a floating composite honeycomb anti-collision structure to reduce damage caused by ship-OWT collisions, applying both analytical and finite element models to analyse the device's performance. These works focused on material modelling and neglected wind and hydrodynamic loads in their analyses. Song et al. (2023) addressed the interactions between a monopile-supported OWT and ice impact, wind loads, and soil contact using nonlinear finite element methods. Both global and local structural dynamic responses were investigated, and the maximum ice loads obtained were consistent with those predicted by ISO standards (see Fig. 9).

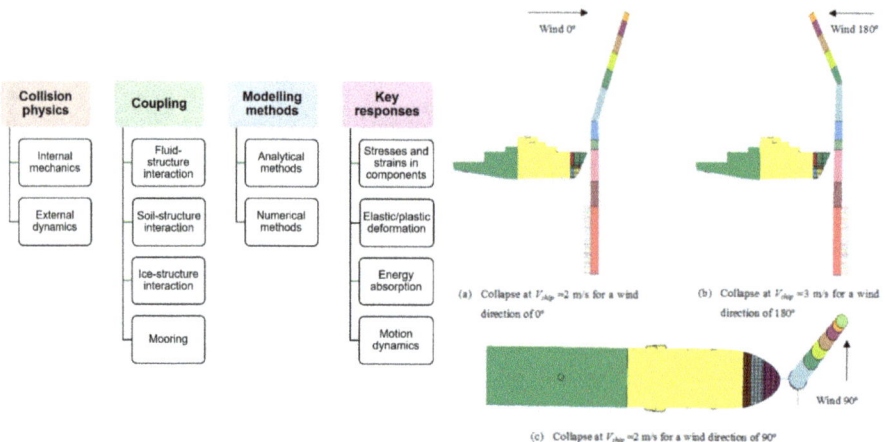

Fig. 8. Classification of ship-OWT collisions with a focus on the physical mechanism, from Song et al. (2021)

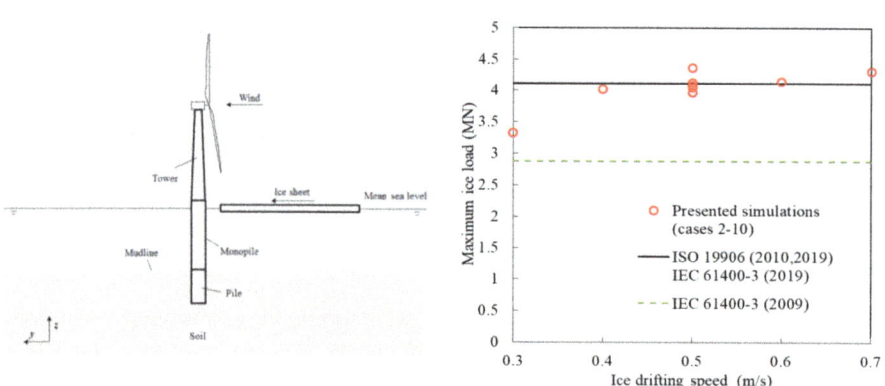

Fig. 9. Comparison between the finite element simulations, ISO, and IEC standards for the maximum ice impact force, from Song et al. (2023)

Ren et al. (2021) numerically examined the dynamic responses of floating wind turbines and the structural damage incurred by a spar floating wind turbine due to ship collision. The results showed that peak collision force, spar platform displacement, and tower top displacement increased almost linearly with the ship's initial velocity. Yu et al. (2022) numerically investigated the ship collision responses of a semi-submersible floating wind turbine, discussing energy absorption and structural responses of both the ship and the OWT. Ladeira et al. (2023) observed that the mooring response in ship-floating wind turbine collision events is understudied, and the limitations of the quasi-static approach compared to the dynamic one need further investigation.

Zhu et al. (2024) investigates experimental testing and evaluation of ship grounding over sharp rocks. The paper presents details of the experimental setup and considers two damage modes: discontinuous fracture/tear and continuous fracture/tear. Horizontal grounding resistance forces, damage extent of ship bottom and ship motion during the tests are presented and compared. The energy dissipation of the grounding process is analysed based on test data. A sensitivity study is carried out on the influence of initial ship velocity, initial relative distance between ship bottom plate and rock eccentricity on ship motion response and structural damage. The study concludes: higher initial velocity, higher initial relative height and sharper rock are more likely to lead to damage mode II, meaning a continuous tear and a more stable grounding force; larger rock eccentricity generally induces larger roll and yaw motion of the ship; for the sharp rock cases considered in this study, the energy dissipated by water resistance force in the grounding process accounts for 4–21% of the initial kinetic energy.

Pineau and Sourne (2023) explore ship grounding accidents, focusing on the interaction between ships and the seabed. The study introduces analytical formulations to calculate the resisting forces on a ship's bottom and the damage caused by combined surge (horizontal) and heave (vertical) motions on shallow and sharp rocks. Using plastic limit analysis, the paper predicts the force required to rupture the bottom plating and transverse floors under these motions. A failure criterion is proposed, and the derived expressions are implemented into an in-house solver. Validated through non-linear finite element simulations, these models show good alignment in predicting damage, especially in cruise ships. The solver models different ship types and assesses how mass and hydrodynamics influence grounding damage. The paper concludes that the methods can efficiently predict grounding responses, useful for pre-design and damage stability analyses, with vertical motions significantly affecting damage extent compared to horizontal movement alone.

Kim et al. (2022) presents a benchmark study comparing different computational methods for analysing ship collisions and groundings. It investigates the structural response of ships under dynamic conditions, particularly focusing on the effects of operational and environmental factors like hydrodynamic restoring forces. The study compares two main computational methods: explicit nonlinear finite element analysis (FEA) and the semi-numerical super-element method. Five grounding scenarios and several ship collision scenarios are simulated, using two different ship types: a passenger ship (Ship A) and a Ro-Pax vessel (Ship B). Key parameters include grounding speed, collision angle, and environmental conditions. The results reveal that accounting for hydrodynamic forces significantly improves the accuracy of predicting damage, especially in collision scenarios. The study also finds that the failure strain used in modelling the ship's bottom plays a crucial role in predicting the extent of damage during grounding.

2.3 Data Driven Model for Load Estimation

Surrogate models that often relate the environmental conditions or load parameters to the structural response parameters in a statistical way, are essential in estimating environmental loads on ships and offshore structures due to their efficiency and accuracy. These models, which use machine learning and data-driven approaches, significantly reduce computation time and can be integrated with high-fidelity simulation tools for real-time predictions. They facilitate uncertainty quantification and design optimization, making the design process more efficient and robust. Surrogate models are adaptable, continuously improving with new data, and are crucial for predicting wave, wind, and current loads, enhancing the safety and resilience of marine structures.

Surrogate models are popularly used for the short-term prediction of both wave load and the corresponding structural response of floating structure under wave actions. Jiang et al. (2024) developed a data-driven model for predicting motions and loads of moored floating structures in waves. The model, trained on wave elevations, predicts hydrodynamic responses in regular and irregular waves effectively. With long short-term memory networks, the model achieves computational efficiency, enabling potential applications in creating digital twins for offshore. Silva and Maki (2024) developed a novel approach using critical wave groups and computational fluid dynamics to efficiently predict extreme ship responses. By integrating long short-term memory neural networks, the method drastically reduces computational costs compared to traditional methods. H. Wang and Ti (2024) used machine learning to predict wave forces on truncated cylinders with various cross-sections like piers and bridge caps. Testing demonstrated high accuracy ($R2 > 0.999$), with over 90% predictions within 5% error, showcasing its efficiency and precision. Yang et al. (2024) focused on predicting wave loads on jack-up offshore platforms, which face harsh marine conditions. Using Stokes wave theory and Morison's formula, a numerical simulation method in ANSYS APDL constructs a sample database with inputs like wave height, period, and flow velocity. A Bayesian regularized neural network then predicts wave loads.

In recent years, some researchers focused their efforts on accurately estimating direction wave spectra around the floating structures using the so called "wave-buoy-analogy" concept. This concept treats a vessel as a wave buoy to estimate wave parameters from vessel motion data. Quasi-steady state has to be assumed when deriving the wave spectrum from vessel motion data. Wave estimation, as an inverse problem to estimate the wave conditions from measured ship motions responses, is also one of the most interesting research topics drawing a lot of research efforts especially in terms of surrogate models.

Park and Kim (2023) developed a method to estimate directional wave spectra crucial for assessing seagoing vessels' structural integrity. Using a 10-parameter bimodal wave spectrum model, they employed Bayesian statistics and the adaptive Metropolis-Hastings algorithm for parameter estimation. Using the "wave-buoy analogy," the ship is treated like a wave buoy. By solving a non-linear least squares problem, the parameters are estimated based on sensor data from the ship. A probabilistic method using Markov-Chain Monte Carlo (MCMC) simulations, specifically the Metropolis-Hastings algorithm, helps identify the best-fit parameters. The method was validated using simulated data, showing it accurately predicts wave responses even where sensors aren't present.

Majidian et al. (2022) examined the feasibility of an onboard instantaneous wave information system by reviewing the concept of real-time sea-state estimation using the "wave-buoy analogy." The study highlights two main methods: model-based approaches, which rely on transfer functions (TFs) that relate wave excitation to vessel responses, and data-driven models that leverage machine learning techniques. While model-based methods have been researched for decades, they still face challenges with accuracy due to transfer function uncertainties. Data-driven approaches, though promising, are sensitive to data quality and characteristics. The paper provides a perspective for future studies, identifying existing challenges and potential solutions.

J. Zhang et al. (2024) introduced an ANN-based model to estimate wave elevation using motion and air-gap responses of floating structures. Compared to traditional methods, the ANN model robustly reconstructs wave-surface elevation, reducing errors from platform motion and wave-structure interaction across domains. Mittendorf et al. (2022) used machine learning to estimate sea state parameters on a medium-sized container vessel. They compared time and frequency domain models for accuracy and efficiency. Various deep neural networks were trained, with an adapted Inception Architecture showing superior performance. Huang et al. (2022) developed a CNN-based surrogate model for regional wind wave prediction, improving computational efficiency over traditional SWAN methods. Demonstrating robustness across computational regions, predictions achieved < 5% relative error compared to SWAN simulations. Den Bieman et al. (2023) introduced an innovative hybrid model combining an operational wave model with a machine learning model trained on wave measurements in the model domain, applied to the Dutch North Sea. The hybrid model significantly improves forecast accuracy, reducing average errors by 21.7% for wave energy density and 25.3% for wave direction.

Chen et al. (2023) introduced an Artificial Neural Network model for predicting moderate directional spreading waves. A quantile loss function is introduced to quantify uncertainty, improving the model's practical value for decision-making and engineering applications in offshore renewable energy systems. Liu et al. (2022) developed a LSSVM model to predict ocean wave elevations in random sea states, integrating both frequency and time domain features of JONSWAP spectrum wave data from indoor experiments. This approach extended prediction timeframes effectively for real-time applications, addressing both wave elevations and associated forces. Ko et al. (2024) proposed a new hybrid method for estimating the wave spectrum of an advancing ship using motion response spectra. It uses motion response spectrum features to determine wave direction via semi-supervised learning and t-SNE visualization. Model tests in various sea states were used to validate the accuracy of both conventional and proposed methods. Li et al. (2024) studied the performance of the Bayesian wave buoy analogy method, Improved CGAN without physics guidance, Improved CGAN with physics guidance, and direct machine learning mapping on the ITTC spectrum. The results show that the Physics Guided Improved CGAN outperforms in prediction error, stability, and generalization ability, including zero-sample predictions of complex wave spectra.

2.4 Load Estimation by Experiments and Full-Scale Measurements

Full scale sea trials are an important way of assessing the real performance and behaviour of marine structures and are of great value for structural design and weight optimization. The idea situation is to have both input environmental conditions and key response parameters measured, so that they can be used for validation of numerical simulations or models. However, it is often difficult or even not possible to directly measure the surrounding environments. Therefore, reconstructions of the hydrodynamic loads or environments from the measured responses are necessary. Gebrezgabir et al. (2023) investigated the reconstruction of the slamming and wave load response of a high-speed catamaran using the transmissibility concept. The reconstruction of the response for different conditions was achieved by deriving the transmissibility function and matrix from the full scale sea trial data. Reconstructed strains from a transmissibility matrix derived from different sea states are shown in Fig. 10. Almallah et al. (2023) investigated the torsional and global loads of 98 m Incat wave-piercing catamaran in irregular slanting waves by full-scale CFD simulations. Internal loads were calculated based on rigid body dynamics and the CFD model was validated and setup. The relationship between the loads and various factors as well as the slamming load characteristics were analysed and compared with the sea trial data.

Offshore platforms and flexible pipes are the key equipment in the offshore oil and gas development system. Wu et al. (2022) reviews the recent works of the theoretical model, numerical simulation, and experimental test in three research areas: hydrodynamic analysis of offshore platforms, structural mechanics analysis of flexible pipe and cable, and monitoring technology of offshore floating structures under marine loads. The field monitoring projects of offshore platforms in China are shown in Fig. 11. Xu et al. (2024) take the "Deep Sea No.1" energy station as an example, evaluate the environmental characteristics and platform service dynamic response state based on the monitoring data of wind, wave and current on site, and indicate the reasons for the difference between the design model and the real motion by constructing the load vector time course and the full-coupled dynamics calculation.

Table 2
Sensor type and location on HSV 2 Swift.

Sensor	Location	Distance from transom (m)	Measured signal type
T1 - 5	Port	31.45	longitudinal bending strain
T1 - 6	Port	55.41	longitudinal bending strain
T1 - 8	Starboard	31.45	longitudinal bending strain
T1 - 9	Starboard	55.41	longitudinal bending strain
G1	LCG	37.14	vertical, transverse and longitudinal acceleration
G2	LCG	37.14	pitch, roll and yaw rate
B1	Bow	87.46	vertical, transverse and longitudinal acceleration

Fig. 1. HSV 2 Swift 98 m Incat catamaran used in sea trials.

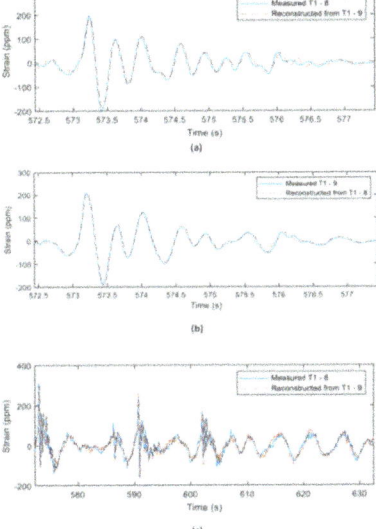

Fig. 10. Reconstructed strain using 8 measured responses and a transmissibility matrix derived from different sea states. (a & c) Run 71 and (b) Run 70 both at 20 knots in starboard bow and head sea, respectively, 2.44 m Hs, from Gebrezgabir et al. (2023)

Fig. 12. Field monitoring system in China (LH11-1 Nanhai Tiaozhan FPS).

Table 1 Field monitoring projects of offshore platforms in China

Year	Platform	Main monitoring content
2021	"Deep Sea No.1" energy station	Environmental data, float response, etc.
2014	LF 7-2 jacket platform	Strain monitoring
2013	PY 34-1 jacket platform	Barge rocker arm stress during jacket launching
2012	LW 3-1 jacket platform	Attitude during jacket launching
2011	LH11-1 FPS (Du et al., 2013)	Environmental data, float response, typhoon data, etc.
2010	BZ28-1 FPSO (Wang et al., 2015)	Environmental data, hull and mooring system response, etc.
2010	BZ28-2S FPSO (Wang et al., 2019)	Environmental data, hull and mooring system response, etc.
2010	HYSY 981 drilling platform	Environmental data, air gap, structural stress characteristics, etc.

Fig. 11. Field monitoring projects of offshore platforms in China, from Wu et al. (2022)

Offshore wind turbines (OWT) perform an essential role in the expansion of clean and renewable energy sources, and the research on their structural safety is of great value. A microwave-based structural health monitoring (SHM) method was proposed by Maetz et al. (2023). The actual working conditions are simulated and fatigue tests are conducted by constructing a scaled laboratory demonstrator. Using a specific radar sensor system and multiple auxiliary sensors for simultaneous measurements, the method can effectively detect grout joint damage. Liu et al. (2023) established a real-time multi-source monitoring system to analyse the impacts of Typhoon "In-fa" on a 4.0MW monopile OWT in Rudong, Jiangsu Province, including environmental, operational conditions and structural responses, as shown in Fig. 12. The above investigations provide a basis for the design, safety monitoring and damage diagnosis of OWT.

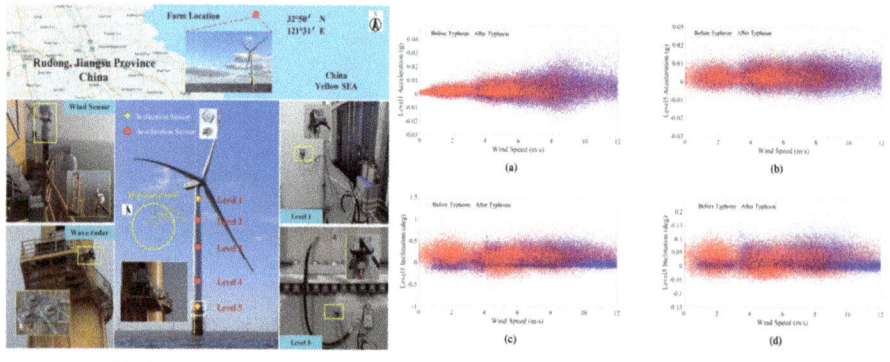

Fig. 12. Wind farm location and monitoring system and structural response comparison before and after typhoon, from Liu et al. (2023)

The ice load is an environmental load that dominates the ice-resistant design of polar ship structures. Research on ice loading is important for polar ship safety. To address the problem that the monitoring area and loading area need to be separated in the indirect measurement of ice loading on polar ship structures, J. Wang et al. (2023) proposed a far-field identification method of ice loading based on RBFNN. Through numerical analysis, scale model test and field application, The Structure of the generalised RBFNN for ice load identification and the testing results of the far-field identification of the total ice load on each frame after noise injection are shown in Fig. 13, it is verified that the method can effectively establish the relationship between ice load and far-field strain, and has the potential for engineering application. To clarify the dynamic ice load of conical offshore wind turbines in cold regions, G. Wang et al. (2022) installed a monitoring system on an OWT in cold region of China to obtain data. A formula for calculating the ice force period and a random dynamic ice force model were established, showing a good agreement with the measured data.

To determine the river ice load characteristics of inland icebreakers during icebreaking operations, a full-scale measurement was carried out by He et al. (2022). The results showed the form and distribution pattern of structural ice load. The river ice load is mainly in the form of a single-peaked triangle, the peak value obeys the Weibull distribution and the peak value is slightly higher than the same thickness of sea ice. To overcome the issue that there is no direct relationship between the RIO value and the probability of ice-induced hull structural damage in POLARIS, Suominen et al. (2024) utilized the full-scale measurement data from the ship PSRV S.A. Agulhas II. After calculation and analysis, the results show that the POLARIS guideline is reasonable and gives the probability of hull structural damage and related influencing factors under different ice levels and operating conditions, as shown in Fig. 14.

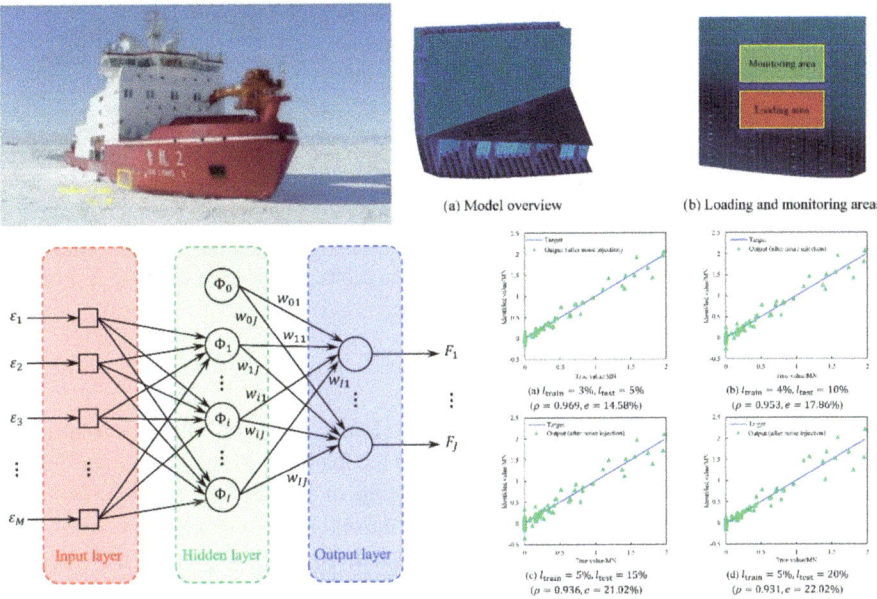

Fig. 13. Structure of the generalised RBFNN for ice load identification and the testing results of the far-field identification of the total ice load, from J. Wang et al. (2023)

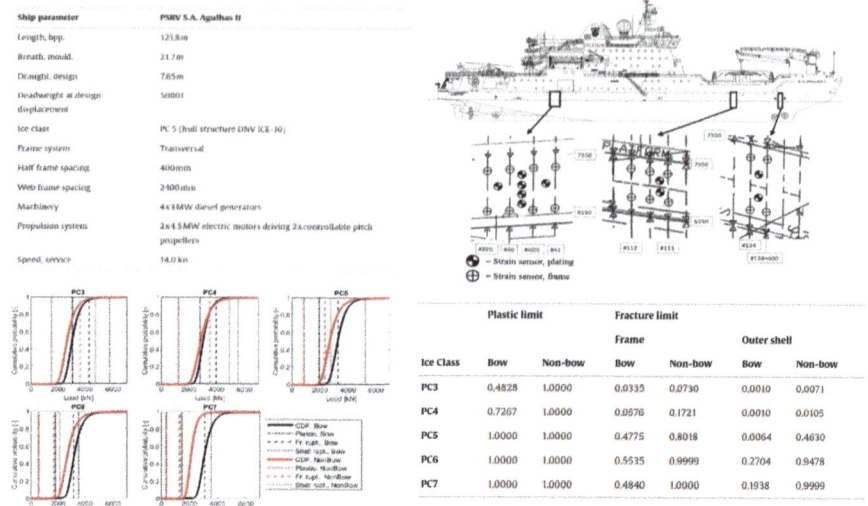

Fig. 14. Cumulative distribution function, CDF, of annual maximum ice-induced loads with structural limit states for the bow and non-bow area, and the probability to exceed different structural capacity limits, from Suominen et al. (2024)

2.5 Concluding Remarks

This section provides an overview of recent developments in modeling environmental loads, with a particular focus on wave, wind and ice loads, as well as accidental forces, such as sloshing and slamming loads, culminating in several key insights. With the growing scale of marine structures and the increasingly severe ocean conditions they face, nonlinear methods are becoming more prevalent for analysing complicated quasi-static load on ships and offshore structures. Recent studies have emphasized the dynamic, stochastic, and probabilistic nature of environmental forces, including those from waves, wind, currents, and ice along with some accidental loads.

Special focuses were given to data-driven modelling which is gathering attention from many researchers to overcome current methodologies. These models leverage machine learning and data-driven techniques to drastically cut down computation time and seamlessly integrate with high-accuracy simulation tools for real-time forecasting. The importance of full-scale measurements in ships lies in their ability to provide accurate, real-world data on structural responses and environmental loads. This data is crucial for validating theoretical models, improving design accuracy, and ensuring the safety and reliability of marine structures under actual operating conditions.

3 Structural Modelling and Response Analysis

In recent years, significant progress has been made in analyzing structural responses of ships, offshore structures, for example floating wind turbine (FWT) structures, particularly with quasi-static methods. These methods offer designers simplified and computationally efficient approaches to estimate load effects and structural responses in the early stages of design, allowing rapid iterations while maintaining accuracy. The quasi-static approach is especially valuable for analysing internal stresses, and fatigue in ships and offshore structures under wind and wave loads with relatively long carrying frequencies, where dynamic effects can often be approximated or ignored in the initial phases. This chapter reviews the major developments in quasi-static methods and their application to different type of ships and offshore structures, including FWTs. The chapter has been divided into several sub-sections to address simplified analysis, direct calculation, optimisation-based analysis, new materials, corrosion and advanced numerical tools.

3.1 Simplified Structural Analysis

While high-fidelity numerical simulations are increasingly used, simplified methods remain essential for initial design assessments. These methods often rely on quasi-static load assumptions, which significantly reduce computational complexity. Recent research has expanded on the use of simplified formulations for ultimate strength, buckling, and local structural responses in floating wind turbine hulls.

S. Wang and Moan (2024) propose a systematic methodology for the load effect analysis and ultimate limit state (ULS) design of semi-submersible hulls of FWTs, focusing on the column design of the floater. As shown in Fig. 15, a multi-body floater model within a fully coupled aero-hydro-servo-elastic numerical system can be developed to calculate the internal loads of the floater, including the cross-sectional forces and moments. The environmental contour method was employed to select critical load cases based on wind speeds and wave periods. The study evaluated von Mises stress in the columns to screen important load cases and investigate lateral pressures on the columns. A detailed ULS check was conducted by combining global loads and lateral pressures to determine design parameters such as plate thickness and ring stiffener dimensions. A sensitivity analysis was also performed to explore the influence of column design parameters on buckling performance, optimizing the stiffener dimensions based on panel buckling criteria. This methodology contributes to improving the detailed structural design of semi-submersible hulls, providing a comprehensive approach to ensuring the reliability and safety of FWTs.

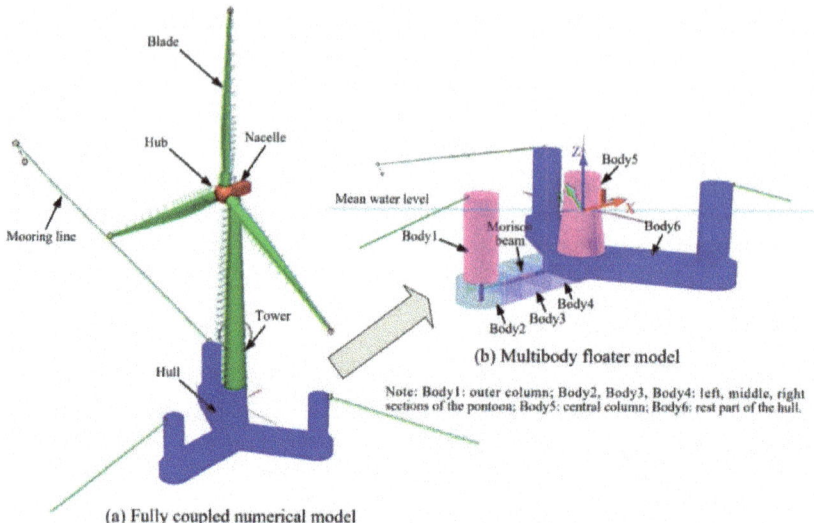

Fig. 15. Numerical model of the fully coupled FWT system with multi-body hull for time domain load effect analysis, from S. Wang & Moan (2024)

S. Wang et al. (2023b) conducted a detailed structural and dynamic performance analysis of a steel semi-submersible platform for a 10-MW FWT. The study outlines a comprehensive design methodology, incorporating time-domain coupled analyses of the wind turbine under wind, wave, and still water conditions. Using advanced hydrodynamic solvers, the pressure distributions on the platform were calculated and then mapped to finite element models (FEM) for stress analysis. The stress calculations were primarily based on a quasi-static approach, meaning that the analysis focused on long-term, steady loads rather than dynamic or transient effects. This method allowed the researchers to evaluate how the structure responds to combined wind and wave loads in a stable manner, without considering the rapid load changes that would occur in a dynamic environment. Buckling analyses were performed in vulnerable areas such as the upper surfaces of pontoons, which experienced significant thrust and hydrostatic pressures. The study concluded that the steel platform design satisfies the structural and stability requirements under both operational and extreme environmental conditions, demonstrating that quasi-static load effects are sufficient for the long-term evaluation of the platform's structural integrity.

Martins et al. (2024) conducted a comprehensive parametric study assessing the accuracy of the Eurocode 3 Part 1–3 (EN 1993-1-3) simplified formulas for calculating the distortional buckling loads of cold-formed steel lipped channels subjected to axial forces and bending about the major and minor axes. The research compared these simplified formulas with numerical methods, specifically using Generalized Beam Theory (GBT). Over 24,000 cases were analysed, covering a wide range of geometric parameters and material properties. Error isoline maps were generated to visualize discrepancies between the simplified and numerical methods. The study found that while the simplified method can yield satisfactory results for certain parameter ranges, significant errors can occur, particularly for more complex geometries and higher width-to-thickness ratios. The authors highlighted that using the simplified method may lead to both unsafe and overly conservative designs, depending on the case. The paper underscores the potential need for more accurate numerical methods, especially in cases where distortional buckling is a critical design consideration.

Beier et al. (2023) conducted a detailed fatigue analysis of inter-array power cables between two floating offshore wind turbines (FOWTs), presenting a simplified method to estimate stress factors for early design stages. They focused on the nonlinear bending behavior of the power cable, considering typical environmental conditions in the North Sea. By using the software OrcaFlex for simulations, the study compared results with UFLEX finite element software, validating that their proposed method provides conservative results for bending loads while maintaining similar accuracy for axial tension loads. The critical fatigue locations were identified near the hang-off points, where curvature variation contributed significantly to fatigue damage. This work offers valuable insights for the design optimization of inter-array power cables, emphasizing the importance of stress factors in determining fatigue life and ensuring the reliability of power cables in floating offshore wind farms.

Lerch et al. (2023) developed a simplified model for the dynamic analysis and power generation of a floating offshore wind turbine (FOWT), aiming at early feasibility studies and pre-engineering phases. The model solves the equation of motion in the time domain and incorporates Morison's equation to calculate hydrodynamic loads, wind thrust at hub height for aerodynamic loads, and a nonlinear model for mooring system loads. The model was validated by comparing its results to the more complex FAST model under two different load cases. The simplified model successfully captured the key motions (surge, heave, and pitch) with acceptable accuracy, showing less than a 1.1% deviation in power generation compared to a bottom-fixed offshore wind turbine (BOWT). The authors conclude that their simplified approach is effective for studying the main behavior of FOWTs, especially for early-stage design, with the potential for further development to include energy generation and cost estimations.

Liang et al. (2023) developed two reduced-order models (ROMs) to improve the efficiency of dynamic analysis for monopile-supported offshore wind turbines (OWTs). The first approach simplifies the OWT structure using substructure-based finite element modeling, dividing the turbine into key components such as blades, the nacelle-hub assembly, and the tower, and then updates the model using an optimization algorithm. The second approach applies proper orthogonal decomposition (POD) and modal truncation to derive a simplified state-space model (SSM), retaining only the most critical modes of vibration. Both methods significantly reduce computational costs while maintaining accuracy within a 5% deviation from full-order models. The study validated these models using a 10 MW reference wind turbine, showing their utility in both frequency and time-domain analyses. The authors emphasize the practical benefits of these ROMs in real-time dynamic analysis, particularly for the design, operation, and maintenance of large-scale OWTs, where computational resources are often limited. Future work could extend these methods to other offshore structural types.

Ali and De Risi (2021) present a simplified blade model for the seismic assessment of wind turbines, addressing the complexities inherent in the dynamic response of wind turbine blades during earthquakes. Given the challenge of using detailed high-fidelity finite element (FE) models due to the composite material layup and computational inefficiency, the authors proposed a genetic algorithm (GA)-based approach for optimizing blade properties like mass, stiffness, and modal response. Their model was validated using a 5 MW reference wind turbine blade (61.5 m long), and the optimized properties were implemented into an FE model for seismic analysis. The study demonstrated that the simplified blade model effectively captures essential dynamic characteristics while reducing computational costs, especially when compared to more detailed models that are difficult to apply in large-scale seismic assessments. Importantly, the results showed that conventional lumped mass approaches could miss the impact of higher blade modes on global system dynamics, emphasizing the need for a more nuanced approach in seismic assessments of wind turbines.

Slagstad et al. (2023) proposed a simplified analytical model for estimating the axial force in the ropes supporting the fish net, utilizing a quasi-static solution based on the principle of virtual displacements. They compared the results between the proposed simplified method and numerical simulations conducted using RIFLEX. The study finds that while the simplified method is highly effective during the initial design phases—particularly for ropes with typical pretension levels—it tends to be conservative and less accurate in scenarios with lower pretension. Additionally, the results emphasize the significant impact of current velocities on fatigue life. Th study indicates that the simplified method can substantially reduce computation time while maintaining reasonable accuracy for preliminary designs and verification purposes.

3.2 First-Principle-Based Direct Calculations

Time-domain direct load and response simulations become popular in the design phase for complex structures under dynamic environmental conditions, due to the maturity of the simulation tools and the increased computational capacity. However, it is still not possible and in many cases not necessary to simulate at once all the global load effects at kilo-meter length scale (for global dynamics) and the local effects at micro-/milli-meter scale (for fatigue cracking). Understanding the effect of global and local dynamics and the limitation of a quasi-static method becomes very important for choosing a right numerical model resolution (both in space and in time) for design analysis.

The algorithm to adjust shear force and bending moment in the FE analysis of shuttle tanker model in global structural analysis was proposed by Lim et al. (2023). The problem of global loads adjustment in global FE analysis was undertaken in this work. As the results, the new algorithm was found to be very accurate and applied global loads were very close to those given in the ship loading conditions.

Gao et al. (2023) present a methodology for time-domain stress analysis of a semi-submersible floater supporting a 15 MW floating wind turbine. As demonstrated in Fig. 16, this study uses a time-domain approach to calculate floater stresses by regenerating hydrodynamic pressure loads on the external surfaces of the floater. These loads are combined with time-varying gravity, inertial forces, and forces at key points like the tower bottom and mooring line fairleads to perform a quasi-static stress analysis. The methodology allows for the evaluation of structural stress without re-solving global motions. The analysis is implemented in a customised MATLAB code and verified against frequency-domain methods, showing good agreement. The study highlights the efficiency of this time-domain approach for structural design, reducing computational time while accurately predicting stresses under wave-induced loading.

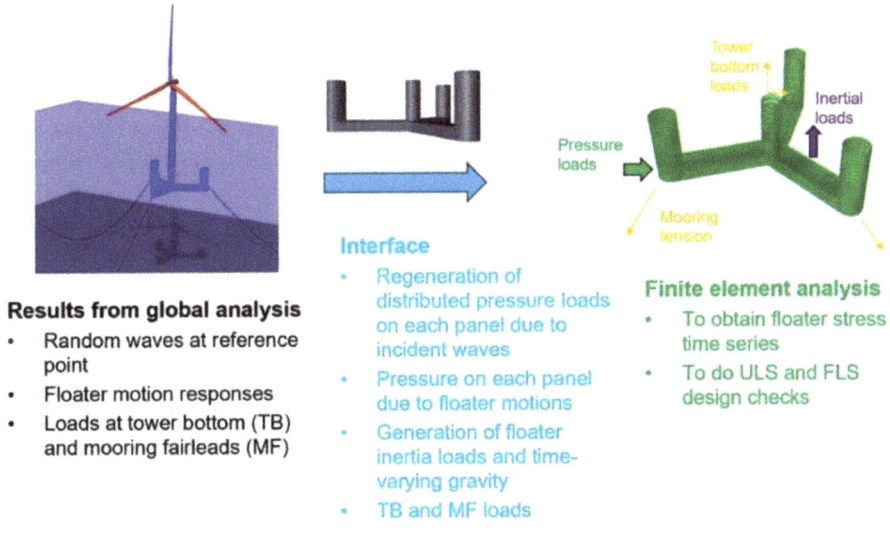

Fig. 16. Analysis procedure for reconstruction of the applied loads for floater stress analysis, from Gao et al. (2024)

S. Wang et al. (2023a) present a comprehensive methodology for assessing the global structural load effects on semi-submersible floating wind turbine (FWT) hulls, focusing on the internal stress and force distribution across the structure. The study performs time-domain simulations to capture internal forces and moments under various environmental conditions, including wind, waves, and still water. Stress calculations in key sections of the floater are based on the quasi-static approach, where static global loads (such as buoyancy and gravity) and time-varying loads (e.g., wind and wave-induced forces) are combined. However, the analysis does not include dynamic stress calculations and transient effects, such as collision or grounding. The study also shows significant cancellation effects between various load components, resulting in reduced total stress values compared to individual contributions. The absence of dynamic stress analysis suggests the findings are primarily applicable for understanding long-term, steady load effects on FWT structures, while instantaneous impacts are not considered.

Fowai et al. (2021) conducted a structural analysis of jacket foundations for offshore wind turbines (OWTs) installed in transitional water depths. The study compared traditional four-legged jacket foundations with X and K joints to newly patented twisted three-legged jackets, evaluating their performance under static and dynamic loading conditions. Static analysis was performed to understand global load-bearing capacity, and dynamic analysis investigated the natural frequencies and response under wind and wave loads. The results indicated that twisted jackets, with a twist angle of 30 or 60 degrees, provided superior structural behaviour while using less material and having fewer nodes compared to traditional designs. The modal analysis showed that the twisted jackets fell under the soft-stiff category, making them less prone to resonance under operational conditions, whereas traditional jackets were classified as stiff-stiff. The study concluded that twisted jackets are cost-effective and efficient alternatives for OWT foundations in transitional water, offering both structural and economic advantages.

Nybø et al. (2021) investigate the quasi-static response of bottom-fixed offshore wind turbines, focusing on how different wind field models, such as Kaimal, Mann, LES, and TIMESR, affect the structural response. The study emphasizes the significance of low-frequency wind loads (below 0.1 Hz), which strongly influence the tower bottom and blade root moments, contributing to fatigue damage. Differences in turbulence intensity and wind coherence across models lead to substantial variations in Damage Equivalent Moments (DEM), especially under stable atmospheric conditions. The study highlights the need for longer simulations than those recommended by standards (e.g., IEC) to reduce uncertainty in estimating the structural response, and it suggests that simplified treatments for transient phenomena, such as reductions in cross-sectional areas for collisions or grounding, can be effectively incorporated into such analyses.

In Li et al. (2021), the quasi-static response of moored floating structures is analysed using a method that minimizes the system's mechanical energy. This method integrates gravitational and buoyancy potential energy, drag-induced kinetic energy, and spring potential energy from line tension, providing an efficient alternative to the traditional catenary equation for calculating mooring system responses. The proposed approach proves more accurate for predicting quasi-static behaviour by incorporating seabed interactions and hydrodynamic effects, which are typically neglected in catenary-based methods. Through comparisons with the AQWA commercial software and ABAQUS, the study demonstrates that the new method is not only computationally efficient but also suitable for early-stage design of moored structures like floating wind turbines, as exemplified in the WindFloat 2 platform case study. This work highlights the method's advantages in handling simplified transient phenomena, such as the impact of ocean currents on mooring tension and equilibrium positions, making it an effective tool for offshore wind energy applications.

Prabowo et al. (2019) investigated the structural crashworthiness of a ship hull under grounding conditions, focusing on the impact of different mesh sizes (element-length-to-thickness, ELT ratios) in finite element modelling. The study simulated the grounding of a chemical tanker on a conical seabed rock, using ANSYS LS-DYNA for explicit dynamic analysis. The authors examined the effect of expanding the ELT ratio from the traditional 10 to 11, 12, and 13, aiming to optimize the balance between computational efficiency and accuracy. The study found that while expanding the ELT ratio from 10 to 13 reduced simulation time by up to 45%, it also increased the likelihood of structural failure, especially in the ship's bottom shell. Despite this, ELT ratios of 12 and 13 provided similar results to the traditional ratio of 10 in terms of internal energy and crushing force, making them a viable option for improving computational efficiency in complex simulations.

Wen et al. (2024) presents a comparative analysis between numerical simulations and experimental tests for floating dock operations with a vessel on board. The study aims to validate a numerical method designed to simulate vessel-docking processes based on a quasi-static assumption. Both model-scale and full-scale experimental tests were conducted to measure the draughts, floating positions, and bending of floating docks during docking operations. The numerical model is built using a hydrostatic force model combined with a Newton-Raphson method to predict the static response of floating docks under various ballast water distributions. A bending model is also used to calculate dock deflection along its longitudinal axis. The results show good agreement between the experimental and numerical predictions, proving the accuracy of the numerical model.

S. Zhang et al. (2024) presents a two-step method for analysing hydro-elastic stresses in marine structures in the frequency domain, focusing on global hydroelasticity and local bending effects. The structure is discretized into rigid modules connected by elastic beams, known as the beam-connected-discrete-modules (BCDM) method, to evaluate global hydro-elastic responses. Then the hydrodynamic pressure, hydrostatic pressure, and inertia forces calculated in the first step is applied to a detailed finite element model (FEM) to assess the stress distribution. The analysis incorporates a quasi-static method for stress calculations, which means the dynamic effects of the structure's flexibility are considered, but local bending-induced inertia forces are neglected. The study compares the proposed method with a modal-based quasi-static approach, validating its accuracy through a flexible barge model. Results indicate that local bending effects increase stresses, particularly at non-resonant frequencies. Furthermore, global flexibility significantly impacts stresses near resonance conditions, driven by associated inertia forces.

Understanding structural and motion responses in engineering analyses requires distinguishing between transient and steady-state behaviours, both of which are crucial for assessing system stability and performance over time. Wu et al. (2022) investigated the collapse load, collapse mode, and implosion effects on adjacent structures using transient dynamic and quasi-static simulations, based on experimental processes with a scale model of a deep-sea pressure hull designed for a service depth of 4500 m. Figure 17 illustrates the determination of collapse pressure and mode through experimental testing under quasi-static conditions, followed by an analysis of the transient dynamic process of implosion and the subsequent fluid-structure coupling, using the collapse load as the initial pressure.

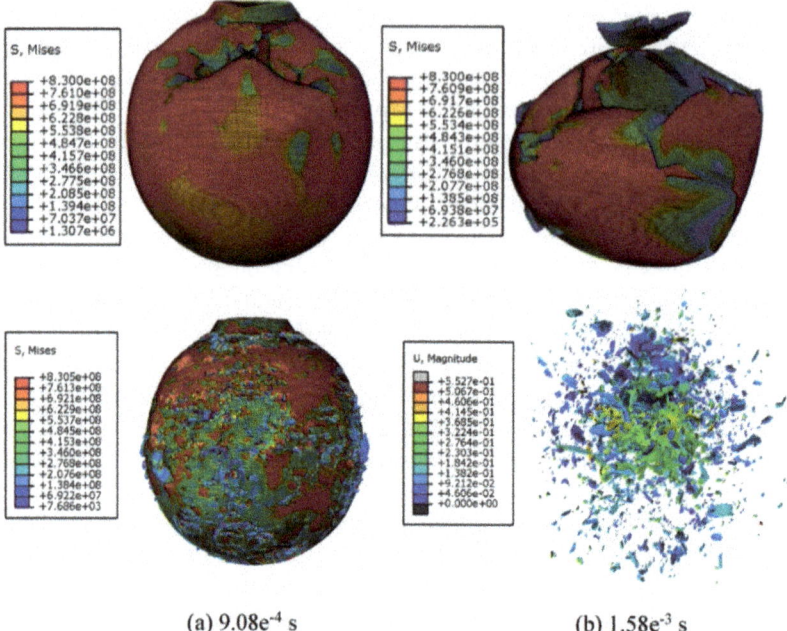

(a) 9.08e^{-4} s (b) 1.58e^{-3} s

Fig. 17. Collapse and implosion process of the spherical hull, from Wu et al. (2022)

3.3 Optimization Analysis

In this section, recent development of optimisation procedures using quasi-static response analysis is reported. Because of the significant computational efforts that are required for optimization analysis, simplified but accurate load and response analyses are necessary.

Aguiari et al. (2022) presented a semi-automatic design procedure for the concept and preliminary phases of hull structural design based on conventional analytical structural models (i.e. beam theory) and rule requirements, which are often under a quasi-static assumption. The proposed scantling design approach is practical and takes advantage of the available computation means to update the traditional sequence of the scantling checks more rationally and to select the optimal layout scantling solutions among truly feasible ones. The idea is to avoid any computation burden for the optimization of the scantling design relying on the fact that the problem is rather over-constrained in practice and feasible solutions are relatively limited in number.

The design of the insulating glass units (IGU) type of windows has become more critical in recent years as large glass-covered areas increase the cruise ship's weight and center of gravity. Heiskari et al. (2023 and 2024) published several papers on the thickness optimization of glass panes to achieve lightweight (IGU) in cruise ships. The papers aim to improve the IGU design framework using different engineering strategies to overcome the shortcomings of rule-based design. These include numerical design methods connected with optimization algorithms and more advanced theoretical assumptions than provided in the existing Rules. The optimized results are compared to the relevant classification rules, and the results show that it is possible to reduce the glass weight up to 50% in monolithic IGUs, see Heiskari et al (2023) and up to 39% for laminated IGUs, see Heiskari et al (2024). However, the percentage depends on the chosen classification society, window shape, design constraints, and level of idealization.

Kang et al. (2023) presented reliability-based design optimisation (RBDO) of a river-sea-going ship based on agent model technology. Figure 18 shows the three steps necessary for RBDO: agent model construction, reliability calculation, and optimisation design. In this study, the high-precision agent model for the limit state of ship hold structure is established based on agent model technology, including BP neural network, Radial Basis Function neural network, and Support Vector Machine combined with SMOTE oversampling algorithm. Furthermore, the reliability computation program is developed using Monte Carlo Simulation Method. The simulated annealing algorithm constructs the RBDO system to investigate the lightweight structure. The weight was reduced by 3.2% compared with the initial design scheme. The RBDO effect is not as good as the deterministic optimization design, but its failure probability is smaller.

The design of cruise ship bulkheads faces is a challenging balance of lightness, load-bearing capacity and fireproof performance. F. Zhang et al. (2022) proposed a lightweight design method based on the topography optimization theory, which can adaptively determine locations, numbers and topography of the bidirectional bead of sandwich bulkhead for cruise ships. Mathematical optimization formulations for bead topography design of sandwich bulkheads consider weight as an objective while strength and buckling criteria define structural constraints. A full-scale lightweight bidirectional bead steel sandwich bulkhead (the specimen) is selected and manufactured by comparing with the stiffened bulkhead. Then a load-bearing fire-resisting test of the specimen is carried out. The test results show a weight reduction of 20.56% for the new designed bulkhead compared with the conventional one and reached the fire-resisting divisions 60 class meeting the requirements of integrity, insulation and load-bearing capacity.

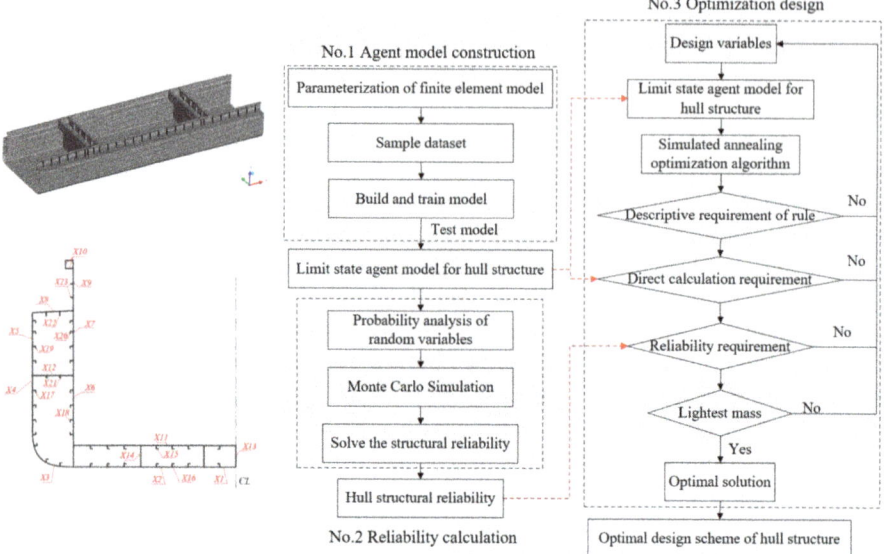

Fig. 18. Strategy of reliability-based design optimisation for hull structures, from Kang et al. (2024)

In the study made by Kendibilir and Kefal (2023), peridynamics topology optimization (PD-TO) framework is implemented to reduce the total weight of ship structures and increase their structural endurance against cracks. To analyse ship structures, both the optimality criteria (OC) algorithm and the proportional (PROP) approach are implemented into the PD-TO solver. As the initial design domain, a reference bulkhead geometry of trailing suction hopper dredger (ship) is modelled. This web frame is optimized according to critical wave/loading conditions such as hogging and sagging. Different volume fractions are investigated to obtain lighter designs based on OC and PROP methods. To prove the efficiency of the proposed method, maximum displacements and compliance of the reference and optimized structures are compared. The results prove that topologically optimized designs offered stiffer structures as compared to conventional designs under the same loading conditions. Overall, it is revealed that the PD-TO can be effectively utilized to design marine structures to achieve a higher strength/weight ratio.

Abedin et al (2024) proposed a two-stage optimization of ship hull structure combining the Fractional Factorial Design technique and Non-dominated Sorting Genetic Algorithm II (NSGA-II) algorithm focusing on a three-cargo hold model for a multipurpose cargo ship. To address this challenge, the study utilises FEMAP-integrated NX NASTRAN software to assess hull girder stress. Furthermore, a novel approach is introduced, integrating the Design of Experiments principles within Minitab 21.4.1 software to identify critical parameters affecting hull girder stress and production costs. This method determined the top five key parameters influencing hull girder stress while also highlighting key parameters that impact production costs. Ship design optimisation is then carried out by incorporating regression equations from Minitab software into the NSGA-II, which is managed using Python software (PyCharm Community Editon 2020.3.1). This optimisation process yields a significant 10% reduction in both ship weight and production costs compared to the previous design, achieved through adjustments in plate thickness, web frame positioning, and stiffener arrangement.

Li et al. (2024) presents a global design methodology for semi-submersible hulls supporting large floating wind turbines (FWTs), using a 10-MW turbine as a case study. The study outlines a step-by-step design process divided into three phases: concept selection, determination of global dimensions, and detailed hull design. The global design process focuses on determining optimal hull dimensions, such as column spacing and outer column radius, which significantly impact the structural response under wind and wave loads. The methodology includes quasi-static analyses to evaluate serviceability criteria like the static heeling angle and intact stability. The study emphasizes the importance of adjusting global parameters to improve intact stability, particularly through increasing outer column radius and freeboard. Additionally, wave-induced response amplitude operators (RAOs) for internal forces and moments are used to assess structural load effects. The findings indicate that critical wave periods can lead to splitting forces on the hull columns, making them a key factor in the structural design.

S. Wang et al. (2024) introduce a multi-objective optimization framework for the design of a 15 MW floating offshore wind turbine (FOWT) supported by a semisubmersible platform. The optimization, aimed at minimizing structural steel mass and enhancing motion performance, utilizes an evolutionary algorithm combined with frequency domain (FD) and equilibrium analysis. The FD simulation evaluates motion response amplitude operators (RAOs), focusing on surge, heave, and pitch under wind and wave loads, while equilibrium analysis assesses pitch heel angles to ensure stability. The optimization framework iteratively updates platform and mooring system parameters, adjusting ballast distribution, pontoon dimensions, and mooring configurations to achieve an optimal design. The results demonstrate a significant reduction in steel usage (up to 11.34%) and improvements in pitch motion performance (up to 20.63%), validated through time-domain simulations using OpenFAST. This method provides a viable approach for balancing cost and structural integrity in large-scale FOWT designs.

H. Zhang et al. (2022b) propose a novel optimization method for floating offshore wind turbine (FOWT) platforms, focusing on improving platform stability and reducing material costs. The study utilizes a fully coupled numerical model to account for the interaction between aerodynamic and hydrodynamic loads, combining design concepts from both semi-submersible and spar platforms. A novel platform with inclined side columns was designed based on the NREL 5 MW wind turbine, optimizing structural parameters such as column inclination, pontoon dimensions, and ballast distribution to enhance hydrodynamic performance. The study revealed that the novel platform demonstrates superior performance compared to traditional designs, reducing material costs by 12.8% compared to a semi-submersible platform while maintaining favourable dynamic responses under wave-induced loads. Additionally, the optimization process for the mooring system and additional structures (e.g., heave plates) further enhanced platform performance, reducing surge motion and cable tension. The novel approach offers a viable solution for deep-sea wind energy exploitation, combining economic feasibility with structural safety.

Kamel et al. (2022) present a modified hybrid method (MHM) for RBDO applied to offshore wind turbines (OWTs) considering soil-structure interaction (SSI). The study addresses the high cost and complexity of designing OWTs, focusing on the monopile foundation and incorporating dynamic analysis. The authors propose an extension of existing RBDO methods, such as the optimum safety factor (OSF) and robust hybrid method (RHM), to improve design reliability. Their MHM approach enhances the efficiency of the design process by reducing computational time while ensuring that the optimized design remains within the safe zone, addressing the limitations of earlier methods that could result in unsafe solutions. This methodology is validated through finite element modeling (FEM) and highlights its potential for reducing material costs while maintaining structural reliability under uncertain environmental conditions.

Stieng and Muskulus (2020) present a reliability-based design optimization (RBDO) framework for the support structures of offshore wind turbines (OWTs), particularly focusing on quasi-static load effects in combination with probabilistic constraints. The study integrates gradient-based optimization methods with analytical sensitivities to enhance the efficiency and accuracy of the design process under uncertain environmental conditions. The quasi-static analysis is applied to examine the load response without considering dynamic effects, and surrogate models are introduced to handle the probabilistic variation of loads. This methodology offers a balance between computational efficiency and accuracy by decoupling the reliability analysis from the design process, making it a viable solution for the optimization of OWT support structures, especially when subjected to extreme load conditions.

Kim and Kim (2020) present a structural analysis and design optimization of a conical concrete support structure (CCSS) for a 3-MW-class offshore wind turbine. The study incorporates fluid-soil-structure interaction (FSI) through an added mass method and a soil spring model to simulate the effects of water and soil surrounding the structure. A quasi-static analysis was used to evaluate the structural displacements and stresses of the CCSS under various load combinations, including self-weight, turbine loads, wave loads, current loads, and wind loads. The analysis showed that the maximum displacements and stresses were within allowable limits, confirming sufficient structural rigidity. To address high tensile stress in the CCSS, the study also implemented a post-tensioning design using prestressed steel, which successfully reduced the tensile stress to acceptable levels. Additionally, the CCSS was found to have favourable dynamic behaviour with its natural frequency falling outside the resonance range of the rotor and blade passing frequencies, ensuring dynamic stability under operational conditions.

3.4 Response Assessment Involving Complex Structures or Degradation Phenomena

3.4.1 Analysis of New Composites and Sandwich Panels

Wu et al. (2023) present a detailed investigation into the quasi-static mechanical properties of composite lattice sandwich structures with enhanced face panels. The study explores the out-of-plane compressive and shear behavior of sandwich structures with four different pyramidal configurations. Theoretical models are developed to predict the equivalent compressive and shear stiffness and strength, and 3D failure mechanism maps are generated to guide the design of the specimens. Experimental tests on these structures confirm that enhanced ribs and vertical struts significantly improve the adhesive bonding strength and load-bearing capacity. The study also highlights that stress concentration factors should be introduced for more accurate predictions of failure modes, such as debonding and buckling, particularly under shear loading.

Taghipoor and Sefidi (2023) conducted a study on the energy absorption behavior of foam-filled corrugated core sandwich panels under quasi-static compressive loading. The research aimed to investigate how different foam densities and core geometries influence energy absorption and crushing forces. Five different sandwich panels with trapezoidal corrugated cores were tested, both experimentally and through numerical simulations using ABAQUS. The results showed that the foam-filled corrugated cores demonstrated superior specific energy absorption (SEA), with bi-core sandwich panels without foam filling exhibiting a 70% better SEA than single-core foam-filled panels. Additionally, panels filled with high-density polyurethane foam showed the best performance in terms of energy absorption. The study highlighted that foam filling significantly improves the crashworthiness of sandwich panels, making them suitable for applications requiring high energy absorption, such as in the automotive and aerospace industries.

Mohaseb Karimlou et al. (2022) conducted an experimental study investigating the effect of clay nanoparticles on the strength of hybrid sandwich panels with polyurethane foam cores under quasi-static loading. The sandwich panels were composed of E-glass woven fabric, basalt fibers, and polyurethane foam of two different densities (40 kg/m^3 and 140 kg/m^3). The study explored the impact of varying clay nanoparticle concentrations (0.0%, 0.1%, 0.3%, and 0.5%) on the mechanical performance of the panels. The quasi-static tests revealed that increasing the percentage of clay nanoparticles improved the panels' resistance to the indenter force and enhanced their energy absorption capabilities. The polyurethane foam cores also played a significant role in preventing damage propagation, with higher-density foam (140 kg/m^3) showing superior performance. The study concluded that the combination of clay nanoparticles and foam density positively influences the overall crashworthiness and energy absorption capacity of the hybrid sandwich panels.

Liu et al. (2022) studied the structural response of U-type corrugated core sandwich panels, commonly used in ship structures, under quasi-static lateral compression loads. The investigation combined experimental, numerical, and analytical methods to evaluate the deformation and resistance characteristics of these panels. The results from quasi-static experiments and finite element simulations indicated that the deformation of the U-type corrugated core could be divided into two phases, with the core either in contact or not during the compression process. Despite the differences in deformation modes, the structural resistance trend remained similar throughout the tests. The study also developed an analytical formula to predict the deformation resistance of the sandwich panels, which showed good agreement with both experimental and numerical results. This research provides valuable insights into optimizing the design and anti-impact performance of corrugated core sandwich panels, particularly for use in marine engineering.

Liu et al. (2022) developed a high-fidelity finite element analysis (FEA) model to simulate and predict the quasi-static flexural behavior of composite sandwich structures with uniform- and graded-density foam cores. These structures, consisting of glass-fiber reinforced polymer (GFRP) skins and polyvinyl chloride (PVC) foam cores, were subjected to three-point bending tests to assess their mechanical performance under flexural loading. The FEA model incorporated both an elastic-plastic damage model for the composite skins and a crushable foam-core model to predict mechanical responses and damage evolution. Experimental results showed that sandwich panels with uniform-density cores exhibited higher load-bearing capacity, displacement, and energy absorption compared to those with graded-density cores. The study concluded that low-density foam layers in graded cores were particularly susceptible to crack initiation, leading to inferior performance in quasi-static flexural tests.

Yang et al. (2021) investigated the quasi-static and dynamic behaviors of sandwich panels with multilayer gradient lattice cores made of stainless steel. The study fabricated and tested two types of sandwich panels—non-gradient and gradient—using quasi-static compression and impact tests to evaluate their mechanical properties. Quasi-static tests showed that the gradient sandwich panels exhibited superior energy absorption and buckling strength compared to non-gradient panels, with failure progressing layer by layer in the gradient panels, following the sequence of relative density from low to high. The finite element model (FEM) used to simulate the quasi-static and impact tests showed good agreement with experimental results, confirming the accuracy of the model. The research concluded that the gradient configuration enhances the impact strength and energy absorption capacity, making it an optimal design for applications requiring high impact resistance.

Garrido et al. (2021) investigated the quasi-static indentation and low-velocity impact behavior of sandwich panels with glass-fiber reinforced polymer (GFRP) faces and low-density core materials, such as polyurethane (PUR), polyethylene terephthalate (PET) foam, and end-grain balsa (BAL). The study tested different indenter geometries and sizes to analyze the effects on stiffness, energy absorption, and damage initiation. The results showed that flat indenters generated higher loads and absorbed more energy than hemispherical ones, with BAL cores exhibiting the highest stiffness and PUR cores having the greatest energy absorption. The quasi-static indentation tests demonstrated similar behavior to low-velocity impact tests, suggesting that quasi-static tests could be a simpler method for predicting impact responses in such panels.

Taghizadeh et al. (2021) conducted an experimental and numerical investigation into the mechanical behavior of novel multi-layer sandwich panels with corrugated cores under quasi-static indentation loading. The study evaluated the effects of different corrugated core geometries, including rectangular, trapezoidal, and triangular shapes, on energy absorption, peak load, and damage mechanisms. The panels were fabricated using glass/epoxy composite materials, and quasi-static tests were performed using a cylindrical indenter to assess mechanical performance. Numerical simulations were carried out in ABAQUS, employing 3D failure criteria to model damage propagation. The results demonstrated that the multi-layer corrugated panels with rectangular cores exhibited the highest energy absorption and load-bearing capacity, primarily due to their superior structural configuration. The main damage mechanisms identified included matrix cracking, fiber breakage, delamination, and core-face sheet debonding, with the rectangular core panels outperforming the other geometries in overall mechanical properties.

Sayahlatifi et al. (2020) investigate the quasi-static behavior of hybrid corrugated composite/balsa core sandwich structures under four-point bending (FPB) through both experimental and numerical studies. The sandwich structure consists of aluminum face sheets, a balsa core, and a corrugated composite layer made of E-glass woven fabric. The study uses a digital image correlation (DIC) technique to monitor strain distribution and deflection patterns, while finite element modeling (FEM) with continuum damage mechanics (CDM) is employed to predict load-bearing capacity and failure mechanisms. The results show that the inclusion of the corrugated composite significantly improves the structure's weight-specific strength by 38.4%, although it has a negligible effect on weight-specific stiffness. The corrugated composite helps prevent catastrophic shear failure, and the study highlights the benefits of optimizing geometric parameters, such as the corrugation angle and thickness, to further improve mechanical performance.

3.4.2 Analysis Involving Corrosion and Other Degradation Effects

This section reviews recent developments in the consideration of corrosion effects in quasi-static approaches.

The basic and most commonly used way to capture the effect of corrosion in quasi-static analysis is still considering uniform thickness reduction. Such an approach is currently adopted in the Rules of Classification Societies. In this view, different works tend to present the corrosion loss in function of exploitation time of structures, considering the environment (e.g. ballast water, oil, etc.) and position of the structural element. Esfidan and Ranji (2024) proposed a corrosion model based on 66,480 data collected from thickness measurements from different classification societies. The data were sorted based on the position of the plate, ship age, country of build, and ship type, and regression formulas were derived. Further, the results were compared with other existing corrosion models, showing that in some places, they underestimated corrosion loss (e.g. in deck and inner bottom plating) and overestimated in other places. Thus, it is important to constantly update existing corrosion models in view of the newest inspection data. Another corrosion model for the inner bottom plating of the bulk carriers was proposed by Ivosevic and Kovac (2023). They found that corrosion usually starts after 7.5 years and then follows a 1.55% thickness reduction per year.

To predict the corrosion loss, some probabilistic tools were used too. Kim et al. (2022) developed the probabilistic corrosion wastage model by modifying the classical wastage model in view of Bayesian inference. Under such considerations, the wastage model could be updated based on the inspection data, and the uncertainty of estimation could be quantified. The Bayesian inference was also used in the work of Woloszyk and Garbatov (2024) for corrosion estimation of ship structural components. In this work, the corrosion loss in the particular structural component is quantified based on measurement data, in view of measurement error and dependent on a number of measurement points, leading to the probability distribution of possible corrosion loss. This could be further used in, for example, the reliability assessment of ship and offshore structures.

Yang et al. (2023) propose a novel methodology for assessing the yield strength of floater hulls for semi-submersible floating wind turbines throughout their entire lifecycle, with a particular focus on the effects of external and internal corrosion damage. The study presents improved corrosion damage models, considering the different marine corrosion zones, including atmospheric, splash, tidal, and immersion zones. These models are incorporated into a time-variant structural finite element model to simulate the corrosion-induced degradation of mechanical properties such as elasticity modulus and yield stress. The methodology assesses structural stresses under both normal and extreme sea conditions using quasi-static analysis and equivalent design wave methods. The results indicate that external and internal corrosion can significantly increase structural stresses and displacements, especially in high-stress regions like the splash zone. The study recommends increasing the material factor (γ_m) from 1.1 to 1.25–1.3 in high-stress areas to account for the accelerated risk of structural failure due to corrosion after prolonged service life.

Ryan and Mehmanparast (2023) developed a new approach for analyzing corrosion-fatigue in offshore steel structures, addressing limitations in existing models. The study reviewed existing corrosion-fatigue crack growth (CFCG) theories and models and proposed a time-dependent fracture mechanics parameter to better account for frequency effects on crack growth rates in corrosive environments. Using experimental data from fatigue crack growth tests on S355G10+M steel in both air and seawater, the authors identified the limitations of the traditional fracture mechanics parameter, ΔK, especially at low frequencies. A new CFCG model was developed that better predicts crack growth rates in seawater based on short-term test data in air. This model was validated using data from both S355G10+M and S355J2+N steels, demonstrating its applicability for improving life predictions for offshore wind turbine foundations.

Abyani et al. (2022) developed a finite element (FE) algorithm and machine learning models to predict the failure pressure of corroded offshore pipelines. The study used 3D FE modeling in ABAQUS to simulate the impact of internal corrosion on pipeline failure under pressure. To reduce computational cost, the researchers implemented various machine learning techniques such as Gaussian Process Regression (GPR) and Multilayer Perceptron (MLP). The models were trained using 1815 realizations of defect parameters, including defect depth, length, and water depth. The study concluded that the GPR and MLP models offered the most accurate predictions, with a strong correlation between predicted and simulated failure pressures. Additionally, the results showed that maximum von Mises stress (MVMS) increased with water depth at lower internal pressure levels, while it decreased at higher pressures due to the opposing effects of internal and external pressures.

Li et al. (2022) examined the impact of corrosion on the fatigue limit capacities of offshore wind turbine (OWT) substructures, particularly focusing on typical tubular joints subjected to wave and operational loads over a long-term service period. The study utilized a two-parameter Weibull distribution to model corrosion damage and applied the Miner linear fatigue accumulation criterion to assess fatigue damage at different stages (10, 20, and 30 years). The findings showed a significant decrease in fatigue capacity, particularly at critical joints such as the connection between pile foundations and braces, due to increasing corrosion over time. The study emphasizes the need to account for corrosion effects in the long-term design and safety assessment of OWT foundation structures to mitigate potential hazards.

Nataraj et al. (2021) present a simplified mechanics-based approach for assessing the seismic capacity of corroded reinforced concrete (RC) structures. The study focuses on developing a methodology to account for corrosion's impact on the residual strength and displacement capacity of corroded RC members. The researchers verified the proposed approach using a comprehensive database of experimental results, and the method was applied to a case study involving an older, severely corroded RC building in New Zealand. The findings show that long-term corrosion significantly reduces displacement capacity, potentially leading to collapse after 30 years without intervention. The methodology incorporates key factors such as reduction in bar cross-sections and bond degradation, making it a practical tool for seismic assessments in aging infrastructure. This approach enhances current seismic assessment guidelines by providing a mechanics-based alternative to regression models, allowing for more accurate predictions of corroded RC structure performance under seismic loading.

Melchers (2021) investigates the reliability of structures affected by pitting corrosion using probabilistic modeling techniques. The study focuses on modeling the maximum depth of corrosion pits, which are critical for assessing the long-term safety of infrastructure like water injection pipelines and oil production pipelines. The Gumbel extreme value distribution is traditionally used for modeling maximum pit depth, but this study finds that it may not be entirely appropriate due to non-linear trends in the data. Melchers demonstrates that pitting corrosion follows a stepwise progression, influenced by environmental factors such as oxygen availability, which varies over time. This stepwise behavior results in multi-modal probability distributions, which provide more accurate insights for predicting long-term deterioration. The research highlights the importance of considering piecewise trends and localized conditions when applying probabilistic models for structural reliability, particularly for offshore structures exposed to aggressive environmental conditions.

3.5 Advanced Numerical Methods

Iso-geometric analysis (IGA) is an innovative numerical approach that bridges the gap between CAD (Computer-Aided Design) and CAE (Computer-Aided Engineering) by using a unified representation—namely, splines—for both geometric modelling and analysis. The primary advantage of IGA is its use of B-splines and NURBS (Non-Uniform Rational B-Splines), which provide a smooth and precise representation of complex geometries while also approximating unknown variables in simulations. Unlike traditional finite element methods (FEM), which rely on polynomial approximations that may need additional refinement, IGA's use of splines improves geometry accuracy, supports efficient data exchange, and enhances computational stability (Cottrell et al., 2006).

In recent years, IGA has been increasingly applied in structural responses of offshore structures. Yildizdag et al. (2022) presented an iso-geometric finite element-boundary element method (FE-BE) to calculate fluid structure interaction effects for vibrating surface-piercing elastic structures or fully submerged structures vibrating close to the free surface. The calculations based on the proposed iso-geometric FE-BE approach show a good agreement with the predictions based on the three-dimensional hydroelasticity theory; see Fig. 19 for a comparison of the non-dimensional added mass and radiation damping in heave and sway, respectively. There is also potential to extend the method to determine the time-domain responses of floating bodies.

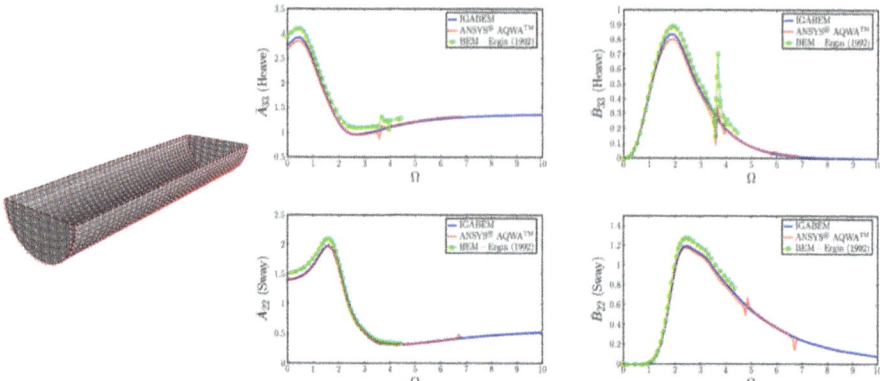

Fig. 19. Comparison of added mass and radiation damping for rigid body motions of a cylindrical shell on free surface, from Yildizdag et al. (2022)

To deal with complex ship hull structures that consist of a large number of plates and stiffeners, Yu et al. (2023) proposed to use the 3 degree-of-freedom Kirchhoff-Love shell elements to model the hull structures along with a coupling method (Nitsche-type) to connect patches. The method is shown to be robust and accurate for the cases of static bending and torsion loads. A key function of structural health monitoring of ships and offshore structures is its capability for real-time shape sensing, which involves reconstructing the complete field of structural displacements.

Dirik et al. (2024) introduced a new method for performing shape sensing of complex stiffened shell structures using a multi-patch iso-geometric inverse finite element method formulation. The aim is to accurately reconstruct the complex geometry of the structure without requiring a fine mesh and to achieve smoother shape sensing even when using fewer strain sensors. The proposed method is validated by solving problems involving simple plates, tee junctions, and partly clamped stiffened panels representing ship structures.

3.6 Concluding Remarks

This chapter provides advancements in the use of quasi-static approaches for evaluating the structural response of different kinds of ship and offshore structures. It is noted that although computational methods development and computer power are still growing, fully dynamic approaches are still rarely used, especially considering initial design and situations where inertia effects could be neglected. This is especially valid when quasi-static approaches are combined with optimization analysis, investigation of new sandwich structural arrangements and modelling of degradation effects (such as corrosion). In all of those cases, the quasi-static approaches simplify the problem and reduce computational time. Some efforts were also made to introduce more advanced numerical techniques (e.g. FEM-based simulations with isogeometric elements) for q-s approaches.

In the case of the application area, the major development is made for FOWT (floating offshore wind turbine) due to growing international attention to replacing classical means of energy production with renewable ones. In the case of ship structures, some progress is noted, too. For initial design and class approval, the quasi-static approaches are still mainly used, and this probably will not change suddenly due to their simplicity and good accuracy at the same time.

4 Structural Strength Assessment

This chapter examines applications of quasi-static loading and response methods in the evaluation of Strength Criteria which are utilized for assessment of failure conditions. The focus of topics presented here is the implementation of quasi-static methods as applied to areas including Ultimate Strength and Fatigue. These topic areas are mainly covered in other committees that focus primarily on the topics themselves (III.1 Ultimate Strength and III.2 Fatigue and Fracture), rather than the application of this specific method of solution. Among the myriad reasons for choosing to implement quasi-static methods in strength assessment are to allow for calculation of solutions with reduced cost where computing and personnel resources, and time are limited, and the accuracies of such solutions are acceptable given the scenario parameters. They also allow for parametric studies and optimization of strength of marine structures at the design phase, as well as to evaluate changes in expected loadings or mission type.

4.1 Buckling and Ultimate Strength

4.1.1 Stiffened Plates and Panels

Buckling and ultimate strength of stiffened plates and panels are critical considerations in the structural design of ships and offshore platforms. Subjected to complex loading conditions, such as axial compression, shear forces, and lateral pressure, their response is influenced by geometric configurations, material properties, and initial imperfections. Recent advancements in empirical modeling and data-driven approaches combined with nonlinear finite element analysis (NLFEM) are continuously improving the ability to predict buckling behavior and ultimate strength with higher accuracy and computational efficiency. These approaches may effectively incorporate factors such as slenderness ratios, imperfection amplitudes, and residual stresses, providing robust tools for assessing structural integrity. This subsection summarizes state-of-the-art methodologies and findings in this area.

Li et al. (2021) present an algorithm to predict load-shortening curves (LSC) across the full strain range for progressive collapse analysis of stiffened panels under longitudinal compression, extending existing empirical methods limited to ultimate compressive strength. The algorithm divides the compressive response into three phases: linear elastic, arc-shaped nonlinear collapse, and exponential post-collapse decay. It focuses on panels under uniform compression, excluding lateral pressure and biaxial loading and assuming elastic-perfectly plastic material behavior without hardening effects. Nonlinear finite element analysis (NLFEM) is used to calibrate empirical constants and ultimate strain for varying slenderness ratios. Initial imperfections, including local plate deflections, column-type bending, and stiffener sideway bending are modelled to capture their impact on elastic stiffness and structural weakening. The algorithm is validated against

NLFEM results, empirical methods, and benchmark studies, demonstrating its accuracy and applicability. The authors recommend expanding the NLFEM dataset, refining calibration parameters, and integrating advanced techniques like machine learning to enhance predictive capabilities, particularly for complex structural scenarios.

Kim et al. (2022a) derived a simplified empirical formula to predict the ultimate strength of plates with imperfections subjected to longitudinal compression based on two critical parameters: i) initial deflection coefficient and ii) plate slenderness ratio and four coefficients for curve-fitting. The methodology involves a parametric study based on NLFEM using 700 plate scenarios that account for geometric properties, material properties, and various levels of initial deflections represented by a buckling mode shape. Two types of steel are considered: i) Mild Steel (MS24) with a yield strength of 235 MPa and ii) High-Tensile Steel (HT32) with a yield strength of 315 MPa. The case study results show excellent agreement with NLFEM results, achieving an $R^2 = 0.98$–0.99, and demonstrate that the effect of initial deflection decreases for very small or very large plate slenderness ratios. The proposed empirical formula is shown to be reliable for predicting ultimate compressive strength and suggest the potential for extension to other conditions such as multi-axial loads, different boundary conditions and material properties.

Kim et al. (2022b) expand their methodology to address combined longitudinal compression and lateral pressure, analyzing 5,600 plate scenarios using the ALPS/ULSAP method. The study examines variations in geometric parameters (plate length, breadth, and thickness), material properties (yield strength: 235–355 MPa), initial deflection levels (seven severities), and lateral pressure (up to 0.2 MPa). An empirical formula is developed with four key coefficients representing relationships between ultimate strength and factors such as plate slenderness, initial deflection coefficient, lateral pressure and yield strength. These coefficients, expressed through sub-coefficients derived from regression analysis, achieved high accuracy ($R^2 > 0.99$). The results highlight the detrimental effects of lateral pressure and initial deflection on ultimate strength, particularly for thinner plates and lower-grade materials. The formula's computational efficiency makes it ideal for pre-FEED and conceptual design stages, providing rapid and accurate predictions.

Li et al. (2022) investigate the reduction in ultimate strength of stiffened panels caused by welding residual stress using a combination of nonlinear finite element modeling and data-driven methods (DDMs). A dataset of 136 NLFEM simulations, incorporating geometric variations, initial imperfections and residual stress distributions, were used to train four predictive models: Support Vector Regression (SVR), Random Forests (RF), Multilayer Perceptron (MLP) and Gaussian Process Regression (GPR). SVR, with a Gaussian kernel, excelled in capturing nonlinear relationships, while RF aggregated decision tree predictions. MLP used neural network techniques like adaptive learning and dropout regularization, and GPR offered both predictions and uncertainty quantification. SVR outperformed others, particularly in extrapolation tasks, achieving accuracy rates of $\approx 99\%$ in interpolation and $\approx 96\%$ in extrapolation scenarios. These models provide an efficient alternative to computationally intensive NLFEM simulations for predicting ultimate strength reductions, with potential for integration into structural codes. However, limitations include the computational effort required for dataset generation and the need for more comprehensive data to improve generalization.

Li et al. (2023) present an empirical formula for predicting the ultimate strength of continuous hull plates under combined biaxial compression and lateral pressure. Using 4,800 nonlinear finite element simulations, the study analyzes 2,800 scenarios for longitudinal ultimate strength (LUS) and 2,000 for transverse ultimate strength (TUS), considering variations in slenderness ratios, aspect ratios, and lateral pressures. Material behavior is modeled as elastic-perfectly plastic with a yield strength of 315 MPa, excluding strain hardening and residual stresses. The empirical formulae are derived using 16 primary coefficients, split equally between LUS and TUS calculations, with each coefficient expressed through sub-formulae optimized by the Levenberg-Marquardt algorithm. A sequential loading method is employed, applying lateral pressure and secondary compressive loads first, followed by primary loads until failure. This approach decouples the interaction between loads, improving computational efficiency and achieving high accuracy ($R = 0.993$ and $R = 0.996$). The results highlight the impact of lateral pressure, slenderness, and aspect ratios on ultimate strength, with thinner plates being more vulnerable. The study focus on the mid-ship bottom plate under biaxial compression and lateral pressure, emphasizing the need to consider shear force effects for side shells in ships with large deck openings. Future research is suggested to decouple interactions between longitudinal and shear strengths, enabling the development of explicit design formulas for such scenarios.

Z. Wang et al. (2023) develop an empirical formula to predict the ultimate strength of passenger ship side shells with openings under combined axial compression and shear loads. The study integrates experimental testing, nonlinear finite element modeling, and parametric analysis of 5,048 scenarios, varying parameters such as plate thickness, opening size, and reinforcement configurations. The finite element model incorporates elastic-plastic material properties and measured initial imperfections. The empirical formula is a polynomial equation with 41 coefficients divided into weakening and strengthening factors. The weakening factor captures the effects of opening size, shear force, and plate thickness, while the strengthening factor represents the influence of reinforcement stiffeners. These factors are expressed as polynomial functions of parameters like the height and breadth ratios of openings, normalized shear force, and normalized plate thickness. Derived through regression analysis, the formula achieves high accuracy ($R^2 = 0.94$–0.98) and shows that reinforcement stiffeners can mitigate the strength reductions caused by large openings, increasing load capacity by up to 60%. The study concludes with guidelines for using the formula in practical design but acknowledges limitations, including the need for expanded validation through additional scenarios to ensure broader applicability and experimental validation using isolated side shell test models.

Hanif et al. (2023) examine the ultimate strength of stiffened panels under uniform axial compression, focusing on uncertainties in geometric parameters. Key factors include plate slenderness relative to plating breadth, column slenderness tied to stiffener dimensions, web slenderness affecting web thickness and height, and span-to-bay ratio, which influences buckling behaviors. Initial imperfections such as local, torsional, and column buckling modes are also incorporated, with amplitudes ranging from 2.5% to 100% of plate thickness. Using data from 750 nonlinear finite element analysis simulations, the study derives a cubic polynomial empirical formula to predict normalized ultimate strength, achieving high correlation ($R^2 = 0.973$). Lower-order models were insufficient to capture complex interactions between geometric parameters, while higher-order models risked overfitting without improving accuracy. The formula demonstrates adaptability to geometric uncertainties and is validated against existing semi-empirical methods, offering accuracy and computational simplicity. However, it is tailored for T-bar stiffened panels under uniaxial compression, excluding other stiffener types and biaxial loading. Future research is suggested to expand its applicability to additional imperfection modes and stiffener configurations (e.g., flat-bar or bulb stiffeners).

Kim et al. (2024) present an empirical formula to predict the ultimate compressive strength of unstiffened cylindrically curved plates subjected to longitudinal compression. The derivation of the empirical formula employed a methodology starting with the collection and definition of 400 curved plate scenarios using ANSYS. These scenarios considered variation of parameters such as plate slenderness ratio (β), aspect ratio (a/b), and flank angle (θ). The numerical simulations assessed the plates' ultimate compressive strength under longitudinal compression, with initial imperfections and simply-supported boundary conditions considered for conservative results. Regression analysis was then employed to establish a mathematical relationship between normalized ultimate strength and the variables, leading to the formulation of a polynomial-based model. The derived formula was validated against experimental data, numerical simulations and existing empirical models. High coefficients of determination ($R^2 > 0.99$) indicated its accuracy in predicting ultimate strength of curved plates. The proposed formula is useful for early design phases and the limitations include the neglection of residual stresses and the need for further studies on small flank angles and aspect ratios to enhance the formula's generality.

S. Wang et al. (2024) explore the impact of transverse loading on the ultimate strength of stiffened panels under biaxial loads, addressing potential limitations of the IACS-UR-S35 rule. Using the large deflection equivalent orthotropic plate theory, they employ membrane stiffness from cross-sectional area ratios and bending stiffness from moments of inertia in longitudinal and transverse directions. Their results show that the IACS beam-column analogy overestimates bending moments under transverse-dominated loads, leading to conservative strength predictions. In general, stiffener yielding is mainly influenced by axial and bending stresses, with torsional stresses playing a minor role. Panels with flat-bar and T-bar stiffeners are more affected by torsional stresses, and higher transverse-to-longitudinal stress ratios increase bending stress contributions while reducing axial and torsional contributions. To enhance prediction accuracy, the study proposes a correction factor, expressed as a first-order polynomial, accounting for stress ratios and buckling utilization factors. Verification of the analytical model is conducted using nonlinear finite element models: detailed stiffened panels, equivalent orthotropic plates, and multi-span models for assessing boundary condition effects. The proposed modified rule formulation may reduce usage factors and aligns more closely with numerical results, accurately addressing failure modes like stiffener yielding and local plate buckling.

4.1.2 Hull Girder Ultimate Strength

Several research groups conduct experimental tests of hull girder ultimate strength and compare them with numerical/analytical results.

Zhao et al. (2022) present an investigation on the ultimate strength of a ship hull girder with relatively large deck openings by experiment, NLFEM and the Smith method. A model with relatively large openings is designed and tested up to the ultimate bearing capacity under a pure sagging bending moment. The nonlinear finite element method analyses the failure mode, the load-displacement curve and the ultimate load of the experimental model. The results of the numerical calculations are compared with the model experiment results and found to be in good agreement. The relative error between the numerical method and the experimental result is up to 3.9%. The study of this type of structure with two decks and openings in both decks provides the basis for the validation of numerical simulations and theoretical methods, extending earlier studies of this type made with simpler structural configurations.

Quispe et al. (2022) conducted a four-point bending test of a small-scale hull box girder. The dimensions of the box girder model are defined from geometric slenderness ratios representing a double bottom panel at the midship of a typical Suezmax tanker. Numerical FE models have been developed, where the friction coefficient between the model and its supports is determined based on experimental correlation. The numerical models with different initial imperfection distributions are considered to perform ultimate strength analyses. In general, this study clearly shows that all cases have an excellent numerical-experimental correlation except for the case with Smith's severe level of imperfection and numerical model without imperfections. Both experimental and numerical models showed that the upper panel of the box girder presented initial structural failure in the plating between stiffeners, which evolved to the stiffeners, following the expected buckling sequence of a reinforced panel.

Cui and Wang (2024) designed experiments and investigated the collapse modes and ultimate strength of corrugated bulkheads and plane bulkheads under lateral pressure. In the experiment, pure air pressure was used to apply uniformly distributed lateral pressure. The results show that the ultimate capacity of a corrugated bulkhead is significantly larger than that of a traditional T-shaped stiffened plane bulkhead under the same weight of materials. The corrugated bulkhead reaches the ultimate limit state after three plastic hinges formed on the corrugation, and experimental results agree well with the nonlinear finite element analysis. The results show that the discrepancy in the ultimate pressure of corrugated bulkhead under clamped and simply supported boundary conditions is very small. Only when the in-plane displacement is fully restrained at the boundaries the clamped supported boundary conditions have better robustness and matches well with the experimental results.

Y. Wang et al. (2024) proposed a new similarity method for box girders subjected to the combined load of bending moment and lateral pressure. Traditional similarity methods consider similarity criteria of cross-section properties, and modern similarity methods introduce some strength-related similarity criteria to consider the nonlinear collapse procedure. However, these two kinds of criteria may be conflicting and are hard to meet simultaneously. The proposed method considers only the strength-related similarity criteria, ignoring the criteria of cross-section properties. The column slenderness, the plate slenderness, and the torsional slenderness are selected as the similarity parameters for every stiffened plate in the box girder. The new similarity relations of the ultimate moment and the applied lateral pressure are derived according to the proposed criteria. A structured step-by-step scale model design procedure is proposed to design corresponding scale models. Four numerical examples are presented to show the correctness and advantages of the proposed method.

SMITH-Based Approach

The simplified progressive collapse method, Smith's method, for hull girders is one of the important methods for the rapid assessment of hull girder ultimate strength. Several studies proposed different improvements and extensions.

The simplified progressive collapse method is a commonly used iterative method for obtaining the accurate ultimate strength of ships. Since the accuracy of the neutral axis position directly affects the accuracy of the ultimate strength, the force equilibrium criterion and the force vector equilibrium criterion are adopted to search for the height and angle of the neutral axis, especially for damaged ships. However, the search for the neutral axis position based on the two criteria requires iterative computation, and it decreases the calculation efficiency. In Zhu et al. (2024), the relationship between the criterion results and the neutral axis position is studied, and it is found that the relationship is approximately linear. Then a new iterative method based on the linear equation is proposed to obtain the neutral axis position and it is adopted to improve the simplified progressive collapse method. Finally, the new method is used to calculate the damaged VLCC. The comparison of the ultimate strength results shows that the improved simplified progressive collapse method based on the linear equation has satisfying efficiency and accuracy.

Woloszyk et al. (2024) proposed a new methodology for the ultimate strength assessment of a ship hull, considering enhanced corrosion modeling. The approach is based on the classical Smith method. However, the recent findings regarding the impact of corrosion degradation on ultimate strength have been proposed. To this end, the stress-strain relationships for particular elements composing ship hull cross-section are modified using a specially developed correction factor. The proposed approach is validated with experimental results of the corroded box girders available in the literature, showing very good agreement. Further, a case study of a VLCC tanker ship is presented, and a comparison between contemporary and enhanced corrosion degradation modeling in terms of resulting ultimate strength is presented. The results indicate that the currently used method may significantly overestimate the hull's structure capacity, especially considering the long exploitation period. Thus, current approaches lead to a non-conservative assessment of the ship hull girder's ultimate strength, potentially increasing the risk of failure.

Cui et al. (2024a) investigate the ultimate strength assessment of hull girders based on Smith's method considering elastic shakedown. A novel average stress–average strain relationship for hull-stiffened plates considering the elastic shakedown limit state was developed by combining the nonlinear finite element method with beam-column theory. Furthermore, the obtained average stress–average strain relationship was introduced into Smith's method to evaluate the ultimate strengths of four hull girders and a scaled test model considering the elastic shakedown limit state. The results showed that the developed average stress–average strain relationship could accurately describe the elastic shakedown characteristics of hull-stiffened plates. Proposed extensions of Smith's method based on the above relationship could effectively evaluate hull girder ultimate strengths under an elastic shakedown limit state. Some limitations of the proposed approach have been underlined.

FEM-Based Approach

Cui et al. (2024b) studied the ultimate strength of hull girder considering the influence of initial imperfections under monotonic/cyclic bending moments which provides a more precise representation of the operational realities faced by vessels in service. When subjected to cyclic bending moments of varying amplitudes, the hull girder may either achieve an elastic shakedown condition or experience progressive plastic accumulation. The nonlinear finite element approach is employed to evaluate the ultimate strength of a bulk carrier girder subjected to monotonic/cyclic bending moments. The study conducts a detailed investigation into the elastic shakedown and plastic accumulation behaviors of the hull girder under various cyclic loading amplitudes. The findings reveal that the bulk carrier hull girder's ultimate strength will be weakened by the welding residual stress, and the extent of this weakening is most noticeable while the hull girder is sagging. At its maximum, the bulk carrier hull girder's ultimate strength is 8.22% less. When considering both the initial deformation and welding residual stresses, the bulk carrier hull girder exhibits the widest elastic shakedown interval. Under the influence of extreme cyclic bending moments, the bulk carrier hull girder ultimately fractures as a result of the ongoing accumulation of plastic deformation.

Jagite et al. (2022) analyzed the dynamic ultimate capacity of ultra-large container ships under realistic loading scenarios. The hull-girder is subjected to a combined bending moment, resulting from a long-term hydro-elastic analysis, with the lateral loads and to local loads given by different cargo loading cases. The numerical results are discussed, and the dynamic load factors are derived as the ratio between the dynamic capacity and the quasi-static one. For the considered 16 000 TEU container ship, the dynamic collapse effect obtained for simple half-sine pure bending moments on a three-frame bay model varies from 1.8% to 2.2%. It is shown that the strain rate effect is negligible in the analysis of the ultimate strength of ship structures subjected to whipping-induced stresses.

Putranto (2024) studied the ultimate strength of a box girder using an equivalent single layer (ESL) approach. The results of ESL were validated using FEM and experiments. Modeling of ESL includes a single plate with the same stiffness as the actual stiffened panel topology, as opposed to traditional FEM modeling and test specimen, where the stiffener is explicitly modeled. For ESL approach, VUGENS subroutine in Abaqus software is used to define the non-linear stiffness properties. Material and geometric non-linearities were considered during simulations. The explicit dynamic analysis is used to allow the large deformation and non-linear material behavior on stiffened panels. The effect of initial imperfection, residual stress, and stiffener slenderness on the ultimate strength was studied. According to the results, the ESL model achieves results similar to the traditional FEM model and experiment. In general, the maximum difference in the ultimate bending moment between ESL and FEM is 12%. One significant advantage of employing the ESL approach over traditional FEM is the substantial reduction in computational time.

Reyes-Casimiro et al. (2023) present a method for optimizing the design of stiffened panels in semi-submersible pontoons using a genetic algorithm (GA) approach. The optimization focuses on minimizing weight while ensuring structural integrity under combined biaxial and lateral pressure loads. The study integrates genetic algorithms with finite element method (FEM) simulations to evaluate the panels' ultimate strength, optimizing variables such as plate thickness and stiffener dimensions. The results show that Tee and angle stiffeners provide superior load-carrying capacity and stiffness compared to plate stiffeners, making them ideal for pontoon design. This approach significantly reduces computational time compared to traditional FEM-only methods, providing an efficient tool for offshore structural design.

4.2 Residual Strength

Residual strength of marine structure's components or system is a quantification of the structure's ability to support loading following a change in resistance relative to its original (as-built or as-designed) state. The change in strength can be attributed to structural fatigue, corrosion or other ageing processes, or due to an unintended overload events including collisions, weather, sea state, fire, explosion, or other means of excess loading. Residual strength evaluation is an important consideration in ships and offshore structures, with long expected life spans and variabilities in loading due to the aforementioned hazards that could be reasonably encountered at sea. Additionally, vessels and platforms often experience extensions of usage beyond their original design life, and beyond incidents of accidental overload, due to economic or other factors.

Despite the common usage of the corrosion model as uniform thickness loss in quasi-static approaches, there were works which tend to consider corrosion as a more complex phenomenon that has a more profound impact on the reduction of the strength of structural members. The evaluation of compressive properties of ship plating subjected to seawater corrosion based on 3D corrosion morphology was presented in Qiu et al. (2022). In this case, the series of nonlinear FE simulations of typical ship stiffened panels were conducted, considering corrosion as non-uniform thickness loss. A useful formulation was proposed, showing the relationship between the loss of bearing capacity of stiffened panels and the time of exploitation.

A key measure of residual strength of damaged vessels is the limit state function, which provides assessment of the longitudinal strength of the post-damaged ship condition under multi-axial bending loads. As shown in Fig. 20, Zhu et al. (2022) performed a comparison study of four fitting methods which are commonly utilized to approximate the limit state function. The methods were least squares including quadratic (LS-Q), cubic (LS-C), and nonlinear fitting (LS-N), moving least squares (MLS), radial bias function neural network (RBFNN), and the weighted piecewise (WP) fitting methods. They were evaluated on four typical closed-curve sample distributions including circular and vertical-, horizontal-, and diagonal-elongated elliptical shaped-distributions. Additionally, the authors calculated distributions on two damaged hull cross-sections: a single-hull bulk carrier and a double-hull oil tanker. The improved Smith Method was utilized for residual strength calculation, including neutral axis rotation accounting. The Smith method enjoys a computational efficiency advantage over finite element methods, and the improved method allows for some improvement of accuracy in asymmetric cross-section cases such as multi-axial bending, as shown in Fig. 20.

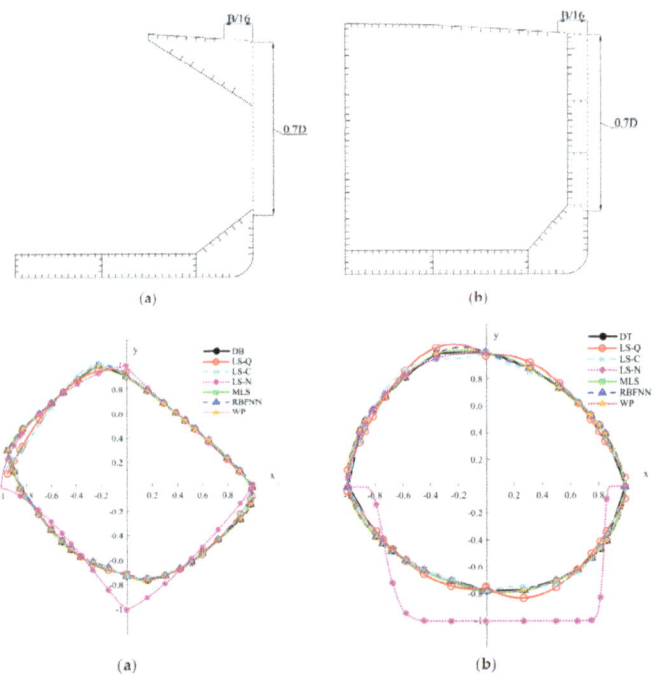

Fig. 20. Cross-sections of Damaged Bulk Carrier (DB, upper plot a) and Damaged Oil Tanker (DT, upper plot - b) and Fitting results of residual strength by different methods for DB (bottom plot - a) and DT (bottom plot -b), from Zhu et al. (2022)

It was noted by Zhu et al (2022) that sample distribution affected the fitting accuracy, especially when curvature changes occurred in a given quadrant, and while the least squares methods are able to fit all distributions, their accuracies are less satisfactory than the more complex MLS, RBFNN, and WP methods which delivered smooth and continuous fitment.

While the Smith method provides for reduced computational cost of residual strength analysis for simple damage cases, finite element analysis is particularly useful in the case of localized damage such as that caused by collisions. Residual strength of a post-ship-collision 5 MW Spar Floating Offshore Wind Turbine (FOWT) was evaluated by Ha and Kim (2022), as shown in Fig. 21. A shell finite element model of a circular tube section was constructed in ABAQUS commercial software, and collision analysis was validated with experimental results of a smaller-diameter cylinder impacted by a striker indenter. The validated collision damage software model was then applied to a collision analysis of the 5 MW FOWT spar structure with cases of athwartship collision of a

4000-ton FOWT maintenance ship with the spar at two different impact angles relative to the internal reinforcements.

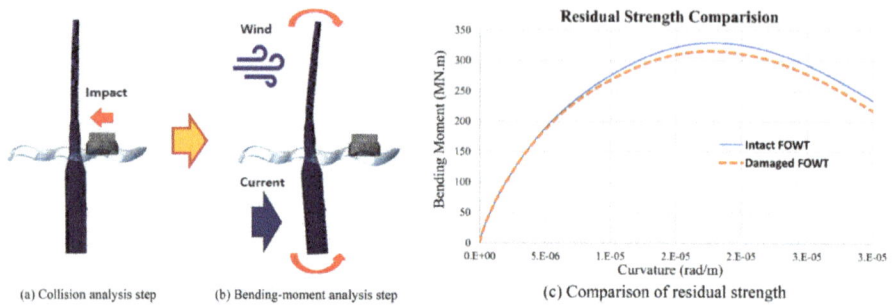

Fig. 21. 5 MW FOWT residual strength analysis, from Ha and Kim (2022)

The resultant damaged structure was evaluated for residual strength using wind and current loadings to create a moment-curvature relation for the damaged FOWT, which when compared to the undamaged FOWT calculation indicated a 4.1% reduction for the residual strength. These residual strength calculations could be utilized to assist FOWT operations and maintenance service providers determine the feasibility of safely transporting and/or repairing collision-damaged units.

The effects of local dent and fracture damage types on residual strength of box girders was investigated by Park et al. (2023), via experimental test data and corresponding numerical modelling, in Fig. 22. Denting damage conditions were created with a knife-edge striker, while a conical striker was used to facilitate fracture damage. Following validation of the numerical model of the damaged states with the test results, a four-point bending load test scenario was utilized to validate calculated residual strength ultimate moments. Strength reductions of ultimate bending moment between 1.6 and 12.9% were calculated for the dent-damaged cases versus undamaged, while fracture-damaged cases exhibited a 5.4–14.9% reduction in strength, and varying substantially with location of damage and damage-deformed shape.

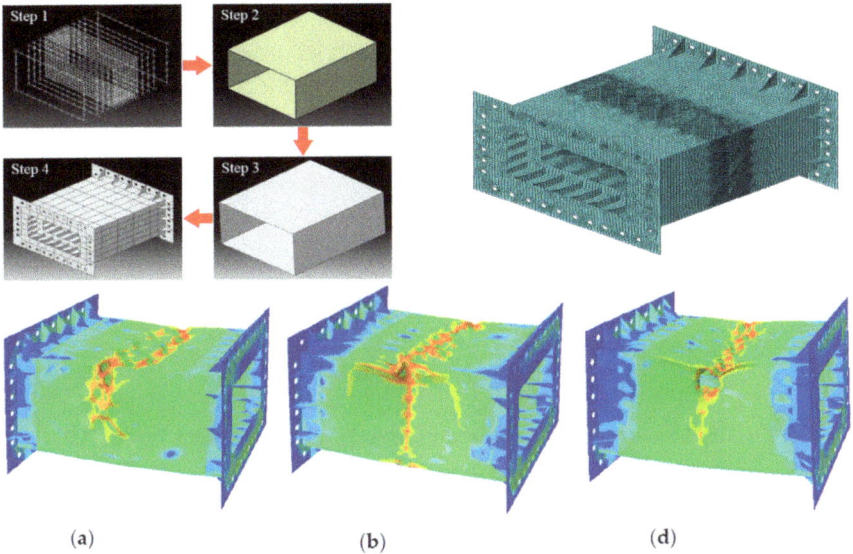

Fig. 22. Numerical models for (a) Undamaged (b) Dented (d) Fractured bending simulations, from Park et al. (2023)

Tools that allow for a rapid analysis of residual strength by leveraging prior results are of key importance to help organizations estimate remaining life of marine components. Gunn, Dimopoulos and Carra (2023) developed their Corroded Chain Analysis and Structural Evaluation (CCASE) tool to assess residual strength of corrosion- and wear-degraded mooring chains. By leveraging existing FE results from the Chain FEA Residual Strength Joint Industry Project (FEARS JIP), and combining them with new models of seabed, hawsepipe, pitting and interlink wear, the authors developed a response surface model (RSM) trained by the various degradation types modelled with FEA. The addition of future FEA modeling allows the RSM to be re-calibrated or re-trained to incorporate new wear scenarios. Users would input photogrammetry scans of degraded chain links to determine a cross-sectional area (CSA) reduction. Residual breaking strength is then rapidly calculated via the RSM fitting. When validating the CCASE tool, several model cases which were not included in the RSM fitment training were evaluated, resulting in predictions within 5% of both their actual tested breaking strength and FE-predicted strength.

Service-life is of paramount importance to ship operators and designers and requires estimation of residual strength due to various form of degradation. In Fig. 23, Tuyen (2023) conducted a study of the effects of initial imperfections and corrosion wastage on residual hull girder ultimate strength, with an example scenario of an aged single-hull bulk carrier of size 56,000 DWT. Initial imperfections, which could represent machining and fabrication anomalies, or the results of residual welding stresses, were modelled as variations in flatness of panels or straightness of stiffeners. Corrosion wastage was quantified as thickness reduction based on variable rates of corrosion for individual structural members as a function of years of service. An incremental-iterative method was employed by the authors for eleven cases including individual cases of slight, average and severe imperfections, average and severe corrosion states, and combinations of the imperfections and corrosion states. Ultimate bending moment (UBM) reductions were quantified versus initial imperfection cases, and ship age for corrosion cases.

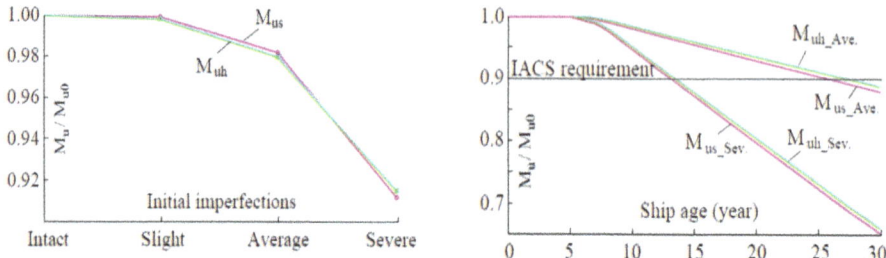

Fig. 23. UBM reductions for imperfection and corrosion cases, from Tuyen (2023)

Corrosion is also a critical source of structural degradation for ring-stiffened cylindrical structures, including semi-submersible platform legs, submarine pressure hulls, and FOWT spars. Park et al. (2024) presented a study of the effects of corrosion damage on hydrostatic collapse of cylindrical test articles and corresponding numerical model simulations. Test articles of varying length, radius, and shell thickness were constructed, both intact and with simulated corrosion via machined 13–25% thickness reductions of discrete areas of shells. The authors observed a reduction in collapse pressure for the shorter and longer articles of 6.3% and 12.1%, respectively. Furthermore, local and overall interactive buckling and stiffener tripping were observed as failure modes in the shorter samples, while overall buckling governed the failure of the longer samples. The numerical FEA modeling of various cases exhibited good correlation with less than 3% differential.

As shown in Fig. 24, a trend analysis indicates that for sample group RSC-CD-2, a reduction in volume due to corrosion areas of 2% can result in a residual strength reduction of 30%. The authors' findings illustrate, via their parametric study, a need for development of a simplified model for quantifying the significant effects of corrosion damage on collapse strength of pressure hull structures.

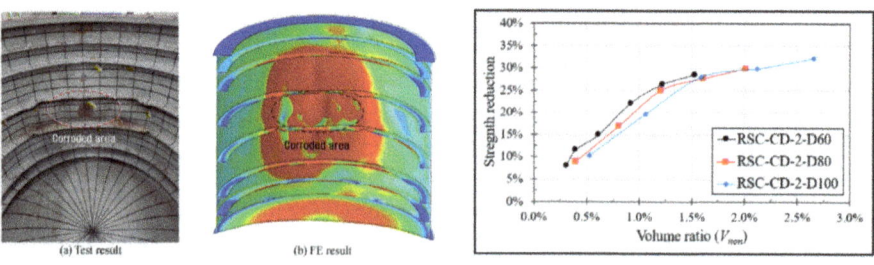

Fig. 24. Test, FEA and residual strength trends for corrosion simulated samples in group RSC-CD-2, from Park et al. (2024)

Explosive events aboard ship, due to both accidental and adversarial sources, are of concern to the strength of the structure post-explosion. Wu et al. (2024) propose the Explosion Smith Method (ESM) as a simplified means of calculating post-explosion residual strength. The location effects of the explosion were studied via 16 discrete cases in a compartmented ship model, simulated by the ESM and compared with FEA results. The ESM utilizes as a basis the Smith method for ultimate strength calculation, with the addition of an explosion damage range calculation that classifies structural compartments and components as either intact, deformed, or broken. The deformed components are assigned a zero stiffness in compression, and an elastic-perfectly plastic behavior in tension. The ESM provided result trends consistent with the finite element simulations for residual strength, with slightly conservative values when compared to FEA of the same cases, and at a reduced computational cost. The authors also noted that the more severe explosive damage cases occurred in compartments inboard and just below the main deck, resulting in more broken and deformed components at the locations of maximum component moment of inertia contribution. Less severe explosion cases were for outboard compartments closer to the neutral axis, where quantity of damage was limited due to outward discharge of explosive energy at the hull edge, and damaged component moment of inertia contributions were minimal.

4.3 Fatigue Strength Analysis

Shamir et al (2023) investigate Offshore wind turbine support structures. These structures connect the wind turbine transition piece and/ or tower to the seabed, are located below the sea level and are in direct contact with seawater during the entire lifespan; therefore, they are highly susceptible to corrosion damage and cracking. In particular, the pitting corrosion is very crucial in these support structures, as it leads to local stress concentrations and thus affects the fatigue life. Although corrosion protection mechanisms are commonly implemented in offshore wind turbines, they have a finite life and therefore corrosion damage cannot be completely avoided during the entire life cycle, and this can lead to pitting corrosion on the steel surface. The investigation of the impact of pitting corrosion on fatigue durability of steel structures is made performing tests on lab-scale coupons, made of S355 structural steel which is widely employed in fabrication of offshore wind support structures. For this purpose, cross-weld uniaxial samples were initially exposed to seawater for different time durations and then tested under cyclic loading condition. The influence of pitting corrosion on the S–N fatigue life of S355 cross-weld specimens was experimentally determined and a predictive fracture mechanics-based model was developed to carry out the durability analysis. Moreover, the influence of seawater exposure time on the fatigue life of the material was critically evaluated at different stress levels. The fatigue samples were initially immersed in artificial seawater for different durations of 0, 2 and 4 months and subsequently tested in air at different stress levels with the load ratio of 0.1 and frequency of 20 Hz. The results from this study show that pitting corrosion significantly decreases the fatigue life of the material and for a given value of stress range the level of fatigue life reduction increases by increasing the seawater exposure time. Moreover, the lines of best fit made to the S–N fatigue data show that similar slopes are obtained for corroded and uncorroded data sets implying that the indicative fatigue endurance limit would continuously decrease as the seawater exposure time and population of corrosion pits increase in the material. The analysis of the corroded samples showed that when Log(N) data is plotted against the seawater exposure time, a linear trend can be observed at high stress levels which will gradually transform into a non-linear trend at lower stress levels. Finally, it was demonstrated that the durability of the corroded material can be predicted in the form of mean S–N curve using the fracture mechanics approach by employing the modified Hartman-Schijve equation and the equivalent initial flaw size (EIFS) as the initial crack length.

Kristiansen et al. (2023) investigate offshore foundations subject to fatigue due to wind and wave loads during their lifetime. Post-weld treatment is a group of techniques well known to improve the fatigue life. This study investigates how this is applicable for large structures, i.e. a 2.4 m jacket T-node, as shown in Fig. 25. A large-scale empirical study was performed and results were obtained from a SN-diagram based on three repetitions and supported by a theoretical background. The as welded samples performed as expected and on the safe side of the category 56 line. Unexpectedly, the burr grinded samples showed no improvement and had an evidently larger variance. Both ultrasonic impact treatment and especially pneumatic impact treatment showed improvements, but did not reach the expected category 125 line. The results were substantiated with strain-gauge measurements for hot spot analysis. The obtained results on fatigue life improvements give rise to apply post weld treatment in an industrial scale for offshore construction. However, it also brings up the discussion of the necessity to redesign structures in order to automate the treatment in a cost-efficient manner.

Fig. 25. The sketch and dimensions in millimeters of the T-nodes used in large-scale tests, from Kristiansen et al. (2023)

This experimental test series show that the fatigue life can be improved by using PWT (Post Weld Treatment) techniques. The results from BG (Burr Grinding), however, are not as convincing as expected. This technique mainly removes the notch effect in the weld toe area, whereas PIT (Pneumatic Impact Treatment) and UIT (Ultrasonic Impact Treatment) also work in the weld toe area. Besides removing the notch effect, these methods also produce compressive stresses in the material in the weld toe area. The UIT and especially the PIT technique showed significant fatigue improvements, but also a lot more scatter in the experimental results.

The FE simulations were in good agreement with the published experimental results, where the AW (As Welded) weld seam is dominated by tension in the range of 300 to 400 MPa and the impact treated weld toe is dominated by compression in the range of − 300 to − 600MPa. The high compressive stresses in the impact treated region is only possible due to strain hardening, where the yield stress is increased from approximately 400 to 800 MPa. PWT could benefit from automation to improve the quality stability and assurance. Automation of PWT would further lead to reduced costs but would require a redesign of the current jacket structure.

Xing et al (2023) investigated that in typical welded ship structures, the fillet welded connections or often referred to as fillet welded joints for connecting secondary to main structures, are very common and about 75% in the percentage of all the welds. Based on whether the loads transfer through welds, fillet welded joints are typically categorized into two different types i.e. load-carrying and non-load carrying. Weld root fatigue cracking has always been a concern in the design and analysis of welded ship structures since the fatigue lives of weld root cracking tend to be significantly lower than that of weld toe cracking. Aluminum 6082 sheets typical for marine structures was selected as the base plate, and two kinds of welding wires AL5183 and AL4943 were included for study of welding wire effects. As shown in Fig. 26, comprehensive fatigue testing had been conducted using load-carrying aluminum cruciform joints with different fillet weld sizes, base plate thicknesses, and materials of weld wires. In addition, the fatigue life improvement technique i.e., UIP (Ultrasonic impact peening) was also investigated. The fatigue failure mode transition behavior of aluminum fillet welded connections had been analyzed systematically by taking advantage of the traction-based weld throat stress solutions and EETS-based (Fatigue failure mode transition) FFMT (Fatigue failure mode transition) criterion. It was found that the test data from shop floor aluminum weldments exhibit a wider scatter band than that seen in steel ones, indicating the importance of relative fillet size over plate thickness (i.e., s/t) in controlling fatigue failure mode transition.

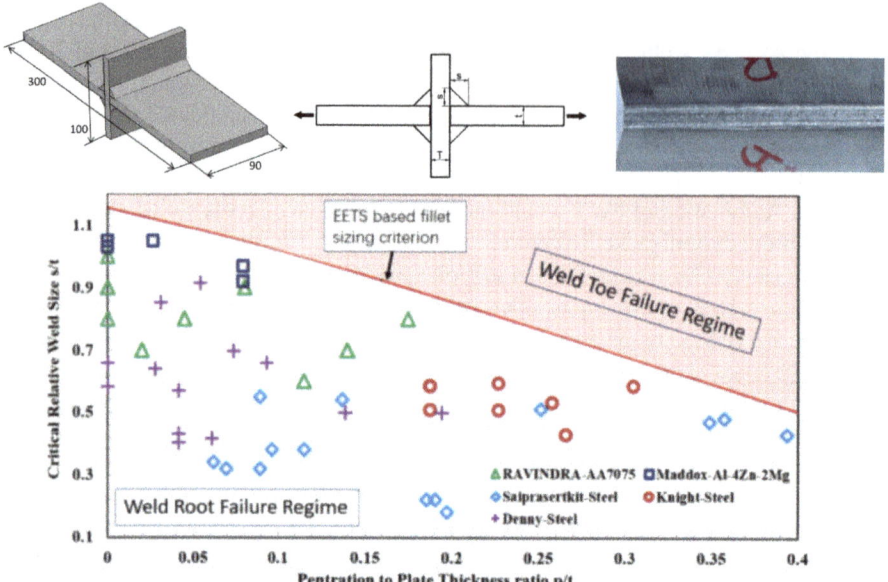

Fig. 26. Geometry, weld and results of cruciform aluminum specimen fatigue tests, from Xing et al. (2023)

5 Uncertainty in Use of Quasi-Static Approaches

5.1 Overview

This chapter mainly introduces the uncertainty assessment when using the quasi-static approaches. Typically, to quantify the accuracy (or the uncertainty) of a method/model is to compare it with more advanced models (for example dynamic approaches) or the direct measurements. In this sense, the literatures reviewed here focus on the uncertainty assessment and differs from those reviewed in the previous chapters.

5.2 Uncertainties in Theoretical and Numerical Models for Environmental Conditions

Most ships and offshore structures are affected by environmental loads, making this a critical factor to consider during the design and operation process. Theoretical and numerical environmental models including both long-term and short-term scenarios, inherently contain uncertainties. These uncertainties stem from various sources, including data set, model assumptions, simplifications, and parameter estimations. Long-term models refer to statistical distributions of major environmental parameters, while short-term models refer to either forecasting of environmental parameters in hours or days or phase-resolved environmental conditions. Forecasting with extremely short periods of seconds or minutes rely on the direct measurement of surrounding environmental conditions and the propagation theory and will not be discussed here.

5.2.1 Long-Term Environmental Models

Long-term environmental models describe environmental conditions over extended periods, often spanning years or decades. They are used to estimate the statistical properties of the environmental conditions and the likelihood of extreme environmental conditions. The long-term models help in understanding climate variability, long-term trends, and extreme events. It is crucial for assessing the long-term safety and durability of marine structures. The inherent unpredictability of climate systems—including wind, waves, currents, ice, and earthquakes—along with their extensive time and spatial frames, presents a significant challenge for establishing long-term environment models. While historical data can serve as a reference for validating model performance, the ability to forecast accurately over long durations is inevitably accompanied by uncertainty.

Formulating a long-term wave model, for example, involves gathering long-term wave data, analyzing the statistical properties of waves, selecting appropriate statistical models to predict the occurrence of rare and unusually large waves that can pose significant risks to marine structures. The Weibull distribution is often used for significant wave heights and the Gumbel distribution is used for extreme wave height, and they jointly formulate the common statistical distributions for extreme wave events. The Peak-Over-Threshold (POT) method focuses on waves exceeding a certain threshold (e.g., top 5% of significant wave heights) and uses these excesses to model extreme events. This approach is better suited for fitting the tails of the distribution, which is important for extreme event prediction. This approach however requires sufficient wave data exceeding the threshold to ensure a reliable statistical model for extremes. Initial Distribution Method (IDM) uses all available wave data to estimate the distribution of wave heights, periods and extreme wave events. This method may struggle to accurately represent the tail of the distribution, which is crucial when predicting rare, extreme events (e.g., 100-year or 1000-year waves), leading to under- or overestimation of extreme wave heights, which are critical in engineering design for safety.

Several key long-term numerical wave models are widely used for predicting wave behaviour and studying wave climate over extended periods. ERA5 (ECMWF Reanalysis 5) is a reanalysis dataset produced by ECMWF (European Centre for Medium-Range Weather Forecasts) that provides global climate and weather data from 1950 to the present. It includes wind and wave parameters at high temporal and spatial resolution. ERA5 is highly reliable for long term wave studies, extreme wave predictions and climate change assessments. Other models include: WAVEWATCH III, WAM, IOWAGA, HIPOCAS.

Amarouche et al. (2023) examines the accuracy and variability of extreme wave height estimates using different data sources and time periods. The study uses four primary data sources: ECMWF Reanalysis v5 (ERA5), SWAN-JRA55 wave hindcast, satellite altimeter observations, and long-term wave buoy measurements. Two statistical models - Annual Maxima (AM) fitted to the Generalized Extreme Value (GEV) distribution and Peak-Over-Threshold (POT) fitted to the Generalized Pareto Distribution (GPD) - are used to estimate extreme significant wave heights (SWH). The authors find that extreme SWH estimates are highly sensitive to both the data source and the analysed time period, with relative differences exceeding 20% in many parts of the world. Longer data periods reduce variability in extreme wave estimates, but significant differences of up to 30% remain, especially with the AM-GEV method. In some cases, extreme wave heights were underestimated by up to 2 m for a 100-year return period in regions like the North-West Atlantic and North-East Pacific. The paper emphasizes the need for careful consideration of data sources and period lengths when estimating global extreme wave conditions for marine structure design and coastal engineering.

Kodaira et al. (2023) investigates the uncertainty in wave hindcast programs used for ship design, focusing on significant wave height (Hs) and mean wave period (Tm02). The authors compare several wave hindcast products - TodaiWW3-NK, ERA5, and IOWAGA (using both CFSR and ERA5 wind fields) - with wave buoy observations from the North Atlantic, evaluating their performance under extreme wave conditions. The study highlights that while all models show good correlation with observational data (correlation coefficients exceeding 0.9), their accuracy varies significantly during high sea states (Hs > 10 m). The analysis also examines the sensitivity of Tm02 to frequency range selection, revealing that reducing the upper frequency limit from 1.0 Hz to 0.4 Hz causes a 10% change in mean values, with less sensitivity in higher wave heights. The study's findings emphasize that different models yield variations in the joint probability of significant wave height and mean wave period under extreme conditions, affecting the estimation of extreme wave loads on ships. The authors suggests that further calibration of numerical models using observational data could improve the reliability of wave hindcasts for extreme wave conditions.

Bitner-Gregersen et al. (2022) provides a comprehensive analysis of the uncertainties associated with wave data and models used in the design and operation of marine structures. Uncertainties are categorized into two types: aleatory (inherent) and epistemic (knowledge-based). For waves, epistemic uncertainty is further divided into data uncertainty, statistical uncertainty, model uncertainty, and uncertainty due to natural climatic variability. The paper focuses on discussing the challenges and limitations of wave data sources, including visual observations, buoy measurements, numerical models, and satellite data, particularly in terms of accuracy and variability. It also addresses the uncertainties in long-term statistical descriptions of wave climate, emphasizing the role of long-term wave data and probabilistic models in marine structure design. The paper calls for further research into uncertainty quantification to enhance these models and methods.

The uncertainties of the long-term wave description lie in the accuracy of applied wave data sets, the rationality of the selected distribution function of main wave parameters and join distributions of wind and wave; accuracy of the distribution parameters; and the influence of climate change on the long-term wave description as stated in Bitner-Gregersen et al. (2022).

5.2.2 Short-Term Environmental Models

In contrast, a short-term wave model is used to predict wave conditions over relatively short time periods, typically ranging from hours to a few days. These models are primarily used for real-time or near-future forecasting, allowing marine operators to make immediate decisions regarding navigation, offshore operations, and safety. Short-term wave models simulate the ocean's surface wave field based on current meteorological conditions, particularly wind speed and direction.

Despite the difficulties in accurately predicting rapid changes or extreme sea events, short-term environmental models can capture the statistical characteristic of the environmental condition. Short-term wave models often provide wave data with high temporal and spatial resolution, making them suitable for localized regions or specific operational tasks. For instance, this type of model may predict significant wave height, spectral wave period, and direction every 1 to 3 h at specific locations. Examples of short-term wave models include: WAVEWATCH III (NOAA), SWAN (Simulating Waves Nearshore), WAM (wave model) and ECMWF Wave Forecasts.

van Essen et al. (2023) investigates the convergence of short-term wave impact extreme value statistics. The study focuses on understanding how long experiments need to be conducted to achieve reliable statistical estimates of wave crest heights and green water impact forces on a sailing ferry. Due to the stochastic nature of ocean waves, estimating extreme values for design and reliability purposes requires tests of sufficient duration to minimize uncertainty. The authors perform long-duration experiments and analyse various test durations, such as 3 h, 1 h, and 30 min, using multiple wave seeds (realizations of sea states), to explore the number of seeds required for the convergence of extreme value predictions and how fitting procedures can impact the results. The paper also highlights the variability in wave-induced loads and stresses the need for proper experimental design to reduce uncertainty in maritime structure assessments.

Kordatos et al. (2024) presents a machine learning framework for predicting significant wave heights (SWH) in nested fine-resolution domains. The study employs radial basis function neural networks (RBFNNs) combined with 2D geospatial interpolation techniques to reduce computational intensity while maintaining high accuracy. The authors first use the WAM model to generate ocean wave data in both coarse and dense grids. The RBFNN, trained using the fuzzy means algorithm, approximates the high-resolution SWH based on low-resolution grid data, wind, and bathymetry inputs. The framework's performance is tested in two Mediterranean case study sites, demonstrating high accuracy. This framework offers a computationally efficient alternative to traditional numerical models while still providing reliable predictions for real-time applications in ocean engineering. The study suggests that integrating machine learning with numerical wave models can improve both accuracy and efficiency in wave forecasting.

Xu et al. (2024) proposes an efficient and precise hybrid method for predicting significant wave height (SWH) in the short term. The approach uses wavelet decomposition (WD) to break down the original wave signal into different components, each of which is forecasted separately using appropriate models. The study uses real wave data from three measurement stations, with 80% of the data used for training and 20% for testing. The hybrid model integrates multiple machine learning techniques, including extreme learning machines (ELM), nonlinear autoregressive networks (NARN), and backpropagation neural networks (BPNN), to improve prediction accuracy. The proposed method is evaluated against other benchmark models for one-hour and multi-hour forecasting. The results show that the hybrid model achieves high accuracy, with correlation coefficients greater than 0.99 for one- and three-step predictions and over 0.96 for five- and eight-step forecasts.

Wu et al. (2024) introduces a hybrid model integrating the Rime Optimization Algorithm (RIME), Convolutional Neural Networks (CNN), and Bidirectional Long Short-Term Memory (BiLSTM) for significant wave height (SWH) prediction. The model is designed to improve the accuracy of short- and long-term SWH forecasting at three buoy stations in the Gulf of Mexico. The RIME algorithm optimizes CNN and BiLSTM parameters to capture nonlinear ocean dynamics more effectively. The study compares the hybrid model to other benchmark models, demonstrating improved performance, particularly in high-energy sea states and outlier predictions. Wu and Gao (2021) investigated the effect of weather forecast uncertainty on the responses of offshore structures and found that allowable sea states of blade installation will be overestimated if the sea state forecast uncertainty is not considered.

Cheynet et al. (2024) analyzed 40 years of hindcast numerical data from two offshore Norwegian areas, revealing significant spatial and temporal variations in environmental parameters that impact the design, operation, and maintenance of offshore wind farms. Using the NORA3 dataset, which provides high-resolution hindcast data from 1982 to 2022, the study examines wind and wave conditions essential for offshore wind farm development. The paper explores wind speed profiles, extreme wave heights, and wind-wave misalignment. The study also discusses joint distribution models of wind and wave parameters, emphasizing their importance in optimizing wind turbine design, layout, and maintenance operations.

5.2.3 Field Measurements and Comparison

In the past, environmental data like wave data were typically gathered through observations by sailors and ship officers, leading to the development of Global Wave Statistics (GWS) atlases. While these datasets offer a long-term record of wave conditions, they are prone to have biases and inaccuracies, especially in remote ocean regions. In contrast, wave buoys and radars deployed at sea provide direct measurements of wave height, period, and direction, commonly used at fixed coastal locations and offshore platforms. However, their global coverage is limited, and accuracy can vary based on the technology used and the maintenance of the instruments. Satellites equipped with altimeters, scatterometers, and synthetic aperture radar (SAR) provide global wave data, capturing sea surface heights and wind speeds. While satellite data offers comprehensive coverage and is invaluable for monitoring long-term trends and extreme events, it is limited by spatial and temporal resolution, making it less effective for short-term variability. Global hindcast and reanalysis datasets, such as those from WAVEWATCH III, ERA5, and WAM, simulate wave conditions based on atmospheric inputs like wind fields, providing continuous global coverage at varying resolutions. These models are validated against in-situ and satellite data to ensure accuracy.

Yu et al. (2023) conducted a comprehensive review of advanced technologies using deep learning algorithms for fluid motion measurement and estimation, particularly focusing on fluid motion estimation and high-resolution reconstruction of velocity fields. It categorizes the discussion into two main areas: fluid motion estimation and velocity field Super-Resolution reconstruction. Initially, the principles of deep learning and its application in fluid motion estimation, emphasizing particle image velocimetry (PIV) algorithms are outlined. The benefits of deep learning over traditional methods are discussed, such as improved accuracy and efficiency, while acknowledging ongoing challenges like model robustness and the need for better temporal resolution. The shift from 2D to 3D fluid motion estimation and the importance of integrating physical knowledge into neural network designs are also discussed.

McCauley et al. (2024) examines the accuracy of wave measurements from a Radac wave radar installed on the TetraSpar floating wind turbine. The study compares the wave radar data with measurements from a nearby Fugro Midi wave buoy and model forecasts from MET Norway. Overall, the wave radar was found to provide reliable wave height, period, and direction estimates, with a good match to buoy measurements and model forecasts. However, some discrepancies were noted in the directional spreading measurements and high-frequency components of the wave spectra due to wave diffraction by the structure. The study also identified occasional spurious large wave crest measurements, likely caused by sea spray interacting with the radar during specific wave conditions. These spurious measurements were filtered by the radar's software, ensuring the accuracy of bulk wave parameters.

Ølberg et al. (2024) reviews the design and signal processing of a ship-borne ultrasonic altimeter system used to measure sea surface elevation. The system integrates a downward-facing ultrasonic altimeter with an inertial measurement unit (IMU) to compensate for the ship's motion. During a 20-month One Ocean Expedition, the system continuously recorded data, with results from a one-month crossing of the Tropical Atlantic highlighted. The study presents a one-dimensional wave spectrum and related wave parameters derived from the sea surface elevation time series, showing significant wave heights that align well with satellite altimetry and spectral wave models. Doppler shift corrections were applied to improve wave period estimates, enhancing the accuracy of mean periods but not peak periods. Future improvements may include directional wave measurements using marine X-band radar. The findings emphasize the utility of the ship-mounted system for real-time data broadcasting and validating satellite observations, while advocating for a collaborative approach in sharing sensor technology within the scientific community to enhance in situ observations.

5.3 Environmental Load Modelling and Their Uncertainties

Wave, current and wind loads applied on marine structures can be treated as quasi-static load when the time scale of the load changes is much longer than the time scale of the structure's dynamic response. In such a case, the dynamic effects like inertia and damping are not very important. The marine structure is assumed to be in a state of static equilibrium at every time instant. The complex dynamic problem is simplified and broken down into a series of static equilibrium problems along the time. Slowly varying wave/wind load and current load can typically be taken as quasi-static load.

Modeling uncertainties in wave loads are essential for formulating rules governing their impact, facilitating efficient structural reliability analysis, and supporting decision-making tools for structural health monitoring and ship digital twins. Parunov et al. (2022a) provided a literature review on modelling wave load uncertainties, which presents general concepts and modelling of the uncertainty in linear and non-linear low-frequency wave-induced loads. The paper categorizes wave load uncertainties into those of wave loads calculated under linear assumptions and those of nonlinear effects, including slamming, sloshing, green water loads, and more. It first investigates the uncertainties in the linear wave load transfer functions, which connect wave characteristics to ship responses and loads. It explores how different modelling methods, such as strip theory and three-dimensional (3D) panel techniques, introduce uncertainties due to factors like numerical approximations and hull geometry. The influence of ship speed, heading, and sea conditions on wave load predictions is also analysed. The paper also lists several studies on uncertainties related to vessel vertical bending moments and the differences between sagging and hogging. The paper emphasizes that understanding and modelling wave load uncertainties are essential for ship design, risk-based decision support systems, and structural reliability analysis. The authors advocate for improved methods to account for uncertainties, particularly in long-term predictions, and emphasize the importance of probabilistic design approaches.

Perunov et al. (2022a) explored the uncertainties in wave-induced rigid-body ship responses, emphasizing global wave loads. The study aims to foster a shared understanding of uncertainty modelling in waves and wave-induced responses for marine structures, offering insights and recommendations for future research directions. Moreover, Perunov et al. (2022b) summarized a benchmark study by the ISSC-ITTC Joint Committee, focusing on global linear wave loads on a container ship with forward speed. The research evaluated the uncertainties in linear transfer functions across multiple seakeeping codes from seven institutes. Results indicated that strip theory codes showed lower uncertainty and better agreement with experiments compared to other methods.

Wang and Guedes Soares (2022) investigated experimental uncertainties in an LNG carrier's 1:70 scale model tests. They focused on heave, pitch, vertical bending moments, and hull slamming pressures in irregular waves of Hs 11.5m and Tp 12s, generated using the JONSWAP spectrum. The study applied the Generalized Extreme Value (GEV) distribution to assess uncertainty, showing that motion and bending moment uncertainties were low (below 5%), while slamming pressure uncertainties were slightly higher, particularly at higher forward speeds.

Huang and Chen (2022) thoroughly examined the effects of wave impacts on marine structures. Their study reveals that wave conditions have a limited influence on the probability of different impact types, whereas structural geometry—such as the length of the overhanging slab and structural clearance—significantly affects the occurrence of impulsive wave impacts. Experiments were conducted in a wave flume with various configurations of overhanging slab lengths and structural clearances, while wave heights and periods were varied to assess their effects on impact types. As shown in Fig. 27, a comparison of the experimental data with the predicted values indicates that most data points align closely with the fitted curve, suggesting that the method provides satisfactory predictions for various structures, including the overhanging horizontal slab and the vertical wall. Additionally, the study quantified the impact of wave conditions and structural geometry on the magnitude of wave impacts and proposed empirical formulas.

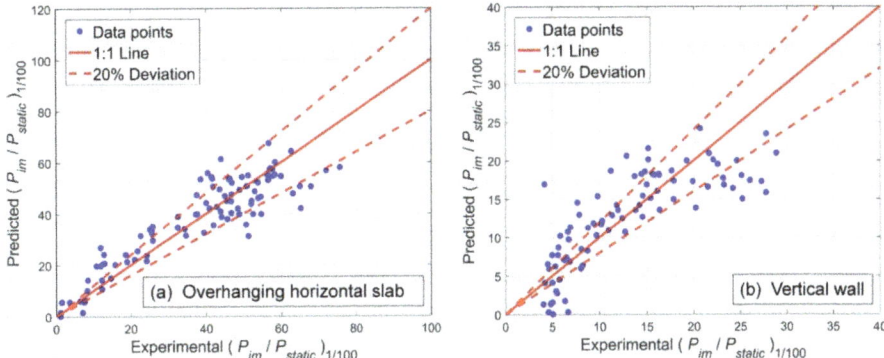

Fig. 27. Comparison of predicted $(P_{im}/P_{static})_{1/100}$ with experimental $(P_{im}/P_{static})_{1/100}$ for wave impact on the overhanging horizontal slab (left) and the vertical wall (right), from Huang & Chen (2022)

Simulation of flooding is an important approach to illustrate the survivability of marine structures in case of a flooding accident. Simulation methods stretching from quasi-static calculations to more complex CFD solutions have been used to perform this type of simulations. Quasi-static methods are often applied for onboard monitoring and decision-making support system due to its computational efficiency and capability of capturing the physics of a slow flooding accidents. More advanced flooding simulations have been proposed as well to address the dynamics of the overall flooding process, especially the initial transient part. Ruponen et al. (2022) presented an international benchmark study on simulation of flooding and motion of damaged ships. Results show there are variations in the simulation of internal flooding and capsize mechanisms including capsize rate, time-to-capsize, etc.

Moreover, in a study on operational loads, Braidotti et al. (2021) utilized sensor data to perform progressive flooding simulations based on three different linearized formulations: Linearized Ordinary Differential Equation (LODE), Linearized Ordinary Differential Equation with Grouping of Completely Filled Rooms (LODEG), and Linearized Differential-Algebraic Equations (LDAE). These methods are all quasi-static. Application of these methods on a large cruise ship revealed that the simplistic assumptions of the LODE formulation can result in overly optimistic time-to-flood estimates or even entirely erroneous predictions when compared to the LODEG and LDAE formulations.

In another paper by Dafermos and Zaraphonitis (2024), a novel flooding simulation tool was developed by combining the smoothed particle hydrodynamics method with a ship motion solver. This methodology examined the interaction between the hull and floodwater, highlighting its advantages in addressing complex ship-floodwater interaction phenomena compared to the quasi-static method.

5.4 Response Modelling and Their Uncertainties

The uncertainties in environmental factors, such as wind and waves, and their forecasts will impact the environmental loads and responses of offshore structures subjected to marine conditions. The dynamic responses of ships and offshore structures are crucial for safety and are of interest to designers and practitioners. These dynamic responses, which include platform motions and structural reactions, must be evaluated according to design guidelines, such as DNV (2010). Depending on the type of loading and system properties, dynamic responses can be steady-state or transient. Although generalizations are challenging, quasi-static methods can be applied when inertial effects are minimal. For steady-state line profiles and tensions in mooring systems, the modelling errors of quasi-static formulations are relatively small compared to finite element or lumped mass approaches, as noted by Liang et al. (2021) and Hall (2024). In contrast to steady-state responses, transient responses in offshore structures may occur over short durations and are often associated with large-cycle oscillations that can compromise structural safety. An example of this is the emergency shutdown of offshore wind turbines (Jiang et al., 2014) with rapid blade pitching. For such transient events, the control strategy (Jiang and Xing, 2022) or damping mechanisms (Liu et al., 2024) are crucial in shaping the responses. To date, the uncertainties of the quasi-static approach have not been systematically quantified.

Focusing on epistemic uncertainty, Ramezani et al. (2023) reviewed the uncertainty models related to the structural design of floating offshore wind turbines, addressing environmental, material, mechanical, and geotechnical aspects. This work highlights the growing uncertainties associated with time-dependent phenomena such as fatigue and corrosion damage, as well as the existing knowledge gaps in this research area. For long-term environmental conditions at offshore sites, uncertainties in wind and wave data are particularly significant.

Furthermore, Li et al. (2023) compared the one-way hydro-structural coupling model with key findings from physical model tests of a jack-up conducted in the TCOMS Ocean Basin. In particular, a single set of hydrodynamic coefficients is consistently employed for deriving forces and additional fluid damping across all cases, eliminating the need for adjusting coefficients as required by the standard Morison equation in different wave conditions. The study demonstrates that accurate dynamic structural responses can be derived for any sea state once the hydrodynamic inputs are appropriately represented.

Liu et al. (2022) examined the structural deformation mechanism of U-type corrugated core sandwich panels used in ship structures under lateral quasi-static compression loads, utilizing experimental, numerical, and analytical methods. Based on the results from quasi-static experiments and finite element simulations, they developed an analytical formula for the deformation resistance of the U-type corrugated core sandwich panel. The performance of this analytical formula has been validated by comparing it to experimental and numerical results, as illustrated in Fig. 28.

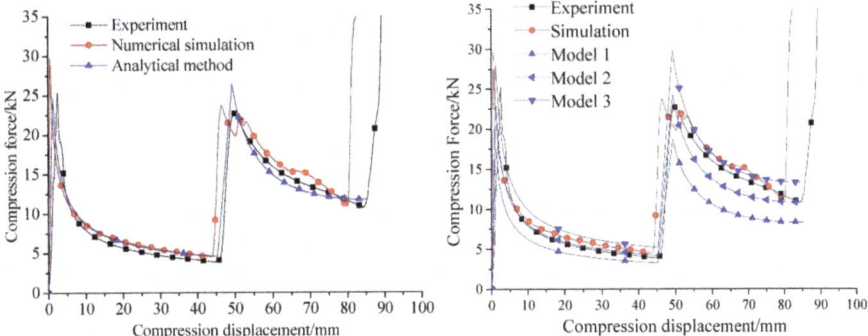

Fig. 28. Comparation results of the compression forces from different methods or different material models (one of the specimens), from Liu et al. (2022)

In a recent paper by Trubat et al. (2022), a quasi-dynamic mooring model was introduced and validated through a parametric study and comparison with experimental results. It is a new approach to assess the mooring line tension that considers the static solution of the catenary shape but updates the distributed vertical force along the mooring line as a function of the external forces, drag and inertial, acting on the line. An updated apparent weight is applied to the static solution of the catenary shape and it improves significantly the solution obtained by the quasi-static approach. The findings indicate that the error of the quasi-static tension within one cycle is less than 20% in only 18% of simulations, while the quasi-dynamic model achieves an error below 20% in 84% of simulations.

Ullah and Choi (2024) presented an economical solution of Submerged Floating Tunnels (SFTs) for crossing deep and wide water bodies. Mooring cables, crucial for the stability of SFTs under hydrodynamic loads, are typically modelled using a quasi-static catenary approach. However, this study introduces a discrete catenary element for fully coupled dynamic analysis, examining the effects of cable dynamics. Using the Qiandao Lake SFT as a prototype, a comparison between quasi-static and dynamic models reveals significant differences in displacements, cable tensions, shear forces, and bending moments. The quasi-static model underestimates these forces, making it unsuitable for structural design. In contrast, the fully coupled dynamic model offers more accurate predictions, especially at natural frequencies. Although the quasi-static model can estimate tunnel displacements, the dynamic model is necessary for precise internal force calculations, with an easy-to-implement methodology.

5.5 Strength Assessment and Their Uncertainties

The ultimate strength of stiffened steel plates, sections and systems for ships and offshore structures are commonly determined based on applicable rules / standards or using a quasi-static nonlinear finite element analysis. Similar approaches can also be used to estimate the ultimate strength of ship hull girders. In these approaches, the dynamic effects are often simplified or neglected. Significant research is still required to quantify the uncertainty associated with these assumptions.

In fatigue analysis, local dynamic effects are often neglected when estimating the stress distributions at hot-spots for fatigue analysis. This is acceptable for most cases. The uncertainties in nominal fatigue stress analysis will be discussed considering the simplifications in modelling, i.e. shell modelling, neglecting of actual shape of weld, imperfections, etc.

An experimental investigation with modelling uncertainties is conducted on single-sided stiffened and double-skinned steel sandwich panels under quasi-static penetration by Romanoff et al. (2022). Through both experiments and numerical simulations, this study explores the effects of collision parameters, the sequence of integration point failures, modelling of weld seams, mesh resolution, and failure criteria on the outcomes of numerical simulations.

In fatigue analysis, local dynamic effects are frequently disregarded when estimating stress distributions at hot spots, which is acceptable in most studies. In order to perform more precise stress analysis, Liu and Ren (2022) introduced a local stress correlation method aimed at obtaining structural stress data for strength evaluations in digital twin models of ship structures monitoring. The proposed method provides a relatively simple approach to identifying the external loads acting on the ship hull structure without the performing load inversion.

In another paper by Liu and Ren (2022), a novel method for evaluating the structural strength with ship digital twin models was presented, which enables rapid acquisition of yield strength evaluation at both monitored and non-monitored points. The results indicate that the maximum error in the calculation results obtained by the proposed method is within 0.0005% for determining the structural yield strength at the monitoring point, and within 0.032% at the non-monitoring point.

5.6 Probabilistic Modelling of Loads, Responses and Strengths and Reliability Analysis

When ships and offshore structures are subjected to various sources of uncertainty, structural reliability is often expressed in terms of probability of failure. Probabilistic models are essential in structural reliability analysis, offering a systematic approach to assess the safety and performance of structures under uncertainty. To calculate the probability of failure, a limit-state function must be mathematically formulated which is defined as a differentiable function g ($\mathbf{X}$) where $\mathbf{X} = [X1, X2,..., Xn]$ are independent variables that concern uncertain factors like mechanical properties, geometry, and load actions (Ditlevsen and Madsen, 1996). The limit state is given as the set of values for which the limit-state function becomes zero. $g(\mathbf{X}) < 0$ denotes the failure domain. The probability of failure is calculated by integrating the joint probability density function of $\mathbf{X}$ over the failure domain. Due to the complexity of practicable problems, there exist few closed-form solutions in general (Jiang et al., 2017).

Several approximate methods are commonly employed to calculate the probability of failure, including first- and second-order reliability methods (FORM and SORM) and simulation methods such as Monte Carlo simulation and importance sampling (Rubinstein and Kroese, 2016). For high-dimensional problems with small failure problems, the subset simulation techniques (Au and Beck, 2001) can be adopted to improve computational efficiency. The first-order reliability method approximates failure probability through linearization, making it efficient for large problems, while the second-order reliability method improves accuracy by considering nonlinearities. These methods collectively enhance our ability to predict structural failures and ensure safety. Monte Carlo Simulation uses random sampling to estimate failure probabilities, providing detailed insights but with high computational demands.

Fang et al. (2022) presented a fatigue crack growth prediction approach for offshore platforms using digital twin technology, which integrates Gaussian process and dynamic Bayesian network algorithms. To improve the computational efficiency of the crack growth prediction process, a Gaussian process-based surrogate model was designed. The superiority was demonstrated through experiments conducted on 7075-T6 aluminium alloy. Specifically, uncertainty analysis was performed on model parameters across various specimens, which indicated that the parameter uncertainty will decrease with the increase of the number of inspections, as shown in Fig. 29.

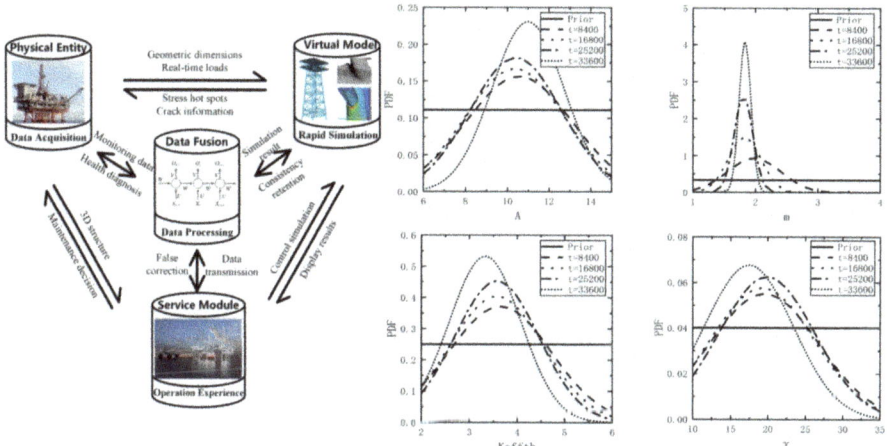

Fig. 29. The overall digital twin concept and the probability density function of the parameters after each inspection, from Fang et al. (2022)

In order to enhance structural integrity management for marine structures, Li & Brennan (2024) explored digital twin technology to virtually monitor and assess the condition of structures subjected to cyclic wave loads, which can lead to fatigue cracks. By integrating virtually monitored data into inspection planning, a more accurate and cost-effective approach by prioritizing inspections based on the probability of failure was provided. This paper also involves Bayesian updating with Markov Chain Monte Carlo simulation (MCMC in figures) to refine stress range distributions based on monitored data, improving fatigue load predictions. The reliability index is calculated using a first-order reliability method, guiding inspection planning by balancing structural safety and inspection costs. Figure 30 illustrates the enhanced accuracy in time-variant reliability indices achieved through this approach.

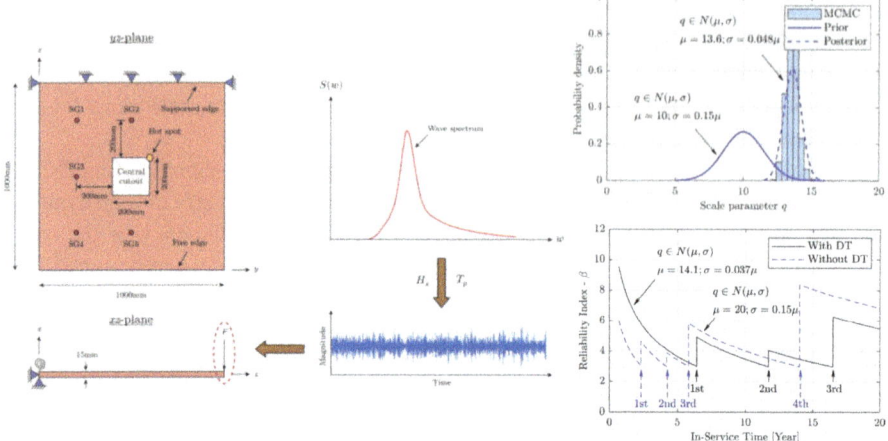

Fig. 30. Comparison of time-variant reliability index (right) evaluated with and without digital twin enabled virtually monitored data (Scenario 1 of the case study model, left), from Li & Brennan (2024)

Shittu et al. (2020) present a structural reliability assessment of offshore wind turbine (OWT) support structures subjected to pitting corrosion-fatigue using a damage tolerance modeling approach. The study integrates probabilistic finite element analysis (FEA) with a stochastic model, focusing on crack growth due to corrosion pits under cyclic loading. The analysis includes the use of artificial neural networks (ANN) and the First Order Reliability Method (FORM) to calculate the reliability indices of the structure over time. The results indicate that the OWT support structure becomes unsafe after approximately 18 years, prior to its designed 20-year life span. The sensitivity analysis highlights that the aspect ratio of corrosion pits is critical to the reliability of the structure. The study provides a comprehensive methodology for assessing the long-term reliability of offshore wind turbine support structures, particularly under the combined effects of corrosion and fatigue.

5.7 Concluding Remarks

In this chapter, we have reviewed the state-of-the-art in quasi-static response analysis for ships and offshore structures, highlighting the necessity of systematic assessments of the uncertainties associated with these methods. It is crucial to compare quasi-static methods against more dynamic and advanced simulations or experimental data to quantify their accuracy. While quasi-static methods remain useful for initial designs, optimization, and scenarios with limited dynamic influences, there is a growing need to refine these methods to better account for complex environmental loads and structural responses.

The ongoing development of computational tools and increased computational power is driving a shift towards more integrated and dynamic analysis approaches. However, for many practical applications, quasi-static methods offer a balance between accuracy and efficiency, particularly when combined with probabilistic models to account for uncertainties. Future research should focus on enhancing the reliability and accuracy of quasi-static approaches, expanding their applicability through improved understanding and quantification of the associated uncertainties.

6 Conclusions and Recommendations

6.1 Conclusions

During this ISSC term, the Quasi-static Response committee has tried to review the recent publications that are related to quasi-static approaches for load and response analysis of ships and marine structures. Discussions have been made in Sects. 2–4 with respect to load modelling, response analyses and strength assessments that may have an element of quasi-static and/or quasi-steady state assumptions. This is purposely done in order to have adequate references for this committee, since papers purely on quasi-static response analysis methods are very limited. In Sect. 5, the aspects related to the uncertainty assessment of quasi-static approaches as well as probabilistic modelling of quasi-static responses are discussed. Although it was very difficult to find highly relevant publications for this committee, we didn't mean that quasi-static approaches are not used by the research community or by the industry. In fact, the boundaries within which a quasi-static response analysis approach can be applied are well understood and their engineering applications are well documented in structural design rules, especially for traditional ships and offshore structures. In that sense, there are not so many new publications on quasi-static methodologies in recent years.

From the load modelling perspective, the review covers different types of ships (in particular large container vessels, FPSOs, LNG vessels and high-speed vessels) and marine structures (like floating wind turbines), under various loads from wind, waves and ice interactions. Similarly as in the last term, researchers are more interested in and therefore reported more cases of nonlinear hydrodynamic loads in extreme waves and due to sloshing and slamming, and quasi-static structural response analysis as long as the local hydroelasticity effect is not profound. Wind load assessment research focused on wind turbines and windsails. A particular interest is to develop databases for wind load coefficients for constant wind speeds based on CFD simulations so that they can be used in long-term simulations with more complicated environmental conditions. This implies a quasi-static or quasi-steady state assumption of the wind loads. There are increasing research interest and activities on ice-structure interactions and collision/grounding of ships, which are highly dynamic and involve a direct assessment of structural failures (of ice or ship hulls). In such cases, global dynamic features need to be captured, while structural failure criteria might be derived under a quasi-static assumption of failure mechanisms, neglecting the local dynamics.

Traditional and well-established methods for ship or floater structural analysis due to wave loads, separate stress analysis (which is typically quasi-static) from motion analysis (which is a dynamic analysis but often performed in frequency domain). Such methods are commonly used as a simplified structural analysis for example in ship design rules or offshore standards. They are often combined with formulae-based design checks for engineering design of ships and marine structures. Moreover, with the increasing interest in novel floating structures, such as floating wind turbines, coupling the combined effects of wind and wave induced responses requires a design analysis based on the first principles. This often demands a direct load and response analysis in time domain. Such applications are also found in recent publications for analysing stress distributions in floaters for floating wind turbines using a time-domain method. Challenges for engineering design analysis of such structures are related to the significant number of load cases of wind and wave combinations that need to be considered for long-term fatigue assessments and therefore the tremendous computation efforts. Quasi-static floater stress analysis based on shell FEMs is therefore necessary to apply. On the other hand, slender elements in floating wind turbines, such as blades, tower, mooring lines and power cables, are analyzed based on dynamic beam FEMs in a global coupled analysis.

Other situations in which a quasi-static structural analysis can be applied and have been applied are also reviewed. These include analysis of complex structures, for example composite structures or sandwich structures (such as steel plate-concrete-steel plate combinations). For such structures, local structural dynamics require a composite layer-to-layer modelling, but the overall structures are often simplified as an integrated structural element when assessing the global structural dynamics. Marine structures under degrading mechanisms (for example corrosion) represent another challenging feature to fully consider the dynamic interaction between the corrosion phenomena and the structural global/local dynamics. Piece-wise constant corrosion rate models or end-of-life corrosion allowance models might be used to simplify the analysis procedures. In structural optimization analysis, it is not affordable to run all cases using advanced load and structural analysis methods for all possible design parameters. It is therefore a full dynamic load analysis with the original given structural design parameters is first applied, followed by a series of quasi-static structural analyses with varying design parameters, so that an optimization can be achieved.

Iso-geometric analysis has been developed in other industries (for example automobile industry) to more precisely match the complex geometry of 3D structures represented in a CAD design and the numerical model for structural load and response analysis, as compared to the traditional finite element-based approach. High accuracy is achieved at a cost of high computational efforts using such modelling methods. Such methods have also been applied to structural elements (such as panels, stiffened plates) in ships and offshore structures for strong fluid-structure interaction problems (such as hydroelastic slamming) to achieve more accurate results. However, it still requires further development for the applications to a complete ship or offshore structure model. In general, the benefits of using such methods still need to be demonstrated.

A chapter on quasi-static analysis for structural strength problems is provided. Since strength issues are mainly discussed in ISSC III.1 Ultimate Strength Committee, a brief review with focus on quasi-static approach development or validation is made. It covers from rule-based approaches to FEM-based assessments, from stiffened plates to complete hull structures and from numerical analysis to large-scale test validations. Discussions about residual strength of offshore structures (for example floating wind turbines) due to external collisions or ship hull structures due to corrosion are also introduced. Extensive research has been conducted for welded aluminum ship structure elements with respect to fatigue strength.

With the increasing availability of the data from model tests and more importantly from field measurements as well as the booming of AI (Artificial Intelligence) and ML

(Machine Learning) in our society, the use of AI becomes more and more popular in the marine technology field. The importance of full-scale measurements from operating ships or offshore structures lies in their ability to provide accurate and real-world data on structural responses, for the validation of theoretical and numerical models. However, due to the limited coverage of direct measurements, machine learning based approaches are developed to extend the measurements from local positions to full structures. Surrogate models are developed to cut down computation time and seamlessly integrate with high-accuracy simulations to consider varying environmental conditions so that they can be easily applied for long-term assessment or for real-time forecasting. The integration and applications of the AI and ML models in our field still need time to be developed, with the joint efforts from the research community and the industry.

Finally, we still keep a chapter on systematic assessment of the uncertainties associated with quasi-static approaches, for modelling of environmental conditions, loads and structural responses. In recent years, providing an uncertainty assessment of either developed theories and methods or measurements from tests or field campaigns become necessary for scientific publications. It was found that the research purely on the uncertainty assessment of quasi-static approaches is limited and they are typically published as a part of the study using more advanced methods. The use of uncertainty assessment in probabilistic modelling and structural reliability assessment is also discussed. Further research in this direction is still needed.

6.2 Recommendations

As given in the mandate to review quasi-static response assessment for ships and offshore structures, with respect to the method developments, validations and applications, it is this committee's view that, most of the recent research developments in numerical methods and computational tools in this field aim for more advanced, more accurate, more integrated dynamic analysis approaches. In general, this is requested by the fact that it is the responsibility of the research community to develop, validate and apply new methods to new problems. Therefore, from the methodology point of view, quasi-static response analysis methods, as traditional simplified methods, do not gain much of the research interest in recent years.

However, the community requires a series of numerical methods/models with different fidelities, accuracies and efficiencies, for different purposes, ranging from conceptual study to engineering design and from design analysis to real-time applications. The applications of quasi-static methods to novel offshore structures under complex life-time environmental loads, still need to be investigated. The emerging areas are, but not limited to, LNG, FPSO, high-speed and Arctic vessels, autonomous vehicles, deepwater platforms, offshore renewable energies, aquaculture plants and novel coastal infrastructures. In addition, some of these structures, such as offshore wind turbines and aquaculture plants, are deployed in a farm configuration. They require a cross-dimensional load and response analysis, in which a hierarchical analysis procedure should be implemented and in the final step, local stress analysis is typically conducted using quasi-static approaches. The focus of such research should be given to justifying their suitability to these new structures/problems and especially quantifying the uncertainties of such methods by comparing with the more advanced dynamic analysis methods and the measurements from the model tests or field campaigns.

Assessments with high-computational efforts, for example analysis involving multiple environmental actions, long-term fatigue analysis, optimization, structural reliability analysis, etc., demand efficient response analysis methods. In such cases, research are still needed to clarify the suitable quasi-static analysis procedures and to benchmark their accuracies and limitations.

Development or implementation of new numerical algorithms/methods/tools for either separate load/response analysis and strength analysis or integrated load/response/failure analysis using for example iso-geometric analysis and model-reduction schemes, and for strong fluid-structure interactions, are also of high importance and interest.

From the practical applications point of view, due to their maturity and simplifications, quasi-static methods have been successfully implemented in design rules for ships and offshore structures for a long time, offering a balance between accuracy and efficiency in engineering designs. Although such design rules and standards are not reviewed by the committee in this term, in general it requires clarifications about the limitations of such methods for different problems. It is also the responsibility of the rule/standard developers to provide multiple choices of analysis methods to the engineers with clear indications of assumptions, simplifications and limitations.

On one hand, the enabling technologies, such as AI, machines learning, sensing technology, digital twin, etc., gains rapid development and broad applications in a very short period in recent years and deserve further attentions and investigations in our field for the benefit of our domain knowledge development. On the other hand, physics-based theories, methods and models in our field have been widely developed and applied, which in fact is an advantage, as compared to the fields with only data and empirical judgements. A good balance or a full integration between data-driven methods and physics-based approaches needs to be considered in the future development. In order to fully utilize the AI technology, it is important and urgent to increase the research activities to collect sufficient data from real-time operations of ships and offshore platforms and to formulate a sharing mechanism so that our field can benefit from the totality of the domain knowledge development.

As already mentioned in the introduction, due to the maturity of the quasi-static approaches and the scarcity of the related publications, it is the current committee's view that the quasi-static response committee could be combined with the dynamic response committee in the future ISSC framework. In general, as most of the discipline-based theoretical frameworks and numerical/experimental methods have been gradually developed and become mature, most of the research in the field of ship and offshore technologies focus on novel engineering structures and the applications, extensions or refinements of these developed methodologies. It is therefore more straightforward to conduct the ISSC work if more application-based committees are formulated.

Acknowledgements. The committee is thankful to Dr. Shuaishuai Wang from Norwegian University of Science and Technology for his contributions to Chapter 3 and Mr. Tianqi Pei from Shanghai Jiao Tong University for his work on the reference list.

References

Abedin, J., Franklin, F., Mahmud, S.M.I.: A two-stage optimisation of ship hull structure combining fractional factorial design technique and NSGA-II algorithm. J. Marine Sci. Eng. **12**(3), 411 (2024)

Abyani, M., Bahaari, M.R., Zarrin, M., Nasseri, M.: Predicting failure pressure of the corroded offshore pipelines using an efficient finite element based algorithm and machine learning techniques. Ocean Eng. **254**, 111382 (2022)

Aguiari, M., Gaiotti, M., Rizzo, C.M.: Ship weight reduction by parametric design of hull scantling. Ocean Eng. **263**, 112370 (2022)

Ahn, Y., Lee, J., Park, T., Kim, Y.: Long-term approach for assessment of sloshing loads in LNG carrier, Part II: grouping method. Mar. Struct. **89**, 103398 (2023)

Ahn, Y., et al.: Long-term approach for assessment of sloshing loads in LNG carrier, Part I: Comparison of Short- and Long-term approaches. Marine Struct. **89**, 103381 (2023)

Ali, A., De Risi, R.: A simplified blade model for reliable seismic assessments of wind turbines. In: Proceedings of the 14th world congress in computational mechanics (WCCM) and ECCOMAS Congress 2020, Paris, France (2021)

Almallah, I., Ali-Lavroff, J., Holloway, D.S., Davis, M.R.: Estimation of torsional and global loads for a wave-piercing high-speed catamaran at full-scale in irregular bow quartering seas using CFD simulation. Ocean Eng. **266**, 113006 (2023)

Amarouche, K., Akpınar, A., Kamranzad, B., Khames, G.E.Y.: Global extreme wave estimates and their sensitivity to the analysed data period and data sources. Mar. Struct. **92**, 103494 (2023)

Arabgolarcheh, A., Micallef, D., Rezaeiha, A., Benini, E.: Modelling of two tandem floating offshore wind turbines using an actuator line model. Renewable Energy **216**, 119067 (2023)

Arai, Y., Katagiri, K., Ikeda, S.: CFD assessment of the wind forces and moments of superstructures through RANS. Appl. Ocean Res. **129**, 103364 (2022)

Au, S.K., Beck, J.L.: Estimation of small failure probabilities in high dimensions by subset simulation. Probab. Eng. Mech. **16**(4), 263–277 (2001)

Beier, D., Schnepf, A., Van Steel, S., Ye, N., Ong, M.C.: Fatigue analysis of inter-array power cables between two floating offshore wind turbines including a simplified method to estimate stress factors. J. Mar. Sci. Eng. **11**(6), 1254 (2023)

Bitner-Gregersen, E.M., et al.: Uncertainties in long-term wave modelling. Mar. Struct. **84**, 103217 (2022)

Carmo, L., Jonkman, J., Thedin, R.: Investigating the interactions between wakes and floating wind turbines using FAST. Farm. Wind Energy Sci. Discuss. **9**(9), 1827–1847 (2024)

Chen, J., Taylor, P.H., Milne, I.A., Gunawan, D., Zhao, W.: Wave-by-wave prediction for spread seas using a machine learning model with physical understanding. Ocean Eng. **285**, 115450 (2023)

Chen, Z., He, Y., Ren, Y., Liu, Y.: Analysis and realization of the influence of sea ice flexural strength on ice resistance in numerical simulation of icebreaking by icebreaker. Ocean Eng. **273**, 113995 (2023)

Chen, Z.W., Jiao, J., Wang, S., Guedes Soares, C.: CFD-FEM simulation of water entry of a wedged grillage structure into Stokes waves. Ocean Eng. **275**, 114159 (2023)

Cottrell, J.A., Reali, A., Bazilevs, Y., Hughes, T.J.: Isogeometric analysis of structural vibrations. Comput. Methods Appl. Mech. Eng. **195**(41–43), 5257–5296 (2006)

Cui, J., Wang, D.: An experimental and numerical investigation on ultimate strength of corrugated bulkheads and plane bulkheads subjected to lateral pressure. Ocean Eng. **295**, 116895 (2024)

Cui, H., Chen, Z., Hu, R., Ding, Q.: Ultimate strength assessment of hull girders considering elastic shakedown based on Smith's method. Ocean Eng. **293**, 116695 (2024)

Cui, H., Chen, Z., Hu, R., Zheng, C.: Research on ultimate strength of hull girder considering initial imperfections under monotonic/cyclic bending moments - A bulk carrier case. Ocean Eng. **311**, 118862 (2024)

Dao, M.H., et al.: Wind tunnel and CFD studies of wind loadings on topsides of offshore structures. Ocean Eng. **285**, 115310 (2023)

Den Bieman, J.P., De Ridder, M.P., Mata, M.I., Van Nieuwkoop, J.C.C.: Hybrid modelling to improve operational wave forecasts by combining process-based and machine learning models. Appl. Ocean Res. **136**, 103583 (2023)

Dirik, Y., Oterkus, S., Oterkus, E.: Isogeometric mindlin-reissner inverse-shell element formulation for complex stiffened shell structures. Ocean Eng. **305**, 118028 (2024)

Erceg, S., Erceg, B., von Bock und Polach, F., Ehlers, S.: A simulation approach for local ice loads on ship structures in level ice. Mar. Struct. **81**, 103117 (2022)

Esfidan, M.R., Ranji, A.R.: Corrosion models for steel plates in ship structure based on statistical data. Proc. Inst. Mech. Eng. Part M: J. Eng. Marit. Environ. (2024). https://doi.org/10.1177/14750902241253319

Fang, X., Wang, H., Li, W., Liu, G., Cai, B.: Fatigue crack growth prediction method for offshore platform based on digital twin. Ocean Eng. **244**, 110320 (2022)

Fowai, I., Zhang, J., Sun, K., Wang, B.: Structural analysis of jacket foundations for offshore wind turbines in transitional water. Brodogradnja **72**(1), 109–125 (2021)

Fujiwara, T., Michio, U., Ikeda, Y.: A new estimation method of wind forces and moments acting on ships on the basis of physical components models. J. Jpn. Soc. Naval Arch. Ocean Eng. **2**, 243–255 (2006)

Gao, Z., Merino, D., Han, K.J., Li, H., Fiskvik, S.: Time-domain floater stress analysis for a floating wind turbine. J. Ocean Eng. Sci. **8**(4), 435–445 (2023)

Garrido, M., Teixeira, R., Correia, J.R., Sutherland, L.S.: Quasi-static indentation and impact in glass-fibre reinforced polymer sandwich panels for civil and ocean engineering applications. J. Sandwich Struct. Mater. **23**(1), 194–221 (2021)

Gebrezgabir, S., Holloway, D.S., Ali-Lavroff, J.: Slam and wave load response reconstruction in high-speed catamarans using transmissibility on full-scale sea trials. Ocean Eng. **271**, 113822 (2023)

Gu, Y., Zhou, L., Ding, S., Tan, X., Gao, J., Zhang, M.: Numerical simulation of ship maneuverability in level ice considering ice crushing failure. Ocean Eng. **251**, 111110 (2022)

Gunn, D., Dimopoulos, S., Carra, C.: Rapid residual strength assessment of degraded mooring chains using an FEA based response surface modeling tool. In: Proceedings of the ASME 2023 42nd International Conference on Ocean, Offshore and Arctic Engineering (2023)

Ha, K., Kim, J.B.: Collision analysis and residual longitudinal strength evaluation of a 5 MW spar floating offshore wind turbine impacted by a ship. Int. J. Precision Eng. Manuf.-Green Technol. **9**, 841–858 (2022)

Han, B., Li, H., Zhang, B., Zou, J., Zhao, W.: Experimental investigation of slamming characteristics of stiffened elastic wedges with different deadrise angles: Part I-Slamming pressure. Ocean Eng. **303**, 117728 (2024)

Han, B., Qu, X., Zou, L., Liu, S., Li, H., Chen, Z.: Experimental investigation of slamming characteristics of stiffened elastic wedges with different deadrise angles: Part II–Structural response. Ocean Eng. **308**, 118288 (2024)

Han, S., Yang, B., Yang, B., Zhang, G.: Numerical simulation of heterogeneous ice sheet-structure interaction based on cohesive element method. Appl. Ocean Res. **145**, 103942 (2024)

Hanif, M.I., Adiputra, R., Prabowo, A.R., Yamada, Y., Firdaus, N.: Assessment of the ultimate strength of stiffened panels of ships considering uncertainties in geometrical aspects: finite element approach and simplified formula. Ocean Eng. **286**, 115522 (2023)

He, S., Chen, X., Zhang, B., Wang, J., Xie, Z., Ji, S.: Study on river ice load characteristics based on field measurements. Shipbuilding China **63**(4), 46–59 (2022)

Heiskari, J., Romanoff, J., Laakso, A., Ringsberg, J.W.: Influence of the design constraints on the thickness optimization of glass panes to achieve lightweight insulating glass units in cruise ships. Mar. Struct. **89**, 103409 (2023)

Heiskari, J., Romanoff, J., Laakso, A., Ringsberg, J.W.: On the lightweight design of laminated insulating glass units in cruise ships. Ships and Offshore Structures, Ahead-of-print, 1–19 (2024)

Hosseinzadeh, S., Tabri, K., Hirdaris, S., Sahk, T.: Slamming loads and responses on a non-prismatic stiffened aluminium wedge: Part I. Exprimental study. Ocean Eng. **279**, 114510 (2023)

Hosseinzadeh, S., Tabri, K., Topa, A., Hirdaris, S.: Slamming loads and responses on a non-prismatic stiffened aluminium wedge—Part II. Numerical simulation. Ocean Eng. **279**, 114309 (2023)

Huang, L., Jing, Y., Chen, H., Zhang, L., Liu, Y.: A regional wind wave prediction surrogate model based on CNN deep learning network. Appl. Ocean Res. **126**, 103287 (2022)

ITTC. Guideline on the CFD-based determination of wind resistance coefficients. In: Specialist Committee on Ships in Operation at Sea of the 29th ITTC (2021)

Ivošević, Š, Kovač, N.: The analyses of the failures of hull structure plating caused by corrosion. Trans. Maritime Sci. **12**(02), 1–10 (2023)

Jagite, G., Bigot, F., Malenica, S., Derbanne, Q., Sourne, H.L., Cartraud, P.: Dynamic ultimate strength of an ultra-large container ship subjected to realistic loading scenarios. Mar. Struct. **84**, 103197 (2022)

Jeon, S., Kim, Y.: Fatigue damage estimation of icebreaker ARAON colliding with level ice. Ocean Eng. **257**, 111707 (2022)

Ji, S., Yang, D.: Ice loads and ice-induced vibrations of offshore wind turbine based on coupled DEM-FEM simulations. Ocean Eng. **243**, 110197 (2022)

Jiang, C., Xu, P., El Moctar, O., Zhang, G.: Analysis of a moored and articulated multibody offshore system in steep waves. J. Offshore Mech. Arct. Eng. **145**(4), 041401 (2023)

Jiang, C., Zhang, Q., El Moctar, O., Xu, P., Iseki, T., Zhang, G.: Data-driven modelling of wave-structure interaction for a moored floating structure. Ocean Eng. **300**, 117522 (2024)

Jiao, J., Chen, Z., Xu, W., Bu, S., Zhang, P.: Asymmetric water entry of a wedged grillage structure investigated by CFD-FEM co-simulation. Ocean Eng. **302**, 117612 (2024)

Jonkman, J., Shaler, K.: FAST.farm user's guide and theory manual. golden, CO: National Renewable Energy Laboratory. NREL/TP-5000–78785 (2021). https://www.nrel.gov/docs/fy21osti/78485.pdf

Kamel, A., Dammak, K., El Hami, A., Ben Jdidia, M., Hammami, L., Haddar, M.: A modified hybrid method for a reliability-based design optimization applied to an offshore wind turbine. Mech. Adv. Mater. Struct. **29**(9), 1229–1242 (2022)

Kang, Y., Pei, Z., Ao, L., Wu, W.: Reliability-based design optimization of river-sea-going ship based on agent model technology. Mar. Struct. **94**, 103561 (2023)

Kendibilir, A., Kefal, A.: Enhanced ship cross-section design methodology using peridynamics topology optimization. Ocean Eng. **286**, 115531 (2023)

Kim, C., Oterkus, S., Oterkus, E., Kim, Y.: Probabilistic ship corrosion wastage model with Bayesian inference. Ocean Eng. **246**, 110571 (2022)

Kim, D.K., Li, S., Lee, J.R., Poh, B.Y., Benson, S., Cho, N.-K.: An empirical formula to assess ultimate strength of initially deflected plate: Part 1 = propose the general shape and application to longitudinal compression. Ocean Eng. **252**, 111151 (2022)

Kim, D.K., Li, S., Yoo, K., Danasakaran, K., Cho, N.K.: An empirical formula to assess ultimate strength of initially deflected plate: Part 2 = combined longitudinal compression and lateral pressure. Ocean Eng. **252**, 111112 (2022)

Kim, D.K., Wong, A.M.K., Hwang, J., Li, S., Cho, N.-K.: A novel formula for predicting the ultimate compressive strength of the cylindrically curved plates. Int. J. Naval Archit. Ocean Eng. **16**, 100562 (2024)

Kim, H.G., Kim, B.J.: Design optimization of conical concrete support structure for offshore wind turbine. Energies **13**(18), 4876 (2020)

Kim, S.J., et al.: Comparison of numerical approaches for structural response analysis of passenger ships in collisions and groundings. Mar. Struct. **81**, 103125 (2022)

Kim, S., Bouscasse, B., Ducrozet, G., Delacroix, S., De Hauteclocque, G., Ferrant, P.: Experimental investigation on wave-induced bending moments of a 6,750-TEU containership in oblique waves. Ocean Eng. **284**, 115161 (2023)

Kim, S., De Hauteclocque, G., Bouscasse, B., Lasbleis, M., Ducrozet, G.: Experimental analysis of extreme wave loads on a containership. Ocean Eng. **306**, 118031 (2024)

Ko, D.H., Nam, B.W., Choi, J.H.: A study on dimensionality reduction technique-based wave estimation using the motion response spectra of the advancing ship. In: Proceedings of the ASME 2024 43rd International Conference on Ocean, Offshore and Arctic Engineering. Volume 5B: Ocean Engineering, Singapore (2024)

Kozmar, H., Hadžić, N., Čatipović, I., Rudan, S.: Wind load assessment in marine and offshore engineering standards. Ocean Eng. **252**, 110872 (2022)

Kramer, J.V., Steen, S.: Sail-induced resistance on a wind-powered cargo ship. Ocean Eng. **261**, 111688 (2022)

Kristiansen, M., et al.: Improving the fatigue life of large offshore foundations. Mar. Struct. **87**, 103314 (2023)

Kume, K., Hamada, T., Kobayashi, H., Yamanaka, S.: Evaluation of aerodynamic characteristics of a ship with flettner rotors by wind tunnel tests and RANS-based CFD. Ocean Eng. **254**, 111345 (2022)

Ladeira, I., Márquez, L., Echeverry, S., Le Sourne, H., Rigo, P.: Review of methods to assess the structural response of offshore wind turbines subjected to ship impacts. Ships Offshore Struct. **18**(6), 755–774 (2023)

Lee, J., Ahn, Y., Kim, Y., Jung, J.: Sloshing phenomenon occurring in the foremost two tanks of LNG carrier. Int. J. Offshore Polar Eng. **32**(4), 434–439 (2022)

Lemström, I., Polojärvi, A., Tuhkuri, J.: Model-scale tests on ice-structure interaction in shallow water: Global ice loads and the ice loading process. Mar. Struct. **81**, 103106 (2022)

Lemström, I., Polojärvi, A., Puolakka, O., Tuhkuri, J.: Load distributions in the ice-structure interaction process in shallow water. Ocean Eng. **258**, 111730 (2022)

Lerch, M., De-Prada-Gil, M., Molins, C.: A simplified model for the dynamic analysis and power generation of a floating offshore wind turbine. Ocean Eng. **271**, 113785 (2023)

Li, C.B., Chen, M., Choung, J.: The quasi-static response of moored floating structures based on minimization of mechanical energy. J. Mar. Sci. Eng. **9**(9), 960 (2021)

Li, D., Chen, Z.: Advanced empirical formulae for the ultimate strength assessment of continuous hull plate under combined biaxial compression and lateral pressure. Eng. Struct. **285**, 116041 (2023)

Li, S., Brennan, F.: Implementation of digital twin-enabled virtually monitored data in inspection planning. Appl. Ocean Res. **144**, 103903 (2024)

Li, S., Coraddu, A., Oneto, L.: Computationally aware estimation of ultimate strength reduction of stiffened panels caused by welding residual stress: From finite element to data-driven methods. Eng. Struct. **264**, 114423 (2022)

Li, S., Kim, D.K., Benson, S.: An adaptable algorithm to predict the load-shortening curves of stiffened panels in compression. Ships Offshore Struct. **16**, 122–139 (2021)

Li, W., Wang, S., Moan, T., Gao, Z., Gao, S.: Global design methodology for semi-submersible hulls of floating wind turbines. Renewable Energy **225**, 120291 (2024)

Li, X., Ma, N., Shi, Q., Gu, X.: Directional wave Spectrum estimation using ship motion data by improved CGAN with physics guided. In: Proceedings of the ASME 2024 43rd International Conference on Ocean, Offshore and Arctic Engineering. Volume 5B: Ocean Engineering, Singapore (2024)

Li, Y., Zhang, S., Zhang, C., Law, Y.Z., Cai, M., Santo, H.: Wave-current-structure blockage: Validation of one-way hydro-structural coupling with large-scale model tests of a stiffness-similar jack-up in elevated condition. J. Fluids Struct. **121**, 103960 (2023)

Li, Y., Zhang, Y., Wang, W., Li, X., Wang, B.: Influence of corrosion damage on fatigue limit capacities of offshore wind turbine substructure. J. Mar. Sci. Eng. **10**(8), 1011 (2022)

Liang, J., Kato, B., Wang, Y.: Constructing simplified models for dynamic analysis of monopile-supported offshore wind turbines. Ocean Eng. **271**, 113785 (2023)

Lim, C., et al.: Shear force and bending moment tuning algorithm of shuttle tanker model for global structural analysis. J. Mar. Sci. Eng. **11**(10), 1900 (2023)

Liu, F., Yu, Q., Li, H., Zhan, Y., Zhou, H., Liu, D.: Field data observations for monitoring the impact of typhoon "In-fa" on dynamic performances of mono-pile offshore wind turbines: A novel systematic study. Mar. Struct. **89**, 103373 (2023)

Liu, H., et al.: Modelling the quasi-static flexural behaviour of composite sandwich structures with uniform- and graded-density foam cores. Eng. Fract. Mech. **259**, 108121 (2022)

Liu, K., Zong, S., Li, Y., Wang, Z., Hu, Z.: Structural response of the U-type corrugated core sandwich panel used in ship structures under the lateral quasi-static compression load. Mar. Struct. **84**, 103198 (2022)

Liu, L., Ji, S.: Comparison of sphere-based and dilated-polyhedron-based discrete element methods for the analysis of ship–ice interactions in level ice. Ocean Eng. **244**, 110364 (2022)

Liu, Q., Feng, X., Tang, T.: A machine learning model for wave prediction based on support vector machine. Int. J. Offshore Polar Eng. **32**, 394–401 (2022)

Liu, X., Jiang, D., Liufu, K., Fu, J., Liu, Q., Li, Q.: Numerical investigation into impact responses of an offshore wind turbine jacket foundation subjected to ship collision. Ocean Eng. **248**, 110825 (2022)

Liu, Y., Ren, H.: Acquisition method of evaluation stress for the digital twin model of ship monitoring structure. Appl. Ocean Res. **129**, 103368 (2022)

Liu, Y., Ren, H.: Rapid acquisition method for structural strength evaluation stresses of the ship digital twin model. Ocean Eng. **285**, 115323 (2023)

Ma, S., Zhu, M., Liu, D., Liu, J., Wang, W.: Experimental study of wet deck slamming for a SWATH in regular waves. Ocean Eng. **288**, 115996 (2023)

Maetz, T., et al.: Microwave structural health monitoring of the grouted connection of a monopile-based offshore wind turbine: fatigue testing using a scaled laboratory demonstrator. Struct. Control. Health Monit. **2023**(1), 1981892 (2023)

Majidian, H., Wang, L., Enshaei, H.: Part A: a review of the real-time sea-state estimation, using wave buoy analogy. Ocean Eng. **266**, 111684 (2022)

Martins, A.D., Peres, N., Jacinto, P., Gonçalves, R.: An assessment of the eurocode 3 simplified formulas for distortional buckling of cold-formed steel lipped channels. Appl. Sci. **14**(6), 4924 (2024)

Matsui, S., Sugimoto, K., Shinomoto, K.: Study on simplified estimation formula for long-term predictions of ship response in waves – application to vertical bending moment. Mar. Struct. **92**, 103480 (2023)

Melchers, R.E.: New insights from probabilistic modelling of corrosion in structural reliability analysis. Struct. Saf. **88**, 102034 (2021)

Mittendorf, M., Nielsen, U.D., Bingham, H.B., Storhaug, G.: Sea state identification using machine learning - A comparative study based on in-service data from a container vessel. Mar. Struct. **85**, 103274 (2022)

Mohaseb Karimlou, M.R., Rezvani Tavakol, M., Yarmohammad Tooski, M.: Experimental study of the effect of clay nanoparticles on the strength of hybrid sandwich panels under quasi-static loading. Aerosp. Sci. Technol. **121**, 107313 (2022)

Nataraj, S., Hogan, L., Scott, A., Ingham, J.: Simplified mechanics-based approach for the seismic assessment of corroded reinforced concrete structures. J. Struct. Eng. **148**(3), 04021296 (2021)

Nguyen, M.Q., De Hauteclocque, G., De Lauzon, J., Malenica, S., Chen, X.B.: Quasi-Static fluid structure interactions for ship advancing in waves with constant forward speed. In: Proceedings of the ASME 2024 43rd International Conference on Ocean, Offshore and Arctic Engineering. Volume 6: Polar and Arctic Sciences and Technology; CFD, FSI, and AI, Singapore (2024)

Nybø, A., Bachynski, E.E., Jacobsen, A.K., Moan, T.: Quasi-static response of a bottom-fixed wind turbine subject to various incident wind fields. Wind Energy **24**(7), 781–798 (2021)

Pan, J., Zhang, W.Z., Sun, Z.M., Qu, X., Xu, M.C.: Experimental study on the dynamical response of elastic trimaran model under slamming load. J. Mar. Sci. Technol. **29**(1), 20–35 (2024)

Park, M.J., Kim, Y.: Probabilistic estimation of directional wave spectrum using onboard measurement data. J. Mar. Sci. Technol. **29**(1), 200–220 (2023)

Park, S.H., Yoon, S.H., Muttaqie, T., Do, Q.T., Cho, S.R.: Effects of local denting and fracture damage on the residual longitudinal strength of box girders. J. Mar. Sci. Eng. **2023**(11), 76 (2023)

Park, S.H., Lee, S.M., Yu, Y., Cho, S.R.: Residual strength of corroded ring-stiffened cylinder structures under external hydrostatic pressure. Int. J. Naval Archit. Ocean Eng. **16**, 100590 (2024)

Parunov, J., Guedes Soares, C., Hirdaris, S., Wang, X.: Uncertainties in modelling the low-frequency wave-induced global loads in ships. Mar. Struct. **86**, 103307 (2022)

Parunov, J., et al.: Benchmark study of global linear wave loads on a container ship with forward speed. Mar. Struct. **84**, 103162 (2022)

Parunov, J., Soares, C.G., Hirdaris, S., Wang, X.L.: Uncertainties in modelling the low-frequency wave-induced global loads in ships. Mar. Struct. **86**, 103307 (2022)

Pineau, J.P., Le Sourne, H.: Analytical modelling of ship bottom grounding considering combined surge and heave motions. Mar. Struct. **88**, 103364 (2023)

Prabowo, A.R., Putranto, T., Sohn, J.M.: Simulation of the behavior of a ship hull under grounding: effect of applied element size on structural crashworthiness. J. Mar. Sci. Eng. **7**(8), 270 (2019)

Putranto, T.: Equivalent single layer approach for ultimate strength analysis of box girder under bending load. Ocean Eng. **292**, 116535 (2024)

Qiu, F., Wang, H., Qian, H., Hu, H., Jin, X., Fan, F.: Evaluation of compressive properties of the ship plate after seawater corrosion based on 3D evolution prediction. Ocean Eng. **266**, 112561 (2022)

Quispe, J.P., Estefen, S.F., Lourenço de Souza, M.I., Chujutalli, J.H., Amante, D.A.M., Gurova, T.: Numerical and experimental analyses of ultimate longitudinal strength of a small-scale hull box girder. Mar. Struct. **85**, 103273 (2022)

Shamir, M., Braithwaite, J., Mehmanparast, A.: Fatigue life assessment of offshore wind support structures in the presence of corrosion pits. Mar. Struct. **92**, 103505 (2023)

Ren, D., Park, J.C.: Particle-based numerical simulation of continuous ice-breaking process by an icebreaker. Ocean Eng. **270**, 113478 (2023)

Ren, Y., Meng, Q., Chen, C., Hua, X., Zhang, Z., Chen, Z.: Dynamic behavior and damage analysis of a spar-type floating offshore wind turbine under ship collision. Eng. Struct. **272**, 114815 (2022)

Reyes-Casimiro, M., Felix-Gonzalez, I., Perea, T.: Design optimization for production semi-submersible pontoons based on genetic algorithms and finite element analysis. Ocean Eng. **268**, 113291 (2023)

Romanoff, J.K., Berntsson, K., Remes, H.: Experimental investigations on stiffened and web-core sandwich panels made for steel under quasi-static penetration. In: Proceedings of the ICSI 2021 the 4th International Conference on Structural Integrity, Madeira, Portugal (2022)

Ryan, H., Mehmanparast, A.: Development of a new approach for corrosion-fatigue analysis of offshore steel structures. Mech. Mater. **176**, 104526 (2023)

Sayahlatifi, S., Rahimi, G.H., Bokaei, A.: The quasi-static behavior of hybrid corrugated composite/balsa core sandwich structures in four-point bending: experimental study and numerical simulation. Eng. Struct. **210**, 110361 (2020)

Shan, Y., et al.: Experimental investigation of slamming pressure on 3D bow flare. Ocean Eng. **312**, 118898 (2024)

Shittu, A.A., Mehmanparast, A., Shafiee, M., Kolios, A., Hart, P., Pilario, K.: Structural reliability assessment of offshore wind turbine support structures subjected to pitting corrosion-fatigue: a damage tolerance modelling approach. Wind Energy **23**(11), 2004–2026 (2020)

Silva, K.M., Maki, K.J.: Implementation of the critical wave groups method with computational fluid dynamics and neural networks. Ocean Eng. **292**, 116468 (2024)

Song, M., Jiang, Z., Yuan, W.: Numerical and analytical analysis of a monopile-supported offshore wind turbine under ship impacts. Renewable Energy **167**, 457–472 (2021)

Song, M., Jiang, Z., Liu, K., Han, Y., Liu, R.: Dynamic response analysis of a monopile-supported offshore wind turbine under the combined effect of sea ice impact and wind load. Ocean Eng. **286**, 115587 (2023)

Stieng, L.E.S., Muskulus, M.: Reliability-based design optimization of offshore wind turbine support structures using analytical sensitivities and factorized uncertainty modeling. Wind Energy Sci. **5**(1), 171–198 (2020)

Sun, J., Fang, H.: The analysis of crashworthiness and dissipation mechanism of novel floating composite honeycomb structure against ship-OWT collision. Ocean Eng. **287**, 115819 (2023)

Suominen, M., Korgesaar, M., Taylor, R., Bergstrom, M.: Probabilistic analysis of operational ice damage for Polar class vessels using full-scale data. Struct. Saf. **107**, 102423 (2024)

Taghipoor, H., Sefidi, M.: Energy absorption of foam-filled corrugated core sandwich panels under quasi-static loading. J. Mater. Des. Appl. **237**(1), 234–246 (2023)

Taghizadeh, S.A., Naghdinasab, M., Madadi, H., Farrokhabadi, A.: Investigation of novel multilayer sandwich panels under quasi-static indentation loading using experimental and numerical analyses. Thin-Walled Struct. **160**, 107326 (2021)

Takami, T., Fujimoto, W., Houtani, H., Matsui, S.: Extreme wave and vertical bending moment predictions by higher order spectrum method and form. In: Proceedings of the ASME 2023 42nd International Conference on Ocean, Offshore and Arctic Engineering. Volume 2: Structures, Safety, and Reliability, Melbourne, Australia (2023)

Thies, F., Fakiolas, K.: Chapter 8 - Wind propulsion. In: Sustainable Energy Systems on Ships, pp. 353–402. Elsevier (2022)

Toedter, S., Neugebauer, J., El Moctar, O.: Experimental investigation of the influence of short-crested seas and swell on sloshing-induced impact loads. Int. J. Offshore Polar Eng. **34**(1), 47–55 (2023)

Truong, D.D., Jang, B.S.: Estimation of ice loads on offshore structures using simulations of level ice-structure collisions with an influence coefficient method. Appl. Ocean Res. **125**, 103235 (2022)

Tuyen, V.V.: Ultimate strength of aged ships under hull structure's imperfections. In: International Conference on Marine Sustainable Development and Innovation, vol. 1278, p. 012018 (2023)

Wang, G., Zhang, D., Yue, Q., Yu, S.: Study on the dynamic ice load of offshore wind turbines with installed ice-breaking cones in cold regions. Energies **15**(9), 3357 (2022)

Wang, H., Ti, Z.: Wave force prediction on truncated cylinders with arbitrary symmetric cross-sections using machine learning. Ocean Eng. **295**, 116716 (2024)

Wang, J., Chen, X., Sun, K., Ji, S.: Far-field identification of ice loads on ship structures by radial basis function neural network. Ocean Eng. **282**, 115072 (2023)

Wang, J., Ren, Y., Shi, W., Collu, M., Venugopal, V., Li, X.: Multi-objective optimization design for a 15 MW semisubmersible floating offshore wind turbine using evolutionary algorithm. Appl. Energy **377**, 124533 (2024)

Wang, P., Yu, W., Zhao, M., Du, X.: Effects of wind-wave-current-earthquake interaction on the wave height and hydrodynamic pressure based on CFD method. Ocean Eng. **305**, 117909 (2024)

Wang, S., Guedes Soares, C.: Random experimental uncertainty analysis on the model tests of an LNG carrier in extreme seas. In: Proceedings of the ASME 2022 41st International Conference on Ocean, Offshore and Arctic Engineering. Volume 5B: Ocean Engineering, Hamburg, Germany (2022)

Wang, S., Moan, T.: Methodology of load effect analysis and ultimate limit state design of semi-submersible hulls of floating wind turbines: with a focus on floater column design. Mar. Struct. **93**, 103526 (2024)

Wang, S., Moan, T., Gao, Z.: Methodology for global structural load effect analysis of the semi-submersible hull of floating wind turbines under still water, wind, and wave loads. Mar. Struct. **91**, 103463 (2023)

Wang, S., Xing, Y., Balakrishna, R., Shi, W., Xu, X.: Design, local structural stress, and global dynamic response analysis of a steel semi-submersible hull for a 10-MW floating wind turbine. Eng. Struct. **291**, 116474 (2023)

Wang, X., Liu, L., Wang, S., Ji, S.: Interaction between ice cover and floating platform simulated by dilated-polyhedron-based DEM. Ocean Eng. **301**, 117413 (2024)

Wen, X., Zhang, J., Ong, M.C., Kniat, A.: Comparative study of numerical modelling and experimental investigation for vessel-docking operations. Mar. Struct. **98**, 103680 (2024)

Wang, X., Yu, Z., Amdahl, J.: Effects of transverse loading on the ultimate strength prediction of stiffened panels under biaxial loads: potential improvements to the IACS rule formulation, Ocean Eng. **313**, 119642 (2024)

Wang, Y., Wei, P., Wang, Q., Dai, Z., Wang, D.: A new similarity method for the ultimate strength of box girders subjected to the combined load of bending and lateral pressure. Ocean Eng. **313**(2024), 119270 (2024)

Wang, Z., Kong, X., Wu, W.: Empirical formula to assess ultimate strength of side shell on passenger ship under combined axial and shear load based on experimental and numerical analysis. Ocean Eng. **288**, 116149 (2023)

Woloszyk, K., Garbatov, Y.: A probabilistic-driven framework for enhanced corrosion estimation of ship structural components. Reliab. Eng. Syst. Saf. **242**, 109721 (2024)

Woloszyk, K., Goerlandt, F., Montewka, J.: A methodology for ultimate strength assessment of ship hull girder accounting for enhanced corrosion degradation modelling. Mar. Struct. **93**, 103530 (2024)

Wu, J., Yan, M., Sun, X.: Simplified approach for assessing the explosive residual strength of ship hulls. Latin Am. J. Solids Struct. **21**(10), e567 (2024)

Wu, Q., Gao, Y., Xiong, J.: Quasi-static mechanical properties of composite lattice sandwich structures with enhanced face panels. Eur. J. Mech. A. Solids **97**, 104808 (2023)

Wu, W., Zhao, Y., Gou, Y., et al.: An overview of structural design, analysis and common monitoring technologies for floating platform and flexible cable and riser. China Ocean Eng. **36**(4), 511–531 (2022)

Wu, Y., Ding, J., Wang, F., Sun, Z.Z., Zhao, M., Wang, Y.M.: Research on the quasi-static collapse and instantaneous implosion of the deep-sea spherical pressure hull. Mar. Struct. **83**, 103191 (2022)

Xia, J., Chen, Z., Zhao, N., Zhao, W., Tang, Q., Cai, S.: Free-drop experimental and simulation study on the ultimate bearing capacity of stiffened plates with different stiffnesses under slamming loads. J. Mar. Sci. Eng. **12**(8), 1291 (2024)

Xiao, J.W., Liu, C., Wang, J.H., et al.: A two-way coupled fluid-structure interaction method for predicting the slamming loads and structural responses on a stiffened wedge. Phys. Fluids **36**(7), 077123 (2024)

Xie, H., Dai, X., Ren, H., Liu, F.: Experimental characterization on slamming loads of a truncated ship bow under asymmetrical impact. Ocean Eng. **284**, 115195 (2023)

Xing, S.Z., Pei, X.J., Mei, J.F., Dong, P.S., Su, S.J., Zhen, C.B.: Weld toe versus root fatigue failure mode and governing parameters: a study of aluminum alloy load-carrying fillet joints. Mar. Struct. **88**, 103344 (2023)

Xu, H., Lyu, B., Li, D.: Assessment of in-service dynamic response status of Deep Sea No.1 energy station based on field monitoring. China Offshore Oil Gas **36**(3), 190–197 (2024)

Xu, Y., Wu, J., Li, P., Kujala, P., Hu, Z., Chen, G.: Investigation of the fracture mechanism of level ice with extended finite element method. Ocean Eng. **260**, 112048 (2022)

Xue, S., Xu, G., Xie, W., Xu, L., Jiang, Z.: Characteristics of freak wave and its interaction with marine structures: a review. Ocean Eng. **287**, 115764 (2023)

Yang, B., Liu, J.: Research on the construction method of neural network-based surrogate model for wave load prediction of offshore platforms. In: Proceedings of the 34th International Ocean and Polar Engineering Conference, Rhodes, Greece (2024)

Yang, L., et al.: Quasi-static and dynamic behavior of sandwich panels with multilayer gradient lattice cores. Compos. Struct. **255**, 112970 (2021)

Yang, Y., Chen, C., Zhao, W., Fan, T.: An innovative method of assessing yield strength of floater hull for semi-submersible floating wind turbine in whole life period. Ocean Eng. **270**, 113679 (2023)

Yildizdag, M.E., Ardic, I.T., Ergin, A.: An isogeometric FE-BE method to investigate fluid–structure interaction effects for an elastic cylindrical shell vibrating near a free surface. Ocean Eng. **251**, 111065 (2022)

Yu, C.D., Bi, X.J., Fan, Y.W.: Deep learning for fluid velocity field estimation: a review. Ocean Eng. **271**, 113693 (2023)

Yu, Y., Wang, Y., Lin, Y.: IsoGeometric Analysis with non-conforming multi-patches for the hull structural mechanical analysis. Thin-Walled Struct. **187**, 110757 (2023)

Yu, Z., Amdahl, J., Rypestøl, M., Cheng, Z.: Numerical modelling and dynamic response analysis of a 10 MW semi-submersible floating offshore wind turbine subjected to ship collision loads. Renewable Energy **184**, 677–699 (2022)

Zhang, F., Yang, D., Qiu, W.: Design and experimental study of lightweight fireproof bidirectional bead bulkheads in a cruise ship. Mar. Struct. **83**, 103180 (2022)

Zhang, H., Cui, J., Liao, X., Shi, H., Guedes Soares, C.: Numerical study on the vertical response of LNG carrier in abnormal waves generated with different mechanisms. Ocean Eng. **262**, 112090 (2022)

Zhang, H., Wang, H., Cai, X., Xie, J., Wang, Y.: Novel method for designing and optimising the floating platforms of offshore wind turbines. Ocean Eng. **266**, 112781 (2022)

Zhang, J., Lu, W., Li, J., Li, X., Cheng, Z.: A data-driven methodology for wave time-series measurement on floating structures. Ocean Eng. **303**, 117629 (2024)

Zhang, S., Fu, S., Li, S., Moan, T., Xu, Y., Pan, Z.: Frequency-domain hydroelastic stress analysis considering local bending effect based on a two-step procedure. Mar. Struct. **95**, 103580 (2024)

Zhao, N., Chen, B.Q., Zhou, Y.Q., Li, Z.J., Hu, J.J., Guedes Soares, C.: Experimental and numerical investigation on the ultimate strength of a ship hull girder model with deck openings. Mar. Struct. **83**, 103175 (2022)

Zhong, K., et al.: Direct measurements and CFD simulations on ice-induced hull pressure of a ship in floe ice fields. Ocean Eng. **272**, 113523 (2023)

Zhou, L., Zheng, S., Ding, S., Xie, C., Liu, R.: Influence of propeller on brash ice loads and pressure fluctuation for a reversing polar ship. Ocean Eng. **280**, 114624 (2023)

Zhu, L., Zhou, Z., Pedersen, P.T.: Ship grounding model tests in a water tank: an experimental study. Mar. Struct. **93**, 103529 (2024)

Zhu, Z., Ren, H., Wang, X., Zhao, N., Li, C.: Methods for fitting the limit state function of the residual strength of damaged ships. J. Mar. Sci. Eng. **10**, 102 (2022)

Zhu, Z., Ren, H., Incecik, A., Lin, T., Li, C., Zhou, X.: A novel method for determining the neutral axis position of the asymmetric cross section and its application in the simplified progressive collapse method for damaged ships. Ocean Eng. **301**, 117390 (2024)

Open Access This chapter is licensed under the terms of the Creative Commons Attribution-NonCommercial-NoDerivatives 4.0 International License (http://creativecommons.org/licenses/by-nc-nd/4.0/), which permits any noncommercial use, sharing, distribution and reproduction in any medium or format, as long as you give appropriate credit to the original author(s) and the source, provide a link to the Creative Commons license and indicate if you modified the licensed material. You do not have permission under this license to share adapted material derived from this chapter or parts of it.

The images or other third party material in this chapter are included in the chapter's Creative Commons license, unless indicated otherwise in a credit line to the material. If material is not included in the chapter's Creative Commons license and your intended use is not permitted by statutory regulation or exceeds the permitted use, you will need to obtain permission directly from the copyright holder.

Committee II.2: Dynamic Response

D. Dessi[1(✉)], V. De Diego[2(✉)], S. Dhavalikar[3(✉)], M. Holtmann[4(✉)], S. J. Kim[5(✉)],
L. Kaydihan[6(✉)], L. Moro[7(✉)], T. Pais[8(✉)], A. Paiva[9(✉)], G. Storhaug[10(✉)],
H. Takahashi[11(✉)], S. Tavakoli[12(✉)], S. Wang[13(✉)], and B. Zao[14(✉)]

[1] CNR-INM, Rome, Italy
daniele.dessi@cnr.it
[2] Barcelona, Spain
[3] Indian Register of Shipping, Mumbai, India
[4] DNV, Hamburg, Germany
[5] Seoul, Republic of Korea
[6] Maritime Research Institute Netherlands (MARIN), Wageningen, Netherlands
[7] Memorial University of Newfoundland, St. John's, Canada
[8] Genoa, Italy
[9] Liege, Belgium
[10] DNV, Oslo, Norway
[11] Tokyo, Japan
[12] Aalto, Finland
[13] Lisboa, Portugal
[14] Harbin, China

Committee Mandate. Concern for the dynamic response of marine structures as required for safety, serviceability, habitability, and sustainability. This should include steady state, transient and random responses, including noise. Attention shall be given to dynamic responses resulting from environmental, machinery, propeller, and accidental loads. Uncertainties should be highlighted in the report.

Keywords: Dynamic response · whipping · springing · hydroelasticity · vibration · noise · blast · shock · underwater · wind · wave · ice · current · internal flow · propeller · machinery · equipment · pile-driving · vortex-induced · underwater · airborne · model tests · full-scale · measurement · experimental · statistical · method · system · monitoring · condition · structural · uncertainty · fatigue · extreme · accidental · comfort · damping · countermeasures · control · active · passive · acceptance criteria · standards · rules · digitalization · sensor · numerical · analysis

1 Introduction

The Dynamic Response Committee operated within a mandate that remained largely consistent with previous iterations. This continuity reflects the long-established structure of Technical Committees, which ensures a clear delineation between thematic areas, and avoids issues related to redefining the general framework. This Committee has principally focused on the dynamic response of structures in which inertial forces play a significant role, particularly in transient phenomena. In such instances, the velocity of elastic motions renders structural dissipative forces non-negligible in determining the peak amplitudes under resonance conditions. However, strict boundaries between ISSC committees are impractical, as interdisciplinarity is essential, and overlapping topics exist. For instance, there is a natural contiguity with the "Quasi-static Response" Committee when analysing time-varying forced responses. Similarly, while the "Loads" Committee focuses on recent improvements in modelling of unsteady forces, related advancements have significant implications for predicting vibrations and noise. Mitigation of noise and vibrations, whether passive or via control feedback, remains a distinctive topic of the "Dynamic Response" Committee. A long-standing scope has also been the review of literature related to noise generation and propagation, both internal and external, encompassing vibroacoustics and hydroacoustics. Additionally, the literature review specifically included the dynamic response of marine structures, such as offshore wind turbines (OWTs), given the increasing importance of vibration-related challenges and the rapid advancement of modelling techniques and technical solutions in this field. Structural monitoring, particularly in the context of dynamic measurements, was also considered. This serves as a gateway for data integration into structural health monitoring procedures (SHM) and emerging digital twin technologies. Although the present Committee was not specifically tasked with surveying digital twin applications in general (as was the case of ISSC 2022), the connection to data science is increasingly relevant. Table 1 reports the most frequently used abbreviations within the document and implicitly delineates some of the topics they represent.

The strength of this Committee, like others, lies in the diversity of perspectives, balancing academic research, which focuses on methodology, with industrial expertise, which seeks practical solutions to specific problems and enables technology transfer. In past ISSC work, this diversity has also influenced access to literature sources and the approaches followed in reviewing them. To harmonize the members' activities, the initial step was to compare the traditional ISSC review process with more structured methodologies, such as systematic review procedures. This effort benefited from the Committee's members experienced in these approaches. The guiding principle is that the Committee's work should be transparent and retraceable, achieved using well-defined keyword combinations in search engines and the adoption of common selection criteria. This structured approach aims to improve the overall quality, coverage, and thematic organization of ISSC reports. For instance, a consequence of this harmonized process has been a more extensive use of journal references and a greater openness towards scientific contributions beyond the core ship and offshore structure field. A detailed description of the implemented methodology would take up space allocated for the literature review, but the Committee is open to further communicating the lessons learned in this regard.

Committee II.2: Dynamic Response   339

Table. 1. General and most used acronyms

ABS	American Bureau of Shipping	LES	Large Eddy Simulation
AI	Artificial Intelligence	LR	Lloyds Register
(A/D) NN	(Artificial/Deep) Neural Network	OMA	Operational Modal Analysis
BEM	Boundary Element Method	RINA	Registro Italiano Navale
BPF	Blade Passing Frequency	RMS	Root Mean Square
CFD	Computer Fluid Dynamics	RPM	Revolutions Per Minute
CSD	Computer Structural Dynamics	SCR	Steel Catenary Risers
DNV	Det Norske Veritas	SHM	Structural Health Monitoring
DOF	Degree-of-Freedom	TEU	Twenty-foot Equivalent Unit
FE (A/M)	Finite Element (Analysis / Method)	TMD	Tuned Mass Damper
FFT	Fast Fourier Transform	VBM	Vertical Bending Moment
(F)OWT	(Floating) Offshore Wind Turbine	VDV	Vibration Dose Value
FSI	Fluid Structure Interaction	TTR	Top-Tensioned Risers
GHG	Greenhouse Gas	VIV	Vortex Induced Vibrations
HBM	Horizontal Bending Moment	VLFS	Very Large Floating Structure
IACS	International Association of Classification Societies	UNDEX	Underwater Explosion
IR	Indian Register of Shipping	URN	Underwater Radiated Noise
ISO	International Organization for Standardization	LR	Lloyds Register

2 Ship structures

2.1 Wave Induced Vibrations

2.1.1 Full Scale Measurements

Full-scale measurements can be used to understand the significance of wave-induced vibrations, for comparison with numerical calculations, for safety assessment, confirmation of fit for purpose, and for input to maintenance and inspection planning. Standardized installations based on hull monitoring rules from class societies can be used. While recent ISSC reports 2015/2018/2022 focused on container ships and the effect of springing/whipping on fatigue and extreme loading, with some exceptions, the current report differs. New measurement campaigns on fatigue and extreme loading have declined, while a newly emerging topic related to comfort requires new specific measurements and processing.

Recent studies have addressed measurements relative to the following ship types:

- 8600 TEU container ship (Storhaug and Jagite, 2024)
- Polar supply and research vessel, SA Agulhas II, 122 m long (e.g., Engelbrecht and Becker, 2023)
- Research icebreaker FS Polarstern, 118 m long (Soal et al. 2019)
- High-speed catamaran, HSV 2 Swift, Incat 98m long (e.g., Gebrezgabir et al., 2023)
- High-speed catamaran, Volcan De Tagoro, a 111 m Incat (Shabani et al., 2023)

Comfort papers address the topic of vibration dose value (VDV), as described by ISO 2631–1 (ISO, 1997), where the vibration pass-filtered signal of the acceleration is taken to a power of four and integrated over time, implying that whipping now and then with higher peak amplitudes may be more important than resonant springing. Dividing springing from whipping is uncommon, but root mean square assessment is an alternative for a more constant amplitude response that could result from springing. With respect to the calculation of the VDV, the acceleration in the horizontal plane is weighted differently than that in the vertical plane because of human perception. Griffin (1990) indicated a 1-h VDV threshold of 0.4 m/s above which complaints are triggered. The concept behind comfort is that, for transient response issues, it may take some time before annoyance develops. Very short time periods show large variations, whereas very long periods are not necessarily stationary. A period of 1-h is therefore regarded as a practical compromise (Griffin, 1990), but sometimes it can be reduced to 0.5 h, as adopted in some EU and UK health and safety guidance documents. ISO (1997) shows how these values can be scaled to other time periods. Engelbrecht and Becker (2023) confirmed this threshold for a polar supply and research vessel when 50% of the respondents onboard confirmed discomfort. Storhaug and Jagite (2024) indicated that in the bow of an extreme design 8600TEU the 1-h VDV level could reach 4.0, which is ten times the threshold as an indication of the upper limit for ships. Another aspect of VDV is related to the long-term exposure of personnel to vibrations, which may cause sleep disturbances. A 24-h VDV measurement primarily highlights the correlation between slamming-induced vibration and sleep disturbances, and secondary to issues related to equipment use, visual tasks, typing/writing, and equipment damage. Daily VDV levels between 8 and 10 for this polar supply and research vessel were found to provide 50% of the recorded responses related to sleep issues (Omer and Bekker, 2018). When divided by 24, this also fits the threshold of 0.4 m/s for hourly VDV levels, but the ISO scaling differs.

The topic of discomfort due to global hull girder vibration is known to cruise ships with confidential reports but lacks publicly available literature. This is also known for container ships and other types of ships during troubleshooting projects. Communication with experts suggests that reading the computer screen may be problematic when vibration levels exceed 45 mm/s as a rule of thumb. Tian et al. (2024) conducted laboratory experiments with a crew working in front of a computer while being exposed to ship vibrations. They evaluated the task performance of different computer equipment such as a mouse, trackball, touchscreen, and leap motion. The ship vibrations affected the performance, especially for leap motion, whereas the mouse and touchscreen performances were the best. Touchscreen and leap motions also resulted in relatively greater muscle fatigue. On the other hand, vibration discomfort perception is also affected by mood, as pointed out by Lorenzino et al. (2023) at low vibration intensities. The crew was placed inside a full-scale mock-up cabin exposed to three levels of vibration intensity and without vibration. The vibration levels were initially obtained from measurements performed onboard a cruise vessel. The results suggest that mood and individual differences and psychological processes are more important at low vibration intensities and that there is a gap between technical guidelines/engineering and theories about the psychological processes related to environmental experience.

Vibration may also affect the performance of the onboard equipment. The equipment includes the weapons' aiming capabilities for the navy (Liu and Chen, 2021). While this may be well known to the industry and handled daily, it has not been frequently covered in recent literature. An example from Kang et al. (2023) illustrates how vibration may affect a naval ship-based electrical switchboard, which by laboratory testing was equipped with a wire rope isolator and exposed to vibration responses in different directions according to the military standard MIL-STD-167-1A. The responses were compared with numerical calculations, demonstrating fair to poor agreement; therefore, even for this simple case, the prediction of wire rope stiffness in different directions and local stress modelling can be challenging. Another example is wire arc additive manufacturing for the repair of marine components onboard under exposure to ship vibrations. Such repair practices are already used on US Navy ships for noncritical components. Shi *et al.* (2023c) tested this technology in a laboratory. The manufacturing equipment under vibration exposure and resonance provided components with an appearance that suggested reduced quality and irregular shapes; however, the interior was dense and defect-free. This suggests that onboard additive manufacturing is a promising solution.

Regarding other uses of onboard measurements, Van Zijl et al. (2021) used accelerometers onboard a polar supply and research vessel to assess model properties, such as natural frequencies and damping, for five modes under four different environmental conditions. Frequency domain decomposition (FDD) and data-driven stochastic subspace identification (SSI) were used for the operational modal analysis (OMA). The natural frequency was relatively independent of the condition, but in ice operation and in heavy seas, the damping increased dramatically compared to that in calm water. The uncertainties in the damping estimates were significant; this is not surprising and has motivated this Committee to further investigate it in the benchmark work reported in Sect. 8. The vibration level was relatively low in the ice operation and with a short duration owing to damping. The damping ratio varied from 0.78% to 3.5%, and higher modes did not always exhibit higher damping. Bossau and Bekker (2022) studied slam response detection from the accelerometers on the same vessel based on comparing different methods such as Morlet wavelet transform and analysis of signal peaks and spectrograms and concluded that a threshold was in any case beneficial for consistent detection of slam response in varying conditions, but that Morlet was the most reliable without a threshold. Accelerometer data from another research vessel, the FS Polarstern, were used to study system identification (SID) under changing environmental and operational conditions, which may reduce the reliability of damage identification (Soal et al. 2019). Stochastic Subspace Identification (SSI) was used to obtain eigenvalues and mode shapes. The idea is to reduce the uncertainty of SID using a trained data-driven model based on environmental and operational inputs and a Kalman filter. The Kalman filter was employed to combine the model predictions with those from the SID, resulting in enhanced results.

A predictive sliding model was found to be necessary because of the burning of fuel and ballasting. The Kalman filter can also produce results when SSI is missing. Gebrezgabir et al. (2023) studied a reconstruction technique for missing measurement data using a transmissibility function and matrix concept applied to sea trial data from a 98 m high-speed catamaran. It was based on the linear response theory and needed to be trained in identifying impact events. It was found to better reproduce more intense responses, including whipping, than the lower responses. However, these more intense events are also of greater concern with regard to fatigue and extreme loading. They pointed out that the technique could prove useful in combination with FE analysis to use fewer sensors, improve redundancy, or infer responses anywhere in the structure for similar future designs.

The same high-speed catamaran was studied with respect to ride control to reduce the slamming and whipping responses by Alsalah et al. (2024). Motivation was based on several recorded damages to the bow of catamarans, and the catamaran was deliberately operated under extreme conditions. Empirical mode decomposition (EMD) was used to retain vibration-detected slamming events and was combined with logistic regression to identify extreme slamming events. Activating the ride control reduced the probability of an extreme slamming response but did not fully consider the deliberate extreme conditions to obtain slams. The vessel heading, speed, short-wave period, and high sea state were confirmed to be important, and the benefit of the ride control system was equivalent to a 1 m drop in significant wave height or 15 knots drop in speed. Shabani et al. (2021) developed a procedure to identify slamming events for the same 98 m catamaran for planning a hull monitoring system on a larger 111 m catamaran. Whipping events were implicit in the measured vertical bow acceleration used to detect wet-deck slamming, using unsupervised and supervised ML techniques. The probability of events exceeding 0.5 and 1.0 g was predicted. For the same catamaran, Shabani *et al.* (2023) also focused on motion-sickness incidents that occur at lower frequencies than those related to VDV. Slamming detection was based also on the analysis of bending stresses.

Few full-scale measurements have been carried out recently, and the effect of whipping/springing on fatigue and extreme loading should be studied more for ship types other than container ships. This includes blunt vessels, such as oil tankers and bulk carriers, and intermediate slender ships, such as gas carriers, Ro-Ro, and cruise ships, where the concern for wave-induced vibrations may differ. Concerns may also be related to other equipment, such as the effect of wave-induced vibrations on the support structures of independent fuel-gas tanks, which is a new trend. The focus on comfort is relatively new and should be studied more as it creates passenger discomfort and complaints on cruise vessels. It may also affect the safety and fatigue of crew members trying to execute their work or experiencing sleep disturbances. The correlation between the VDV and voluntary speed reduction requires further clarification. We have seen how to improve measurement data quality and operation, and how special events can be detected. More focus should be placed on how performance and safety can be improved based on decision support from full-scale measurements. This is much related to monitoring, digitalization, and structural health monitoring, which are further discussed in Sect. 5.

2.1.2 Model Tests

Model test data can be used to validate numerical methods and predict the full-scale ship dynamic response. In addition to geometric similarity, the model also needs to be similar in terms of elastic characteristics and mass distribution. Two types of elastic model continue to be commonly used: a segmented elastic model and a fully elastic model. In the first category, Chen et al. (2023d) performed experimental parametric study on the hydroelastic behaviour of a 15000-TEU container ship based on segmented model experiments in regular head waves. They analysed the influences of different test parameters (number of segments, propulsion mode of the model, and backbone shape) on the ship hydroelastic behaviour. Their results show that the propulsion mode has more impact on hydroelastic responses than the number of segments. Si et al. (2024) studied the effects of the design and fabrication of segmented ship models on wave-induced loads. Three different segmented ship models of a 20,000 TEU ultra-large container ship were designed and fabricated with different support frames and backbones. The shell and wooden frames of Model A and B were fabricated using the same method, but with different types of backbones and scales. Model C was a steel framing model. They showed that the uncertainties caused by the repeated tests, moment of inertia, calibration coefficient, and wave height affected significantly the results in all the three model tests, while the uncertainties caused by the ship line, height of the CG, and acquisition system could be ignored. Grammatikopoulos (2023) gave an extended survey about elastically scaled physical models by reviewing almost all the related publications. He classified the flexible models by type and research scope, highlighting the main features and typical layouts (e.g., number of segments, inner structure, segment-backbone connection). Fully elastic models are much less common compared to segmented elastic models. Chen et al. (2024b) designed, constructed and tested a fully elastic ship model to investigate hydro-elastic responses of container ships. The global rigid-body motions and elastic structural responses, including the vertical shear force, vertical bending moment, and torsional moment, were measured. Large-scale model sea trials pose considerable difficulties and are expected to become a prominent focus in the future. Jiao et al. (2021b) analysed the wave-induced VBM and HBM on a large-scale segmented model (1:25 scale), equipped with a rectangular tubular steel backbone. The ship navigation information (speed, heading angle, and position) and ship motions (roll angle, pitch angle, and vertical speed) were measured using a Global Position System/Inertial Navigation System (GPS/INS) device that was installed at the centre of gravity (COG) of the model. The VBM and HBM on the ship model under complex wave conditions could also be accurately measured. Therefore, the measurement data obtained from the sea trials demonstrated the feasibility and usefulness of open-water seakeeping tests by large-scale models. Li et al. (2022a) conducted experimental study on stern slamming and global response of a large cruise ship in regular waves. They found that the longitudinal deadrise angles were usually small in the following waves, which resulted in a large slamming force loaded on the stern in a short time. In addition, the whipping response of the hull girder was excited. Ibrahim and Judge (2024) discussed the slamming effects on global hull girder loads using a hydroelastic model test of a semi-displacement vessel in both regular and irregular waves. Their data show significant increase in global hull girder loads due to slamming at Froude numbers of 0.45 and 0.7.

In recent years, extreme waves have received increasing attention. Tang et al. (2022) studied the nonlinear bending moments of an ultra large container ship in extreme waves based on a segmented model test. In their study, ship responses in harsh waves were analysed and the largest wave height to wavelength ratio exceeded 1/10. They found that the nonlinearities of VBM are mainly caused by high-order harmonic components, and these high-order harmonics affect the sagging VBM more than the hogging VBM. Lu et al. (2022, 2024) chose a 20,000 TEU containership with an overall length of about 400 m as the target ship to investigate the hydroelastic response. A set of systematic model tests was carried out in the seakeeping wave basin of the China Ship Scientific Research Centre (CSSRC). The experiments provided benchmark model test data for the validation of the numerical software and methods. Si et al. (2022) conducted an uncertainty analysis of the linear VBM by using model tests and numerical predictions. Hirdaris et al. (2023) reviewed the uncertainties associated to hull girder hydroelastic response and wave load predictions. In their review, they concluded that challenges in realizing and modelling uncertainties can be attributed to the ambiguity of model tests, systematic use of data emerging from computational, model, or full-scale methods, and limitations of numerical methods.

In the future, more research could be conducted on ship dynamic response in oblique long-crested waves, short crest waves, rogue waves, large-scale model sea trials, fully elastic models, bending-torsion coupled characteristics, and so on.

2.1.3 Analysis Methods

The wave-induced vibrations of a ship hull girder are referred to as springing when they are associated with a resonance phenomenon and whipping when they are caused by transient impact loading. Both phenomena excite the lower global hull vibration mode shapes and high-order modes with respect to the duration of impact loading. Consequently, both phenomena increase fatigue and may cause extreme loading on the hull girder. Analysing these phenomena requires both a hydrodynamic model and a structural model. These models are typically coupled using an algorithm with the level of complexity depending on the specifics of the numerical models involved. The interaction between the fluid and structure is generally obtained by either one-way or two-way coupling methods in numerical FSI problems. In one-way coupling, the hydrodynamic loads are transferred to the structural model but are not affected by the resulting elastic displacements; in two-way coupling, structural deformations update the boundary conditions for the load calculation. Two-way coupling is considered computationally expensive because of mesh deformation needed by CFD codes and use of inner loops to advance the solution at each time step, thus requiring long solution times.

Jiao *et al.* (2021a) adopted a two-way FSI using a CFD–FEA numerical method to predict motions, wave loads and hydroelastic responses in regular waves for a flexible S175 container ship. High-order hull girder vertical vibrations of harmonic springing up to the 12th order and up to 4-node whipping responses were successfully reproduced, indicating that the CFD–FEA co-simulation method is reliable and capable of simulating ship nonlinear hydroelastic responses in waves. While comparing the results of the study with the theoretical and experimental data in the literature, it was observed that the springing and whipping loads obtained by different numerical and experimental methods can differ significantly. The uncertainty and difference in 2-node whipping loads were found to be larger, which is probably due to the difference in the structural parameters of the ship models used in different studies. Therefore, it is stated that structural details of typical hull forms, such as the S175 hull should be made available to the scientific community in a form suitable to carry on benchmark studies for instance by ITTC (International Towing Tank Conference) and ISSC committees. As a continuation of the above study, Huang et al. (2022) performed a similar study to evaluate the global motions, wave loads, and springing and whipping loads on the same S175 containership model using a two-way fluid-structure coupling method. Verification and uncertainty analyses of the influence of the fluid mesh and domain, time step, fluid viscosity, and number of structural elements on the co-simulation results were performed systematically, focusing on ship motions and loads. A validation and benchmark study was also conducted by comparing the wave load responses of the S175 ship (obtained by the present FSI method) with experimental and numerical data in the literature. The authors observed that data or results on springing and whipping loads are not accompanied by an evaluation of uncertainties in the available literature. The whipping responses are affected by a wide variety of factors, including internal influences such as the model's structure, distribution of stiffness, number of segments, modal behaviour, and structural damping ratio, and external influences such as wave type, wave steepness, and modelling of impact loads.

Jagite et al. (2021) presents a new approach developed to compute the nonlinear whipping response using a fully coupled hydro-elasto-plastic model with a one-way fluid structure interaction method. In the proposed method, the structure is modelled as two nonuniform Timoshenko beams connected via a nonlinear hinge, whereas the hydrodynamic part is modelled using the 3D BEM. The time-domain simulation was performed using frequency-dependent hydrodynamic coefficients to calculate the diffracted-incident wave loads. In addition, the radiation force is calculated from the memory-response functions and the past history of the velocities. The nonlinear pressures resulting from slamming were calculated for multiple 2D sections and later integrated over the 3D hydrodynamic mesh. Finally, the hydro-elastoplastic problem was solved in the time domain by numerical integration. This means that when a ship is subjected to an extreme loading scenario, the hull girder may suffer from local buckling or plastic deformation, which is different from the linear elastic assumption. This paper presents a numerical investigation of the nonlinear whipping response on a database of 14 container ships ranging from 160 m to 350 m. The outcome of this study provides useful information regarding the effects of nonlinear structural behaviour on the slamming-induced whipping response of ships. For ultra-large container ships, the dynamic ultimate capacity factor is defined as the ratio between the linear and nonlinear whipping responses, which varies from 1.001 to 1.013. Therefore, it was concluded that nonlinear structural behaviour can be neglected in the analysis of the maximum hydroelastic response. Jagite et al. (2022) also analysed the dynamic ultimate strength capacity of ultra-large container ships under realistic loading scenarios to find out the effect of the whipping-induced stresses on the dynamic capacity and additional stress reserves. The paper is based on a series of dynamic collapse analyses for a 16000 TEU container ship. Two finite element models with different longitudinal extensions were used: a three-frame bay model and two-hold model. The hull girder was subjected to a combined bending moment resulting from a long-term hydroelastic analysis, with lateral and local loads given by different cargo loading cases. The dynamic ultimate strength was calculated using an implicit nonlinear FEM solver, in which both material and geometrical nonlinearities were taken into account. The dynamic capacity, defined as the maximum load that the structure can withstand without collapsing, was determined using an iterative approach, where each iteration required an independent time-domain analysis. The numerical results are discussed, and the dynamic load factors are derived as the ratio between the dynamic and quasistatic capacities. Finally, it was shown that the strain rate effect is negligible in the analysis of the ultimate strength of ship structures subjected to whipping-induced stresses.

Pal *et al.* (2023) presented an extreme value distribution method for the VBM, which was developed using a reduced-order model (ROM) based on CFD-FEM coupling analysis. The reference ship was a post-Panamax containership for which basin tests were performed in a previous study. To build the ROM, the VBM for all components, that is, wave-induced, whipping, and springing components, are calculated separately and superimposed. The vessel motions and wave-induced components were estimated from the transfer functions derived from the nonlinear strip theory and later corrected using CFD-FEM coupling. The whipping component is evaluated using von Karman's momentum theory with proper adjustment to the specific hull parameters and additionally corrected using CFD-FEM coupling. In this way, a new semi-empirical springing ROM was developed by combining coupled CFD–FEM analysis and experimental results. The effects of springing and whipping in calculating the extreme value distribution of the VBM were described for different probability levels. The effect of consecutive bow-stern slamming was also discussed. The development of this method will aid in attaining real-time shipload prediction without expensive computational costs owing to the coupled CFD-FEM.

Parunov et al. (2024) presented the results of a benchmark study organized by the Marstruct Virtual Institute on the motion and global wave loads on a warship model in regular waves, a Canadian Patrol Frigate, for which experimental results were available for comparison. The aim of this study was to quantify the uncertainty in numerical whipping predictions. Nine institutions participated in the benchmark with six codes to simulate the hydroelastic responses. The numerical methods employed include nonlinear strip theory, 3D BEM formulated in the frequency and time domains, and CFD. Euler and Timoshenko beam theories were used to model the stiffness of the hull girder. Both one- and two-way coupling models were used. To calculate the slamming loads, experimentally based methods, CFD and momentum theories were employed. The comparison between the different approaches includes wet natural frequencies of vertical bending vibrations, vertical ship motions, and both vertical wave and whipping bending moments at midships. Wave-induced and whipping responses were analysed for regular head waves of different steepness and for two ship speeds. The frequency-independent model error, which is commonly used for uncertainty quantification of rigid body seakeeping responses, was extended to quantify uncertainties in whipping bending moments.

It was found that fully coupled CFD-FEM models provide results consistent with measurements, but such simulations are prohibitively computationally expensive, and the interpretation of results can be challenging. The combination of the potential theory seakeeping method with correction based on CFD-FEM simulation for a limited number of cases is a promising alternative.

Lu et al. (2023a) proposed a time-domain 3D nonlinear hydroelastic method considering asymmetric slamming in oblique waves with one-way coupling method. The asymmetric slamming loads and whipping responses of an ultra-large container ship in oblique regular waves were investigated. According to this study, steady time-domain results can be obtained using the developed hydroelasticity algorithm, facilitating the analysis of hydroelastic responses with asymmetric slamming. The numerical algorithm effectively captured the features of the wave-load responses related to slamming and whipping. Asymmetric slamming leads to a significant increase in the high-frequency components. This effect is attributed to the influence of torsional and horizontal vibrations on the system. Higher-order whipping is observed because the natural frequency is close to integer multiples of the wave encounter frequency, leading to significant amplification of the high-frequency components.

Riesner and Moctar (2021a, 2021b) introduced a numerical method to compute the global elastic vibrations of ships advancing at a constant forward speed in regular oblique waves, using a two-way coupling method. The structural solver was based on a beam-element approach that considered vertical and horizontal bending, as well as nonuniform torsion. The beam-element approach was coupled with a weakly nonlinear time-domain BEM to compute the elastic response of ships to regular waves. To validate the new numerical method, the results were compared with results from experimental model tests of a post-Panamax containership at a constant forward speed of 15 knots in regular waves at 150 and 120 degrees heading. In general, the predictions from the new numerical method were in agreement with the model test measurements. Resonant second-, third-, and fourth-order springing vibrations occurred at encounter wave frequencies close to one-half, one-third, and one-fourth of the natural frequency of the two-node bending mode, respectively. It was found that it is important to consider many waves of different frequencies to cover all the higher-order effects. The new method can predict springing-induced higher-order resonant vibrations and the associated VBM for a ship in head waves. It was demonstrated that it is essential to consider a stationary wave system to reliably predict the springing-induced vertical bending moments for ships at a forward speed. Furthermore, it has been shown that torsion and horizontal bending are strongly coupled, and pure torsion or pure horizontal bending is not evident. Coupled, linear, and higher-order springing effects were observed for one- and two-node torsion horizontal bending vibrations.

Li et al. (2024a) present the development of a time-domain hydroelastic numerical model for investigating the slamming and whipping loads imposed on a ship traveling at different forward speeds. The numerical model integrates a beam model, 3-D Rankine panel model, and 2-D modified Logvinovich model. The computational results were compared with those of a segmented model test of a large cruise ship. The convergence of the model with different numbers of elastic modes and choices of the 2D hull profiles to calculate the slamming force was investigated to ensure that the numerical results were stable. The main contributor to the whipping response was the vibration of the first-order elastic mode. The slamming force was sensitive to the longitudinal interval of the slamming profiles, particularly in the stern. The numerical and experimental whipping responses were found to be in good agreement, and the local slamming pressure based on the slamming profile was more accurate when the effect of the ship's forward speed was considered.

Tang et al. (2023) proposed a fully nonlinear hydroelastic numerical method for the one-way fluid structure interactions of the ship navigating in head seas. The fully nonlinear method was first verified to be accurate for wave diffraction around a fixed vertical cylinder and was then applied to ships. The numerical model consisted of a hydrodynamic part and a structural part based on the boundary element method and non-uniform Timoshenko beam theory, respectively. Hydrodynamic and structural models were coupled to calculate the symmetric responses of an ultra-large container ship (ULCS), and comparative investigations against model tests and numerical predictions with two other large container ships were carried out. The numerical heave and pitch motions, time records of the total VBM, amplitude of the first-order harmonic, and phase difference of the first-order and high-frequency VBM were compared with the experimental results and were found to be in good agreement. It was observed that high-order harmonics of the VBM may couple with the whipping responses when their frequencies are close to the first-order natural frequency. One interesting result is that the seventh-order harmonic of the VBM may couple with the whipping load and hence the amplitude is magnified. Vijith and Rajendran (2023) developed a 2D time-domain code based on strip theory to investigate the VBM acting on an ULCS in head and oblique waves with one–way coupling method. Simulations were carried out for both regular and irregular waves, and the results were compared with the experimental results. The proposed numerical method can capture the springing and whipping effects due to the hydroelastic effect. It can be observed that the springing responses increased with the encountering frequency. The contribution of the flexible response to the total VBM response was quantified to be a maximum of 12%. The effect of speed on the springing response was numerically investigated for waves with a length equal to the ship length. The springing component contributed almost 33% of the first harmonic at the highest speed. The VBM in irregular oblique seas was numerically simulated and compared with the experimental results. It was highlighted that the springing and whipping responses were better predicted by the numerical method.

Liu *et al.* (2023c) present a study about a model based on Long Short-Term Memory (LSTM) neural networks for the reconstruction and short-term forecast of global whipping responses using ship motion data. The model was fine-tuned and trained based on a dataset of heave, pitch, surge motion data, vertical acceleration data, and vertical bending moments from a large cruise ship model experiment by utilizing 5-fold cross-validation. The established model was tested with a pre-split test dataset and used to perform the reconstruction and multistep ahead prediction (forecast) of the VBM amidship in the following and head seas. The reconstructed results demonstrated a high degree of fit with the measured VBM amidship under both regular and irregular waves. The prediction accuracy tends to diminish as the forecast horizon increases. Comparisons show that the LSTM model has a strong ability to capture whipping responses, and the proposed model is more accurate when the whipping responses are stronger.

Few papers proposed significant advancements on the calculation of structural responses following water impacts due to sloshing and green water. Ju et al. (2022) proposed a method to predict sloshing pressures and hot-spot stresses in LNG cargo containment systems (CCS) in real time using inner-deck stress measurements (numerically simulated in this study). They introduced an impulse/space superposition method to enhance sloshing pressure estimation accuracy and applied a mode superposition method for stress prediction. An additional mode selection process improved modal amplitude estimation. Numerical simulations validate the methods, demonstrating their potential for real-time structural health monitoring and assessing design sloshing pressures during vessel operation.

Researchers have used complex analysis methods for application and comparison studies of the whipping and springing effects. It is observed that two-way coupling of the full 3D FEM and CFD models is still not feasible for a wide range of applications because of the computationally expensive nature of the problem. Therefore, researchers are still looking for improved methods for analysing this problem. One of the outcomes of these studies is that the combination of fully coupled CFD and the FEM provides consistent results with measurements, but the combination of the potential theory seakeeping method with a correction based on CFD-FEM simulations is a good alternative. It is clear from the papers that the simplicity of the one-way coupling approaches with a beam assumption for the structural model, together with the BEM for the hydrodynamic model, are still attractive to researchers and is worth improving. There is still a need for research on the effects of nonlinearity to obtain a better approximation of two-way coupling models that use full 3D structural and hydrodynamic definitions of the physics involved.

There are not many papers and data on the accuracy and uncertainty of the results, especially for the hydroelastic responses of springing and whipping loads for the S175 benchmark ship. One possible reason for the uncertainty in these responses is that the structural models used in the numerical calculations and experiments developed by different researchers can be quite different. As stated in a previous paper, there is a need for the structural details of typical hull forms, such as the S175 hull, to be uniform, provided as publicly available and generally accepted by the scientific community of ship hydroelasticity. Studies using neural networks for the reconstruction and prediction of global whipping responses using ship motion data may be a good alternative to computationally expensive solutions. They tend to produce good results as long as the model is fine-tuned and trained based on an existing ship dataset of vertical acceleration data, vertical bending moments, and so on. The reliability of these models remains heavily dependent on how well they are trained. It may take some time for these applications to gain thrust from industry. To increase the reliability of these models, there is still a need for additional research showing how the training and fine-tuning of neural network systems should be performed from the perspective of effort, time, and resources for a better approximation.

2.2 Ice Induced Vibrations

Shipping in polar seas has become more frequent, largely because of the reduction in the extent of ice in the Northern Hemisphere. However, polar seas are not completely ice-free; ships may still encounter ice masses ranging from small fragments to large icebergs, or even thin ice covers. This has led to the emergence of various research areas including studies on ship resistance in ice-covered waters, ice loads on ships, and ice-induced vibrations. However, the first two topics are beyond the scope of this section, and we only cover ice-induced vibrations.

Research on the ice-induced dynamic responses of ships can be divided into three main clusters. The first group focuses on advancing our physical, mathematical, and numerical understanding of the dynamic response of ships when they collide with ice floes or icebergs, which is viewed as ice-structure interaction in many different studies (i.e., ice bending or crushing under the ship force along with the excited plate response). In the numerical study of this problem, the interaction between ice and the structure must be modelled. A contact force that lasts for a very short duration emerges, and leads to dynamic responses in the structure, ultimately causing plastic deformation. This force can also occur repeatedly. This numerical simulation involves modelling plastic and large deformations, performing numerical FEA, and exploring various options for representing the mechanical behaviour of both the ice and structure of the ship. An example of such research is the numerical study conducted by Yi et al. (2021). The authors simulated the plastic response of a ship structure under the load caused by an ice plate, modelling the ship using a material model that incorporates plastic deformation with kinematic hardening, and the ice plate using a material model that represents isotropic elastic behaviour until failure. A recent study was conducted by Li *et al.* (2023a), who modelled the responses of a cargo containment system (CCS) under the load caused by icebergs. The authors developed a new numerical approach to first calculate loads and then estimate structural responses. Their results showed how the ice load initially affects the side plate of the ship and is then transferred to the inner hull structure, with the transverse girder plate responsible for this transfer.

They also demonstrated that an ice collision load was transmitted to the bottom of a secondary insulation box by compressing the masticrope. Interestingly, the CCS responses relative to the transverse girder plate were greater than those between the plates. Nonetheless, this investigation overlooked hydrodynamic effects, thereby limiting the scope of its practical applications. Hydrodynamic effects are believed to affect the dynamic responses of structures, as the additional added mass and damping (associated with different structural modes) would lead to fluid stresses, which may couple with structural stress. Such a problem, ship-ice-water interactions, is very complicated and has not received much attention until recently, despite its importance. A simple but valuable numerical model that considers ship-ice-water interactions was recently developed by Shi *et al.* (2023d). In this research, the first steps towards considering hydrodynamic effects on the vibration of structures interacting with level ice were considered. The fluid-structure interaction problem is solved using an arbitrary Lagrangian-Eulerian method.

In the second research area, researchers modelled the response of ice subjected to ice impact, which resembles the impact of ice on a ship plate, a phenomenon occurring quite often in the real maritime world. The impact problem of ice on a plate can be modelled using either a single impact process or repeated impacts, with the latter leading to stress accumulation in the structure. In addition, the ice crushing phenomenon, which occurs during the impact process, can affect the local stiffness and contact forces. Various research teams worldwide have conducted laboratory tests. Cai et al. (2022a) conducted a simple physical test aimed to model the dynamic responses of steel plates subjected to repeated ice loads. The authors also employed nonlinear FEA to model the problem and enhance our understanding of the underlying physics. The plate structure was modelled using a piecewise linear plasticity model and the ice structure was modelled using a soil and concrete plasticity model. Their study reported that it is easier for the plate to reach a pseudo-shakedown state under repeated ice impacts than under repeated rigid mass impacts.

Another interesting experimental study was conducted by Li et al. (2022d), who experimentally modelled the dynamic responses of a sandwich panel subjected to ice wedge and rigid wedge impacts (a single impact, not repeated). It was observed that the peak force and structural deformation under ice-wedge impact were lower than those of a plate impacted by a rigid wedge, highlighting the effects of ice crushing on the load and resulting responses. Other physical and numerical studies simulating dynamic responses to single or repeated ice impacts were reported by Truong et al. (2022), Yu et al. (2022d), Xiao et al. (2023) and Jang et al. (2024). Truong et al. (2022) showed that, under repeated impacts, the stiffness of the plate increases with the number of impacts, leading to a gradual increase in the applied force. Yu et al. (2022d) reported that under repeated loading from ice, a stiffened plate may not experience failure. However, the deformations became more localized, and the plate exhibited buckling. Xiao et al. (2023) compared the responses of a Polyvinyl Cloride (PVC) plate and equivalent weight hull plate under repeated ice loading. They observed that bending was the dominant response mode, with the PVC plate exhibiting better resistance to ice loads. In addition to these impact tests, conducted either numerically or physically, Cai et al. (2022b) developed a new theoretical model for ship-ice collisions using rigid-plate theory. They demonstrated that their method can be applied to extreme events in which ice impact loads are highly significant. The predictions of the analytical model were compared with those obtained from numerical simulations achieved via a nonlinear FEA model.

Another research direction in this field (third cluster), although smaller in terms of the number of publications, focuses on the ice-induced vibrations of the shaft line. One example is the study by Zambon et al. (2022) in which the authors measured and reported ice-induced vibrations in the shaft line of an icebreaker. Later, Zambon et al. (2023) investigated the influence of ice on the transient torsional dynamics of a polar-class propeller shafting system by developing a numerical model of the shaft line for the Canadian Coast Guard icebreaker Terry Fox.

Most studies on ice-induced vibrations in ships have relied on numerical methods or experiments that focus on the repeated loading of ice on plate structures. However, many of these studies overlooked hydrodynamic effects, particularly fluid-structure interactions. Observations in this area have been insightful, offering a significant physical understanding of ice crushing and bending during hull-ice interactions, as well as the hull's response. Ice-induced vibration studies have gained increased attention in recent years, leading to the development of new vibration measurement methodologies and mathematical models with practical applications in polar shipping.

2.3 Machinery and Propeller Induced Vibrations

Through its Greenhouse Gas Strategy towards 2050, the International Maritime Organization (IMO) has set ambitious goals in the global fight against climate change, including a 20% reduction in emissions by 2030, a 70% reduction by 2040 (compared to 2008 levels), and the ultimate goal of achieving net-zero emissions by 2050. The IMO recognizes that research and development activities are essential for achieving these goals. Addressing marine propulsion, alternative low-carbon and zero-carbon fuels, and innovative technologies is crucial. The committee reviewed recent research and development efforts and has provided a high-level overview of how different measures affect the shipboard dynamic response regarding machinery- and propeller-induced vibrations. In the past, research activities primarily focused on vibration control and mitigation of habitability and comfort level onboard, and propulsion efficiency has now emerged as a significant driver for innovation. The propeller and machinery onboard are the main vibration excitation sources. Therefore, any changes in power generation and propulsion can affect onboard vibrations. This aspect is primarily discussed in terms of machinery-induced vibrations in the following subsections. For propeller-induced vibrations, recent research is still predominantly focused on the propeller–rudder interaction, cavitation phenomena, and related numerical methods.

2.3.1 Propeller Induced Vibration

The propeller excitation forces mainly consist of the surface and bearing forces. The surface force acts directly as a pressure fluctuation on the shell plate of the ship structure above the propeller, and the bearing forces act as forces and moments on the ship structure through the propeller shaft. It is important to accurately evaluate the absolute value of each excitation force and its transfer function to a ship structure. These forces may impact accommodation spaces, reduce habitability, and affect the ship structure, causing fatigue damage or mechanical problems to machinery and equipment. In recent years, there has been a trend of increasing the diameter of propellers and reducing their rotational speed to improve their propulsion efficiency and lower CO_2 emissions. Consequently, it is necessary to pay attention to the increase in the surface force owing to the decrease in tip clearance (distance between the hull and tip of the propeller) and the possible excitation of hull resonances at a lower frequency because of the decrease in the rotation speed. Nevertheless, research on propeller excitation forces, particularly concerning the vibrational effects on accommodation spaces, hull structures, machinery, and equipment, remains limited.

Ship rudders are typically placed behind the propeller. This standard layout is effective for rudder performance; however, it causes pressure fluctuations involving the rudder, which lead to vibrations of the rudder itself and steering gear components. Furthermore, these vibrations are transmitted to the hull structure, thereby resulting in structure-borne noise. Regarding the analysis of the flow around the rudder, research mainly dealt with the rudder effectiveness and propulsive performance. In recent years, research has focused on the vibration of the rudder as well. Magionesi et al. (2023) analysed the dynamic response of a marine rudder located in the wake of a propeller (referenced as INSEAN E779A) using a one–way FSI approach. The contribution of the hub and tip vortices to rudder excitation was investigated, and this analysis can be helpful in designing a propeller blade shape capable of weakening or altering the noise and vibratory effects of the ship. The time-dependent pressure distribution on the rudder, obtained through Detached Eddy simulation (DES), was used to evaluate the pressure field, which was the input for a structural solver to determine the resulting deformations and stresses. The evaluations considered the propeller at a neutral rudder deflection angle δ and then equal to ±4°, which are representative of weak manoeuvring conditions. The results indicated that at δ = 4° the hub and tip vortex systems contributed equally to the rudder response, whereas the tip vortex contributed more than 60% to the overall displacement at δ = −4°. The hub vortex contribution was negligible at the neutral position. Lock-in and resonance may occur when the excitation frequencies are close to the natural frequencies, leading to severe vibrations and structural damage. Zhang et al. (2022e) examined the lock-in process of rudder vibrations induced by the propeller wake by considering a two-way coupling FSI. The vibrations of the rudder fixed at one end were analysed under different propeller rotational speeds, and the frequency-domain characteristics and vibration modes under different excitation frequencies were compared. The turbulent wake of the propeller was solved using large-eddy simulation (LES). The frequency range of the lock-in process was quantitatively investigated and specified. The results indicated that the vibrations of the rudder were locked at the first natural frequency (f_1) when the BPF (blade passing frequency) was lower than $1.83 f_1$. Intense vibrations were observed at the BPF when the BPF exceeded $1.83 f_1$. The application of these methodologies is expected to facilitate the identification of the causes of damage in actual ships, which may lead to increased noise, vibrations, and potential blade damage.

Zhang et al. (2023f) investigated the relationship between propeller fatigue life and propulsion efficiency under cavitation conditions using a FSI analysis. The hydrodynamic performance of the propeller under different cavitation numbers was studied using the finite volume method (VOF), and the S-N curve of the blade material was used to evaluate the fatigue life of the propeller model. The findings emphasize the significant impact of cavitation on blade service life and vibration. The vibration amplitude decreases with an increase in the advance coefficient, and the maximum vibration amplitude occurs at the critical point of cavitation. Cavitation appears to induce vibrations in the propeller blade, posing a risk of damage to the blade root. On the other hand, Chen et al. (2022) carried out the investigation on hydrodynamic damping characteristics of elastic rotating propeller blades theoretically and experimentally. First, the modal hydrodynamic damping ratio of a propeller blade rotating in water was theoretically predicted based on cantilever plate theory. Subsequently, a series of experiments was conducted to determine the mechanical, acoustic radiation, and hydrodynamic damping of the first-order bending mode of the rotating blade. The results indicate that the theoretical predictions were in good agreement with the experimental results. Hydrodynamic damping increased linearly with the rotational speed of the propeller. The theoretical analysis also showed that the modal hydrodynamic damping ratio of the first-order bending mode was significantly larger than that of higher-order bending modes. It was concluded that hydrodynamic damping plays a dominant role in the damping of rotating blades; therefore, it cannot be ignored in vibration analysis. For twin-skegs, the wake field behind the tail is nonuniform because of the shielding of the skeg, which causes propeller-induced vibration of the ship. Pan et al. (2022) carried out optimization of the skeg shapes by studying propeller surface force to reduce hull vibration for the twin-skeg type ship. The results showed that the variation in the skeg parameters represented by the exit angle and tail thickness had a significant influence on the unevenness of the wake field and the propeller induced fluctuation pressure. It was possible to reduce the vibration of the ship by modifying the skeg parameters, which provided an optimum direction for the vibration reduction design of a RO-RO ship with twin-skeg.

A twisted submarine rudder was designed to reduce the unsteady force and improve the propulsion efficiency. Ye et al. (2023) studied the resistance, pressure coefficient, and flow field of a fully appended SUBOFF submarine (a kind of hull form developed for submarines) using the shear stress transport (SST) k–ω turbulence model. The design concept of the twisted rudder was comprehensively analysed, and the basic form of the rudder was determined. It was concluded that the circumferential velocity in the propeller wake, and the pulsation amplitude of the unsteady force were significantly reduced. The main BPF amplitudes of the force and torque decreased by more than 40% and 35%, respectively. This type of research provides useful ideas for reducing the vibration response of ship structures.

Dai et al. (2022) investigated the problem of vibration in an axial flow waterjet propeller with high power density and low specific speed. Based on a coupled fluid–structure analysis, vibration characteristics such as the unsteady excitation of the rotor, fluid–structure interaction effects, and fluctuating pressure of the inner flow of the waterjet propeller under different speed conditions were studied. The numerical results for the hydrodynamic performance were compared with the experimental results and showed good agreement with the experimental data. For instance, the results showed that the maximum fluctuating pressure value was in the tip clearance region close to the blade inlet edge.

Vortex-induced vibration (VIV), resulting in the lock-in phenomenon where the vibration frequency of the structure is locked in its resonance frequencies, is a potential cause of vibration fatigue and/or singing of the propellers of large merchant ships. Lin and Kim (2022) investigated a deep-learning-based indirect VIV detection algorithm that used the vibration measurement data of a hull structure instead of a propeller during a sea trial. RPM frequency representations of the vibration signal, the so-called waterfall chart of 2D data, were measured and fed into the proposed convolutional neural network (CNN) architecture for the detection of the VIV frequency and RPM range. To generate a large dataset, a method based on the modal superposition method, instead of a time-consuming fluid-structure interaction analysis, was proposed. To test the proposed VIV detection algorithm, investigations were conducted in a laboratory on a small-scale ship propulsion system designed in such a way that the vortex shedding frequency and underwater natural frequency match each other. The validity of the proposed algorithm was tested using the structural vibration signal measured at the hull structure of a crude oil carrier, in which propeller singing occurred during a sea trial. It was concluded that a detector replacing vibration experts could be developed if sufficient actual ship data were used to train the network model developed through this feasibility study.

As previously discussed, several studies have been conducted on the propeller excitation forces. A numerical analysis of the rudder vibration induced by the propeller excitation force was performed using a commercial software. To clarify the actual phenomenon and verify the validity of the analysis, full-scale measurements of the rudder vibration are expected.

2.3.2 Machinery Induced Vibration

As outlined in the introduction to this topic, the IMO has outlined short to long-term measures to achieve net-zero emissions. As a short-term measure, the IMO aims to enhance energy efficiency in international shipping and sets an ambitious goal: zero or near-zero GHG emission technologies, fuels, and energy sources should represent at least 5%, with a striving target of 10% of the energy used by international shipping by 2030. Thus, the energy-efficiency design index (EEXI) for existing ships came into force on January 1, 2023. Thus, it became mandatory for all ships to calculate their attained EEXI to measure their energy efficiency and to initiate the collection of data to report their annual operational carbon intensity indicators (CII) and CII ratings. The present committee reviewed recent research and development efforts and has provided a high-level overview of the impact of different measures on shipboard dynamic response.

The two main options to reduce carbon intensity in compliance with EEXI and CII regulations are power limitation in both engine and shaft, and energy efficiency improvements. In most cases, limiting the power range is accompanied by limitations in the range of operational revolution. The adjusted revolution rate may cause resonance conditions, resulting in increased structural or machinery vibration. It is recommended to carefully assess the risk of excessive vibrations resulting from power limitation solutions as well as other solutions, such as the deactivation of one or more cylinders. Various approaches can enhance energy efficiency without slowing down ships as well. These measures apply to both existing and new fleets. It is important to note that the machinery onboard alongside the propeller is a significant vibration excitation source. Therefore, any changes in power generation can affect onboard vibrations. In the following four areas of energy efficiency improvement and its implication on vibrations are discussed.

New technologies.

Equipped with new technologies, onboard DC (direct current) grids have proven to have several advantages, including a significant reduction in specific fuel consumption and reduced emissions. This can be achieved by installing an optimized variable-speed diesel generator. The engine speed is adjusted according to the required power, which provides the highest efficiency for the combustion process. However, this also causes the vibration characteristics of the system to vary, resulting in stronger vibrations at specific speeds. Most generators exhibit resonance at lower operational speeds, a phenomenon that is of little concern for fixed-frequency generators that always operate at a constant speed. The effects of variable-speed power generation on vibrations were studied by Vuong et al. (2021). Based on experimental investigations, different solutions were proposed to improve the vibrational behaviour. Redesigning or changing the firing order is a possible solution but at a high cost. Altering the natural frequency by changing the rigidity of the frame is an economical and efficient solution for diesel generator sets. In addition, replacing the rubber vibration isolator with a spring is the best option for experimental systems. While the previous study focused on the gen-set vibration itself, Lekatompessy (2021) simulated the effect of different engine speeds on the vibration amplitude in an engine room of an aluminium catamaran. The investigation aimed to analyse the effect of engine speed with variations of 500, 1000, 1500, and 2300 RPM on structural vibrations without additional load. This study indicated that the highest vibration amplitudes appeared at 1000 and 1500 RPM and that, not surprisingly, the structural response can vary significantly depending on the engine's revolution rate. These two studies underline the need to carefully align the power generation concept, genset, and structural design to avoid excessive machinery-induced vibrations. Carbon capture and storage (CCS) is another new technology used to reduce CO_2 emissions in the shipping industry. Hua et al. (2023) summarized research progress in ship-based carbon capture technology. Noise or vibrations induced by CCS systems have not yet been reported. However, although this technology has been proven for land-based operations, the effect of the shipboard environment, including vibrations, has not yet been studied. In scrubber retrofits, when added to a slender funnel, the additional heavy weight can substantially change the natural frequency and dynamic response of the funnel. In such a case, or if installed on a new filigree framework in the aft ship, the retrofitted scrubber and its supporting structure may exhibit elevated vibrations, whereas permissible vibration levels are seldom defined. It is therefore suggested that same as for scrubbers retrofits, the system integration including the storage tank call for a careful consideration of vibration aspects and should be subject for future research. In addition to the above-described vibration-related challenges of variable-speed generator sets of DC grids, such power generation concepts also allow for the simple, flexible, and functional integration of energy sources, such as shaft generators, batteries, and fuel cells.

Hybrid systems

The combination of a battery energy storage system and conventional engine generates less noise and vibration when the ship is running on battery power. If shoreside power is on offer at the ports you visit, the electricity already comes from up to 50% of renewable sources. Quantitative investigations about the achieved reduction in emitted machinery noise for different power-sources combinations are still missing.

Alternative fuels

To enhance safety beyond that of traditional fuels, storage tanks for alternative fuels (such as types A, B, and C) are designed with careful consideration for vibration resistance. However, the combustion of these alternative fuels, as well as conventional fuel oil containing multifunctional composite additives intended for emission reduction, may influence the vibrations generated by the engines themselves. Xu *et al.* (2024b) studied the effects of fuel composite additives on the vibration, wear, and emission performances of diesel engines. They found that composite additives to diesel tend to affect the operational reliability through the thermal effect, thereby intensifying engine vibration and resulting in localized excessive wear, particularly fuels containing aviation kerosene or 2-Ethyl 1-hexanol, resulting in decreased combustion stability and increased cylinder pressure and heat release rate, which increased the mechanical vibration energy by up to 41.8%. The combustion intensity and mechanical stress of the fuel additives (2-nitropropane and toluene) were moderate, exhibiting an ideal ability for vibration control with a slight decrease of 7.6% in the mechanical vibration level during the running-in period.

Wind assisted propulsion (WASP)

Interest in WASPs to improve propulsion efficiency and reduce greenhouse gas emissions from ships is increasing. A rotor sail, which is a typical WASPs, can provide an auxiliary propulsive force by rotating a cylindrical structure based on the Magnus effect. Although the rotational speed is low, the rotational forces from a massive structure can cause structural damage and rapid wear of the supporting bearings. Therefore, it is necessary to analyse the vibration response of the rotating rotor sail. This includes evaluating the natural frequencies of the rotating structure, assessing the possibility of resonance due to excitation sources such as mass imbalance, and anticipating vibration responses in the rotating structure. Kim et al. (2024) investigated the accuracy of a simplified four-degree-of-freedom idealization of the rotor by comparing it with a FEM. The model accurately estimated the first natural frequency of the cylindrical mode accurately with about 0.3 Hz discrepancy only. This approach is suitable for application to dynamic response analysis under off-resonance conditions. Furthermore, it was found that the bottom supporting bearing significantly affected the dynamic responses of the rotor sail compared with the upper supporting bearings. Furthermore, the maximum vibration velocity occurred in the case of the mid-plate imbalance condition. Mass imbalances can occur in the rotor sail at two main locations. First, at the topmost Thom disk, which has a wide plate-like structure relative to the cylindrical rotor. If the rotation axis and Thom disk are offset during installation, it can result in mass imbalance. Second, at the midplate, which is composed of ship structural steel and has a higher mass concentration than the cylindrical rotor. In addition, Parmar (2024) studied the imbalance of Flettner rotors by analysing the readings from the strain gauges installed on the rotor of the Fehn Pollux vessel. This study utilized a deep learning approach and proposed two methods for predicting rotor vibrations. For prediction, the long short-term memory model (LSTM) demonstrated better accuracy than the gated recurrent unit (GRU) model. Owing to its simplicity and efficiency, the LSTM can be deployed on the system onboard and utilized to predict the vibration, unbalanced forces, and direction of unbalance in the rotor. An automated self-balancing approach is discussed as an alternative to manual balancing, which could be an interesting topic for future work. In this approach, the balancing process can be performed dynamically during operation of the Flettner rotor. Automatic balancing can be enabled when the vibration exceeds predetermined limitations.

2.4 Shock Response

2.4.1 Air Blast

Blasts, whether caused by internal explosions or external attacks, pose a critical challenge in naval engineering with significant implications for mission accomplishment and safety. Research in this field aims to understand the dynamic effects of high-pressure shock waves on ship structures and develop strategies to enhance structural resistance. These efforts include test correlations using similarity laws, small-scale testing with simplified cabin structures, and analyses of material responses to extreme dynamic forces.

The use of similarity laws allows researchers to predict full-scale structural responses based on scaled experiments, thereby simplifying the complexity of such analyses. However, the analysis of underwater explosions involves highly non-linear responses. Qin *et al.* (2022b) modified the charge mass to compensate for the dynamic response distortion caused by the strain-rate effect in the analysis response of cabin structures under internal blast loading. The results of the scale model tests demonstrated that the plastic properties of mild steel are highly sensitive to the strain rate. The transient impact load leads to a high loading acceleration rate and variation range of the strain rate, which causes a large error in the scale model used to predict the prototype. Qin *et al.* (2022a) proposed the dimensionless damage number of stiffened cabin structures under an internal blast, considering the strength of the stiffeners. Several studies focused on testing scaled structures that are subjected to internal blast loads. Ren et al. (2022) analysed three sets of cabin structure models with different scaling factors combined with different explosive masses. A FE model was developed to investigate the modification of the scaling law. The experimental results showed that the final deflection-to-thickness ratio increased with an increase in model size. The reason for this inconsistency was discussed based on the traditional scaling law, and a modified formula considering the effects of size and strain rate was provided. The proposed fitting formula expresses the deflection-to-thickness ratio of the centre of the plate as a function of the scale factor and scaled distance. Lin *et al.* (2024c) carried out a study on the distribution of explosion load in the ship steel cabin and the propagation law of the structural impact. The numerical model of an explosion in a closed cabin using the Arbitrary Lagrangian Eulerian (ALE) method shows a clear convergence phenomenon of shock waves at the corner of the cabin, and this convergence effect is correlated with the length-width ratio of the cabin. It also shows that the main failure mode of the cabin plate structure was tearing failure along the corners and a large deflection. In addition, using an entire ship model shows that the presence of a bulkhead structure has a major impact on peak acceleration. It was verified that in the process of a local internal explosion of the entire ship, there is a limit to the energy generated by the explosion that can be transmitted along the structure. Yao et al. (2024) examined the dynamic response of mild-steel plate-side cabin structures under unconfined (UB) and confined blast (CB) loads. The experimental and numerically calculated curves agreed well for both trends and peak values. The dynamic response of the plate under UB and CB loads can be divided into three phases. In phase I, the plastic hinge started at the centre and propagated to the boundary in the case of the UB condition, whereas in the CB condition, it occurred close to the boundary and propagated in the opposite direction, caused by the converged, superimposed, and intensified blast load in the corners. In phase II, two plastic hinge lines propagated towards each other, and a platform existed between the boundary while the centre remained undeformed in the UB condition, whereas in the CB condition, a unique phenomenon of larger deformation in the peripheral region than in the central area was produced. Phase III is the final central deformation in both UB and CB conditions. In the design of a cabin structure subject to a CB, more attention should be paid to the strength of the connections at the edges, because CB loading may lead to tearing or shear failure at the corners and edges because of higher loads than an unconfined load. Xu et al. (2022) presented a series of experiments and FE simulations relative to welded aluminium-alloy 5083-H111 plates representing a ship superstructure under near-field air-blast loadings to examine their dynamic plastic deformation and failure. They used one bare plate and four butt-welded plates to evaluate the effects of explosive charge-plate distance and weld seam. In the numerical computations, the material properties in the heat-affected zone (HAZ) and the remaining plate regions were defined separately. The simulations agreed with the experimentally obtained deformation and failure shapes. Material degradation occurred in the HAZ, which strongly affected the failure mode of the aluminium plates. Experimental and numerical studies of protective materials provided valuable insights. Gargano and Mouritz (2023) presented a comparative assessment of the dynamic deformation and damage of a mild steel (type G250), an aluminium alloy (AA5083-H116), a glass fibre-vinyl ester laminate and a carbon fibre-vinyl ester laminate under different air blast explosion conditions. When the materials are compared at a similar areal density, which is critical for lightweight design, the blast performance of the composite materials is similar to that of aluminium alloys and superior to that of steel. Gargano et al. (2022) fabricated flat square panels in sandwich composites used in ship construction for blast testing. Digital image correlation (DIC) was used to measure the out-of-plane deformation and back-surface strain of the sandwich panels over the duration of the blast event. In the context of explosive blast testing conditions employed in this study, the glass fibre sandwich panels exhibited greater resistance to out-of-plane deformation than the carbon fibre panels did. The glass fibre facesheets were more resistant to damage because the flexural strength and failure strain of their sandwich panels were higher than those of carbon fibre panels. The balsa core provided the laminate facesheets with higher damage resistance than the polymer foam core, and this was due to its high flexural and shear strengths. Li et al. (2023b) examined the dynamic response of all metallic honeycomb sandwich panels (L907A low-alloy ship steel) under confined blast loads. The simulation results were validated against the experimental measurements for both the sandwich and monolithic target plates. The results demonstrate that the proposed sandwich panel exhibits superior blast resistance compared with its monolithic counterpart with an equal areal density. This is primarily owing to the absorption of the impact energy in out-of-plane bending, in-plane stretching, and localized core crushing. Patel et al. (2023) studied the blast performance of plate models with or without foam-filled honeycomb sandwich structures with the skin of fibre metal laminates (FML) and/or steel. The use of the proposed FML as the front skin of both the foam-filled and non-foam-filled honeycomb sandwich structures significantly improved the blast mitigation performance, particularly in terms of high specific energy absorption and reduced back skin deflection.

The study of blast responses in compartments enables the optimization of ship designs. To further enhance these designs, future work should include the propagation of blasts through ship structures and the analysis of multiple detonations. Further research is required to develop blast-resistant materials that can effectively dissipate blast energy for use on naval ships.

2.4.2 Underwater Explosion

An underwater explosion (UNDEX) generates shockwaves and gas bubble pulsations that propagate through the water, imposing severe hydrodynamic loads on the hull of the ship. The dynamic response of a ship structure involves rapid pressure fluctuations, whereas the oscillation of a gas bubble created by an explosion can induce additional loading cycles. In extreme cases, these effects can result in localized deformation, fracture, and catastrophic failure.

With regard to simulation-testing correlation Huang *et al.* (2024a, 2024b) investigated the response of an initially damaged double-hull structure ship under UNDEX loading. The impact of flooding water through the orifice is the primary factor causing the deformation of the inner plate. Deformation in the outer plate will decrease as the water level in the compartment increases, whereas deformation in the inner plate will increase with increasing water level. Under certain specific damage conditions, the ingress of water into a compartment effectively enhances the explosion resistance of the double-hull plates. Zheng et al. (2023) analysed the implosion caused by the failure of the spherical pressure hull and the interference with the adjacent cylindrical shell in the double-hulled cabin of a submarine. The implosion of the cabin structure was examined under various triggering modes and pressures. Their research showed that the triggering mode affects the deformation, particularly when the pressure is increased. However, the cylindrical shell does not deform vertically.

In the context of testing real-scale or simplified hull girder structures, He *et al.* (2023b) presented a near-field UNDEX test on a full-scale ship. The numerical model was validated through the application of the coupled Eulerian–Lagrangian method. The near-field UNDEX can be divided into three stages: shock waves, after-flows, bubble pulsations, and water jets. The overall strength of the ship is compromised by the combined effects of these three stages. The sagging deformation of the ship is caused by the negative pressure resulting from the bubble pulsation. Zhou et al. (2021) introduced a method for calculating the dynamic response of a simplified hull girder under combined UNDEX and wave-induced loads. The acoustic-structure interaction method was employed to determine the UNDEX load, and the wave-induced response of the hull girder was derived using a hydrodynamic-to-structure calculation procedure. The synergistic effect of the combination of different levels of wave-induced loads and underwater explosions was analysed. The results indicated that the hull girder exhibits increased vulnerability under the combined effects of underwater explosion (UNDEX) and wave-induced loads, especially in severe sea states. This study assumes that the wave-induced load on the hull girder remains unchanged during the UNDEX period.

Regarding the development of new models to study FSI, Sigrist and Broc (2023) presented a new approach for computing the dynamic response of equipment mounted on submarine hulls that are exposed to underwater explosions without performing fluid-structure coupled calculations. This method is based on the calculation of the transfer functions representing the mechanical interactions at the stake (fluid/structure and structure/structure coupling). To validate the new numerical method, the results were compared with those from the FEM for various shock problem scenarios (single mass or multiple masses mounted on a hull). Finally, the numerical algorithm effectively captured the dynamic response of a 3D system representative of the equipment and its supporting structure mounted on the hull and frames. Yin et al. (2023) proposed a 1D model to simulate the underwater shock wave propagation characteristics and cavitation in water-filled double-plate structures typical of submarines. The dynamic responses of the outer and inner plates were compared with those of water-backed and air-backed single plates, respectively. The results showed that the transmitted wave in the gap water was influenced by the area density and supporting spring of the outer plate. A thicker outer plate can reduce the peak value of the transmitted wave, and a stiffer supporting spring can decrease the transmitted impulse. However, if the outer plate is freestanding or softly supported, it has minimal effect on the response of the inner plate. Moreover, water cavitation is determined by the supporting spring of the inner plate in the case of a softly supported outer plate, and cavitation initially develops in the gap water near the inner plate before expanding towards the external water. When the supporting spring of the outer plate is stiff, it significantly reduces the occurrence of cavitation in the external water, whereas cavitation in the gap water is still determined by the support spring of the inner plate. Subsequent studies will further consider real three-dimensional models to investigate the influence of structural deformation on the transmission of shock waves and cavitation. Zhang *et al.* (2022g) explored a method to identify ship damage during an explosion event based on mode curvature shapes (MCS). This damage reduces the stiffness and mass of the hull girder, which is reflected in its vibration signal of the hull girder. After the damage position was correctly identified, the MCS corresponding to the damage position was used to quantify the extent of damage. A damage indicator was defined to accurately locate the damage. The damage indicator can efficiently locate both single and multiple damages of the hull girder. This method requires further investigation, and when the damage extent is small, the quantitative identification error of this method will be large. For damage occurring near the neutral axis of the hull, in the bow or aft of the ship, the identification accuracy is low because such damage will not cause a large change in the modes. To study FSI problems resulting from close-in UNDEX, Löhner et al. (2023) developed a numerical model that coupled an in-house structural code for large deformations and a CFD code, using FEMAP to control the transfer of information between codes. A mathematical model was developed to describe compressible fluids including air, water, and high explosives. The numerical model was validated against experiments or other models for several 1D–3D cases. The final application case considers a generic ship section. The differences in the flow mechanisms between the rigid and deforming targets were quantified and evaluated. Cavitation was modelled only approximately and may require further refinement.

The application of machine learning to the study of UNDEX phenomena and the structural response and damage of structures has been widely studied. Gannon and Marshall (2023) presented an approach to obtain the response of a floating steel plate to an underwater shock. The velocity time series for a rigid floating plate subjected to UNDEX shock was calculated to train the recurrent neural network (RNN) to predict the plate response, including the effect of reloading due to the cavitation closure pulse. The resulting shock-response model demonstrated a high degree of accuracy in predicting the shock response of a floating plate. The model exhibited an average discrepancy of 2.3% between the predicted peak plate velocities and numerical model results. Furthermore, the model predictions of the plate response to the cavitation closure pulse were found to be in close agreement with the numerical model. Liu *et al.* (2022g) proposed two deep neural network (DNN) models for predicting nonlinear elastic-plastic responses of stiffened plates subjected to near-field underwater explosion. Two DNN models, composed of three hidden layers, were trained based on the dynamic results collected using an LS-DYNA simulation containing a stiffened plate of ship structures, water, and air. It was found that the DNN can accurately estimate the dynamic responses of the stiffened plate considering fluid-structure coupling effects and geometric and material nonlinearities. DNN models can capture the strain concentration at the boundary and oscillation of the displacement-time curves.

The topic addressed in this section is well documented, particularly in terms of studying and developing models for fluid-structure interaction analyses. While research continues to use scaled or simplified structures, a notable gap remains in the analyses involving real cases or tests with complex, full-scale naval structures.

2.5 Noise

This section presents the recent advances in design methodologies and technical solutions for controlling ship noise. Noise control has been an essential aspect of ship design since the advent of motor vessels. Underwater noise signatures of ships have been studied primarily for warfare and defence. In contrast, onboard noise has become critical with the increased transportation of passengers and as crew health and safety standards have improved over time. Over the last few decades, ship noise has become even more important, as increasing scientific evidence has highlighted the negative effects of ship noise on passengers and crew, as well as on the surrounding environment. The latter includes underwater radiated noise that affects sensitive marine ecosystems and airborne noise radiated from ships navigating inland waters or manoeuvring in harbours, which affects the surrounding residential neighbourhoods. Consequently, studies on ship noise control have started focusing on strategies to mitigate underwater radiated and airborne external noise via technical solutions and new regulatory frameworks.

Over the reporting period, research has continued to advance our understanding of the complex phenomena that characterize noise generation and propagation on and from ships. Studies have contributed to the development of procedures for characterizing noise sources on ships and simulating and controlling noise propagation on board (airborne and structure-borne), underwater, and externally (airborne). Regarding the analysis of sources, it is important to highlight that the recent delivery and operation of electrified ships has raised interest in understanding how electric propulsion systems may affect radiated noise. A first answer is provided by Andersson et al. (2024). They recently published results of comparative measurements of noise from ships with different propulsion systems, including battery-operated vessels. They examined airborne and underwater radiated noise and found that while battery-operated vessels reduce external airborne noise—eliminating low-frequency noise from diesel engines, which is often the reason for the annoyance of residents close to harbours and inland waterways—the same beneficial effect is not observed for underwater radiated noise. While this is the first important study on this topic, further studies are needed to understand whether the results are also confirmed for other types of battery-operated ships.

The following subsections present recent advances in noise control and design. The results from the reviewed scientific and technical papers show that ship noise is still relevant, and more studies are needed to effectively mitigate its negative impact.

2.5.1 Internal Noise

Internal noise has long posed significant challenges to shipbuilders, prompting national and international standards for regulating noise levels in shipboard spaces. These regulations aim to safeguard crew health and enhance passenger comfort through precise measurement procedures, exposure calculations, and noise reduction guidelines for both new and existing vessels.

Despite ongoing progress in this area, recent studies have cast doubts on the conclusions of the ISSC 2022 report, which asserted that existing regulations and technological advancements have effectively addressed internal noise concerns. New research highlights persistent issues, showing that noise exposure levels aboard vessels frequently exceed the established limits. Burella et al. (2021) demonstrated the presence of hazardous noise levels for fish harvesters operating hydraulic deck equipment and outboard engines, with skippers frequently unaware of these risks. Similarly, Stone and Moro (2022) identified dangerously high noise levels in salmon aquaculture facilities, caused by feed blowers and vessel engines. Despite awareness among facility managers, employees frequently neglect hearing protection, thereby highlighting a gap between awareness and practice. Cui et al. (2022) demonstrated the adverse impact of engine noise on seafarers' sleep quality, affecting safety and performance. Febriyanto *et al.* (2023) further reviewed the broad physical and psychological health issues linked to noise, including hearing loss, stress, and sleep disturbances. Together, these findings underline the pressing need for improved noise control strategies and awareness programs to bridge the gap between knowledge and implementation.

Experimental techniques have significantly improved the accuracy of internal noise measurements in marine environments for assessing and identifying potential design issues. For example, Li *et al.* (2022c) highlighted the limitations of noise prediction using a source–path–receiver model. Although this method yielded acceptable results in rooms with sound sources, it was less reliable in predicting noise at higher frequencies, particularly for structural radiation and penetration sound. Bassetti et al. (2022) explored the applicability of Transfer Path Analysis (TPA), a well-established technique in the automotive industry, for naval applications. TPA was found to be effective in identifying critical noise sources and improving the system characteristics to satisfy vibration and noise requirements. Their work successfully demonstrated the robustness of the technique in naval contexts, particularly when applied to specific ship sections using artificial exciters such as hammers and shakers. Future applications with more detailed identification steps could provide deeper insights into the vibrational dynamics of ships, making TPA a promising tool for enhancing noise control strategies in the maritime sector. Moreover, Prato et al. (2023) introduced a novel modal approach for measuring the transmission loss properties of ship bulkheads and deck treatments at low frequencies (50–80 Hz). This approach addresses the limitations of standard methods, which often yield fluctuating results in non-diffuse sound fields, making them less reliable in marine environments. By focusing on the attenuation of the source room modes and introducing normalization terms related to the receiving room volume and modal damping, their method offers a more accurate and representative measure of sound transmission loss, particularly in low-frequency ranges. This innovation is significant because it aligns more closely with the actual sound field conditions encountered in both laboratory settings and onboard ships, providing results that are more reflective of real-world noise exposure and subjective user perception.

Recent advancements in numerical methods for assessing internal noise have significantly improved predictive accuracy. Liu *et al.* (2022f) introduced a averaged-energy method designed to predict mechanical noise across the whole frequency range under kinematic excitation. Their experimental verification demonstrated the precision of the method, particularly for predicting the structural mechanical noise at both low-mid and high frequencies. Key innovations include a load identification method that minimizes calculation errors even when the acceleration data are incomplete. Chen *et al.* (2023c) developed a unified analysis model for rotary acoustic cavity systems, offering reliable predictions of vibro-acoustic behaviour. Shi *et al.* (2024b) used frequency domain noise analysis to study cabin noise during icebreaking operations, identifying critical areas where noise control is necessary. Kyaw Oo D'Amore *et al.* (2023) presented a combined CFD and FEM framework to optimize marine engine exhaust systems, effectively reducing noise without compromising efficiency. Luan et al. (2024) conducted CFD and hybrid computational aeroacoustics (CAA) simulations to analyse marine gas turbine exhaust systems, introducing design modifications such as protrusions to reduce both low-frequency internal noise and high-frequency external jet noise. Liu *et al.* (2022e) developed a method to accurately predict mechanical noise under diesel engine excitation by combining numerical simulations with hydroacoustic experiments. Their approach, which identifies generalized forces through equivalent load and gradient meshing models, highlights the critical need to account for acoustic cavity resonance, particularly in low-frequency predictions. Louvros et al. (2022) addressed the increasing complexity and costs of late-stage design changes in shipbuilding by proposing an optimization framework that integrates both general and internal design tools. This modular and efficient approach enhances design reliability and minimizes revisions, particularly during the early design stages. Luyun et al. (2023) extended the theory of human-induced vibration and noise, typically applied in buildings and bridges, to ship corridors. Their study highlighted the challenges of accurately modelling human-induced noise, particularly under random walking conditions, and called for further research to address the complexities of low-frequency noise, which are difficult to measure experimentally.

Data-driven methods introduced new approaches for predicting and controlling internal noise in marine environments. Ye et al. (2022) developed an innovative acoustic reconstruction model using neural networks to enhance the accuracy and efficiency of low-frequency vibration predictions in ship hulls. By optimizing the use of limited vibration monitoring points and addressing complex excitation problems, this model reduces both sensor requirements and computational costs, making it a promising tool for real-time noise assessment. Similarly, Bocanegra et al. (2023) conducted a comprehensive analysis of shipborne noise by using a large dataset of noise records from various ships. Their study employed statistical and AI techniques to identify the key characteristics of the noise signals, emphasizing the frequency content and tonal components. By exploring different single-value indicators and spectral shapes, their research highlighted the limitations and strengths of current noise indicators, underscoring the need for more refined classifications and experimental measurements to improve regulatory standards and onboard noise control strategies.

Despite advances in noise prediction and control methods, such as Transfer Path Analysis and improved numerical simulations, noise levels on ships frequently exceed established limits, leading to ongoing health impacts among seafarers. These persistent issues highlight the need for not only refined regulations but also better awareness training for workers and more effective integration of noise control strategies in the early stages of ship design to ensure long-term improvements in onboard conditions.

2.5.2 External Noise

The two most recent ISSC Dynamic Committee reports emphasized the nascent nature of research in the field of external airborne ship noise. Although this is still partially true, over the previous reporting period, there were significant steps forward in research that led to the implementation of voluntary class notations and preliminary measures to control this environmental and community health concern, making this field more understood and mature. External airborne noise is a particular concern when ships are berthed in harbours situated near residential neighbourhoods because its effects on workers and inhabitants were documented. For these reasons, several research projects have been developed to i) understand sources of noise and their contribution to overall port noise, ii) develop standardized methods to quantify noise levels and map them in ports and surrounding areas, iii) conceptualize methodologies for simulating noise propagation, and iv) find solutions to mitigate hazardous noise.

Fredianelli et al. (2021) highlight the complexity of the characterization of noise pollution in ports and propose to divide the sources into the following categories: road sources, railway sources, port sources, and industrial sources. They can be further subdivided based on their operations and positions. In this paper, the authors show how current rules and standards are not sufficient to characterize and regulate harbour noise. Some authors have published important data from noise measurements that characterize noise sources in harbours, including ships and other harbour activities. Schiavoni *et al.* (2022b) reviewed several studies published in recent research projects to provide an overview of the main sound sources commonly found in harbours. In addition to broadband levels, they provided the sound power spectra of the main noise sources, which are important for characterizing acoustic emission. The database created by the authors clearly shows that ships are relevant contributors to overall noise levels; however, other heavy equipment, such as cargo handling facilities, is also a major contributor to overall noise pollution from harbour activities. Fredianelli et al. (2024) presented sound power levels of equipment used for ports operations with the aim of providing information that can be used for noise mapping. In their study, they stressed the importance of defining a clear methodology to assess these sources because their sound levels vary widely depending on their mode of operation.

To address the lack of standardized methods for characterizing sound sources, some classification societies (e.g., ABS, LR, DNV, and RINA) have issued measurement procedures to assess external airborne noise from ships, which can be used to obtain voluntary class notations. These notations may be used by port authorities to develop noise control strategies in ports and by ship owners/operators to obtain reduced fees in port jurisdictions where noise levels are controlled. Fredianelli et al. (2022) argued that the measurement procedure proposed by the classification societies mentioned above and by both ISO 2922:2020 and ISO 14509–1:2008, present practical challenges in their implementation that prevent their global application. The authors proposed a series of feasible and reproducible measurements to overcome these challenges and characterize ships as exterior airborne noise sources while engaging in different operations. While the proposed guidelines are promising, there are still questions regarding the uncertainty of the measurements performed.

Data from characterized sources can be used to simulate noise propagation. Simulations can be used to evaluate alternative mitigation strategies and assess their efficacy. In a review paper, Biot et al. (2024) described commercial software packages that can be used to simulate noise propagation in harbours. The authors presented some simulations of external airborne noise from a case study vessel and discussed in detail the impact of the numerical modelling parameters on the accuracy of the simulation results. For instance, the simulations highlighted that some details of the ship model geometry, including the exact positions of the sources, did not significantly affect the outcomes of the simulations. A key challenge identified by the authors in developing these numerical simulations is the validation of the models, which is not straightforward because of the large number of sources and uncertainty in their characterization. To overcome these challenges, the authors proposed a self-validation method based on the equivalence of the overall sound power calculated using the model in accordance with ISO 8927 and ISO 3744. The uncertainty of the sound power levels calculated using this method has not been discussed and should be the focus of future studies.

Concerning mitigation strategies, one of the proposed advantages of external airborne noise class notations provided by certain classification societies is the potential reduction in port authorities' fees (e.g., ABS, 2024). For instance, the Vancouver Fraser Port Authority offers fee reductions for ships with underwater noise notations. Rising concerns related to external airborne noise may lead port authorities to adopt similar approaches to this pollutant. Some jurisdictions have issued guidelines and protocols for controlling external noise during operation. For instance, Transport Canada (2018) issued an interim protocol that outlines guidelines to mitigate external airborne noise from ships at anchor with the aim of minimizing the noise exposure of residents (TC, 2018). In addition to these operational solutions, other mitigation solutions focusing on the sources have already been presented in the previous ISSC report. An interesting proposal for alternative solutions to reduce environmental noise is the creation of hills used as natural infrastructure for transportation and sound barriers that divide the port from residential areas. Schiavoni *et al.* (2022a) showed in a practical case study how the re-positioning of sources and reorganization of the layout of a port may effectively reduce noise.

2.5.3 Underwater Noise

Marine propulsors are identified as the main contributors to a vessel's underwater radiated noise. Vessel signatures in terms of spectral limits can be experimentally obtained for a complete vessel at full scale; however, the prediction capability of sound sources is still rudimentary at best. The acoustic evaluation of a marine propeller using CFD methods requires an accurate representation of turbulence and cavitation, which are the main sources emitting broadband and high-frequency noise, whereas the water volume displacement by the blades creates only a distinct tonal noise at multiples of the blade-passing frequency. The latter can also be predicted by low-fidelity methods with sufficient accuracy; however, broadband noise sources require high spatial and temporal resolutions of flow features, which necessitates resolving low- and medium-wavelength turbulence with LES.

Kimmerl et al. (2021), evaluated the tip and hub vortex cavitating flows with implicit LES focusing on probable sound source representation. The cavitation structures for free-running propellers were investigated using the open-source code OpenFOAM. To accurately resolve the structure of the tip vortex, a prior mesh refinement was employed during the simulation in regions of high vorticity. Acoustic simulations required fine time steps and a large number of rotations for spectral analysis. The numerical effort was reduced by relaxing the wall modelling associated with LES for tip-vortex flow arising primarily due to flow shear behind sharp edges rather than boundary layer shear at the wall. Empirical modes for the cavitating tip vortex were not observed in the simulations. It might require larger sampling frequencies in the model scale or the Schnerr–Sauer cavitation model and, despite these improvements, results can be insufficient because of the missing higher-order terms of the linearized Rayleigh–Plesset equation. Physical mechanisms, such as a spatially pulsating vapor vortex core along the running length and the conservation of angular momentum around the core, could be reproduced. The authors categorized vortices from the tip as primary (cavitating) or secondary (noncavitating). The primary cavities at the centre are compressed by the secondary vortices, resulting in an elliptical shape. The authors remarked that the primary cavitating core was composed of at least two internal vortices, contributing to the formation of an elliptical cavity shape. Higher mesh resolutions are required to investigate whether these vortices were originated from a different origin in the mixing zone. It was proposed that a tip vortex, like any cavitating vortex, assumes an energetically superior mode (breathing mode) when additional vortices no longer constrain the shape by physical or numerical dissipation or merging with the primary vortex structure. The capability of the method to accurately capture cavitation macrostructures within the margins of feasible mesh refinements was discussed. It is to be noted that the propeller noise is a result of dynamic cavitation structures and bubble collapses. Shear and vortex cavitation structures require further understanding. Presented simulations implemented Schnerr–Sauer cavitation model which cannot capture the effect of bubble collapses. Hence, authors remarked that additional intensive future research efforts are required to investigate these aspects.

Sezen and Atlar (2022) assessed full-scale URN characteristics using CFD tools. Numerical calculations were performed using a hybrid method combining detached eddy simulation (DES) and permeable formulation of the Ffowcs-Williams-Hawkings (FWH) equation. A commercial CFD solver (Star CCM +) was used for the numerical calculations. Two different permeable noise surfaces encapsulating the hull and propeller (PS2) and only the propeller and its slipstream (PS1) were utilised. To accurately solve the tip vortex flow and model the tip vortex cavitation (TVC) in the propeller slipstream, the vorticity-based adaptive mesh refinement (V-AMR) technique was applied in the numerical calculations, along with the Schnerr–Sauer mass transfer model to model the cavitation. URN computations and a comparison with full-scale measurements are presented for the three different RPMs. Foamy and strong bubbles could not be predicted in the CFD compared to the sea trial. The blade-pass frequencies were generally well captured. The numerical results were underpredicted for a lower RPM. The application of the V-AMR technique enabled the TVC observation of the propeller slipstream at full scale, the authors noted variations with respect to sea trials were noted. As the sliding mesh was used in this study, the TVC could not be extended towards the rudder, which inevitably influenced the accurate prediction of the propeller URN. The 'PS2' configuration showed a spectral hump with higher amplitudes, mainly associated with the TVC dynamics, as in the sea trial data. Such a spectral hump was not present in the 'PS1' configuration. The cavitation dynamics and nonlinear noise sources occurring inside 'PS2' might cause a hump compared to 'PS1', even though the hull interference might contribute considerably to the amplitude of this hump, which requires further investigation. The general trend of the predicted noise spectrum encourages the use of the proposed CFD method in full-scale applications.

Tanttari and Hynninen (2024) presented an equivalent source method (ESM) for propeller noise. ESM is a noise radiation prediction tool based on a set of discrete elementary sources that form an equivalent source (ES) model. The accuracy and reliability of ESM prediction depend on the complexity of the real source, acoustic environment, ES model size, geometry, and quantities to be predicted. The Potsdam Propeller Test Case (PPTC) model scale propeller in a cavitation tunnel was used as an example. The authors performed systematic testing and screening of various choices (model size, source, number of sensors, and their locations, ideal vs. practical) for ES models in underwater noise simulation. Both the sound power and sound pressure at individual points were examined. The authors remarked that the performance of the ES models created was highly dependent on the source-sensor distances and shorter distances were better. The exact location of the sources, combined with the acoustic environment, also play a role. Concerning sound power, simple models, even those consisting of a single monopole source, work quite well. Concerning the average sound pressures at the sensor points, a considerable number of sources and sensors, together with a reasonable choice of their locations, is important. However, single-point responses predicted using an ES model are subject to considerable uncertainties. It was concluded that by using a simple ES model, a real source can be transferred to other environments if the sound power level is the main interest. If the details of the sound field, such as the directivity or responses at certain points, are of interest, models with a larger number of sources are needed. Further research is required to apply ESM to real ships at a full scale.

For naval ships, the URN due to waterjet propulsion can be of concern, and it can be broadly categorized as pump noise and nozzle efflux noise. Pump noise mainly includes cavitation noise and tonal mechanical/hydrodynamic noise, whereas efflux noise is caused by the jet impact and bubble entrainment. Dhavalikar et al. (2022) addressed efflux noise in wet and dry transom condition. LES was performed to simulate the hydrodynamics of turbulent multiphase flow behind both the submerged and free-surface water jets. The impact of the water jet on the free surface and the subsequent bubble generation rate was captured using the multiphase volume of fluid (VOF) technique.

The authors used the commercial (CFD) tool StarCCM +. The impact noise generated by the waterjet was attributed to the oscillating pressure in the fluid domain outside the jet-flow region. A basic validation was conducted by examining the oscillating pressure as a function of jet velocity. To compute the bubble generation rate, a blob-detection algorithm was employed. Additionally, a mathematical model for calculating the bubble pulsation noise based on the bubble generation rate was validated and implemented. The authors recommended further validation with the help of experiments, which could be useful in further investigation of URN from water jets of stealth vessels, as well as to investigate bubble noise from propeller cavitation, air lubrication systems (ALS), etc.

In view of the growing concerns over URN due to shipping as well as GHG emissions from shipping, the topics of URN and energy efficiency are being discussed together at the international level, for instance at the IMO. A detailed matrix of measures to control URN and GHG emissions is presented in the Vard Report (Terweij, 2023). Measures to reduce URN may not necessarily reduce GHG emissions. The effectiveness of the noise reduction measure is related to the noise source being treated, so limited to its frequency characterization, and not necessarily to the overall noise signature of the ship. Resilient mounts and rafting systems are primarily intended to block the transmission of energy at the characteristic frequencies of the source, such as engine firing rate and harmonics. Cavitation noise reduction has broad-spectrum benefits, and it also addresses the blade rate effects at lower frequencies. The authors remarked that only a few of the methods listed in the matrix were explored in sufficient detail to define their URN benefits for typical ship applications. Therefore, there is an urgent need for additional measurement campaigns to provide better definitions in this area.

A qualitative comparison of the numerical results with sea trial data is rather difficult. Many factors can influence noise measurements at sea, which are not present in CFD simulations. Moreover, the recorded acoustic signals certainly include several contributions to the measurements. In addition, the cavitation dynamics are certainly different in the CFD predictions compared with the sea trial data, owing to the assumptions in the numerical calculations. Therefore, it is not possible to expect complete agreement between the predictions and full-scale measurements. In addition, to fine-tune the numerical methods, further validation with experimental and full-scale measurements would be required in due course. Numerical methods are also being applied to assess evolving technologies, such as air lubrication system (ALS), wind assisted ship propulsion (WASP) and so on. These technologies are primarily applied for energy efficiency improvement, although with growing global concerns for URN due to ships, they are also being investigated for their noise impact.

It is generally claimed that ALS helps in reducing the URN because of the mismatch in the speed of sound between the bubble curtain and the surrounding water. However, the noise generated by actual bubbles is uncertain, and more work is required to better understand the effects of ALS on the URN generated by a vessel. Techniques such as WASP may result in a reduced URN; however, this is solely because of the reduced propeller loading with such technologies, and URN due to the propeller remains a primary concern.

3 Offshore structures

3.1 Wave Induced Vibrations

Wave-induced vibrations of marine structures can be categorized into two types. In the first category, structures such as a Very Large Floating Structure (VLFS), ice sheet, or wave energy converter (WEC) are assumed to float on the water, and while they can be moored, they are assumed to be exposed to water waves that cause both rigid and elastic responses. This type of study has several applications. In the second category, the structures are placed vertically in water and mounted on the sea bottom. In this case, waves and currents can induce structural vibrations. Addressing problems related to these two categories requires the consideration and coupling of both fluid and solid mechanics. This has led to a large number of scholarly studies, and researchers have attempted to develop new and more robust models that consider different aspects of physics (e.g., nonlinearities associated with waves and plastic deformations in the structure).

The potential-based flow assumption for the fluid component has been the predominant approach for addressing such problems since the 1950s, and significant advancements have continued in this field. The fluid problem can be solved using different methods, from panels methods to matching methods, the latter solving the governing equations in sub-domains and then applying continuity and physical consistency at the domain interfaces. The solid problem is usually solved using FEM. A recent example is the study by Bispo et al. (2022), who developed a model to solve the dynamic response of a VLFS exposed to water waves. The fluid problem was solved for linearized free surface conditions using a matching method, whereas the solid problem was solved using a multi-body method. It was shown that the stiffness of the hinges placed in a VLFS has a minimal impact on the wave transmission coefficients, although they could reduce the vertical displacement. These investigations could also encompass more complex scenarios and incorporate nonlinear boundary conditions for the water surface. An illustrative example is the study by Hartmann *et al.* (2022). They developed a numerical tank that could solve the elastic responses of a large thin structure subjected to water waves by considering nonlinearity. It was concluded that although a nonlinear model would capture the nonlinearities related to displacements and waves, it might still exhibit inaccuracies, as certain artificial damping effects observed in tank tests would not be accounted for in this model.

Another set of studies in this category used CFD to model the fluid-solid interaction of floating structures. In a recent work, Liu *et al.* (2022c) built a new computational tank to numerically simulate the dynamic responses of floating objects that may also exhibit plastic deformation. The model was developed by coupling CFD and nonlinear FEM models. Two coupling approaches, one- and two-way, were used, with the latter demonstrating a higher level of accuracy. Another example of coupling CFD and FEM simulations can be found in the work of Gu et al. (2023), who numerically simulated the dynamic responses of floating offshore structures. They showed that a CFD-based simulation, by considering nonlinearities, would provide a smaller bending moment for cases with an overtopping event compared to a potential analysis. Tavakoli et al. (2022) developed a *numerical* tank to simulate dynamic responses of a floating viscoelastic floater exposed to water waves, which resembles an ice floe. The fluid and solid dynamic responses were solved using the finite volume method (FVM). The authors used *solids4foam,* which is the flexible FSI library of openFOAM. It was demonstrated that the dispersion and dissipation of waves in a viscoelastic structure can be accurately tracked using the Maxwell model. These results can be applied to the design of floating breakwaters.

Although many models use CFD and potential-based flow models to address wave-induced vibrations of floating structures, notable progress has been made in modelling these problems using the Green-Naghdi (GN) method. This advancement is evidenced in research conducted by Kostikov *et al.* (2022, 2024). They initially applied the GN method to solve the drift and dynamic responses of multiple elastic sheets subjected to waves and currents (Kostikov *et al.* 2022). Later, they used this method to analyse the dynamic response of a moored elastic object subjected to water waves (Kostikov et al., 2024). It was found that the wave-induced vibration of a moored plate can occur at a frequency different from that of waves, likely because of the influence of the mooring effects on the stiffness of the structure.

Progress in model development has not been limited to CFD, potential-based flow models, and GN modelling. Researchers have begun to use emerging methods (e.g., AI-based models) to study the interaction of water waves with elastic floating structures. Recent advancements in modelling the dynamic responses of floating offshore structures subjected to water waves can be found in the work of Tay (2023). The author developed a surrogate model and used an ANN to solve the dynamic responses of objects exposed to water waves. Notably, the data for this modelling approach were generated using a model that coupled linear potential flow modelling with the linear FEM. The author also conducted irregular wave test cases, which demonstrated the effectiveness of the model in addressing real-world maritime environments. The study concluded that the model performed well under irregular (random) wave conditions and can be integrated into digital twins that are currently being developed for VLFS. Another intelligent-based model was developed by Rajiv et al. (2023), who developed a metamodel to predict the dynamic responses of floating offshore structures exposed to water waves. The authors generated a set of synthetic data using the OPENFast code.

Finally, it should be noted that studies on the dynamic responses of floating offshore structures mostly present heave and pitch responses (and do not consider elastic responses). The novelty is mostly related to the consideration of other external factors, such as wind load. A successful example of such a study can be seen in the work of Shi *et al.* (2023b). The authors used two different sets of codes to estimate the hydrodynamic and aerodynamic loads acting on the structures and then calculated the responses of the structures for different water depths. Water depth has been reported to reduce motion.

In the second category, as explained previously, the authors studied structures mounted on the seabed. Such studies require a different approach for solving the problem compared with problems related to the first category. Probabilistic and stochastic models are frequently used. For example, Syed Ahmad et al. (2022) modelled the dynamic responses of fixed offshore platforms using efficient-time simulation (ETS) regression models, considering linear, polynomial, and cubic regression approaches. The forces acting on the structure were predicted using the Morison equation, and the wave-current effects were also studied. The results of this study indicate that more accurate structural response can be obtained using the linearized response as a new input variable in the development of regression models. In another study, Gaidai and Xing (2022b) used a statistically based model based on Gaidai *et al.* (2022a), to predict the probability of extreme dynamic responses of jackets by considering non-linearity. Fluid forces were calculated using the Morison method.

Studies in this field are not limited to wave effects, and some also provide new modelling and understanding of the combined effects of waves and wind on the dynamic responses of floating structures. These methods are applied in the design of offshore wind turbines (OWT). A good example of a recent study in this field can be found in the work of Ti et al. (2022). A new model was developed to predict the dynamic responses of a bottom-mounted slender structure using a quasi-stationary method that calculates loads based on instantaneous structural responses, including those caused by waves and winds. The wave loads were predicted using the BEM for linearized free-surface boundary conditions. The authors demonstrated that it is essential to properly account for aeroelastic and hydroelastic effects; failing to do so could lead to an overestimation of the structural responses by as much as 32%.

Another stream of research has also been conducted in which scholars have studied the dynamic response of mounted piles by considering soil effects. These studies have practical applications for the design of fixed floating offshore structures. An illustrative example is the research conducted by Yang *et al.* (2023b), who utilized a stochastic analysis approach employing an integrated FE model of a soil-pile-structure system. This study examines the dynamic responses of an OWT under varying wave conditions by employing both empirical and actual wave spectra. Numerical simulations were conducted to analyse the performance differences across various scenarios.

While most studies in the second category have developed simple models for solving dynamic responses and have avoided using CFD to find hydrodynamic loads, a recent CFD-based study that solved a fully nonlinear fluid dynamic problem by incorporating water surface effects induced by nonlinear waves was conducted by Lin *et al.* (2024a). The authors presented a comprehensive model that predicts the dynamic response of a jack-up structure exposed to water waves, currents, wind loads, and aftershocks. In this research, the authors developed a method capable of analysing the fluid-structure interaction effects and evaluated the impact of seismic ground accelerations using the CFD software FLUENT and a CFD-FEA coupling approach.

3.2 Wind Induced Vibrations

The study of wind-induced vibrations remains a critical research area for ensuring the safety, reliability, and efficiency of offshore structures and pipelines in challenging marine environments. Recent advances in both experimental studies and numerical simulations have provided deeper insights into the fluid-structure interactions involved in these phenomena.

Recent experimental investigations yielded valuable data, particularly for understanding flow- and vortex-induced vibrations (FIV and VIV). Praveen and Avula (2022) conducted a comprehensive study on wind tunnels, focusing on the downstream cylinder behaviour in the presence of an upstream stationary cylinder. Their findings revealed that the reduced velocity increased with the mass-damping ratio across different gap ratios between the cylinders, emphasizing the significant influence of the gap ratios on the reduced velocity and the role of the initial perturbations in inducing galloping vibrations. Domingos et al. (2024) provided insights from Europe's first full-scale measurement campaign for installing an OWT tower using a floating heavy-lift vessel. Their analysis showed that approximately 96% of the tower motion response was due to wave action, with only 3% attributed to the VIVs caused by a passive tugger line, the latter highlighted by a natural frequency aligned with the vortex shedding frequency of the tower.

In numerical simulations, progress has been made in modelling the combined effects of wind, ice loads, and soil-structure interaction on offshore structures. Wu *et al.* (2024b) investigated the dynamic behaviour of monopile OWTs under these combined loads, emphasizing that such interactions can lead to substantial tower-base acceleration and vibration. This poses a potential threat to the personnel on the operating platform, highlighting the critical importance of controlling these combined vibrations to ensure safety. Additionally, Yin et al. (2024) explored the single-blade installation method for large-scale OWTs, which offers advantages in reducing deck and crane requirements on installation vessels. Their study developed a real-time analysis program using the Lagrangian method to simulate the coupled dynamic responses during the entire hoisting process, factoring the aerodynamic loads on the blade. They found that turbulent wind significantly increases the dynamic response complexity, necessitating careful trajectory planning to avoid collisions with surrounding structures. Liu *et al.* (2023e) extended these efforts by investigating the dynamic behaviour of long floating bridges under the combined effects of wind and waves. Using a 1:100 scaled-down model based on Froude similarity criteria, their experimental study revealed that the coupling effect of wind and waves amplifies the motion response amplitudes and broadens the probability distribution of the responses, particularly under high wind velocities. Wind loads primarily influenced the mean response, whereas wave loads significantly affected the standard deviation, with sway, heave, and roll responses showing increased vulnerability in wind-wave environments.

Mitigation strategies for VIV have led to innovative developments. Youssef et al. (2022) examined the use of helical strakes and fin plates to suppress the VIV in OWT towers and submerged cylinders. Although helical strakes provided some benefits, the fin plates were more effective, reducing streamwise and lateral vibrations by 81% and 75%, respectively. These results suggest that fin plates can be a standard solution for VIV mitigation in future offshore structures. Advanced control systems have also been developed to enhance the stability of FOWTs. Mu et al. (2022) introduced a radial basis neural network sliding mode control strategy, which outperformed traditional controllers by reducing crossflow vibration by 94.7% and nearly eliminating downstream pulsation vibrations. This approach, which decouples the lift frequency of the bluff body from its natural frequency, represents a significant advancement in the FOWT stability. Additionally, Qi et al. (2023) proposed a superimposed proportional model-free adaptive control (SP-MFAC) strategy to address the coupling effects between rotor aerodynamics and platform motions in FOWTs. By optimizing the blade root moments, rotor speed, and platform pitch and yaw motions, the SP-MFAC strategy, combined with trailing edge flap control, achieved better overall performance compared to traditional PID-based controls, especially in turbulent and extreme gust conditions. Wang *et al.* (2024a) proposed a combined active torque control algorithm integrating feedback from tower vibrations and feedforward from misaligned wind and incoming waves. This method effectively suppressed dynamic loading in OWTs and increased the fatigue life of monopiles from 19 to 39 years with minimal impact on power production. Furthermore, Elias and Beer (2024) explored the use of a tuned mass damper inerter with multiple electromagnetic motors (TMDI-EM) for wind- and wave-induced vibration suppression and energy harvesting in OWTs. Their findings indicated that TMDI-EMs not only improved vibration control, but also enhanced energy harvesting capabilities, making them a potentially promising solution in future OWT design.

Gusts and extreme conditions pose critical challenges to the stability of OWTs and require careful design and risk assessment. Torrielli and Giusti (2023) emphasized the importance of accounting for uncertainties in wind-induced loads, noting that neglecting this aspect may lead to an underestimation of the risk by up to 20%. In a related study, Liu *et al.* (2022b) examined a 15 MW FOWT under extreme conditions, such as typhoons and abrupt wind changes. Their findings showed that these conditions significantly amplify turbine motion and structural stresses, particularly in mooring lines and tower responses, thereby threatening turbine stability. Similarly, Sun et al. (2024) emphasized the critical role of environmental load charts in ensuring the structural safety of offshore platforms under extreme wind, wave, and current conditions, highlighting how these tools can be used to evaluate and enhance the adaptability of the platform to such conditions, thereby extending its operational lifespan. The reviewed studies underscore the need for advanced design and control strategies to address the increasing severity of the environmental conditions affecting OWTs. Climate change, leading to more frequent high wind gusts, presents a significant challenge to wind turbine performance and safety. Abdelaziz et al. (2023) emphasize the necessity for resilient design standards to cope with future climate risks. Future research should focus on optimizing turbine design for extreme wind conditions, enhancing control strategies, and integrating climate projections into long-term planning. Additionally, exploring the impacts of various turbine sizes, platform types, and mooring systems under different conditions can provide valuable insights into improving the resilience and efficiency of future offshore installations. A critical issue with wind-induced vibrations, particularly in the design of FOWTs, is that the design criteria remain subject to change. Although all OWTs adhere to a design basis, the industry has not yet matured, and the final safety levels and acceptance criteria, including vibration thresholds, are still under development. This lack of standardization significantly affects optimization and economic efficiency, highlighting the need for clear and consistent guidelines as the industry evolves.

3.3 Current Induced Vibrations

Current-induced vibration (CIV) is a significant challenge for marine and offshore structures, impacting critical components, such as risers, pipelines, and floating platforms, including spars and semi-submersibles. These vibrations can cause considerable fatigue damage, compromising both the structural integrity and the operational lifespan of these systems. In recent years, research has increasingly focused on vortex-induced vibrations (VIV) as a major component of the broader subject area.

The ISSC 2018 report highlighted the limited understanding of VIV within prototype Reynolds number flow regimes and underscored the importance of studying the effects of the surface roughness. Building on this, the ISSC 2022 report further emphasized the need for research into Reynolds number scale effects and the continued development of coupled CFD solvers. Recent advancements in analysis methods, particularly in CFD, have been notable, with the integration of machine learning techniques emerging as a promising approach to enhance efficiency. Although some issues have been addressed, many remain unresolved.

3.3.1 Experiments

Recent advancements in experimental research on VIV in marine structures have yielded significant insights into the influence of various configurations and conditions on VIV responses. Li et al. (2023c) conducted model tests to explore the VIV characteristics of a flexible pipe under combined uniform and oscillatory flow conditions, showing that such combined flow conditions can lead to severe fatigue damage. Zhu et al. (2023c) focused on the VIV response of inclined flexible pipes at various oblique angles and evaluated the validity of the independence principle over a range of reduced velocities. This study demonstrates the necessity for a comprehensive consideration of the spatial-temporal evolution of the VIV response, mode transition features, and vortex shedding characteristics. Rueda-Bayona et al. (2023) applied an experimental and analytical approach to measure the FSI near offshore monopiles under wet conditions, quantifying viscous damping and the effects of structural and 3D hydrodynamic accelerations. Their experiments indicated that VIV contributed to a 5% increase in the structural damping coefficient, and extreme wind loading further increased the damping by 74%. Rashki et al. (2022) investigated the effects of soft marine fouling on VIV-based hydrokinetic energy generation using circular cylindrical oscillators in towing-tank tests. Marine fouling significantly reduces the maximum amplitude of crossflow oscillations and affects the synchronization range, which has implications for the efficiency of energy generation systems. Liu et al. (2023b) examined a segmented free-hanging marine riser used for deep-sea mining. At low reduced velocities, the dominant frequency did not follow the expected doubling principle, indicating an anomaly in VIV behaviour. The mass concentration at the bottom disrupted the vortex shedding and reduced the bending response. This study highlighted the significant effects of weakly constrained boundary conditions and segmentation on the VIV response of risers. Zhu et al. (2023d) conducted laboratory experiments on a free-spanning submarine power cable under steady current. Their analysis revealed that even minor geometric variations can lead to substantial differences in VIV responses. Wu et al. (2024a) conducted a model test of a steel catenary riser (SCR) at SINTEF Ocean to investigate the effects of wave and VIV loads. The study analysed selected test cases to understand the impact of wave loads on VIV responses. Nonlinear time-domain simulations using VIVANA-TD were also performed, and they showed good agreement with the model test results. The research highlighted key parameters, such as the current flow velocity ratio α and Keulegan-Carpenter number (KC), as influential factors in riser responses under combined wave and VIV loads that affect the synchronization of vortex shedding. This study represents progress in the understanding of the interaction between wave loads and VIV in SCRs. However, this emphasizes the need for a further analysis of additional test cases and more complex flow conditions. Future studies should include a systematic analysis of field measurements to strengthen the conclusions and enhance the understanding of this phenomenon.

3.3.2 Analysis Method Based on Potential Flow

There have also been developments in numerical models based on potential flow theory for the calculation of hydrodynamic loads. Regarding the structural description, Euler–Bernoulli beam theory and FEM have been widely used. Qu et al. (2023) simulated VIV in top tensioned risers under spanwise varying flow directions. Their findings indicated that the impact of variation in the flow direction on VIVs can generally be ignored at high flow velocities, where the riser response is mainly governed by traveling waves. Liu et al. (2024a) modelled VIVs in steel catenary risers using Euler-Bernoulli beam theory with absolute nodal coordinate formulation. Their model successfully captured synchronized excitation coefficients under oblique flow conditions. Zhao et al. (2023b) created a fluid-structure coupled vibration model for Pipe-in-Pipe systems using a double-beam Euler-Bernoulli model and a wake oscillator model. Their results showed that the slenderness ratio and connection layer stiffness of the system affected the inner and outer pipe displacements, frequency ratios, and traveling wave velocities, illustrating the nonlinear system characteristics. Mao et al. (2023a) formulated a dynamic mathematical model for a suspended riser under different top boundary joint conditions, using finite element and Newmark-β methods for dynamic analysis. Both welded and flexible top joins were considered, and a weight was hanged at the bottom of the riser to simulate the actual withdrawal process This study demonstrates that both structural components and buoyancy modifications play crucial roles in controlling VIV. Guo et al. (2023) developed a VIV model for flexible risers using the Hamilton variational principle and incorporated a wake oscillator model. The impact of the top tension and internal flow on the VIV and the significant frequency-locking effects under various flow conditions were noted. Zhu et al. (2023b) introduced a novel sparse modal decomposition method to reconstruct the VIV in risers, which was validated using data from the Norwegian Deepwater Program. The method exhibited high accuracy and stability, particularly in noisy and data-sparse environments, making it highly suitable for engineering applications. Li et al. (2023d) refined a semi-empirical model to simulate the VIV of flexible risers under parametric excitation in unsteady combined flows. Extensive simulations revealed that the VIV is predominantly influenced by the combined flow conditions and frequency of excitation relative to the natural frequencies of the system. Several analyses were conducted on other marine and offshore structures. Karrech et al. (2024) derived equations for free-spanning pipelines subject to lateral and transversal non-linear deformations caused by VIV. Their research revealed a critical period-doubling bifurcation at high oscillation levels, highlighting a previously overlooked aspect of the VIV predictive tools. Wang et al. (2024c) explored the VIV of an NACA0009 hydrofoil due to trailing edge vortex shedding. Their research analysed the frequency-locking mechanism and introduced a new two-dimensional model tailored for lock-in studies.

3.3.3 Numerical and CFD-Based Methods

As computational capabilities continue to develop, significant advancements in the understanding and mitigation of VIV through high-fidelity CFD modelling and FSI techniques have been observed. The commercial tools ANSYS-FLUENT and Star CCM +, and the open-source code OpenFOAM, are common approaches. Hu et al. (2023) explored VIV suppression in cylindrical risers using the viv3D-FOAM-SJTU solver based on OpenFOAM. They focused on the effectiveness of grooves with specific geometries on the riser surface, identifying 90° angle grooves with depths of 0.12D and 0.16D (where D is the diameter) as most effective in reducing VIV. However, they noted the need for further optimization at higher Reynolds numbers, indicating an ongoing challenge in scaling these solutions for practical applications. Li et al. (2024c) investigated the impact of cavitation on VIV using a FSI solver in OpenFOAM, finding that cavitation significantly extends the lock-in frequency range and reduces vibration amplitude. Karthikeyan and Nallayarasu (2023) utilized ANSYS Multiphysics to study 3D elastic cylinders using two-way FSI coupling, an advanced technique that outperforms traditional one-way coupling by capturing higher-order harmonics in fluid forces, thereby offering a deeper insight into VIV dynamics. Similarly, Liu et al. (2023a) employed Star CCM + to investigate the VIV of elastically coupled cylinders, highlighting the limitations imposed by Re on VIV suppression strategies in coupled systems. Shao et al. (2023) utilized the immersed boundary-lattice Boltzmann method (IB-LBM) to analyse an elastically mounted square cylinder with a detached solid or flexible plate at Reynolds number 150. Their study investigated the effects of the gap distance, plate length, and plate flexibility on VIV responses. Moradi and Mojia (2024) simulated the impact of free-surface proximity on VIV characteristics at high Froude numbers using ANSYS-FLUENT. They noted substantial changes in flow parameters, such as reductions in drag and alterations in lift and Strouhal numbers, emphasizing the influence of environmental factors on the VIV behaviour. Xu et al. (2024a) evaluated the computational demands for simulating the vortex-induced motion (VIM) in a semi-submersible offshore platform using a URANS model in ANSYS-FLUENT, revealing the specific requirements for accurate VIM simulations compared with more static flow simulations around bluff bodies. Pathak and Saha (2024) performed a one-way fluid-structure interaction (FSI) analysis using ANSYS-FLUENT and ANSYS-STRUCTURAL to examine the dynamic behaviour of submerged floating tunnels (SFTs) under current action. Their study concluded that accurate load prediction and structural response require full-scale analysis, including 3D simulations that account for both the inflow and crossflow directions. Additionally, incorporating more advanced structural modelling, complex loading conditions, and surface roughness would further enhance the accuracy of the predictions. Zhang et al. (2023b) developed a comprehensive 3D model to analyse the VIV of flexible risers by considering both internal and external flows. Their findings underscore the significant nonlinear effects of internal flow velocities on VIV responses, illustrating the complexity of fluid-solid interactions in marine structures. Wei et al. (2022) used RANS equations to analyse the VIM of square columns arranged in a square configuration and discovered that such configurations experience galloping at lower critical velocities than circular columns. This study highlights the importance of considering geometry in VIM analysis. The influence of the three-dimensional effect should be considered in future studies. Li *et al.* (2024b) investigated scale effects on VIV of a 2D circular cylinder using CFD solver in OpenFOAM. The results indicated that the drag coefficient, lift coefficient, and vortex shedding frequency exhibited significant variations with scale changes, even at the same Reynolds number. This variation was more pronounced under large-scale and low-flow-velocity conditions.

3.3.4 Machine Learning

In addition, various machine-learning techniques have been applied to analyse CIVs. The VIV of an inclined flexible cylinder was predicted by Xu et al. (2023) using machine learning methods. Three different machine learning algorithms namely, back propagation neural network (BP-NN), support vector machine (SVM), and Gaussian process regression (GP) were employed using previous tests as a database. Within the parameter space of the present study, the SVM is regarded as the most suitable for predicting the VIV of an inclined flexible cylinder. However, more systematic experimental tests are required to improve the model further. Mentzelopoulos et al. (2023) improved the flexible cylinder VIV prediction by machine learning hydrodynamic databases using measurements along the structure. This study demonstrates the effectiveness of the framework for flexible vertical risers in a stepped current and flexible catenary riser.

3.3.5 Conclusion and Further Work

As computational capabilities and theoretical understanding continue to evolve, future research is likely to focus on refining these approaches to enhance the prediction and control of VIV across a broader range of marine applications. While combined uniform and oscillatory flows have been considered in some studies, the impact of more complex loading conditions, such as varying current profiles and the combined effects of waves, has not been thoroughly analysed. As in previous reports, the scaling laws from model tests to full-scale applications (Reynolds number scale effects) remain poorly understood. The limited number of full-scale measurements has contributed to this lack of understanding. To address these gaps, systematic numerical modelling of VIV problems incorporating variations in scale at different Reynolds numbers can be conducted. In addition, 3D effects should be considered in future studies.

3.4 Internal Flow Induced Vibrations

Marine risers, cold-water pipes, and centrifugal turbomachines such as fans, pumps, and compressors are extremely important components in oil exploitation and transportation. Throughout the operational lifespan of a riser or pipeline, multiphase flows comprising crude oil, natural gas, and water in combined oil and gas transportation frequently occur within its internal structure. Consequently, the total fluid density and pressure may be unsteady and inhomogeneous along the pipe, inducing internal flow-induced vibration (FIV) that can cause large oscillations, raising concerns about fatigue.

Though the fluid is typically multiphase, such as gas-liquid or solid–liquid, the internal flow is usually considered as a single-phase flow in the numerical models coupling flow and structural response in VIV investigations. Chang et al. (2023) studied the axial-transverse coupled VIV characteristics of a composite riser under a gas-liquid two-phase internal flow. Based on Hamilton's principle and a semi-empirical wake oscillator model, a predictive model of the response of the composite riser under internal flow excitation was developed. The effects of ocean current velocity, liquid phase velocity, gas volume fraction, and fibre orientation angle on the coupled VIV in the three directions of the riser were discussed. The flow in the riser was assumed to be steady, which is an ideal model. Therefore, it is necessary to further improve the two-phase flow model to describe the fluid characteristics and unsteady variation characteristics of the riser more accurately. Mao et al. (2023b) studied the VIV characteristics of mining risers under the coupling effect of external ocean current and internal gas–liquid two-phase flow. The effects of the discharge rate, mixed fluid density, inlet ratio, and other factors on VIV were investigated. The gas-liquid two-phase flow reduced the natural frequency of the riser. Compared with pure liquid flow, the vibration amplitude and frequency of the riser were larger, and the modal response involved higher-order modes under the same external excitation. Therefore, more attention should be paid to the effect of gas-liquid two-phase flow on the VIV in practical engineering problems.

Edge boundary conditions are important factors in the analysis of internal flow-induced vibrations. Leng et al. (2022) analysed the internal flow effects on the cross-flow VIV of marine risers with different support methods (hinged-hinged, hinged-cantilevered, and fixed at both ends). A numerical model was established in which the internal flow in the riser VIV process was considered. The proposed model was verified using experimental data. The internal flow effects (e.g., internal flow density and internal flow velocity) on the VIV of marine risers with the different support methods were investigated. The dependency of internal flow features on the riser different support methods was sorted in descending order with respect to induced VIV as follows: hinged-cantilevered, hinged-hinged and fixed at both ends. Tan et al. (2024) analysed the vibrations and stability of a cold-water pipe (CWP) for Ocean Thermal Energy Conversion (OTEC) subjected to clump weight under different boundary conditions. The six boundary conditions of the CWP were: Clamped-Clamped (C–C), clamped-clump weight at the bottom (C–W), clamped-free (C–F), simply supported-simply supported (S–S), simply supported-clump weight at the bottom (S–W), and simply supported-free (S–F). They found that the clump weights significantly affected the critical flow velocity of the pipe under six boundary conditions. Augmenting the clump weight decreased the critical flow velocity for the C–C and C–W boundary conditions. Conversely, it elevated the critical flow velocity for the S–S, S–W, and S–F boundary conditions, with no effect observed under the C–F boundary condition. For different boundary conditions, the clump weights had a destabilizing or stabilization effect on the CWP system. Increasing the clump weights reduces the modal amplitudes under the C–C, S–W, and S–F boundary conditions, while it increases them under the C–F and S–S boundary conditions. For the S–S boundary condition, larger clump weights shifted the maximum modal amplitude towards the mid-pipe. However, increasing the clump weights leads to a complex trend in the modal amplitudes under C–W boundary conditions.

The occurrence of rotation during pipe operation can significantly affect the rheological vibration characteristics of pipes. Unlike pipes under external flow, the flow-induced vibration characteristics of pipes undergo complex changes when they appear to rotate. Zhang *et al.* (2024b) conducted a numerical investigation of the flow-induced vibration of a rotation pipe with internal flow. A theoretical vibration model of a three-dimensional rotating pipe with internal flow was constructed, and a rotating pipe with a rotation ratio in the range of 0–1.5 and an internal flow velocity in the range of 0–5 m/s was numerically investigated using the Newmark-β and FE methods. The wake pattern, vibration response, and hydrodynamic characteristics of the rotating pipe were illustrated. With increasing rotation ratio, the vibration amplitude and hydrodynamic force increased, and the motion trajectory of the rotating pipe varied from "8" to "elliptical." In particular, when the rotation ratio reached 1.5, the range of the motion trajectory remained steady.

The internal flow-induced vibration responses of the steel catenary riser (SCR) differ significantly from those of the top-tensioned riser (TTR). Li *et al.* (2022b) studied the VIV response characteristics of a catenary riser conveying two-phase internal flow. Based on the force decomposition semi-empirical time-domain model of VIV for a flexible riser, the VIV responses of the TTR and SCR conveying internal flow were calculated and compared. Under an external shear current, the VIV responses of the TTR and SCR exhibited traveling wave characteristics, which is the result of a multi-frequency response. Under similar external currents and with similar sizes, SCR had smaller tension and lower natural frequency than TTR, the modal orders of SCR were higher, and the fatigue problem of SCR was also more significant. The increase in the internal flow velocity triggered higher-order modes of the VIV response and intensified the fatigue damage of the risers. The stiffness loss of the SCR was more pronounced than that of the TTR with an increase in the internal flow velocity. Therefore, the internal flow effect on SCR appears to be more significant than that on TTR.

In the near future, more studies on internal multiphase flow-induced vibrations should be conducted, especially those related to experimental investigations. Although many simulation tools are available, experimental validation is still lacking in this area. The real boundary conditions and external flow environment are important when studying the internal flow-induced vibration features.

3.5 Equipment Induced Vibrations

An offshore platform comprises numerous critical components essential for marine system operations. In particular, rotating machinery such as pumps, motors, engines, and compressors induce dynamic responses in the platform structure owing to their repetitive actions. These recurring structural vibrations and dynamic behaviours can compromise the stability of both the equipment and the structure itself, directly affecting the safety of personnel residing on the offshore platform. Most research on equipment-induced vibrations in offshore structures has focused on the control and absorption methods for the resulting vibrations rather than on their analysis. Studies have included methods for mitigating equipment vibrations, such as tuned-mass dampers and damping methods. However, detailed case studies of tuned mass dampers or vibration mitigation methods are more appropriately covered in Sect. 4.1.2 dedicated to active control of vibrations, and thus they will be minimally discussed in this section. With the increase in environmental concerns, research predominantly focused on equipment related to OWT systems. As digital twins for simulating the dynamic behaviour of offshore structures are being actively investigated, the application of virtual sensing and digital twin models for core equipment has also been identified as a significant area of research. These studies mainly dealt with updating simulation models based on actual measured data obtained from offshore structures, as well as advanced modelling techniques that address multibody dynamic characteristics, and modal expansion analysis.

Jaen-Sola et al. (2021) proposed an optimization method for the rotor structure of an offshore direct-drive wind turbine to ensure the structural stability of the supporting structure and to prevent resonance caused by vibrations from the generator, which is a rotating machine. To minimize the vibrations and resultant dynamic response from the wind turbine generator, they parameterized the design variables of the generator rotor structure. By varying these parameters, a systematic design approach was proposed to maintain the overall structural rigidity while reducing the mass and natural frequency. The design parameters were optimized using a topology optimization method to enhance the rotor structure and simultaneously reduce its mass and operational energy consumption. Although this study is in its initial stages of research, it represents a valuable example of structural optimization research essential for scaling up OWTs.

Augustyn et al. (2021) reported on modal expansion techniques to improve the accuracy of models for virtually simulating the dynamic responses of offshore wind jacket substructures. They proposed a virtual sensing technique using additional sensor configurations and investigated the feasibility of modal expansion using virtual sensing data. The vibrations occurring in the substructures resulted from a combination of global and local responses, excitation due to persistent winds, and dynamics related to stationary waves. To enhance accuracy of the simulation of the substructure response additional information from the surrounding and interacting substructures was incorporated. Therefore, in this study, two additional points at the bottom of the substructures were measured using the physical sensors, and those were combined with virtual sensors to perform the modal expansion of the structure. This approach allows for a clearer representation of vibrations due to wave loads and local brace modes in the substructures. Although this study focuses more on the vibrations of subsystems of offshore structures rather than the vibrations induced by equipment, it is included in this section as a representative study of virtual sensing techniques for substructure dynamics.

Recently, the expansion of electric drive systems has altered the dynamic response characteristics of offshore structures. Yousri et al. (2023) examined the impact of electric machines on structural reliability. In particular, they analysed the dynamic response caused by torque ripples in offshore crane winch systems. Recently, there has been a trend toward electrification of winch systems, shifting from hydraulic to electric drive systems. Changes in the torque ripple characteristics due to this shift can induce different mechanical stresses on the drivetrain. In this study, torque ripples were simulated and their impact on the drive gearbox was analysed. The results showed that the torque ripples in electric drive systems were smaller than those in hydraulic systems, thus improving the reliability of the drive gearbox. Wan et al. (2023) evaluated the structural vibrations that occur when large rotating machinery is added to an offshore platform. Using SACS (Structural Analysis Computer System) and ANSYS software, they implemented a simulation model and assessed the structural modal analysis and dynamic response of the offshore platform to evaluate stability by adding rotating equipment installation. The frequency margin and vibrational response were evaluated based on ISO standards and CCS guidance, suggesting that this evaluation could aid in the development of guidelines for adding new equipment. As documented by the previous papers, it is evident that research is needed on how the dynamic response of existing structures changes with the addition of new types of power systems and rotating machinery to offshore platforms. The ongoing accumulation of research concerning the excitation properties and structural effects of innovative equipment is expected to facilitate the development of guidelines for integrating such apparatus.

This section has discussed equipment-induced vibrations affecting offshore structures. As rotating machinery for new functions is added to offshore structures, along with the electrification of equipment, studies on the structural dynamic response to various excitation sources are increasingly necessary. Moreover, because offshore systems are typically operated over extended periods, it is essential to focus on the adaptation and updating of descriptive models over time from a dynamic response perspective. Such research requires precise dynamical models of the equipment that influences the dynamic response in offshore systems, and techniques to calibrate these models to match real-time data. Future research should focus on developing high-fidelity models that consider also the measurements taken on-board.

3.6 Ice Induced Vibrations

Studies on ice-induced vibrations in marine structures have predominantly focused on the dynamic responses of OWTs impacted by drifting ice. However, the interactions between pack ice and offshore platforms were also studied in the 2010s (see, for instance, Sun and Shen, 2011). These studies often focus on ice loads rather than ice-induced vibrations and are beyond the scope of the present section.

Researchers have followed two major approaches to study ice-induced vibrations in marine structures. In one approach, to model the structural vibrations caused by ice, the interactions between the ice and the structure were not considered, and the ice loads were treated as a forcing condition. Consequently, neither ice breaking nor crushing was observed. In other words, these studies focused on the vibrations of the structure and overlooked ice-related physics (i.e., mutual ice-structure interactions are neglected). However, this approach generally requires a thorough understanding of the ice loads. An example of scholarly work in this area is the study by Zhu et al. (2021), who investigated ice-induced vibrations of monopile OWTs, taking into account aerodynamic effects. Aerodynamic loads were calculated using the blade element theory, while ice loads were estimated using the Määttänen-Blenkarn model. The structural response of the monopile was simulated using FEA. It was observed that ice-induced frequency lock-in of the monopile occurred when the drift ice velocity was between 0.01 m/s and 0.06 m/s for 7 cm thick ice, and between 0.03 m/s and 0.09 m/s for 32 cm thick ice. Another study by Zhang et al. (2022d) modelled the dynamic responses of a jacket structure using a simplified model by applying a basic equation for ice loads. Subsequently, using synthetic data, they developed an intelligent method, the gated recurrent unit (GRU) network, to forecast the structural responses of jackets subjected to ice loads. The authors demonstrated that by coupling the model with a long short-term memory (LSTM) approach, the active control of structural vibrations can be achieved. In this area of study, while the effects of the structure on ice (breaking and crushing) are not considered, models can be made more comprehensive by incorporating pile-structure interactions. For instance, a recent holistic model capable of solving the dynamic responses of a monopile impacted by drifting ice was developed by Wu et al. (2024b). Aerodynamic loads were calculated using blade element theory, and the soil-structure interaction model was developed within the ABAQUS code, whereas ice loads were implemented using a stochastic model, meaning that ice-structure interactions were not considered. The results revealed that the co-occurrence of ice and aerodynamic loads reduced the bending moment compared with the case in which the cumulative effect of individual ice and aerodynamic load contributions were considered. The model was later used to evaluate the effectiveness of a tuned-mass damper for controlling structural vibrations (Wu et al., 2024c). The authors found that a tuned mass damper specifically tailored for the first mode can effectively reduce the displacement of the OWT. However, to reduce acceleration responses of the structure, a multiple tuned mass damper targeting higher-order modes is required.

In contrast, another stream of studies focuses on models that consider ice-structure interactions. These models not only examine ice-induced vibrations but also study the effects of the structure on ice, leading to more complex physics. Experimental tests are feasible, albeit challenging, as towing ice at a desired drifting speed presents significant difficulties. Alternatively, a monopile can be slowly moved in a tank covered with thin ice to obtain a similar ice-structure interaction. For example, Hammer and Hendrikse (2023) conducted ice tank tests in the Aalto University ice tank to measure how misalignment between the wind and ice loading directions affects ice-induced vibrations in OWTs impacted by drifting ice. Using a hybrid test setup, they observed that misalignment led to sustained vibrations in the direction of the ice load. They also found that, while ice interactions constrained wind-induced motions at low drift velocities, this effect failed rapidly at high velocities. An interesting aspect of these tests is that they considered hydrodynamic and soil effects, whether significant or not. Most studies in this stream have used computational solid mechanics codes. The most challenging task is to choose the proper numerical method to solve the ice motions and their interactions with the structure during collisions. In recent years, researchers have employed methods such as FEA, CEM (cohesive element method), and DEM (discrete element method) to numerically simulate ice motions, with the structure typically modelled using FEA.

Ji and Yang (2022) and Song et al. (2023a) used FEA to model ice motion interacting with structures. Ji and Yang (2022) numerically modelled the mutual interaction between a monopile and drifting ice and evaluated the validity of their full FEA model by comparing its predictions with previous ice tank tests. They concluded that, under different conditions, the vibration responses of the wind turbine varied locally. The vibration frequency in the foundation was observed to be close to the third-order natural frequency of the structure, whereas that of the tower was observed to be close to the first-order natural frequency. In addition, Song et al. (2023a) investigated the dynamic response of monopile-supported OWTs subjected to combined sea-ice impacts and wind loads using nonlinear FEA in LS-DYNA. Similar to the study by Ji and Yang (2022), both the sea ice and wind turbine were modelled with FEA, with the ice represented by a piecewise linear plasticity model. Song et al. (2023a) reported that co-existence of ice and wind loads caused larger dynamic responses in the OWT, including increased tower-top fore-aft (FA) displacement and higher tower bending moments at both the mean sea level and the mudline, compared to cases involving only wind or ice loads.

As mentioned earlier, DEM can also be used to model the motion of the ice layer. Long et al. (2023) employed the DEM method to simulate the motion of drifting ice impacting a monopile. They developed a numerical approach to address this interaction, modelling the ice using DEM and the structural motions of the monopile using the Verlet integration scheme. Interestingly, it was shown that when the breaking process occurs in the ice sheet, the ice undergoes a ductile-brittle transition with periodicity, leading to self-excitation of the OWT. This results in larger dynamic responses in the structure with higher frequencies compared to cases that do not experience self-excitation.

The final approach used for modelling ice behaviour in ice-structure interaction problems is the CEM. The work of Shi *et al.* (2023a) is a good example of such research. The authors developed a fully coupled model to simulate the dynamic response of a monopile impacted by drifting ice, incorporating structure-soil interaction. FEA was used to solve the dynamic responses, and the structure-soil interaction was modelled using the beam theory. It should be noted that ice-bending failure was represented using the Tsai-Wu yield criterion. These findings indicate that structural damage is more sensitive to ice thickness than to ice velocity. Similarly, Wang *et al.* (2022a) utilized CEM to numerically model ice motion and solve the mutual interaction between drifting ice and a monopile. Nonlinear springing was considered in the structural model to account for nonlinearity.

In general, research studies pursue two distinct methodological approaches with the objective of developing more comprehensive models that account for soil-structure interaction and aerodynamic effects on the structural responses of a monopile subjected to drifting ice impact. However, hydrodynamic effects are often neglected even though they may be significant, particularly if accelerations occur, which could involve added mass effects.

3.7 Response to Accidental Loads

Accidents are defined as unexpected events that ships and offshore structures may be subjected to during operation. Although a broad range of potential accidents exists, this section focuses on the dynamic responses caused by these events. Such accidents can be primarily categorized as collisions between vessels and offshore structures, the consequences of falling objects, and natural disasters, such as earthquakes and tsunamis. There is also significant research interest in mitigation measures to absorb collision energy to reduce damage and dynamic response.

Zhang et al. (2021b) studied dynamic responses relative to ship-FOWT collision, namely a 5-MW OC3 Hywind Spar-type FOWT and an offshore service. They first developed a mathematical model for the external mechanism of this collision scenario, which was then combined with an in-house tool to predict the nonlinear dynamic responses of the entire FOWT system in the time domain including aero-hydro-servo-elastic coupling. The tool was then used to simulate different scenarios and investigate the nonlinear dynamic responses of the FOWT system. Collision scenarios included still water, wave-only, and wind-wave conditions. The ship FOWT collision model was then verified by comparing the results of the still water case with an analytical solution for ship-spar collisions given in the literature.

It is shown that in still water conditions, the ship impact affects the responses of the platform and mooring system more than those in wave and wave wind conditions. In the wave-only condition, the motion responses of the platform were suppressed by the wave effect; however, the tower vibration and tower top deformation remained sensitive to ship collisions. For the safety of the FOWT, the acceleration at the nacelle was analysed because some equipment might be sensitive to acceleration. The analysis results indicate that even though the FOWT structure does not suffer critical damage owing to ship impact, the equipment inside may still fail to work because of the high value of acceleration induced by ship impact. Currently, the combined method used in this analysis is only able to address the external dynamic response analysis of FOWT by ignoring the structural damage caused by ship impact. A method that considers both the external and internal structural dynamic responses should be developed in the future.

Yu *et al.* (2022c) performed a study on the dynamic response analysis of a 10 MW semi-submersible FOWT subjected to ship collision loads. The ship collision responses of the OO-STAR floater were investigated using the nonlinear FE in-house code USFOS. The selected striking ships are modern supply vessels of 7,500 tons and shuttle tankers of 150,000 tons, representing service/coastal merchant vessels and large passing vessels, respectively. Modelling of the FOWT is described in detail, including the OO-STAR floater, DTU 10 MW turbine blade, turbine tower, and mooring system. The modelled hydrodynamic loads included buoyancy loads and motion-induced radiation loads using the Morrison equation. An eigenmode analysis of the turbine model was performed to verify the established model and the results agree well with the DTU report. Global collision response analyses of the FOWT were performed under both the parked and operating conditions. Upon examination, the semi-submersible floating offshore wind turbine (FOWT) exhibits a satisfactory level of safety when encountering collisions with modern supply vessels. However, nacelle accelerations may exceed the allowable operational limit and damage the delicate equipment inside the nacelle. For shuttle tanker collisions, the FOWT undergoes large displacements and rotations, and may eventually lose hydrostatic stability and capsize.

Sun *et al.* (2023) proposed a novel floating composite honeycomb anti-collision structure to reduce damage caused by ship-OWT high energy collisions. In this study, a novel UHPC-steel-EPP floating composite anti-collision structure (USEFCAS) was examined. An evaluation and comparison of the performance of the anti-collision structure in terms of the OWT response were performed using ABAQUS/Explicit. This paper also presents the dynamic response of the nacelle after collision in terms of acceleration over time. Both the rubber and USEFCAS honeycomb structures significantly reduced the nacelle accelerations. USEFCAS effectively absorbed 80% of the initial kinetic energy of the ship, significantly reduced the collision force of the ship and the nacelle's maximum acceleration, and prolonged the collision time. Thus, the investigated anti-collision system seems to be a promising alternative for reducing damage to ships and OWT and avoiding secondary disasters.

Underwater explosions produce an impulsive load causing relevant damage, in which the dynamic response of the ship structure plays an important role. Li et al. (2021)

presented a simplified hull-girder model to quickly and accurately predict the overall whipping response of a vessel subjected to an underwater explosion. The load profile of an underwater explosion was first divided into five phases, and a theoretical model of the fully elastic response and elastic-plastic response of the hull girder under the combined effects of an explosion shock wave and pulsating bubble was established and verified using experimental data. The effects of damping on the overall mode, period, and amplitude of the deformation were investigated. The effect of the presence of the damping on the plastic hinge of the girder were also considered. The results show that the proposed model can capture the repetitive loading and unloading processes leading to the plastic deformation of the girder and accurately forecast the period and amplitude of the overall oscillations of the hull girder with an error of less than 10%. Damping suppressed the whipping response, and the suppression effect increased as the distance from the explosion decreased. Damping is found to be an important factor for the plastic hinge rotation angle of the girder that needs careful evaluation.

Ships and offshore structures are naturally exposed to repeated impacts during their lifespan, such as vessel collisions and dropped objects. Wang *et al.* (2024b) studied the energy dissipation of ship plates under repeated impacts by using a FE model which included friction and damping effects. Numerical simulations were performed for a rectangular plate and validated against experimental results of both single and repeated impacts. The numerical analysis shows that elastic energy increases with an increase in plate deflection, regardless of the magnitude of the impact energy. The accumulated energy dissipated through the friction effect increased linearly with the impact number. After considering the energy dissipated by friction, damping, and other effects, a general energy criterion was suggested for further studies.

Research on the critical issue of ship-to-offshore wind turbine (OWT) collisions has indicated that numerical calculation methodologies can generate valuable data regarding the vibratory response of wind turbine structures following an impact event. This response can also magnify the acceleration response and, therefore, can be dangerous for the equipment that works inside. Another interesting topic is the design of anti-collision structures for wind turbines to reduce acceleration responses after collisions.

3.8 Noise

The rapid expansion of offshore energy production, including that of wind farms and oil and gas platforms, has introduced several technical and operational challenges. Noise and vibration management are of particular concern because of their impact on structural integrity, worker safety, and overall operational efficiency. Offshore installations generate significant levels of noise during both the construction and operation phases, which can affect the structural performance and well-being of the personnel. Therefore, understanding and mitigating these noise impacts is crucial for the sustainable development and safe operation of offshore energy infrastructure.

In the following sections, two distinct aspects of noise management in offshore environments are discussed: Sect. 3.8.1 focuses on noise and vibration control on board offshore platforms, exploring innovative solutions and technologies aimed at enhancing occupational safety and comfort. This includes the development of advanced predictive models and noise-reduction devices tailored to the unique conditions of these installations. In contrast, Sect. 3.8.2 examines underwater noise radiation associated with offshore wind farms and marine construction activities, discussing various strategies and technologies designed to minimize its operational and environmental impact.

3.8.1 Noise on Board

Effective management of noise and vibration on offshore platforms and wind farms is essential because of their substantial impact on both structural integrity and human well-being. Despite the critical nature of this issue, current research specifically focused on noise management in offshore environments remains limited, leaving key gaps in our understanding of optimal control strategies for these challenging settings. Nonetheless, recent studies have made significant advancements by introducing sophisticated modelling techniques, innovative noise attenuation solutions, and thorough assessments of occupational health risks. Nektarios et al. (2021) explored the specific hazards faced by workers in the wind-energy sector. As wind farms become a key source of renewable energy, new materials and technologies expose workers to unique occupational risks, such as noise, electromagnetic fields, shadow flicker, epoxy and styrene exposure, and physical stress. This review highlights the lack of comprehensive studies on factors, such as vibrations, welding fumes, harmful substances, weather conditions, and biological hazards. This document calls for further research to manage these risks across the wind energy production lifecycle and stresses the need for collaboration among the government, academia, and industry to enhance knowledge and safety measures. Kim et al. (2023) investigated a silencer designed to reduce noise on offshore platforms without using traditional sound-absorbing materials like glass wool, in response to stricter NORSOK standards. This study introduced a silencer based on the acoustic characteristics of a periodic structure that creates a bandgap to suppress acoustic wave propagation. Modelling and experiments demonstrated the effectiveness of the silencer in noise attenuation, making it suitable for offshore platform requirements and enhancing habitability by reducing noise pollution. Lastly, Chin and Zhang (2021) focused on improving noise prediction on offshore platforms using deep learning techniques. This study employs a Deep Belief Network (DBN) to predict sound pressure levels in various rig compartments, optimizing neuron numbers to minimize errors. The findings indicate that the optimized DBN-DNN model significantly reduces the prediction error compared to traditional methods. The document underscores the potential of deep learning models in enhancing offshore platform design and calls for further research to improve model accuracy.

3.8.2 Underwater Radiated Noise

The development and expansion of OWT are critical components of the global transition to sustainable energy sources. However, the environmental impact of the underwater noise generated during both the construction and operation phases of these projects has become a significant concern. Various studies have been conducted to understand the mechanisms of noise generation and to develop effective mitigation strategies. Peng *et al.* (2021a) defined the use of air bubble curtains as an effective method for reducing underwater noise during pile-driving operations. This technique, as demonstrated in their study, showed that air bubble curtains can significantly scatter and absorb sound waves, leading to noise reduction of up to 20 dB. This method is particularly effective in protecting marine life from the harmful effects of intense noise levels typically associated with pile driving. Additionally, Jestel et al. (2021) with their damped cylindrical-spreading model provide a more accurate prediction of mitigated underwater noise levels by considering various environmental factors. This model improves upon traditional spherical-spreading approaches, facilitating better planning and implementation of noise mitigation strategies during marine construction. To enhance noise prediction capabilities, Peng *et al.* (2021b) developed a fast computational model for near- and far-field noise, which integrates environmental and operational parameters to provide comprehensive predictions of underwater noise propagation. This model is valuable for environmental impact assessments and helps to mitigate noise pollution effectively. Von Pein et al. (2022) investigated how strike energy, pile diameter, ram weight, and water depth affect noise levels, resulting in scaling laws that facilitate accurate predictions based on these variables. This is crucial for designing quieter pile driving operations. To address near-field noise, He *et al.* (2023a) incorporated the interaction between the pile and the surrounding soil. Their model highlighted the significant influence of soil properties on noise levels, emphasizing the need for detailed soil characterization in noise predictions. The study by Molenkamp *et al.* (2024a) on vibratory pile driving with nonlinear frictional pile-soil interaction emphasized the importance of realistic modelling in noise mitigation. Noise fields were found to be highly sensitive to variations in system dynamics and excitation spectra, underscoring the need for sophisticated models to accurately predict noise emissions. Molenkamp *et al.* (2024b) examined the impact of asymmetric impact forces and pile inclination on underwater noise levels during monopile installation. These findings highlight the need for strategies to minimize noise during these operations to reduce environmental impacts. Operational noise from OWTs poses several challenges. Yang *et al.* (2023a) emphasized the importance of understanding the relationship between the wind speed, rotor speed, and noise levels. Using accelerometers and hydrophones, they found a strong correlation between these factors, suggesting that continuous monitoring and analysis is essential for effective noise management in wind farms. In addition, Yoon et al. (2023) conducted a study involving ten days of data collection near a 3-MW wind turbine, revealing that noise levels vary with rotor speed, peaking at approximately 198 Hz. The high correlations between underwater noise, tower vibration, wind speed, and rotor speed underscore the need for integrated noise assessment frameworks. Zhou and Guo (2023) explored innovative mitigation techniques, including the use of compact circular structures. These structures act as barriers to scatter and absorb sound waves, effectively reducing noise levels by up to 10 dB and enhancing the environmental compatibility of OWTs. A significant advancement was presented by Molenkamp et al. (2023), who introduced a model that allows for relative motion between the pile and soil, thereby improving the noise prediction accuracy. This research shows that traditional models that do not account for pile slip are less effective for vibratory installations, emphasizing the need for improved models that consider soil dynamics. Wang *et al.* (2022c) focused on identifying key frequency components of operational noise. They found that dominant frequencies vary with rotor speed, offering insights into noise generation mechanisms and suggesting ways to optimize turbine designs to minimize environmental impacts. This study highlights challenges such as environmental variability and proposes strategies to mitigate underwater noise. Although significant progress has been made in understanding and addressing noise from offshore wind turbines, research gaps persist regarding noise from other renewable energy systems and traditional oil and gas operations. De Jong et al. (2023) provided a comprehensive review of methods for measuring underwater noise from oil and gas activities, including seismic surveys, drilling, production, and transportation. Their work details key sound sources, measurement techniques (e.g., hydrophones and autonomous vehicles), and critical parameters, such as the sound pressure level (SPL) and frequency bands.

Bridging these gaps is essential for the development of robust noise and vibration management strategies for offshore installations. Future research should focus on emerging challenges, such as the acoustic impacts of deeper offshore installations, noise profiles of innovative turbine designs, and the effects of climate change on underwater noise propagation. Addressing these issues is crucial for advancing sustainable and environmentally responsible offshore energy development.

4 Vibrations and Noise Control

Damping and counter measures for vibration and noise may differ and are therefore handled separately in this chapter. Furthermore, active and passive controls can be distinguished. Passive controls rely on the system's fixed damping and stiffness, require less maintenance because they do not rely on external power, and are, in most cases, less complex and more robust structural design features. The main limitation of passive systems is that they are less adaptable to changing conditions. Once installed, their performance is fixed, and they cannot be easily adjusted to cope with different types or levels of vibration and noise. Active controls are characterized by an additional control force generated by a controller; therefore, they are generally more complex and costly. Active control techniques are typically selected for their adaptability and precision in dynamic and high-demand applications. The literature review also revealed that the research is quite separated between the offshore and ship structure segments.

4.1 Vibration

Several literature reviews have been published, for instance, by Tian et al. (2023) and Liu et al. (2022d), which generally categorize structural vibration control methods into four types: passive, active, semi-active, and hybrid. Machado et al. (2024a) provided a good overview of the commercial devices available in the market and used to control vibrations in wind turbines.

4.1.1 Passive Techniques

Analysis and techniques.
Limited research has recently been conducted on damping measurements, vibration treatment, and damping enhancement technologies for ship structures. The fact

that damping data, especially for ship structures, are often lacking in the literature has also motivated the Committee's benchmark study. Nonetheless, some research is ongoing because accurate dynamic models are required to predict structural vibrations and improve the analysis of fatigue damage. Van Zijl et al. (2021) used OMA to identify the vibration characteristics of a polar vessel using full-scale acceleration measurements. Five modes were tracked over a range of operating and environmental conditions, including calm and stormy seas, and two ice cases. Compared to calm open water conditions, vertical bending modes were generally affected the most by the operating environment, with the natural frequency and damping increasing up to 3.7% and 349%, respectively, in ice.

Shrivastava et al. (2022) proposed a methodology to accommodate the welding uncertainties in the strength and damping of a stiffened plate in a theoretical model. Uncertainties were incorporated using coupling springs and dampers. The variation in damping due to welding was determined using a damping identification technique that requires prior accurate knowledge of the stiffness of the weld. Analyses were conducted to investigate the robustness of the proposed algorithm by inducing random noise in modal parameters.

The effects of structural damping and fluid damping on the vessel dynamic response to an explosion were analysed by Li et al. (2021) based on experimental data from model tests and theoretical models. The dimensionless damping ratio corresponding to the first-order wet mode frequency of the girder was determined to be close to 0.3, which is similar to the results previously reported in the literature. It was found that the suppression effects due to damping increased as the explosion distance decreased. Moreover, damping was shown to be an important factor affecting the plastic hinge rotation angle of the girder and should therefore not be ignored in whipping response simulations of hull girders subjected to underwater explosions.

Compared to the shipping sector, significantly more research in the field of damping identification has been reported in the offshore sector. Norén-Cosgriff *et al.* (2021) and Song *et al.* (2023b) identified the modal parameters of an OWT using full-scale monitoring. It was observed that depending on the mode shape, the identified natural frequencies and damping ratios may show significant variability. Considering the difficulties as well as the importance of accurate damping input, Liu *et al.* (2024b) proposed a novel damping separation method based on operating state-measured data collected from an OWT to provide various damping ratios caused by different loads. In addition, Kjeld *et al.* (2023a, 2023b) performed a systematic analysis of the influence of wind and wave characteristics on the modal properties of an idling OWT based on data from a monitoring campaign. In general, they concluded that the wind and wave characteristics had a non-negligible influence on most of the modal parameters of the OWT. The damping ratios tended to have a positive correlation with the wind speed and significant wave height. In contrast, the natural frequencies tended to decrease as the wind speed/significant wave heights decreased. The strong correlation between these variables presents a significant challenge in quantifying which of them is the primary driver of the change in dynamics. In addition, they studied how to minimize the empirical uncertainty bounds of the experimentally derived damping estimates. A closed-form analytical expression for dynamic response was derived by Adhikari *et al.* (2021). The results show that the damping in the wind turbine dynamic analysis is completely captured by seven different physically realistic damping factors and suggests a value that is very useful for benchmarking related experimental and finite element studies at the initial design/analysis stage. Jiang et al. (2023) derived a new theoretical calculation method of the along-wind aerodynamic damping considering separately the contributions resulted from the changes in wind speed, rotor speed and pitch. It is shown that along-wind aerodynamic damping increases as the value of the above influencing factors increases. It was also demonstrated that increasing aerodynamic damping reduces the effectiveness of a TMD.

Foundation damping is considered the second most valuable component of system damping after aerodynamic damping. Malekjafarian et al. (2021) presented a critical review of recently published studies on foundation damping for OWTs on monopiles and explained how soil damping contributes to the total damping of OWTs. The authors also reviewed the main methods used to estimate the foundation damping in numerical and experimental studies. Three years later, Bradshaw et al. (2024) studied foundation damping in a piled offshore wind-jacket structure. As shown by Zhu *et al.* (2023a), the effective foundation damping ratio determined using FEA is not always consistent, reaching only 15%, which is much smaller than the input value. The inclusion of the soil–structure interaction and elastoplastic behaviour of soil only increased the effective damping ratio to 35% of the input value and obfuscated the roles of the various factors in the effective damping. Zhang *et al.* (2021a) presented a simplified model for estimating soil damping due to the nonlinear soil response relatively to pile foundations. The model is linked to the damping response of the soil measured at the element level; therefore, it offers design engineers an efficient and accurate way to estimate soil-pile interaction damping based on site-specific soil data. Lian et al. (2021) developed an estimation approach for the total damping of an OWT supported by a wide-shallow bucket foundation, because its damping characteristics were unknown.

Applications.

Among the control techniques, the passive control system has the advantage that it does not require an external power supply, and probably will become the most widely used control countermeasure in the future. Passive control systems can be subdivided into dynamic vibration absorbers (DVA), energy dissipators, and isolators. In terms of DVAs, two classical devices, namely a tuned mass damper (TMD) and a tuned liquid damper (TLD), are commonly utilized.

The only recent TLD-related research was that of Sardar *et al.* (2023), who explored the prospect of a typical deep-water storage tank. The focus of research in the field of passive vibration control in OWTs is mainly related to TMD, as reported by Villoslada et al. (2022) and Zhang *et al.* (2023c). Gao et al. (2024) and Verma et al. (2022) focused on optimal design of conventional TMDs, while Zhang *et al.* (2023a) provided an experimental investigation. Liu *et al.* (2022d) developed and studied a spring pendulum pounding TMD. Similar to Lin *et al.* (2024b), Dong et al. (2024) proposed new TMD concepts to overcome the issues of the conventional TMD, such as the spatial limits of wind turbines and towers, preventing the installation of excessively large or heavy dampers, and avoiding excessive damper displacements. Dong et al. (2024) investigated a new C-shaped particle damping-tuned mass damper (C-shaped PD-TMD). The C-shape ensures compatibility with cable transmission and unobstructed personnel access, whereas the friction and collision of particles within the muffler are used to dissipate energy and improve damping efficiency. Zhang et al. (2023d) proposed a tuned mass damper inerter to further enhance the classical TMD, which is a two-terminal mechanism that transforms translational motion into rotational motion. Similarly, Zhang and Høeg (2021) investigated an inerter-enhanced vibration absorber, namely the rotational inertia double-tuned mass damper (RIDTMD), to dampen the in-plane vibrations of a FOWT. The RIDTMD consistently outperformed the TMD owing to the extra degrees of freedom (extra resonance) introduced but at the cost of a slightly larger damper stroke.

Wang et al. (2023) introduced an amplifying damping transfer system (ADTS) which transfers the rotational deformation of the turbine tower to the jacket platform in order to control the dynamic response of the OWT. The ADTS significantly amplified the rotational deformation. Therefore, the ADTS can provide considerable additional damping. Unlike traditional TMD, ADTS not only provides additional damping for the fundamental mode vibration but also achieves a considerable damping effect on higher modes. In addition to the aforementioned TMD-related studies, Zhao et al. (2023a) investigated a constrained layer damping (CLD) system, while Liang et al. (2024) and Sarkar et al. (2023) examined a viscoelastic damper to mitigate vibrations in an OWT tower and an offshore jacket platform, respectively. The latter is based on a deck-isolation layer comprising laminated rubber bearings and supplemental damping, in the form of linear viscous dampers. Zhang et al. (2024a) described a bidirectional bistable nonlinear energy sink (BBNES) to mitigate bidirectional vibration (e.g., wind-wave loading) and the resulting OWT vibrations in the fore-aft and side-to-side directions during an emergency shutdown. Traditional TMDs can mitigate emergency shutdowns but can suffer from frequency detuning and stroke issues.

Offshore structures are critical infrastructure that may be under the threat of earthquake shaking. Zuo and Zhu (2022) developed a novel track nonlinear energy sink to improve the seismic performance of OWT towers. Ma et al. (2023) introduced an innovative inerter-based damping isolation system (IDIS) for the seismic protection of offshore platforms. The concept integrates inerter-based dampers with conventional bearings. The considered inerter possesses the following characteristics: a two-terminal mass element, negative stiffness effect, and mass amplification effect. An analytical model of the system was developed to model this system and demonstrate that a properly optimized IDIS is more effective than a conventional DIS.

Passive control is emerging as the preferred choice owing to its ease of manufacture, cost-effectiveness, and simple installation. Nevertheless, its limitations often lie in the substantial size and weight associated with optimal installation positions for achieving good performance. However, recent studies on novel local structural design features and metastructures have demonstrated that these disadvantages can be overcome.

Ruan et al. (2021) studied by numerical simulations and experiments a locally resonant system made of phononic crystals (LRPC) that is composed by a cylinder, a spiral beam, and a matrix. The results show that the spiral phononic crystal has good adaptability for a wide range of low-frequency bands, around 15–45 Hz, by suitably adjusting the cylinder properties. The LRPC has the potential to isolate vibrations to protect electronic devices and precision instruments, despite the complex vibration sources and boundaries on the ship. Shen et al. (2023) designed a LRPC sandwich plate structure that opened two wide band gaps between 60–100 Hz and 250–655 Hz, meaning that propagation of vibration within the band gap is largely suppressed. The band gap concept was also introduced as the basis for the investigation of the vibration isolation characteristics of sandwich plate-type elastic metamaterials conducted by Chen et al. (2021). A steel stub supported by a rubber layer was positioned on the sandwich plate. For full-scale marine sandwiches a bandgap of 80–140 Hz could be achieved experimentally thus confirming numerical studies. Chen *et al.* (2024a) studied the same resonator also for a periodic hull grillage resulting in a vibration attenuation region between 30–170 Hz. An et al. (2024) introduced a sandwich-plate-type metastructure consisting of rubber mass subsystems arranged in the core layer of a sandwich plate. Multiple bandgaps below 200 Hz were identified.

Zhao *et al.* (2021) introduced an L-extension-type ship vibration-isolation pedestal based on the impedance mismatch principle. The results from numerical simulations show that under the premise that the weight of the foundations is like a straight-wall foundation, the total vibration level drops by 40.56% in the whole frequency band of 10–250 Hz.

As previously reported by the ISSC 2022 Dynamic Response Committee, further research has been conducted in the field of quasi-zero stiffness (QZS) isolators, which can provide adequate static stiffness to support heavy machines while causing extremely low natural frequencies. This characteristic enables the suppression of machine vibrations at lower frequencies compared with conventional linear isolators. Research by Sui et al. (2024) revealed that their adaptable beam-type nonlinear QZS isolator achieves an initial isolation frequency of 2.14 Hz, which is 68.5% lower than the 6.79 Hz frequency observed in linear isolators. Shi *et al.* (2024a) introduced a QZS isolator with further reduced resonance frequency and improved load capacity. Yu *et al.* (2022a) designed a QZA isolator for shock isolation under strong acceleration excitation exceeding 1000 g. A similar principle was applied by Kampitsis et al. (2022) for a passive vibration control device of an OWT named Extended K-Damper (EKD). Contrary to the conventional TMD, the EKD can increase its vibration absorption capability by introducing negative stiffness elements instead of increasing the additional mass at the top of the towers. The EKD manages to isolate the vibrations of the nacelle from the tower, retaining the relative displacements of the nacelle and tower within a reasonable range, while increasing the effective damping of the OWT tower.

Other research relative to the offshore segment was also aimed at metamaterials, which have demonstrated advancements in mechanical efficiency regarding band isolation and vibration mitigation features, coupled with enhanced manufacturing possibilities. Machado *et al.* (2024b) studied a novel approach to control the vibration of wind turbines by employing mechanical metamaterials based on the theoretical foundation of infinite resonators. A significant reduction in vibration amplitudes, combined with a wide operational bandwidth and effective frequency attenuation compared to conventional TMD, was assessed. This achievement is made possible by employing smaller and less stiff mass control devices within the turbine structure.

4.1.2 Active Control

Active and semi-active control techniques are effective in controlling vibrations in ships and offshore structures. Active control uses sensors to monitor the vibration response of the structure in real time and actuates devices to generate a counteracting force and suppress vibrations. It can reduce the nonstationary vibration response over a wide frequency range, but it requires external energy, and the system is complicated. In contrast, semi-active control uses sensors to monitor the vibration response of the structure in real time, similarly to active control, and reduces vibrations by controlling the characteristics of the damper (damping device). It consumes less energy and is a highly reliable system; however, there are limitations to the frequency range that can be controlled and its effectiveness. Active and semi-active controls are typically selected based on these advantages and disadvantages, respectively. Recently, there has been little documentation on the application of vibration-active control devices to ship structures.

In recent years, several studies have investigated the application of active control devices to ship propeller shafting systems. The vibration and acoustic characteristics of the stern structure affect hull structure and equipment reliability, crew and passenger comfort, and stealth performance of naval ships. The main engine excitation force, propeller excitation force, and force due to the misalignment of the propeller shaft cause vibrations in the propeller shaft, which are transmitted to the ship hull through the thrust block and bearings. Zhang *et al.* (2022b) conducted numerical analysis and experiments to control the transverse vibration of a propeller shafting system of a ship using active control. Owing to the higher number of radial bearings, the transmission of the transverse vibration of the shafting system between the shaft and hull is more complex than the longitudinal vibration typically characterized by only one bearing. Both external force and equipment excitation were considered, and the electromagnetic inertial actuator was located on the bearing support. An adaptive feedforward control scheme, reported as more reliable than feedback control based on the Filtered-x Least Mean Square (Fx-LMS) algorithm, was used. It reduced the transverse vibration of the shafting system by more than 10 dB at the highest amplitude frequency of 224 Hz, and the total vibration level across the frequency range also decreased. Lu *et al.* (2023b) performed numerical and experimental analyses to control both transverse and longitudinal vibration in the propeller shafting system using active control combined with passive control technology. Applying a hybrid of active and passive controls provides the benefits of passive systems, such as energy savings and high reliability. The Fx-LMS active control algorithm with a large-value low-frequency electromagnetic inertia actuator, combined with passive control based on the longitudinal arrangement of rubber damping plates in the propulsion shaft system, was applied to examine the effectiveness of the combination of active and passive control. The experimental results showed that the active-passive control approach achieved an 8–20 dB control effect on the mainline spectrum of the propulsion shaft system vibration. Xie et al. (2021) demonstrated the effectiveness of an active vibration control method, both numerically and experimentally, in reducing the longitudinal vibration of a ship propulsion shafting system. The investigation of the longitudinal vibration control is more comprehensive than lateral vibration control because the longitudinal excitation force is predominant and significantly greater in magnitude than the lateral excitation force. The main challenge was the variation of the dynamic parameters of the shaft support with rotational speed, particularly the longitudinal stiffness of the thrust bearing. Two inertial actuators were mounted symmetrically on the thrust bearing, and an adaptive algorithm with online auxiliary filter estimation was applied. Numerical and experimental results demonstrated that the active control method could reduce longitudinal vibration by 10–20 dB. The comparison between the adaptive control with online auxiliary filter estimation and the Fx-LMS method shows that the former can adapt to variations in the dynamics of the system and is more robust than the latter.

Sharma et al. (2024) conducted a numerical analysis and experiments to control the longitudinal vibration of a ship propulsion shafting system using a magnetorheological (MR) damper. An adaptive Neuro-Fuzzy Inference System (ANFIS) was applied as a damper controller to achieve the desired damping force, and a Linear Quadratic Regulator (LQR) was applied as a system controller to calculate the active control force. ANFIS was shown to be an effective new control method. Experiments were conducted out on the MR damper and shafting system to verify the accuracy of the numerical models. The proposed system showed significant performance in vibration mitigation of 24.7% to 46.9% for different rotational speeds (200, 300, 400, 500, 600, and 750 r/min) and 68.21% at the response peak of the frequency domain analyses compared with the passive system.

Offshore platforms are usually subjected to environmental loading such as waves, winds, ice, currents, and earthquakes, which may lead to the failure of deck facilities and platforms, inefficiency of operation, and even discomfort of crews. Environmental loads on offshore structures have a wide spectrum, and semi-active control can adapt to a wider frequency range than passive control. Therefore, MR dampers are increasingly used in offshore structures. Hosseini et al. (2023) studied the simultaneous effect of wave load and earthquake on the Ressalat oil platform in the Persian Gulf, and vibrations were reduced using two MR dampers. The fractional-order proportional integral derivative (FOPID) was used to estimate the required voltage of the MR dampers. To enhance the performance of the FOPID, 2-interval type fuzzy systems (IT2FLC) were employed to calculate the proportional, integral, and derivative gains of the FOPID. The observer teacher learner-based optimization (OTLBO) algorithm was also employed to optimize the parameters of the IT2FLC and the integral and derivative operators. To evaluate the effectiveness of the proposed control, the maximum value and root mean square responses of the platform under five waves with different return periods and seven far-fault earthquake records were compared. Modified endurance wave analysis (MEWA) was used to produce waves corresponding to the location of the structure. Therefore, it is worthwhile to conduct a study under conditions closer to the actual load. The proposed control system reduced the maximum displacements of the platform deck under all possible loading scenarios by more than 35%. It was found that the wave load could act as a damping system against an earthquake load, albeit to a small extent.

Leng et al. (2024) proposed a smart variable-stiffness isolation system based by magnetorheological elastomer (MRE) laminated structure. A wave tank test for the scaled offshore platform equipped with an MRE-based isolation system was conducted; regular and random waves were generated to reproduce realistic wave loadings with different dynamic characteristics. In addition, a fuzzy logic controller was proposed to achieve real-time decoupling control. Research findings indicate that under regular wave loads, with the implementation of a semi-active MRE-based system, the RMS values of the deck acceleration were reduced by approximately 28% and 38%, respectively, depending on the wave period. Under random wave loads, reductions of up to 20% and 33% were observed depending on the wave period and length, respectively.

Energy production by OWTs is a key strategy at global scale to reduce carbon dioxide emissions that contribute to global warming. The vibrations caused by environmental loads, such as wind, waves, and earthquakes, may result in a reduction in the power production capacity and fatigue failure of the structure and wind turbine components. Consequently, vibration control of OWTs has become an increasingly important aspect. Jothinathan et al. (2022) developed the feed-forward Neural Network controller (NNC) for MR damper to control the structural vibrations of a fixed wind turbine in a water depth of 64.5m induced by wind and wave excitation. The application of neural networks is effective because the characteristics of MR dampers are nonlinear. The controller was trained for various environmental conditions using the displacement time histories of the structure with the output from the backstepping controller to predict the MR damper voltage. Backstepping control is a widely used approach for synthesizing stabilizing controllers for linear systems and some classes of nonlinear systems. It was reported that the MR damper provides displacement control of 61% at the tower top at the turbine's rated speed and acceleration control of 31% at the tower top and 46% at the tip of the jacket structure.

Lara et al. (2023) studied control strategy by pitch control for the reduction of vibration response of lateral vibration by the tower side loads of an IAE 15 MW OWTs with a monopile structure. Mitigating vibrations using blade pitch control is effective because no additional equipment is required. Active generator torque control (AGTC) was applied to reduce the lateral tower vibrations of the wind turbine. The generator torque affects the lateral motion of the structure through the reaction in the generator stator, which is attached to the main frame at the top of the tower. Therefore, it is possible to increase the damping of the lateral vibration modes of the structure by controlling the generator torque in the phase opposite to the lateral velocity. Multi-objective optimization using genetic algorithms was employed to determine the optimal AGTC parameters based on wind speed and wave height. The results indicate that the AGTC strategy without a filter effectively reduced the lateral tower fatigue load at the expense of higher power signal oscillations and increased fatigue in the low-speed shaft. However, using a filter in the AGTC mitigated these adverse effects, smoothing power oscillations to achieve similar values as the baseline control and significantly reducing the load experienced in the low-speed shaft. The AGTC strategy with a filter effectively obtained acceptable results for the generated power quality while maintaining a significant reduction in tower lateral fatigue loading.

Active and semi-active vibration control are primarily applied to propulsion shafting systems and offshore structures. Verification in an actual structural environment is expected to confirm the effectiveness of control against many exciting force components, such as the main engine and propeller excitation forces for ship propulsion shafting systems, and against wave loads, wind loads, and seismic loads with various frequency components for offshore structures, considering the complex vibration characteristics of actual structures.

4.2 Noise

4.2.1 Passive Techniques

The increasing demand to reduce onboard and underwater noise owing to shipping has challenged the maritime industry to develop new mitigation methods. Passive techniques are one of the countermeasures used to achieve this goal. It emerges from recently published papers that metamaterials, viscous damping applications, acoustic black hole (ABH) techniques, negative stiffness dampers, along with their optimized inclusion in the overall design, are among these mitigation measures.

Zhou et al. (2023) investigated the use of damping materials to reduce underwater noise radiation in OWTs. In this study, an acoustic liner for underwater noise control of OWTs was developed and numerically analysed. Periodic air cavities consisting of a circular truncated cone and a cylinder were embedded into the lined layer. The acoustic performance of the lined layer was analysed using the equivalent medium and transfer matrix methods. Based on the analytical prediction model, an optimal configuration targeting the low-frequency range was designed, and its absorption coefficient under 2500 Hz was as high as 0.96. The FE method was employed to examine the effects of the lined layer on a scaled tower-water model. The presence of the lined layer lowered the sound pressure level by up to 18 dB. Because the speed of sound in the lined layer is lower than that in water, the radiation angle of the sound wavefront was smaller than that of the baseline model. This study can be considered a starting point for the development of viable solutions to control underwater noise related to OWTs. A ring-stiffened cylindrical shell model was presented by Zou et al. (2023) to represent the primary hull. They also provided a theoretical framework combining analytical and numerical methods for calculating the coupled vibration of fluid and structure, as well as underwater acoustic radiation, in underwater vehicle structures. By introducing a high-stiffness and high-damping viscous material into the structure, the damping loss of the entire main hull can be increased, and the methodology described in this study can be utilized to optimize the installation of multiple high-damping cylindrical shells, which can effectively minimize noise emanation from the underwater vehicle structure. However, the method described herein has some limitations. Specifically, the present methodology is applicable solely to cases where the main hull comprises a single ring-stiffened cylindrical shell.

Recently, more studies on acoustic black holes have been conducted in the field of vibration and noise reduction. Gao et al. (2022) proposed a systematic method for the application of the ABH technique in raft structures. The developed approach provided new ideas for improving not only the vibration isolation performance of floating raft systems but also for reducing the level of ship vibration and noise. The influence of each parameter on the structural vibration and the recommended value range of each parameter were illustrated. 1D and 2D ABH applications on flat plates have a considerable effect on reducing vibration levels, whereas application on a floating raft system yields much less reduction. Rognoni et al. (2023) addressed the ABH technique combined with viscoelastic materials (VEMs), as a promising new solution for mitigating underwater radiated noise (URN). In this paper, they presented the characterization of an acoustic black hole as a metamaterial solution for the mitigation of noise radiated by a ship panel. The proposed procedure combines a method for the evaluation of the VEM properties of high-damping materials with an analysis of the parameters that control the ABH performance. The results demonstrate the effectiveness of the ABH in mitigating low-frequency (50 – 200 Hz) ship-radiated noise.

Merino-Martinez et al. (2021) performed an interesting study on a novel approach to estimate the annoyance caused by wind turbine noise and to evaluate the performance of rotor blade trailing-edge add-ons to reduce it. A case study was presented featuring four state-of-the-art noise-reduction measures applied to two full-scale wind turbines at nominal power. Synthetic sound signals were auralized and propagated to three observer locations. The expected annoyance in each case was estimated by employing a combination of psychoacoustic sound quality metrics and a listening experiment with 16 participants. The importance of the sound characteristics of wind turbine noise for perceived annoyance, such as tonality, spectral content, and amplitude modulation, was highlighted. A close relationship was found between the results of the psychoacoustic metrics and listening experiment. In general, this holistic approach provides valuable information for the design of optimal noise-reduction measures applied to wind turbines.

Metamaterials have recently been researched owing to their high performance in reducing vibration and noise. Kyaw Oo D'Amore et al. (2022) presented an eco-sustainable solution to replace traditionally adopted techniques to reduce the noise transmitted through cabin ceilings onboard ships. The investigated metasolution, made only with steel parts, is a valid replacement and a reasonable green alternative to noise-absorbing panels mostly made of mineral wool. A systematic FE analysis was performed to design and optimize the metasolution. This configuration was then tested in a double chamber and showed an advantage of 6 dB in terms of sound reduction index (Rw) compared to the mineral wool solution. The authors also investigated metamaterials from the perspectives of production cost and fire resistance. The authors concluded that metamaterials have suitable characteristics to be a valid eco-sustainable solution for noise control onboard ships as an alternative to traditional mineral wool products.

Azevedo Vasconcelos et al. (2024) proposed an elastic metamaterial-based interface structure to attenuate underwater noise resulting from the OWT monopile driving process. The proposed structure, which is called meta-interface, is introduced between the monopile and hammer, and is used to remove energy from the input signal associated with high noise levels. To this end, they first identified the frequency ranges associated with high sound pressure levels, which were shown to be related to the eigenmodes of the monopile. The periodic unit cells of the metainterface were designed such that the elastic/acoustic waves at the identified frequency ranges were attenuated. A frequency analysis of the pile driving system with the meta-interface showed that the new noise levels attained significant attenuation in frequency ranges below 1000 Hz. This demonstrates a novel solution for the low-frequency underwater noise issue during the hammering of offshore monopiles.

Beside classical damping material layer-based sound transmission mitigation measures, innovative techniques, such as acoustic black holes, negative stiffness, and metamaterials, have been increasingly researched. It has been shown that metamaterials have achieved a high technological readiness level and are ready to be used in industrial applications. Nonetheless, further verification tests for metamaterials are required to convince industry partners about their use. However, acoustic black holes require further research and verification comparisons. There is also a need for more detailed investigations and verification of negative-stiffness techniques for sound transmission mitigation.

4.2.2 Active Control

To date, active techniques for controlling ship noise have been developed theoretically and in prototype form; however, practical applications in real-world scenarios have been limited and confined to specific cases. During the reporting period, not much work has been reported and published on this topic. Regarding onboard ships, active noise control is challenging, especially with regard to low-frequency noise in small environments, such as crews or passenger cabins. Peng et al. (2023) suggested overcoming the technological challenges of controlling global noise in ship cabins by creating localized quiet areas. The authors proposed using the filtered-x normalized least mean square (FxNLMS) algorithm coupled with multichannel virtual sensors as an adaptive technique to control noise and actively create a localized silence area. The authors performed numerical simulations of the effect of the proposed technology on controlling noise in a localized field surrounding a cabin bed. The simulations numerically proved the effectiveness of the algorithm. Yu and Cao (2022) applied the aforementioned algorithm to actively control noise in a marine engine room. Although this study was also confined to simulations, the researchers applied the algorithm to real-life noise signals collected in the engine room of a vessel and demonstrated that the active noise control system could potentially reduce the noise levels by approximately 20 dB within the frequency range of 20 – 500 Hz. Both studies reported that they did not address the ongoing issue of technology readiness, which should be examined further in future research to allow the industry to translate these results into practice.

A similar trend can be observed in underwater radiated noise control. Few studies have been conducted to explore the implementation of traditional active noise control techniques to mitigate underwater radiated noise. Zhang *et al.* (2023e) developed a multichannel adaptive control method based on bilateral secondary sources to control underwater noise radiated from a cylindrical structure. The control system was composed of two secondary sources, two error hydrophones, and a reference hydrophone, all of which were controlled using a multichannel adaptive control system. The authors presented the results of a set of experiments performed in a controlled environment, where a cylindrical shell made of AISI 4340 steel was dynamically excited, and an active noise control system was used to mitigate the radiated noise. The results indicated a noise reduction of 10 dB. The effective application of this setup to actual ships is yet to be demonstrated, and the technology needs to be proven further, but the development of these exists.

Another active control system to mitigate URN is based on the injection of bubbles or water to create a curtain around the hull and propeller, which masks the propeller and hull radiated noise (Smith and Rigby, 2022). This system has proven its effectiveness on a number of vessels, approximately 23 vessels up to 2018, even if it has not been applied on a larger scale. Tests performed before the reporting period (AQUO 2015) showed a reduction of 3–6 dB for a cargo vessel at 14 knots. Few studies have been conducted on this system during the reporting period. As part of their involvement in the SATURN European research project, MARIN performed tests on two injection systems to reduce ship-radiated underwater noise (SATURN 2023). The two systems were designed to control the machinery-radiated noise (Masker) and propeller noise (Prairie). The results showed insertion losses of up to 22 dB and 7 dB, respectively. However, it is important to note that these results were only preliminary and rigorous data that have not yet been published.

5 Monitoring and Digitalization

5.1 Sensor and Monitoring Systems

Sensors and monitoring systems are increasingly utilized for diverse applications, including those that do not account for dynamic responses, and are diffused in other engineering sectors as well. The primary focus of this discussion, however, is on sensors and systems specifically designed for dynamic structural responses, with applicability to ship and offshore structures.

5.1.1 Sensors

Sensors for structural dynamics are used to detect the response, measure noise, or observe vibratory deflections, which are addressed in this review. Two studies specifically addressed the topic of noise sensors. The first study pertains to onboard noise, whereas the second focuses on URN. Li *et al.* (2024c) utilized four fibre optic acoustic sensors of the external Fabry-Perot interferometer type in a 1 m cube configuration to localize sound pressure in air, achieving approximately 1–3% position accuracy under laboratory conditions. This study remains in the experimental phase. Li et al. (2020) developed a prototype low-frequency, low-noise vector sensor for potential application in underwater radiated noise measurement, based on an accelerometer with an integrated preamplifier, designed to minimize self-noise. The operational frequency range was 60–1000 Hz.

Several studies have recently dealt with vision sensors aimed at measuring either onboard mechanical vibrations, or global vibrations of large structures. Na et al. (2023) developed a dynamic vision sensor, also known as event camera, which utilized an event filter-based phase correlation template matching method to measure the microvibrations of mechanical equipment. With some performance improvement techniques, they showed that micro-vibrations were measured and that faults could be detected with subsequent calculation of operational deflection shapes of a rotor system. However, further improvements were still necessary to realize practical applications. Zhong et al. (2022) proposed a vision-based fringe projection vibration measurement system for the radial displacement of shafts, using a high-speed camera. A single fixed camera and projector producing a fringe pattern at a precise angle can reveal multidimensional vibrations. The device sensitivity to angle, fringe intensity, and fringe period was simulated. The comparison with eddy current sensors was good with vibration amplitudes in the micromillimeter range (up to 0.5 mm), suggesting that it is a potentially useful method for real applications.

Ground-based synthetic aperture radar (GBSAR) has a wide range but minute resolution. High-frequency multiple-input multiple-output radars (MIMO) can monitor sub-second structural vibrations but have limited cross-range resolution. Hosseiny et al. (2024) modified GBSAR to evaluate subsecond vibrations with MIMO imaging capabilities. Numerical simulations and experiments confirmed the capabilities with a promising signal-to-noise ratio and peak-to-sidelobe ratio. Further studies should address actual SHM applications.

Yu et al. (2022b) studied the use of high-rate GPS/BDS satellites for vibration monitoring with a sampling frequency of up to 20 Hz. The vibrating structure was a small linear vibrating table. Global navigation satellite system (GNSS) technology uses two positioning strategies: real-time kinematic (RTK) positioning and precise point positioning (PPP). The GNSS receiver using PPP and RTK, Leica total station (laser), and MEMS triaxial accelerometer were compared. The accelerometer was used as a reference for the frequency and total station for displacements. The other receiver was placed 500 m away from each other. The natural frequency (0.62 Hz) and amplitudes (approximately 3.4 cm) were in fair agreement, and the accuracies of PPP and RTK were 7 and 5 mm, respectively, for horizontal motions. For vertical motions, the noise level was approximately 1 cm.

Among the other vibration sensors, which are particularly relevant for mechanical equipment, the following can be mentioned. Wang et al. (2022b) made a double-spring-piece triboelectric sensor (DS-TES) for broad band vibration monitoring. TES technology implies that the sensor is self-powered by a triboelectric nanogenerator (TENG), without the need for cabling. The size of the sensor was approximately 2 cm × 2 cm. Experiments on a vibration table and testing on two motors concluded that high linearity was achieved from 0 to 200 Hz, with an error rate of less than 0.015% and a high signal-to-noise ratio. Another example of TENG sensors was investigated by Zhang et al. (2022f) using one TENG for energy harvesting and another connected TENG for vibration sensing in the range of 6–20 Hz. The device also included automatic wireless transmission to the cloud platform by IoT. The accuracy tested in a laboratory environment with a shaker was fair with a relative standard deviation of 0.09 and a stability of 36 000 cycles, which may be a concern in some cases. The application of the sensor going to sleep and waking up again was also demonstrated, but with a concerning wake-up time of the order of half an hour. Fang et al. (2022) developed another type of vibration energy harvesting sensor having larger dimensions (about 20x20 cm). This sensor included temperature measurements, two types of accelerometers, a GPS, capacitor storage, and an inertial pendulum. This was intended for trains placed on the bogie wheel system to assess wheel/rail-induced vibration tracking rail irregularities. Such a system could be further developed for other applications, such as ship monitoring, as the first tests were promising when applied to an express train.

A traditional telemetry unit, equipped with a strain gauge and fitted to a shaft, requires battery swaps. To avoid this, Lee et al. (2018) developed a battery-free magnetostrictive torsional vibration sensor (MTVS) for rotating shafts. Magnetostrictive patches and small magnets were mounted on the shaft with a solenoid coil to convert the patch deformation due to shaft vibration into electric voltage. It was tested and compared with a traditional telemetry unit on a small test shaft and a real shaft of an LNG carrier. The optimal number of patches must be calculated. Tests on a 25 mm diameter shaft showed small errors for several modes and shaft lengths. The variation in coil spacing has little influence, but a degrading bounding of the patches is an issue (5% increased error in eight weeks for poor bonding). For the real 490 mm diameter shaft, a comparison between the telemetry unit and the corrected MTVS was good, identifying the barred range at the RPM run-up, which in this case was 35 RPM. Correction by calibration was required because the MTVS measures the strain rate and not strain. The results suggest that the MTVS is a promising and cost-effective alternative; however, calibration without strain gauges is a future development task. Pawlenka et al. (2024) developed a capacitive sensor for measuring vibrations and small displacements on high-speed machines with frequencies up to 3.8 MHz and a small relative error of 0.7%, which could measure displacements from 0 to 1200 μm with a resolution down to 715 nm. The focus was on designing it at a low cost while maintaining its performance. Device testing was conducted using a laboratory apparatus that produced vibrations with a shaft position amplitude of 40 μm rather than being subjected to commercial testing procedures.

Arrays of embedded fibre Bragg grating sensors for SHM may be desirable for OWT blades as well as for other equipment or vessels made of composites. Tests of such sensors were conducted on a fast patrol boat based on impact events (Mieloszyk et al., 2021). Results conclude that relative direction between fibre reinforcements and fibre optic is a critical issue, and careful design and processing are necessary to achieve reliable strain results, and more research is needed. Ducoin et al. (2023) conducted a similar study on embedded fibre sensors, focusing on their application to propellers. Model testing with modal vibrations and associated frequencies based on initial and corrected values deviated from the numerical predictions, but the flow detachment generated a Strouhal frequency that was visible in the spectra of the drag force. More work is necessary to improve the evaluation of the hydroelastic response and compare it with CFD. These two papers exemplify the trend of incorporating sensors into composites and equipment, as also demonstrated by sensors in composite pressure vessels (tanks), where optical fibre sensors may more effectively sense vibrations (Meemary et al., 2025). The concept of distributed sensing in SHM is not novel and has been implemented in civil engineering and aviation through the utilization of accelerometer arrays, as demonstrated by Johnson et al. (2004).

We have seen continued developments in various sensors for vibration and noise monitoring, although less for the latter. GPS sensors may be useful for horizontal motions from 5 cm upward and for vertical motion from 50 cm upward, and radars may also be used for sub-second vibrations. Optical sound pressure sensors may be used for the localization of noise and accelerometers for URN up to 1000 Hz. Self-powered small TES sensors may be used for vibration monitoring up to 200 Hz, but both the sampling frequency and wake-up time may be an issue, as well as endurance. Battery-free MTVS for rotating shafts are promising but lack improved calibration procedures. Self-powered inertial pendulums may also be used but not applied in the maritime industry, and capacitive sensors may measure vibration amplitudes down to approximately 0.01 mm. Cameras can be used for the detection of micro-vibrations and mode shapes. This remains promising but is still in the development stage, has not yet reached practical applications, and requires success stories. It is necessary to investigate practical systems that are used or developed where the sensor is more off-the-shelf.

5.1.2 Review Papers and Reports

A monitoring system may not yield a useful outcome without further processing, and this principle applies even to an SHM system designed for a specific purpose. Starting with review papers on SHM, Jayawardena *et al.* (2022) applied deep learning to data from fibre-optic sensors for civil engineering structures with a focus on detecting damage and flaws. Three types of systems were used: fibre-Bragg grating interrogators, optical distributed sensor interrogators and optical backscatter reflectometers, and the last is less but recently used. The fibre Bragg grating reflects part of the light at a certain Bragg wavelength and offers high sensitivity. Distributed optical fibre sensors are newer for more spatial distribution along the fibre for a predefined path and may cover thousands of points. Three different scattering processes may be used, but only Brillouin and Rayleigh scattering can measure strain and vibration, while Raman sensors only measure temperature. Rayleigh scattering with a millimetre-scale spatial resolution is behind the optical backscattered reflectometer. Several systems are used in pipelines, and most of these systems measure strain and vibration. Pipeline systems focusing on vibration use Rayleigh optical time-domain reflectometry with pulsed lasers. For deep learning, a range of different supervised and unsupervised learning methods are currently used in combination with optical fibres, but they are mostly supervised for dynamic applications. However, industrial applications have not yet been reached. These optical fibres are not only used for fatigue but also for corrosion and leakage monitoring recently, while the use of fibre-Bragg grating for strain and temperature monitoring is declining despite its high sensitivity. This may not be true for the maritime industry as one of the leading suppliers of fibre-Bragg gratings has installed systems on 300 + assets, some of which are used for SHM.

The SHM review paper by Zhang *et al.* (2022a) focuses on vibration extraction in civil engineering using signal processing techniques in the time and frequency domains with respective categorization. The categorization of the six methods in the time domain and five methods in the frequency domain can be useful, focusing on the advantages and disadvantages of the different methods. Time domain techniques are mainly used for linear systems and are sensitive to noise; however, autoregressive moving average with exogenous excitation (ARMAX) is a more useful method for nonlinear and non-stationary signals with noise. For the frequency domain, multiple signal classification (MUSIC) is promising because it has a high frequency resolution, can estimate closely spaced modes, and employs a high signal-to-noise ratio. The SHM review paper by Pozo et al. (2021) examines the utilization of sensors, encompassing 19 papers and 7 special issues related to the journal "Sensors". Regarding vibration, it is noteworthy that the Savitzky-Golay filter was employed for damage localization and quantification, and it was identified as insensitive to noise. A laser Doppler vibrometer was used for non-stationary conditions, and a method to minimize the number of sensors was related to minimizing the nondiagonal entries of the modal assurance criterion matrix.

The SHM paper by Vieira et al. (2022) is not a classical review. It focuses on the maintenance and operation of offshore wind support structures and illustrates the levelized costs of energy from three projects with an operational cost of approximately 20%. The authors listed 17 failures that are mostly related to grouting and scour, with the possibility of changing natural frequencies. The study also considered the potential failure modes of the support structure, including buckling as a high-consequence failure, but with cracking as noncritical. The latter may be questioned for non-redundant structural details. This study primarily employed a stochastic Monte Carlo model to analyse the impact of SHM on health assessment. The results suggest that SHM was useful for reducing the number of critical events and the interval between inspections, the fraction of monitored turbines within the farm, and the damage detection rate. The use of SHM in 10% of the structures did not provide a competitive advantage. It criticized the fact that the mean time between failures and other data are not publicly available to improve learning by the scientific community.

A review by Silva-Campillo et al. (2023) focused on different types of sensors for different applications related to marine structures, with a special focus on the optimal localization of the different sensors. The latter is covered by the inverse FEM (iFEM), Fisher Information Matrix (FIM), Information Entropy (IE), and Modal Assurance Criterion (MAC). One observation is that the study did not address the relationship between sensor placement and inspection requirements, nor did it discuss the implications for the measurement accuracy. It briefly overviews data processing, but with more examples of practical installations on ships and offshore structures. They emphasize the importance of having SHM onboard and point to the industry focus on combining fibre Bragg grating sensors (FBG) with digital twins based on IoT. The review paper by Pezeshki et al. (2023) covers ship and offshore SHM methods, with an overview of model-based, vibration-based, and digital twin methods. Vision and population-based approaches are briefly discussed. It provides an overview of the advantages and disadvantages of the methods and suggests that novel signal processing and ML should be used more frequently, especially in relation to vibration-based SHM. Chen *et al.* (2023b) made a SHM review paper on optical fibre sensors. As previously mentioned, they may be useful in the case of vibrations, even with arrays, to cover the spatial distribution. This is covered, for instance, in relation to submarine cables, but also in relation to earthquakes, tunnels, and hydropower dams, but mentioned as being under development in the marine field. They point to several development areas such as smart materials, AI, installation techniques, and damage protection.

Only a few reports have provided an overview of the framework and damage sources for the offshore sector with reference to Svendsen *et al.* (2022a, 2022b) and Horn et al. (2024). Svendsen *et al.* (2022a) included a sensor overview and sensing parameters for different failures, reference standards for digital solutions and structural integrity management and a framework for digital twins and SHM. The latter refers to 5 levels: 1) screening and diagnosis, 2) FE model updating, 3) load model updating, 4) quantification of uncertainties, and 5) detection of changes (damage detection). Level 0 is a phase with a pre-study and an SHM system design. Svendsen *et al.* (2022b) provided results from the damage database for the Norwegian continental shelf (NCS) pointing to 10 major damages where excessive vibration and VIV issues are directly related to dynamic response while cracking, member separation, missing members and overloading may have contribution from dynamic response. Approximately 90% of the damages reported was related to cracking and those damages are regarded as mainly minor; however, it must be considered that the NCS is a harsh environment where cracking is regarded as relatively more probable compared to other damages. The report criticizes the energy sector for lagging on SHM, not understanding the value and benefits, and requesting standardization. Guidance was also included in this implementation. This work was continued by Horn et al. (2024), who extended the framework beyond the model focus, giving an overall picture related to structural integrity management and barrier management. In this general framework, SHM becomes one of the key elements in addition to data-driven inspection planning, data quality, and security management. They also extended Rytter's definition of SHM from 1993, which included levels of damage detection, such as existence, localization, size assessment, and consequence evaluation. Horn et al. (2024) included also a prognostic approach (damage may occur), fit for purpose confirmation (damage is not likely) and diagnostic approach (looking back to what happened). They criticize standards, documents, and references dealing with SHM for not defining SHM properly and provide a proposal for a better definition. Furthermore, they provided an overview of SHM with examples and highlighted an almost exponential increase in research for all industries, but with more attention within the renewable sector than the oil, gas, and shipping sectors. A fourth PSA study (https://www.havtil.no/en/) addressed SHM for topside piping systems and equipment for fatigue, corrosion, and moisture penetration through insulation with vibration as the main source of fatigue but with different excitation sources (Horn *et al.*, 2022 and Horn and Wiggen, 2021). It is concluded that the use of sensors is essential and useful as part of hybrid solutions using design models because proper modelling of excitation and response may be missing in the design, including degradation and changed conditions.

Another review supporting the technical reports above was provided by Li and Brennan (2024). The focus is shifted from the digital twin to the application of structural integrity management. Although FEM remains central, model updating, real-time simulation, and data-driven forecasting are mentioned as enablers for diagnostics, prognosis, and decision-making. They highlighted some specific needs and targets ahead, such as improvements and focus on standardization, redundancy, data quality and with data mining capability as a value releaser. Although a dynamic response was mentioned, this paper provides a more general review.

SHM can be used in combination with inspection, which is still regarded as an essential part of structural integrity management. One purpose is to detect the damage due to vibrations. However, vibration analysis and vibration excitation may also be non-destructive inspection techniques for revealing the change in state. For composite and sandwich structures, a review study on non-destructive testing (NDT) and evaluation methods was conducted by Ibrahim (2014) and repeated by Nsengiyumva et al. (2021). Vibration analysis or vibration testing according to ASTM E756–05 is just one of NDT methods that can provide an alternative to visual inspection and strain monitoring for assessing material properties. It may be more suitable for specimens and equipment on a smaller scale, but potentially applicable to the whole structure, which is known as a method before and after storms on fixed platforms in the Gulf of Mexico. Another survey-related study was conducted by Lin and Dong (2023). They focused on ship hull inspection, including vibration-based monitoring, and also referred to Ibrahim (2014). Among health-monitoring approaches, this technique may be complimentary to more manual inspection techniques in-dock and during underwater inspections.

5.1.3 Asset Specific Monitoring Systems in Research

More specific SHM papers address various asset types and develop concepts that are increasingly being tested for use in the maritime sector. A prominent terminology is "population based SHM" related to the wind offshore industry and missioned by the University of Sheffield. The term is attracting attention, but for the sake of completeness, it is worth noting that past papers have also considered the idea of using information across several similar assets within the same geographic area. Many of these papers are reviewed by Horn et al. (2024). It is not surprising that vibration is an issue for these tall and slender structures, and even more so when the foundation floats, with about 20 FOWTs existing per 2024 (13 on NCS and 3 in China).

Other asset types are also relevant. Studies indicate potential use cases, such as model updating, early warning, damage localization, sensor localization, and optimization, although in many cases, the value is not necessarily clear; that is, studies are more research-related than giving the owner or operator a direct value or benefit. For ships, a good example of the efforts spent on the assessment of ship responses by model tests, full-scale measurements, and numerical predictions is the Japanese R&D program (Fujikubo et al., 2024). They provided an overview of the achievements and challenges related to digital twins for ship structures intended to reduce uncertainties in design and operation. This framework extends beyond ships and may be applicable to other assets.

The modal decomposition for a monopile OWT was studied by Iliopoulos et al. (2016) using FE and accelerometers from real-time data from Vestas V90 3 MW. They identified various natural frequencies and fair agreement with FE modal analysis having an average 4% error but varying between 1 and 8% for the different modes. An expansion approach was used to estimate the response at unmeasured positions. Changing the operational and ambient conditions may affect the modes and frequencies, suggesting that the future use needs to tune the FE model for specific conditions and extend it to strain. A monopile OWT was also studied by Ren et al. (2023) based on experiments with varying layered-soil conditions. Model updating of the pile-soil interaction is considered essential for refining the frequency estimation, which also influences operation and maintenance. The model updating focused on monitoring the stiffness and damping based on the measured vibrations of rigid and slender piles. The results show that natural frequencies change in layered soil compared to pure sand conditions, and the foundation stiffness depends on the burial depth under cyclic loading (opposite to the initial static tests), not showing a linear behaviour with time and depending on the stiffening of the sand layer and softening of the soil layer. Damping may change by either reducing or increasing. The behaviour may differ for slender and rigid piles; thus, model updating is essential. Wind turbines are also associated with various issues. De Novaes Pires Leite *et al.* (2021) focused on the fault detection and diagnostics of gears and bearings based on the time, frequency, and time-frequency domains utilizing the Bandt and Pompe approach, power spectrum, and wavelet decomposition. The results suggest that alternatives to signal processing have the potential to detect faults, but there are still challenges related to the complexity of environmental data and conditions or data quality from real measurements. Bandt and Pompe probability mass functions were regarded as the most sensitive to changes in structural conditions, but the change indicating a fault was different for gears and bearings. Gearbox modifications were associated with abrupt changes. Future work will focus on fault patterns and an appropriate threshold.

A riser is a critical component exposed to fatigue loading from floater motion and vortex-induced vibration at the connectors, and the logistics of the riser elements is relevant. Yan et al. (2022) proposed a data-driven system integrity management of riser connectors with respect to fatigue risk using a set of methods arranged in a novel manner, including FEA, rain flow counting, Hilbert-Huang transform, power spectral density, empirical mode decomposition, intrinsic mode functions, artificial neural network (ANN), risk-based inspection, and risk management measures. The intention was to provide an early warning. A case study using three years of data was conducted to demonstrate the methodology suggesting low risk, but with events reaching medium risk, which may be the basis for inspection decisions. This risk is demonstrated by displaying the transverse deformation to diameter ratio which is correlated to fatigue damages by training the ANN based on past data and simulations. Future developments are regarded necessary to bring it into practical use.

Pipelines may also be exposed to dynamic response and cracking; however, the main challenge is to detect the location of the damage. Kong Chen et al. (2022)et al. (2022 studied an improved damage features extraction method based on guided ultrasonic wave (GUW) monitoring system to more reliably detect small damages like cracks. The performance is affected by environmental and operational conditions, which is why detection of defects needs to be more robust to improve reliability. This method is based on the residual reliability criterion (RRC). Based on simulated GUW data, it was shown to be more successful than the three conventional damage-detection schemes evaluated using a receiver operating characteristic (ROC) performance evaluation method. Future developments may include the use of artificial neural networks (AAN) in an automatic damage monitoring system and testing of more realistic pipeline features and different environmental and operational conditions to determine if it can be useful in real applications.

Jacket structures are used both in the oil and gas industry and for fixed OWTs up to approximately 80–90 m water depth. Many joints in the splash zone are difficult to inspect. Jahangiri et al. (2023) studied how small damage can be detected in a vibration based SHM for a truss structure. The minimum observable damage (MOD) was defined as the smallest damage detected by SHM with a given probability of detection (POD). MOD was computed using ROC analysis based on a damage index. Mode shapes and natural frequencies are considered, as well as environmental uncertainties and measurement noise, to build the damage index based on damaged and undamaged FE. The simulated demonstration cases are academic, and only one damage is considered as a time, but the idea appears promising for the initial design phase of vibration-based SHM and should be tested in real cases. Eichner et al. (2023) investigated an academic jacket structure and optimal vibration sensor localization relative to SHM and fatigue inspection and maintenance, with the aim of maximizing the beneficial outcomes relative to installation costs.

Operational modal analysis, a genetic algorithm for the stochastic optimization of sensor locations, and Bayesian techniques related to updated reliability estimates were used. The results indicate that SHM was effective, only few sensors were needed in the upper part of the structure, and the method should be further tested in real applications. Another method for placing sensors was demonstrated by Martarella de Souza Mello et al. (2024) for a long slender structure that twists and bends; however, the application is a rotor blade. The modal information under the effect of noise in the sensor data was studied using Kriging interpolation to establish mode shapes based on FEA. A multi-objective Lichterberg algorithm was used in the optimization process. This method is robust and effective for distributing a reduced number of sensors, and the locations are not trivial and symmetrical. However, this method is computationally expensive, and other, more efficient algorithms should be tested.

SHM for aging platform structures approaching their design lifetime is highly relevant, but the retrofitting of systems as well as data storage and transfer come at a cost. Tang et al. (2020) studied the effect of sampling rate on the damage identification. A nonlinear 2-DOF simulation model, a model experiment of an academic structure, and full-scale data from a soft yoke single-line mooring system related to an FPSO were used to demonstrate the methodology. Random decrements and spectrum characteristics were used to identify system parameters from accelerations and displacements, and a support vector machine was used to identify the rule of change. The results suggest that a low sampling rate, lower than common sense (10 times eigenfrequency), is still useful for damage identification, and displacement data at a low sampling rate can still be used and show better identification capabilities than low sampling rate acceleration data. Even sampling rates lower than the highest natural frequencies may be used, which is not common sense.

For rotating equipment such as shafts or rotor sails, mass imbalance may occur because of construction tolerances or degradation, which may lead to lateral vibration in the service range. Puerto-Santana et al. (2022) investigated a non-intrusive monitoring system (not using additional weights) to reveal such an imbalance by system identification and signal processing using the subspace system identification (SSI) technique and Hilbert transformation in the time domain based on vibration and rotation sensors. A calibration method using speed run-up was used, and a speed range with associated modal parameters was chosen outside the resonance. The method was shown to be successful in two rotating experiments to identify unbalanced conditions without FE. This confirms the improvement achieved through balancing. The method should be applied to real applications with multiple disks and check if it works at very high speeds and if several speed ranges are necessary. It is unclear whether it is useful for a tall rotating sail that may bend owing to wind forces that introduce an imbalance.

The optimal localization of sensors for different purposes is a hot topic, although it has not been well covered in previous ISSC reports. Nonetheless, the examples included above are still at a research level. For suppliers, this has not been a hot topic, which may be the reason for unnecessary extensive instrumentation. For daily troubleshooting projects in the industry, the location of sensors is driven by practical engineering and installation practices, such as access, availability, and planning time. Numerous challenges arise in this situation: for example, the ability of a sensor to be immersed, its suitability for use in dangerous environments, the possibility of extracting information from alternative systems, the timely availability of the sensor, the extent of power (battery) capacity, the expert's ability to board the vessel the following day, the duration of time available on board, whether the tank is empty or occupied, the feasibility of erecting scaffolding, and the potential for installation using rafts, among other considerations. Standard services for noise and vibration exist as well, and in many cases, spot checks are performed using handheld equipment such as velocimeters. This is also a primitive and inexpensive form of SHM, but it is often used for purpose confirmation or to detect damages. The excitation source, consequence, inspection techniques, repair, and mitigation actions may still need to be determined. The overall SHM methodology may need to incorporate all these aspects.

An alternative to vibration testing is strain sensing. Aravanis et al. (2023) considered an array sensor setup with multiple sensors on a plate element. While plate fatigue after a short time is often caused by vibration issues, the focus of this study was buckling collapse inspired by the recorded hull girder collapse of two container ships with potential whipping contribution. However, it is important to maximize the benefits of the SHM system by using an optimal sensor placement framework. There are potentially many plate fields of interest that comply with optimal sensor positions, but not all of them are practically available and finding the critical weak spot may be difficult. The authors recognized that more work needs to be done.

The research focus on SHM is high, but the improvement of real application SHM is limited; thus, researchers should cooperate with the leading suppliers of SHM. The research papers extend over a broad area from different asset types to the optimization of sensor locations, improvement in small damage detection, non-intrusive unbalance monitoring in rotating systems, minimum sampling rate, and early warning for risk exposure. On the regulatory side, the term SHM is not well defined, and guidelines and standards are lacking, with a few exceptions such as fixed OWTs (BSSE, 2017). SHM appears relevant to the topic of fatigue, but it also needs to cover major damage risks and show potential for model parameter changes. There is great potential for learning from other industries where this topic is much more mature, but it also indicates that vibration is a leading parameter. Regarding noise, there are few SHM related publications. A final general statement independent of the topic is that without efficient data sharing and high-quality data, progress will remain slow. The industry is responsible for facilitating the turning of research into practical use. In this respect, data platforms could be a meeting point.

5.2 Use Cases

Maintaining the continuous operation of ships and offshore platforms is beneficial from an economic and safety perspective. Clearly identifying the operational status is essential for establishing efficient maintenance plans and ensure stable system operation. While Sect. 5.1 synthesized research on sensors and equipment systems for monitoring, this section analyses several monitoring applications in the fields of ships and offshore platforms. The use cases are organized into separate subsections that focus on specific aspects related to monitoring applications, i.e., system cost reduction through monitoring, solutions to address the lack of data in monitoring, and approaches that combine simulations with actual measured signals for monitoring.

5.2.1 Use Cases to Reduce System Costs

Maintenance costs for ships and offshore platforms have been reported to account for a significant portion of the total lifecycle cost. For example, maintenance costs for OWTs account for approximately 20–35% of the total lifecycle. To minimize the maintenance and total lifecycle costs, an effective monitoring system and optimal planning are necessary. The main objective of condition monitoring is to carry out pro-active maintenance, based on the physical condition of the equipment, before failure occurs. This helps ensure continuous operation and full functionality of the equipment by enabling the preventive replacement of critical parts. Although various condition monitoring systems and methods for ships and offshore structures have been studied, research on their overall impact on the cost and revenue remains limited.

Recent studies on monitoring applications have quantitatively analysed the economic effects of follow-up actions, including the impact of maintenance errors (Koukoura et al. 2021). These studies incorporated specific maintenance -related costs, such as repair costs, failure costs, and power production losses, to support monitoring and preventive planning (see Toftaker et al., 2022 and Zhang et al., 2022c). Additionally, to apply economical monitoring technologies, research has been conducted using IoT-aided SCADA (Supervisory Control and Data Acquisition) systems with low-cost electronics (Qays et al., 2022). Koukoura et al. (2021) developed a condition monitoring technology for OWTs, which are not easily accessible at sea, and analysed the impact of the warning time before actual failure occurred. In contrast to other studies on condition monitoring of OWTs, this study analysed the correlation between longer warning times and asset availability. Based on the warning time defined by the P-F interval (potential-to-functional failure), this study demonstrated that proactive warnings through condition monitoring can increase the availability of offshore wind farms and help reduce costs for preventive maintenance activities.

5.2.2 Methods to Overcome Lack of Database

Condition monitoring and fault diagnosis algorithms are generally classified into model-based, signal-based, data-driven, and hybrid methods that combine various approaches. Recently, several studies have focused on data-driven methods that leverage machine learning algorithms. Data-driven methods do not require expert knowledge but instead require a large quantity of high-quality labelled data to identify relationships among signals for condition monitoring and fault diagnosis. This approach requires high-quality labelled data. Although SCADA systems can accumulate historical data for OWTs, obtaining high-quality labelled data remains challenging because of cost issues for high-level systems. Additionally, because fault situations represent a small proportion of the entire life cycle, acquiring sufficient fault data is limited. To address this issue, few-shot learning (FSL) techniques have been investigated to perform fault diagnosis using a small amount of high-quality labelled data (Siraj et al. 2024).

Jin et al. (2023) proposed a novel ordinal classification prototypical network (OCPN) model that considers the absence of high-quality labelled data and fault severity for OWTs. The OCPN algorithm was validated using data from several wind turbines in China. This primarily involved identifying the types and severity of bearing faults in OWTs by measuring vibrational features in both time and frequency domains. This study demonstrated the effectiveness of a monitoring system based on processing WT vibration data.

5.2.3 Real-World Application Comparing with Simulations

Most research on condition monitoring typically relies on existing data to train algorithms and validate them using previously acquired datasets. However, studies that evaluate performance in real-world systems or analyse the limitations of real-world applications remain limited.

Kim et al. (2022) applied structural condition monitoring technology to an actual offshore structure investigating both its applicability and limitations. A newly developed cosine similarity-based structural condition monitoring algorithm was adopted for the offshore structure of Gageocho in South Korea. This algorithm fundamentally relies on a FE model to pre-calculate changes in natural frequencies under various damage scenarios. SHM was conducted by quantitatively comparing these pre-calculated natural frequencies with observed data from the actual structure. The monitoring system implemented in this study consists of three sub-processes: damage individualization, recognition, and identification. In the damage individualization phase, a damage estimation vector (DEV) was used to quantify the difference between the pre- and post-damage states based on a pre-updated FE model. During the recognition stage, a normalized warning index (NWI) was calculated using the measured natural frequencies; damage was detected if this index exceeded a predefined threshold. Finally, in the damage identification stage, the cosine similarity between the predefined damage scenarios from the FE model and the damage reflection vector were computed to identify the most likely damage condition. The analysis procedure described above was applied in particular to the actual damage monitoring for 36 joints of the Gageocho offshore structure, and this is indeed one of the strengths of this research. However, the study also highlights a key limitation: accurate damage assessment is only feasible when the damage scenarios closely match those pre-analysed. Thus, developing efficient models that ensure sufficient data acquisition through pre-analysis using FEA remains a challenge.

Sakaris et al. (2021) also presented a case study on SHM for FOWT. In their study, a monitoring technology was developed for the tendons of a 10 M-class FOWT, stabilized by two rigid-body tanks and twelve tendons. The monitoring process involved three stages: damage detection, damaged tendon identification, and precise damage quantification. It was assumed that damage to FOWT results in a reduction in tendon stiffness. Therefore, a database of the dynamic response of the platform under various tendon stiffness reduction scenarios was generated through simulations. Damage diagnosis was conducted using a functional model-based method. The system achieved a 95% accuracy rate in identifying damaged tendons, and even with a single accelerometer, the mean error in damage quantification was only 4%, demonstrating excellent monitoring performance.

5.2.4 Condition Monitoring Applications for Environmental Sustainability

Condition monitoring technology for equipment on ships can impact not only maintenance costs but also environmental and energy efficiency. Zamorano et al. (2022) applied condition monitoring using wavelet packet transform in the time-frequency domain to shipboard centrifuge lube oil separation systems. The goal was to maintain the cleanliness of lubricating oil and reduce energy loss in ship engines. In this study, the optimal mother wavelet and patterns were selected to determine the cleanliness state of marine lubrication oils. An algorithm was developed to classify the measured wavelet transformation features. Vibration signals were collected from four lube-oil centrifugal separator systems on a Ro-Pax merchant ship. These systems were powered by 7.5 kW electric motors operating at 3000 RPM. The mother wavelets used - dbN, coifN, and symN- were selected based on the DEV value, which indicates energy variation. The states of the lube oil separator were then classified using a linear support vector machine (SVM) algorithm. This study presents a significant real-world application of condition monitoring using dynamic response, wavelet transform and machine learning-based classification. It also demonstrates how condition monitoring can contribute to system efficiency and environmental sustainability.

To summarize, recent studies on condition monitoring increasingly focus on offshore wind turbine (OWT) structures, aligning with recent eco-friendly trends. While past research emphasized damage classification and severity estimation, recent studies have also considered the uncertainty of monitoring results and the economic implications of early warnings. Additionally, practical applications have advanced toward algorithms that enable few-shot learning, even when databases for structural vibrations, natural frequencies, and time-frequency features are limited.

6 Standards and Acceptance Criteria

6.1 Vibration

When evaluating the vibration response of ships or marine structures, it is important that the measurement methods, analysis methods, and acceptance criteria are consistent to enable reliable evaluations. This consistency allows for the comparison of measurement data with specifications, comparisons with other ships, and further development and improvement of vibration regulations. These methods are specified in ISO standards and classification society guidelines, helping to ensure consistency in the data and evaluation results.

ISO 21984:2018 (ISO, 2018) has been published as a guideline for the measurement, evaluation, and reporting of vibrations regarding the habitability of specific ships. ISO 20283–5:2016 (ISO, 2016a), which has been published to replace the former standard ISO 6954:2000 (ISO, 2000), is generally applicable to all passenger and merchant ships. ISO 21984:2018 is not intended to complement or add to ISO 20283–5:2016, but rather to serve as alternative. The shipbuilder may choose either ISO 21984:2018 or ISO 20283–5:2016 for a specific ship, taking into account the ship's individual design conditions of the ship and, where applicable, experience with sister or similar ships. The selected standards should be agreed upon with the shipowner.

ISO 21984:2018 provides guidelines regarding habitability for all persons on board ships that meet one or both of the following conditions:

- 2-stroke cycle, long-stroke, low-speed diesel engine directly coupled to a fixed-pitch propulsion propeller is installed.
- The length of the deckhouse (L) is limited to its height (H)

A comparison of the acceptable vibration guidance values of ISO 21984:2018 and ISO 20283-5:2016 is presented in Table 2. The guidance values of the accommodation space and wheelhouse of ISO 21984:2018 were smaller than those of ISO 20283-5:2016 because the guideline values of ISO 21984:2018 were set based on the vibration response conditions of actual merchant ships.

Table 2. Comparison of Guidance values of acceptable vibration

Type of occupied space (Crew spaces)	Guidance value (Velocity)	
	ISO 21984	ISO 20283-5
Accommodation space	5,0 mm/s	3,5 mm/s
Open-deck recreation space	4,5 mm/s	4,5 mm/s
Office	4,5 mm/s	4,5 mm/s
Engine control room	5,0 mm/s	5,0 mm/s
Wheel house excluding bridge wings	6,0 mm/s	5,0 mm/s
Other workspaces	6,0 mm/s	6,0 mm/s

In recent years, research has been conducted on comfort in the accommodation area of ships with respect to whole-body vibration caused by whipping or springing. The long duration of serious vibrations induced by slamming or springing may affect human comfort and performance on board the vessel. Such vibrations cause discomfort, sleep disturbance, and interfere with activities and performance on-board, such as the operation of equipment, writing, typing, and visual tasks. Researchers refer to ISO 2631-1:1997/AMD 1:2010 (ISO, 1997) for evaluating such vibration phenomena. It is expected that research on this topic will continue especially from a medical perspective.

In general, shipyards seldom disclose design methods for controlling structural and machinery vibrations, or case studies of related problems, as such information is typically considered confidential. However, classification societies have published these antivibration design methods and case studies in their guidelines, which provide valuable insights for shipyards and shipowners.

Information on methods relative to vibration analysis and mitigation measures is very useful during the design stage and when problems occur. ABS (2023) has provided general guidance for ship vibration analysis, including the effectiveness of top bracing for the main engine, the use of vibration compensators to reduce superstructure response, and modelling techniques for fluid behaviour in cargo and ballast tanks.

DNV (2021b) has provided methods for assessing the effects of vibration on the fatigue of lashing bridges of container ships. The excessive global vibration of lashing bridges may cause significant stress in the connections to the deck structure. Methods for natural vibration analysis, forced vibration analysis, and subsequent fatigue strength assessment have been introduced to support comprehensive design verification of lashing bridges, aiming to prevent resonance and fatigue-related issues..

In addition, case studies on vibration problems offer valuable lessons for both designers and operators. ABS (2022) presented a case study of flare-tower resonance on an FPSO vessel. Fatigue cracks appeared at the flare tower foundation after a short period of operation. The excessive vibrations were mainly caused by resonance, as the flare tower's natural frequency closely matched that of the hull girder's two-node vertical mode. The total fatigue damage was calculated by combining the effects of both low- and high-frequency responses. A fatigue analysis of the repaired structure confirmed that all critical locations met the fatigue criteria.

ABS (2022) also presented a case study of container stack resonance. Stack snapping occurs when the twist locks between the containers suddenly open during extreme roll motions and starts to oscillate at its natural frequency. Stack resonance occurs when the natural frequency of one of the hull girder natural vibration modes coincides with that of the stack. An infinite number of combinations of stack height, mass distribution, lashing pattern, lashing bridge height, container size, and stiffness can influence the natural frequency of the container stack. A container stack behaves as a highly nonlinear system, with its natural frequency depending on the magnitude of the excitation force.

Since the publication of ISO 20283–5:2016, there have been relatively few new ISO standards or revisions related to ship and offshore structure vibrations. Under these circumstances, ISO 21984:2018 was introduced to address specific aspects of habitability. It is anticipated that an ISO standard will be established to evaluate the effects of vibrations on the human body, specifically those caused by slamming and whipping, from a medical perspective. In contrast to ISO, classification societies update and publish their guidelines more frequently. The vibration-related guidelines and case studies introduced by these societies are particularly valuable to the maritime industry.

6.2 Noise

Underwater-radiated noise (URN) from ships continues to be a current topic of discussion at international level. The IMO guidelines for underwater noise - MEPC.1/ Circ. 833 (IMO, 2014) - recognized that commercial shipping is one of the main contributors to URN. With adverse effects on critical life functions for a wide range of marine life, including marine mammals, fish, and invertebrate species, upon which many indigenous coastal communities depend on their food, livelihoods, and cultures. The revised guidelines - MEPC.1/ Circ. 906, Oct 2023 (IMO, 2023) - present an overview of mitigations measures to reduce the underwater radiated noise of any given ship. They are intended to assist relevant stakeholders such as designers, shipbuilders, and ship operators in establishing mechanisms and programs through which noise reduction efforts can be realized. The guidelines include updated technical knowledge, with references to international measurement standards, recommendations, and classification society rules. They also provide sample templates to assist shipowners in the development of an underwater radiated noise management plan. Furthermore, the working group of the IMO's sub-committee on ship design and construction at its 10th session (SDC, 2024) suggested that the URN management planning chart should be included in future revised guidelines. IMO working group also discussed the experience matured building phase of three years to gain experience and develop best practices in the use of the revised guidelines.

The tripartite working group (MEPC, 2024) comprising IACS, Shipbuilding associations, and industry trade associations has suggested various actions, including the following:

- Track and log the number of ships that have adopted the IMO URN Guidelines and/or quiet ship notations over time.
- Collect noise measurements and demonstrate the number of co-benefits obtained from energy efficiency.
- Develop one or more case studies comparing the costs of adopting quiet ship notations/IMO URN Guidelines with the benefits provided by various environmental award schemes.

The IMO guidelines (IMO, 2023) define the baseline URN as the ship's source level – along with associated depth sources -- that follows from initial predictions and trials under typical operational conditions, preferably using standardized measurements. The ISO 17208 series outlines procedures for measuring and analysing underwater sound from ships. Part 1 (ISO, 2016b) of the 17208 series covers the measurements of radiated noise level in deep water, while Part 2 (ISO, 2019) includes methods for calculating source levels from deep-water data. Part 3 standard (ISO, 2023b) is currently under development and is in the enquiry phase among ISO member bodies. This part aims to address the need for source-level measurements in shallow water, recognizing that deep-water measurement locations are often far from the operational zones of many ships.

The various Classification societies have provided notations for URN using on-field measurements. However, there are differences in the measurement procedures followed by each classification society. IACS has published Recommendation No. 181 (IACS, 2024) on the measurement of URN from ships, which aims to harmonize the methods used among IACS members, ensuring consistency and comparability across different class notations. The new recommendation establishes common definitions and terminologies, relevant measurement procedures based also on the latest industry and ISO developments, and appropriate data post-processing techniques. The considered Recommendation refers to ISO/DIS17208-3 (ISO, 2023b) and will therefore be revisited upon the finalization of the ISO standard. It can be noted that various notations indicate compliance with URN levels for certain purposes or standards, and the notations of each class will continue to vary. Only URN measurement procedures are stably targeted for harmonization.

The uncertainty in associated with URN measurement is discussed in ISO 17208–1 (ISO, 2016b). When the methodology described in this standard is carefully implemented, an overall uncertainty of less than 5 dB across the entire frequency range is considered achievable. In this context, actual measurement uncertainties from URN tests conducted south of Izu-Oshima Island, Japan, between 2020 to 2022, are presented in SDC 9/INF 10 (SDC, 2022b). Based on the findings in this document, it was concluded that managing URN reduction through direct measurement is impractical, as the uncertainty in the absolute value of the URN exceeds 5 dB. Further research is required to evaluate the impact of slow steaming, which is one of the primary operational strategies proposed for URN reduction. The study concluded that it is difficult to estimate the effect of slow steaming on URN reduction on ship speed alone.

Due to challenges in identifying and monitoring URN from ships using the current standards, the Republic of Korea introduced a novel URN monitoring method based on onboard measurement in SDC9/INF 9 (SDC, 2022a). To verify this technology, a sea trial was conducted to simultaneously measure the URN based on onboard measurements and the standard ISO measurement methods. The results demonstrate the feasibility of the URN prediction methodology based on onboard sensors. Research is still under development and more full-scale trials are planned to further ensure the reliability of URN predictions. The present study offer some advantages compared to the conventional measurement approach because self-monitoring of URN can be conducted for a long time period through the onboard measurement system.

The SATURN project (de Diego et al., 2023) attempted to compare various measurement standards and classification society procedures by measuring the URN due to two vessels at different sites. For the first vessel, detailed measurements were performed for over 190 passages. A thorough comparison of reported vessel signatures from the different methods was provided, proving the need for unifying methodologies as most of them returned lower levels than those obtained under ISO 17208–1 procedure, with differences ranging from -1.5 dB to -12 dB for the same vessel.

ISO/DIS 17208–3 (ISO, 2023b) has been identified as a consistent procedure for characterizing vessel sound in shallow water, showing low deviations in reported levels against the consolidated international standard for deep-water URN measurements, ISO 17208–1 (ISO, 2016b). Its analytical formulation of propagation loss (PL), used to estimate the vessel URN in such complex environments, presents a good trade-off between usability and accuracy. During the development of this standard, details of the testing procedure and post-processing methods, lessons learned, and suggestions were shared with the ISO working group (ISO TC 43/SC 3/WG 1). For the second vessel, only limited data on vessel passages were recorded; however, additional onboard measurements were also collected. The gathered data are intended for use in a follow-up correlation study under project task T2.2.4, which aims to estimate vessel URN based on onboard measurements as a cost-effective alternative.

7 Benchmark Study

7.1 Introduction

The concept of uncertainty is often neglected when discussing research or industrial work. Understanding uncertainties may be essential. It is crucial to avoid making decisions based on unreliable results, and when the uncertainties are known to be high, it may be necessary to find a way to reduce them. In simple statistical terms, it is often desirable to determine both the mean value and the standard deviation or the confidence interval. The confidence interval can be specified for both the sample uncertainty and mean value.

This study does not deal in depth with the concept of uncertainty but rather defines uncertainty in terms of bias and variability. ISO 5725–1:2023 (ISO, 2023) classifies accuracy into trueness and precision. Trueness is regarded as bias, precision is regarded as variability, and uncertainty is represented by the level of accuracy. Measurements are affected by uncertainties arising from both bias and variability. When these measurements are utilized within a model to predict a specific quantity, additional uncertainties are introduced, which are inherent to the model itself due their bias and variability. Beyond this, the assessment is conducted by an individual whose level of experience also contributes to the overall uncertainty. At the end, there are many sources of uncertainty that progress into the overall result. It may be difficult to determine exactly how to reduce the uncertainties without knowing the different processes: measurements, models, and user experience.

In the context of dynamic response, there are several parameters and quantities which may be related to different levels of uncertainty. Some key parameters are considered in this benchmark. A literature study on the topic of uncertainty in dynamic response quantification is not included herein but is considered by Storhaug et al. (2024).

7.2 Objective

The objective of the benchmark is to examine the uncertainties related to post-processing of experimental data. These uncertainties primarily arise from the individuals conducting the assessment although they are also slightly influenced by the intrinsic performances of the numerical models employed for data processing. Though inevitably present, the present analysis does not scrutinize the model uncertainty associated with the limitations inherent in each model. The numerical procedures used to predict quantities are considered as relatively simple for some quantities and slightly more complicated for others. The measurements which were provided to all participants are to be regarded as accurate. However, in conjunction with the assumptions and decisions from each participant, the uncertainties are mainly related to the varying level of user expertise.

7.3 Measurement Data

The measurement data were obtained from the 8600 TEU container vessel investigated by Heggelund et al. (2011). A deck strain sensor located at midship on the starboard side is considered, providing a 24-h time series of nominal measured stress in N/mm^2. The sampling frequency is approximately 42.37 Hz with a total of 3661049 samples for the one-day period. No other information was provided to the benchmark participants. The mean stress value is non-zero and positive, indicating tension (i.e., hogging); however, it cannot be considered reliable, as it appears to correspond to the design-level stress in hogging, which is unlikely. In fact, the initial "zero" setting in port at the time of installation was probably inaccurate, a common issue in such procedures. The entire time series, from 13:00 (computer time) on 17th of November 2010 to 13:00 on 18th of November 2010, is shown in the left plot of Fig. 1. The dynamic stress range exhibits significant amplitudes near the design level, indicating a relevant storm event with contributing whipping effects. Storhaug & Jagite (2024), who analysed the vibration dose value during the same storm using a bow accelerometer, characterized the storm as moderate, with a significant wave height of 5–6 m. Based on operational expertise, the vessel speed is assumed to be about 17 knots. The spectrum in the right plot of Fig. 1 reveals a wave-induced peak at 21 s, likely due to swell, and a dominant peak at approximately 10 s, attributed to the wind sea. In addition, vibratory responses are observed around 0.50 Hz (2.0 s), 0.98 Hz (1.02 s) and 1.43 Hz (0.7 s), before reaching an apparently flat "noise level". Considering the position of the sensor at midship, the first and stronger (vibratory) peak is likely associated to the vertical 2-node vibration mode. The subsequent weaker peaks may correspond to the horizontal 2-node mode and the vertical 4-node mode. If there is torsional vibration, its frequency may be close to that of the vertical 2-node, rendering it indistinct or negligible.

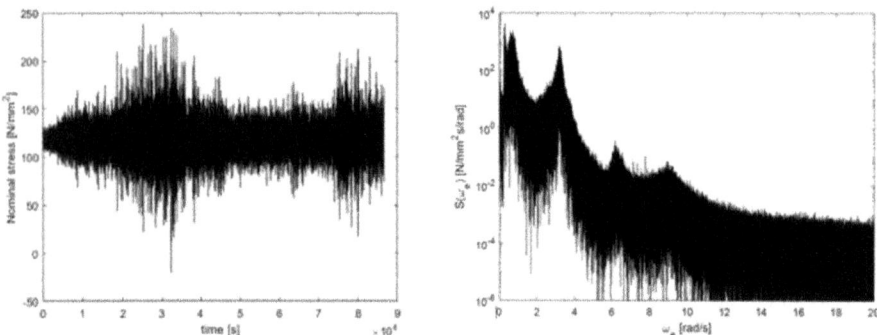

Fig. 1. 24-h time series from a midship deck sensor (left) and a 24-h spectrum (right)

7.4 Benchmark Tasks

The benchmark is divided into four tasks. In each task, the one-day time series is divided into 48 half-hours, and quantities are assessed within each half-hour. Each half-hour may then be represented by 76271 samples rounded down. The target quantities that have to be determined across the tasks are:

- Dynamic maximum and minimum values with vibration
- Natural frequency
- Damping ratio
- Fatigue damage with and without vibration

All these quantities are regarded relevant in relation to the dynamic response.

7.5 Benchmark Results

7.5.1 Evaluation of Results for Each Task

Before describing the tasks in detail, it is worth to explain the procedure employed to evaluate the uncertainties:

- The quantity x_i is predicted for a specific half hour by all participants with N the number of participants who provided an answer.

- The mean value is predicted for that half hour as: $\bar{x} = \frac{1}{N}\sum_i x_i$.
- The standard deviation is predicted for the same half hour as: $s_x = \sqrt{\left(\frac{1}{N-1}\right)\sum_i (x_i - \bar{x})^2}$.
- The 2.5% confidence interval of the Student's t-distribution with N-1 degrees is freedom is calculated as: $\hat{x}_{0.025} = \bar{x} + t_{0.025, N-1}\frac{s_x}{\sqrt{N}}$.
- The 97.5% confidence interval of the Student's t-distribution with N-1 degrees of freedom is calculated as: $\hat{x}_{0.975} = \bar{x} + t_{0.975, N-1}\frac{s_x}{\sqrt{N}}$.
- The 95% confidence interval of the mean is thereby defined by the interval from $\hat{x}_{0.025}$ to $\hat{x}_{0.975}$.
- The ratio of the confidence interval and the mean value quantifies the relative uncertainty for that half hour (this is necessary because a predicted quantity may be high or low).
- The average of the relative uncertainty for all the 48 half-hours intervals is regarded as the final uncertainty of the measured quantity and will be reported. The value is multiplied by 100 to express it as a percentage. The 95% confidence interval is thereby considered to be the mean value ± 0.5 times the relative uncertainty.

Although alternative procedures may exist, this method is considered effective for quantifying and representing the uncertainty associated with the 95% confidence interval of both the mean value and width of the confidence interval of the dynamic quantity. It should be noted that in Storhaug et al. (2024) a factor of 2.3 was used to define the 90% confidence interval based on the standard deviation; however, that approach is not statistically accurate.

7.5.2 Task 1: Maximum and Minimum Dynamic Stress Per Half Hour

In the first task, the dynamic maximum (indicative of hogging) and minimum (indicative of sagging) values are determined. While extracting a peak value is straightforward, defining the signal mean is essential due to its non-zero nature, which influences the peaks. Physically, the mean value is affected by nonlinearities in the dynamic signal, temperature effects, and the vessel's forward speed, which establishes a sagging moment and modifies the loading condition. Additionally, the burning of fuel changes the loading condition and trim of the vessel, as does the potential variation of filling levels in the ballast tank. It is therefore not so easy to determine exactly what the mean value should be when identifying the dynamic part, but the mean value should be extracted anyway from the dynamic component. This can be done in different ways like:

- The mean value for the whole day is taken simply as the average of all the samples, and the dynamic values are taken relative to that.
- The mean value for the half hour is taken simple as the average of all the samples, and the dynamic value is taken relative to that.
- The mean value for the whole day is taken based on a filtering process to remove drift by suppressing low cycle variations; this can be optionally performed by setting a cut-off frequency in the signal spectrum provided by FFT.

- The mean value for the half hour is taken based on a filtering process that eliminates drift by removing low-cycle variations, using the same approach described previously with a specified cut-off frequency.

Variations exist in this context, including the possibility for the cut-off frequency to vary from one half-hour interval to another. Thus, different participants have chosen different approaches, which inherently introduce uncertainties. In the spectrum to the right in Fig. 1, the mean value for the whole day was deducted from the time series before the FFT filtering was done. However, non-negligible spectrum values can be identified at low frequencies below 0.2 rad/s (or below 0.032 Hz or above 31 s) suggesting that there is some slowly varying mean value.

The maximum and minimum dynamic stresses were calculated by each participant for each half hour. The results are shown in Fig. 2 and Fig. 3. Storhaug et al. (2024) also included tabulated values. The relative uncertainties are 2.7% for hogging and 2.4% for sagging responses including the whipping contribution. It should be considered that conducting this analysis with or without whipping introduces additional methodological choices, potentially increasing the uncertainties, particularly regarding the whipping amplitudes. This implies that inherent uncertainties exist in the estimation of whipping amplification across different numerical hydroelastic tools, depending on how each method accounts for whipping effects. This variability may be relevant for rule development to ensure that the total load is correctly combined from the different components.

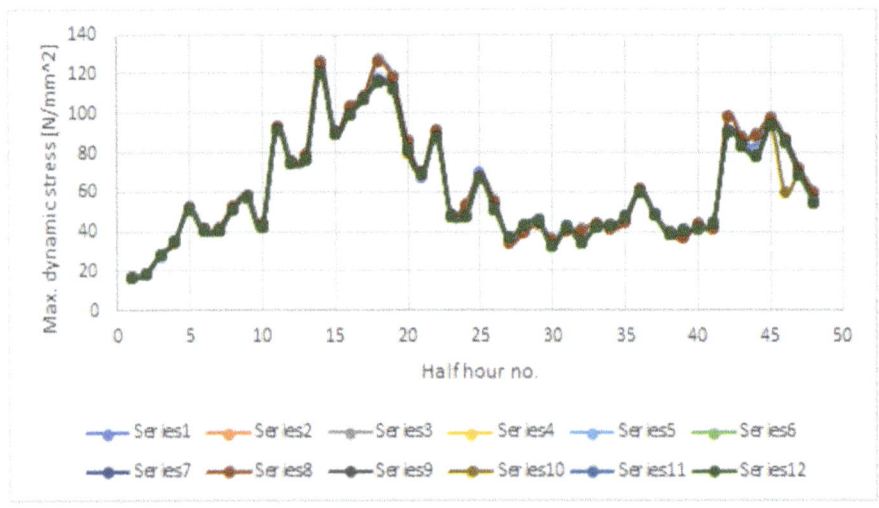

Fig. 2. Predicted maxima for all half-hours for all participants.

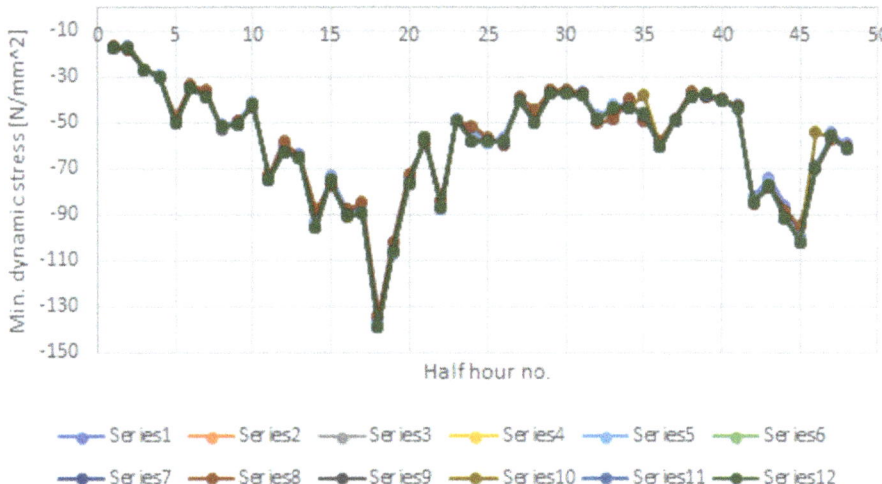

Fig. 3. Predicted minima for all half-hours for all participants.

7.5.3 Task 2: Determination of Natural Frequency

The vertical 2-node vibration mode was present in all the 48 half-hour records. The participants were asked to determine its natural frequency for each half hour. However, this value may be subject to variation due to added mass along the vessel, water depth, and variations in the loading condition, such as fuel consumption. The location of the vessel suggests that for practical purposes, the water depth may be regarded as infinite. Consequently, the added mass effect is likely to be more pronounced at extreme wave heights when the bow and stern experience significant vertical motion relative to the water surface (the natural frequency may change in a single wave event). The change of the loading condition for one day is also limited.

The natural frequency can be identified through the FFT spectrum, but the spectrum may be spiked or smoothed; a better evaluation can be obtained by calculating and averaging multiple spectra. Similarly to the mean static stress value, the frequency can be calculated from the measurements during the entire day. In Fig. 4, it is worth noting that the estimation of one participant differs significantly from the others, frequently being much lower. In this case, the number of vibration cycles was counted, and the natural frequency was estimated by dividing the number of identified vibration cycles by 1800 s. Though this method is not accurate, it provides additional information when the natural frequency is known. This procedure implicitly provides the fraction of the signal that contains vibration cycles; therefore, if the natural frequency is 0.50 Hz, and this method gives 0.40 Hz, then vibration is present in the response 80% of the time. This confirms that for the half-hours, the response contains vibration more than 80% of the time. This is often a sign of springing, but it is well known from this storm that many whipping events occurred.

The uncertainty of the natural frequency estimate is 3.2%; if Participant 3 is regarded as an outlier, it drops to 1.7%, which is better than the dynamic peak values. Participant 8 has the lowest uncertainty of 0.4% but also the highest mean estimate of 0.52 Hz. In general, the uncertainty for the 48 half hours is in average 0.8%, so the uncertainty of the average natural frequency for each method is less than the uncertainty between the methods. This suggests that considering uncertainty using just one method may fail to address the uncertainty of the method itself. Two participants also showed a frequency that increased with time, which is acceptable if related to the progressive burning of fuel; however, the increase is within the uncertainty of the estimates, and thus, it is not regarded as methodologically sound to conclude that there is an increasing trend. Such a conclusion is a typical example of misuse of statistics (Storhaug et al., 2024).

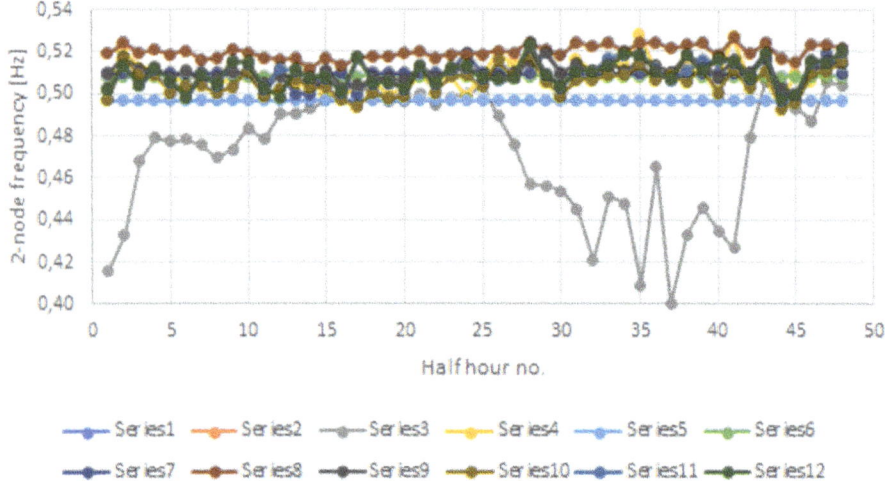

Fig. 4. Predicted 2-node natural frequency for all half-hours for all participants.

7.5.4 Task 3: Estimate of Damping Ratio

Since there are vibrations in all 48 half-hour records, it is possible to estimate the damping. Different approaches exist for damping identification, a process known to be challenging; some of these methods are covered by Storhaug et al. (2017). In this benchmark the following methods were used:

- Spectral 1-DoF fit to the smoothed FFT spectrum using a Hanning window with 4096 samples and with a lower and upper cut-off frequency 1 rad/s different than the peak frequency.
- Half band width method on the band-pass filtered probability density function based on the natural frequency obtained from FFT.
- Random decrement method #1 based on time series made by inverse FFT between 0.3 and 1.0 Hz.
- Random decrement method # 2 using a positive trigger function, with a decay length of 11 s, and a logarithmic decrement method which is applied to extract the damping ratio as average of the peak-to-peak log decrements. The same band-pass filtering window is adopted for all the 48 half-hours.
- Divided half-hour into segments with maxima identified. Starting from the maximum value, a logarithmic approximation of the envelope of the time series was performed

for a time period shorter than the pitching period. The damping ratio was derived from a logarithmic approximation. The mean damping ratio was then calculated from all the small segments.
- Enhanced Frequency Domain Decomposition (EFDD) which first computes the singular value decomposition (SVD) of the spectrum matrix for each frequency, decoupling the structural system into a series of 1-DoF systems. The Modal Assurance Criterion between the peak singular vector and the others is calculated to help the user choose the boundary frequencies for each peak. Finally, the damping ratio is computed using the curve-fitting method on the autocorrelation function determined using the inverse FFT of the autospectral density. It can be estimated by checking the decay process cycle by cycle or over many cycles; however, other techniques can be applied, for instance fitting a response process from a 1-DoF system to the signal, which is assumed to be a clear modal response. Two other methods employed a basic frequency domain (BDF) either with half-power band or with curve fitting method on the autocorrelation function of the equivalent single degree of freedom system; they provided a lower mean damping ratio with significantly more spread.
- Half-power bandwidth method. Spectrum based on the average of periodograms with a length of 4096 samples. Fitting the spectrum was done between 0.3 and 1.0 Hz. The logarithmic decrement method was also used but it was only partially effective.
- Damping-ratio identification based on the half-quadratic gain method (half-power bandwidth method was considered as well but without significant change in the results). This method is regarded as valid for damping ratio less than 5%.
- Half-power bandwidth method using the envelope of the FFT spectrum.
- Half-power bandwidth method using a spectrum smoothed from FFT.

Several participants used the half-power bandwidth method but with some adjustments. The basics behind the method are simple. The results are displayed in Fig. 5. Participants 2, 6 and 11 show results considerably higher for certain half-hour intervals. Results from participant 6 are higher all the time and relate to the half-quadratic gain method. Participant 5 values have the lowest time average with the least spread. The global uncertainty of all the participants' estimations is 108%, i.e., as high as the mean damping ratio of 2.61%. If Participant 6, with significantly higher damping ratio values, is excluded in the statistics, the overall average among the remaining participants drops to 2.41% with an uncertainty of 87%, which is still high. Participant 5 employs a methodology that is considered cumbersome and may require refinement for general use but is anyway optimal in this context, while the random decrement method, although less effective in this instance, has previously demonstrated superior performance (Storhaug et al., 2017). Again, the average uncertainty of the individual methods is well below 15%, suggesting potential bias in the estimates compared to uncertainties between methods. The uncertainties of the individual methods are listed in Table 3. The half-power bandwidth method exhibits a tendency toward higher uncertainty.

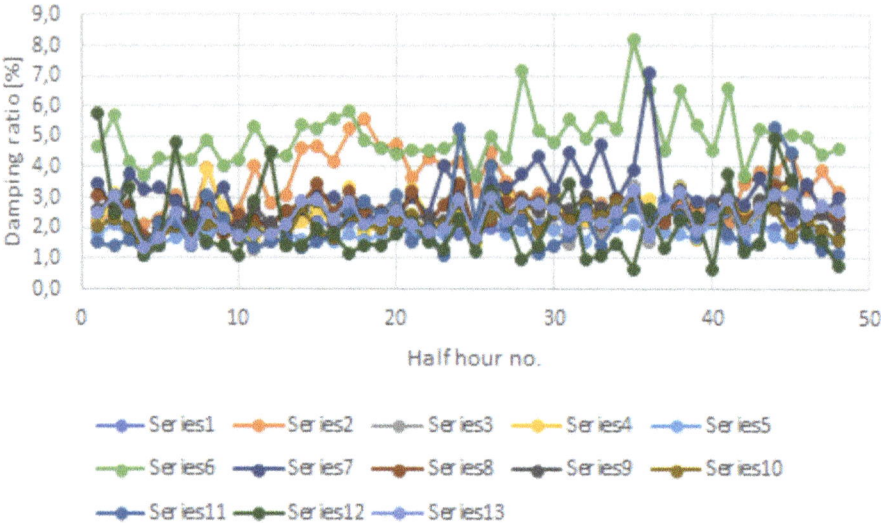

Fig. 5. Predicted damping ratio for all half-hours for all participants.

Table 3. Uncertainty of individual methods

Participant	Method	Mean damping ratio	Uncertainty in %
1	Smoothed spectrum fit	2.15	11.1
2	Spectrum half bandwidth method	3.31	16.5
9b	Basic frequency domain method with spectrum half bandwidth method	2.23	15.2
9c	Basic frequency domain method with curve fitting on the autocorrelation function of single DOF system	2.38	12.3
4	Random decrement method between 0.3 and 1.0 Hz	2.43	14.3
5	Mean of all fits of envelope decay curve for whipping time series with period less than pitch period	1.83	8.3
6	Half quadratic gain method	4.99	10.3
7	Likely half power bandwidth method	3.12	16.1
8a	Half power bandwidth method	2.58	10.7
8b	Logarithmic method	1.80	8.3
9a	EFDD with curve fitting on the autocorrelation function using inverse FFT of the autospectral density	2.51	9.3

(*continued*)

Committee II.2: Dynamic Response   425

Table 3. (*continued*)

Participant	Method	Mean damping ratio	Uncertainty in %
10	Random decrement method	2.16	10.5
11	Half power bandwidth method	2.14	26.5
12	Half power bandwidth method	2.08	33.0

Participant 1 investigated also the sensitivity of the estimations on the number of samples in the Hanning window and width of the spectrum fit in rad/s about the natural frequency, based on a robust method that is not necessarily accurate. The related results are provided in Table 4. Result No. 8 is reported as final with low uncertainty and low damping combination, considering that the employed method is known to overestimate the damping slightly. The results suggest that such an exercise is necessary for this method to ensure the best results. This approach may also be applicable to several other methods. However, it is not easy to select the best result, as the criterion for making this choice is somewhat arbitrary and often user dependent. If the ratio of the uncertainty to damping ratio is minimized, then result No. 6 has the lowest ratio, with an average damping value of 2.54. If maximizing this value, then result No. 9 would be chosen with an average damping value of 1.97. The value selected by Participant 1 as final (No. 8) was close to minimizing this ratio; however, this choice was based on engineering judgement, resulting in an average damping ratio of 2.15. The uncertainty of a random choice in resolution (number of samples) and width based on the figures in Table 4 would be 25%. This uncertainty is reduced when a rational initial setting is first established; therefore, this choice should not be made arbitrarily. Ensuring both low damping value and low uncertainty appears to be relevant for this method, despite the absence of a scientific basis for making such a selection. Hence, it can be considered as a user uncertainty related both to the method chosen and to the parameters used within the method.

Table 4. Uncertainty of spectrum fit methods depending on number of samples in the Hanning window and the width of the fit about the natural frequency

No	Samples in Hanning Window	Width of fit in ± rad/s about the natural frequency	Average damping ratio	Uncertainty in %
1	1024	± 1.0	3.86	7.1
2	2048	1.0	2.64	9.1
3	4096	1.0	2.28	10.6
4	8192	1.0	2.19	11.6
5	16384	1.0	2.22	12.3
6	4096	0.6	2.54	10.3
7	4096	0.8	2.42	10.2
8	4096	1.2	2.15	11.1

(*continued*)

Table 4. (*continued*)

No	Samples in Hanning Window	Width of fit in ± rad/s about the natural frequency	Average damping ratio	Uncertainty in %
9	4096	1.4	1.97	19.8
10	4096	1.6	1.95	18.6
11	8192	1.4	1.93	18.3
12	16384	1.4	2.01	16.6
13	8192	1.2	2.05	14.7

7.5.5 Task 4: Estimate of the Fatigue Damage and the Vibration Damage

In this case, the two slope S-N curve for welded details in air by DNV (2021a) was provided, along with a stress concentration factor of 1.33. The fatigue rate was calculated by multiplying the fatigue damage by the factor of 25·365.25·48, which defines the number of half-hours within 25 years design life. This factor relates to the budget damage per half hour. However, the fatigue damage should be calculated as the total fatigue damage, including vibration, followed by the fatigue damage due to the vibration contribution alone. Several participants erroneously calculated the fatigue damage from the vibration cycles though they had been instructed to separately calculate the fatigue damage for the total signal and for the wave-response band-pass filtered time series, and then to obtain the vibration damage as their difference. This misinterpretation has been the reason for underestimating the vibration contribution to fatigue damage for decades; this issue was first addressed properly when rainflow-counting was developed in the seventies but remained frequently misunderstood until about year 2000. The choice of the cut-off frequency between static and wave response, between wave response and vibration, and between vibration and noise, was left to the participants as well as the size and number of bins and the other method-dependent parameters.

Most participants employed rainflow-counting, with the exception of one participant who used a closed form formulation based on narrow-band assumption, calculating standard deviations between 0.03 Hz and 0.15 Hz and between 0.3 Hz and 1.0 Hz. Participant 2 employed a wave-pass frequency band from 0.02 Hz to 0.3 Hz, while Participant 1 utilized a band from 0.024 Hz to 0.318 Hz. Other participants applied a frequency cut-off between wave and vibration at 0.25 Hz with 80 bins, with some opting for cut-offs at 0.2 Hz and others at 0.3 Hz. One participant used a cut-off frequency for noise at 12 Hz, another at 7.8 Hz and two employed 1.0 Hz, which neglected higher vibration modes. One participant used the old S-N curve with a weld factor of 1.5 excluded and a stress concentration factor of 2.0. This was assessed to cause within 10% difference by Participant 1 and 9. Participant 9 used both an algorithm based on ASTM and AFNOR with 2.5% difference in average on the total damage, representing the difference between two rainflow-counting algorithms.

The total fatigue rate including vibration is shown in Fig. 6, and the vibration fatigue rate alone is illustrated in Fig. 7. The uncertainty is 31.6% for the total fatigue rate and

75.9% for the vibration fatigue rate. These are significant uncertainties, and upon examination of Fig. 6 and 7, interpretation may be challenging as the data appear to closely align, with the notable exception that Participant 3 and 4 exhibit distinct differences. It is important to note that most of the time the fatigue rates are low and not at peak levels as observed during this storm. The expected average should be less than a fatigue rate of 1 over 25 years. The smaller fatigue rates show significant relative differences especially during the half-hours from 27 to 41 on the vibration fatigue rate (Fig. 7). This difference is revealed by this approach; however, if one examines the aggregated fatigue rates with the peak values dominating, then the difference may be smaller. Removing Participant 3 and 4 reduces the uncertainty to 16.6% for the total fatigue rate and 62.0% for the vibration fatigue rate, which are still significant.

The vibration contribution varies between 59% and 68% for the different participants on the aggregated (average fatigue rates). The uncertainty in the total average fatigue rate is 5.6%, while for vibration the average fatigue rate is 7.0%. This confirms that during this extreme storm the uncertainties related to the vibration part are reduced compared to what should be expected in less severe storms. This may be attributed to the selection of bin sizes and frequency bands. However, the latter uncertainties should be disregarded in the bigger picture as they are more storm specific. This serves as another illustration that uncertainties may be misquantified when the uncertainty is not normalized.

7.6 Summary

A single day in an extreme storm has been examined focusing on the stress response level for a deck sensor located amidship on an 8600TEU container vessel with pronounced bow flare. The time-history has been divided into 48 half-hours and for each half hour several quantities were computed such as the maximum dynamic value, minimum dynamic value, natural frequency for the vertical 2-node vibration, damping ratio, total fatigue rate and vibration fatigue rate. Relative uncertainty has been calculated as the average relative uncertainty of the 48 half hours, where the relative uncertainty for each half hour is estimated as the 95% confidence interval by the t-distribution. The relative uncertainties regarding the quantities considered are:

- 2.7% for the maximum wave induced hogging values with whipping
- 2.4% for the minimum wave induced sagging values with whipping
- 1.7% for the natural 2-node vertical vibration frequency
- 108% on the damping ratio
- 17% for the total fatigue rate
- 62% for the vibration fatigue rate

The results are largely influenced by user choices. The associated uncertainty can be considered low for the maximum and minimum values, as well as for the determination of natural frequencies. However, the uncertainty in damping ratio is high and at the edge of acceptability. The uncertainty affecting the vibration damage is also relatively high. For the total fatigue damage, it is moderate and falls within the typical range observed in fatigue-related quantities when using the rainflow-counting method.

The benchmark results indicate that the user-related uncertainties cannot be neglected in the estimation of the damping ratio and the assessment of vibration contributions to fatigue damage. This is consistent also considering the different choices made in the numerical analysis of the damping ratio, without being supported by any sensitivity analysis to determine their impact on the dynamic response.

Future work should be focused on standardizing methods for the damping ratio estimation and establishing best practice procedures. It is recommended to use simulation of dynamic systems with known damping to tune and validate estimation methods. This should be done before applying a certain method and its input parameters for evaluating the damping ratio, especially when such a value may serve as a reference value in future numerical studies, potentially influencing results related to damping.

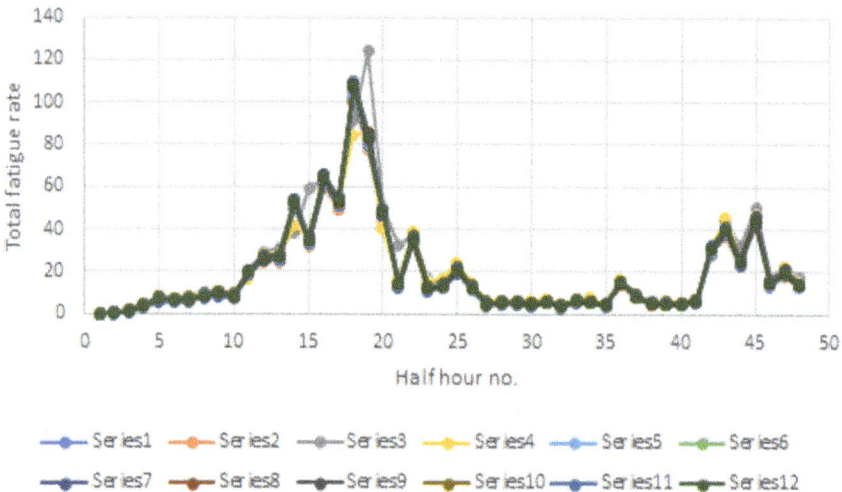

Fig. 6. Predicted total fatigue rate for all half-hours for all participants.

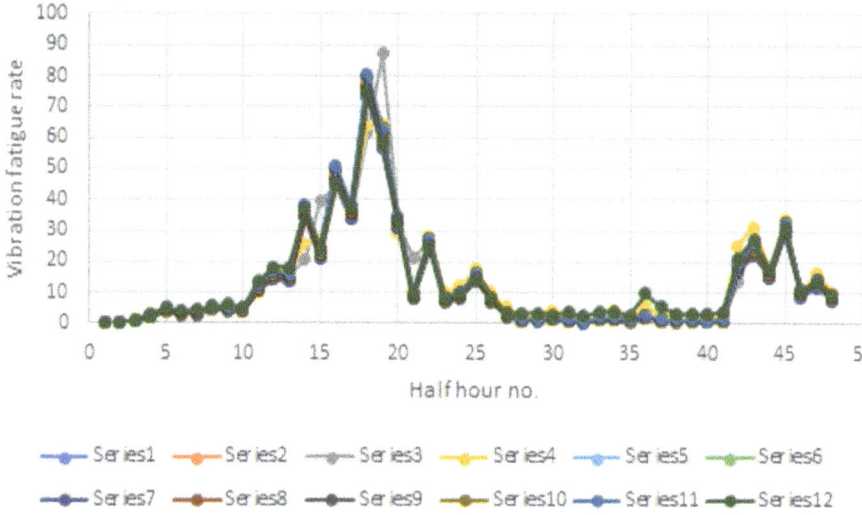

Fig. 7. Predicted vibration fatigue rate for all-half hours for all participants.

7.7 Acknowledgements

The committee would like to thank the ship operator, owner, yard and DNV for arranging the measurements and sharing the data and, in particular, G. Storhaug for organizing the benchmark. Further, the committee thanks the following participants contributing to this benchmark study as well as their companies/institutions for the support given:

- Centec Technico Ulisboa University: S. Wang
- Chungnam National University: S.J. Kim
- Det Norske Veritas: G. Storhaug, M. Holtmann, U. Behrens, G. Jagite
- Indian Register of Shipping: S. Dhavalikar
- Light Structures AS: G. Sagvolden
- National Research Council of Italy: D. Dessi
- Navantia: V. De Diego
- Nihon Shipyards Co., Ltd.: H. Takahashi
- University of Geneva, DINAV: T. Pais

8 Conclusions

"*Panta rhei*"- "Everything flows". This ancient wisdom from Heraclitus (6th century BC) aptly captures the essence of dynamical problems in engineering. Whether analysing transient responses in marine structures, vibrations in on-board mechanical systems, or hydroelastic coupling in fluid-structure interactions, engineers must grapple with constantly changing states. Understanding and controlling these dynamic behaviours is crucial for designing stable, efficient, and resilient systems. "Everything flows," but much faster than it did even a few decades ago. The continuous and overwhelming flow of information pervading today's design and management of structural systems presents both an opportunity and risk. One can easily become lost in it. Far from witnessing the conclusion of the mechatronic revolution, this Committee has aimed to document the evolution and impact of both consolidated and emerging trends in structural dynamics applied to ships and marine structures.

Enhancing the prediction of wave-induced vibrations and their impact on structural reliability has remained a key focus during this period. These efforts are supported by numerical investigations and model experiments, driven by safety regulations, structural lightening, and the challenges of climate change. High-fidelity simulations, while valuable in capturing nonlinear behaviours, are computationally demanding, Thus, alternative approaches that balance accuracy and computational efficiency remain relevant. Advanced reduced-order models of dynamic responses are still an option for deriving reliable and comprehensive statistical insights. A major challenge lies in assessing the reliability of predictive models. Recent benchmark studies revisiting historical databases (like those relative to S175 scale-model) have addressed this issue. The Committee believes that more efforts should be made to validate and compare the prediction capabilities of wave-induced responses under uncertainty. It encourages future research, including by upcoming Dynamic Response Committees, to pursue this direction. In scaled testing with elastic models, emerging technologies like 3D printing, inclusion of nonlinear effects, and distributed sensing should be further explored to improve hydroelastic scaling and fidelity.

Full-scale measurements are increasingly focused on the impact of vibrations on passengers, crew, and onboard equipment, extending beyond the pure structural integrity. This shift highlights the importance of incorporating human factors into predictive modelling. Simultaneously, real-time monitoring of elastic responses is advancing to support vibration mitigation strategies via both informed decision-making and automated ride control. Ice-induced vibrations in ships are gaining attention, with a significant emphasis on modelling ice-structure interactions. While laboratory experiments remain necessary to validate numerical models, full-scale observations are critical for understanding real-world effects. Greater integration of scale-model testing and full-scale measurements into predictive models also deserves attention. The use of data fusion techniques that merge models and observations could accelerate this process, particularly from a virtual sensing perspective.

Propeller-induced vibrations remain a key research area, now expanded to include rudder vibrations and blade flexibility. Advances in propulsion, hull design, and operational regimes require assessments of transmitted vibration levels, especially as composite materials and bio-inspired solutions become more prevalent. The push toward net-zero emissions by 2050 brings sustainable fuels, new onboard equipment, and alternative energy sources, requiring vibration assessment. These introduce new vibration-related structural challenges, such as those related to wind-assisted propulsion. Although improved efficiency is expected to reduce overall vibration levels, new installation layouts may introduce unforeseen issues requiring detailed analysis.

Shock responses continue to be a priority, particularly for internal explosions affecting passenger safety and shipping regulations. External air blast and underwater explosions, critical in naval applications, demand time-resolved nonlinear simulations and also increasingly drive innovations in materials and countermeasure development.

Internal noise remains a concern for onboard comfort and crew health. Recent research, mainly based on empirical observations, have revealed potentially underestimated effects and issues. Advances in noise prediction, including transfer path analysis, hybrid aeroacoustic models, and AI-based techniques, offer promising improvements. Airborne noise pollution in ports, especially near residential zones, demands modelling approaches and standardized measurement to assess noise levels aligned with evolving regulations. Predicting and mitigating underwater noise to reduce environmental pollution and comply with class notation, remains challenging, particularly due to turbulence and cavitation modelling. Hybrid modelling techniques show promise for bridging the gap between computational models and on-field observations, provided that extensive data collection is conducted. The evaluation of underwater radiated noise through onboard vibration measurements is an emerging approach applicable to both cavitating and non-cavitating propeller regimes. However, simultaneous and extensive data collection, both onboard and underwater, is still lacking, limiting the validation of models in both direct and inverse analyses.

In offshore structures, research emphasis is increasingly shifting from traditional oil and gas platforms or very large floating structures to offshore wind turbines (OWTs). These systems face complex interactions with waves, wind, and currents, as well as growing concern over ice-induced vibrations. With increasing global deployment, floating OWTs encounter significant challenges related to accidental loads, structural damage, and survivability, necessitating nonlinear analyses.

Vortex-induced vibrations in secondary structures, such as cables and risers, remain a research priority. To achieve a balance between accuracy and computational efficiency, it is essential to consider the integration of low- and high-fidelity modelling approaches. Interest is growing in off-design conditions and mitigation strategies. Despite the availability of commercial software, emerging numerical methods like the Green-Naghdi theory and immersed boundary-lattice Boltzmann methods continue to be explored to manage these complex applications. Offshore structures' inherent complexity - topological, physical, and operational - demands sophisticated data-driven models leveraging AI. Furthermore, as previously noted, increased attention to worker safety and health, structural integrity, and operational efficiency in offshore environments has renewed interest in vibroacoustic and hydroacoustic research.

Vibration control strategies depend on application-specific requirements, whether for offshore platforms, wind turbines, or ship propulsion systems. Consequently, the choice among active, semi-active, and passive control methods is context-dependent. Active systems offer greater flexibility but at a higher cost, while passive solutions are simpler and require less maintenance. Research into smart materials is expanding, but long-term economic viability remains uncertain and warrants further analysis.

Noise mitigation technologies, including damping materials and metamaterials, are promising for reducing both airborne and structure-borne noise. Innovative techniques such as acoustic black holes, negative stiffness, and metamaterials (which are maturing toward higher technology readiness levels) are increasingly being explored. They should be increasingly combined with low-noise structural design practices. Active noise control, particularly for underwater noise, is still in early development. For instance, bubble curtains around hulls and propellers are being studied for underwater noise mitigation, but further validation is needed. A key point is determining control effectiveness in terms of damping changes before and after mitigation, and understanding the environmental and operational conditions that affect damping features. While studies, especially in offshore contexts, have addressed this issue, uncertainties persist that hinder the reliability of damping estimations.

The benchmark conducted by this Committee has further underlined this evidence. A 24-h acceleration signal relative to onboard measurements on a containership was processed by several users to identify dynamic parameters such as signal amplitude, frequency and damping. The spread of damping estimations points to a pressing need for reliability assessment, best practices, and eventual standardization of methodologies.

While ship vibration regulations remain largely unchanged, recent research has focused on less explored issues, such as whole-body vibration exposure, fatigue effects on lashing bridges, and container stack resonances. In October 2023, the IMO released updated guidelines on underwater noise (IMO, 2023), outlining mitigation strategies for ship designers, builders, and operators. A tripartite working group (MEPC, 2024), involving IACS, shipbuilding associations, and industry trade bodies, is developing further initiatives. A major challenge remains the lack of extensive URN measurement campaigns to support regulatory progress. Correlating onboard vibration measurements with URN assessments is an emerging method that may enhance rule compliance and real-time monitoring in the future. In general, updating of regulation and standards to new technologies and materials should be a permanent effort.

The digitalization of structural systems and digital twin technology is gaining interest. Applications of data-driven models, especially for OWTs, are growing, though hybrid digital twins, combining sensor data and numerical models, remain rare for dynamic response applications. Sensor networks and real-time monitoring systems are improving structural health assessment and enabling data-driven design updates via extensive and continuous data collection. Advances in imaging, remote sensing, fibre-optics, and data fusion algorithms are enhancing dynamic data reconstruction across spatial domains. Technologies such as edge computing, IoT, self-powering and modular digital infrastructure are expected to accelerate the digital twin adoption for structural health monitoring.

Data-driven methods are expected to shape future modelling of complex dynamic systems. Although their application to some problems like wave-induced responses, remains limited, several ongoing research projects point to increasingly explore their use. In this respect, a clear understanding of their cost-effectiveness, accuracy, and data requirements compared to physics-based approaches should be pursued in the future. While complexity drives interest in AI-based approaches, it also raises questions about users' ability to evaluate the result robustness. Best practices and guidelines are needed to help researchers and engineers use pure or hybrid data-driven methods with greater confidence.

These final considerations serve as a reminder that, despite growing accuracy, our predictions remain inherently affected by uncertainties at various levels. In this context, the proposed benchmark study highlights the crucial link between the quality of analysis and the depth of understanding and expertise in applying the chosen modelling tools.

References

Abdelaziz, S., Sparrow, S.N., Hua, W., Wallom, D.: Future Climate change impact on wind gust for offshore wind Turbines cut-out within the UK exclusive economic zone. In: 2023 IEEE Green Technologies Conference (GreenTech), IEEE, pp. 10–14 (2023)

ABS: Insights into Ship Vibration Analysis (2022)

ABS: Guidance Notes on Ship Vibration (2023)

ABS: Guide for the Classification Notation: Underwater Noise and External Airborne Noise. ABS-295 (2024)

Adhikari, S., Bhattacharya, S.: A general frequency adaptive framework for damped response analysis of wind Turbines. Soil Dyn. Earthq. Eng. **143**, 106605 (2021)

Alsalah, A., Holloway, D., Ali-Lavroff, J.: Reducing wave impacts on high-speed catamarans through deployment of ride control: analysis of full-scale measurements. Ocean Eng. **292**, 116581 (2024)

An, X., et al.: Sandwich plate-type metastructures with periodic graded resonators for low-frequency and broadband vibration attenuation. Ocean Eng. **298**, 117229 (2024)

Andersson, C., Johansson, A.T., Genell, A., Winroth, J.: Fully electric ship propulsion reduces airborne noise but not underwater noise. Ocean Eng. **302**, 117616 (2024)

AQUO, 2015. Practical guidelines Task 5.3: Assessment of the Solutions to Reduce Underwater Radiated Noise. WP5 Report, AQUO Project

Aravanis, G.I., Silionis, N.E., Anyfantis, K.N.: Damage detection in ship hull structures under operational variability through strain sensing. Ocean Eng. **286**, 115537 (2023)

Augustyn, D., Pedersen, R.R., Tygesen, U.T., Ulriksen, M.D., Sørensen, J.D.: Feasibility of modal expansion for virtual sensing in offshore wind jacket substructures. Mar. Struct. **79**, 103019 (2021)

Azevedo Vasconcelos, A.C., Valiya Valappil, S., Schott, D., Jovanova, J., Aragón, A.M.: A metamaterial-based interface for the structural resonance shielding of impact-driven offshore monopiles. Eng. Struct. **300**, 117261 (2024)

Bassetti, M., Tonna, R., Pais, T., Silvestri, P., Lembo, E., Iuliano, A.: Application of Transfer Path Analysis Technique to Cruise Ships. Progress in Marine Science and Technology, 397–407 (2022)

Bindingsbø, O.T., Singh, M., Øvsthus, K., Keprate, A.: Fault Detection of a Wind Turbine Generator Bearing Using Interpretable Machine Learning. Frontiers in Energy Research 11 (2023)

Biot, M., et al.: Quantification of airborne noise emitted by ships based on class notation. Ocean Eng. **296**, 117085 (2024)

Bispo, I.B.S., Mohapatra, S.C., Guedes Soares, C.: Numerical analysis of a moored very large floating structure composed by a set of hinged plates. Ocean Eng. **253**, 110785 (2022)

Bocanegra, J.A., Borelli, D., Gaggero, T., Rizzuto, E., Schenone, C.: Characterizing onboard noise in ships: insights from statistical, machine learning and advanced noise index analyses. Ocean Eng. **285**, 115273 (2023)

Bossau, J.C., Bekker, A.: Line detection techniques to pinpoint slamming impulses in time-frequency images of hull acceleration measurements. Ocean Eng. **249**, 110841 (2022)

Bradshaw, A.S., et al.: A case study of foundation damping in a piled offshore wind jacket structure. Soil Dyn. Earthq. Eng. **180**, 108605 (2024)

BSSE: Offshore Wind Recommendations - Guidelines for Structural Health Monitoring for Offshore Wind Turbine Towers & Foundations. Doc.no. 20160190 16–1036 rev. 3 (2017)

Burella, G., Moro, L., Neis, B.: Is on-board noise putting fish harvesters' hearing at risk? a study of noise exposures in small-scale fisheries in newfoundland and labrador. Saf. Sci. **140**, 105325 (2021)

Cai, W., Zhu, L., Qian, X.: Dynamic responses of steel plates under repeated ice impacts. Int. J. Impact Eng **162**, 104129 (2022)

Cai, W., Zhu, L., Gudmestad, O.T., Guo, K.: Application of rigid-plastic theory method in ship-ice collision. Ocean Eng. **253**, 111237 (2022)

Chang, X., Song, Q., Qu, C., Li, Y., Liu, J.: Axial-transverse coupled vortex-induced vibration characteristics of composite riser under gas-liquid two-phase internal flow. Ocean Eng. **277**, 114163 (2023)

Chen, D., Li, Y., Gong, Y., Li, X., Ouyang, W., Li, X.: Low frequency vibration isolation characteristics and intelligent design method of hull grillage metastructures. Mar. Struct. **94**, 103572 (2024)

Chen, D., Zi, H., Li, Y., Li, X.: Low frequency ship vibration isolation using the band gap concept of sandwich plate-type elastic metastructures. Ocean Eng. **235**, 109460 (2021)

Chen, H., Tong, X., Chen, Y., He, J.: Theoretical and experimental studies on the hydrodynamic damping of elastic rotating propeller blades. Int. J. Naval Architecture Ocean Eng. **14**, 100446 (2022)

Chen, S., et al.: Marine structural health monitoring with optical fiber sensors: a review. Sensors **23**, 1877 (2023)

Chen, T., Zhang, H., Huang, B.: Acoustic field characteristic analysis of the rotary acoustic cavity coupled system. J. Low Frequency Noise Vibration Active Control **42**, 3–24 (2023)

Chen, Y., Zhang, S., Chen, W.K., Magee, A.: Design, Construction and Testing of a Fully Elastic Ship Model for Investigating Hydro-Elastic Responses of Container Ships. Mar. Struct. **98**, 103663 (2024)

Chen, Z., Zhao, N., Zhao, H., Guo, R., Zhao, W.: Experimental parametric study on hydroelastic behaviour of a 15000-TEU container ship based on segmented ship model. Ships Offshore Struct. **18**(7/9), 948–959 (2023)

Chin, C.S., Zhang, R.: Noise modeling of offshore platform using progressive normalized distance from worst-case error for optimal neuron numbers in deep belief network. Soft. Comput. **25**, 495–515 (2021)

Cui, R., Liu, Z., Wang, X., Yang, Z., Fan, S., Shu, Y.: The impact of marine engine noise exposure on seafarer fatigue: a China case. Ocean Eng. **266**, 112943 (2022)

Dai, Y., Liu, Z., Zhang, W., Chen, J., Liu, J.: CFD-FEM analysis of flow-induced vibrations in waterjet propulsion unit. J. Marine Sci. Eng. **10**, 1032 (2022)

de Diego, R.Y., Ghasemi, M., Pettersen, O.S.: Deliverable D2.1, Final Report on Vessel URN Measurements, SATURN Project (2023)

de Jong, C.A.F., Halvorsen, M.B., Hannay, D.E., Ainslie, M.A.: Methods to Measure Underwater Sound Sources from Oil and Gas Activities, pp. 1–17. The Effects of Noise on Aquatic Life. Springer International Publishing, Cham (2023)

de Novaes Pires Leite, G., et al.: Alternative fault detection and diagnostic using information theory quantifiers based on vibration time-waveforms from condition monitoring systems: application to operational wind Turbines. Renewable Energy **164**, 1183–1194 (2021)

Dhavalikar, S., Ramkumar, J., Kumar C.H., Chaturvedi, A.: Investigation of Waterjet Efflux Noise with Numerical Techniques. Waterjet 2022 (2022)

DNV: Fatigue Assessment of Ship Structures. DNV class guideline DNV-CG-0129, edition October 2021 (2021a)

DNV: Lashing Bridge Vibration. DNV Class Guideline, DNV-CG-0447, edition August 2021 (2021b)

Domingos, D.F., et al.: Full-scale measurements and analysis of the floating installation of an offshore wind Turbine tower. Ocean Eng. **310**, 118670 (2024)

Dong, X., Ren, S., Jia, Y., Yu, T.: Parameter optimization and vibration reduction effect of one C-shaped particle damping-tuned mass damper for offshore wind Turbine. Ocean Eng. **305**, 117856 (2024)

Ducoin, A., Barber, R.B., Wildy, S.J., Codrington, J.D., Baker, A.: Experimental evaluation of the use of embedded fiber bragg gratings to measure steady and unsteady flow-induced marine propeller blade deformation. Ocean Eng. **281**, 114889 (2023)

Eichner, L., Schneider, R., Baeßler, M.: Optimal vibration sensor placement for jacket support structures of offshore wind Turbines based on value of information analysis. Ocean Eng. **288**, 115407 (2023)

Elias, S., Beer, M.: Vibration control and energy harvesting of offshore wind Turbines installed with TMDI under dynamical loading. Eng. Struct. **315**, 118459 (2024)

Engelbrecht, M., Bekker, A.: A discomfort threshold for impulsive whole-body vibration on a slamming-prone vessel. Appl. Ergon. **109**, 103992 (2023)

Fang, Z., et al.: A novel vibration energy harvesting system integrated with an inertial pendulum for zero-energy sensor applications in freight trains. Appl. Energy **318**, 119197 (2022)

Febriyanto, K., Rahman, F.F., Guedes, J.C.C.: The physical and psychological effects of occupational noise among seafarers: a systematic review. Int. J. Environ. Health Res. **34**, 2674–2686 (2024)

Fredianelli, L., Bernardini, M., D'Alessandro, F., Licitra, G.: Sound power level and spectrum of port sources for environmental noise mapping. Ocean Eng. **306**, 118094 (2024)

Fredianelli, L., Bolognese, M., Fidecaro, F., Licitra, G.: Classification of noise sources for port area noise mapping. Environments **8**, 1 (2021)

Fredianelli, L., et al.: Source characterization guidelines for noise mapping of port areas. Heliyon **8**, e09021 (2022)

Fujikubo, M., et al.: A Digital Twin for Ship Structures - R&D Project in Japan. Data-Centric Eng. **5**, e7 (2024)

Gaidai, O., Fu, S., Xing, Y.: Novel reliability method for multidimensional nonlinear dynamic systems. Mar. Struct. **86**, 103278 (2022)

Gaidai, O., Xing, Y.: Novel reliability method validation for offshore structural dynamic response. Ocean Eng. **266**, 113016 (2022)

Gannon, L.G., Marshall, C.R.: A recurrent neural network model for structural response to underwater shock. Ocean Eng. **287**, 115898 (2023)

Gao, S., Tao, Z., Li, Y., Pang, F.: Application research of acoustic black hole in floating raft vibration isolation system. Rev. Adv. Mater. Sci. **61**, 888–900 (2022)

Gao, Y., et al.: Integrated design and real-world application of a Tuned Mass Damper (TMD) with displacement constraints for large offshore monopile wind Turbines. Ocean Eng. **292**, 116568 (2024)

Gargano, A., Das, R., Mouritz, A.P.: Comparative experimental study into the explosive blast response of sandwich structures used in naval ships. Composites Commun. **30**, 101072 (2022)

Gargano, A., Mouritz, A.: Comparative study of the explosive blast resistance of metal and composite materials used in defence platforms. Composites Part C: Open Access **10**, 100345 (2023)

Gebrezgabir, S., Holloway, D.S., Ali-Lavroff, J.: Slam and wave load response reconstruction in high speed catamarans using transmissibility on full scale sea trials. Ocean Eng. **271**, 113822 (2023)

Grammatikopoulos, A.: A review of physical flexible ship models used for hydroelastic experiments. Mar. Struct. **90**, 103436 (2023)

Griffin, M.J.: Handbook of Human Vibration. Academic Press, Elsevier (1990)

Gu, N., Liang, D., Zhou, X., Ren, H.: A CFD-FEA method for hydroelastic analysis of floating structures. J. Marine Sci. Eng. **11**, 737 (2023)

Guo, X., Wang, P., Liu, J., Nie, Y., Dai, L., Zeng, L.: Investigation on vortex induced vibration characteristics of three-dimensional marine riser considering cross and inline flows coupling effect. Aust. J. Mech. Eng. **21**, 1335–1349 (2023)

Hammer, T.C., Hendrikse, H.: Experimental study into the effect of wind-ice misalignment on the development of ice-induced vibrations of offshore wind Turbines. Eng. Struct. **286**, 116106 (2023)

Hartmann, M.C.N., et al.: Hydroelastic Potential Flow Solver Suited for Nonlinear Wave Dynamics in Ice-Covered Waters. Ocean Eng. **259**, 111756 (2022)

He, R., Xiang, Y., Guo, Z.: A poroelastic model for near-field underwater noise caused by offshore monopile driving. J. Sound Vib. **564**, 117878 (2023)

He, Z., Du, Z., Zhang, L., Li, Y.: Damage mechanisms of full-scale ship under near-field underwater explosion. Thin-Walled Struct. **189**, 110872 (2023)

Heggelund, S.E., Storhaug, G., Choi, B.-K.: Full Scale Measurements of Fatigue and Extreme Loading Including Whipping on an 8600TEU Post Panamax Container Vessel in the Asia to Europe Trade. 30th International Conference on Ocean, Offshore and Arctic Engineering, vol. 2: Structures, Safety and Reliability, 273–282, Rotterdam, Netherlands (2011)

Hirdaris, S., et al.: Review of the uncertainties associated to hull girder hydroelastic response and wave load predictions. Mar. Struct. **89**, 103383 (2023)

Horn, A.M., Wiggen, F. (2021). Structural Health Monitoring of topside piping systems and equipment. Report No.: 2021–3187, Rev. 1, PSA study

Horn, A.M., Wiggen, F., Gustafsson, S.: Sensor Data for Maintenance Planning of Topside Piping. 41st International Conference on Ocean, Offshore and Arctic Engineering, vol. 3: Materials Technology; Pipelines, Risers, and Subsea Systems, virtual conference (2022)

Horn, A.M., Storhaug, G., Håland, J., Jagite, G.: Structural Health Monitoring by Use of Sensor Data. Report no.: 2023–1241, Rev. 1. PSA study (2024)

Hosseini Lavassani, S.H., Mousavi Gavgani, S.A., Doroudi, R.: Optimal control of jacket platforms vibrations under the simultaneous effect of waves and earthquakes considering fluid-structure interaction. Ocean Eng. **280**, 114593 (2023)

Hosseiny, B., Amini, J., Aghababaei, H.: Spectral estimation model for linear displacement and vibration monitoring with GBSAR system. Mech. Syst. Signal Process. **208**, 110916 (2024)

Hu, H., Pan, Z., Zhao, W., Wan, D.: Numerical investigation of vortex-induced vibrational responses to a flexible tensioned riser with symmetric grooves in uniform currents. Ocean Eng. **271**, 113780 (2023)

Hua, W., Sha, Y., Zhang, X., Cao, H.: Research Progress of Carbon Capture and Storage (CCS) technology based on the shipping industry. Ocean Eng. **281**, 114929 (2023)

Huang, S., Jiao, J., Guedes Soares, C.: Uncertainty analyses on the CFD-FEA co-simulations of ship wave loads and whipping responses. Mar. Struct. **82**, 103129 (2022)

Huang, X., Mao, J.-W., Li, Q., Wang, Z., Pan, G., Hu, H.-B.: On the Interaction Between the Underwater Explosion and the Double-Layer Structure with an Orifice on the Outer Plate. Ocean Eng. **306**, 118050 (2024)

Huang, X., Mao, J.-W., Luo, X., Du, P., Ouahsine, A.: Dynamic response of a warship's metal-jet-damaged double-layer plates subjected to the subsequent underwater explosion. J. Marine Sci. Eng. **12**, 854 (2024)

IACS: Measurement of Underwater Radiated Noise from Ships. IACS Rec. No 181 (2024)

Ibrahim, M.E.: Nondestructive evaluation of thick-section composites and sandwich structures: a review. Compos. A Appl. Sci. Manuf. **64**, 36–48 (2014)

Ibrahim, A.M., Judge, C.Q.: Investigations of hull girder slamming factor for a semi-displacement vessel using model testing. Appl. Ocean Res. **150**, 104084 (2024)

Iliopoulos, A., Shirzadeh, R., Weijtjens, W., Guillaume, P., Hemelrijck, D.V., Devriendt, C.: A modal decomposition and expansion approach for prediction of dynamic responses on a monopile offshore wind Turbine using a limited number of vibration sensors. Mech. Syst. Signal Process. **68–69**, 84–104 (2016)

IMO: Guidelines for the Reduction of Underwater Radiated Noise from Shipping to Address Adverse Impacts on Marine Life. MEPC.1/Circ.833 (2014)

IMO: Revised Guidelines for the Reduction of Underwater Radiated Noise from Shipping to Address Adverse Impacts on Marine Life. MEPC.1/Circ.906 (2023)
ISO: Accuracy (Trueness and Precision) of Measurement Methods and Results – Part1: General Principles and Definitions. ISO **5725–1**, 2023 (2023)
ISO: Underwater Acoustics - Quantities and Procedures for Description and Measurement of Underwater Sound from Ships - Part 3: Requirements for Measurements in Shallow Water. ISO/DIS **17208–3**, 2023 (2023)
ISO: Underwater Acoustics - Quantities and Procedures for Description and Measurement of Underwater Sound from Ships - Part 2 Determination of Source Levels from Deep Water Measurements. ISO **17208–2**, 2019 (2019)
ISO: Ships and Marine Technology - Guidelines for Measurement, Evaluation and Reporting of Vibration with Regard to Habitability on Specific Ships. ISO **21984**, 2018 (2018)
ISO: Mechanical vibration - Measurement of Vibration on Ships - Part 5: Guidelines for Measurement, Evaluation and Reporting of Vibration with Regard to Habitability on Passenger and Merchant Ships. ISO **20283–5**, 2016 (2016)
ISO: Underwater acoustics - Quantities and Procedures for Description and Measurement of Underwater Sound from Ships - Part 1: Requirements for Precision Measurements in Deep Water Used for Comparison Purposes. ISO **17208–1**, 2016 (2016)
ISO: Mechanical Vibration and Shock - Evaluation of Human Exposure to Whole-Body Vibration - Part 1: General Requirements. ISO **2631–1**, 1997 (1997)
ISO: Mechanical Vibration – Guidelines for the Measurement, Reporting and Evaluation of Vibration with regard to Habitability on Passenger and Merchant Ships. ISO **6954**, 2000 (2000)
Jaen-Sola, P., Oterkus, E., McDonald, A.S.: Parametric lightweight design of a direct-drive wind Turbine electrical generator supporting structure for minimising dynamic response. Ships Offshore Struct. **16**, 266–274 (2021)
Jagite, G., Bigot, F., Malenica, S., Derbanne, Q., Le Sourne, H., Cartraud, P.: Dynamic ultimate strength of a ultra-large container ship subjected to realistic loading scenarios. Mar. Struct. **84**, 103197 (2022)
Jagite, G., Derbanne, Q., Malenica, S., Bigot, F., Le Sourne, H., Cartraud, P.: Investigation of the nonlinear slamming-induced whipping response of ships using a fully-coupled hydroelastoplastic method. Ocean Eng. **238**, 109751 (2021)
Jahangiri, M., Palermo, A., Kamali, S., Hadianfard, M.A., Marzani, A.: A procedure to estimate the minimum observable damage in truss structures using vibration-based structural health monitoring systems. Probab. Eng. Mech. **73**, 103451 (2023)
Jang, H.S., Hwang, S., Yoon, J., Lee, J.H.: Numerical analysis of ice-structure impact: validating material models and yield criteria for prediction of impact pressure. J. Marine Sci. Eng. **10**(4), 504 (2024). https://doi.org/10.3390/jmse10040504
Jayawickrema, U.M.N., Herath, H.M.C.M., Hettiarachchi, N.K., Sooriyaarachchi, H.P., Epaarachchi, J.A.: Fibre-optic sensor and deep learning-based structural health monitoring systems for civil structures: a review. Measurement **199**, 111543 (2022)
Jestel, J., von Pein, J., Lippert, T., von Estorff, O.: Damped cylindrical spreading model: estimation of mitigated pile driving noise levels. Appl. Acoust. **184**, 108350 (2021)
Ji, S., Yang, D.: Ice loads and ice-induced vibrations of offshore wind Turbine based on coupled DEM-FEM simulations. Ocean Eng. **243**, 110197 (2022)
Jiang, J., Lian, J., Dong, X., Zhou, H.: Research on the along-wind aerodynamic damping and its effect on vibration control of offshore wind Turbine. Ocean Eng. **274**, 113993 (2023)
Jiao, J., Huang, S., Guedes Soares, C.: Viscous fluid-flexible structure interaction analysis on ship springing and whipping responses in regular waves. J. Fluids Struct. **106**, 103354 (2021)
Jiao, J., Ren, H., Soares, C.G.: Vertical and horizontal bending moments on the hydroelastic response of a large-scale segmented model in a seaway. Mar. Struct. **79**, 103060 (2021)

Jin, Z., Xu, Q., Jiang, C., Wang, X., Chen, H.: Ordinal few-shot learning with applications to fault diagnosis of offshore wind Turbines. Renewable Energy **206**, 1158–1169 (2023)

Johnson, T.J., Brown, R.L., Adams, D.E., Schiefer, M.: Distributed structural health monitoring with a smart sensor array. Mech. Syst. Signal Process. **18**, 555–572 (2004)

Jothinathan, S., Kashyap, S., Kumar, D., Saha, N.: Response Control of Fixed Offshore Structure with Wind Turbine Using MR Damper. OCEANS 2022 - Chennai, India, 1–8 (2022)

Ju, H.B., Jong, B.S., Yim, K.H.: Prediction of sloshing pressure and structural response of LNG CCS. Ocean Eng. **266**(3), 112298 (2022)

Kampitsis, A., Kapasakalis, K., Via-Estrem, L.: An Integrated FEA-CFD SIMULATION OF OFFSHORE Wind Turbines with vibration control systems. Eng. Struct. **254**, 113859 (2022)

Kang, M.-S., Kim, J.-H., Kim, M.-H.: Experimental and numerical study on the vibration characteristics of an electric switchboard with wire rope isolators in naval ships. Ocean Eng. **283**, 115172 (2023)

Karrech, A., et al.: Non-linear vibration of free spanning subsea pipelines with multi-dimensional mid-plane stretching. Eng. Struct. **301**, 117265 (2024)

Karthikeyan, S., Nallayarasu, S.: CFD simulation of vortex-induced vibration of an elastic cylinder in subcritical flow regime using a two-way coupled model validated by experiment. Ocean Eng. **273**, 113956 (2023)

Kim, B., Oh, J., Min, C.: Investigation on applicability and limitation of cosine similarity-based structural condition monitoring for gageocho offshore structure. Sensors **22**, 663 (2022)

Kim, D.M., Hong, S.H., Jeong, S.H., Kim, S.J.: Analysis of dynamic characteristics of rotor sail using a 4DOF rotor model and finite element model. J. Marine Sci. Eng. **12**, 335 (2024)

Kim, J.-Y., Hong, S.-Y., Song, J.-H., Min, D.-K.: Analysis of noise attenuation for offshore platform silencer in duct with flow. Proceedings of the Institution of Mechanical Engineers, Part M: Journal of Engineering for the Maritime Environment **237**, 344–356 (2023)

Kjeld, J.G., Avendaño-Valencia, L.D., Brandt, A., Christensen, S.S., Andersen, J.K.F.: Towards minimal empirical uncertainty bounds of damping estimates of an offshore wind turbine in idling conditions. Mech. Syst. Signal Process. **191**, 110180 (2023)

Kjeld, J.G., Avendaño-Valencia, L.D., Vestermark, J.: Effect of wind and wave properties in modal parameter estimates of an idling offshore wind Turbine from long-term monitoring data. Mech. Syst. Signal Process. **187**, 109934 (2023)

Kong Chen, Y., et al.: Efficient residual reliability criterion index in a permanent guided wave monitoring system. Measurement **197**, 111292 (2022)

Kostikov, V.K., Hayatdavoodi, M., Ertekin, R.C.: Moored elastic sheets under the action of nonlinear waves and current. Mar. Struct. **93**, 103542 (2024)

Kostikov, V.K., Hayatdavoodi, M., Ertekin, R.C.: Drift of Elastic floating ice sheets by waves and current: multiple sheets. Phys. Fluids **34** (2022)

Koukoura, S., Scheu, M.N., Kolios, A.: Influence of extended potential-to-functional failure intervals through condition monitoring systems on offshore wind Turbine availability. Reliab. Eng. Syst. Saf. **208**, 107404 (2021)

Kyaw Oo D'Amore, G., Caverni, S., Biot, M., Rognoni, G., D'Alessandro, L.: A Metamaterial Solution for Soundproofing on Board Ship. Appl. Sci. **12**, 6372 (2022)

Kyaw Oo D'Amore, G., Morgut, M., Biot, M., Mauro, F., Kašpar, J.: Integration and Optimization of the After-Treatments Systems to Reduce the Acoustic Footprint of the Ships. Applied Acoustics **213**, 109625 (2023)

Lara, M., Vázquez, F., Sandua-Fernández, I., Garrido, J.: Adaptive active generator torque controller design using multi-objective optimization for tower lateral load reduction in monopile offshore wind Turbines. IEEE Access **11**, 115894–115910 (2023)

Lee, J.K., Seung, H.M., Park, C.I., Lee, J.K., Lim, D.H., Kim, Y.Y.: Magnetostrictive patch sensor system for battery-less real-time measurement of torsional vibrations of rotating shafts. J. Sound Vib. **414**, 245–258 (2018)

Lekatompessy, D.R.: Analysis of the effect of engine rotation on vibration amplitude in the engine room of aluminum ship. Presented at the 7th International Conference on Basic Science 2021, Ambon, Indonesia, 040014 (2023)

Leng, D., Liu, D., Li, H., Jin, B., Liu, G.: Internal flow effect on the cross-flow vortex-induced vibration of marine risers with different support methods. Ocean Eng. **257**, 111487 (2022)

Leng, D., Lv, P., Zhu, Z., Li, Y., Liu, G.: Experimental study on semi-active magnetorheological elastomer based isolation system for offshore platform using wave tank. Ocean Eng. **292**, 116467 (2024)

Li, H., Zou, J., Deng, B., Liu, R., Sun, S.: Experimental study of stern slamming and global response of a large cruise ship in regular waves. Mar. Struct. **86**, 103294 (2022)

Li, H., Zhang, C., Zheng, X., Mei, Z., Bai, X.: A simplified theoretical model of the whipping response of a hull girder subjected to underwater explosion considering the damping effect. Ocean Eng. **239**, 109831 (2021)

Li, H., Zou, J., Peng, Y., Zhou, X., Lu, L., Sun, S.: Numerical study of slamming and whipping loads in moderate and large regular waves for different forward speeds. Mar. Struct. **94**, 103563 (2024)

Li, H.S., Wang, S. and Guedes Soares, C.: Uncertainty assessment of the scale effects on a submerged cylinder. Advances in Maritime Technology and Engineering, Guedes Soares, C. & Santos T.A. (eds.) Taylor and Francis Group, London, UK, 139–148 (2024b)

Li, J., Hu, J., Zhao, Z., Duan, C., Li, F.: Effect of cavitation on vortex-induced vibration of hydrofoil. J. Phys. Conf. Ser. **2709**, 012007 (2024)

Li, M., Wan, Z., Yuan, Y., Tang, W.: A numerical method for analysing the responses of cargo containment system of an LNGC under iceberg collision. Ships and Offshore Structures, 1–14 (2023a)

Li, S., Brennan, F.: Digital Twin Enabled Structural Integrity Management: Critical Review and Framework Development. Proceedings of the Institution of Mechanical Engineers, Part M: Journal of Engineering for the Maritime Environment **238**, 707–727 (2024)

Li, X., et al.: Three-dimensional sound source localization system based on fiber optic sensor array with an adaptive algorithm. Optics Commun. **559**, 130383 (2024)

Li, X., et al.: Dynamic responses of ultralight all-metallic honeycomb sandwich panels under fully confined blast loading. Compos. Struct. **311**, 116791 (2023)

Li, X., Yuan, Y., Duan, Z., Xue, H., Tang, W.: Experimental investigation on vortex-induced vibration of a flexible pipe in combined uniform and oscillatory flow. Ocean Eng. **285**, 115375 (2023)

Li, X., Yuan, Y., Duan, Z., Xue, H., Tang, W.: Coupling effect of time-varying axial tension and combined unsteady flow on vortex-induced vibration of flexible risers. Ocean Eng. **285**, 115358 (2023)

Li, X., Yuan, Y., Xue, H., Tang, W.: Vortex-induced vibration response characteristics of catenary riser conveying two-phase internal flow. Ocean Eng. **257**, 111617 (2022)

Li, Y., Jiang, S., Feng, S., Cao, W., Zhang, K.: Experimental study of an engineering prediction approach for shipboard cabin noise. J. Phys: Conf. Ser. **2383**, 012092 (2022)

Li, Y., Wu, X., Xiao, W., Wang, S., Zhu, L.: Experimental study on the dynamic behaviour of aluminium honeycomb sandwich panel subjected to ice wedge impact. Composite Struct. **282**, 115092 (2022d)

Li, Z., Yang, S., Wang, S., Xu, Z., Zhang, Q.: Development of a new kind low frequency low-noise vector sensor. Sens. Actuators, A **301**, 111743 (2020)

Lian, J., Jiang, J., Dong, X., Wang, H., Zhou, H.: One damping estimation approach of the parked offshore wind Turbine supported by wide-shallow bucket foundation. Ocean Eng. **235**, 109387 (2021)

Liang, C., Yuan, Y., Yu, X.: Applying new high damping viscoelastic dampers to mitigate the structural vibrations of monopile-supported offshore Wind Turbine. Ocean Eng. **295**, 11691 (2024)

Lin, D.H., Kim, K.S.: Development of deep learning-based detection technology for vortex-induced vibration of a ship's propeller. J. Sound Vib. **520**, 116629 (2022)

Lin, B., Dong, X.: Ship hull inspection: a survey. Ocean Eng. **289**, 116281 (2023)

Lin, H., Luan, H., Uzdin, A.M., Zhang, S., Wei, L., Yang, L.: A CFD-FEA coupled model for simulating dynamic response of offshore jacket platform under earthquake considering wind, wave. Current and Aftershock Loads. Ocean Engineering **300**, 117481 (2024)

Lin, J., Wang, Y., Zhang, G., Liu, Y., Zhang, J.: Novel tuned mass dampers installed inside tower of spar offshore floating wind Turbines. Ocean Eng. **301**, 117412 (2024)

Lin, X., Wang, S., Zhang, L., Xu, S., Hu, Y.: Study on load characteristics and structural impact response propagation law of cabin internal explosion. Ocean Eng. **301**, 117582 (2024)

Liu, D., Ai, S., Sun, L., Guedes Soares, C.: Vortex-induced vibrations of catenary risers in varied flow angles. Int. J. Mech. Sci. **269**, 109086 (2024)

Liu, G., Luo, J., Huang, G., Li, R., Ji, W., Long, T.: Data driven damping separation method for the mega-offshore wind Turbine tower. Ocean Eng. **306**, 118158 (2024)

Liu, G., Song, Z., Xu, W., Sha, M.: Numerical study on the VIVs of two side-by-side elastically coupled cylinders under different re and natural frequencies. Ocean Eng. **284**, 115261 (2023)

Liu, J.H., Corbita, N.T., Lee, R.M., Wang, C.C.: Wind Turbine anomaly detection using mahalanobis distance and SCADA alarm data. Appl. Sci. **12**, 8661 (2022)

Liu, R., Li, H., Zou, J., Ong, M.C.: Reconstruction and prediction of global whipping responses on a large cruise ship based on LSTM neural networks. Ocean Eng. **285**, 115393 (2023)

Liu, S., Chuang, Z., Qu, Y., Li, X., Li, C., He, Z.: Dynamic performance evaluation of an integrated 15 MW floating offshore wind turbine under typhoon and ECD conditions. Front. Energy Res. **10** (2022b)

Liu, W., et al.: Development of a fully coupled numerical hydroelasto-plastic approach for offshore structure. Ocean Eng. **258**, 111713 (2022)

Liu, X., Xu, J., He, G., Chen, C.: Lateral vibration mitigation of monopile offshore wind Turbines with a spring pendulum pounding tuned mass damper. Ocean Eng. **266**, 112954 (2022)

Liu, X., Yang, D., Li, Q., Liu, J.: Energy-averaged method for ship mechanical noise prediction under kinematic excitation in the full frequency domain. Ocean Eng. **246**, 110615 (2022)

Liu, X., Yang, D., Li, Q., Liu, J.: Fast average prediction method based on the upper -lower limit theory for ship's mechanical noise. Appl. Acoust. **197**, 108894 (2022)

Liu, Y., Jiang, Y., Zhao, H., Wang, S., Han, J.: Experimental investigation on vortex-induced vibration characteristics of a segmented free-hanging flexible riser. Ocean Eng. **281**, 115032 (2023)

Liu, Y.Z., Ren, S.F., Zhao, P.-F.: Application of the deep neural network to predict dynamic responses of stiffened plates subjected to near-field underwater explosion. Ocean Eng. **247**, 110537 (2022)

Liu, Z., Chen, W.: Research and analysis on firing accuracy of naval gun. J. Phys: Conf. Ser. **1948**, 012081 (2021)

Liu, Z., Liu, J., Yan, J., Guo, A., Li, H.: Dynamic responses of the end-anchored floating bridge under the combined action of wind and waves. Ocean Eng. **288**, 115907 (2023)

Löhner, R., Li, L., Soto, O.A., Baum, J.D.: An arbitrary lagrangian-eulerian method for fluid-structure interactions due to underwater explosions. Int. J. Numer. Meth. Heat Fluid Flow **33**, 2308–2349 (2023)

Long, X., Liu, L., Ji, S.: Discrete element analysis of ice-induced vibrations of offshore wind turbines in level ice. J. Marine Sci. Eng. **11**(11), 2153 (2023)

Lorenzino, M., D'Agostin, F., Rigutti, S., Bovenzi, M., Fantoni, C., Bregant, L.: Mood regulates the physiological response to whole-body vibration at low intensity. Appl. Ergon. **108**, 103956 (2023)

Louvros, P., Boulougouris, E., Coraddu, A., Vassalos, D., Theotokatos, G.: Multi-Objective Optimisation as an Early Design Tool for Smart Ship Internal Arrangement. Ships and Offshore Structures **17**, 1392–1402 (2022)

Lu, L., Ren, H., Li, H., Zou, J., Chen, S., Liu, R.: Numerical method for whipping response of ultra large container ships under asymmetric slamming in regular waves. Ocean Eng. **287**, 115830 (2023)

Lu, Y., et al.: Modal investigation on a large-scale containership model for hydroelastic analysis. Shock. Vib. **1**, 2539870 (2022)

Lu, Y., Jin, Q., Yang, J., Tian, C., Si, H., Zhou, Q.: The effect of forward speed on hydrodynamic response of a containership in head waves. Ocean Eng. **304**, 117835 (2024)

Lu, L., Zhang, L., Zhang, C., Li, H-S., Wang, G.: Experimental Study on the Active-Passive Approach of Vibration of Ship Propulsion Shaft System. In: Presented at the 28th International Congress on Sound and Vibration (ICSV28), Singapore (2023b)

Luan, Y., Yan, L., Sun, T., Zunino, P.: Aerodynamic noise and its reduction of the marine gas Turbine air exhaust system. J. Acoustical Soc. Am. **155**, 2728–2740 (2024)

Luyun, C., Yingying, Z., Deqing, Y.: Human-induced noise study for cruise cabin: numerical analysis and experimental validation. Ocean Eng. **286**, 115498 (2023)

Ma, R., Bi, K., Zuo, H., Du, X.: Inerter-based damping isolation system for vibration control of offshore platforms subjected to ground motions. Ocean Eng. **280**, 114726 (2023)

Machado, M.R., Dutkiewicz, M.: Wind Turbine vibration management: an integrated analysis of existing solutions, products, and open-source developments. Energy Rep. **11**, 3756–3791 (2024)

Machado, M.R., Dutkiewicz, M., Colherinhas, G.B.: Metamaterial-based vibration control for offshore wind turbines operating under multiple hazard excitation forces. Renewable Energy **223**, 120056 (2024)

Magionesi, F., Dubbioso, G., Muscari, R.: FSI analysis of a marine rudder behind a propeller. Int. J. Offshore Polar Eng. **34**, 291–298 (2024)

Malekjafarian, A., Jalilvand, S., Doherty, P., Igoe, D.: Foundation damping for monopile supported offshore wind turbines: a review. Mar. Struct. **77**, 102937 (2021)

Mao, L., Wu, M., Zhang, W., Guo, C., Zeng, S.: Analysis of large deformation of deep water drilling riser considering vortex-induced vibration. Appl. Ocean Res. **133**, 103484 (2023)

Mao, L., Yan, J., Zeng, S., Cai, M.: Vortex-induced vibration characteristics of mining riser under the coupling effect of external ocean current and internal multiphase flow. Appl. Ocean Res. **140**, 103747 (2023)

Meemary, B., Vasiukov, D., Deléglise-Lagardère, M., Chaki, S.: Sensors integration for structural health monitoring in composite pressure vessels: a review. Compos. Struct. **351**, 118546 (2025)

de Souza, M., Mello, F., Pereira, J.L.J., Gomes, G.F.: Multi-objective sensor placement optimization in shm systems with kriging-based mode shape interpolation. J. Sound Vib. **568**, 118050 (2024)

Mentzelopoulos, A.P., del Águila Ferrandis, J., Rudy, S., Sapsis, T., Triantafyllou, M.S., Fan, D.: Physics-based data-informed prediction of vertical, catenary, and stepped riser vortex-induced vibrations. Int. J. Offshore Polar Eng. **33**, 367–379 (2023)

MEPC: The Tripartite Working Group on Underwater Radiated Noise, MEPC-82/9/2 (2024)

Merino-Martínez, R., Pieren, R., Schäffer, B.: Holistic approach to wind Turbine noise: from blade trailing-edge modifications to annoyance estimation. Renew. Sustain. Energy Rev. **148**, 111285 (2021)

Mieloszyk, M., Majewska, K., Ostachowicz, W.: Application of embedded fibre bragg grating sensors for structural health monitoring of complex composite structures for marine applications. Mar. Struct. **76**, 102903 (2021)

Molenkamp, T., Tsetas, A., Tsouvalas, A., Metrikine, A.: Underwater noise from vibratory pile driving with non-linear frictional pile-soil interaction. J. Sound Vib. **576**, 118298 (2024)

Molenkamp, T., Tsouvalas, A., Metrikine, A.: A numerical study on the effect of asymmetry on underwater noise emission in offshore monopile installation. Ocean Eng. **299**, 117351 (2024)

Molenkamp, T., Tsouvalas, A., Metrikine, A.: The Influence of Contact Relaxation on Underwater Noise Emission and Seabed Vibrations due to Offshore Vibratory Pile Installation. Frontiers in Marine Science 10 (2023)

Moradi, M.A., Mojra, A.: Free-surface flow past a circular cylinder at high froude numbers. Ocean Eng. **295**, 116804 (2024)

Mu, A., et al.: Bluff body vortex-induced vibration control of floating wind Turbines based on a novel intelligent robust control algorithm. Phys. Fluids **34**, 114103 (2022)

Na, W., Sun, K.H., Jeon, B.C., Lee, J., Shin, Y.: Event-based micro vibration measurement using phase correlation template matching with event filter optimization. Measurement **215**, 112867 (2023)

Nektarios, K., et al.: Occupational health hazards and risks in the wind industry. Energy Rep. **7**, 3750–3759 (2021)

Norén-Cosgriff, K., Kaynia, A.M.: Estimation of natural frequencies and damping using dynamic field data from an offshore wind Turbine. Mar. Struct. **76**, 102915 (2021)

Nsengiyumva, W., Zhong, S., Lin, J., Zhang, Q., Zhong, J., Huang, Y.: Advances, limitations and prospects of nondestructive testing and evaluation of thick composites and sandwich structures: a state-of-the-art review. Compos. Struct. **256**, 112951 (2021)

Omer, H., Bekker, A.: Human responses to wave slamming vibration on a polar supply and research vessel. Appl. Ergon. **67**, 71–82 (2018)

Kimmerl, J., Mertes, P., Abdel-Maksoud, M.: Application of Large Eddy Simulation to Predict Underwater Noise of Marine Propulsors. Part 1: Cavitation Dynamics. J. Marine Sci. Eng. **9**(8), 792 (2021)

Pal, S.K., Ono, T., Takami, T., Tatsumi, A., Iijima, K.: Effect of springing and whipping on exceedance probability of vertical bending moment of a ship. Ocean Eng. **266**, 112600 (2022)

Pan, L.-F., Zhou, H.-W., Mou, L.-W., Mao, X.-Q., Ma, C.-F.: Study on Vibration Reduction Design of a Ro-Ro Ship with Twin-Skeg. The International Society of Offshore and Polar Engineers, Shanghai, China (2022)

Parmar, C., Wings, E., Nourmohammadi, F.: Identification of vibration for balancing in fehn pollux ship with ECO flettner rotor. J. Energy Syst. **8**, 1–10 (2024)

Parunov, J., et al.: Benchmark on the prediction of whipping response of a warship model in regular waves. Mar. Struct. **94**, 103549 (2024)

Patel, M., Patel, S., Ahmad, S.: Blast analysis of efficient honeycomb sandwich structures with CFRP/steel FML skins. Int. J. Impact Eng **178**, 104609 (2023)

Pathak, K., Saha, N.: One-way fluid-structure interaction of submerged floating tunnel under current action. In: 43rd International Conference on Ocean, Offshore and Arctic Engineering, Vol. 6: Polar and Arctic Sciences and Technology; CFD, FSI, and AI, Singapore (2024)

Pawlenka, T., Škuta, J., Tůma, J., Juránek, M.: Development of capacitive sensors for measuring vibrations and small displacements of a high-speed rotating machines for use in active vibration control systems. Sens. Actuators, A **365**, 114902 (2024)

Peng, A., Wu, H., Li, X., Tong, Z., Yin, Z., Hu, J.: Research on application technology of local sound field virtual control technology in ship cabin noise reduction. In: IEEE 11th International Conference on Computer Science and Network Technology (ICCSNT), IEEE, pp. 165–168 (2023)

Peng, Y., Tsouvalas, A., Stampoultzoglou, T., Metrikine, A.: Study of the sound escape with the use of an air bubble curtain in offshore pile driving. J. Marine Sci. Eng. **9**, 232 (2021)

Peng, Y., Tsouvalas, A., Stampoultzoglou, T., Metrikine, A.: A fast computational model for near- and far-field noise prediction due to offshore pile driving. J. Acoustical Soc. Am. **149**, 1772–1790 (2021)

Pezeshki, H., Adeli, H., Pavlou, D., Siriwardane, S.C.: State of the art in structural health monitoring of offshore and marine structures. Proceedings of the Institution of Civil Engineers - Maritime Engineering **176**(2), 89–108 (2023)

Pozo, F., Tibaduiza, D.A., Vidal, Y.: Sensors for structural health monitoring and condition monitoring. Sensors **21**, 1558 (2021)

Vijith, P.P., Rajendran, S.: Hydroelastic effects on the vertical bending moment of a container ship in head and oblique seas. Ocean Eng. **285**, 115385 (2023)

Prato, A., Silvestri, P., Pais, T., Gaggero, F., Schiavi, A.: Airborne sound transmission loss of ship bulkheads: a new methodological approach for in-laboratory assessment by including low frequencies. Ocean Eng. **285**, 115428 (2023)

Praveen, B.M.N., Avula, V.R.: Experimental Investigation of Downstream Cylinder Vibrations Caused by FIV in the Presence of Upstream Stationary Cylinder. Marine Structures **82**, 103137 (2022)

Puerto-Santana, C., Ocampo-Martinez, C., Diaz-Rozo, J.: Mechanical rotor unbalance monitoring based on system identification and signal processing approaches. J. Sound Vib. **541**, 117313 (2022)

Qays, M.O., et al.: Monitoring of renewable energy systems by IoT-aided scada system. Energy Sci. Eng. **10**, 1874–1885 (2022)

Qi, L., et al.: Regulating rotor aerodynamics and platform motions for a semi-submersible floating wind turbine with trailing edge flaps. Ocean Eng. **286**, 115629 (2023)

Qin, Y., Wang, Y., Wang, Z., Yao, X.: Investigation on similarity laws of cabin structure under internal blast loading. Ocean Eng. **260**, 111998 (2022)

Qin, Y., Yao, X., Wang, Z., Wang, Y.: Experimental investigation on damage features of stiffened cabin structures subjected to internal blast loading. Ocean Eng. **265**, 112639 (2022)

Qu, Y., Wang, P., Fu, S., Zhao, M.: Vortex-induced vibrations of a top tensioned riser subjected to flows with spanwise varying directions. Int. J. Mech. Sci. **242**, 107954 (2023)

Rajiv, G., Verma, M., Subbulakshmi, A.: Gaussian process metamodels for floating offshore wind turbine platforms. Ocean Eng. **267**, 113206 (2023)

Rashki, M.R., Hejazi, K., Tamimi, V., Zeinoddini, M., Harandi, M.M.A.: Impacts of soft marine fouling on the hydrokinetic energy harvesting from one-degree-of-freedom vortex-induced vibrations. Sustainable Energy Technol. Assess. **54**, 102881 (2022)

Ren, X., et al.: The scaling laws of cabin structures subjected to internal blast loading: experimental and numerical studies. Defence Technol. **18**, 811–822 (2022)

Ren, X., Xu, Y., Shen, T., Wang, Y., Bhattacharya, S.: Support condition monitoring of monopile-supported offshore wind Turbines in layered soil based on model updating. Mar. Struct. **87**, 103342 (2023)

Riesner, M., el Moctar, O.: A numerical method to compute global resonant vibrations of ships at forward speed in oblique waves. Appl. Ocean Res. **108**, 102520 (2021)

Riesner, M., el Moctar, O.: Assessment of wave induced higher order resonant vibrations of ships at forward speed. J. Fluids Struct. **103**, 103262 (2021)

Rinaldi, G., Thies, P.R., Johanning, L.: Current status and future trends in the operation and maintenance of offshore wind turbines: a review. Energies **14**, 2484 (2021)

Rognoni, G., Kyaw Oo D'Amore, G., Brocco, E., Moro, L., Biot, M.: Investigation on the impact of a metamaterial solution for the mitigation of noise radiated by a ship panel. In: 33rd International Ocean and Polar Engineering Conference, Ottawa, Canada, ISOPE-I-23-162 (2023)

Ruan, Y., Liang, X., Hua, X., Zhang, C., Xia, H., Li, C.: Isolating low-frequency vibration from power systems on a ship using spiral phononic crystals. Ocean Eng. **225**, 108804 (2021)

Rueda-Bayona, J.G., Guzmán, A., Eras, J.J.C.: Vortex-induced vibration effect of extreme sea states over the structural dynamics of a scaled monopile offshore wind Turbine. J. Ocean Eng. Marine Energy **9**, 359–376 (2023)

Sakaris, C.S., et al.: Structural health monitoring of tendons in a multibody floating offshore wind turbine under varying environmental and operating conditions. Renewable Energy **179**, 1897–1914 (2021)

Sardar, R., Chakraborty, S.: Wave induced vibration control of offshore jacket platform by tuned liquid damper with floating base. Ocean Eng. **273**, 113948 (2023)

Sarkar, N., Ghosh, A.: A frequency domain study on deck isolation effectiveness in control of wave-induced vibration of offshore jacket platform. Ocean Eng. **270**, 113682 (2023)

SATURN: Mitigating Ship Noise Using Bubble Injection. Website Post (2023). https://www.saturnh2020.eu/post/mitigating-ship-noise-using-bubble-injection. Accessed 31 Jan 2025

Schiavoni, S., Baldinelli, G., Presciutti, A., D'Alessandro, F.: Acoustic mitigation of noise in ports: an original methodology for the identification of intervention priorities. Noise Mapping **9**, 211–226 (2022)

Schiavoni, S., et al.: Airborne sound power levels and spectra of noise sources in port areas. Int. J. Environ. Res. Public Health **19**, 10996 (2022)

SDC: Report of the Working Group. IMO Sub-Committee on Ship Design and Construction, SDC 10/WP.3 (2024)

SDC: Monitoring Technology of Underwater Radiated Noise from Ships Using Onboard Noise Measurement. IMO Sub-Committee on Ship Design and Construction, SDC 9/INF.9 (2022a)

SDC: Report on Underwater Sound Measurements in Japan and Discussion on Estimating Source Levels of Underwater Radiated Noise from a Ship. IMO Sub-Committee on Ship Design and Construction, SDC 9/INF.10 (2022b)

Sezen, S., Atlar, M.: Marine propeller underwater radiated noise prediction with the FWH acoustic analogy part 3: assessment of full-scale propeller hydroacoustic performance versus sea trial data. Ocean Eng. **266**(2), 112712 (2022)

Shabani, B.: Intelligent monitoring of a large catamaran ferry. Int. J. Maritime Eng. **165**, 11–22 (2023)

Shabani, B., Ali-Lavroff, J., Holloway, D.S., Penev, S., Dessi, D., Thomas, G.: Using remote monitoring and machine learning to classify slam events of wave piercing catamarans. Int. J. Maritime Eng. **163**(A3) (2021)

Shao, J.Y., Zhang, L., Wen, J.D.: Vortex induced coupled vibration of an elastically mounted square cylinder with a detached solid and flexible plate. Ocean Eng. **283**, 115092 (2023)

Sharma, S., Nava, V.: Condition monitoring of mooring systems for floating offshore wind turbines using convolutional neural network framework coupled with autoregressive coefficients. Ocean Eng. **302**, 117650 (2024)

Shen, C., Huang, J., Zhang, Z., Xue, J., Qian, D.: Sandwich plate structure periodically attached by s-shaped oscillators for low frequency ship vibration isolation. Materials **16**, 2467 (2023)

Shi, W., et al.: Wide quasi-zero stiffness region isolator with decoupled high static and low dynamic stiffness. Mech. Syst. Signal Process. **215**, 111452 (2024)

Shi, W., Liu, Y., Wang, W., Cui, L., Li, X.: Numerical study of an ice-offshore wind Turbine structure interaction with the pile-soil interaction under stochastic wind loads. Ocean Eng. **273**, 113984 (2023)

Shi, W., Zhang, L., Karimirad, M., Michailides, C., Jiang, Z., Li, X.: Combined effects of aerodynamic and second-order hydrodynamic loads for floating wind turbines at different water depths. Appl. Ocean Res. **130**, 103416 (2023)

Shi, X., Cai, C., Bao, P., Li, Z.: Influence of ship-based vibration on characteristics of arc and droplet and morphology in wire arc additive manufacturing. Chinese J. Mech. Eng. Additive Manuf. Front. **2**, 100067 (2023)

Shi, Y., Yang, D., Li, Q., Qin, J.: Cabin noise analysis of polar transport vessels under ship-ice-water-air coupling continuous icebreaking based on the S-ALE algorithm. Mar. Struct. **95**, 103601 (2024)

Shi, Y.H., Yang, D.Q., Wu, W.W.: Numerical analysis method of ship-ice collision-induced vibration of the polar transport vessel based on the full coupling of ship-ice-water-air. J. Ocean Eng. Sci. **8**(4), 323–335 (2023)

Shrivastava, K., Vijayan, K., Arora, V.: Identification of stiffness and damping of the weld in stiffened plates using model updating. Mar. Struct. **82**, 103140 (2022)

Si, H., et al.: Uncertainty analysis of linear vertical bending moment in model tests and numerical prediction. Mech. Syst. Signal Process. **178**, 109331 (2022)

Si, H., Zhao, N., Tian, C., Zhou, Q., Geng, Y., Hu, J., Yang, P.: Effects of Design and Fabrication of Segmented Ship Models on Test Results of Wave-Induced Loads. Ocean Engineering 305 (2024)

Sigrist, J.-F., Broc, D.: A versatile method to calculate the response of equipment mounted on ship hulls subjected to underwater shock waves. Finite Elem. Anal. Des. **218**, 103917 (2023)

Silva-Campillo, A., Pérez-Arribas, F., Suárez-Bermejo, J.C.: Health-monitoring systems for marine structures: a review. Sensors **23**, 2099 (2023)

Siraj, F.M., Ayon, S.T.K., Samad, M.A., Uddin, J., Choi, K.: Few-shot lightweight SqueezeNet architecture for induction motor fault diagnosis using limited thermal image dataset. IEEE Access **12**, 50986–50997 (2024)

Smith, T.A., Rigby, J.: Underwater radiated noise from marine vessels: a review of noise reduction methods and technology. Ocean Eng. **266**, 112863 (2022)

Soal, K., Govers, Y., Bienert, J., Bekker, A.: System Identification and tracking using a statistical model and a Kalman filter. Mech. Syst. Signal Process. **133**, 106127 (2019)

Song, M., Jiang, Z., Liu, K., Han, Y., Liu, R.: Dynamic response analysis of a monopile-supported offshore wind Turbine under the combined effect of sea ice impact and wind load. Ocean Eng. **286**, 115587 (2023)

Song, M., Partovi Mehr, N., Moaveni, B., Hines, E., Ebrahimian, H., Bajric, A.: One year monitoring of an offshore wind turbine: variability of modal parameters to ambient and operational conditions. Eng. Struct. **297**, 117022 (2023)

Stone, J.K., Moro, L.: Occupational noise exposure in Canada's salmonid aquaculture industry. Aquaculture **550**, 737831 (2022)

Storhaug, G., et al.: Uncertainties in predicted quantities based on measured time series of dynamic response. In: 34th International Ocean and Polar Engineering Conference, Rhodes. Greece (2024)

Storhaug, G., Jagite, G.: Vibration Dose Value Assessment from Full Scale Measurements. Advances in Maritime Technology and Engineering. CRC Press, London, pp. 493–500 (2024)

Storhaug, G., Laanemets, K., Edin, I., Ringsberg, J.W.: Estimation of Damping from Wave Induced Vibrations in Ships. Progress in the Analysis and Design of Marine Structures. CRC Press, 121–130 (2017)

Sui, G., et al.: Research on flexible beam-type nonlinear vibration isolators suitable for low frequencies. Ocean Eng. **293**, 116652 (2024)

Sun, C., Lin, H., Zhang, Z., Sun, H., Zhou, L.: New approach for environmental load charts of jack-ups considering actual preloading operations. J. Mar. Sci. Appl. **23**, 209–221 (2024)

Sun, J., Fang, H.: The analysis of crashworthiness and dissipation mechanism of novel floating composite honeycomb structure against ship-OWT collision. Ocean Eng. **287**, 115819 (2023)

Sun, S., Shen, H.H.: Simulation of pancake ice load on a circular cylinder in a wave and current field. Cold Reg. Sci. Technol. **78**, 31–39 (2012)

Svendsen, B.T, Tygesen, U.T., Kelly-Rosenville, J., Azam, N., Dollerup, N., Jønsson, B.V.: The use of digital solutions and structural health monitoring for integrity management of offshore structures. Report REN2021N00099-RAM-RP-00004, Petroleum Safety Authority (PSA) Norway (2022a)

Svendsen, B.T., Dollerup. N., Azam, N., Tygesen, U.T., Smith, A., Jønsson, B.V.: The Evaluation of Damage Detection and Structural Health Monitoring for the Integrity Management of Offshore Structures. Report REN2022N00396-RAM-RP-00001, Petroleum Safety Authority (PSA) Norway (2022b)

Syed Ahmad, S.Z.A., Abu Husain, M.K., Mohd Zaki, N.I., Mukhlas, N.A., Najafian, G.: An improved version of ETS-regression models in calculating the fixed offshore platform responses. J. Marine Sci. Eng. **10**, 1727 (2022)

Tan, J., An, C., Zhang, Y., Zhang, Y., Duan, Q., Duan, M.: Vibrational and stability analysis of internal flow-induced cold-water pipe for OTEC subjected to the clump weight under different boundary conditions. Ocean Eng. **296**, 116849 (2024)

Tang, D., et al.: Research on sampling rate selection of sensors in offshore platform SHM based on vibration. Appl. Ocean Res. **101**, 102192 (2020)

Tang, Y., Sun, S.L., Yang, R.S., Ren, H.L., Zhao, X., Jiao, J.L.: Nonlinear bending moments of an ultra large container ship in extreme waves based on a segmented model test. Ocean Eng. **243**, 110335 (2022)

Tang, Y., Sun, S.-L., Abbasnia, A., Guedes Soares, C., Ren, H.-L.: A fully nonlinear BEM-beam coupled solver for fluid-structure interactions of flexible ships in waves. J. Fluids Struct. **121**, 103922 (2023)

Tanttari, J., Hynninen, A.: Marine propulsor underwater radiated noise emission characterization using sensor arrays. Appl. Acoust. **221**, 110021 (2024)

Tavakoli, S., Huang, L., Azhari, F., Babanin, A.V.: Viscoelastic wave-ice interactions: a computational fluid-solid dynamic approach. J. Marine Sci. Eng. **10**, 1220 (2022)

Tay, Z.Y.: Artificial neural network framework for prediction of hydroelastic response of very large floating structure. Appl. Ocean Res. **139**, 103701 (2023)

Terweij, R.: Ship Energy Efficiency and Underwater Radiated Noise - Report 545-000-01, Rev. 3, Vard Marine Inc. (2023)

Ti, Z., Wang, Y., Song, Y.: Frequency-domain approach of aero-hydro-elastic response for offshore bottom-mounted slender structures under wind and wave. Ocean Eng. **260**, 111795 (2022)

Tian, H., Soltani, M.N., Nielsen, M.E.: Review of floating wind turbine damping technology. Ocean Eng. **278**, 114365 (2023)

Tian, Y., Shi, Y., Wu, Y., He, W., Liu, S., Tao, D.: Assessing mouse, trackball, touchscreen and leap motion in ship vibration conditions: a comparison of task performance, upper limb muscle activity and perceived fatigue and usability. Int. J. Ind. Ergon. **101**, 103585 (2024)

Toftaker, H., Bødal, E.F., Sperstad, I.B.: Joint optimization of preventive and condition-based maintenance for offshore wind farms. J. Phys: Conf. Ser. **2362**, 012041 (2022)

Torrielli, A., Giusti, A.: Uncertainties in the response of tower-like structures to wind gust buffeting. J. Wind Eng. Ind. Aerodyn. **236**, 105404 (2023)

Transport Canada: Interim Protocol for the Use of Southern B.C. Anchorages. Transport Canada - Interim Protocol (2018). https://tc.canada.ca/en/marine-transportation/ports-harbours-anchorages/interim-protocol-use-southern-bc-anchorages (last visited 31 Jan. 2025)

Truong, D.D., Cho, S.R., Huynh, V.V., Dang, X.P., Duong, H.D., Tran, T.H.: A study on dynamic response of steel plates under repeated impacts. In: Long, B.T., Kim, H.S., Ishizaki, K., Toan, N.D., Parinov, I.A., Kim, Y.H. (eds.) Proceedings of the International Conference on Advanced Mechanical Engineering, Automation, and Sustainable Development 2021 (AMAS2021). Lecture Notes in Mechanical Engineering. Springer, Cham (2022) van Zijl, C., Soal, K., Volkmar, R., Govers, Y., Böswald, M., Bekker, A.: The use of operational modal analysis and mode tracking for insight into polar vessel operations. Mar. Struct. **79**, 103043 (2021)

Verma, M., Nartu, M.K., Subbulakshmi, A.: Optimal TMD design for floating offshore wind Turbines considering model uncertainties and physical constraints. Ocean Eng. **243**, 110236 (2022)

Vieira, M., Henriques, E., Snyder, B., Reis, L.: Insights on the impact of structural health monitoring systems on the operation and maintenance of offshore wind support structures. Struct. Saf. **94**, 102154 (2022)

Villoslada, D., Santos, M., Tomás-Rodríguez, M.: TMD stroke limiting influence on barge-type floating wind Turbines. Ocean Eng. **248**, 110781 (2022)

von Pein, J., Lippert, T., Lippert, S., von Estorff, O.: Scaling laws for unmitigated pile driving: dependence of underwater noise on strike energy, pile diameter, ram weight, and water depth. Appl. Acoust. **198**, 108986 (2022)

Vuong, Q.D., et al.: Study on the variable speed diesel generator and effects on structure vibration behavior in the DC grid. Appl. Sci. **11**, 12049 (2021)

Wan, Y., Wang, L., Chen, D.: Analysis of vibration of in-service offshore platform induced by addition of large rotating machinery. Ship Building China **64**, 86–95 (2023)

Wang, B., Liu, Y., Zhang, J., Shi, W., Li, X., Li, Y.: Dynamic Analysis of Offshore Wind Turbines Subjected to the Combined Wind and Ice Loads Based on the Cohesive Element Method. Frontiers in Marine Science 9 (2022a)

Wang, C., et al.: Double-spring-piece structured triboelectric sensor for broadband vibration monitoring and warning. Mech. Syst. Signal Process. **166**, 108429 (2022)

Wang, L., Wang, L., Hong, Y.: Mitigation of side-to-side vibration of a 10MW monopile offshore wind Turbine under misaligned wind and wave conditions by an active torque control. J. Sound Vib. **574**, 118225 (2024)

Wang, M., Zhang, W.-Q., Wang, P.-G., Du, X.-L.: Multiple hazards vibration control of jacket offshore wind turbines equipped with amplifying damping transfer systems: winds, waves, and earthquakes. Ocean Eng. **285**, 115355 (2023)

Wang, R., Xu, X., Zou, Z., Huang, L., Tao, Y.: Dominant frequency extraction for operational underwater sound of offshore wind turbines using adaptive stochastic resonance. J. Marine Sci. Eng. **10**, 1517 (2022)

Wang, X.G., Zhu, L., He, X., Soares, C.G.: Numerical study on energy dissipation of ship plates under repeated impacts. Advances in Maritime Technology and Engineering. CRC Press (2024b)

Wang, Z., Hu, J., Li, F., Ning, X., Sun, S.: Fluid and Structure Coupling Analysis on Frequency Locking of an Elastic Hydrofoil. Appl. Ocean Res. **143**, 103853 (2024)

Wei, D., Chang, S., Bai, X., Pan, Y., Zhou, Y.: Investigation on vortex-induced motions of four-square columns in a square configuration considering galloping under different current velocity and incident angle. Ocean Eng. **266**, 112653 (2022)

Wu, J., Yin, D., Lie, H., Grytøyr, G., Hogg, B.F.: Investigation of VIV Responses of Slender Structures under Waves and Currents. 43rd International Conference on Ocean, Offshore and Arctic Engineering, vol. 6: Polar and Arctic Sciences and Technology; CFD, FSI, and AI, Singapore (2024a)

Wu, T., Zhang, C., Guo, X.: Dynamic responses of monopile offshore wind Turbines in cold sea regions: ice and aerodynamic loads with soil-structure interaction. Ocean Eng. **292**, 116536 (2024)

Wu, T., Zhang, M., Peng, R., Yu, H. (2024c). Numerical Study on Ice-Induced Vibration Control of a 10-MW Monopile Offshore Wind Turbine Using Tuned Mass Damper. Ocean Engineering 311, 118909.Xiao, D., Hao, Z., Zhou, T., Zhu, H: Experimental Studies on Vortex-Induced Vibration of a Piggyback Pipeline. Fluids **9**, 39 (2024)

Xiao, W., Hu, Y., Li, Y.: Ice impact response and energy dissipation characteristics of PVC foam core sandwich plates: Experimental and numerical study. Mar. Struct. **89**, 103407 (2023)

Xie, X., Ren, M., Zheng, H., Zhang, Z.: Active vibration control of a time-varying shafting system using an adaptive algorithm with online auxiliary filter estimation. J. Sound Vib. **513**, 116430 (2021)

Xu, L., Wang, J., Triantafyllou, M.S., Fan, D.: Predictions of multi-scale vortex-induced vibrations based on a multi-fidelity data assimilation method. Mar. Struct. **93**, 103539 (2024)

Xu, S., Wen, H., Liu, B., Guedes Soares, C.: Experimental and numerical analysis of dynamic failure of welded aluminium alloy plates under air blast loading. Ships Offshore Struct. **17**, 531–540 (2022)

Xu, W., He, Z., Zhai, L., Wang, E.: Vortex-induced vibration prediction of an inclined flexible cylinder based on machine learning methods. Ocean Eng. **282**, 114956 (2023)

Xu, Y., et al.: Effects of fuel composite additives on the vibration, wear and emission performances of diesel engines under hot engine tests. Eng. Fail. Anal. **160**, 108156 (2024)

Yan, Y., Zhang, S., Jin, X., Xu, L., Yan, X.: Applications of continuum fatigue risk monitoring in riser connectors system integrity management. Ocean Eng. **245**, 110540 (2022)

Yang, C., Li, R., Lü, L., Liu, Z., Jiang, Y., Xu, Z.: Vibration mechanism and noise characterization of offshore wind Turbines. Acoustics Australia **52**, 69–76 (2023)

Yang, S., Deng, X., Zhang, M., Xu, Y.: Effect of Wave Spectral Variability on the Dynamic Response of Offshore Wind Turbine Considering Soil-Pile-Structure Interaction. Ocean Eng. **267**, 113222 (2023)

Yao, S., Chen, Y., Sun, C., Zhao, N., Wang, Z., Zhang, D.: Dynamic response mechanism of thin-walled plate under confined and unconfined blast loads. J. Marine Sci. Eng. **12**, 224 (2024)

Ye, J., Zhang, D., Zheng, Z., Yang, W., Ke, L.: Numerical analysis of wake field and unsteady forces on submarine propeller with twisted rudders. Ocean Eng. **287**, 115798 (2023)

Ye, L., Shen, J., Tong, Z., Liu, Y.: Research on acoustic reconstruction methods of the hull vibration based on the limited vibration monitor data. Ocean Eng. **266**, 112886 (2022)

Yi, D., He, Z., Zhao, Y.: Structural response of ship-ice collision based on finite element method. In: Proceedings of the 2021 IEEE 11th Annual International Conference on CYBER Technology in Automation, Control, and Intelligent Systems (CYBER), 136–140 (2021)

Yin, C., Yu, H., Jin, Z., Liu, J., Huang, W., Wu, S.: Investigation of shock wave propagation and water cavitation in a water-filled double plate subjected to underwater blast. Int. J. Mech. Sci. **253**, 108400 (2023)

Yin, L., Qiao, D., Tang, G., Yan, J., Lu, L., Ou, J.: Dynamic responses analysis of crane-blade coupling system for the single blade installation of offshore wind Turbine considering the wind effect. Mar. Struct. **94**, 103570 (2024)

Yoon, Y.G., Han, D.-G., Choi, J.W.: Measurements of Underwater Operational Noise Caused by Offshore Wind Turbine Off the Southwest Coast of Korea. Frontiers in Marine Science 10 (2023)

Yousri, M., Jacobs, G., Neumann, S.: Effect of electrification on the quantitative reliability of an offshore crane winch in terms of drive-induced torque ripples. Modeling, Identification and Control: A Norwegian Research Bulletin **44**, 1–16 (2023)

Youssef, M., el Moctar, O., el Sheshtawy, H., Tödter, S., Schellin, T.E.: Passive flow control of vortex-induced vibrations of a low mass ratio circular cylinder oscillating in two degrees-of-freedom. Ocean Eng. **254**, 111366 (2022)

Yu, B., Liu, H., Fan, D., Xie, X.: Design of quasi-zero stiffness compliant shock isolator under strong shock excitation. Precis. Eng. **78**, 47–59 (2022)

Yu, Q., Cao, E.: Active control for marine engine room noise using an FXLMS algorithm. Sci. Program. **2022**, 1–11 (2022)

Yu, T., Wang, J., Liu, J., Liu, K.: Experimental and numerical simulation of the dynamic response of a stiffened panel suffering the impact of an ice indenter. Metals **12**(3), 505 (2022)

Yu, W., Peng, H., Pan, L., Dai, W., Qu, X., Ren, Z.: Performance assessment of high-rate GPS/BDS precise point positioning for vibration monitoring based on shaking table tests. Adv. Space Res. **69**, 2362–2375 (2022)

Yu, Z., Amdahl, J., Rypestøl, M., Cheng, Z.: Numerical modelling and dynamic response analysis of a 10 MW semi-submersible floating offshore wind turbine subjected to ship collision loads. Renewable Energy **184**, 677–699 (2022)

Zambon, A., Moro, L., Kennedy, A., Oldford, D.: Torsional vibrations of Polar-Class shaftlines: Correlating ice–propeller interaction torque to sea ice thickness. Ocean Eng. **267**, 113250 (2023)

Zambon, A., Moro, L., Oldford, D.: Impact of different characteristics of the ice–propeller interaction torque on the torsional vibration response of a Polar-Class shaftline. Ocean Eng. **266**, 112630 (2022)

Zamorano, M., Avila, D., Marichal, G.N., Castejon, C.: Data preprocessing for vibration analysis: application in indirect monitoring of 'ship centrifuge lube oil separation systems.' J. Marine Sci. Eng. **10**, 1199 (2022)

Zhang, C., Mousavi, A.A., Masri, S.F., Gholipour, G., Yan, K., Li, X.: Vibration feature extraction using signal processing techniques for structural health monitoring: a review. Mech. Syst. Signal Process. **177**, 109175 (2022)

Zhang, C., Wang, G., Wei, D., Tian, Y., Yang, L.: The research on the transverse vibration active control model of ship propulsion shaft with the active control force on the bearing support. Ocean Eng. **266**, 112722 (2022)

Zhang, H., et al.: Experimental study on mitigating vibration of floating offshore wind Turbine using tuned mass damper. Ocean Eng. **288**, 115974 (2023)

Zhang, J., Chen, N.-Z., Shen, C.: Fluid-structure interaction vibration characteristics of flexible riser transporting high-speed spiral flow in deep-sea mining. In: 42nd International Conference on Ocean, Offshore and Arctic Engineering, Vol. 7: CFD & FSI, Melbourne, Australia (2023b)

Zhang, J., Liang, X., Wang, B., You, P., Yun, L.: Vibration control and stroke evaluation of bi-directional bistable nonlinear energy sinks applied in monopile offshore wind Turbines during emergency shutdown. Appl. Ocean Res. **148**, 104044 (2024)

Zhang, J.W., Liang, X., Wang, L.Z., Wang, B.X., Wang, L.L.: The influence of tuned mass dampers on vibration control of monopile offshore wind Turbines under wind-wave loadings. Ocean Eng. **278**, 114394 (2023)

Zhang, P., et al.: Marine systems and equipment prognostics and health management: a systematic review from health condition monitoring to maintenance strategy. Machines **10**, 72 (2022)

Zhang, P., Wu, Z., Cui, C., Yao, R.: Mitigation of ice-induced vibration of offshore platform based on gated recurrent neural network. J. Marine Sci. Eng. **10**, 967 (2022)

Zhang, T., Wang, W., Li, X., Wang, B.: Vibration mitigation in offshore wind Turbine under combined wind-wave-earthquake loads using the tuned mass damper inerter. Renewable Energy **216**, 119050 (2023)

Zhang, W., Li, F., Ma, J., Ning, X., Sun, S., Hu, Y.: Fluid-structure interaction analysis of the rudder vibrations in propeller wake. Ocean Eng. **265**, 112673 (2022)

Zhang, X., et al.: Numerical investigation of the flow-induced vibration of a rotation pipe with internal flow. Ocean Eng. **294**, 116809 (2024)

Zhang, X., Li, Z., Li, Y., Wang, S., Li, W., Xiao, Y.: Multi-channel adaptive control of underwater structure radiated sound field based on bilateral secondary sources. In: 11th International Conference on Computer Science and Network Technology (ICCSNT), pp. 422–427. IEEE (2023e)

Zhang, X., Xu, Q., Zhang, M., Xie, Z.: Numerical prediction of cavitation fatigue life and hydrodynamic performance of marine propellers. J. Marine Sci. Eng. **12**, 74 (2023)

Zhang, X., et al.: Broadband vibration energy powered autonomous wireless frequency monitoring system based on triboelectric nanogenerators. Nano Energy **98**, 107209 (2022)

Zhang, Y., Aamodt, K.K., Kaynia, A.M.: Hysteretic damping model for laterally loaded piles. Mar. Struct. **76**, 102896 (2021)

Zhang, Y., Guo, J., Xie, Y., Xu, J.: Warship damage identification using mode curvature shapes method. Appl. Ocean Res. **129**, 103396 (2022)

Zhang, Y., Hu, Z., Ng, C., Jia, C., Jiang, Z.: Dynamic responses analysis of a 5 MW spar-type floating wind Turbine under accidental ship-impact scenario. Mar. Struct. **75**, 102885 (2021)

Zhang, Z., Høeg, C.: Inerter-enhanced tuned mass damper for vibration damping of floating offshore wind Turbines. Ocean Eng. **223**, 108663 (2021)

Zhao, J., et al.: Damping performance and its influencing factors of wind Turbine towers with constrained layer damping treatment. Structures **57**, 105322 (2023)

Zhao, X., Tan, M., Zhu, W., Shao, Y., Li, Y.: Study of vortex-induced vibration of a pipe-in-pipe system by using a wake oscillator model. J. Environ. Eng. **149**, 4023007 (2023)

Zhao, X.H.: Study of L-Extension Type Ship Vibration Isolation Pedestal Based on Impedance Mismatch Principle. Chinese Journal of Ship Research 16(3) (2021)

Zheng, J., He, Y., Zhao, M., Xia, J.: Dynamic response analysis of spherical pressure hull implosion inside adjacent underwater structure. Ocean Eng. **283**, 115169 (2023)

Zhong, J., Liu, D., Chi, S., Tu, Z., Zhong, S.: Vision-based fringe projection measurement system for radial vibration monitoring of rotating shafts. Mech. Syst. Signal Process. **181**, 109467 (2022)

Zhou, H., Kong, X., Wang, Y., Zheng, C., Pei, Z., Wu, W.: Dynamic response of hull girder subjected to combined underwater explosion and wave induced load. Ocean Eng. **235**, 109436 (2021)

Zhou, T., Guo, J.: Underwater noise reduction of offshore wind Turbine using compact circular liner. Appl. Energy **329**, 120271 (2023)

Zhu, B., Sun, C., Jahangiri, V.: Characterizing and mitigating ice-induced vibration of monopile offshore wind Turbines. Ocean Eng. **219**, 108406 (2021)

Zhu, H., Su, K., Luo, G.: Effective foundation damping prediction of monopile-supported offshore wind Turbines based on integrated fitting equation and PSO–SVM algorithm. Ocean Eng. **285**, 115306 (2023)

Zhu, H., Wu, J., Du, Z.: Application of regularization method with sparsity prior in riser vortex-induced vibration reconstruction. Ocean Eng. **287**, 115833 (2023)

Zhu, H., Zhang, X., Zhao, H., Xie, Y., Tang, T., Zhou, T.: Experimental investigation on the vortex-induced vibration of an inclined flexible pipe and the evaluation of the independence principle. Phys. Fluids **35**, 037115 (2023)

Zhu, J., Ren, B., Dong, P., Chen, W.: Vortex-induced vibrations of a free spanning submarine power cable. Ocean Eng. **272**, 113792 (2023)

Zou, M.-S., Tang, H.-C., Liu, S.-X.: Modeling and calculation of acoustic radiation of underwater stiffened cylindrical shells treated with local damping. Mar. Struct. **88**, 103366 (2023)

Zuo, H., Zhu, S.: Development of novel track nonlinear energy sinks for seismic performance improvement of offshore wind turbine towers. Mech. Syst. Signal Process. **172**, 108975 (2022)

Committee II.2: Dynamic Response 451

Open Access This chapter is licensed under the terms of the Creative Commons Attribution-NonCommercial-NoDerivatives 4.0 International License (http://creativecommons.org/licenses/by-nc-nd/4.0/), which permits any noncommercial use, sharing, distribution and reproduction in any medium or format, as long as you give appropriate credit to the original author(s) and the source, provide a link to the Creative Commons license and indicate if you modified the licensed material. You do not have permission under this license to share adapted material derived from this chapter or parts of it.

The images or other third party material in this chapter are included in the chapter's Creative Commons license, unless indicated otherwise in a credit line to the material. If material is not included in the chapter's Creative Commons license and your intended use is not permitted by statutory regulation or exceeds the permitted use, you will need to obtain permission directly from the copyright holder.

Committee III.1: Ultimate Strength

J. W. Ringsberg[1(✉)], L. Brubak[2], B.-Q. Chen[3], X. Chen[4], M. Chun[5], I. Darie[6], M. I. L. de Souza[7], M. Gaiotti[8], D. Georgiadis[9], M. Kõrgesaar[10], T. Magoga[11], A. M. Mohammad Zubar[12], K. Nahshon[13], T. Okafuji[14], M. Paredes[15], J. Romanoff[16], I. Schipperen[17], Y. Wang[18], A. Zamarin[19], and Z. Zhan[20]

[1] Chalmers University of Technology, Gothenburg, Sweden
jonas.ringsberg@chalmers.se
[2] DNV, Oslo, Norway
[3] Instituto Superior Técnico, Lisbon, Portugal
[4] Technical University of Denmark (DTU), Lyngby, Denmark
[5] KRISO, Daejeon, Republic of Korea
[6] University of Rostock, Rostock, Germany
[7] Rio de Janeiro, Brazil
[8] University of Genoa, Genoa, Italy
[9] Athens, Greece
[10] Tallinn University of Technology, Tallinn, Estonia
[11] Fishermans Bend, Australia
[12] Makassar, Indonesia
[13] Naval Surface Warfare Center, Carderock, USA
[14] Tokyo, Japan
[15] Texas A&M University, College Station, USA
[16] Aalto University, Espoo, Finland
[17] TNO, Delft, The Netherlands
[18] Southampton, UK
[19] University of Rijeka, Rijeka, Croatia
[20] Shanghai, China

Committee Mandate. Concern for the collapse behaviour of ships and offshore structures and their structural components under ultimate conditions. Uncertainties in strength assessment shall be highlighted. Attention shall be given to the influence of load combinations, fabrication-induced imperfections, life-cycle effects, damage, and user approach. Consideration shall be given to the practical application of methods.

Keywords: Buckling · Collapse · Corrosion · Fabrication-induced imperfections · In-service damage and degradation · Load-carrying capacity · Offshore structures · Rules and standards · Ships · Ship structures · Ultimate limit states · Ultimate strength · Uncertainty

1 Introduction

Ultimate strength analysis is one of the most fundamental and critical assessments performed for ships or offshore structures and is required to ensure safe and reliable marine structures. This analysis is initially performed in the design phase to ensure that expected load conditions do not exceed the maximum load-bearing capacity of a structure. This assessment may also be revisited post-design if structural capacity diminishes due to damage from accidents, for example, collisions or groundings, or in-service degradation, including corrosion and fatigue. There is growing interest in predicting diminished load-carrying capacity during normal to extreme operating conditions and understanding its impact on global load-bearing capacity.

This report presents a summary and discussion of an extensive literature review of work published primarily from 2021 to 2024 that was performed by members of ISSC Committee III.1. The committee's mandate, which was unchanged from that of ISSC 2022 (ISSC, 2022a), guided this work. The effort began with a careful review of the three most recent committee reports from 2015, 2018, and 2022 to identify subjects that had been unaddressed or minimally discussed and to review the former committees' recommendations for future work, as well as discussions and questions raised during ISSC 2022 congress sessions. Subject areas covered extensively in the past reports have deliberately received less attention in this report. Subjects with low or moderate levels of progress during the committee's mandate period are specifically highlighted.

The report is organized into seven sections. Section 2 presents updates to fundamental topics related to ultimate strength analysis, including recent developments in computational tools and examples of experimental procedures and investigations. Uncertainties related to material and fabrication and how they affect the ultimate strength are presented in Section 3, including in-service degradation effects, such as corrosion and mechanical damage, and life-cycle management analysis. Sections 4 and 5 focus on the ultimate strength aspects particular to ships and offshore structures, respectively. They describe relevant load cases, structural elements, structural systems, and associated rules and regulations. Section 6 presents the report from a benchmark study in which the committee and additional invited participants simulated, analysed, and predicted in "blind" the failure progression of a transversely stiffened thin-plated structure. As in the benchmark study presented in the 2022 committee report (ISSC, 2022a), the current benchmark was also divided into three phases. Finally, in Section 7, the major conclusions, trends, and recommendations of the committee are presented.

2 Fundamentals

This section collates the main definitions and assumptions, first principles and theories, and methods, tools, and approaches for the assessment of ultimate strength for both ships and offshore structures. Ultimate strength is defined as the maximum resistance that a material, structural element, or larger structure can carry when exposed to external loading with a certain loading pattern. Ultimate strength is commonly defined as an ultimate stress, ultimate force, or ultimate moment. Examples include the maximum moment a hull girder can experience, the maximum axial load of a monopile, and the maximum pressure a subsea structure can endure before hydrostatic collapse. An excellent review of this definition for the static and dynamic situations in a maritime context is provided in a recent article by Jagite et al. (2022) who discuss the roles of the inertia and strain rate as dynamic effects that differentiate the two situations. The inertia rate primarily affects the structural response, i.e., the amplitude of the internal load, which can differ from the applied external loading, unlike in the static case. In contrast, the strain rate affects the material level responses through changes in yield strength. Nevertheless, most current analyses are conducted statically, neglecting dynamic effects (inertia, damping, and strain rate).

2.1 Definitions, Assumptions, and Uncertainties

It is often desirable for the structural analyst to not only know the ultimate strength but also the shape of the deformation-loading curve, where deformation is quantified as either strain, displacement, or curvature. This provides insight into the stiffness characteristics of the structure as well as the energy absorbed in both elastic and inelastic deformation. In most cases, monotonic loading is considered. However, in other cases, loading path dependency may be important. This issue was also highlighted by the Official Discusser of the previous committee (ISSC, 2022b) and is accounted for in this report.

Loading on ships and offshore structural systems consists of environmental loads during normal operations, such as waves, currents, wind, and ice, as well as damage events such as collisions and groundings, dropped objects, fires, and explosions. Structural systems and sub-systems experience load effects at different times and in different ways. Quite often, this modelling is simplified from dynamic to static, in which case these phase differences of load actions are lost, along with those of inertia and strain rate effects. For example, global wave-induced hull-girder bending moments typically manifest as tensile and compressive loads at the panel level, and augmented local hydrostatic and dynamic pressures. Other loads are highly localised, such as ice impact. Wave-induced loads tend to transfer to maximum tensile and compressive stresses at the extreme "fibres" of thin-walled structures, away from the neutral axis, and their application is often based on deformation (curvature) for which the internal reactions in terms of stresses are integrated to the section moments acting on the hull girder. Wave-induced loads tend to transfer to maximum tensile and compressive stresses at the extreme "fibres" of thin-walled structures, away from the neutral axis. In comparison, ice loads tend to cause the greatest load effects close to the neutral axis of the ship hull girder, generating stronger interactions with the hull girder shear loads. In contrast, offshore applications, such as windmills, may experience interactions between local ice-induced loads and the maximum tensile and compressive loads caused by primary responses of the entire structural system. These fundamentally different load-carrying mechanisms require careful consideration in ultimate strength assessments. In addition, Fig. 1 presents the ultimate strength paths of the (a) material, (b) the sub-structure (e.g., stiffened panel), and (c) the structural system (e.g., the hull girder); and in the case of ships and offshore structures these three different length scales are coupled. Moving upwards across these scales, the effects at lower scales are "smeared" (or *averaged*) over the volume of the structure. Conversely, moving downwards in the length scales concentrates or *localises* the damage.

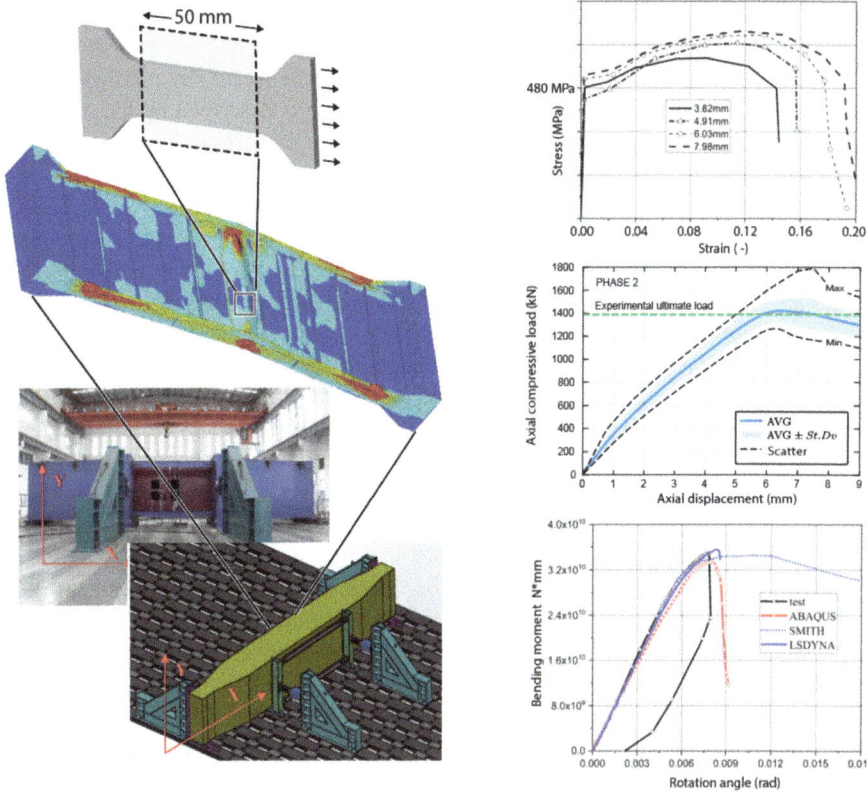

Fig. 1. Illustration of the ultimate strength progression across three hierarchical levels: (a) the material level, (b) the sub-structure level (e.g., a stiffened panel), and (c) the structural system level (e.g., the hull girder). In ships and offshore structures, these length scales are interconnected. Adopted from Ringsberg et al. (2021) and Zhao et al. (2022).

Structural elements or sub-structures are characterised by load-shortening, pressure-displacement, or other similar relationships that are analogous to the stress-strain curve at the material level. The shape and characteristics of these curves are affected by the confluence of material behaviour, structural geometry, initial imperfections, residual stresses, and flaws, such as cracks induced by manufacturing or corrosion induced by the maritime environment, and operating conditions. In the analysis, the question is whether the ultimate strength should be determined either using (a) nominal, (b) actual (which also includes life-cycle effects such as dents) or (c) anticipated values of parameters to determine the future ultimate strength or the residual strength, for example, after an accident. The use of ductile materials does not preclude quasi-brittle failure where structures fail under relatively low tensile loads. This occurs in cases of high levels of local corrosion/degradation as well as highly undermatched weld joints that appear in extruded aluminium structures. An additional feature at this level is the notable observable differences in structural behaviour when the direction of loading is changed.

Under tension, straightening leads to fracture, whereas under compression, it results in progressive buckling failure.

Larger structural systems consist of a series of interconnected structural elements. In the case of the hull girder of a ship, ultimate strength is described by the moment-curvature relationship of a cross-section. This is affected by the resistance of the structural system components, redundancy of the load-carrying mechanism, and interaction with environments, such as waves and ice, seafloor in groundings, and other structures in collisions. Therefore, in this report, attention is given to the influence of load combinations, fabrication-induced imperfections, life-cycle effects, damage from accidents, and operational practice on the estimation of resistance and ultimate strength.

The most important mathematical parameters that influence the shape of these load-deformation relationships are the initial and tangent stiffnesses of the curves (these must equate initially; for example Young's modulus in materials, and bending stiffness in beams), as well as the area below the curve, which is generally related to the energy or energy per volume of the material or structural system when loaded statically. The design point is typically much lower than the ultimate strength, and safety factors are used to create a buffer between the design values and the ultimate strength. These safety factors are usually set based on consideration of the behaviour of the material or structure beyond the ultimate strength. Due to variations in plastic behaviour, safety factors can substantially differ in ductile and brittle materials. Defining the ultimate strength of a complete structure through full-scale experimentation over its lifetime is practically infeasible; therefore, computational methods and various levels of simplification are necessary, which influence the failure sequence from the material level to the hull girder level. This issue is discussed further in Sect. 3.

2.1.1 Assumptions

All structures are fabricated based on a design that specifies materials, fabrication processes, thicknesses, dimensions, and geometrical locations for all structural elements. In the design phase, ultimate strength assessment must utilise either assumed nominal characteristics or historical ranges for deviations from nominal values, such as actual versus nominal material properties, and nominal dimensions versus dimensional tolerances. At present, the in-service condition of ships and offshore structures is minimally tracked throughout the lifecycle from the time of manufacturing to the end-of-life. There are currently efforts to accomplish this via the use of digital twins in which an analysable digital version of a structure in its existing condition is maintained allowing designers, shipbuilders, operators, and authorities to continuously monitor the real-time condition of the structural system until dismantling. However, in practice, this is a very difficult and expensive endeavour for most ships and offshore structures, as the collection and maintenance of required actual data are cumbersome within complex structural systems. Due to this, estimates are often exploited.

As a result, ultimate strength analysis of in-service structures requires assumptions about the current state of a structure, which creates uncertainties in ultimate strength predictions; these are reviewed in detail in Sect. 3. Even in instances where the current condition is described, such as the amount of wastage due to corrosion, it is rarely in the form of full-field measurements. Therefore, assumptions must be made regarding conditions in unmeasured locations. The load history of the structure may also involve numerous assumptions. Although the instantaneous load from cargo can generally be determined with high accuracy, several assumptions are required to estimate environmental loads, such as those from waves, wind, and ice, during ultimate strength simulations. These will manifest in the ship dynamics and load distribution as local hydrodynamic pressures and hull girder loads and their phase differences; see the following sections and other committee works on direct methods for loads. If the load history has created significant fatigue damage or accumulation of plastic damage in the structure, the ultimate strength will be affected, and the way plastic damage accumulates at the material level has a substantial influence on the ultimate strength of the panel (Cui and Ding, 2022). Accidents, such as collisions, groundings and explosions, lead to sudden reduction of the capacity of the structural system. Several investigations have also focused on the dynamics of ultimate strength, i.e., where the load is not introduced quasi-statically as has been done traditionally. Load and strength may also interact dynamically.

Ultimate strength simulations require assumptions within the computational models at the material, sub-structural and structural system levels. In the context of finite element (FE) analysis, solid elements produce the most accurate mathematical descriptions of material behaviour. However, shell and beam elements are most commonly used due to the impracticality of modelling a ship or offshore structure with solid elements as a result of the number of required elements, small timesteps, and resulting computational complexity. Recent research efforts have focused on bridging beam and shell structural elements with solid elements that can capture realistic material behaviour, including multi-scale and multi-physics analyses in which the material microstructure and mechanisms are modelled and simulated explicitly and not solely from material tests. The aerospace industry has developed the NASMAT platform (Arnold et al., 2023), which aims to create an integrated framework for multi-scale and multi-physics materials and structural analyses within a seamless modelling package. This platform builds on advances in the modelling of multi-scale structures, allowing two-way coupling between material properties and structural behaviour, thereby enabling updates to material properties based on load actions at the structural level. Similar modelling approaches have been explored in maritime applications, such as multi-scale and multi-physics modelling to determine the effects of corrosion damage on ultimate strength damage (Ilman et al., 2021) by directly accounting for the interaction between corrosion and mechanical stresses. Additionally, research by Peng et al. (2022) demonstrated how the combined effects of periodic tidal movement and the harsh environment created by the tide, seawater, air temperature, and sea temperature can be incorporated into models that can analyse the degradation of marine reinforced concrete in cold regions resulting in significant stiffness reduction after only tens of load cycles. Structural elements in these models are represented by shell, plate, and beam formulations, with decreasing mathematical complexity moving from the shell towards the beams. Materials are typically considered anisotropic allowing the use of the same elements to model both composites and metals. The Carrera unified formulation (CUF) allows systematic selection of the shell, plate, and beam kinematics as a function of the thickness coordinate of the structural element, and "Best Theory Diagrams" (see Fig. 2) are used to identify the optimal kinematic assumptions for specific structural problems, such as boundary and loading conditions or material distribution throughout the thickness (Petrolo and Carrera, 2019; Petrolo and Iannotti, 2023). Recent advancements in CUF also extend to non-classical continuum mechanics, allowing material points to possess finite sizes with defined microstructures (Carrera and Zozulya, 2020). These are important in modelling the strain fields, which may differ from those generally adopted (e.g., linear); the issue is especially important in composite structures and near the free edges. In hull girder modelling, FE models are widely used due to their flexibility and accuracy. However, due to computational costs, the application of model reduction techniques, such as the Smith method, to beam models is often preferred in practical design work, e.g., the Common Structural Rules (CSR).

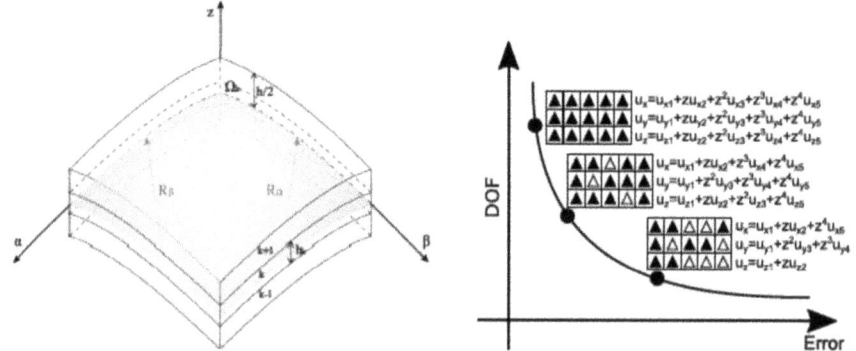

Fig. 2. Derivation of the Best Theory Diagram of a shell element based on the 4th order polynomial expansion of the in-plane displacement field throughout the thickness coordinate (Petrolo and Carrera, 2019).

2.1.2 Uncertainties

In this report, the classical definitions of uncertainty are used (epistemic and aleatoric), both individually and in combination. Section 2, which addresses the fundamentals, focuses primarily on epistemic uncertainty, while Sect. 3 considers aleatoric uncertainty. The choice is made this way as the fundamentals section focuses on the modelling aspects, while Sect. 3 on life-cycle aspects. The effects of uncertainty on hull girder and offshore structural strength are of particular interest, as demonstrated by a study on the impact of uncertainty on load-shortening curves of a panel as part of the ISSC 2022 Ultimate Strength committee's benchmark study (ISSC, 2022a). Input data accuracy, particularly in the design phase, significantly affects results, with uncertainties arising from geometry, material properties, load data, and modelling strategies. Woloszyk *et al.* (2021a) investigated the use of photogrammetry to generate meshable geometry for conducting ultimate strength calculations using the true shape of a stiffened plate of three different thicknesses. Comparisons with Smith's (1977) closed-form formulae showed similar ultimate strength values for both approaches, although the use of the measured shape moderately influenced pre- and post-ultimate load-displacement behaviour, with this effect increasing for thinner sections, as shown in Fig. 3.

In their work, Georgiadis and Samuelides (2021a) applied random field theory to describe stochastic geometric imperfections on plates and their effects on hull girder strength, modelling initial imperfections using mean values and variances, and using artificial neural networks (ANNs) to reduce the number of non-linear FE analyses. While an ANN is trained on a dataset to provide results for arbitrary input, applying Bayesian statistical interference provides an update of parameters of the probability density functions based on a number of observations. In contrast, Chen et al. (2023) showed that treating initial imperfections solely as sinusoidal distortions was insufficient, and longitudinal stiffener misalignment (LSM) due to welding must also be considered. Their work highlighted the importance of symmetric and antisymmetric modes and emphasised that LSM accelerates the reduction of load-carrying capacity.

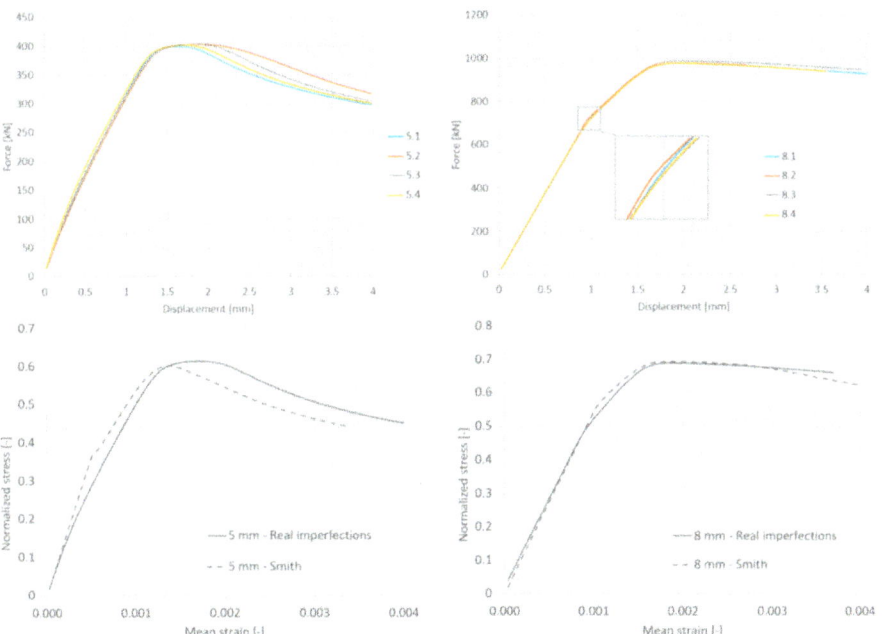

Fig. 3. Comparison of the load-displacement (upper figures), and the normalized stress average strain (lower figures) of 5 mm and 8 mm thick panels with flat bar stiffeners as predicted using the true geometry and the imperfection equations by Smith (Woloszyk *et al.*, 2021a).

The quantification of model uncertainty for hull girder ultimate strength prediction is another important aspect that has been raised in previous ISSC reports (see, e.g., ISSC, 2018a, 2022a). As prescribed by the CSR, the primary methods for assessing hull collapse are the Smith method and FE analysis. Li et al. (2021b) proposed a probabilistic approach to evaluate the uncertainty in estimates of hull girder ultimate strength resulting from variations in the load-end shortening curves (LSCs) by modelling the critical characteristics of the LSCs in probabilistic terms. The probability distributions of the critical LSCs characteristics of the stiffened panels were developed based on a dataset generated using empirical formulae and FE analysis results. The impact on the variability of the ultimate strength of the hull girder was assessed for four merchant ships and four naval vessels. A key finding from this study was that the mean ultimate strength predicted for all case study ships in hogging closely matched the CSR baseline, whereas for sagging the mean was approximately 95–97% of the CSR baseline. Georgiadis *et al.* (2023b) applied Bayesian statistical inference to estimate the uncertainty when using the Smith method. Engineering judgment and FE analysis data were formally combined through Bayesian analysis, while the uncertainty in the FE analysis results was explicitly accounted for through the likelihood formulation. Their work produced new probabilistic models for container ships to estimate uncertainty when using the Smith method. The authors applied the same approach in their study of double-hull oil tankers (Georgiadis and Samuelides, 2023a).

While design calculations must use specified minimum material values, analyses seeking to capture true ultimate strength behaviour require a decision regarding the choice of material properties. The choice between using the design or actual values is a key consideration. The actual properties of normal-strength steel typically exceed those provided by material manufacturers, implying that the ultimate strength of the actual structure could be higher than calculated. However, this safety margin tends to decrease for higher-strength steels. The same consideration applies to the assumed state of the structure, such as whether it is considered damaged or undamaged. Do et al. (2021) investigated the effects of local damage in stiffened cylinders after impact loads. In their simulations, the impact analysis defined geometric changes due to impact, including contact and strain rate effects. The resulting new nodal positions were mapped onto the structure for buckling analysis, thereby accounting for the effects of residual stresses. At this stage, the material was adjusted to elastic-perfectly plastic, neglecting material memory effects. This type of numerical analysis showed a reasonably good correlation with the experimental results, although some differences were observed in the pre- and post-ultimate strength behaviour between load (pressure) and the axial shortening (see Fig. 4).

Committee III.1: Ultimate Strength   461

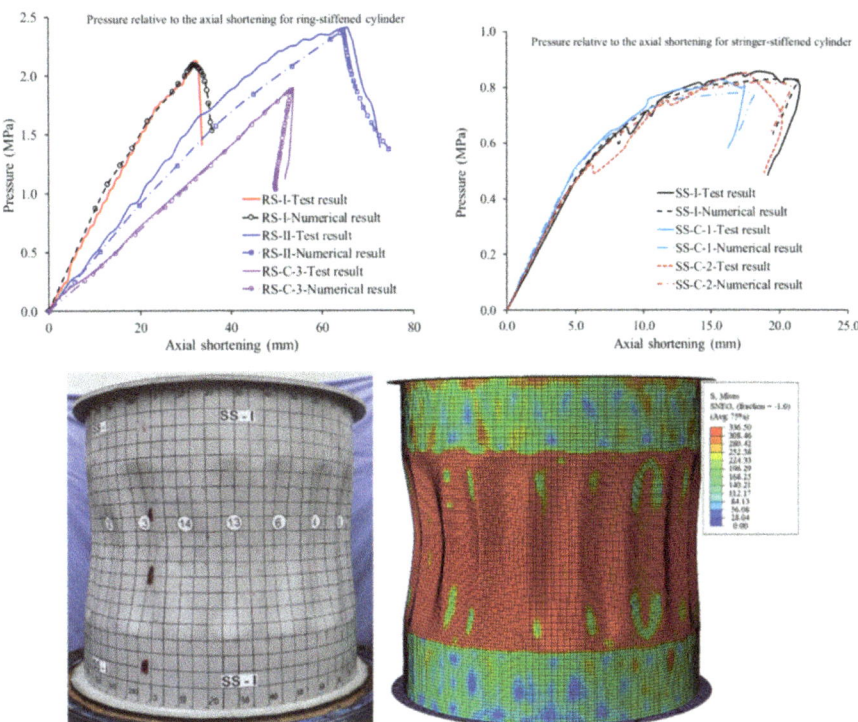

Fig. 4. Comparison of simulations and experiments on pre-damaged cylindrical shells (Do et al., 2021).

Kuznecovs et al. (2021) used FE analyses to investigate the effect of ship collisions on ultimate strength, predicting residual strength using the Smith method for tankers, with and without corrosion degradation (both general and pitting corrosion) included. They demonstrated the effects of collision damages along the hull girder length by utilising 3D demand-capacity plots to describe the relationship between horizontal and vertical bending moments at different sections along the prismatic body of the ship.

Li et al. (2021a) used a shell-element-based FE analysis to study the individual, combined, and superimposed effects of pitting corrosion and cracks in various orientations on the ultimate strength of box girders. Their findings confirmed the expected reduction in ultimate moment due to both types of damage but also highlighted the important role of structural integrity in ensuring that the connections of stiffeners and plating were appropriate for carrying loads and supporting the plate against buckling. Investigations by Piscopo and Scamardella (2023) utilised Monte Carlo simulations to estimate the statistical properties of plate ultimate strength of pitted simply supported plates. Similarly, the importance of stiffener integrity was demonstrated by Li et al. (2021c) who investigated the influence of residual stresses on the ultimate strength of ship panels under compression and noted that the residual stresses in the stiffeners may lead to beam-column type failure and thus a significant reduction in ultimate strength due to loss of support for the plate. Zhong and Wang (2021a) investigated the ultimate strength of laser-welded web-core sandwich plates, demonstrating the combined effects of geometrical and material non-linearities on the resulting load-end-shortening path when compared to models solely considering geometrical non-linearity. Their results indicate that highly localized plasticity affects the onset of local buckling, which subsequently leads to overall buckling of the panel.

A quasi-static approach is typically assumed to be appropriate for loading; for instance, experiments and simulations are conducted such that strain rate and kinetic energy effects are negligible. Kong et al. (2021) investigated the dynamic buckling of a deck grillage with an opening subjected to in-plane loads, using both experimental and numerical methods. Their findings indicated that the dynamic buckling mode differed significantly from the failure mode of the structure subjected to quasi-static in-plane compressive load. Specifically, they observed that vertical displacements at the deck edge and axial displacements increased as impact frequency decreased. Their study also demonstrated the effects of dynamic buckling on hull girder failure. In their study, Jagite and Bigot (2023) further explored the evaluation of the ultimate strength of a hull girder when loaded dynamically by wave loads.

When choosing between solid, shell, plate, and beam structural models, as well as between analytical and numerical solutions, the accuracy of the stress field versus computational cost is one of the primary considerations in ultimate strength analyses. It is widely known that material failure is dependent on stress triaxiality, and single-layer shell and line-type beam elements cannot model stresses in all three directions directly. Approaches exist to approximate these omitted stresses in post-processing with certain assumptions. However, uncertainties may be introduced particularly in problems where damage localizes quickly, affecting the load-carrying mechanism of the structural system due to interdependent length scales. Another significant source of uncertainty lies in the level of idealization of the loads and boundary conditions used in simulations and their practical realism. When verifying and validating models, uncertainties can arise from the reproducibility of results under seemingly similar conditions, as well as from systematic and random errors in experimental data. This aspect is discussed in detail in the benchmark study presented in Sect. 6.

Wang et al. (2022b) investigated the buckling of aluminium extruded panels using FE analysis with both static and dynamic solvers. They discovered that the static solver, which is commonly used in buckling simulations, may not accurately model the experimentally observed bucking and post-buckling behaviour, as it neglects inertia forces. These forces were found to be crucial in driving the buckling process forward, especially when the stiffener-tripping mode occurred. Therefore, they recommended using the dynamic solver, as the inclusion of inertia effects improves numerical stability over time steps, thus avoiding severe overprediction of the ultimate load. Additionally, they noted that the residual stress model should be carefully selected since including the residual stress may strengthen the panels, resulting in an overestimation of the ultimate load. Consequently, they advised against using excessive residual stress levels. As noted by the ISSC 2022 Ultimate Strength committee (ISSC, 2022a), the focus of the post-processing of theoretically or experimentally obtained data is often the actual ultimate strength rather than residual capacity, resulting in an emphasis on the energy of the structural system beyond the ultimate strength. Therefore, in many comparative studies, the focus is not on the tail parts of the load-end-shortening or moment-curvature curves, and thus the uncertainties are fundamentally related to the fact that these parts of the curves are overlooked or the communication in scientific papers is not highlighting this region carefully enough. This concern will be further examined in the benchmark study in Sect. 6. Advances in technology, such as digital image correlation (DIC), provide a valuable opportunity to address these gaps. With access to large datasets from both numerical and experimental sources, techniques such as DIC, which enables precise strain field measurements on actual structures, can be powerful. These approaches may enhance our understanding of uncertainties not only in the context of ultimate strength but also in the case of residual capacity following ultimate load. Such approaches could also be expected to enable a better explanation of variability across repeated experiments.

2.2 Computational Tools

2.2.1 Numerical Methods

Material modelling serves as input for all FE simulations, regardless of the selected element type. The manner in which load effects are incorporated into plasticity models influences the computational prediction of ultimate strength. The stress-strain curve consists of several regions: the linear-elastic region, the yielding plateau, the strain-hardening region modelling (which models plasticity), and the softening region (which models the reduction of material resistance to zero as strain increases). This monotonic representation of material stress-strain behaviour is a simplification that neglects the accumulation of damage due to cyclic loading, such as high- and low-cycle fatigue. These effects are always present in ships and offshore structures, but their effects over the life cycle of the modelled structure are difficult to reliably assess in practice. How these aspects are modelled will influence the failure behaviour of the material, influencing the structural response at larger scales. It is recognised that strain rate and temperature affect material properties and ultimately the capacity of the structural system. As noted, multi-scale and multi-physics approaches to materials and structural modelling (e.g., with computational tools like NASMAT) enable these complex simulations. Structures composed of different materials have unique characteristics that must be considered in numerical assessments of ultimate strength.

Cui and Ding (2022) used FE analysis to study the effect of various material models on the ultimate strength and fracture failure of hull-stiffened panels made from *steel* under cyclic loads. They varied the material modelling strategy among elastic-perfectly plastic, isotropic hardening, and Chaboche models (i.e., combined isotropic and kinematic hardening accounting for both monotonic and cyclic effects) to examine plastic accumulation during cyclic loading and its impact on ultimate strength under compression at the structural level and combined compression-tension at the material level. Plastic accumulation was highest in the initial cycles and decreased rapidly with constant amplitude loading in subsequent cycles, particularly in the weakest region of the structure where fracture typically occurs, and at higher load levels. Both stiffness and ultimate strength were reduced after unloading. The choice of hardening model significantly influenced results, with the Chaboche model providing the most stable accumulation of plastic strain. When plastic strain accumulation reached a specific threshold, it did not increase further, and the point of final failure was governed by low-cycle fatigue. One key aspect in this type of plasticity modelling is the determination of material parameters experimentally, see for example Shi et al. (2013) and Ma et al. (2018), and the effects of the material model on the hull-girder ultimate strength under several load cycles, see for example, Preventas et al. (2021). In a series of studies, Barsotti and Gaiotti (2022; 2023a,b) analysed the cumulative process of buckling deformation on a stiffened panel subjected to uniaxial cyclic loading. They demonstrated that pure plane stress modelling with shell elements is insufficient to accurately capture the accumulation of plastic strain. Instead, solid elements were required in areas with regions of high plastic strain, such as stiffener intersections, where constraint effects resulted in a triaxial stress state. Kinematic compatibility between shell and solid elements was modelled with rigid links, and they applied a plastic bilinear model that was in accordance with the isotropic hardening law. Over one thousand load cycles with 90% of the structure's ultimate monotonic load, deformation accumulation rates were observed to increase with the number of load cycles.

Heat-affected zones (HAZ) significantly influence the welding connections of structures made from aluminium and high-strength steels; thus, their modelling should not be neglected when ultimate strength is to be assessed, even when it typically complicates the numerical modelling substantially. Wang *et al.* (2022b; 2024) investigated the effects of the HAZ, residual stresses, and initial imperfections on the numerical modelling of welded aluminium panels and considered inertia effects and compared the results from static and dynamic solvers. They concluded that inertia effects are important as they stabilise the numerical solution, particularly in the case of the rapid collapse of stiffeners. After comparing their results with those obtained through approaches guided by IACS and the computational tool developed by DNV (the panel ultimate limit state, PULS), they concluded that these methods yielded satisfactory results in cases where welding effects can be neglected. In *reinforced concrete structures*, durability degradation modelling is essential due to corrosive ion penetration and freeze-thaw cycles (FTCs), as reported by Peng et al. (2022). Modelling damage in concrete requires accounting for temperature distribution, pore water crystallization laws, and critical saturation under seawater FTCs to predict durability degradation accurately.

In *composite* materials and structures, material performance is strongly correlated with the structural model (beams, plates, or shells). Peeters et al. (2022) demonstrated the impact of high-fidelity kinematics on predictions of the buckling strength of composite aeronautical stiffened panels, where the composite layup was designed to trigger multiple buckling and failure modes within 10% of the lowest buckling mode. The Carrera unified formulation (CUF) was used, enabling natural activation of the relevant high-order kinematic terms without significantly increasing computational load. The model was validated through repeated experiments, and experimental and modelled results showed excellent agreement. Augello et al. (2022) expanded on this work, applying a refined equivalent single-layer model in which the critical point of buckling and post-buckling behaviour was modelled with excellent accuracy by using fewer degrees of freedom (DOFs). CUF enabled the development of equivalent 1D models, incorporating complicated stacking sequences through refined expansion functions (Lagrange polynomials). In cases where the shear transfer between material layers is incomplete or layers are allowed to deflect differently to different extents, these kinematical changes become very important when assessing deformations, strains, and stresses. Insulated Glass Units (IGU) represent an extreme case in this respect. Heiskari et al. (2022; 2023) conducted numerical investigations on insulated *glass* units used on passenger ships; these units consist of a hermetically sealed void between two glass panes and their study focused on the impact load-sharing due to the void and the influence of von Karman strains on glass pane thickness. They assumed no influence of structural strength on glass strength and linear-elastic material behaviour. Results indicated a significant increase in predicted ultimate strength. These theoretical findings are currently being further explored in an experimental study, the results of which are expected to be published soon. The main idea is to learn about shear transfer, load-sharing effects, and geometrical non-linearities, and how these interact when glass structures are loaded to their ultimate limit.

2.2.2 Reduced-Order and Analytical Models

Dimension reduction is often necessary in engineering to reduce the computational effort required for assessments of ultimate limit state. In the ISSC 2022 Ultimate Strength committee report (ISSC, 2022a), dimension reduction strategies were characterised as presented in Fig. 5. At the highest level, reduction involves simplifying the physical problem size, for example, by isolating a panel from a larger structural assembly (such as a hull girder) and limiting the analysis to that panel. The second level of reduction involves reducing the degrees of freedom, often achieved by applying kinematic assumptions inherent in beam, plate, and shell theories. In the third level, problem size may be reduced through closed-form solutions by using semi-analytical, multi-scale numerical methods, or advanced structure-specific methods, such as the Smith method, which uses numerical methods with reduced degrees of freedom. Machine learning (ML) methods are emerging as a fourth model reduction method that can be used to explore the design space effectively without requiring exhaustive simulations or experiments, enabling faster predictions of structural behaviour. In the future, when computational resources increase, it is worth considering whether these computationally cheaper models with substantial idealisations and simplifications are used over more exact models with fewer simplifications, but higher computational costs. Such choices require substantial theoretical knowledge, practical experience, and skills from the analyst in balancing between accuracy and speed.

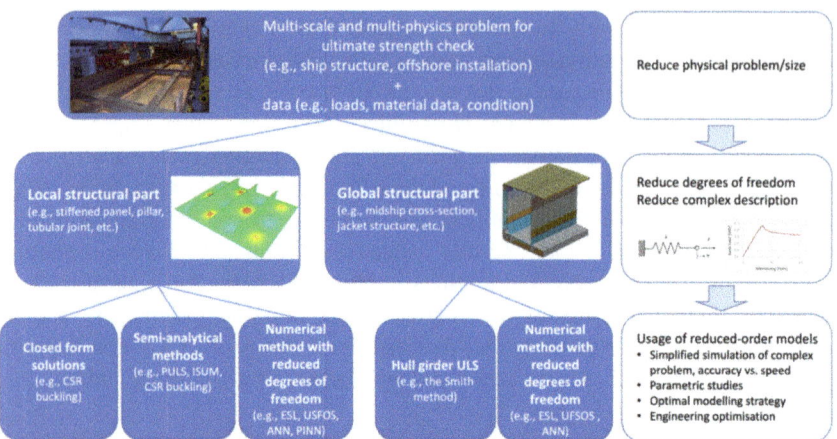

Fig. 5. Dimension reduction in ultimate strength assessment (updated from ISSC, 2022a).

Ma et al. (2022) extended the idealised structural unit method (ISUM) to the analysis of the dynamics of the ultimate strength of stiffened plates. In their approach, the Newmark method to solve the equations of motion, and the ISUM model comprised an ISUM plate and ISUM beam-column elements. The static solution was expanded to include mass and damping matrices (Rayleigh damping). They concluded that the effects of the strain rate and inertia of the structure increased ultimate strength. The computational tool for panel ultimate limit state developed by DNV (PULS, see Byklum and Amdahl, 2002), which is based on analytical, formulations, has been used as the basis for recent research seeking to incorporate the FE method. Assuming first-order shear deformation theory (FSDT) for plates, Putranto and Kõrgesaar (2021) and Putranto et al. (2021; 2022a,2022b) employed parametric FE analysis to derive non-linear stiffness models for unit cells in ABAQUS, which serve as input for non-linear FSDT plates with orthotropic material definitions. This approach, which accounts for both geometric and material non-linearities, demonstrated good agreement with high-fidelity 3D FE models of the actual geometry, both at the sub-structural level (e.g., stiffened panels) and at the structural system level (e.g., ships). However, recent investigations have highlighted challenges in addressing loads in the weaker direction, i.e., the direction normal to the stiffeners (Kõrgesaar et al., 2023). The prevailing differential equations are the root cause since they cannot distinguish between antisymmetric and symmetric out-of-plane shear strains, which negatively affect buckling strength (Romanoff et al., 2020). In FE analyses, this limitation can lead to mesh size-dependent problems in defining the bifurcation buckling load, which serves as input for the non-linear solution phase. The benchmark study presented in Sect. 6 will address this topic in further detail.

Ishibashi et al. (2024) developed a simplified method grounded in elastic large deflection plate theory to estimate the ultimate strength of simply supported rectangular plates under combined biaxial and shear loads, without requiring numerical iterations. Yielding was verified analytically in pre-defined locations identified through extensive non-linear FE analyses of steel plates representative of typical ship structure designs. While generally accurate, the method tended to overestimate the ultimate strength in thinner plates subjected to certain tension/compression loading scenarios. Similarly, Li and Chen (2023a) developed explicit empirical formulae to estimate longitudinal and transverse ultimate strengths, based on non-linear FE analyses. These formulae share a common general format, providing practical tools for structural assessment.

Li et al. (2022a) estimated the ultimate strength of cracked plates using finite element analysis and ANNs. Their work focused on parametrising the problem through the load-end-shortening curve and key input parameters, such as plate slenderness, crack orientation, length and number of cracks. The results indicated that the ultimate strength, expressed as the peak stress of the plated element, can be predicted with relatively high accuracy ($\pm 5\%$ for 87% of the cases) for designs outside of the training dataset. However, the output data did not include details about the tangent modulus or energy absorption of the plated element. Although this was not discussed in this study, in principle if the ANN is trained to account for these factors, its predictive capabilities could be improved, allowing it to capture additional features related to load-end-shortening. Park and Kim (2022) investigated the sloshing-induced buckling failure of LNG cargo containment systems to investigate the efficacy of an ANN-based approach for multi-material systems under complex loading. Given the complexity of the problem, their ANN model utilised two hidden layers, with layer sizes optimised through trial and error to balance between under- and over-fitting. Both numerical and categorical variables were used to improve regression results; 85% of the total data was used for the development of the model and 15% for testing.

Ishibashi et al. (2024) investigated the ultimate strength of rectangular plates in the longitudinal and transverse directions through statistical modelling of the initial imperfections of 120 measurements of bulk carrier and pure car carrier deck and bottom plates. They assumed a double trigonometric shape model and estimated the component wave amplitudes through Bayesian linear regression. They concluded that the correlations between trigonometric component waves should be considered when deriving shapes for strength analyses. The true ultimate strength was estimated through non-linear FE analysis, linear-elastic perfectly plastic material models, and by estimating the ultimate strength numerically without accounting for residual stress effects. They concluded that (particularly in the case of thin plates) bimodal strength probability distributions can occur below the yield point due to mode shifts in plate shape from the initial imperfection to ultimate strength failure. In thicker plates, the shape remains unchanged, and the strength approaches that of the yield point. Lima et al. (2023) proposed a bi-fidelity Kriging model for reliability analysis of the ultimate strength of stiffened panels that leverages the accuracy of high-fidelity (HF) simulations and the coverage and efficiency of low-fidelity (LF) simulations by utilising distance correlations between the two sets of results. This multi-fidelity framework provided similar accuracy to computationally intensive HF modelling but with considerable computational time savings. Coraddu et al. (2023) developed a surrogate model approach based on non-linear FE analysis that enables the determination of the optimal design of the ultimate strength of stiffened panels in a given design space while accounting for residual stresses. Hernandez Ramos et al. (2024) integrated the non-dominated sorting genetic algorithm II (NSGA-II) and the original Smith method, which employs genetic operators, such as crossover, mutation, non-dominated sorting, and population diversification to explore the connection between reduced-order models and their associated modelling errors. They used a Sigma function to evaluate force equilibrium errors and force vector equilibrium errors. The applicability and accuracy of their proposed approach were validated on a double-hull VLCC and demonstrated good agreement with methods proposed in ISSC reports on ultimate strength 2000 and 2012 (ISSC, 2000; 2012).

Physics-informed neural networks (PINNs) are an emerging machine learning technique for problems with established physical models, which can be used to augment data obtained from simulations and/or experiments. Incorporating these laws into the training process enables a reduction in the amount of learning data required, resulting in faster convergence rates. Several recent advancements have been made that could benefit future ultimate strength analyses. Zhuang et al. (2021) proposed a deep autoencoder-based energy method for addressing linear bending, vibration, and buckling of Kirchhoff plates, utilising the minimum potential energy principle to model the plate bending effectively in their selected benchmark cases. In turn, Yan et al. (2022) investigated shell structures made from composite materials and introduced the concept of an "extreme learning machine" in the context of PINNs for linear-elastic analysis of isotropic plates with a cutout and free vibration analysis of stiffened composite panels. Li *et al.* (2021d) highlighted that energy-based training is more efficient than PDE-based training due to the lower-order approximations of the displacement fields, while Petrolo and Carrera (2021) proposed the use of ANNs as an effective means for selecting the element kinematics in finite element meshes. Bastek and Kochmann (2023) explored the use of PINNs in the analysis of thin shells, specifically investigating shell element behaviour against locking. They considered three well-known benchmark case studies involving doubly-curved shells subjected to gravity and point loads under conditions of linear elasticity and low strain. Their findings suggest that the weak form, which is commonly used in FE analysis, performs better in the context of PINN than the strong form. They also noted that scaling differences in membrane, bending, and shear energies become particularly pronounced in the thin-plate limit. These investigations have implications for the fundamental underlying assumptions of ultimate strength analyses, particularly in the selection of the initial conditions of simulations, and simulation accuracy in general, highlighting the need for care to ensure realistic estimates.

Some studies have addressed the prediction of strength or failure. Huang et al. (2022) used physics-guided deep neural networks (PGDNN) to identify structural damage in large structures. Their approach focused on addressing two challenges: the large datasets required for pure data-based machine learning methods, which can be computationally infeasible, and the difficulty of achieving high precision in FE analysis-based damage identification. A loss function was developed to minimise the discrepancy between the simulation results and measured data, enabling the model to learn damage features from actual structures. The accuracy of the PGDNN was substantially better than that of conventional neural networks (CNN). Tran et al. (2024) also adopted a machine-learning approach in their investigation of the crack-tip cohesive law. They used PINNs to construct a fully functional form of the cohesive traction separation relationship at the crack tip based on far-field stresses and displacements. Leveraging Maxwell-Betti's reciprocal theorem and small-scale yielding enabled successful modelling of fatigue crack propagation. This study demonstrated the efficacy of the PINN in stable cohesive zone extraction, suggesting the potential for extension to higher plasticity levels required for ultimate strength predictions.

2.3 Experimental Investigations

2.3.1 Full- and Model-Scale Experiments

Full-scale observations of the failure of built structures are vital when the design principles for ultimate strength are to be derived or updated, particularly for consideration of the interplay between environmental and strength modelling. The Norwegian Safety Investigation Authority (NSIA, 2023) reported on Viking Polaris superstructure window failures at the southeast of Cape Horn, November 22nd, 2022. The ship hit a breaking wave, which resulted in the breakage of the windows of seven staterooms of the cruise ship, with severe damage to both people (1 died, 8 injured) and materials. The crew were unable to predict the risk associated with the breaking wave, particularly the effect of the height of the wave on the hull or the force that might be exerted. The brittleness of the material led the structure to shatter after the ultimate strength was reached. This case motivated recommendations for updates to the current design rules.

Another branch of the literature has focused on investigating ultimate strength in carefully controlled laboratory settings. Zhao et al. (2022) conducted a series of experiments on the midship section of a fast ferry featuring two decks, multiple materials, and an opening in the deck, to assess the ultimate strength of large structural systems containing several connected plate fields and intermediate frames. Testing was conducted using a 1:5 scale model subjected to pure bending, and comparisons were made against FE simulations carried out using the ABAQUS and LS-Dyna software tools and the Smith method. Similar studies have been conducted for box-girders (e.g., Quispe et al., 2022). These results indicated good agreement between the simulation results and the experimental results, particularly in terms of the moment-curvature relationship. In another study, Shi and Gao (2021) explored the impact of a large superstructure on the load-carrying capacity of a passenger ship hull girder. While their model was simplified, their study demonstrated the effect of the interaction between bending moments and shear forces on the ultimate strength of the complex hull girder.

2.3.2 Experiments on Panels and Grillages

Understanding the behaviour of structural sub-systems behaviour is essential for validating numerical simulations and reduced-order models, which are used to accelerate the design process. Experiments play a critical role in testing the validity of the assumptions used in model development, thereby improving the accuracy and efficiency of the methods used to assess ultimate strength. While laboratory experiments offer well-controlled conditions, they often require idealized loading conditions and scaled-down specimens, which may not fully replicate real-world scenarios. Ideally, experiments would be repeated on similar specimens to increase confidence in the results; however, practical and economic constraints often limit the feasibility of multiple tests.

Xu and Guedes Soares (2021) investigated the impact of the size of the test unit, along with the applied loads and boundary conditions, on structural performance by conducting experiments on small-scale panels. They relied on validated FE analysis to enable scale transitions between model-scale and full-scale specimens. They emphasised the importance of precise specimen geometry and support system design for accurate test results. Their experiments included stiffened panels with both two-span and three-span configurations, each with identical end-edge support; the two-span models demonstrated greater strength, attributed to increased rotational stiffness in the central span compared to the three-span setup. In $1/2 + 1 + 1/2$ span configurations, failure typically occurred in the central span, aligning with expected outcomes, while in the three-span models, failure often began in the outer spans.

Several studies have sought to address the discrepancies arising from scale differences. Xu et al. (2021b) explored *partial* or *response* similarity methods in relation to the more complete *geometric* similarity methods. They found that the primary limitation of partial similarity methods is the alteration of collapse modes. Ma et al. (2023) developed a general scaled model technique for stiffened panels to preserve the similarity between full and model scale models, by focusing on plate and column slenderness, as well as non-dimensional lateral pressure. Their method involves a two-step process: first, the number of stiffeners needed to ensure plate slenderness is consistent between scales is determined, second, the dimensions of both the plate and stiffener are adjusted to maintain consistency in the sectional area. Liu et al. (2021b) conducted experiments to evaluate the compressive strength of long-span stiffened panels with large openings in the web frames and girders of passenger ship decks supported by tension pillars under combined uniaxial and lateral loads. The scaling factor for the lengths was 1:10, while plate thickness was scaled 1:2.33. Irregular shapes were observed in the scanned imperfections, and the ultimate strength results showed good agreement with FE simulations. However, due to the experimental challenges in applying loads and boundary conditions, the shapes of the stress-strain curves differed significantly between experimental and FE results. This discrepancy was attributed to the difficulty in retraining unloaded edges in physical experiments, whereas restraints are easily applied in FE analysis. Liu et al. (2021a) expanded their prior investigation to grillages with large openings typical of those found on the top decks of large passenger ships. Similar challenges with unloaded edges were observed experimentally. Their findings indicated that the complex shape of the grillage openings and the presence of reinforcements altered the overall buckling and collapse behaviour of the stiffened panel under compressive loads. This investigation also highlights the effect of edge stiffeners on the ultimate strength. Ma et al. (2021) performed experiments on biaxially loaded panels under lateral loads and used laser scanning to capture the shapes of initial imperfections. They concluded that incorporating these measured imperfection shapes into FE simulations led to significantly better agreement with experimental results compared to relying on the standard design equations. Xu et al. (2022b) used scaled experiments to validate their FE analyses on perforated rectangular specimens, finding good agreement in terms of ultimate strength, though initial stiffness alignment was less precise.

Guo et al. (2021) performed tensile experiments on laser-welded aluminium panels using DIC instrumentation to capture the strain field evolution and damage formation. This was complemented with a thermal-elastic-plastic analysis to assess the temperature field, welding deformations, and residual stresses. These data were combined to assess the ultimate strength of the panels using non-linear FE analyses. This hybrid modelling approach, which combines measured strain fields with computed stress values (since these cannot be measured), effectively accounted for all significant effects and therefore resulted in excellent agreement in the force-displacement relationships of the tested panels.

Barsotti et al. (2025) performed experiments on a full-scale stiffened panel loaded in the transverse direction towards the deck stiffeners to study the failure mode differences for a scenario where the load acts in the weaker direction of the panel. This case, identified by Kõrgesaar et al. (2023) as challenging for reduced-order models, provided insights into multiple sources of uncertainty affecting agreement between experimental results and FE simulations. Due to these complexities, this experiment was selected as a benchmark for the current committee's study.

2.3.3 Experiments at the Component Level

Experiments may also focus on component and material specimen testing to better understand the effects of various environmental conditions, such as temperature, corrosion, or moisture, on material properties under applied loads, including strain or stress. These studies often examine factors such as strain rate, stress triaxiality, and material inhomogeneity, which may arise due to the material manufacturing process or joining methods that localise damage. Additionally, they may investigate how different material systems influence structural failure modes, particularly in composite materials where operational damage, moisture, temperature variations, and delamination significantly impact strength.

3 Construction, in-Service, and Life-Cycle Effects

The multivariate nature of ultimate strength prediction is reflected in both the inherent complications associated with prediction methods (as discussed in Sect. 2) and the very nature of the life cycle of a given structure. Aspects such as material properties, fabrication techniques, environmental effects, and structural maintenance all contribute to the ultimate strength prediction, which has further implications for design optimisation, service management, and potential life extension. Because there are uncertainties in all of these aspects, which are often difficult to quantify across structural scales, there is an uncertainty in the ultimate strength prediction. Research shows that a 10% uncertainty in the ultimate strength of plating may translate to a 20% variation in the ultimate strength capacity of a ship hull (Li et al., 2021b). In the pursuit of more insightful analysis, research has been focused on the application of more realistic material models, more representative loads, improved geometry characterisation, and improved modelling formulations to account for uncertainties. Compared to the current industrial practice and classification society requirements, many of these applications are achieved at an increased computational cost. However, researchers have been adopting machine learning methods, such as ANNs, as a simpler tool to infer the current serviceability and remaining service life of a given asset, which would not be possible with an idealised model (Graves et al., 2023). Before delving into ship and offshore-specific topics, this section reviews the latest literature on fabrication, as well as in-service and life-cycle effects, covering both metallic and non-metallic constructions.

3.1 Material Uncertainties

The uncertainties associated with material inputs strongly affect the ultimate strength prediction of both metallic and non-metallic structures. This section discusses key research advances in the past three years relating to this aspect.

3.1.1 Steel

The role of mechanical properties in the ultimate strength characteristics of steel structures is generally well established. The peak load prediction at the design stage is generally acceptable even when the minimum rule-required properties and a simplified bi-linear stress-strain relationship are utilised (Ringsberg et al., 2021). However, the true ultimate strength highly depends on the actual mechanical properties of individual structural members.

Zhao et al. (2022) conducted both experimental and numerical investigations of the ultimate strength of a ship hull girder model with large openings. As part of their experimental work, tensile tests were carried out on DH40 and DH36 grade steels. The tensile test coupons were made of different thicknesses, ranging from approximately 3 mm to 8 mm. The results were then fed into an FE model using a simplified stress-strain relationship (elastic-perfectly-plastic model). It was found that the failure modes of the strong girder and the first deck structure were relatively consistent between the simulations and model tests. However, this was not the case for the deformation direction of the panel centre of the second deck. The ultimate strength prediction error of the FE model was as low as 1.6%, implying that the second deck deformation effect was minimal. These results were fairly consistent with the findings from the ISSC 2022 Ultimate Strength benchmark study (Ringsberg et al., 2021); when the actual material properties were used in modelling, even with a simplified bi-linear stress-strain relationship, the ultimate strength prediction error was only 3%. However, the failure mode or plasticity location predictions at both the onset of yielding and the ultimate strength points proved to be a more difficult task. This implies that even though it may be possible to confidently estimate the magnitude of the ultimate strength with a small error, it is difficult to precisely predict the location of damage. Note that failure mode validation is more challenging than validating the ultimate strength value, as in most large-scale experiments in the literature, the failure mode is typically measured by strain gauges at discrete locations rather than through a full-field measurement. It is challenging to identify any onset of plasticity that can be compared with FE results.

Pan et al. (2023) addressed a gap in the applicability of existing load-end shortening formulae in the IACS Common Structural Rules for Bulk Carriers and Oil Tankers (CSR-H) in the case of hard corner elements. The assumption that hard corner elements follow elastic-perfectly-plastic collapse may lead to an overestimation of the ultimate load-carrying capacity due to the local buckling of plates. To more realistically reflect the progressive collapse behaviour under longitudinal compression of structural elements, the authors proposed an improvement of the average load-end shortening curves in CSR-H. This was achieved via experimental results and non-linear FE analyses of hard corner elements to modify the CSR-H prescriptive load-end shortening curve formula, resulting in more realistic collapse behaviour and increased accuracy of the hull girder ultimate strength assessment. Therefore, it may be beneficial to assign more realistic material properties to hard corner elements in reduced-order/analytical methods, such as the Smith method. The authors also investigated the effect of steel grade, finding that steel grades with higher yield strength had lower plastic flows that may affect the equivalent stress distribution of the structural element; the lower the yield strength of the material, the more likely the structural element's response will be in the plastic region and the larger the increase in the errors of load-end shortening curves because of elastic-plastic buckling.

The role of yield strength in ultimate strength was further demonstrated during the ISSC 2022 Ultimate Strength benchmark study (Ringsberg et al., 2021). It was shown that the assumption of nominal versus actual material strength had the largest impact regardless of the numerical method selected. Also note that the examination of the M/V MOL Comfort accident conducted by ClassNK (ClassNK, 2014) found an 8% higher girder ultimate capacity when replacing the minimum mandated yield strength with more realistic values.

The effect of yield strength uncertainty manifests itself when considering cyclic loading. Jagite and Bigot (2023) analysed the ultimate strength of a container ship using a hydro-elastic-plastic model. The model permitted the application of time-dependent cyclic loads, determined from a long-term direct hydro-elastic analysis, rather than monotonic increasing loads. Interestingly, the authors concluded that the reduction of the ultimate strength of the hull girder due to cyclic loading was minor relative to the uncertainty in the material yield strength or the uncertainty in the effects of residual stresses and geometric imperfections. The authors recommended that such studies should be conducted on different ships of varying sizes and functions to gain a better understanding of the influence of cyclic loading on the ultimate strength of hull girders, with consideration of the probability of occurrence of "significant loading cycles". In addition, cyclic testing of steel materials typically used in ship construction was suggested to calibrate material models, such as the Chaboche model.

For cases where material coupons may be retained for testing, uncertainties surrounding mechanical properties could be addressed through testing efforts. However, in real-world structures or design-related analysis, this is not an option. In such cases, prior investigations into the uncertainty of steel properties must be examined. High-strength steels for marine use generally exhibit higher material property variation than mild steels (Storheim et al., 2018). Additionally, the former does not typically display a clear yield plateau, posing a challenge for modelling efforts that must consider using the 0.2% proof stress or proportional limit. Nahshon et al. (2018) applied various uncertainty quantification methods to examine the role of material uncertainties in the ultimate strength of a stiffened panel using available material and geometrical variability data from Hess et al. (2002). Using various statistical sampling methods to generate FE analysis input models that incorporated the statistical nature of steel material properties, as well as geometric imperfections and fabrication tolerances, they found a large range of possible load-shortening curves, as shown in Fig. 6. It can be expected that such variation will manifest itself when a structural system is considered.

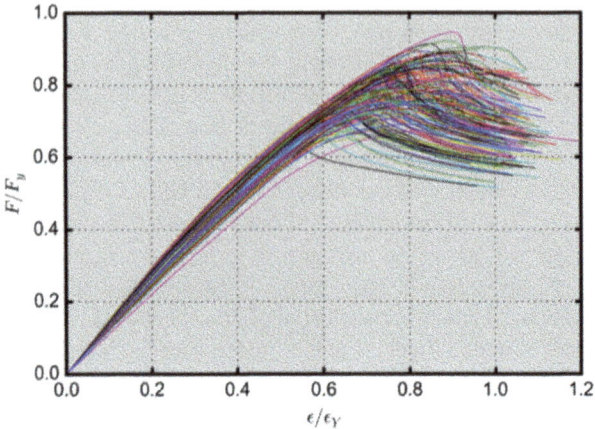

Fig. 6. Range of load shortening curves for typical stiffened panels, accounting for steel material, thickness, initial imperfections, and fabrication uncertainties (Nahshon et al., 2018).

Currently, steel grades for shipbuilding are capped at S460 (460 MPa minimum yield strength for container ships) by classification societies, although higher-grade steels are used extensively in naval, offshore and lifting appliance sectors. Ultra-high strength steels are often used for the construction of key sections of ice breakers and other specialised ships since they allow for a reduction in scantlings while maintaining the same design stress level. However, high-strength steels, while advantageous for their strength, present challenges, such as reduced toughness, high yield-to-tensile strength ratio (currently limited to 0.94 by classification societies), more complex welding processes, and lower ductility. These effects, combined with possible low operational temperatures, raise concerns regarding ultimate strength performance. Wong and Walters (2023) studied the decreased strain-hardening ability of high-strength steels and its effect on the elastic-plastic rotation capacity and local buckling of a stocky I-beam (with an overall cross-sectional slenderness below 0.51). The rotation capacity of a plastic hinge needs to be large enough so that the plastic hinge can maintain its strength until complete collapse. Both geometric and welding-induced residual stresses were included in the numerical models. A total of 217 beams with various yield strengths (from 460 to 960 MPa), yield-to-tensile ratios (from 0.80 to 0.99), and beam geometries were analysed. It was found that the rotation capacity of the beam was simultaneously dependent on all of the investigated parameters. Based on their findings, Wong and Walters suggested that a cross-sectional slenderness limit could replace the yield-to-tensile ratio limit to enable broader use of high-strength steels. However, their analysis appears to have been primarily informed by the rotational requirements for civil engineering applications, along with yield-to-tensile limits from classification societies, and slenderness requirements from IACS requirements for Polar Class ships. The differences in the assumed safety margins and design approaches in these requirements were not considered.

3.1.2 Aluminium Alloys

Liu *et al.* (2020a), Hosseinabadi and Khedmati (2021), and Wang *et al.* (2022d) published comprehensive reviews examining various factors and their interactions that influence the ultimate strength of aluminium structures. These included material properties, investigation methods, structural geometry/arrangement, initial imperfections, boundary conditions, loading conditions, and in-service degradations. Unlike steel structures, where an elastic-perfectly-plastic material model is commonly used, the Ramberg-Osgood approximation model is generally adopted for aluminium alloys (also required by classification societies). Aluminium structures exhibit a high level of sensitivity to temperature with rapid loss of stiffness above 100 °C. As such, it is crucial to prioritise the design at the member, joint, and structural levels, along with ensuring effective fire protection for the structural systems. They are also most likely to contain under matched heat-affected zones (HAZ) in weld areas, as opposed to overmatching in most steel welds. For example, Liu *et al.* (2020b) reported such characteristics for AA5083-H116 coupons (see Fig. 7), which were fed into their LS-DYNA model of butt weld and fillet weld joints. It was found the difficulties in predicting plastic responses were primarily due to the use of simplified shell elements and material properties; the actual material properties would vary with distance to the weld seams. In addition, ultimate strength prediction methods, such as ISUM and ISFEM, may require further evaluation for their applicability to aluminium vessels (Hosseinabadi and Khedmati, 2021).

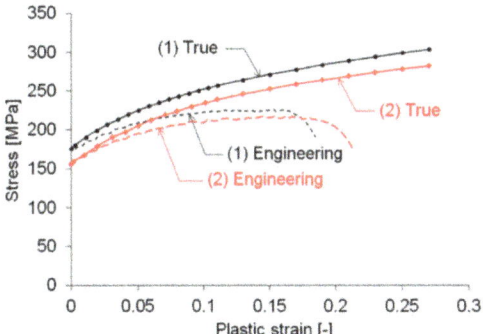

Fig. 7. Material properties of AA5083-H116: (1) base and welded materials, and (2) HAZ material (Liu et al., 2020b).

3.1.3 Composites and Sandwich Structures

The application of composites in marine environments has steadily expanded, encompassing applications from low/high-performance vessels and offshore renewable energy systems to wind-assisted propulsion systems and dock infrastructure. Given the diversity of factors, such as composite materials, layup configurations, and manufacturing techniques, it is essential to accurately characterise and correctly apply material properties in any structural analysis, especially when predicting strength capacity and failure modes. Lowde et al. (2022) conducted an in-depth review focused on the prospect/feasibility of building over 100 m long composite ships. Their review included manufacturing processes, material properties, and economic and life cycle considerations. One of the key obstacles preventing scaling up to longer composite ships is the lack of a comprehensive database of material properties. In addition, new materials, such as 2D monolayers and nanotubes, are often costly and are available in limited quantities. Nevertheless, research continues, and novel methods are being used to improve methods for accurately predicting material properties and failure modes.

Barsotti et al. (2020) reviewed industrial developments in limit state assessment and design approaches for marine composites, with a focus on pleasure yachts/crafts and naval ships. They concluded that experimental testing was the most reliable method for determining material properties and failure modes. However, when experiments are infeasible, designers often rely on established rules, the research literature, and manufacturer reports. The use of cohesive elements in FE modelling represents the state-of-the-art approach in the assessment of delamination, which is one of the most critical failure modes. Despite the effectiveness of this approach, cohesive elements are limited to small-scale applications due to computational demands and the complexity of calibration. Additionally, the joining of composites with other materials (for example, for repair) has become an increasingly explored area of research in recent years.

Liu et al. (2023) performed a multi-scale analysis on glass-fibre-reinforced polymer (GFRP) materials. The micro- and mesoscale representative volume element (RVE) models of components used in GFRP materials were established, incorporating failure criteria and stiffness degradation models through a user-defined subroutine (VUMAT) in ABAQUS. The equivalent material properties at the micro-scale (mesoscale) obtained by a homogenisation method were used to define the mesoscale (macro-scale) mechanical properties in the FE analyses. It was shown that the ultimate strength of the GFRP stiffened panels was primarily determined by the failure of chopped strand mat (CSM) fibre bundles and woven roving (WR) yarns. The findings from this work suggest that the parametric study of meso-mechanics could be an effective guiding method in the optimisation of macro ultimate strength capacity in stiffened panels.

Jafarzadeh and Khedmati (2020) modelled the midship section of a 50 m long WR/polyester GFRP vessel subjected to four-point bending to assess its ultimate strength behaviour. Due to the 45-degree fibre orientation, symmetry conditions could not be applied, necessitating a full model of the section. Modelling was conducted (in ANSYS using SHELL281) based on first-order shear deformation theory. No initial imperfection was included. The Tsai-Wu failure criterion was employed, and bending moment versus displacement was analysed for both hogging and sagging conditions. In both cases, failure was observed to initiate in the deck structure. Validation was performed through a comparison with numerical results from the literature, showing relatively good agreement.

Based on the active learning Kriging (ALK) model and the Hashin failure criterion, Wang et al. (2023a) proposed a new reliability evaluation model for composite stiffened panels and conducted a reliability analysis of their ultimate bearing capacity. They also studied the importance ranking of the input variables. Comparing the results of the ALK model with those from the Monte Carlo method demonstrated the model's accuracy and efficiency in various cases. Their discussion of the effects of longitudinal elastic modulus and fibre-direction tensile strength on post-buckling failure probability provides valuable insights and references for the optimisation and design of composite stiffened plates. An ANN was used by Sun *et al.* (2021b) to predict the ultimate strength of composite hat-stiffened panels under in-plane shear. The database for training the ANN was generated using FE models, which considered several material parameters. The ANN model achieved over 98% accuracy in predicting the buckling load and the ultimate load. However, its performance with input parameters outside its training domain (i.e., its generalisability) remains uncertain.

Aluminium alloy honeycomb sandwich (AHS) structures are known for their excellent impact-absorption capabilities and high stiffness-to-weight ratio. Compared to composites, they also offer improved recyclability without sacrificing their lightweight character. Research has explored hybrid lightweight ship designs that combine steels with AHS structures to improve energy efficiency. For example, Garbatov et al. (2023) introduced a hybrid bulk carrier design in which part of the cargo holds was replaced with AHS panels. The ultimate strength of the hull girder was incorporated into a multi-objective optimisation design process, and two designs were selected for further reliability analysis, which will be discussed in detail in Sect. 3.4. Corigliano et al. (2023) conducted both experimental and numerical studies of an AHS panel subjected to uniaxial compressive loads. The numerical studies were based on a modified Smith method and the Johnson-Ostenfeld formulation. Relatively good agreement of the load-shortening curves was obtained between the two methods. However, the numerical method did require calibration based on experimental results.

Honeycomb sandwich structures produced by additive manufacturing were investigated by Garbatov et al. (2024). The core material consisted of thermo-plastic nylon reinforced with 10% chopped carbon fibres, manufactured using the classical fused deposition modelling technique. The skins were made from thermoplastic nylon reinforced with continuous carbon fibres, produced via continuous fibre co-extrusion technology. Printed samples were subjected to 3-point bending tests, which fed into a subsequent numerical approach for the prediction of the progressive compressive failure modes. The response of the compressed face (global and local buckling followed by delamination) governed the flexural response of the structure. On a larger scale, Zhong and Wang (2022) conducted a numerical analysis to study the ultimate strength of sandwich box girders. The analytical results were validated through comparison with experimental results. Their findings indicated that increasing face plate thickness while reducing web plate thickness led to higher warping stress in the box girder, ultimately decreasing the ultimate bearing capacity. However, further research is needed to understand these structures' performance under ship-specific design conditions.

3.2 Fabrication Uncertainties

Consideration of initial imperfections has always been a key part of ultimate strength assessment. Most imperfections stem from structural connections (often seen as weak spots). The quality of both traditional welding and non-metallic bonding is influenced by the materials used, fabrication procedures, and workmanship. This section reviews the recent literature that discusses the consideration of these uncertainties when predicting ultimate strength.

3.2.1 Welding-Induced Distortions and Residual Stresses

Steel

Welding of steel structures introduces distortions and residual stresses that have a profound effect on ultimate strength at the structural element, grillage, and system levels. As such, approaches to account for these effects have been a longstanding research topic. Within the context of FE analysis, efforts generally focus on direct or implicit simulation of distortions and residual stresses through imperfection generations in a deterministic, statistical, or stochastic manner. In contrast, reduced-order models such as the Smith method consider these effects through modification of the underlying load-shortening curves of structural elements to perform progressive collapse analyses.

Given the large temperature variation during a welding process, temperature-dependent material properties play an essential role in welding-induced imperfections. However, there are limited material data, especially at high temperatures available in the public domain. This research gap motivated Chen and Guedes Soares (2021) to conduct tests with metal inert gas T-joint fillet welding tests on rectangular steel plates. Subsequently, a 3D sequentially coupled thermo-elastic-plastic modelling approach was developed, with both welding-induced distortion and residual stresses validated against experiments. A simplified temperature-based material model was then proposed which could be used in welded structure models.

The uncertainties associated with welding-induced distortions and their effect on ultimate strength were also investigated by Li et al. (2022c). The authors conducted FE analysis case studies on a steel grillage model and tested both deterministic and probabilistic geometric imperfections. This work emphasised that, beyond influencing the magnitude of the ultimate strength, the selection of a particular distortion mode also has implications for model convergence and hence prediction robustness. It was also concluded that adopting a stochastic modelling approach could provide more realistic assessments of structural responses.

A new probabilistic-based imperfection model was introduced by Georgiadis and Samuelides (2021a) via a spectral representation method. The method was based on random field theory and actual measurements of ship structures. The impact of stochastic geometric imperfections on the ultimate strength of plates was assessed using Monte Carlo simulations. The proposed imperfection model was then applied to a tanker under extreme sagging conditions to predict the stochastic ultimate strength of the hull-girder by training an ANN. Figure 8 shows the comparison with conventional imperfection models and the Smith method. Notably, the proposed model resulted in a reduction of ultimate strength by up to 4.5% compared to the commonly used "hungry horse" imperfection mode. The use of an ANN makes it feasible to incorporate the full spectrum of input uncertainties in ultimate strength assessments.

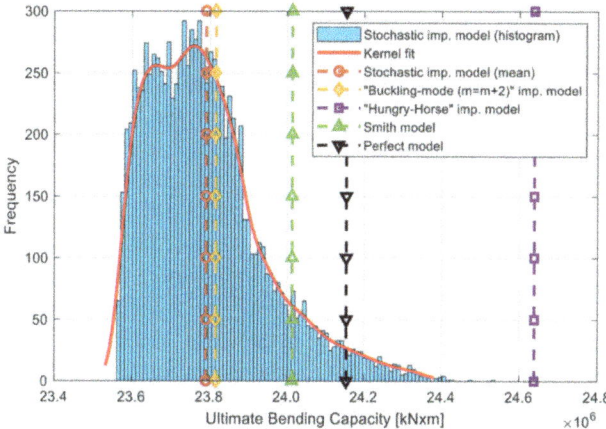

Fig. 8. Histogram of stochastic ultimate strength of a VLCC in sagging compared to conventional imperfection models and the Smith method (Georgiadis and Samuelides, 2021a).

Graves et al. (2023) proposed a new method for mapping point cloud data obtained from laser scanning or photogrammetry to FE models to characterise geometric imperfections. Mapping was performed using Kriging methods that inherently smooth out noise in the point cloud data and effectively fill data gaps. The method was then applied to FE analysis calculations of a typical tee-stiffened steel grillage. Results were compared to those performed using a typical spectral-distortion approach. The proposed method provides a robust approach for incorporating measured in-service distortions when commonly assumed welding-induced distortion shapes (e.g., eigenvalue buckling mode or hungry horse) are unsuitable. Such in-service distortions could be located where deck repairs are made, or impact damage has occurred.

Residual stress can be redistributed or relaxed under applied loads during service. Elastic shakedown involves initial cycles of plastic deformation that adjust plasticity-induced residual stresses, redistributing them to achieve internal equilibrium after unloading. Consequently, the original welding residual stress state is modified. Gadallah et al. (2024) developed a numerical procedure to analyse the relaxation of welding residual stress due to elastic shakedown in various welded components. They found that early cyclic loading, especially near weld toes, significantly altered the initial welding residual stresses, highlighting the importance of accounting for this redistribution. Hayama et al. (2024) conducted *in situ* X-ray stress measurements to assess fatigue properties and relaxation behaviour throughout the fatigue process. They found a linear correlation between the relaxation threshold stress and initial compressive residual stress with specimen hardness, which aligned with the compressive yield strength. This underscores the importance of using relaxation threshold stress for quantitative evaluations to predict fatigue limits, particularly considering the compressive residual stress induced by surface treatments.

Aluminium

Ultimate strength analyses of aluminium structures present additional challenges compared to similar analyses of steel structures due to (1) the significant impact of welding on the reduction of local material properties (softening in the HAZ), (2) the use of extruded structures, and (3) the generally lightweight scantling of these structures. Having reviewed the welding-affected material properties in Sect. 3.1.2, it is notable that understanding and data on initial deflections in welded aluminium structures lag somewhat behind those of steel structures. The majority of aluminium stiffened panel models in the literature apply similar methods as for steel structures, e.g., using Fourier series to generate initial deflections. Typical amplitudes are summarised in the review by Hosseinabadi and Khedmati (2021). Soleimani et al. (2020) carried out a numerical buckling analysis of a midship section of a real high-aspect-ratio twin hull (HARTH) vessel, made of AA5083-H321 plating and AA6062-T6 stiffeners. A HARTH vessel has a smaller waterplane area than conventional catamarans, and experiences reduced hull drag, allowing it to more easily break through the water during movement. However, it also means that the vessel is more susceptible to buckling in rough seas. Initial geometric imperfections in plates were considered using a half-wave distortion mode as described by Smith et al. (1987), incorporating plate aspect ratio, plate thickness, and an average distortion factor for aluminium plates, as recommended by Paik et al. (2008). These initial imperfections play a critical role in the buckling behaviour of thin plates (particularly those less than 8 mm thick). While the plate aspect ratio has minimal effect on overall strength, it significantly influences the failure mechanism. Welding residual stresses and heat-affected zone properties were not included in this analysis.

3.2.2 Non-welded Connections

The impact of manufacturing technology (welding versus extrusion) on the ultimate compressive strength of aluminium-alloy stiffened panels has been explored through a series of FE analyses (Liu et al., 2020a). Assuming similar levels of initial distortion, the load-carrying capacity of extruded panels was reported to be 4.6% higher than welded panels due to the consideration of HAZ properties. The use of large aluminium extrusions is common in high-speed craft such as passenger ferries. However, research on extruded structures is still limited (Hosseinabadi and Khedmati, 2021).

L-joints, such as those typically found between the hull's outer shell and the upper deck, are a common feature in ship structures. Kai et al. (2020) showed that changes in the structural transitional area of such joints can change the failure mechanisms under tensile loads (see Fig. 9). Joints with a radius of 45, 90, and 180 mm were tested and analysed using FE analyses with a new damage criterion, which differentiated between matrix and fibre response under both tension and compression. The foam core was modelled using an ideal elastic-plastic model, and debonding between the stiffener and the panels was included via a cohesive zone approach. The authors subsequently proposed an improved L-joint, where the common bonding mechanisms at the end of the joint were replaced by integrated-moulding T structures with rounded corners. This design incorporated a larger no-damage zone and had a higher ultimate strength than conventional designs. It must be noted that this improved design was only studied numerically. Although no experimental validation was provided, the numerical model was validated against the original tested geometries. The damage and failure prediction maps presented in the paper (Fig. 9c) aid in understanding the failure mechanisms and optimising joint design.

3.3 In-Service Effects: Degradation and Damage

This section focuses on the effect of material degradation and mechanical damage on ultimate strength. Both aspects have been intensively investigated in the past few years to better estimate the remaining strength capacities of structures, which has key implications for the optimisation of the design and life-cycle management of a given asset. The effects of asset-specific accidental damage, such as collision and grounding, are outside the scope of this section and will be covered in Sects. 4 and 5.

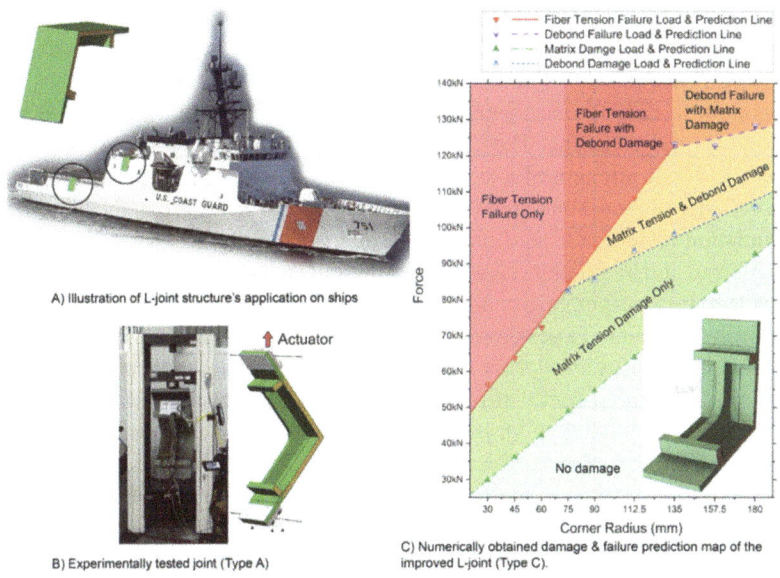

Fig. 9. (a) L-joint application case in ship structures, (b) experimental testing of the L-joint, and (c) damage and failure prediction map of the improved L-joint (Kai et al., 2020).

3.3.1 Corrosion and Material Degradation

Degradation of Metallics

Requirements for and advancements in coating applications for metallic structures exposed in marine environments highlight the significant threat posed to ultimate strength capacity by localised corrosion in the form of pitting or crevice corrosion, as it results in localised thinning, eventually leading to perforation (often covered by rust and/or coating) that is difficult to detect. Understanding how and when this process takes place, what the controlling parameters are, and how they correlate with one another in the context of ultimate strength analysis is essential for developing more robust and reliable evaluation procedures and fitness-for-service practice codes. Consequently, there has been a continuous research effort on this topic in recent decades.

In their consideration of structural members, Feng et al. (2020) focused on the ultimate strength characteristics of three stiffened panels with pitting corrosion distributed on both plates and webs. A random as well as uniform distribution of circular through-thickness pits was employed, and their influences on the ultimate strength were investigated. Their findings indicated that pitting distribution and pitting severity were less harmful in the context of ultimate strength reduction when they occurred on the webs rather than on the panels. These results support previous findings related to corrosion wastage effects where web pitting was found to be less influential on ultimate strength. Based on these results, the authors proposed an empirical formula for predicting the ultimate strength of stiffened panels as a function of the ratio of the corroded volume to the undamaged volume of the reinforced panel. The robustness of this formula was verified against 243 simulated cases with random pitting distributions and degrees of severity. These findings may provide valuable information for supporting maintenance decisions and emergency stop protocols.

Mursid et al. (2023) investigated the effect of pitting corrosion on a ship bottom plate damaged by grounding. They found that the pitting location played a significant role in the ultimate strength of a structure by conducting a very detailed FE analysis, where 14 scenarios were simulated using different pitting positions on the bottom plate. The simulation results indicated that the location of pitting corrosion impacts stress concentrations, crack initiations, load re-distribution, and penetrating position when the crack nucleates. For example, a configuration comprising four small pits near the centre of the indenter will exhibit a peak force that is approximately twice the force that would be experienced if occurring away from the indenter. These findings indicate that the critical position of pitting significantly affects the load-carrying capacity of the bottom plate, especially as the relative spacing between pits increases radially. Zhang et al. (2022) also investigated the combined effects of pitting and cracking on the ultimate strength of stiffened panels under compression, addressing a gap in research on how these defects impact load-carrying capacity. Their study aimed to confirm the interactive effects of cracks and pitting with plate and column slenderness ratios on ultimate strength.

A more general framework for assessment of the ultimate strength of a bulk carrier subject to pitting corrosion was developed by Piscopo and Scamardella (2021). They modified the well-established incremental-iterative method (IACS, 2020a,2020b) to study the ultimate strength of a hull girder and the combined effects of random pitting corrosion wastage. In their approach, the authors introduced a slight modification to the existing expression using an extended version of the Faulkner equation (Faulkner, 1975). This method incorporates plate panels as part of grillage structures, whose ultimate capacity is provided by the Faulkner formula, which enables allowance for a certain degree of rotational restraint along the longitudinal edges of the plate panels. This correction provides an explicit strength check criterion for the ultimate strength assessment of pitted plating, which is not considered by current rules and guidelines. The reliability of the proposed method was evaluated against results obtained from a benchmark study using four reference scenarios, characterised by different locations and the extent of pitting corrosion wastage. In their study that also focused on ship structures, Shi et al. (2021) proposed a new method to study the ultimate strength of a pitted ship hull through a series of numerical analyses with various levels of corrosion damage and damage locations. Using a series of Monte Carlo simulations, they demonstrated a strong relationship between the criticality of corroded structures and the sensitivity of related factors impacting hull collapse.

In their work related to general/uniform corrosion, Kim *et al.* (2022a) introduced a Bayesian inference model to address uncertainties in corrosion prediction. They proposed a probabilistic model in which the parameters introduced in the corrosion wastage model were treated as random variables and the probability distribution of these random variables was updated accordingly when measurement data were obtained. This iterative updating approach estimated corrosion depth distributions with approximately 95% and 99% reliability. The predicted corrosion depth from posterior results decreased at each time point, reflecting the incorporation of new data through Bayesian inference. This model could support health monitoring and assist in determining optimal times for periodic inspections. Similarly, Garbatov (2020) developed closed-form solutions for determining the corrosion allowance of ship structures based on probabilistic and risk principles. It is worth mentioning that the proposed set of equations takes into consideration the time-dependent degradation factor for wastage estimation in cargo ships coupling with the Two-Areas-At-Time (TAAT) analysis to determine corrosion allowance in different environments. Although this model assumes uniform thickness reduction without local pitting, it offers a practical probability-based approach for predicting general corrosion wastage.

General corrosion wastage was also indirectly studied by Liu (2021) through FE modelling of lattice corrugated panels made of E690 for offshore applications. The panels were subjected to quasi-static and low-speed impact loading. Uniform thickness reduction representing general corrosion was assumed to circumvent complex material dissolution modelling. It was found that corrosion in the inner part of the structure had a more severe effect on the structural strength than the outer surface corrosion. In a similar analysis using uniform thinning to represent corrosion, Zhang et al. (2024) conducted experimental and numerical analyses of "corroded" concrete-filled double-skin steel (CFDST) subjected to eccentric compression. CFDST is increasingly being used as an alternative to circular steel tubes (CST) for offshore structures, such as wind turbine jackets, for enhanced structural resilience. Partial uniform thinning of the outer steel skin of the tube was considered. Four levels of corrosion and two types of cross-sections were analysed. It was found that CFDST exhibits enhanced robustness and energy dissipation capacity against sudden damage due to local buckling compared to CST. The authors also recommended further research on CFDST structures, considering both dynamic loading and the effects of corrosion.

To simulate more realistic corrosion morphologies on steel structures, Georgiadis and Samuelides (2021b) studied the effect of non-uniform thickness reduction on stiffened panels subjected to compressive loads using random field theory. Their findings indicated that thickness data collected from inspections could be used to inform spatial variations in the material, enabling more reliable vessel-specific predictions of ultimate strength. However, this conclusion cannot be a priori extrapolated in the case of pitting corrosion in areas with pronounced stress concentrations. Zhang and Zaghi (2023) used 3D scanning of heavily corroded steel bridge structures to estimate their residual strength. They generated FE models directly from 3D point cloud data of the damaged structures, incorporating perforations and thickness reductions. It was demonstrated that corrosion substantially changed the failure mode and encouraged failure to occur at the damaged location where longitudinal and transverse structures intersected, as illustrated in Fig. 10. Although this study focused on bridges, the corrosion features and resulting analysis framework could be equally valuable for understanding failure modes in ship and offshore structures under similar conditions.

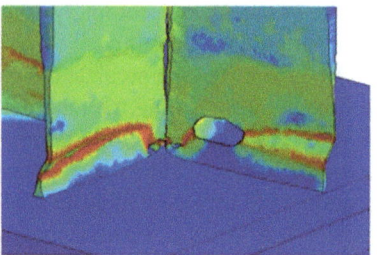

Fig. 10. Failure mode of corroded girder (Zhang and Zaghi, 2023).

Tekgoz et al. (2020) summarised key factors influencing changes in the ultimate strength of ship structures that have aged or experienced damage during their operational lifetime. They showed that the ultimate strength of corroded structures is influenced not only by thickness wastage but also by microstructural changes that may occur during the environmentally assisted degradation processes, which eventually undermine mechanical performance. In ship design regulations, the net scantling approach is applied to oil tankers and bulk carriers covered by the CSR, as well as to other ship types as directed by some classification societies. The full corrosion addition is deducted from the scantling for buckling analyses, whereas for hull girder analyses, only 50% of the corrosion addition is removed. IACS specifies corrosion additions in the CSR for tankers and bulk carriers, UR-S11a for container ships, and UR-S21 for hatch covers/coamings. Based solely on general corrosion, these guidelines were derived from extensive survey thickness measurements collected by classification societies over several years, followed by statistical analysis. Final corrosion additions also account for coating life, coating failure period, and survey intervals.

Structural integrity assessment standards for the offshore sector, such as API 579 (2021), define corrosion degradation as either local metal loss (LML) or general metal loss (GML) for in-service structures. As specified in API 579, every structure is designed with a corrosion allowance (CA). Inspections that find metal loss exceeding the specified CA necessitate the evaluation of the suitability of the structure for continued use, as well as predictions of future corrosion progression to guide inspection frequencies and protocols.

Assessment procedures are based on point thickness readings and thickness profiles for a prescribed grid. For GML, the minimum thickness measurement is often used as a baseline across the plate or structure, providing a conservative estimate. As in the procedure for LML, if the GML assessment does not meet basic criteria, more advanced assessment methods are available. When calculating the minimum required wall thickness, including the CA, engineers often assume that the steel is purchased at a standard thickness, which is usually greater than needed. This practice introduces an additional safety margin beyond the design requirements due to the excess material. These considerations are particularly important for evaluating the ultimate strength behaviour of offshore structures, as corrosion progression may bring the structure closer to its limit state over time.

Degradation of Composites

The growing use of composites has raised concerns about their structural integrity, particularly under marine environmental degradation, where the underlying mechanisms are not fully understood. This has spurred intensive research efforts to gain further understanding of the interaction between the matrix material and surrounding conditions with applied stress. Recently, Vizentin and Vukelic (2022) investigated composite degradation experimentally. Coupons made of glass fibre with epoxy and polyester resin were immersed in real sea environments for up to 12 months. Material characterisation and tensile tests were subsequently conducted. It was found that the 0/45/90 layout for the polyester/glass combination showed the greatest resilience to seawater. However, the polyester coupons were produced using a hand-layup process, and the impact of this method was not assessed. It was reported that microorganism growth resulted in voids in the resin matrix and affected the mechanical properties. However, this mechanism must be further researched.

Accelerated ageing tests were conducted by Ghabezi and Harrison (2022) on glass/epoxy and carbon/epoxy samples in artificial seawater at room and elevated temperatures for over 180 days. Failure modes and locations were recorded using the 3-part failure mode code from ASTM D3039. Experiments showed that mechanical-based damage, such as softening and swelling, is the primary source of physical degradation of polymer resins. Even after water penetration reaches the saturation point, water molecules still react with the epoxy over time, causing irreversible chemical changes (chain scission) and reduced material properties. Bonsu et al. (2022) also performed tests in artificial seawater and evaluated the failure mechanisms of plain glass and basalt fibre-reinforced composites, as well as a selected glass/basalt hybrid composite sequence. The results showed that some hybrid laminates with sandwich-like and alternating sequencing exhibited superior mechanical properties and ageing resistance than plain laminates. For instance, a $[B_2G_2]_S$ (where B stands for basalt fibre and G stands for glass fibre) hybrid composite with basalt fibre outer plies retained 100% tensile strength and 86.6% flexural strength after ageing, which was the highest among all the laminates. However, $[BGBG]_S$ specimens with alternating sequencing retained the highest residual impact strength after ageing. Scanning electron microscopy (SEM) analysis of the failed specimens showed fibre breakage, matrix cracking, and debonding caused by fibre-matrix interface degradation due to seawater exposure. However, various hybrid configurations significantly hindered crack propagation across specimens, thereby altering their overall damage morphology.

Wang et al. (2022c) developed a progressive damage model that included the effect of hygrothermal conditioning on both the strength and stiffness of glass fibre composite structures. The model was developed based on quasi-static tensile and compressive tests on open-holed specimens under both dry-state room temperature and hygrothermal conditions. Failure models, including both the Hashin and maximum stress criteria, were adjusted to match the experimental data. While the authors reported good agreement between numerical analyses and experimental results, discrepancies were noted in the displacement and strain values at which the ultimate load was reached. Based on these findings, the authors concluded that the ultimate strength of these high-strength quasi-isotropic panels decreased by approximately 35–40% for both tension and compression when tested at 70 °C with equilibrium moisture absorption. For compressive loading, the failure mode remained consistent, but for tensile loading, the failure morphology shifted from an X-shape around the hole to an I-shape. This study underscores the importance of accounting for hygrothermal effects when evaluating composite materials for ship and offshore applications.

3.3.2 Mechanical Damage

Cracks in metallic marine structures are commonly observed along weld lines and at the intersection of stiffening elements. Understanding the impact of cracks on structural strength is necessary for preventing complete failure and informing maintenance strategies, yet studies on cracked aluminium panels are limited despite extensive literature on the residual strength of aluminium residual strength. Attia et al. (2023) conducted a series of FE analyses on cracked AA5083 H116 panels, varying crack length, location, and orientation to evaluate their effects on ultimate strength. It was found that the location of the crack significantly impacted the ultimate capacity, equivalent stress distributions, and deformation shapes for transverse and inclined cracks, whereas this was not the case for longitudinal cracks. As the crack length increased, the ultimate capacity of the panel decreased. Taking the loading rate into account, Shi *et al.* (2023a) analysed the dynamic ultimate strength of cracked, thin steel plates under compression. Steel with 315 MPa yield strength was modelled using non-linear FE analysis. An empirical formula was then derived accounting for loading rate, crack, and plate geometry. It was reported that cracks caused up to a 61% reduction in the dynamic ultimate strength of plates when compared to uncracked specimens. Both a longer crack and a larger crack angle reduced the ultimate strength. Similar to the findings by Attia et al. (2023), the transverse location of the crack also influenced the dynamic ultimate strength more than the longitudinal location, with a decreased effect observed as the crack approached the unloaded side of the plate. The length and aspect ratio of the plate were found to influence the dynamic ultimate strength, while the slenderness ratio had little impact at a high strain rate.

Through a combination of non-linear FE analysis and the use of an ANN, Li *et al.* (2022a) found that the ultimate strength of plating is significantly influenced by crack orientations and configuration features. The combined effect of these influential factors was captured through an empirical mathematical expression derived by the ANN. The machine learning algorithm was trained and validated with an independent database to evaluate the accuracy and applicability of the suggested approach. The impact of the crack location on the ultimate strength of the hull girder was examined by Babazadeh and Khedmati (2021) for a bulk carrier. Their study indicated that the presence of cracks in deck structures and the side structures near the decks has the most significant effect on ultimate strength under both sagging and hogging conditions.

Low cycle loading is another key source of damage affecting the ultimate strength of marine structures during service. Li *et al.* (2019a) presented an analytical method to predict the collapse behaviour of plates and stiffened panels under severe low-cycle loading scenarios where loading magnitudes approach or exceed ultimate strength. The method was developed by performing parametric FE studies. This study included an investigation of the effects of kinematic versus isotropic hardening. The model was then formulated in a manner suitable for introduction into an ultimate strength framework based on the Smith method, thereby facilitating rapid predictions of potential hull disintegration. Deng et al. (2022) conducted experiments as well as FE calculations on single and double-hulled girder test sections with large deck openings to investigate severe cycling loading beyond the ultimate strength of the girder. This work included numerous details related to the tested structures, such as material stress-strain curves, residual stress measurements, and initial distortions. Additionally, 3D scanning was performed after every load cycle to characterise the damage progression. The corresponding FE simulations demonstrated the accumulation of residual stress, damage, and reduction in structural capacity as a result of cyclic-loading-induced damage, see Fig. 11.

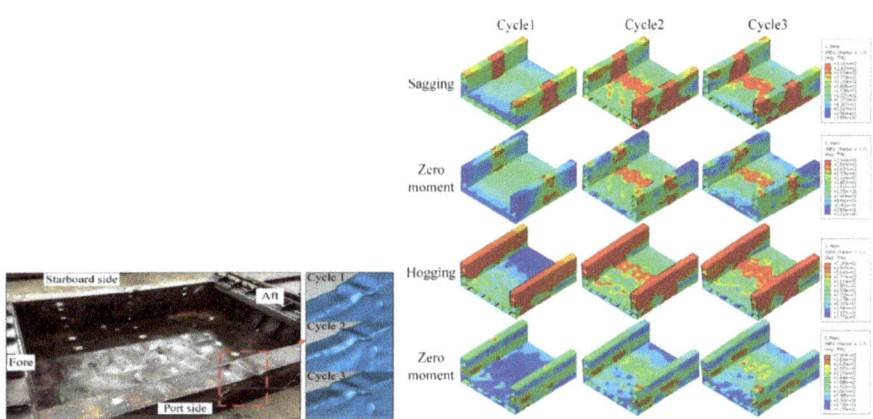

Fig. 11. Experimental setup and corresponding FE analyses for a double-hulled case showing residual stress accumulation during severe hogging/sagging cycles (Deng et al., 2022).

Studies have also been conducted to investigate the combined effects of cracks and cyclic loading. Hu et al. (2022) studied the failure mechanism and behaviour of cracked panels under cyclic loading. They concluded that when the buckling half-wave number is odd, the crack location at the trough/crest of the buckling mode results in larger deformation in the adjacent area around the crack. This accelerates strain accumulation near the crack tip, promoting crack propagation and leading to greater ultimate strength reduction. Conversely, when the buckling half-wave number is even, its effect is insignificant. This was found to be true for cracks either located in the centre or at the edge of the panel. Kang et al. (2022) investigated cracked steel panels under extreme cyclic loading and developed a formula to predict the ultimate strength for different types of crack propagation. FE analysis revealed that cyclic loading induces smaller deformations than monotonic loading due to lower load amplitudes. Hence the ultimate strength of a non-cracked panel remains relatively unaffected by an increase in cycles. However, increasing the load accelerates crack propagation and decreases ultimate strength for all crack positions. Double-edged cracks exhibited the fastest crack growth and ultimate strength reduction, while thicker panels facilitated greater crack extension. The study conducted by Song et al. (2022) focused on the effects of pre-existing and low-cycle fatigue (LCF) cracks on the residual ultimate strength of steel plates. An elastic-plastic analysis was carried out to determine the bearing capacity of plates under various loading conditions for cracks with different lengths and locations. FE analysis was then used to replicate buckling and collapse behaviour. A lower bearing capacity was observed as the crack propagated when compared to static cracks. Additionally, cyclic tensile loading produced deflection in the opposite direction of initial deflection, which improved compressive bearing capacity. However, higher cyclic load amplitudes increased deflection and reduced ultimate strength. Future improvements to the proposed LCF model may include consideration of the crack surface contact effect.

Openings in composite stiffened panels can cause cracking. Liu et al. (2022b) investigated the effect of open cracks on the residual bearing capacity of a single-hat-stiffened composite panel, using an ANN model. The model input parameters included crack orientation, length, width, and angle. The main emphasis of the study was on producing an ANN with a satisfactory learning rate without overfitting. The network was trained using approximately 90% of the 3830 FE analyses that included a Hashin criterion for the in-plane failure, and a cohesive zone approach for the possible debonding between the stiffener and panel. The use of the ANN allowed the authors to study the effect of the input parameters of the crack on the residual buckling strength.

The buckling and post-buckling behaviour of J-type composite stiffened panels were theoretically predicted and tested both numerically and experimentally by Wu et al. (2023). The authors also investigated the influences of adhesive layers, the layup configurations of both the stiffened rib and skins, geometric dimensions, and accidental impact load positions of the J-type ribs on the structural buckling responses. It was found that the failure modes included delamination, debonding, skin crushing, and rib fracture. The skin between the ribs was the first buckling location. The load-bearing capacity was found to be sensitive to the impact energy when the back of the rib was impacted.

3.4 Life-Cycle Management Analysis

Life-cycle analysis can provide a rational basis for the management of ageing ships and offshore structures with reduced ultimate strength. Uncertainties that may arise from time-varying deterioration phenomena (e.g., corrosion, crack, delamination), loads, or load effects, and modelling/prediction techniques dictate the use of probabilistic approaches for life-cycle management. Structural reliability and risk-based methods constitute important performance indicators for structural safety in the presence of uncertainties, based on which key actions and decisions related to life-cycle aspects (e.g., maintenance, inspection, repair, service life extension) can be obtained. This section discusses the state-of-the-art research related to the life-cycle analysis of ships and offshore structures, including inspection, maintenance, service life extension, damage detection, and monitoring. Moderate progress has been made in the field. Risk-based analysis tools are the most commonly used analysis method, whereas digital twin and machine learning applications are emerging as viable solutions to facilitate more effective life-cycle management.

Inspection planning is an important aspect of ship life-cycle management. Unlike standard periodic class surveys, optimal inspection planning seeks to reduce costs and minimise unnecessary downtime while maintaining structural safety. Machine learning techniques offer a promising approach for optimising inspection decisions. Cheng et al. (2022) proposed a reinforcement learning method to optimise the dynamic inspection policy of ship structures. Unlike fixed inspection policy, dynamic inspection policy allows inspections to be based on the actual state and can avoid unnecessary or insufficient inspections. Results indicated that dynamic inspections can effectively reduce the expected life-cycle costs of a vessel. Acquiring information from inspections can assist in future inspection planning (e.g., selecting the next inspection time) and life-cycle management by reducing the uncertainties in the prediction models (e.g., crack propagation or corrosion wastage models). Bayesian inference is another powerful mathematical tool for this purpose. Liu *et al.* (2021c) used a Bayesian inference technique to reduce the uncertainties related to crack propagation and refine the prediction of fatigue-induced structural deterioration. Kim *et al.* (2022a) also employed a Bayesian inference tool in an analysis of inspection data to update corrosion wastage model predictions. Other relevant studies have been published over the past decade, see ISSC (2022a).

In the context of material/structural degradation, recent work has focused on optimising corrosion addition. Gong et al. (2020) introduced a risk-based framework that allows decision-makers to determine the economically optimal corrosion addition and the benefits realisable through reducing expected life-cycle costs. This framework incorporates reliability analysis in ultimate limit state (ULS) assessments and considers the impact of periodic dry-docking maintenance on the life-cycle cost estimates of an oil tanker. The study found that the expected life-cycle cost of a hull girder with added corrosion allowance is significantly lower than that of a net scantling design.

A first-order reliability analysis (ultimate limit state) was carried out by Garbatov et al. (2023) on two lightweight hybrid bulk carrier designs containing aluminium honeycomb sandwich panels in the inner hull over their service lives. A two-stage corrosion model was used for the honeycomb structures, assuming a coating life of 5 years and a coating-break-down transition time of 6–7 years (as opposed to 4 years for steels). This was also combined with a cost-benefit analysis including factors such as loss of cargo, human life, and cost of CO_2 emission. However, only general corrosion was considered, whereas corrosion of aluminium alloys in marine environments is primarily pitting/localised.

Recent research has also examined the impact of corrosion on the operational life of traditional offshore platforms. Mohd et al. (2022) applied Melcher's two-stage empirical corrosion model to assess the global ultimate strength and reserve strength ratio (RSR) (the ratio of base shear ultimate strength versus its design strength with a 100-year return period) of a platform through push-over analysis, considering both average and severe corrosion levels. The PETRONAS guideline requires an RSR of 1.5 to be adopted for manned platforms and 1.32 for unmanned platforms. The study found that, with an RSR of 1.5, the platform could safely operate for up to 50 years under average corrosion conditions, but only 35 years under severe corrosion. Furthermore, the authors concluded that the corrosion in the atmospheric zone did not affect the global strength of the structure (Othman et al., 2023). Corrosion in the immersion zone was found to be more critical than that in the splash zone for platform life extension. The corrosion model led to a 6-year extended life span for all the studied jacket offshore platforms. However, only uniform corrosion (thickness reduction) was considered in these studies and the sensitivity of the corrosion model to life span estimation was not assessed.

The offshore wind sector has widely adopted supervisory control, data acquisition, and condition monitoring systems during operations (Yeter *et al.*, 2022b). However, as the industry pushes to reduce costs further, there is a growing need for "smarter" life cycle management and life extension strategies. These strategies should not only draw on experience from conventional offshore platforms but also integrate "big data", AI-based predictive models, and intelligent decision-support tools. Yeter *et al.* (2022b) published a comprehensive review of AI-aided life extension studies of offshore wind structures. They discussed key challenges, including data acquisition, preprocessing (such as noise filtering algorithms), and risk/reliability-based machine learning approaches. The authors highlighted the importance of reassessing structural capacity for various failure mechanisms (linked to different levels of risks), such as the ultimate strength of corroded leg components, to inform the decision-making process. Additionally, they explored a cycle-by-cycle approach for structural integrity assessment to extend the life of wind turbine support structures (Yeter *et al.*, 2022a). Although this research primarily focused on the operational stage, the interconnected nature of design, manufacturing, life extension, and decommissioning stages suggests that AI-based approaches will see increasing applications across the offshore wind sector shortly.

The final key aspect of a life-cycle analysis is determining an optimal end-of-life strategy, i.e., service life extension or decommissioning. Service life extension of ship structures has become increasingly important due to sustainability goals and the significant costs associated with new ship construction (Liu *et al.*, 2020c). Risk-based analysis is a viable tool to assess the performance of hull structures beyond their intended design life and for planning necessary maintenance actions. Liu *et al.* (2021c) presented a risk-based cost-benefit analysis for extending ship service life by addressing hull structural failure. Two cost-benefit indicators, benefit-cost ratio and net present value, were used as financial performance indicators to guide optimal end-of-life decision-making. This framework provides a structured approach for determining optimal strategies across various oil tanker sizes.

4 Ships

Section 4 is focused on the ultimate strength of ship structures. Unlike fixed offshore structures, ships are inherently meant for transit operations. As the world becomes increasingly focused on addressing climate change and reducing emissions, ship operations need to become more environmentally sustainable. Naval architects can contribute to resolving the ongoing energy crisis by designing lightweight ship structural systems or developing tools that enable the design of such systems. These novel systems should facilitate increasing the amount of cargo transported per unit of energy spent. This must be balanced with the need for ship structures to withstand imposed loads over their lifetimes with a sufficient margin of safety. Optimising the design and construction of ship structures can enable weight reductions and increased efficiency, which can ensure significant fuel savings and a corresponding reduction in emissions. However, the design of such lightweight structures is a compromise between efficiency and safety. In other words, lightweight efficient designs must not compromise the safety of structures, operators, or people at large. Therefore, we need to be more aware and knowledgeable of the ultimate limit states of ship structures, and phenomena that affect ultimate strength. With this aim, this section presents the most relevant findings from recent years.

4.1 Loads

This section provides an overview of the loads (or load effects) acting on ship structures and their effect on ship response. Broadly speaking, these loads include (i) static loads and (ii) low- and high-frequency dynamic loads caused by waves. Static loads are caused by the uneven distribution of weight and buoyancy, while dynamic loads result from various forces, such as wave-induced pressure on a hull's wetted surface, inertial force, sloshing, slamming, whipping, and green water on deck.

4.1.1 Non-linear Wave Load Effects

The assessment of the ultimate strength of hull girders is a central element in ship structural design. Ensuring the safety of the ship structure from an ultimate strength perspective necessitates designing hull structures that have a load-carrying capacity that surpasses applied loads. This underscores the importance of investigating the loads applied to hull girders, particularly focusing on the vertical wave bending moment, to assess the safety of the ship structure in terms of its ultimate strength.

The response of a ship structure to time-varying loads is inherently a dynamic event. However, it is convenient to analyse this response as a quasi-static problem. The frequency domain approach for determining design load in ship structural design has been extensively employed. However, this approach inevitably introduces approximation errors resulting from the linearisation of complex problems. For instance, the frequency domain-based analysis method assumes identical hogging and sagging wave bending moment magnitudes, while in reality, large waves acting on a container ship can produce a significantly larger sagging bending moment. To address these errors, attempts are being made to directly calculate irregular wave loads using computational fluid dynamics (CFD). Choi et al. (2023) proposed a method that applies high-order spectral (HOS) to model non-linear ocean waves, and rapidly generate the necessary irregular waves for structural analysis by interpolating the fluid domain information of the modelled non-linear ocean waves using 3rd and 4th order B-spline interpolation. The method is computationally more efficient than CFD-based wave generation using linear superposition and provides a more accurate non-linear description of waves.

Takami et al. (2023) introduced a method for predicting the extreme distribution of the vertical wave bending moment in ships considering non-linear wave fields that include freak waves by combining the higher order spectrum method (HOSM) and the first order reliability method (FORM). They scaled down a 6,600 TEU container ship to create two hypothetical vessels with lengths of 100 m and 200 m, respectively, and estimated their extreme hogging vertical bending moments. They found that the shorter vessel was more sensitive to the effects of wave shape non-linearity. Therefore, traditional linear methods may underestimate the wave-induced vertical bending moment (VBM) because they exclude non-linear wave shape effects. The influence of wave non-linearity on ship motion and vertical wave bending moment was also investigated numerically by Houtani et al. (2023). The simulations involved varying the wave height and wave non-linearity order, specifically considering the first and fifth orders, for a wall-sided ship with the same dimensions (length, width, and depth) as a 6,600 TEU container ship with a large flare. They found that both the shape of the ship and the non-linearity of the wave shape can significantly affect the wave-induced VBM. These studies demonstrate the importance of non-linear wave shapes on wave-induced VBM, which becomes more pronounced in large waves. Moreover, the incorporation of these methods in the design for ultimate strength allows consideration of more realistic loading scenarios (dynamic and non-linear effects) and scenario-based modelling as advocated in the ISSC 2018 Ultimate Strength committee report (ISSC, 2018a), which should result in higher fidelity ultimate strength analysis.

4.1.2 Dynamic Load Effects

Dynamic load effects on ultimate strength have historically received limited attention, as noted in the ISSC 2018 Ultimate Strength committee report (ISSC, 2018a). However, recent research shows growing interest in this area. Xiong et al. (2023) developed an empirical formula to predict the dynamic ultimate strength of stiffened panels under longitudinal compressive impact loads, addressing the scarcity of studies on rapid dynamic loads in extreme wave conditions. Using FE analysis, they evaluated steel panels with varying yield strengths (315–340 MPa) and analysed parameters like impact duration, shape, and plate geometry. The findings indicate that impact duration significantly affects dynamic ultimate strength, with longer durations reducing strength, while factors like amplitude, impulse, and impact shape have minimal influence. The study emphasizes the need to further investigate in-plane impact loads.

Zhong and Wang (2021b) studied the dynamic ultimate strength of steel-stiffened panels under in-plane impact load and lateral pressure, focusing on rapid dynamic loads like explosions. Using non-linear FE analysis, they found lateral pressure minimally affects dynamic ultimate strength for short impacts but becomes significant for longer durations. Short-duration impacts exhibited higher dynamic ultimate strength, which was influenced by the model range. The strength was notably impacted by local plate deflection, showing an inverse correlation with impact load impulse. Additionally, increased lateral pressure led to a reduction in quasi-static ultimate strength.

Guo et al. (2024) examined the dynamic ultimate strengths and deformation behaviours of stiffened plates under various loading cases using non-linear FE analysis. The study confirmed that dynamic ultimate strength decreases with longer slamming load durations, with collapse modes transitioning from local buckling to edge collapse. Lateral loads significantly reduced strength after 20 ms of slamming duration, while transversal in-plane compressive loads further decreased strength due to global buckling.

Jagite et al. (2022) used FE analysis to investigate the ultimate strength of large containerships under dynamic loading scenario due to slamming induced whipping. Novelty of the work relates to realistic loading conditions and the way dynamic effects are calculated. Namely, dynamic effects are quantified by considering two scenarios: a wave scenario, and a wave-whipping scenario wherein high-frequency whipping-induced stresses are superimposed to the low-frequency wave-induced stresses. The dynamic collapse effect was calculated as the relative value between whipping and wave scenarios, and it quantifies the increase of structural capacity due to slamming-induced whipping. In considered case, the effect varied from 1.05% to 2.26%. This was considerably less than obtained with traditional approach 4.8% to 8.4%, which quantifies the effect as the ratio between the dynamic ultimate strength and the quasi-static one. The authors stress that the obtained results with new calculation method align well with long-established industry practice which considers the wave loads as quasi-static and disregards any dynamic effects associated with the wave loads.

Kong et al. (2021) investigated the ultimate strength and damage mechanisms of a steel deck under quasi-static and dynamic in-plane compression loading through experimental and numerical methods. A structural model was used for quasi-static testing, while simulations analysed the effects of load amplitude, frequency, initial stress, and deflection on strength and failure modes. Dynamic buckling modes differed from quasi-static ones, with buckling strength decreasing proportionally to pre-applied quasi-static loads. Certain load frequencies induced dynamic buckling even at amplitudes near static load capacity. Geometry, boundary conditions, and material properties influenced buckling, with dynamic loads causing more damage than static ones. The study emphasized further testing to understand post-buckling behaviour under in-plane impact loads.

Shi and Gao (2020) analysed two container ships (14,500 TEU and 19,000 TEU) that failed due to whipping loads to assess the impact of whipping on hull girder reliability. Using the first order reliability method (FORM), they identified significant variability in whipping moments, model tests, and calculations. It was also found that evaluation criteria for hull ultimate strength are not the same across all classification guidelines. Therefore, large container ships should have their minimum ultimate capacity increased to maintain a safe failure probability. The authors recommended that more case studies be carried out on other ships to improve their analysis.

4.1.3 Other Load Effects and Load Combinations

Empirical models, FE analysis, and machine learning based approaches have been employed to improve the accuracy of strength predictions under complex loading conditions. These approaches aim to enhance the safety and reliability of structures by addressing factors such as external pressure, cyclic loading, and temperature effects.

Seyffert et al. (2019) presented an interesting approach that employs the non-linear Design Loads Generator (NL-DLG) process to efficiently estimate the lifetime performance of stiffened panels under combined loading. The approach allows for a direct comparison between panel options and highlights critical panel parameters that are strongly related to design robustness. The approach was used to compare the failure probabilities of six different stiffened panel designs for a specific vessel in a long exposure to harsh ocean excitation. The comparison was especially interesting since some class society had previously vetted all designs. It was demonstrated that the NL-DLG process yields similar statistical characteristics to those of brute-force Monte Carlo simulations, while reducing simulation time by a factor of 87,000.

Ma and Wang (2021) investigated the effect of lateral pressure on the collapse behaviour of stiffened plates under combined loading, developing an empirical formula verified through numerical analysis to provide a faster, reliable method for estimating ultimate strength. They found that lateral pressure influences initial deflection and pre-buckling strength, significantly reducing ultimate strength, especially for thinner plates. Longitudinal ultimate strength may increase for column slenderness above 0.7 with periodic symmetrical collapse modes, but lateral pressure impacts longitudinal and transverse strength differently. Thicker plates or those with larger stiffeners were less affected. The authors recommended refining the empirical formula by analysing parameters influencing ultimate strength and validating it with additional data.

Li and Chen (2023b) investigated the effects of extreme cyclic loading and lateral pressure on the collapse behaviour and ultimate strength of continuously stiffened panels using non-linear FE analysis. The study focused on parameters such as slenderness, stiffener geometry, and load characteristics under combined loading scenarios, addressing gaps in prior monotonic loading studies. Findings showed that cyclic loading can induce buckling, especially with smaller stiffeners, and strength reductions increase with panel thickness, stiffener size, and cycle numbers. Weak stiffeners combined with thick panels were identified as particularly detrimental. Under combined loads, lateral pressure can prevent overall buckling, but this effect decreases with an infinite number of stiffeners and is more significant for slender panels. The authors recommended further study of complex loading parameters to refine collapse and strength predictions for stiffened panels.

Hu-Wei and Ping (2018) proposed a simplified approach to represent the stress-strain relationship of plates under monotonic uniaxial compression using the envelope of the stress-strain curve from cyclic uniaxial compression. The approach was validated against previous tests and analyses. Tests were conducted on six square-column models to simulate the collapse behaviour of hull plates under cyclic compression, and the results were compared with the predictions from the simple approach. The results from simple approach showed good agreement with experimental results (see Fig. 12). The study concluded that the method's accuracy could be enhanced by incorporating residual stress and uneven plastic deformation effects.

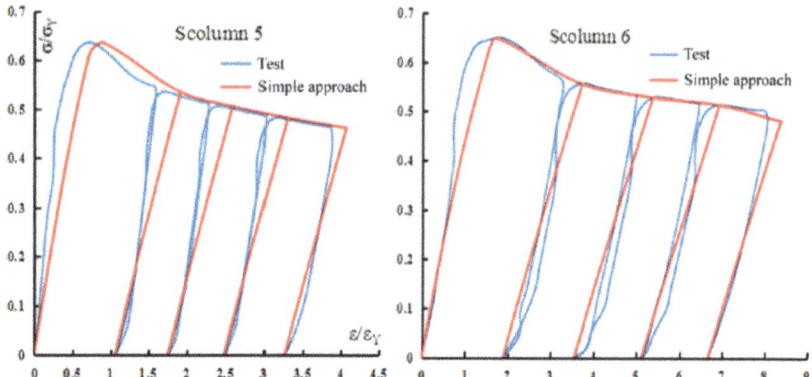

Fig. 12. Curve comparison between the simple approach and experiment (Hu-Wei and Ping, 2018).

Park and Kim (2022) utilized an ANN to predict the ultimate strength of a GTT LNG cargo containment system under sloshing impact, reducing reliance on complex numerical analyses during design. The ANN, with two hidden layers and 64 neurons per layer, was trained on 24191 FE analyses, achieving high accuracy. The study also evaluated the influence of including more realistic onboard boundary conditions, sloshing rise time, load area, and temperature. The onboard conditions significantly affected ultimate buckling strength compared to rigid boundaries. A key limitation was that the ANN is applicable only within the scope of the training data.

4.2 Structural Elements

Stiffened panels are the fundamental building block and load-carrying component of ship structures. This section reviews recent developments in the ultimate strength analysis of stiffened panels, considering aspects such as material properties, geometric imperfections, residual stresses, and load combinations. With improved access to computational resources, these models have become increasingly detailed, which can influence the final collapse mode of stiffened panels. Additionally, there is a clear trend toward developing computationally efficient, machine-learning-based methods to incorporate all these complex details.

4.2.1 Stiffened Panels

Various influences on the ultimate strength of the stiffened panel were investigated, such as initial imperfections of the actual panel, stiffener misalignment (Chen et al., 2023), small and large openings (Liu *et al.*, 2021b), residual stresses (Guo et al., 2021), welding techniques (Mun and Ri, 2022), and corrosion (Zhu et al., 2021; Mokhtari et al., 2023).

Different buckling failure modes in stiffened panels lead to variations in ultimate strength. To address this, Zhang *et al.* (2023a) performed a series of non-linear finite element analyses with a stiffened panel benchmark model, including appropriate initial imperfections, boundary conditions, and mesh size. Compared with earlier closed-form analytical expressions for ultimate strength (Paik and Thayamballi, 1997; Zhang and Khan, 2009), which considered column slenderness λ, plate slenderness β, web height to thickness ratio h_w/t_w, their evaluation additionally accounts stiffener tripping slenderness λ_e parameter. Thereby, the fitting equation accounts for beam-column buckling, local plate buckling, stiffener web buckling, and stiffener tripping. Comparison with earlier closed-form expressions and FE results demonstrates the increased accuracy of the fitting expression especially in the range where stiffener failure is expected (see Fig. 13).

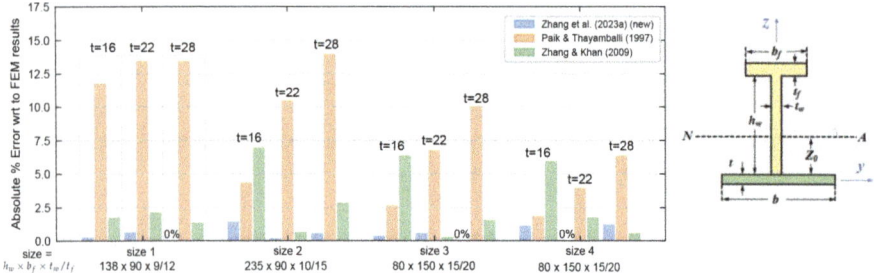

Fig. 13. Table 8 from Zhang *et al.* (2023a) converted to a figure showing the error of different fitting formulas with respect to FE analysis results.

Another fitting formula for the ultimate strength of stiffened panels was derived by Hanif et al. (2023). However, compared to Zhang *et al.* (2023a), their fitting equation focused on the imperfection mode, with separate parameters for column, torsional, and local imperfections. A total of 750 stiffened panel configurations were investigated using a non-linear FE analysis. Regression analyses performed across studied cases showed that plate slenderness and local imperfection modes had the largest influence on strength. Furthermore, their sensitivity analysis revealed that, among the imperfection modes considered, the local imperfection mode had the strongest influence on the normalized strength, confirming the results of Li *et al.* (2022c). The derived formula demonstrated good accuracy when compared with the FE analysis results, with a mean absolute percentage error (MAPE) of 1.9%. The model was also compared against six other fitting equations, with the best showing a 6.89% MAPE. This confirms its validity as a reference for evaluating the ultimate strength value in scenarios where an initial geometric imperfection is present.

Li *et al.* (2022c) provide recommendations for selecting an imperfection model for buckling analysis of ship-type stiffened plated structures based on their comparative study regarding the effects of different imperfection modelling approaches on the ultimate compressive strength. Models were categorized into deterministic and probabilistic approaches. Deterministic models included hungry-horse (HH), Admiralty Research Establishment (ARE), and critical buckling (CM) mode, which correspond better to as-built panels rather than in-service panels. In contrast, the probabilistic approach treats initial imperfections as a random field generated from specified statistical parameters and is thus believed to be applicable for both as-built and in-service panels. The authors conclude that it is difficult to conclusively suggest the best imperfection model for collapse analysis. Different models have different advantages, but recommendation given by the

Committee III.1: Ultimate Strength 497

authors take into account the convergence of numerical simulations and the effect on accuracy. Considering the randomness of imperfections authors advocate the use of probabilistic imperfection model but concede that it is more complicated than deterministic approaches.

One approach to account for mechanical damage to a stiffened panel is to consider the damaged geometry as a form or type of imperfection. Chujutalli et al. (2020) investigated the ultimate strength reduction of stiffened panels under compression due to indentation damage. The ultimate strength, failure mode, and deformation field were measured experimentally for selected scaled stiffened panels, which were used to validate the simulation model. The validated FE model was subsequently used for parametric study to gauge the effects of indentation dent depth, indenter diameter, indentation location and stiffener slenderness. It was found that depending on the stiffener type used (flat bar or T-stiffener), the indentation parameters have a different effect on the response.

Xu et al. (2021b) proposed a partial similarity method to account for buckling and distortion when assessing the ultimate strength of small-scale stiffened panel models. Complete similarity methods are unsuitable for use on slender/thin elements, therefore, highlighting the need for a more accurate similarity method that can predict the ultimate strength of a full-scale ship from a smaller model, as the proportions of members on a small-scale model are not necessarily reflective of those on a full-scale model. This study was also driven by a need to consider plate buckling on the scaled model. AH32 steel coupons were tested and used in the verification of the FE model. It was concluded that for partial similarity, plate and column slenderness govern beam-column buckling, while plate torsional slenderness governs tripping buckling. Partial similarity methods can lead to dimensional distortion across different buckling modes, potentially altering the collapse mode and load-carrying capacity of stiffened panels. The collapse mode and governing parameters should be carefully considered during model scaling to minimise dimensional distortion. Complete geometric similarity may be applied to small-scale models for validation. Based on results for collapse mode pre- and post-buckling average stresses, partial similarity can be used to guide the design of test models if key parameters specific to the collapse mode are considered.

It is well established that distortions in a structure's geometry resulting from fabrication and manufacturing processes significantly influence the ultimate strength and structural stability of the system. However, existing methods for quantifying these distortions heavily rely on assumptions that limit their practicality. With the growing use of full-field 3D measurement and inspection techniques such as light detection and ranging (LiDAR) and photogrammetry, it is now possible to inspect and quantify distortions during inspection activities or even continuously during operations. The capability, use, and availability of such data will only increase over time. This trend has sparked interest in the use of quantitative 3D inspection data to create digital twins of structures for the production of data that can be used to update FE models so that they accurately reflect the actual state of the structure. As noted in the ISSC 2018 Ultimate Strength committee report (ISSC, 2018a), these data are also essential for reliability analysis, this data is crucial for reliability analysis, though the problem was previously unaddressed in the literature at that time. By integrating inspection data with FE models, strength reductions can be assessed, enabling more accurate insights into current serviceability and remaining life than an idealised model could provide. For instance, Graves et al. (2023) presented a process for incorporating point cloud measurements of distortions into FE models through a Gaussian process interpolation that inherently filled in gaps in data, eliminated outliers, and provided quantitative measures of uncertainty. They demonstrated the utility of this approach for the provision of strength estimates that were comparable to established methods and its wider applicability to a range of cases where the quantification of a performance envelope for risk analysis would be useful.

As large amounts of data become available and data analysis techniques evolve, data-driven approaches are expanding to the field of structural engineering. One such data-driven approach was implemented by Li *et al.* (2022b) to estimate the ultimate strength reduction factor of stiffened panels caused by welding residual stress. Various machine learning (ML) methods were compared based on their ability to predict this reduction factor. The data consisted of 136 simulated scenarios of simply-supported two-bay and two-span stiffened panel models under uniaxial compression. The simulation model had been previously introduced by the same author (Li *et al.*, 2021a). The authors reported that the best ML method had a 99% prediction accuracy for panels within the limits of the dataset and 96% accuracy for panels with configurations which deviated from those used for learning by up to 15%. The approach demonstrates the potential of data-driven approaches in the realm of ship structural design, which often requires expensive non-linear numerical simulations, but as shown here, can be replaced by a computationally inexpensive data-driven method.

Hosseinpour et al. (2022) used FE models to study the behaviour of steel plates with a central circular hole subjected to compressive axial loading. A total of 270 holed steel plate configurations were modelled and analysed using the ABAQUS software. The effects of four main variables including plate length, hole diameter, plate thickness, and yield stress were discussed. Using the database provided by the FE analyses, the ANN method was used to develop a predictive model to estimate the ultimate strength of steel plates with a circular hole in the centre. The accuracy of the ANN-based formula was compared and confirmed with formulations presented in previous studies. Lima et al. (2023) also introduced a method to replace computationally expensive structural reliability models with a surrogate model based on a bi-fidelity Kriging model. The results suggest that the proposed bi-fidelity framework, which used the correlation between high-fidelity (HF) and low-fidelity (LF) non-linear FE analysis models (as shown in Fig. 14), can provide an accurate failure probability estimation with less computational cost compared to solely using HF non-linear FE analysis.

Most numerical studies exploring the ultimate strength of ship structures utilise a large number of calculations to explore the parameter space. Papanikolaou and Anyfantis (2022) employed a design of experiments (DoE) and response surface methodology (RSM) combined with numerical non-linear FE analysis to develop a surrogate model for the ultimate strength of stiffened panels. It was demonstrated that the buckling response with respect to the nondimensional slenderness ratio could be fitted with nine runs per stiffener geometry. The T-bar stiffener surrogate model was found to be sufficiently accurate for ultimate stress prediction in the practical design space, while the surrogate models for angle bars and flat bars demonstrated a difference between 10 and 30% from common structural rules (CSR).

The equivalent single-layer (ESL) homogenisation method was presented by Putranto et al. (2021). ESL transforms a 3D stiffened panel into a plate with equivalent stiffness with comparable mechanical behaviour. ESL stiffness was obtained through unit-cell analyses based on the stiffened panel where periodicity was imposed with boundary conditions based on the first-order shear deformation theory (FSDT). Stiffnesses were determined from the first derivative of the membrane force and bending moment obtained through numerical simulations. The effect of the initial imperfection shape was included in the analysis to account for local and global buckling behaviour. ESL with non-linear stiffness was implemented in the ABAQUS UGENS subroutine, allowing incremental evaluation of stiffness. The key aspect of this analysis was the use of non-linear ESL stiffness, as linear analysis was unable to detect the point at which the grillage reached its ultimate strength capacity. The method was computationally efficient and allowed for considerable reduction in modelling effort by smearing of stiffeners. The method was further validated by authors in Putranto *et al.* (2022a) for combined compression and in-plane shear.

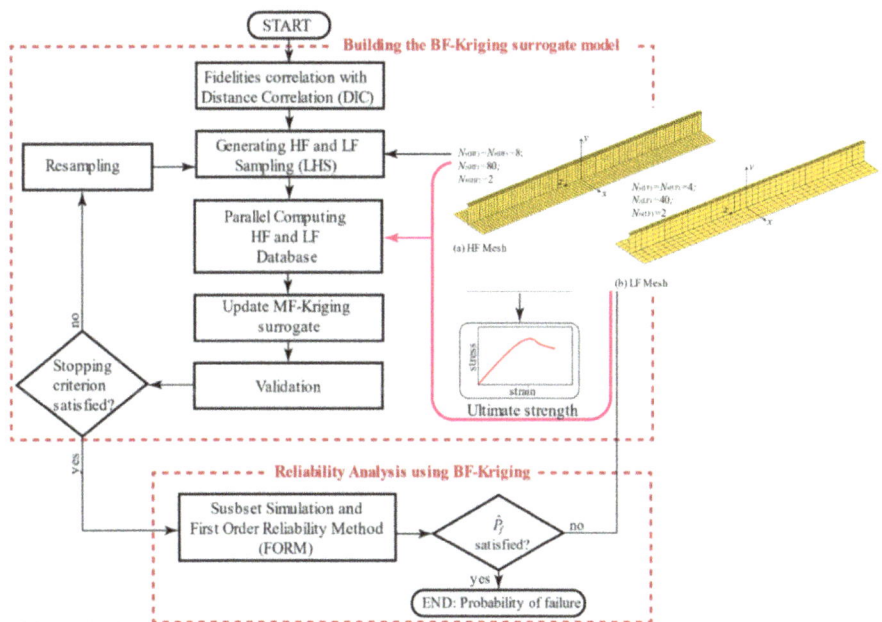

Fig. 14. Flowchart of the bi-fidelity Kriging framework.

Kõrgesaar et al. (2023) demonstrated the applicability of the ESL approach for the ultimate strength assessment of stiffened panels under various loading combinations of uniaxial (longitudinal and transverse), biaxial and shear. Using the ESL approach, the 3D stiffened panel was replaced with its ESL representation, and the tertiary stiffening elements were excluded from the FE model, as shown in Fig. 15. This approach offers significant time savings in modelling the final structure. ESL analyses were conducted using both the ABAQUS implicit and explicit subroutines to compare numerical efficiency and accuracy between the two schemes.

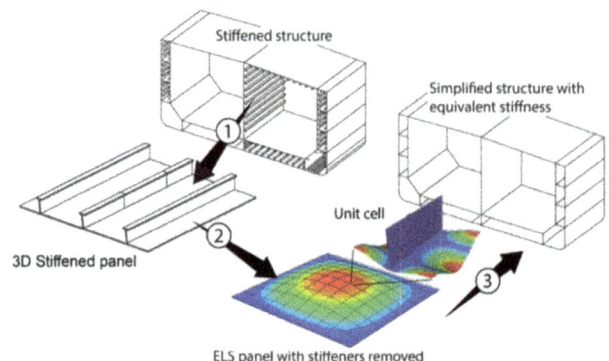

Fig. 15. FE model simplification using ESL approach (Kõrgesaar et al., 2023).

Long-span stiffened panels, which often utilise girders with deep webs and multi-web openings, are commonly employed in open spaces without stanchions to create large open spaces, such as theatres and lounges in large passenger ships. These panels are often supported with in-tension pillars to decrease deck deflection and out-of-plane deformations, but their effect on the compressive strength of supported panels is unknown. Liu *et al.* (2021b) presented experimental and FE analyses of the ultimate compressive strength of two long-span stiffened panels subjected to combined uniaxial and lateral loads. The comparison with experimental results demonstrated that the in-tension pillars only marginally increased the ultimate compressive strength of the stiffened panels. The experimental results aligned with the numerical simulations, and this study provides insights into the buckling and ultimate behaviour of long-span stiffened panels, including their progressive collapse processes.

Extruded aluminium structures are increasingly being used for high-speed craft. Such structures are often stiffened using floating girders to achieve lightweight and affordable ship hulls. Sun *et al.* (2021a) conducted uniaxial compression experiments with simply supported boundary conditions to investigate the differences in ultimate strength and collapse behaviour between a novel aluminium stiffened panel with a floating girder (SPT) and a fixed girder (SPX) (see Fig. 16). These experiments showed that a novel aluminium SPT had a higher ultimate strength than an SPX. The former experienced web tripping, while the floating girder panel exhibited girder buckling.

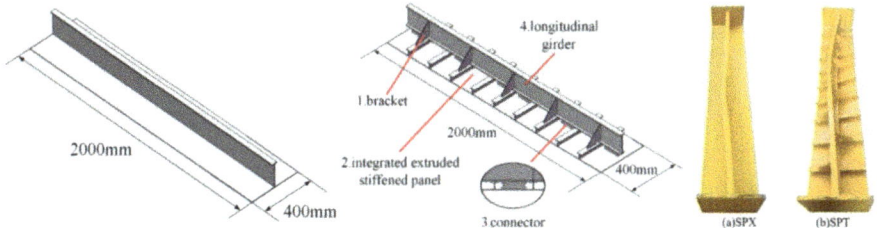

Fig. 16. Appearance of stiffened panel with fixed girder (SPX) and floating girder (SPT), and comparison of collapse deformation.

4.2.2 Curved Stiffened Panels

Unstiffened and stiffened cylindrically curved plates are often used in ship structures, such as in cambered decks, side shells at the fore and aft sections, and the circular bilge area. Park et al. (2020) conducted numerical analyses to investigate the ultimate collapse response of stiffened curved plates commonly used in ship structures. Various stiffener geometries shown in Fig. 17 were evaluated under axial compressive loading, considering quasi-static and cyclic loads, with initial imperfections and geometric non-linearity included. Results showed that increased stiffener height and plate curvature improved the elastic and elastic-plastic buckling strengths. Furthermore, elastic buckling exhibited complex secondary behaviour, absent in elastic-plastic buckling due to plasticity effects. The study emphasized designing for elastic-plastic buckling to account for the geometric instability and material non-linearity of curved plates compared to flat plates.

Alinia et al. (2019) studied the shear buckling and post-buckling behaviour of thin curved panels under lateral pressure and increasing in-plane shear forces, considering lateral pressure magnitude, the radius of curvature, and the panel aspect ratio. Theoretical buckling load predictions were compared with experimental results. The results showed that inward pressure eliminated the snap-through phenomenon and softening stage in shallow panels but had minimal effect in on moderately curved panels under low pressures. Increased inward pressures significantly reduced the ultimate shear capacity of highly curved panels, leading to unstable buckling and suppressed hardening stages due to released strain energy.

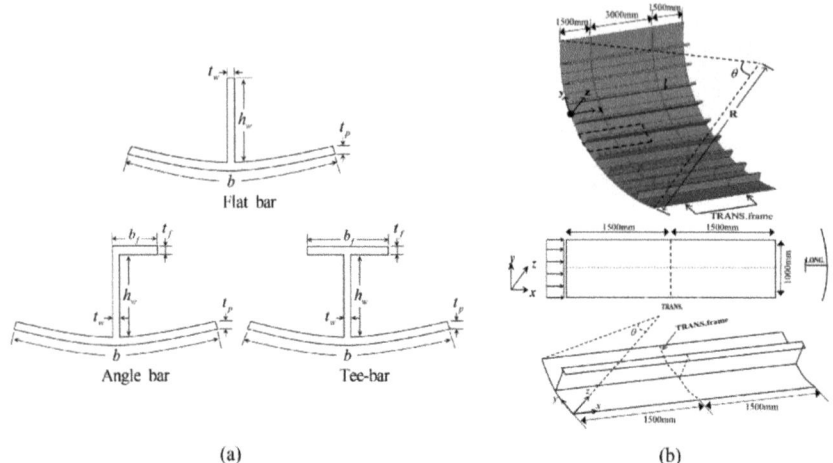

Fig. 17. (a) Typical shapes of curved plate and stiffener, and (b) double span/double bay model of a stiffened curved plate (Park et al., 2020).

Kim et al. (2021) developed empirical design equations to predict the ultimate strength of curved plates under longitudinal compressive loads. Their study included experiments on buckling collapse, considering factors like flank angle, slenderness ratio, and aspect ratio. Detailed finite element analyses were conducted to investigate buckling and collapse behaviour of curved plates under elastic-plastic large deflection conditions. The proposed approach allows for estimating collapse using these geometric characteristics.

Cho et al. (2022) employed a deep learning model to predict the ultimate strength of curved plates under compression. While the fully connected network provided satisfactory estimations, discrepancies in mean squared error results between training and validation sets emerged post the 500th epoch. Exploring alternative regularization techniques or learning rate schedulers could have mitigated this issue. Furthermore, due to the non-linear input-output vector relationship in one-dimensional space, the use of a "tanh" activation function might enhance performance. Despite these challenges, the deep learning-based predictions for ultimate strength showed notably superior accuracy compared to empirical formulae.

Shiomitsu et al. (2023) found that critical yielding locations impacted the ultimate strength of curved plates under axial compression. Yielding occurred along the longitudinal edges when dominated by membrane action, while it manifested at locations of maximum curvature when influenced by combined bending and membrane actions. The study proposed a straightforward approach for predicting the ultimate strength of simply supported curved plates under axial compression, which was verified through nonlinear finite element analysis outcomes for various flank angles.

4.2.3 Composite Elements

This section explores various methods for analysing and predicting the behaviour of composite materials under compressive loading, with a particular focus on delamination and failure mechanisms. Techniques such as FE analysis, neural networks (NN), and progressive damage models are typically utilised to assess the strength and structural integrity of laminated strips. Experimental studies complement these models by examining the effects of material properties, temperature, and moisture on the mechanical behaviour of composites.

Alizadeh et al. (2022) conducted FE analyses to evaluate the residual ultimate strength of composite strips with different layup configurations and a single through-the-width delamination under compressive loading. Using solid 8-noded layered elements delamination growth was modelled via the strain energy release rate calculated with the virtual crack closing technique, offering an efficient method to predict delamination regions. Two layups, $[0/90/90/0]_4$ and $[90/0/0/90]_4$ were analysed, with delamination introduced between the 5th and 6th plies. The first layup exhibited stable load stages, while the second showed a sudden capacity drop. This approach provides a straightforward assessment of delamination effects on compressive strength of laminated strips.

Liu *et al.* (2022b) built neural networks (NN) to examine the effect of crack characteristics – length, orientation, and width – on the compressive residual strength of single hat-stiffened composite panels. A network with five hidden layers, each containing 5 to 30 neurons, achieved optimal accuracy without overfitting. The model was trained on 90% of 3,830 FE analyses, which utilized the Hashin criterion for in-plane failure and a cohesive zone model for stiffener-panel debonding. This NN effectively predicted the influence of crack parameters on residual buckling strength.

Wang *et al.* (2022c) developed a progressive damage model to evaluate hygrothermal effects on strength and stiffness, based on tensile and compressive tests of open-holed specimens under dry and hygrothermal conditions. Failure models incorporating Hashin and maximum stress criterion were updated with the fitted relationships. While numerical predictions aligned well with experimental ultimate loads, discrepancies occurred in displacement or strain levels at failure. Results showed a 35–40% reduction in ultimate strength for both tension and compression under 70 °C with equilibrium moisture absorption. The compressive failure mode remained unchanged, but tensile failure shifted from an X-shape to an I-shape. This study emphasizes accounting for hygrothermal effects in composite material evaluations for marine applications.

Xu *et al.* (2022a) conducted experimental and numerical studies to investigate the failure behaviour of GFRP L-joints with a PVC core. They used a material model based on Shokrieh et al. (1996), a 3D adaptation of the Hashin criterion, to model fibre and matrix failure, shear failure, and delamination. Cohesive zone elements were employed to simulate core-laminate debonding, with a Maxwell element added to capture loading rate effects from the viscoelastic behaviour of the adhesive. FE analyses, conducted in ABAQUS with 8-noded linear brick elements, were compared to tensile ultimate strength tests. Incorporating viscoelastic adhesive behaviour provided better agreement with experimental results, especially in capturing post-linear behaviour and failure modes.

4.2.4 Complex Loading

Li and Chen (2023a) proposed empirical formulae for assessing the ultimate strength of continuous hull plates under combined biaxial compression and lateral pressure. Based on 4,800 numerical analyses using non-linear FE methods, the formulas treat longitudinal and transverse strength separately, enhancing applicability and accuracy. Anyfantis (2020) evaluated the ultimate strength of stiffened panels under combined uniaxial thrust and bending moments. Using design of experiments and non-linear FE analyses, the study found that non-uniform thrust significantly impacts ultimate strength of stiffened panels.

Woloszyk et al. (2021b) analysed the effect of boundary conditions on stiffened plates under compressive loads. The study, comparing FE analyses with experimental tests, showed similar ultimate strength and post-collapse shapes across different boundary conditions, though discrepancies were noted for fully clamped setups. Barsotti and Gaiotti (2024) studied residual displacements accumulation in HSLA steel stiffened panel under cyclic loading. They found a decrease in ultimate strength under alternating tension-compression loads compared to compression-only loading, with material hardening law significantly influencing results under alternating loading.

4.2.5 Empirical Formulae

This section explores methods for predicting the ultimate compressive strength of steel-stiffened panels using physical testing and empirical formulae. New empirical models are being developed to improve structural design and safety predictions under various loading conditions.

Useful empirical formulae have historically been developed by fitting curves to data from relevant testing databases. An example of this is the Paik-Thayamballi (Paik and Thayamballi, 1997) formula which provides a closed-form function of plate and column slenderness ratios that was developed by fitting to available data. Since 1997, high-precision data-acquisition equipment has been used to gain insights from large-scale physical models built from modern steel types (e.g., AH32) using modern manufacturing technologies (e.g., flux-cored arc welding technique). It is essential to assess the compatibility of these advanced testing data with established empirical formulae. In a study conducted by Lee et al. (2023b), benchmark studies were conducted to evaluate this compatibility, focusing specifically on the Paik-Thayamballi formula. This study offers valuable insights into the alignment between the advanced testing data and the established empirical formulae.

Li and Chen (2023a) focused on the development of empirical formulae for hull structural design and safety estimation. They verified and adopted an equivalent sequential loading method for biaxial compression to decouple the conventional implicit strength relationship of plates expressed by the elliptic function. A total of 2,737 and 1,944 scenarios, involving different combinations of slenderness ratio, aspect ratio, and applied loads, were selected for non-linear FE analysis to assess longitudinal and transverse ultimate strength, respectively. Based on their numerical findings, the authors proposed two empirical formulas with the same general explicit form using the response surface method. These formulas are versatile, accommodating various load scenarios, including combined biaxial compression, lateral pressure, and other load combinations. Statistical analysis showed that the equations aligned closely with ABAQUS simulation results ($R^2 = 0.993$ and 0.996), validating their effectiveness as a practical approach for ultimate strength assessment in design.

Kim *et al.* (2022b) proposed a simplified empirical formulation to predict the ultimate strength of initially deflected plates subjected to longitudinal compression. The applicability and accuracy of the empirical formulation were verified through comparison with a non-linear FE analysis. A total of 700 plate scenarios with initial deflections, modelled based on assumed buckling mode shapes, were used as input data. A generalised empirical formula was developed from the input data to simplify the plate design process, yielding a reliable closed-form expression that aged well with results from the non-linear FE analysis ($R^2 = 0.98$ to 0.99). Additionally, Kim *et al.* (2022c) proposed an empirical formula to predict the ultimate strength of initially deflected plates subjected to combined longitudinal compression and lateral pressure. This formulation was validated using the ALPS/ULSAP method, analysing a total of 5,600 plate scenarios that included variations in geometry, material properties, and applied loads. These examples, along with a detailed user guide, provide valuable resources for structural analysis.

4.3 Structural Systems

In this section, hull girder, intact and damaged residual strength, superstructure inclusion, and their effects on the ultimate strength of ships are discussed by reviewing recent advances in the structural systems of ships. Emphasis is placed on numerical investigations using FE simulations and combined loading effects.

4.3.1 Intact Hull Girder

Combined loading effects have become an increasingly important topic in ULS assessment because of the M/V MOL Comfort and M/V MSC Napoli accidents. Therefore, this sub-section explores various methods, including FE analysis, incremental-iterative techniques, and empirical models, to assess the impact of combined loads, initial imperfections, residual stresses, and corrosion on the ultimate strength of hull girders. The findings emphasise the need for advanced models and surrogate techniques to improve the accuracy and efficiency of predicting the structural behaviour of ships under complex loading conditions while highlighting areas for further research, such as fatigue in sandwich decks.

The combined effect of hogging and bottom local loads (double-bottom effect) on the hull girder's ultimate strength prediction and annual failure probability in container ships was investigated in Georgiadis *et al.* (2023b). The estimate of uncertainty based on predictions from the conventional Smith model was decreased by including a model correction factor that combined engineering judgment and FE analysis data, which included the double-bottom bending effect. A formal method based on Bayesian analysis was developed to explicitly account for the uncertainties in FE analysis results. The findings show that a significant increase in the failure probability of container ships was due to the double-bottom effect in hogging conditions.

Tatsumi and Fujikubo (2020) evaluated the ultimate hull girder strength of container ships subjected to combined hogging moment and bottom local loads using FE analysis. They concluded that there was a significant reduction in hull girder strength due to the influence of bottom local loads. These results were used in a companion study by Tatsumi et al. (2020) to develop a simplified method for the analysis of the progressive collapse of container ships under these combined loading conditions. The authors proposed an extension to the conventional Smith method, to account for the effects of bottom local loads. The extended Smith method was validated through a series of FE analysis results showing good agreement between the two approaches.

Lindemann et al. (2022) investigated the ultimate strength of a double-hull VLCC using non-linear FE analysis under pure vertical, horizontal, and biaxial bending, considering both initial imperfections and residual stresses due to welding. They concluded that welding residual stresses had a negligible effect on the progressive collapse of the hull girder. In a subsequent study, Lindemann et al. (2023) examined the impact of various methods and model parameters on the ultimate strength of ships in bending, comparing results from a non-linear FE analysis (frame space model), application of the Smith method, and a deep neural network (DNN) model trained using results from the application of the Smith method. Consistent with their earlier findings, they concluded that initial deflections have a greater impact than welding residual stresses. All methods showed good agreement in terms of moment-curvature and neutral axis shift for an intact hull girder; however, the DNN tended to overestimate the residual hull girder strength for scenarios that included damage. It is the opinion of the committee that while the application of surrogate models, such as DNN, is certainly attractive from the perspective of computational efficiency, the assumptions and limitations of these methods should be well communicated, and training data should be free of any bias or modelling errors. Furthermore, surrogate models inherit the limitations of the models used in training, raising the question of which base model (a simplified Smith method or more detailed FE analysis results) should be used as a standard. Both have their advantages, which need careful consensus within the community to ensure the advancement of the field.

Lindemann et al. (2024) also investigated the applicability of the ISUM method to determine the ultimate strength of ships under bending. They found that ISUM predicted slightly higher final bending moments than the Smith method. Consistent with their previous findings, they noted that welding residual stresses had a negligible effect on ultimate strength in horizontal bending. In a parallel study, Lindemann and Kaeding (2024) further explored the applicability of ISUM for performing progressive collapse analyses of welded box girders in bending. They discussed the versions of ISUM and proposed a combined ISUM plate/beam-column element to model the progressive collapse behaviour of welded box girders in four-point bending. The results were compared to experimental data, with additional comparisons made using non-linear finite element analyses.

Faqih et al. (2023) evaluated the ultimate strength of the hull girder of a bulk carrier exposed to various corrosion damage rates. Two corrosion conditions, standard and severe, were considered, and the effect of corrosion was modelled as thickness reductions over various service times (0, 10, 15, and 20 years). Ultimate strength was assessed using incremental-iterative methods based on IACS-CSR, both with and without accounting for the inclination of stiffened panels relative to thrust load (referred to as non-uniform uniaxial thrust). The findings from the study suggest that considering non-uniform uniaxial thrust reduces the ultimate bending moment, and severe corrosion has a pronounced negative impact on the ULS of the structure.

Xu *et al.* (2022c) analysed the collapse severity of a steel hull girder under extreme loading, along with and its influencing parameters. Due to the limitations of experimental setups in replicating the extreme wave loads encountered at sea, this study aimed to assess the magnitude of collapse events and post-ultimate strength behaviour through analytical methods. Experiments typically cannot produce sufficiently large loads to induce collapse or reliably elicit dynamic responses, and post-ultimate strength behaviour is challenging to observe in tank experiments. An analytical approach was proposed to predict the collapse severity of a box girder under extreme loading across all unloading phases to address these gaps. The study found that collapse severity is reduced when there is a gradual capacity drop in bending moment-curvature, whereas a rapid drop in capacity increases the severity of collapse. While collapse modes and behaviours have been explored in other studies, specific focus on collapse severity under extreme loading remains limited.

Recent studies of the ultimate strength of hull girders with sandwich decks have shown promising results. In a comparative study, Zhong and Wang (2022) numerically analysed a hull girder with an upper laser-welded web-core sandwich under various loading conditions, including vertical bending, horizontal bending, torsion, and combined loads. The results indicated that hull girders with sandwich decks offer substantial advantages in ultimate strength over traditional hull girders. The stiffness of the hull girder under sagging was improved by using the laser-welded web-core sandwich deck. The sagging ultimate strength of the hull girder was increased by 15%, and the horizontal bending ultimate strength was increased by 6.2%. Interestingly, the finite laser weld rotation stiffness was found to have no significant effect on the ultimate strength of the hull girder with the upper sandwich deck, which suggests that the welding technique used for the sandwich deck does not compromise the overall strength of the hull girder. This conflicts with earlier findings by Jelovica et al. (2012) who observed that laser welds affect buckling sandwich plates. It is possible that the global analysis conducted by Zhong and Wang (2022) may have smoothed out local effects from the laser welds, warranting further investigation. The study also established high-accuracy ultimate strength interaction relationships for hull girders under combined two- or three-moment loads, providing a valuable predictive tool for practical applications.

The failure modes of the hull girder with the upper sandwich deck were also investigated. Under sagging conditions, failure occurred through global buckling and local buckling of the sandwich deck. Horizontal bending led to buckling of the outer side plates as well as the deck and bottom plates on the compression side. Under torsion, shear buckling of the outer side plates and the bottom plates was more pronounced compared to the prototype hull girder. Overall, this study highlights the potential utility of laser-welded web-core sandwich decks for the improvement of the ultimate strength of hull girders. However, future investigations should clarify the applicability and limitations of sandwich panels from other limit state perspectives (e.g., fatigue).

4.3.2 Residual Strength of Damaged Hull Structures

This sub-section focuses on advances in the assessment of the residual strength of damaged ships, particularly using methods, such as the Smith method and non-linear FE analysis. Studies show that asymmetrical bending, damage from collisions or groundings, and corrosion significantly impact the residual strength of ships, with variations between different vessel types like bulk carriers and oil tankers. Recent work includes the development of rapid assessment tools and empirical formulae for estimating strength loss after damage, providing critical insights for emergency response and structural optimisation.

Residual strength assessment has progressed significantly over the past decade. Cerik and Choung (2020) investigated the hull girder strength of intact and damaged ships using the incremental-iterative method developed by Smith for progressive collapse analysis. The authors extended the Smith method to address the asymmetrical bending of beams with arbitrary cross-sections by determining the translation and rotation of the instantaneous neutral axis at each curvature increment. They conducted a series of analyses on a double-hull VLCC and a bulk carrier, considering various loading plane angles and damage conditions. These analyses confirmed the earlier findings by Fujikubo et al. (2012), which indicated that the rotation of the neutral axis had minimal impact on the strength of oil tankers but significantly affected bulk carriers, where it can lead to a faster loss of strength after reaching the collapse load.

Kuznecovs et al. (2020) compared the ultimate and residual strength of a vessel using non-linear FE analysis and the Smith method proposed by Fujikubo et al. (2012). Their case study examined both an intact hull and a hull damaged by collision, under two conditions: newly built (non-corroded) and aged due to corrosion. The calculation of a residual strength index showed a greater reduction in strength in the FE analysis compared to the Smith method. The difference in residual strength index was more pronounced in the damaged and corroded cases, as the Smith method simplified the effects of damage and corrosion.

Li and Kim (2022) evaluated the residual ultimate strength of grounded ships using numerical and Smith-based methods. A residual strength versus damage index (R-D) diagram was developed, which accounts for varying extents of damage across various types of double-hull oil tankers. This R-D diagram serves as a valuable tool for quickly assessing the residual ultimate strength in emergency response situations.

Kuznecovs et al. (2021; 2022) introduced a methodology called SHARC for simulating and analysing a ship's damage stability and ULS conditions following a collision. SHARC integrates three methods: advanced non-linear finite element simulations that simulate the collision scenario, a dynamic damage stability simulation tool called SIMCAP, and a modified Smith method for the ULS analysis of a collision-damaged ship structure. They demonstrated the capabilities of the SHARC analysis approach by applying the methodology to a case of an intact and a damaged oil tanker under both non-corroded and corroded structural conditions across various sea states. Corroded structures resulted in larger damage opening and penetration depth of the striking bow, but the damage was limited to the same internal compartments. Consequently, this resulted in similar floodwater volumes and distributions, with the corroded tanker reaching the steady state faster due to the larger damage opening area.

Komoriyama et al. (2024) investigated the collapse behaviour and ultimate longitudinal bending strength of damaged box girders in upright and inclined conditions through experimental and numerical analyses. These experiments involved applying four-point bending loads to box girders with elongated holes. The authors concluded that the Smith method can estimate the ultimate strength of damaged box girders similarly to the FE model. They suggested further research to investigate different hole sizes, shapes, and locations, as well as real ship structures.

Understanding the remaining/residual ultimate strength of a ship post-collision or grounding is key for deciding on appropriate remedial actions. However, such studies have been relatively scarce in the literature. Do et al. (2024) conducted a comprehensive numerical analysis of the post-collision ultimate strength of a handy-size container ship mid-section. The non-linear FE method using the Hosford-Coulomb crack model was first validated against a previously published box girder impact and bending test, with up to 9% difference in the residual ultimate strength between the simulation and test. This was followed by a parametric study of a container ship mid-section with classic T-bone damage from a bulbous bow, see Fig. 18. The influence of impact velocity (1 m/s to 11 m/s), bow shape of the striking vessel, and boundary conditions of the FE model were investigated. The results were used to develop an empirical formula for rapid estimation of the ultimate strength reduction. Similar studies on different ship types would be valuable not only for optimising scantling designs but also for informing emergency responses upon damage occurrence.

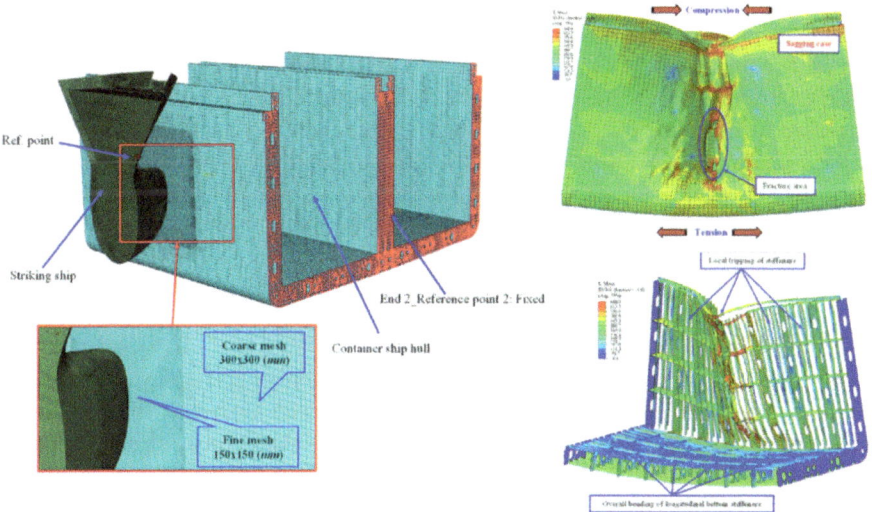

Fig. 18. Container ship mid-section FE model subject to collision damage and sagging bending moment (Do et al. 2024).

4.3.3 Load Combinations

Previous ISSC Ultimate Strength committee reports, e.g., the ISSC (2018a) report, have identified the need for investigation of the combined load effects on ULS. Zhang *et al.*

(2023a) investigated structural dynamic ultimate strength under combined bending, torsion, and lateral pressure (slamming) using non-linear FE analysis with ABAQUS as a solver. The analyses were performed with a partial length model of 26 m even though the Official Discusser in the ISSC 2018 Ultimate Strength committee report (ISSC, 2018b) suggested that torsion cannot be adequately assessed using partial length models. Their analyses showed that when slamming duration was on the order of seconds there was little difference between the dynamic and the static ultimate strength. For shorter events, the dynamic ultimate strength of the hull girder was higher than the static ultimate strength. Although not explicitly noted by the authors, this suggests that modelling slamming events as quasi-static is a conservative approach. However, these results require further scrutiny.

Lee and Paik (2020) investigated the ultimate strength characteristics of as-built ultra-large containership hull structures under combined vertical bending and torsional moments. The analysis was conducted using the intelligent supersize finite element method (ISFEM) and the ALPS/HULL3D program. Their results showed that the ultimate strength of the hull structures was influenced by ship size and the combination of vertical bending and torsional moments. This study provided insights and structural design recommendations for enhancing the ULS capacity of containership hull structures and achieving robust structural design for ultra-larger containerships. Containership hulls with large deck openings may be vulnerable to structural failure due to torsional moments combined with vertical bending, and this vulnerability becomes more significant with an increase in containership size.

Dynamic loads play an important role in structural safety analysis. Shi *et al.* (2023a) investigated the dynamic elastic-plastic response and ultimate strength of a box girder under a bending moment. One set of results was validated with previously conducted experiments, which were then used as a basis for six different box girder models that were used to gauge the effect of model length, plate thickness, mass density, and load excitation period. Based on the structural dynamic response, they proposed an evaluation criterion of the dynamic limit state for the box girder under a bending moment. However, it is left to the reader to interpret how this can be exploited in an actual ship analysis procedure.

4.3.4 Progression Sequence of Failure

Deng et al. (2022) presented numerical and experimental findings for the ultimate bearing capacity and collapse behaviour of ship hull girders with large deck openings under cyclic ultimate bending moments. These results provide valuable insights into structural behaviour, validate numerical models, and support previous findings that cyclic plastic loading reduces the capacity of the hull girder in subsequent load cycles. Notably, the double-hull structure exhibited a higher margin of residual strength in subsequent loading cycles than the single-hull structure, which the authors attributed to better stiffness distribution and load redistribution capability. Liu and Guedes Soares (2020) similarly demonstrated that the ultimate strength reduction of hull girders due to cyclic loading can be captured through numerical simulations. They used a kinematic hardening material model to account for both the ratcheting and Bauschinger effects. While these results demonstrate the capabilities of numerical simulations, the question remains of how to integrate these relatively costly simulations into ship structure analysis and rule development. Simplified approaches developed for this purpose would be beneficial for future investigations (see, e.g., Preventas et al., 2021), as ULS analysis under monotonically increasing bending moments is already complex and numerical simulations under cycling loading further increase simulation time and complexity.

Jagite and Bigot (2023) questioned the accuracy of the approach proposed by Liu and Guedes Soares (2020) due to the effects of load amplitude, period, and cyclic load magnitude. Motivated by the lack of research on the cyclic ultimate strength of ship hull girders, they analysed the influence of cyclic loads on structural capacity by considering realistic loading scenarios through a long-term hydro-elastic analysis. The ultimate strength of a containership was analysed numerically on a partial structural model using non-linear FE analysis. Their study clearly explains the assumptions and modelling principles, providing a solid foundation for future research. They compared cyclic ultimate strength to traditional ultimate strength obtained with monotonically increasing loads, finding that cyclic loading had an unfavourable, but negligible 1.2% effect on the ultimate strength, which is considerably lower than the 10% reduction reported earlier by Liu and Guedes Soares (2020). Jagite and Bigot (2023) attributed this difference to the more realistic loading assumptions of their approach and called for further experimental studies on the cyclic loading behaviour of typical marine structural steels to advance the field.

Hu et al. (2022) investigated the ultimate bending moment of a cracked stiffened box girder using a series of FE analyses, focusing on the effects of crack length and location under cyclic loads. The study found that crack location had minimal impact on the, focusing on the effects of crack length and location under cyclic loads. The study found that crack location had minimal impact on the ultimate bending moment of stiffened box girders under monotonic bending and bi-directional cyclic bending, but it did have some influence under unidirectional cyclic bending. However, crack length played a more significant role in affecting the ultimate bending moment under both unidirectional and bi-directional cyclic bending. The authors concluded that the ultimate strength reduction under extreme cyclic bending was attributable to the coupling effect of fatigue crack damage and accumulated plastic damage.

Zhou et al. (2023) modelled the progressive collapse behaviour of ship structures considering fluid-structure interaction effects using ABAQUS StarCCM coupling. The difference between ship structure collapse under pure bending and ship structure progressive collapse under different wave conditions was compared. They found that wave load significantly influenced both the ultimate strength and deformation mode of the ship structure. Additionally, wave parameters, such as wave height and wavelength, had a notable impact on the structural response of the ship.

4.4 Rules and Guidelines

Merchant ships have historically been primarily designed and sized following the rules prescribed by ship class societies, while naval ships have historically been designed and sized based on naval methods and guidelines. These ship design rules and guidelines were largely derived from the principles of structural mechanics as well as from the wide experience of individual ships in operation over prior decades. This section focuses on current methods and new developments given by class societies and IACS, concerning buckling and the ultimate strength assessment for ships.

The global ship fleet continues to evolve with increasing vessel sizes, greater cargo capacities, and specialised vessel types for various purposes. Different approaches to buckling requirements exist across ship types, and classification societies have specific buckling procedures tailored to different vessel designs and structural arrangements. During the design phase of a ship structure, ultimate strength checks are conducted on both a global and local level: (i) a global ultimate hull girder strength check (e.g., the Smith method), and (ii) a local buckling check for individual structural members (e.g., explicit design rules).

4.4.1 Development of Rules and Regulations

A ship's structure must be designed to withstand severe wave environmental loads, and for global operations, the North Atlantic scatter diagram is typically used in these structural designs. The International Association of Classification Societies (IACS) has recently undertaken substantial work to update the wave scatter diagram currently provided in the IACS "Rec. No. 34" (IACS, 2022b), which is used as a basis for the designs of nearly all commercial shipping vessels worldwide. Austefjord et al. (2023) presented some of the background theory and discussed the recent IACS updates. The updated wave scatter diagram was derived from state-of-the-art hindcast datasets, incorporating bad-weather avoidance patterns observed in real ship voyages, using a comprehensive position database from automatic identification system (AIS) data. The spectral shape and spreading were adjusted based on these datasets, resulting in a slightly narrower spectrum. Additionally, AIS data were post-processed to derive statistics of ship speed and heading and their correlation with significant wave height. The updated wave data generally result in reduced loads; this reduction is consistent with the calibration factors currently used in the design rules.

In response to the IMO Goal-Based Standards (GBS) observation on requirement FR1–8/OB/02, and to ensure that the wave data used in ship design rules accurately reflect North Atlantic conditions, IACS (2022a, 2022b) has allocated significant technical resources and applied modern methodologies to update Recommendation No. 34, "Standard Wave Data". Revision 2 now offers an extensive, comprehensive dataset that provides a more accurate representation of the North Atlantic Sea state (IACS, 2022b). Building on the work in Recommendation 34, IACS has also initiated a large-scale project to develop design wave loads based on the updated wave scatter diagram. This project will re-evaluate all wave loads in the Common Structural Rules (CSR), including, hull girder loads, sea pressures, and accelerations, to establish a detailed technical foundation and implement necessary rule changes. Additionally, for other ship types, hull girder loads such as wave bending moments and shear forces will be reviewed, and if necessary, several IACS Unified Requirements may be updated.

4.4.2 Class Requirements: Ultimate Hull Girder Strength Assessment for Ships

Assessments of ultimate hull girder strength commonly use the Smith method (Smith, 1977), which is an incremental procedure to compute the ultimate of a ship's hull girder structure. A summary and a more detailed review of this approach are presented in ISSC (2018a; 2022a). The Smith method, which is used by most class societies, is applied to incrementally calculate the hull girder bending capacity for each cross-section of the ship, under hogging and sagging conditions. This method has not been further developed since its first implementation in the CSR for bulk carriers and oil tankers (IACS, 2023a).

Hull girder ultimate strength assessment is currently required by most class societies for ships over 150.0 m in length. However, this mandatory assessment typically only considers global vertical bending moments. Assessment of hull girder ultimate strength capacity for ships under horizontal bending moment or torsion is not required by class societies. Darie and Rörup (2017) showed that large containerships operating in oblique seas experience maximum horizontal and torsional loads that are as relevant to the ship's integrity in terms of hull girder strength capacity as the vertical bending moment (head sea conditions). The effect of torsion and horizontal bending moment on hull girder ultimate strength capacity, particularly for ships with large deck openings requires further exploration.

Local loads, such as external pressure, cargo load, and ballast loads are not directly considered in the ultimate strength assessments mandated in class requirements. These local loads, which have a notable effect on the ultimate strength capacity of a ship, are currently only addressed through partial safety factors. For example, to account for double-bottom bending due to local loads on bulk carriers in hogging conditions, IACS (2023a) requirements specify a partial safety factor of 1.25 to reduce the hull girder ultimate strength capacity of the cross-section. Similarly, the mandatory IACS (2020a) requirements for container ships include a partial safety factor of 1.15 to address the local bending effect on the double-bottom in hogging conditions. This factor applies specifically to the midship region, and it may be adjusted in narrower double-bottom areas, such as the engine room, where it can be reduced to 1.0. However, recent increases in the size and capacity of certain ship types, particularly very large containerships with greater width and draught, have resulted in significantly increased local loads on the double-bottom structure. This suggests that the partial safety factor for double-bottom bending effects on hull girder ultimate strength may need to be re-evaluated, potentially using finite element calculations, to ensure accuracy for these larger vessels.

The approach for calculating the hull girder ultimate strength implemented in the requirements presented in IACS (2020a, 2020b; 2023b) is based on a net scantling concept. This concept involves deducting half of the corrosion addition values, as specified by the rule's scantling requirements, from all structural elements of the cross-section under consideration. Since all IACS-classed ships undergo regular surveys by classification societies to examine their corrosion status, the net scantling concept provides a safety margin for hull girder ultimate strength. IACS also requires a partial safety factor of 1.05 (IACS, 2020a, 2020b) or 1.1 (IACS, 2023a) for hull girder ultimate strength, which is intended to account for uncertainties in the construction materials used. An additional partial safety factor, specifically applied to the vertical wave bending moment, is set at 1.2. This factor further enhances the safety margin in assessing hull girder ultimate strength for ships under the IACS classification.

The whipping effect is another important factor influencing the hull girder's ultimate strength of the ships. Whipping is often experienced on large container ships, commonly occurring in head sea conditions, and can reduce the hull girder's ultimate strength capacity, as evidenced in accident analyses of the container ships M/V MSC Napoli and M/V MOL Comfort. Currently, IACS (2020a, 2020b) advises that classification societies consider the whipping effect in hull girder ultimate strength assessments for container ships; however, it does not provide a specific calculation method. In developing these rules, classification societies have conducted extensive research—including theoretical studies, model testing, and numerical verification across various ship types. This research led to the development of hydro-elastic load calculation software and established a technical approach for evaluating hull girder ultimate strength under whipping, characterized through specific class notations. These specific technical requirements are presented in Table 1.

IACS established working group PH38 to produce the study "Whipping on Container Ship" (IACS, 2022a). After theoretical analysis and comparison with real ships, the first draft of Annex 4 of UR-S11A "Functional Requirements for Ultimate Strength Assessment of Hull Beams Under Whipping Effect" was developed (to be released), which provides a foundational method for evaluating hull girder ultimate strength considering the whipping effect in large container ships. It includes core requirements for hydro-elastic analysis, a long-term method for analysing whipping loads, and guidance on the application of whipping factors.

Table 1. Class notations for ultimate strength assessments including whipping.

Classification (Acronym)	Class Notation	Reference
ABS	WIP	ABS (2025)
BV	WhiSp2, WhiSp3	BV (2024)
CCS	WAU	CCS (2018; 2022)
DNV	WIV	DNV (2021; 2022a)
KR	WHIP	KR (2022)
LR	WDA1, WDA2	LR (2022)

4.4.3 Local Buckling Assessment for Individual Structural Members

Local buckling assessments are conducted for individual structural members, such as stiffened panels, plates, and pillars, under various load conditions. Given the complexity of a ship structure, where multiple load cases must be evaluated for each stiffened plate, it is essential to use computationally efficient tools. These include explicit design rules employing closed-form methods or semi-analytical tools. For specific cases not covered by existing design rules, other more advanced methods, such as FE methods can be used. However, due to the high number of structural elements and load cases, using non-linear FE analysis extensively would be impractical.

Traditional methods for local buckling assessments, such as CSR and UR-S35 (IACS, 2024), often rely on empirical approaches. These methods typically involve plastic corrections to the elastic buckling strength or curve fitting based on non-linear FE analyses.

Semi-analytical approaches offer an alternative to these conventional methods. A recent study by Ishibashi et al. (2024) introduced a semi-analytical method that employed elastic large-deflection theory combined with trigonometric functions to derive analytical solutions describing the elastic post-buckling behaviour. The ultimate strength was then predicted by assessing the yield at critical locations within the plate. This approach is akin to the PULS buckling procedure (DNV, 2023). A key distinction between the two methods is that PULS employs an iterative process with a series of trigonometric functions to compute response and ultimate strength.

4.4.4 Harmonisation of IACS Buckling Procedures

IACS is responsible for numerous ship strength regulations, including both generally applicable Unified Requirements and specific rules tailored to different ship types. There are varying approaches to buckling requirements across vessel types, with rule sets, such as UR-S11, UR-S11A, UR-S21, UR-S21A (IACS, 2010; 2015a; 2015b; 2020a), and CSR (IACS, 2023a). These rule sets collectively cover 90% of ships worldwide. Currently, IACS is working towards harmonising these procedures to create a single, unified buckling standard.

Based on insights gained over the past several years gained through the application of the buckling methodology in CSR for bulk carriers and oil tankers, IACS decided to adapt the CSR buckling assessment procedure for use across all ship types. The objective is to establish a buckling procedure accepted by most IACS members, applicable to all relevant Unified Requirements related to ship strength without reservations. Following this decision, a dedicated project team was established by IACS and a new set of unified buckling strength requirements UR-S35 (IACS, 2024) was issued in 2024. In addition, some improvements to the existing CSR were introduced, as presented in Sect. 4.4.3. The new UR-S35 is a separate document and will function as a toolbox for the various unified requirements for buckling, as shown in Fig. 19. This new rule set of buckling formulae is already in effect for hatch covers, replacing buckling formulae in the previous rules (IACS, 2010; 2015b). The longitudinal strength standards, UR-S11 and UR-S11a, stipulate that the longitudinal strength members of a ship hull must be checked for buckling. However, before incorporating the new buckling requirements, updates to the wave loads are needed.

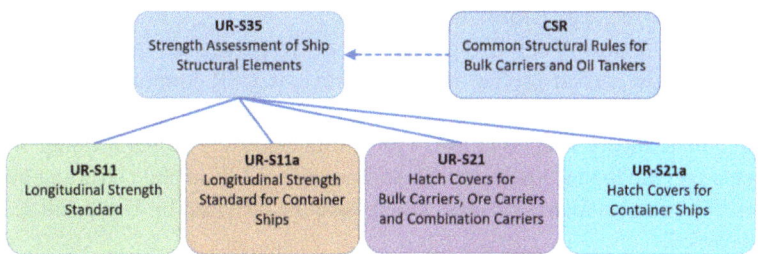

Fig. 19. UR-S35 as a buckling toolbox for all unified requirements in IACS.

Alongside efforts to harmonise buckling strength formulations, another IACS expert team has been updating the wave loads mentioned above. In addition to the harmonisation of buckling strength assessments and updating wave loads, a consequence assessment must be completed before the rule changes for longitudinal strength can be put into force. This involves evaluation of various aspects, such as corrosion additions, slenderness requirements. IACS is actively working on these aspects as part of the ongoing effort.

4.4.5 Recent Updates to the CSR

IACS has focused on the improvement and harmonisation of buckling assessment methods, with recent advancements led by an IACS project team established in 2018. The buckling procedure within the CSR has been incorporated into the new unified requirements, UR-S35 (IACS, 2024). As noted, one objective is to adopt this approach for longitudinal strength checks across all vessel types, which would encompass over 90% of the global fleet. Some of the improvements implemented by this project team, particularly in the context of the stiffener buckling approach, have been presented in Brubak et al. (2023). Additional background theory for these updates can be found in the technical background for CSR (IACS, 2023b). Issues identified with recent updates in the CSR buckling approach are summarised below:

- Buckling of global stiffeners
- Buckling of torsional stiffeners
- U-type stiffeners
- Sniped stiffeners
- Plates with openings
- Plates with one free edge

Only minor corrections have been implemented for the first two cases since the publishing of ISSC (2022a), and a short summary is presented here for the completeness of this report. In global stiffener buckling, out-of-plane deflections are computed using orthotropic plate theory rather than an equivalent beam, which has improved accuracy in certain cases, such as for plates with long, slender stiffeners under transverse loads. The formula for torsional elastic buckling was updated, and a load-dependent expression for torsional capacity was introduced. Further details, including comparisons with CSR and FE analysis, are available in the CSR technical background documents (IACS, 2023b) and the literature (see, e.g., Brubak et al., 2023).

The updated approach for U-type stiffeners incorporates the effect of rotational stiffness due to the closed profile, replacing previous CSR formulae that treated U-type stiffeners as equivalent T-stiffeners using beam theory. In the updated formulae, orthotropic plate theory is applied to U-type stiffeners, using global bending stiffness coefficients specific to the closed profile. Additionally, for local plate buckling between stiffeners, the extra rotational stiffness along the edge provided by U-type stiffeners is now considered.

A new approach was recently introduced in the CSR buckling procedure for sniped stiffeners, incorporating an additional bending moment due to the eccentricity of axial loads, as shown in Fig. 20. In the original formula, the impact of sniped stiffeners was accounted for by including an additional imperfection in the calculation of the bending moment for second-order effects. This may be realistic for plates with pure axial loads in the stiffener direction. However, for combined loads, including transverse stress and shear stress, the original formula can produce unrealistic results. For example, under the original approach, increasing plate thickness could appear more effective than increasing stiffener size to prevent stiffener buckling, since the increased plate thickness would reduce the initial imperfection. This is counterintuitive, as increasing stiffener size is generally more efficient in preventing stiffener buckling. Additionally, the CSR buckling formulation has been updated for plates with a free edge, which is particularly relevant for plates with openings. In these cases, the buckling procedure for plates with a free edge applies to the individual plate segments on either side of the opening, as illustrated for the plates P1 and P2 in Fig. 21.

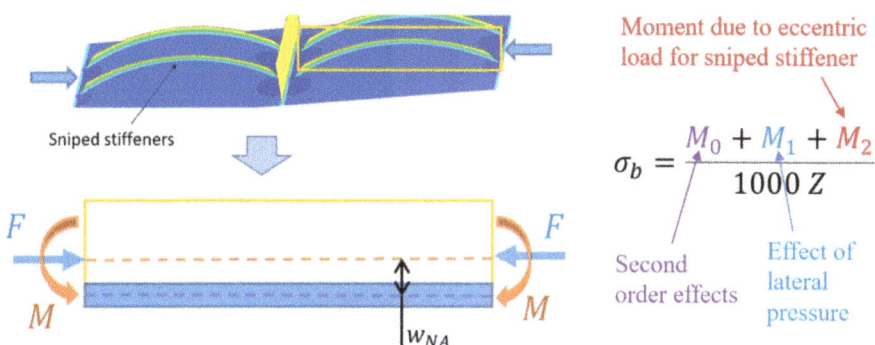

Fig. 20. Bending stress for sniped stiffeners with additional bending moment M_2 to account for the load eccentricity.

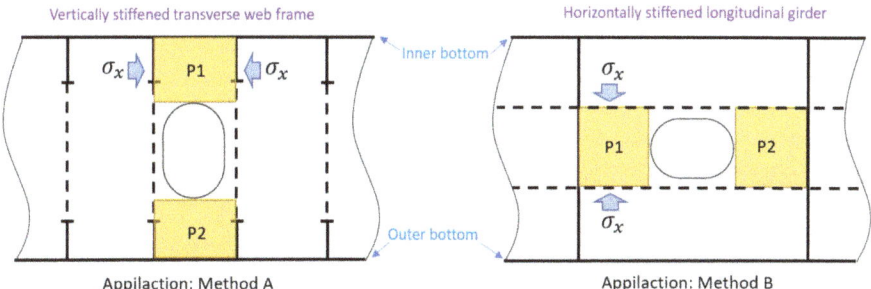

Fig. 21. Plates with opening where buckling formulae with a free edge are used for two different applications (Method A and B).

This update was prompted by industry feedback indicating that the existing formulae could be overly conservative, as demonstrated in the technical buckling documentation for CSR (IACS, 2023b). An additional application was introduced to address this, reducing conservatism in certain special cases. Consequently, there are now two applications for a plate with a free edge, as illustrated in Fig. 22. In the left figure, the loaded edges are forced to remain straight and parallel (i.e., displacement-controlled), which is referred to as Method A in CSR. This application is suitable for plates where the neighbouring plates in the load direction are continuous, as illustrated to the left in Fig. 21. In such cases, the plate can re-distribute stresses, allowing it to retain reserve strength beyond its elastic buckling limit. Method B is applied in other applications, as illustrated on the right in Fig. 22. In this case, there is less in-plane support along the loaded edges, which are free to move in-plane. This is approach is equivalent to the method outlined in the original CSR and is more conservative. Method B is applied to plates without a continuous structure in the load direction, as illustrated to the right in Fig. 21. In these scenarios, the plate lacks reserve strength beyond elastic buckling because the loads are transferred directly into the plate, preventing stress redistribution.

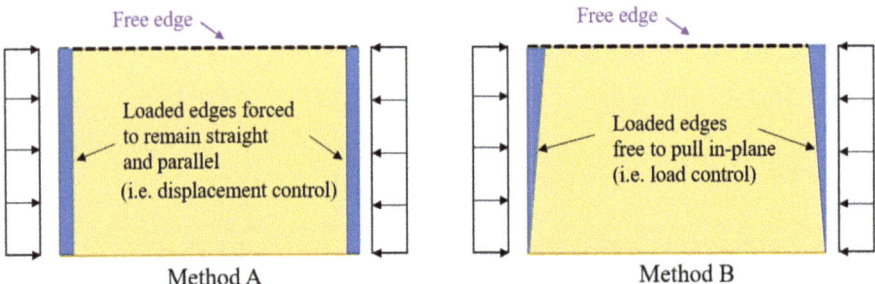

Fig. 22. Two different applications for a plate with one free edge.

5 Offshore Structures

Offshore structures, including oil and gas platforms and wind farms, are essential infrastructures in marine environments for harnessing valuable resources and energy. These structures must withstand the harsh conditions of the open sea, including extreme weather events, waves, and corrosive saltwater. Recent advances and future research needs are presented in this section, along with a summary of the latest updates in relevant design standards and regulations. Research on the ultimate strength of oil and gas offshore structures appears to be in decline, likely due to the global transition from fossil fuels to renewable energy. Furthermore, there is limited literature on the ultimate strength of other offshore structures, such as very large floating offshore structures, floating production storage and offload systems (FPSOs), and floating solar panel structures.

5.1 Loads

Extensive research has been conducted on the impact of load effects on the ultimate strength of traditional offshore structures. However, there has been a growing interest in relatively new structures, such as offshore wind turbines, arctic offshore structures, and composite risers.

5.1.1 Load Effects on Offshore Wind Turbines

Offshore wind is now recognised globally as one of the principal energy sources for combatting climate change. The development of offshore wind energy is forecasted to increase globally to over 290 GW by 2030 (Musial et al., 2023). Zavvar et al. (2022) presented a numerical study of the wave-induced motions and loads on the free-float capable tension leg platform (TLP) named CENTEC-TLP with a 10 MW floating wind turbine (FWT). The study analysed the response of the TLP to various wave conditions and wind speeds in ANSYS® AQWA. A 1:50 scale model of a TLP FOWT supporting a 5 MW wind turbine was designed and tested under combined wind and wave conditions by Ren et al. (2024). The tests conducted included free-decay tests, regular and random wave tests, as well as tendon-broken tests. The experimental results indicated that following tendon failure, the tendons adjacent to the broken tendon experienced a significant increase in tension force.

Li *et al.* (2024b) analysed the tower and blade responses of a spar-type floating offshore wind turbine (FOWT) to the original freak wave crest and subsequent changes. The study concluded that significant blade edgewise deformations occurred, and the tower resonated with the floating foundation. The most noticeable influences were observed in the fluctuation degree changing waves. Ding et al. (2024) described an experimental investigation into the non-linear wave forces acting on a monopile offshore wind turbine foundation under various wave conditions. Utilising a phase-based harmonic separation method, the study explored multi-directional and bi-directional wave interactions, achieving a clear separation of harmonic components even in complex wave scenarios. The results highlighted that non-linear loading could account for up to 40% of the total under certain conditions, with a reduction in non-linear high-order harmonics observed in wider wave spreading. A study by Wang and Moan (2024) investigated the influence of load effects on ultimate strength in the case of semi-submersible hulls for FWTs. A multi-body floater model integrated within a fully coupled aero-hydro-servo-elastic system was developed, enabling precise calculation of cross-sectional forces and moments under complex environmental conditions. This approach combined global load effects with lateral pressures to assess ULS, providing insights into critical design parameters and safety factors, which are essential for optimising hull and column structures in FWTs.

The impact of second-order wave hydrodynamic loads combined with aerodynamic loads on the dynamic responses of semisubmersible FOWTs was investigated by Shi *et al.* (2023b). Three water depths (50, 100, and 200 m) were examined to understand the depth effect. Simulation results demonstrated that second-order hydrodynamic loads were significant factors in motion and structural responses under extreme sea conditions, regardless of whether aerodynamic loads were included in the analysis. Additionally, aerodynamic loads were observed to substantially impact the dynamic responses of all semi-FOWTs during normal operational conditions. Among the three FOWTs analysed, the V-shaped semi-FOWT exhibited the highest pitch motion responses in extreme sea conditions.

The ship collision responses of a semi-submersible FOWT, i.e., the OO-STAR floater with DTU 10 MW blades, were investigated by Yu et al. (2022) using USFOS. It was revealed that the FWTs exhibit a higher capacity to withstand higher energy impacts without collapse compared to bottom fixed installations. Notably, collisions with operating FOWTs tend to have more severe consequences, with the worst-case scenario occurring when the vessel strikes from the opposite direction of the wind. An aero-hydro coupled method was proposed by Zhang and Hu (2022) to analyse the dynamic response of FOWT under ship-FOWT collision under wind-wave conditions. Though the structure can withstand the ship collision at the beginning, the residual strength of FOWT may not be enough to withstand the continuous wind-wave loads. Therefore, a lower impact velocity could cause the tower to collapse in wind-wave conditions, in contrast to still-wave conditions.

The presence of energetic steep or breaking waves (ESBWs) in the targeted locations for OWTs can significantly contribute to the overall loading of the structure, impacting the ultimate limit state (ULS). Martin et al. (2022) utilised the computational fluid dynamics (CFD) solver Code Saturne to model the wave-breaking slamming phenomenon on a cylinder. Figure 23 illustrates the test model of a 1:55 scale representation of a 31m diameter rigid vertical column situated at a water depth of 121 m. The comprehensive test program encompassed 354 irregular wave tests lasting 3 h each in full-scale time in the SINTEF Ocean Basin laboratory. Slamming panels were installed on the area of the column facing the incoming waves.

Among the various load scenarios, the most critical for ULS assessment in FOWTs are those involving combined extreme wind and wave conditions, particularly during parked or shutdown states. These can induce significant tower base bending moments and tendon overloads. Additionally, tendon failure scenarios, as observed in experimental studies (Ren et al., 2024), can lead to load redistribution and localized overstressing, making them critical for collapse assessment.

Fig. 23. Test model of a vertical column: (left) the full model, (middle) the lower column, and (right) the upper column (Økland et al., 2022).

5.1.2 Ice Load on Arctic Offshore Structures

The Arctic area is known for its abundant oil and gas resources, accounting for 13% of the world's unproved reserves and 30% of the world's total reserves (Zhang et al., 2023b). Truong and Jang (2022) presented a FE analysis on the interaction between a moving offshore structure and fixed-level ice to estimate ice-load distributions and structural response characteristics, where the sophisticated procedure of ice load estimation using the influence coefficient method (ICM) was derived. Braun et al. (2022) introduced an approach to determine combined load spectra for offshore structures and standardised time series for wind, wave, and ice action. The load interaction of a monopile support structure in the Baltic Sea was numerically assessed based on over 3000 simulations of different load conditions, each lasting 600 s, for four design load cases (DLCs). It was concluded that combined stress spectra for wind, wave, and ice action on an OWT differed significantly from standard load spectra without ice contribution. This can be attributed to the occurrence of very high ice-related load cycles, which can lead to overload effects similar to those seen in storm events. Therefore, applicable damage accumulation sums for such spectra should be determined experimentally.

The definition of ice loads varies based on measurement techniques, encompassing terms such as contact pressure, contact forces, and line load, which depend on whether pressure gauges or shear strain gauges are employed. Li et al. (2024a) delineated ice load measurements, discussing direct methods using pressure panels and indirect methods that infer loads from structural responses. This included the use of strain gauges which, despite introducing uncertainties due to assumptions about load locations and heights, remain a prevalent practice for quantifying ice loads due to their practical applicability in field conditions.

The construction of OWTs in sub-arctic areas, such as the Baltic Sea and the Bohai Bay, presents a unique challenge, primarily due to the presence of sea ice (Hammer et al., 2023). In the numerical assessment conducted by Tabri et al. (2022), the ice involved in the interaction with a wind turbine foundation in the Baltic Sea was treated as an isotropic, brittle material. The behaviour of the ice was described using stress-strain curves to analyse its response during the interaction process. Barooni et al. (2022) also presented a suite of ice load models using a coupled aero-hydro-servo-elastic numerical model to study the dynamic response of FOWTs in regions with cold climates and the presence of sea ice. The findings of this study suggest that aerodynamic loads from wind inflow primarily control the dynamic response of wind turbines. When operating at the rated wind speeds, ice loads have little impact on power output and are statistically insignificant for floating spar-buoy wind turbines. Future work may pursue a further investigation of the effect of stochastic ice loads on the dynamic response of floating offshore wind turbines across various operational modes.

An experimental study was conducted to estimate the effects of hydrodynamics on the ice impact rate and load between an approaching iceberg and a moored FPSO by Huang et al. (2023). A probabilistic analysis was initially conducted to identify potential scenarios contributing to the design of iceberg loads, focusing on the iceberg population and environmental conditions on the Grand Banks and a standard FPSO. Three iceberg sizes and their corresponding environmental and test conditions were selected for the study. The experiments indicated that the hydrodynamic effect led to a decreased probability of impact for smaller icebergs compared to the platform. Further experiments are suggested to investigate the influence of hydrodynamics on larger icebergs. The influence of hydrodynamic effects might be significant in some scenarios and should not be ignored for estimating iceberg design loads. Gu et al. (2023) conducted a comprehensive review of ice detection and mitigation technology for bottom-fixed offshore wind turbines deployed in the Arctic. The ice mitigation technologies were divided into two categories: passive technologies, including ice-phobic coating and black paint, and active technologies, such as heating resistance, pneumatic, ultrasonic, and hot air injection techniques.

For Arctic offshore structures, the most critical ice load patterns for ULS assessment are those involving high-velocity ice crushing and bending failure against vertical or inclined structural members. These loads are particularly severe at the waterline and can lead to local or global structural failure. The interaction of ice with compliant structures, such as monopiles or jackets, under dynamic conditions can also result in significant stress concentrations that govern ULS.

5.1.3 Load on Composite Risers

In response to the increasing need for sustainable and versatile materials, the offshore industry is progressively adopting composite materials. Their unique combination of lightweight properties, corrosion resistance, thermal insulation, and fatigue resistance makes composite risers an attractive choice for deep-sea operations. Chang et al. (2022) concluded that the internal flow velocity, external ocean current velocity, fibre orientation angle and top axial tension have a substantial impact on the coupled vibration stability and frequency-locking characteristics of composite risers.

Amaechi (2022) presented the locally tailored design of deep-water composite risers subjected to burst, collapse, tension, and combined loads. ANSYS ACP was used for modelling composite risers with different material configurations. Figure 24 illustrates the structure of the risers and their common loadings. Internal pressure was identified as the most critical factor of all the loadings investigated since it imposed the highest level of stress on the riser layers.

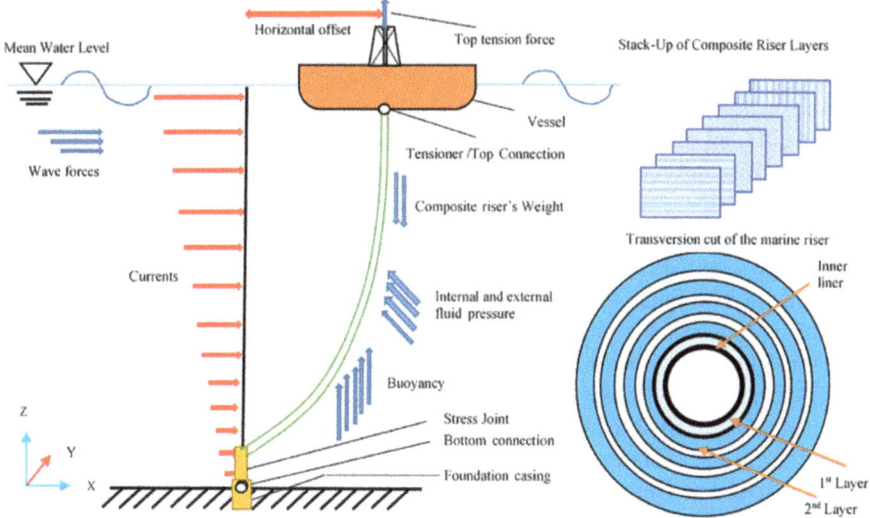

Fig. 24. Schematic of a composite marine riser highlighting loads, stack-up and transversion-cut of plies (Amaechi, 2022).

Hastie et al. (2022) conducted a failure analysis for a thermoplastic composite pipe (TCP) under loads illustrative of deepwater riser operation by FE analysis using ABAQUS. The analysis considered temperature-dependent material properties, examining various laminate stacking sequences. A multi-angle stack was found to be effective for both small and large tension operating scenarios. They concluded that TCP designed for extreme operating conditions will inevitably necessitate the use of large-diameter storage spools. He et al. (2023) conducted both theoretical and FE analysis to investigate the behaviour of thermoplastic composite pipes subjected to axisymmetric loadings. Their study found that the axial stiffness of the pipes remained constant regardless of the internal and external pressures. However, there was a slight decrease in stiffness as the prescribed temperature increased.

Theoretical and numerical models of composite flexible risers under axisymmetric loads were proposed by Sun et al. (2024). It was found that the structural radial stiffness of high-modulus glass fibre is better than that of high-strength glass fibre. Specifically, when the fibre volume fraction (FVF) of high-modulus fibreglass reaches 53.3%, it can achieve an equivalent stiffness to high-strength fibreglass with an FVF of 80%. An overview of the developed modelling approaches and the CFD results from verifiers in blind tests were provided by Yeon et al. (2022), and CFD was employed to analyse the model of an offshore floating structure. Results from these approaches were investigated in a comparison benchmark study, which included test results from existing models for both a semi-submersible rig and an FPSO. Through the Joint Industry Project (JIP), it was determined that CFD readiness for estimating wind loads on offshore floating structures was quite high. However, a major obstacle to achieving accurate CFD modelling was the need to repair geometry to ensure a watertight surface from CAD, especially when numerous defects were present.

5.2 Structural Elements

5.2.1 Tubular Joints

Research on tubular joints has recently focused on analysis methods and local strengthening of different joint types. Saberi and Fantuzzi (2022) analysed results from prior experiments on reinforced/unreinforced tubular T-joint components subjected to tension-compression loading conditions via the open-source finite element package Code_Aster. The code produced accurate predictions for ultimate load resistance against ovalisation for large deformation settings. Configurations with a brace-to-chord diameter ratio greater than 0.5 exhibited higher load-bearing capacity compared to those with a lower diameter ratio, where the failure mode was local yielding, causing the brace to yield before the joint. Nassiraei and Yara (2023) investigated the local joint flexibility of K-joints reinforced with external plates subjected to axial loads. An FE model was developed and validated against experimental results, with 150 K-joints simulated to analyse the effects of reinforcing plate size and joint geometry. Key findings include that external plates effectively reduce local joint flexibility in K-joints and the creation of a design formula to predict the local joint flexibility factor for reinforced K-joints under axial loads.

Kadry et al. (2022) presented a mathematical formula to calculate the capacity of a symmetrically loaded multi-planar KK joint. A verified ANSYS model was used to generate 172 records for parametric study with different geometries, member sections, and material properties. The novelty of the developed model lies in its ability to predict joint capacity as a true 3D joint, rather than an approximated 2D joint, as in prior studies. The results of the developed formula were compared with experimental test results and showed an average accuracy of 92.5%.

Feng et al. (2021) conducted a numerical study on the ultimate strength of stainless-steel hybrid tubular X-, T-, and Y-joints with square braces and circular chords (SHS-to-CHS). A parametric study using 288 FE models was performed, with half representing T/Y joints and the other half X joints. Various geometric properties were adjusted, revealing that the load-carrying capacity of the joints was particularly sensitive to the brace width-to-chord diameter ratio. A comparison of the results with current design guidelines indicated existing design codes inaccurately predict the strength of stainless steel SHS-to-CHS hybrid tubular T/Y- and X-joints. Consequently, improved design equations were proposed.

The effects of FRP on SCFs in uniplanar KT-joints reinforced with different fibre-reinforced polymers Glass/vinyl ester, Glass/epoxy (Scotchply 1002), S-glass/epoxy, Aramid/epoxy (Kevlar9/Epoxy), Carbon/Epoxy (AS/3501), and Carbon/epoxy (T300/5208) under five axial loading conditions were numerically investigated by Zavvar et al. (2023). The commercial FE software ANSYS was used to create 5184 finite element models with different geometries defined by dimensionless parameters, FRP materials, and several FRP layers/orientations. This range of dimensionless parameters effectively covers a broad spectrum of tubular connections used in offshore structures. In related work, Pandey and Young (2021) conducted experiments on novel joint configurations, including brace-rotated (BR), square bird-beak (SBB), and diamond bird-beak (DBB) T-joints, to assess their load versus deformation behaviour and static joint resistances, see Fig. 25. The joint failure resistances of BR, SBB, and DBB

T-joints were 1.53, 1.34, and 1.53 times the joint failure resistances of identical traditional RHS-to-RHS (rectangular hollow section) T-joints, respectively. Moreover, the joint ultimate capacities of BR, SBB, and DBB T-joints were 1.84, 2.12, and 2.24 times the joint failure resistances of identical traditional RHS- to-RHS T-joints, respectively. The authors also performed numerical FE simulations of the SBB joint and proposed accurate and reliable design equations for predicting the static strength of the SBB T- and X-joints.

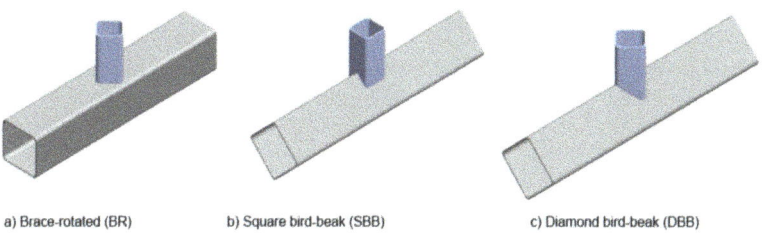

Fig. 25. Three-dimensional models of BR, SBB, and DBB T-joints (Pandey and Young, 2021).

5.2.2 Tubular Members

Tubular structural members, from pressure vessels and line pipes to piping and fittings at smaller scales, are often subjected to stresses and loads that can exceed design limits. These extreme conditions can bring components to a critical state, potentially compromising structural integrity. There are two distinct cases when referring to extreme loading conditions. One is when the extreme load causes actual damage to the structure, thereby reducing its residual or ultimate strength. The other is when the extreme load is used as a design scenario to assess the structure's capacity under ULS or ALS without prior damage. The former affects the remaining strength and may require damage-tolerant design or repair strategies, while the latter is used to verify that the structure can withstand extreme but undamaged conditions.

While large-scale experiments provide valuable insights, they can be costly and time-consuming, especially for certain structural components. As a result, computational tools are highly valued for their cost-effectiveness, speed, and ability to conduct parametric analyses. However, these benefits come with certain drawbacks, particularly in terms of reliability and accuracy. Despite extensive research over recent decades to address these issues, challenges remain that warrant further improvement.

Li *et al.* (2023a) conducted comprehensive computational analyses on multiaxially loaded pipe components with defects, made from 316L stainless steel, to assess plastic collapse and compare results with safety limit load requirements. The modelling included low-temperature behaviour, incorporating kinetic phase transformation and temperature-dependent yield strength formulations to capture mechanical responses at near-cryogenic temperatures (Paredes et al., 2020). This study was largely motivated by the need for long-distance modular transportation of liquid natural gas (LNG), where components are subjected to complex loading conditions, such as combined torsion, bending, and internal pressure, in Arctic-like environments. The findings were compared with analytical solutions from standards, such as ASME (2017), Rules for in-service inspection of nuclear power plant components: boiler and pressure vessel code, section XI; and ASME (2017), boiler and pressure vessel code; see Fig. 26.

Yuan et al. (2022) focused on the plastic collapse behaviour of subsea-lined pipes deployed using the S-lay installation method. During this process, the lined pipe was subjected to combined tension-bending loading which typically leads the liner material to detach from the pipeline due to severe ovalisation. Using a custom numerical framework with quasi-two-dimensional and three-dimensional models, they investigated two loading sequences of bending and tension on lined pipes with both perfect and imperfect geometries. The results indicated that tension, either applied before or after bending, affects the growth of ovalisation and liner collapse, the extent of which depends on the load path followed. The authors also demonstrated that bending with modest levels of internal pressure delays the onset of liner collapse, but the threshold pressure varies for different load paths.

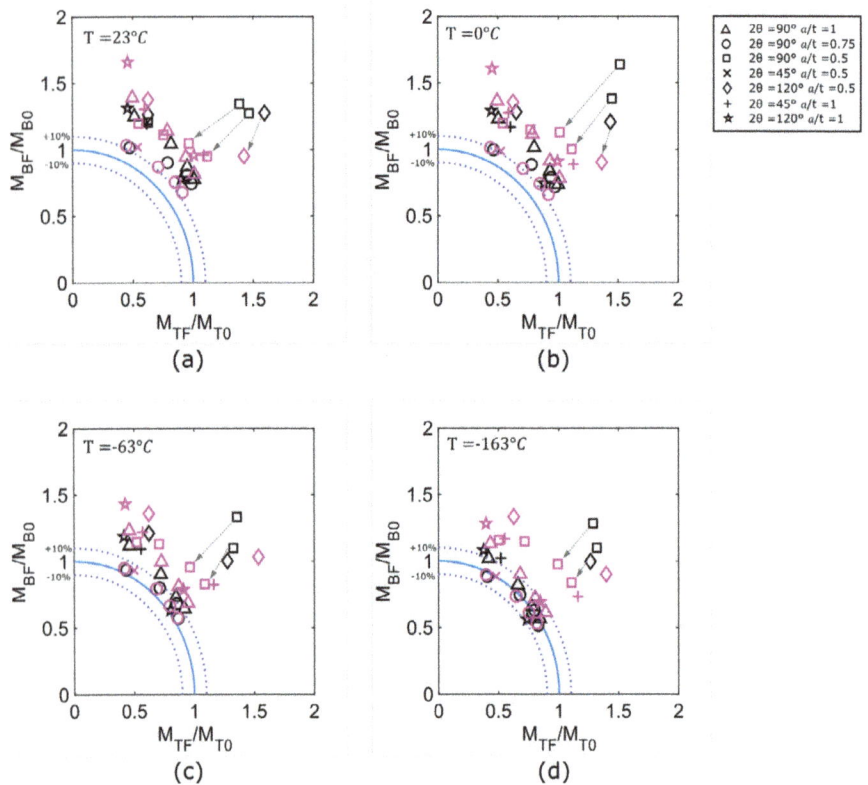

Fig. 26. Failure interaction curves for distinct temperatures: (a) 23 °C, (b) 0 °C, (c) − 63 °C, and (d) − 163 °C. The black open symbols represent simulations without internal pressure ($p_i = 0$ MPa), while the magenta colour marks depict pressurized pipe configurations ($p_i = 20$ MPa); Li et al. (2023a).

5.2.3 Reinforced Tubular Members

With the increase in the rated power of offshore wind turbines, the size of support structures has also grown, posing new challenges in the design of monopile tower barrels and support systems. To address these requirements, tapered concrete-filled double-skin steel tubular (CFDST) structures have recently been proposed by Wang *et al.* (2023b) (see Fig. 27). They developed an analysis tool for the evaluation of axial compression strength and torsional capacity for tapered CFDST members. Their investigation addressed the torque versus deformation response, interaction behaviour, and component contributions, as well as the influence of various parameters on the torsional capacity.

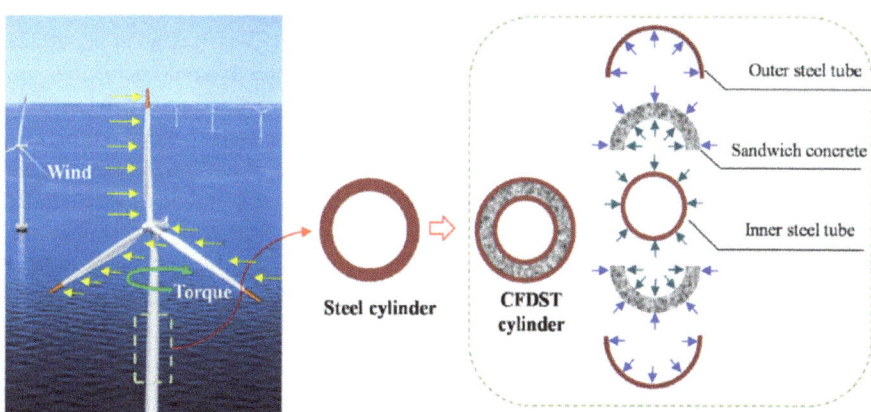

Fig. 27. Potential application of CFDST members in support structures (Wang *et al.*, 2023b).

Sundarraja and Prabhu (2011) demonstrated experimentally that the external bonding of the carbon-fibre reinforced polymer (CFRP) to concrete-filled steel tubular (CFST) structures provides effective confinement pressure and delays the local buckling of the steel tuber under axial compression. It is now well established that the presence of CFRP sheets considerably enhances the ultimate strength and deformation of traditional CFST columns. Therefore, different approaches have been proposed to predict the ultimate strength of FRP-CFST columns (e.g., Güneyisi and Nour, 2019) and their performance has been investigated under different combinations of variables (e.g., Shen et al., 2019). Wang *et al.* (2022a) conducted a comprehensive FE analysis, offering insights into the axial compression behaviour of circular FRP-CFST columns. They investigated the parameters affecting the axial compression behaviour of stub columns and a new regression-based formula for ultimate bearing capacity was proposed.

In a broader context, Duong et al. (2023) proposed a machine learning-based approach for the prediction of the ultimate capacity of CFST columns with different cross-sectional shapes: circular, elliptical, square, and rectangular. The training database was gathered from various studies available in the literature, and the results are accessible through a freely available MATLAB-based graphical user interface. The effects of cross-sectional shapes on the axial performance of CFST were also investigated numerically by Ayough et al. (2021) who concluded that hexagonal and octagonal CFST columns behaved similarly to square columns, whereas circular CFST columns demonstrated superior structural performance.

5.3 Structural Systems

5.3.1 Offshore Aquaculture Platforms

Recent developments in aquaculture fish farms have included a shift towards offshore farming in response to increasing seafood demand and limited space in coastal areas. The size of fish farms is expected to increase, and new designs are being developed to withstand high-energy waves and stronger currents in the open ocean. There is growing interest in combining floating offshore wind turbines with aquaculture cages, creating a new concept called the floating wind-aquaculture platform (FWAP) (Cao et al., 2022). Experimental investigations are being conducted to study the dynamic behaviour and effects of aquaculture nets on the FWAP. Another example is vessel-shaped fish farms (Li et al., 2019b) to improve the longevity of aquaculture systems and the welfare of farmed fish. Furthermore, there is a focus on improving the strength of net structures under the action of flow to ensure the survival and growth of farmed fish. Regulations and standards, such as the Norwegian Standard NS 9415, have been introduced to address technological investments and reduce the number of escaped fish. A study by Chu et al. (2023) presented a review of the most recent developments related to standards and guidelines for the design and analysis of offshore fish farms. Their work covered various crucial aspects for consideration when developing offshore fish farming facilities, such as design criteria, global performance analysis procedures, and decision-making methods. The authors also discussed the limitations and considerations associated with the references used in the review.

A novel BT-FOWT-AC integrated structure designed for a water depth of 100 m was proposed by Zhai et al. (2024). Their study assessed the impact of integrating an aquaculture cage and netting on the structures and responses of three different configurations: Stru1, which included an aquaculture cage and netting; Stru2 without netting; and Stru3 without an aquaculture cage., The study evaluated these configurations under various environmental conditions, including wave and current loads, as well as combined wind, wave, and current forces during both normal operations and shutdown scenarios. Findings revealed that the netting's inherent damping properties significantly diminished structural fluctuations, thereby enhancing the stability and safety of the integrated system.

5.3.2 Offshore Wind Farms

Offshore wind turbines (OWTs) are subjected to various environmental loads such as wind, waves, and currents, which can cause fatigue and ultimate strength failure. Various numerical simulation tools have been developed to predict the dynamic response of OWTs under various loading conditions (Wang et al., 2021). Due to the combined effect of various factors, the support structure may fail and the wind turbine may collapse. Nevertheless, the reliability analysis of OWT trends towards simplified load analysis. The reliability evaluation of these structure is used as a basis for the estimation of safety levels. For example, Ivanhoe et al. (2020) developed a reliability assessment framework for OWT jacket support structures considering five limit states (including the ULS) using a parameterised finite element model. Wilkie and Galasso (2020) proposed a probabilistic risk modelling framework to assess the structural risk posed by extreme weather conditions to OWTs. Specifically, they formulated a fragility function, which quantified the probability of different damage levels occurring across various wind and wave intensities. Through analysis of two case study sites, they found that the tower generally had a higher probability of failure at the ultimate limit state (ULS) than other structural components, such as the monopile, transition piece, and blades. Notably, the monopile was typically the safest component under these conditions. Moreover, they highlighted that ULS assessment of components is more relevant to sites exposed to hurricane-level conditions (East Coast of the USA), while for milder European conditions fatigue limit states are expected to be more relevant.

Yeter et al. (2019) presented an uncertainty analysis of soil-pile interactions for monopile offshore wind turbine support structures. Their study examined the sensitivity of pile design parameters to variations in soil properties, such as strength, stiffness, and damping. They employed a probabilistic approach in their analysis to assess the uncertainty in the response of the pile, including the ultimate capacity and the natural frequency. Their results suggested that soil properties have a significant impact on pile response, and the level of uncertainty in the design parameters may be relatively high. The authors concluded that a comprehensive understanding of soil properties is essential for the accurate and reliable design of OWT support structures.

In recent years, machine learning has been applied to predict the ultimate strength of offshore structures. For instance, Muhaimin Ishak et al. (2022) developed a machine learning-based tool, IGUSA (Intelligent Global Ultimate Strength Analysis), which was used in a case study to predict the ultimate strength of a fixed offshore jacket platform installed in Malaysian waters. The input parameters for their model included water depth, leg strength, number of risers, number of conductors, and number of legs, as well as response parameters, such as end-on base shear at collapse, broadside base shear at collapse, and diagonal base shear at collapse. They concluded that their model was suitable for rapid assessments of ultimate strength for fixed offshore structures, reducing time and cost in structural integrity management.

Offshore structures are likely to be exposed to corrosive environments, consequently affecting their ultimate strength. Therefore, it is important to investigate the improvement in ultimate load capacity after the application of strengthening and/or repairing methods. George et al. (2022a) investigated the flexural response of tubular steel members retrofitted by carbon fibre-reinforced polymer (CFRP) wraps. Two specimen sets underwent a four-point bending test. The ultimate strength of the specimens showed improvement in the range of 18.6 to 22.3% for the "Intact and Strengthened", which were insensitive to the number of CFRP layers, and 14.9 to 32.1% for the "Corroded and Repaired" specimens, which increased as the number of CFRP layers increased. However, the ductility index of the tested specimens fell as the number of CFRP layers increased. They also noted that the ultimate strength of the retrofitted "Intact and Strengthened" specimens was reduced by a maximum of 2.2% after one year of exposure to saline water. Nevertheless, the ultimate strength of "Corroded and Repaired" specimens was at least 4% higher than the corresponding values of their counterparts before exposure.

George et al. (2022b) also investigated the ultimate strength of retrofitted tubular steel members with CFRP, with different retrofitted lengths under axial eccentric compression loading. Their investigation revealed that partial-length CFRP retrofitting did not enhance the ultimate strength of leg specimens, likely due to local yielding that occurred outside the retrofitted zone, resulting in earlier failure compared to non-retrofitted specimens. While leg specimen failure was predominantly due to cross-section yielding, brace specimens—designed to be governed by Euler buckling—showed an average strength improvement of 19.1% due to CFRP retrofitting. Additionally, corroded and subsequently repaired leg and brace specimens demonstrated significant strength gains, with improvements of up to 40.2% and 22.1%, respectively. Full-length CFRP retrofitting of leg specimens led to a substantial increase in ultimate load capacity, reaching up to 53.8% with an increasing number of CFRP layers.

A review of the recent worldwide reports of failure incidents of offshore wind farms highlighted an incident that occurred at the Anholt offshore wind farm in Denmark in April 2022, where a rotor, including three blades, detached from the nacelle of an offshore wind turbine and fell into the sea (Ørsted Media Relations, 2022). This wind farm, commissioned in 2013, comprises 111 S-Gamesa turbines, each with a capacity of 3.6 MW (Rapacka, 2023). Following a thorough investigation, the incident was attributed to a production fault in the affected rotor (Energy Watch, 2023). Although the rotor detachment incident was attributed to a production fault rather than an external load, it highlights the importance of considering manufacturing-induced uncertainties in ULS assessments. Such uncertainties can significantly reduce the actual capacity of structural components and should be incorporated into probabilistic design frameworks.

With the growing number of floating offshore wind turbines, the importance of time-dependent structural assessments has increased due to the unsteady, turbulent wind, and ocean environments that induce a coupled, time-varying structural response. Lee et al. (2023a) proposed a time-domain structural analysis method, applied to the substructure of a 15 MW floating offshore wind turbine. This method was tested under two critical load cases: extreme operational conditions and parked states in high wind and wave environments, using two distinct turbulence models. Extreme Von Mises stress values were computed from a failure distribution function, based on three hours of data from the actual structure. The study concluded that the method was computationally efficient and capable of capturing the coupled effects between transient wind loads on the turbine and irregular wave loads on the floating structure.

Xu et al. (2021a) conducted a comparative study of the ultimate limit state design of different mooring systems for floating wind turbines in shallow water. Seven mooring concepts designed for a 5 MW semi-submersible floating wind turbine at 50 m water depth (see Fig. 28) were compared to identify solutions that were structurally reliable and economically attractive. This study introduced the concept of mooring system design for further strength assessment.

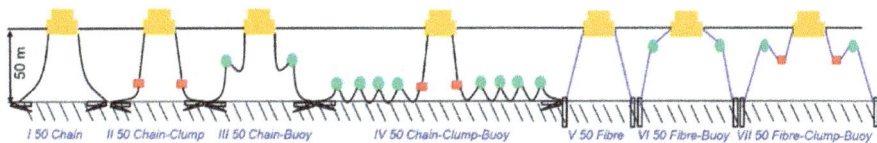

Fig. 28. Seven mooring concepts designed for a 5 MW semi-submersible floating wind turbine at 50 m water depth (Xu et al., 2021a).

An incident involving a ship collision with an offshore wind turbine substructure has raised important considerations for structural resilience (Lewis, 2023). The cargo ship, transporting 1,500 tons of grain from Poland to Belgium, collided with an offshore wind turbine in Germany. Following a post-accident inspection, the turbine was reactivated within 24 h. However, this event highlights the need for further research into the effects of ship collisions on the ultimate load capacity of offshore wind turbine structures. Ladeira et al. (2023) provided a review of structural assessment methods for offshore wind turbines under ship impact. Analytical models, experimental studies, and numerical simulations can play a critical role in predicting the structural response to such collisions.

Jia et al. (2020) conducted an experimental study in China to quantify the effects of ship collision on the structural load of an offshore wind turbine. The collision generated impact forces ranging from 2.4 to 25 kN, and strain measurements were made at the blades of the turbine and tower during the event. The researchers calculated the resulting bending moments in these structural components and observed a slight increase in peak bending moments on both the tower and the blades due to the collision. Despite this increase, the collision did not compromise the structural integrity of either component. Nevertheless, there remains a need for more extensive experimental research on the impact of ship collisions on offshore wind turbine structures, as this area is currently underexplored.

Liu et al. (2022a) conducted a three-point bending test and finite element simulation to investigate the effect of impact location, collision angles, and impact velocity on crashing response, see Fig. 29. They designed a down-scaled tubular member of an offshore jacket platform made of DH-36 steel for testing, and modelled plastic deformation using a power law hardening approach. After validating the FE model, they simulated an offshore jacket foundation and considered 27 accident cases with different impact locations, angles, and velocities in the range of 0.5 to 4 m/s applied to the legs and braces. They found that the most severe scenario led to a local indentation of 916 mm on the leg. Additionally, they noted that as the angle of impact increased, the severity of local indentation on the structural members decreased.

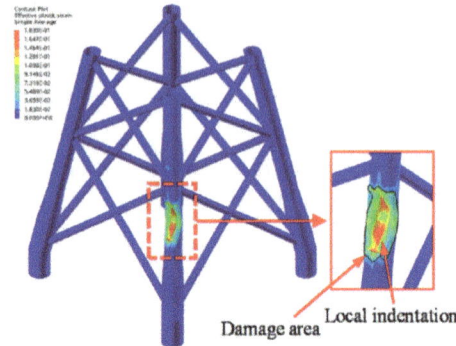

(a) Damage in an actual ship-jacket collision   (b) Simulated plastic strain field of the jacket (case ILL3)

Fig. 29. Comparison of actual and simulated damage to the jacket (Liu *et al.*, 2022a).

5.3.3 Other Floating Structures

To address the coupling between hull motion and turbine loading, class societies recommend fully coupled time-domain analyses of floating offshore wind turbines, followed by response-based structural assessments to evaluate structural strength. Recognizing the impracticality of analysing a large number of design load cases with lengthy simulation times using current tools, Moon et al. (2023) proposed an efficient time-domain structural analysis for buckling and ultimate strength assessment. For various design scenarios, they performed time-domain structural analyses by mapping aero-elastic, hydrodynamic, hydrostatic, inertial, and mooring loads onto a FE structural model. To enhance computational efficiency, they introduced a pseudo-spectral stress synthesizer based on "lodal" response analysis, which combines linear frequency-domain loads with non-linear time-domain loads. This lodal response concept enables the method to generate stress component time histories that account for transient wind loading, irregular wave loading, and the resulting motion response of the floating hull.

To address the coupling between hull motion and turbine loading, class societies recommend fully coupled time-domain analyses of floating offshore wind turbines, followed by response-based structural assessments to evaluate structural strength. Recognizing the impracticality of analysing a large number of design load cases with lengthy simulation times using current tools, Moon et al. (2023) proposed an efficient time-domain structural analysis for buckling and ultimate strength assessment. For various design scenarios, they performed time-domain structural analyses by mapping aero-elastic, hydrodynamic, hydrostatic, inertial, and mooring loads onto a FE structural model. To enhance computational efficiency, they introduced a pseudo-spectral stress synthesizer based on "lodal" response analysis, which combines linear frequency-domain loads with non-linear time-domain loads. This lodal response concept enables the method to generate stress component time histories that account for transient wind loading, irregular wave loading, and the resulting motion response of the floating hull.

Mooring system design for floating structures is often governed by ultimate limit state (ULS) conditions. Cai et al. (2024) numerically investigated the ultimate strength of the mooring lines and suspension ropes of a novel semi-spar floating wind turbine platform under extreme conditions, as well as the relationship between the ropes' tension and the wave direction. Simulation revealed that the lateral suspension ropes parallel to the propagation direction were sensitive to winds and waves.

The FPSO is one of the most common platform types for offshore oil production. Silva et al. (2022) investigated the ultimate strength of FPSO hulls designed with SPS sandwich plates according to the DNV rules. Numerical simulations showed that SPS design provided a reduction of 2.8% of the total weight and a better overall structural performance (an increase of 26% in the magnitude of the ultimate strength of the hull). The authors recommended further research regarding the local ultimate strength and fabrication process of SPS panels.

Fonseca et al. (2024) conducted a mooring analysis of an FPSO in ballast conditions, focusing on horizontal wave drift loads and associated slow drift oscillations. These factors represent the largest contributors to extreme line loads and also carry the greatest uncertainty in numerical predictions. This study highlighted a limitation in state-of-the-art radiation/diffraction potential flow codes, which often underestimate wave drift loads in severe sea states for large floating structures like FPSOs with substantial waterplane areas. To address this, the authors applied empirical quadratic transfer functions (QTFs) of horizontal wave drift loads directly within time-domain simulations for the mooring analysis. The methodology was validated through a case study of an FPSO operating in the North Sea, demonstrating improved accuracy in predicting extreme mooring loads.

Floating photovoltaic (FPV) systems are drawing considerable interest in solar energy generation because of their unique benefits, including space efficiency and cooling effects from water, which improve panel efficiency. A study by Li *et al.* (2023b) assessed the ultimate strength of a modular FPV system designed with ultra-high-performance concrete (UHPC) and fibre-reinforced polymer (FRP). Their study included hydro-elastic and structural analyses focusing on the influence of hinge connections and module configurations on bending capacity under wave loads. Such evaluations are critical for ensuring structural resilience in offshore settings, where wave action can impose significant stresses on the system.

5.4 Rules and Guidelines

Offshore rules and regulations for ultimate strength are essential for ensuring the safety and reliability of offshore structures. These regulations guide the design, construction, and maintenance of offshore structures, ensuring that they can withstand the harsh environmental conditions and extreme loads they are subjected to. Design standards vary depending on the application and are generally divided into two main types: *fixed structures* and *floating production units* (FPUs). Fixed structures, often comprising tubular members and connections, must account for seismic response among other environmental loads, while FPUs, built primarily of plated structures and stiffened panels, face different challenges. Moreover, structural materials and inspection criteria also differ, with tailored standards for each type. International standards organisations also play a significant role in establishing these standards, which are essential for uniformity and reliability across the offshore industry, which are as follows:

- International Organization for Standardization, ISO 19900, Series of standards for offshore structures (ISO, 2019).

- API, Offshore standard, American Petroleum Institute (API, 2021).

Classification societies, which are non-governmental organisations, establish rules, which are typically linked to international standards. The rules and regulations governing the ultimate strength of offshore structures vary across regions. There are also national standardisation organisations that develop standards, such as NORSOK Standards N-series in Norway, the British Standards Institute (BSI) in the UK, Deutsches Institut für Normung (DIN) in Germany, and Danish Standards (DS) in Denmark. Many of these national standards closely relate to, and sometimes incorporate, parts of international standards (such as those of the International Organization for Standardization (ISO)), thus maintaining a level of alignment with global regulations while addressing specific regional needs.

Over the past decade, the offshore industry has increasingly focused on harmonising standards to ensure that offshore structures are designed and maintained to the same standard across different regions. Although there are some differences between standards, such as NORSOK and ISO, these are gradually becoming more aligned. One key difference lies in the frequency of updates. NORSOK standards are typically revised annually to incorporate minor adjustments, whereas ISO standards are updated less frequently, leading to longer intervals between revisions.

Differences in the standards for ultimate strength assessment also vary based on the application. For example, while ISO, NORSOK, and API standards are closely aligned regarding connection requirements, column buckling provisions are very similar in ISO and NORSOK. However, lateral pressure effects differ slightly, resulting in a higher computed capacity in ISO due to a larger safety factor. The use of safety factors also differs across these standards, with ISO applying varied material factors depending on the specific failure modes. NORSOK has recently updated its N-004 standard, adding supplementary material to address topside structures, bolted connections, and corrosion.

Design standards are continuously refined to enhance accuracy and address special cases or innovative designs. There is a growing trend towards advanced methods, such as non-linear finite element (FE) analysis. In support of this, a new Eurocode standard (CEN, 2023) is being developed, which outlines principles and requirements for numerical methods in steel structure design, particularly for ultimate limit state, fatigue, and serviceability limit state verifications. Additionally, classification societies have introduced detailed standards for FE analysis, such as DNV-RP-C208 (DNV, 2022b), a widely recognized guideline covering non-linear FE analyses for ultimate strength, accidental scenarios, and low-cycle fatigue. This guideline provides comprehensive instructions for FE modelling, including aspects such as material properties, mesh configurations, boundary conditions, analysis procedures, and post-processing.

In addition to the requirements mentioned above, offshore operators also have supplementary specifications for offshore structures, which vary depending on the company. The oil industry has a long tradition of cooperation between oil companies in the context of engineering standards, and the industry has recently developed common supplementary specifications for topside structures (IOGP, 2020) and for *Fixed Steel Offshore Structures* (IOGP, 2021). One of the objectives of this undertaking was to limit substantial budget and schedule overruns for oil and gas projects. The oil industry community have implemented an initiative which seeks to drive a reduction in upstream project costs with a focus on industry-wide, non-competitive collaboration and standardization. It must be noted that these common supplementary specifications are not mandatory, and oil companies in IOGP may still use their own supplementary specifications for specific projects and offshore installations.

DNV provides a set of rules for assessing the strength of offshore structures (DNV-OS-C101, OS-C102, OS-C103, OS-C104, OS-C105, OS-C106, RP-0286, and RP-C208). Buckling assessments are covered in both the load resistance factor design (LRFD) format in OS-C101 and the working stress design (WSD) format in OS-C201; however, the WSD format (OS-C201) was retired in July 2023. Recently, DNV introduced a new voluntary class notation for hull girder strength in OS-C102 (Structural Design of Offshore Ship-Shaped and Cylindrical Units). This addition provides specific hull girder strength capacity checks, addressing the requirements of NORSOK N-004 Annex C that extend beyond the standard's basic requirements.

6 Benchmark Study

The previous benchmark study by the ISSC 2022 Ultimate Strength committee (ISSC, 2022a) highlighted the importance of assessing numerical methods based on the entire end-shortening curve of ship structures, rather than focusing solely on the ultimate capacity point. This is particularly crucial in scenarios where the dissipation of energy in a structure plays a significant role, such as hull girder failure under wave loading (Li and Benson, 2019). The maritime industry's shift towards high-strength steels, though beneficial for reducing hull weight, poses challenges in designing slender structures susceptible to elastic buckling, since while the yield stress is increased, the Young's modulus does not change significantly, complicating the design of such structures. While transverse loads may not be highly intense in ships with longitudinal structures, the low resistance to elastic buckling due to the aspect ratio of plating panels remains a concern. Ongoing international debates focus on the regulatory framework for buckling verification with transverse stress components (IACS, 2023a). A research project funded by Fincantieri in Italy (NBA - Non-linear Buckling Assessment, 2021) produced a full-scale test geometry to assess the structural response of slender steel ship structures to transverse loads. The structure was designed to induce local elastic instability at lower thresholds than its ultimate capacity to provide valuable insights into load redistribution across structural components.

The committee's benchmark study was designed to test researchers' predictive capabilities by having them generate end-shortening curves based on identical information provided to all participants. During the first two phases of the study, no experimental data were made available, allowing researchers to rely solely on theoretical models. In the third and final phase, participants proposed hypotheses to explain any numerical-experimental discrepancies observed.

6.1 Description of the Benchmark Study

The committee-coordinated blind benchmark study evaluated researchers' ability to predict the ultimate strength and non-linear response up to the collapse of a panel structure.

Participants were not provided with experimental results, highlighting the challenge of making accurate predictions based solely on limited data. The study was divided into three phases to assess how additional data and a more in-depth understanding of specific information could affect and potentially improve prediction accuracy. Participants were encouraged to apply their knowledge and expertise, with flexibility in selecting methods, finite element software solvers, numerical damping, element types, formulations, mesh sizes, imperfection modelling, material properties, boundary conditions, and other factors to assess ultimate strength capacity and post-buckling behaviour.

6.1.1 Benchmark Phases

Figure 30 outlines the workflow of the benchmark study, focusing on a full-scale physical model (see Fig. 31) manufactured by Fincantieri and tested at the University of Genoa, Italy. Following the approach of the ISSC 2022 Ultimate Strength committee (ISSC, 2022a), the study was divided into three phases (Phases 1 to 3). Building on the ISSC 2022 Ultimate Strength committee benchmark, the current study incorporated the actual specimen geometry, including imperfections, in the initial phase, with laser scans provided. In Phase 2, more information regarding material properties was disclosed, while in Phase 3, the participants fitted the actual experimental curve using their unique skills.

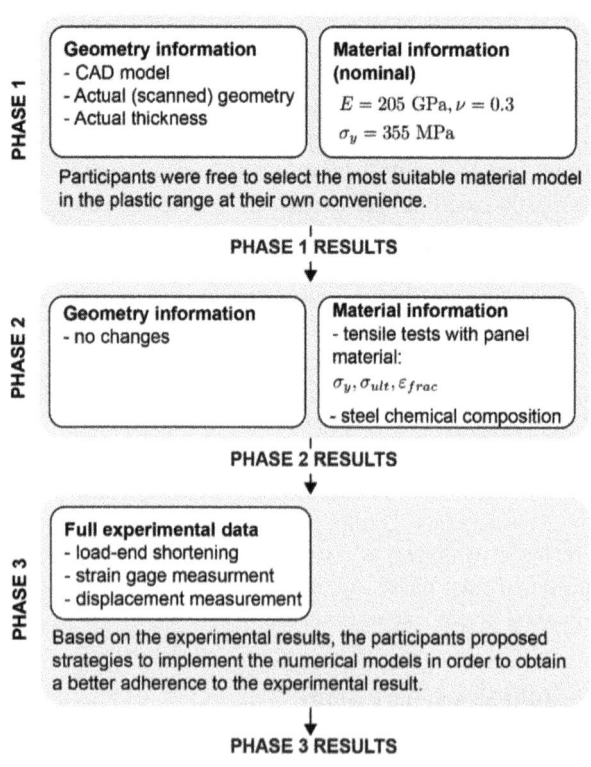

Fig. 30. Benchmark flowchart.

6.2 Description of the Experiment

6.2.1 Setup of the Experiment

Thanks to a collaborative partnership with Fincantieri, the committee was given access to the results of physical experiments on a large-scale structural component. The test structure, depicted in Fig. 31, underwent longitudinal compression until collapse using two 150-ton capacity actuators. These actuators were linked to a shared hydraulic system to ensure uniform internal pressure, with the opposite end of the structure entirely fixed. The setup was designed to ensure equal transmission of external forces to the specimen. Custom jack heads were used to fit precisely within the reinforced structure at the specimen's edge, minimising tolerances and enabling the external edge to withstand concentrated forces. However, machine compliance was not measured, and all results exclude this factor.

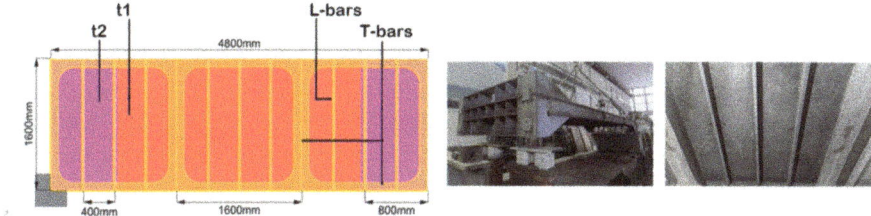

Fig. 31. Panel geometry and detail of the central bay of the as-built panel.

The experimental test, conducted by MaSTeL (Marine Structures Testing Lab) at the University of Genoa, involved several measurements, including:

- Axial displacement of the loaded edge.
- Axial displacement of the fixed edge to evaluate potential deformation of the supporting structure.
- Data collected by four load cells on the fixed edge, providing vertical support to the panel and addressing uncertainties from the friction of supporting elements.
- Active loads applied by the actuators, specifically two on the loaded edge, for comparison with the axial load measured by the load cells.
- 40 measurements of different strain components obtained through 16 strain gauges (12 rosettes and 4 linear gauges).
- Two high-definition videos of the shell plating captured from different angles, featuring markers on the plating suitable for 3D reconstruction.

6.2.2 Panel Geometry

The designed test panel, divided into three bays with transverse stiffening and frames, aimed to induce early local elastic buckling and achieve a substantial post-buckling range before the ultimate load. This design considered both the structural aspects and the dimensions of the laboratory test bench. The panel's slenderness was determined based on Fincantieri's construction geometries, addressing regulatory challenges for local instability criteria. While these geometries are uncommon in cargo ships, they establish standards for weight-minimising constructions. The dimensions and thicknesses of various structural elements of the panel were as follows:

- Plates: t1 = 4 mm, t2 = 8 mm, where t2 represents the thickness of plating extending longitudinally 850 mm from the ends and t1 the one of the central parts of the panel (see Fig. 31: the red area represents the t1 area which is 4 mm thick).
- L-bars: 60 × 5 × 30 × 5 mm (hw × tw × bf × tf), transverse ordinary stiffeners.
- T-bars: 150 × 6 × 100 × 6 mm, extending longitudinally along the longer edge of the panel and transversally at the end of each bay.

The use of varying plating thicknesses was intentional to prevent collapse at the panel ends, effectively defining a clearer experimental focus area. The panel material was DH36 steel.

6.2.3 Geometrical Imperfections

The shipbuilding standard IACS Rec 47 (IACS, 2021) was considered in the panel fabrication, with a maximum limit of 8 mm for plate imperfections between frames and a standard of 4 mm. The panel manufactured by the shipyard exhibited significant geometric imperfections, with deviations from the average plane of the plating reaching up to 6 mm in the most deformed areas, particularly in the two end intervals where a greater plating thickness was chosen to mitigate edge effects. Hence, the panel exceeded the IACS standard for fairness of plating between frames but still complied with the maximum imperfection limit. Additionally, the common stiffeners displayed a noticeable tilt relative to the welding angle onto the plating. In response to these imperfections, a decision was made to conduct a laser scan of the panel geometry from its flat side to accurately document the deviations of the plating from its mid-plane, see Fig. 32. Furthermore, 3D reconstructions were carried out based on local scans to ascertain the tilt angle of the common beams, as illustrated in Fig. 32. Unfortunately, a complete 3D reconstruction of the full panel specimen was not available as alignment of scans was not maintained during the scanning process.

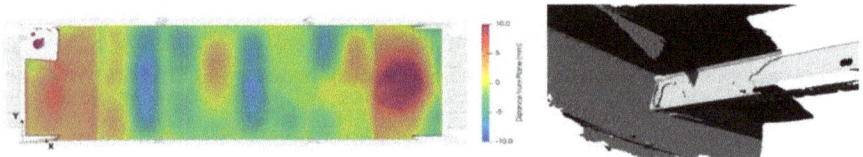

Fig. 32. Deviation from the best-fit plane (left) and example of 3D reconstruction of some details (right).

6.3 Phase 1 Results: Numerical Assumptions

The participants of the benchmark study were provided with the following information:

- An IGES format CAD model of the test specimen.
- Laser scan data file capturing the flat surface of the specimen.

- 3D reconstructions of all connections between common reinforcements and primary beams, along with mid-span areas of individual reinforcements.
- Map of thickness measurements obtained using an ultrasound probe in various areas of the specimen.
- Limited information on material properties: Young's modulus ($E = 205$ GPa), Poisson's ratio ($v = 0.3$), and yield stress ($\sigma_y = 355$ MPa).
- Fully clamped boundary conditions at the panel's ends.
- No residual stresses were considered.

The participants were encouraged to utilise engineering judgement and expertise when data were not explicitly provided. This included devising strategies for incorporating initial geometrical imperfections and selecting an appropriate non-linear material model. Four FE software packages were used, with ABAQUS being the most popular (9 out of 16 groups). Other software used were LS-Dyna, NX-Nastran and Marc. All participants chose a 4-node shell element, except for one who preferred an 8-node element. The preferred mesh size was 20 × 20 mm, but ranged from 12.5 × 12.5 mm (used by one participant for specific details) to a larger mesh size of 50 × 50 mm for the entire group. Three participants opted for an explicit analysis. The interpretation of initial imperfections is of interest, as seven out of fourteen participants decided to disregard the shape provided through laser scanning, opting for alternative strategies based on their experience.

One participant among the sixteen who submitted results in Phase 1 had significantly different values, and this outlier was excluded from the current analysis. The end-shortening curves from the remaining fifteen participants are shown in Fig. 33 (left). Different modelling strategies for initial imperfections are represented by red for eigenmode imperfections, black for laser scanning-based geometry approximations, and blue for alternative representations. Four teams treated the model as perfectly plastic, while the remaining twelve assumed a work-hardening behaviour, applying a modulus of 1000 MPa with minimal variations. Table 2 presents a summary of the modelling assumptions of each participant and the deviations from the mean values of ultimate strength (K and n are material parameters in the Ramberg-Osgood relationship).

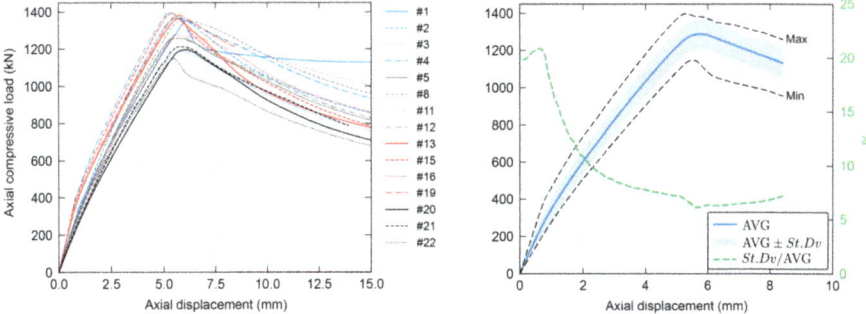

Fig. 33. (Left) The resulting end-shortening curve after Phase 1 of the benchmark study, and (right) average values and dispersion.

Table 2. Salient modelling assumptions of participants and resulting ultimate strength (I: implicit solver; E: explicit solver; E.M. #: eigenmode, # buckling shape number).

Participant	Software	Mesh size [mm]	Element: nodes, int. Points, sect. Points	Solver	Imperfection model Shape	Imperfection model Amplitude [mm]	Ultimate strength [kN]	Deviation from average [%]	Material hardening	Tan. Mod., or $(K; n)$ [MPa]; [-]
#1	NX-Nastran	50 × 50	QUAD4: 4, 4, 6	I	E.M. 1	2.0	1348.3	4.1	Yes	1000
#2	LS-Dyna	20 × 20	Type 16: 4, 4, 5	I	Mapped	6.0	1366.3	5.4	Yes	900
#3	ABAQUS	10 × 10	S4R: 4, 1, 5	E	Mapped	6.0	1188.6	-8.3	Yes (NL)	770; 0.161
#4	LS-Dyna	20 × 20	Type 16: 4, 4, 5	I	E.M. 1	1.5	1398.2	7.9	Yes	1000
#5	ABAQUS	20 × 20	S4R: 4, 1, 5	I	Sinusoidal	6.0	1256.7	-3.0	No	0
#8	ABAQUS	20 × 20	S4R: 4, 1, 5	I	Mapped	6.0	1351.8	4.3	Yes	1000
#11	ABAQUS	20 × 20	S4: 4, 4, 5	I	Mapped	6.0	1296.6	0.1	Yes (NL)	822; 0.164
#12	ABAQUS	19 × 19	S4: 4, 4, 11	I	Mapped	6.0	1258.0	-2.9	Yes (NL)	Swift-Voce model
#13	ABAQUS	12.5 - 50	S4R: 4, 1, 5	E	Only tilt	-	1362.7	5.2	Yes (NL)	790; 0.01
#15	LS-Dyna	20 × 20	ELFORM 2: 4, 1, 5	E	Not used	-	1393.8	7.6	No	0
#16	ABAQUS	25 × 25	S8R: 8, 5, 5	I	Mapped	6.0	1276.7	-1.5	Yes (NL)	606; 0.479
#19	ABAQUS	15 × 15	S4R: 4, 1, 5	I	E.M. 1–5	0.8	1382.6	6.7	Yes (NL)	750; 0.154
#20	LS-Dyna	20 × 20	ELFORM 2: 4, 1, 5	E	Mapped	6.0	1195.6	-7.7	Yes	
#21	Marc	20 × 20	ELEMENT 75: 4, 4, 5	I	Mapped with tilt	6.0	1212.2	-6.4	Yes (NL)	799; 0.171
#22	ABAQUS	20 × 20	S4: 4, 4, 5	I	E.M. 1	2.0	1147.9	-11.4	Yes (NL)	794; 0.11
					Average value		1295.7			
					Standard deviation		80.3			

Averaging and dispersion calculations were applied to the results from different participants, as illustrated in Fig. 33 (right). The average value curve (along with curves derived by adding and subtracting the standard deviation) is presented. The absolute maximum and minimum curves are also depicted, along with a curve representing the ratio of standard deviation over the average value, indicated on the right vertical axis. The maximum relative dispersion occurs in the initial phase of the curve for small applied loads, stabilising at approximately 6% of the average load until reaching the ultimate load. A possible explanation for this phenomenon is discussed in the following paragraph. The increase in dispersion slightly increases again in the post-collapse region after the ultimate load is reached.

Committee III.1: Ultimate Strength   541

6.3.1 Effects of Initial Imperfections

The numerical curves in Fig. 33 reveal a key characteristic of the early stages of the behaviour of the specimen. The elastic buckling load threshold is low, as planned in the design phase of the test, primarily due to local instability in the elementary plating panels. This unique characteristic leads to a sudden change in stiffness and the slope of the end-shortening curve upon reaching the threshold. Initial stiffness varies significantly among different models, with the stiffest showing a pronounced change in stiffness at the first critical load threshold. However, some models with lower initial stiffness do not exhibit a clear transition to post-buckling behaviour due to instability in the plating panels.

The variation in initial stiffness is linked to how geometric imperfections were interpreted. Models without shell plating imperfections (#13 and #15), the one featuring combined eigenmode-type imperfections (#19), and the model incorporating a mode-one eigenmode imperfection with a 1.5 mm amplitude (#4) demonstrate higher initial stiffness and exhibit a noticeable shift due to local elastic instability. Models using a mode-one eigenmode imperfection with a larger amplitude of up to 2.0 mm (#1 and #22) display behaviour closer to models with actual measured imperfections, though they remain slightly stiffer. A comparable response is seen in the model with sinusoidal imperfections (#5), where the amplitude closely matches that observed on the actual specimen. It was noted that in the current study, no significant differences were observed in the results between those who used the stiffeners' tilt and those who ignored it in the imperfection model.

6.3.2 Ultimate Strength and Failure Mode

The average predicted ultimate strength was 1295.7 kN, with a standard deviation of 80.3 kN, equal to 6.20% of the average value. The average axial displacement at which the ultimate response occurred was estimated at 5.77 mm, with a standard deviation of 0.26 mm equal to 4.54% of the average value. Results ±one standard deviation are presented in Fig. 33 (right). The maximum value found exceeds the average ultimate load value by 7.9%, while the minimum value is 11.4% lower. It was difficult to find a certain correlation that links the effect of the salient modelling characteristics with the relative ultimate load value. In the first instance, the effect of the initial imperfections on the value of the ultimate load can be excluded, since the average value between only the results obtained from the curves with eigenmode type imperfection and the average value of the results obtained from the curves with real imperfection modelled according to laser scanning is very close (1255.5 vs. 1251.7 kN). Furthermore, the average curves ± one standard deviation are also very similar in the two cases.

6.3.3 Post Ultimate Response

There seems to be no correlation between predicting material model selection and the predicted peak load of the structure. Figure 34 (left) illustrates that averaging the results of perfectly plastic material models closely matches those incorporating hardening up to ultimate strength, with similar error bars. Most participants considered the work-hardening effect and selected a hardening modulus of around 1000 MPa. Each participant described an identical collapse scenario, where a plastic hinge forms roughly at the midpoint of the specimen. This hinge corresponds to significant deformations concentrated on the beams' web aligned with the applied load, as shown in Fig. 34 (right). Unlike the observations related to ultimate strength, the behaviour post-ultimate is influenced by the material model, as depicted in Fig. 34 (left). Models assuming perfect plasticity show a converging post-ultimate response with minimal deviation. Conversely, models incorporating hardening exhibit a more dispersed pattern, with the lower error band closely aligning with results from perfect plasticity assumptions.

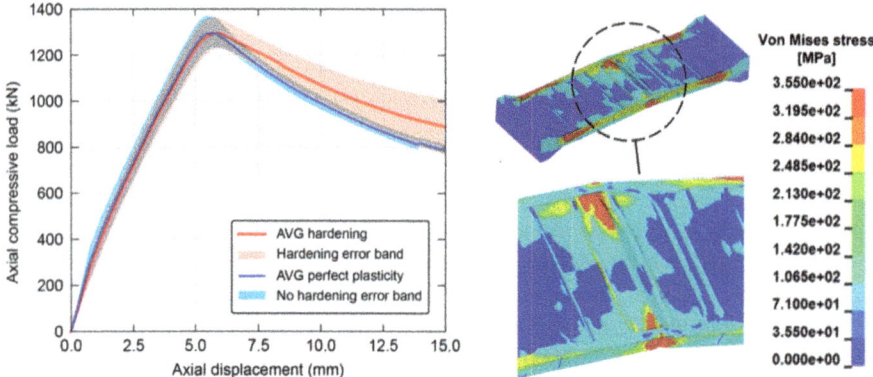

Fig. 34. (Left) Resulting end-shortening curves with and without hardening, and (right) collapse mode, as depicted by all participants with minimum variations.

A key takeaway from Phase 1 is the challenge in linking specific modelling parameters to their influence on predicting the structure's ultimate strength. However, the impact of initial imperfections on local buckling response in the shell plating, as well as the effect of the work hardening model on post-ultimate response, is more evident.

6.4 Phase 2 Results: Material Modelling

Phase 2 involved the sharing of information regarding the characteristics of the material used to construct the test panel. Information was provided about the chemical composition of the alloy used, which was obtained from a hot chemical analysis of samples collected during the production phase. Furthermore, the salient data of tensile tests on specimens extracted from the sheets with which the specimen was built, having different thicknesses, were shared, such as:

- yield strength,
- maximum engineering strength, and
- elongation at break.

The participants were asked to modify their numerical model from Phase 1 taking into account the newly disclosed material properties only, leaving the remaining parameters unaltered. In total, sixteen specimens were tested, the data of which were provided to the participants via the table presented in Table 3. The table also reports the average values and standard deviations of the specimens derived from plating having thicknesses of 4 and 10 mm, as well as the total average and standard deviation of the whole set. In Phase 2, the participants were also asked to modify only the material properties, leaving all the other input parameters of the numerical model unchanged. In total, fifteen participants completed Phase 2.

In Phase 1, participants uniformly applied the same yield strength, making only minor adjustments to the work-hardening model. However, in Phase 2, with more detailed material information, participants interpreted the plastic behaviour of the material differently. Some tailored the material model to account for the varying sheet thicknesses. Notably, the 4 mm plating model saw substantial diversity in approaches, as illustrated in Fig. 35 (left). Participants adopted various hardening models, including:

- exponential model following the yield stress.
- an initial yield plateau followed by an exponential model.
- bilinear model.
- multi-linear model.

Table 3. Material data for steel DH36 (Young's modulus, $E = 205$ GPa) provided to the participants in Phase 2.

Specimen no	Thickness [mm]	Yield strength [MPa]	Engineering strength [MPa]	Elongation [%]
1	4	417	505	38.2
2	4	394	495	40.2
3	4	416	499	34.1
4	4	382	502	34.7
5	4	385	493	33.5
6	4	412	495	36.9
Average value		401	498	36.3
Standard deviation		15	4	2.4
7	5	398	494	39.4
8	6	387	497	38.3
9	8	399	529	31.8
10	10	395	546	
11	10	406	532	
12	10	397	529	
13	10	396	526	
14	10	389	533	
Average value		397	533	31.3
Standard deviation		6	7	2.9
15	12	425		

(*continued*)

Table 3. (*continued*)

Specimen no	Thickness [mm]	Yield strength [MPa]	Engineering strength [MPa]	Elongation [%]
16	12	392		

Such variations have implications for the structural response near and beyond the ultimate load, with a clear increase in the dispersion of the results in this phase, as highlighted by Fig. 35 (right). The slight deviation in the dispersion in the first section of the curve is not immediately explainable, since the differences compared to the first phase should only concern the structural response once the plasticisation has occurred in the model.

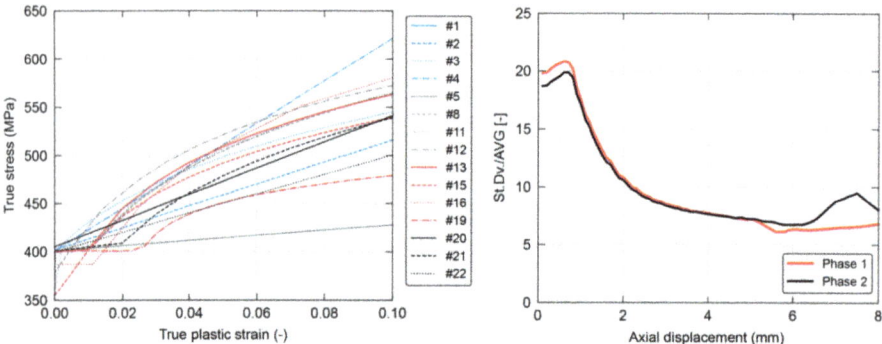

Fig. 35. (Left) True plastic strain vs. true plastic stress as interpreted by different participants in Phase 2, following the material data provided. (Right) Standard deviation of the end shortening curve over the average values observed in Phase 1 and 2.

Figure 36 shows the average of the end shortening curves ± the standard deviation obtained in Phases 1 and 2. The minimum and maximum values for the two phases are also shown. It can be clearly observed that, as expected, the results tend to differ only for high compressive load values capable of inducing a plastic response in the material.

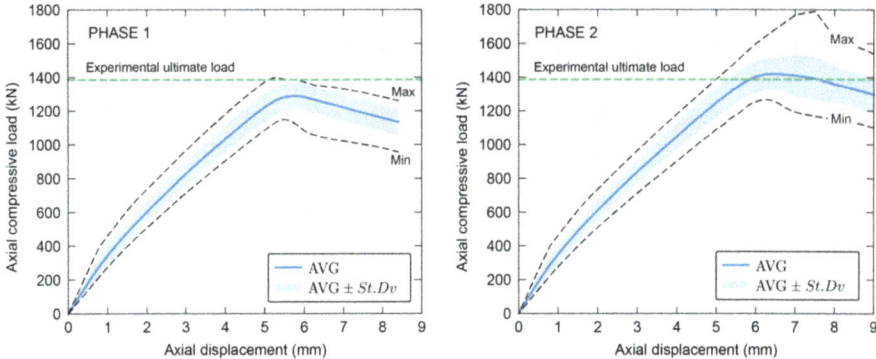

Fig. 36. The resulting average end-shortening curves after Phase 1 and 2, and standard deviation.

At the conclusion of Phase 2, participants received the experimental results, including the measured ultimate compressive force, which is shown on the graph. This additional material data led to numerical results that aligned more closely with the experimental findings, though the flexibility in interpreting the material model contributed to a wider spread in numerical outcomes. Specifically, the average ultimate strength from Phase 1 underestimated the experimental value by 7.47%, whereas in Phase 2, it was overestimated by 1.98%. This progression illustrates how detailed material information can improve accuracy, while also highlighting the impact of varied modelling assumptions on result dispersion.

A key takeaway from Phase 2 is that the additional material data provided to participants generally enhanced the accuracy of the numerical predictions. However, varied interpretations of the plasticity model led to increased result dispersion, underscoring the influence of modelling assumptions on prediction consistency.

6.5 Phase 3 Results: Interpretation of Experimental Results

In Phase 3, after sharing the experimental data, the participants were asked to formulate hypotheses about the modelling parameters to explain the numerical/experimental discrepancies. Following a collegial discussion, it was agreed that the experimental curve shows that boundary conditions changed during the experiment, as shown by the four distinct changes in the experimental curve in Fig. 37. It can also be seen that there is an initial settlement in the experiment followed by 3 drops in load corresponding to support failures. This was not considered in Phase 1 and 2 efforts. It was decided to proceed in separate working groups, each of which with the task of investigating a specific task and its effect on the load-end shortening curve. Three participants focused on the effect of welding residual stresses while two participants investigated constraint conditions that best represented the reality of the experiment without adding residual stresses to their models. Finally, a participant attempted to introduce a different modelling strategy, representing the shell plating through three layers of solid elements, connected by constraint equations to the shell elements still used for the representation of the beams. The solid mesh was created by mapping the actual imperfections of the panel. The participant observed an increase in the ultimate load of less than 0.5% for Phase 2, with no significant effects observed in the post-ultimate stage. Therefore, for analyses involving thin plating, using solid modelling is generally not recommended due to its complexity and time requirements. Instead, shell or plate elements are often more efficient and sufficient to accurately capture the structural response without the added computational burden of solid modelling.

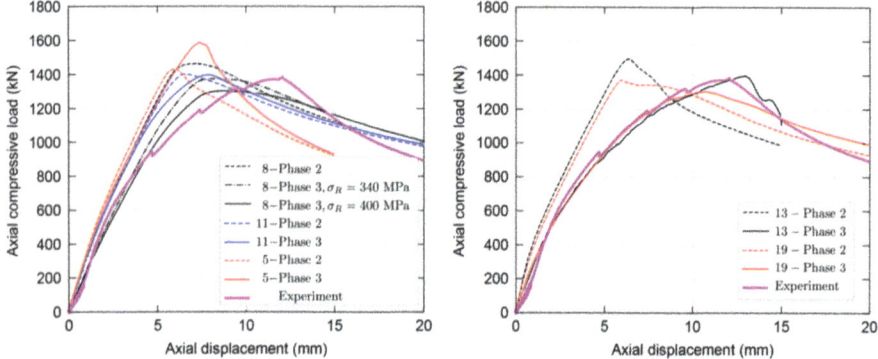

Fig. 37. (Left) Effect of residual stresses. End shortening curve after introducing residual stresses and the Phase 2 curve from the same participants. (Right) The effect of modified boundary conditions.

6.5.1 Effect of Residual Stresses

Three participants incorporated residual stresses due to welding in their models. Participant #8 applied a simplified thermal model targeting residual stress levels of 340 MPa and 400 MPa, as per the Eurocode (CEN, 2023). Participants #5 and #11 followed the approach by Yi et al. (2018), detailed in Paik (2018), directly inputting the stress field and verifying that the residual values matched their target after convergence. Participant #11 maintained a residual-to-yield stress ratio of 0.85 across all materials and thicknesses throughout the structure's welds, yielding a residual stress of 340 MPa in the same areas as Participant #8. Participant #5, however, limited residual stresses to the main parts of the test specimen, excluding the surrounding loading frame.

Participant #8's results indicate a clear effect of residual stresses, showing a reduced ultimate load compared to Phase 2. Participant #11's results also displayed a slight reduction in ultimate load, accompanied by lowered stiffness and greater displacement at peak load. When a residual stress of 400 MPa was simulated, similar to Participant #8, the ultimate load reduction was more pronounced. Interestingly, Participant #5's model showed an increase in ultimate load. A plausible hypothesis is that the residual stress field applied in this model may not be fully self-equilibrated across the shell panel, resulting in residual tensile stress on the beams, which could enhance the ultimate strength of the specimen, as the collapse mode in compression involves beam buckling. Despite these variations, all models demonstrated a stiffness reduction more consistent with experimental observations. However, residual stresses alone do not fully account for the numerical and experimental discrepancies.

6.5.2 Effect of Modified Boundary Conditions

Further investigation of end conditions was conducted. Using axial displacements evaluated at multiple points, it was possible to develop an estimate of the rotations at the ends of the panel about the transversal y-axis. The evaluation is presented in Fig. 38. It is observed that while the rotation is negligible at the loaded edge (actuator side), there is significant edge rotation at the fixed side. After stabilising at 0.1 degrees, the rotation increases dramatically following failure at the end support at 930 kN, after which the rotation increases linearly until the next support failure begins just before 1200 kN.

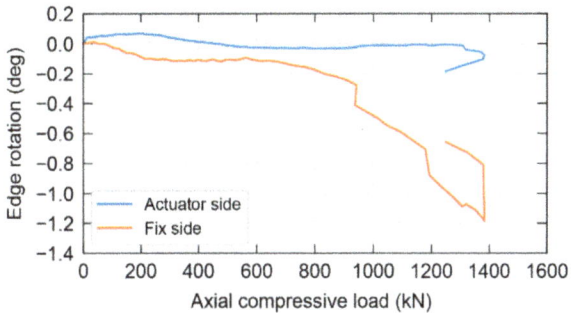

Fig. 38. Evaluation of rotation at edges during the experiment about the transversal y-axis.

Motivated by this finding, two Phase 3 participants investigated the role of boundary conditions and machine compliance on the predicted results. Participant #19 analysed the data from the displacement sensors placed at the two ends of the test panel and hypothesised different constraint conditions at the two ends; see Fig. 39. In particular, the absence of rotation was noted at the loaded end, while on the side in contact with the load cells, the boundary conditions were modified according to the following scheme:

- The load cell that consistently indicated a compressive load equal to zero was considered not in contact and consequently the area was left completely free.
- One support was considered perfectly clamped.
- The other two were left free to rotate about an axis transverse to the direction of application of the load.

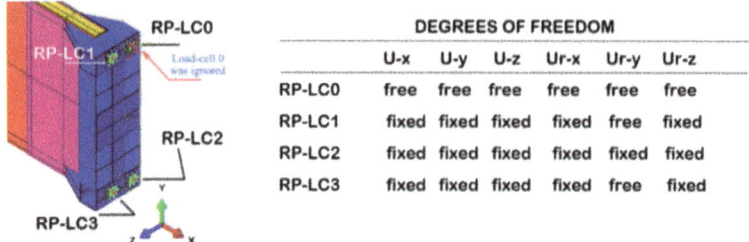

Fig. 39. Boundary conditions simulated by participant #19.

The correspondence with the experimental curve is excellent up to approximately 1300 kN, as seen in Fig. 37 (Right). However, after this point, the model tends to slightly underestimate the load compared to the experimental curve.

In Phase 3, Participant #13 explored the impact of the loading frame's elastic compliance on structural response, as illustrated in Fig. 37 (right). This analysis employed load pads at the loaded end of the frame to replicate the spring-like behaviour of the elastic compliance system. The load pads were modelled as $100 \times 100 \times 2.5$ mm^3 thin square plates and were positioned at one end of the frame. Normal-hard contact and surface-to-surface contact between the load cell and frame were defined in ABAQUS/Explicit. A reverse-engineering iterative approach determined that a load pad with an elastic modulus of 300 GPa and elastic-perfect plastic behaviour with a yield stress of 355 MPa produced the best fit to the experimental data. Although this setup improved correlation with experimental results, it failed to account for support failures and sharp load drops in the load-shortening curve. This suggests that the observed numerical/experimental discrepancies may stem not only from the elastic compliance system of the load frame but also from complex constraint conditions that are challenging to model accurately.

6.6 Comparison with CSR and UR-S35

One motivation for the experimental test in the benchmark study was to demonstrate that a stiffened plate has reserve strength beyond elastic buckling. In the current formulae for buckling in CSR and UR-S35 (IACS, 2024), local elastic plate buckling is accepted, and the plate can have a reserve above local elastic buckling; however, global elastic buckling is not accepted since this will result in stress redistribution over a larger area, and global elastic buckling is a rather unstable failure mode. In this section, the ultimate load computed by CSR is compared with the non-linear FE analyses calculated for the benchmark study. As mentioned in Sect. 4, these rules are applicable to more than 90% of the ships all over the world.

The computed ultimate stress by CSR for the stiffened plate is equal to 45 MPa, and this corresponds to a force equal to 568 kN which is rather conservative compared to the results obtained by the experimental test and non-linear FE analyses. The reason for the conservative results computed by CSR is that global elastic buckling (i.e., stiffener and plate buckles together out-of-plane) is the critical failure mode and this is a cut-off limit in the rules. This is a sound design principle to prevent very slender stiffeners. It can also be mentioned that the yield strength will not affect the computed strength by the rules, since global elastic buckling is independent material yield strength as studied in the different phases of the benchmark study.

In some rule frameworks, such as in the DNV Ship Rules (DNV, 2023), the use of non-linear FE analysis is permitted. However, these rules stipulate that global elastic stiffener buckling must be verified separately through eigenvalue analysis to ensure it does not occur. The ultimate capacity is defined as the lower of the global elastic buckling load or the load determined by non-linear FE analysis. In this case, the global eigenmode obtained through FE analysis yields a buckling load of 891 kN, with the associated buckling mode (eigenmode 13) shown in Fig. 40. Thus, within this rule context, global elastic buckling becomes the critical factor since it occurs at a lower load than the maximum achieved in non-linear FE analysis. The computed buckling load from the FE analysis is notably higher than that from the CSR approach. This difference is partially due to boundary conditions: in CSR, the stiffeners are modelled as simply supported, whereas in the FE model, they are welded at the ends, introducing some rotational restraint. This boundary condition significantly influences the global elastic buckling load, making it more sensitive. When plates are simply supported, the results align more closely with those from CSR.

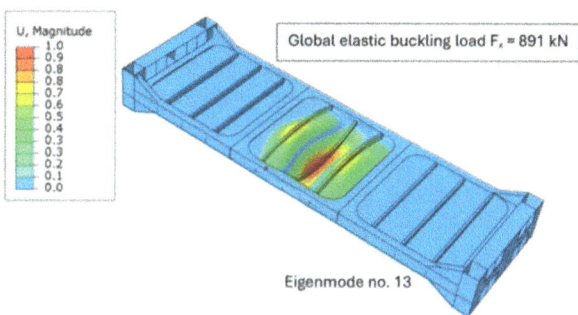

Fig. 40. Global elastic buckling load computed using finite element analysis.

6.7 Concluding Remarks

The results obtained from the committee's benchmark can be summarised as follows.

- Despite all participants using the same input geometry and being given limited modelling parameters in Phase 1, the results displayed considerable variation. This variation underscores the extent to which non-linear FE analysis outcomes are still heavily influenced by user decisions.
- It appears difficult to find a correlation between the effect of the salient modelling characteristics and the relative ultimate load value.
- Variations in the choice of initial imperfection models had a significant impact on the early part of the end-shortening curve, particularly affecting the local buckling response of the shell plating. However, this did not substantially influence the ultimate strength. Nonetheless, differing initial stiffness may lead to varying load redistributions across structural components. It is recommended, therefore, to use real imperfections where possible or to apply an initial imperfection pattern derived from

an eigenvalue analysis with a sufficient amplitude (at least 1.5 mm in this study) to prevent abrupt stiffness changes due to local buckling of the plating.
- Providing data on the yield and engineering ultimate strength of the material is not sufficient to reduce the modelling uncertainties. The participants of this benchmark study formulated different assumptions on the work hardening model, which increased the dispersion of the results.
- Welding-induced residual stresses show a non-negligible, effect on the end-shortening curves. However, the effect on the ultimate strength is still unclear and doubts remain about the correct modelling strategy to introduce this stress field into the model.
- The correct set of experimental boundary conditions appears difficult to deduce. However, the availability of displacement data from multiple measurement points in the experiments enabled the formulation of hypotheses regarding the actual boundary conditions, which were essential to refine the numerical/experimental comparisons. It is therefore recommended that experiments should acquire the axial displacements at intermediate frames as well as panel ends as well as monitor all degrees of freedom at the panel ends to be able to more effectively reconstruct the exact boundary conditions.
- Overall, the Phase 3 simulations and analyses indicate that including welding residual stresses in the model achieves an excellent match with the experimentally measured specimen stiffness, as particularly demonstrated by participant #8 in Fig. 37 (left). However, the numerical model begins to diverge from the experimental data at the first fluctuation in the experimental load, around 950 kN of compressive force. This fluctuation is attributed to the failure of one of the panel supports, leading to a subsequent change in the boundary conditions. To accurately replicate the experimental observations from this point onward, it is necessary to adjust the boundary conditions in the numerical model (see Fig. 37, right).

7 Conclusions, Trends and Recommendations

The committee's report presented a review of published work relevant to the committee's mandate along with results from a benchmark study with submissions from researchers inside and outside the committee. This section summarises the committee's essential findings/trends and offers recommendations for the community to further improve its understanding of the ultimate strength of ships and offshore structures.

7.1 Conclusions and Trends

The committee has identified four areas where the reviewed literature shows important progress during the mandate period: uncertainties/scale effects, numerical methods, multi-scale and multi-physics simulations, and the need for life-cycle management procedures.

Sections 2 and 3 focus on the aspects fundamental to ultimate strength prediction and uncertainties. The influence of scale effects and factors on the material level and the geometry's size (compared to, e.g., sheet thickness) have received much attention during the reporting period. Structures are characterised by load-end-shortening, pressure-displacement, or other similar relationships that are analogous to the stress-strain curve at the material level. However, the shape and characteristics of these relationships are much more complex due to various uncertainties.

Epistemic and aleatory uncertainties remain a central challenge in determining the ultimate strength of ships and offshore structures. The roles of various factors, such as geometry, imperfections, material properties, residual stresses, boundary conditions, loads, flaws, and their interdependence are best understood through continued physical testing, monitoring of real-world structures, and the development and benchmarking of numerical methods. Furthermore, the role the analyst plays in decisions of what to include in the analysis and how to conduct it remains a significant factor in the outcome of non-linear FE analysis predictions.

The increasing shift from traditional steel structures to lightweight constructions, e.g., aluminium alloys and composite materials is anticipated to continue, and this necessitates continued research. For example, further study is needed regarding initial imperfections of extruded aluminium structures, particularly where friction stir welding or other novel welding techniques are utilised. For composites, major challenges exist in the effective incorporation of material defects, such as delamination, in large-scale structural models. Features, such as welding-induced distortions and corrosion damage, tend to be more realistically simulated in modelling work using the 3D profilometry technique. However, scaling up from the structural element level to the system level remains challenging.

Research efforts focusing on the role of dynamic and cyclic effects are particularly noteworthy, as this area has been under-explored until recently. Recent studies have examined the role of inertia, structural damping, and strain rate effects as well as low-cycle loads in association with crack simulations. Analyses showing that the ultimate strength capacity can be reduced highlight the need for further efforts on these topics, including testing to validate numerical simulations.

Various experimental studies have been conducted since the last report examining the ultimate strength of model and full-scale structures. These efforts provided valuable data but also highlighted the challenge in controlling tests of large-scale objects with regard to control of load application and the boundary conditions. This is particularly noteworthy in the committee's benchmark study where changes in boundary conditions during the experiment were not captured in numerical simulations. Control of the initial stiffness and the boundary conditions, which can greatly affect the collapse sequence and behaviour, and the post ultimate characteristics, is crucial to the value of such efforts.

The effects of various environmental conditions, such as temperature, corrosiveness, or moisture on the material properties in combination with loading effects have been explored via experiments on either a material or component length scale in order to understand the effects of these conditions on ultimate strength. Other experiments have been presented from model to full scale for panels and grillage structures. These have often been conducted to validate numerical simulation procedures.

The committee has seen progress in the development of numerical methods divided into different categories. Fast-running or practical codes such as the Smith method are still used. However, their developments have been minimal due to faster computation resources and easy-to-use CAD and computation tools. Conventional FE methods are the predominate numerical method, most likely due to the readily available nature of commercial software, whereas other methods, such as the ESL, ISUM and ISFEM, have been reported in the literature as promising alternatives for large-scale and complex structure elements but appear to receive less use. The use of machine learning (ML) and physics-informed neural network (PINN) methods is increasing, especially in studies related to a large number of input parameters with uncertainties, such as structural health monitoring and diagnostics of aged ships and offshore structures where the ultimate strength capacity is reduced due to, e.g., corrosion and accumulated plastic deformation. The most exciting development originates from the aerospace industry, which has developed a framework (the NASMAT software) in which multi-physics, multi-scale materials, and structural analyses are coupled into one seamless modelling package, building on multi-scale structures modelling research.

The ISSC 2022 Ultimate Strength committee (ISSC, 2022a) reported that life-cycle management methodology had become more widely adopted in the literature. This trend has continued during the mandate period of the current committee, and there are examples of studies emphasising the prediction of (residual/effective) ultimate strength capacity. The progress identified by the committee originates from the fact that it is practically impossible to define the ultimate strength capacity at full scale during the lifetime, and thus, computational methods and various level simplifications are needed that affect the sequence of failure from the material to the hull girder level. The development of AI-based tools and methods to create digital twin models, which could include ML-based data or a PINN-based approach, offers the opportunity to more effectively utilise large databases of sensing/inspection records and hence more realistically monitor a structure's structural health as well as the ultimate strength capacity to assist life-cycle management/decision making of the structure.

Section 4 presents the influence of structural configuration, components, and loads on ultimate strength. Compared to previous committee reports, our review notes that dynamic effects and complicated loading scenarios are receiving increasing attention. Recent studies have demonstrated the importance of non-linear wave shapes on wave-induced VBM, which becomes more pronounced in large waves. Traditional linear methods may underestimate the wave-induced VBM because they do not account for the non-linear effects of wave shape. Moreover, incorporating complicated loading scenarios, and non-linear wave shapes, in the design for ultimate strength allows consideration of more realistic loading scenarios (dynamic and non-linear effects) and scenario-based modelling, which should result in higher fidelity ultimate strength analysis. These should also consider dynamic load effects on the ultimate strength from, e.g., whipping and sloshing loads. Another load-related topic is cyclic loading and its accumulation of residual strains and displacements (local and global levels) that affect the ultimate strength capacity in the long term. This effect, together with the influence of corrosion and the possibility of reaching a plastic shakedown response, is recommended to be investigated in more detail.

Empirical formulae are still being developed, particularly to account for different loading conditions and applications. In the shipping industry, there is also a significant focus on harmonising the buckling formulae to obtain one common buckling procedure. For instance, IACS is working on harmonising the procedures for local buckling checks and the new set of rules will cover 90% of the global fleet. The empirical formulae are also increasingly being replaced by machine learning models that demonstrate better accuracy. Future work may focus on the development of more advanced machine learning models for strength assessment, integrating inspection data into finite element models, and exploring the application of data-driven approaches in ship structural design to reduce reliance on expensive numerical simulations.

The committee found that there is an increasing trend towards accounting for realistic loading scenarios as well as the in-service condition of ships as it relates to ultimate strength analysis. With the increased proliferation of full-field 3D measurement and inspection techniques, inspecting and quantifying distortions during inspection activities or even continuously during operations is possible. This has led to the development of structural digital twins, with inspections, 3D measurements, and sensors providing inputs to model updates. These digital twins can be used as the basis for real-time evaluation of ultimate strength, providing safer ship operations and timely warnings of potentially dangerous combinations of loading and structural conditions.

FE analysis results that account for realistic loading and structural condition inputs are needed to accomplish this goal. These results must then be transitioned to user-friendly reduced-order models, often using ML methods, that are suitable for integration into ship operations. Therefore, the committee recommends continued development of ML-based and other reduced-order models based on high-fidelity computations.

Section 5 presents the recent development of the knowledge and methods on the ultimate strength of various offshore structures and related components. In general, studies on the ultimate strength of oil and gas offshore structures have declined since the committee's last report, which is attributable to the ongoing global green energy transition from fossil fuels to renewables. The global shift towards cleaner and more sustainable energy sources has driven unprecedented growth in offshore structures throughout the reporting period. These structures play a central role in the global energy infrastructure, serving as a cornerstone for the swift advancement of the offshore industry, notably in the renewable energy sector. They play a pivotal role in harnessing the vast potential of offshore wind and tidal energy, ensuring the resilience and durability of these installations in challenging marine environments. The significance of comprehending and enhancing the ultimate strength of these structures cannot be overstated, as it directly correlates with the reliability, safety, and longevity of our offshore energy endeavours, ultimately contributing to a more sustainable and environmentally responsible future.

The rapid expansion of offshore wind energy has introduced new technical challenges, particularly in understanding ultimate strength under extreme operating conditions and unusual load cases. For instance, ship collisions with operational floating wind turbines can severely impact the residual strength of these structures, necessitating detailed research to enhance resilience and safety. Hydrodynamic loads caused by breaking waves can significantly increase the structural demand and may contribute to conditions approaching the ultimate limit state, particularly in shallow water or nearshore environments. The construction of offshore wind turbines in sub-Arctic areas, such as the Baltic Sea and the Bohai Bay, presents a unique challenge, primarily due to sea ice.

Recent trends in the design and longevity assessments of offshore structures emphasise advanced, integrated approaches that enhance safety, reliability, and sustainability in marine environments. Extensive research on the ultimate strength of structural elements has been conducted over the past decades; however, further advancements could be achieved by leveraging machine learning and digital twin technologies to improve predictions of buckling and post-buckling behaviour. Recent publications have shown that life-extension decisions could significantly benefit from big data analysis using machine learning to establish precise relationships between variables influencing the life-cycle performance of, e.g., offshore wind turbines.

The variety of offshore structures is increasing, expanding to include structures such as oil and gas platforms, floating offshore wind, wave energy converters, photovoltaic platforms, and aquaculture installations. During the reporting period, there has been an increasing interest in mooring designs and ultimate strength assessment of the moorings of these structures. Additional research is essential to ensure safe and reliable mooring designs that account for ultimate strength, particularly for dynamic power cables, which in some structures are suspended freely in the ocean. At a system level, studies on the structural integrity and failure mechanisms of offshore wind farms have also become a key focus area. In addition to conventional research areas, such as system reliability analysis and corrosion-affected components, a few real case failure incidents of offshore wind farms, such as blade breakage from the wind turbines and falling into the sea, highlight new technical challenges in the design of large offshore wind farms and how failure incidents impact ships and other offshore structures or infrastructures. A holistic research approach, which considers technical, economic, societal and environmental aspects, is crucial for the large-scale deployment of offshore wind farms. From a structural perspective, this includes ensuring the reliability and safety of support structures through a comprehensive assessment of ultimate strength, fatigue performance, and degradation mechanisms under long-term environmental exposure.

One evident observation regarding guidelines and regulations is that, while well-established for the oil and gas industry—despite occasional improvements or adjustments—they remain limited for offshore renewable energy exploration. Anticipated guidelines should address specific aspects such as site-specific environmental loading, fatigue and ultimate strength criteria for novel structural configurations, material degradation in marine environments, and design provisions for modularity, maintenance, and decommissioning. These elements are critical to ensure the safety, reliability, and long-term performance of offshore renewable energy structures. Furthermore, many models and methods proposed in academic research hold potential for practical application, though only some have been implemented in the field. Therefore, increased collaboration between academia and industry is needed to ensure that these innovative approaches are effectively implemented in real-world offshore engineering projects. This collaboration can bridge the gap between theoretical advancements and practical applications, ultimately enhancing the safety, reliability, and efficiency of offshore structures.

Section 6 presents the benchmark study coordinated by the committee, involving participants who were internal and external to the committee. As in the benchmark study reported by the committee in the ISSC 2022 report (ISSC, 2022a), the benchmark for this report was divided into three phases. Participants were first furnished identical information related to the modelling task, without access to experimental data in the first two phases. They were then required to propose a series of hypotheses to explain the

Committee III.1: Ultimate Strength 555

observed discrepancies between the numerical and experimental results. Following the recommendation by the Official Discusser in 2022, the current benchmark study aimed to investigate the ability of analysts to predict the end-shortening curve, i.e., the entire load-response path, including post-collapse behaviour. The results obtained from this benchmark can be summarised as follows.

Despite the growing reliability of non-linear numerical models and the availability of increasingly precise experimental data (from multi-point displacement sensors, strain gauges, and digital image correlation, for example), tracing the sources of discrepancies between numerical and experimental approaches has become more complex as data volume and precision increase. This challenge arises partly because, even with ample knowledge, it is easy to focus on detailed analysis rather than the overall system response.

- The results from Phase 1 were rather scattered despite little freedom for an analyst to make assumptions related to the boundary conditions, load, or material modelling; such a varied interpretation of the geometric imperfections by the participants was not foreseen, having been provided by the actual geometric imperfections of the structure. It was not possible to identify a correlation linking the effect of the salient modelling characteristics (such as the FE solver selected, mesh size, and element type) with the relative ultimate load value and post-collapse behaviour. Hence, the results in this phase show, for the case studied, that the skill and overall judgement of analysts are uncertainty factors that are challenging to control or reduce without having strict modelling and simulation guidelines and procedures. This underscores the crucial need for such guidelines in our community. Existing guidelines provide limited practical advice on defining non-linear FE analysis. For example, DNV (2019) highlights the importance of accurately or conservatively representing real conditions, suggesting the use of sensitivity studies when uncertainty exists. ABS (2025) stresses the crucial role of boundary conditions in non-linear FE analysis and cautions against relying on assumptions from linear models. However, no practical examples specifically useful for the analyst are provided. While supplementary materials offer examples, these are often highly idealized.
- The modelling of geometrical imperfections greatly influenced the panel's stiffness in the initial loading phase, affecting load redistribution among the various structural components, as well as the failure sequence and post-collapse behaviour. The participants of the benchmark study compared different alternatives to model geometrical imperfections. The committee recommends modelling the structure using measured structural imperfections, when available, or using an initial imperfection obtained from modal analysis, preferably including the first three modal shapes. Analysts should take particular care in selecting appropriate amplitude magnitudes for these modal shapes, ensuring an adequate amplitude magnitude to correctly represent the initial stiffness (for the current specimen, the threshold identified was 2.0 mm for mode 1).
- Providing the participants with actual material data in Phase 2 did not reduce the scatter in results. The diverse selection of material models (such as yield plateau and work hardening) led to even greater variability than observed in Phase 1. This outcome reinforces the committee's recommendation to develop unified modelling and

simulation guidelines, ideally under IACS leadership with ISSC support. Nonetheless, addressing the dispersions in experimental data remains an open topic for further discussion.
- This committee did not work on defining an acceptable level of deviation in the results. However, it limited the work to photographing the current situation, where significant deviations were observed even when working with a relatively limited number of input parameters. Unfortunately, apart from the clear indications that emerged in the first phase regarding the initial geometric imperfection model, no trends were observed regarding the solver, element, or mesh size.
- Prior to the start of Phase 3, it was identified that the experimental data were significantly influenced by uncertainties related to the intended boundary conditions for the panel. Unfortunately, the designed boundary conditions were not consistently maintained during testing, while numerical models in Phases 1 and 2 adhered to the original boundary conditions as planned. Some participants incorporated the updated information about these shifting boundary conditions into their simulations and successfully achieved a close match with the entire experimental end-shortening curve.
- FE simulations were conducted to investigate the influence of welding-induced residual stresses on the end-shortening curve. The results show that, as expected, they have a non-negligible effect and should preferably be considered.

7.2 Recommendations

The committee presented an extensive literature review in this report and conducted a benchmark study. Several of the committee's members were members of the ISSC 2022 Ultimate Strength committee. Many of the recommendations made in the previous committee's report still apply, since a relatively long time is required to conduct studies and publish findings, develop and implement new methods, and revise guidelines and recommended practices. In addition to the committee's new recommendations, we echo some of the ISSC 2022 committee's recommendations (ISSC, 2022a) to ensure continuity in the development of specific topics and progress in broader research areas:

- Future studies conducted by academia, research institutes, and the industry should focus on developing more advanced machine learning models for strength assessment, integrating traditional inspection as well as LiDAR/3D geospatial data into FE models, and exploring the application of data-driven approaches in ship and offshore structural design to reduce reliance on expensive numerical simulations.
- Distortions resulting from fabrication processes should be incorporated into structural analysis to improve the accuracy of strength predictions. Depending on the stage of the design/assessment, these can be included from actual measurements, idealised distortions, or from modal analysis, where several modes are mapped on the geometry.
- This report emphasises the importance of utilising advanced monitoring and inspection techniques to update FE models and create digital twins that reflect the actual state of the structure. User-friendly practical tools and interfaces need to be developed to minimise uncertainty as a result of the skill level of the analyst.
- Once the failure mode has been validated via a comparison of results from numerical simulation and experiment, any discrepancy in terms of ultimate capacity can

Committee III.1: Ultimate Strength 557

be evaluated by investigating aleatoric and epistemic uncertainties in the numerical model and the experiments.
- Shell elements currently provide a well-balanced approach for modelling large structures. Future advances in computational power could make large-scale models using solid elements more feasible. The development of ESL (Equivalent Single Layer), ISUM (Idealised Structural Unit Method), and ISFEM (Incremental Structural Finite Element Method) should also be pursued for large-scale applications. Reduced-order models are increasingly proposed in the literature but often lack direct comparisons with new or established models. The next committee is advised to include these comparisons in a benchmark study to assess the strengths and limitations of each model.
- Multi-scale and multi-physics frameworks, such as the NASMAT software originally designed for aerospace applications hold significant promise for transforming design methodologies for ships and offshore structures, promoting sustainability throughout their service lives. Although this may extend beyond the current committee's direct focus, it is highly pertinent to the ISSC and aligns with the committee's mandate, as it intersects with ultimate strength, life-cycle management, and life-cycle engineering principles.
- Future research is needed to better predict/quantify the material degradation processes (e.g., corrosion, repeated plastic deformation, development of cracks), especially when new materials are increasingly used for ship and offshore constructions, and to better incorporate the human element (from fabrication to operation) into the analysis.
- The committee did not find any specific studies on the ultimate strength of wind assisted propulsion systems (WAPS), or indeed vessels with WAPS installed. However, the WAPS technology is becoming increasing popular driven by the IMO net zero target set for 2050 in the shipping sector. As these systems, such as Flettner rotors and wingsails, can be quite large (currently up to 50 m tall and 200 tonnes per unit), the structural failure of WAPS or WAPS-ship interface may significantly affect the integrity/ultimate strength of hulls. Therefore, the committee recommends future research into the structural integrity of these systems and their hull-supporting structures.
- Standards and guidelines specifically for offshore renewable energy structures, such as floating wind turbines, should be developed and consolidated, as current regulations are still largely based on oil and gas structures.
- There should be stronger collaboration between research institutions and industry to ensure that innovations in predicting and improving ultimate strength are applied practically in the design and maintenance of offshore structures.
- Continued research into the effects of the corrosive marine environment on the ultimate strength of structures throughout their operational lifespan is essential. Investigating advanced materials, such as composites, along with effective strengthening and retrofitting techniques, will be critical for extending the service life and resilience of these structures. In addition, uncertainties arising from operational conditions, such as variable loading, maintenance practices, and accidental events, should also be considered in the assessment of ultimate strength.

Acknowledgements. The committee extends its gratitude to Dr Beatrice Barsotti for her assistance, coordination, and support in the benchmark study. The committee also thanks Fincantieri (Italy) for financing the experimental test that formed the foundation of the benchmark study and for making the results accessible to all participants.

References

ABS: ABS notations and symbols (2025). https://ww2.eagle.org/content/dam/eagle/rules-and-resources/RuleManager2/class-notations-table.pdf. Accessed 11 July 2025

Alinia, M.M., Saeidpour, A., Amani, M.: The shear buckling and post-buckling behavior of laterally pressured curved panels. Eng. Archive (engrxiv), 1–48 (2019). https://doi.org/10.31224/osf.io/jg27s

Alizadeh, F., Mazruee Sebdani, R., Guedes Soares, C.: Numerical analysis of the residual ultimate strength of composite laminates under uniaxial compressive load. Compos. Struct. **300**, 116161 (2022). https://doi.org/10.1016/j.compstruct.2022.116161

Amaechi, C.V.: Local tailored design of deep water composite risers subjected to burst, collapse and tension loads. Ocean Eng. **250**, 110196 (2022). https://doi.org/10.1016/j.oceaneng.2021.110196

Anyfantis, K.N.: Ultimate strength of stiffened panels subjected to non-uniform thrust. Int. J. Naval Archit. Ocean Eng. **12**, 325–342 (2020). https://doi.org/10.1016/j.ijnaoe.2020.03.003

API: Recommended practice for fitness-for-service, API 579–1/ASME FFS-1. Standard by American Petroleum Institute, Washington, D.C., USA (2021)

Arnold, S.M., Ricks, T., Pineda, E., Bednarcyk, B.: An enabling platform for achieving multiscale multiphysics analysis of multiphase materials. NASA, Glendale Research Center, Cleveland, Ohio, USA (2023). https://ntrs.nasa.gov/citations/20230015286. Accessed 11 July 2025

ASME: BPVC.XI - BPVC Section XI-Rules for inservice inspection of nuclear power plant components. American Society of Mechanical Engineers, New York, NY, USA. ISBN: 9780791871027 (2017)

Attia, A., El Kilani, H.S., El Afandy, M.M., Saad-Eldeen, S.: Numerical simulation of intact and cracked aluminium-alloy stiffened panels: calibration based on experiment. Ocean Eng. **269**, 113474 (2023). https://doi.org/10.1016/j.oceaneng.2022.113474

Augello, R., et al.: Buckling test of stiffened panels: Evaluation of post-buckling and failure by testing and layerwise models. In: Vassilopoulos, A.P., Michaud, V. (eds.), Composites Meet Sustainability; Proceedings of the 20th European Conference on Composite Materials (ECCM 20), Lausanne, Switzerland, 25-30 June 2022. Lausanne: EPFL Lausanne, Composite Construction Laboratory, pp. 459–467 (2022)

Austefjord, H.N., de Hauteclocque, G., Johnson, M.C., Zhu, T.: Update of wave statistics standards for classification rules. In: J.W. Ringsberg & C. Guedes Soares (eds.), Advances in the Analysis and Design of Marine Structures; Proceedings of the 9th International Conference on Marine Structures (MARSTRUCT 2023), Gothenburg, Sweden, 3–5 April 2023. London: CRC Press, pp. 43–52 (2023). https://doi.org/10.1201/9781003399759-5

Ayough, P., Ibrahim, Z., Sulong, N.H.R., Hsiao, P.C.: The effects of cross-sectional shapes on the axial performance of concrete-filled steel tube columns. J. Constr. Steel Res. **176**, 106424 (2021). https://doi.org/10.1016/j.jcsr.2020.106424

Babazadeh, A., Khedmati, M.R.: Progressive collapse analysis of a bulk carrier hull girder under longitudinal vertical bending moment considering cracking damage. Ocean Eng. **242**, 110140 (2021). https://doi.org/10.1016/j.oceaneng.2021.110140

Barooni, M., Nezhad, S.K., Ali, N.A., Ashuri, T., Sogut, D.V.: Numerical study of ice-induced loads and dynamic response analysis for floating offshore wind turbines. Mar. Struct. **86**, 103300 (2022). https://doi.org/10.1016/j.marstruc.2022.103300

Barsotti, B., Gaiotti, M.: Cumulative buckling deformation of stiffened panel under cyclic loading. In: Ergin, S., Guedes Soares, C. (eds.), Sustainable Development and Innovations in Marine Technologies; Proceedings of the 19th International Congress of the International Maritime Association of the Mediterranean (IMAM 2022), Istanbul, Turkey, 26–29 September 2022. London: CRC Press, pp. 109–113 (2022). https://doi.org/10.1201/9781003358961

Barsotti, B., Gaiotti, M.: Evaluation of residual plastic strain on a stiffened panel subjected to compression and tension-compression cyclic load. In: Ringsberg, J.W., Guedes Soares, C. (eds.), Advances in the Analysis and Design of Marine Structures; Proceedings of the 9th International Conference on Marine Structures (MARSTRUCT 2023), Gothenburg, Sweden, 3-5 April 2023. London: CRC Press, pp. 403–409 (2023a). https://doi.org/10.1201/9781003399759-44

Barsotti, B., Gaiotti, M.: FEM numerical strategies for the evaluation of the accumulated plastic strain due to cyclic load condition. In: J.W. Ringsberg & C. Guedes Soares (eds.), Advances in the Analysis and Design of Marine Structures; Proceedings of the 9th International Conference on Marine Structures (MARSTRUCT 2023), Gothenburg, Sweden, 3-5 April 2023. London: CRC Press, pp. 411–417 (2023b). https://doi.org/10.1201/9781003399759-45

Barsotti, B., Gaiotti, M.: Residual plastic strain effects on the ultimate capacity for a HSLA stiffened panel subjected to two different cyclic load conditions. In: Proceedings of the ASME 2024 43rd International Conference on Ocean, Offshore and Arctic Engineering (OMAE 2024), Singapore, 9–14 June 2024. (OMAE2024-126720) (2024)

Barsotti, B., Gaiotti, M., Rizzo, C.M.: Recent industrial developments of marine composites limit states and design approaches on strength. J. Mar. Sci. Appl. **19**, 553–566 (2020). https://doi.org/10.1007/s11804-020-00171-1

Barsotti, B., Battini, C., Gaiotti, M., Rizzo, C.M., Vergassola, G.: Experimental assessment of ultimate strength of a transversally loaded thin-walled structure and comparison with numerical models at different levels of complexity. Mar. Struct. **103**, 103793 (2025). https://doi.org/10.1016/j.marstruc.2025.103793

Bastek, J.H., Kochmann, D.M.: Physics-informed neural networks for shell structures. Eur. J. Mech. A/Solids **97**, 104849 (2023). https://doi.org/10.1016/j.euromechsol.2022.104849

Bonsu, A.O., Mensah, C., Liang, W., Yang, B., Ma, Y.: Mechanical degradation and failure analysis of different glass/basalt hybrid composite configuration in simulated marine condition. Polymers **14**(17), 3480 (2022). https://doi.org/10.3390/polym14173480

Braun, M., et al.: Development of combined load spectra for offshore structures subjected to wind, wave, and ice loading. Energies **15**(2), 559 (2022). https://doi.org/10.3390/en15020559

Brubak, L., Lv, Y., Ishibashi, K., Bollero, A., Bøe, Å.: Rule formulation updates on buckling strength requirements in Common Structural Rules. In: Ringsberg, J.W., Guedes Soares, C. (eds.), Advances in the Analysis and Design of Marine Structures; Proceedings of the 9th International Conference on Marine Structures (MARSTRUCT 2023), Gothenburg, Sweden, 3-5 April 2023. London: CRC Press, pp. 339–346 (2023). https://doi.org/10.1201/9781003399759-37

BV: Rules for the classification of steel ships (2024). https://marine-offshore.bureauveritas.com/nr467-rules-classification-steel-ships. Accessed 11 July 2025

Byklum, E., Amdahl, J.: A simplified method for elastic large deflection analysis of plates and stiffened panels due to local buckling. Thin-Walled Struct. **40**(11), 925–953 (2002). https://doi.org/10.1016/S0263-8231(02)00042-3

Cai, Q., Chen, D., Yang, N., Li, W.: A novel semi-spar floating wind turbine platform applied for intermediate water depth. Sustainability **16**(4), 1663 (2024). https://doi.org/10.3390/su16041663

Cao, S., Cheng, Y., Duan, J., Fan, X.: Experimental investigation on the dynamic response of an innovative semi-submersible floating wind turbine with aquaculture cages. Renew. Energy **200**, 1393–1415 (2022). https://doi.org/10.1016/j.renene.2022.10.072

Carrera, E., Zozulya, V.V.: Carrera unified formulation (CUF) for the micropolar plates and shells. I. High order theory. Mech. Adv. Mater. Struct. **29**(6), 773–795 (2020). https://doi.org/10.1080/15376494.2020.1793241

CCS: Guidelines for direct calculation assessment of hull structure including springing and whipping (2018). https://www.ccs.org.cn. Accessed 11 July 2025

CCS: Rules for structures of container ships (2022). https://www.ccs.org.cn. Accessed 11 July 2025

CEN: Eurocode 3, prEN 1993-1-14: Design of steel structures. Part 1.14: Design assisted by finite element analysis, European Committee for Standardisation, Brussels, 2023 (issued for hearing 2023, planned for publication in April 2025) (2023)

Cerik, B.C., Choung, J.: Progressive collapse analysis of intact and damaged ships under unsymmetrical bending. J. Mar. Sci. Eng. **8**(12), 988 (2020). https://doi.org/10.3390/jmse8120988

Chang, X.P., Qu, C.J., Song, Q., Li, Y.H., Liu, J.: Coupled cross-flow and in-line vibration characteristics of frequency-locking of marine composite riser subjected to gas-liquid multiphase internal flow. Ocean Eng. **266**, 113019 (2022). https://doi.org/10.1016/j.oceaneng.2022.113019

Chen, B.Q., Guedes Soares, C.: Experimental and numerical investigation on welding simulation of long stiffened steel plate specimen. Mar. Struct. **75**, 102824 (2021). https://doi.org/10.1016/j.marstruc.2020.102824

Chen, X., Chen, Z., Li, D.: Misalignment effect on the ultimate strength of welded stiffened panels under uniaxial compression. Ocean Eng. **284**, 115190 (2023). https://doi.org/10.1016/j.oceaneng.2023.115190

Cheng, J., Liu, Y., Cheng, M., Li, W., Li, T.: Optimum condition-based maintenance policy with dynamic inspections based on reinforcement learning. Ocean Eng. **261**, 112058 (2022). https://doi.org/10.1016/j.oceaneng.2022.112058

Cho, S.J., Ban, I.J., Shin, S.C.: A novel deep learning model to predict ultimate strength of ship plates under compression. Appl. Sci. **12**(5), 2522 (2022). https://doi.org/10.3390/app12052522

Choi, Y.M., et al.: An efficient methodology for the simulation of nonlinear irregular waves in computational fluid dynamics solvers based on the high order spectral method with an application with OpenFOAM. Int. J. Naval Archit. Ocean Eng. **15**, 100510 (2023). https://doi.org/10.1016/j.ijnaoe.2022.100510

Chu, Y.I., et al.: Offshore fish farms: a review of standards and guidelines for design and analysis. J. Mar. Sci. Eng. **11**(4), 762 (2023). https://doi.org/10.3390/jmse11040762

Chujutalli, J.H., Estefen, S.F., Guedes Soares, C.: Indentation parameters influence on the ultimate strength of panels for different stiffeners. J. Constr. Steel Res. **170**, 106097 (2020). https://doi.org/10.1016/j.jcsr.2020.106097

ClassNK: Investigation report on structural safety of large container ships. Investigative panel on large container ship safety (2014). https://www.classnk.or.jp/hp/pdf/news/Investigation_Report_on_Structural_Safety_of_Large_Container_Ships_EN_ClassNK.pdf. Accessed 11 July 2025

Coraddu, A., Oneto, L., Li, S., Kalikatzarakis, M., Karpenko, O.: Surrogate models to unlock the optimal design of stiffened panels accounting for ultimate strength reduction due to welding residual stress. Eng. Struct. **293**, 116645 (2023). https://doi.org/10.1016/j.engstruct.2023.116645

Corigliano, P., Palomba, G., Crupi, V., Garbatov, Y.: Stress-strain assessment of honeycomb sandwich panel subjected to uniaxial compressive load. J. Mar. Sci. Eng. **11**(2), 365 (2023). https://doi.org/10.3390/jmse11020365

Cui, H., Ding, Q.: Ultimate strength and fracture failure of hull stiffened plates based on plastic accumulation under cyclic loading. Ocean Eng. **261**, 112016 (2022). https://doi.org/10.1016/j.oceaneng.2022.112016

Darie, I., Rörup, J.: Hull girder ultimate strength of container ships in oblique sea. In: Guedes Soares, C., Garbatov, Y. (eds.), Progress in the Analysis and Design of Marine Structures; Proceedings of the 6th International Conference on Marine Structures (MARSTRUCT2017), Lisbon, Portugal, 8-10 May 2017. London: CRC Press, pp. 225–233 (2017). https://doi.org/10.1201/9781315157368

Deng, H., Yuan, T., Gan, J., Liu, B., Wu, W.: Experimental and numerical investigations on the collapse behavior of box type hull girder subjected to cyclic ultimate bending moment. Thin-Walled Struct. **175**, 109204 (2022). https://doi.org/10.1016/J.TWS.2022.109204

Ding, H., et al.: Experimental investigation of nonlinear forces on a monopile offshore wind turbine foundation under directionally spread waves. In: Proceedings of the ASME 2024 43rd International Conference on Ocean, Offshore and Arctic Engineering (OMAE 2024), Singapore, 9-14 June 2024. (OMAE2024-125160) (2024)

DNV: Fatigue and ultimate strength assessment of container ships including whipping and springing. Class guideline DNV-CG-0153. DNV AS, Høvik, Norway (2021). www.dnv.com. Accessed 11 July 2025

DNV: Rules for classification: Pt.5 Ch. 2. Class Rules DNVGL-RU-SHIP. DNV AS, Høvik, Norway (2022a). www.dnv.com. Accessed 11 July 2025

DNV: Determination of structural capacity by non-linear FE analysis methods. Recommended practice DNV-RP-C208. DNV AS, Høvik, Norway (2022b). www.dnv.com. Accessed 11 July 2025

DNV: Buckling. Class guideline DNV-CG-0128. DNV AS, Høvik, Norway (2023). www.dnv.com. Accessed 11 July 2025

Do, Q.T., Huynh, V.V., Cho, S.R., Vu, M.T., Vu, Q.T., Thai, D.K.: Residual ultimate strength formulations of locally damaged steel stiffened cylinders under combined loads. Ocean Eng. **225**, 108802 (2021). https://doi.org/10.1016/j.oceaneng.2021.108802

Do, Q.T., Xuan-Phuong, D., Tra, T.H., Tuyen, V.V., Prabowo, A.R., Hung, T.D.: Parametric study of side collision-induced denting failures on the ultimate strength of a handy-size containership under vertical bending. Ocean Eng. **309**, 118534 (2024). https://doi.org/10.1016/j.oceaneng.2024.118534

Duong, T.H., Le, T.T., Le, M.V.: Practical machine learning application for predicting axial capacity of composite concrete-filled steel tube columns considering effect of cross-sectional shapes. Int. J. Steel Struct. **23**, 263–278 (2023). https://doi.org/10.1007/s13296-022-00693-0

Energy Watch: Ørsted CEO says confidence not shaken by malfunction boom in Danish offshore wind (2023). https://energywatch.com/EnergyNews/Renewables/article14304136.ece. Accessed 11 July 2025

Faqih, I., Adiputra, R., Prabowo, A.R., Muhayat, N., Ehlers, S., Braun, M.: Hull girder ultimate strength of bulk carrier (HGUS-BC) evaluation: structural performances subjected to true inclination conditions of stiffened panel members. Results Eng. **18**, 101076 (2023). https://doi.org/10.1016/j.rineng.2023.101076

Faulkner, D.: A review of effective plating for use in the analysis of stiffened plating in bending and compression. J. Ship Res. **19**(1), 1–17 (1975). https://doi.org/10.5957/jsr.1975.19.1.1

Feng, L., Hu, L., Chen, X., Shi, H.: A parametric study on effects of pitting corrosion on stiffened panels' ultimate strength. Int. J. Naval Archit. Ocean Eng. **12**, 699–710 (2020). https://doi.org/10.1016/j.ijnaoe.2020.08.001

Feng, R., Wu, C., Chen, Z., Roy, K., Chen, B., Lim, J.B.P.: Finite element modeling and proposed design rules of stainless steel hybrid tubular joints with square braces and circular chord. J. Construct. Steel Res. **179**, 106557 (2021). https://doi.org/10.1016/j.jcsr.2021.106557

Fonseca, N., Tahchiev, G., Ygre Rogne, O.: Mooring system ULS analysis based on empirical QTFs of wave drift loads. In: Proceedings of the ASME 2024 43rd International Conference on Ocean, Offshore and Arctic Engineering (OMAE 2024), Singapore, 9-14 June 2024. (OMAE2024-127825) (2024)

Fujikubo, M., Alie, M.Z.M., Takemura, K., Iijima, K., Oka, S.: Residual hull girder strength of asymmetrically damaged ships: influence of rotation of neutral axis due to damages. J. Japan Soc. Naval Archit. Ocean Eng. **16**, 131–140 (2012). https://doi.org/10.2534/jjasnaoe.16.131

Gadallah, R., Murakawa, H., Shibahara, M.: Study of various parameters on residual stress relaxation for different welded components. J. Constr. Steel Res. **214**, 108503 (2024). https://doi.org/10.1016/j.jcsr.2024.108503

Garbatov, Y.: Risk-based corrosion allowance of oil tankers. Ocean Eng. **213**, 107753 (2020). https://doi.org/10.1016/j.oceaneng.2020.107753

Garbatov, Y., Palomba, G., Crupi, V.: Risk-based hybrid light-weight ship structural design accounting for carbon footprint. Appl. Sci. **13**(6), 3583 (2023). https://doi.org/10.3390/app13063583

Garbatov, Y., Marchese, S.S., Epasto, G., Grupi, V.: Flexural response of additive-manufactured honeycomb sandwiches for marine structural applications. Ocean Eng. **302**, 117732 (2024). https://doi.org/10.1016/j.oceaneng.2024.117732

George, J.M., Kimiaei, M., Elchalakani, M., Efthymiou, M.: Flexural response of underwater offshore structural members retrofitted with CFRP wraps and their performance after exposure to real marine conditions. Structures **43**, 559–573 (2022). https://doi.org/10.1016/j.istruc.2022.06.075

George, J.M., Kimiaei, M., Elchalakani, M., Fawzia, S.: Underwater strengthening and repairing of tubular offshore structural members using carbon fibre reinforced polymers with different consolidation methods. Thin-Walled Struct. **174**, 109090 (2022). https://doi.org/10.1016/j.tws.2022.109090

Georgiadis, D.G., Samuelides, M.S.: Stochastic geometric imperfections of plate elements and their impact on the probabilistic ultimate strength assessment of plates and hull-girders. Mar. Struct. **76**, 102920 (2021). https://doi.org/10.1016/j.marstruc.2020.102920

Georgiadis, D.G., Samuelides, M.S.: The effect of corrosion spatial randomness and model selection on the ultimate strength of stiffened panels. Ships Offshore Struct. **16**(sup1), 140–152 (2021). https://doi.org/10.1080/17445302.2021.1907063

Georgiadis, D.G., Samuelides, E.S.: A Bayesian approach for the quantification of strength model uncertainty factor in ultimate limit state. In: Ringsberg, J.W., Guedes Soares, C. (eds.), Advances in the Analysis and Design of Marine Structures; Proceedings of the 9th International Conference on Marine Structures (MARSTRUCT 2023), Gothenburg, Sweden, 3-5 April 2023. London: CRC Press, pp. 871–876 (2023a). https://doi.org/10.1201/9781003399759-96

Georgiadis, D.G., Samuelides, M.S., Straub, D.: A Bayesian analysis for the quantification of strength model uncertainty factor of ship structures in ultimate limit state. Mar. Struct. **92**, 103495 (2023). https://doi.org/10.1016/j.marstruc.2023.103495

Ghabezi, P., Harrison, N.M.: Hygrothermal deterioration in carbon/epoxy and glass/epoxy composite laminates aged in marine-based environment (degradation mechanism, mechanical and physicochemical properties). J. Mater. Sci. **57**, 4239–4254 (2022). https://doi.org/10.1007/s10853-022-06917-2

Gong, C., Frangopol, D.M., Cheng, M.: Risk-based decision-making on corrosion delay for ship hull tankers. Eng. Struct. **212**, 110455 (2020). https://doi.org/10.1016/j.engstruct.2020.110455

Graves, W., Nahshon, K., Aminfar, K., Lattanzi, D.: Finite element model updating with quantified uncertainties using point cloud data. Data-Centric Eng. **4**, e16 (2023). https://doi.org/10.1017/dce.2023.7

Gu, Y., Ong, M.C., Janocha, M.J., Blomgren, A.: Review on feasibility of bottom-fixed offshore wind development in the arctic focusing on icing problems. In: Proceedings of the ASME 2023 42nd International Conference on Ocean, Offshore and Arctic Engineering (OMAE 2023), Melbourne, Australia, 11-16 June 2023. (OMAE2023-102822) (2023). https://doi.org/10.1115/OMAE2023-102822

Güneyisi, E.M., Nour, A.I.: Axial compression capacity of circular CFST columns transversely strengthened by FRP. Eng. Struct. **191**, 417–431 (2019). https://doi.org/10.1016/j.engstruct.2019.04.056

Guo, G., Cui, J., Wang, D.: A numerical investigation on collapse modes of stiffened plates under combined dynamic loads. In: Proceedings of the ASME 2024 43rd International Conference on Ocean, Offshore and Arctic Engineering (OMAE 2024), Singapore, 9–14 June 2024. (OMAE2024-126571) (2024)

Guo, Z., Bai, R., Lei, Z., Jiang, H., Zou, J., Yan, C.: Experimental and numerical investigation on ultimate strength of laser-welded stiffened plates considering welding deformation and residual stresses. Ocean Eng. **234**, 109239 (2021). https://doi.org/10.1016/j.oceaneng.2021.109239

Hammer, T.C., Willems, T., Hendrikse, H.: Dynamic ice loads for offshore wind support structure design. Mar. Struct. **87**, 103335 (2023). https://doi.org/10.1016/j.marstruc.2022.103335

Hanif, M.I., Adiputra, R., Prabowo, A.R., Yamada, Y., Firdaus, N.: Assessment of the ultimate strength of stiffened panels of ships considering uncertainties in geometrical aspects: finite element approach and simplified formula. Ocean Eng. **286**, 115522 (2023). https://doi.org/10.1016/j.oceaneng.2023.115522

Hastie, J.C., Guz, I.A., Kashtalyan, M.: Failure analysis of a composite riser pipe under operational and spooling loads. Procedia Struct. Integr. **42**, 614–622 (2022). https://doi.org/10.1016/j.prostr.2022.12.078

Hayama, M., Kikuchi, S., Tsukahara, M., Misaka, Y., Komotori, J.: Estimation of residual stress relaxation in low alloy steel with different hardness during fatigue by in situ X-ray measurement. Int. J. Fatigue **178**, 107989 (2024). https://doi.org/10.1016/j.ijfatigue.2023.107989

He, Y., Vaz, M.A., Caire, M.: Thermoplastic composite pipe failure envelopes under axisymmetric and thermomechanical loading. Eng. Struct. **278**, 115499 (2023). https://doi.org/10.1016/j.engstruct.2022.115499

Heiskari, J., Romanoff, J., Laakso, A., Ringsberg, J.W.: On the thickness determination of rectangular glass panes in insulating glass units considering the load sharing and geometrically nonlinear bending. Thin-Walled Structures **171**, 108774 (2022). https://doi.org/10.1016/j.tws.2021.108774

Heiskari, J., Romanoff, J., Laakso, A., Ringsberg, J.W.: Influence of the design constraints on the thickness optimization of glass panes to achieve lightweight insulating glass units in cruise ships. Mar. Struct. **89**, 103409 (2023). https://doi.org/10.1016/j.marstruc.2023.103409

Hernandez Ramos, M., Zhou, X., Ding, Z.: NSGA-II Enhanced Smith method for evaluation of the ultimate bending strength of damaged ships. In: Proceedings of the 2nd International Conference on the Stability and Safety of Ships and Ocean Vehicles, Wuxi, China, 14-18 October 2024, 895–905 (2024)

Hess, P.E., Bruchman, D., Assakkaf, I.A., Ayyub, B.M.: Uncertainties in material and geometric strength and load variables. Nav. Eng. J. **114**(2), 139–166 (2002). https://doi.org/10.1111/j.1559-3584.2002.tb00128.x

Hosseinabadi, O.F., Khedmati, M.R.: A review on ultimate strength of aluminium structural elements and systems for marine applications. Ocean Eng. **232**, 109153 (2021). https://doi.org/10.1016/j.oceaneng.2021.109153

Hosseinpour, P., Hosseinpour, M., Sharifi, Y.: Artificial neural networks for predicting ultimate strength of steel plates with a single circular opening under axial compression. Ships Offshore Struct. **17**(11), 2454–2469 (2022). https://doi.org/10.1080/17445302.2021.2000265

Houtani, H., Matsui, S., Fujimoto, W.: Numerical investigation of the statistics of vertical bending moments of ships in nonlinearly evolving irregular waves. In: Proceedings of the ASME 2023 42nd International Conference on Ocean, Offshore and Arctic Engineering (OMAE 2023), Melbourne, Australia, 11-16 June 2023. (OMAE2023-104733) (2023). https://doi.org/10.1115/OMAE2023-104733

Hu-Wei, C., Ping, Y.: Ultimate strength of hull plates under monotonic and cyclic uniaxial compression. J. Ship Res. **62**(3), 156–165 (2018). https://doi.org/10.5957/josr.170080

Hu, K., Yang, P., Xia, T.: Ultimate strength prediction of cracked panels under extreme cyclic loads considering crack propagation. Ocean Eng. **266**, 112948 (2022). https://doi.org/10.1016/j.oceaneng.2022.112948

Huang, Y., Wang, J., Taylor, R., Talimi, V., Fuglem, M.: An experimental study of iceberg hydrodynamic interactions with a generic floater on the grand banks NL. In: Proceedings of the 33rd International Ocean and Polar Engineering Conference (ISOPE 2023), Ottawa, Canada, 19–23 June 2023. (ISOPE-I-23-282) (2023)

Huang, Z., Yin, X., Liu, Y.: Physics-guided deep neural network for structural damage identification. Ocean Eng. **260**, 112073 (2022). https://doi.org/10.1016/j.oceaneng.2022.112073

IACS: Evaluation of scantlings of hatch covers and hatch coamings of cargo holds of bulk carriers, ore carriers and combination carriers. Unified Requirements Strength UR-S21 2010. In: International Association of Classification Societies, London, UK (2010)

IACS: Longitudinal strength standard. Unified Requirements Strength UR-S11, June 2015. In: International Association of Classification Societies, London, UK (2015a)

IACS: Evaluation of scantlings of hatch covers and hatch coamings and closing arrangements of cargo holds of ships. Unified requirements strength UR-S21A, May 2015. In: International Association of Classification Societies, London, UK (2015b)

IACS: Requirements concerning strength of ships. Unified Requirement S11A: Longitudinal strength standard for container ships. In: International Association of Classification Societies. London, UK (2020a)

IACS: Technical background for rule change notice for CSR, Pt 1, Ch 8, version 1 (2020). In: International Association of Classification Societies. London, UK (2020b)

IACS: Rec 47 Shipbuilding and repair quality standard, September 2021. In: International Association of Classification Societies, London, England (2021)

IACS: Annual review 2022. In: International Association of Classification Societies. London, UK (2022a). https://iacs.org.uk/about-us/annual-review. Accessed 11 July 2025

IACS: Standard wave data international association of classification societies. London, UK (2022b). https://iacs.org.uk/resolutions/recommendations/21-40. Accessed 11 July 2025

IACS: Common structural rules for bulk carriers and oil tankers (CSR-H). In: International Association of Classification Societies, London, England (2023a)

IACS: Technical background for CSR, 2023. In: International Association of Classification Societies, London, England (2023b)

IACS: Requirements concerning strength of ships. Unified Requirements Strength UR-S35: buckling strength assessment of ship structural elements. In: International Association of Classification Societies, London, England (2024)

Ilman, E.C., Wang, Y., Wharton, J.A., Sobey, A.J.: A hybrid corrosion-structural model for simulating realistic corrosion topography of maritime structures. Thin-Walled Struct. **169**, 108481 (2021). https://doi.org/10.1016/j.tws.2021.10848

IOGP: Supplementary specification for offshore topside structures (S-631–04) 2020. In: The International Association of Oil & Gas Producers (2020)

IOGP: Supplementary specification for fixed steel offshore structures (S-631-11) (2021). In: The International Association of Oil & Gas Producers (2021)

Ishibashi, K., Shiomitsu, D., Tatsumi, A., Fujikubo, M.: Simplified ultimate strength estimation method of rectangular plates under combined loads. Mar. Struct. **95**, 103592 (2024). https://doi.org/10.1016/j.marstruc.2024.103592

ISO: Petroleum and natural gas industries - general requirements for offshore structures, ISO 19900. International Organization for Standardization (2019). https://www.iso.org/standard/69761.html. Accessed 11 July 2025

ISSC: Report of specialist committee IV.2: Ultimate Hull Girder Strength. In H. Ohtsubo & Y. Sumi (eds.), Proceedings of the 14th International Ship and Offshore Structures Congress (ISSC 2000), vol. 2, Nagasaki (Japan), 2-6 October 2000. Elsevier, Oxford, UK, pp. 321–391. ISBN: 0080436021 (2000)

ISSC: Technical committee iii.1: ultimate strength. In: Fricke, W., Bronsart, R. (eds.) Proceedings of the 18th International Ship and Offshore Structures Congress (ISSC 2012), vol. 1, Rostock (Germany), 9–13 September 2012. Schiffbautechnische Gesellschaft, Hamburg, Germany, pp. 285–363 (2012). ISBN: 978-3-87700-131-8

ISSC: Technical Committee III.1: Ultimate Strength. In: Kaminski, M.L., Rigo, P. (eds.), Proceedings of the 20th International Ship and Offshore Structures Congress (ISSC 2018), vol. 1, Liège (Belgium) and Amsterdam (The Netherlands), 9–14 September 2018. IOS Press, Amsterdam, The Netherlands, pp. 335–439 (2018a). https://doi.org/10.3233/978-1-61499-862-4-335

ISSC: Discussion of committee iii.1: ultimate strength. In: Kaminski, M.L., Rigo, P. (eds.), Proceedings of the 20th International Ship and Offshore Structures Congress (ISSC 2018), vol. 3, Liège (Belgium) and Amsterdam (The Netherlands), 9-14 September 2018. IOS Press, Amsterdam, The Netherlands, pp. 61–84 (2018b). https://doi.org/10.3233/PMST200006

ISSC: Technical committee iii.1: ultimate strength. In: Wang, X., Pegg, N., (eds.), Proceedings of the 21st International Ship and Offshore Structures Congress (ISSC 2022), vol. 1, Vancouver, Canada, 11-15 September 2022. The Society of Naval Architects & Marine Engineers, pp. 395–500 (2022a). https://doi.org/10.5957/ISSC-2022-COMMITTEE-III-1

ISSC: Discussion of committee III.1: ultimate strength. In: Wang, X., Pegg, N. (eds.), Proceedings of the 21st International Ship and Offshore Structures Congress (ISSC 2022), vol. 3, Vancouver, Canada, 11-15 September 2022. The Society of Naval Architects & Marine Engineers, pp. 57–72 (2022b). https://doi.org/10.5957/ISSC-2022-DISCUSSION-III-1

Ivanhoe, R.O., Wang, L., Kolios, A.: Generic framework for reliability assessment of offshore wind turbine jacket support structures under stochastic and time dependent variables. Ocean Eng. **216**, 107691 (2020). https://doi.org/10.1016/j.oceaneng.2020.107691

Jafarzadeh, S., Khedmati, M.R.: Progressive collapse analysis of a composite ship hull girder under vertical bending using finite element method. Internal J. Maritime Technol. **14**, 21–32 (2020). http://ijmt.ir/article-1-684-en.html

Jagite, G., Bigot, F.: Numerical investigation of the hull girder ultimate strength under realistic cyclic loading derived from long-term hydroelastic analysis. Ships Offshore Struct. **18**(4), 515–528 (2023). https://doi.org/10.1080/17445302.2022.2035566

Jagite, G., Bigot, F., Malenica, S., Derbanne, Q., Le Sourne, H., Cartraud, P.: Dynamic ultimate strength of a ultra-large container ship subjected to realistic loading scenarios. Mar. Struct. **84**, 103197 (2022). https://doi.org/10.1016/j.marstruc.2022.103197

Jelovica, J., Romanoff, J., Ehlers, S., Varsta, P.: Influence of weld stiffness on buckling strength of laser-welded web-core sandwich plates. J. Constr. Steel Res. **77**, 12–18 (2012). https://doi.org/10.1016/j.jcsr.2012.05.001

Jia, H., Qin, S., Wang, R., Xue, Y., Fu, D., Wang, A.: Ship collision impact on the structural load of an offshore wind turbine. Glob. Energy Interconnect. **3**(1), 43–50 (2020). https://doi.org/10.1016/j.gloei.2020.03.009

Kadry, A.A., Ebid, A.M., Mokhtar, A.S.A., El-Ganzoury, E.N., Haggag, S.A.: Parametric study of unstiffened multi-planar tubular KK-Joints. Results Eng. **14**, 100400 (2022). https://doi.org/10.1016/j.rineng.2022.100400

Kai, Q., Renjun, Y., Mingen, C., Haiyan, Z.: Failure mode shift of sandwich composite L-Joint for ship structures under tension load. Ocean Eng. **214**, 107863 (2020). https://doi.org/10.1016/j.oceaneng.2020.107863

Kang, H., Yang, P., Xia, T.: Ultimate strength prediction of cracked panels under extreme cyclic loads considering crack propagation. Ocean Eng. **266**(3), 112948 (2022). https://doi.org/10.1016/j.oceaneng.2022.112948

Kim, C., Oterkus, S., Oterkus, E., Kim, Y.: Probabilistic ship corrosion wastage model with Bayesian inference. Ocean Eng. **246**, 110571 (2022). https://doi.org/10.1016/j.oceaneng.2022.110571

Kim, D.K., Li, S., Lee, J.R., Poh, B.Y., Benson, S., Cho, N.K.: An empirical formula to assess ultimate strength of initially deflected plate: part 1 - propose the general shape and application to longitudinal compression. Ocean Eng. **252**, 111151 (2022). https://doi.org/10.1016/j.oceaneng.2022.111151

Kim, D.K., Li, S., Yoo, K., Danasakaran, K., Cho, N.K.: An empirical formula to assess ultimate strength of initially deflected plate: part 2 - combined longitudinal compression and lateral pressure. Ocean Eng. **252**, 111112 (2022). https://doi.org/10.1016/j.oceaneng.2022.111112

Kim, J.H., Park, D.H., Kim, S.K., Kim, M.S., Lee, J.M.: Experimental study and development of design formula for estimating the ultimate strength of curved plates. Appl. Sci. **11**(5), 2379 (2021). https://doi.org/10.3390/app11052379

Komoriyama, Y., et al.: Ultimate longitudinal bending strength of damaged box girder in upright and inclined conditions - Model experiment and numerical analysis. In: Le Sourne, H., Guedes Soares, C. (eds.), Advances in the Collision and Grounding of Ships and Offshore Structures, Proceedings of the 9th International Conference on Collision and Grounding of Ship and Offshore Structures (ICCGS 2023), Nantes, France, 11–13 September 2023. London: CRC Press, pp. 377–385 (2024). https://doi.org/10.1201/9781003462170-46

Kong, X., Zhou, H., Zheng, C., Pei, Z., Yuan, T., Wu, W.: Research on the dynamic buckling of a typical deck grillage structure subjected to in-plane impact load. Mar. Struct. **78**, 103003 (2021). https://doi.org/10.1016/j.marstruc.2021.103003

Kõrgesaar, M., Putranto, T., Jelovica, J.: Equivalent single layer approach for predicting ultimate strength of stiffened panel under different load combinations. In: J.W. Ringsberg & C. Guedes Soares (eds.), Advances in the Analysis and Design of Marine Structures; Proceedings of the 9th International Conference on Marine Structures (MARSTRUCT 2023), Gothenburg, Sweden, 3-5 April 2023. London: CRC Press, pp. 347–354. https://doi.org/10.1201/9781003399759-38

KR.: Guidance on strength assessment of container ships considering the whipping effect (2022). https://www.krs.co.kr/KRRules/KRRules2022/KRRulesE.html. Accessed 11 July 2025

Kuznecovs, A., Ringsberg, J.W., Johnson, E., Yamada, Y.: Ultimate limit state analysis of a double-hull tanker subjected to biaxial bending in intact and collision-damaged conditions. Ocean Eng. **209**, 107519 (2020). https://doi.org/10.1016/j.oceaneng.2020.107519

Kuznecovs, A., Ringsberg, J.W., Mallaya Ullal, A., Janardhana Bangera, P., Johnson, E.: Consequence analyses of collision-damaged ships—damage stability, structural adequacy and oil spills. Ships Offshore Struct. **18**, 1–15 (2022). https://doi.org/10.1080/17445302.2022.2071014

Kuznecovs, A., Schreuder, M., Ringsberg, J.W.: Methodology for the simulation of a ship's damage stability and ultimate strength conditions following a collision. Mar. Struct. **79**, 103027 (2021). https://doi.org/10.1016/j.marstruc.2021.103027

Ladeira, I., Márquez, L., Echeverry, S., Le Sourne, H., Rigo, P.: Review of methods to assess the structural response of offshore wind turbines subjected to ship impacts. Ships Offshore Struct. **18**(6), 755–774 (2023). https://doi.org/10.1080/17445302.2022.2072583

Lee, D.H., Paik, J.K.: Ultimate strength characteristics of as-built ultra-large containership hull structures under combined vertical bending and torsion. Ships Offshore Struct. **15**(sup1), S143–S160 (2020). https://doi.org/10.1080/17445302.2020.1747829

Lee, H., et al.: Time-domain response-based structural analysis on a floating offshore wind turbine. J. Mar. Sci. Appl. **22**(1), 75–83 (2023). https://doi.org/10.1007/s11804-023-00322-0

Lee, H.H., Kim, H.J., Paik, J.K.: Use of physical testing data for the accurate prediction of the ultimate compressive strength of steel stiffened panels. Ships Offshore Struct. **18**(4), 609–623 (2023). https://doi.org/10.1080/17445302.2022.2087358

Lewis, M.: In a first, cargo ship strikes an offshore wind turbine (2023). https://electrek.co/2023/06/04/cargo-ship-offshore-wind-turbine/. Accessed 11 July 2025

Li, D., Chen, Z.: Advanced empirical formulae for the ultimate strength assessment of continuous hull plate under combined biaxial compression and lateral pressure. Eng. Struct. **285**, 116041 (2023). https://doi.org/10.1016/j.engstruct.2023.116041

Li, D., Chen, Z.: Progressive collapse analysis and ultimate strength estimation of continuous stiffened panel under longitudinal extreme cyclic load and lateral pressure. Ocean Eng. **285**, 115340 (2023). https://doi.org/10.1016/j.oceaneng.2023.115340

Li, D., Chen, Z., Li, J., Yi, J.: Ultimate strength assessment of ship hull plate with multiple cracks under axial compression using artificial neural networks. Ocean Eng. **263**, 112438 (2022a). https://doi.org/10.1016/j.oceaneng.2022.112438

Li, D., Feng, L., Huang, D., Shi, H., Wang, S.: Residual ultimate strength of stiffened box girder with coupled damage of pitting corrosion and a crack under vertical bending moment. Ocean Eng. **235**, 109341 (2021a). https://doi.org/10.1016/j.oceaneng.2021.109341

Li, F., Suominen, M., Kujala, P., Zhou, L.: A literature survey of probabilistic modelling methods of local ice loads on marine structures. In: Proceedings of the ASME 2024 43rd International Conference on Ocean, Offshore and Arctic Engineering (OMAE 2024), Singapore, 9-14 June 2024. (OMAE2024-127097) (2024a)

Li, H., Li, Y., Li, G., Zhu, Q., Wang, B., Tang, Y.: Transient tower and blade deformations of a Spar-type floating wind turbine in freak waves. Ocean Eng. **294**, 116801 (2024). https://doi.org/10.1016/j.oceaneng.2024.116801

Li, S., Benson, S.D.: A re-evaluation of the hull girder shakedown limit states. Ships Offshore Struct. **14**(sup1), 239–250 (2019). https://doi.org/10.1080/17445302.2019.1573872

Li, S., Kim, D.K.: A comparison of numerical methods for damage index based residual ultimate limit state assessment of grounded ship hulls. Thin-Walled Struct. **172**, 108854 (2022). https://doi.org/10.1016/j.tws.2021.108854

Li, S., Coraddu, A., Oneto, L.: Computationally aware estimation of ultimate strength reduction of stiffened panels caused by welding residual stress: from finite element to data-driven methods. Eng. Struct. **264**, 114423 (2022b). https://doi.org/10.1016/j.engstruct.2022.114423

Li, S., Georgiadis, D.G., Kim, D.K., Samuelides, M.S.: A comparison of geometric imperfection models for collapse analysis of ship-type stiffened plated grillages. Eng. Struct. **250**, 113480 (2022c). https://doi.org/10.1016/j.engstruct.2021.113480

Li, S., Hu, Z., Benson, S.: An analytical method to predict the buckling and collapse behavior of plates and stiffened panels under cyclic load. Eng. Struct. **199**, 109267 (2019). https://doi.org/10.1016/j.engstruct.2019.109627

Li, L., Jiang, Z., Ong, M.C., Hu, W.: Design optimization of mooring system: an application to a vessel-shaped offshore fish farm. Eng. Struct. **197**, 109363 (2019). https://doi.org/10.1016/j.engstruct.2019.109363

Li, S., Kim, D.K., Benson, S.: A probabilistic approach to assess the computational uncertainty of ultimate strength of hull girders. Reliab. Eng. Syst. Saf. **213**, 107688 (2021b). https://doi.org/10.1016/j.ress.2021.107688

Li, S., Kim, D.K., Benson, S.: The influence of residual stress on the ultimate strength of longitudinally compressed stiffened panels. Ocean Eng. **231**, 108839 (2021c). https://doi.org/10.1016/j.oceaneng.2021.108839

Li, W., Bazant, M.Z., Zhu, J.: A physics-guided neural network framework for elastic plates: comparison of governing equations-based and energy-based approaches. Comput. Methods Appl. Mech. Eng. **383**, 113933 (2021d). https://doi.org/10.1016/j.cma.2021.113933

Li, Y., Sakonder, C., Paredes, M.: Plastic collapse analysis in multiaxially loaded defective pipe specimens at different temperatures. J. Pipeline Sci. Eng. **3**(1), 100092 (2023). https://doi.org/10.1016/j.jpse.2022.100092

Li, Z., Chen, D., Feng, X., Chen, J.F.: Hydroelastic analysis and structural design of a modular floating structure applying ultra-high performance fiber-reinforced concrete. Ocean Eng. **277**, 114266 (2023). https://doi.org/10.1016/j.oceaneng.2023.114266

Lima, J.P.S., Evangelista, F., Guedes Soares, C.: Bi-fidelity kriging model for reliability analysis of the ultimate strength of stiffened panels. Mar. Struct. **91**, 103464 (2023). https://doi.org/10.1016/j.marstruc.2023.103464

Lindemann, T., Kaeding, P.: Formulation of idealized structural unit method to perform progressive collapse analyses of welded box girders in bending. In: Proceedings of the ASME 2024 43rd International Conference on Ocean, Offshore and Arctic Engineering (OMAE 2024), Singapore, 9–14 June 2024. (OMAE2024-125828) (2024)

Lindemann, T., La Ferlita, A., Di Nardo, E., Kaeding, P.: Application of different methods to determine the ultimate strength of ships in bending. In: Proceedings of the ASME 2023 42nd International Conference on Ocean, Offshore and Arctic Engineering (OMAE 2023), Melbourne, Australia, 11–16 June 2023. (OMAE2023-103731) (2023). https://doi.org/10.1115/OMAE2023-103731

Lindemann, T., La Ferliata, A., Kaeding, P.: Application of idealized structural unit method to determine the ultimate strength of ship model in bending. In: Proceedings of the ASME 2024 43rd International Conference on Ocean, Offshore and Arctic Engineering (OMAE 2024), Singapore, 9–14 June 2024. (OMAE2024-125843) (2024)

Lindemann, T., Okpeke, B.E., La Ferlita, A., Mühmer, M., Kaeding, P.: Numerical investigations on ultimate strength of a double hull VLCC under combined loads and initial imperfections. In: Proceedings of the ASME 2022 41st International Conference on Ocean, Offshore and Arctic Engineering (OMAE 2022), Hamburg, Germany, 5–10 June 2022. (OMAE2022-80204) (2022). https://doi.org/10.1115/OMAE2022-80204

Liu, B., Guedes Soares, C.: Ultimate strength assessment of ship hull structures subjected to cyclic bending moments. Ocean Eng. **215**, 107685 (2020). https://doi.org/10.1016/j.oceaneng.2020.107685

Liu, B., Doan, Y.T., Garbatov, Y., Wu, W., Guedes Soares, C.: Study on ultimate compressive strength of aluminium-alloy plates and stiffened panels. J. Mar. Sci. Appl. **19**, 534–552 (2020). https://doi.org/10.1007/s11804-020-00170-2

Liu, B., Gao, L., Ao, L., Wu, W.: Experimental and numerical analysis of ultimate compressive strength of stiffened panel with openings. Ocean Eng. **220**, 108453 (2021). https://doi.org/10.1016/j.oceaneng.2020.108453

Liu, B., Liu, K., Villavicencio, R., Dong, A., Guedes Soares, C.: Experimental and numerical analysis of the penetration of welded aluminium alloy panels. Ships Offshore Struct. **16**(5), 492–504 (2020). https://doi.org/10.1080/17445302.2020.1736856

Liu, B., Yao, X., Lin, Y., Wu, W., Guedes Soares, C.: Experimental and numerical analysis of ultimate compressive strength of long-span stiffened panels. Ocean Eng. **237**, 109633 (2021). https://doi.org/10.1016/j.oceaneng.2021.109633

Liu, B., Zhang, X., Garbatov, Y.: Multi-scale analysis for assessing the impact of material composition and weave on the ultimate strength of GFRP stiffened panels. J. Mar. Sci. Eng. **11**(1), 108 (2023). https://doi.org/10.3390/jmse11010108

Liu, L., Yang, D.Y., Frangopol, D.M.: Probabilistic cost-benefit analysis for service life extension of ships. Ocean Eng. **201**, 107084 (2020). https://doi.org/10.1016/j.oceaneng.2020.107094

Liu, L., Yang, D.Y., Frangopol, D.M.: Ship service life extension considering ship condition and remaining design life. Mar. Struct. **78**, 102940 (2021). https://doi.org/10.1016/j.marstruc.2021.102940

Liu, M.: Effect of uniform corrosion on the mechanical behavior of E690 high-strength steel lattice corrugated panel in the marine environment: a finite element analysis. Mater. Res. Express **8**, 066510 (2021). https://doi.org/10.1088/2053-1591/ac0655

Liu, X., Jiang, D., Liufu, K., Fu, J., Liu, Q., Li, Q.: Numerical investigation into impact responses of an offshore wind turbine jacket foundation subjected to ship collision. Ocean Eng. **248**, 110825 (2022). https://doi.org/10.1016/j.oceaneng.2022.110825

Liu, Y., Lei, Z., Zhu, R., Shang, Y., Bai, R.: Artificial neural network prediction of residual compressive strength of composite stiffened panels with open crack. Ocean Eng. **266**, 112771 (2022). https://doi.org/10.1016/j.oceaneng.2022.112771

Lowde, M.J., Peters, H.G.A., Geraghty, R., Graham-Jones, J., Pemberton, R., Summerscales, J.: The 100 m composite ship? J. Mar. Sci. Eng. **10**(3), 408 (2022). https://doi.org/10.3390/jmse10030408

LR: Global Design Loads of Container Ships and Other Ships Prone to Whipping and Springing. Lloyd's Register, London (2022)

Ma, H., Wang, D.: Lateral pressure effect on the ultimate strength of the stiffened plate subjected to combined loads. Ocean Eng. **239**, 109926 (2021). https://doi.org/10.1016/j.oceaneng.2021.109926

Ma, H., Xiong, Q., Wang, D.: Experimental and numerical study on the ultimate strength of stiffened plates subjected to combined biaxial compression and lateral loads. Ocean Eng. **228**, 108928 (2021). https://doi.org/10.1016/j.oceaneng.2021.108928

Ma, H., Yang, Y., He, Z., Jia, Z., Zhang, Y.: Experimental study on constitutive relation of the high performance marine structural steel under extreme cyclic loads. Ocean Eng. **168**, 204–215 (2018). https://doi.org/10.1016/j.oceaneng.2018.09.003

Ma, Z., Pei, Z., Wu, W.: Dynamic ultimate strength analysis of stiffened plate based on Idealized Structural Unit Method. Mar. Struct. **84**, 103203 (2022). https://doi.org/10.1016/j.marstruc.2022.103203

Ma., H., Kawamura, Y., Okada, T., Wang, D., Hayakawa, G.: A general scaled model design method of stiffened plate subjected to combined longitudinal compression and lateral pressure considering the ultimate strength and collapse modes. Marine Struct. **90**, 103435 (2023). https://doi.org/10.1016/j.marstruc.2023.103435

Martin, M.B., Harris, J.C., Renaud, P., Hulin, F., Filipot, J.F.: Numerical investigation of slamming loads on floating offshore wind turbines. In: Proceedings of the 32nd International Ocean and Polar Engineering Conference (ISOPE 2022), Shanghai, China, 5–10 June 2022. (ISOPE-I-22-031) (2022)

Mohd, M.H., et al.: The effect of corrosion depth on the ultimate strength of an aging fixed offshore structure. In: Ismail, A., Dahalan, W.M., Öchsner, A. (eds.), Design in Maritime Engineering, Proceedings of the 2nd International Conference on Marine and Advanced Technologies (ICMAT 2021), Kuala Lumpur, Malaysia, 24 August 2021. Springer, pp. 271–286 (2022). https://doi.org/10.1007/978-3-030-89988-2_21

Mokhtari, M., Wang, X., Amdahl, J.: Buckling and post-buckling behaviour of extruded aluminium panels subject to the combined effects of welding and pitting corrosion. In: Proceedings of the ASME 2023 42nd International Conference on Ocean, Offshore and Arctic Engineering (OMAE 2023), Melbourne, Australia, 11–16 June 2023. (OMAE2023-105048) (2023). https://doi.org/10.1115/OMAE2023-105048

Moon, W., et al.: Time-domain response-based structural assessment of a FOWT – buckling and ultimate strength assessment. In: ASME 2022 4th International Offshore Wind Technical Conference (IOWTC 2022), Boston, Massachusetts, USA, 7–8 December 2022. (IOWTC2022-96497) (2023). https://doi.org/10.1115/IOWTC2022-96497

Muhaimin Ishak, M.I., et al.: IGUSA: prediction of ultimate strength of fixed offshore structures in Malaysian waters using machine learning techniques. In: Proceedings of the Offshore Technology Conference Asia (OTC 2022), Virtual and Kuala Lumpur, Malaysia, 22–25 March 2022. (OTC-31493-MS) (2022). https://doi.org/10.4043/31493-MS

Mun, J.S., Ri, Y.H.: Study on ultimate strength estimation of intermittently welded stiffened plates under uniaxial compression. Mar. Struct. **84**, 103163 (2022). https://doi.org/10.1016/j.marstruc.2022.103163

Mursid, O., et al.: Effect of pitting corrosion position to the strength of ship bottom plate in a grounding incident. Curved Layered Struct. **10**, 20220199 (2023). https://doi.org/10.1515/cls-2022-0199

Musial, W., et al.: Offshore wind market report: 2023 edition. U.S. Department of Energy (2023). https://www.energy.gov/eere/wind/articles/offshore-wind-market-report-2023-edition. Accessed 11 July 2025

Nahshon, K., Rynolds, N., Shields, M.D.: Efficient uncertainty propagation for high-fidelity simulations with large parameter spaces: application to stiffened plate buckling. J. Verification, Validation and Uncertainty Quant. **3**(1), 011003 (16 pages) (2018). https://doi.org/10.1115/1.4039836

Nassiraei, H., Yara, A.: Numerical analysis of local joint flexibility of K-joints with external plates under axial loads in offshore tubular structures. J. Mar. Sci. Appl. **21**(4), 134–144 (2023). https://doi.org/10.1007/s11804-022-00302-w

NSIA: Marine casualty involving the cruise ship "Viking Polaris" south-east of Cape Horn, 29 November 2022. Report MARINE 2023/06. Norwegian Safety Investigation Authority, Lillestrøm, Norway (2023)

Othman, N.A., Mohd, M.H., Rahman, M.A.A., Musa, M.A., Fitriadhy, A.: Investigation of the corrosion factor to the global strength of aging offshore jacket platforms under different marine zones. Int. J. Naval Archit. Ocean Eng. **15**, 100496 (2023). https://doi.org/10.1016/j.ijnaoe.2022.100496

Paik, J.K.: Ultimate Limit State Analysis and Design of Plated Structures (2nd edn.) Wiley, Hoboken (2018). https://doi.org/10.1002/9781119367758

Paik, J.K., Thayamballi, A.K.: An empirical formulation for predicting the ultimate compressive strength of stiffened panels. In: Proceedings of the 7th International Ocean and Polar Engineering Conference (ISOPE 1997), Honolulu, Hawaii, USA, 25–30 May 1997. (ISOPE-I-97-444) (1997)

Paik, J.K., Andrieu, C., Cojeen, H.P.: Mechanical collapse testing on aluminum stiffened plate structures for marine applications. Marine Technol. SNAME News **45**(4), 228–240 (2008). https://doi.org/10.5957/mtl.2008.45.4.228

Pan, J., Xu, R.J., Song, Z.J., Wan, Q., Li, X.B.: Study on the load-end shortening formulae of structural elements including hard corner element considering collapse modes based on test and numerical modeling. Thin-Walled Struct. **185**, 110624 (2023). https://doi.org/10.1016/j.tws.2023.110624

Pandey, M., Young, B.: Ultimate resistances of member-rotated cold-formed high strength steel tubular T-joints under compression loads. Eng. Struct. **244**, 112601 (2021). https://doi.org/10.1016/j.engstruct.2021.112601

Papanikolaou, N., Anyfantis, K.: Construction of surrogate models for predicting the buckling strength of stiffened panels through DoE and RSM methods. Eng. Comput. **39**(4), 1374–1406 (2022). https://doi.org/10.1108/EC-03-2021-0176

Paredes, M., Grolleau, V., Wierzbicki, T.: On ductile fracture of 316L stainless steels at room and cryogenic temperature level: an engineering approach to determine material parameters. Materialia **10**, 100624 (2020). https://doi.org/10.1016/j.mtla.2020.100624

Park, J.S., Ha, Y.C., Seo, J.K.: Estimation of buckling and ultimate collapse behavior of stiffened curved plates under compressive load. J. Ocean Eng. Technol. **34**(1), 37–45 (2020). https://doi.org/10.26748/ksoe.2019.108

Park, Y.I., Kim, J.H.: Artificial neural network based prediction of ultimate buckling strength of liquid natural gas cargo containment system under sloshing loads considering onboard boundary conditions. Ocean Eng. **249**, 110981 (2022). https://doi.org/10.1016/j.oceaneng.2022.110981

Peeters, D.M.J., et al.: Buckling test of stiffened panels: Modeling and vibrational correlation testing. In: Vassilopoulos, A.P., Michaud, V., (eds.), Composites Meet Sustainability; Proceedings of the 20th European Conference on Composite Materials (ECCM 20), Lausanne, Switzerland, 25-30 June 2022. Lausanne: EPFL Lausanne, Composite Construction Laboratory, pp. 763–770 (2022)

Peng, R.X., Qiu, W.L., Teng, F.: Investigation on seawater freeze-thaw damage deterioration of marine concrete structures in cold regions from multi-scale. Ocean Eng. **248**, 110867 (2022). https://doi.org/10.1016/j.oceaneng.2022.110867

Petrolo, M., Carrera, E.: Best theory diagram for multi-layered structures via shell finite elements. Adv. Model. Simul. Eng. Sci. **6**, 4 (2019). https://doi.org/10.1186/s40323-019-0129-8

Petrolo, M., Carrera, E.: Selection of element-wise shell kinematics using neural networks. Comput. Struct. **244**, 106425 (2021). https://doi.org/10.1016/j.compstruc.2020.106425

Petrolo, M., Iannotti, P.: Best theory diagrams for laminated composite shells based on failure indexes. Aerotecn. Missili Spazio **102**, 199–218 (2023). https://doi.org/10.1007/s42496-023-00158-5

Piscopo, V., Scamardella, A.: Incidence of pitting corrosion wastage on the hull girder ultimate strength. J. Mar. Sci. Appl. **20**, 477–490 (2021). https://doi.org/10.1007/s11804-021-00218-x

Piscopo, V., Scamardella, A.: Ultimate strength assessment of simply supported pitted platings: a new stochastic approach based on Monte Carlo simulation. Mar. Struct. **87**, 103312 (2023). https://doi.org/10.1016/j.marstruc.2022.103312

Preventas, M., Fanourgakis, S., Samuelides, M.S.: Effect of low cycle/high amplitude loads on the moment carrying capacity of a ship's hull. In: Amdahl, J., Guedes Soares, C. (eds.), Developments in the Analysis and Design of Marine Structures; Proceedings of the 8th International Conference on Marine Structures (MARSTRUCT 2021), Trondheim, Norway, 7-9 June 2021. London: CRC Press, pp. 137–146 (2021). https://doi.org/10.1201/9781003230373-16

Putranto, T., Kõrgesaar, M.: Numerical investigation on the buckling response of stiffened panel subjected to biaxial compression with non-linear equivalent single layer approach. In: Proceedings of the 31st International Ocean and Polar Engineering Conference (ISOPE 2021), Rhodes, Greece, 20–25 June 2022 (2021). (ISOPE-I-21-4206)

Putranto, T., Kõrgesaar, M., Jelovica, J.: Ultimate strength assessment of stiffened panels using equivalent single layer approach under combined in-plane compression and shear. Thin-Walled Struct. **180**, 109943 (2022). https://doi.org/10.1016/j.tws.2022.109943

Putranto, T., Kõrgesaar, M., Tabri, K.: Application of equivalent single layer approach for ultimate strength analyses of ship hull girder. J. Mar. Sci. Eng. **10**(10), 1530 (2022). https://doi.org/10.3390/jmse10101530

Putranto, T., Kõrgesaar, M., Jelovica, J., Tabri, K., Naar, H.: Ultimate strength assessment of stiffened panel under uni-axial compression with non-linear equivalent single layer approach. Mar. Struct. **78**, 103004 (2021). https://doi.org/10.1016/j.marstruc.2021.103004

Quispe, J.P., et al.: Numerical and experimental analyses of the ultimate strength of a small-scale hull box girder. Marine Struct. **85**, 103273 (2022). https://doi.org/10.1016/j.marstruc.2022.103273

Rapacka, P.: Anholt wind farm incident. Developer asks to establish 'no sail zones' (2023). https://balticwind.eu/anholt-wind-farm-incident-developer-asks-to-establish-no-sail-zones/. Accessed 11 July 2025

Ren, Y., Shi, W., Venugopal, V., Zhang, L., Li, X.: Experimental study of tendon failure analysis for a TLP floating offshore wind turbine. Appl. Energy **358**, 122633 (2024). https://doi.org/10.1016/j.apenergy.2024.122633

Ringsberg, J.W., et al.: The ISSC 2022 committee III.1-Ultimate strength benchmark study on the ultimate limit state analysis of a stiffened plate structure subjected to uniaxial compressive loads. Marine Struct. **79**, 103026 (2021). https://doi.org/10.1016/j.marstruc.2021.103026

Romanoff, J., Karttunen, A., Varsta, P.: Design space for bifurcation buckling of laser-welded web-core sandwich plates as predicted by classical and micropolar plate theories. Ann. Solid Struct. Mech. **12**, 73–87 (2020). https://doi.org/10.1007/s12356-020-00064-6

Saberi, S., Fantuzzi, N.: Analysis of unreinforced and reinforced tubular T-joints structures with open source finite element software. Mech. Adv. Mater. Struct. **30**(4), 912–922 (2022). https://doi.org/10.1080/15376494.2022.2028043

Seyffert, H.C., Troesch, A.W., Collette, M.D.: Combined stochastic lateral and in-plane loading of a stiffened ship panel leading to collapse. Mar. Struct. **67**, 102620 (2019). https://doi.org/10.1016/j.marstruc.2019.04.008

Shen, Q., Wang, J., Wang, J., Ding, Z.: Axial compressive performance of circular CFST columns partially wrapped by carbon FRP. J. Constr. Steel Res. **155**, 90–106 (2019). https://doi.org/10.1016/j.jcsr.2018.12.017

Shi, G., Wang, M., Wang, Y., Wang, F.: Cyclic behavior of 460 MPa high strength structural steel and welded connection under earthquake loading. Adv. Struct. Eng. **16**(3), 451–466 (2013). https://doi.org/10.1260/1369-4332.16.3.451

Shi, G.J., Gao, D.W.: Reliability analysis of hull girder ultimate strength for large container ships under whipping loads. Struct. Infrastruct. Eng. **17**(3), 319–330 (2020). https://doi.org/10.1080/15732479.2020.1744023

Shi, G.J., Gao, D.W.: Model experiment of large superstructures' influence on hull girder ultimate strength for cruise ships. Ocean Eng. **222**, 108626 (2021). https://doi.org/10.1016/j.oceaneng.2021.108626

Shi, G.J., Wang, D.Y., Wang, F.H., Cai, S.J.: Analysis of dynamic response and ultimate strength for box girder under bending moment. J. Mar. Sci. Eng. **11**(2), 373 (2023). https://doi.org/10.3390/jmse11020373

Shi, W., Zhang, L., Karimirad, M., Michailides, C., Jiang, Z., Li, X.: Combined effects of aerodynamic and second-order hydrodynamic loads for floating wind turbines at different water depths. Appl. Ocean Res. **130**, 103416 (2023). https://doi.org/10.1016/j.apor.2022.103416

Shi, X.H., Shen, H., Zhang, J., Guedes Soares, C.: Uncertainty of ultimate strength of ship hull with pits. In: Guedes Soares, C. (ed.), Maritime Technology and Engineering 5 Volume 1; Proceedings of the 5th International Conference on Maritime Technology and Engineering (MARTECH 2020), Lisbon, Portugal, 16-19 November 2020. London: CRC Press, pp. 583–590 (2021). https://doi.org/10.1201/9781003216582-65

Shiomitsu, D., Ishibashi, K., Yanagimoto, F., Fujikubo, M.: A simple estimation method for ultimate strength of curved plates under axial compression. In: Proceedings of the ASME 2023 42nd International Conference on Ocean, Offshore and Arctic Engineering (OMAE 2023), Melbourne, Australia, 11-16 June 2023. (OMAE2023-102688) (2023). https://doi.org/10.1115/OMAE2023-102688

Shokrieh, M., Poon, C., Lessard, L.: Three-dimensional progressive failure analysis of pin/bolt loaded composite laminates. In: Proceedings of the 83rd Meeting of the AGARD Structures and Materials Panel (AGARD CP-590), Florence, Italy, 2-3 September 1996 (1996)

Silva, J.P., Chen, B.Q., Videiro, P.M.: FPSO hull structures with sandwich plate system in cargo tanks. Appl. Sci. **12**(19), 9628 (2022). https://doi.org/10.3390/app12199628

Smith, C.S.: Influence of local compressive failure on ultimate longitudinal strength of ship's hull. In: Proceedings of the International Symposium on Practical Design in Shipbuilding (PRADS), Tokyo, Japan, 17-21 October 1977, pp. 73–79 (1977)
Smith, C.S., Davidson, P., Chapman, J.C.: Strength and stiffness of ship's plating under in-plane compression and tension. Royal Inst. Naval Archit. Trans. **130** (1987). https://api.semantics cholar.org/CorpusID:134846745
Soleimani, E., Tabeshpour, M.R., Seif, M.S.: Parametric study of buckling and post-buckling behavior for an aluminium hull structure of a high-aspect-ratio twin hull vessel. Proc. IMechE Part M, J. Eng. Maritime Environ. **234**(1), 15–25 (2020). https://doi.org/10.1177/147509021 9868635
Song, Y., Yang, P., Peng, Z., Xia, T.: Residual ultimate strength of ship cracked plates considering fatigue crack propagation under cyclic loads. Ships Offshore Struct. **17**(6), 1403–1412 (2022). https://doi.org/10.1080/17445302.2021.1926143
Storheim, M., Alsos, H.S., Amdahl, J.: Evaluation of nonlinear material behavior for offshore structures subjected to accidental actions. J. Offshore Mech. Arctic Eng. **140**(4), 041401 (7 pages) (2018). https://doi.org/10.1115/1.4038585
Sun, H., Wang, Y., Jia, L., Lin, Z., Yu, H.: Theoretical and numerical methods for predicting the structural stiffness of unbonded flexible riser for deep-sea mining under axial tension and internal pressure. Ocean Eng. **310**, 118672 (2024). https://doi.org/10.1016/j.oceaneng.2024.118672
Sun, K., et al.: Experimental study on ultimate strength of novel aluminum alloy stiffened panel with floating girder. In: Proceedings of the ASME 2021 40th International Conference on Ocean, Offshore and Arctic Engineering (OMAE 2021), Virtual, Online, 21-30 June 2021. (OMAE2021-63761) (2021a). https://doi.org/10.1115/OMAE2021-63761
Sun, Z., Lei, Z., Zou, J., Bai, R., Jiang, H., Yan, C.: Prediction of failure behavior of composite hat-stiffened panels under in-plane shear using artificial neural network. Compos. Struct. **272**, 114238 (2021). https://doi.org/10.1016/j.compstruct.2021.114238
Sundarraja, M.C., Prabhu, G.G.: Investigation on strengthening of CFST members under compression using CFRP composites. J. Reinf. Plast. Compos. **30**(15), 1251–1264 (2011). https://doi.org/10.1177/0731684411418018
Tabri, K., Tõns, T., Suominen, M., Kõrgesaar, M.: Ice-induced loads on offshore wind turbines in the Baltic Sea. In: Proceedings of the ASME 2022 41st International Conference on Ocean, Offshore and Arctic Engineering (OMAE 2022), Hamburg, Germany, 5-10 June 2022. (OMAE2022-79035) (2022). https://doi.org/10.1115/OMAE2022-79035
Takami, T., Fujimoto, W., Houtani, H., Matsui, S.: Extreme wave and vertical bending moment predictions by higher order spectrum method and FORM. In: Proceedings of the ASME 2023 42nd International Conference on Ocean, Offshore and Arctic Engineering (OMAE 2023), Melbourne, Australia, 11-16 June 2023. (OMAE2023-101876) (2023). https://doi.org/10.1115/OMAE2023-101876
Tatsumi, A., Fujikubo, M.: Ultimate strength of container ships subjected to combined hogging moment and bottom local loads, part 1: nonlinear finite element analysis. Mar. Struct. **69**, 102683 (2020). https://doi.org/10.1016/j.marstruc.2019.102683
Tatsumi, A., Ko, H.H.H., Fujikubo, M.: Ultimate strength of container ships subjected to combined hogging moment and bottom local loads, part 2: an extension of Smith's method. Mar. Struct. **71**, 102738 (2020). https://doi.org/10.1016/j.marstruc.2020.102738
Tekgoz, M., Garbatov, Y., Guedes Soares, C.: Review of ultimate strength assessment of ageing and damaged ship structures. J. Mar. Sci. Appl. **19**, 512–533 (2020). https://doi.org/10.1007/s11804-020-00179-7
Tran, H., Gao, Y.F., Chew, H.B.: Numerical and experimental crack-tip cohesive zone laws with physics-informed neural networks. J. Mech. Phys. Solids **193**, 105866 (2024). https://doi.org/10.1016/j.jmps.2024.105866

Truong, D.D., Jang, B.S.: Estimation of ice loads on offshore structures using simulations of level ice-structure collisions with an influence coefficient method. Appl. Ocean Res. **125**, 103235 (2022). https://doi.org/10.1016/j.apor.2022.103235

Vizentin, G., Vukelic, G.: Failure analysis of FRP composites exposed to real marine environment. Procedia Struct. Integr. **37**, 233–240 (2022). https://doi.org/10.1016/j.prostr.2022.01.079

Wang, B., Luo, L., Nie, X., Duan, S., Wang, L.: Post-buckling reliability and sensitivity analysis of composite stiffened plates based on adaptive Kriging method. Acta Mech. Solida Sin. **36**, 340–348 (2023). https://doi.org/10.1007/s10338-022-00366-9

Wang, J.T., Liu, X.H., Sun, Q., Li, Y.W.: Analytical behavior and bearing capacity research on out-of-code tapered CFDST members under pure torsion and compression-torsion combination. Ocean Eng. **284**, 115324 (2023). https://doi.org/10.1016/j.oceaneng.2023.115324

Wang, M., Wang, C., Hnydiuk-Stefan, A., Feng, S., Atilla, I., Li, Z.: Recent progress on reliability analysis of offshore wind turbine support structures considering digital twin solutions. Ocean Eng. **232**, 109168 (2021). https://doi.org/10.1016/j.oceaneng.2021.109168

Wang, Q., Liu, K., Zhang, M.: Numerical studies on the performance of the circular fibre reinforced plastics-concrete filled steel tubes stub column under axial compression. J. Reinf. Plast. Compos. **41**(9–10), 383–398 (2022). https://doi.org/10.1177/07316844211051707

Wang, S., Moan, T.: Methodology of load effect analysis and ultimate limit state de-sign of semi-submersible hulls of floating wind turbines: with a focus on floater column design. Mar. Struct. **93**, 103526 (2024). https://doi.org/10.1016/j.marstruc.2023.103526

Wang, X., Amdahl, J., Egeland, O.: Numerical study on buckling of aluminum extruded panels considering welding effects. Mar. Struct. **84**, 103230 (2022). https://doi.org/10.1016/j.marstruc.2022.103230

Wang, X., Jia, P., Wang, B.: Progressive failure model of high strength glass fiber composite structure in hygrothermal environment. Compos. Struct. **280**, 114932 (2022). https://doi.org/10.1016/j.compstruct.2021.114932

Wang, Z., et al.: Structural fire behavior of aluminium alloy structures: review and outlook. Eng. Struct. **268**, 114746 (2022). https://doi.org/10.1016/j.engstruct.2022.114746

Wang, X., Yu, Z., Amdahl, J.: Ultimate strength of welded aluminium stiffened panels under combined biaxial and lateral loads: a numerical investigation. Mar. Struct. **97**, 103654 (2024). https://doi.org/10.1016/j.marstruc.2024.103654

Wilkie, D., Galasso, C.: Site-specific ultimate limit state fragility of offshore wind turbines on monopile substructures. Eng. Struct. **204**, 109903 (2020). https://doi.org/10.1016/j.engstruct.2019.109903

Woloszyk, K., Bielski, P.W., Garbatov, Y., Mikulski, T.: Photogrammetry image-based approach for imperfect structure modeling and FE analysis. Ocean Eng. **223**, 108665 (2021). https://doi.org/10.1016/j.oceaneng.2021.108665

Woloszyk, K., Garbatov, Y., Kowalski, J., Samson, L.: Numerical and experimental study on effect of boundary conditions during testing of stiffened plates subjected to compressive loads. Eng. Struct. **235**, 112027 (2021). https://doi.org/10.1016/j.engstruct.2021.112027

Wong, W.J., Walters, C.L.: Effect of strain hardening on the bending capacity of high-strength welded I-section beams. Ships Offshore Struct., 1–16 (2023).https://doi.org/10.1080/17445302.2023.2195727

Wu, Q., Hu, S., Tang, X., Liu, X., Chen, Z., Xiong, J.: Compressive buckling and post-buckling behaviors of J-type composite stiffened panel before and after impact load. Compos. Struct. **304**, 116339 (2023). https://doi.org/10.1016/j.compstruct.2022.116339

Xiong, Y., Li, C., Cai, S., Wang, D.: The dynamic ultimate strength of stiffened panels under axial impact loading. Ships Offshore Struct. **18**(5), 707–720 (2023). https://doi.org/10.1080/17445302.2022.2067417

Xu, G., Qin, K., Yan, R., Dong, Q.: Research on failure modes and ultimate strength behavior of typical sandwich composite joints for ship structures. Int. J. Naval Archit. Ocean Eng. **14**, 100428 (2022). https://doi.org/10.1016/j.ijnaoe.2021.100428

Xu, K., Larsen, K., Shao, Y., Zhang, M., Gao, Z., Moan, T.: Design and comparative analysis of alternative mooring systems for floating wind turbines in shallow water with emphasis on ultimate limit state design. Ocean Eng. **219**, 108377 (2021). https://doi.org/10.1016/j.oceaneng.2020.108377

Xu, M.C., Guedes Soares, C.: Experimental evaluation of the ultimate strength of stiffened panels under longitudinal compression. Ocean Eng. **220**, 108496 (2021). https://doi.org/10.1016/j.oceaneng.2020.108496

Xu, M.C., Song, Z.J., Pan, J.: Study on the similarity methods for the assessment of ultimate strength of stiffened panels under axial load based on tests and numerical simulations. Ocean Eng. **219**, 108294 (2021). https://doi.org/10.1016/j.oceaneng.2020.108294

Xu, M.C., Song, Z.J., Wang, T., Zhang, W.Z., Pan, J.: Empirical formulae assessment of ultimate strength for perforated stiffened panels under longitudinal compression. Ocean Eng. **264**, 112445 (2022). https://doi.org/10.1016/j.oceaneng.2022.112445

Xu, W., Zhou, X., Li, C., Ren, H.: Post-ultimate strength behavior and collapse severity of ship hull girder under extreme wave load by an analytical method. Ships Offshore Struct. **17**(2), 410–424 (2022). https://doi.org/10.1080/17445302.2020.1834753

Yan, C., Vescovini, R., Dozio, L.: A framework based on physics-informed neural networks and extreme learning for the analysis of composite structures. Comput. Struct. **265**, 106761 (2022). https://doi.org/10.1016/j.compstruc.2022.106761

Yeon, S.M., et al.: Development and verification of modeling practice for numerical estimation of wind loads on offshore floating structures. Int. J. Naval Archit. Ocean Eng. **14**, 100434 (2022). https://doi.org/10.1016/j.ijnaoe.2021.100434

Yeter, B., Garbatov, Y., Guedes Soares, C.: Uncertainty analysis of soil-pile interactions of monopile offshore wind turbine support structures. Appl. Ocean Res. **82**, 74–88 (2019). https://doi.org/10.1016/j.apor.2018.10.014

Yeter, B., Garbatov, Y., Guedes Soares, C.: Life-extension classification of offshore wind assets using unsupervised machine learning. Reliab. Eng. Syst. Saf. **219**, 108229 (2022). https://doi.org/10.1016/j.ress.2021.108229

Yeter, B., Garbatov, Y., Guedes Soares, C.: Review on artificial intelligence-aided life extension assessment of offshore wind support structures. J. Mar. Sci. Appl. **21**, 26–54 (2022). https://doi.org/10.1007/s11804-022-00298-3

Yi, M.S., Hyun, C.M., Paik, J.K.: Three-dimensional thermo-elastic-plastic finite element method modeling for predicting weld-induced residual stresses and distortions in steel stiffened-plate structures. World J. Eng. Technol. **6**(1), 176–200 (2018). https://doi.org/10.4236/wjet.2018.61010

Yu, Z., Amdahl, J., Rypestøl, M., Cheng, Z.: Numerical modeling and dynamic response analysis of a 10 MW semi-submersible floating offshore wind turbine subjected to ship collision loads. Renew. Energy **184**, 677–699 (2022). https://doi.org/10.1016/j.renene.2021.12.002

Yuan, L., Liu, Z., Chen, N.: On the buckling of mechanically lined pipes under combined tension and bending. Ocean Eng. **262**, 111991 (2022). https://doi.org/10.1016/j.oceaneng.2022.111991

Zavvar, E., Abdelwahab, H.S., Uzunoglu, E., Chen, B.Q., Guedes Soares, C.: Numerical study of the wave induced motions and loads on the CENTEC-TLP floating wind turbine. In: Guedes Soares, C. (ed.), Trends in Renewable Energies Offshore; Proceedings of the 5th International Conference on Renewable Energies Offshore (RENEW 2022), Lisbon, Portugal, 8-10 November 2022. London: CRC Press, pp. 567–573 (2022). https://doi.org/10.1201/9781003360773-65

Zavvar, E., Henneberg, J., Guedes Soares, C.: Stress concentration factors in FRP-reinforced tubular DKT joints under axial loads. Mar. Struct. **90**, 103429 (2023). https://doi.org/10.1016/j.marstruc.2023.103429

Zhai, Y., Zhao, H., Li, X., Feng, M., Zhou, Y.: Effects of aquaculture cage and netting on dynamic responses of novel 10 MW barge-type floating offshore wind turbine. Ocean Eng. **295**, 116896 (2024). https://doi.org/10.1016/j.oceaneng.2024.116896

Zhang, J., Shi, X.H., Guedes Soares, C., Liu, J.: Ultimate strength of stiffened panels with a crack and pits under uni-axial longitudinal compression. Ships Offshore Struct. **17**(2), 319–338 (2022). https://doi.org/10.1080/17445302.2020.1827805

Zhang, Q., Yang, H., Wu, S., Cheng, W., Liang, Y., Huang, Y.: A study on the ultimate strength and failure mode of stiffened panels. J. Marine Sci. Eng. **11**(6), 1214 (2023). https://doi.org/10.3390/jmse11061214

Zhang, T., Zaghi, A.E.: Estimation of the residual bearing strength of corroded bridge deck girders using 3D scan data. Thin-Walled Struct. **188**, 110798 (2023). https://doi.org/10.1016/j.tws.2023.110798

Zhang, X., Wang, Y., Chemori, A.: Structural reliability based energy-efficient arctic position mooring control of moored offshore structures under ice loads. Ocean Eng. **268**, 113435 (2023). https://doi.org/10.1016/j.oceaneng.2022.113435

Zhang, Y., Hu, Z.: An aero-hydro coupled method for investigating ship collision against a floating offshore wind turbine. Mar. Struct. **83**, 103177 (2022). https://doi.org/10.1016/j.marstruc.2022.103177

Zhang, Y., et al.: Seismic performance evaluation and numerical analysis of CFDST long columns with local corrosion under eccentric compression. Ocean Eng. **306**, 118006 (2024). https://doi.org/10.1016/j.oceaneng.2024.118006

Zhao, N., Chen, B.Q., Zhou, Y.Q., Li, Z.J., Hu, J.J., Guedes Soares, C.: Experimental and numerical investigation on the ultimate strength of a ship hull girder model with deck openings. Mar. Struct. **83**, 103175 (2022). https://doi.org/10.1016/j.marstruc.2022.103175

Zhong, Q., Wang, D.Y.: Ultimate strength behavior of laser-welded web-core sandwich plates under in-plane compression. Ocean Eng. **238**, 109685 (2021). https://doi.org/10.1016/j.oceaneng.2021.109685

Zhong, Q., Wang, D.Y.: Dynamic ultimate strength characteristics of stiffened plates subjected to the in-plane impact load and lateral pressure. In: Proceedings of the ASME 2021 40th International Conference on Ocean, Offshore and Arctic Engineering (OMAE 2021), Virtual, Online, 21-30 June 2021. (OMAE2021-62663) (2021b). https://doi.org/10.1115/OMAE2021-62663

Zhong, Q., Wang, D.: Numerical investigation on ultimate strength of sandwich box girders under vertical bending and torsion. Ocean Eng. **254**, 111338 (2022). https://doi.org/10.1016/j.oceaneng.2022.111338

Zhou, J., Pei, Z., Wu, W., Ding, J.: Progressive collapse behavior of ship structures considering fluid-structure interaction effect. In: Proceedings of the ASME 2023 42nd International Conference on Ocean, Offshore and Arctic Engineering (OMAE 2023), Melbourne, Australia, 11-16 June 2023. (OMAE2023-102755) (2023). https://doi.org/10.1115/OMAE2023-102755

Zhu, Y., Zhang, Y., Du, F.: Ultimate strength of hull structural stiffened plate with grooving corrosion damage under uniaxial compression. J. Ship Res. **65**(4), 309–319 (2021). https://doi.org/10.5957/JOSR.03200014

Zhuang, X., Guo, H., Alajlan, N., Zhu, H., Rabczuk, T.: Deep autoencoder based energy method for the bending, vibration, and buckling analysis of Kirchhoff plates with transfer learning. Eur. J. Mech. A/Solids **87**, 104225 (2021). https://doi.org/10.1016/j.euromechsol.2021.104225

Økland, O.D., Lian, G., Vestbøstad, T.: Experimental investigation of slamming loads on vertical column exposed to short and long crested waves. In: Proceedings of the ASME 2022 41st

International Conference on Ocean, Offshore and Arctic Engineering (OMAE 2022), Hamburg, Germany, 5–10 June 2022. (OMAE2022–79076) (2022). https://doi.org/10.1115/OMAE2022-79076

Ørsted Media Relations: Incident at Anholt offshore wind farm (2022). https://orsted.com/en/media/newsroom/news/2022/04/incident-at-anholt-offshore-wind-farm. Accessed 11 July 2025

Open Access This chapter is licensed under the terms of the Creative Commons Attribution-NonCommercial-NoDerivatives 4.0 International License (http://creativecommons.org/licenses/by-nc-nd/4.0/), which permits any noncommercial use, sharing, distribution and reproduction in any medium or format, as long as you give appropriate credit to the original author(s) and the source, provide a link to the Creative Commons license and indicate if you modified the licensed material. You do not have permission under this license to share adapted material derived from this chapter or parts of it.

The images or other third party material in this chapter are included in the chapter's Creative Commons license, unless indicated otherwise in a credit line to the material. If material is not included in the chapter's Creative Commons license and your intended use is not permitted by statutory regulation or exceeds the permitted use, you will need to obtain permission directly from the copyright holder.

Committee III.2: Fatigue and Fracture

H. Remes[1(✉)], G. An[2], M. Deul[3], P. Dong[4], P. Haselbach[5], S. Heggelund[6], P. Jurisic[7], A. Kahl[8], T. Kawabata[9], J. Liu[10], F. Prasetyo[11], M. Sicchiero[12], M. Soliman[13], M. Vicente del Amo[14], B. Yeter[15], J. Maljaars[16], S. Nakayama[17], A. Niraula[18], M. Ozdemir[19], J. Rodenburg[16], J. Roh[20], X. Song[21], F. Yanagimoto[17], and S. Wu[22]

[1] Aalto University, Espoo, Finland
  `heikki.remes@aalto.fi`
[2] Seoul National University, Seoul, Republic of Korea
[3] Delft, The Netherlands
[4] University of Michigan, Ann Arbor, USA
[5] Technical University of Denmark (DTU), Lyngby, Denmark
[6] DNV, Oslo, Norway
[7] Split, Croatia
[8] Hamburg, Germany
[9] University of Tokyo, Tokyo, Japan
[10] Wuxi, China
[11] Jakarta, Indonesia
[12] Trieste, Italy
[13] Oklahoma State University, Stillwater, USA
[14] Southampton, UK
[15] University of Strathclyde, Glasgow, UK
[16] Aagtekerke, Netherlands
[17] Kita, Japan
[18] Espoo, Finland
[19] Cambridge, UK
[20] Seoul, Korea
[21] Guanzhong, China
[22] Thousand Oaks, USA

Committee Mandate. Concern for crack initiation and growth under cyclic loading and unstable crack propagation and tearing in the ship and offshore structures. Due attention shall be paid to the suitability and uncertainty of physical models and testing. Consideration is to be given to the practical application, statistical description, and fracture control methods in design, fabrication, and service.

Keywords: Fatigue · Fracture · Damage modeling · Crack growth · Steel · Composites · Fabrication · Verification · Reliability · Standards · Rules

---

External Contributors—J. Maljaars, S. Nakayama, A. Niraula, M. Ozdemir, J. Rodenburg, J. Roh, X. Song, F. Yanagimoto, S. Wu.

© The Author(s) 2026
W. Wu and J. Ding (Eds.): ISSC 2025, LNME, pp. 578–727, 2026.
https://doi.org/10.1007/978-981-95-2668-0_6

1 Introduction

Understanding of fatigue and fracture in engineering structures has significantly evolved over the past 150 years. Despite extensive prior advancements, it has remained a critical developing discipline in nearly all engineering sectors, particularly maritime and offshore industries. One specific example is that progress in computational mechanics and advancements in computing power have enabled more sophisticated analysis methods. Still, continuous improvement is essential. The global push for sustainability and carbon neutrality presents new demands on fatigue and fracture analyses. Furthermore, emerging applications, such as floating offshore wind turbines and tanks for alternative fuels, require advanced modeling under diverse conditions. This emphasizes the necessity for specialized fatigue and fracture assessment methods. Also, new lightweight and sustainable materials such as carbon-neutral high-strength steels and composites are deemed necessary to reduce greenhouse gas emissions and achieve sustainability objectives.

To address emerging development needs, the ISSC2025 fatigue and fracture committee has concentrated on recent advancements in load scenarios, fatigue and fracture fundamentals, and modeling techniques for assessing fatigue life, fracture, and reliability. Emphasizing sustainability challenges, climate change, and new materials, the report reviews the state-of-the-art strategies and methods to enhance assessment protocols. To support this review, the committee has conducted benchmark studies on fatigue and fracture assessments for selected relevant case structures, including a floating offshore wind turbine structure and an LNG fuel tank. Targeted toward structural analysts, experts, and researchers in the maritime industry, the report aims to provide insights into advancements for designing new structures, assessing existing ones, and exploring innovative solutions. The report also presents forward-looking strategies and highlights future research needs to improve the resilience and sustainability of ships and offshore structures.

The report structure is aligned with the committee's objectives. Section 2 is dedicated to load modeling, addressing combined loading scenarios typical of ships and offshore structures while also considering the impact of climate change. Section 3 focuses on the fatigue and fracture fundamentals of modern materials, emphasizing the influence of innovative manufacturing techniques like additive manufacturing and the importance of material testing and characterization methods. Section 4 discusses recent developments and trends in computational fracture and damage modeling, particularly in crack growth and fatigue life prediction, with special emphasis on model validations and structural modeling approaches. Section 5 emphasizes reliability-based approaches for fatigue and fracture analysis, showing increasing importance for fatigue design and structural integrity assessment. Section 6 reviews recent developments in rules and standards, highlighting ongoing standardization efforts and providing practical insights obtained from the benchmark studies of floating wind turbine structures (Appendix I) and the LNG fuel tank of an ultra-large container carrier (Appendix II).

2 Loads for Fatigue Assessment

The fatigue assessment of the ships and offshore structures is conducted by applying analysis based on S-N curves or fracture mechanics. The fatigue life of these structures largely depends on internal and external cyclic loads. The external loads are mainly caused by environmental factors, such as waves, wind, and current, and in some cases, including the extreme conditions on them. Simplified environmental loads are typically assumed for fatigue assessment.

This chapter addresses recent issues of fatigue loading from the environment (wave, wind, current, and ice), as well as the statistical load modeling based on the history of environmental data and in-service measurements from different structures. This chapter mostly reports the fatigue load on offshore structures, such as offshore wind and subsea structures, based on journal articles published between 2021 and 2024. Additionally, the chapter discusses fatigue loads on ship structures, e.g., referencing the new wave statistics data revision in IACS (2022). Operational issues and aspects of climate change are also discussed.

2.1 Fatigue Loads

2.1.1 Waves

To predict fatigue damage in marine structures, accurate wave data is of paramount importance. IACS Recommendation 34, which was recently updated, hereafter Rec. 34 (IACS, 2022), provides wave statistics intended for the design of sea-going ships above 90 m, including the effect of bad weather avoidance. It is based on North Atlantic trade, which represents the most severe conditions in which ships tend to operate. The standard includes advice on sea states, wave spectrum, spreading, heading distribution, and vessel speed. The sea state is given by a scatter diagram, providing a joint probability distribution for the significant wave heights (H_s) and the mean up-crossing period (T_z). The latest, i.e., the second revision of Rec. 34, incorporates significant updates, utilizing hindcast data validated by buoys and altimeters for more realistic sea state information compared to the previous version, which relied on shipboard visual observations. In addition, AIS position data are used to determine actual shipping routes for the North-Atlantic scatter diagram. Technical justification for these changes is given in (Austefjord et al., 2023). The most significant changes are found in the scatter diagram, where the wave spectrum is slightly adjusted. The update from revision 1 to revision 2 is expected to change design loads such as pressures, motions, accelerations, hull girder, and extreme and fatigue loads. More detailed information about Rec. 34 is given in Chapter 6.

Wave load analysis often interacts with wind load analysis. For instance, tall offshore wind turbine (OWT) structures, consisting of the foundation and wind turbine tower, are subjected to stochastic wind and wave loads, which are exciting rigid-body and flexural responses. Zhang and Sclavounos (2021) developed an analytical non-linear wave load model for monopiles supporting multi-megawatt OWTs in deep and finite water depth. The non-linear wave model is developed on fluid impulse theory (FIT) and accounts consistently for all second-order and quadratic effects in long waves compared to the cylinder diameter. Explicit expressions for the sum- and difference-frequency Quadratic

Force Transfer Functions (QTFs) in deep and finite water depth in unidirectional and directional seas were derived. The wave load model is found to be very well in agreement with experimental measurements. Nonlinear load components are found to be significant relative to the linear wave load. Mei and Xiong (2021) studied the effect of second-order wave loads on a 15 MW floating offshore wind turbine. A fully coupled simulation of wind and wave loads was performed. The sum- and difference-frequency QTFs were calculated for a wide range of wave frequencies. The fatigue damage over 25 years at the tower base is overestimated by around 17% without the contribution of second-order wave excitations. Therefore, non-linear hydrodynamics must be considered in the design of a floating offshore wind turbine (FOWT) tower to reduce the cost. Saenz-Aguirre et al. (2022) presented a new method for predicting energy generation and fatigue loads. The method combines cluster analysis for wind and wave data, the statistical distribution of the environmental data, and their effect in a highly detailed aeroelastic wind turbine model. The results showed that the waves had a limited effect on the operation and power production of a FOWT.

Recent studies also include wave load analysis for mooring systems, which are important structural components of very long floating bridges. The mooring systems effectively limit the transverse motions of the bridge under environmental loads and add viscous hydrodynamic damping to the entire bridge system. For bridges with large spans, complex topology can create inhomogeneous wave conditions (wave height, wave period, and wave direction varying along the length of the bridge). Dai et al. (2021a) investigated a 4.6 km long straight and side-anchored floating pontoon bridge. It was found that the wave load inhomogeneity had a considerable effect on the standard deviations of mooring line tension, which characterize the dynamic components of the responses. The correlation of sea states along the length of the floating bridge also affects the fatigue damage in the mooring lines. The applicability of different spectral methods (i.e., Dirlik's method for the wide band Gaussian process and the narrow band based on the Rayleigh model) was also investigated by comparing the results with the conventional Rainflow cycle counting algorithm. Both studied spectra provided reasonable estimates for the studied load histories, with errors within 6% compared to rain flow counting. Calculation of mooring line response and fatigue damage is typically done using a large number of structural time-domain simulations. As an alternative, Xie et al. (2023c) presented a deep-learning approach for these calculations. The proposed method was based on a GRU neural network. It was found that the technique could predict fatigue damage with high accuracy and efficiency, even when the wave direction was included as a variable. A sensitivity analysis revealed that the wave direction had a significant influence on the structural fatigue damage.

2.1.2 Wave-Induced Vibrations

Wave-induced vibrations can have a significant influence on the fatigue performance of marine structures such as ship hulls, rudder blades, and subsea wellhead connectors. Osawa et al. (2021) performed fatigue tests of the longitudinal deck of a 6500TEUs Container ship. The occurrence stress of whipping vibration and slamming impact load are determined and characterized using waveform data histories of ships in real operation conditions. Both stresses that result from whipping vibration and slamming impact load were simplified to be used in the high-frequency effect (HFE) fatigue test. The simplified whipping superimposed loading can be applied to the HFE fatigue testing, assuming linearity between load and strain at high-speed rotation frequency or intermittent hammering of a vibrator aid. Hageman et al. (2022a) reviewed the whipping factor (the ratio between the total and wave-induced fatigue damage) for a Coast Guard cutter. It was found that the whipping factor was most influenced by bow flare and its wave steepness. The whipping factor was assessed for various operation areas, revealing its contribution to fatigue damage in the ship's structure across different ocean regions. The findings indicate that in the Bering Sea, whipping accounts for 21% of the long-term fatigue damage. In the Pacific Ocean and Caribbean Sea, whipping contributes 5%, while in the North Atlantic Ocean, it accounts for 10% of the long-term fatigue damage.

In recent years, advancements have also been made in fatigue analysis for oil and gas infrastructure, addressing the complex dynamic environments in which these systems operate. Zhang et al. (2022b) presented a numerical model for the fatigue analysis of oil offloading lines (OOLs) in the Calm Buoy oil offloading line system of a floating production, storage, and offloading (FPSO). Specifically, the effect of the vortex-induced vibrations due to current was investigated. The results show that vortex-induced vibrations cause significant fatigue damage in the middle part of OOL by increasing it by 5–10 times its fatigue damage. Li et al. (2024) proposed an assessment method to determine the fatigue life of subsea wellhead connectors by considering riser wave-induced vibration. The non-linear dynamic load was determined using a global coupling analysis model of the drilling system. The time history of the bending moment of the wellhead connectors was found, and the fatigue loading spectrum was established using the Rainflow method. The most accurate fatigue life predictions were obtained by the Dowling and Huffman-Beckman theories.

Recently, the fatigue failure of ship rudders owing to vortex-induced vibration has increased as commercial ships become faster and larger. Jang et al. (2023) presented a fatigue damage prediction method that can be used in the design stage. The fatigue damage was estimated by a fluid-structure interaction (FSI) model implementing orthonormal mode shapes to the hybrid coupling method. It was verified that the new calculation method agreed well with the experimental results. The potential fatigue damage was analyzed by comparing the stress distribution for a flow velocity range of 18.5 m/s to 20 m/s with the S-N curves provided by the classification society.

2.1.3 Wind

Fatigue loading due to wind is predominantly relevant for OWTs but could also become more relevant for ships with the introduction of wind-assisted ship propulsion (WASP). During the last committee period (between 2019 and 2022), a lot of work was performed regarding OWTs. Wind loads have naturally been an important part of this. Fatigue is a common and critical source among typical failure modes, and fatigue damage due to wind loads is significant. Therefore, many papers have been found on this topic, including wind and wave interaction.

Recent research has studied critical factors influencing fatigue damage analysis of FOWTs. The study carried out by Hegseth et al. (2021) confirmed that fatigue damage was sensitive to wind turbulence, wind direction, and especially wind-wave misalignment for the Spar-type substructure; however, it is more so for the tower than the platform. Furthermore, Lizarraga-Saenz et al. (2022) studied the importance of the correlation between wind speed and wave data in the calculated fatigue damage for FOWT. It was concluded that simplified metocean data without correlation between wind and wave measurements could be a reasonable approach for early fatigue lifetime load estimations.

The fatigue assessment of OWT, considering full-directional wind inflow under the coupling effects of non-Gaussian wind fields and waves, has been studied by Li et al. (2021a). The work results showed that the crack initiation life is greatly affected by the non-Gaussian wind field, while the crack propagation life is not that sensitive. The fatigue life of OWT under a full-directional wind inflow is approximately three times longer than that under a single-directional wind inflow. It is recommended to consider full-directional wind inflow in the non-Gaussian wind field for the OWT design It was also found that wind-induced loading is a significant source of uncertainty in assessing fatigue damage in floating offshore wind turbines, primarily due to spatial-temporal variability and local speed-up factors. This issue is further exacerbated by the inability of physics-based models to keep pace with the increasing demand for larger wind turbines, as highlighted by Veers et al. (2023).

One crucial advantage of OWT deployed further offshore is that they tend to have higher wind speed but less fluctuation in wind speed, leading to higher energy production efficiency. In this regard, Ramezani et al. (2023) referred to the Weibull distribution as the most suitable probability density function to describe wind speed, whereas the Gamma distribution was found to be appropriate for wave height and period. Natarajan (2022) investigated a procedure to determine long-term fatigue damage and remaining fatigue life from a combination of stochastic extrapolation and computationally fast synthesis. The stochastic extrapolation damage equivalent load due to the wind load effect is based on the three-parameter Weibull distribution. The results are validated using measured loads from a wind farm.

Fatigue analysis of monopile OWTs involves dynamic analyses, coupling aerodynamics, hydrodynamics, soil-structure interaction, and the control system for wind turbines. However, a fully integrated analysis is time-consuming. Katsikogiannis et al. (2021) investigated a procedure to reduce computational time. The site-specific environmental parameters were condensed by a lumping process. The lumping process condensed each scatter diagram associated with a wind speed class to a single sea-state load case. The lumping was done both in the time and frequency domain based on damage-equivalent contour lines. A significant reduction of computational effort (93%) was achieved compared to a full long-term assessment. The lumping method had an accuracy of 94–98%. Yeter and Garbatov (2022) discussed how the fatigue damage approaches for wide-band load processes, such as the Dirlik and Benasciutti-Tovo models, could be alternatives to computationally expensive time-domain fatigue damage analysis if the response amplitude operator of the support structure is well understood.

Bladed, OpenFast, and HAWT are the multi-physics simulation tools that have been widely used by academia and industry to perform fully-couple dynamic simulations of offshore wind turbines in the time domain. However, there is still significant room for improvement in terms of implementing these high-fidelity numerical tools into design optimization, as there is a trade-off between accuracy and computational effort for aero-hydro-servo-elastic simulations (Otter et al., 2022), which is an obstacle to the full-scatter fatigue analysis. In this regard, Lemmer et al. (2020) developed a low-fidelity multi-physical model of a multi-body system to represent complex interactions between aerodynamics, hydrodynamics, and structural dynamics in FOWT. The model was found to be quite promising as it could perform analysis in less than a minute for a 1-h load case with a 6% difference in fatigue load.

2.1.4 Current

In marine and offshore engineering, the influence of current forces can play an important role in fatigue assessment. Mullings and Stallard (2021) investigated the variation in the fatigue loading on a tidal turbine. The loading was due to tidal current. Calculations were based on an efficient blade element model. A frozen turbulence field was used as inflow. Fatigue assessment was done using damage equivalent loads (DELs), i.e., the single amplitude load that gives the same fatigue damage as a variable amplitude loading time series. Fatigue loads at different depths and locations (within a given tidal site) were investigated. The calculations were based on measured shear profile and turbulence. In addition, the effect of waves was investigated. For both locations, the measured and predicted loads with the flood tide were within 5%. For the ebb tide, there was more significant load variation between different locations, with the near bed turbine having greater variation between the predicted and measured profiles, consistent with the variation of shear. For the near-bed turbine, fluctuation intensity also contributed to the difference between measured and calculated values for the ebb tide. The presence of waves was found to increase the fatigue loads for the near-surface position significantly. For the near bed position, the effect of waves was small. The range of predicted and measured profiles was averaged to give a single defining profile, combined with the average fluctuation intensity for each position and location. Averaging the profiles tends to give lower fatigue loads than using the whole range of profiles, i.e., averaging profiles is a non-conservative approach.

Zheng and Chen (2022) did a time-domain fatigue assessment for preloaded blade root bolts in an FOWT. The current was included as a drag force calculated using the Morrisons equation. The current itself was considered as a steady flow. The variability in current load was caused by the relative velocity between the current and platform. The calculations focused on bolt strength, and the paper did not contain many results showing the effect of current. Zhang et al. (2022b) investigated fatigue damage in oil offloading lines (OOLs) in the FPSO catenary anchor leg mooring (CALM) buoy offloading system under wave and current loads in the West Africa Sea area. Numerical simulations and comparison with experimental results were carried out. The numerical model was used to study various parameters, including current. The current flow velocity is in the form of shear flow. The current loads were found using the Morison equation and the relative velocity between the OOL and the current flow. The lift force was calculated to predict the effect of vortex-induced vibrations. The fatigue damage along the OOL was predicted. It was found that fatigue damage due to the current mainly occurred near the CALM buoy.

The effect of vortex-induced vibration was concentrated primarily on the middle part of the OOL and could increase its fatigue damage by 5–10 times but had little influence on the fatigue damage of the parts near the CALM buoy.

Slagstad et al. (2023) presented a simplified analytical model for estimating the ropes' axial forces for the fish net fatigue life prediction. Assumed displacement shape under combined wave and current loading given the high natural frequencies of pre-tensioned ropes compared to wave frequencies. The fatigue life was also numerically performed by combining wave periods and heights for different current velocities. Results show that this simplified method is extremely efficient in computation time compared to numerical simulations. The fatigue life of both the rope and the connection point is strongly affected by loading-induced current.

2.1.5 Operation

For ships navigating in ice-covered waters, fatigue damage due to ice loads can also be an important issue from the fatigue perspective. The ice load is generated from a subset of operational loads during ice-covered water navigation. Until now, studies of ice loads have often relied on field measurements. Recently, several numerical studies have also been performed. Zhao et al. (2021c) provided a simulation-based procedure for probabilistic fatigue damage assessment of ships in level ice fields. Jeon and Kim (2022) proposed a methodology for calculating the fatigue damage of the icebreaker ARAON during level ice collisions, factoring in ice load and ship speed. Gaidai et al. (2023b) used a route-specific ice thickness distribution to estimate ice loads on an oil tanker's bow, noting a reduced probability of encountering thick ice due to route optimization. Shin et al. (2023) studied ice-induced fatigue loads on semi-submersible platforms, comparing finite element analysis results with simplified methods. Han et al. (2024) performed numerical fatigue calculations for ships in broken sea ice, examining variables like speed and load magnitude in relation to ice thickness and concentration.

Braun et al. (2022b) highlighted the need for better consideration of ice loads in fatigue strength assessments as offshore wind turbines become more adopted in harsh northern environments. Considering the ice-induced vibrations, they developed a combined load spectrum for wind, wave, and ice actions. Their follow-up study (Braun et al., 2022c) applied this spectrum to fatigue test butt-welded joints, and the fatigue test results were compared against the fatigue damage sums from other load spectrums. Hammer et al. (2023) experimentally investigated ice-induced vibrations on offshore wind turbines on monopile foundations, considering 50-year Southern Baltic Sea conditions. Cai et al. (2024) reviewed the dynamic response of marine structures under repeated mass impacts, including the ice-structure interactions. The ice-structure interactions were considered as deformable bodies. Also, the extreme mooring loads and accumulated fatigue damage corresponding to different ice management schemes have been investigated based on the full-scale tests of station-keeping trials (SKT) in drifting ice (Sinsabvarodom et al., 2021).

Recent studies also focused on the fatigue life of marine propulsion shafts. An original parametric method has been introduced for the assessment of the fatigue life of marine shaft lines, which introduces a relevant load (the loads are rotary bending, torque loading, and propeller thrust) modeling around a limited number of parameters specific to marine shaft lines while depicting the loading complexity (multiaxiality, mean stress effect, non-proportionality of the loading path) (Guellec et al., 2023). Purcell et al. (2021) calculated the short-term fatigue damage on a propulsion shaft during ice impacts. Different sources of uncertainties were investigated. The uncertainties related to the ice load calculation method and the dynamic analysis model significantly affected the fatigue life.

In recent advancements in structural safety assessments, researchers have also focused on evaluating the fatigue strength and remaining life of critical components subjected to cyclical loading conditions. During multiple submergences, the evaluation of the pressure hull's fatigue strength and remaining life has become crucial for the safety of deep submersibles. The fatigue life of a 7000 m class submersible pressure hull subject to 10–70 MPa hydrostatic pressure during submergences has been evaluated with the equivalent structural stress method considering the heat-affected zone (HAZ) of the weld seam (Yu et al., 2022a). Additionally, the high demand for transshipment operations worldwide, which often involve mooring buoys, has prompted the development of a method for fatigue analysis of these structures. This method uses computational modeling and simulation via the FEM, focusing on the points of greatest cyclical loading of the pad eye. The main tensions in the buoy are obtained from the application of calculated forces for traction on a mooring cable at the bit of the buoy (Cabral et al., 2023).

2.2 Extreme Loads

Generally, fatigue load is considered the environmental condition experienced by the structure. While the structure poses extreme environmental phenomena, such as typhoons or storms, the extreme structural response should be considered in investigating the safety aspect. Extreme load is mostly modeled by extrapolating short-term extreme loading into long-term distribution for fatigue assessment.

Qin et al. (2023) investigated the effect of typhoon conditions on a 10 MW large-scale monopole offshore wind turbine that developed in the southern China area. The typhoon conditions are modeled as a coupling of the typhoon wind field and the typhoon wave field. Qin et al. (2023) proposed to generate the statistical characteristics of response value and short-term extreme response from the history of typhoon conditions in the chosen area. Furthermore, the extreme load is estimated using extrapolation, and the global maxima method is specifically used. The extreme stress of extreme typhoon environmental conditions is used to determine fatigue damage based on a combination of the rain flow counting for process time series of extreme stress in the low-cycle fatigue loading regime, S-N curve for the er of cycles to failure, and Miner's rule cumulative fatigue damage.

Song et al. (2023) did multi-parameter full probabilistic modeling of long-term joint wind-wave actions using multi-source data to analyze fatigue in floating offshore wind turbines. A full probabilistic model of the multiple environmental variables has been established using multi-source data and a physically based approach. This full probabilistic model can be applied in the fatigue analysis of offshore structures to consider the long-term combined wind and wave excitations reasonably. The long-term metocean data at the South China Sea region are affected by typhoons, which should be considered in the joint distribution modeling of environmental variables. The proposed method using the multi-source data and physically based approach can effectively separate the typhoon data from the normal wind data.

The work from Sinsabvarodom et al. (2021) proposed the prediction method of short-term extreme mooring load for fatigue damage assessment of station keeping in ice. The load is estimated from the full-scale measurement data during the station-keeping trials in drifting ice. Ship-ice interaction frameworks are used to determine the ice-induced loads on a ship's bow or hull due to drifting sea ice by application of the resistance method. The extreme load is derived from the time history of ice-induced load by employing the peaks over the threshold and the block maxima method as well as the average conditional exceedance rate. Thus, the fatigue damage calculation for the mooring line of station keeping is calculated using the Palmgren Miner rules by applying the short-term extreme mooring load estimated using the Gumbel distribution.

2.3 Statistical Modeling of Loads

The fatigue assessment analysis is proceeded by assuming the set of fatigue load blocks representing the experimental load cycle, actual loading condition, and the environmental load. In general, most of them are simplified by using statistical models. The statistical modeling of fatigue load could be defined using short-term and long-term load modeling.

In Subchapter 2.1, fatigue loading was applied to the ship, and FOWT was reported due to the wave, wind, as well as current and ice load operation. Mao et al. (2024) reported the study for the applicability of the spectral fatigue assessment framework of a FOWT system in the early stage of the design process. The spectral fatigue model is based on the stress-transferred functions calculated using FAST simulation. The structural stress response is estimated from FAST simulation from the extracted environmental load, and it is assumed to be a highly wide band non-Gaussian process with a white noise approach. To determine the fatigue life of a wind turbine yaw bearing with different support foundations, Xu et al. (2021) developed a time-domain method to estimate the fatigue damage. Using three wind velocities, a time-domain method is applied to predict the wind and hydrodynamic loads on the yaw bearings. A Gumbel distribution, Rainflow-counting algorithm, linear cumulative damage model, and S–N curve theory generate the lifetime damage equivalent loads.

Zhao et al. (2021c) presented a novel procedure for fatigue assessment for ships navigating in ice-covered waters. Simulation-based procedures are developed for a probabilistic fatigue damage assessment of ships moving through a level ice field. The simulated long-duration ice load record is proposed for the fatigue analysis. The extreme value statistics for the load peak of ice load are estimated based on the average conditional exceedance rate (ACER) method and are converted to the structure stresses by utilizing the beam theory. The probabilistic model of stress distribution, the Weibull distribution, the lognormal distribution, and the Gumbel distribution are applied to derive the stress distribution due to ice load interaction. The proposed method was applied to the transit of the chosen icebreaker in first-year sea ice. In another study Zhao et al. (2021b) conducted a study to estimate fatigue life based on fatigue crack propagation of the polar vessels. The stress range distribution due to ice load at the level ice was determined based on the maximum likelihood estimation method and fitted with the Weibull distribution.

Purcell et al. (2021) investigated the sources of uncertainty to assess fatigue damage of propulsion under ice load. The long-term stress distribution due to ice load response was modeled using a two-parameter Weibull distribution. The long-term fatigue damage was determined using a combination of Palmgren Miner fatigue damage and the Rainflow counting method. The short-term fatigue model is determined by considering several sources of uncertainties. The sources of uncertainties are lumped mass model parameters, discretized temporal simulation, the inverse calculation of ice load, and the fatigue damage method, which are qualified and are expressed as normal, log-normal, and Weibull distribution.

Lone et al. (2022) introduce a probabilistic fatigue model concept for analysis of the design and life extension of the mooring chain. The probabilistic fatigue model is calculated considering the short-term and long-term environmental load. The long-term environmental load is assumed to represent a stationary process and is independently distributed from year to year. Furthermore, the fatigue load is assumed by considering annual distribution to account for the seasonal variation of environmental load. Thus, the annual fatigue load is modeled based on the lognormal distribution.

2.4 Operation and Climate Change Induced Environmental Loads

2.4.1 Ship Routing

Wave height and mean wave period are two important sea state characteristics in ship routing. Hildeman et al. (2021) developed a non-stationary Gaussian random field model for significant wave heights. The model parameters are estimated sea state data from the ERA-Interim dataset at the North Atlantic Ocean using a maximum likelihood approach. The fitted model is used to compute wave height exceedance probabilities and the distribution of accumulated fatigue damage for ships traveling a popular shipping route. The model results agree well with the data, indicating that it could be used for route optimization in naval logistics.

A joint spatial model of wave height and mean wave period on the North Atlantic Ocean has been proposed by Hildeman et al. (2022). The model describes the distribution of the logarithm of the two quantities as a bivariate Gaussian random field, modeled as a solution to a system of coupled fractional stochastic partial differential equations. The bivariate random field is non-stationary and allows for arbitrary and different smoothness for the two marginal fields. The model's parameters are estimated from data using a stepwise maximum likelihood method. The fitted model is used to derive the distribution of accumulated fatigue damage for a ship sailing a transatlantic route.

A new database of wave scatter diagrams describing the joint probability of wave height and wave period worldwide has been developed, which can provide a practical and representative wave input for the design of structures subjected to waves. Accounting for the weather routing effect in estimating wave conditions, an additional database of weather encountered by ships has been built that would result in more realistic modeling of long-term ship response by comparing with wave scatter diagrams of Rec. 34 rev.1 (de Hauteclocque et al., 2023).

2.4.2 Climate Change

Tamimi et al. (2022) showed the simulation-based framework for predicting the probability of ship structure failure due to changes in climate by considering wave parameters, storm characteristics, and ship routes. The work considered 8 different routes across the North Atlantic Ocean. The proposed framework was for quantifying the long-term impact of climate change on the vertical bending moment and subsequent fatigue crack propagation in ship hulls. Firstly, Global Climate Models (GCMs) are used to quantify the long-term effects of climate change on the sea conditions (wave and wind as well as information corresponding to its speed and direction) and the resulting ship loading time histories. Thus, a probabilistic fatigue crack propagation approach is developed to account for uncertainties associated with material properties and loading conditions, specifically those affected by climate change. The proposed approach is applied to a tanker ship operating within predefined routes in the North Atlantic Ocean.

Zou et al. (2023b) conducted work to explore the projection of climate change's impact on the fatigue damage of offshore floating photovoltaic structures. Based on the prediction of wave and wind parameters in the North Sea, the annual fatigue damage due to climate change and the high emission of greenhouse gases in the area was evaluated. They simulated the projected specific climate scenario, sea state, on 2011–2020, 2051–2060, and 2091–2100, and showed that the significant wave height declined by 15% maximum in the North Sea from 2011–2100. The decreasing significant wave height affected annual fatigue damage by 19% - 28%.

2.4.3 Operation

Dong et al. (2022b) assessed the impact of start-up and shut-down cycles on the fatigue life of single-sided girth welds at the offshore pipelines. A local notch strain approach was employed for the fatigue life calculations. It was then shown that the number of start-up and shut-down cycles be limited to 150 if the maximum allowed fatigue damage index is 0.1, which is the value usually adopted for pipeline fatigue assessments.

Low cycle fatigue (LCF) cracks may be seen in ship structural details in the early stages, even if the calculated fatigue life based on the dynamic wave stresses is sufficiently long. In this case, the static still water stresses due to the loading/unloading cycles are to be combined with the dynamic stress ranges due to the wave loads for reliable fatigue life estimation. Dong et al. (2021a) carried out strain-based fatigue reliability assessment of welded joints for ship structures. A probabilistic distribution for the low cycle fatigue stress range was developed considering the uncertainties in the static and dynamic stresses in the LCF assessment.

2.5 In-Service Measurement for Fatigue Loads

Gaidai et al. (2023a) used 2800TEUs Panamax Containership, equipped with onboard motion sensors, during her crossing at the trans-Atlantic voyage routes in actual and rarely bad weather. The motion sensor is to handle the complexity of high-dimensional dynamic systems with nonlinear cross-correlation of all data components. The data components consist of ship roll angle and mid and rear longitudinal stress as a three-dimensional (3D) dynamical system. The onboard data was used to evaluate the multidimensional reliability and determine the lifetime of container vessels.

Hageman and Thompson (2022) used full-scale measurements and hindcast data as input to numerical analysis (virtual hull monitoring) of frigate-size vessels. The measurement was a local strain measurement at a longitudinal bracket on the portside near midship to filter high-frequency components for comparing estimated fatigue damage from linear wave load. Both estimated fatigue damage based on strain measurement and hind-cast wave data were compared. The agreement with measurement data is found to be good when using hindcast data for lower sea states. For higher sea states, the fatigue accumulation seemed to be underpredicted.

Takeuchi et al. (2023) proposed a long-term fatigue assessment method based on EWP (Equivalent Wave Probability). The model enables multi-position long-term fatigue assessment solely through hull monitoring data at a few reference sensors. The statistical model is calibrated by Bayesian inference from measured data. The validity of this method is verified by using the long-term (about four years) multi (12)-position hull monitoring data of an 8,600TEU container ship. The estimated fatigue damage agreed with measurements, independent of the chosen data assimilation period.

2.6 Summary

The fatigue loads, which are mainly considered for the fatigue assessment, are a derivative of the external environmental loads. These fatigue loads for ships and floating offshore structures arise from similar sources, such as waves and wind, current, and operational conditions. This chapter reports the developed models that could represent the fatigue load under various external environmental conditions. Thus, the chapter mostly discusses floating offshore structures as objects based on collected reference articles. It seems that the fatigue load models vary depending on the floating structure type. The approach and methodology to model the fatigue load may differ for each target structure. There may be a need for harmonization of the fatigue load approach and methodology to ensure consistency across different types of structures experiencing the same external conditions. Developing this harmonization procedure is a crucial future research task. Moreover, future research should pay special attention to how different loads interact, including wave, wind, and current, as they play a crucial role in fatigue damage assessment of marine structures. Wave loads, particularly non-linear ones, can have a significant impact on fatigue damage, making accurate wave data essential for predicting fatigue damage. Wind loads also contribute to fatigue damage, especially in offshore wind turbines, where wind-wave misalignment can worsen fatigue damage. Similarly, current-induced forces can have a significant influence on fatigue damage accumulation, particularly for tidal turbines and oil offloading lines. Also, the influence of extreme load frequency distribution on fatigue damage and service life estimation for ships, floating offshore structures, and oil offloading lines is an important area for future research.

3 Fatigue and Fracture Fundamentals

Fatigue and fracture analysis of ships and offshore structures relies primarily on experimentally determined material properties. Therefore, test methods and interpretation of material properties are essential for the fatigue and fracture design of ships and offshore structures. In addition, numerical analysis has been used in design in recent years, and much attention has been given to modeling materials and quantifying fracture driving forces under the operating conditions of the structure. This chapter addresses recent issues and advances in the field of fatigue and fracture assessment, with particular emphasis on the fundamental areas of material property characterization and modeling, as well as assessment test methods. In addition to traditional topics, recent research trends in fatigue and fracture of structures, including cyclic material behavior, non-metallic materials, materials for decarbonized infrastructure, and advanced joining methods, are thoroughly reviewed. The review evaluation focuses on publications from the period 2021–2024. However, for areas not covered in previous ISSC reports, a more comprehensive review is conducted, and older key references are also included.

3.1 Resistance

3.1.1 Parent Materials

Common Structural Steels Due to the potential for structural weight reduction, there is still strong momentum for applying high-tensile steels (HSS) in the marine field. Several reports have been published on the application of high-tensile steels to the marine industry, including crack arresting and fatigue properties. Whereas the yield stress is shown to have no noticeable influence on the fatigue strength of welded joints (Qin et al., 2019) the introduction of HSS in the maritime industry increased fatigue cracks observed as the material allowed for smaller plate thicknesses and, thereby, higher applied stresses. The YP47 development and its CAT (Crack Arrest Temperature) evaluation report from Hyundai Steel was presented in (An et al., 2024), and a study on improving fatigue crack propagation properties in low-carbon steel through microstructural control was discussed in (Hyodo et al., 2023).

Titanium and Aluminum Alloys Titanium alloys are widely used as pressure shells of deep-diving submersibles, the excellent balance of strength and ductility. The fatigue life of a 7,000 m class submersible pressure hull titanium manufactured with alloy TC4 subject to 10–70 MPa hydrostatic pressure has been evaluated, and results show the fatigue life decreases sharply in the stress range of 10–20 MPa and inclines to converge when the pressure reaches 40 MPa (Yu et al., 2022a). The LCF behavior of Ti-6Al-4V-0.55Fe alloy has been investigated, and results indicate its LCF life is similar to Ti-6Al-4V alloy at high strain amplitudes ($\Delta\varepsilon_t/2 > 1.0\%$), while much higher at low strain amplitudes ($\Delta\varepsilon_t/2 < 1.0\%$), which could be attributed to the effect of Fe microalloying (Sun et al., 2021b). The LCF properties of Ti-6Al-3Nb-2Zr-1Mo and Ti-6Al-4V ELI titanium alloys developed for deep-diving submersibles have been studied. The results show that their LCF properties clearly depend on the strain amplitudes ($\Delta\varepsilon_t/2$). i.e., when $\Delta\varepsilon_t/2$ is greater than 0.7%, the fatigue life of Ti-6Al-3Nb-2Zr-1Mo is slightly longer than that of Ti-6Al-4V ELI, and when $\Delta\varepsilon_t/2$ is less than 0.7%, the former fatigue life becomes shorter than that of the latter (Wang et al., 2023b). The cyclic working state of diving, operating, and rising may induce a creep-fatigue problem in the pressure structures, which has become the research area of the deep-sea manned submersibles. Researchers have performed a structures-grade dwell fatigue test for the pressure hull and a variety of specimen-grade compressive dwell fatigue tests for Ti-6Al-4V ELI alloys, and proposed fatigue crack growth rate models considering the effects of dwell time and short cracks (Wang et al., 2024) (Wang et al., 2023b).

Although the carbon footprint in the manufacturing process of aluminum alloys (primary aluminum) is much higher than that of steel, aluminum alloys have been applied in marine engineering due to its remarkable mechanical properties, such as high strength over density ratio, excellent formability, and high recyclability. However, aluminum alloy welded structure has some prominent shortcomings: large welding deformation, low fatigue strength, reduced strength of HAZ compared to base metal (other than annealed), and worse fire resistance compared to steel. To tackle these challenges, the temperature field simulation program of a 5059 aluminum alloy thin plate welding structure is designed to predict deformation and residual stress. The simulation results are in good agreement with the test results (Xu et al., 2022). Also, the fatigue properties of 5086 aluminum alloy and its weld joints have been investigated. Test results show the difference in fatigue lives of base alloy and its weld joint. Mean stress has a detrimental effect on fatigue properties. The main factor determining the fatigue life of a welded joint is the delay in crack growth during intergranular fatigue crack propagation due to frequent deflection of crack path, which is absent in the base alloys (Jaisawal et al., 2023).

Nickel Alloys and High Manganese Steels Nickel alloys have become the most widely used cryogenic material for LNG tanks due to their high strength, good corrosion resistance, and good cryogenic toughness. It has been observed from the fatigue test that the S-N curve of 9% Ni steel butt joints with plasma arc welding confirms the IIW FAT 80 class, and its fatigue strength at cryogenic temperature is about 20% higher than that of high Mn steel (Kim et al., 2021). Fatigue life for 9% Ni steel butt welded joints has been predicted using the proposed fatigue life prediction method based on the Coffin-Manson approach, which considered the geometric effect, the difference of cyclic stress-strain properties between parent material, the HAZ, and weld metal, and the mean stress effect by welding residual stress and applied stress (Shiratsuchi and Osawa, 2021). Also, the fatigue crack growth (FCG) mechanism of Ni-based weld metal in a 9% Ni steel joint has been investigated at different temperatures (room temperature and 77 K) and driving force ΔK levels. It can be found that the FCG rate decreased with decreasing temperature, and FCG behavior transformed with increasing ΔK or temperature (Li et al., 2022a). Due to its good corrosion resistance, the nickel-based alloy has also been used in the rotating components of marine compressors. The corrosion fatigue (CF) behavior of nickel-based alloy A286 has been studied using a peridynamic mechanical-chemical coupled model. The damage evolution process during corrosion fatigue has been captured, and the precision of this method has been verified by CF lifetimes tests (Wang et al., 2023a). In the construction material field, Ni-bearing steels have mainly been used for above-ground storage tanks at LNG temperatures since the 1960s. The 9% Ni steel was developed by Brophy and Miller (1948) as a reflection of the Cleveland accident in 1944 (Kawabata and Hirose, 2017) and has been applied in many LNG tanks worldwide. Against a background of rising Ni prices, 7% nickel steel (Kubo et al., 2010) was developed in the 2010s and has become a new standard material for LNG tanks. On the other hand, the tanks in carrier ships are manufactured mainly by aluminum alloy. However, high manganese steel has recently been actively developed, and its properties, both in the parent metal and in the welded joint, are suitable for cryogenic use. As a result, high manganese steel has been applied to actual ship cargo tanks and above-ground tanks (Lee et al., 2023) in the LNG field.

Carbon Fiber Reinforced Thermoplastics Thermoplastic composite use is steadily increasing because of its potential for cost-effective, automated, and rapid manufacturing. These materials generally entail a high specific stiffness and strength, which are natural properties of fiber composite materials. Additionally, an increased fracture toughness and indefinite shelf-life are inherent to the thermoplastic matrix. Compared to thermosetting and elastomer polymers, carbon fiber-reinforced thermoplastic (CFRTP) is gaining popularity in many industrial sectors due to its recyclability, simplicity of processing, less likelihood of change in its properties, freedom in shaping, and shorter production time. Often, CFRTP is used as a structural material, where CFRTP is joined to metallic alloys. Fatigue crack growth resistance can be improved by silane coupling treatment, which utilizes the anchoring effect of the nanostructure on the aluminum penetrating the resin of CFRTP (Saito et al., 2021). Also, for other hybrid materials, applying a suitable resin layer and chemically treating the surfaces improves joint strength (Hu and Li, 2024). Even so, CFRTP requires care when used, as experiments on submerged specimens have shown increased mass due to water absorption and the growth of adhering algae and marine microorganisms. Consequently, it can reduce the ultimate tensile strength and fatigue strength, depending on the fiber layout configurations (Vizentin, 2022). A review of the failure mechanisms and changes in the mechanical performance of FRP-metal adhesive joints under various temperature and humidity conditions highlights the deterioration of joint performance. The internal stress, caused by differing thermal expansion coefficients of the adjacent materials, affects interface stability. Additionally, water molecules can infiltrate and reduce interfacial adhesion due to corrosion, leading to molecular chain breakage or flow under environmental influences (Guo et al., 2023). Also, the durability and resistance to impact damage for underwater composite structures as marine propellers are studied by Islam et al. (2022). Here, influencing factors such as laminate curvature, thickness, impact angle, inter-ply stacking sequence, constituent materials, water diffusion, and fluid-structure interaction, among others, are reviewed to improve marine propeller design and its impact behavior.

3.1.2 As Manufactured Materials

Additive Manufacturing There have been a large number of publications in metal additive manufacturing (AM) over the last few years, among which some of the representative work on major issues for fatigue critical application are given in (Solberg and Berto, 2019, Dong, 2020, Shrestha et al., 2021, Li et al., 2022b) for powder-bed AM processes. As far as fatigue is concerned, one critical issue is how to qualify and certify metal AM parts for safe-critical applications due to randomly distributed defects or discontinuities (see Fig. 1a) due to rapid heating and solidification associated with fusion-based AM processes (Solberg and Berto, 2019, Dong, 2020, Shrestha et al., 2021, Li et al., 2022b, Dong et al., 2023a, Wu et al., 2023, Wu and Dong, 2023). In this regard, some of the most recent studies have shown that rapid fracture propagation modeling techniques can effectively establish defects' interaction criteria by introducing some quality acceptance criteria for supporting NDE (Non-Destructive Evaluation) certification requirements. It has also been shown that the fatigue behavior of metallic AM parts is largely like that of welded components in terms of the S-N curve slope in the log-log plot. As a result, a structural strain (ΔEs) based master E-N curve method can be used to correlate a large amount of metal AM fatigue test data among various AM materials, as illustrated in Fig. 1b.

Due to the existing limitations in build volume associated with powder-bed AM processes, there has been increased attention on wire-arc additive manufacturing (WAAM) processes, which can take advantage of existing welding wire and well-accepted gas metal arc welding processes for large-volume AM applications, which are of particular interest for applications ship and offshore structures for part replacements, particularly as an effective means of on-site or on-board fabrication capabilities for structural maintenance. The most recent developments in WAAM processes and their qualification and certification issues can be found in some recent publications, e.g., (Svetlizky et al., 2021, Kladovasilakis et al., 2021, Sherman et al., 2023). Again, the effects of distributed defects on fatigue performance remain to be a major challenge for deploying WAAM components for fatigue-critical applications. In addition to distributed defects like those in powder-fusion processes (e.g., see Fig. 1a), rough surface conditions, without machining, are the additional issues that must be considered for some components for which some as-built surface conditions may not be avoidable (Leonetti et al., 2024).

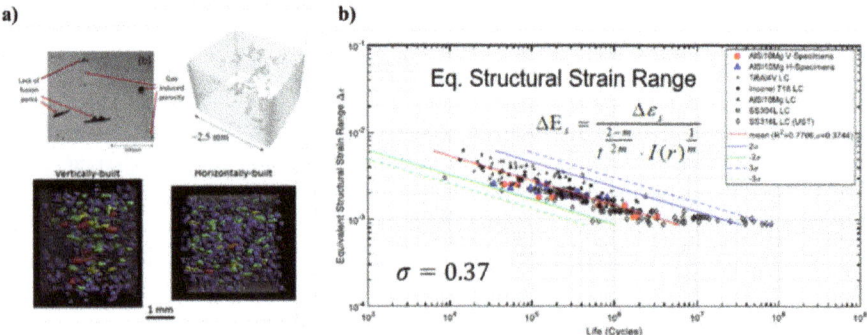

Fig. 1. Example of distributed defects (a) and AM material fatigue test results (b).

Friction Stir Welding Friction stir welding (FSW) as a solid-state welding technology has the advantages of low residual stresses and small deformations, which are primarily set for applications on aluminum and its alloys. The fracture and fatigue characteristics of aluminum alloy FSW joints and the importance of optimizing FSW process characteristics such as tool rotational speed, travel speed, and tool offset have been reviewed

(Kumar Yadav et al., 2023). The effect of microstructure and residual stresses on fracture and fatigue have been emphasized. The fatigue failure behavior of FSW dissimilar T-lap joints between AA7075 and AA5083 has been investigated, and results show that its fatigue life is significantly dependent on interface morphology. The geometry and orientation of kissing bonds played an important role in the fatigue failure behavior of the joints (Duong et al., 2021). The influence of Cu-reinforcement on fatigue properties of AA5086 friction stir weld joints was studied. Results show that compared to the simple weld joints without any reinforcement, the fatigue strength of Cu-reinforced weld joints was improved, and the location of failures in the samples tested under fatigue loading was shifted from the thermo-mechanical affected zone in the simple weld to the base material in the double-pass Cu-reinforced welded joints (Choudhary and Gaur, 2023). The FSW technique also provides a reliable welding method for high-quality and low-deformation titanium alloy, which is required for marine applications. Two different FSW sequences (single weld and two welds) have been investigated for their effect on the low-cycle fatigue performance of Ti-4Al-0.005B titanium alloy T-joints. Results show that with increasing applied strain amplitude, fatigue damage gradually increased. The fatigue life of single-weld T-joints is close to that of the double-weld T-joints. The single-weld T-joint breaks at the HAZ on the advancing side, while the double-weld T-joint breaks at the HAZ of the second weld (Su et al., 2022).

High Heat Input Welding The high heat input welding process, especially electron gas welding (EGW), is a very useful process for joining thick steel plates up to 100 mm in shipbuilding, bridge construction, and similar heavy fabrication industries. Recently, a few publications have been published on the EGW process to evaluate fatigue and fracture mechanisms. The higher heat input is associated with lower brittle crack arrestability in thick steel plates such as 80 mm because the fracture toughness decreases with increased welding heat input, especially in HAZ in EH47 grade steel plates (An et al., 2021a). The stress concentration factor was used to analyze the influence of heat input on the formation of surface roughness and surface cracks. Higher heat input leads to a longer cooling time and the formation of the weld surface, thus giving more time for the molten metal of the welded joint to form a favorable surface of the welded joint, reducing cracks. The occurrence of fatigue cracks could be minimized by increasing the arc voltage and the arc current or by reducing the welding speed to cooling speed (Randić et al., 2022). Also, in laser welding, the weld beads produced with high-heat inputs exhibit significant sensitivity in CTOD (crack tip opening displacement) fracture toughness and hardness due to the microstructural changes. The weld beads containing martensite and bainite showed lower fracture toughness and brittle fracture characteristics with no significant plastic deformation and shallow cavities. The FCG was less sensitive to microstructural alterations than the CTOD fracture toughness. The weld beads showed higher FCG resistance than the base metal due to the mechanical properties and morphologies of the microstructures. There was no significant difference in FCG among the weld beads (Ribeiro et al., 2021).

Laser and Electron-Beam Welding Laser welding (LW) has gained popularity over conventional techniques due to its high weld quality, high energy density, high flexibility, and better specific heat input, so it is widely used for high-strength dissimilar metal welding in marine structures. The deformation behavior and microstructure characteristics of laser-welded Ti-6Al-4V titanium alloy joints under variable amplitude fatigue ranging from 320 to 380 MPa have been investigated, and results show that the fatigue limit was 339 ± 5 MPa (Chen et al., 2023a). Research has shown that treatment via ultrasonic rolling is an effective method for improving fatigue life and fracture toughness in Ti-6Al-4V titanium alloy laser-welded joints (Chen et al., 2023a). The fatigue behavior of Ti-6Al-4V/Inconel 625 (Nickel alloy 625) dissimilar metals laser-welded joints have been analyzed, and the experimental results in terms of fatigue strength demonstrate a good quality of the obtained joints (Corigliano and Crupi, 2021). The fatigue performance of the butt joint of Q345 steel with 20 mm thickness has been studied by comparing high-power LW with narrow gap hybrid laser-arc welding. Results show that the fatigue property (S-N curve) of the LW is better than that of the hybrid laser-arc weld, which was thought to be strongly associated with the fine multidirectional grains within the laser weld and the precipitation of particle phases inside the hybrid weld (Wang et al., 2021a).

Electron beam welding has the advantage of strong penetration ability with small HAZ, and thus, it is often used to weld or repair structural components of large thickness and large equipment. The low-cycle fatigue properties, high-cycle fatigue properties, and fatigue crack propagation rates of the materials in different zones in the thickness direction of a double-sided electron beam welded joint of TC4 titanium alloys with a thickness of 140 mm have been investigated, and it was revealed that the overlap region of two weld seams was the weakest region of the entire joints in terms of fatigue performance (Long et al., 2021a, Long et al., 2021b, Long et al., 2022). The microstructure evolution of the nail-shaped weld on electron-beam welded joint of Ti–6Al–4V alloy has also been studied, and the fracture mechanism has been explored through in situ observation of tension and fractography. It was revealed that the aggregation and slip of dislocations along needle α′ phases trigger crack initiation (Sun et al., 2021a).

3.1.3 Special Purpose Materials

Liquefied Hydrogen Storage Liquefied hydrogen was first obtained by Dewar, who developed a technique for cooling gases using the Joule-Thomson effect. Using this technique, he liquefied hydrogen in minute quantities in 1895 (Kapitza, 1997). He built a larger facility and was able to produce liquid hydrogen in large quantities in 1898. Since then, it has been used primarily for the following applications: Hydrogen bubble box for particle research, Space Development, and Study of application to aircraft fuel due to the oil crisis. Regarding materials, stainless steel, aluminum alloys, and titanium alloys have been mainly used. Considering the size of the storage tank scale, it is not easy to apply titanium alloys or aluminum alloys. More recently, the application of carbon steel as an extension of 9%Ni steel for LNG tanks has also been considered a challenging issue. The following is a summary of the application of each material to liquid hydrogen storage tanks to date.

Austenitic stainless steels have good low-temperature performance, making them the first choice of material for low-temperature working conditions and the most widely used material for liquid hydrogen storage and transport containers. The first attempts to use liquid hydrogen as an aviation fuel were made in 1955. Two stainless steel wing-tip liquid hydrogen tanks were developed (Qiu et al., 2021). The 300 series have been widely used for cryogenic liquid storage and transport containers due to their excellent overall performance, 304, 304L, 316, 316L, 321, 347, etc. In 1961, tensile properties were investigated using liquid helium. The tensile properties have been actively evaluated from that time to the present. In 1970, hydrogen fuel was applied to commercial aircraft and liquid hydrogen storage tanks, which were made of 2219 aluminum alloy and located at the front and rear of the fuselage, respectively. The liquid hydrogen tanks of the Space Shuttle (1981-) were made of 2195 alloy and 2090 alloy (both aluminum-lithium alloys). Also, titanium and titanium alloys (α-Ti, α-β duplex Ti) were used for containers and structural pipes for liquid helium and liquid hydrogen in the Apollo space rocket launched in 1981 (Qiu et al., 2021). The biggest problem with titanium alloys in low-temperature applications is that their elongation, impact toughness, and fracture toughness decrease with decreasing temperature. However, materials have been developed to overcome these problems. Based on recent research, Ni-bearing steel is a strong candidate for liquefied hydrogen storage tanks. Figure 2 shows the trend of the minimum service temperature based on the toughness improvement effect obtained with increasing Ni content. Considering this relationship and the boiling point of liquid hydrogen is -253 °C, it is possible to use steel with about 12% Ni added. Furthermore, considering the recent application of 7% Ni steel for LNG tanks (Kubo et al., 2010), it is also possible to consider further Ni reduction at the liquefied hydrogen temperature. In addition, the low cost of high manganese steel makes it a viable option, and its application is being actively considered (Lee et al., 2023).

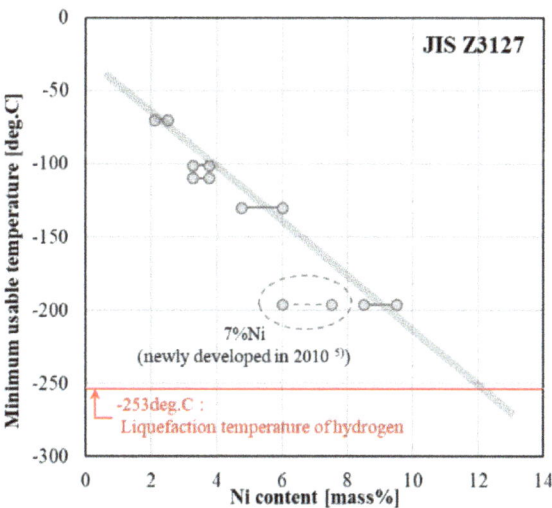

Fig. 2. Relationship of Ni content and minimum usable temperature (JIS, 2021).

Transformation-Induced Plasticity Behavior To use liquefied hydrogen in a stable manner, it is essential to construct unprecedentedly large storage tanks. However, the temperature of liquefied hydrogen is 20 K, an extremely low temperature. Currently, austenitic stainless steel is considered the material of choice for liquefied hydrogen storage tanks.

The mechanical properties of metastable materials, particularly Fe-Cr-Ni steels, have long been the subject of experimental and theoretical research. It is known that the initial microstructure of this type of material can be altered by phase transformation depending on the service environment, such as temperature and applied strain, which is problematic in some cases.

In metastable austenitic stainless steels, plastic deformation below the Md temperature can produce body-centered cubic α' and hexagonal closed-shell ε-martensite, resulting in significant strength. Md is the maximum temperature at which martensitic transformation can occur by plastic deformation. The transformation from face-centered cubic austenite γ to α' and ε-martensite is known to depend on alloy composition, stacking defect energy, deformation degree, and temperature. Early studies (Eichelman and Hull, 1953, Bannykh and Kovneristyi, 1969, Hirayama, 1970) investigated the phase transformation of several pure and commercial Fe-Cr-Ni alloys by plastic deformation at room temperature and various low temperatures. One of the main conclusions of these studies was that the rate of martensitic transformation is highly dependent on the chemical composition of the steel, with Ni, Cr, and C having a particularly strong influence on the phase transformation process. The total effect of the main alloying elements on martensitic transformation is assessed from the nickel equivalent (Hirayama, 1970) and Md30 (Angel, 1954, Nohara et al., 1977). Martensitic transformation of austenitic stainless steels has been the focus of research in uniaxial tensile at low strain rates (static loading) (Angel, 1954, Lagneborgj, 1964, Wigley, 2012, Olson and Cohen, 1975, Tamura, 1982). Tsuchida and Tomota (2000) also developed a method to accurately obtain stress-strain relationships in the phase-mixed state during transformation using a homogenization method called the secant method proposed by Weng (Tsuchida and Tomota, 2000, Tsuchida et al., 2011, Tsuchida et al., 2021b, Tsuchida et al., 2021a). In these studies, little attention was paid to the influence of stress state on phase transformation, but in (Lebedev and Kosarchuk, 2000, Polatidis et al., 2021, Beese and Mohr, 2011, Morohoshi and Kawabata, 2022) it has been shown through systematic experiments and discussion that the influence of the stress field is very significant. Figure 3 shows an example of an investigation of the effect of changes in the stress field from uniaxial tension to uniaxial compression on the relationship between plastic strain and the amount of transformation. Considering that the range of changes in these stress fields is much milder than that of the strongly constrained areas in notches and crack tips in real structures, structural or fracture mechanics analyses that take these transformation behaviors into account are necessary to predict deformation and stress concentrations accurately. Tsuda et al. (2023) developed a numerical module that can reproduce the TRIP (Transformation induced plasticity) effect in arbitrary stress fields and incorporated it into the 3-D elastoplastic finite element method. This enables accurate consideration of TRIP effects in complex geometries and under high stress-triaxialities, such as at crack tips. This will contribute to the safety assessment of cryogenic facilities such as liquefied hydrogen storage tanks.

The fatigue properties of metastable austenite have been investigated in relation to martensitic transformation in the crack tip region during fatigue crack growth. Pineau and Pelloux (1974) and Mei and Morris (1990) showed that the transformation significantly reduces the crack propagation rate, which is an advantageous property. It is also pointed out that the fatigue crack propagation rate is significantly degraded when the material is subjected to plastic deformation in front of the crack propagation path due to earthquakes. Martelo et al. (2015) have shown a unified equation developed by Kujawski (2001) for the difference in fatigue crack growth rate depending on the stress ratio where the coefficient alpha in the Walker equation should be increased to 0.6 to fit all data of TRIP steel.

$$\frac{da}{dN} = C\left(\Delta K^{1-\alpha} K_{max}^{\alpha}\right)^m \quad (1)$$

Mechanical simulation incorporated with TRIP behavior has been attempted using approaches such as the phase field method (Yeddu and Somers, 2021, Liu et al., 2021), the crystal plasticity method (Liu et al., 2021, Lindroos et al., 2022), and the strain gradient plasticity method (Sedaghat and Abdolvand, 2021), and will be implemented in structural analysis tools such as the finite element method to contribute to high-precision fracture analysis of structures and specimens.

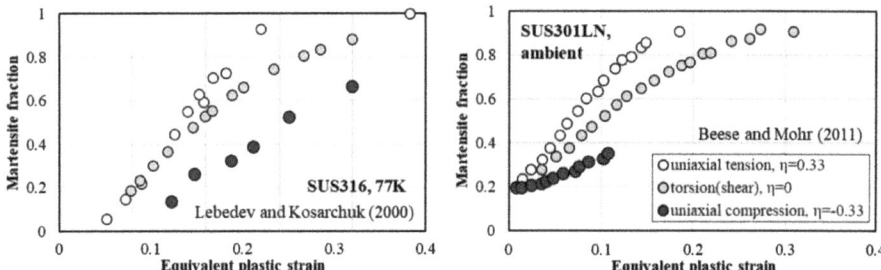

Fig. 3. Correlation of equivalent plastic strain and martensite transformation rate for stainless steel (Taken from (Lebedev and Kosarchuk, 2000) and (Beese and Mohr, 2011)).

Hydrogen Embrittlement Hydrogen embrittlement (HE) is the mechanical damage mechanism caused by the penetration of hydrogen into the metal, causing a reduction in tensile strength and ductility, followed by deformation of larger grains, cracks, and blisters. Furthermore, the resulting hydrogen embrittlement of the material may accelerate the initiation of surface flaws under cyclic loading (Wasim and Djukic, 2020). The hydrogen embrittlement originates from manufacturing processes as well as the byproduct of the electrochemical reactions occurring on the steel surface (corrosion, H2S reaction). The Hydrogen embrittlement mechanism depends on metal susceptibility to a specific corrosive environment and the yield strength of the used material. In this regard, mild steel is more prone to it in alkali environments, whereas high-strength and stainless steel are more prone to it in environments with hydrogen sulfide, carbonite, and chloride in more moderated temperatures (Campari et al., 2023).

Regardless of hydrogen ingress occurring during manufacturing or operation due to a corrosive environment, hydrogen embrittlement is an important issue that needs to be considered for offshore steel, especially high-strength offshore steel. For instance, steel can be strengthened by adding Mo in bainitic steel. Nam et al. (2023) investigated the impact of adding Molybdenum (Mo) up to 0.2 wt% to high-strength steel. The study concluded that with the addition of Mo, the strength of 0.2 Mo steel increased; however, HE resistance (crack initiation) was almost similar to that of 0.02 Mo steel. Nam et al.

(2023) stated that the more Mo content the bainitic steel has, the more the martensite-austenite (MA) constituents become sensitive to HE. However, the Mo solutes, serving as trap sites for hydrogen, increase as well, which results in hydrogen diffusion being delayed by the Mo solutes and alleviating the degradation in the HE resistance.

Choi et al. (2023) analyzed austenitic stainless steel subjected to hydrogen under various temperature conditions. The study reported a decrease in the tensile strength and elongation, whereas an increase was observed in the yield strength in all cases. Especially, the tensile test at room temperature for 20% pre-strain at cryogenic temperature after subjecting hydrogen demonstrated fracture in the elastic region. Another type of steel commonly used in offshore oil platforms is 21Cr2NiMo steel. Liu et al. (2022a) examined the stress-corrosion cracking behavior of steel in SO_2-polluted coastal environments. Their findings showed that cracks began at corrosion pits, with crack propagation driven by mechanisms of anodic dissolution and hydrogen embrittlement. Similarly, a study by Yan et al. (2023) confirmed that H2S corrosion fatigue was controlled by the combined effect of anodic dissolution and hydrogen embrittlement for X65 welded joints. Furthermore, Zhou et al. (2022) studied the hydrogen embrittlement on the fatigue performance behavior of pure iron in different environments through fatigue crack growth tests. The results showed that the fatigue crack growth rate in H_2 is much higher than in N_2, and NG and CO_2 exacerbated the crack growth rate in an H_2 environment. This is because the presence of CO2 reduced hydrogen diffusion activation energy.

Besides the corrosive environment reaction, manufacturing and large residual tensile stresses near welds can facilitate hydrogen diffusion and accumulation in high-strength steel plates. Jiang et al. (2022) analyzed the distribution of hydrogen diffusion under the influence of residual stress using numerical simulations validated by related experimental research and tests. The weak area of the welding joint was found to be near the weld toe, which exhibited high hydrostatic stress and hydrogen concentration, and the hydrostatic tensile stress in the vertical weld path was maximized (~345 MPa), degrading the material properties and causing hydrogen-related cracking. Fernando Maia de Almeida et al. (2023) presented a methodology to assess the hydrogen embrittlement threshold force (P_{th}) in high-strength steels using the Small Punch Test (SPT) technique based on material specimen stiffness. The incremental SPT and the incremental step loading (ISL) techniques were applied to validate the methodology by testing notched specimens manufactured from high-strength steels. A systematic error of -4.0% and 2.2% was found for the Pth-SPT values with respect to the ISL method.

Zhu et al. (2022a) investigated welded titanium alloy joints used in hydrogen environments in aerospace, shipbuilding, and chemical industries. To do so, an ultrasonic surface rolling process is used to treat TC4 titanium alloy joints welded by laser welding method, followed by slow-rate tensile tests under electrochemical hydrogen charging conditions. The mechanical properties of the TC4 laser welded joints before and after the ultrasonic surface rolling process (USRP) were compared. The study concluded that USRP could significantly improve the anti-hydrogen embrittlement behavior of the TC4 laser welded joint due to grain refinement and USRP-induced residual compressive stress in the deformation layer. The impact of microstructure heterogeneity on hydrogen-assisted fracture was studied by Li et al. (2023) for X80 weld metal. To this end, CTOD tests in air and an H2S-saturated solution were carried out, and the tests revealed that hydrogen embrittlement was the primary cause of the degradation of fracture toughness associated with microstructure heterogeneity, especially the amount of inclusions. This is because hydrogen-dislocation interactions make it easier for dislocations to follow the slip plane, resulting in void formation and crack initiation. San Marchi et al. (2021) showed that hydrogen suppressed the strain-induced martensitic transformation when the α transformation fraction is more than 20%. These observations are hypothesized as hydrogen-promoting nucleation of strain-induced martensite while suppressing the growth of these martensite nuclei.

3.1.4 Testing Aspects

Development and Modification of Fracture Toughness Evaluation Standards Crack Tip Opening Displacement (CTOD) is understood to be an elasto-plastic fracture mechanics parameter compatible with the J-integral. CTOD is still more commonly used in design codes for welded steel structures or in toughness evaluations of steel plates because the J-integral was not applied to unstable brittle fractures until 1996, and CTOD was mostly used in the field of welded steel structures. Tagawa et al. (2010) and Tagawa et al. (2014) pointed out the large differences in CTOD calculation values between BS7448-part1 and ASTME1290. In particular, the BS did not consider work hardening properties of materials, which was considered problematic, and activities to improve the CTOD calculation formula were initiated. The CTOD was calculated comprehensively under a wide range of conditions while observing the nature of rotational deformation in detail through numerical analysis using a mesh division that can well reproduce the deformation of the crack tip well. Kawabata et al. (2016), Kawabata et al. (2017a), and Kawabata et al. (2017b) developed a CTOD calculation formula that enables a highly accurate estimation of CTOD as a driving force under a wide range of conditions. The equations for compact specimens and for a wide range of a0/W conditions were further investigated, and after discussions in ISO-TC164-SC4-WG3, the equations were incorporated into ISO15653 -2018 (ISO, 2018) and ISO12135-2021 (ISO, 2021) as shown in Table 1. In addition, Kawabata et al. (2018) have proposed a conversion formula between CTOD and J based on this new CTOD definition for J users.

Symbols: a_0: Initial crack length, W: Width, δ_0: CTOD, S: Span, F: Force, B: Thickness, B_N: Thickness for side-groove position, g_1. g_2: SIF function determined by specimen shape, $R_{p0.2}$: Yield strength, R_m: Tensile stress, V_p: Plastic component of clip gage displacement at fracture, E: Young's modulus.

In the CTOD evaluation, there is the problem of uniformity of fatigue pre-crack tip straightness, which is often a difficulty in the evaluation of welded joints, especially important in CTOD critical value evaluation. The evaluator's specific work still required a great deal of experience. The 2018 revision of ISO 15653 will include detailed information on the procedure for performing reverse bending methods (Kawabata et al., 2024a, Kawabata et al., 2024b), and the new edition, which has now been completed at the ISO meeting and is undergoing revision, will include more effective local compression methods to modify the residual stress field (Ozawa et al., 2021, Ozawa et al., 2022, Ozawa et al., 2023), which will greatly reduce the differences in the possibility of success of CTOD tests on welded joints due to differences in user experience. A schematic diagram of the newly specified local compression method is shown in Fig. 4.

Table 1. CTOD equation newly employed into ISO12135-2021 (ISO, 2021) and ISO15653-2018 (ISO, 2018).

a_0/W	Specimen	Calculation formulae and coefficients
$0.45 \leq \frac{a_0}{W} \leq 0.70$	Bend specimen	$\delta_0 = \left[\left(\frac{S}{W}\right)\frac{F}{(BB_N)^{0.5}} \times g_1\left(\frac{a_0}{W}\right)\right]^2 \left[\frac{(1-v^2)}{mR_{p0,2}E}\right] + \tau \bullet \frac{0{,}43(W-a_0)V_p}{0{,}43(W-a_0)+a_0}$
	Compact specimen	$\delta_0 = \left[\left(\frac{S}{W}\right)\frac{F}{(BB_N)^{0.5}} \times g_2\left(\frac{a_0}{W}\right)\right]^2 \left[\frac{(1-v^2)}{mR_{p0,2}E}\right] + \tau \bullet \frac{0{,}52(W-a_0)V_p}{0{,}52(W-a_0)+a_0}$
$0.10 \leq \frac{a_0}{W} \leq 0.45$ (ISO15653 AnnexE)	Bend specimen	$\delta_0 = \left[\left(\frac{S}{W}\right)\frac{F}{(BB_N)^{0.5}} \times g_1\left(\frac{a_0}{W}\right)\right]^2 \left[\frac{(1-v^2)}{mR_{p0,2}E}\right] + \tau \bullet C_{Vp}\frac{0{,}43(W-a_0)V_p}{0{,}43(W-a_0)+a_0}$
Constraint factor		$m = 4{,}9 - 3{,}5\frac{R_{p0,2}}{R_m}$
Plastic term correction factor		$\tau = \left\{-1{,}4\left(\frac{R_{p0,2}}{R_m}\right)^2 + 2{,}8\left(\frac{R_{p0,2}}{R_m}\right) - 0{,}35\right\}[0{,}8 + 0{,}2\exp\{-0{,}019(B-25)\}]$
a_0/W correction factor		$C_{Vp} = -1{,}74\left\{\left(\frac{a_0}{W}\right) - 0{,}45\right\}^2 - 1$

Parameter	Recommendation
Platen contact area, S	$0{,}24B^2$
Distance from the edge of the platen to the expected fatigue crack tip, d, mm	$0{,}1B$–$0{,}2B$
Total strain in the thickness direction, ε_p, %	Up to 4,0%

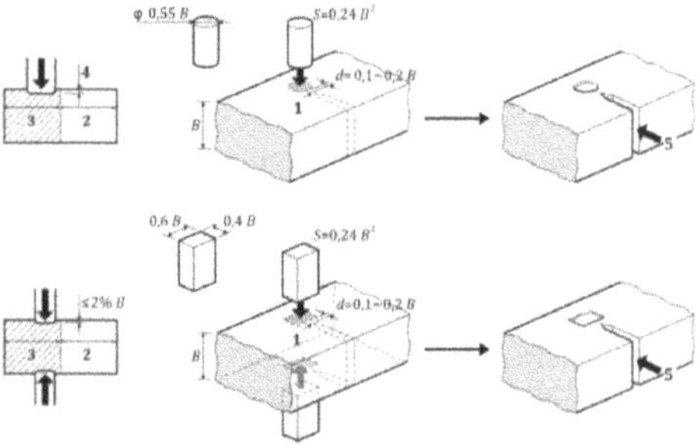

Fig. 4. Recommended parameters for local compression that will be printed in the new version of ISO15653 (ISO, 2018).

Furthermore, in the field of CTOD-R curves, the conversion method from the ASTM J-integral, which was previously used in ISO 12135 (ISO, 2021), has been updated and replaced with a calculation formula that can be derived from macro load to clip gauge displacement from CTOD measured by the silicone rubber casting method (Khor et al., 2016, Khor et al., 2018a, Khor et al., 2018b, Khor, 2019, ISO, 2021). The new method is expected to improve the accuracy of CTOD as a driving force, especially when determining the R-curve of FCC-based metallic materials such as stainless steel.

In December 2018, there was the publication of a new version of the fracture toughness testing standard BS 8571 (BS, 2018) for carrying out Single Edge Notch Tension (SENT) tests (Moore and Hutchison, 2016). The main differences in the 2018 version in comparison to the 2014 version are that in BS 8571 (BS, 2018), the calculation of J for clamped specimens is now the same for specimens with all W/B ratios and included as informative annexes on 'Guidance for stable crack path deviation during ductile crack extension' and 'Guidance for using non-fatigue pre-cracked specimens'. The formula to calculate the elastic component of CTOD has been updated. The crack length to specimen width ratio range is from 0.3 to 0.5 (but shallower notches can be used on agreement of all parties). Another main difference is that the validity of the shape of the initial and final crack shapes is now the same as those given in ISO 15653 (ISO, 2018), and other typographical and formatting changes, additional clarifications, and guidance notes have been made. The BS 8571 standard (BS, 2018) has become the industry-accepted method for performing SENT testing, being required by DNV for subsea pipeline design and pipeline girth weld assessment according to DNVGL-RP-F108 (DNVGL, 2017a) and DNVGL-ST-F101 (DNVGL, 2017b).

Size Effect and Transferability Issue It is known that fatigue and fracture characteristics depend on the specimen size. Specimen size is often significantly smaller than the size of a real structural component. This is referred to as the transferability issue. An overview of recent developments related to the size effect is given by Zhu et al. (2022b), Braun et al. (2022a). The typical approach to cope with the transferability issue is the application of a safety factor. However, for the interpretation of (e.g., monitoring) data, this safety factor approach does not suffice, and more understanding of the transferability issue is required. The size effect that is considered in standards e.g. (DNV, 2021e) is predominantly the thickness effect (part of the geometrical effect). An example is in section F.5 of DNV-RP-C203 (DNV, 2021e) which provides guidance on accounting for the geometrical (weld thickness) and system effects (weld length). The statistical size effect and residual stress differences are not explicitly covered.

The transferability issue is typically considered in three domains, which can occur separately or in combination:

Statistical aspect

This covers the amount and distribution of flaws over the weld volume. "Research on this topic is relatively mature" (Zhu et al., 2022b).

Geometrical aspect

This covers changes and differences in stress gradient. Next to differences in the geometry, this also includes residual stress differences, the influence of the number of welding passes, and potentially parallel load paths. However, research on the last two points is scarce. Full-scale fatigue test data (refer to Chapter 4 of this ISSC 2025 committee report) can be used to assess these differences with small-scale specimens. The interpretation in literature is typically focused on the thickness effect (Corigliano and Crupi, 2022), whilst acknowledging other influence factors. This focus is due to the increased interest in thin and lightweight structures. On the other hand, there is a need for increased thickness for special marine structures such as offshore wind monopiles. This thickness effect is well known and already covered with correction factors in class documents. The influence of residual stress is less straightforward, and literature has shown strong influences of the context of the specimens. It is observed that residual stress differences do not always lead to a relief in the relevant direction, as FE results validated with strain gauge measurements indicate (Deul et al., 2022). This is attributed to the net effect of the relief of boundary conditions. Next to these context-specific results, most research on the geometrical size effect is performed for uniaxial fatigue test data (Zhu et al., 2022b). On the other hand, differences in the initiation properties of brittle fracture due to specimen geometry have been explained by the equivalent Weibull stress concept; see (Beremin et al., 1983) and (ISO, 2016). Recently, Oie et al. (2021) showed that a more accurate explanation is possible by considering the plastic strain contribution to micro-crack formation.

Technological aspect

This covers differences in production methods and production quality. Compared with the other two, this topic has not been well investigated yet. Surface finish is identified as a key influential parameter of fatigue strength (Zhu et al., 2022b). The recent developments in relation to the transferability issue are on the influence of production processes on specific materials (e.g. EDM and grinding of AISI D2 tool steel in (de Jesus et al., 2020) and laser shock peening on HCF properties of aluminized AISI 321 stainless steel in (Li et al., 2021b). Research that was found is typically very specific to one application, making generalization difficult. The literature is mostly focused on novel manufacturing equipment but not on generic and readily used equipment. As long as fatigue design is based on the S-N curves, an important challenge will be the available data on the fatigue test database used to infer the design curves. Information on residual stresses and production quality is to be included to enable meaningful analyses. With all uncertainties, generalization is difficult, and design standards have to ensure high reliability, as highlighted in (Braun et al., 2022a). This is why fatigue design guidelines propose a safety factor which often leads to conservative results (Sonsino, 2009). Next to that, the transferability issue limits the amount of data points that can be combined to infer the S-N curves. The current state-of-the-art literature, as analyzed in this section, does, in the committee's opinion, not allow for the reduction of these safety factors.

3.2 Driving Forces

3.2.1 Fatigue and Fracture Parameters

There have been no noticeable developments in fatigue and fracture parameters over the 3 year evaluation period, except for new testing methods for developing certification requirements for extreme service environments, e.g., arctic operations, LNG tankers, hydrogen transportation, etc. These include fatigue damage and life evaluation methods (Braun et al., 2020, An et al., 2021b, Feng et al., 2021, Zhao et al., 2021b, Braun, 2022, Korotygin et al., 2023, Qiao et al., 2023) and fracture toughness requirements (Gook et al., 2020, Kim et al., 2020a, Kim et al., 2021, Ying et al., 2024), and specific fracture mechanics solutions (Aubert et al., 2020) for supporting fitness for service or engineering criticality assessment.

3.2.2 Multiaxiality

Welds in ships and offshore structures are typically subjected to multiaxial loading conditions in which fatigue damage at a given critical location or hot spot has more than one operative stress component, e.g., normal stress and shear stress. The two stress components can be in phase (i.e., proportional loading) or out-of-phase (non-proportional). It has been shown that out-of-phase loading can be 4 to 5 times more damaging in terms of fatigue lives than proportional loading. This is mainly because fatigue damage is load-path dependent (Jimenez-Martinez, 2020, Yang et al., 2020, Haselibozchaloee et al., 2022). It is essential to develop structural fatigue evaluation techniques that can effectively consider both proportional and non-proportional loading spectrums relevant to offshore and ship structures. Such a technique must consider two issues: (1) how to determine a fatigue damage cycle; (2) how to establish a fatigue damage parameter in terms of stress or strain or other related parameter that can be cycle-counted in the time domain in a consistent manner. Classical methods, e.g., the critical plane method, have not been effective nor practical for applications in ship and offshore structures (Wei et al., 2023c). The newest developments along this line are the path-dependent maximum range (PDMR) cycle counting method (Mei et al., 2021, Wei et al., 2023b, Wei et al., 2023d), as illustrated in Fig. 5 in the normal stress and shear stress (σ-τ) plane. In this method, load path length (Fig. 5(a)) and load-path moments (1st or 2nd moment, see Fig. 5(b)) can be used as a multiaxial fatigue damage assessment. The damage parameter can be its cycle (one-half cycle in this case). The effectiveness in treating variable-amplitude proportional and non-proportional multiaxial loading histories in the time domain can be found in (Mei et al., 2021, Wei et al., 2022, Wei et al., 2023b, Wei et al., 2023d) and in the frequency domain in (Ravi and Dong, 2020, Pei et al., 2022a, Ravi et al., 2022). The most recent investigations on using the effective notch stress recommended by IIW for treating Modes I and III multiaxial fatigue loading were documented in (Bufalari et al., 2024b). They concluded that the data analyzed seems to fit a uniaxial Mode I reference data. The same authors (Bufalari et al., 2024a) also investigated how a total stress method based on the critical plane concept can be used to correlate the same test data and showed that both the total stress and effective notch stress methods yield similar data correlation.

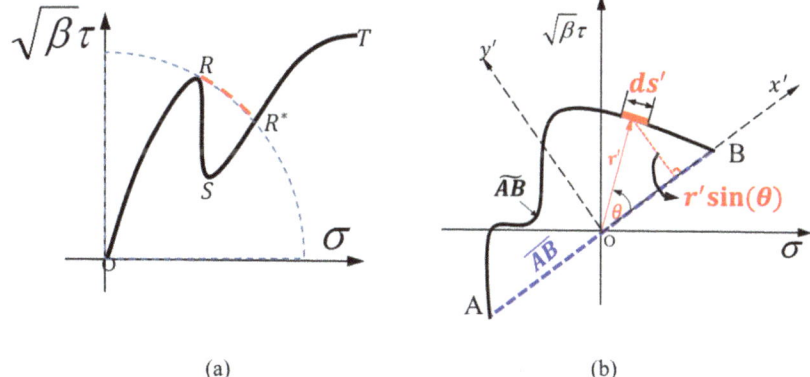

Fig. 5. Concept of path-dependent maximum range (PDMR) cycle counting method (Mei et al., 2021)

3.2.3 Damage Accumulation and Cyclic Loading

The limitations of the Palmgren-Miner linear damage accumulation rule are well-known and studied. The most important challenge of this rule is that the non-linear nature of damage accumulation under random loads is not covered, also called the load sequence effect. Many alternative accumulation models have been presented over the years, whereas recent developments have moved towards a more practical implementation of advanced theoretical models. Alternative damage accumulation models are typically more time-consuming, which might hinder rapid implementation in engineering processes. It is observed that the recent focus is on developments towards empirical-based methods to cover the underlying load sequence and load interaction effects instead of extensive theoretical models that can be too elaborate for the design practice. Next to that, probabilistic-based approaches are increasing in interest.

Amplitude Variability New variable amplitude (VA) fatigue test results of S355 and S1100 are presented and evaluated using the 4R approach in (Grönlund et al., 2024). The 4R approach poses a local mean stress correction as a function of the weld toe quality, material strength, stress ratio, and residual stress. This approach is used by Grönlund et al. (2024) to incorporate the influence of an overload and variable amplitude loads. VA fatigue test results are published for high-strength mooring lines in (Aursand et al., 2024), for which the VA characteristic load sequence is based on time domain simulations of the loads over a 62-year period.

Another noted development is the application of the Peak Stress Method (PSM) to the evaluation of VA uniaxial and multiaxial loadings. It should be noted that the PSM is merely to determine the local stress criterion, it is not an accumulation model. It is, however, able to combine multi-axial loads. The PSM is combined with the damage accumulation model by Palmgren-Miner to validate against test data from literature (Campagnolo et al., 2022) and against newly generated uniaxial and multiaxial fatigue test data (Vecchiato et al., 2023). The PSM proved to predict the crack initiation point accurately and compared well to experimental fatigue test data of steel welded joints.

Damage Envelopes and Isodamage Curves Originally, the idea of damage envelopes started with the definition of straight isodamage curves, converging at the knee point of the Wöhler curve (Rege and Pavlou, 2017). The slope of the S-N curve above the fatigue limit is a function of the damage. However, as written in (Pavlou, 2018), this straight nature of the isodamage curve does not fulfill the boundary conditions at both the axes of the S-N curve (zero cycles and zero stress), which should result in zero damage. As an alternative, the definition of curved isodamage lines is proposed by Pavlou (2018). These lines form the so-called 'damage envelope'. Isodamage curves are a material property that will be derived once (Pavlou uses an ANSYS module). Further verification and implementation of the damage envelope concept is given in (Bjørheim et al., 2022a). Verification is performed for various materials for two-stage and multi-stage loading. The model is, to date, not yet verified for irregular loads and multi-stage loads on steel but can provide a relatively easy-to-adopt alternative for the Palmgren-Miner rule.

S-N Curve Slope Modifications Below the Fatigue Limit The 6-parameter random fatigue limit model as a continuous formulation of the S-N curve (Leonetti et al., 2017) along with the non-linear damage accumulation model that modifies the slope below the fatigue limit as damage progresses (Leonetti et al., 2020) are already considered in the 2022 report of ISSC. The required fitting parameters are available for the considered datasets in the respective papers, and more generically applicable parameters have been derived recently (Qin et al., 2021). A recent use of the model is the derivation of partial safety factors for the fatigue design of steel bridges (Maljaars et al., 2022), which is an approach that can be extended to welded ship and offshore structures. The application of the random fatigue-limit based models results in higher estimated reliability levels than a conventional probabilistic S-N model. However, compared to the Eurocode model, the partial safety factor is still to be increased to reach the required reliability level (Maljaars et al., 2022, Kahl et al., 2023) experimentally revealed the effects of the thermal cut plate edges and proposed a modification method.

3.2.4 Welding-Induced Residual Stress

The presence of residual stresses from welding can lead to brittle fracture, corrosion cracking, and deformations. These issues collectively weaken the material's resistance to fatigue and fracture. There is a strong correlation between residual stress and fatigue, as the fatigue potential can be inferred by the level of residual stress.

Generally, fatigue life has been improved, such as changing the weld bead shape using grinding, peening, and thermal treatment together with favorable compressive residual stress. The degree of shot peening exerted exhibits a significant role in elevating fatigue life and enhancing the material's ductility (Qu et al., 2021, Maleki et al., 2021). Additionally, controlling welding residual stress can lead to an improvement in fatigue life. Improving the bead shape to enhance fatigue life may also influence changes in welding residual stress depending on the connection type. Enhancing material fatigue resistance necessitates implementing treatments aimed at mitigating residual stresses. These stresses can be alleviated by focusing on three distinct conditions: preprocessing, in situ processing, and post-processing. Preprocessing entails actions like preheating and optimizing the weld groove's shape to counteract uneven temperature dispersion around the welding zone (Zhao et al., 2021a).

The magnetic-vibration stress relief treatment is an effective method for reducing residual stress without using temperature treatments. However, exceeding fatigue limits can compromise material integrity, causing crack formation and eventual fracture. Huang et al. (2023) provide new insights into these mechanical treatments for reducing residual stress in structural and functional materials. The treatment reduced residual stresses of the studied high-strength steel in both longitudinal and transverse directions, being higher for the longitudinal direction. The residual stress homogenization mechanism results from the combined effect of material softening and deformation superposition, which leads to local micro-plastic deformation The experimental and numerical influence of residual stresses on the fatigue life of welded joints and notched specimens were reported. The implemented thermal-structural model, although simplified, revealed that residual stresses have a larger effect on fatigue life under pure torsion loading compared to pure bending loading (Chiocca et al., 2022).

Zhang et al. (2024) studied the effect of welding residual stress redistribution on crack closure and its impact on FCG, particularly in relation to the crack wake and applied load. FCG rates are reduced at the beginning of crack growth due to compressive RS at the crack tip and the crack closure effect. RS redistribution causes variations in FCG rates, with RS relaxation leading to the gradual recovery of crack growth rates as the crack propagates through different zones (parent material, HAZ, and weld material). During fatigue, the total amount of residual stress remains constant, while the peak residual stress increases when the crack tip is far from the weld center line. The total and peak residual stresses gradually decrease as the crack tip approaches the weld center line.

Dai et al. (2023) purposed a practical and efficient numerical simulation procedure for high-frequency mechanical impact treatment of welded joints. The recommendations of the FE modeling and determination of displacement-controlled simulation parameters were presented in detail. The validity of the high-frequency mechanical impact simulation was studied numerically and experimentally in an out-of-plate gusset welded joint. The radius of the peening tool has been modified to enable more accurate reproduction of the treated local profile in the experiment. This facilitates the consideration of local stress concentrations and RS relaxation in subsequent fatigue assessment. To accurately evaluate the improvement by the high-frequency mechanical impact process on the fatigue life of the structures, it is necessary to perform stress relaxation analyses after high-frequency mechanical impact simulation to consider the residual stress relaxation and its influence on crack initiation and short crack growth rate (Ono et al., 2024) and (Ono and Remes, 2025).

Residual stress, inherently generated during welding, varies based on the welding process, the number of passes, and plate thickness. An et al. (2020) studied the impact of residual stress on the unstable fracture of thick, high-strength steel plates used in structures like large container ships. The study examined unstable crack propagation under low temperatures using three specimen types: no stress, edge tensile stress, and edge compressive stress. The experimental results were verified through full-scale brittle crack arrest tests. Results showed that tensile welding residual stress at the crack tip accelerated brittle crack propagation, while compressive stress mitigated it, suggesting that applying compressive stress to the weld joint may prevent complete brittle fracture.

3.3 Testing Method Developments for Decarbonization

This section discusses the developments regarding testing methods for evaluating the safety and efficiency of transporting liquid hydrogen, including transfer from production sites to end-users. This section also considers issues such as temperature control and insulation, given the cryogenic condition of liquid hydrogen.

3.3.1 Hydrogen

As hydrogen is increasingly used for energy storage and transportation, the need for reliable test methods continues to grow. The evaluation of hydrogen embrittlement has received considerable attention due to its impact on the mechanical performance of materials. Strength, ductility, and fracture toughness at cryogenic temperatures are also in focus as liquefaction (20K) technology is considered a realistic transportation and storage method.

Kim et al. (2022a) studied the effects of hydrogen at 300K and 20K on mechanical behavior and fracture morphology using hydrogen-charged specimens. Tensile and fracture-toughness tests were conducted at room temperature and low temperatures of 77 K and 20 K to analyze how the mechanical behavior at low temperatures differs from that at room temperature. Weeks et al. (2022) performed tensile tests for weldment in liquid helium (4 K). As the testing temperature decreases from 77 K to 4 K, all welds exhibit a rise in yield strength, plus a decrease in total elongation and reduction in area. Benzing et al. (2022) investigated ductile crack resistance using Charpy-type single-edge bend, SE(B). Fracture toughness specimens were extracted from welds in four welded 316L stainless steel plate samples. As expected, decreasing the test temperature from 77 K to 4 K caused a reduction of fracture toughness (critical J-integral and tearing modulus) for all investigated welds. Unlike the specimens tested at 77 K, those tested at 4 K showed less overall plastic deformation.

Lucon and Benzing (2021) highlight that ample evidence is available in the literature that Charpy testing below 77 K is not technically feasible due to both temperatures rising during the transfer of the specimens from the cooling medium to the impact position and adiabatic heating. While temperature rise during transfer can be minimized by appropriately modifying the test setup, adiabatic heating at elevated strain rates is unavoidable. Adiabatic heating occurs when the energy generated during plastic deformation is converted into heat but cannot dissipate fast enough to the surrounding environment, causing the temperature of the material to increase. Regardless of the approach taken to keep Charpy specimens fully immersed in liquid helium at the time of impact, all methods required a correction factor to account for the influence of fracturing the container on the energy spent to break the specimens (Lucon and Benzing, 2021). Slow Strain Rate Tensile (SSRT) tests are sometimes used to evaluate the fracture resistance of materials used for high-pressure hydrogen and liquefied hydrogen. For example, Boot et al. (2021) used SSRT to evaluate the hydrogen embrittlement properties of pipelines that have already been utilized for natural gas transportation. Conducting SSRT tests under high-pressure hydrogen is challenging in terms of cost and time required for delivery, but as shown in (Ogata, 2022), an idea to simplify the facility using hollow specimens has been developed and is expected to become ISO-approved.

Although there has been progress in testing methods, the evaluation of fatigue and fracture of applied steel for storage of liquefied hydrogen is very limited. Additionally, it is important to establish clear experimental methods and requirements in the future to realize decarbonization.

3.3.2 Ammonia

In the 1950s, a small leak was found in a large amount (3%) of a small agricultural ammonia tank that had been used for three years, leading to the establishment of the "Agricultural Ammonia Association Research Special Committee. Air and CO2 are harmful. If 0.1% or more of water is added, it acts as an inhibitor. Based on these findings, the Association recommended the addition of 0.2% or more water, the removal of air, and the use of stress relief annealed (SR) or hot-formed for tanks larger than 36 in (910 mm) in diameter (Dawson, 1956, Loginow and Phelps, 2013). As a result, stress corrosion cracking (SCC) problems in agricultural tanks have since decreased. For many years, it was believed that SCC occurred in ambient temperature ammonia tanks only and not in low-temperature storage tanks operating at - 33 °C (- 27°F) and atmospheric pressure. In the 1980s and 90's surveys, stress corrosion cracks have also been reported in low-temperature tanks (Nyborg and Lunde, 1996). Figure 6 shows the experimental results of (Nyborg and Lunde, 1996) using ASTM A537 Grade 1 and compact specimen.

At -33 °C, the crack propagation rate is somewhat lower than at room temperature but not zero.

Since around 2015, ammonia has been considered one of the leading transport methods (carriers) for hydrogen, the next-generation energy, and various studies have opened the way for its "direct use" as a fuel, making it a carbon-free fuel alongside hydrogen. The most promising applications of ammonia as a fuel are "marine use" and "power generation". Kawarasaki et al. (2023) present Ammonia SCC properties of steel focusing on low pressure and temperature conditions. In addition, in 2023, Korea submitted a document to IMO on the applicability of high Mn steel to ammonia storage tanks (IMO, 2023).

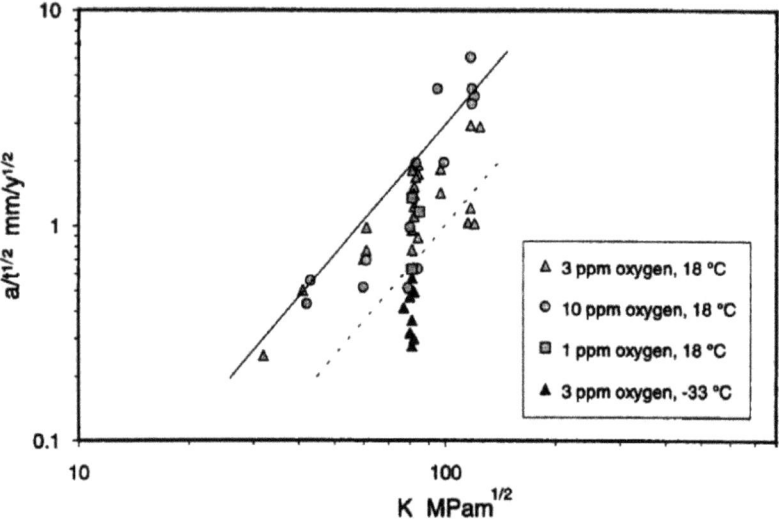

Fig. 6. Stress corrosion crack divided by the square root of exposure time as a function of stress intensity factor (Nyborg and Lunde, 1996).

3.3.3 Co_2

To promote carbon capture and storage (CCS) on the scale necessary to achieve greenhouse gas reduction targets, it is necessary to utilize sites with sufficient capacity for stable storage and effective utilization of the CO_2 generated, along with efforts to control CO_2 emissions themselves. The optimal method of transportation to the storage site should be selected considering the location of the CO_2 emission source and the storage site, the state of infrastructure development between them, and the amount of CO_2 transported. If the storage site is on land, transportation methods include pipelines, tank trucks (automobiles), and railroads. If the storage site is in a sea area, there are means of transportation by submarine pipelines and ships. The most important effect of impurities on CO_2 pipeline transport is an increase in corrosion rate and changes in the decompression curve of CO_2 (Bilio et al., 2009). Also, ISO (2024) showed a new evaluation method for unstable ductile crack arrestability, as shown in Fig. 7. The A-B-C line is the boundary where the crack stops. The crack can be stopped below this line. This approach is based on DNV-RP-F104 (DNV, 2021a). However, it is important to note that this is only a statistical finding based on experimental data and cannot be applied to steel pipe types that do not meet the experimental assumptions.

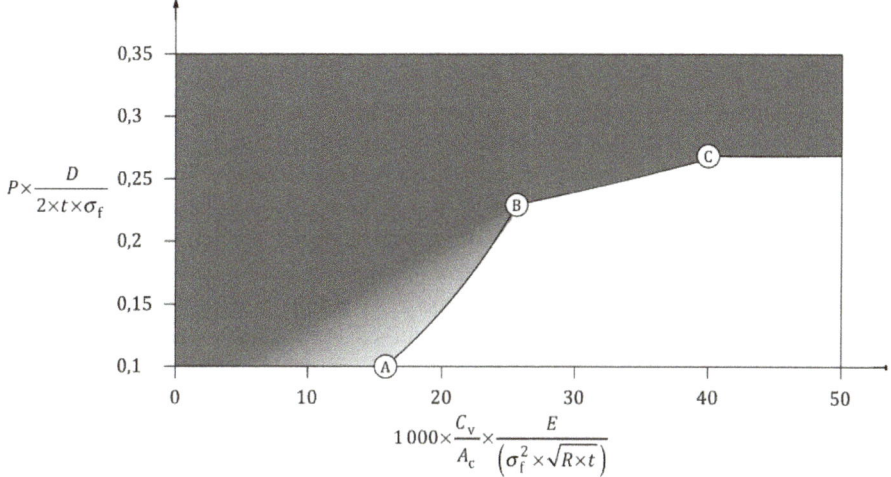

Fig. 7. Evaluation of fracture arrest in dense CO2 stream pipelines.

For pipeline materials, Bilio et al. (2009) recommended that, as with natural gas transportation, materials should be selected with consideration for brittle and ductile fracture. For casings, Millet et al. (2021) recommend carbon steel, mainly 13Cr steel, etc. For ships, Tanaka et al. (2022) discussed the concept of cryogenic transportation and the importance of computational fluid dynamics (CFD), as shown in Fig. 8. To quantify the fracture strength required in the material, it is not enough to assume that the contents are subject to inertial forces, and it becomes more important to consider complex fluid forces using CFD. Lee et al. (2023) presented an analytical approach to evaluate the effect of mechanical stress relief (MSR) on relieving residual stress, where loading conditions were simulated from an operation condition to calculate required fracture toughness.

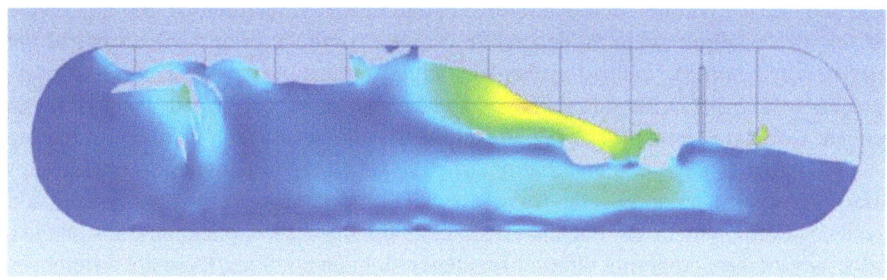

Fig. 8. CFD sloshing simulation of a liquefied CO_2 cargo tank.

3.4 Summary

This chapter reviewed recent advancements in fatigue and fracture analysis, focusing on material properties, numerical modeling, and assessment methods. It examined high-tensile steels, titanium, aluminum, nickel alloys, and composites, highlighting their fatigue behavior, welding characteristics, and structural applications. Developments in additive manufacturing techniques, such as WAAM and FSW, were analyzed for their effects on fatigue resistance. The chapter also discussed fracture toughness testing, CTOD evaluation, and fatigue crack growth mechanisms. Special materials, including TRIP steels and hydrogen embrittlement-resistant alloys, were evaluated, particularly for cryogenic applications such as LNG and hydrogen storage. The fatigue performance and challenges of carbon fiber-reinforced polymers in marine environments were also explored. Furthermore, multiaxial fatigue was analyzed, focusing on proportional and non-proportional loading effects. Alternative damage accumulation models to the Palmgren-Miner rule were reviewed. The impact of welding residual stresses on fatigue life and brittle fracture resistance was also examined. Additionally, emerging trends in testing methods for decarbonization-related materials, including hydrogen, ammonia, and CO_2 storage and transport solutions, were highlighted.

4 Computational Modeling of Fracture and Damage Mechanics

Fatigue life modeling has evolved from traditional S-N curve methods to more sophisticated computational techniques integrating fracture and damage mechanics. This chapter explores the latest advancements in predicting fatigue life and fracture for ship and offshore structures. The key concepts and state-of-the-art techniques are outlined, emphasizing the integration of computational fracture and damage mechanics.

4.1 Fatigue Life Modeling

Fatigue life modeling begins with understanding the cyclic nature of loading, which leads to progressive material degradation. The fatigue life of a component can be divided into two stages: fatigue crack initiation and propagation. For ship and offshore structures, fatigue life prediction must account for the variability of cyclic loads, complex geometries, and harsh environmental conditions. Fatigue life is often divided into high-cycle and low-cycle fatigue regimes, depending on the magnitude of applied stress and the corresponding number of load cycles for fatigue failure. As emphasized in (Muñiz-Calvente et al., 2022), fatigue damage assessment can be performed using time- or frequency-domain methods; see Fig. 9. Fatigue damage assessment involves evaluating the cumulative damage in a material or structure subjected to fluctuating stresses over time. The two primary approaches for assessing fatigue damage are the time-domain and frequency-domain methods. Each has distinct methodologies, applications, advantages, and disadvantages. Applying time- or frequency-domain methods, there are differences in the description of the input parameters and methods, such as material characterization, the definition of the reference parameter, and the treatment of loading history with different cycle counting algorithms.

Fatigue assessment in the time-domain analyzes stress or strain signals directly in the time domain, typically obtained from measurements or simulations. A fatigue damage model, such as Miner's Rule, is applied after extracting stress cycles using a technique like Rainflow Counting. The advantage of this method is its accuracy, as it accounts for the actual sequence and amplitude of load cycles. Thus, it is ideal for irregular or non-stationary signals, especially applicable for transient, non-linear, or time-varying loads. Moreover, the method can directly incorporate advanced material models and damage accumulation theories. The disadvantages of the fatigue assessment in the time domain lay in its computational cost, as it requires large data sets and higher computational power for long-duration signals, the signal preprocessing, as it demands signal filtering and cycle-counting techniques, which can introduce errors, and thus its complexity, as processing real-world signals with noise or irregularity can complicate the analysis.

The fatigue assessment based on the frequency domain evaluates fatigue damage by analyzing the frequency content of the stress signal using the Fourier transform or power spectral density. Statistical properties of the stress signal (e.g., root mean square stress, spectral moments) are used with models like Dirlik's, Wisrching-Light's, Zhao-Baker's, or Bendat's approach to estimating fatigue damage. The advantage of the fatigue assessment based on the frequency domain is its efficiency due to faster computation compared to time-domain analysis, especially for long signals. Thus, the method is well-suited for stationary and stochastic signals where stress amplitudes follow statistical distributions such as wave loading, wind loads, or vibration-induced fatigue. Its compact representation, which reduces data to its spectral characteristics and lowers storage and computational requirements, is advantageous. However, the disadvantages of the fatigue assessment based on the frequency domain are rooted in the assumptions, where linearity and stationarity of the load are often assumed, which may not reflect real-world scenarios. In consequence, it may show lower accuracy because it can fail to capture transient or non-stationary phenomena accurately. Moreover, with this method, it can be difficult to incorporate non-linear material properties or sequence effects directly. In addition, the interpretation of multi-modal or complex spectra can be challenging.

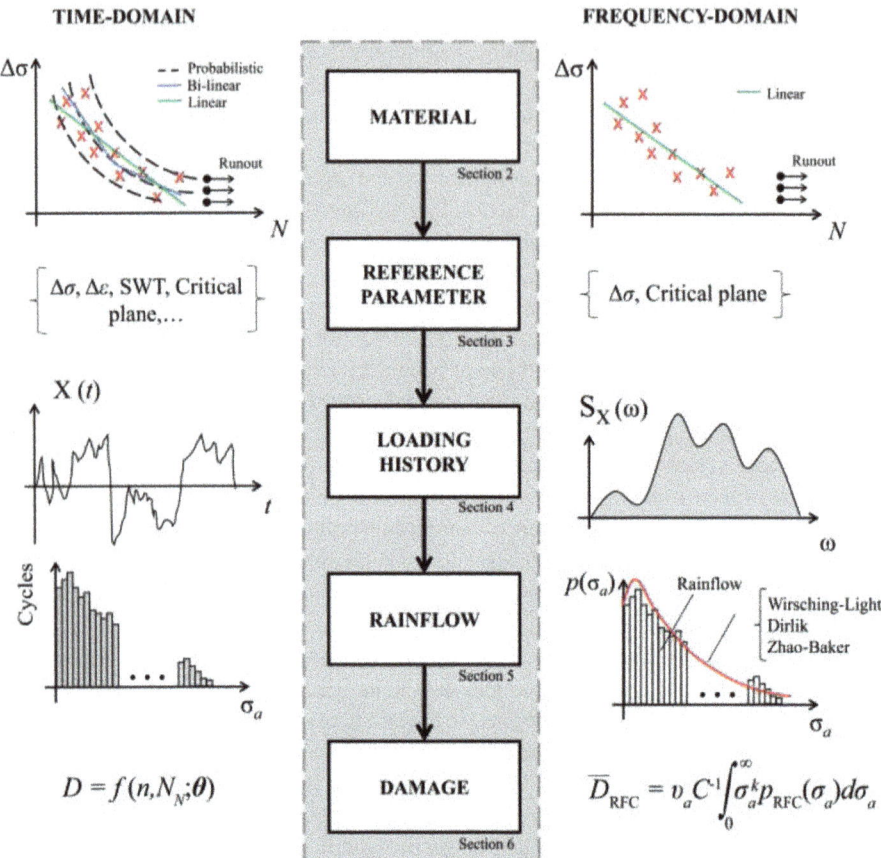

Fig. 9. General flowchart illustrating the fatigue damage assessment from both time- and frequency-domain approaches (Muñiz-Calvente et al., 2022).

4.1.1 Continuum Damage Mechanics

Continuum damage-based approaches rely on modeling the progressive accumulation of microstructural defects, which reduces the material's stiffness and load-carrying capacity. Damage mechanics are often implemented as cohesive zone models or continuum damage mechanics (CDM). CDM models introduce a scalar damage variable ranging from 0 (undamaged) to 1 (wholly damaged).

The main two branches of continuum damage-based approaches for fatigue are based on the work by Chaboche and Lemaitre. Chaboche models the deterioration before macrocrack initiation by considering the degradation of the modulus of elasticity. It can be used as a simple engineering tool that does account for non-linear damage accumulation. A recent example of applying the Chaboche model to offshore support substructure can be found in (Muzaffer et al., 2024). The model by Chaboche can be modified to incorporate residual stresses by incorporating the hydrostatic mean stress in the formulation, as shown by Zhang et al. (2012). This modification is applied to A516 pressure vessel steel, yielding a successful prediction of the fatigue life of notched specimens (Attarha and Sattari-Far, 2023). More advanced and often applied is the two-scale fatigue model by Lemaitre, which accounts for micro-plasticity at the crack tip. Lemaitre applies a micro-element embedded in an elastic representative volume element (RVE) to model the microcrack initiation and propagation. This model by Lemaitre is based on physics at the crack tip and can, in general, be considered more advanced and accurate than that of Chaboche. A recent development is the application of the constitutive equation by Lemaitre to the probabilistic fatigue life prediction of an aluminium alloy 2024-T3 base material (Liu et al., 2022b). First, macro-meso two-scale stress-strain equations are established by combining the Eshelby-Kroner localization law and classic elastic-plastic constitutive equation. Second, two parameters (S and s) in Lemaitre's fatigue damage evolution model are randomized. The parameters (S and s) are regarded as random variables that present the randomness of fatigue life defined by fitting experimental data. The probabilistic distribution of S and s over the background sample dataset is assumed to represent the scatter in the fatigue life. This fitted model is validated using test data of a full-scale panel structure with a stress ratio $R = -1$, of which all three results are within the 95% confidence bounds of the fitted model.

CDM can be applied to simulate the influence of microstructural instabilities and inhomogeneities on low-cycle fatigue (LCF) resistance. An overview of developed models for LCF is given in (Abdul-Latif, 2021). Recent developments are the application of damage-mechanics-based fatigue initiation predictions to compression-compression LCF with validation using experimental data for 7057-T6 aluminium alloy (Dong et al., 2023b) and to LCF for marine high-strength steels (Chen et al., 2023b). Based on the CDM, a damage-coupled cyclic elastoplastic constitutive model is proposed to describe the hysteresis behaviors of marine high-strength steels (Chen et al., 2023b). The CDM approach is based on experimental data, using the Ramberg-Osgood model and Manson-Coffin equation to fit and characterize the cyclic stress-strain relation and low cycle fatigue damage behaviors of marine HSSs. Petry et al. (2022) study the Voce–Chaboche (V-C) material model parameter optimization for welded high-strength steel joints. The Newton trust region method with an accumulated true strain approach successfully determined the V–C model parameters for different material zones in the welded joint and closely estimated the strain range and the fatigue life for a variable amplitude load history.

A strain-based damage model accounting for the effect of different strain ratios on fatigue failure is proposed in (Pandey et al., 2021). Low-cycle fatigue crack growth based on CDM and the extended finite element method was simulated using the Chaboche mixed hardening model to determine the stress and strain in the cracked domain accurately. Attarha and Sattari-Far (2024) studied the fatigue crack growth life of standard and smooth notched specimens containing residual stresses, which are evaluated based on the continuum damage and fracture mechanics approaches. It is indicated that fatigue crack growth lives assessed by the continuum damage model are closer to the experimental results, with a difference of $< 10\%$ in the specimens containing residual stresses. It is found that in residual stresses, the plastic area at the crack tip is relatively large. Consequently, the stress intensity factor K is not a proper parameter to evaluate the stress state near the crack tip. CDM was also used to model short crack initiation and growth using material microstructures-dependent RVE. Ono and Remes (2025) examine how surface integrity characteristics affect fatigue damage accumulation in high-frequency mechanical impact-treated welded joints. Using a non-local continuum damage mechanics model combined with elastic-plastic finite element simulations, the study investigates the effects of geometry, compressive residual stress, and work hardening on the initiation and growth of short cracks in high-strength steel joints. The results demonstrate that compressive residual stress significantly extends the growth time of short cracks (up to 0.2 mm), while the work-hardened layer primarily delays crack initiation and early growth phases. However, high-peak loads reduce the beneficial effects of residual stress, though its influence persists during short crack propagation. The study also demonstrates that most fatigue life is consumed during crack initiation and early growth stages, highlighting the critical role of surface integrity in improving fatigue resistance. Validated against experimental data, the model provides a tool for predicting the fatigue life of high-performing structures under variable surface conditions and loading histories.

CDM has been recently applied to understand fatigue behavior in offshore structures to ensure their longevity and operational safety. Beier et al. (2024) investigated the fatigue behavior of suspended inter-array power cables connecting two floating offshore wind turbines. Two spars and two semi-submersible FOWTs interconnected with dynamic power cables are chosen, respectively, for the present investigation. A parametric study is performed on the effect of the number of buoys attached to the cable on the fatigue life. For the specific design, the power cable fatigue life is higher for the semi-submersible floaters than the spar floaters under the same met-ocean conditions. The critical component prone to fatigue failure for all evaluated cases is the armor layer. The largest fatigue damage is observed at buoy-cable connection points. Okenyi et al. (2023) employed a novel corrosion-based fatigue model to determine the fatigue life of offshore welded structures. In addition to the material's ultimate strength, endurance limit, and stress ratio (mean stress effect), the model includes a corrosion factor concept to account for the impact of corrosion pits on the fatigue performance of welded S355 steel.

4.1.2 Fracture Mechanics

As a result of sustainability concerns, there has been renewed interest in reliable life extension methodologies for existing marine structures (both offshore installations and transport systems) and fixed and floating wind energy support structures (Fajuyigbe and Brennan, 2021, Zhao et al., 2021b, Wang et al., 2021b, Fang et al., 2022, Yeter and Garbatov, 2022, Yeter et al., 2022b, Yeter et al., 2022a, Yeter and Brennan, 2024, Zhu et al., 2024) over the last few years. Fracture mechanics-based assessment methods, often referred to as Fitness for Service (FFS) or Engineering Criticality Assessment (ECA), have become indispensable, particularly for implementing an effective digital twin (Fang et al., 2022). In this regard, a Paris-type crack growth model is typically used to interpret application-specific data, e.g., addressing low-temperature service environments in polar regions (Zhao et al., 2021b) or corrosion environments (Jacob and Mehmanparast, 2021, Mehmanparast and Vidament, 2021). An emphasis has been placed on a quantitative determination of risk level for continued operation without or with remediation after detecting structural damage or anomalies (Hlaing et al., 2021, O'Donnell et al., 2021). To support such risk-informed FFS or ECA assessment, reliable structural health monitoring methods become essential, which are increasingly relying on the recent advances in machining learning (ML) or artificial intelligence (AI) algorithms, e.g., as discussed recently in (Yeter et al., 2022b, Yeter et al., 2022c).

Lee and Kim (2022b) proposed a new crack growth model specifically to be applied to ship structures, including LNG vessels. The model was formulated and improved by considering complex environmental conditions experienced by ships, such as under varying loads, including storms. The new crack growth formula combined the existing Huang's model and Porter's model for the effect of stress ratio and load history factor with an additional number of variables like the constant parameter of load history parameter and exponential weighting factor to the load history and stress ratio parameters. Aursand et al. (2024) proposed a crack growth model and method for establishing equivalent cracks from surface scans of corrosion conditions to estimate the remaining fatigue life of a corroded mooring chain. The crack growth model assumes the prediction of fatigue crack growth rate based on linear elastic fracture mechanics together with the Paris-Erdogan law. Besides estimating the fatigue life, the crack propagation model is also used to identify the effect of structural degradation. Elahi et al. (2023) investigate the effect of pitting corrosion by using a short fatigue crack propagation model. Liao et al. (2022b) proposed an improved crack growth model based on the Paris equation and continuum damage mechanics by introducing corrosion damage into the Paris equation. The model demonstrates good accuracy and predicts the corrosion fatigue life of T-welded joint specimens in artificial seawater. Feng et al. (2023) conducted high-cycle fatigue tests of welded plate joints and proposed a new and enhanced constitutive damage model, incorporating a new fatigue indicator involving the plastic strain energy density, positive elastic strain energy density, and mean stress effect. The proposed constitutive damage model successfully predicts the fatigue life of welded plate joints under different levels of cyclic loadings. The remaining fatigue life of welded plate joints is updated through adsorbing periodic crack measurement information and achieves good agreement with the experimental results.

A new simple nonlinear creep-fatigue interaction (CFI) damage model coupling CDM and extended finite element method (XFEM) based methodology is developed by Pandey et al. (2023) to predict the crack growth life under a creep-fatigue environment. The proposed model computes stresses in the cracked domain and determines fatigue life. Ragab et al. (2022) studied creep crack growth modeling of weldments at high temperatures using a modified ductility-based CDM model, which considers ductility exhaustion, i.e., lost ability to undergo plastic deformation. This phenomenon typically occurs in ductile materials under certain loading conditions, such as cyclic loading, high-stress concentrations, or after prolonged exposure to environmental factors like high temperatures. The proposed model holds a key advantage over existing models in that it requires fewer material constants to be identified and calibrated. The modified model was implemented into a user-defined subroutine in ABAQUS and then used to predict the creep crack growth of Grade 91 vessel weldments. Creep crack initiation and growth in the vessels were predicted to occur in the HAZ of the weldments, which showed good agreement with the corresponding vessel tests. Further, the predicted creep rupture lives of the notched bars and the vessels using the proposed model correlated reasonably well with the experimental results. This suggests that the modified ductility-based damage model can be used for creep crack growth and life prediction for high-temperature structures.

4.1.3 Influence of Imperfections

Imperfections such as voids, inclusions, surface roughness, and weld discontinuities play a significant role in fatigue life prediction. The size, distribution, and nature of imperfections affect the fatigue life and contribute to a considerable scatter of fatigue properties. He et al. (2022) proposed a probabilistic fatigue life prediction model considering imperfection location and size sensitivity. The model is based on the calibrated weakest-link theory. Specifically, a surface factor was defined to consider the dominance of the surface defect over the interior one. In addition, a size sensitivity factor was developed to capture the sensitivity of fatigue strength to the volume and size change. Their model validation and comparison using two types of metallic materials indicated that the developed surface factor and size sensitivity factor improve the predictive ability of the traditional weakest-link theory model.

Local undercut imperfections at the welded toe are critical sites for fatigue crack initiation, significantly affecting the fatigue strength of welded structures. While the impact of continuous undercut depth on fatigue performance is well-documented, research on the influence of localized undercut geometry remains limited. Niraula et al. (2024) investigate the effects of three-dimensional undercut geometry on the fatigue strength of welded joints through elasto-plastic finite element analysis, incorporating both geometric and plasticity effects. A parametric model derived from high-resolution measurement data is employed to represent realistic undercut geometries. A full-factorial design of six geometric parameters (depth, radius, width, length, and others) enables systematic exploration of their individual and combined effects on fatigue performance. The study proposes a novel three-dimensional undercut geometry index validated for estimating fatigue crack initiation life. It provides valuable insights into fatigue strength-based quality criteria for welds with local undercut defects.

4.2 Brittle and Ductile Fracture Modeling

Offshore and ship structures are often subjected to extreme loads, which may cause brittle or ductile fractures, depending on the material properties and operating conditions. Brittle fractures are characterized by sudden failure with minimal plastic deformation, while ductile fractures involve significant plastic deformation before rupture. Predicting these failure modes is critical for ensuring the safety and longevity of marine structures.

4.2.1 Continuum Damage-Based Approaches

Besides fatigue life modeling, CDM offers a powerful framework for modeling brittle and ductile fracture behavior by simulating the progressive degradation of materials under stress. In this approach, a damage variable is introduced to represent the internal material degradation evolving under loading conditions. For brittle materials, the damage evolution is rapid, leading to sudden fracture, while for ductile materials, damage accumulates more gradually due to plastic deformation. However, the numerical solution of CDM problems can suffer from convergence-related challenges during the material softening stage, and consequently, existing iterative solvers are subject to a trade-off between computational expense and solution accuracy. Saji et al. (2024) present a novel unified arc-length (UAL) method, where they formulated the analytical tangent matrix and governing system of equations for both local and non-local continuum damage problems. This approach renders the proposed solver substantially more efficient and robust than existing solvers used for CDM problems. The proposed UAL approach exhibits a superior ability to overcome critical increments along the equilibrium path and is 1–2 orders of magnitude faster than force-controlled arc-length and monolithic Newton-Raphson solvers.

Regarding ductile fracture modeling, Park et al. (2024) propose the Gurson-Cohesive model to investigate 3D ductile fracture accounting for void growth, coalescence, and complete failure phenomena. The proposed model idealizes the ductile fracture process as continuum damage evolution, cohesive crack initiation, nonlinear softening along the crack surface, and complete failure and describes continuum damage with void growth. In contrast, the cohesive zone model is utilized to introduce discontinuous cracks and represent nonlinear softening behavior. The transition from the continuum damage to discontinuous crack is systematically considered using a porosity-based crack initiation criterion considering stress triaxiality. Ziccarelli et al. (2023) extended a micromechanics-based adaptive cohesive zone model for simulating ductile crack propagation under monotonic loading to handle cyclic loading. The proposed model adaptively modifies the cohesive traction–separation relationship for crack opening and closure, as the loading reverses between tension and compression. They implemented their approach into the finite element analysis platform WARP3D. Validations between simulations and coupon tests demonstrate that the proposed model can accurately simulate the effect of crack propagation on specimen response, as well as other key aspects of observed behavior, including crack face closure and crack tunneling. Padilla-Llano et al. (2022) present a finite element-based model for direct simulation of cyclic fracture in steel components and connections with application to predicting failure and collapse of steel structures. The formulation couples a plasticity model for large deformations that captures plastic work stagnation and the Bauschinger effect with a two-stage damage model to simulate fracture initiation, propagation, and failure through an element deletion strategy. Their model includes the non-proportional loading and load-history effects in the fracture initiation and propagation. Fracture initiation and propagation are controlled using new fracture initiation strain and fracture energy surfaces with stress triaxiality and Lode angle dependence that can represent different fracture behaviors.

4.2.2 Fracture Mechanics -Based Approaches

Fracture mechanics-based approaches offer a robust framework for understanding and predicting both brittle and ductile fractures by focusing on the behavior of cracks and defects. For brittle materials, the propagation of cracks occurs rapidly, often driven by the stress intensity at the crack tip, which is analyzed using linear elastic fracture mechanics. In contrast, the ductile fracture is typically modeled using elastic-plastic fracture mechanics, which considers plastic deformation around the crack tip, often using parameters such as the J-integral or crack tip opening displacement. Al-Hagri et al. (2024) investigated how different modeling choices may influence the calculation of the stress intensity factors (SIF) in the joints of the steel jacket structures. The different fidelity approaches were especially studied in terms of their influence on the SIFs at the crack fronts. Moreover, this study elaborated on the sensitivity of the SIFs by considering the flexibility of the joints. The findings emphasized the viability of employing multi-fidelity modeling, as this approach demonstrated a favorable compromise between computational efficiency and the precision of SIF calculations. Moreover, the incorporation of the joints' flexibility revealed a corresponding rise in the computed SIFs, reaching up to 7%, attributed to the resultant reduction in the overall stiffness of the steel jacket structure.

4.2.3 Recent Advances in Predictive Fracture Modeling

Tlatlik et al. (2023) explored the initiation of brittle fracture in safety-critical components as recent research has shown the limitations of macroscopic methods, such as the Master Curve concept. These traditional macroscopic approaches are inadequate for reliably assessing brittle failure under dynamic loading conditions, particularly due to their severe shortcomings in evaluating cleavage fractures at elevated loading rates (crack tip loading rate of about 10^3 to 10^5 MPa$\sqrt{m}$/s). The study utilized local cleavage fracture models to assess brittle fracture at elevated loading rates, proposing adjustments to enhance the analysis. The adjusted "dynamic" Master Curve improved the accuracy of predicting brittle fracture initiation. Additionally, Shen et al. (2023) investigated cryogenic fracture properties of a bcc steel, providing insights into the transition from brittle to ductile behavior at $-196\,°C$. Their study introduced an anisotropic unified fracture criterion that demonstrated remarkable predictive capability and enhanced the understanding of cryogenic fracture initiation mechanisms. Ruggieri and Jivkov (2022) presented a probabilistic framework for characterizing cleavage fracture initiation in structural ferritic steels, showing promise as an engineering-level procedure for predicting brittle fracture initiation behavior across temperatures and crack-tip constraints. Addressing the applicability of the Weibull stress method in predicting brittle fracture initiation at beam-to-diaphragm connections, Shimizu et al. (2023) proposed a fracture evaluation method considering mixed-mode loading conditions, improving the accuracy of predicting brittle fracture initiation, particularly at singularities subjected to the weld toes of cope holes.

In (Shibanuma et al., 2022), a novel approach utilizing the s-version finite element method was introduced to accurately assess local stress during dynamic brittle crack propagation in 3D solids. This strategy defines a local, structured mesh covering the crack front and aligned with the crack propagation direction, demonstrating exceptional accuracy compared to standard FEM, adaptive re-meshing, or even XFEM simulations without requiring intricate re-meshing procedures. Validated through analytical closed-form solutions for stationary and dynamically propagating circular crack problems, this approach offered a promising foundation for numerically analyzing dynamic brittle crack propagation issues. As a separate investigation, Dai et al. (2021b) explored fracture parameters in cracked shells subjected to out-of-plane loading using ordinary state-based peridynamics. The authors employed adaptive dynamic relaxation for steady-state solutions using a non-local deformation gradient and equivalent domain integral. The proposed peridynamics shell model successfully evaluated fracture parameters under single- and mixed-mode loading conditions, contributing valuable insights into brittle fracture propagation behavior in shells and enhancing the understanding of structural responses under diverse loading scenarios.

Several studies contributed valuable insights into modeling and simulating the intricate ductile fracture process. Bergo et al. (2021) presented a non-local Gurson-Tvergaard-Needleman (GTN) model for explicit finite element analysis, incorporating an integral condition on the rate of change of porosity. This non-local approach demonstrated mesh independence during ductile damage and fracture stages, offering a unique perspective on strain localization and crack propagation. Chen et al. (2022) investigated ductile tearing in a full-size pre-cracked pipe using a non-local GTN model, successfully reproducing experimentally observed crack branching, a challenging feat for local GTN models. Nkoumbou Kaptchouang et al. (2021) adopted a cohesive GTN model for simulating crack initiation and propagation in ductile materials, demonstrating efficacy in simulating crack growth in standard ferritic steel. Seo et al. (2022) focused on simulating ductile crack growth in thin plates using a simplified strain-based damage model. The model, dependent on stress triaxiality with two parameters determined from tensile test results, exhibited good agreement between simulated and experimental results, emphasizing the importance of in-plane element size in crack growth simulations. Wang et al. (2020) proposed a material model based on the Johnson-Cook (J-C) and GTN models for ship collision and grounding simulations, addressing the limitations of the GTN model and demonstrating applicability for structural failure response prediction in maritime scenarios. Huang et al. (2020) introduced a continuous damage model for Q690D steel, considering the influence of the Lode parameter. The Lode parameter is a scalar value used in material mechanics and plasticity theory to characterize the state of stress in a material, particularly in three-dimensional stress states. It provides a measure of the deviation of the stress state from purely uniaxial tension, pure shear, or equibiaxial tension and compression, which are important in understanding material behavior under different loading conditions. The model, calibrated with experimental results, accurately predicted crack initiation and evolution for Q690D steel, showcasing improved accuracy compared to models neglecting Lode parameter effects. Two additional studies by Cerik et al. (2021) evaluated localized necking models for fracture prediction in punch-loaded steel panels, comparing experimental results with numerical predictions using different localized necking modeling approaches, highlighting the mesh size sensitivity and the improved estimations achieved with the Bressan-Williams-Hill criterion. Furthermore, Lee and Basaran (2021) reviewed degradation and damage evolution models in the literature for various engineering materials, primarily metals and composites. Empirical models established under the framework of Newtonian mechanics, GNT, J-T, microplasticity, and other micro-mechanism-based damage models are thoroughly discussed. Also, and models using irreversible entropy as a metric with an empirical evolution function are included. Together, all these recent studies contributed diverse perspectives and methodologies, enriching our understanding of ductile crack growth and providing valuable tools for predictive modeling in various structural scenarios.

Beyond traditional methods, phase-field and peridynamic models offer powerful non-local frameworks for simulating complex fractures, each addressing limitations of classical fracture mechanics in different ways. Phase-field models employ a continuous field variable to represent fracture, allowing the simulation of intricate crack paths, branching, and fatigue behavior without explicitly tracking crack surfaces. Kalina et al. (2023) Kalina et al. (2023) present an overview of phase-field models for fatigue fracture in a unified framework, categorizing them into two main types: (1) those that model fatigue through gradual degradation of fracture toughness, and (2) those that incorporate an additional fatigue term into the crack-driving force. Their work aims to guide in selecting the most suitable model for different applications. Despite their versatility, phase-field approaches are computationally intensive and require careful parameter calibration, especially under cyclic loading. While phase-field models rely on a diffusive representation of cracks, peridynamics takes a different route. It replaces the partial differential equations of continuum mechanics with integral equations that capture long-range interactions between material points. This formulation naturally accommodates crack initiation, branching, and interaction without remeshing, making it particularly effective for simulating multiple interacting cracks. Bachimanchi and Saha (2023) model tubular joints having T, Y, K, and X configurations in the peridynamic framework by implementing peridynamic shell governing equations. A comparative study is performed among the offshore tubular joint configurations. The analysis considered the linear displacement variation and critical loads for the unstable deformation due to damaged material points at the joint intersection. However, like phase-field methods, peridynamics also faces challenges in notably high computational cost, complex parameter definition, and ongoing research needs for robust application to elastoplastic and anisotropic materials.

4.3 Structural Modeling Approaches and Guidelines

4.3.1 Global and Local Models with Sub-Structuring Techniques

Structural modeling techniques are essential in engineering analysis for predicting the fatigue and fracture behavior of structures under various loads. These techniques can be broadly categorized into global and local models, each serving different aspects of the design and analysis process. Global models represent the entire structure, capturing components' overall behavior and interactions. They provide a broad understanding of how a structure responds to external forces. However, they often lack the precision needed for localized areas of high stress or complexity due to the limitations in computational resources or insufficient levels of discretization. Local models, on the other hand, focus on specific regions of interest, typically areas prone to high stress, deformation, or failure (e.g., joints, notches, or interfaces). By creating a refined mesh in these regions, local models provide detailed insights into the structural response. Local models often use data from global models to define boundary conditions, ensuring the consistency of analysis. In Finite Element Analysis (FEA), both submodeling and substructure approaches are used to deal with complex systems, but they serve different purposes and are applied in distinct ways. Substructuring is a technique used to simplify analysis by breaking a structure into smaller, more manageable components (substructures). These components can be analyzed independently and then assembled into the global model. Thus, it enhances efficiency by reducing the size of the global problem, allowing smaller, condensed models to be solved and assembled faster than analyzing the entire structure at once. This technique often relies on creating super-elements, condensed representations of the substructures, capturing their essential stiffness and mass characteristics. Submodeling helps balance computational efficiency, analyzing local parts of a structure or component using a refined mesh. Two different submodel approaches are typically applied. The classical submodel is a local model where the loads and boundary conditions are based on the interpolation of the solution from an initial global model (usually with a coarser mesh). It projects these onto the nodes on the appropriate parts of the boundary of the submodel, as shown in Fig. 10. The classical submodeling approach is a widely used technique that integrates global and local modeling. General commercial finite element codes support sub-modeling procedures. Also, integrated CAD/CAE software allows it (Narvydas et al., 2021). Moreover, more specialist software codes like, e.g., Sesam HydroD with its Submod software module, allow a closer look at details without refining the entire model by taking global model results and a local model as input to compute the boundary loads (DNV, 2025).

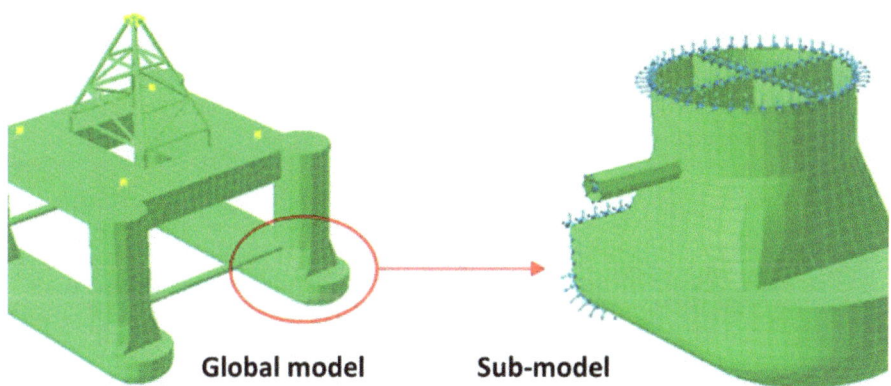

Fig. 10. Global-local model of an offshore platform illustrating the sub-modeling approach and its boundary conditions. The figure is taken from (DNV, 2025).

Creating a classical submodel involves several steps. The global model is created and analyzed, simulating the entire structure or domain and capturing the overall response of the structure. Once results from the global model are obtained, the region of interest is isolated, and a local model (submodel) is developed with a finer mesh for increased accuracy. Then, the results (boundary conditions and displacements) from the global analysis are used to refine and analyze a local sub-model. Time-dependent variables saved from the global model's analysis drive the submodel's boundaries. These boundaries can be driven either by applying boundary conditions or, in certain cases, using stresses from the global model. The submodel can be run separately from the global analysis. The only link between the submodel and the global model is the transfer of the time-dependent and time-independent values of variables saved in the global analysis to the relevant boundary nodes of the submodel. Consequently, the submodel response does not influence the global response, and thus, it is crucial to check that the submodel response has an insignificant impact on the global response. This is the fundamental assumption of submodeling. If the global response is affected by the submodel response, a multi-fidelity submodeling approach is recommended, linking the global model and the submodel. In multi-fidelity sub-modeling, meshes of different element types and discretization, e.g., beam-to-solid or shell-to-solid coupling, can be used. Unlike classical submodeling, multi-fidelity simulation approaches typically link meshes of different element types and can be run as a single analysis. In this way, the response of the high-fidelity local model is directly linked to the global low-fidelity model, and its structural behavior can affect the global response. Another possibility that really distinguishes between a global model and a submodel (without integration) is a "two-way coupled" analysis, where the response of the local model is fed to the global model. Al-Hagri et al. (2024) demonstrate a multi-fidelity submodel approach, where a low-fidelity jacket offshore structure model is combined with a high-fidelity fracture mechanics model with cracks in the welds, as shown in Fig. 11. Different submodeling approaches and their impact on the stress-intensity factor (SIF) were studied. The findings emphasize the viability of multi-fidelity modeling as a compromise between computational efficiency and the precision of SIF calculations. A similar study was conducted by Peng et al. (2023), who compared the fatigue damage data of a jacket model calculated using the multi-scale finite element model with those of the conventional beam model. Their multi-scale finite element model results show that the difference in uniaxial fatigue damage can be about 15% larger than the beam model.

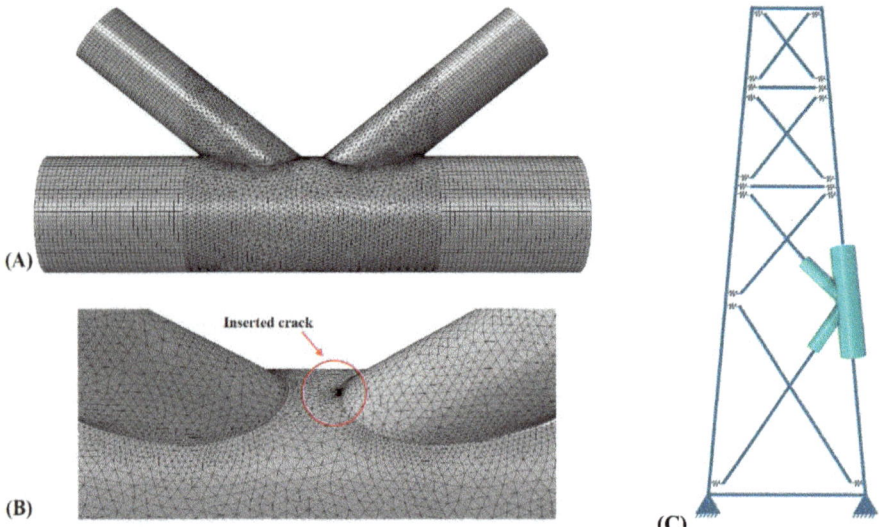

Fig. 11. (A) Meshing pattern of the K-joint and the localized region, and (B) Close-up on the mesh surrounding the inserted crack, (C) Representation of the multi-fidelity FE model of the jacket structure (pictures are taken from (Al-Hagri et al., 2024)).

Lu et al. (2023) proposed an efficient time-domain fatigue damage assessment approach for tripod structures by establishing a global-local FE model with detailed 3D solid element modeling for critical zones and a simplified beam model for the remaining regions and came to similar conclusions. Important component of their methodology was the Stress Influence Matrix (SIM), which mathematically represents how an external load or force applied at one point in a structure influences the stresses at other points or elements. The SIM, formed by multiple finite element analyses, allows for the efficient calculation of time-variant stress based on the equivalent static loads and the linear superposition principle. All publications suggest using multi-scale/fidelity finite element models for better accuracy in predicting jacket-type offshore structures exposed to multi-axial stress states.

Zhao et al. (2022) compared flexible and rigid large-volume substructure models to highlight the effect of substructural flexibility on the hydrodynamic loads and dynamic responses of FOWT by implementing a fully coupled simulation analysis in the time domain. They also examined how flexible and rigid large-volume substructure models impact the structural fatigue behavior of a novel semi-submersible floating offshore wind turbine (FOWT). The results show that substructural flexibility significantly impacts fatigue damage of the integrated FOWT system in operating sea states rather than in extreme sea states. Marjan and Hart (2022) investigated different parameters and design modifications that can impact the design life of an offshore jacket foundation by substructuring using a super-element model. The super-element model connected to an NREL 5-MW wind turbine is used to compute the fatigue damages during the service life. The impact of soil non-linearity, marine growth, scour size, the mass of the transition piece, and the grouted connection's design on the dynamic response and fatigue damages are compared. A 30% increase in fatigue life for tubular joints based on geometry and butt welds was observed by replacing the concrete transition piece with a lightweight steel configuration. Augustyn et al. (2021) presented a probabilistic framework for updating the structural reliability of offshore wind turbine substructures using digital twins to quantify and update the structural and loading uncertainties associated with the structural dynamics and load modeling parameters in fatigue damage accumulation. For this purpose, a numerical model of a 7 MW jacket-supported turbine using a substructure model is reduced to a Craig–Bampton supplement with 30 internal modes. The quantification and update of the uncertainties associated with the structural dynamics and load modeling parameters in fatigue damage accumulation are included in a probabilistic model for fatigue damage accumulation to update the structural reliability. It is found that updating the soil stiffness significantly affects the reliability of the joints close to the midline while updating the wave loading substantially affects the reliability of the joints localized in the splash zone.

Toor and Lotsberg (2024) performed fatigue calculations for geometric singularities at welded components using a sub-model shell-to-solid interface to model welded components in FE analysis and to calculate the hot spot stress at the fatigue critical location; see Fig. 12. Their study concluded that the weld geometry and stiffness significantly influence the hot spot stress calculation at the considered cope hole. Thus, the weld geometry and stiffness must be included in the finite element model for an accurate fatigue damage calculation of such details. Chang et al. (2022) demonstrated that submodeling can also be applied to plane frame models using a large bridge model as a case study. They presented an example of a plane frame model developed using the submodeling technique based on nodal displacements. The results were analyzed and compared regarding the stress field and the plastic damage observed in the concrete.

The structural modeling with welding distortion is explored by Mancini et al. (2024). They develop modeling strategies for thin-walled stiffened panels with welding-induced distortions. Thin stiffened panels, common in lightweight ship superstructures, suffer from complex distortion shapes that complicate traditional structural stress analysis methods. The study evaluates gradual scale reductions, transitioning from 3D nonlinear FEA to simplified 2D and 1D models, using experimental distortion data from full-scale ship-deck panels. A 2D analytical plate model accurately predicts structural stress distributions when distortion curvature slopes are below 0.02 radians and amplitudes smaller than the plate thickness. For localized stresses near welds, a simplified non-linear 1D beam model effectively estimates structural hot-spot stress when the maximum distortion amplitude is less than 0.6 times the plate thickness. These models significantly reduce computational complexity while maintaining acceptable error margins (below 10%) in most cases. The study bridges the gap between computationally intensive 3D models and industry-standard simplified methods, providing tools for early fatigue design stages.

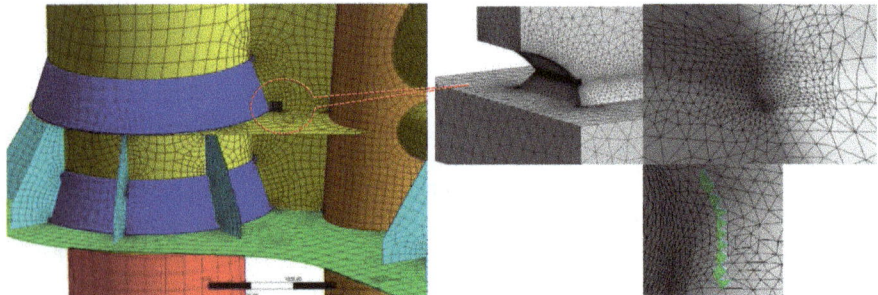

Fig. 12. Global and sub-model mesh using a shell-to-solid coupling to calculate the hot spot stress at the fatigue critical location (Toor and Lotsberg, 2024).

4.3.2 Structural Modeling Techniques and Approaches

DNV (2021d) developed a classification guideline finite element analysis tool for the fatigue assessment of ship structures to get the best accuracy that supports fatigue fracture analysis. DNV guideline categorizes the FE modeling technique on three levels. There is a global model, a partial ship model, and a local model. According to DNV (2021f), the global model represents a model of the entire ship, including all ship structural elements (longitudinal and transverse structural members). Within the global model, the element mesh size is standardized so that the element size will have a proper aspect ratio of the elements and adequate mesh arrangement. Generally, the element size is between the spacing of longitudinal girders, transverse webs, or between stringers and decks. All primary longitudinal and transverse structural members for the global model are modeled by membrane or shell elements. In the partial ship model technique, finite element (FE) analysis is conducted on the cargo hold region, which includes three cargo holds. The model uses a mesh size of s × ss × ss × s (where sss represents the stiffener spacing). Shell elements are employed to represent the plates, while beam elements are used for the stiffeners. The local model is utilized to perform finite element (FE) analysis for specific details, including bracket toes and flange terminations of stiffeners, slots, and lugs in web frames at their intersections with stiffeners, bracket and flange terminations in girder systems, and panel knuckles. DNV (2021d) describes the element model and meshing size of the local model based on the considered structural detail. In general, shell elements are used.

It has been well established that finite element stress analysis of welded structures must address mesh sensitivity because of stress/strain singularity at the weld toe or root (Xing et al., 2023, Yu et al., 2022c, Zhang et al., 2022a). This was the primary reason the extrapolation-based hot spot stress (HSS) method was introduced in the 1980s for tubular structures and in 1990 for ship structures. Although widely adopted for applications in offshore and ship structures [e.g., (Heyraud et al., 2021, Bao et al., 2023, Chen et al., 2023c, Alencar et al., 2021)], it should be noted that HSS methods still suffer a certain degree of mesh-size sensitivity, mainly when dealing with complex stress states, e.g., multiaxial stress states in which shear stress cannot be ignored (Pei et al., 2022a, Mei et al., 2021). To address such mesh-size sensitivities, a traction-based structural stress method using nodal forces/moment from finite element model output has been developed, which has led to the development of a single master S-N curve and its scatter band (Wei et al., 2024a)). Along this line, related approaches can be found in several recent publications, e.g., (Alencar et al., 2021, Rautiainen et al., 2021). Since its adoption by ASME Sec III Div 2 BP&V Code and Standard, the method has recently been further enhanced as the structural strain method (also known as the master E-N curve method by analytically converting elastically computed traction structural stress to a structural strain parameter for accommodating local plastic deformation in low cycle fatigue regime, e.g., (Rautiainen et al., 2021, Pei et al., 2022b). As such, the structural strain method takes advantage of the mesh-size insensitivity in linear elastic structural stress calculation while offering an analytical approach to treating plastic deformation effects on low cycle fatigue at hot spot locations if required.

Peng et al. (2023) developed a multi-scale modeling method for jacket-type offshore wind turbines, in which the local joints of a jacket are modeled in detail using solid elements, and other components are modeled via the common beam element. Considering the multiaxial stress state of the local joint, multi-axial fatigue damage analysis based on the multiaxial S-N curve is performed using two equivalent stress (von Mises and Lemaitre) methods. Their results indicated that the tubular joint between the jacket leg and brace connections can be modeled using the multi-scale method. Specifically, a local model utilizing solid elements for the typical joint is coupled with beam elements through constraint equations, enabling the integration of the higher-fidelity joint discretization. Furthermore, a comparison of uniaxial and multiaxial fatigue results from the multi-scale finite element model revealed that the difference could be approximately 15% larger than results from a simpler beam element model. Xu et al. (2020) investigated tubular K-joints, frequently employed in offshore structures, subjected to multi-axial stresses, proposing a multiaxial fatigue life prediction method for tubular K-joints, which is based on an improved zero-point structural stress approach that accounts for the stress gradient in the thickness direction to quantify the structure stress at the weld toe of a tubular K-joint.

4.3.3 Screening of Fatigue and Fracture Critical Areas

Fatigue-critical areas are usually the locations where multiple structural elements relate to different angles. Screening fatigue and critical areas is a special field that has shown only limited recent research activities in terms of scientific publications. The current design guidelines provide approaches for identifying fatigue-critical locations; see, e.g., (ABS, 2020). The course finite element model can used to identify highly stressed areas, and then a detailed fatigue assessment of identified locations is performed with a fine element mesh model. Also, the predefined structural list of potential critical locations can be utilized. However, it is worth mentioning that special attention should be paid partially when new structural solutions with advanced geometry topologies are utilized. Thus, further research in that field will be beneficial in the future. Also, screening automation is an area of interest to the industry.

In the predefined structural list, fatigue-critical areas are usually locations where multiple structural elements relate to different angles. In this regard, hopper knuckle connections are usually prone to fatigue failures regardless of ship type. For instance, Lloyd's Register (LR) generally recommends checking the following details (LR, 2009): 1) Hopper knuckle connections, where misalignments may exist 2) Bulkhead stool to inner bottom connections 3) End connections of longitudinal; see Fig. 13. Misalignments are advised to be modeled explicitly by FEM if the generic SCF formulas do not apply to the complex details. Like the hopper knuckle connections, bulkhead stool to inner bottom connections are prone to misalignments, and careful screening is required. The stresses at the stool sloping plate, inner bottom plate, and double bottom floor web may generate a complex stress state at the connections, and detailed FEM modeling, including misalignments, may be required to calculate the hotspot stresses. Longitudinal stiffener end connections are susceptible to fatigue failure due to the SCFs in the transverse frames. The fatigue life of the stiffener end connections may be obtained by simplified stress analysis based on beam theory and suitable SCFs from IACS CSR-H (IACS, 2023b). Stress analysis of the free edges at the stiffener passage slots must be undertaken by dummy rod elements for accurate predictions in the fatigue assessment.

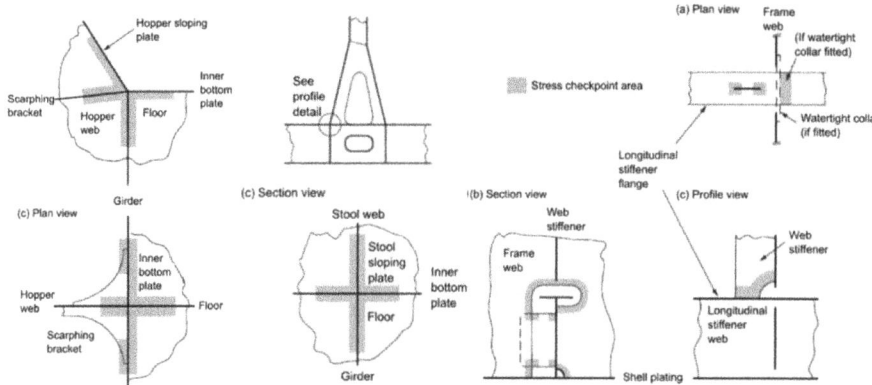

Fig. 13. Examples of fatigue-critical areas for hopper knuckle connections, bulkhead stool, and the end connections of longitudinal stiffeners.

There are more specific recommendations depending on the relevant ship type and size. For instance, considering LNG and LPG carriers with IMO Type-A and Type-B tanks constructed using flat surfaces, special consideration is needed for fatigue screening. LR recommends conducting screening analyses based on the nominal stress and using appropriate design S-N curves with geometric stress concentration factors as necessary for several specific details (LR, 2022c). These details include the primary and secondary member face plates at attachment points such as brackets or webs, stiffener heel and toe boxing fillet weld connections, and tank skin fillet weld toes at T-joints with primary and secondary members. It's important to note that the locations to be assessed are not limited to these details, and each site may need to be considered on a case-by-case basis. For fatigue screening of an LNG carrier with an IMO Type-C tank, the following locations should be considered (LR, 2024b): the tank shell at support locations, tank joint connections between cylinders and their longitudinal bulkheads, support ring frames, and high-stressed locations identified from ultimate limit state assessments. Detailed FEA analyses are required at locations identified in the screening assessment, such as cruciform joints, tank corner skin butts, seats and chocks, primary member butt joints, and bracket toes. When assessing the tank and hull structures, particularly around supports and chocks, special attention should be given to the vertical supports at the aft and forward ends of the independent tank, as well as the supports along the mid-length of the tank, due to hull girder bending issues such as sagging and hogging.

4.3.4 Overview of Software and Applications

This section presents selected examples of widely used software and subroutines for fatigue and fracture analysis, with emphasis on their strengths and limitations in computational modeling of fracture and damage mechanics for marine structures.

Software examples for fatigue and fracture analysis:

- CrackWISE: An engineering critical assessment (ECA) tool developed by TWI, based on BS 7910. The latest release, CrackWISE 6, supports various structural and

flaw geometries and includes an Annex T toolkit to guide NDT capability. It largely complies with BS 7910:2019, except for Annex V (strain-based flaw assessment).
- FRANC3D: The FRacture ANalysis Code 3D is designed to simulate three-dimensional crack growth in complex structures, accommodating realistic geometry, loading, and evolving crack shapes. It operates in conjunction with general-purpose FE solvers.
- IWM VERB: A PC-based tool for assessing fractures in components with crack-like defects, using both elastic and elastic-plastic fracture mechanics. It adheres to international guidelines and standards (FITNET, SINTAP, R6, BS 7910, API 579, FKM Guideline, ASME BPVC Section XI). While primarily for metallic components under static or cyclic loading, it can also be applied to non-metallic materials describable via conventional fracture mechanics.
- Verity: A module within FEA-SAFE (3DS) for predicting failure locations and fatigue lives in welded joints and structures, including spot welds. Its nodal-force–based approach reduces mesh sensitivity compared to stress-based methods.
- CAE Fatigue: Developed by Hexagon, this software provides random response and vibration fatigue solvers in both frequency and time domains. It can assess fatigue life for spot and seam welds (2D and 3D) according to BS 7608.

In addition to these specialized codes, several major FE platforms integrate fatigue and fracture capabilities:

- ANSYS Mechanical: Includes a fatigue module (strain-life ε-N and stress-life S-N) and SMART crack growth analysis. SMART (Separating Morphing and Adaptive Remeshing Technology) automatically remeshes only the crack-tip region each iteration, improving efficiency for large models.
- MSC Nastran: Provides embedded fatigue tools for time- and frequency-domain analyses, with S-N and ε-N approaches applied via static, modal, or transient simulations.
- MSC Fatigue: Developed jointly by nCode International and MSC Software, integrated into MSC Patran. Supports S-N, ε-N, and LEFM-based approaches, including welded-structure life estimation per BS 7608.
- 3DS fe-safe: Interfaces directly with Abaqus, ANSYS, and Nastran. Predicts fatigue crack initiation and propagation locations and timelines.

An applied example of commercial software for stress concentration factor (SCF) analysis is provided in Appendix A, featuring an SCF assessment of a K-joint in a floating wind turbine support structure.

While commercial packages offer robust built-in capabilities, advanced or experimental methods often require tailored computational strategies. Two primary approaches are common:

- Secondary Development within Commercial Software: Extends existing FEA platforms via user-defined subroutines. For instance, ABAQUS provides UMAT/VUMAT for custom constitutive laws and USDFLD/VUSDFLD for user-defined field variables. This approach leverages established pre- and post-processing environments while enabling integration of research-level models.

– Fully Autonomous Code Development: Suitable for methodologies fundamentally different from classical FEA, such as peridynamics. A stand-alone implementation allows full control, optimization for method-specific requirements (e.g., non-local interactions), and customization for high-performance computing environments.

4.4 Validation Based on Full- and Large-Scale Tests

Full-scale tests provide essential real-world data for validating predictive models, helping to identify failure modes and develop improved structures for engineering applications. This chapter considers full-scale structure specimens with representative manufacturing quality and structural redundancy in the possibility of multiple load paths and redistribution.

4.4.1 Experimental Studies on Marine Structures

In recent years, significant research has been conducted to understand the impact of various factors on the fatigue resistance of marine structures. Mendoza et al. (2022) studied the fatigue of catenary-type mooring lines and empirically estimated the effect of pitting corrosion on the fatigue resistance of mooring lines. Data from fatigue testing of both new and used chain links are considered. The used chain link samples were retrieved from several offshore floating units. A hierarchical statistical analysis is proposed to use the available information effectively. The mean stress effect is taken into account in the analysis of the data. Results show that pitting corrosion's effect on the mooring lines' structural reliability is significant. Kristiansen et al. (2023) conducted an empirical study obtaining an S-N curve for post-weld treatment (PWT) based on a large-scale test. The results of the PWT and its mechanical mechanisms involved in the impact treatment techniques are supported and explained by a theoretical background. The authors concluded that both ultrasonic impact treatment and especially pneumatic impact treatment showed improved fatigue life compared to untreated welded joints but did not reach the expected detail category of 125. Tamimi and Soliman (2024) studied the influence of geometry and material input parameters on the crack propagation and fatigue service life of welded stiffened panels. Three-dimensional finite element analysis, artificial neural networks, and an elastic-plastic crack advancement rule are integrated to predict the crack propagation. Sensitivity analysis is conducted to evaluate the effect of variability in the considered input parameters on the variance of the fatigue service life of these panels. It was found that neglecting the uncertainties associated with the geometric parameters can lead to unconservative estimates of fatigue reliability.

To carry out advanced experimental investigations, robust measuring and monitoring techniques for structural response and damage detection are important. In this regard, Haselbach et al. (2024) developed a new method to determine blade properties based on photogrammetry and strain measurement data. Their so-called blade property extractions method allows for calculating the bending stiffness, the orientation of the principal axes, and the elastic center of the wind turbine blade using standalone and combinations of different load scenarios. This method enables the comparison of numerical and experimental responses more accurately compared to pure strain and displacement field measurements. The work by Bjørheim et al. (2022b) contains a comprehensive review of fatigue damage detection and measurement techniques, specifically when applied to fatigue testing. The work concludes that fatigue damage measurement during fatigue testing is time-consuming, not in the least, due to the continuous monitoring and amount of data gathered for each test. They, therefore, advocate for the provision of extensive descriptions of experimental work so as to be able to compare and combine fatigue test data that includes monitoring efforts.

4.4.2 Steel Structure Tests

For full- and large-scale specimens, one can distinguish between specimens with specific complex full-scale geometries and specimens with simplified structures to reduce computational complexity in large-scale structural analysis while retaining the essential features of the physical problem. The latter builds upon the idea of simplified structures by expanding their scope to include additional critical features or effects necessary for accurate analysis and decision-making. The difference between both categories lies in defining failure and testing termination. For a specific complex geometry, the fatigue life is typically expressed by the crack length at a certain number of cycles (Rautiainen et al., 2023) or, more precisely, at a 20 mm crack in the bulb surface (Fricke and Paetzold, 2010). For an extended simplified structure, the fatigue life reported indicates the number of cycles to a through-thickness crack (Lillemäe et al., 2017) or penetrating the weld surface (Fricke et al., 2006). Residual stress differences between small- and full-scale specimens are only explicitly considered (Fricke and Paetzold, 2010, Lillemäe et al., 2017). From the above-listed database, it can be concluded that tests of full-scale specimens are mostly aimed at obtaining data for specific typical connections with limited sample size, as shown in Table 2. However, for more generic specimens, there could be great value in obtaining more data to validate the current fatigue life prediction based on empirical data from small-scale specimens. A significant set of standardized small-scale tests together with full-scale tests are very valuable in validating advanced analysis methods, as demonstrated, e.g., in (Lillemäe et al., 2017).

4.4.3 Composite Structure Tests

Currently, the limited understanding and predictive capabilities regarding the fatigue and fracture behavior of composite materials are often mitigated through conservative design practices. These approaches involve the application of partial safety factors to account for manufacturing defects, variations in composite material quality, and undetected impact damage, aiming to prevent crack propagation and catastrophic failure. However, this overdesign approach prevents the optimization of weight and material usage. Due to limited access to modern fatigue models for composites, manufacturers still rely heavily on costly and time-consuming testing campaigns to validate their designs. Especially under extreme conditions but also for fatigue, testing is necessary due to insufficient prediction capabilities and material models.

Elamin et al. (2018) investigated the impact response and damage mechanisms of composite sandwich structures in arctic conditions. Carbon fiber-reinforced composite sandwich panels with polyvinyl chloride (PVC) foam core are subjected to low-velocity

Table 2. Summary of full and large-scale fatigue tests on steel structures.

Type of connection	Material	Load type	Number of tests	Failure location	Reference
Web frame corner	Steel S355	CA	3	Weld toe	(Fricke and Paetzold, 2010)
Web frame corner	Steel S355	VA	2	Weld toe	(Fricke and Paetzold, 2010)
Stiffened panel	Seel S235	CA	9	Weld toe	(Lillemäe et al., 2017)
Rectangular hollow section (RHS) axial load	Steel S235	CA	9	Weld root	(Fricke et al., 2006)
RHS joint in bending	Steel S235	CA	13	Weld root	(Fricke et al., 2006)
RHS pillar connection	Steel S235	CA	8	Weld root	(Rautiainen et al., 2023)

impact at extremely low temperatures, representing the harsh arctic environmental conditions. Force-time history curves show that test temperature significantly influences the impact damage behavior, showing lower impact strength and reduced penetration energy required to perforate the top face sheets. Specimens impacted at extremely low temperatures ($-70\ °C$) exhibit less impact strength and higher susceptibility to damage, resulting in severe penetration by the impactor. X-ray micro-computed tomography technique is employed to reveal multiple complex impact damage modes. Specifically, results from this work elucidate arctic temperature influence on detrimental failure mechanisms: large face sheet-core debonding, extensive composite face sheet delamination, significant core shearing and crushing, and severe face sheet fiber fracture.

Fatigue testing excitedly on natural frequencies, especially for huge composite structures, becomes time-consuming and costly. Therefore, new standards and approaches aim at subcomponent testing, allowing higher frequencies. Different concepts and methods are developed to address the challenges posed by parts of the loading spectrum or of the structures that are "truncated" or omitted during testing, often due to time or equipment limitations, such as compensating for omitted cycles through adjusted load magnitudes or applying scaling factors to preserve damage equivalence. Also, hybrid testing can be applied to combining high-frequency testing with selected, slower low-frequency cycles to retain key damage effects. The study "A novel rotor blade fatigue test setup with elliptical biaxial resonant excitation" (Melcher et al., 2020, Melcher et al., 2022) describes a new method to design suitable test setups. A parameterized finite element (FE) model of the test with beam elements for the blade represents the test setup. A harmonic analysis of the FE model can identify the load distribution and the test conditions of a specific test setup within seconds. An optimization algorithm that varies the parameters of the model and searches for the optimal setup is then applied to the analysis. This approach efficiently determines a test setup suited to the predefined requirements. The method is validated by using it in three different test scenarios for a modern rotor blade: (a) state-of-the-art uniaxial setups, (b) uniaxial setups including springs (mining 1-D loading), and (c) a biaxial setup.

4.5 Summary

This chapter explores advanced computational techniques for predicting fatigue life and fracture in ship and offshore structures, integrating modeling approaches, assessment methods, and practical applications. Computational modeling of fracture and damage mechanics primarily follows two continuum damage-based approaches for fatigue, originally developed by Chaboche and Lemaitre and later refined through various modifications. While the peridynamic framework remains significant, hybrid simulations, particularly those leveraging sub-modeling techniques, are gaining prominence. Research in submodeling and substructuring highlights the importance of global-local models in accurately simulating fracture and damage mechanics, especially in offshore structures where structural reliability is crucial, and a single model often lacks the necessary level of detail while being computationally expensive.

Fatigue life modeling is essential for understanding crack initiation and propagation under cyclic loading. Time-domain methods, such as Rainflow Counting, offer high accuracy by capturing load cycle sequences, while frequency-domain methods, like Fourier analysis, provide computational efficiency but may be less precise for non-stationary signals. Complementing this, continuum damage mechanics models, particularly those developed by Chaboche and Lemaitre, simulate microstructural defect accumulation, residual stress effects, and material degradation under cyclic loading. Fracture mechanics-based methods, including Fitness for Service (FFS) and Engineering Criticality Assessment (ECA), employ Paris-type crack growth models to assess structural integrity in challenging environments, such as low temperatures and corrosive conditions.

The chapter also examines the impact of imperfections, such as voids, inclusions, and surface roughness, on fatigue life. Probabilistic models enhance predictive accuracy by accounting for the sensitivity of fatigue strength to these imperfections. Furthermore, it differentiates between brittle and ductile fractures, introducing advanced models such as the Gurson-Cohesive approach to simulate failure mechanisms and understand material behavior under extreme loads. Experimental validation remains crucial, with large-scale and full-scale testing providing insights into fatigue resistance and structural integrity. Studies on composite structures highlight the challenges and methodologies for testing under extreme conditions. Structural modeling techniques, including global-local modeling, submodeling, substructuring, and finite element analysis, balance computational efficiency with detailed analysis, enabling precise fatigue assessments.

Finally, the chapter reviews commercially available software such as CrackWISE, FRANC3D, and ANSYS Mechanical, discussing their applications in computational modeling for marine structures. By integrating advanced modeling techniques with experimental validation, this chapter provides a comprehensive discussion of fatigue life prediction, damage assessment, and fracture analysis of marine structures.

5 Fatigue and Fracture Reliability Analysis

The fatigue phenomenon of welded structural components is a complex process influenced by several variables related to loading and resistance. These variables have an intrinsic uncertainty. Hence, fatigue damage assessment and life prediction are subject to a great deal of uncertainty. To account for these uncertainties, reliability theory, and probabilistic methods have been essential in aiding the structural design and structural integrity management, resulting in more accurate and precise preventative and corrective remedial actions ensuring structural safety above the permissible limits for marine structures. This chapter reviews recent scientific work on fatigue and fracture reliability of marine structures, focusing on uncertainties and reliability-based design and integrity management.

5.1 Uncertainties

Uncertainty is the central element in fatigue-related structural safety assessments, regardless of whether they are addressed through safety factors in deterministic fatigue life calculation or fully probabilistic design approaches. These uncertainties can stem from randomness found in the nature of physical processes and imprecision in measuring techniques (aleatory uncertainties), or they can stem from the models developed to explain natural phenomena and mechanisms (epistemic uncertainties). The fundamental difference between these two types of uncertainty is that aleatory uncertainty cannot be reduced by acquiring more information, whereas epistemic uncertainty can be reduced by more information and experience.

5.1.1 Load Modeling

Wang et al. (2022b) examined the variability in environmental conditions and its impact on the fatigue design of floating offshore wind turbines. The study considered mean wind speed, turbulence intensity, wind shear, wave height, and period. Aero-hydro-servo-elastic simulations were used to assess the global dynamic responses of floating structures. Primarily, the focus was on understanding the sensitivity of the main bearings in the drivetrain to these environmental variables, providing insights into which factors were most critical in contributing to fatigue damage. The study revealed that fatigue damage in the main bearings increased approximately linearly with the turbulence intensity within a certain range in all environmental conditions. Also, the difference in fatigue damage can go up to 8%, 15%, and 30% with a change of 10% turbulence intensity variation under below-rated, rated, and above-rated conditions. Further, uncertainty in bending moment increased nonlinearly with the increasing mean wind speed, and different components, such as bearings and gearbox, were affected differently. Whilst the variability in wave height and peak period had negligible influence on the fatigue damage of the bearings, a stronger influence should be expected for the floating substructure. The study recommended 5 (up to 10) independent 1-h simulations with random wind and wave seeds to achieve high accuracy of the drivetrain short-term fatigue damage. Although the study was performed for drivetrain bearings, the conclusions should shed some light on the fatigue damage assessment of fatigue reliability of the structural components, such as the tower and substructure.

There is a gap in the literature addressing quantitatively the trade-off between accuracy and computational effort for aero-hydro-servo-elastic simulations, which can be essential to the fatigue reliability analysis of these structures subjected to complex loading. In this regard, the review is given by Otter et al. (2022) on the modeling approaches for floating offshore wind turbine design, and the analysis pointed out the trade-off between simulation fidelity and computational feasibility. The review also pointed out the need for further investigation into how to account for modeling uncertainty and biases across different modeling techniques. Ćorak et al. (2022) investigated uncertainties of wave data of the Adriatic Sea collected from different sources and analyzed the impact of the difference on the design of marine structures using two databases based on in-situ measurements and two databases based on hindcasting. Between the in-situ and hindcast data fit the Weibull Distribution for long-term wave height distribution, the study found up to a 6% and 40% deviation for shape and scale parameters, respectively. Such a discrepancy led to fatigue life prediction differences by a factor larger than three. The outcome of this study underscored two important aspects regarding environmental load modeling: duration and time resolution of the data and geographical discretization.

Liao et al. (2022a) reviewed the recent developments and prospects of fatigue reliability of wind turbines, which revealed wind load modeling uncertainty as the most significant. This can be explained by ever-increasing turbine size pushing the boundaries of models developed based on physical experiments, as explained by Veers et al. (2023). Although Liao et al. (2022a) did not provide any quantitative analysis of the literature, it listed the sources of uncertainty in terms of load and environment as such statistical distribution of wind speed (10-min. Mean) and turbulence intensity, local wind measurements, and speed-up factors. The study indicated that wind load models are subject to great uncertainty because load distribution variation under wide spatiotemporal variability does not fully represent the load characteristics, and the impact of uncertainty in the site-specific wind climate parameters on total uncertainty involved in load analysis is currently based on engineering judgment and awaits further research. However, the review suggested that the reliability analysis results could be used to set partial safety factors to account for factors affecting fatigue performance, such as mean wind load and turbulence under large spatiotemporal variability. Otter et al. (2022) and Liao et al. (2022a) recommended further development of high-precision simulation tools considering the interaction of different fields and components, such as complex inflow, wind turbine control, elastic structural behavior, hydrodynamic loads, and coupled dynamic response. Nevertheless, the review also suggested considering the joint action of multisource uncertainties in addition to reducing computational costs. Hegseth et al. (2021) showed the contribution of different parameters influencing the environmental load on the overall cost of a spar-type floating structure, considering both design and inspection. The probabilistic load analysis was conducted using Monte Carlo Simulation (MCS); however, the statistical descriptors of these parameters were not reported. An important result of this study was that fatigue reliability is sensitive to wind turbulence, wind direction, and especially wind-wave misalignment. This sensitivity is higher for the tower than the platform, and it strongly indicates that the aleatoric and epistemic uncertainties related to loading must be considered to reach optimal support structural design. Such studies pave the way for a holistic approach to investigating loading-related uncertainties.

Han et al. (2021) studied the stochastic structural dynamic response of the tension-leg platform for offshore wind turbines, and they quantitatively demonstrated how the mean value and standard deviation of the dynamic response change depending on the operational mode, which depends on the wind speed. Low-frequency components (wind) showed a higher standard deviation than high-frequency components in tower-base bending moment and tension in mooring during the load cases with wind speed close to the rated wind speed of the wind turbine (Coefficient of variation CV = 0.25–0.35). As the sea states get harsher (higher wind speed and wave height), the mean value of tower-base bending moment and tension in mooring decreases; however, the overall standard deviation increases due to the contribution from the wave-induced loading and high-frequency components (CV = 0.50–0.66). The study discussed above did not note the random seeds used in the simulations. Thus, it is challenging to identify to what extent the randomness of the signal causes the variation. To date, there is not a peer-reviewed study exploring the physical and modeling uncertainty associated with the multi-modal dynamic behavior of complex marine structures.

Ramezani et al. (2023) gave an extensive review of various sources, types, and models of uncertainties affecting the structural reliability of floating offshore wind turbines. The review presented a meta-analysis of the statistical descriptors associated with the wind speed for different wind turbines, which indicated that offshore wind turbines in deeper waters demonstrated higher wind speed with less fluctuation than those in shallow water. The review also stated that the Gamma, Lognormal, and Weibull distributions are three widely used PDFs for modeling environmental loads; it is found that the Weibull distribution is a well-accepted representation of the wind speed and wave height, while the Gamma distribution is mostly used for the wave height and period. The lack of existing knowledge on the nature of uncertainties over time is also recognized. CV for wind speed is nearly 0.49–0.55, which is relatively low compared to the wave loads and current loads. The modeling uncertainty related to the environmental loading (CV ~ 0.1–0.2) is much higher than that related to the material strength (0.02–0.10). The review in (Ramezani et al., 2023) presents an extensive collection of aleatory and epistemic uncertainty that have been used in the literature on offshore structures, including both floating oil and gas platforms and floating offshore wind turbines. Dong et al. (2022a) gave an extensive review of uncertainties in fatigue loads, material properties, manufacturing process, damage accumulation, crack growth, and fatigue life of ships and offshore structures.

There has been considerable effort with respect to developing surrogate models for the effect of complex dynamic loading within the fatigue damage assessment. The common purpose for developing these surrogate models is to improve efficiency whilst maintaining accuracy, which is ultimately very valuable for comprehensive fatigue damage assessment of marine structures subjected to multiple loading sources, such as environmental, operational, and environment-structure interaction. Recently, an example of such development was given by Lim et al. (2022) to address the fatigue damage in marine risers caused by vortex-induced vibrations using polynomial chaos expansion, a statistical method for approximating the output of a complex system as a sum of orthogonal polynomials with random coefficients. In addition, as an alternative to the time-domain and frequency-domain approaches, Li and Zhang (2020) and Zhao and Dong (2021) showed that with sufficient data, surrogate models can be used to perform probabilistic fatigue damage assessment, which is significant for reliability-based design optimization for marine structures requiring tremendous computational effort. Multi-layer Artificial Neural Networks (ANN) and kriging models stood out as common tools to develop efficient surrogate models. For such surrogate models, it is important to quantify the modeling uncertainty prior to performing fatigue reliability analysis to avoid misleading the design recommendations.

There have been long-standing assumptions about how waves and wind should be mathematically modeled. The long-term fatigue damage assessment framework is generally employed, considering that the load process is Gaussian, stationary, and ergodic. Several studies have tackled the inaccuracy and imprecision caused by these underlying assumptions. Kim et al. (2020b) addressed computational accuracy limitations with respect to the wide-banded Gaussian random process in the first part of a two-part paper. Alternative to the commonly used Dirlik and Benasciutti-Tovo fatigue damage models, a joint probability density function of the cycle mean value and amplitude of stress cycles was devised using the nonlinear regression analysis of spectral characteristics (moments) of the power spectrum of stress amplitude. In a subsequent paper, the model for the Gaussian process of stress ranges was improved using a nonlinear transformation function by Kim and Jang (2021) to deal with non-Gaussian random loading processes. The study concluded that when strong non-Gaussian characteristics are involved in a process, even well-established wide-band models can underestimate the fatigue damage (12%), and the developed model lowers this underestimation significantly (1%). The models developed to address wide-band and non-Gaussian loading processes are rather overlooked research areas. The fatigue reliability analysis tends not to account for uncertainties associated with the loading process modeling.

In addition to the loading process modeling, the generation of generalizable load-time history is important for simulating realistic loading conditions because such load-time histories can be employed for uncertainty analysis in crack growth analysis. In this regard, Li et al. (2020) presented an approach to generate design load-time histories for assessing the fatigue strength of offshore structures, integrating storm models and seasonal waves. The study pointed out significant uncertainties with respect to crack size, which can be obtained depending on sea state sequence, including calm and storm conditions. This highlighted the importance of considering the load sequence effect in the damage assessment. Apart from this, developing a generalizable load-time history for long-term fatigue loading is far from plausible. Incorporating the climate change model into the long-term fatigue loading modeling is expected to gain more importance. Although the impact might be very small (~1–2%) in long-term fatigue damage, the frameworks introduced by Tamimi et al. (2022) and Hübler and Rolfes (2021) can be adapted to expand the investigation of the impact of changing climate on the long-term fatigue damage analysis and design of marine structures.

Fatigue design recommended practices (e.g., DNV-RP-C203 and JCSS) indicate the same set of uncertainties for various structures, covering ships, offshore structures, and land-based infrastructures. According to the recommendations in these documents, the loading uncertainty is the same for fixed and floating structures, which in itself indicates that differentiation between the two support structure types could result in different uncertainty levels. In the 2021–2024 timeframe, no literature was found that explicitly considers this difference, although its influence on resulting reliability factors is significant.

5.1.2 Structure-Environment Interactions

In recent years, there has been considerable interest in structure-environment interaction, such as ice-structure, wave-structure, and pile-soil interactions. A probabilistic fatigue damage assessment approach was investigated by Zhao et al. (2021c) for ships navigating through ice fields. The complex nature of ship-ice interactions was modeled using statistical distributions within a probabilistic numerical simulation. The appropriateness of Weibull, Lognormal, and Gumbel distributions were compared for stress amplitude distributions due to ice load actions. The Weibull distribution was found to be a better fit for the probability distribution of stress amplitude, and its parameters converged within a numerical simulation of 1200 s. Shi et al. (2023) reported one of the few studies regarding the ice-induced collision and vibration mechanisms in OWTs, emphasizing the importance of considering pile-soil interaction effects. The stochastic simulation revealed the influence of cone angles and ice conditions on OWT fatigue damage, which was particularly significant. The comparison between the simulated dynamic ice force and the measured ice force in the model test indicated that the difference in the maximum ice force was 6%, and the average horizontal ice force was 26%.

Mujeeb-Ahmed et al. (2021) studied the impact of mooring line layout on the loads of ship-shaped offshore installations (FPSO) through a probabilistic analysis. The sensitivity analysis showed that chain diameter, current speed, and wave direction contribute significantly to the variance in fatigue damage of mooring lines. Like Mujeeb-Ahmed et al. (2021), Chen and Low (2021a) trained artificial neural networks using outputs of Monte Carlo to develop a computationally efficient method for long-term fatigue damage analysis of deepwater risers, emphasizing the importance of including wave directionality, which is prone to discretization error while analyzing the metocean data. The outcomes of these studies are far from conclusive; however, they show the importance of accounting for the uncertainty associated with parameters of environmental loading, such as wave direction and current speed, together with the geometrical properties of the mooring lines and risers. Furthermore, Hageman et al. (2022b) and Hirdaris et al. (2023) highlighted the importance of accurately modeling springing, whipping, and slamming effects and the challenges in doing so due to limitations in numerical methods, ambiguities in modeling and testing, and the use of computational and full-scale data. The studies also emphasize the need for structural reliability methods incorporating non-linearity in ship design, especially since springing can contribute to fatigue damage significantly in unfavorable sea conditions.

Apart from the environmental loading-structure interaction, the soil-structure interaction is also subjected to significant modeling uncertainty. In this regard, Sørum et al. (2022) performed aero-hydro-servo-elastic simulations using probabilistic variables to study the dynamic response of monopiles, and they concluded that the uncertainties in the wave and soil models dominated the fatigue damage at the tower base and the monopile of 5, 10, and 15 MW wind turbines. Consequently, due focus must be given to reducing uncertainty in environmental conditions and soil using as accurate models as possible. The importance of the wind-related parameters (e.g., wind speed, turbulence, etc.) on the total fatigue damage for the monopile increased with increasing turbine size, suggesting that uncertainties in the wind parameters become more important for the emerging large turbines. Among the environmental variables, it was recommended to focus on wave loading. The most important ones were found to be the coherence model in the tower top and the soil model for the monopile at the seafloor and tower base. Wave loading had the highest influence on the maximum fatigue damage in the monopile.

Yu et al. (2022b) modeled soil-related parameters, such as undrained shear strength and unit weight, as random variables within a multi-objective design optimization framework. Lognormal distributions with a CV of 0.3 and 0.05 were used to model the independent random variables "undrained shear strength" and "unit weight", respectively. When the correlation between these random variables was considered, the design robustness (rotation, displacement, axial stress, and fatigue) decreased, and in turn, the feasible design space shrunk.

5.1.3 Fatigue and Fracture Strength

The fabrication of welded structural details comprises a series of processes that are difficult to control fully. Consequently, fatigue strength characteristics of structural details are subject to a great deal of uncertainty. In addition, the tests performed for material characterization are usually performed for constant amplitude fatigue loading, meaning that there is natural variability concerning response to the loading and damage accumulation. Besides that, the dataset used to derive the S-N curves can have some deviations from actual structures in terms of residual stresses, large-scale effects, production tolerances, and potential deviations in production, by which the actual scatter in fatigue strength in practice is even larger. It is for this reason that design fatigue factors (DFF) are devised to account for these uncertainties by adding a conservative safety factor to the fatigue life.

Velarde et al. (2020) presented a demonstration of the application of reliability analyses to a large monopile, incorporating a reliability-based calibration of the DFF to account for the underlying uncertainties related to the stress concentration factor (SCF), damage accumulation, the dynamic model, and the wave load model. The formulation of the set of uncertainties was in itself subject to uncertainties and assumptions; next to that, the set of uncertainties differs between applications. Furthermore, they applied a calibration of the DFF by also considering the statistical distribution of turbulence effects and wave-induced resonance on top of the structural response uncertainties. As a result, it was concluded that accounting for the uncertainties, a DFF larger than 3 is recommended to have a sufficiently large annual reliability index after 25 years.

Zhao (2021) reanalyzed and calibrated DFFs used for the oil and gas structures for the offshore wind turbine structures through fatigue reliability analysis. This work is based on the recalibration on modification of the probability values of given stress ranges. In addition to the DFF calibration, several studies have examined the factors affecting the fatigue life of ships and offshore structures. Dong et al. (2021b) employed an analytical notch stress-strain estimation approach consisting of several steps to consider the material memory effect, overcoming some limitations of the existing methods with respect to including the load sequence effect. In addition to the load sequence effect, weld-induced tensile residual stresses and misalignments were shown as the factors contributing the most to the uncertainty regarding fatigue damage accumulation. Whilst the weld-induced tensile residual stresses, surface roughness, HAZ, and internal defects and flaws can bring weld- and material-related uncertainty, weld shape, and size, misalignments, and stress concentrations factors are geometry-related uncertainty affecting fatigue life of welded marine structures. In this regard, Larsen et al. (2022) accounted for the statistical weld size effect within a design optimization problem using Karhunen-Loéve expansion, where the results revealed that the use of high-quality welds (welding by robots) resulted in a higher reliability level, in turn, a lower mass of the K-type welded-tubular. Hariprasath et al. (2023) confirmed that the weld (GTAW) had a comparably higher notch sensitivity rate than the base material, which amplified with the increasing stress ratio and mean stress. Mikulski and Lassen (2020) showed that there is a higher variability in the number of cycles to reach crack initiation as the initial crack size is smaller, especially for the member loading model. Furthermore, the variation in the toe radius had a significant impact on the crack initiation and early crack growth. Regarding structural analysis, Tamimi et al. (2023) developed a probabilistic approach based on the Weibull stress criterion for characterizing the fracture strength of stiffened ship hulls. The approach was based on high-fidelity finite element modeling of a stiffened ship hull that accounted for welding residual stresses. The analysis shows that the fracture strength of a ship hull is significantly higher than that of standard fatigue test coupons, necessitating similar analyses to compute the fracture strength in reliability quantification studies.

There is continuous development with respect to adjusting the material's fatigue strength characteristics to account for the uncertainty from manufacturing and the temporal development of corrosion. In this regard, Lone et al. (2022) developed a new probabilistic model for the design and life extension of mooring chains, including mean load and corrosion effects. The case study presented showed that the coefficient of variation of accumulated fatigue damage due to corrosion development can be up to 31%, considering a fast initial corrosion grade development with an uncertain final value. Shittu et al. (2021) compared the fatigue reliability using the S-N curve approach with the fracture mechanics approach for a typical welded jacket support structure for OWTs. Whilst for the S-N approach, the loading variable was the one that contributed the most to the fatigue damage variability, the reliability was most sensitive to the uncertainties of the crack growth law constants for the fracture mechanics. For a case concerning pitting corrosion fatigue, the sensitivity analysis showed that the aspect ratio of pits at critical size plays a significant role in the reliability of the structure, such that the structural integrity could be compromised before the end of the design life (20 years). For the additively manufactured materials, the size effect on the fatigue performance variation was investigated by Niu et al. (2022) based on the computed tomography (CT) scan of defects. The study presented a methodology whereby the extreme value statistics theory was used to extrapolate the maximum defects of additively manufactured materials under size effect. Subsequently, the study proposed a strength-altering factor to characterize the different defects, originating from process and volume differences, on fatigue strength scatter using the Kitagawa–Takahashi diagram.

5.1.4 Damage Accumulation and Crack Growth Assessment

Mell et al. (2022) presented a method to estimate discretization error for time-domain structural response obtained by numerical methods such as the finite element method, which propagates through time. Though the method introduces additional computational effort due to using signal processing procedures, e.g. signal smoothening techniques, it can be used to reduce the uncertainty in the accumulated fatigue damage, or it can provide discretization error bounds to be used in the structural reliability. Although to what extent the discretization error affects the accumulated fatigue damage has not been quantitatively shown, the study certainly raises questions similar to the review reported by Dong et al. (2022a) and Yeter and Garbatov (2022) for ship and offshore structures. Explicit time-domain fatigue damage assessment is required due to nonlinear interaction between the multi-physics nature of the overall system, and it can be very useful for mesh size sensitivity analysis. In addition to the fatigue damage accumulation process, the statistical signal process for the load spectrum generation for variable amplitude fatigue testing is important. In this regard, the derivation of the fatigue strength characteristics is shown to be subjected to uncertainty, as discussed by Jimenez-Martinez (2020), based on the analysis performed for accelerated fatigue testing.

Chen and Low (2021b) explored the potential of a new approach using Artificial Neural Networks (ANN) to reduce the variance in Monte Carlo Simulation performed time-domain fatigue damage assessment. This new variance reduction approach can be quite useful, especially when there is no prior information on the system response, e.g. importance sampling. The approach, whilst more effective in linear cases, shows promise for wider application in offshore structural engineering, offering a computationally efficient and unbiased alternative for fatigue damage estimation. As also foreseen by the study, machine learning techniques will continue to be integrated into the already existing fatigue damage assessment framework to improve the computational efficiency and accuracy of the fatigue damage predictions. Wavelet ANN, long short-term memory networks, and nonlinear autoregressive models with exogenous input are among the methods that can be implemented in highly non-linear cases such as fatigue crack growth.

In recent years, several studies have also been conducted to develop more efficient calculation methods. Wilkie and Galasso (2021) introduced a Gaussian Process regression-based framework for efficient fatigue damage assessment of offshore wind turbines based on Aero-elastic simulations. Whilst the approach offers a generalizable fatigue damage assessment tool for European waters, further confirmation is needed. The uncertainty associated with the developed surrogate model of the dynamic response of an offshore wind turbine resulted in a coefficient of variation of 30% in fatigue damage. Zhu et al. (2024) developed a cost-efficient computational framework to perform reliability-based fatigue analysis in the design of FOWT substructures. The developed framework combined a smart sea state selection algorithm with the MSC-based reliability analysis using bootstrapping from the selected sea states, contributing to the most accumulated fatigue damage. The proposed time-domain fatigue reliability analysis framework promised to significantly reduce the computational effort needed to obtain the dynamic responses in the time domain and the fatigue reliability analysis using MCS. Another surrogate model example was given by Ibarra et al. (2022), incorporating a probabilistic domain transformation and a multi-dimensional mathematical approximation for the function, resulting in a univariate dimension-reduction method. The effectiveness of the developed approach was confirmed by benchmarking against the crude MSC, which required a smaller number of simulations to achieve accurate long-term fatigue damage estimates. Dabetwar et al. (2021) and Yeter et al. (2022c) showed the feasibility of using advanced signal processing techniques and machine learning algorithms to reduce the dimensionality and assess the randomness of the fatigue damage classification in composites and steel structures, which can have a significant impact on the early warning system for offshore wind turbine components (blades and support structures) under complex loading for proactive maintenance strategy (Cai et al., 2020, Maleki et al., 2021, Wei et al., 2023a).

5.2 Reliability-Based Design and Structural Integrity Management

Structural reliability measures the probability of a structure fulfilling its intended purpose for a given environmental and operational condition during its service life. All the important uncertainties related to load and resistance must be accounted for, and the estimated structural reliability must be checked against a reliability level resulting in an acceptable risk status. There are various approaches to achieve satisfactory structural reliability and risk status.

5.2.1 Reliability Methods and Approaches

Wang et al. (2022a) grouped the structural reliability methods according to the degree of sophistication (Level I-IV). Level I reliability methods account for uncertainties employing global and partial safety factors. In Level II, the first-order second-moment method is used to estimate reliability, whilst the first- and second-order reliability methods (FORM/SORM) and Monte Carlo Simulation (enhanced with sampling methods) are considered Level III. Finally, Level IV can be deemed an improved FORM/SORM with an optimizer. Level IV can also be considered part of the risk-based assessment as it incorporates the probability of failure and costs associated with the failure, such as initial construction, inspection, repair, and failure costs, as well as the benefits obtained from its operation. As also argued by Wang et al. (2022a), the quantitative reliability analysis has become an integral part of the design, code calibration, multi-hazard analysis, monitoring-based updating, and inspection planning for offshore renewable structures. Although the structural reliability methods are well-established, some dispersion might be found in the application of risk-based methodologies for the inspection and maintenance planning of marine structures. This is due mainly to the assumption associated with the construction cost, operational cost, and discount rate applied in the calculation, which is rather overlooked.

Despite many successful implementations of reliability analysis on prominent failure mechanisms, there are challenges with the reliability methods, such as defining dynamic system characteristics, nonlinearities related to load and failure mechanism, and site-specific environmental conditions leading to higher uncertainties, data confidentiality, and computational effort. To address some of these concerns, Shittu et al.

(2021) recommended developing machine learning-assisted (e.g., ANN) structural reliability analysis (e.g., FORM) to deal with complex, nonlinear, and time-dependent structural problems, such as fracture mechanics-based structural integrity assessment. More technical details on the machine learning-based methods applied in the structural reliability analysis can be found in a review given by Saraygord Afshari et al. (2022). Also, dynamic target reliability was recommended for the structural integrity management of multi-unit offshore structures in a review presented by Yeter and Garbatov (2022). The review highlighted the continuous advancement in asset management, such as effective structural health monitoring, reliable remote inspection techniques, and condition-based maintenance strategies that consider the structural asset with other non-structural assets holistically. The review also highlighted the fact that there is no common data usage when it comes to the cost definition, which makes the conclusions regarding the inspection and maintenance planning difficult to compare. A dynamic target reliability index that changes depending on the economic conjuncture and fatigue life improvements (repair and structural health monitoring) can help build flexible inspection and maintenance policies, reducing structural risk. Interdisciplinary research is the key to achieving such a complex structural integrity management system of marine structures. The introduction of advanced remotely operated underwater vehicles, inspection techniques, and digital twins empowered by recent developments in automation and machine learning would be an integral part of the risk-based methodologies.

5.2.2 Digital Twins and Machine Learning in Reality Analysis

Digital twins are expected to pave the way for innovative structural designs and structural integrity management. The continuous learning and updating features of digital twins have already started to replace some of the risk mitigation strategies in the design stage. According to Wagg et al. (2020), the digital twin concept with a real-time representation of the overall marine structure helps remove some of the uncertainties linked to the structural response; therefore, a more transparent and realistic remaining life estimation can be performed. Such capabilities from the design state facilitate novel structural management techniques but still require a long-term technology qualification process. The issues regarding the reliability of such technologies should be further analyzed, and the cost associated with such capabilities should also be part of the risk-based decision-making process.

Wang et al. (2021b) reviewed progress in emerging digital twins within the reliability analysis framework for structures supporting offshore wind turbines (OWT). The review illustrated a comprehensive digital twin framework encompassing real-time monitoring, fault diagnosis, crack detection, and reliability-based remaining life prediction and operation optimization of OWT support structures. Another review by Amiri et al. (2021) delineated the latest design and monitoring methods developments. The review examined probabilistic dynamic response analysis, resulting in an optimal seismic demand model and fragility curves. Furthermore, robust fault-tolerant control and multi-objective optimization were discussed within the scope of asset integrity management. In this regard, Augustyn et al. (2021) presented the structural reliability updating framework for OWT support structures using the digital twin generated by operational data. In this framework, the digital twin serves to quantify uncertainties associated with the structural dynamics and load modeling parameters in fatigue damage accumulation, and it is used to update the structural reliability accordingly. One caveat the study urged the uncertainty propagation regarding virtual sensing over time, which could reduce structural reliability. Thus, there will likely be a trade-off between the benefit obtained from virtual sensing and structural health monitoring that must be examined carefully. Different sensors would have different specifications such as measurement range, sampling rate, uncertainty tolerance, size, etc. In an uncontrolled environment, such as offshore, the data is subjected to a great deal of noise. Yeter et al. (2022c) argued that dedicated data quality assurance is necessary to de-risk the offshore application; thus, prior to training supervised and unsupervised machine learning models, big data must be cleansed from noise and outliers using digital signal processing techniques. As of now, no guidance is available for the specification of monitoring units applied to digital twin-based structural integrity management. Furthermore, Yeter et al. (2022b) demonstrated a risk-based framework to classify offshore wind turbines considered for life extension using an unsupervised machine learning technique. Although the study addressed structural risk management during the life extension, such a risk-based framework can be applied to manage structural risks from the design stage as well. However, such a dynamic risk management framework requires revisiting well-established design philosophies for conventional marine structures, considering the developments in fatigue life improvement methods.

The digital twin implementation must be comprehended within a risk-based methodology that can be used either for the design, operation management, or both. The risk-based methodology must be combined with an optimization process where either life-cycle cost can be an appropriate objective function, or a multi-objective optimization can be developed. In this regard, Kim et al. (2022b) modeled an optimization process that sets reliability-index and life-cycle cost as objective functions considering life-cycle cost, extended service life, maintenance delay, damage detection delay, and time-based probability of failure, forming a Pareto front. Machine learning-based surrogate models are a step forward for the structural reliability analysis for a complex structure that an analytical solution cannot approximate. Moreover, reliability updating is a very important aspect of structural integrity management of ship and offshore structures, and Hlaing et al. (2022) demonstrated that Bayesian Networks and Markov Decision Processes have become widely employed tools to help stochastic decision-making processes that require sequential decisions over time, such as monitoring and inspections for OWT support structures.

The digital twin representation of the structural asset can also be developed by training the data obtained using a high-fidelity numerical model. An example of such an approach was given by Hejazi et al. (2021) for aging catenary risers. The study applied the Bayesian machine learning technique, adopting Gaussian Processes for regression to a predictive model. However, the data used to train this predictive model was based on probabilistic numerical simulations, which are still computationally expensive. Another example of virtual monitoring of the structural response was presented by Hageman and Thompson (2022) for a frigate-type hull. The stress response assessment was done using the numerical model of the vessel, and the three-wave datasets were compared to strain measurements. The assessment was expanded to the accumulated fatigue damage assessment at a midship bracket, resulting in the calculated fatigue damage being in good agreement between the wave hindcast data and strain measurements, provided that the wave spectrum shape parameter was updated appropriately. Xu et al. (2023) gave an example of such a procedure for deepwater risers where the annual crack growth model of the riser based on fracture mechanics was established by considering the crack inspection data as a factor, and the crack growth dynamic Bayesian network was established to evaluate and update the fatigue reliability of the riser.

Virtual sensing is particularly important for structural components of marine structures that are impractical to physical instrumentation, such as fixed-bottom offshore wind turbines' structural details at midline. In this regard, Zou et al. (2023a) applied a generalized Bayesian virtual sensing technique, "Gaussian process latent force model", merging physics-driven and data-driven models of latent variables of the system to extrapolate unmeasured strain state. The study evaluated the performance of the applied model by comparing it to in-situ data collected from an operating OWT. Such a technique is particularly important for modeling and analyzing physical uncertainties across different parts of the support structures and implementation for digital twin-enabled asset integrity management. Hübler and Rolfes (2022) focused on the spatial and temporal extrapolation methods where the data regarding structural condition state was acquired based on simulations. In addition to Gaussian process regression and artificial neural networks, a binning approach was introduced to generate a functional relationship between short-term fatigue damage and environmental and operational conditions, yielding a very low error rate. Xie et al. (2023b) demonstrated machine learning application for damage detection by combining an artificial neural network with a genetic algorithm to correct the uncertainty involved in the damage location. The dataset was generated using modal analysis results based on the Finite Element Method. The study reported an up to 11% improvement in fatigue crack localization under noisy environments when an artificial neural network was combined with the genetic algorithm. Regarding damage detection under uncertainty and noise, Xie et al. (2023a) showed that empirical mode decomposition with a probabilistic neural network can successfully separate the field noise and localize the fatigue crack through acoustic emission with less than 2% error.

5.2.3 System-Level Analysis and Practical Applications

In recent years, many studies have also presented structural integrity assessment frameworks with a focus on damage-tolerant philosophy for various engineering problems of marine structures, such as (Amirafshari et al., 2021, Fajuyigbe and Brennan, 2021, Yeter and Garbatov, 2022, Wang et al., 2021b, Amiri et al., 2021, Wahab et al., 2020, Wang and Cui, 2020). These were significant overlaps regarding the best practices in structural integrity assessment and the definition of failure and its consequences. However, Amirafshari et al. (2021) gave more emphasis on the design and inspection planning optimization, Fajuyigbe and Brennan (2021) put more effort into flaw acceptability assessment and sensitivity of failure mechanism related to the crack depth-to-thickness ratio and Yeter and Garbatov (2022) focused more on the retardation effect and probabilistic structural integrity assessment. Wang and Cui (2020) gave a historical review of the unified fatigue life prediction method for marine structures, in which the importance of cycle-by-cycle crack growth simulation accounting for load sequence and overload-induced plasticity was emphasized. Such a method requires a significant computational effort; thus, the fatigue reliability analysis might not be possible without any help of the response surface method and machine learning techniques to replace or at least assist the physics-based models. Along these lines, Tamimi et al. (2023) utilized ANN and a cycle-by-cycle crack propagation approach to quantify the reliability of ship hulls under random sea loading. The study utilized numerical finite element analysis that was surrogated with ANN to quantify the crack-driving measures, thereby achieving a considerable reduction in the computational cost.

Another important development in recent years is the analysis of the structural safety of a marine structure system, covering multiple failure mechanisms. In this regard, Hameed et al. (2021) applied a risk-based inspection methodology to certify the integrity management of a pipeline system, considering different failure mechanisms like external/internal sheath damage, fatigue damage, and corrosion, resulting in a system-level structural safety measure. Similarly, Garbatov and Huang (2020) performed a reliability-based design optimization considering multiple failure mechanisms for ship structural components. The multi-objective optimization problem was solved using the non-dominated sorting genetic algorithm (NSGA-II), where the reliability index was used as a design constraint; consequently, a nearly 10% reduction was achieved in the net cross-section compared to the original design. A case-specific yet relevant study combining the multiple failure modes was presented by Ruiz Muñoz and Sørensen (2020), where the probabilistic fracture mechanics-based assessments are performed for a plate structure accounting for the corrosion-free phase to the corrosive environment phase. Such an approach allows for novel inspection planning for both new designs and existing structures with no inspection data.

There is also an emerging research direction that looks at structural integrity management in a holistic way, adopting damage-tolerant design principles. Zou et al. (2020), Yeter and Garbatov (2022), and Nielsen and Sørensen (2021) showed the risk-based holistic decision-making approach encompassing the design, inspection, and maintenance decisions against fatigue cracking, taking into account the uncertainties influencing the failure and the consequences of failure (including different quality of inspection). The main feature of this approach is that it captures the trade-off between design (capital cost) and maintenance (operation cost) costs, leading to a more realistic optimal decision with respect to the asset. This approach is more compatible with the desired digital twin implementation as the holistic approach can take into account the benefits and costs of information obtained from such advanced and costly condition monitoring capability. Ship structures are typically one of their own kind, so the time available for the fatigue analysis in the design stage is limited. Advanced probabilistic analysis methods can be time-consuming to implement in the design of ship structures with multiple fatigue-prone welded structural details. To address this issue, Deul et al. (2023) proposed tailoring the fatigue detail category class to represent the uncertainties in a design-practice-friendly manner. The tailored FAT classes were derived based on a priori estimates of the uncertainties of both the load and the strength. Cai et al. (2020) presented a hybrid framework where the physics-based and data-driven models were merged to assess the remaining useful life of the structural system of subsea pipelines. As for the degradation processes with fatigue, corrosion, sand erosion, and internal waves were modeled using dynamic Bayesian networks, and sensor and expert knowledge were used to update the remaining life.

5.3 Summary and Outlook

This chapter discussed the recent developments in probabilistic fatigue and fracture analysis for marine and offshore structures. Studies covering the uncertainties in load prediction and modeling were first discussed, followed by those discussing the fatigue strength and damage accumulation and propagation models. The chapter also covered recent developments in reliability- and risk-based management techniques for marine and offshore structures. This analysis of the literature shows that most studies in this field are dedicated to probabilistic structural integrity management.

The recent studies mainly focused on 1) implementing more accurate models that can achieve better prediction of the damage propagation and structural reliability, 2) integration of sensor data to update and improve the digital twin representation of the structure, and 3) incorporation of machine learning to reduce the computational cost and allow integration of high-fidelity, complex models into the probabilistic management framework. The high level of uncertainty related to predicting fatigue damage propagation still represents a major challenge. The randomness in marine loading and the complexity of properly modeling its effects are still the main sources of uncertainty related to fatigue life prediction and risk-based management. In this regard, meticulous round-robin benchmark studies, like the ones presented in this committee report, can provide valuable insight into the sources of physical and model (global and local) uncertainties related to fatigue life and strength predictions, as well as how to reduce these uncertainties. It is also worth noting that such benchmark studies should be followed by uncertainty analysis and quantification.

For OWTs, many recent publications offer quite academic approaches to the probabilistic analysis that are well-suited for offshore structures, yet they are not always applicable to ship structures. The main reason for the different perspectives between OWTs and ship structures is argued to be a combination of structural complexity and the number of structures made of one design. For ships, this number is typically one, making optimization difficult due to large changes that can occur during the design process (sometimes when the keel is already on the slipway). Conservative measures are thereby preferred for the fatigue design of ship structures at the design stage. Despite these challenges, the recent developments in computational approaches offer several opportunities that can address the fatigue damage assessment and management processes. For instance, high-fidelity digital twins can offer a better representation of the structural behavior than numerical models, allow for considering complex loading conditions, and result in a more accurate prediction of crack growth and damage propagation. The advanced capabilities of surrogate models based on machine learning allow for integrating these high-fidelity models into the probabilistic analysis for a more comprehensive treatment of uncertainties with feasible computational costs. The utilization of sensor data to continuously update the digital twins and, subsequently, the probabilistic prediction models can lead to significant enhancement in the accuracy of these fatigue damage assessment processes and improve the quality of the management planning process under uncertainty.

6 Rules and Standards for Fatigue and Fracture Assessment

This chapter focuses on recent changes in rules and standards relevant to fatigue and fracture assessment of ships and offshore structures. Changes and updates in the period 2021–2024 are discussed.

6.1 Ship and Offshore Structures Design Rules and Guidelines

Table 3 summarizes recent updates in classification society rules and guidelines. Although there have been no fundamental changes, there has been significant activity on improving and expanding the existing approaches. Recent updates enhance structural integrity assessment, fatigue analysis, and regulatory alignment. Key changes include revised S-N curves, stress concentration factors, and expanded use of finite element analysis for critical components. Also, life extension, crack propagation, fatigue assessments for wind-assisted propulsion, and rules alignments with IACS and CSR are addressed.

During the last committee period, there has been extensive activity related to LNG as cargo or fuel and floating offshore wind turbines. These topics are discussed separately in Sect. 6.4 and Sect. 6.5. Also, the DNV RP-C203 (Fatigue design of offshore steel structures) was updated in October 2024. The most significant changes are related to S-N curves:

- For S-N curves B1 to C2 in air, the slope transition point for variable amplitude loading is moved from 1×10^7 cycles to 5×10^6 cycles.
- For S-N curves in seawater with cathodic protection, the fatigue limit is increased.

The change in the slope transition point is justified by the recent review of fatigue test data. DNV and TWI are starting (kick-off early 2025) up a JIP on Fatigue Integrity, where one aim is to see if the slope transition point can be harmonized between DNV and BS7608. The change in slope transition point is also expected to be implemented in the new edition of DNV-CG-0129 (to be published in 2025).

6.1.1 IMO and IACS Work on the Common Structural Rules

The IMO conducted a Goal Based Standard (GBS) verification audit of the IACS Common Structural Rules for bulk carriers and oil tankers (CSR) in 2015 (IMO, 2016). Based on this audit, changes were implemented in the January 2017 version of the CSR (IACS, 2017), summarized in (Garbatov et al., 2022). A further change was implemented in the January 2020 version of the CSR (IACS, 2021a). This was initiated by the auditors' observation that a "corrosion-free condition" as a prerequisite for the effectiveness of post-weld treatment cannot be ensured in practice and the concern that the calculated fatigue life of a post-weld treated joint may become less than that of an as-welded joint unless corrosion protection is maintained to a higher standard. In response, IACS issued a technical background report (IACS, 2021b) showing that the latter is not the case, mainly due to the provision that for a post-weld treated joint, a minimum of 17 years of as-welded fatigue life shall be predicted. To alleviate the auditors' concern further, the rule has been changed so that a corrosion-free condition or a protective coating

Table 3. Recent updates in classification society rules and guidelines.

Document	Comment
LR Guidance note for the life extension of floating offshore installations at a fixed location (March 2023)	Life extension assessment based on FDA level 2 and level 3 methods
LR Recommended Practice for Subsea Pipelines (January 2023)	Supplementary guidelines, in addition to the industry codes and standards. The document addresses the selection of S-N curves, maximum allowed damage parameters, and loads to be considered if these issues are not clearly defined in the main design code
LR Recommended Practice for Fixed Offshore Installations (July 2022)	Replaces the former Rules and Regulations for the Classification of Fixed Offshore Installations. Contains, among other things, the identification of critical locations, description of analysis methods, selection of S-N curves, and design fatigue factors
ABS Guide for spectral-based fatigue analysis for vessels (July 2022)	Providing S-N data for aluminium, Ref. *Rules for Building and Classing Light Warships, Patrol and High-Speed Naval Vessels*
ABS Guidance Notes on Fracture Analysis for Marine and Offshore Structures February 2022)	The first edition includes the general procedure for crack propagation analysis and structural integrity assessment for marine and offshore structures with a defect, a discontinuity, or a stress riser. Handy for evaluating the impact of a defect on structural integrity during the service life
DNV Rules for Classification of Ships (January 2022)	The welding quality level for the bottom plating of container ships has been reduced to ISO 5817 level C. Internal fatigue assessments show that ISO 5817 level C is sufficient for bottom plating
DNV standard for wind-assisted propulsion systems (December 2023)	This standard contains an updated description of fatigue assessment. It describes the calculation of wind loads, inertia loads, and their combination. Also, high-frequency cyclic loads for rotor sails are covered
DNV Recommended Practice for Fatigue Design of Offshore Structures (October 2024)	Changed S-N curves B2 to C2 in air, Updated SCFs for tubular joints, Other changes related to S-N curves, and More detailed methods for calculating fatigue damage in weld improved details

(*continued*)

Table 3. (continued)

Document	Comment
CCS Guidelines for fatigue strength of ship structures (August 2021)	
CCS Guidelines for Fatigue Strength Assessment Based on Fracture Mechanics Methodology (January 2024)	This new guideline from CCS addresses issues related to fatigue strength assessment of marine structures. Developed based on theoretical analysis, numerical analyses, and experiments
LR Guidance Notes for the Calculation of Stress Concentration Factors, Fatigue Enhancement Methods and Evaluation of Fatigue Tests for Crankshafts (July2024)	Guidance notes on the evaluation of SCFs for the crankshaft fillets based on FEAs as an alternative to the analytical formulas, which are limited to certain geometries. See also LR 2024c
LR Calculation Procedures for Stress Concentration Factor and Strength Assessment of Hatch Corner (August 2023)	Guidance on determining Stress Concentration Factor (SCF) on feeder container ship hatch corners for peak stress and stress range analyses
LR Guidelines for the Finite Element Analysis of Type C Tank (January 2024)	New document involving FEA procedures for the strength and fatigue assessment of Type-C tanks. This document involves recommendations on the fatigue critical locations and guidance for the hot-spot stress evaluation
IRCLASS Guidelines on Fatigue Design Assessment of Ship Structures (May 2021)	First edition Basis of class notation FDA
BV Guidelines for fatigue assessment of ships and offshore structures (November 2021)	Now also applicable for aluminium and stainless steel: section on *Fatigue strength of aluminium alloys plated welded details* Residual stresses included in the formula for mean stress factor
ABS Fatigue assessment of offshore structures (June 2020)	Updated S-N curves for tubular joints Updated sections on time domain analysis methods and fatigue strength based on fracture mechanics approach
LR ruleset	Ongoing process to align LR ruleset with IACS CSR. In this regard, extensive research on the consistency between the new rulesets and ShipRight FDAs is ongoing
RINA Amendments to the "Guide on Complete Ship Model Calculation of Passenger Ships" (Effective from 1/1/2022)	Material factor f is introduced for the calculation of the elementary notch stress range. This change significantly benefits assessing fatigue life when using high-strength material for free edge openings

maintained during the design lifetime is not a prerequisite for applying the benefit from post-weld treatment.

The IMO also conducted a Goal-Based Standard (GBS) maintenance verification audit of IACS Common Structural Rules for bulk carriers and oil tankers (CSR) in 2018 (IMO, 2018). Based on this audit, a non-conformity was identified. The observation was made regarding the functional requirement to fatigue life, related to the S-N curve for base material at free pate edges in a corrosive environment and to hatch corners of bulk carriers. The auditors claimed that there was not enough technical background provided. To rectify this, IACS revised the technical background report (IACS, 2020), and CSR was updated, requiring protection of cargo hatch opening corners against mechanical damage (IACS, 2021a). Based on this, IMO considered the non-conformity as rectified (IMO, 2021). There are no remaining open non-conformities and observations related to fatigue.

6.1.2 IACS Work on Wave Loads

During the last period, IACS revised Recommendation No. 34, Standard Wave Data (IACS, 2022). The recommendation gives an updated scatter diagram describing the wave data for the North Atlantic. It is intended for sea-going ships with a length of 90 m and larger operating in unrestricted service, excluding vessels that operate at a fixed location. The main reason for revising the recommendation was the observation from IMO GBS verification that *"modern data show both an increase in mean significant wave height for the North Atlantic and that more extreme weather is being experienced in recent years, including the existence of rogue waves and the possible effect of climate change"*. Revision 1 of Rec. 34 was based on old wave statistics from visual eyeball observations. Revision 2 of the wave standard is defined using state-of-the-art wave data sources (numerical hindcast) combined with a ship position dataset (AIS); see Austefjord et al. (2023). The wave scatter diagram is significantly modified and includes the effect of bad-weather avoidance. Furthermore, the spectrum and spreading shapes are slightly narrower than in Rec. 34, Rev. 1. When comparing the fatigue loads based on Rev. 2 with the loads based on Rev. 1, it is found that the new fatigue loads are generally significantly lower. The reduction ranges from 5% to 50%. The most significant reduction is seen for longer vessels. Even if the updated scatter diagram as such gives reduced wave loads, it will not necessarily translate directly into a scantling reduction. For instance, the current IACS unified requirement S11A for container vessels considers a routing factor to correct the Rec. No. 34, as Rev. 1 does not account appropriately for bad weather avoidance. This factor will thus be adjusted when Rec. No. 34, Rev. 2 is implemented. Similar updates are expected for the worldwide scatter diagram.

6.2 International Standards and Recommendations

Several international standards and recommendations serve as the basis for the rules and guidelines. This section contains a summary of the key developments according to Table 4.

Committee III.2: Fatigue and Fracture 653

Table 4. Review summary of international standards and recommendations.

Section	Standard or recommendation	Content
6.2.1	British Standard for assessing flaws (BS 7910)	Assessment of the acceptability of flaws in all types of structures and components
6.2.2	British Standard for fatigue design and assessment (BS 7608)	Fatigue design and assessment in general
6.2.3	EN ISO 5817	Quality levels of imperfections in fusion-welded joints in all types of steel, nickel, titanium, and their alloys for material thickness ≥ 0.5 mm
6.2.4	Eurocode 3, EN 1993 1.9	Fatigue assessment in general: • Nominal, hot-spot, and notch stress approaches • Stress concentration factors for nominal stress approach
6.2.5	FKM Guideline	Fatigue and strength assessment in general: • Valid for components of iron or aluminium, materials including cases of elevated temperatures
6.2.6	IIW Recommendations for fatigue design	Fatigue assessment in general: • Nominal, structural, and notch stress approaches • Fracture mechanics approach • Mean stress effect • Post-weld treatment, residual stresses

6.2.1 British Standard for Assessing Flaws (BS 7910)

BS 7910 gives guidance and recommendations for assessing the acceptability of flaws in all types of structures and components. BS 7910:2019 (BS, 2019) supersedes the withdrawn BS7910:2013 + A1:2015. Selected changes are (a full overview of the revisions is given on page ix of the standard):

- Annexes have been designated as "informative", whereas they were classified as "normative" in the earlier editions.
- A new Annex V has been added for strain-based assessment and design, i.e., the failure assessment diagrams for the engineering critical assessment (ECA) of flaws can now be established using strain data.
- The leak-before-break assessment procedure has been simplified. Leak-before-break assessment is a part of ECA of flaws in, e.g., LNG containment systems.
- More guidance has been added in Annex J on the Master Curve approach using data from Charpy V-notch impact tests.

- More resources on the distinction of local and global solutions for limit loads have been added. The limit load solutions for offshore tubular joints and clad plates containing a repair weld have been removed.

The new Annex V (strain-based assessment) has been added as the significant recent interest in the resistance of structures to higher nominal loads makes it necessary to use engineering fracture mechanics to determine the flaw tolerance of welds subjected to high strain levels. E.g., offshore pipelines are subjected to a high strain level and sometimes yield. Annex V assumes that the baseline loading might be easily expressed as a displacement or rotation, but the system performance might display a load-controlled behavior. Annex V considers only the assessment of planar flaws under tensile applied strains. The derivation of the methods given in this annex is based on uniaxial primary stresses. Still, recent work suggests that the methods may be conservatively applied to cases of biaxial loading. For more information, see (McCaig and Wang, 2023).

6.2.2 British Standard for Fatigue Design and Assessment (BS 7608)

BS 7608:2014 + A1:2015 (BS, 2015) supersedes BS 7608:2014, which is withdrawn. BS 7608 + A1 is a full revision of the standard, introducing the following main changes:

- Introduction of the hot-spot stress method with guidance on finite element stress analysis
- New correction for both plate thickness and applied bending with allowance for welded joint proportions
- Additional weld details: some details have been reclassified
- Comprehensive guidance on the use of weld-toe improvement methods
- New guidance on acceptance of fatigue testing and statistical analysis of results
- Revised seawater corrosion fatigue data
- Design data to resist shear fatigue failure
- Guidance on stress calculation for combined loading
- Revised cumulative damage rules.

6.2.3 En ISO 5817

This International Standard provides quality levels of imperfections in fusion-welded joints (except for beam welding) in all types of steel, nickel, titanium, and their alloys for material thickness ≥ 0.5 mm. It covers fully penetrated butt welds and all fillet welds. Its principles can also be used to partial-penetration butt welds. Three quality levels are used to permit application to a wide range of welded fabrication. They are designated by symbols B, C, and D. Quality level B corresponds to the highest requirement on the finished weld. Annex B (formerly Annex C in Version 2014) gives additional criteria to meet fatigue class (FAT) levels.

The additional requirements are given for quality levels C and B to adjust the limits for imperfections to the fatigue class FAT 63 for quality level C, giving C63, and FAT 90 for quality level B, giving B90. A quality level B125 corresponds to fatigue level FAT 125, represented by additional requirements to level B for some imperfections. Level C63 covers FAT 63 and lower, level B90 covers FAT 90 and lower, and level B125 covers FAT 125 and lower. B125 may not be practically achieved for welded joints.

Table 5 summarizes the requirements of construction tolerances for the relevant fatigue classes, where e is the eccentricity value, and the β stands for angular distortion of the joints. These requirements must be strictly fulfilled if the nominal stress approach (without explicitly introducing SCFs due to the construction tolerances) is employed for the considered locations. In the case of the hotspot stress approach, the SCFs due to the misalignments should be explicitly calculated (e.g., according to BS7910) and introduced in the fatigue assessment.

Table 5. Requirements of construction tolerances for the relevant fatigue classes.

Imperfection	C63	B90	B125
Linear misalignment	$e \leq 0.15 t_{min}$ Max. 4 mm	$e \leq 0.1 t_{min}$ Max. 3 mm	$e \leq 0.05 t_{min}$ Max. 1.5 mm
Angular misalignment	$\beta \leq 2°$	$\beta \leq 1°$	$\beta \leq 1°$

6.2.4 Eurocode 3, EN 1993.1.9

The next generation of Eurocode 3, i.e., EN 1993, "Design of Steel Structures," is being developed as part of the whole development of the 2nd Generation of Eurocodes. Kuhlmann et al. (2021) and Lukić et al. (2024) give an overview of updates to the fatigue-related parts (EN 1993–1-9) as follows.

Without changing the basic rules of the existing EN 1993-1-9 Fatigue, but with the aim of clarification, the future EN 1993-1-9 will distinguish between fatigue design concepts representing the design philosophies (such as damage-tolerant and safe-life concepts) and the different fatigue design methods that are the tools used for the design concepts. One major change is the introduction of specific recommendations for other stress-based design methods, particularly the hot-spot stress method and the notch stress method, besides the well-known nominal stress method. To distinguish between the different stress methods, a more precise definition has been added to clarify how hot-spot and notch stresses should be computed.

As before, the main document of prEN 1993-1-9:2020 focuses on fatigue verification based on the nominal stress method because of its great practical importance. Specific annexes are additionally provided for the hot-spot stress and the notch stress methods. Another significant change compared to the existing EN 1993–1-9 concerns the detail tables, which are the heart of the nominal stress method and have been completely revised. The user gets better illustrations and an improved and clarified compilation of different execution qualities and associated detail categories for each detail. Moreover, a "symbol" column has been added for welded details to indicate the appropriate weld quality compatible with the detail category considered. In general, the use of the detail tables requires a weld quality level B according to EN ISO 5817, an accredited assignment of personnel, and an extent of non-destructive testing (NDT) as specified by EN 1090-2. Consequently, the last column in each table only contains supplementary requirements beyond the specifications of EN 1090-2.

A separate annex is also included to address stress concentration factors due to the transition of plate width or plate thickness. The effect of misalignments beyond 5% of plate thickness is also addressed. Also, an annex defining the requirements for post-weld treatment by high-frequency impact treatment (HFMI) is included. Only constructional details with potential fatigue cracking from the weld toe and where extensive research has been conducted are appropriate to benefit from HFMI treatment effects. The annex is, among others, based on the work summarized in (Marquis and Barsoum, 2016) (Fig. 14).

Detail category	Constructional detail	Symbol	Description	Supplementary Requirements
112	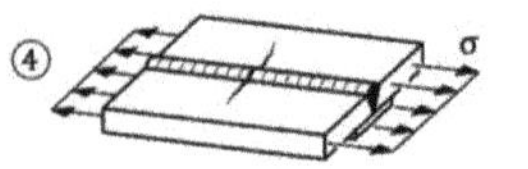 ④	⍙	④ Automatic or fully mechanised butt welds, welded from one side, with a continuous root backing, without stop-starts	
100			as forementioned, but with stop-starts or manual butt welds	
125	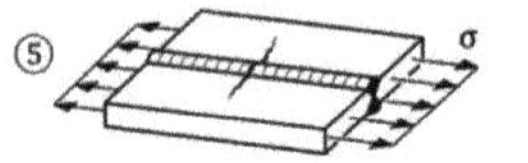 ⑤	⍙⍙	⑤ Butt welds, welded from both sides, ground flush parallel to load direction, without stop-starts	Extent of NDT according to EN 1090-2: 100%.
112		⍙X	as forementioned, but no grinding	
90			as forementioned, but with stop-starts	

Fig. 14. Extract from revised detail table for built-up members (Kuhlmann et al., 2021).

6.2.5 FKM Guideline

The guideline was first published in 1994 by the Forschungskuratorium Maschinenbau e.V. (FKM). The current 7th edition was published in December 2020 and is a complete revision (FKM, 2020). The Guideline is valid for mechanical engineering and related fields. It allows an analytical assessment of the static and fatigue strength, the latter in the form of an assessment of the fatigue limit, the fatigue strength for finite life, or the variable amplitude fatigue strength, depending on the service stress conditions. The guideline is valid for components of iron or aluminum materials, including cases of elevated temperatures, produced by milling, forging, casting, with or without machining, or welding. The maximum material tensile strength is Rm = 1400 MPa for steel and cast iron and 610 MPa for aluminum.

6.2.6 IIW Recommendations for Fatigue Design

IIW Recommendation for fatigue design of welded joints and components (Hobbacher, 2016) has recently been revised. The work on the update started in 2017. The publication of the new IIW recommendations is expected in 2025. The recommendations contain information on the design of welded structures under cyclic loading. Within the recommendations, all relevant assessment approaches, i.e., nominal, structural, and notch stresses, as well as fracture mechanics, are included. In addition, the relevant influences are covered, such as the effect of post-weld treatment, mean respectively residual stresses, and wall thickness. In the current update, the structure is enhanced and straightened, and with it, the applicability is improved. As entirely new content, the assessment of welded thin joints with the notch stress approach and the impact of post-weld treatment procedures on the endurable stresses and the slope of the S-N curve is added. Baumgartner et al. (2024) provide an overview of the most relevant changes, and references to the source of the changes are given.

IIW Recommendations for the high-frequency mechanical impact (HFMI) treatment for improving the fatigue strength of welded joints (Marquis and Barsoum, 2016) are currently being revised. New developments and guideline updates are summarized in Marquis et al. (2024), noting that new research and test data confirm the important elements of that guideline. However, it also demonstrated that certain portions of those recommendations need to be clarified, expanded, and improved, including:

- the scope of the guidelines for steels with yield strengths from 960–1300 MPa,
- relaxing the fatigue geometry requirement of the weld profile prior to HFMI treatment and the groove dimension following treatment,
- presenting fatigue strength modification factors also numerically as an alternative to modifying the fatigue class,
- incorporating the updated thickness correction factors from the newest IIW Fatigue Design Recommendations,
- adding a series of tables that summarize the rules for defining the appropriate fatigue strength for different combinations of material strength, load ratio, stress range, and assessment method,
- provide guidelines on HFMI treatment of prefatigued structures, and
- present a methodology for verifying the effectiveness of HFMI devices based on documentation and experimental data.

Marquis et al. (2024) presents details of this new information that will be incorporated into revised IIW recommendations for HFMI treatment to improve the fatigue strength of welded joints.

6.3 Design Rules and Guidelines for Carriage of LNG as Cargo or Fuel

To effectively reduce CO_2 emissions, alternative fuels, such as LNG, hydrogen, methanol, ammonia, and electricity, can be used. Of these, LNG has been adopted at the largest scale so far. The recent surge in the number of larger vessels using LNG as fuel has led to some activity in the development of rules and regulations. These rules address the challenges of significant temperature variations and cryogenic materials used in liquefied natural gas (LNG) fuel storage tanks. In addition, cargo/fuel loading/unloading-induced low cycle fatigue is relevant for these vessels.

The structural integrity of the LNG tanks in either gas carriers or LNG-fueled ships is crucial, and LR has recently updated gas ship rules (LR, 2022b) and rules for ships using low-flashpoint fuels (LR, 2022a). In these rules, IGC code requirements in terms of fatigue & fracture assessment of containment systems have been fully integrated along with further clarifications. LR also published a guideline document for the FEA procedure of Type-C LNG tanks (LR, 2024a). The document describes the procedures for a fine mesh application for strength and fatigue assessments. Fatigue-critical locations are also concisely summarized.

DNV introduced requirements for FE calculations of independent prismatic gas fuel tanks (DNV, 2022). The purpose of the FE analyses is to assess the structural adequacy of structural members of the gas fuel tank, tank supports, and associated hull structure under consideration of all relevant loads. The analysis requirements are described for container ships with independent prismatic tanks of Type-A or Type-B. An additional class guideline (DNV, 2021g) provides acceptable procedures for strength, fatigue, crack propagation, and fracture analyses. In addition, ABS published guidance notes on the Strength Assessment of Independent Type C Tanks (ABS, 2022), addressing both high- and low-cycle fatigue assessment for bilobe tanks and their supporting structures.

CCS has comprehensively specified descriptive and finite element strength requirements for independent liquid cargo tanks in the Rules for Construction and Equipment of Ships Carrying Liquefied Gases in Bulk (CCS, 2023). It provides requirements for fatigue strength assessment based on S-N curves for independent cargo tanks and their supporting structures. The requirements for fatigue strength assessment based on fracture mechanics are also given. The requirements for fatigue crack propagation and leakage analysis should be carried out for Type B liquid cargo tanks according to the IGC rules, and the methodology applies to other types of liquid cargo tanks.

KR (2018) has recently published a guideline related to fatigue and crack propagation in high manganese austenitic steel for low-temperature application. The guideline provides information such as S-N curves and procedures for fatigue strength assessment and crack growth tests. Also, CCS (2021) and LR (2023a) have developed guidelines for high manganese austenitic cryogenic steels. These guidelines mainly formulate the corresponding technical standards and inspection requirements in the aspects of high manganese austenitic steel, welding materials, application of high manganese austenitic steel, welding of high manganese austenitic steel, etc. The fracture toughness simulation of high manganese austenitic steel and fatigue crack propagation evaluation process based on CTOD method was also proposed.

A benchmark study has been carried out to establish a common framework for the fatigue, crack growth, and leak assessments of Type B prismatic LNG tanks. This is described in Appendix II. It was found that there are significant variations in how procedures from the classification societies are implemented. These variations were mostly observed regarding the following process steps:

- Selection of S-N curves and combination of high cycle and low cycle fatigue damage,
- Spectrum implementation and consideration of failure modes in the crack growth assessment of a surface defect,
- SIF solutions for through-thickness cracks,
- Consideration of equivalent crack opening stress in the leak assessment.

It is recommended to establish a common framework, particularly for crack growth and leak assessment to reduce the uncertainties in these calculations.

6.4 Design Rules and Guidelines for Floating Offshore Wind Turbines

In recent years, the activity on offshore floating wind turbines (FOWTs) has increased significantly. Rules and regulations for OWTs have not been discussed in previous ISSC reports. Therefore, a separate section is devoted to this.

Several rules and regulations are in place for the fatigue assessment of offshore wind turbines. Some of these include:

- Standards developed by international or national consensus-based groups such as IEC, ISO, or API. An example is IEC 61400-3, which provides guidelines for the design of offshore wind turbines, including fatigue assessment.
- Classification societies such as DNV, ABS, LR, BV, and NK have specific rules and regulations for the design and assessment of offshore wind turbines, including fatigue assessment.
- National regulations may also apply, depending on the location of the wind farm. For example, in the UK, the Health and Safety Executive (HSE) provides guidelines for the design and assessment of offshore wind turbines, including fatigue assessment. It is important to note that the specific rules and regulations may vary depending on the location and the organization responsible for the assessment.
- Internal standards by larger companies.

The intention of the ISO standards is to achieve reliability levels appropriate for manned and unmanned offshore structures, whatever the type of structure. Thus, they do not contain specific requirements for the fatigue design of FOWTs. IEC is the internationally recognized main standardization body for wind energy, including offshore floating wind. Design requirements for FOWTs are described in IEC (2019). This document contains some descriptions of loading and load calculation (wind and wave loads and the combination of these). It is emphasized that non-linear load effects (e.g. from wave loads) need to be incorporated through non-linear time domain analyses.

Structural design of floating wind turbine structures is covered in depth in DNV-ST-0119 Floating wind turbine structures. This standard is applicable to all types of support structures for floating wind turbines. The standard is applicable to the design of complete structures, including substructures, but excludes wind turbine components such as nacelles and rotors. It contains a description of environmental conditions, loads, and fatigue design. Reference is made to the following standards for more detailed descriptions: DNV-RP-0286: coupled analysis of floating wind turbines (guidance for modeling, load analysis, and model testing of FOWTs), DNV-ST-0437: principles, technical requirements and guidance for loads and site conditions for wind turbines, DNV-RP-C203: general description of fatigue design of steel structures, DNV-RP-C205: General guidance regarding environmental conditions and environmental loads. The DNV standard DNV-ST-0376 Rotor blades for wind turbines focus particularly on blades made from fiber-reinforced plastics. The standard was updated in April 2024 and now contains a separate section on fatigue models. It also contains descriptions of fatigue calculation and fatigue testing.

Guidance notes from various classifications societies for floating offshore wind turbines are listed in Table 6. Jiang (2021) conducted a comprehensive state-of-the-art review of the technical aspects, rules, and standards for installing offshore wind turbines. The following standards were listed:

- DNVGL-ST-N001 (DNV, 2023b) provides guidance on marine operations for various offshore structures, with Chapter 8 specifying general requirements for OWT installation. Fatigue during transport and installation is discussed in Chapters 11 and 13.
- The ISO 29400 (ISO, 2020) is similar to the aforementioned but applies to port and marine operations involving offshore structures like subsea templates, foundations, and OWTs. Some advice related to the fatigue assessment is given.
- The DNV-RP-N103 (DNV, 2017) offers simplified formulas for determining design loads, which is crucial for planning and executing marine operations. While it covers towing, weather, and lifting operations, it does not explicitly address OWTs. Some advice is given related to the fatigue assessment, especially due to vortex-induced vibrations (VIV).
- The DNVGL-ST-0378 (DNV, 2021c) defines requirements for lifting appliances and deck equipment for offshore and platform cranes, whereas the DNVGL-ST-0054 (DNV, 2021b) provides safety principles and guidance for transporting and installing onshore and OWT components. Some advice related to the fatigue assessment is given.
- Guidance on loads and site conditions for wind turbines, emphasizing its application in design load cases for transport, installation, maintenance, and repair, are specified in DNVGL-ST-0437 (DNV, 2024). Load cases for fatigue analysis are described in detail.

Despite these standards and guidelines reflecting contemporary best practices, continuous improvement is essential. Guachamin Acero et al. (2016) proposed a method for evaluating marine operations' operational limits and operability. Noble et al. (2021) suggest standardizing Marine Renewable Energy (MRE) technology due to the broad spectrum of disciplines, necessitating more comprehensive guidance to address its diverse facets. They reviewed recently published guidance, identified perceived gaps, and proposed a pathway toward a standardized approach to MRE technology and development. Among other things, they pointed out a lack of guidance on dealing with combinations of effects, e.g., composite aging and fatigue or wear combined with corrosion or fatigue. Wang et al. (2022a) reviewed a state-of-the-art reliability assessment of OWT support structures, providing a comprehensive review of the structural reliability, reliability-based calibration of codes, fatigue reliability, and the implementation of reliability assessment.

For a floating offshore wind turbine, an important part of the design is the K-joint of tubular members connecting the columns of the floating structure. The complexity of the calculations required for this type of structure makes simplified screening methods for preliminary fatigue life assessment of real interest. At present, the effective application of design standards provided by various classification societies cannot be considered consolidated. The committee, therefore, performed a benchmark study dedicated to the stress concentration factor on a K-joint with the aim of investigating the applicability and limitations of known approaches in this field. This benchmark study is reported in Appendix I. In general, there was good consistency among the results regarding

Table 6. Selected design rules and guidelines for floating offshore wind turbines.

Body/Publisher	Document	Edition
IEC	IEC TS 61400–3-2 Wind energy generation systems – Part 3–2: Design requirements for floating offshore wind turbines	April 2019
DNV	DNV-ST-0119 Floating wind turbine structures	June, 2021
DNV	DNV-RU-OU-0152 Floating wind installations	July, 2023
BV	NI 572 DT R02 E, Classification and Certification of Floating Offshore Wind Turbines	January 2019
ABS	Guide for Building and Classing Floating Offshore Wind Turbines	July 2020
RINA	Guide for Certification of Floating Offshore Wind Turbine Installations	August 2021
NK	NKRE—GL-FOWT01, Guidelines for Floating Offshore Wind Turbines	December 2021
LR	LR GN22 Guidance Notes for Offshore Wind Farm Project Certification	May 2022
LR	Recommended practice for Floating Offshore Wind Turbine Support Structures	January 2024

the qualitative distribution of SCFs along the joint and in terms of absolute values. The scatter of the results is generally small when braces are the loaded members. If the chord is the loaded member, the scatter increases, especially on the chord itself. The regression equations provided by the classification societies do not always seem conservative compared to participants' results.

6.5 Summary

Recent changes in rules and standards relevant to fatigue and fracture assessment of ships and offshore structures have been discussed. During the last committee period, there has been significant activity on improving and expanding the existing approaches. This applies to classification society rules and guidelines as well as international standards and design recommendations. There has been a particularly large activity related to LNG as cargo or fuel and to floating offshore wind turbines. Therefore, an in-depth discussion has been provided on these topics.

7 Conclusions

This ISSC Committee III.2 report reviews the latest advancements for addressing fatigue and fracture challenges in the ship and offshore structures. The work is aligned with the committee's mandate to improve understanding of fatigue and fracture assessment to control crack initiation and growth under cyclic loading.

Recent advancements have enhanced fatigue load modeling to represent environmental sources, such as waves and wind, considering climate change and extreme conditions. Despite similar environmental loads, the modeling approaches often vary depending on structure type. Future research should harmonize these methods for consistency across different structures and focus on modeling load interactions and integrating real-time environmental data.

Progress in fatigue- and fracture-resistant materials and their characterization methods are also apparent, resulting in, for instance, improved crack tip opening displacement formulas. The developments also enhanced mitigating welding residual stresses and multiaxial fatigue evaluation. Advanced metallic materials and manufacturing processes can improve fatigue and fracture resistance. However, challenges in quality assurance and standardization remain. Future research should also include bio-based, recyclable, and hybrid materials to address fatigue and fracture in demanding environments, considering sustainability issues.

Recent advancements in computational modeling of fracture and damage mechanics have led to improved methods for fatigue life prediction. For instance, continuum damage-based approaches have been extended through various modifications. Hybrid simulations, such as multi-fidelity sub-modeling and -structuring, have increased the accuracy of simulating fracture and damage mechanics. Future research should address localized imperfections, enhance multiscale models, and develop efficient computational techniques validated through large-scale testing.

Reliability-based approaches have become crucial for managing uncertainties in material behavior, operational conditions, and long-term degradation. The recent advancements in probabilistic analysis have focused on uncertainties in load prediction, modeling, and damage accumulation. Key developments include integrating more accurate models, sensor data, and machine learning. Despite good progress, challenges remain in predicting fatigue damage due to random marine loading and modeling complexities. Another significant challenge is the accurate prediction of damage under variable amplitude loading conditions, where accurate models and sufficient tests with realistic load spectra are lacking. Future research should also emphasize uncertainty quantification, high-fidelity digital twins, and advanced surrogate models.

Recent changes in rules for fatigue and fracture assessment highlight significant improvements in existing approaches. Key updates from classification societies include enhanced fatigue analysis, S-N curves, and stress concentration factor assessments. Advancements particularly address LNG carriage and floating offshore wind turbines with new guidelines for fatigue strength and structural requirements. International standards have also evolved, focusing on flaw assessment, fatigue design, and post-weld treatments. Future research will include harmonizing fatigue data, addressing high-strain resistance, and standardizing fatigue and crack assessments.

Committee III.2: Fatigue and Fracture   663

Regarding future research needs, there is also a pressing demand for simplified screening tools to enhance early-stage fatigue design and computational efficiency in optimization algorithms. Also, while much recent research has been dedicated to offshore wind structures, many of these findings are applicable to ships, which face additional challenges due to their structural complexity and unique designs. Furthermore, focusing more research on real-world applications in marine structures could significantly increase the impact of recent advancements in fatigue and fracture engineering.

Appendix I: Benchmark Study on a k-joint of Floating Structure

A stress concentration factor study of a K-joint of the floating structure supporting a wind turbine

Objectives

The fixed installations of wind turbines are widespread. Still, the growing demand for green energy has initiated interest in expanding to new areas, i.e., floating offshore wind turbine installations in deep water. As design concepts for these new areas vary significantly from existing structures in terms of geometry and operational profile, the applicability of the existing design standards and rules requires further evaluation. Unlike conventional offshore structures, where environmental wave loads are typically considered the dominant loads from a global strength point of view, floating wind turbines present unique challenges. In addition to wave loads, operational loads (turbine thrust, strictly related to wind loads) play a crucial role and must be properly combined with the wave loads. Furthermore, the flexibility of the offshore wind turbine support structures makes simple rigid body analysis approaches unsuitable for estimating the loads and corresponding structural response.

The commonly utilized approach in engineering practice involves the fully coupled time domain analysis of the system, which consists of a floater and turbine, often conducted by specialized software such as Orcaflex. This approach yields large sets of complex load cases and uses one-dimensional structural modeling with beam elements. Since beam elements provide results on the forces and moments acting on them, it is necessary to correlate the beam element loads with the structural hot spot stresses at fatigue-critical locations such as joints of trusses to perform a fatigue assessment. Although numerous design standards offer formulae for this purpose, they may not cover all the possible load scenarios across the entire perimeter of the joint.

To facilitate the development of effective fatigue design approaches for floating offshore wind turbine installations, this benchmark study carries out structural stress analyses of a large wind turbine installation as to compare its results and derive potential trends and recommendations. The studied structure is a semi-submersible column stabilized design with overall dimensions that can reach up to 100 m to install large wind turbines, such as those in the 15–18 MW range; see Fig. 15. This study focuses on a fatigue-critical K-joint of tubular members that connect the columns of the floating structure supporting a wind turbine. The primary objective of this study is to compare different state-of-the-art approaches for calculating stress concentration factors to identify suitable methods for applications in floating wind turbine structures. Additionally, the study evaluates the applicability of existing class rule SCF formulas for applications in tubular joints, some of which may be outside their parameter ranges.

Fig. 15. The studied floating offshore wind turbine installations (Floater shape).

Review of Codes and Standards

The calculation of the Stress Concentration Factor (SCF) values and subsequent application in fatigue life prediction of marine structures is commonly based on the hot spot structural stress. The structural stress method can be used for plated or tubular structures at a stress concentration location corresponding to fatigue cracking at the weld toe. It is, however, not applicable to weld root details in fillet or partial penetration welds or cracks initiating and propagating perpendicular to the weld (for the tubular joint in this study: e.g., location c in the brace and chord for a load normal to the chord shown in Fig. 20), or other locations without a stress concentration occurring in shell elements. For these details, a separate and different type of fatigue assessment (e.g., nominal stress, effective notch stress) needs to be performed. It should be noted that the hot spot structural stress method through surface stress extrapolation is sensitive to the mesh size and mesh quality. Therefore, DNV (2021d) and IIW (Hobbacher and Baumgartner, 2024) provide guidelines on mesh sizes in combination with extrapolation distances. These documents provide advice for general purposes, particularly for planar surface joints. More specific guidance for tubular joints is given in DNV-RP-C203 (DNV, 2021e), which also considers curvature. A summary is given in the Fig. 16 DNV (2021e) also presents recommendations for mesh quality; for example, corner angles between 60° and 120° and length/breadth ratio less than 5.

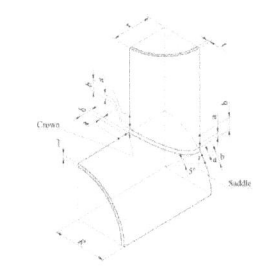	Extrapolation of stress along the brace surface normal to the weld toe: $a = 0.2\sqrt{rt}; b = 0.65\sqrt{rt}$ Extrapolation of stress along the chord surface normal to the weld toe at the crown position: $a = 0.2\sqrt{rt}; b = 0.4\sqrt[4]{rtRT}$ Extrapolation of stress along the chord surface normal to the weld toe at the saddle position: $a = 0.2\sqrt{rt} ; b = \frac{\pi R}{36}$
Alternative analysis approach: use the stresses at the Gaussian points directly in the fatigue assessment if placed at $0.1\sqrt{rt}$ from the weld toe (r = radius of considered tubular and t = thickness).	

Fig. 16. Recommendation for hot spot calculation in tubular joints. From DNV (2021e).

In the current study and often in current industrial practice, FE models with shell elements are used to model a structure for fatigue assessment using the hot spot method without modeling the weld. Neglecting the weld in shell element FE models can result in additional limitations of the hot spot structural stress method. Figure 17 presents a plate with two plates attached in the same plane. For hot spot location c, in combination with loading direction I, the hot spot stress will equal the nominal stress using shell elements that neglect the welds because shell elements are infinitely thin, and stress is not distributed through the vertical plate or welds. Therefore, for this location, the nominal stress is recommended for fatigue assessment. For the tubular connection regarded in the current study, this effect is visible in locations a and e (see Fig. 20) in the chord for an applied load normal to the chord. This effect will also occur in other load cases and locations where the load direction is not clear due to the complex load and curved weld line. Most probably for this reason, DNV (2021e) recommends using solid elements, including weld, for fatigue assessment of tubular joints using the hot spot method.

DNV (2021e) includes, in addition to the hot spot stress evaluation indication, regression formulae to calculate the stress concentration factors for simple tubular joints and overlap joints under the most significant loading conditions. These and similar formulae are the result of studies begun in the 1960s and developed over the following decades based on regression of FE model results. The widespread use of this type of joint and its recent application in wind turbine installations is keeping interest alive. To give just a few examples, increasing geometric complexity is studied by considering multiplanar configurations from the theoretical point of view (Zavvar et al., 2021) or with the support of experimental data (Bao et al., 2023). From the point of view of building regressions on the SCF, the use of machine learning techniques, such as neural network models (Mohammed et al., 2024), seems particularly promising.

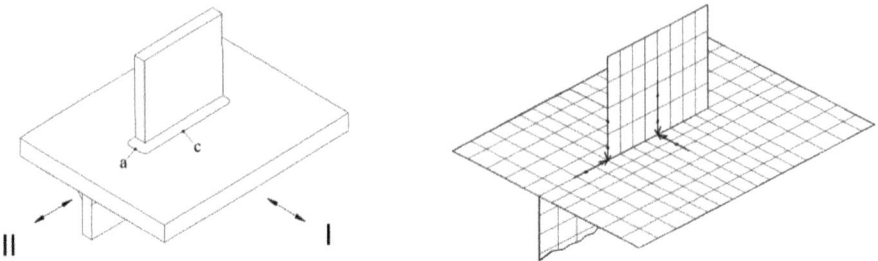

Fig. 17. Example of hot-spot limitations. From DNV (2021e).

Benchmark Case Study Presentation

Provided with details of the K-joint and the loading conditions to be analyzed, the case study participants were asked to evaluate the stress concentration factors at the welded connection. No specific guidance has been given on the methodology used to assess how individual choices might affect the values obtained and the overall spread of results, except that a shell element FE model was provided.

Details of K-Joint

The K-joint that is the subject of this study is part of a floater of considerable size, approximately 100m long and wide, supporting a 15MW wind turbine, and identifies the connection of two diagonals (braces) to the main chord.

Figure 18 below shows the overall dimensions and scantlings intended for the beams. The length of the pipes in relation to the diameters can be considered sufficient to justify the use of the 1D element global F.E. model to calculate the forces acting on the members.

The main horizontal chord is continuous without internal reinforcements, and the inclined braces are interrupted at the connection. A full penetration weld is present at the joint. This work focuses on determining hot spot stress at the joint. Therefore, the geometry of the weld, referenced only for the definition of the calculation point, may be disregarded considering the geometrical intersection of the external surfaces of the pipes.

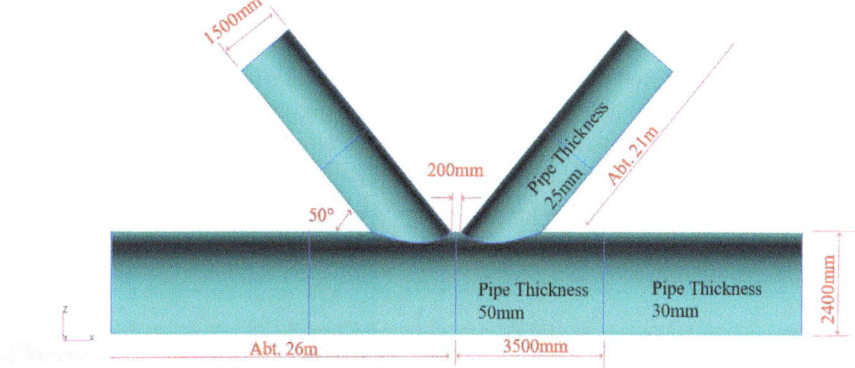

Fig. 18. K-joint details.

Load Cases

The load cases considered in the benchmark study were limited to a set of elementary load cases that could be appropriately combined to describe more complex loads and load histories. In addition, for some load cases considered, stress concentration factor values are formulated in the regulations and standards, which would be a starting point for the initial analysis of the results obtained.

The finite element model was constrained at the rear end of the main chord, and the other ends have been provided with rigid multi-point constraints (e.g., Nastran MPC, see Fig. 19) with an independent node at the center of the profile on which the loads have been applied.

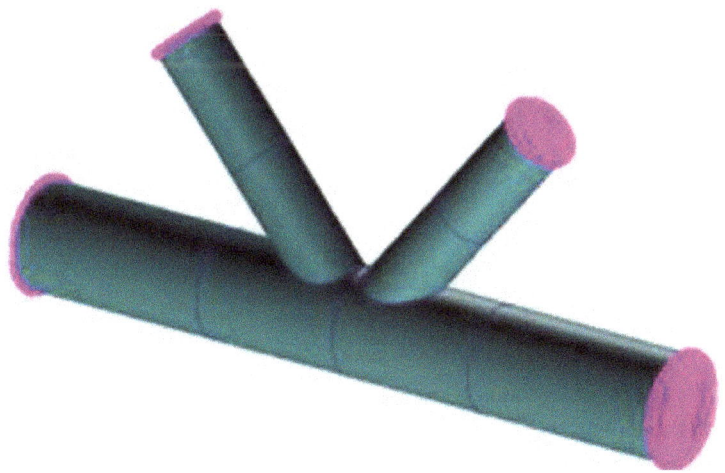

Fig. 19. Multi-point constraints.

A single nominal stress has been calculated for each load case as the stress acting in the loaded member (e.g. if the chord is the loaded member, the nominal stress is the nominal stress calculated on the chord, and the stress concentration factor on one brace is the ratio between the hot spot stress on the brace and the nominal stress calculated on the chord). This definition is consistent with the closed-form formulae available in the literature, codes, and standards and has the advantage of ensuring a non-zero value. This also provides the opportunity to define a stress concentration factor for all structural members.

$$\sigma_n = \frac{N_x}{A_x} + \frac{\sqrt{M_y^2 + M_z^2}}{W}$$

where:

- σ_n: Nominal stress
- N_x: Axial force acting on the loaded member
- $\sqrt{M_y^2 + M_z^2}$: In-plane or out-of-plane bending moment acting on the loaded member
- A_x: Section area of the loaded member
- W: Section modulus of the loaded member

As the load cases are based on unitary loads, the nominal stress is inversely proportional to one of the section properties, section area, or section modulus of the loaded member (Table 7).

Due to the simplification introduced in the definition of the load cases and the future interest in their combination, for each load case, not only the hot spot stress at the worst location has been considered, but the stress concentration factors have been evaluated for each member all along the connection, concentrating on five specific locations shown and defined in Fig. 20.

SCF locations:
- a: crown heel
- b: in between
- c: saddle
- d: in between
- e: crown toe

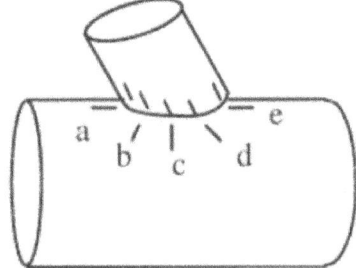

Fig. 20. SCF location definition.

Modeling and Calculations Approaches

The approaches used by the participants were generally quite similar but with some notable differences, which are indicated below.

A commonly used standard is the International Institute of Welding Guidelines, which is similar to DNV (2021e) used by one of the participants. In the latter case, the

Table 7. Load Cases.

Pure unitary axial load on the main chord	Pure unitary balanced axial load on the braces
$\sigma_n \propto 1/A_{Chord}$	$\sigma_n \propto 1/A_{Brace}$
Pure unitary in-plane bending moment on the chord	Pure unitary in-plane bending moment on the brace 1
$\sigma_n \propto 1/W_{Chord}$	$\sigma_n \propto 1/W_{Brace}$
Pure unitary in-plane bending moment on the brace 2	Pure unitary out-of-plane bending moment on the chord
$\sigma_n \propto 1/W_{Brace}$	$\sigma_n \propto 1/W_{Chord}$
Pure unitary out-of-plane bending moment on the brace 1	Pure unitary out-of-plane bending moment on the brace 2
$\sigma_n \propto 1/W_{Brace}$	$\sigma_n \propto 1/W_{Brace}$

participant used the most straightforward approach, avoiding extrapolation, and read the stress at the Gaussian point at $0.1\sqrt{rt}$ from the weld line (r = radius of considered tubular and t = thickness). Another participant also provided an example of the application of the ASME PVcode2007.

Depending on availability at the participants' organization, two commercial finite element software packages were used: Nastran and ABAQUS. The nodal force/moment-based structural stress method commercially available in the fe-safe (Verity Module) was also used since it has been adopted for the specific application of the ASME B&PV Sec. VIII Div 2 International Code for which the technical basis is provided in (Dong et al., 2010) since 2007. Note that the nodal force/moment-based method is mesh-insensitive in structural stress calculation and applicable for both weld toe and throat fatigue failure modes (Dong et al., 2010). It also enables a direct treatment of low cycle fatigue for which closed forms have been recently developed for adoption by the same ASME Code in its 2025 revision (Wei et al., 2024a, Wei et al., 2024b).

Common to all results is the mid-plane surface idealization associated with the shell element mesh. Only one participant used 8-node quadratic quadrilateral elements, while others preferred linear quadrilateral elements. The mesh size has been selected according to the reference standard. Only the participant using the mesh-insensitive method adopted ASME B&PV Sec. VIII Div 2 Code (Dong et al., 2010) considered three element sizes in the FE model: $0.5t_{max} \times 0.5t_{max}$, $t_{max} \times t_{max}$, $2t_{max} \times 2t_{max}$. As the results obtained were almost identical, as shown in Fig. 21, only the results obtained with the mesh size $t_{max} \times t_{max}$ are presented later in this report. For the rest of the participants, t x t mesh size is adopted, with minor adjustments for defining stress extrapolation positions. According to the idealization adopted, the results have been read on element nodes or centroids, always on the outer layers.

Significant differences were noted in the choice of reference stress and extrapolation method. Three of the participants used the component of the stress perpendicular to the weld line, while one referred to the principal stresses with a specified range angle with the normal to the weld line direction.

Another important aspect is the definition of the extrapolation direction; see Fig. 22. The normal to the weld direction is defined by the symmetry plane at the crown heel and the crown toe (locations a and e). Still, at the saddles and the points in between (locations b, c, and d), it is easy to confuse it with the neutral axis of one of the joint members. Participants paid attention to this aspect to avoid the approximation leading to different results, especially when using methods based on longer extrapolation distances (e.g. 0.5t-1.5t).

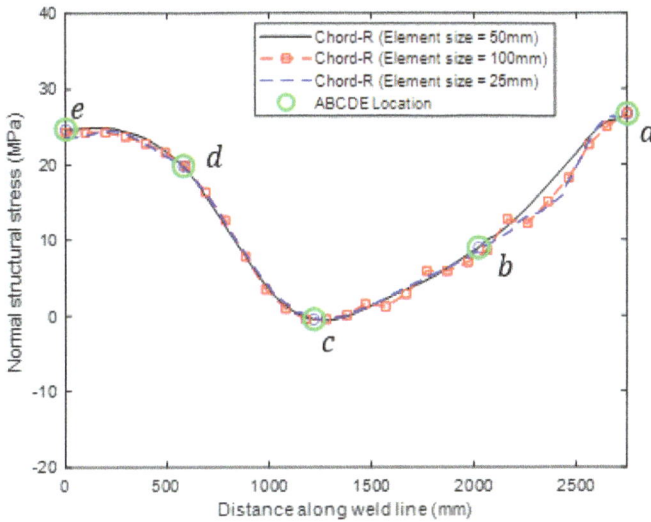

Fig. 21. Mesh insensitive method test on three different mesh sizes.

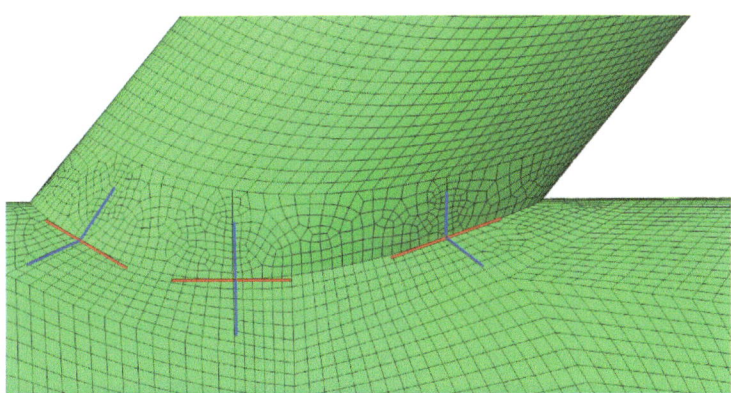

Fig. 22. Extrapolation direction definition.

It should be noted that when using the mesh-insensitive structural stress method (adopted by ASME B&PV 2007 Code), nodal forces and moments for each structural member (e.g., chord) are determined with respect to the ring of elements. Then, the structural stress along the node line at the brace-chord intersection was computed through a matrix equation.

The resulting structural stresses are expressed in terms of both membranes plus bending parts perpendicular to the weld line. In this method, the element set is highlighted in Fig. 23 for collecting the nodal forces and moments at the nodes representing the weld toe line (inner edge of these elements).

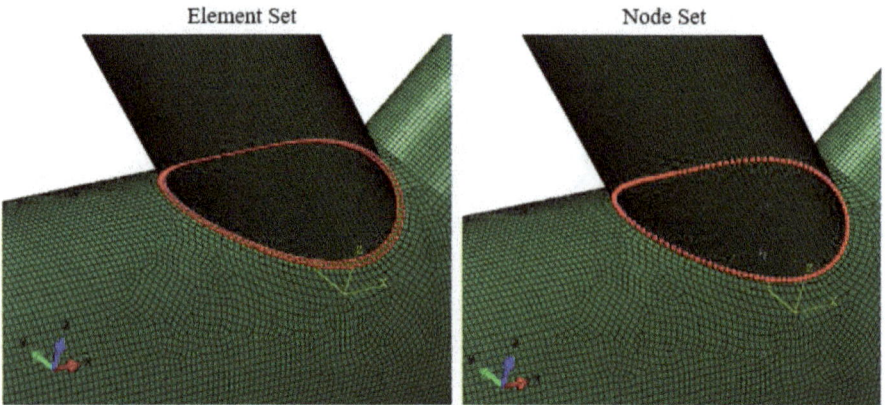

Fig. 23. The element and node sets around the joint in Abaqus CAE.

FE-safe/verity has been used to convert the nodal forces to structural stress, considering only the components normal to the weld line. The equations used are presented as follows:

$$\begin{Bmatrix} F_1 \\ F_2 \\ F_3 \\ . \\ . \\ F_{n-1} \end{Bmatrix} = \begin{bmatrix} \frac{(l_1+l_{n-1})}{3} & \frac{l_1}{6} & 0 & 0 & .. & \frac{l_{n-1}}{6} \\ \frac{l_1}{6} & \frac{(l_1+l_2)}{3} & \frac{l_2}{6} & 0 & .. & .. \\ 0 & \frac{l_2}{6} & \frac{(l_2+l_3)}{3} & \frac{l_3}{6} & .. & .. \\ .. & .. & .. & .. & .. & .. \\ .. & .. & .. & .. & .. & .. \\ \frac{l_{n-1}}{6} & 0 & 0 & 0 & \frac{l_{n-2}}{6} & \frac{(l_{n-2}+l_{n-1})}{3} \end{bmatrix} \begin{Bmatrix} f_1 \\ f_2 \\ f_3 \\ . \\ . \\ f_{n-1} \end{Bmatrix}$$

$$\sigma_s = \sigma_m + \sigma_b = \frac{f_r}{t} - \frac{6m_\theta}{t^2}$$

where:

- $f_1, f_2, f_3, \ldots$ Are the local line forces at nodes around the weld,
- $F_1, F_2, F_3, \ldots$ Are local nodal forces in local coordinate systems at nodes around the weld,
- $l_1, l_2, l_3, \ldots$ Are the element lengths between corresponding nodes.

Figure 24 shows the normal and tangential directions around the weld, respectively. The stress distribution around the weld is a direct output from FE-safe/verity, from which the stress values at the hot-spot locations can be extracted for comparison with those from other methods. It was only in this specific application that the participant carried out the analysis with different mesh sizes, demonstrating the insensitivity of the results.

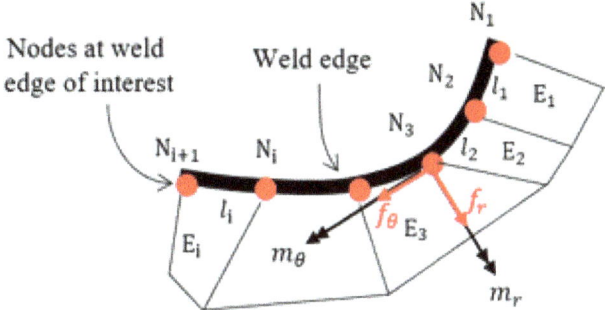

Fig. 24. The definitions of local nodal forces/moments and local line forces/moments around the weld.

The order of calculations was also monitored, highlighting whether the participant performed the derivation of the result first (e.g., calculation of the principal stress) and then the extrapolation or vice versa.

To address potential inconsistencies in defining the exact locations of the hotspot points (i.e., points a, b, c, and d) among participants, particularly for the intermediate points b and c, their coordinates were carefully monitored. All participants defined these as the intersection of the surfaces with the symmetry plane rotated by 45° around the axis of the braces.

For ease of comparison, the above has been summarized in Table 8, using different grey scale backgrounds to highlight differences and similarities.

Results

The following results plots show the calculated SCFs at all locations for each loading condition and member. To assist the reader in data interpretation, each plot is arranged to reflect joint geometry. SCFs on the braces are shown at the top, while SCFs on the chord are shown at the bottom in the corresponding locations. The sides of the plots highlight the maximum stress concentration for each member of the joint, with evidence of the spread of the results and the detailed values provided by the participants. These values are summarized in terms of average value and standard deviation (Figs. 25, 26, 27, 28, 29, 30, 31 and 32).

Table 8. Summary of participants approaches.

	Participant 1	Participant 2	Participant 3	Participant 4
Reference standard	DNV-RP-C203	IIW	ASME B&PV Sec. VIII Div 2 (2007)	IIW
Software	MSC Nastran 2018.2	ABAQUS 2023	ABAQUS 2021+ Fe-Safe/Verity	ABAQUS 2023
Geometrical idealization	Midplane Surfaces	Midplane Surfaces	Midplane Surfaces	Midplane Surfaces
Element Type	Linear quadrilateral shell elements (QUAD4)	Linear, finite-membrane-strain, fully integrated, quadrilateral shell element (S4)	Linear, finite-membrane-strain, fully integrated, quadrilateral shell element (S4)	Quadratic quadrilateral shell elements with reduced integration (S8R)
Mesh size	$\frac{2}{3} 0.1\sqrt{r\,t}$	$0.4t - 1.0t$	$t_{max} \times t_{max}$ where t_{max}: $\max(t_{chord}, t_{brace})$	$t \times t$
Read out point	Element centroids; external layer	Element nodes; external layer	Element nodes; external layer	Element nodes; external layer

(*continued*)

Table 8. (*continued*)

	Participant 1	Participant 2	Participant 3	Participant 4
Reference stress	Stress normal to the weld line	Stress normal to the weld line	Stress normal to the weld line	Principal stress within ± 60° from the weld normal direction or stress normal to the weld 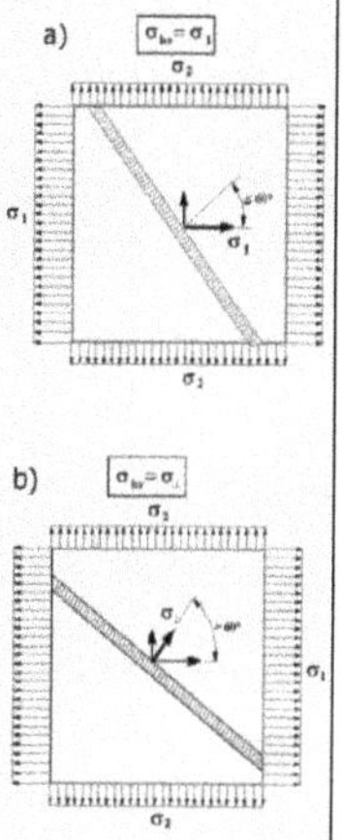
Extrapolation method	No Extrapolation, Stress directly read at $0.1\sqrt{rt}$	Linear 0.4t; 1.0t 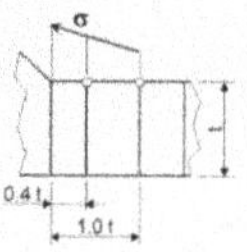	Mesh Insensitive, No extrapolation. Stress directly calculated at weld toe line 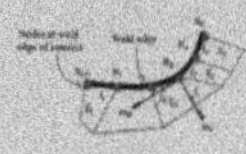	Linear 0.5t; 1.5t 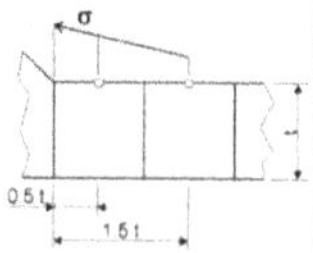
Extrapolation direction	No Extrapolation	Perpendicular to the weld line	No Extrapolation	Perpendicular to the weld line

(*continued*)

Table 8. (continued)

		Participant 1	Participant 2	Participant 3	Participant 4
Calculation sequence		Shape and orientation of elements created to read the stress normal to weld line directly as one of the components in the local coordinate system	Derive and Extrapolate Derive reference stresses at 0.4t and 1.0t. Extrapolate stress, calculated at 0.4t and 1.0t, to the hot spot location	Directly Derive Calculate forces and moments at nodes along the weld line and derivation of the stresses.	Extrapolate and Derive Extrapolate normal parallel and shear stress components, calculated at 0.5t and 1.5t, to the hot spot location. Derive the principal stress or the normal stress from the values extrapolated at the hot spot.
Hot Spot Location (x, y, z)	a	(2021,0,1175)	(2021,0,1175)	(2018,0,1175)	(2021,0,1175)
	b	(1636,521,1053)	(1636,522,1053)	(1635,520,1054)	(1636,522,1053)
	c	(840,738,915)	(840,738,915)	(841,736,916)	(840,738,915)
	d	(275,522,1053)	(275,522,1053)	(277,520,1054)	(275,522,1053)
	e	(95,0,1175)	(95,0,1175)	(98,0,1175)	(95,0,1175)

Committee III.2: Fatigue and Fracture 677

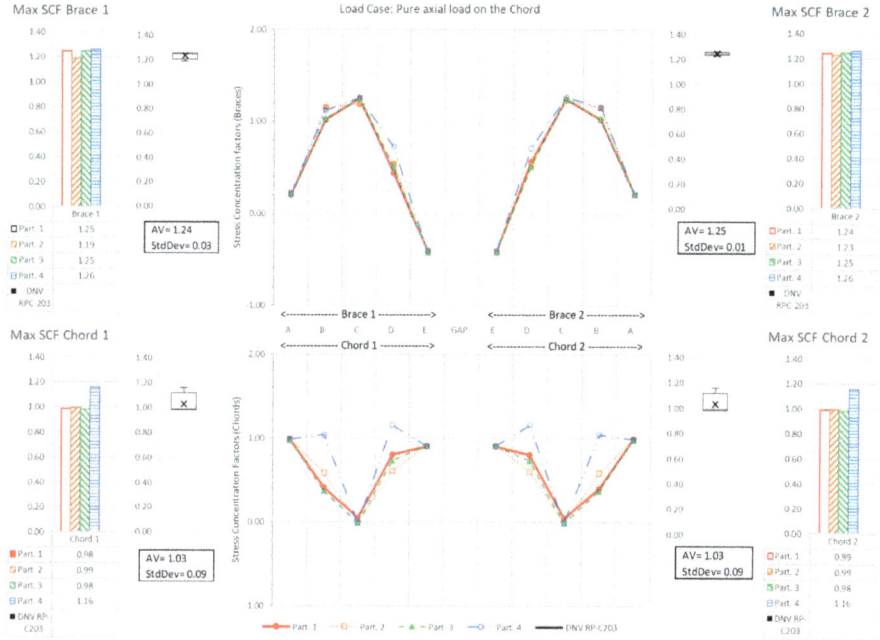

Fig. 25. SCF in Pure axial load on the Chord load case.

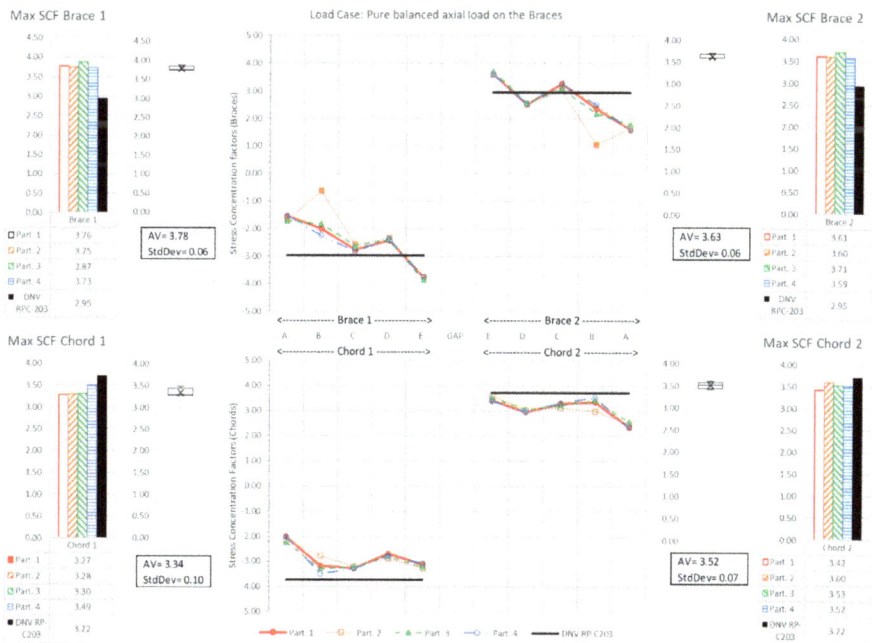

Fig. 26. SCF in Pure balanced load on the Braces load case.

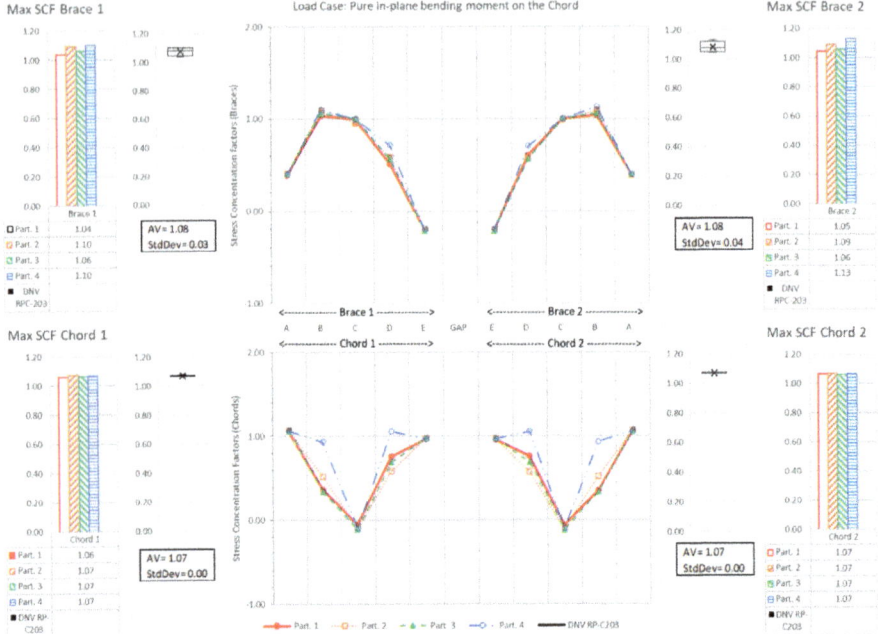

Fig. 27. SCF in Pure in-plane bending moment on the Chord load case.

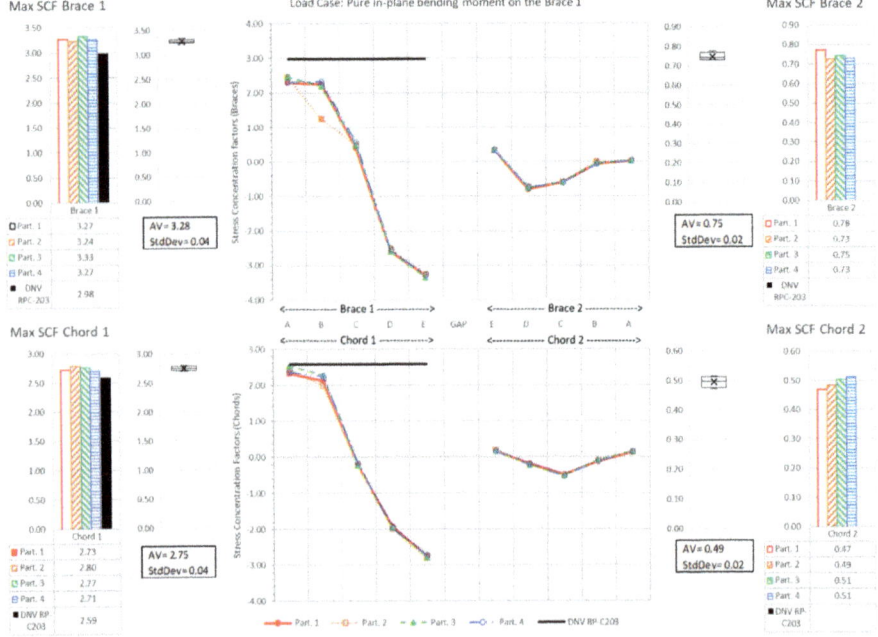

Fig. 28. SCF in Pure in-plane bending moment on the Brace 1 load case.

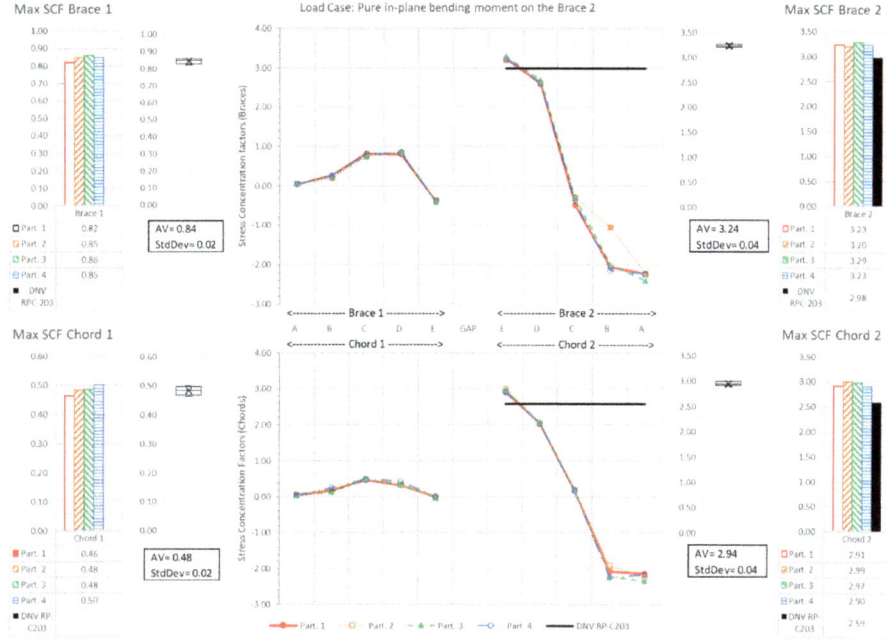

Fig. 29. SCF in Pure in-plane bending moment on the Brace 2 load case.

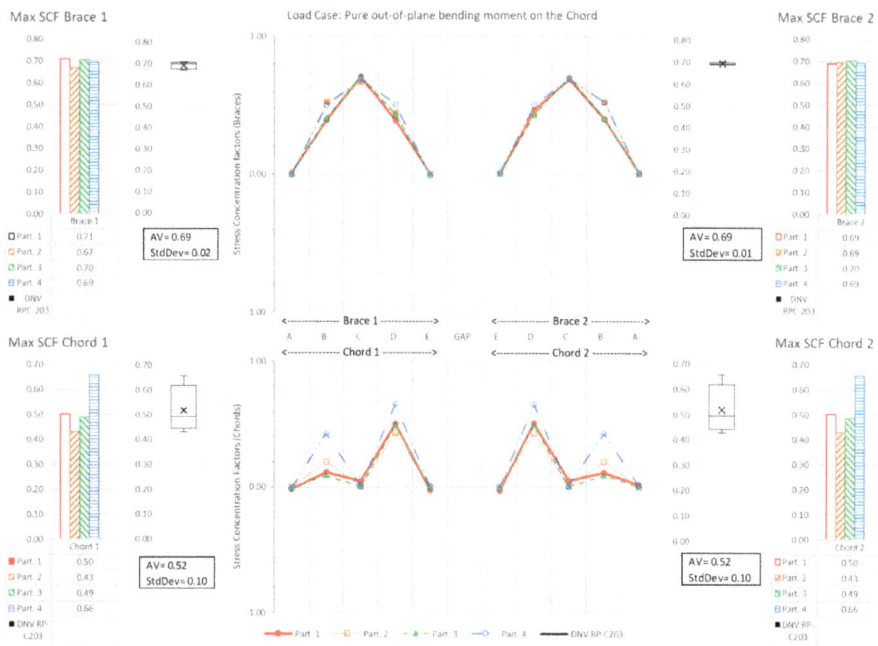

Fig. 30. SCF in Pure out-of-plane bending moment on the Chord load case.

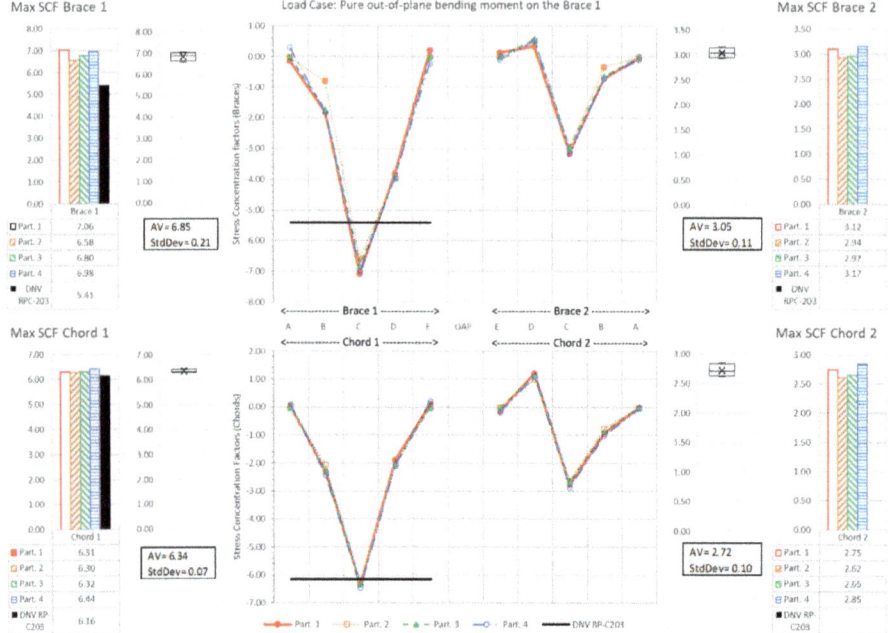

Fig. 31. SCF in Pure out-of-plane bending moment on the Brace 1 load case.

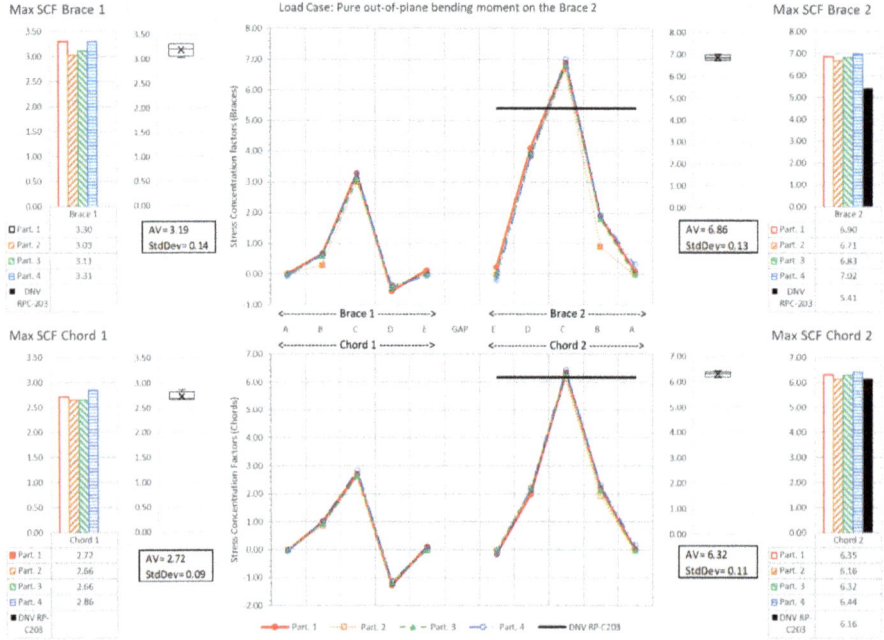

Fig. 32. SCF in Pure out-of-plane bending moment on the Brace 2 load case.

The scatter of results is present in all load cases, but it is more pronounced in some of them. To highlight this, Fig. 33 shows the standard deviation divided by the average value for all structural members in all load cases.

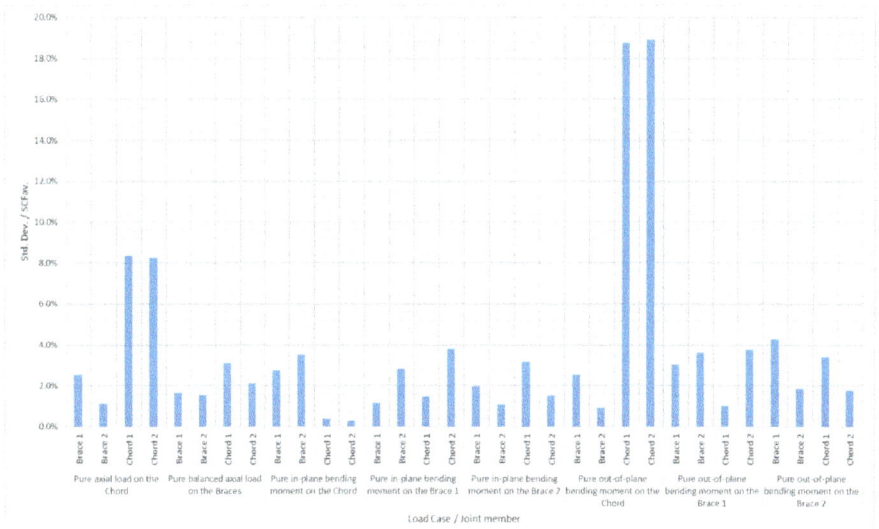

Fig. 33. Scatters of the results.

Results Analysis

Before presenting the reflections on the results, it is important to stress that what is presented in the previous section is already the result of an adjustment to correct errors and misalignments in the assumptions made. The aspect where this need for clarification was most evident was the definition of the position of the calculation points on the saddles and the intermediate points (i.e., points b, c, and d).

It could be argued that methods based on solid finite elements or more complex extrapolation methods (e.g. $0.2\sqrt{rt} - 0.65\sqrt{rt}$ proposed by DNV (2021e)) have not been investigated in this work. The need for simplification was highlighted from the beginning due to the complexity of the geometry and the number of results processed. This was, therefore, a deliberate choice made by the committee members, who were aware that some locations might be underestimated (unconservative) because no weld was included in the model and who wanted to focus on the most straightforward approaches suitable for large-scale industrial applications.

Table 9 summarizes the comparison of the results obtained. It highlights the agreement between the participants in predicting the maximum stress concentration factors. For each load case, even where minor differences in the maximum stress concentration factor were found, the intermediate points where the results diverged have been indicated. It also shows for which load case or analyzed member the rule formulae (when available) are conservative or not with respect to the participants' results.

In general, there was good consistency among the results regarding the qualitative distribution of SCFs along the joint and absolute values. There was unanimous agreement in identifying the most critical point (a, b, c, d, or e) for each loading condition.

The scatter of the results is generally small when braces are the loaded members. If the chord is the loaded member, the scatter increases, especially on the chord itself. It should be noted that in these cases, there is no real stress concentration (i.e., SCF < 1), and the limitations of the hot spot method presented in Section I.2 become more apparent, but ignoring these values completely or approximating them to 1.0 will result in an inappropriate approximation for complex load combinations.

The principal stress approach (participant 4) has the most significant discrepancy. While, in general, it does not differ much from the others in predicting the maximum SCF, it shows differences in the evaluation of intermediate points (i.e., locations b and d) on the chord in the load cases where the chord is the loaded member. This effect can be considered predictable, as the principal stress calculation introduces a portion of the stress component parallel to the weld line into the results, which may not be negligible in certain load cases and locations.

The approaches used by participants 1, 2, and 3, although quite different, give very similar results, clearly demonstrating that once the reference stress is clearly defined as the stress perpendicular to the weld line, the other aspects of the approach used become secondary.

Somewhat surprising is the comparison with SCF values obtained from regression formulae (e.g. (DNV, 2021e)) which, contrary to expectations, do not always seem conservative if compared with participants' results.

Conclusions and Recommendations

This annex presents a benchmark study on structural stress assessment of large floating wind turbine installation. The study focused on a single geometry of K-joint to enable systematic analysis of different evaluation methods and to explore and reflect on several aspects of the analysis. In this benchmark study, different geometric idealizations, mesh sizes, FEA readout points, different types of structural stresses (principal and normal to the weld line), and extrapolation methods were compared. The key findings of the study can be summarized as follows:

- Although different structural stress approaches were used, the variation in the SCFs is mostly very small. This confirms the general reliability of the most used engineering practice, FEA-based direct calculations.
- Classification society regression equations may not always be conservative for specific applications like large pipes. Using FEA-based direct calculations, even in simplified form, is recommended. Standard formulas do not cover all cases and overlook the signs of stress, limiting their use for cases requiring the superposition of effects.
- In surface extrapolation approaches, the choice of reference stress plays an important role. Using stress perpendicular to the weld line simplifies post-processing and permits immediate application of superposition of effects in complex loading conditions. However, principal stress, while increasing calculation complexity, captures local criticalities related to stress parallel to the weld line.

Table 9. Summary of the comparison of results.

Load Case		Max SCF	Divergent Locations	Rule formula	Rule Formula	Divergent Locations	Max SCF	
Pure axial load on the Chord	Brace 1	✓	B, D	✗	✗	B, D	✓	Brace 2
	Chord 1	▮	B, D	✗	✗	B, D	▮	Chord 2
Pure balanced load on the Braces	Brace 1	✓	B	⊖	⊖	B	✓	Brace 2
	Chord 1	✓	B	⊕	⊕	B	✓	Chord 2
Pure in-plane bending moment on the Chord load case	Brace 1	✓	D	✗	✗	B, D	✓	Brace 2
	Chord 1	✓	B, D	✗	✗	B, D	✓	Chord 2
Pure in-plane bending moment on the Brace 1	Brace 1	✓	B	⊕	✗	-	✓	Brace 2
	Chord 1	✓	-	⊕	✗	-	✓	Chord 2
Pure in-plane bending moment on the Brace 2	Brace 1	✓	-	✗	⊖	B	✓	Brace 2
	Chord 1	✓	-	✗	⊖	-	✓	Chord 2
Pure out-of-plane bending moment on the Chord	Brace 1	✓	B, D	✗	✗	B, D	✓	Brace 2
	Chord 1	▮	B, D	✗	✗	B, D	▮	Chord 2
Pure out-of-plane bending moment on the Brace 1	Brace 1	✓	B	⊖	✗	B	✓	Brace 2
	Chord 1	✓	-	✓	✗	-	✓	Chord 2
Pure out-of-plane bending moment on the Brace 2	Brace 1	✓	B	✗	⊖	B	✓	Brace 2
	Chord 1	✓	-	✗	✓	-	✓	Chord 2

List of symbols:
- ✓ Good match between participants' results
- ▮ Differences found between participants' results
- ⊕ Conservative Rule Formula (compared to participants' results)
- ⊖ Not conservative Rule Formula (compared to participants' results)
- ✗ Rule Formula not available for the specific load case

- The structural stress approach with single-point reference stress at $0.1\sqrt{rt}$ (Participant 1) has special advantages for engineering practice as it avoids stress extrapolation. The single-point reference stress approach requires no post-processing and eventually produces results comparable to those of more complex methods.
- The traction-based mesh-insensitive structural stress method (Participant 3) provides a promising alternative to commonly used surface extrapolation approaches. The traction stress approach uses nodal forces and moments to compute line forces and moments, and thus, structural stress is mesh-size insensitive, unlike other approaches. Although calculating the extracting nodal forces and moments to transform them into structural stresses seems demanding, it can be automated, as done in Fe-safe Verity. It makes the approach efficient for calculating the SCFs over an entire weld line in space.

Although tubular joints have been studied extensively over the last few decades, especially in complex engineering applications like offshore wind turbine support structures, there is still a need for future research to develop robust and efficient approaches for welded structures with complex load cases and load histories. For example, the effect of the weld using solid elements has not been investigated in this study. Thus, it would be essential to investigate how the impact of weld geometry could be added to a simple shell element model. Also, the limitation of using the hot spot stress method is only marginally discussed in this study, highlighting the need to clarify how different methods, such as hot spot stress and nominal stress, can be efficiently combined. Future studies must also consider both the reliability of the intermediate results and the actual impact on the structure assessment (e.g., fatigue life estimation) to balance the complexity of the proposed method against the required accuracy. Thus, future research should pay attention to the fundamental aspects and the theoretical developments arising from them without neglecting the practical aspects of industrial implementation.

Appendix II: Benchmark Study on an Lng Tank

Study on fatigue and fracture engineering critical assessment of an IMO Type-B prismatic LNG fuel tank

Introduction

The demand for Liquefied Natural Gas (LNG) carriers has been increasing worldwide since the beginning of the war between Ukraine and Russia. Additionally, Regulation 14 in MARPOL Annex VI (1997) mandates the reduction of sulphur emissions, driving the shipping industry to explore alternative fuels. LNG emerges as a viable low-sulphur fuel option in response to these regulations. The independent IMO Type-B prismatic LNG tanks are regarded as a promising choice for LNG fuel tanks on ships.

LNG is stored at cryogenic temperatures, for which significant changes in material properties are expected to occur. Therefore, it is crucial to evaluate the risk of fracture and leaks from LNG tanks to prevent the catastrophic failure of the hull structure, which could potentially occur after coming into contact with leaked cryogenic liquid.

The common materials used for cryogenic applications are 9% Ni and high-manganese steels. It is important to properly obtain the mechanical properties of those materials at cryogenic temperatures. Mechanical properties of metals and welded joints at cryogenic temperatures have been addressed extensively by Choi et al. (2012), Jeong et al. (2016), and Kim et al. (2021). In summary, it was reported in those works that 9% Ni steel has improved fatigue and fracture resistance at cryogenic temperatures.

Kim et al. (2021) compared the fatigue and fracture performances of automatic plasma arc welded high-manganese and 9% Ni steels at room (298 K, 25 °C) and cryogenic (110 K, −163 °C) temperatures. This work reported a significant improvement in the crack growth performance of the 9% Ni steel at cryogenic temperatures, while the high-manganese steel showed a consistent crack growth performance at room and cryogenic temperatures. However, both materials showed better crack growth performance in room and cryogenic temperatures than the proposed crack growth parameters of BS 7910:2019 (BS, 2019).

Although Type-B prismatic tanks are designed based on the model tests, advanced analytical tools and analysis methods to assess the stresses, fatigue life, crack propagation, and leakage according to the IGC Code (IMO, 2014), the fatigue and crack growth analysis procedures for LNG tanks have rarely been reported, e.g., (Suga et al., 2014), (Lee and Kim, 2022a).

Classification societies have released rules, procedures, and guidelines for the fatigue and crack propagation assessment of Type-B prismatic tanks, as outlined in (ABS, 2023), (CCS, 2023), (ClassNK, 2018), (DNV, 2023a), (KR, 2022), and (LR, 2012). It is worth noting that the procedures of these classification societies may vary in the implementation.

In this context, a benchmark study was proposed with the objective of establishing a common framework to reduce uncertainties and mitigate over-conservatism in the procedures and guidance notes. The study aimed to compare the practices of the participants, as well as the selected methods, procedures, and parameters used for fatigue, crack growth, and leak assessments of an LNG fuel tank installed on a container ship.

The benchmark study has been designated to have 4 phases:

(1) Phase 1: Estimation of stresses on the tank boundary plates using simplified formulas,
(2) Phase 2: Fatigue assessment of critical locations using S-N curves,
(3) Phase 3: Crack growth assessment of a critical location,
(4) Phase 4: Leak assessment for a location where a full crack penetration may occur.

To examine possible sources of differences in the results, the parameters after each phase are unified, and all participants are advised to use those parameters in the subsequent phase.

- The stress results after Phase 1 are unified. All participants are then advised to employ the identical data in the fatigue, crack growth, and leak assessment.
- The dimensions of the through-thickness crack are unified after Phase 3, and all participants then adopt the identical crack dimensions in the leak assessments.

The benchmark studies were carried out by participants from the alphabetically listed countries as below. Later, each of the contributing groups will be assigned a number, which is not necessarily in the same order as the below list.

- China: China Ship Scientific Research Centre (CSSRC) and China Classification Society (CCS)
- Germany: Det Norske Veritas (DNV)
- Japan: University of Tokyo and ClassNK
- Korea: Chosun University and Samsung Heavy Industries (SHI)
- United Kingdom: Lloyd's Register (LR)

The organization of this work is given as follows. Firstly, the tank geometry, loads, and the assessed structural details will be presented in Sect. II.2. The results and discussion of the stress estimation (Phase 1) will be given in Sect. II.3. Section II.4 will cover the results and discussion of fatigue assessments based on S-N curves (Phase 2). The crack propagation assessments (Phase 3) will be presented in Section II.5. The leak assessment (Phase 4) will be described in Sect. II.6. Finally, concluding remarks will be drawn in Sect. II.7.

Tank Geometry, Loading, and Structural Details

The benchmark study deals with the Type B prismatic LNG-fuel tank of a post-Panamax ultra-large container carrier. The principal particulars of the ship are given in Table 10.

Table 10. Principal particulars of the ship.

Length O.A	Length B.P	Breadth (Moulded)	Depth (Moulded)	Design Draught (Moulded)	Scantling Draught (Moulded)	Air draft (Above baseline)
366.0 m	350.5 m	51.0 m	29.3 m	14.5 m	16.0 m	Less than 67.2 m

Tank Data

The approximate volume of the tank is 12,000 m^3 with the dimensions of $L \times W \times D = 23.8 \times 20.4 \times 25.78$ m. The tank material is 9% Ni steel with the following properties in Table 11.

Table 11. Material properties for the tank structure.

Tank material	Young's Modulus (E)	Poisson's Ratio	Base material yield stress	Base material tensile stress	Welded mat. Yield stress	Welded mat. Tensile stress
9% Ni	206,800 N/mm^2	0.3	490 N/mm^2	690 N/mm^2	420 N/mm^2	655 N/mm^2

Loads

Loads acting on the tank are determined from a hydrodynamic analysis. The accelerations at the tank center of gravity were determined with the probability of exceedance of 10^{-8}. The hydrodynamic analyses and pressure calculations were performed by Participant #1.

The ship is assumed to be fully loaded during 85% of its lifetime, and the remaining 15% of its lifetime is assumed to be spent in ballast condition. The time spent in the maintenance and docking was omitted for the sake of conservatism (Table 12).

Table 12. Long-term accelerations at the tank center of gravity with a probability level of 10^{-8}

Load condition	Ax [m/s^2]	Ay [m/s^2]	Az [m/s^2]
Fully loaded ship	0.26	2.80	3.28
Ballast condition	0.31	4.89	3.33

The dynamic pressure acting on the critical locations is to be derived according to the IGC code (IMO, 2014) using pressure ellipsoid. In this regard, the following formula, derived by Sung and Randall (2015), was employed in the pressure calculations.

$$P_{tot} = \rho g \left[(z_0 - z_c) + \sqrt{(a_x(x_c - x_0))^2 + (a_y(y_c - y_0))^2 + (a_z(z_0 - z_c))^2} \right] + P_0 \quad (2)$$

This expression represents the total pressure acting on a critical location with the coordinates x_c, y_c and z_c in local tank coordinates. The coordinates of the reference point that generates maximum total pressure on a critical location are x_0, y_0 and z_0. The above pressure formula was expressed in terms of non-dimensional accelerations, i.e., $a_x = A_x/g, a_y = A_y/g$ and $a_z = A_z/g$, where Earth's gravity is $g = 9.81$ m/s^2.
The tank overpressure is taken as $P_0 = 25$ kPa. The ullage factor in the pressure calculation is omitted, i.e., set to zero. The density of LNG is taken as $\rho = 0.5$ ton/m^3.

The static and dynamic pressures for each location were determined with respect to a reference corner of the tank so that the total pressure becomes maximum. The pressure data were derived and distributed by Participant #1.

Once the pressure components were determined, each participant calculated the corresponding stresses, i.e., static and dynamic stresses, using simplified stress formulas.

Critical Locations for Investigation

The critical locations were selected from the tank boundary plates, especially in the tank corners and primary member-boundary plate joints. The critical locations and geometric representations are listed in Table 13.

The parameters in Table 14 are:

- t, t_1: the thickness of the plate accommodating the flaw,

Table 13. Critical locations.

Location	Geometric Representation	Dimensions	Structural Detail
Loc #1: Bottom plate – CL connection Dist. from tank bottom: 0 mm		$t = 16$ mm $s = 875$ mm $S = 3400$ mm $L = 14$ mm	As-weld condition
Loc. #2: Bottom corner plate i.w.o Transverse and Long. Bhds. Dist. from tank bottom: 0 mm		$t = t_1 = 13$ mm $t_2 = 20$ mm $s = 840$ mm $S = 1600$ mm $L = 10$ mm	Ground-flushed weld-toe
Loc. #3: Bottom plate connection to the aft end Trans. Bhd. Dist. from tank bottom: 0 mm		$t = 16$ mm $s = 875$ mm $S = 1600$ mm $L = 8$ mm	Ground-flushed weld toe
Loc. #4: Bottom plate connection to the Long. Bhd. Dist. from tank bottom: 0 mm		$t = 13$ mm $s = 840$ mm $S = 3400$ mm $t_{bhd} = 12$ mm $L = 7$ mm	Ground-flushed weld-toe
Loc. #5: Long. stringer connection to the long. Bhd. Dist. from tank bottom: 8690 mm		$t = 10.5$ mm $s = 790$ mm $S = 3400$ mm $L = 14$ mm	As-weld condition
Loc. #6: Tank corner of the transverse Bhd. and long. Bhd. Dist. from tank bottom: 8295 mm		$t = t_1 = 11$ mm $t_2 = 10$ mm $s = 790$ mm $S = 3400$ mm $L = 6$ mm	Ground-flushed weld-toe
Loc. #7: Horizontal stringer connection to the Trans. Bhd. i.w.o. CL. Dist. from tank bottom: 8295 mm		$t = 10.5$ mm $s = 790$ mm $S = 3400$ mm $L = 12$ mm	As-weld condition

- s: the short edge of the elementary plate panel (EPP),
- S: the long edge of the elementary plate panel (EPP).
- L: weld attachment length
- The distance between the tank bottom and the hotspot location is required in the leak assessment to calculate the pressure heads.

All locations were determined by Participant #1 and distributed to the other participants with the relevant geometric and pressure data. The pressures were derived for the load calculation points of EPPs as per CSR – Part 1, Ch. 3, Sect. 7 (IACS, 2023a). The calculated pressures and other relevant details for the critical locations are listed in Table 14.

Table 14. Pressure data for the critical locations.

Loc	s [mm]	S [mm]	t [mm]	P_{st} [kPa]	P_{dyn} [kPa] – Full load cond	P_{dyn} [kPa] – Ballast cond
1	875	3,400	16.0	151.5	53.8	60.2
2	840	1,600	13.0	151.5	60.5	77.2
3	875	1,600	16.0	151.5	53.9	60.3
4	840	3,400	13.0	151.5	60.4	77.1
5	790	3,400	10.5	112.7	48.1	67.6
6	790	3,400	11.0	112.7	48.2	67.7
7	790	3,400	10.5	112.7	41.6	52.5

Although the exact locations for hotspots #5, #6, and #7 are not the same, the vertical coordinates of the pressure calculation points are identical. This is why the static pressures (P_{st}) are identical for Locs. #5, #6 and #7.

The dynamic pressures (P_{dyn}) in Table 14 represent the pressure amplitude. So, the dynamic stresses calculated by the dynamic pressures will be multiplied by a factor of two to obtain the stress ranges for fatigue and crack propagation calculations.

Phase 1: Stress assessments

The participants were asked to obtain the stress data for the locations based on the data provided in Table 14. As a guidance, the plate scantling formula of CSR - Part 1, Ch. 6, Sect. 3 (IACS, 2023a) under local pressure was recommended. However, the participants were also advised that they were free to choose any convenient simplified approach for deriving stress data.

Participant #1 has picked the plate scantling formula from the LR Gas Ship Rules (LR, 2023b) and converted that expression into the stress estimation formula with certain assumptions. The plate scantling formula of LR Gas Ship Rules (LR, 2023b) is given as:

$$t = 0.035 sf \sqrt{P_{eq} \cdot k} \qquad (3)$$

The plate scantling formula of CSR (IACS, 2023a) is given below.

$$t = 0.0158 \alpha_p s \sqrt{\frac{|P|}{\chi C_a R_{eH}}} \quad (4)$$

The correction factor for the plate aspect ratio is represented by "f" in the LR formula, while this factor is "α_p" in the CSR formula. Applied pressure on the plate surface is represented as P_{eq} in (LR, 2023b) and as P in the CSR formula. The parameter k is a material correction factor, coefficient C_a the permissible bending stress ratio, R_{eH} yield stress, and for the material is The parameter, χ location of the plate under intact and flooded conditions; see CSR (IACS, 2023a).

The above formulae are converted to stress estimation formulas with the following substitutions:

- $k = \frac{235}{\sigma}$ for LR formula
- $R_{eH} = \sigma$ for CSR formula

It must be noted that this study focuses on fatigue and crack propagation analysis procedures, not stress prediction. For this reason, the stresses are estimated using simplified approaches. All participants have calculated static stresses from the given static pressure and geometric data in Table 14.

Table 15 presents membrane (σ_m) and bending (σ_{bend}) components of static stresses derived by each participant. In the given table, the bending stress components were directly derived from the approximate expressions converted from the plate scantling formulas, while the membrane stress components represent predictions from the inherent safety factor of the formulas. This inherent safety factor allows for the addition of a proportion of the bending stress as membrane stress on top of the bending stress. Participants #3 and #4 have adopted 13% of the bending stress as the membrane stress component, whereas other participants have used 15% of the bending stress as the membrane stress component.

The dynamic stresses and the stress ranges were calculated using the same logic and are presented in Table 16 and Table 17.

The stress estimations generally align well; however, there is a noticeable difference between the stress results of Participant #1 and other stress data for Locations #2 and #3. This variance arises from the implementation of aspect ratio correction. At Locations #2 and #3, the aspect ratio of the local plate is smaller than 2.0, resulting in the correction factor f being less than 1.0. Participant #5 also explicitly calculated the aspect ratio correction factor, α_p in the CSR formula. Other participants have assumed this factor to be 1.0 for a more conservative stress prediction.

In the subsequent phases, the stresses were unified, and the stress data obtained by Participant #1 were selected for the fatigue and crack growth assessments.

Phase 2: Fatigue Assessment Based on S-N Curves

In the fatigue assessments, certain assumptions are made to determine if locations meet the criteria set by the IGC code (IMO, 2014) and IGF code (IMO, 2017). These codes limit the fatigue damage index to 0.1 for locations that cannot be inspected. For locations

Committee III.2: Fatigue and Fracture 691

Table 15. Static stress [MPa] predictions based on the plate scantling formulas.

Loc	Participant #1		Participant #2		Participant #3		Participant #4		Participant #5	
	σ_m	σ_{bend}	σ_m	σ_{bend}	σ_m	σ_{bend}	σ_m	σ_{bend}	σ_m	σ_{bend}
1	17	112	17	113	15	113	15	113	17	113
2	19	125	24	158	21	158	21	158	21	142
3	13	88	17	113	15	113	15	113	15	100
4	24	158	24	158	21	158	21	158	24	158
5	24	159	24	159	21	159	21	159	24	159
6	22	145	22	145	19	145	19	145	22	145
7	24	159	24	159	21	159	21	159	24	159

Table 16. Stress ranges [MPa] for the fully loaded condition.

Loc	Participant #1		Participant #2		Participant #3		Participant #4		Participant #5	
	$\Delta\sigma_m$	$\Delta\sigma_{bend}$	$\Delta\sigma_m$	$\Delta\sigma_{bend}$	$\Delta\sigma_m$	$\Delta\sigma_{bend}$	$\Delta\sigma_m$	$\Delta\sigma_{bend}$	$\Delta\sigma_m$	$\Delta\sigma_{bend}$
1	12	80	12	80	10	80	10	80	12	80
2	15	100	19	126	16	126	16	126	17	114
3	10	62	12	80	10	80	10	80	10	71
4	19	126	19	126	16	126	16	126	19	126
5	21	136	20	136	18	136	18	136	20	136
6	19	124	19	124	16	124	16	124	19	124
7	18	118	18	117	15	118	15	117	18	117

Table 17. Stress ranges [MPa] for ballast condition.

Loc	Participant #1		Participant #2		Participant #3		Participant #4		Participant #5	
	$\Delta\sigma_m$	$\Delta\sigma_{bend}$	$\Delta\sigma_m$	$\Delta\sigma_{bend}$	$\Delta\sigma_m$	$\Delta\sigma_{bend}$	$\Delta\sigma_m$	$\Delta\sigma_{bend}$	$\Delta\sigma_m$	$\Delta\sigma_{bend}$
1	14	89	13	90	12	90	12	90	13	90
2	20	128	24	161	21	161	21	161	22	145
3	11	70	14	90	12	90	12	90	12	80
4	25	161	24	161	21	161	21	161	24	161
5	29	191	29	191	25	191	25	191	29	191
6	27	174	26	174	23	175	23	174	26	174
7	23	148	22	148	19	149	19	148	22	148

inspectable by appropriate means and leak detection systems, the fatigue damage index is limited to 0.5. The design life of the ship and the tank is taken as 25 years in the present calculations.

In this study, it is assumed that all locations are inspectable. Therefore, for compliance with the IGC/IGF code requirements, the fatigue damage indexes are not to exceed 0.5, i.e., the minimum fatigue life must be 50 years.

The modified Miner-Palmgren rule from IGC Code (IMO, 2014), which includes the fatigue damage caused by loading/unloading cycles (low cycle fatigue- LCF), is used to evaluate the fatigue life of the critical locations.

Preliminary fatigue calculations for all locations were undertaken by Participant #1 using the stress data from Sect. 3, and the assessment results are given in Table 18. The results indicate that the maximum damage occurs at Loc. #5, also susceptible to through-thickness crack penetration and leakage.

Table 18. Preliminary fatigue assessment results by Participant #1 for all locations.

Loc	K_t	S-N Class	K_{ft}	m_{ft}	Knuckle Point for S-N Curves	HCF	LCF	Combined Damage Index	Fatigue Life [Years]
1	1.20	90	1.46E+12	3	1.0E+07	0.014	0.002	0.016	>250
2	1.31	100	2.00E+12	3	1.0E+07	0.032	0.002	0.034	>250
3	1.35	100	2.00E+12	3	1.0E+07	0.003	0.001	0.003	>250
4	1.42	100	2.00E+12	3	1.0E+07	0.098	0.004	0.103	243.1
5	1.20	90	1.46E+12	3	1.0E+07	0.222	0.006	0.228	109.7
6	1.45	100	2.00E+12	3	1.0E+07	0.110	0.003	0.113	221.4
7	1.20	90	1.46E+12	3	1.0E+07	0.101	0.006	0.106	235.1

Consequently, it was decided to continue with only Loc. #5 to facilitate more qualitative discussions. The fatigue assessment results and relevant analysis parameters for Loc. #5 from all participants are listed in Table 19.

Participants #1 and #5 adopted the FAT 90 class from IIW (Hobbacher and Baumgartner, 2024) for the fatigue assessments. Participant #2, however, utilized the FAT 96 class S-N curve for the considered detail. This would result in the FAT 80 class if the stress concentration factor $K_t = 1.2$ was considered on the resistance side, and for pure membrane stress, the same as used by participant #3. FAT 96 is a higher fatigue class, and this is one of the reasons why Participant #2 has achieved the longest fatigue life in Table 19. Another reason, the most influential parameter, is the total number of wave load cycles throughout the ship's lifetime. Participants #1, #3, and #5 have assumed 1.25E+8 cycles for 25 years of design life, whereas Participants #2 and #4 have assumed 1.0 E+8 over the design life of the ship. The latter assumption results in a longer fatigue life. The third factor affecting the estimated fatigue life is the assumption of the combined

Table 19. Fatigue assessment parameters and calculated damage indexes for Loc. #5.

Participant	K_T	S-N Class	K_{ft}	m_{ft}	Knuckle Point for S-N Curves	HCF	LCF	Combined Damage Index	Fatigue Life [Years]
#1	1.20	90	1.46E+12	3	1.0E+07	0.222	0.006	0.228	109.7
#2	1.20	96	1.77E+12	3	1.0E+07	0.128	0.022	0.128	195.3
#3	1.00	80	1.02E+12	3	1.0E+07	0.233	0.006	0.239	104.6
#4	1.00	71	7.05E+11	3	2.0E+06	0.227	0.011	0.238	105.0
#5	1.20	90	1.46E+12	3	1.0E+07	0.224	0.005	0.229	109.1

fatigue damage. Since the LCF damage index is lower than 0.25, Participant #2 did not consider it in the combined fatigue damage index.

Participant #4 has adopted a categorized S-N curve for each type of welded joint based on the experimental results organized by a domestic committee. This curve assumes a lower fatigue class for the considered location, which is reasonable when no stress magnification factor is introduced explicitly. This approach might be considered equivalent to the nominal stress approach of IIW(Hobbacher and Baumgartner, 2024); however, it is more conservative than the IIW recommendation of FAT 80 in the case of the nominal stress approach for T-joints under as-weld conditions. Another point to note about the S-N curve of Participant #4 is the knuckle point, which is at 2.0 E+06 cycles. This means that the Haibach correction was applied less conservatively compared to the implementation of other participants. The slope of the S-N curves beyond the knuckle points was the same for all participants, i.e., $m_{ft2} = m_{ft} + 2 = 5$.

The considered location is a T-joint, and the stress magnification factor, K_t is related to the angular distortion of the plating. The axial misalignments are not applicable to this location. Participants #1, #2, and #5 have adopted $K_t = 1.2$ in line with the recommendation of IIW (Hobbacher and Baumgartner, 2024), and hotspot stress S-N curves have been employed. Despite IIW (Hobbacher and Baumgartner, 2024) recommends FAT 100 hotspot S-N curve for the considered location, Participants #1 and #5 remained conservative, considering the uncertainty in the stress estimation process.

Participant #3 has assumed that the construction tolerances were within the limit of the inherent value for the considered nominal stress S-N curve and has adopted the FAT 80 class without introducing any stress magnification factor.

In the LCF assessment, the static stress components for the empty tank condition were taken as $\sigma^e_{bend} = \sigma^e_m = 0$, while the static stresses for the fully loaded condition are given in Table 15. All participants have taken 1,250 loading/unloading cycles for the considered tank through 25 years of lifetime.

Participants #2 and #4 calculated the LCF damage by combining the maximum stress ranges from the wave loads with the static stress cycles of full and empty tank conditions. This explains why the LCF damages predicted by these participants are higher than others. The fatigue lives for Loc. #5, estimated by all participants, is visualized in Fig. 34.

Fundamentally, all the approaches here in the fatigue assessment are consistent. The number of wave cycles, in general, depends on the ship's length and its operating condition. In this benchmark study, however, the difference comes from the interpretation of whether the total number of cycles relates to 20 or 25 years of operation time.

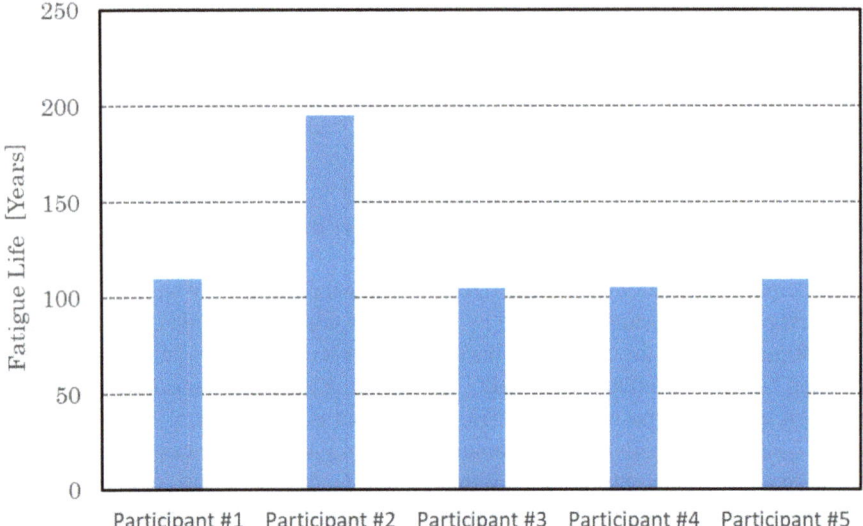

Fig. 34. Predicted fatigue lives by all participants for Loc. #5.

Phase 3: Crack Growth Assessments

Using the unified stress data and the stress concentration factors, the same as for the fatigue assessments in Phase 2, crack growth assessments for a surface defect at Loc. #5 were undertaken. The participants were asked to freely choose any suitable crack growth parameters and spectrum applications.

Spectrum Implementation

In this phase of the work, it was foreseen that one of the important factors influencing the crack growth rates and the fatigue crack growth lives would be the spectrum implementation. In this regard, the analysis setup parameters are given in Table 20.

In Table 20, N_R represents the reference number of cycles to generate the spectrum, which corresponds to the long-term probability level of accelerations, 10^{-8}. The number of stress cycles in a spectrum block is represented by P_{SB}.

The length of the spectrum is represented by t_{rep}, which may also be called a repeat period. The spectrum is then repeated until the full penetration of the crack or for 100 years, whichever occurs first.

Table 20. Analysis setup parameters for the crack growth assessment.

Participant	N_R[cycles]	P_{SB}	t_{rep}	Spectrum sequence	dN_{max}[cycles]	Q_m[MPa]
#1	1.0 E+8	200[a]	30 days	Ballast cond.+full load	1000	420
#2	1.0 E+8	61[b]	90 days	Full load+ballast cond.+loading/unloading	100	490
#3	1.0 E+8	100[c]	25 years[d]	Full load+ballast cond.+loading/unloading	N.A	420
#4	1.0 E+8	500[e]	35 days	4 x Full load+ballast cond	100	N.A
#5	1.0 E+8	100[c]	30 days	Full load + ballast cond	100	420

[a] *100 stress range blocks for ballast condition + 100 stress range blocks for full load condition*
[b] *30 stress range blocks for full load condition + 30 stress range blocks for ballast condition + 1 loading/unloading*
[c] *50 stress range blocks for full load condition + 50 stress range blocks for ballast condition*
[d] *Spectrum was applied as full ship condition + ballast condition + loading/unloading cycles at once for 25 years*
[e] *The number of stress range blocks for each loading condition (block) is 100. A spectrum unit has 5 blocks (4 x Full load + 1 x Ballast)*

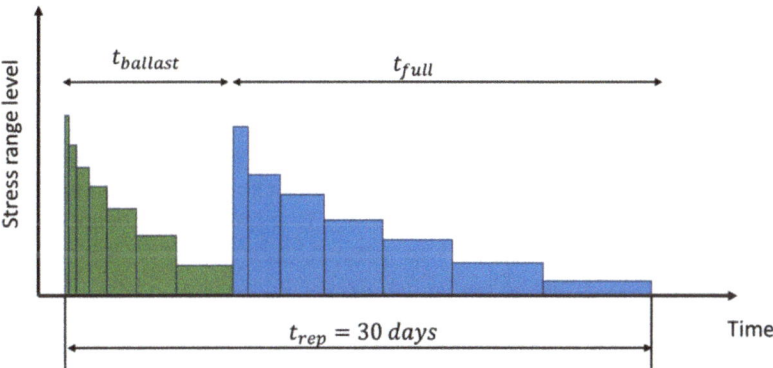

Fig. 35. Schematic implementation of the spectrum sequence by Participant #1.

The implementation of the spectrum sequence may be explained by the help of Fig. 35. The length of each spectrum sequence is adjusted considering the operation data, i.e., 15% of lifetime in ballast condition and 85% in full load condition. In this regard, $t_{ballast} = 4.5$ days and $t_{full} = 25.5$ days were adopted by Participant #1. The number of stress range blocks (P_{SB}) for each loading condition was taken as 100 by Participant #1, regardless of the spectrum duration, i.e., $P_{SB-ball} = P_{SB-full} = 100$.

The approach of Participant #2 is like that of Participant #1, but the full load condition is applied first. The length of each loading condition is adjusted accordingly, i.e., $t_{ballast} = 13.5$ days and $t_{full} = 76.5$ days. The number of stress range blocks in each loading condition is taken equal as $P_{SB-ball} = P_{SB-full} = 30$. Moreover, one load cycle for loading/unloading operation was considered by Participant #2 in the crack growth assessment stress spectrum.

Participant #3 generated the spectrum for 25 years and applied it once rather than repeating it for shorter periods. This implementation simplifies the procedure, but the results are more susceptible to being affected by the spectrum implementation sequence.

The approach by Participant #4 is relatively different from the others, as shown in Fig. 36. The ballast condition spectrum block is applied after 4× full load condition spectrum block to adjust the overall operation condition of 15% ballast duration. The length of each spectrum block is approximately the same, i.e., 7 days. Then, the repeat period for each spectrum unit (5 blocks) becomes $t_{rep} = 7 \times 5 = 35$ days.

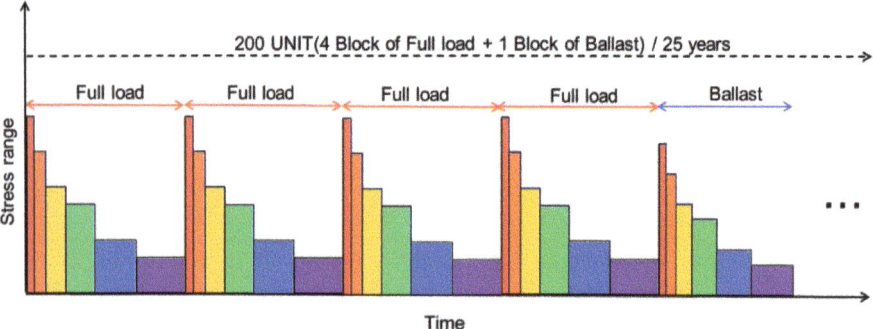

Fig. 36. Schematic implementation of the spectrum sequence by Participant #4.

The approach of Participant #5 is like that of Participant #1, but the full load condition was applied first, and the number of stress range blocks for each loading condition was taken as 50, i.e., $P_{SB-ball} = P_{SB-full} = 50$. To meet the operational conditions, $t_{full} = 25.5$ days and $t_{ballast} = 4.5$ days are adopted by Participant #5.

In Table 20, the parameter dN_{max} is the resolution parameter for the crack growth integration. The larger the value, the quicker the crack growth integration. The influence of this parameter on the crack growth rate is small, particularly for the surface crack propagation. A special consideration might need to be given to this parameter during the through-thickness crack growth and leak assessments.

Finally, the parameter Q_m stands for the membrane residual stress (secondary stress) to be used in the criticality assessment of the crack, which is usually taken as the yield stress of the material. This parameter is considered for fracture limit load (criticality) assessments. Participant #4 assumed that there would be no sudden breakage of the plating, so the secondary stress is not applicable in the assumption of Participant #4.

Crack Growth Parameters and Methodology

The participants were asked to choose any available stress intensity factor (SIF) solution method, software package, or applicable standards in the crack growth analysis and criticality assessment. Table 21 shows the related crack growth parameters, applied standards, and the employed software packages in this study stage. The constants are

Committee III.2: Fatigue and Fracture 697

denoted by C_1 and C_2 for the two-stage Paris law. Similarly, the slope constants are m_1 and m_2. If a participant assumes single-slope Paris law, the related constants are given as C_1 and m_1. The unit in this table for K_{mat} (critical SIF) and ΔK_{th} (threshold value) is $N \cdot mm^{-1.5}$.

Table 21. Crack growth analysis parameters, solution methods, and software.

Participant	C_1	m_1	C_2	m_2	ΔK_{th}	K_{mat}	SIF solution	Criticality Assessment	Software
#1	2.10E−17	5.1	1.29E−12	2.88	63	5776.30	BS 7910	BS 7910 – Option 1	In-house code
#2	2.10E−17	5.1	1.29E−12	2.88	63	5351.06	API 579/ ASME FFS-1	BS 7910 – Option 1	IWM - Verb
#3	2.10E−17	5.1	1.29E−12	2.88	63	2200.31	BS 7910	BS 7910 – Option 1	CrackWise
#4	1.04E−11	2.64	–	–	40.1	N.A	Newman-Raju Eqs	N.A	In-house code
#5	2.10E−17	5.1	1.29E−12	2.88	63	5776.30	BS 7910	BS 7910 – Option 1	CrackWise

All participants, except #4, have used the two-stage Paris law with the crack growth parameters from BS 7910:2019 (BS, 2019). For comparison with other results, participant #4 adopted the crack growth parameters obtained from the welding joints of 9% Ni steel, organized by a domestic committee. The comparison of two-stage crack growth parameters from BS 7910:2019 and the test results of the domestic committee is given in Fig. 37. This figure suggests that the crack growth parameters of the domestic committee are more conservative than those of BS 7910:2019.

The initial crack sizes were selected by each participant based on their internal procedures or best practices. The crack dimensions at the beginning and the end of assessments are given in Table 22. The initial and final crack depths are denoted by a_0 and a_{final}, respectively. The initial and the final full crack lengths are represented by $2c_0$ and $2c_{final}$ in the given table.

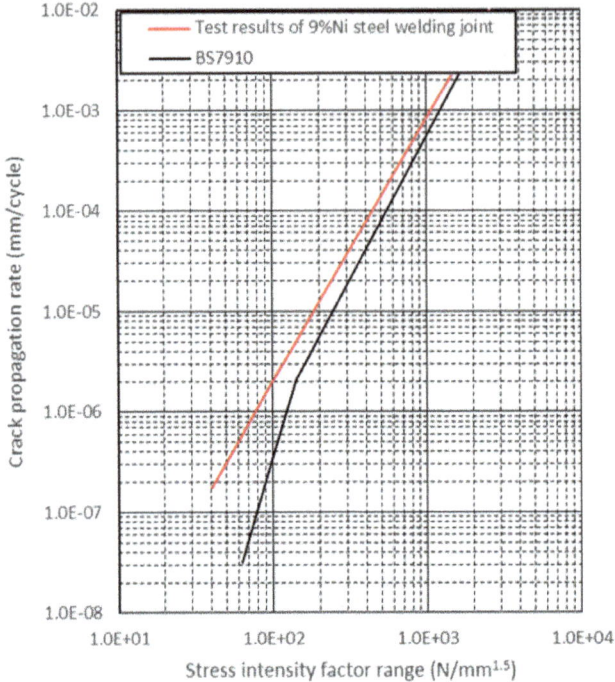

Fig. 37. Comparison of crack growth parameters from BS 7910 and the domestic committee.

Table 22. The summary of the surface crack growth assessments.

Participant	Initial crack dimensions		Surface crack growth			
	a_0[mm]	$2c_0$[mm]	t_{final}[years]	a_{final}[mm]	$2c_{final}$[mm]	
#1	1.0	5.0	60.75	5.47*	43.00	
#2	0.5	5.0	91.20	4.45*	38.55	
#3	1.0	5.0	65.70	3.15*	20.95	
#4	1.0	5.0	100.00	6.61	17.41	
#5	0.5	5.0	77.60	5.55*	42.54	

The participants have carried out the crack growth assessments until the full penetration or 100 years, whichever occurs first. All participants, except #4, have undertaken criticality assessments for the crack depth according to Clause 7 of BS 7910:2019 (BS, 2019). If the crack depth reaches the critical value, as shown by asterisk (*) in Table 22, it is assumed that there is a ligament instability ahead of the crack front, and the crack penetrates through the thickness. Then, the surface crack growth assessment is terminated. Participant #4 assumed that the crack penetrates through the thickness once the crack depth reaches 80% of the thickness.

The difference between the critical crack depths may be attributed to the differences in the SIF solutions, fracture toughness values, and the welding residual stress (Q_m) magnitudes. Another parameter affecting the critical size could be the peak stresses employed. Participant #3 adopted only static (mean) stress as the peak stress, while Participants #1, #2, and #5 considered the summation of static stress and the maximum stress amplitudes from the ballast condition. The assumption of Participant #3 would normally result in a higher final crack size; however, it is the opposite case because of the smallest fracture toughness value, which cancels out the less conservative impact of the peak stress assumption.

It was reported that all participants had applied the weld notch effect factors (M_k) in the SIF solutions. Finally, the crack growth curves in the thickness and length direction are respectively given in Fig. 38 and Fig. 39. To eliminate the influence of load cycle number over time, the crack growth curves were also plotted as a function of load cycles. In this way, the gaps between the curves are reduced noticeably.

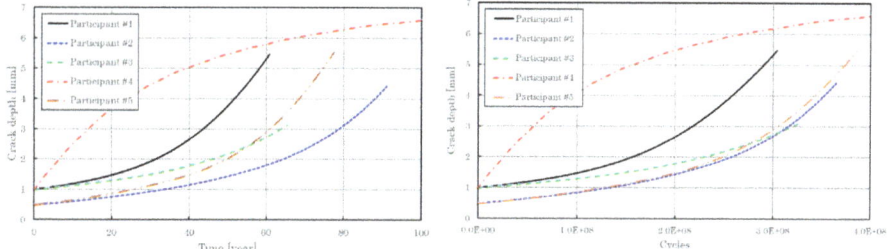

Fig. 38. Crack growth curves in thickness direction.

The characteristics of the crack growth curves by Participant #4 are rather different from the ones obtained by other participants. This may be explained by the fact that the design code BS 7910:2019 (BS, 2019) and API579-1 (API, 2021), treat surface cracks as equivalent through-thickness cracks in the SIF solutions.

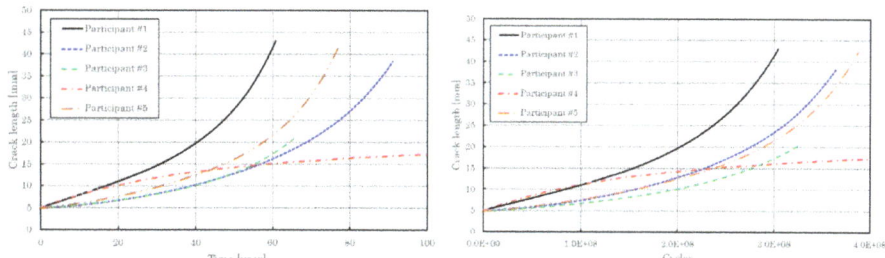

Fig. 39. Crack growth curves in the length direction.

The scatter in the solutions of Participants #1, #2, #3, and #5 is related to the assumption on the initial crack depth, spectrum implementation, and the number of cycles over the considered period of time, i.e., 10^8 cycles were considered over 25 years by Participants #2 and #4, while 1.25×10^8 cycles were considered over the same period by Participants #1, #3 and #5.

Phase 4: Leak Assessments

The philosophy behind the design of IMO Type B tanks is the occurrence of the leak before a crack reaches the critical, i.e., unstable state, called "Leak before Failure (LbF)". This concept applies to tank locations where a failure would lead to immediate leakage. For other locations, e.g., internal structures, criteria related to inspection intervals, or the lifetime of the tanks need to be applied.

Since the tank material, 9% Ni steel, keeps its ductility at cryogenic temperatures, the same logic from Annex F – BS 7910:2019 (BS, 2019) maybe applied here for the LbF assessment.

The LbF assessment process may be described as follows, considering the requirements of IGC/IGF codes.

- Crack depth reaches a critical value according to the criticality assessment; a ligament instability occurs at this point and the crack penetrates through the thickness.
- The surface crack is now to be recharacterized. This may be achieved by following the guidance from Annex E – BS 7910:2019 (BS, 2019).
- A through-thickness crack growth analysis is to be performed. The analysis is required for at least $15 + t_{det}$ days duration. The number of days until the leakage volume triggers the leak detection system is represented by t_{det}.
- The 15-day duration assumes remedial actions are taken once the leakage is detected.
- Check whether the crack growth is stable after $15 + t_{det}$ days, i.e., no failure.
- Calculate the total leak amount and average leak rate, which may be used for the leak protection system and drip tray design.

In this stage of the benchmark study, all participants adopted the crack dimensions obtained by Participant #1 to start the leak assessment and through-thickness crack propagation.

Crack Recharacterization

Participant #1 follows Annex E – BS 7910:2019 for the crack recharacterization once the crack's full-penetration has been achieved. The recharacterization process may be explained as follows.

The crack length on the initiation side is to be updated as $2c_1 = 2c_{ls} + t$, where t is the plate thickness; c_{ls} represents the half crack length on the initiation side of the plate immediately before the full penetration, see Fig. 40.

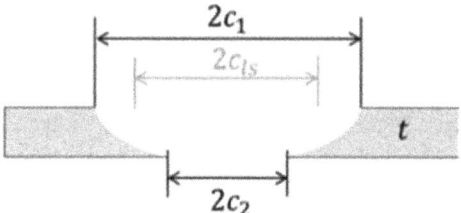

Fig. 40. Crack recharacterization.

The crack length ($2c_2$) on the penetration side can be determined as follows, assuming the flaw keeps the same aspect ratio of the elliptical shape.

$$c_2 = \sqrt{c_1^2 - c_{ls}^2}$$

The unified crack data, according to the calculations by Participant #1, are given below to start the leak assessment.

- $2c_{ls} = 43.00$ mm
- $2c_1 = 53.50$ mm

If there is no specific requirement for the minimum amount of leak that triggers the detection system, it may be assumed as $V_{det} = 5$ litre.

Stress Spectrum

IGC code (IMO, 2014) specifies the simplified 15-day spectrum using the most probable maximum stress over the ship's lifetime. According to this definition, Participant #1 recommended to assume the LNG fuel tank as full, but the ship is in the ballast condition as the stress ranges for ballast condition are higher than those of the full ship condition. The total number of cycles for 15-day spectrum is taken as $N_{leak} = 2.05 \times 10^5$.

Leak Assessment Methodologies

Each participant has chosen a suitable methodology and formulation to calculate effective crack opening area and leak rate. The formulations are given in Table 23.

The coefficients in the last column are designated for unit conversion. Otherwise, the leak rate formulas are identical except for the one adopted by Participant #4. This participant applied the Hagen-Poiseuille equation, which describes the steady laminar flow of an incompressible viscous fluid through a cylindrical pipe of constant diameter, for comparison with other results.

The parameters in Table 23 are not explained here in detail. However, the main differences in the assumptions are discussed as follows.

Participants #1 and #4 have adopted the membrane component of static/mean stresses (σ_{sm}) as the effective crack opening stress, while the other participants have considered the dynamic bending stress component (σ_{db}) as well. This assumption makes a significant difference in the calculated results as the bending stresses are dominant in the considered location.

Table 23. Summary of leak assessment methodologies.

Participant	Effective crack opening area	Effective crack opening stress	Leak rate
#1	$A = \frac{2\pi\sigma_{eqv}c_2^2}{E}$	$\sigma_{eqv} = \sigma_{sm}$	$Q = C_d A \sqrt{2g\left(h + \frac{P_0}{\gamma}\right)}$
#2	$A = \frac{2\pi\sigma_{eqv}c_2^2}{E}$	$\sigma_{eqv} = \sigma_{sm} + \sigma_{db}/3$ for $\sigma_{sm} > \sigma_{dm}$	$Q = 3.6 C_d A \sqrt{2g\left(h + 100\frac{P_0}{\gamma}\right)}$
#3	$A = \frac{2\pi\sigma_{eqv}c_2^2}{E}$	$\sigma_{eqv} = \sigma_{sm} + \sigma_{db}/3$ for $\sigma_{sm} > \sigma_{dm}$	$Q = 3.6 C_d A \sqrt{2g\left(h + 100\frac{P_0}{\gamma}\right)}$
#4	$A = \frac{2\pi\sigma_{eqv}c_2^2}{E}$	$\sigma_{eqv} = \sigma_{sm}$	$Q = \frac{\pi P_0 c_2^3 b^3}{4\mu t(c_2^2 + b^2)}$
#5	$A = \frac{2\pi\sigma_{eqv}c_2^2}{E}$	$\sigma_{eqv} = \sigma_{sm} + \sigma_{db}/3$ for $\sigma_{sm} > \sigma_{dm}$	$Q = 3.6 C_d A \sqrt{2g\left(h + 100\frac{P_0}{\gamma}\right)}$

Parameters to be Calculated

The following parameters were calculated:

- t_{det}: time until detection of the leakage, i.e., the time until the total leak amount reaches V_{det}. It is conveniently given in "days" rather than years.
- t_{leak}: time until the crack becomes unstable from the onset of the leak. It may be taken as $t_{leak} = t_{det} + 15$ days if the crack growth rate is small and the time until unstable crack growth is long.
- V_{leak}: total amount of leak in "liter" after t_{leak} days.
- Leakrate: average leak rate during t_{leak} days in "liter/day".
- $2c_1$ and $2c_2$ after t_{leak} days of crack propagation.

Assessment Results

The outcomes of the leak assessment are given in Table 24. The most important finding from the results in the table is the variation in leak rate and total amount of leaks. One potential reason for this could be the different equivalent stress assumptions for the crack opening area. The final crack dimensions after 15 days are also different between the participants. Participants #1, #3, and #5 obtained similar crack sizes, which may be attributed to the fact that the participants have used the SIF solutions from BS 7910:2019 (BS, 2019). On the other hand, Participant #2 adopted SIF solutions from API 579-1 (API, 2021); while Participant #4 adopted the well-known Newman-Raju equations.

Apart from the differences in Sect. 6.3, there are also differences in the implementation of these formulas. For instance, Participants #1 and #4 updated the leak rate after every crack growth incrementation with the new c_2 value. On the other hand, other participants adopted the c_2 value after 15 days of crack propagation and the leakage was calculated using that value. Updating the crack opening area after each crack growth

Table 24. Summary of leak assessment results.

Participant	$2c_1$[mm]	$2c_2$[mm]	t_{det}[days]	t_{leak}[days]	V_{leak}[litre]	Leak rate [liter/day]
#1	65.0	48.8	0.14	15.25	951	62.3
#2	53.7	32.2	0.06	15.00	1215	81.0
#3	60.6	42.7	0.02	15.00	2097	139.8
#4	53.5	31.9	0.00	15.00	7367	491.1
#5	62.1	44.8	0.01	15.00	3889	259.3

incrementation is rather complex but will provide a more accurate/less conservative leak rate.

Another point to highlight is the crack length in the SIF calculation for through-thickness crack growth assessments. Participant #4 considered the equivalent crack length for an elliptical shape through-thickness crack in this case, while others adopted the crack length on the initiation side as a more conservative approach for the crack growth assessment. The shape of the through-thickness crack and its impact on the crack growth under various loading combinations require further examination. The present design codes are based on rectangular-shaped through-thickness cracks. However, elliptical-shaped through-thickness cracks are relatively common, and more precise SIF solutions for various crack shapes would produce less conservative crack growth rates.

The leak rate and the total leak obtained by Participant #4 sets the highest values. This could be explained by the flow characterization in the Hagen-Poiseuille equation, which assumes a laminar flow in a long cylindrical pipe; however, the crack size is quite large compared to the plate thickness in the present problem. For this specific case, the Hagen-Poiseuille equation is considered less accurate.

The scatter in the total leak and leak rates suggests that more work is required in identifying the most suitable method and the implementation of the current practices.

Neither of the participants reached the critical condition in terms of the unstable crack growth after 15-day of leak assessment. Therefore, no further details were given regarding the criticality of the locations.

Conclusion and Final Remarks

In the past, limited work has been done to establish fatigue and crack growth analysis procedures for LNG tanks. Thus, due to the increasing demand for these tanks, a benchmark study has been conducted to create a common framework and to reduce uncertainties and over-conservatism in the existing procedures.

A type-B prismatic LNG fuel tank for a post-Panamax container vessel was considered. Each participant was asked to carry out stress estimation (Phase 1), fatigue life estimation based on S-N curves (Phase 2), crack growth assessment for a surface flaw (Phase 3), and leak assessment along with a through-thickness crack growth (Phase 4). Phase 1 was undertaken for seven different locations. However, only one location was considered from Phase 2 onwards (Loc. #5).

The predicted stresses in Phase 1 were found to be mostly consistent. A slight difference was noted between the results of Participant #1 and other participants' results due to the use of the plate aspect ratio correction factor. Participants #1 and #5 explicitly calculated this factor, while the others assumed a value of 1.0 for all locations, which is a more conservative approach. Otherwise, the stress estimation formulas derived from the LR gas ship rules (LR, 2023b) and CSR (IACS, 2023a) were consistent. It is also believed that calculating the correction factor, rather than assuming it, will help reduce over-conservatism in the stress prediction phase.

The fatigue lives based on S-N curves for Loc #5 were found to be longer than 100 years by all participants. Participant #2 estimated the fatigue life around 195 years, while others estimated it in the 100–110 years range. This is because of the number of cycles assumed for the design life (25 years). It was taken as 1.0 E+8 cycles by Participants #2 and #4, while other participants assumed the value as 1.25 E+8 cycles. Another reason is the higher-class S-N curve (FAT 96) adopted by Participant #2.

Participant #2 did not combine the HCF damage and LCF damage since the LCF damage was lower than the threshold value of 0.25. However, the impact of the LCF damage on fatigue life is negligible for the case considered here.

The crack growth assessments for the surface defect at Loc #5 demonstrated variation in time until full penetration and in the crack dimensions at the time of full penetration. Potential reasons for the observed scatter are:

- Variations in the implementation of stress spectrums,
- Number of load cycles over the analysis duration,
- Different initial defect sizes,
- Different interpretations of failure mechanisms.

It was observed that when the surface crack growth curves are plotted against the number of cycles, the SIF solutions from BS 7910:2019 and API 579-1:2021 produced similar results, provided that all other parameters are consistent (see the results of Participants #2 and #5 in Fig. 38 and Fig. 39).

The selection of initial defect sizes depends on the yard's capability in non-destructive testing and visual inspection. While all participants selected an initial crack length of 5 mm, the initial crack depths varied. Although conservative, an initial crack depth of 1.0 mm is more likely to be detected during the inspection and may be used for crack growth assessments.

The leak rate estimations have shown significant variation. This is mainly because of the several parameters depending on the interpretation of the participants and different effective crack-opening stress assumptions.

Participant #4 adopted equivalent crack length in the through-thickness crack growth assessment for SIF calculations. On the other hand, other participants employed the crack length on the initiation side for the SIF calculations.

The leak rate formula of Participants #2, #3, and #5 considers both membrane and dynamic bending stress components in the calculation of effective crack opening area, while the formulas by Participants #1 and #4 consider only membrane stress component. Introducing the dynamic bending stresses in the leak rate assessment increased the leak rate noticeably as the bending stresses dominate in the present location.

Committee III.2: Fatigue and Fracture   705

The leak rate formula of Participant #4 is based on the Hagen-Poiseuille equation, which is less accurate in the case of larger cracks on relatively thin plates. The variation in the leak rate results of Participants #1, #2, #3, and #5 are related to the effective crack opening area and the crack dimensions; the leak rate formulas themselves (last column in Table 23) are, however, expected to produce similar results.

The findings from the benchmark study suggest that the fatigue and crack growth assessment process of LNG tanks are subjected to significant uncertainties. In this regard, a reliability analysis perspective may be required to quantify how reliable the assessment results are. Reliability assessment methods and recent developments in the reliability of fatigue and fracture of marine structures are examined comprehensively in Chapter 5 of the committee report.

The following recommendations may reduce the uncertainties in the assessment results and provide more accurate results.

- For stress predictions:

  – Considering the plate aspect ratio explicitly.

- For fatigue assessment with S-N curves:

  – Using hotspot stress approach with explicitly calculated or obtained SCFs. In this way, the impact of the SCFs due to the construction tolerances may be represented well, which will be less in bending dominant cases.
  – Proper estimation of the total number of load cycles throughout the ship's lifetime.

- For surface crack growth assessment:

  – Proper estimation of the total number of load cycles throughout the ship's lifetime.
  – The application of stress spectra for each loading condition should be performed repetitively at short intervals rather than at all conditions at once during the crack growth assessment. Otherwise, significant variations in results may occur depending on whether the stress spectrum is applied from highest to lowest or vice versa
  – Clear definition of the material failure mechanisms, i.e., whether unstable crack growth failure is applicable.

- For through-thickness crack growth and leak assessment:

  – Updating the crack opening area after each crack growth increment will result in a more accurate leak rate estimation than using the crack size only at the end of the 15-day crack propagation period.
  – Using a more representative leak rate formula considering the shape of the crack and the plate thickness.

The recommendations for the future are:

- To reduce the over-conservatism in the fatigue and crack growth properties of materials at cryogenic temperatures, these properties should be sourced from well-established references specifically developed for materials under such conditions.
- A surface crack growth under bending dominant case needs more examination and understanding.
- The modeling of the through-thickness cracks in the design codes is idealized to be rectangular. This assumption may result in sufficiently accurate results under membrane-dominant cases; however, for bending-dominant cases, a more accurate crack modeling may be required.

References

Abdul-Latif, A.: Continuum damage model for low-cycle fatigue of metals: an overview. Int. J. Damage Mech **30**, 1036–1078 (2021)

ABS: Guide for fatigue assessment of offshore structures (2020). https://ww2.eagle.org/content/dam/eagle/rules-and-guides/current/offshore/115_fatigueassessmentofoffshorestructures/offshore-fatigue-guide-jun20.pdf. American Bureau of Shipping

ABS: Guidance Notes on Strength Assessment of Independent Type C Tanks. American Bureau of Shipping (2022)

ABS: Guide for Building and Classing - Liquified gas carriers with independent tanks. Am. Bureau Shipp. (2023)

Al-Hagri, A., Paamand, J., Haselbach, P.U., Stang, H., Kolios, A., Katsanos, E.: On the sensitivity of stress intensity factors to modelling choices for steel K joints. In: The 34th International Ocean and Polar Engineering Conference (2024)

Alencar, G., de Jesus, A., da Silva, J.G.S., Calçada, R.: A finite element post-processor for fatigue assessment of welded structures based on the Master S-N curve method. Int. J. Fatigue **153**, 106482 (2021)

Amirafshari, P., Brennan, F., Kolios, A.: A fracture mechanics framework for optimising design and inspection of offshore wind turbine support structures against fatigue failure. Wind Energ. Sci. **6**, 677–699 (2021)

Amiri, N., Shaterabadi, M., Reza Kashyzadeh, K., Chizari, M.: A comprehensive review on design, monitoring, and failure in fixed offshore platforms. J. Marine Sci. Eng. **9** (2021)

An, G., Park, J., Bae, H.: Brittle fracture avoidance technology in large structures with thick steel plates. J. Nanosci. Nanotechnol. **21**, 4926–4930 (2021a)

An, G., Park, J., Han, I., Woo, W.: Unstable fracture phenomenon of welded joints with weld residual stresses. Theoret. Appl. Fract. Mech. **109**, 102747 (2020)

An, G., Park, J., Park, H., Han, I.: Fracture toughness characteristics of high-manganese austenitic steel plate for application in a liquefied natural gas carrier. Metals (2021b)

An, G., Park, J., Seong, D., Seo, J.: Brittle crack arrest temperature estimation method utilizing a small-scale test with a thick steel plate for shipbuilding. Metals (2024)

Angel, T.: Formation of martensite in austenitic stainless steels effects of deformation, temperature, and composition. J. Iron and Steel Inst. **177**, 165–174 (1954)

API. API 579–1/ASME FFS-1: Fitness-For-Service. American Petroleum Institute (2021)

Attarha, M.J., Sattari-Far, I.: Application of continuum damage mechanics in fatigue assessment of A516 steel specimens considering residual stresses. Int. J. Damage Mech **32**, 1057–1076 (2023)

Attarha, M.J., Sattari-Far, I.: Comparison of the continuum damage and fracture mechanics in fatigue assessment of components containing residual stresses. Mech. Based Des. Struct. Mach. **52**, 5518–5535 (2024)

Aubert, J.-M., Dong, P., Sauvage, J.-P.: A Comprehensive Set of Round-Bar Stress Intensity Factor Solutions for ECA of Mooring Shackle and Chain Components. Offshore Technolo. Conf. (2020)

Augustyn, D., Ulriksen, M.D., Sørensen, J.D.: Reliability updating of offshore wind substructures by use of digital twin information. Energies (2021)

Aursand, M., Frøseth, G.T., Haagensen, P.J., Skallerud, B.H.: Crack growth in high strength mooring line steel under variable amplitude loading. Mar. Struct. **93**, 103534 (2024)

Austefjord, H.N., de Hauteclocque, G., Johnson, M.C., Zhu, T.Y.: Update of wave statistics standards for classification rules. In Ringsberg, J.W., Guedes Soares, C. (eds.) Advances in the Analysis and Design of Marine Structures: Proceedings of the 9th International Conference on Marine Structures (MARSTRUCT 2023, Gothenburg, Sweden, 3–5 April 2023). 1st ed. ed. London, CRC Press (2023)

Bachimanchi, P., Saha, N.: Peridynamic analysis of tubular joints of the offshore jacket structure. J Offshore Mech. Arctic Eng. **146** (2023)

Bannykh, O.A., Kovneristyi, Y.Y.: Steels for low-temperature application. Metallurgiya 13–41 (1969)

Bao, S., Wang, W., Zhou, J., Li, X.: Study on hot spot stress distribution of three-planar tubular Y-joints subjected to in-plane bending moment. Mar. Struct. **87**, 103326 (2023)

Baumgartner, J., Hobbacher, A., Levebvre, F.: Recent update of the IIW-recommendations for fatigue assessment of welded joints and components. Procedia Struct. Integr. **57**, 618–624 (2024)

Beese, A.M., Mohr, D.: Effect of stress triaxiality and Lode angle on the kinetics of strain-induced austenite-to-martensite transformation. Acta Mater. **59**, 2589–2600 (2011)

Beier, D., Janocha, M.J., Ye, N., Ong, M.C.: Fatigue assessment of suspended inter-array power cables for floating offshore wind turbines. Eng. Struct. **308**, 118007 (2024)

Benzing, J., Derimow, N., Lucon, E., Weeks, T.: Fracture Toughness Tests at 77 K and 4 K on 316L Stainless Steel Welded Plates. Technical Note, NIST 2230 (2022)

Beremin, F.M., Pineau, A., Mudry, F., Devaux, J.-C., D'Escatha, Y., Ledermann, P.: A local criterion for cleavage fracture of a nuclear pressure vessel steel. Metall. Trans. A **14**, 2277–2287 (1983)

Bergo, S., Morin, D., Sture Hopperstad, O.: Numerical implementation of a non-local GTN model for explicit FE simulation of ductile damage and fracture. Int. J. Solids Struct. **219–220**, 134–150 (2021)

Bilio, M., Brown, S., Fairweather, M., Mahgerefteh, H.: CO2 pipelines material and safety considerations. Sympos. Ser. Hazards XXI **155**, 423–429 (2009)

Bjørheim, F., Pavlou, D.G., Siriwardane, S.C.: Nonlinear fatigue life prediction model based on the theory of the S-N fatigue damage envelope. Fatigue Fract. Eng. Mater. Struct. **45**, 1480–1493 (2022a)

Bjørheim, F., Siriwardane, S.C., Pavlou, D.: A review of fatigue damage detection and measurement techniques. Int. J. Fatigue **154**, 106556 (2022b)

Boot, T., et al.: Assessing the susceptibility of existing pipelines to hydrogen embrittlement. In: TMS 2021 150th Annual Meeting & Exhibition Supplemental Proceedings. Springer, Cham (2021)

Braun, M.: Statistical analysis of sub-zero temperature effects on fatigue strength of welded joints. Welding World **66**, 159–172 (2022)

Braun, M., Ahola, A., Milaković, A.-S., Ehlers, S.: Comparison of local fatigue assessment methods for high-quality butt-welded joints made of high-strength steel. Forces Mech. **6**, 100056 (2022a)

Braun, M., et al.: Development of combined load spectra for offshore structures subjected to wind, wave, and ice loading. Energies (2022b)

Braun, M., et al.: Fatigue strength of fixed offshore structures under variable amplitude loading due to wind, wave, and ice action (2022c)

Braun, M., et al.: Sub-zero temperature fatigue strength of butt-welded normal and high-strength steel joints for ships and offshore structures in arctic regions. In: ASME 2020 39th International Conference on Ocean, Offshore and Arctic Engineering, OMAE2020. Virtual, Online (2020)

Brophy, G.R., Miller, A.J.: The Metallography and heat treatment of 8 to 10% Nickel steel. Trans. A.S.M., **41**, 1185–1203 (1948)

BS: Code of practice for fatigue design and assessment of steel structures. BS 7608:2014+A1:2015. British Standard (2015)

BS: Method of test for determination of fracture toughness in metallic materials using single edge notched tension (SENT) specimens. BS 8571:2018. British Standards (2018)

BS: Guide to methods for assessing the acceptability of flaws in metallic structures, BS 7910:2019. British Standard Institution (2019)

Bufalari, G., den Besten, H., Hong, J.K., Kaminski, M.L.: Mode-{I, III} multiaxial fatigue of welded joints in steel maritime structures: total stress based resistance incorporating strength and mechanism contributions. Int. J. Fatigue **188**, 108499 (2024a)

Bufalari, G., den Besten, H., Kaminski, M.L.: Mode-{I, III} multiaxial fatigue of welded joints in steel maritime structures: effective notch stress based resistance incorporating strength and mechanism contributions. Int. J. Fatigue **180**, 108067 (2024b)

Cabral, W., Lameira, P., Araújo, A.: Floating buoy fatigue analysis for barge-to-ship operations. Ocean Eng. **278**, 114479 (2023)

Cai, B., et al.: Remaining useful life estimation of structure systems under the influence of multiple causes: subsea pipelines as a case study. IEEE Trans. Industr. Electron. **67**, 5737–5747 (2020)

Cai, W., Li, S., Zhu, L., Cao, D., Guo, K., Li, Y.: A systematic review on dynamic responses of marine structures under repeated mass impacts. Ocean Eng. **294**, 116790 (2024)

Campagnolo, A., Vecchiato, L., Meneghetti, G.: Multiaxial variable amplitude fatigue strength assessment of steel welded joints using the peak stress method. Int. J. Fatigue **163**, 107089 (2022)

Campari, A., Ustolin, F., Alvaro, A., Paltrinieri, N.: A review on hydrogen embrittlement and risk-based inspection of hydrogen technologies. Int. J. Hydrogen Energy **48**, 35316–35346 (2023)

CCS: Guidelines for high manganese austenitic cryogenic steels (in Chinese). Beijing, China, China Classification Society (2021)

CCS: Rules for construction and equipment of ships carrying liquefied gases in bulk. China Classification Society (2023)

Cerik, B.C., Lee, K., Choung, J.: Evaluation of localized necking models for fracture prediction in punch-loaded steel panels. J. Marine Sci. Eng. (2021)

Chang, S., Liu, K., Yang, M., Yuan, L.: Theory and implementation of sub-model method in finite element analysis. Heliyon **8** (2022)

Chen, J., et al.: Deformation behavior and microstructure characteristics of the laser-welded Ti-6Al-4V joint under variable amplitude fatigue. Mater Charact **196**, 112606 (2023a)

Chen, R., Low, Y.M.: Efficient long-term fatigue analysis of deepwater risers in the time domain including wave directionality. Mar. Struct. **78**, 103002 (2021a)

Chen, R., Low, Y.M.: Reducing uncertainty in time domain fatigue analysis of offshore structures using control variates. Mech. Syst. Signal Process. **149**, 107192 (2021b)

Chen, X., Yue, J., Wu, X., Lei, J., Fang, X.: A damage coupled elasto-plastic constitutive model of marine high-strength steels under low cycle fatigue loadings. Int. J. Press. Vessels Pip. **205**, 104982 (2023b)

Chen, Y., Lorentz, E., Dahl, A., Besson, J.: Simulation of ductile tearing during a full size test using a non local Gurson–Tvergaard–Needleman (GTN) model. Eng. Fract. Mech. **261**, 108226 (2022)

Chen, Z., Yu, B., Wang, P., Qian, H.: Fatigue properties evaluation of fillet weld joints in full-scale steel marine structures. Ocean Eng. **270**, 113651 (2023c)

Chiocca, A., Frendo, F., Bertini, L.: Residual stresses influence on the fatigue strength of structural components. Procedia Struct. Integr. **38**, 447–456 (2022)

Choi, J.K., Lee, S.-G., Park, Y.-H., Han, I.-W., Morris Jr., J.W.: High manganese austenitic steel for cryogenic applications. In: The Twenty-second International Offshore and Polar Engineering Conference (2012)

Choi, Y.-H., et al.: Temperature-dependent hydrogen embrittlement of austenitic stainless steel on phase transformation. Metals **13** (2023)

Choudhary, S., Gaur, V.: Enhanced fatigue properties of AA5086 friction stir weld joints by Cu-reinforcement. Mater. Sci. Eng. A **869**, 144778 (2023)

Class, N.K.: Guidelines for Liquefied Gas Carrier Structures – Independent prismatic tanks (2018)

Ćorak, M., Mikulić, A., Katalinić, M., Parunov, J.: Uncertainties of wave data collected from different sources in the Adriatic Sea and consequences on the design of marine structures. Ocean Eng. **266**, 112738 (2022)

Corigliano, P., Crupi, V.: Fatigue analysis of TI6AL4V/INCONEL 625 dissimilar welded joints. Ocean Eng. **221**, 108582 (2021)

Corigliano, P., Crupi, V.: Review of fatigue assessment approaches for welded marine joints and structures. Metals (2022)

Dabetwar, S., Ekwaro-Osire, S., Dias, J. P.: Damage classification of composites based on analysis of lamb wave signals using machine learning. ASCE-ASME J. Risk and Uncert Engrg. Syst. Part B Mech Engrg. **7** (2021)

Dai, J., Leira, B.J., Moan, T., Alsos, H.S.: Effect of wave inhomogeneity on fatigue damage of mooring lines of a side-anchored floating bridge. Ocean Eng. **219**, 108304 (2021a)

Dai, M.-J., Tanaka, S., Bui, T.Q., Oterkus, S., Oterkus, E.: Fracture parameter analysis of flat shells under out-of-plane loading using ordinary state-based peridynamics. Eng. Fract. Mech. **244**, 107560 (2021b)

Dai, P., et al.: Numerical study on local residual stresses induced by high frequency mechanical impact post-weld treatment using the optimized displacement-controlled simulation method. J. Manuf. Process. **92**, 262–271 (2023)

Dawson, T.J.: Behavior of welded pressure vessels in agricultural ammonia service. Welding J **35**, 568–574 (1956)

de Hauteclocque, G., Maretic, N.V., Derbanne, Q.: Hindcast based global wave statistics. Appl. Ocean Res. **130**, 103438 (2023)

de Jesus, A.M.P., Ramos, G.F.S., Gomes, V.M.G., Marques, M.J., de Figueiredo, M.A.V., Marafona, J.D.R.: Comparison between EDM and grinding machining on fatigue behaviour of AISI D2 tool steel. Int. J. Fatigue **139**, 105742 (2020)

Deul, M., van Lieshout, P., Werter, N.: On the validity of using small-scale fatigue data to design full-scale steel welded structures: testing assumptions on residual stress relief. Dubrovnik, Croatia (2022)

Deul, M.L., van Battum, C.H.H., Hoogeland, M.: Tailoring the fatigue detail category class: a deterministic implementation of a probabilistic-based approach to consequence- and uncertainty-informed fatigue life prediction of ships. J. Mar. Sci. Eng. (2023)

DNV: DNV-RP-N103, Modelling and analysis of marine operations (2017)

DNV: DNV-RP-F104, Design and operation of carbon dioxide pipelines (2021a)

DNV: DNV-ST-0054, Transport and installation of wind power plants (2021b)

DNV: DNV-ST-0378, Offshore and platform lifting appliances (2021c)

DNV: Fatigue assessment of ship structures. Dnv-CG-0129, Edition October 2021. Oslo, DNV (2021d)

DNV: Fatigue design of offshore steel structures. DNV-RP-C203 (2021e)

DNV: Finite element analysis, DnV-CG-0127, Edition August 2021. DNV, Oslo (2021f)
DNV: Gas fueled container ship with independent prismatic tanks type-A and type-B, DNVGL-CG-0554. Det Norske Veritas, Oslo (2021g)
DNV: Rules for classification: Ships, Ship types, Container ships, DNVGL-RU-SHIP Pt.5 Ch.2, . Oslo, Det Norske Veritas (2022)
DNV: Class Guideline CG-0133 – Liquified gas carriers with Independent prismatic tanks of type-A and B (2023a)
DNV: Wind Assisted propulsion systems. DNV-ST-0511. Oslo, Det Norske Veritas (2023b)
DNV: DNV-ST-0437, Loads and site conditions for wind turbines (2024)
DNV (2025). https://www.dnv.com/services/offshore-and-marine-structural-engineering-sesam-for-fixed-structures-1096/
DNVGL: Assessment of flaws in pipeline and riser girth welds. Recommended Practice, DNVGL-RP-F108 (2017a)
DNVGL: Submarine pipeline systems. Standard DNVGL-ST-F101 (2017b)
Dong, P.: Quantitative weld quality acceptance criteria: an enabler for structural lightweighting and additive manufacturing. Weld. J. **99**, 39S-51S (2020)
Dong, P., Hong, J., Osage, D., Dewees, D., Prager, M.: The master SN curve method an implementation for fatigue evaluation of welded components in the ASME B&PV Code. Sect. VIII, Div. **2**, 579–581 (2010)
Dong, P., Rawls, G., Krentz, T.: Fatigue Property Evaluation of Metallic AM Parts and Determination of Design Stress Allowables (2023a)
Dong, Q., Shi, X., Tong, D., Liu, F., Wang, L., Zhao, L.: Crack initiation life model for compression-compression low cycle fatigue based on damage mechanics. Int. J. Fatigue **169**, 107495 (2023b)
Dong, Y., Garbatov, Y., Guedes Soares, C.: Strain-based fatigue reliability assessment of welded joints in ship structures. Mar. Struct. **75**, 102878 (2021a)
Dong, Y., Garbatov, Y., Guedes Soares, C.: Review on uncertainties in fatigue loads and fatigue life of ships and offshore structures. Ocean Eng. **264**, 112514 (2022a)
Dong, Y., Garbatov, Y., Soares, C.G.: Fatigue strength assessment of a butt-welded joint in ship structures based on time-domain strain approach. J. Ship Res. **65**, 123–138 (2021b)
Dong, Y., Kong, X., An, G., Kang, J.: Fatigue reliability of single-sided girth welds in offshore pipelines and risers accounting for non-destructive inspection. Mar. Struct. **86**, 103268 (2022b)
Duong, H.D., Okazaki, M., Tran, T.H.: Fatigue behavior of dissimilar friction stir welded T-lap joints between AA5083 and AA7075. Int. J. Fatigue **145**, 106090 (2021)
Eichelman, G.H., Hull, F.C.: The effect of composition on the temperature of spontaneous transformation of austenite to martensite in 18-8-type stainless steel. Trans. Am. Soc. Metals **45**, 77–104 (1953)
Elahi, S.A., Sofiani, F.M., Chaudhuri, S., Balbín, J.A., Larrosa, N.O., De Waele, W.: Investigation of the effect of pitting corrosion on the fatigue strength degradation of structural steel using a short crack model. Procedia Struct. Integrity **51**, 30–36 (2023)
Elamin, M., Li, B., Tan, K.T.: Impact damage of composite sandwich structures in arctic condition. Compos. Struct. **192**, 422–433 (2018)
Fajuyigbe, A., Brennan, F.: Fitness-for-purpose assessment of cracked offshore wind turbine monopile. Mar. Struct. **77**, 102965 (2021)
Fang, X., Wang, H., Li, W., Liu, G., Cai, B.: Fatigue crack growth prediction method for offshore platform based on digital twin. Ocean Eng. **244**, 110320 (2022)
Feng, L., Qian, X., Zhang, W.: Adaptive fatigue assessment of welded plate joints based on crack measurements. In: Geng, G., Qian, X., Poh, L.H., Pang, S.D. (eds.) Proceedings of The 17th East Asian-Pacific Conference on Structural Engineering and Construction, 2022. Singapore, Springer Nature Singapore (2023)
Feng, L., Zhang, L., Liao, X., Zhang, W.: Probabilistic fatigue life of welded plate joints under uncertainty in Arctic areas. J. Constr. Steel Res. **176**, 106412 (2021)

Maia, F., et al.: The evaluation of hydrogen embrittlement threshold force using the Small punch test. Theoret. Appl. Fract. Mech. **125**, 103673 (2023)

FKM: Analytical strength assessment of components made of steel, cast iron and aluminum material, FKM Guideline, 7. Ed.,. VDMA Verlag (2020)

Fricke, W., Kahl, A., Paetzold, H.: Fatigue assessment of root cracking of fillet welds subject to throat bending using the structural stress approach. Weld. World **50**, 64–74 (2006)

Fricke, W., Paetzold, H.: Full-scale fatigue tests of ship structures to validate the S-N approaches for fatigue strength assessment. Mar. Struct. **23**, 115–130 (2010)

Gaidai, O., Yakimov, V., Wang, F., Hu, Q., Storhaug, G., Wang, K.: Lifetime assessment for container vessels. Appl. Ocean Res. **139**, 103708 (2023a)

Gaidai, O., Yan, P., Xing, Y., Xu, J., Zhang, F., Wu, Y.: Oil tanker under ice loadings. Sci. Rep. **13**, 8670 (2023b)

Garbatov, Y., Huang, Y.C.: Multiobjective reliability-based design of ship structures subjected to fatigue damage and compressive collapse. J. Offshore Mech. Arctic Eng. 142 (2020)

Garbatov, Y., et al.: Committee III.2: fatigue and fracture. In: 21st International Ship and Offshore Structures Congress, Volume 1 (2022)

Gook, S., Krieger, S., Gumenyuk, A., El-Batahgy, A.M., Rethmeier, M.: Notch impact toughness of laser beam welded thick sheets of cryogenic nickel alloyed steel X8Ni9. Procedia CIRP **94**, 627–631 (2020)

Grönlund, K., Ahola, A., Riski, J., Pesonen, T., Lipiäinen, K., Björk, T.: Overload and variable amplitude load effects on the fatigue strength of welded joints. Weld. World **68**, 411–425 (2024)

Guachamin Acero, W., Li, L., Gao, Z., Moan, T.: Methodology for assessment of the operational limits and operability of marine operations. Ocean Eng. **125**, 308–327 (2016)

Guellec, C., Doudard, C., Levieil, B., Jian, L., Ezanno, A., Calloch, S.: Parametric method for the assessment of fatigue damage for marine shaft lines. Mar. Struct. **87**, 103325 (2023)

Guo, J., et al.: A review on failure mechanism and mechanical performance improvement of FRP-metal adhesive joints under different temperature-humidity. Thin-Walled Struct. **188**, 110788 (2023)

Hageman, R., Drummen, I., Thompson, I., Stambaugh, K.: Fleet Structural Integrity through Monitoring and Data Fusion. IN Vladimir, N., Malenica, Š. & Senjanović, I. (Eds.) Proceedings of the 15th International Symposium on Practical Design of Ships and Other Floating Structures. Dubrovnik, Croatia (2022a)

Hageman, R.B., Thompson, I.: Virtual hull monitoring using hindcast and motion data to assess frigate-size vessel stress response. Ocean Eng. **245**, 110338 (2022)

Hageman, R.B., van der Meulen, F.H., Rouhan, A., Kaminski, M.L.: Quantifying uncertainties for Risk-Based Inspection planning using in-service Hull Structure Monitoring of FPSO hulls. Mar. Struct. **81**, 103100 (2022)

Hameed, H., Bai, Y., Ali, L.: A risk-based inspection planning methodology for integrity management of subsea oil and gas pipelines. Ships Offshore Struct. **16**, 687–699 (2021)

Hammer, T.C., Willems, T., Hendrikse, H.: Dynamic ice loads for offshore wind support structure design. Mar. Struct. **87**, 103335 (2023)

Han, Y.-Q., Le, C.-H., Zhang, P.-Y., Dang, L., Fan, Q.-L.: Stochastic analysis of short-term structural responses and fatigue damages of a submerged tension leg platform wind turbine in wind and waves. China Ocean Eng. **35**, 566–577 (2021)

Han, Y., Zhu, X., Song, M.: Fatigue damage calculation of ship hulls caused by ice loads in broken ice fields. Ships Offshore Struct. **19**, 310–322 (2024)

Hariprasath, P., Sivaraj, P., Balasubramanian, V., Pilli, S., Sridhar, K.: Influence of stress ratio on fatigue behaviour of gas metal arc welded naval grade HSLA steel joints: assessment of safe and unsafe region for ship hull fabrication. Eng. Fail. Anal. **148**, 107216 (2023)

Haselbach, P.U., et al.: Quantifying the accuracy of different modelling techniques and element types to predict the structural performance of the DTU 12.6 m wind turbine blade. In: 21st European Conference on Composite Materials. European Society for Composite Materials (2024)

Haselibozchaloee, D., Correia, J., Mendes, P., de Jesus, A., Berto, F.: A review of fatigue damage assessment in offshore wind turbine support structure. Int. J. Fatigue **164**, 107145 (2022)

He, J.-C., Zhu, S.-P., Luo, C., Niu, X., Wang, Q.: Size effect in fatigue modelling of defective materials: application of the calibrated weakest-link theory. Int. J. Fatigue **165**, 107213 (2022)

Hegseth, J.M., Bachynski, E.E., Leira, B.J.: Effect of environmental modelling and inspection strategy on the optimal design of floating wind turbines. Reliab. Eng. Syst. Saf. **214**, 107706 (2021)

Hejazi, R., Grime, A., Randolph, M., Efthymiou, M.: A Bayesian machine learning approach to rapidly quantifying the fatigue probability of failure for steel catenary risers. Ocean Eng. **235**, 109353 (2021)

Heyraud, H., et al.: A two-scale finite element model for the fatigue design of large welded structures. Eng. Fail. Anal. **124**, 105280 (2021)

Hildeman, A., Bolin, D., Rychlik, I.: Deformed SPDE models with an application to spatial modeling of significant wave height. Spat. Stat. **42**, 100449 (2021)

Hildeman, A., Bolin, D., Rychlik, I.: Joint spatial modeling of significant wave height and wave period using the SPDE approach. Probab. Eng. Mech. **68**, 103203 (2022)

Hirayama, T.: Influence of chemical composition on martensitic transformation in Fe-Cr-Ni stainless steel. J. Jpn. Inst. Met. **34**, 507–510 (1970)

Hirdaris, S., et al.: Review of the uncertainties associated to hull girder hydroelastic response and wave load predictions. Mar. Struct. **89**, 103383 (2023)

Hlaing, N., Morato, P.G., Nielsen, J.S., Amirafshari, P., Kolios, A., Rigo, P.: Inspection and maintenance planning for offshore wind structural components: integrating fatigue failure criteria with Bayesian networks and Markov decision processes. Struct. Infrastruct. Eng. **18**, 983–1001 (2022)

Hlaing, N., Morato, P.G., Rigo, P., Amirafshari, P., Kolios, A., Nielsen, J.S.: The effect of failure criteria on risk-based inspection planning of offshore wind support structures. In: Life-Cycle Civil Engineering: Innovation, Theory and Practice. CRC Press (2021)

Hobbacher, A.F.: Fatigue actions (Loading). In: Hobbacher, A.F. (ed.) Recommendations for Fatigue Design of Welded Joints and Components. Springer, Cham (2016)

Hobbacher, A.F., Baumgartner, J.: Recommendations for Fatigue Design of Welded Joints and Components. Springer, Cham (2024)

Hu, S., Li, F.: Laser joining of CFRTP to metal: a review on welding parameters, joint enhancement, and numerical simulation. Polym. Compos. **45**, 1931–1955 (2024)

Huang, G., Liu, R., Hu, S.: Investigation of the mechanism for reduction of residual stress through magnetic-vibration stress relief treatment. J. Magn. Magn. Mater. **582**, 171041 (2023)

Huang, X., Ge, J., Zhao, J., Zhao, W.: A continuous damage model of Q690D steel considering the influence of Lode parameter and its application. Constr. Build. Mater. **262**, 120067 (2020)

Hübler, C., Rolfes, R.: Analysis of the influence of climate change on the fatigue lifetime of offshore wind turbines using imprecise probabilities. Wind Energy **24**, 275–289 (2021)

Hübler, C., Rolfes, R.: Probabilistic temporal extrapolation of fatigue damage of offshore wind turbine substructures based on strain measurements. Wind Energ. Sci. **7**, 1919–1940 (2022)

Hyodo, Y., et al.: Improvement of fatigue crack propagation property in low carbon steel by microstructural control and an investigation of its practical benefit. ISIJ Int. **63**, 1738–1746 (2023)

IACS: Common Structural Rules for Bulk Carriers and Oil Tankers. London, International Association of Classification Societies (2017)

IACS: Fatigue Assessment of Free Edges (former Fatigue Assessment on Hatch Corner). TB Report Pt 1, Ch 9, Sec 5. Rev. 2. London, International Association of Classification Societies (2020)
IACS: Common Structural Rules for Bulk Carriers and Oil Tankers. London, International Association of Classification Societies (2021a)
IACS: Post Weld Treatment (2015 GBS audit IACS/2015/FR1–8/OB 17). TB Report Pt 1 and Pt 2. Rev.1. London, International Association of Classification Societies (2021b)
IACS: Recommendation No. 34. Standard Wave Data. Rev. 2. London, International Association of Classification Societies (2022)
IACS: Common Structural Rules for Bulk Carriers and Oil Tankers (2023a)
IACS: Common Structural Rules for Bulk Carriers and Oil Tankers, Edition 01 January 2023. London, IACS (2023b)
Ibarra, M.A.C., Simão, M.L., Videiro, P.M., Sagrilo, L.V.S.: Long-term fatigue analysis of mooring lines considering wind-sea and swell waves using the Univariate Dimension-Reduction Method. Appl. Ocean Res. **118**, 102997 (2022)
IEC: 61400 - Wind turbines - Part 1: Design requirements International Electrotechnical Commission (2019)
IMO: IGC Code - International code for the construction and equipment of ships carrying liquefied gases in bulk (2014)
IMO: GBS Audit Team 5 Report on IACS Common Packages Regarding GBS Functional Requirements 1–8. MSC 96/5 Annex 13. International Maritime Organization (2016)
IMO: IGF Code - International code of safety for ships using gases or other low-flashpoint fuels (2017)
IMO: Goal-Based New Ship Construction Standards - Final report of the GBS maintenance of verification audit of 12 recognized organizations and IACS' common structural rules for bulk carriers and oil tankers (CSR). Note by the Secretary-General. MSC 100/6/5. International Maritime Organization (2018)
IMO: Goal-Based New Ship Construction Standards - Final report of the combined GBS audit on the rectification of non-conformities of IACS and DNV-GL ship construction rules. Note by the Secretary-General. MSC 103/7/1. International Maritime Organization (2021)
IMO: Technical information on the ammonia compatibility test of high manganese austenitic steel, submitted by the Republic of Korea (2023)
Islam, F., Caldwell, R., Phillips, A.W., John, N.A.S., Prusty, B.G.: A review of relevant impact behaviour for improved durability of marine composite propellers. Composites Part C: Open Access **8**, 100251 (2022)
ISO: Metallic materials - Method of constraint loss correction of CTOD fracture toughness for fracture assessment of steel components. ISO 27306:2016 (2016)
ISO: Method of test for the determination of quasistatic fracture toughness of welds. ISO15653 (2018)
ISO: Ships and marine technology — Offshore wind energy — Port and marine operations. ISO 29400 (2020)
ISO: Metallic materials — Unified method of test for the determination of quasistatic fracture toughness, ISO12135 (2021)
ISO: ISO 27913:2024, Carbon dioxide capture, transportation and geological storage — Pipeline transportation systems (2024)
Jacob, A., Mehmanparast, A.: Crack growth direction effects on corrosion-fatigue behaviour of offshore wind turbine steel weldments. Mar. Struct. **75**, 102881 (2021)
Jaisawal, R., Gaur, V., Ahmed, S.: On improved fatigue properties of aluminum alloy 5086 weld joints. Int. J. Fatigue **174**, 107712 (2023)

Jang, W.-S., Choi, W.-S., Choi, H.-G., Hong, S.-Y., Song, J.-H.: Fatigue damage prediction of ship rudders under vortex-induced vibration using orthonormal modal FSI analysis. Mar. Struct. **88**, 103376 (2023)

Jeon, S., Kim, Y.: Fatigue damage estimation of icebreaker ARAON colliding with level ice. Ocean Eng. **257**, 111707 (2022)

Jeong, D., Sung, H., Park, T., Lee, J., Kim, S.: Fatigue crack propagation behavior of Fe25Mn and Fe16Mn2Al steels at room and cryogenic temperatures. Met. Mater. Int. **22**, 601–608 (2016)

Jiang, J., Zeng, W., Li, L.: Effect of residual stress on hydrogen diffusion in thick butt-welded high-strength steel plates. Metals (2022)

Jiang, Z.: Installation of offshore wind turbines: a technical review. Renew. Sustain. Energy Rev. **139**, 110576 (2021)

Jimenez-Martinez, M.: Fatigue of offshore structures: a review of statistical fatigue damage assessment for stochastic loadings. Int. J. Fatigue **132**, 105327 (2020)

JIS: JIS G3127:2021, Nickel steel plates for pressure vessels for low temperature services (2021)

Kahl, A., von Bock und Polach, F., von Selle, H., Braun, M., Grimm, J.-H.: Yield strength and thickness effect on fatigue strength of thermal cut plate edges. In: The 33rd International Ocean and Polar Engineering Conference (2023)

Kalina, M., Schneider, T., Brummund, J., Kästner, M.: Overview of phase-field models for fatigue fracture in a unified framework. Eng. Fract. Mech. **288**, 109318 (2023)

Kapitza, P.L.: The liquefaction of helium by an adiabatic method. Proc. Roy. Soc. Lond. Ser. A – Math. Phys. Sci. **147**, 189–211 (1997)

Katsikogiannis, G., Sørum, S.H., Bachynski, E.E., Amdahl, J.: Environmental lumping for efficient fatigue assessment of large-diameter monopile wind turbines. Mar. Struct. **77**, 102939 (2021)

Kawabata, T.: Hirose, H.: A transition of philosophies of fracture safety and materials in liquefied natural gas storage tanks. In: 27th International Ocean and Polar Engineering Conference, ISOPE. San Francisco, California, USA (2017)

Kawabata, T., Kitano, H., Ozawa, T.. Mikami, Y.: Effect of Reverse bend process on CTOD toughness evaluation and understanding of its mechanisms (Part1: Base plate experiments and appropriate reverse bend amount for precrack straightness). Eng. Fract. Mech. (under review) (2024a)

Kawabata, T., Kitano, H., Ozawa, T., Mikami, Y.: Effect of Reverse bend process on CTOD toughness evaluation and understanding of its mechanisms (Part2: Brittle fracture criterion for prestrained material and reverse bend recommended conditions). Eng. Fract. Mech. (under review) (2024b)

Kawabata, T., et al.: Investigation on η and m factors for J integral in SE(B) specimens. Theoret. Appl. Fract. Mech. **97**, 224–235 (2018)

Kawabata, T., et al.: Applicability of new CTOD calculation formula to various a0/W conditions and B×B configuration. Eng. Fract. Mech. **179**, 375–390 (2017a)

Kawabata, T., et al.: Plastic deformation behavior in SEB specimens with various crack length to width ratios. Eng. Fract. Mech. **178**, 301–317 (2017b)

Kawabata, T., et al.: Proposal for a new CTOD calculation formula. Eng. Fract. Mech. **159**, 16–34 (2016)

Kawarasaki, T., Sakakibara, Y., Shinozaki, I., Nakayama, G.: Evaluation of liquid ammonia SCC susceptibility of steels by low-temperature, low-pressure test. Zairyo-to-Kankyo **72**, 312–317 (2023)

Khor, W.: A CTOD equation based on the rigid rotational factor with the consideration of crack tip blunting due to strain hardening for SEN(B). Fatigue Fract. Eng. Mater. Struct. **42**, 1622–1630 (2019)

Khor, W., Moore, P.L., Pisarski, H.G., Brown, C.J.: Determination of the rigid rotation plastic hinge point in SENB specimens in different strain hardening steels. J. Phys. Conf. Ser. **1106**, 012015 (2018)

Khor, W., Moore, P.L., Pisarski, H.G., Haslett, M., Brown, C.J.: Measurement and prediction of CTOD in austenitic stainless steel. Fatigue Fract. Eng. Mater. Struct. **39**, 1433–1442 (2016)

Khor, W.L., Moore, P., Pisarski, H., Brown, C.: Comparison of methods to determine CTOD for SENB specimens in different strain hardening steels. Fatigue Fract. Eng. Mater. Struct. **41**, 551–564 (2018)

Kim, B.E., Park, J.Y., Lee, J.S., Lee, J.I., Kim, M.H.: Effects of the welding process and consumables on the fracture behavior of 9 Wt.% nickel steel. Exp. Tech. **44**, 175–186 (2020a)

Kim, H.-J., Jang, B.-S.: Fatigue life prediction of ship and offshore structures under wide-banded non-Gaussian random loadings: Part II: extension to wide-banded non-Gaussian random processes. Appl. Ocean Res. **106**, 102480 (2021)

Kim, H.-J., Jang, B.-S., Kim, J.D.: Fatigue-damage prediction for ship and offshore structures under wide-banded non-Gaussian random loadings Part I: approximation of cycle distribution in wide-banded gaussian random processes. Appl. Ocean Res. **101**, 102294 (2020b)

Kim, M.-S., et al.: Metallic material evaluation of liquid hydrogen storage tank for marine application using a tensile cryostat for 20 K and electrochemical cell. Processes (2022a)

Kim, S., Frangopol, D.M., Ge, B.: Probabilistic multi-objective optimum combined inspection and monitoring planning and decision making with updating. Struct. Infrastruct. Eng. **18**, 1487–1505 (2022b)

Kim, T.-Y., Yoon, S.-W., Kim, J.-H., Kim, M.-H.: Fatigue and fracture behavior of cryogenic materials applied to lng fuel storage tanks for coastal ships. Metals (2021)

Kladovasilakis, N., Charalampous, P., Kostavelis, I., Tzetzis, D., Tzovaras, D.: Impact of metal additive manufacturing parameters on the powder bed fusion and direct energy deposition processes: a comprehensive review. Prog. Add. Manuf. **6**, 349–365 (2021)

Korotygin, D., Nammi, S.K., Pancholi, K.: The effect of ice floe on the strength, stability, and fatigue of hybrid flexible risers in the arctic sea. J. Compos. Sci. (2023)

KR: Guidelines on Assessment of Fatigue and Crack Propagation for High Manganese Steel. Busan, Korean Register (2018)

KR: Guidelines on Assessment of Fatigue and Crack Propagation. Korean Register (2022)

Kristiansen, M., et al.: Improving the fatigue life of large offshore foundations. Mar. Struct. **87**, 103314 (2023)

Kubo, N., et al.: Development of 7%Ni-TMCP steel plate for LNG storage tanks (concept of development and properties of 10, 25 and 40 mm thick 7%Ni steel plate). Quart. J. Jpn. Weld. Soc. **28**, 130–140 (2010)

Kuhlmann, U., Schmidt-Rasche, C., Jörg, F., Pourostad, V., Spiegler, J., Euler, M.: Update on the revision of Eurocode 3. Steel Construct. **14**, 2–13 (2021)

Kujawski, D.: A fatigue crack driving force parameter with load ratio effects. Int. J. Fatigue **23**, 239–246 (2001)

Kumar Yadav, B., Singh Bhadauria, S., Sharma, V.: A review on fracture and fatigue behaviour of FSW joints of Al alloys. Mater. Today Proc. (2023)

Lagneborgj, R.: The martensite transformation in 18% Cr-8% Ni steels. Acta Metall. **12**, 823–843 (1964)

Larsen, M.L., Arora, V., Adhikari, S., Clausen, H.B.: Optimization of welded K-node in offshore jacket structure including the stochastic size effect. Mar. Struct. **82**, 103128 (2022)

Lebedev, A.A., Kosarchuk, V.V.: Influence of phase transformations on the mechanical properties of austenitic stainless steels. Int. J. Plast. **16**, 749–767 (2000)

Lee et al.: High mn steel for storage and transportation of liquefied hydrogen. In: Proceedings of the 33rd International Ocean and Polar Engineering Conference, MaTCH Symposium Canada (2023)

Lee, H.W., Basaran, C.: A review of damage, void evolution, and fatigue life prediction models. Metals (2021)

Lee, M.-S., Kim, M.-H.: Fatigue crack growth evaluation of IMO Type B Spherical LNG cargo tank considering the effect of stress ratio and load history. J Weld Join **40**, 40–47 (2022a)

Lee, M.-S., Kim, M.-H.: A new fatigue crack growth model considering underloads and overloads history together with stress ratio. Int. J. Naval Arch. Ocean Eng. **14**, 100481 (2022b)

Lemmer, F., Yu, W., Luhmann, B., Schlipf, D., Cheng, P.W.: Multibody modeling for concept-level floating offshore wind turbine design. Multibody Sys.Dyn. **49**, 203–236 (2020)

Leonetti, D., Maljaars, J., Snijder, H.H.: Fitting fatigue test data with a novel S-N curve using frequentist and Bayesian inference. Int. J. Fatigue **105**, 128–143 (2017)

Leonetti, D., Maljaars, J., Snijder, H.H.: Probabilistic fatigue resistance model for steel welded details under variable amplitude loading – Inference and uncertainty estimation. Int. J. Fatigue **135**, 105515 (2020)

Leonetti, D., Zancato, E., Meneghetti, G., Maljaars, J.: Experimental and numerical determination of the fatigue notch factor in as-built wire arc additive manufacturing steel components. Fatigue Fract. Eng. Mater. Struct. **47**, 4372–4389 (2024)

Li, B., Rong, K., Cheng, H., Wu, Y.: Fatigue assessment of monopile supported offshore wind turbine under non-gaussian wind field. Shock. Vib. **2021**, 6467617 (2021a)

Li, K., et al.: Fatigue crack growth mechanism of Ni-based weld metal in a 9% Ni steel joint. Mater. Sci. Eng. A **832**, 142485 (2022a)

Li, Q., et al.: Role of microstructure heterogeneity on hydrogen-assisted fracture toughness degradation of X80 weld metal in H2S-saturated solution. Corros. Sci. **224**, 111467 (2023)

Li, S., Dong, Y., Guedes Soares, C.: A procedure to generate design load-time histories for fatigue strength assessment of offshore structures. Ocean Eng. **213**, 107707 (2020)

Li, W., et al.: Effect of laser shock peening on high cycle fatigue properties of aluminized AISI 321 stainless steel. Int. J. Fatigue **147**, 106180 (2021b)

Li, X., Zhang, W.: Long-term fatigue damage assessment for a floating offshore wind turbine under realistic environmental conditions. Renewable Energy **159**, 570–584 (2020)

Li, Y., et al.: Low cycle fatigue behavior of wire arc additive manufactured and solution annealed 308 L stainless steel. Addit. Manuf. **52**, 102688 (2022b)

Li, Z., et al.: A method for the fatigue-life assessment of subsea wellhead connectors considering riser wave-induced vibration. Ocean Eng. **306**, 118044 (2024)

Liao, D., Zhu, S.-P., Correia, J.A.F.O., De Jesus, A.M.P., Veljkovic, M., Berto, F.: Fatigue reliability of wind turbines: historical perspectives, recent developments and future prospects. Renewable Energy **200**, 724–742 (2022a)

Liao, X., Li, Y., Qiang, B., Wu, J., Yao, C., Wei, X.: An improved crack growth model of corrosion fatigue for steel in artificial seawater. Int. J. Fatigue **160**, 106882 (2022b)

Lillemäe, I., Liinalampi, S., Remes, H., Itävuo, A., Niemelä, A.: Fatigue strength of thin laser-hybrid welded full-scale deck structure. Int. J. Fatigue **95**, 282–292 (2017)

Lim, H., Manuel, L., Low, Y.M., Srinil, N.: A surrogate model for estimating uncertainty in marine riser fatigue damage resulting from vortex-induced vibration. Eng. Struct. **254**, 113796 (2022)

Lindroos, M., Isakov, M., Laukkanen, A.: Crystal plasticity modeling of transformation plasticity and adiabatic heating effects of metastable austenitic stainless steels. Int. J. Solids Struct. **236–237**, 111322 (2022)

Liu, C., et al.: CALPHAD-informed phase-field modeling of grain boundary microchemistry and precipitation in Al-Zn-Mg-Cu alloys. Acta Mater. **214**, 116966 (2021)

Liu, M., Liu, Z., Du, C., Zhan, X., Yang, X., Li, X.: Stress corrosion cracking behavior of high-strength mooring-chain steel in the SO2-polluted atmosphere. Int. J. Miner. Metall. Mater. **29**, 1186–1196 (2022a)

Liu, X., Wang, X., Liu, Z., Chen, Z., Sun, Q.: Continuum damage mechanics based probabilistic fatigue life prediction for metallic material. J. Market. Res. **18**, 75–84 (2022b)

Lizarraga-Saenz, I., Artal-García, J., Martín-San-Román, R., Vittori, F., Azcona-Armendáriz, J.: Study of the influence of met-ocean data in fatigue loads calculations of a floating offshore wind turbine. J. Phys. Conf. Ser. **2265**, 042014 (2022)

Loginow, A.W., Phelps, E.H.: Stress-corrosion cracking of steels in agricultural ammonia. Corrosion **18**, 299t–309t (2013)

Lone, E.N., Sauder, T., Larsen, K., Leira, B.J.: Probabilistic fatigue model for design and life extension of mooring chains, including mean load and corrosion effects. Ocean Eng. **245**, 110396 (2022)

Long, J., Jia, J.-L., Zhang, L.-J., Zhuang, M.-X., Wu, J.-H.: Fatigue inhomogeneity of 140 mm thick TC4 titanium alloy double-sided electron beam welded joints. Int. J. Fatigue **165**, 107214 (2022)

Long, J., Zhang, L.-J., Ning, J., Ma, Z.-X., Zang, S.-L.: Zoning study on the fatigue crack propagation behaviors of a double-sided electron beam welded joint of TC4 titanium alloy with the thickness of 140 mm. Int. J. Fatigue **146**, 106145 (2021a)

Long, J., et al.: Analysis of heterogeneity of fatigue properties of double-sided electron beam welded 140-mm thick TC4 titanium alloy joints. Int. J. Fatigue **142**, 105942 (2021b)

LR: ShipRight Desing and Construction, Fatigue Design Assessment, Level 1: Structural Detail Design Guide, Edition February. London, Lloyd's Register (2009)

LR: ShipRight ADP - Guidance notes for liquefied gas carriers adopting IMO Type-B independent tanks primarily constructed of plane surfaces. Lloyd's Register (2012)

LR: Rules and Regulations for the Classification of Ships using Gases or other Low-flashpoint Fuels. Lloyd's Register (2022a)

LR: Rules and Regulations for the Construction and Classification of Ships for the Carriage of Liquefied Gases in Bulk. Lloyd's Register (2022b)

LR: ShipRight Design and Construction, Procedure for Ship Units, Edition February 2022. London, Lloyd's Register (2022c)

LR: Guidance for Approval, Manufacture, Testing and Certification of High Manganese Austenitic Steel for Low Temperature Service. Lloyd's Register (2023a)

LR: Rules and regulation for the construction and classification of ships for the carriage of liquefied gases in bulk. Lloyd's Register (2023b)

LR: Guidelines for the Finite Element Analysis of Type C Tank. Lloyd's Register (2024a)

LR: ShipRight Design and Construction, Additional Design and Construction Procedures: Guidelines for the Finite Element Analysis of Type C Tank, Edition January 2024. London, Lloyd's Register (2024b)

Lu, F., Long, K., Diaeldin, Y., Saeed, A., Zhang, J., Tao, T.: A time-domain fatigue damage assessment approach for the tripod structure of offshore wind turbines. Sustainable Energy Technol. Assess. **60**, 103450 (2023)

Lucon, E., Benzing, J.: Instrumented charpy tests at 77 K on 316L stainless steel welded plates. Technical Note, NIST, 2196 (2021)

Lukić, M., Euler, M., Nussbaumer, A., Rauch, M., Pope, D.: New developments on fatigue in Eurocode on steel structures. Procedia Struct. Integrity **57**, 550–559 (2024)

Maleki, E., Farrahi, G.H., Reza Kashyzadeh, K., Unal, O., Gugaliano, M., Bagherifard, S.: Effects of conventional and severe shot peening on residual stress and fatigue strength of steel AISI 1060 and residual stress relaxation due to fatigue loading: experimental and numerical simulation. Met. Mater. Int. **27**, 2575–2591 (2021)

Maljaars, J., Leonetti, D., Hashemi, B., Snijder, H.H.: Systematic derivation of safety factors for the fatigue design of steel bridges. Struct. Saf. **97**, 102229 (2022)

Mancini, F., Remes, H., Romanoff, J.: On the modelling of distorted thin-walled stiffened panels via a scale reduction approach for a simplified structural stress analysis. Thin-Walled Struct. **197**, 111637 (2024)

Mao, W., Nembach, B., Tian, W., Wu, D.: A spectral fatigue assessment framework for floating offshore wind turbines. Ships Offshore Struct. 1–10 (2024)

Marjan, A., Hart, P.: Impact of design parameters on the dynamic response and fatigue of offshore jacket foundations. J. Mar. Sci. Eng. (2022)

Marquis, G.B., Barsoum, Z.: IIW recommendations on High Frequency Mechanical Impact (HFMI) treatment for improving the fatigue strength of welded joints. In: Marquis, G.B., Barsoum, Z. (eds.) IIW Recommendations for the HFMI Treatment: For Improving the Fatigue Strength of Welded Joints. Singapore, Springer Singapore (2016)

Marquis, G.B., Barsoum, Z., Leitner, M.: New developments and guideline updates for HFMI treatment for improving the fatigue strength of welded joints, XIII-2032-2024. In: 77th IIW Annual Assembly. Rhodes, Greece (2024)

Martelo, D.F., Mateo, A.M., Chapetti, M.D.: Fatigue crack growth of a metastable austenitic stainless steel. Int. J. Fatigue **80**, 406–416 (2015)

McCaig, E., Wang, Y.: On the application of Engineering Critical Assessment specified in BS 7910 to ship structures. In: Ringsberg, J., Guedes Soares, C. (eds.) Advances in the Analysis and Design of Marine Structures: Proceedings of the 9th International Conference on Marine Structures (MARSTRUCT 2023, Gothenburg, Sweden, 3–5 April 2023). CRC Press (2023)

Mehmanparast, A., Vidament, A.: An accelerated corrosion-fatigue testing methodology for offshore wind applications. Eng. Struct. **240**, 112414 (2021)

Mei, J., et al.: An overview and comparative assessment of approaches to multi-axial fatigue of welded components in codes and standards. Int. J. Fatigue **146**, 106144 (2021)

Mei, X., Xiong, M.: Effects of second-order hydrodynamics on the dynamic responses and fatigue damage of a 15 MW floating offshore wind turbine. J. Mar. Sci. Eng. (2021)

Mei, Z., Morris, J.W.: Influence of deformation-induced martensite on fatigue crack propagation in 304-type steels. Metall. Trans. A **21**, 3137–3152 (1990)

Melcher, D., Bätge, M., Neßlinger, S.: A novel rotor blade fatigue test setup with elliptical biaxial resonant excitation. Wind Energ. Sci. **5**, 675–684 (2020)

Melcher, D., Petersen, E., McInnes, M.: Potential of damage accumulation during segmented rotor blade fatigue tests. J. Phys. Conf. Ser. **2265**, 032061 (2022)

Mell, L., Rey, V., Schoefs, F., Rocher, B.: Uncertainty propagation of structural computation for fatigue assessment. J. Mar. Sci. Appl. 21, 55–66 (2022)

Mendoza, J., Haagensen, P.J., Köhler, J.: Analysis of fatigue test data of retrieved mooring chain links subject to pitting corrosion. Mar. Struct. **81**, 103119 (2022)

Mikulski, Z., Lassen, T.: Crack growth in fillet welded steel joints subjected to membrane and bending loading modes. Eng. Fract. Mech. **235**, 107190 (2020)

Millet, C., Mallick, D., Néel, G., Faramarzi, L., Amoli, A.M.: Material integrity aspects of CCS: an overview for CO2 transport and storage. In: TCCS–11. CO2 Capture, Transport and Storage. Trondheim 22nd–23rd June 2021. Short Papers from the 11th International Trondheim CCS Conference. SINTEF Academic Press (2021)

Mohammed, A., Dasari, S.R., Khandelwal, S.K., Desai, Y.M.: Advancements in stress concentration factor computation for tubular X joints through Bayesian-optimized neural networks. Structures **67**, 106962 (2024)

Moore, P., Hutchison, E.: Comparison of J equations for SENT specimens. Procedia Struct. Integrity **2**, 3743–3751 (2016)

Morohoshi, R., Kawabata, T.: Effect of stress field on TRIP behavior and its influence on fracture behavior of commercial stainless steels at cryogenic temperature. In: Proceedings of IIW2022 Conference. Tokyo (2022)

Mujeeb-Ahmed, M.P., Cabrera, J., Kim, H.J., Paik, J.K.: Effect of mooring line layout on the loads of ship-shaped offshore installations. Ocean Eng. **241**, 110071 (2021)

Mullings, H., Stallard, T.: Assessment of dependency of unsteady onset flow and resultant tidal turbine fatigue loads on measurement position at a tidal site. Energies (2021)

Muñiz-Calvente, M., Álvarez-Vázquez, A., Pelayo, F., Aenlle, M., García-Fernández, N., Lamela-Rey, M.J.: A comparative review of time- and frequency-domain methods for fatigue damage assessment. Int. J. Fatigue **163**, 107069 (2022)

Muzaffer, S., Chang, K.-H., Hirohata, M.: Fatigue life comparison of tubular joints in tripod and jacket offshore support substructures using 3D fatigue FE analysis. KSCE J. Civ. Eng. **28**, 849–859 (2024)

Nam, J., et al.: Alleviation of hydrogen embrittlement to delay effective hydrogen diffusion against ma constituent formation due to mo solutes for 420 mpa grade offshore steels. Met. Mater. Int. **29**, 1625–1636 (2023)

Narvydas, E., Puodziuniene, N., khan Thorappa, A.: Application of finite element sub-modeling techniques in structural mechanics. Mechanics **27**, 459–464 (2021)

Natarajan, A.: Damage equivalent load synthesis and stochastic extrapolation for fatigue life validation. Wind Energ. Sci. **7**, 1171–1181 (2022)

Nielsen, J.S., Sørensen, J.D.: Risk-based derivation of target reliability levels for life extension of wind turbine structural components. Wind Energy **24**, 939–956 (2021)

Niraula, A., Remes, H., Nussbaumer, A.: Strain-based analysis on the influence of local undercut geometry on fatigue crack initiation life. Int. J. Fatigue **186**, 108392 (2024)

Niu, X., et al.: Defect tolerant fatigue assessment of AM materials: size effect and probabilistic prospects. Int. J. Fatigue **160**, 106884 (2022)

Nkoumbou Kaptchouang, N.B., Monerie, Y., Perales, F., Vincent, P.-G.: Cohesive GTN model for ductile fracture simulation. Eng. Fract. Mech. **242**, 107437 (2021)

Noble, D.R., et al.: Standardising marine renewable energy testing: gap analysis and recommendations for development of standards. J. Mar. Sci. Eng. (2021)

Nohara, K., Ono, Y., Ohashi, N.: Composition and grain size dependencies of strain-induced martensitic transformation in metastable austenitic stainless steels. Tetsu-to-Hagane **63**, 772–782 (1977)

Nyborg, R., Lunde, L.: Measures for reducing SCC in anhydrous ammonia storage tanks. Process. Saf. Prog. **15**, 32–41 (1996)

O'Donnell, J., Kyoung, J., Samaria, S., Sablok, A.: Engineering criticality assessments of floating offshore platforms based on time domain structural response analysis (2021)

Ogata, T.: Development of a method for testing materials in high-pressure hydrogen using hollow specimens and acquisition of external funding and ISO Standardization History and Status, NIMS Materials Standardization Activities Report (2022)

Oie, N., Kawabata, T., Kishiki, S., Nakagomi, T.: Specimen size effect on the brittle fracture properties of steel for buildings subjected to large earthquakes. Procedia Struct. Integrity **33**, 586–597 (2021)

Okenyi, V., et al.: Corrosion surface morphology-based methodology for fatigue assessment of offshore welded structures. Fatigue Fract. Eng. Mater. Struct. **46**, 4663–4677 (2023)

Olson, G., Cohen, M.: Kinetics of strain-induced martensitic nucleation. Metall. Trans. A **6**, 791–795 (1975)

Ono, Y., Remes, H.: Influence of surface integrity on short crack growth behavior in HFMI-treated welded joints. Weld. World **69**, 227–243 (2025)

Ono, Y., Remes, H., Kinoshita, K., Yıldırım, H.C., Nussbaumer, A.: Local relaxation of residual stress in high-strength steel welded joints treated by HFMI. Welding in the World **68**, 2187–2202 (2024)

Osawa, N., De Gracia, L., Iijima, K., Yamamoto, N., Matsumoto, K.: Study on fatigue strength of welded joints subject to intermittently whipping superimposed wave load. In: Okada, T., Suzuki, K., Kawamura, Y. (eds.) Practical Design of Ships and Other Floating Structures. Springer, Singapore (2021)

Otter, A., Murphy, J., Pakrashi, V., Robertson, A., Desmond, C.: A review of modelling techniques for floating offshore wind turbines. Wind Energy **25**, 831–857 (2022)

Ozawa, A.T., Kosuge, B.H., Mikami, C.Y., Kawabata, D.T.: Typical local compression effect on crack front straightness and fracture toughness. Weld. World **65**, 1777–1790 (2021)

Ozawa, A.T., Kosuge, B.H., Mikami, C.Y., Kawabata, D.T.: Local compression process avoiding toughness change. Welding in the World **67**, 607–615 (2023)

Ozawa, T., Kawabata, T., Mikami, Y.: Quantitative evaluation of fracture toughness deterioration due to pre-strain. Eng. Fract. Mech. **272**, 108683 (2022)

Padilla-Llano, D.A., Schafer, B.W., Hajjar, J.F.: Cyclic fracture simulation through element deletion in structural steel systems. J. Constr. Steel Res. **189**, 107082 (2022)

Pandey, V.B., Singh, I.V., Mishra, B.K.: A Strain-based continuum damage model for low cycle fatigue under different strain ratios. Eng. Fract. Mech. **242**, 107479 (2021)

Pandey, V.B., Singh, I.V., Mishra, B.K.: A new creep-fatigue interaction damage model and CDM-XFEM framework for creep-fatigue crack growth simulations. Theoret. Appl. Fract. Mech. **124**, 103740 (2023)

Park, J., Kweon, S., Park, K.: Gurson-Cohesive modeling (GCM) for 3D ductile fracture simulation. Int. J. Plast. **175**, 103914 (2024)

Pavlou, D.G.: The theory of the S-N fatigue damage envelope: generalization of linear, double-linear, and non-linear fatigue damage models. Int. J. Fatigue **110**, 204–214 (2018)

Pei, X., Krishnan Ravi, S., Dong, P., Li, X., Zhou, X.: A multi-axial vibration fatigue evaluation procedure for welded structures in frequency domain. Mech. Syst. Signal Process. **167**, 108516 (2022a)

Pei, X., Li, X., Zhao, S., Dong, P., Liu, X., Xie, M.: Low cycle fatigue evaluation of welded structures with arbitrary stress-strain curve considering stress triaxiality effect. Int. J. Fatigue **162**, 106969 (2022b)

Peng, M., Liu, M., Gu, S., Nie, S.: Multiaxial fatigue analysis of jacket-type offshore wind turbine based on multi-scale finite element model. Materials (2023)

Petry, A., Gallo, P., Remes, H., Niemelä, A.: Optimizing the voce–chaboche model parameters for fatigue life estimation of welded joints in high-strength marine structures. J. Mar. Sci. Eng. (2022)

Pineau, A.G., Pelloux, R.M.: Influence of strain-induced martensitic transformations on fatigue crack growth rates in stainless steels. Metallurgical Trans. **5**, 1103–1112 (1974)

Polatidis, E., et al.: The effect of stress triaxiality on the phase transformation in transformation induced plasticity steels: experimental investigation and modelling the transformation kinetics. Mater. Sci. Eng. A **800**, 140321 (2021)

Purcell, E., Nejad, A.R., Valavi, M., Bekker, A.: On uncertainty assessment of fatigue damage of propulsion shaft under ice impact (2021)

Qiao, K., Liu, Z., Sun, Z., Wang, X.: Effects of low temperature overload and cycling temperature on fatigue crack growth behavior of ship steels in Arctic environments. Ocean Eng. **288**, 116090 (2023)

Qin, M., Shi, W., Chai, W., Fu, X., Li, L., Li, X.: Extreme structural response prediction and fatigue damage evaluation for large-scale monopile offshore wind turbines subject to typhoon conditions. Renewable Energy **208**, 450–464 (2023)

Qin, Y., den Besten, H., Palkar, S., Kaminski, M.L.: Fatigue design of welded double-sided T-joints and double-sided cruciform joints in steel marine structures: a total stress concept. Fatigue Fract. Eng. Mater. Struct. **42**, 2674–2693 (2019)

Qin, Y., den Besten, H., Palkar, S., Kaminski, M.L.: Mid- and high-cycle fatigue of welded joints in steel marine structures: effective notch stress and total stress concept evaluations. Int. J. Fatigue **142**, 105822 (2021)

Qiu, Y., Yang, H., Tong, L., Wang, L.: Research progress of cryogenic materials for storage and transportation of liquid hydrogen. Metals (2021)

Qu, S., Duan, C., Hu, X., Jia, S., Li, X.: Effect of shot peening on microstructure and contact fatigue crack growth mechanism of shaft steel. Mater. Chem. Phys. **274**, 125116 (2021)

Ragab, R., Parker, J., Li, M., Liu, T., Sun, W.: Creep crack growth modelling of Grade 91 vessel weldments using a modified ductility based damage model. Eur. J. Mech. A. Solids **91**, 104424 (2022)
Ramezani, M., Choe, D.-E., Heydarpour, K., Koo, B.: Uncertainty models for the structural design of floating offshore wind turbines: a review. Renew. Sustain. Energy Rev. **185**, 113610 (2023)
Randić, M., Pavletić, D., Potkonjak, Ž.: The influence of heat input on the formation of fatigue cracks for high-strength steels resistant to low temperatures. Metals (2022)
Rautiainen, M., Remes, H., Niemelä, A.: A traction force approach for fatigue assessment of complex welded structures. Fatigue Fract. Eng. Mater. Struct. **44**, 3056–3076 (2021)
Rautiainen, M., Remes, H., Niemelä, A., Romanoff, J.: Fatigue strength assessment of complex welded structures with severe force concentrations along a weld seam. Int. J. Fatigue **167**, 107321 (2023)
Ravi, S.K., Dong, P.: A spectral fatigue method incorporating non-proportional multiaxial loading. Int. J. Fatigue **131**, 105300 (2020)
Ravi, S.K., Dong, P., Wei, Z.: Data-driven modeling of multiaxial fatigue in frequency domain. Mar. Struct. **84**, 103201 (2022)
Rege, K., Pavlou, D.G.: A one-parameter nonlinear fatigue damage accumulation model. Int. J. Fatigue **98**, 234–246 (2017)
Ribeiro, H.V., Reis Pereira Baptista, C.A., Fernandes Lima, M.S., Santos Torres, M.A., Marcomini, J.B.: Effect of laser welding heat input on fatigue crack growth and CTOD fracture toughness of HSLA steel joints. J. Mater. Res. Technol. **11**, 801–810 (2021)
Ruggieri, C., Jivkov, A.P.: A probabilistic approach for cleavage fracture including the statistics of microcracks: application to a reactor pressure vessel steel. Eng. Fract. Mech. **272**, 108702 (2022)
Ruiz Muñoz, G.A., Sørensen, J.D.: Probabilistic inspection planning of offshore welds subject to the transition from protected to corrosive environment. Reliab. Eng. Syst. Saf. **202**, 107009 (2020)
Saenz-Aguirre, A., Ulazia, A., Ibarra-Berastegi, G., Saenz, J.: Floating wind turbine energy and fatigue loads estimation according to climate period scaled wind and waves. Energy Convers. Manage. **271**, 116303 (2022)
Saito, K., Jespersen, K.M., Ota, H., Wada, K., Hosoi, A., Kawada, H.: Fatigue delamination growth characterization of a directly bonded carbon-fiber-reinforced thermoplastic laminates and aluminum alloys with surface nanostructure using DCB test. J. Compos. Mater. **55**, 3131–3140 (2021)
Saji, R.P., Pantidis, P., Mobasher, M.E.: A new unified arc-length method for damage mechanics problems. Comput. Mech. (2024)
San Marchi, C., Ronevich, J.A., Sabisch, J.E.C., Sugar, J.D., Medlin, D.L., Somerday, B.P.: Effect of microstructural and environmental variables on ductility of austenitic stainless steels. Int. J. Hydrogen Energy **46**, 12338–12347 (2021)
Saraygord Afshari, S., Enayatollahi, F., Xu, X., Liang, X.: Machine learning-based methods in structural reliability analysis: a review. Reliab. Eng. Syst. Saf. **219**, 108223 (2022)
Sedaghat, O., Abdolvand, H.: Strain-gradient crystal plasticity finite element modeling of slip band formation in α-Zirconium. Crystals (2021)
Seo, J.-M., Kim, Y.-J., Omiya, M.: Crack growth simulation in thin plate using simplified strain based damage model. Eng. Fract. Mech. **260**, 108188 (2022)
Shen, F., Münstermann, S., Lian, J.: Cryogenic ductile and cleavage fracture of bcc metallic structures – influence of anisotropy and stress states. J. Mech. Phys. Solids **176**, 105299 (2023)
Sherman, R.J., Kessler, H.D., Frank, K.H., Medlock, R.: Evaluation of Large-Format Metallic Additive Manufacturing (AM) for Steel Bridge Applications: Final Report of Tensile, Impact, and Fatigue Testing Results. No. GT-SEMM-23-01. Georgia Tech Research Corporation (2023)

Shi, W., Liu, Y., Wang, W., Cui, L., Li, X.: Numerical study of an ice-offshore wind turbine structure interaction with the pile-soil interaction under stochastic wind loads. Ocean Eng. **273**, 113984 (2023)

Shibanuma, K., Kishi, K., He, T., Morita, N., Mitsume, N., Fukui, T.: S-version finite element strategy for accurately evaluating local stress in the vicinity of dynamically propagating crack front in 3D solid. Comput. Methods Appl. Mech. Eng. **399**, 115374 (2022)

Shimizu, M., Akahoshi, T., Azuma, K., Iwashita, T.: Effect of different modes on fracture prediction using weibull stress to predict brittle fracture from defects at beam-to-column connections. In: The 33rd International Ocean and Polar Engineering Conference (2023)

Shin, Y.-C., Kim, J.-H., Kim, Y.: Estimation of the ice-induced fatigue damage to a semi-submersible platform under level ice conditions. Appl. Sci. (2023)

Shiratsuchi, T., Osawa, N.: Fatigue life prediction for 9%Ni steel butt welded joints. Int. J. Fatigue **142**, 105925 (2021)

Shittu, A.A., Mehmanparast, A., Hart, P., Kolios, A.: Comparative study between S-N and fracture mechanics approach on reliability assessment of offshore wind turbine jacket foundations. Reliab. Eng. Syst. Saf. **215**, 107838 (2021)

Shrestha, R., Simsiriwong, J., Shamsaei, N.: Fatigue behavior of additive manufactured 316L stainless steel under axial versus rotating-bending loading: synergistic effects of stress gradient, surface roughness, and volumetric defects. Int. J. Fatigue **144**, 106063 (2021)

Sinsabvarodom, C., Leira, B.J., Chai, W., Næss, A.: Short-term extreme mooring loads prediction and fatigue damage evaluation for station-keeping trials in ice. Ocean Eng. **242**, 109930 (2021)

Slagstad, M., Yu, Z., Amdahl, J.: A simplified approach for fatigue life prediction of aquaculture nets under waves and currents. Ocean Eng. **272**, 113859 (2023)

Solberg, K., Berto, F.: Notch-defect interaction in additively manufactured Inconel 718. Int. J. Fatigue **122**, 35–45 (2019)

Song, Y., Sun, T., Zhang, Z.: Fatigue reliability analysis of floating offshore wind turbines considering the uncertainty due to finite sampling of load conditions. Renewable Energy **212**, 570–588 (2023)

Sonsino, C.M.: Effect of residual stresses on the fatigue behaviour of welded joints depending on loading conditions and weld geometry. Int. J. Fatigue **31**, 88–101 (2009)

Su, Y., et al.: Comparing the fatigue performance of Ti-4Al-0.005B titanium alloy T-joints, welded via different friction stir welding sequences. Mater. Sci. Eng. A **859**, 144227 (2022)

Suga, K., Endo, T., Kikuchi, M.: Crack growth simulation at welded part of LNG tank. J. Ocean, Mech. Aerospace **3**, 1–7 (2014)

Sun, W., et al.: Revealing tensile behaviors and fracture mechanism of Ti–6Al–4V titanium alloy electron-beam-welded joints using microstructure evolution and in situ tension observation. Mater. Sci. Eng. A **824**, 141811 (2021a)

Sun, Y., Alexandrov, I.V., Dong, Y., Valiev, R.Z., Chang, H., Zhou, L.: Optimized low-cycle fatigue behavior and fracture characteristics of Ti–6Al–4V alloy by Fe microalloying. J. Market. Res. **15**, 5277–5287 (2021b)

Sung, C.H., Randall, C.: An analytical approach for determining the design pressures of tank structures based on the idealised acceleration ellipsoid distribution described in the International Code for the Construction and Equipment of Ships Carrying Liquefied Gases in Bulk MSC 370(93). Lloyd's Register Technical Association, Session: 2014–15 (2015)

Svetlizky, D., et al.: Directed energy deposition (DED) additive manufacturing: physical characteristics, defects, challenges and applications. Mater. Today **49**, 271–295 (2021)

Sørum, S.H., Katsikogiannis, G., Bachynski-Polić, E.E., Amdahl, J., Page, A.M., Klinkvort, R.T.: Fatigue design sensitivities of large monopile offshore wind turbines. Wind Energy **25**, 1684–1709 (2022)

Tagawa, T., et al.: Experimental measurements of deformed crack tips in different yield-to-tensile ratio steels. Eng. Fract. Mech. **128**, 157–170 (2014)

Tagawa, T., et al.: Comparison of CTOD standards: BS 7448-Part 1 and revised ASTM E1290. Eng. Fract. Mech. **77**, 327–336 (2010)

Takeuchi, T., et al.: Fatigue assessment of ship structures based on equivalent wave probability (EWP) concept (1st report): proposal of EWP concept and its verification by 8600TEU container ship's onboard hull monitoring. Mar. Struct. **91**, 103476 (2023)

Tamimi, M.F., Khandel, O., Soliman, M.: A framework for quantifying fatigue deterioration of ship structures under changing climate conditions. Ships Offshore Struct. **17**, 2745–2760 (2022)

Tamimi, M.F., Soliman, M.: Sensitivity assessment of the crack propagation behavior in welded stiffened panels. Ships Offshore Struct. 1–16 (2024)

Tamimi, M.F., Soliman, M., Khandel, O.: A comprehensive approach for quantifying the reliability of ship hulls under propagating fatigue cracks. Ocean Eng. **279**, 114488 (2023)

Tamura, I.: Deformation-induced martensitic transformation and transformation-induced plasticity in steels. Metal Sci. **16**, 245–253 (1982)

Tanaka, T., Terada, S., Tanaka, H., Kawamata, S., Abe, K., Watanabe, M.: Development of liquefied CO_2 carriers and onboard CO_2 capture equipment to realize a carbon neutral society. Mitsubishi Heavy Industries technical review (2022)

Tlatlik, J., Hohe, J., Münstermann, S.: Analysis and improvement of local cleavage fracture models at elevated loading rates. Theoret. Appl. Fract. Mech. **125**, 103857 (2023)

Toor, K.K., Lotsberg, I.: Fatigue calculation at hot spot in cope hole welded details using finite element analysis. Procedia Struct. Integrity **57**, 772–784 (2024)

Tsuchida, N., Ishimaru, E., Kawa, M.: Role of deformation-induced martensite in TRIP effect of metastable austenitic steels. ISIJ Int. **61**, 556–563 (2021a)

Tsuchida, N., Morimoto, Y., Tonan, T., Shibata, Y., Fukaura, K., Ueji, R.: Stress-induced martensitic transformation behaviors at various temperatures and their TRIP effects in SUS304 metastable austenitic stainless steel. ISIJ Int. **51**, 124–129 (2011)

Tsuchida, N., Tomota, Y.: A micromechanic modeling for transformation induced plasticity in steels. Mater. Sci. Eng. A **285**, 346–352 (2000)

Tsuchida, N., Ueji, R., Inoue, T.: Effect of temperature on stress–strain curve in SUS316L metastable austenitic stainless steel studied by in situ neutron diffraction experiments. ISIJ Int. **61**, 632–640 (2021b)

Tsuda, R., Morohoshi, R., Tsuchida, N., Kawabata, T.: FEM module development considering TRIP for evaluating liquefied hydrogen storage tank deformation under huge earthquake. In: ISOPE International Ocean and Polar Engineering Conference. Canada, ISOPE (2023)

Vecchiato, L., Campagnolo, A., Meneghetti, G.: The Peak Stress Method for fatigue lifetime assessment of fillet-welded attachments in steel subjected to variable amplitude in-phase multiaxial local stresses. Int. J. Fatigue **169**, 107482 (2023)

Veers, P., et al.: Grand challenges in the design, manufacture, and operation of future wind turbine systems. Wind Energ. Sci. **8**, 1071–1131 (2023)

Velarde, J., Kramhøft, C., Sørensen, J.D., Zorzi, G.: Fatigue reliability of large monopiles for offshore wind turbines. Int. J. Fatigue **134**, 105487 (2020)

Vizentin, G.: Durability modeling and analysis of composite structures exposed to the marine environment. Maritime Faculty in Rijeka. Rijeka (2022)

Wagg, D.J., Worden, K., Barthorpe, R.J., Gardner, P.: Digital twins: state-of-the-art and future directions for modeling and simulation in engineering dynamics applications. ASCE-ASME J. Risk Uncert Engrg Syst. Part B Mech. Engrg. **6** (2020)

Wahab, M.M.A., Kurian, V.J., Liew, M.S., Kim, D.K.: Condition assessment techniques for aged fixed-type offshore platforms considering decommissioning: a historical review. J. Mar. Sci. Appl. **19**, 584–614 (2020)

Wang, C., Mi, G., Zhang, X.: Welding stability and fatigue performance of laser welded low alloy high strength steel with 20 mm thickness. Opt. Laser Technol. **139**, 106941 (2021a)

Wang, F., Cui, W.: Recent developments on the unified fatigue life prediction method based on fracture mechanics and its applications. J. Mar. Sci. Eng. **8** (2020)

Wang, H., Cai, Z., Dong, H., Liu, Y., Wang, W.: Mechanical-chemical-coupled peridynamic model for the corrosion fatigue behavior of a nickel-based alloy. Int. J. Fatigue **168**, 107400 (2023a)

Wang, L., Kolios, A., Liu, X., Venetsanos, D., Cai, R.: Reliability of offshore wind turbine support structures: a state-of-the-art review. Renew. Sustain. Energy Rev. **161**, 112250 (2022a)

Wang, L., et al.: Experimental investigation on compressive dwell fatigue behavior of titanium alloy pressure hull for deep-sea manned submersibles. Ocean Eng. **303**, 117646 (2024)

Wang, M., Wang, C., Hnydiuk-Stefan, A., Feng, S., Atilla, I., Li, Z.: Recent progress on reliability analysis of offshore wind turbine support structures considering digital twin solutions. Ocean Eng. **232**, 109168 (2021b)

Wang, Q., et al.: Low cycle fatigue behavior of near-alpha titanium alloys used in deep-driving submersible: Ti-6Al-3 Nb-2Zr-1Mo vs. Ti-6Al-4 V ELI. J. Alloys Compoun. **934**, 167856 (2023b)

Wang, S., Moan, T., Jiang, Z.: Influence of variability and uncertainty of wind and waves on fatigue damage of a floating wind turbine drivetrain. Renewable Energy **181**, 870–897 (2022b)

Wang, Z., Hu, Z., Liu, K., Chen, G.: Application of a material model based on the Johnson-Cook and Gurson-Tvergaard-Needleman model in ship collision and grounding simulations. Ocean Eng. **205**, 106768 (2020)

Wasim, M., Djukic, M.B.: Hydrogen embrittlement of low carbon structural steel at macro-, micro-and nano-levels. Int. J. Hydrogen Energy **45**, 2145–2156 (2020)

Weeks, D., Derimow, N., Benzing, J.: Tensile Tests at 77 K and 4 K on 316L Stainless Steel Welded Plates, Technical Note (NIST TN). Gaithersburg, MD, [online], National Institute of Standards and Technology (2022)

Wei, S., et al.: Fatigue performance assessment of thick TIG-Dressing cruciform welded joints made by Q355D structural steel. J. Market. Res. **27**, 5977–5993 (2023a)

Wei, Z., Dong, P., Mei, J., Pei, X., Ravi, S.K.: A moment of load path-based parameter for modeling multiaxial fatigue damage of welded structures. Int. J. Fatigue **171**, 107575 (2023b)

Wei, Z., Dong, P., Mei, J., Ravi, S.: An improved path-dependent multi-axial fatigue cycle counting and fatigue life assessment method. In: 13th International Conference on Multiaxial Fatigue and Fracture (ICMFF13) (2022)

Wei, Z., Dong, P., Mei, J., Ravi, S.K.: Comparison of two damage parameters for nonproportional multiaxial fatigue assessment of welded structures. In: Advances in Accelerated Testing and Predictive Methods in Creep, Fatigue, and Environmental Cracking, pp. 239–261 (2023c)

Wei, Z., Dong, P., Pei, X.: Effective second moment of load path (ESMLP) method for multiaxial fatigue damage and life assessment. SAE International (2023d)

Wei, Z., Dong, P., Pei, X.: The structural strain method for fatigue evaluation of welded components: closed-form solutions. Int. J. Fatigue **180**, 108119 (2024a)

Wei, Z., Lei, L., Pei, X., Dong, P.: The structural strain method for fatigue evaluation of welded components: analytical treatment of reversed plasticity. Int. J. Press. Vessels Pip. **210**, 105249 (2024b)

Wigley, D.: Mechanical Properties of Materials at Low Temperatures. Springer (2012)

Wilkie, D., Galasso, C.: Gaussian process regression for fatigue reliability analysis of offshore wind turbines. Struct. Saf. **88**, 102020 (2021)

Wu, S., Cheng, J., Dong, P., Zhang, Y., Zhang, L.: Defect interactions and effects on fatigue behaviors of additively manufactured parts (2023)

Wu, S., Dong, P.: Modeling of distributed defects and a proposed fatigue interaction criterion for additively manufactured metal parts. Int. J. Press. Vessels Pip. **206**, 105048 (2023)

Xie, Y., Gao, C., Wang, P., Qu, X., Cui, H.: Research on vibration fatigue damage identification of oil and gas pipeline under the condition of measured noise injection. Appl. Ocean Res. **134**, 103512 (2023a)

Xie, Y., Gao, C., Wang, P., Zhou, L., Zhang, C., Qu, X.: Research on vibration fatigue damage locations of offshore oil and gas pipelines based on the GA-improved BP neural network. Shock. Vib. **2023**, 2530651 (2023b)

Xie, Y., Tang, H., Low, Y.M.: Deep gated recurrent unit networks for time-domain long-term fatigue analysis of mooring lines considering wave directionality. Ocean Eng. **284**, 115244 (2023c)

Xing, S., Pei, X., Mei, J., Dong, P., Su, S., Zhen, C.: Weld toe versus root fatigue failure mode and governing parameters: A study of aluminum alloy load-carrying fillet joints. Mar. Struct. **88**, 103344 (2023)

Xu, C., Liu, G., Li, Z., Huang, Y.: Multiaxial fatigue life prediction of tubular K-joints using an alternative structural stress approach. Ocean Eng. **212**, 107598 (2020)

Xu, J., Benson, S., Wetenhall, B.: Comparative analysis of fatigue life of a wind turbine yaw bearing with different support foundations. Ocean Eng. **235**, 109293 (2021)

Xu, L., Hu, P., Li, Y., Qiu, N., Chen, G., Liu, X.: Improved fatigue reliability analysis of deepwater risers based on RSM and DBN. J. Mar. Sci. Eng. **11** (2023)

Xu, S., Chen, J., Shen, W., Hou, R., Wu, Y.: Fatigue strength evaluation of 5059 aluminum alloy welded joints Considering welding deformation and residual stress. Int. J. Fatigue **162**, 106988 (2022)

Yan, Y., et al.: Corrosion fatigue behavior of X65 pipeline steel welded joints prepared by CMT/GMAW backing process. Corros. Sci. **225**, 111568 (2023)

Yang, H., Wang, P., Qian, H.: Fatigue behavior of typical details of orthotropic steel bridges in multiaxial stress states using traction structural stress. Int. J. Fatigue **141**, 105862 (2020)

Yeddu, H.K., Somers, M.A.J.: Martensite formation during heating from cryogenic temperatures – a phase-field study. Comput. Mater. Sci. **196**, 110529 (2021)

Yeter, B., Brennan, F.: Probabilistic structural integrity assessment of a floating offshore wind turbine under variable amplitude loading. Procedia Struct. Integrity **57**, 133–143 (2024)

Yeter, B., Garbatov, Y.: Structural integrity assessment of fixed support structures for offshore wind turbines: a review. Ocean Eng. **244**, 110271 (2022)

Yeter, B., Garbatov, Y., Guedes Soares, C.: Analysis of life extension performance metrics for optimal management of offshore wind assets1. J. Offshore Mech. Arctic Eng. **144** (2022a)

Yeter, B., Garbatov, Y., Guedes Soares, C.: Life-extension classification of offshore wind assets using unsupervised machine learning. Reliab. Eng. Syst. Saf. **219**, 108229 (2022b)

Yeter, B., Garbatov, Y., Guedes Soares, C.: Review on artificial intelligence-aided life extension assessment of offshore wind support structures. J. Mar. Sci. Appl. **21**, 26–54 (2022c)

Ying, G., Shangyu, Y., Cong, Z., Fulong, X., Jianfeng, R.: Low-temperature toughness enhancement of 9% ni steel girth welds in LNG storage tanks via a TIP–TIG welding process. J. Mater. Eng. Perform. **33**, 7098–7110 (2024)

Yu, C., Guo, Q., Gong, X., Yang, Y., Zhang, J.: Fatigue life assessment of pressure hull of deep-sea submergence vehicle. Ocean Eng. **245**, 110528 (2022a)

Yu, Y., Chen, X., Guo, Z., Zhang, J., Lü, Q.: Robust design of monopiles for offshore wind turbines considering uncertainties in dynamic loads and soil parameters. Ocean Eng. **266**, 112822 (2022b)

Yu, Y., Pei, X., Wang, P., Dong, P., Fang, H.: A structural stress approach accounting for notch effects on fatigue propagation life: Part I theory. Int. J. Fatigue **159**, 106793 (2022c)

Zavvar, E., Hectors, K., De Waele, W.: Stress concentration factors of multi-planar tubular KT-joints subjected to in-plane bending moments. Mar. Struct. **78**, 103000 (2021)

Zhang, L., Dong, P., Wang, Y., Mei, J.: A Coarse-Mesh hybrid structural stress method for fatigue evaluation of Spot-Welded structures. Int. J. Fatigue **164**, 107109 (2022a)

Zhang, T., McHugh, P.E., Leen, S.B.: Finite element implementation of multiaxial continuum damage mechanics for plain and fretting fatigue. Int. J. Fatigue **44**, 260–272 (2012)

Zhang, X., Ni, W., Sun, L.: Fatigue analysis of the oil offloading lines in FPSO system under wave and current loads. J. Mar. Sci. Eng. (2022b)

Zhang, Y., Sclavounos, P.D.: Nonlinear wave loads on offshore wind turbines: extreme statistics and fatigue. J. Offshore Mech. Arctic Eng. **143** (2021)

Zhang, Z., Yang, B., James, M.N., Xiao, S.: Evolution of residual stress at a fatigue crack tip and its influence on crack tip shielding and plasticity. J. Market. Res. **32**, 1749–1760 (2024)

Zhao, W.: Calibration of design fatigue factors for offshore structures based on fatigue test database. Int. J. Fatigue **145**, 106075 (2021)

Zhao, W., Jiang, W., Zhang, H., Han, B., Jin, H., Gao, Q.: 3D finite element analysis and optimization of welding residual stress in the girth joints of X80 steel pipeline. J. Manuf. Process. **66**, 166–178 (2021a)

Zhao, W., Leira, B.J., Feng, G., Gao, C., Cui, T.: A reliability approach to fatigue crack propagation analysis of ship structures in polar regions. Mar. Struct. **80**, 103075 (2021b)

Zhao, W., Leira, B.J., Kim, E., Feng, G., Sinsabvarodom, C.: A probabilistic framework for the fatigue damage assessment of ships navigating through level ice fields. Appl. Ocean Res. **111**, 102624 (2021c)

Zhao, Y., Dong, S.: Probabilistic fatigue surrogate model of bimodal tension process for a semi-submersible platform. Ocean Eng. **220**, 108501 (2021)

Zhao, Z., Wang, W., Shi, W., Qi, S., Li, X.: Effect of floating substructure flexibility of large-volume 10 MW offshore wind turbine semi-submersible platforms on dynamic response. Ocean Eng. **259**, 111934 (2022)

Zheng, T., Chen, N.-Z.: Time-domain fatigue assessment for blade root bolts of floating offshore wind turbine (FOWT). Ocean Eng. **262**, 112201 (2022)

Zhou, C., et al.: Hydrogen uptake induced by CO_2 enhances hydrogen embrittlement of iron in hydrogen blended natural gas. Corros. Sci. **207**, 110594 (2022)

Zhu, F., Yeter, B., Brennan, F., Collu, M.: Time-domain fatigue reliability analysis for floating offshore wind turbine substructures using coupled nonlinear aero-hydro-servo-elastic simulations. Eng. Struct. **318**, 118759 (2024)

Zhu, R., Ma, S., Wang, X., Chen, J., Huang, P., Wang, Y.: Effect of ultrasonic surface rolling process on hydrogen embrittlement behavior of TC4 laser welded joints. J. Mater. Sci. **57**, 11997–12011 (2022a)

Zhu, S.-P., Ai, Y., Liao, D., Correia, J.A.F.O., De Jesus, A.M.P., Wang, Q.: Recent advances on size effect in metal fatigue under defects: a review. Int. J. Fract. **234**, 21–43 (2022b)

Ziccarelli, A., Kanvinde, A., Deierlein, G.: Cyclic adaptive cohesive zone model to simulate ductile crack propagation in steel structures due to ultra-low cycle fatigue. Fatigue Fract. Eng. Mater. Struct. **46**, 1821–1836 (2023)

Zou, G., Faber, M.H., González, A., Banisoleiman, K.: A holistic approach to risk-based decision on inspection and design of fatigue-sensitive structures. Eng. Struct. **221**, 110949 (2020)

Zou, J., Lourens, E.-M., Cicirello, A.: Virtual sensing of subsoil strain response in monopile-based offshore wind turbines via Gaussian process latent force models. Mech. Syst. Signal Process. **200**, 110488 (2023a)

Zou, T., Niu, X., Ji, X., Chen, X., Tao, L.: The projection of climate change impact on the fatigue damage of offshore floating photovoltaic structures. Front. Mar. Sci. **10** (2023b)

Open Access This chapter is licensed under the terms of the Creative Commons Attribution-NonCommercial-NoDerivatives 4.0 International License (http://creativecommons.org/licenses/by-nc-nd/4.0/), which permits any noncommercial use, sharing, distribution and reproduction in any medium or format, as long as you give appropriate credit to the original author(s) and the source, provide a link to the Creative Commons license and indicate if you modified the licensed material. You do not have permission under this license to share adapted material derived from this chapter or parts of it.

The images or other third party material in this chapter are included in the chapter's Creative Commons license, unless indicated otherwise in a credit line to the material. If material is not included in the chapter's Creative Commons license and your intended use is not permitted by statutory regulation or exceeds the permitted use, you will need to obtain permission directly from the copyright holder.

Committee IV.1: Design Methods, Principles and Criteria

Y. Kawamura[1(✉)], J. Sales[2], P. Georgiev[3], J. Jelovica[4], W. Tang[5], A. Kolios[6], S. Vhanmane[7], M. Sidari[8], H. Amlashi[9], J. M. Kwon[10], A. Sobey[11], Y. Yan[12], and J. Sirkar[13]

[1] Yokohama National University, Yokohama, Japan
kawamura-yasumi-zx@ynu.ac.jp
[2] Federal University of Rio de Janeiro, Rio de Janeiro, Brazil
[3] Technical University of Varna, Varna, Bulgaria
[4] University of British Columbia, Vancouver, Canada
[5] China Ship Scientific Research Center, Wuxi, China
[6] Technical University of Denmark (DTU), Roskilde, Denmark
[7] Indian Register of Shipping, Mumbai, India
[8] Fincantieri, Trieste, Italy
[9] University of South-Eastern Norway, Kongsberg, Norway
[10] Seoul, Republic of Korea
[11] University of Southampton, Southampton, United Kingdom
[12] Houston, USA
[13] Washington D.C, USA

Committee Mandate. Concern for the overall design process for ships and offshore structural safety and performance, including unmanned, and its integration with production, maintenance and repair. The development of appropriate principles for rational life-cycle design using general sustainability criteria in economic, societal and environmental terms shall be addressed. Particular attention shall be given to the roles and requirements of computer-based design and production, and to the utilization of information technology. Possible differences with the safety requirements in existing standards within the regulatory frameworks, including comparisons with other relevant industries and Goal-Based Standards, shall be considered. The role of reliability-based design codes and requirements shall be treated as well as their calibration to established safety levels.

Keywords: Digital Twin · Autonomous Ships · Update of Regulations · Machine Learning based Design · AR & VR · Optimization · Risk-based Design · Structural Reliability · Integrated Approach · Data Exchange

---

J. Sirkar has passed away during the term of ISSC activity. We would like to send our condolences, and we appreciate his effort to the activity of this committee IV.1.

Committee IV.1: Design Methods, Principles and Criteria 729

1 Introduction

In the previous ISSCs, there were two committees related to design of ships and offshore structures, "IV.1 Design Principle and Criteria" and "IV.2 Design Methods". On the other hand, this committee of ISSC 2025, "IV.1 Design Methods, Principles and Criteria", is a new committee created by merging these two committees of the previous ISSCs. In the previous report of IV.1, the framework for assessing marine structures against societal sustainability goals for economic, social, and environmental performance was discussed. Especially, an extensive discussion was made for the sustainability criteria, including the growing push to decarbonize the maritime industry. The report also focused on digital twin monitoring systems, accidental limit state, human factors, and structural reliability. Also, the previous committee, IV.2, mainly focused on the state-of-the-art design related issues, such as the design tools, structural optimization, lifecycle data management, or integrated design concept. Considering the topics treated in the previous report of IV.1 and IV.2, this Committee decided the structure of this report so that the report covers wide range of design related topics reported in the past report as follows.

Section 2 presents overview of design methods, principles and criteria. The chapter introduces Design Principles and Criteria before discussing changes in design processes required by environmental changes and arctic operations. Design methods are then introduced, which includes an overview of Design for X (DfX) approaches, Risk-based Design and Autonomous Vessels. Finally, a section on Digitalization is included, due to the growing importance of these approaches to structural design.

Section 3 presents overview of updates in governing rules and regulations which influence ship and offshore structure designs. This chapter holds the latest updates from International Maritime Organization (IMO) regulations and goal-based standards as well as the recent changes and information from International Association of Class Societies (IACS) and classification societies. It also addresses the current interest in offshore winds, alternative fuels and autonomous ships.

Section 4 presents the latest developments in tools for structural design. Advancements in software tools from various classification societies are reviewed, along with recently proposed Computer Aided Design (CAD) exchange formats for design software in the marine industry. Structural optimization methods are discussed, including the recent proliferation of machine learning approaches in design. Virtual and augmented reality, beyond aiding in production, are also considered as tools for conceptual development. The chapter reviews the role of digital twins in both design and production. Finally, an overview of design methods for offshore structures is provided.

Section 5 presents recent advancement related to lifecycle management of ships and offshore structures, which is important for efficient and sustainable development of ships and offshore structures. Especially, data exchange and standards, digital twin in the lifecycle, design and operation tools, etc. is discussed.

Section 6 focuses on the critical developments in Reliability and Risk Assessment for ships and offshore structures. This section reviews the evolution of structural reliability analysis (SRA) techniques, moving from traditional statistical methods to advanced multi-dimensional and probabilistic approaches. The integration of digital technologies like Structural Health Monitoring (SHM) and Digital Twin (DT) systems has greatly enhanced the ability to predict and mitigate structural failures under complex marine environments. Specific challenges such as ice accumulation, corrosion, and vibration-induced stresses have been addressed using advanced computational and probabilistic models. The chapter emphasizes the importance of incorporating lifecycle considerations into the design and maintenance of marine structures, making them more resilient and efficient in the face of evolving global challenges.

Section 7 aims to bridge the gap between research and practice in the maritime industry. Coordination and communication between distinctive design tools are crucial, especially considering emerging Information and Communication Technology (ICT) technologies like AI, big data, cloud computing, etc. These technologies offer new opportunities and challenges, and their successful implementation is a significant focus. Emerging ICT technologies for ships and offshore structures include AI for design optimisation, decision-making, and autonomous navigation, big data for monitoring, fault detection, and safety enhancement. This chapter deals with the recent state-of-the-art and state-of-practice in design methodology and tools, the development of design optimisation and decision-making tools, and advancements in lifecycle management and integrated design approaches.

In Sect. 8, benchmark study about topology optimization for cross-section of the tanker under multiple loading conditions is reported. As concluded in the previous IV.2 report, topology optimization has seen an increase in popularity in the marine industry. In recent years, many commercial finite element software become to provide topology optimization tool which can be used without detailed knowledge of the theory of topology optimization. In this benchmark study, we used three commercial software and two in-house programs to solve the benchmark problem. By applying topology optimization to ship structures, applicability of topology optimization to practical design is discussed.

Finally, in Sect. 9, we concluded the discussion of each chapter, and presented future challenges and recommendations, which might have an impact in the field of design of marine structures.

It is noted that some of the topics included in the previous reports of IV.1 and IV.2 are not addressed in this report. For example, the topic of Human factor is not addressed in this report, because of limited length of the report and the limited number of committee members. However, this Committee included recent important topics such as autonomous ships and machine learning approach for design.

2 Overview of Design Methods, Principles and Criteria

This section follows a new structure for the ISSC committees, with the amalgamation of Design Principles and Criteria and Design Methods. The structure for the overview seeks to take the core elements of each of the previous overview chapters. It was decided to keep the two previous chapters separate, with Principles and Criteria and Methods having different focuses. However, a new section on Digitalization is included to focus on the importance of innovative approaches in this area and combines literature on both Principles and Criteria and Design Methods.

2.1 Principles and Criteria

2.1.1 Introduction to Principles and Criteria

Design Principles are the intent for judging the worthiness of a design. They result from accumulated wisdom over time and are applied during the design process in conjunction with engineering principles: fundamental concepts, regulations and standards. In Ship and Offshore Structures, the general design principle is defined by IMO which has moved away from prescriptive rules towards Goal-Based Standards. These Goal-Based Standards are described as: "Ships shall be designed and constructed for a specified design life to be safe and environmentally friendly, when properly operated and maintained under the specified operating and environmental conditions, in intact and specified damage conditions, throughout their life (SOLAS, 2010). It is the design principles that seek to judge the worthiness of a design next to these criteria.

The IACS (2024) gives the clearest definition of a design principle. In summary the safety of the structure is assessed using the potential failure mode(s). The ship should be designed using consistent design load sets that cover the operating mode and is designed so that it has redundancy. The ship structure should have a hierarchy, where elements lower in the hierarchy do not result in immediate failure of those higher in the hierarchy. The rules are based on limit state design, for each failure mode one or more limit states might be relevant, these are divided into four categories: Serviceability Limit State (SLS), Ultimate Limit State (ULS), Fatigue Limit State (FLS) and Accidental Limit State (ALS).

A systems engineering approach (Crawley et al., 2015; NASA, 2016) is increasingly employed to analyse stakeholder values and system functions, as ships are large and complex systems involving multiple stakeholders. The design increasingly requires a sociotechnical evaluation. Castaneda et al. (2020) used the systems approach to better assess the long-term impact of a feed-in tariff reduction on households' photovoltaic (PV) adoption, utilities, and on solar companies. It is reported that such an approach is important to ascertain counterintuitive effects. System Dynamics simulation developed in this study established that cutting subsidies to PV users could result in the reduction of PV installations and lower benefits for solar companies. However, this study found that the market would still have potential for solar energy-related technology growth in the UK. Ichinose et al. (2022) designed and proposed concepts for data centre systems as integrated infrastructure consisting of both data centre and green power plant in response to rapid regional growth and potential of offshore and nearshore infrastructure in Southeast Asia. Using the systems approach, the proposed system was carefully analysed as a sociotechnical system to effectively select multiple evaluation criteria and define various design parameters. It is concluded that Singapore is an attractive location for investment given an offshore data centre with an offshore fixed wind power plant.

Previously Committee Design Principles has covered a range of topics, in ISSC 2018 these included Sustainability and Lifecycle Principles, Goal-Based Approaches, Reassessment for Life Extensions, Human Performance in Engineering and Criteria Evaluation, The Challenge of Human Performance in Engineering and Inland and Coastal Vessels in ISSC 2022 this included Design Criteria, Design Philosophy, Design criteria and ship life cycle, world environmental indicators, ship energy efficiency measures, ship particulars optimization and lightening the hull structure. The current overview will focus on two key topics: sustainability in the Marine Industry and Artic Operations.

2.1.2 Sustainability in the Marine Industry

Shipping is currently a prominent topic in discussions on sustainability. Like other industries, shipping produces greenhouse gas (GHG) emissions and needs to decrease its impact on the environment. With international shipping accounting for more than 80 percent of global trade volume, it is accountable for nearly 3 percent of total global GHG emissions (UN, 2023). The aviation and maritime transport account 14.4% and 13.5% of EU transport emissions, respectively (EU, 2023). Although maritime transport has a small share without additional actions, emissions from the shipping sector will continue to increase.

The activities of the IMO as a UN agency responsible for safe, secure and efficient shipping and the prevention of pollution from ship started 20 years ago with resolution A.963(23) (IMO, 2004). The large-scale development of new legislation began with the adoption of Annex VI of MARPOL which became effective on 19 May 2005. Annex VI aims to reduce air emissions from ships, including sulfur oxides (SOx), nitrogen oxides (NOx), ozone-depleting substances (ODS), volatile organic compounds (VOCs), and shipboard incineration. The goal is to significantly decrease the carbon intensity of global shipping, ultimately eliminating its impact on both local and global air pollution as well as environmental issues (IMO, 2023a).

The Initial IMO Strategy on reduction of GHG emissions from ships adopted on 13 April 2018 was revised by Resolution MEPC.377(80) (MEPC, 2023). By preserving the goals set out in the initial strategy for further improving the Energy Efficiency Design Index (EEDI) for new ships and reducing CO_2 emissions by at least 40% by 2030 compared to 2008, more ambitious goals have been established. The new goals are "*...uptake of zero or near zero GHG emission technologies, fuels and/or energy sources to represent at least 5%, striving for 10%, of the energy used by international shipping by 2030*", and "*… to peak GHG emissions from international shipping as soon as possible and to reach net zero GHG emissions by or around, i.e. close to 2050*" (MEPC, 2023).

To achieve these ambitious goals, numerous rules and new requirements have been issued in the last more than 10 years, both for new and existing ships. An overview of the currently effective rules is presented in Table 1.

The main principle embedded in these requirements is the implement different measures to reduce the CO2 emissions per unit of transport work expressed as the product of capacity and speed, or capacity and distance covered in one year as it is for CII.

2.1.2.1. Structures Related to Energy Efficiency

The influence of structural design is important for energy efficiency and the environment, considering the life cycle of the ship (Fig. 1).

A hybrid structure of 116,226 tDW bulk carrier's cargo hold by replacing the steel plate of the inner side shell with an aluminium honeycomb sandwich (AHS) was created to find the optimal structural solution. The goal was to maximize the strength of the ship hull while minimizing the cost of the ship, lightship weight, transportation expenses, and maximizing the annual cargo capacity (Garbatov et al., 2022). The hybrid solution

Table 1: Summary of the actual requirements concerning reduction of emissions in shipping

Applicable to	Index/Indicator	Entered into force	Rules/Req.
New Ships	**EEDI** – Energy Efficiency Design Index	on 1 January 2013	Res. MEPC.203(62), Adopted on 15 July 2011
Existing Ships	**EEXI** – Energy Efficiency Existing Ships Index	on 1 November 2022	Res. MEPC.333(76) Adopted on 17 June 2021
	CII – Carbon Intensity Indicator	on 1 January 2023	MEPC.352(78);MEPC.353(78) MEPC.338(76);MEPC.354(78) MEPC.355(78)

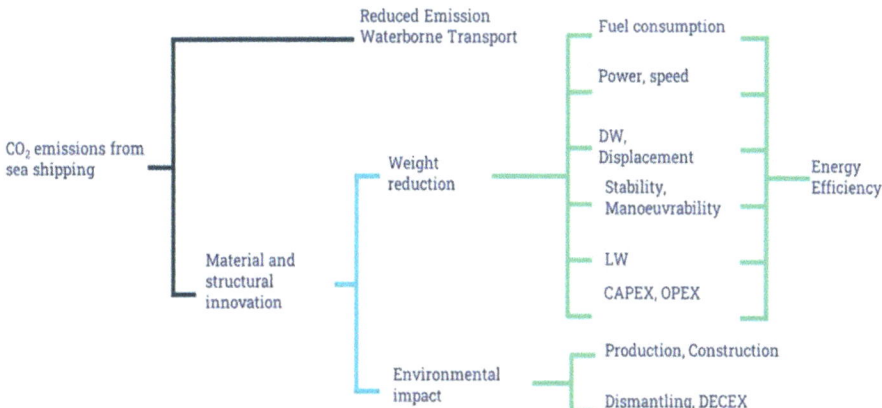

Fig. 1. Key points to address to achieve environmental impact reduction in the shipping industry (Reused from Garbatov, Palomba, et al. (2023b) under CC BY 4.0)

offers several advantages, including approximately 13% reduction in weight and around 11% reduction in ship cost. However, there are no significant improvements observed in terms of the attained EEDI (Energy Efficiency Design Index). The hybrid solution has a higher beta-reliability index of 1.7 compared to the steel solution at the time to the first repair. Additionally, the hybrid solution has a 30% longer period to the first repair and a 55% lower repair cost per year (Garbatov, Palomba, et al., 2023).

A review of green composites (Crupi et al., 2023) presents a selection of bio-based materials that exhibit favourable mechanical and environmental characteristics making them potentially well-suited for use in various marine applications after incorporation of life cycle assessment to the current standards and eco-design methodologies.

The survey analysis (Dolz et al., 2023) indicates that nearly 95% of EU shipyards either presently utilize or have intentions to utilize composite materials. The most used composites consist of fiberglass and polyester resin.

Recently, DNV issued Class Guideline that establishes the necessary requirements and suggestions for constructing sandwich plate systems (SPS) (DNV, 2022).

2.1.2.2. Alternative Fuels

An analysis (Tadros et al., 2023a) of all papers concerning current regulations, available technologies, and future trends in green shipping industry reveals that out of the 6,247 papers, 2,170 titles are dedicated to engine technologies, followed by 776 titles focused on propeller selection and 687 titles focused on alternative fuel.

The databases of scientific literature can provide a significant number of reviews on the use of various alternative fuels such as: Ammonia (Balci et al., 2024); E-fuels (Nemmour et al., 2023) Hydrogen (Martin et al., 2023). The current state of alternative fuels can be judged by the number of ships of different types that are in operation or on order (Table 2) (DNV, 2024).

Table 2. Number of vessels by fuel type

	LNG	Methanol	LPG	H_2 Fuel Cell	H_2 ICE	Ammonia
In operation	509	31	107	1	9	0
On order	524	236	75	20	6	19
Total	1033	267	182	21	15	19

In addition to overview articles, the rules of classification organizations for the introduction of alternative fuels into operation are particularly important. Recently, one can use the rules of classification societies for example for methanol & ethanol fuelled ships (BV, 2022a), ammonia (ABS, 2020b) and hydrogen (ABS, 2021), (BV, 2023b).

2.1.3 Arctic Operation

The Arctic shipping fleet expansion is anticipated in the future. To ensure safe operation of vessels and allow structural weight optimization, reliable prediction techniques are needed in design. Their precision hinges on the accurate representation of ice-ship interactions. These methods are based on fundamental laws of physics and serve as a viable alternative to full-scale ice field trials and experiments for forecasting ice actions on ships navigating in ice-covered waters. It is crucial to validate these methods against experimental results, measurements, and observations from ice field trials to ensure their reliability. Research efforts have been directed toward enhancing understanding of ice-ship structure interactions, predominantly through numerical simulation and experimental methodologies. The following sections outline some notable contributions in recent years.

Accurate estimation of ice loads is necessary before conducting structural analysis. Kim et al. (2021) investigated uncertainty in rule-derived ice loads using an ice mechanics point of view. The focus was on local ice crushing loads and the vessel speeds in the Unified Requirements for Polar Class Ships of the International Association of Classification

Societies (IACS). The results showed that the choice of parameters in IACS ice crushing loads for larger vessels with higher ice classes (PC1) has high uncertainty and the upper limit values are greater than those predicted by the ice mechanic's approach. Lemström et al. (2022) conducted laboratory-scale experiments investigating ice-structure interaction in shallow water. Ten-meter-wide ice sheet was pushed against an inclined structure of the same width. The ice thickness remained constant at about 50 mm across seven experiments. In one series, both compressive and flexural strengths were approximately 50 kPa, while in the other two series, the ice strength was two and four times higher. The loading process exhibited two phases: initially, the ice load on the structure increased linearly at a constant rate for all experiments, followed by a steady-state phase with a nearly constant load. Interestingly, the magnitude of ice loads did not directly correlate with ice strength; weaker ice yielded higher loads than ice twice its strength. Although ice rubble grounded in all experiments, the bottom bore only a fraction of the load. Load records were normalized using a factor combining ice weight and characteristic length. This normalization facilitated the derivation of a model elucidating the loading process: the weight of incoming ice predominates during phase one, while buckling governs the transition to phase two when the ice reaches sufficient strength. Notably, the loading process for the weakest ice differed from the other two types, forming a dense pile of slush instead of distinct ice blocks.

Significant efforts are being directed towards estimating ice loads on offshore structures, including offshore wind farms, and analysing their effects on structural response. Hammer et al. (2023) conducted scale-model tests for offshore wind turbines interacting with drifting sea ice using a real-time hybrid test setup. The experiments aimed to reproduce at scale the interaction of an idling and operational 14 MW turbine with ice representative of 50-year return period Southern Baltic Sea conditions. The effect of turbine operation on the development of ice-induced vibrations was found to be minor at low ice speeds. A new finding is the development of an interaction regime with a strongly amplified non-harmonic first-mode response of the structure, combined with higher modes after moments of global ice failure. The regime develops between speeds where intermittent crushing and frequency lock-in in the second global bending mode develop. Numerical simulations with a phenomenological ice model coupled to a full wind turbine model show that intermittent crushing and the new regime result in the largest bending moments for a large part of the support structure. P. Zou et al. (2024a) studied the dynamic behaviour of offshore wind turbine (OWT) systems under ice-structure-soil inter-action. They employed a coupled approach using cohesive element method (CEM) and finite element method (FEM) to simulate level ice sheets. The Mohr–Coulomb (M-C) model was utilized for soil-structure interaction (SSI), focusing on glacial soils. Their three-dimensional model in LS-DYNA systematically examined the effects of numerous factors such as structure geometry, ice loading conditions, and soil characteristics on ice actions and OWT displacement. Results indicate that conical structures reduce horizontal ice forces, enhancing OWT stability. Variations in soil properties, particularly elastic shear modulus, cohesion, and friction angle, significantly influence OWT dynamics. Glacial soil's elastic shear modulus affects structural displacement, potentially compromising stability, while reduced cohesion and friction angle contribute to greater displacement.

Data-driven models are being increasingly explored as design tools and design load estimators. Meng et al. (2023) developed a recurrent neural network (RNN) model to predict the flexural and uniaxial compressive strengths of sea ice. This model can estimate sea ice mechanical properties solely from physical parameters like sea ice temperature, salinity, density, and loading rate. The average prediction error is less than 20%. Comparative analysis with an empirical formula, back-propagation neural network, and genetic algorithm back-propagation neural network revealed that the optimized RNN is better suited for this prediction task. Q. Sun, J. Chen, et al. (2024) introduced a deep learning model for predicting ice resistance, which involved selecting suitable ship and ice parameters and discussing the impact of various data preprocessing methods on model outcomes. They augmented processed data using Deep Convolutional Generative Adversarial Network (DCGAN) and constructed a Graph Neural Network (GNN) to elucidate network weight distribution and model generalization. Comparisons were made between the established model (GNN-DCGAN) and several other ice resistance prediction methods. Furthermore, considering the dynamic changes in ship design parameters and ice mechanical properties, they conducted a parameter sensitivity analysis using MT UIKKU as a case study. Results indicated that the DCGAN method enhanced prediction accuracy and applicability of the GNN model. Predictions closely matched experimental data, offering valuable insights for ship design in ice-covered waters.

Several notable structural analysis advancements and reviews can be outlined. Mokhtari et al. (2023) introduced an elastoplastic material model for freshwater polycrystalline ice, incorporating pressure and rate-dependent yield criteria. This model was applied to simulate structural damage in a full-scale AA5083-H116 aluminium panel subjected to ice impact in a drop test. Employing the Tsai-Wu yield surface, the model adapts differently in ductile and brittle ice regimes to accommodate the ductile-to-brittle transition. Constitutive laws were implemented using the VUMAT material subroutine for the Abaqus Explicit solver. The mechanical behaviour of the AA5083-H116 alloy was simulated using the Johnson-Cook plasticity model. The simulation results of the ice drop test on the aluminium panel showed excellent agreement with experimental data using the proposed ice material model and the Johnson-Cook model for the alloy. W. Zhang, J. Li, et al. (2022) reviewed ice failure criteria under complex stress state. The theories behind various criteria, experimental verifications and engineering applications are presented. Recommendations on post-yielding behaviours, damage mechanics, multiscale evaluations, etc. are provided. Braun and Ehlers (2022) reviewed high-cycle fatigue behaviour of steel structures under sub-zero conditions, including innovative approaches for local fatigue assessment. The review provides insights into integrating temperature considerations into fatigue design practices. Bobeldijk et al. (2021) devised a technique for determining the technically safe speed for a single ice floe impact on a standard non-ice-strengthened naval vessel. They evaluated three structural configurations for the side shell structure, including a reference design and two alternative Lean Duplex designs, using a representative ice floe impact load scenario. Practical visual and structural acceptance criteria were established. Implementing suggested modifications to the structural layout and materials would enhance resistance to floe impacts at higher speeds. However, even with these enhancements, some permanent deformations are anticipated when such non-ice-strengthened ships navigate ice-infested waters.

Novel approaches for operation of marine vessels in ice-infested waters were proposed. C. Zhang, D. Zhang, et al. (2022) introduced a dynamic ice routing model using time-varying maps. Expanding the Ant Colony Algorithm into three-dimensional space, their heuristic 3D-ACA method accommodates voluntary and involuntary speed changes for ships, effectively addressing dynamic route planning in the Arctic. Safety and fuel saving were integrated as objective functions, acknowledging the importance of risk reduction and carbon oxide emission reduction in Arctic navigation. Utilizing results from a quantified risk assessment model, the multi-objective route planning approach allows for Pareto solutions tailored to end-users' requirements. Analysing the algorithm's key parameters, a case study demonstrated the model's efficacy in route search scenarios, particularly in the Northeast Passage. Focus of another study was decreasing damage accumulation in propulsion machinery during ice navigation. Purcell et al. (2023) explored existing methods and applied them to automatically detect sea ice using high-frequency shaft response measurements, low-frequency propulsion control measurements, and navigation data. These methods include rule-based approaches, machine learning-based classification, and statistical change detection, evaluated based on misclassification errors during cross-validation. Results demonstrate the feasibility of detection using the selected data with satisfactory prediction accuracy for a case-study vessel. Machine learning approaches offer a balanced solution between prediction accuracy and the rate of false negative predictions, with support vector machine classifiers using a radial basis function kernel being the recommended method.

Prominent levels of noise during ice breaking can negatively affect the crew. Shi et al. (2024) conducted cabin noise analysis of polar transport vessel during continuous icebreaking, accounting for ship-ice-water-air coupling. They numerically studied ice loads, compared resistance values using experimental and empirical formulas, and analysed sound source load excitation. Frequency domain acoustic analysis was performed using acoustic structure coupling and statistical energy analysis models. Transient cabin noise effects on personnel comfort were assessed according to ship noise standards. Results indicated that transient cabin noise disperses from the collision position to the stern and attenuates with increasing frequency. Icebreaking excitation mainly affects noise below 1000 Hz, exhibiting strong nonlinearity at low frequencies. Sound source excitation notably impacts middle to high-frequency bands. A-weighted sound pressure levels in accommodation cabins and wheelhouses exceeded standard limits by 5–20 dB(A) during icebreaking, significantly affecting onboard comfort. These findings offer insights for studying ship-ice collision-induced transient cabin noise and developing relevant standards.

2.2 Design Methods

2.2.1 Introduction to Design Methods

Following the work of the preceding ISSC IV.2 Committee the review is mostly oriented toward the future aspects of ship structural design methodologies including:

- Design for X, where "X" represents a specific goal such as operability, environment, or production; and

- Risk Based Design, as a specific type of Design for X that is widely used as part of a goal-based design approach.

However, an additional category of Autonomous Ships has been added, as this seems to be an important topic moving forwards.

The first aspect is covered in Subsect. 2.2.2, Structural Design for X, while Subsect. 2.2.3 covers the Risk Based Design and Subsect. 2.2.4 covers the Autonomous Ships.

2.2.2 Structural Design for X

The 'Design for X' methodology represents a multi-dimensional approach to ship design, with the aim of concurrently optimizing various objectives. This simultaneous optimization is a key feature of the methodology, giving priority, for instance, to safety measures to safeguard the life, property and environment, improving efficiency to reduce fuel consumption and environmental impact, considering the entire lifecycle of the vessel to minimize costs and environmental footprint, streamlining operational processes to enhance performance, and optimizing production methods to ensure the construction of a high-quality and reliable ship.

Erikstad and Lagemann (2022) discuss recent progress in marine systems design methodologies, outlining four critical pathways: model-based systems engineering, set-based strategies, holistic optimisation, and modular architectures with configuration-based design approaches. They emphasise the importance of design-for-sustainability, operational simulation, and the ability to manage uncertainties with flexibility as crucial focus areas. Additionally, they highlight the potential of emerging technologies, such as artificial intelligence, machine learning, and digital twin technologies, while noting that further development is necessary for these technologies to impact the industry significantly.

Kondratenko et al. (2023) discuss a framework for holistic and sustainable design optimisation of Arctic ships, addressing challenges in shipping in icy conditions and promoting eco-friendly energy sources. Tadros et al. (2023b) systematically analyse decision support methods to optimise ship hulls for enhanced energy efficiency, offering practical tools for ship designers. They emphasise the significance of leveraging approaches such as operational research and machine learning across four research domains: hull form, structure, cleaning, and lubrication. These methodologies are designed to advance maritime decarbonisation efforts. Trivyza et al. (2018) propose a method to optimise ship energy systems, balancing economic and environmental goals. The approach integrates economic and environmental objectives using the NSGA-II algorithm to visualise trade-offs through a Pareto front. It considers innovative technologies like fuel cells and carbon capture, highlighting the need to assess trade-offs between environmental benefits and economic viability. This approach ensures a more realistic assessment of energy system performance over the ship's lifecycle than traditional design-point approaches. Nikolopoulos and Boulougouris (2020) propose a simulation-driven methodology that offers a practical solution for optimising ship design under uncertain conditions. This approach, which considers dynamic market and environmental factors, enhances life-cycle performance and efficiency, making it a valuable tool for addressing real-world challenges in ship design.

Sassanelli et al. (2020) examine how Design for X approaches can be incorporated into the Circular Economy (CE) framework. They identify key objectives and research gaps and propose methods for improving product design. The research primarily focuses on theoretical evaluations, indicating that this area is specialised but growing. Sullivan et al. (2021) develop a customisable lean design methodology for the maritime industry, integrating IoT and real-time data to enhance design efficiency, sustainability, and vessel performance. The Customizable Lean Design Methodology (LDM) enables better decision-making through a structured approach involving real-time data and user feedback. Pereira and Garbatov (2022) present a multi-attribute decision-making procedure for optimising ship structural design. The method focuses on risk, cost, and efficiency to achieve an optimum design solution for safety and environmental impact. It employs multi-attribute decision-making techniques, particularly the TOPSIS method, to evaluate design alternatives. Zaraphonitis et al. (2021) examine hydrodynamic optimisation strategies for a container ship and a bulk carrier. They focus on making trim adjustments and retrofitting bulbous bows to improve fuel efficiency and decrease greenhouse gas emissions. As a result of these optimisation efforts, the propulsion power required for both types of vessels was significantly reduced, particularly after the bulbous bow modifications. Papanikolaou et al. (2022) discuss a holistic ship design approach that integrates various design tools to enhance efficiency and sustainability. This approach emphasises multi-objective optimisation, facilitating exploring a vast design space through parametric modelling.

2.2.3 Risk-Based Design

Risk-based design integrates risk and reliability analysis. It is increasingly being used to support novel designs and to prepare for modern technologies. It allows for designs that are outside of the prescriptive constraints with an equivalent safety level by providing evidence of the structure's safety equivalence. In this approach, safety requirements are no longer constraints but are defined as design objectives. In a risk-based design approach, the first step is to set the design safety goals. Hazards are determined and critical design scenarios are generated. Functional requirements can then be generated, and the risks evaluated. The derived safety levels can then be compared to a traditional design. Some excellent examples of the benefits of this approach are given in Papanikolaou (2009) and Garbatov and Sisci (2018). In these examples several potential benefits are assessed, some of which are focused on structural improvements, but others relate to the operating profit or environmental impact of the design. For example, in Papanikolaou (2009) the analysis of an onion shaped structural arrangement allows for an improvement in revenue without the need to reduce safety, but this would be outside the prescriptive design rules.

Several IACS classification societies have published guidelines on how to perform a risk-based design these include Lloyd's Register, ClassNK, Bureau Veritas and DNV. These elements focus on technical aspects of risk, with limited inclusion of human factors (Ventikos et al., 2021). Ito (2022) highlights the importance of storage and sharing of data alongside Uncertainty Quantification. High quality data is needed for the analysis in the risk-based design process. The data needs to be captured and stored for use in new analysis. Historically much of the data was stored with too little detail to inform this stage and it has been reliant on the opinions of experts, which is expensive and can be prone to bias. Uncertainty Quantification is also an important part of the risk-based design process, as shipping is inherently stochastic, to the weather that a vessel will encounter and to the loading of the vessel. It's important for decision makers not to just have the potential quantification of the risk, but also to understand the confidence in that calculation when making design decisions. Increasing quantities of data being produced, through pervasive sensors, AIS (Automatic Identification System) and oceanographic conditions, and the ability to store and analyse this data, will provide increased confidence in the risk-based design approach moving forwards.

2.2.4 Autonomous Ships

Autonomous ships are a growing area of importance. IMO is developing a MASS (Marine Maritime Autonomous Surface Ships) code (MSC, 2023) that will be non-mandatory in 2025, and this will form the basis for a goal-based MASS code in 2028. Currently some autonomous and remote-controlled ships are being travelled in some sea areas. It is predicted that the initial use for these vessels will be short voyages (IMO, 2024a). Lloyd's Register has defined 6 levels of autonomy.

- AL0 – manual with no autonomous function,
- AL1 – on-Ship decision support,
- AL2 – on and off- ship decision support,
- AL3 – 'Active' human in the loop,
- AL4 – human on the loop – operator/supervisory,
- AL5 – fully Autonomous (& rarely supervised),
- AL6 – fully autonomous (& with no supervision).

At the current state of the maritime industry, it is likely that from a structural standpoint that the structural design of a ship will need to have limited changes until AL6 with humans likely to be required on-board until that point although with remote operation this could be as early as AL4. As the vessels become more autonomous and the size of the vessel grows, it will become important to consider both changes to the structural design of the vessel and to consider how the ship can continue to be maintained. Ventikos et al. (2020) perform an analysis of the hazards of a system with respect to changes in the autonomy level. The results show that as the autonomous system is given more authority, the number of mitigation measures decreases. This is because the system can be designed to reduce the number of hazards. However, after an accident there were no mitigation measures that could be used to reduce damage. There is increasing published guidelines in this domain: BV (2017), ClassNK (2018) and DNV-GL (2018). These refer to risk assessment for proof of a safe design as there is limited operational experience to draw upon in this domain and empirical adjustments are not yet possible.

Alongside these new technical challenges, will be new legislative issues. A key area will focus on responsibility and liability, with Maalouf and Norsetter (2024) illustrating the concept with the notion of seaworthiness. In the case of autonomous vessel, it can be considered whether the responsibility for seaworthiness shifts to the manufacturer from the shipowner, as they might struggle to guarantee this. Seaworthiness is often guaranteed in contracts and therefore raises issues for liability. A shipowner is currently liable for any damage in operation if they fail to use due diligence if they fail to make the ship worthy before a voyage. This concept is also relevant in terms of insurance. The shipowner has no cover in an instance where there is a knowingly unseaworthy vessel. However, a shipowner might not be responsible for faults made before the ship came into their owner's possession. It might be possible to claim that a defect in an autonomous system occurred before they took possession and therefore transfer liability to the manufacturer or software developer. In a structural sense, this will make it difficult to diagnose the cause of a specific incident, due to the increased complexity and larger number of suppliers that will be required to ensure fault detection and maintenance occur. Human error incidences will be replaced by faults in the underlying software, error in the data processing, sensor breakdowns or remote operator issues.

2.3 Digitalization

Future ship designs will need to meet demanding environmental standards with the inclusion of new fuels and a greater rate of change to the design and operation of ships than the industry has traditionally been used to. To safely meet these new innovations and technologies will be challenging. Models, simulations and monitoring tools will therefore likely have a place in ensuring the safety of ship structures. Digitalization is therefore considered explicitly for the first time, due to its increasing significance. Current topics of interest include Generative AI, Digital Twins, Pervasive Sensing, Cloud Computing and Machine Vision. Many of these techniques are finding their way into academic literature. However, there are still a few issues with putting them into practice. Gartner Inc. predicted that across all sectors that half of CIOs (Chief Information Officers) are planning an AI project but that 85% of these projects will fail (Gartner, 2018) this is almost double the corporate IT failures from a decade before (Bojinov, 2023).

To collect the data novel sensors will allow the collection of data at scale. There needs to be an increased focus on how to efficiently and effectively store and retrieve this data. The data needs to be fed back into the design process to ensure safety and will support the move to-wards decarbonization, as revolutionary new structures will be required to support the movement to net-zero fuels. There will need to be an increase in the density of the sensors being installed, with current sensors on large ships normally measuring: vertical accelerations at the bow, transverse accelerations at amidships, ship motions at the centre of gravity, global longitudinal stress amidships, global longitudinal stress amidships (port and starboard side), global longitudinal stress at quarter length fore and aft of midship (port or starboard side), local transverse stress at transverse deck strip amidships, global longitudinal stress below neutral axis amidships (port and starboard), double bottom bending stress, bending/shear stress in pillar bulkheads, lateral loads at bow flare or bottom near forward perpendicular (slamming pressure) and lateral loads at side shell (wave pressure). However, there is a lack of understanding of the response at a local level, which will be required on autonomous vessels without crews and on new fuel storage structures (ABS, 2020a).

To analyse this data, engineers will need to use statistics, scientific computing and scientific methods to extract knowledge and insights from data. These approaches have a long-standing history of use in Engineering and beyond and there is a growing literature in ship structures. These tools are often focused on regression analysis. This has been supported with a move towards physics-based Machine Learning, to account for the generally small datasets and a bias in our analysis away from areas of interest, as we often have limited data of extreme events. The focus of this physics-based Machine Learning has been predominantly towards fluids and energy systems, but there is increasing literature focused on structural analysis. For example, in a recent review in Civil Engineering there are 23 examples of CFD/Turbulence modelling using physics-based Machine Learning and 10 in structural mechanics/Structural Health Monitoring (Vadyala et al., 2022).

To interpret these tools requires an ability to define the uncertainty in a scenario. An increased knowledge of Bayesian approaches and how to derive confidence intervals is required at a higher level. However, uncertainties obtained from in-service measurements are rarely published, (Hageman et al., 2022).

Unsupervised learning is when the machine learning tools are given unlabelled data with the aim to infer the structure of the data. There are three main approaches: clustering, association and dimensionality reduction. Clustering breaks the data into groups and allows a determination of what class a new data point fit to. Dimensionality reduction is used to reduce the number of features, or dimensions, in a dataset and to understand which features are most important.

Moving forwards, we are likely to see more Artificial Intelligence approaches, the ability for a computer to perform tasks at an equivalent level to a human. Much of the focus has been on deep learning where we are using larger networks on bigger datasets than we were previously capable of. This includes Machine Vision, Large Language Models, Natural Language Processing and Speech Recognition which will support users to interact with AI tools, recognise damage scenarios in predictive maintenance (Y. Zou, K. Zhang, et al., 2024) using generative AI to design structures (Ao et al., 2023), structural health monitoring (C. Sun et al., 2024) or summarise large documents (Uddin et al., 2022).

In reinforcement learning we use a reward system to teach a learner. We release an agent into an environment; it sees an observation and selects an action based on this observation. The action then leads to a reward based on the outcome and this is used to train a policy. In ship structures we are already seeing these techniques starting to be used in scheduling of maintenance (J. Cheng et al., 2022a, 2022). Going beyond this Strong AI, or Artificial General Intelligence, will allow our AI to learn many different tasks. However, it is likely that we are many years away from these approaches and these are unlikely to impact on structural design, operations and end-of-life for some years.

However, among the many benefits of digitalization will be new challenges. The Maritime Industry is one in which profit margins are tight and companies have traditionally shied away from being development leaders. Especially in ship structures, there are seen to be more limited benefits with requirements for ship surveys still being necessary. This limits the financial rewards. In addition to this, there are issues around the generation, ownership and sharing of data. In many scenarios the individual shipping companies can't afford to generate the quantity of data required to provide improved analysis. With more stored data and greater connectivity there are also a growing number of potential security risks.

2.4 Conclusions

There are increasing pressures on the structural design of ships. These come in the form of new operating environments, for example the arctic and climate change, using new fuels and through increased pressure to digitalize. IMO are introducing regulation changes within the industry to reduce emissions; in the short term this will require minimal changes to structural design and operation but in the long-term there will need to be substantial differences to the design of ships to contain these new fuels safely. Autonomous vessels will also provide new challenges. In the short-term, we will see limited changes to the design of vessels, but as we move through the levels of autonomy, and the removal of humans, it is likely that this will drive a substantial change to the loads that the vessel experiences and to the structural design. It will also complicate the monitoring and maintenance of these structures. To support a safe transition, we will need to look to Risk-Based Design approaches, which will allow us to assess an equivalent safety for these new vessel types. We can also use new digital approaches, where new sensors will allow us to better understand vessels in real time and where Machine Learning and AI will allow us to monitor the health of these structures.

3 Regulatory Framework for Ships and Offshore Structures

Rule and regulation are in general governing ship and offshore structure's design principle and criteria and those are also influencing on design method. This chapter is to introduce the latest update from regulatory framework including relevant ship and offshore structure standard to investigate the link to design principles and criteria.

3.1 Introduction (Framework)

IMO provides regulations which is prominent level of design principle to ensure the safety of human, property and environment. In general, International Association of Classification Societies (IACS) specifies the requirements as rules to follow the principle of goal how to fulfil the regulation given from IMO. It is researched that the latest discussion in Maritime Safety Committee (MSC) of IMO within 2021 to 2024 and follow up the update in IACS Unified Requirement (UR) how the discussion in MSC is adopted into. In addition, this chapter will include the significant changes in individual classification societies rules. The relevant changes in offshore structure rules and update in international standard will be also discussed.

3.2 *IMO Regulations and Goal Based Standards* (IACS Update)

3.2.1 Update from IMO Regulation

IMO's Maritime Safety Committee (MSC) had held 104th to 108th session from 2021 to 2024. The major topics related to ship structure safety and design principle are as shown in Table 3. (See IMO (2021, 2022b, 2022c, 2023c, 2024b) for details.)

Table 3. Summary of technical topics related to ship structure design in IMO MSC's sessions

Topics	Relevant regulation/code	Enter into force
104th session (Oct. 2021) Many proposals for new work items were not considered due to time constraints during the remote meeting as it was held within COVID pandemic. Thus, no structure related items had been discussed.		
105th session (Apr. 2022) Topics were focused mainly on adopts amendments to the 2024 update of SOLAS and related mandatory codes.		
Design, construction, inspection and testing of portable tanks with shells made of fibre-reinforced plastics materials		
106th Session		
Use of high manganese austenitic steel for type A, B and C tanks for among others butane and methane (LNG) as cargo or fuel	IGC and IGF codes	1st/Jan./2026
Harmonization of technical/operational requirement for watertight doors on cargo ships	MARPOL Annex I, Load line, IBC code, IGC code, SOLAS	1st/Jul./2024
107th session (Jun. 2023)		
Onboard lifting appliances and anchor handling winches	SOLAS II-1/3-13	1st/Jan./2026
Goal-based new ship construction standard	MSC.454(100)	1st/Jan./2025
Emergency towing equipment	SOLAS II-1/3-4	1st/Jan./2028 subject to adoption by MSC
Unified interpretation to SOLAS Ch. II-1	SOLAS II-1, MSC.1/Circ.1362/Rev.1	
108th session (May 2024)		
Emergency towing equipment to all ships over 20,000 DWT	SOLAS II-1/3-4	1st/Jan./2026
High-manganese austenitic steel	MSC.1/Circ.1599/Rev.2, MSC.1/Circ.1622	

3.2.2 Goal Based Standard

As mentioned in Sect. 3.2.1, goal based standard target has been discussed through several MSC session and finally MSC.454(100) (IMO, 2018) has been revised in 107th session. Goal-based standards (GBS) for new ship construction of bulk carriers and oil tankers are, conceptually, the IMO's rules for class rules. Under GBS, IMO auditors use guidelines to verify the construction rules for bulk carriers and oil tankers of class societies acting as Recognized Organizations (Resolution MSC.454(100)), which is to consider experience gained with their application and the recommendations made by the GBS audit and to support their implementation.

3.3 IACS and Classification Rules

3.3.1 IACS Technical Work and Update in Unified Requirement

This sub-chapter includes the major technical work and update of Unified Requirement/Interpretation in IACS as well as CSR BC&OT from 2021 to 2024.

3.3.2 IACS Technical Work in Annual Review

• The crest of accurate wave data (IACS, 2019)

The report explained the background of study that IACS Rec.34 (rev.1) was questioned from GBS audit on CSR since huge progress has been made over the two decades regarding several wave data by available sources including altimetry (measured from satellite), hindcast model (re-analysis of past weather) and wave buoys (see Fig. 2). Hence, IACS has started investigation to improve IACS Rec.34.

Extreme wave height is shown as similar level between sources, however the overall shape of IACS Rec.34(rev.1) sample singular such as wave period and its dependence. The study was focusing on the optimum distribution corresponding to wavelength and height. Actual measurement from buoys and satellites was used for calibration of scatter diagram. This study has been basis for updating IACS Rec.34 (rev.2) which has been published in Dec.2022 (IACS, 2022b). It mentioned that IACS is expected to result in more accurate rule loads and improved standardisation of safety levels of the fleet.

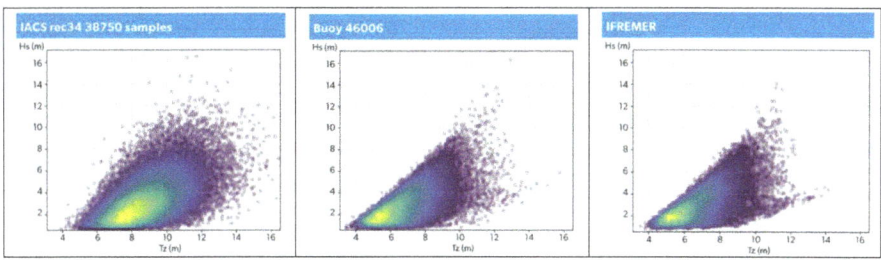

Fig. 2. Hs/Tz scatter plot from various sources (IACS, 2019)

- Monitoring the success of corrosion work (IACS, 2020)

    For the corrosion protection of ballast tanks of bulk carriers and oil tankers, historically IMO issued several resolutions to enhance safety level for the last 30 years, these include the Enhanced Survey Programme (ESP) as well as SOLAS regulations for coating of ballast tanks.

    IACS has conducted three comprehensive statistical analyses of corrosion data obtained from the thickness measurements from tankers and bulk carriers. Statistic results have obtained for oil tankers and bulk carrier by 3 different categories, e.g. phase 1 to 3, as shown Table 4.

Table 4. Corrosion diminution in mm (both sides) of 90% cumulative probability, for web frames in ballast water side tanks – application of ESP and coating requirements for ballast water tanks are indicated

Vessel type		Phase 1	Phase 2	Phase 3
Bulk carrier		2.3 mm	0.82 mm	0.53 mm
Oil tanker		1.1 mm	0.68 mm	0.5 mm
Requirements	Enhanced Survey Program (ESP)	No	Yes	Yes
	Ballast tank coating requirement	No	Yes	Yes

In summary, the typical corrosion in BW tanks significantly reduced between Phase 1 and Phase 2. This concludes ESP implementation following improved and more systematic survey regime has enormous influence on condition of BW tank especially in bulk carrier in operation. When comparing the results from Phase 2 and Phase 3, there is clear reduction in the corrosion level, which means new coating standard (PSPC) has improvement in protection of BW tanks. The report mentioned that an outcome of this comprehensive corrosion analyses will be basis for updating and developing technical background for corrosion margins used in CSR, and corrosion additions for CSR may be re-considered in the future.

- Buckling strength updates (IACS, 2021)

    Buckling strength assessment method is important to ensure ship structure's safety. However, many different assessment methods have been used by class societies and industries. This report has introduced the history and background of IACS Unified Requirement (UR), UR35 Buckling which is based on the current closed form method (CFM) used in CSR BC&OT. This is intended to use a unified buckling capacity toolbox for the design and approval of non-CSR ships as well as CSR ships. The main improvements on buckling capacity methods have reflected based on non-linear finite element analysis (NLFEA). This has been also improved with change of stiffener buckling formula. As a result, the new IACS UR S35 has been published in Feb. 2023 and planned to come into force on 1st/July/2024.

- Improved and enriched wave data (IACS, 2022a)

    IACS Recommendation No.34 Standard Wave Data (Rec.34) is used as a basis for the longitudinal strength of almost all the world's commercial ships as well as for all dynamic loads/motions in the IACS CSR BS&OT. It is also a commonly used reference for direct wave load analysis of ships. As IACS received comments on Rec.34 Rev.1(2001) related to the underlying statistical data which were demonstrated inaccuracies in human estimates, embedded the effect of weather avoidance and the last observation included dated back to 1984, it was necessary to update Rec.34 with reliable statistics. As explained above IACS Annual Report 2019, GBS requests evidence that the wave data used in the rules (Rec.34 Rev.1) accurately represent North Atlantic conditions, inclusive of the possible effects of climate change. An updated scatter diagram of wave height and wave period has been published in Rec.34 Rev.2 (IACS, 2022b).

This report explains the reflection of wave encounters on ships in North Atlantic is to consider realistic combinations of routes and wave data. The study was done from voyages of over 20,000 vessels of AIS data to the same temporal resolution as the hindcast wave data.

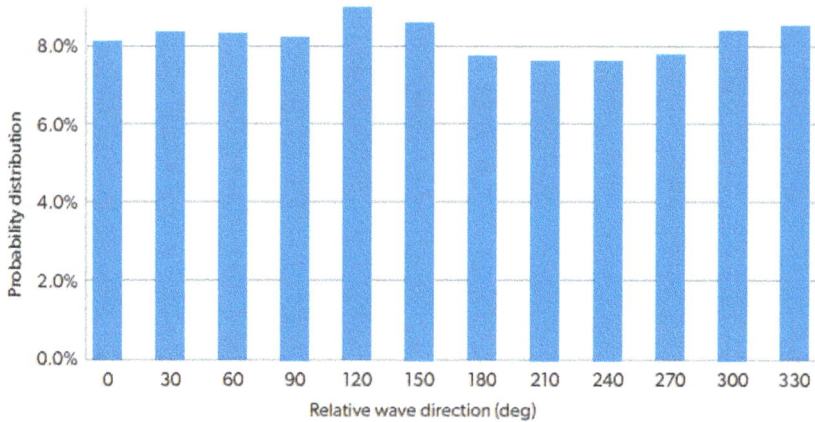

Fig. 3. Heading histogram, all data (IACS, 2022a)

Figures. 3 and 4 present two factors able to be explained.

1. Ships' captains avoid beam seas in harsh weather, to limit roll motion and to avoid stability problems.
2. Harsh weather happens in locations where routes are mostly east-west, with the dominant wave direction from the west.

When the relationship between ship speed and heading has been investigated, the two most plausible reasons are:

1. Voluntary speed reduction to limit ship motions.
2. Involuntary speed reduction due to added resistance in waves.

IACS is considering continuing investigation that all wave loads, e.g. hull girder loads, sea pressures and acceleration will be checked in CSR. Comprehensive technical background for IACS Rec. 34 is available on the IACS home page IACS(HF&TB) (2023) and it may result the relevant rule changes in IACS UR and CSR.

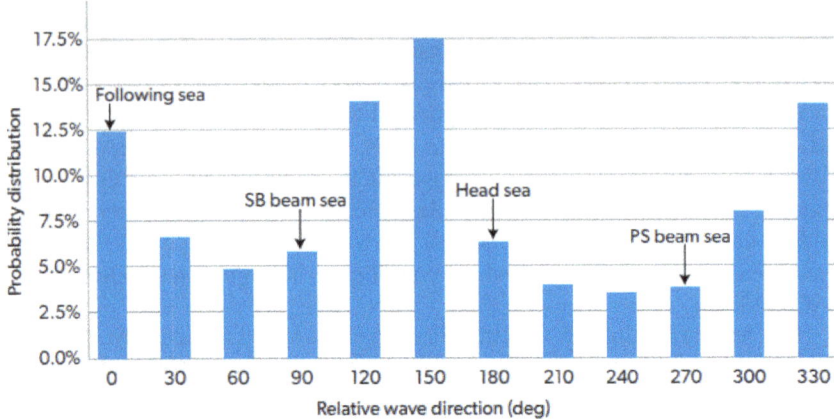

Fig. 4. Heading histogram, Hs > 10 m (IACS, 2022a)

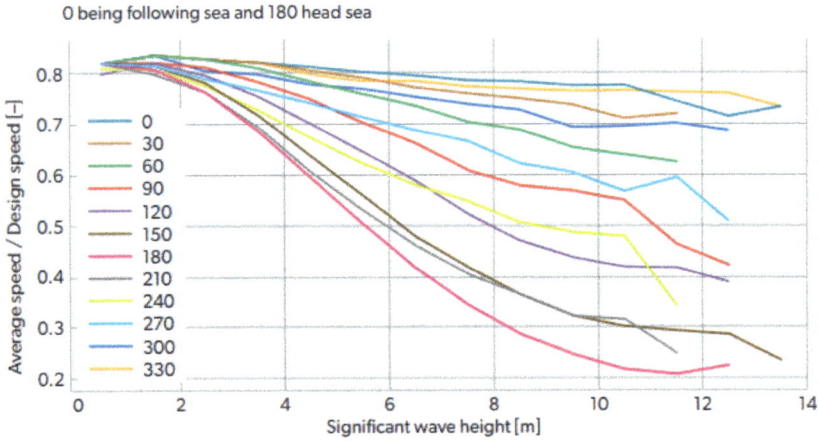

Fig. 5. Average ship speed as function of Hs and relative wave heading (IACS, 2022a)

- Preventing the loss of seaborne containers (IACS, 2023a)

  A significant increase in the number of transported containers has reaffirmed concerns about the loss of containers, despite considerable efforts and technical attention being paid to this issue. IACS assessed the safety issues and research has been taken with respect to operation, design and maintenance. It is noted that IACS is developing UR C6, C7 and C8 which are planned for completion in 2024 with classification societies' implementation in 2025. This will be intended to harmonize the design loads on container stacks and will also provide strength assessment methodology including possibility of unification of calculation procedure and acceptance criteria.

3.3.2.1. New/Revisions to IACS Unified Requirements

Table 5 includes the list of rule changes in UR for structure and welding requirements (IACS(UR), 2023).

3.3.3 IACS CSR BC&OT Rule Update

In Rule Change Notice (RCN) to CSR 1 Jan 2022 (IACS(RCN), 2022), the ship length used in CSR (rule length) is harmonised with that in GBS (freeboard length); buckling strength requirements are enhanced; upper limit of intermittent weld leg length and corrosion application applied to superstructure are defined. Application of rules requirements and associated loads for various structural elements is also defined.

In RCN 2023 to CSR (IACS(RCN), 2023), rule changes are made to clarify the method to get the number of EPP load points, the application of span and the criteria for distributed loads for steel coil calculation; changes are also made, among others, for the bilge plating minimum requirement, hot spot location and stress calculation in way of corrugated bulkhead to lower stool connection, etc. fThe Table 6 summarizes IACS CSR BC&OT Rule update.

3.3.4 Major Changes in Classification Rules

The major changes and updates of classification societies' rules are researched between 2022 and 2024. As these rules are being updated every year, the background and motivation of rule changes in each classification society are focused. Since there are many changes in IACS rules, e.g. UR and CSR as summarized in Sects. 3.3.2.1 and 3.3.3, the rule update in classification societies' rules had been commonly considered to reflect them. The following is the summary of major rule changes for ship hull structure from each classification societies among IACS members. This focuses on major update in rule principles such as the equivalent design wave concept for prescriptive loads and criteria corresponding to goal-based standard.

ABS has combined five rules set into new two sets to simplify in Jan. 2020 as Rules for building and classing Marine Vessels and Rules for building and classing Mobile Offshore Units. The goal-based standards and necessary clarification have been implemented in rule edition Jul. 2024 which describes the definition of goals and functional requirements (ABS, 2024). The necessary goals and functional requirements are defined corresponding to each structure element in existing rule requirements how to utilize and achieve the goals depending on the scope (Table 7 and Table 8).

BV had a full revision of steel ship structure rule NR467 in July 2022 (BV, 2022b). This had been considered to adopt goal-based standard. New equivalent design wave load concept and capacity formula similar with IACS CSR BC&OT had been introduced in this edition. The latest update in 2024 is for updating standard wave data in accordance with IACS Rec. 34 update (BV, 2024).

CCS had maintenance modifications for adjustment of requirements between the years 2022 and 2024 (CCS, 2024). Those are mostly related to dry cargo ships such as bulk carriers, container ships and car carriers.

Table 5. Summary of revision in IACS Unified Requirements between 2021 and 2023

Resolution no.	Revision	Adoption	Title	Implementation Date
2021				
UR W1	Rev.4	Apr 2021	Material and welding for ships carrying liquefied gases in bulk and ships using gases or other low-flashpoint fuels	01 Jul 22
UR W2	Rev.3	Sep 2021	Test specimens and mechanical testing procedures for materials	01 Jan 23
UR W13	Rev.7	Sep 2021	Thickness tolerances of steel plates and wide flats	01 Jan 23
UR W14	Rev.3	Sep 2021	Steel plates and wide flats with specified minimum through thickness properties ("Z" quality)	01 Jan 23
UR W17	Rev.6	Sep 2021	Approval of consumables for welding normal and higher strength hull structural steels	01 Jan 23
UR W18	Rev.6	Sep 2021	Anchor chain cables and accessories including chafing chain for emergency towing arrangements	01 Jan 23
UR W25	Rev.6	Sep 2021	Aluminium alloys for hull construction and marine structures	01 Jan 23
UR W26	Rev.2	Sep 2021	Requirements for welding consumables for aluminium alloys	01 Jan 23
CSR 2021	URCN 1	Aug 2021	Urgent rule change notice 1 to CSR 01 Jan 2021 version	01 Jan 22
CSR 2021	RCN 1	Dec 2021	Rule change notice 1 to CSR 01 Jan 2021 version	01 Jul 22
2022				
UR S14	Rev 7	Dec 2022	Testing procedures of watertight compartments	

(*continued*)

Committee IV.1: Design Methods, Principles and Criteria 751

Table 5. (*continued*)

Resolution no.	Revision	Adoption	Title	Implementation Date
2023				
UR S21	Rev 6	Jan 2023	Requirements concerning strength of Ships	01 Jul 24
UR S21A	Deleted	Jan 2023	Requirements concerning strength of Ships	01 Jul 24
UR S35	New	Feb 2023	Buckling Strength Assessment of Ship Structured Elements	01 Jul 24
UR S10	Rev 7	Feb 2023	Rudders, Side Pieces and Rudder Horns	01 Jul 24
UR W31	Rev 3	Mar 2023	YP47 Steels and Brittle Crack Arrest Steels	01 Jul 24
UR S26	Rev 5	May 2023	Strength and Securing of Small Hatches on the Exposed Fore Deck	01 Jul 24
UR A1	Rev 8	Jun 2023	Anchoring Equipment	01 Jul 24
UR S3	Rev 2	Jun 2023	Strength of End Bulkheads of Superstructures and Deckhouses	01 Jul 24

DNV had minor updates for maintenance purpose including clarification and editorial changes between year 2022 and 2024 as the principle of rules has not been changed since a full revision according to goal-based standard was published in Oct. 2015 (DNV, 2015).

IRS (Indian Register of Shipping) completely enhanced its rules for steel ships in July 2024 (IRS, 2024). The newer version of rules is based on feedback from plan approval and survey functions and follows net scantling approach, EDW loads concept with specific chapter on fatigue and finite element analysis requirements. Also, NovaHULL rule-based scantling calculation software that helps users in verifying the compliance of ship structures against the July 2024 version released for industry use.

KR had no main change in general hull structure rules but new equivalent design wave and capacity formula similar rule concept with IACS OT&BC had been adopted for container ship rules separately (KR, 2023a). Similarly in 2021 the first edition of separate hull structure rules for membrane LNG carrier had been published (KR, 2023b). It is noticed that special attention has been paid on guidelines for ballast loading conditions of cargo vessels involving partially filled ballast tank in 2022. There are also updates of load correction in structural rules for container ships and membrane LNG carrier between 2022 and 2024.

LR's rules have several changes for maintenance both reflecting industries feedback and IACS implementation, but this is similar level with other class societies. However

Table 6. Summary of revision in IACS CSR BC&OT between 2021 and 2023

Resolution no.	Revision	Adoption	Title	Implementation Date
CSR 2021	URCN 1	Aug 2021	Urgent rule change notice 1 to CSR 01 Jan 2021 version	01 Jan 22
CSR 2021	RCN 1	Dec 2021	Rule change notice 1 to CSR 01 Jan 2021 version	01 Jul 22
CSR 2021	Corr.1	Jan 2022	Corrigenda 1 to CSR 01 Jan 2021 Version	01 Jul 21
CSR 2022	2022RCN1	Dec 2022	IACS CSR for bulk carrier and oil tankers	01 Jul 23
CSR 2023	2023		Common Structural Rules – Consolidated 01 Jan 2023	01 Jul 23
CSR 2023	2023 RCN1	Dec 2023	IACS CSR for Bulk Carriers and Oil Tankers	01 Jul 24

Table 7. Goals for structural requirements (ABS, 2024)

Goal No.	Goals
STRU 1	In the intact condition, have sufficient structural strength to withstand the environmental conditions, loading conditions, and operational loads anticipated during the design life.
STRU 2	Resist structural failure associated with accidental conditions.
STRU 3	Provide protection to persons on board, the environment and required safety services.
STRU 3.1	Maintain mechanical properties during extreme temperatures.
MAT 1	The selected materials' physical, mechanical, and chemical properties are to meet the design requirements appropriate for the application, operating conditions, and environment.

major changes are found in class notation ShipRight procedure (LR, 2023) for strength calculation tailored in each ship type, especially the effort is paid continuously in gas carrier and container ships between 2022 and 2024.

ClassNK has done comprehensive rule revision in July 2022 (ClassNK, 2022). In this revision, the equivalent design wave load concept and capacity formula similar with IACS CSR OT&BC has been adopted for direct strength assessment by cargo hold finite element analysis.

Table 8. Functional requirements (ABS, 2024)

Functional Requirement No.	Functional Requirements
Structure (STRU)	
STRU-FR1	Scantlings are to have sufficient strength to resist failure and excessive deformation associated with buckling and yielding when subjected to the loads anticipated throughout the service life, including hull girder loads, deck cargo loads, green water loads, hydrostatic loads (tanks and watertight boundaries), etc.
STRU-FR2	Limit crack propagation from the deck to side shell, side shell to bottom, and vice versa.
STRU-FR3	Limit crack propagation at locations susceptible to fractures.
STRU-FR4	Prevent buckling of structural members.
STRU-FR5	The design of the termination of structural members is to prevent yielding or fatigue cracks.
STRU-FR6	Avoid discontinuities, misalignment, and other harmful details that may lead to failure in service.
STRU-FR7	Design bolted connections to transmit load between structural elements
STRU-FR8	Provide sound quality for all workmanship and structural fabrication.
Materials (MAT)	
MAT-FR1	Provide materials with properties and quality appropriate for the location in the hull based on the criticality of the structure and thickness of material.
MAT-FR2	Provide higher toughness materials in locations subject to low temperatures to prevent brittle fractures.
MAT-FR3	Avoid galvanic corrosion due to dissimilar metals.

3.4 Offshore Structures

Offshore wind energy, as one of the most credible sources for increasing renewable energy production, has the potential to contribute to achieving the United Nations Sustainable Development Goal (SDG) 7, Affordable and Clean Energy (Galparsoro et al., 2022).

G. Wang et al. (2022) compared the offshore wind turbine design requirements between International Electromechanical Commission (IEC), classification societies and institutes, as shown in Table 9. The IEC requirements (IEC 61400-3-1:2019, IEC TS 61400-3-2:2019) start from onshore wind practice, and cover turbine, floater, station keeping, and dynamic cable for floating offshore wind turbine (FOWT). Classification society requirements and guidance are also trusted and used extensively in FOWT design and analyses.

3.5 Other Special Topics

Since CII (Carbon Intensity Indicator) has regulated to apply all ships above 5,000 GT of all cargo, Ro Pax and cruise ships from 1st Jan. 2023, the ship owners have strong demand to improve their ships' efficiency. One of idea to improve energy efficiency is to be assisted by wind power. Most classification societies have developed specific guidelines for design, installation and operation of WAPS (Wind Assisted Propulsion System).

There are a lot of discussion of future alternative fuels. Lately strong interest on ammonia, methanol and hydrogen has been increased rather than LNG or LPG as fuel in long term view. In addition, the interest of liquefied CO_2 is being increased for carbon capture storage or transportation. Design of such tanks have higher technical challenge as IMO IGC code has lot of limitation on those liquids, for instance material yield strength limitation 410 N/mm^2, whilst industries are seeking higher strength material for enlarging tank volume (IMO, 2022a). The material issue is one of specific area to overcome for this decarbonization topic.

Table 9: Comparison of design requirements between IEC and Classes/Institutes (G. Wang et al., 2022)

	IEC	DNV	ABS	API* (for O&G)	NORSOK * (for O&G)
Publication	IEC 61400-3 series	DNV-ST-0119, DNV-ST-0437	ABS FOWT Guide	No API rule for FOWT	No NORSOK rule for FOWT
Coverage	RNA, tower, floater, mooring, dynamic cable	tower, floater, mooring, dynamic cable	tower, floater, mooring	floater, mooring, SURF	floater, mooring, SURF
Design conditions	10 design situations, 42 DLCs	8 design situations, 39 DLCs	10 design situations, 35 DLCs, 2 SLCs for mooring	Conditions for operation, survival, transportation, inspection/maintenance, accidents	Conditions for operation, survival, transportation, inspection/maintenance, accidents
Format	LRFD	LRFD	LRFD or WSD	WSD (mostly)	More LRFD
Analysis method	Time-domain	Time- and frequency-domain **	Time- and frequency-domain **	Mostly frequency-domain	Mostly frequency-domain

*API and NORSOK do not have requirements specifically for offshore wind; they are quoted here for reference only.
**Frequency-domain analysis should only be applied in the preliminary design phases.

It was pointed out that FOWT technologies are still developing, and the rules and requirements will keep evolving.

In 2021 IMO conducted a regulatory scoping exercise on Maritime Autonomous Surface Ships that was designed to assess existing IMO instruments to see how they might apply to ships that use varying degrees of automation. In 2024 at MSC 108 committee agreed to the revised road map for development MASS code. This will come with more detailed regulations which may have a link to consideration of ship's design.

4 Design Tools

Tools for design of ship and offshore structures keep advancing in terms of their computational efficiency and accuracy. This chapter reviews recent advancements in CAD tools, classification societies' software, and developments in optimization approaches, machine learning for structural design and virtual/augmented reality for design and production. The chapter in the end gives overview of recent development in design of offshore structures.

4.1 CAD Systems

Computer-Aided Design (CAD) continues to play a key role in naval architecture. CAD models are usually at the centric position and often treated as the final 'true' source of information in the integration of CAD and computer-aided engineering (CAE), product lifecycle management (PLM). Researchers and software developers have been working to improve efficiency about the information exchanges in various tasks and different software packages.

In ISSC 2022 report, the data exchange and integration of CAD tools including the monolithic software and modular systems were discussed. Progress has been made from that report and some of the efforts and advances are discussed below.

4.1.1 CAD Exchange Format

Started by eight partners from the APPROVED project, OCX standard has gained traction with more than 30 members (https://3docx.org/members) in the maritime industry, including classification societies and ship-building relevant CAD vendors (DELTAMARIN, 2022; Zerbst, 2023). DELTAMARIN (2022) used OCX to streamline the approval processes of classification materials from drawings to 3D models. BV (2023a) presents completed vessel design that was created, reviewed and class-approved using 3D models with OCX file format.

Astrup et al. (2023) show how to use 3D model in the 3D review and approval process by the classification society, where appliance positions and attributes were exported from OCX format and further reviewed and verified according to regulatory requirements. The study pointed out that the current model-based review process is manual but potentially many tasks can be automated and be implemented directly in the design tools.

The quality and required post-processing time of a 3D model transfer between NAPA and CATIA SFD using OCX standard was evaluated by Gušani et al. (2023) and the team sees great potential of OCX in theory and practice for managing structural transfer needs.

Han et al. (2022) presented a code that allows conversion of structural CAD models between AVEVA MARINE and Smart 3D. They explained the database structure of both software packages that allow the transformation of the models. The script was proved on a ship hull and several structural components.

C. Li et al. (2023) proposed a unified ship structure design tool using Multi-Domain Feature Mapping (MDFM) and Extensible Markup Language (XML). It integrates CAD, CAE, and VR into a cohesive entity, employing a visual expression strategy based on the Unified Mesh Model (UMM). The tool enables intelligent finite element (FE) meshing and analysis, offering advantages in model generation, virtual visualization, and evaluation for diverse topologies. Its structure and data flow management align with a common design process, easing the linkage of scenarios with engineering changes.

4.1.2 3D Model-Based Approach

As said in ISSC 2022 report, the 3D model-based approach to the conceptual design of ships has become increasingly common and has significantly picked up speed. Model-based product development is deemed as major precondition to establish product information models for manufacturing and operational purposes, or digital twins (Bitomsky et al., 2022).

Amano et al. (2022) used a 3D product model to streamline the ship structural optimization process, which combines the necessary modelling work, prescriptive rule check, FEM-based rule check and further weight and other calculations.

A model-based manufacturing approach was adopted to replace the traditional document-based processes at some leading shipyards to leverage ship virtual twins from design to production and achieve the strategic move towards "Smart Digital Shipyards" (Chouche, 2022).

Kim et al. (2022) presents an interface between 3D CAD software, NAPA Designer and classification's structural strength assessment software to improve design productivity. Sakagami et al. (2022) conducted a project to integrate engineering process for rule and direct strength calculations into 3D CAD for efficient plan approval and hence realized an efficient 3D model-based approval (3DMBA).

Pérez Fernández (2022) discussed CAD and PLM integration at various levels in the context of FORAN system and Teamcenter PLM system and pointed out that the integration between CAD and PLM has clear advantages such as time saving and user friendliness.

4.2 Software Tools

4.2.1 Classification Societies Software Tools

Classification societies offer software tools that are essential for the shipbuilding industry, surveyors, and consultants. These tools can enhance design accuracy and help the automation of production processes. The proceedings of ISSC 2015 (Collette et al., 2015) presented an overview of the software offered by classification societies. A comparison of classification society tools for Harmonized Common Structural Rules (H-CSR) was conducted in the proceedings of ISSC 2018 (Lazakis et al., 2018). The proceedings of ISSC 2022 (Ivaldi et al., 2022) outlined the classification society tools used for rule-based ship design. This section provides an updated review of the functions and features of the latest version of these calculation tools released in recent years as illustrated in Table 10.

Committee IV.1: Design Methods, Principles and Criteria   757

Table 10. Review of the classification society rule-based calculation tools

CS	Rule-based calculation tools (Latest release date)	Functions & Key Features
ABS	CSR Prescriptive Analysis (2023.01)	• Assess ship structures for compliance CSR January 2015, 2017, 2018, 2019, 2020, 2021, 2022 and 2023 Rules • Track down Rule assessment failures • Rapidly identify areas of concern and the design modifications that might be required
	CSR FE Analysis (2023.01)	• Use FE analysis for direct strength and fatigue assessment complying with CSR January 2015, 2017, 2018, 2019, 2020, 2021, 2022 and 2023 Rules • An interface with the Nastran and Patran FEM tools • Minimum data input and accurate output results • Automated loading generation • A system to rapidly identify areas of concern and the design modifications that may be required
BV	MARS (2023.11)	• Calculation of geometric properties of ship's sections • Hull Girder Strength and Ultimate Strength criteria • Local strength criteria of plates and ordinary stiffeners: Yielding, Buckling, Fatigue, Minimum thickness • Assessment of cross-sections and transverse bulkheads along ship length
	Veristar Hull (2023.11)	• Global strength assessment of cargo hold structures (yielding and bucking criteria) • Stress assessment of structural details • Fatigue strength assessment • Generation of 3D finite element model facilitated by the import/export from various CAD formats • Automatic buckling panel analysis • Full integration of the sub-modelling approach for local and fatigue analysis

(*continued*)

Table 10. (*continued*)

CS	Rule-based calculation tools (Latest release date)	Functions & Key Features
	StarBoat (2023.06)	• Perform comprehensive structural assessments of small cargo and non-cargo ships with regards to the latest regulatory requirements
	ComposeIT (2023.05)	• Enable users to perform detailed strength analyses of composite panels and stiffeners for yacht structure
	Steel Coil (2023.09)	• Assess the local strength of a cargo hold for a foreseen steel coil loading arrangement
	Veristar Lashing (2023.06)	• Assess container securing and locking devices for containers stowed on exposed decks
CCS	COMPASS-HCSR-SDP (2023.10)	• Complete coverage of CSR BC & OT prescriptive requirements • Check hull girder yielding, ultimate and residual strength, scantling requirement, buckling strength, fatigue strength, bow impact, tank sloshing…
	COMPASS-HCSR-DSA (2023.06)	• Direct strength analysis software with full coverage of HCSR direct strength analysis requirements • Perform HCSR rule required FE analysis based on MSC.Patran • Automatic Loads, boundary constraints, corrosion deduction according to CSR BC & OT requirements
DNV	Nauticus Hull (2023.10)	• Complete module-based structural analysis package • Efficient environment for strength assessment, design and verification • Supports the latest updates of CSR BC & OT for prescriptive and Finite Element Method calculations. • Integrated with Sesam GeniE for FE modelling, post-processing and code check. • Import of FE models from common 3D design and FE systems such as MSC Patran/Nastran, ANSYS, AVEVA Marine, NAPA • Proven software used by more than 275 shipyards and ship design offices worldwide. • High quality technical support.

(*continued*)

Table 10. (continued)

CS	Rule-based calculation tools (Latest release date)	Functions & Key Features
	POSEIDON	• Full integration of complete prescriptive checks based on local and longitudinal strength calculations. • Full integration of cargo holds FE analysis • Fully automatic determination of initial scantlings • All rule sets and ship types relevant for 100A5 hull approval (including CSR) are covered. • Full model transfer from NAPA Steel implemented
LR	RulesCalc (2018.07)	• Cover all ship types in the Rules and Regulations, as well as LNG and LPG carriers and bulk carriers. • Can be used as a standalone system or in conjunction with NAPA, Tribon and Lloyd's Register's ShipRight SDA
	ShipRight (2020.08)	• Check your hull structure design for structural and fatigue compliance with the ShipRight Procedures within LR Rules • Automatically generate load cases. Import data from other finite element modelling tools like Hyperworks, FEMAP, and PATRAN
KR	Sea Trust-RuleScant (2023.05)	• Structural strength analysis of bulk carriers, double hull oil tankers, containers ships based on the latest HCSR and KR Rules • Creation of Section-by-Section Wizard, Manual Input and Import CAD Data • Support to Optimized Calculation regarding Longitudinal Strength Calculation and Auto Fine Mesh and Screening
	SeaTrust-Shaft (2023.07)	• Torsional Vibration analysis

(continued)

Table 10. (*continued*)

CS	Rule-based calculation tools (Latest release date)	Functions & Key Features
Class NK	PrimeShip-HULL (HCSR) (2023.09)	• Contain rule calculation tool that allows all the calculations required by the CSR BC & OT • Allows users to conduct finite element analysis and perform strength assessments. Using Altair's HyperWorks as its platform • Allow MSC. Patran users to conduct Finite Element Analysis (FEA) and perform strength assessments. • Support structural optimization through functions such as "Case Study Tool," "Automatic Calculation of Required Scantlings"
RINA	Leonardo Hull 2-D/3-D	• Direct linkage between the software and RINA scantling verification Rules and IACS Common Structural Rules • 2-D analysis tool helps the designer from the initial stages of hull structure definition, and performs fatigue checks • 3-D module performs FEA of ship structures, including a fatigue check of structural details

4.2.2 Other Tools

Beyond the tools offered by classification societies, a variety of calculation tools have been introduced by researchers and institutes for ship design and structure assessment in the last few years.

The HOLISHIP project, as described by Papanikolaou et al. (2022), is focused on developing ship design methods that are more environmentally friendly and cost-effective. This project uses the CAESES optimization platform to automate the exploration within the design space of parametric ship models. CAESES is a software tool based on parametric modelling, which aids naval architects in better understanding the impact of various ship designs on over-all ship performance. This understanding is crucial for meeting the economic and environmental performance requirements of modern ships. In the context of the HOLISHIP project, the CAESES platform is used to navigate through the design space of ship models, aiming to identify the most optimal ship design solutions.

MetOcean is an open-source web service that aims to provide maritime engineers and designers with access to oceanographic and meteorological data for evaluating the performance and safety of ship designs, as Dæhlen et al. (2023) presented. It retrieves historical observed data from multiple meteorological services and combines it with forecast models to provide continuous geospatial data.

4.3 Structural Optimization

Computational structural optimization can be broadly divided into sizing and topology optimization. Additionally, changing the external or internal shape of a part or structure entails shape optimization, which typically concerns with a fluid flow. For effective optimization with higher mathematical rigor, local optimization algorithms are used. If true optima are looked for, metaheuristic algorithms are employed. Topology optimization (TO) aims to provide novel designs by optimally arranging material within a target domain. This framework has the potential to surpass traditional ship structural designs, which are largely based on experience. Note that in this Committee IV.1 report, the benchmark study on topology optimization is also presented in Sect. 8.

Recent years have seen both fundamental developments and practical applications of these frameworks, which are briefly reviewed in this sub-section. Machine/deep learning (ML/DL) is being increasingly used to enhance efficiency of the entire optimization framework. Only a few works on ML/DL are reviewed here; some more can be found below in dedicated sub-section.

4.3.1 Optimization of Ship Structures

Several research can be highlighted for using and advancing metaheuristic optimization algorithms to optimize scantlings in various marine vessels. Silva-Campillo et al. (2022) optimized a corner bracket for converting a conventional to an inverted bow hull of a platform supply vessel under bow flare impact loads. They set up the most effective bracket geometry and stiffening method, proposing alternative geometries to optimize local weight. This resulted in an innovative "omega bracket" that minimizes weight while keeping buckling strength. Putra and Kitamura (2021) proposed a three-stage optimization method to minimize material costs for hatch cover design, adhering to IACS Common Structural Rules. The first two stages, termed hybrid GAs, combine GA optimization for selecting plate material and stiffener types with size optimization to decide optimal plate thicknesses. The third stage, layout optimization, adjusts the welding line position to achieve an optimal plate arrangement. Jiang, Yang, et al. (2023) proposed the Dominating Artificial Bee Colony Algorithm (D-ABC) for multi-objective optimization of ship structures, incorporating Pareto domination and individual crowding in design hyperspace. They also introduced the adaptive ABC algorithm with modified chaotic mapping to improve nectar source diversity. River boat case study showed a 5.08% reduction in main frame cross-sectional area and 7.43% reduction in vertical centre of gravity.

4.3.2 Topology Optimization

Applications of topology optimization (TO) to achieve practical structural designs are becoming more common. Chu et al. (2021) presented the level-set-based optimization method for both critical failure criteria, stress, and buckling, enabling the weight minimization of stiffened panels, simultaneously optimizing size, layout, and topology. The level-set method is used for TO of the stiffeners, as well as the thicknesses of the skin and stiffeners. The p-norm function is used as a stress aggregation to approximate the maximum stress that is considered as a constraint in the optimization problem. Bakker et al. (2021) simultaneously optimized the topology of stiffeners and their layout on a base shell or plate. This is accomplished by introducing a fixed number of modular stiffeners, which are subject to density-based topology optimization (SIMP). Tyflopoulos et al. (2021) applied conventional TO to brake callipers on a student race car. Their process involved design concept, material selection, and topology optimization of the calliper housing, the latter performed using Ansys and Abaqus. Chen et al. (2021) conducted TO on a bracket to enhance its fatigue performance. They used a multiple load-case topology optimization approach to minimize compliance under various asymmetric extreme loading conditions. A new bracket was fabricated using SLM, and fatigue tests verified the design's reliability. C. Zhang, K. Long, et al. (2022) proposed a TO methodology for designing offshore wind turbine (OWT) jacket structures in the concept phase. A weighted normalized compliance objective was minimized under practical aerodynamic and hydrodynamic loads. Using peridynamics topology optimization (PD-TO), Kendibilir and Kefal (2023) optimized the web frame of a bulk carrier. PD-TO is a general nonlocal optimization strategy that minimizes strain energy density as the objective function and uses a predefined volume fraction as a constraint. Several damage (crack) scenarios were considered, showing that the framework can yield viable topologies that are stiffer compared to conventional designs under the same loading conditions.

Topology optimization is also used to develop multi-scale structures or lattice structures with advanced properties which cannot be obtained with a single material. Zheng et al. (2020) proposed a TO method to obtain the orthotropic materials with auxetic property (negative Poisson ratio) using bi-directional evolutionary structural optimization (BESO) in which the variables are not continuous but discrete. For conceptual simplicity, this paper adopts the energy-based homogenization method (EBHM) to evaluate the equivalent elastic properties of the material microstructures. Xiao and Cirak (2022) proposed a shape and topology optimization approach by combining lattice topology optimization with shape optimization of the entire structure. The resulting structure forms a pin-jointed lattice and a thin-shell skin. The pin-jointed lattice is optimized using the SIMP method. For the shape optimization of both the shell and the lattice, the geometry of the lattice-skin structure is parameterized using the free-form deformation technique. Akamatsu et al. (2023) proposed a topology optimization method based on homogenization and level set methods to generate a 3D structure without cavities showing a negative Poisson's ratio, which can be fabricated using a 3D printer. In the first optimization step, a material structure forming a single material, and a cavity region was optimized. In the second step, they used the optimized structure as the first design and optimized a material structure comprising two materials without any cavity region.

Additive manufacturing (AM) is rapidly becoming widely accepted in several industries. TO typically proposes novel structures that can be produced using AM. Zou et al. (2021) presented a TO for AM where a structural self-supporting mathematical model identifies the overhang regions, which is used as a constraint in optimization. Amir and

Amir (2021) proposed a TO methodology for AM using a layered construction approach. Overhang limitations and support structure requirements are embedded in the TO. The method minimizes compliance with a two-material scheme, distributing one phase as a continuous solid-void and the other as a homogenized lattice for additional support. Liu et al. (2021) presented a systematic design optimization approach for shells with self-supporting infills for AM. The workflow includes concurrent TO of the shells and infills using a density based TO. A new overhang constraint, based on the AM filter, ensures the infills are self-supporting in the specified manufacturing direction and provide necessary support to the external shell for successful production. Xu et al. (2022) proposed a TO method to design self-support structures, focusing on controlling residual stress. The goal was to minimize structural compliance while meeting constraints on maximum residual stress and mass fraction. The method was applied to several 2D benchmark examples, showing its effectiveness in controlling residual stress and distortion.

As a form of fundamental development, H. Deng et al. (2022) presented an efficient and compact MATLAB code for three-dimensional stress-based sensitivity analysis. They calculated the maximum von-Mises stress using the concept of p-norm, termed p-norm global stress. The sensitivity analysis for p-norm global stress measure is also derived and verified by comparison with finite difference approximation.

4.3.3 Global Optimization

Goal of the global optimization is to find truly optimal (non-dominated) designs or solutions. This can be achieved using specific algorithms that often require numerous iterations, sometimes numbering in the thousands, to converge. This computational demand is especially pronounced when numerical simulations relying on CFD or FEM are used to evaluate the objective function or constraints. Efficient Global Optimization (EGO) methods utilizing surrogate models like the Kriging model have long been studied (Forrester et al., 2008; Jones, 2001). Wei et al. (2022) introduced a Parallel Efficient Global Optimization (PEGO) algorithm targeting ship hull form optimization. This work addresses computational challenges associated with using Computational Fluid Dynamics (CFD) for evaluating hydrodynamic performance in ship design. The authors extend the EGO algorithm to a parallel version, enabling multiple sample points per iteration and enhancing optimization efficiency. In a related study, S. Zhang et al. (2021) focused on optimizing ship hull forms using CFD methods. They introduced a surrogate model based on the Deep Belief Network (DBN) to predict the wave-making resistance of the Wigley ship. Comparing the DBN method with traditional surrogate models, they claim superior performance for the DBN-based approach.

Research on fundamental development of global optimization algorithms continues. Yuen et al. (2023) examine the evolutionary mechanisms that have been explored and those that have been overlooked in metaheuristic algorithms, using a framework based on the extended evolutionary synthesis (EES), an extension of the genetics-focused modern synthesis. The analysis reveals that while Darwinism and the modern synthesis are incorporated into evolutionary computation, EES is largely ignored except for cultural inheritance (in swarm intelligence algorithms), evolvability (in covariance matrix adaptation evolution strategy), and multilevel selection (in multilevel selection genetic algorithm). A gap in epigenetic inheritance is identified, presenting untapped potential for biologically inspired mechanisms in evolutionary computation. Jelovica and Cai (2022) introduced a novel constraint-handling technique (CHT) for evolutionary algorithm MOEA/D that can improve its speed of convergence to Pareto front and the spread of non-dominated solutions. The approach was proved to have superior performance over several state-of-art optimization algorithms and CHTs on a few engineering problems, including structural optimization of a chemical tanker. The approach relies on variable-constraint mapping which can be determined automatically using neural networks (Cai & Jelovica, 2023). Additional advantage of such approach is that the trained neural network can be used as surrogate model in optimization, replacing the computationally expensive assessment of buckling and yielding constraints. Kupwiwat et al. (2024) introduce a novel multi-objective optimization method using multi-agent reinforcement learning and graph representation. Agents iteratively change solutions to increase hypervolume and improve the distribution of non-dominated solutions, modelled as a Markov game. Each agent uses a policy function and a value function to predict actions and estimate rewards based on observations. The method is evaluated on three optimization problems, proving its effectiveness for small-scale structures.

4.4 Machine Learning Approaches for Design

Deep learning, a subset and more advanced form of machine learning, has seen tremendous developments in recent years (Goodfellow et al., 2016). These tools are part of Artificial Intelligence (AI), which is defined as the ability of a machine to mimic intelligent human behaviour, using human-inspired algorithms to approximate conventionally complicated problems.

Tezzele et al. (2023) present an approach for structural optimization for modern passenger ship hulls using advanced model order reduction techniques to streamline both input parameters and outputs of interest. Their approach integrates parameter space reduction via active subspaces into the proper orthogonal decomposition with interpolation method, running within a multi-fidelity framework, see Fig. 6(a). They evaluated their framework on simplified and full models of a midship section and a passenger ship, respectively, controlled by 20 and 16 parameters. Results demonstrate the effectiveness of these methods, particularly during the preliminary design phase, revealing previously unexplored designs while managing high-dimensional parameterizations. Cai and Jelovica (2024) propose an efficient representation of graphs in a graph neural network for predicting stress distribution across stiffened panels. Instead of considering finite elements as graph vertices, as is typical, structural units (stiffeners and plates in between) are used as vertices (graph nodes). This approach saves a large amount of time during training and reduces GPU memory usage. The effectiveness is proved on a wide range of stiffened panels, including double-curved panels with material nonlinearity.

Yüksel et al. (2023) reviewed recent progress and future research trends in AI applications in design and engineering design concepts. Methods such as machine learning, optimization algorithms, and fuzzy logic have been discussed from an engineering design perspective. Additionally, AI-based design studies are reviewed for various design stages, including inspiration, idea and concept generation, evaluation, optimization, decision-making, and modelling. The review concludes that interest in data-based design methods and Explainable Artificial Intelligence (XAI) has increased in recent years and that the use of AI methods in engineering design applications helps to achieve efficient, fast, accurate, and comprehensive results. Bagazinski and Ahmed (2023b) propose a creation of parametric ship hull designs using a parametric diffusion model. This model generates the tabular parametric design vectors of a ship hull, which are then constructed into a point cloud and mesh for performance evaluation. Parametric ship hulls produced using performance guidance saw an average 91.4% reduction in wave drag coefficients and an average 47.9× relative increase in the total displaced volume of the hulls compared to the mean performance of the hulls in the training dataset.

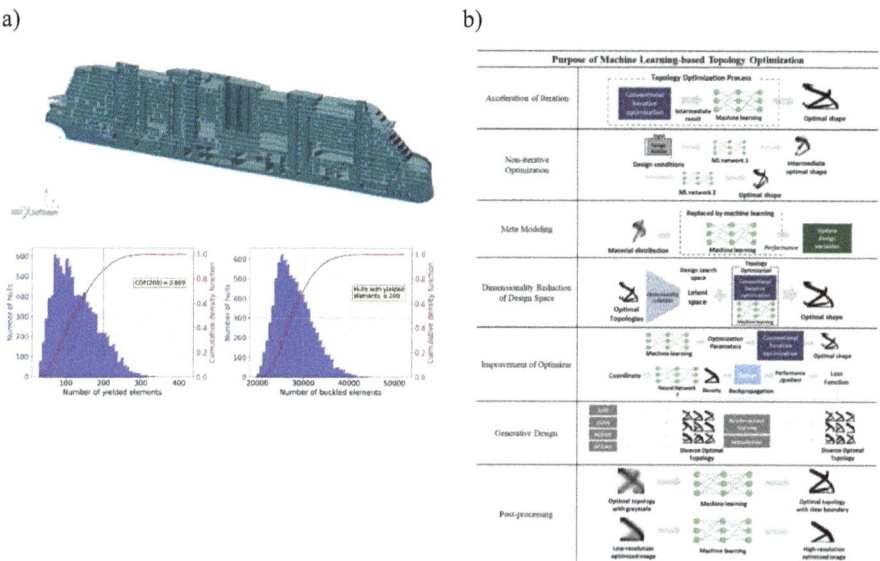

Fig. 6. (a) Structural optimization of passenger ship hulls using proper orthogonal decomposition technique (Reused from Tezzele et al. (2023) under CC BY 4.0); (b) Use of ML to accelerate TO (Reused from Shin et al. (2023) under CC BY 4.0)

One of the interesting applications of machine learning is generating innovative designs. Oh et al. (2019) proposed the deep generative design framework that can generate numerous design options which are not only aesthetic but also optimized for engineering performance. The proposed framework integrates topology optimization and generative models (e.g., Generative Adversarial Networks (GANs)) in an iterative manner to explore modern design options, thus generating many designs starting from limited earlier design data. The 2D wheel design problem is considered as a case study for validation of the proposed framework. Shu et al. (2020) explored a novel approach to 3D design using Generative Adversarial Networks (GANs). They generate 3D models and evaluate them in a physics-based simulation, iteratively improving design quality by incorporating high-performing designs back into the training dataset. This method bridges visual design and functional performance, enhancing the creation of feasible design concepts. For optimization of 2D airfoil design, Y. Wang et al. (2023) proposed a data-driven shape encoding and generating method, which automatically learns representations from existing airfoils. The representations are then used in the optimization of synthesized airfoil shapes based on their aerodynamic performance. The proposed method is built upon VAEGAN, a neural network that combines Variational Autoencoder with Generative Adversarial Network. Yüksel and Börklü (2024) proposed a generative deep learning approach to improve the mechanical properties of 3D concept designs of the component used in aerospace industry by combining the design capability of a generative adversarial network with finite element analysis. An innovative design and evaluation framework has been developed for GAN models to generate 3D models with improved mechanical properties. Generation of innovative design using GANs is also applied to ship design. Khan et al. (2023) introduce a generic parametric modeler built using GANs for the versatile representation and generation of ship hulls. They trained the model on a dataset of 52,591 physically validated designs from a wide range of existing ship types, including container ships, tankers, bulk carriers, tugboats, and crew supply vessels. A new shape extraction and representation strategy was developed to convert all training designs into a common geometric representation of the same resolution. They showed that traditional and novel designs with geometrically valid and practically feasible shapes can be created.

Another recent trend is the use of machine learning approaches in TO development. Qiu et al. (2021) proposed a deep learning model that automatically generates structural topology configurations with minimal compliance and deformation under various load conditions and volume fraction limits. The model combines Convolutional Neural Network (CNN) with U-net architecture and Recurrent Neural Network (RNN) with Long-Short Term Memory (LSTM) architecture. C. Deng et al. (2022) proposed an algorithm called Self-directed Online Learning Optimization (SOLO), integrating Deep Neural Networks (DNN) with FEM calculations to accelerate non-gradient topology optimization. The DNN maps designs to objectives, serving as a surrogate model to approximate and replace the expensive original function. A heuristic optimization algorithm then identifies the optimal design based on the DNN's predictions (Shin et al., 2023) reviewed the development of TO via ML. They concluded that TO benefits from ML in several ways shown in Fig. 6(b). ML can improve each procedure of TO or replacing multiple procedures with one neural network. The most computationally expensive step in TO is solving the linear system, particularly the finite element module. Jokar and Semperlotti (2021) introduced the finite element network analysis (FENA) to predict the response of physical systems using bidirectional recurrent neural networks (BRNN). Jeong et al. (2023) presented a physics-informed neural network (PINN)-based TO framework that employs PINN to predict the structural response in each iteration, fully substituting the original FE module.

Though a variety of neural networks (NNs) can be used in TO, a disadvantage is that most learning techniques rely on a database for training or pre-training and are challenging to extrapolate to larger-scale problems (Jeong et al., 2023; Jokar & Semperlotti, 2021). High-cost FE modules are still required to build the initial training dataset, and the predicted optimal structures are not as effective as those obtained through conventional methods. Inspired by the inverting representation of the image technique, Deep Image Prior (DIP) (Lempitsky et al., 2018), and PINN (Raissi et al., 2019), Z.

Zhang et al. (2021) proposed a topology optimization via neural reparameterization framework (TONR). This framework substitutes the TO process with a convolutional neural network (CNN). The main ideas are (1) to make the design variables in the optimization problem the parameters in the NN, which are updated by the NN, and (2) to make the objective function in the optimization problem the loss function. The complex sensitivity analysis is conducted through automated differentiation in ML. This approach shifts the computational burden from sensitivity analysis to NN updates, but for the first time, it generates better solutions than conventional methods. Zehnder et al. (2021) employed implicit neural representations with periodic activation functions for mesh-free TO and developed NTopo. They critiqued the commonly used ReLU activation functions for leading to oversimplified solutions, and in their study, TO was reconceptualized as a stochastic optimization problem using Monte Carlo sampling with the design objective as the loss function. Although their computational efficiency lags state-of-the-art numerical methods (OC and SIMP), they provided an innovative mesh-free learning method. ML has also been applied to multi-objective TO by Doi et al. (2019). They solved a simple TO problem using a non-gradient optimization algorithm, and then employed a CNN to predict further evaluations, accelerating TO by learning from initial optimizer results. This study, though focused on a small-scale problem, unveils the potential of integrating ML with classic non-gradient optimization algorithms, which are not originally designed for large-scale TO.

4.5 Data Collection and Storage

With the increased adoption of sensing, digitalization, machine/deep learning and digital twins in maritime industry, issues arise with various aspects of data collection, storage, processing, analysis, and utilization. Ensuring data quality is crucial, as incomplete or inaccurate data can lead to erroneous decisions. Data privacy and protection against security breaches are critical. Big data systems must scale efficiently to manage increasing data volumes and user demands. Effective integration of diverse data sources is necessary to maintain data integrity. Analysing vast datasets requires sophisticated processing techniques and efficient resource management. Establishing data governance policies, including regulatory compliance, is essential. Effective data visualization is key to conveying insights and facilitating decision-making. Addressing ethical concerns, such as data ownership, consent, and algorithmic bias, is vital to maintain trust in data-driven processes.

Li et al. (2018) discuss the challenges of managing and utilizing the vast and growing volume of marine wireless big data. They focus on fast, reliable, and sustainable data delivery in hostile marine environments and propose an architecture for heterogeneous marine networks to enhance data transmission. They also explore energy-efficient, reliable undersea transmission schemes and examine applications for sea-surface object detection and marine object recognition. The article concludes by discussing open issues and challenges in data transmission and detection/recognition. Bakdi et al. (2021) introduce a method for designing realistic testbed scenarios for autonomy tests using historical data and high-performance computational techniques. Their approach combines AIS traffic data with digital maps, vessel information, and nautical charts to analyse vessel interactions for collision and grounding conflicts within a 15-min prediction window.

They accurately evaluate relative risk by considering physical constraints and navigational rules. Using directed graphs, they model spatial and temporal dependencies among conflicts, capturing risky traffic situations and vessel behaviours over time. The computationally efficient algorithms can scale to large datasets, enabling the extraction, classification, and scoring of millions of scenarios for comprehensive post-test assessments. Wang (2020) proposes a distributed storage management method for ship big data based on HDFS. The proposed method handles high data volume and fast query requirements, demonstrating high reliability, fast read/write speeds, and support for large-scale data storage and queries through experimental validation. C. Wang et al. (2023) propose a combat methods verification system for surface ships using big data technology. This system integrates big data analysis, modelling, simulation, and performance analysis to tackle combat methods evaluation challenges. The framework and architecture are outlined, with detailed explanations for each phase. The discussion also includes ideas for optimizing combat methods based on big data, providing valuable insights for surface ship commanders to evaluate and enhance combat plans. Fonseca and Gaspar (2021) discuss digital twin applications in the maritime industry, focusing on data handling. Drawing on product data modelling dimensions and efforts to standardize ship data, the authors outline requirements for effective data modelling in digital twin contexts. They offer insights into challenges and future directions for digital twin implementation, proposing an open standardization for digital twin data with potential application to research vessels.

4.6 Virtual Reality, Augmented Reality

Tools for virtual reality (VR) and augmented reality (AR) are starting to be more widely adopted in the shipbuilding industry and better tailored for the specifics of the community. Garcia Agis et al. (2020) examine the use of advanced VR tools to simulate critical ship operations during the conceptual design stage. They emphasize evaluating design parameters and understanding the impact of external factors. The study compares various free-source VR software, concluding that some recent tools, despite being visually impressive, are too complex for effective use in conceptual ship design. The paper calls for a balance between display quality and resource efficiency, focusing on practicality and efficient tool use within time-to-market deadlines and resource constraints. Gernez et al. (2023) investigate the integration of VR in maritime design processes. Through mapping and interviews in two design projects, they explored how VR functioned as a collaborative medium. Despite access to VR, most design conversations primarily relied on emails with screenshots, visual interpretations, and text annotations. VR was occasionally used along with executable files, flythrough videos from real-time rendered models in a game engine, and a web-based 3D model-sharing tool. VR's role in collaboration was supplementary, integrated with various design activities like field studies, 2D sketching, and 3D modelling. The cases highlighted the need for VR to complement other communication channels, with the creation pipeline enabling collaboration through diverse mediums, such as web-based online tools.

AR systems blend real world environment with virtual information and provide an expanded and information-rich view of the user surroundings, generating a greater understanding of—and interaction with—datasets, manufacturing processes, and abstract concepts, as defined by Molina Vargas et al. (2020). This paper reviews concrete AR applications in maritime engineering, production, design, operation, and maintenance. It explains integration of AR in learning factory at NTNU for diverse industrial applications, focusing on its potential in the shipbuilding industry. The findings offer insights into the current state of research and outline opportunities for further exploration aligned with the development of the learning factory concept. Choi and Park (2021) address the challenges posed by the enormous scale of offshore structures by proposing a new way for marker (location) tracking and updating. The method enables construction workers to navigate efficiently and intuitively install and inspect outfitting parts, eliminating the need for paper drawings. Moreover, they show several examples how AR can aid with installation and maintenance of equipment, see examples in Fig. 7. Röltgen and Dumitrescu (2020) discussed the evolving role of AR in industrial applications, emphasizing its potential for performance enhancement. The paper aims to bridge the gap in understanding AR's strategic impact by introducing a classification scheme based on eight criteria tailored to industrial use. Through a literature review, 75 industrial AR applications are classified into 16 clusters using cluster analysis and multidimensional scaling, each representing a generic use case. The paper provides transparency on AR applications and benefits, offering companies insights on leveraging AR for their business. Fernández-Caramés and Fraga-Lamas (2023) conducted a comprehensive review on the applications of AR and mixed reality (MR) in shipbuilding within the context of Industry 4.0. They emphasize the potential of these technologies to enhance various shipbuilding tasks, providing efficient visual interfaces for operators to interact with physical and virtual elements. This paper also covers the state of the art in both commercial and academic AR/MR solutions, analyses challenges in enhancing shipbuilding tasks, and explores the latest hardware, software, and communication architectures for deployment in shipyard workshops and on ships. The review offers valuable insights for future developers in the field.

Fig. 7. a) Example of AR-based self-navigation; (b) Intuitive viewing of engineering points within parts; (c) Capturing an issue and instantly sharing via email (Reused from Choi and Park (2021) under CC BY 4.0).

4.7 Digital Twins for Design and Production

The concept of the digital twin has existed since the beginning of space explorations when NASA implemented similar concepts to a digital twin in the 1960's (Ibrion et al., 2019). Grieves (2014) developed a framework of the digital twin where the information of the physical entity and the virtual entity are synchronized, making digital twin a widespread concept. He provides a clear definition of a digital twin that is composed of three main parts: a physical product in the real environment composed of information about itself, a virtual product in a virtual environment representing the physical product, and a data connection between these two products actively flowing in both ways as so-called mirroring or twinning.

The digital twin technique is envisioned to assist in all phases of the lifecycle but up until now research has primarily focused on the manufacturing and operation phase and has been specifically lacking in the design and retire phase (Mauro & Kana, 2023). Currently, most research publications containing conceptual digital twin applications focus on parts of a ship instead of the total ship itself. A systematic review of the literature finds only two available publications which address a digital twin theoretical framework considering new-build methods (Papanikolaou et al., 2024), and of which only one is associated with the design of the whole ship (Xiao et al., 2022). A digital twin framework is proposed with the use of a vertical-horizontal design method regarding the total ship throughout its lifecycle phases. Even though it provides promising conclusions, it is still a theoretical framework with in-depth research still being conducted as mentioned by the authors.

Madusanka et al. (2023) reviewed the development of DT technology and its admittance to a variety of applications within the maritime domain in general and surface ships in particular. The conceptual theory behind the evolution of DT is highlighted along with the development of the technology and current progress in practical applications with an exploration of the key milestones in the extension from the electrification of the shipping sector towards the realization of a definitive DT-based system. Existing DT-based applications within the maritime sector are surveyed along with the comprehension of ongoing research work. The development strategy for a formidable DT architecture is discussed, culminating in a proposal of a four-layered DT framework.

Several initiatives have been undertaken to advance digital twin-enabled ship design. The application and advantages of digital twin technology in ship design is discussed by Harries et al. (2022) through the example of a tanker in the German MariData project. The researchers used suitable parametric models within CAESES® to redesign the ship hull at various levels of accuracy and analysed the different hull form approximations hydrodynamically using HSVA's database of model tests for similar vessels and high-fidelity CFD simulations. By using digital twin modelling and hydrodynamic analysis, researchers can more accurately capture the impact of the ship's hull shape, appendages, and propulsion system on energy consumption, thereby improving ship performance. Fonseca et al. (2022) offer an open framework that interfaces with services like visualization, simulation, and remote control. The study features an experimental case of a scaled model ship equipped with a dynamic positioning system, assessed under artificial wave conditions. The digital twin prototype highlights real-time mirroring and control of the model's position and utilizes a web application to predict motion responses in various wave scenarios. This approach facilitates closed-loop testing and design by allowing for comparisons with actual data, thereby addressing lifecycle issues.

The digital twin model can also be specifically applied to naval vessels by providing a virtual asset that gathers accurate digital models and simulators within a common environment (Gualeni et al., 2023), which includes a network of sensors deployed on board and a live connection between the real and virtual space. Benefits of the digital twin model for naval vessels include improved design optimization, proactive maintenance, and enhanced lifecycle management. The digital twin model can also support the implementation of a lifecycle approach within the naval industry, enabling cost-effective decision making throughout the entire lifecycle of the ship.

Respected to the digital twin applications for ship production, Burggräf et al. (2022) discusses the use of digital twin technology in a German shipyard in the ProProS (Zerbst et al., 2022) research project, which can improve production planning and control by providing a more detailed and accurate forecast of production times for individual parts. The software enables planners to run a complete Deming process repeatedly with reasonable effort and uses track-and-trace data collected in production to perform an automated check if the developed plan is still valid or needs adjusting.

Another potential application of digital twin technology is in the realm of digital twin-based retrofit design. H. Zhang et al. (2023) propose the construction of a digital twin for an already existing research vessel 'Gunnerus'. The article provides a digital twin architecture including the data-driven design building block method using the Open Simulation Platform, an open-source simulation platform. Even though a unique vessel is being considered, the authors aim is to provide a standardized digital twin concept for the maritime industry. Mouzakitis et al. (2023) address the importance of using high-performance computing together with big data analytics to develop and therefore contribute to high-level digital products for the maritime industry. The digital twin architecture includes a proposed data integration into the existing unified system within the 'VesselAI' project. Additionally, the Digital Twin for Green Shipping (DT4GS, 2024) project aims to develop digital twin for existing commercial vessels, with applications in container ships, bulk carriers, tankers, and RO-PAX ferries.

Compared to the application of digital twin in a total ship itself, more research focused on parts of a ship or the ship subsystems. DeNucci and Brahan (2022) employ the digital twin methodology to simulate the process of ice accumulation on ships. By modelling the physical characteristics of the ship and environmental conditions into a digital twin, real-time predictions and analyses of ice build-up can be achieved. The application of this method lies in monitoring and predicting the ice build-up on ships in real-time, assisting operators in making informed decisions such as route adjustments and implementing de-icing measures to ensure the vessel's safety and operational integrity.

A review of the state-of-the-art of digital twins in the design and retrofit of ships (Papanikolaou et al., 2024) concludes that there is currently no standardized design method or approach for DT-based design or retrofit within the ship industry, or even in other relevant industries. The report identifies several key challenges that need to be addressed to fully leverage the potential of DTs in these phases, including standardized digital twin framework, model-based systems engineering (MBSE) application, and operational phase data collection and utilization.

4.8 Specific Topics Related to Offshore Structures

New techniques for design and optimization of offshore structures have been proposed. Shabakhty et al. (2024) proposed a novel metaheuristic algorithm, Enhanced Colliding Bodies Optimization (ECBO), for optimizing a jacket platform. The study considers constraints on member stress, buckling, horizontal displacements, and connection adequacy. ECBO improved the design by 15%, outperforming GA's 11% improvement, demonstrating ECBO's superior efficiency and ability to avoid local optima. Ojo et al. (2022) presented a review of multi-disciplinary design optimization (MDO) research for floating offshore wind turbine (FOWT) substructures, covering concepts, design techniques using geometric shape parameterization, and analysis methods (time and frequency domain) for assessing system response. They also examined various optimization algorithms, from global search heuristics to local gradient-based methods, used in FOWT substructure design. The review identifies research gaps in MDAO for FOWTs and suggests future research areas: enhancing design space with advanced parameterization techniques, using surrogate models for quick sensitivity studies, and scaling up geometric designs for new, heavier turbines. Zheng et al. (2023) proposed a method using Kriging surrogate models for efficient and rapid optimization of lightweight jackets for offshore wind turbines. The design considers structural responses under extreme loads, with key variables identified through parameter sensitivity analysis. Finite element models are converted into surrogate models, optimized using a genetic algorithm, and verified through dynamic analysis. The method demonstrated high efficiency, reducing optimization time by 98.61% and proving effective for jackets at 30 m, 50 m, and 70 m water depths. Wang, Mantey, et al. (2024) optimized jacket substructure of an offshore wind turbine (OWT) considering self-weight, wind, wave loads, and soil-pile interaction, achieving significant weight reductions. Dyer et al. (2022) assessed platform risk by using two machine learning models, a gradient boosted regression tree (GBRT) and an artificial neural network (ANN), to predict the removal age of platforms in the Gulf of Mexico. The models were trained on an extensive dataset representing the natural and engineered offshore system. Both models provided highly accurate predictions, with 95–97% accuracy and forecasts within 1.42–2.04 years of the actual removal age during validation. da Cunha Jácome Vidal et al. (2022) proposed a conceptual framework for the decommissioning process of offshore platforms, highlighting the interconnected dimensions involved. Using a systematic literature review of 86 articles from five databases, the study developed a model with five key dimensions: (i) Relationship with stakeholders, identifying major parties involved; (ii) Technological solutions, covering technical aspects; (iii) Decommissioning options, outlining alternatives; (iv) Operationalization, focusing on execution; and (v) Impacts, including environmental, financial, socioeconomic, health, and safety effects. Larsen et al. (2022) presented an optimization framework for welded joints in offshore jacket structures, focusing on fatigue damage. The framework optimizes the orientation and location of welds in a welded K-node, considering that longer welds are more prone to fatigue failure. The statistical size effect modelled using the Karhunen-Loéve expansion, accounts for weld length in damage assessment. The framework simulates different welding qualities by adjusting the correlation length and coefficient of variation. Validation on a simple plate structure shows the approach is effective. Additionally, a full-scale welded K-node is optimized for mass, revealing that high-quality welds lead to a lighter structure.

Kapasakalis et al. (2024) presented a multi-objective optimization approach for designing vibration control systems for monopile OWT subjected to wind and wave forces. An Extended K Damper (EKD) is used, incorporating tower dynamics and soil–structure interactions. The design process considers competing objectives like engineering needs, manufacturing specs, budget constraints, and uncertainties. A global sensitivity analysis identifies key parameters, followed by multi-objective optimization to explore trade-offs (Pareto front) and prove the EKD's effectiveness, even without certain components. The robust analysis accounts for uncertainties, leading to EKD designs with superior performance. Results show that the EKD improves peak tower response and damping more effectively than a conventional Tuned Mass Damper with 20 times the mass. Yu and Amdahl (2023) presented a Rayleigh-Ritz solution to predict high-order natural frequencies and eigenmodes for monopile-supported turbines, accounting for tapered towers and soil-pile interactions. The method is applied to a 10 MW turbine, with results showing excellent agreement with nonlinear finite element simulations for the first three eigenmodes.

4.9 Conclusion

Several advancements in tools for design of ship and offshore structures have been proposed in recent years. Notably, advanced machine learning techniques are increasingly proposed in scientific literature for various design scenarios, staring to be used in practice. AR and VR techniques have found their use for production and are increasingly being holistically investigated for concept design. Optimization is advancing with more effective and robust algorithms, including topology optimization that could generate more efficient structural designs than traditionally used in industry.

5 Life Cycle Management

In the maritime industry, it is considered that a more holistic approach for not only the design process but also the entire lifecycle of the structures, which includes design, manufacturing, operation, and dismantling, is necessary to satisfy the demand for efficient and sustainable development of ships and offshore structures. This chapter focuses on the recent advancement in the field of lifecycle management of ships and offshore structures, especially related to data exchange and standards, digital twin, design and operation tools, etc.

5.1 Data Exchange and Standards

The shipbuilding industry can benefit greatly by using digitalization and computational tools throughout the entire life cycle of a ship. This can be done through the creation of cyber-physical systems or digital twins, which have the potential to generate value for stakeholders. However, to successfully take advantage of these technologies, there needs to be a significant reshaping of current tools and practices in the shipbuilding industry.

According to a recent study (Koelman et al., 2024), there has been a lack of updates regarding new standards for information exchange. Among the existing ones, the most discussed are DXF, IGES, and ISO 10303, which is commonly referred to as STEP (STandard for the Exchange of Product model data). The primary goal of these standards is to establish a framework for defining shape, with the potential to expand into the realm of product modelling.

The ISO 15926 series (ISO, 2023) provides guidelines for representing information throughout the life cycle of process industries facilities. This representation is defined by a generic data model that can be used to create a shared database or data warehouse. Another use of this model is to generate handover files that contain clear and consistent life-cycle data, based on a commonly shared data model and reference data library. This choice of terminology suggests that this standard is primarily focused on process plants, which may restrict its suitability for ship design applications.

ISO 81346 (ISO, 2022) is an intriguing industrial standard that goes beyond the mere product and its physical form, offering a broader framework. The standard presents fundamental principles for organizing systems, including the organization of information about systems. These principles serve as a foundation for establishing clear and precise reference designations for objects in any given system and are accompanied by rules and guidelines to facilitate this process. The universality of the standard refers to the assignment of various aspects of the product, where each aspect may have its own hierarchy and taxonomy. Aspects can be the constructional relations with the components, the functional relations and the spatial relations (e.g. the location). Additional aspects that can be considered include financial and logistical.

The mentioned standards pertain to the characteristics of a product during its design and construction stages. However, there is a growing field that also incorporates operational data, particularly through sensor data, which makes a significant contribution.

In this direction, the aim of the published at the beginning of 2024 ISO 19848 (ISO, 2024) standard is to offer instructions and specifications for exchanging data on board a ship. However, in the future, it is conceivable that ship machinery and equipment will be linked to the Internet. The standard outlines two principles and their models for data exchange: Data Channel and Time Series Data. A Data Channel List includes the essential meta-data, while a Time Series Data format is used for measurements. To ensure dependable data exchange, this standard proposes the use of XML (Extensible Markup Language) and XML Schema for encoding and defining the structure of the data.

The shipbuilding industry lacks computational tools to support the digital thread and life cycle process, which is why they do not practice maturity management for used parts. Unlike discrete manufacturing industries, the shipbuilding industry does not have the means to ensure functional safety, traceability, and compliance in their processes. An attempt to overcome this is made through the ongoing (2023–2026) Smart European Shipyard (SEUS) project (Gaspar et al., 2023; Koelman et al., 2024). The goal of SEUS is to develop a technologically advanced solution that offers a range of computational tools for the shipbuilding industry. These tools will integrate data streams as the design progresses, offering a comprehensive set of tools for various stakeholders involved in the shipbuilding process. This will give them access to a centralized and reliable source of data. The project is based on the software systems PIAS (by SARC) and CADMATIC.

Originally defined in a research project, the Open Class 3D Exchange (OCX) standard has swiftly garnered attention from users and tool vendors within the shipbuilding sector (Zerbst, 2023). The OCX is a standard that is specific to vessels and addresses the information requirements of classification societies. It plays a crucial role in replacing traditional 2D class drawings with a 3D model. See Fig. 2 of Zerbst (2023) for the summary of the use cases and their current state of readiness.

5.2 Digital Twin for Lifecycle

Applying digital twin technology to the ship lifecycle effectively enhances design efficiency, boosts digital operation and maintenance management, and advances the development of supporting industries (Lv et al., 2023; Mauro & Kana, 2023). Though recent advancements of digital twin technology for design and production have already reported in Sect. 4 (Sect. 4.7), here again recent topics of digital twin are reported from the viewpoint of lifecycle.

For constructing the digital twin model for ship lifecycle, Xiao et al. (2022) presents an architecture named SWLC-DT (Ship Whole Life Cycle Digital Twin) based on a vertical-horizontal (V-H) design approach. This framework integrates historical data from previous product generations (vertical) with real-time data from current products (horizontal) to refine digital twin models across a product's entire lifecycle, encompassing design, construction, operation, and maintenance. Emphasizing the elimination of information silos, the SWLC-DT framework aims to enhance the digital and intelligent transformation of the shipbuilding industry by providing intelligent services and enabling predictive and optimized lifecycle management. The architecture includes physical, virtual, and knowledge spaces along with a virtual control platform to ensure effective data interaction and system operation. Liang et al. (2023) proposes a method for constructing a digital twin model for ships, which includes functional models like the virtual scene generation system, sensor system, hull model, power system, and security system, comprising data, mechanism, and functional models. Data models are refined using historical and current voyage data and mechanism model simulations, while mechanism models focus on ship equipment and system mechanics, and functional models cater to the ship's function system requirements. The model processes real-time environmental and ship status data from the sensor system for simulations, providing results to support monitoring, task planning, and decision-making within the ship's functional system.

During the ship design stage, dynamic data-driven simulation is pivotal for digital twin technology. Advanced sensor and network communication technology enables real-time collection of observation data from physical systems, which is then incorporated into the simulation process, allowing for dynamic corrections to the simulation model and enhancing the accuracy and credibility of simulation results (Wei et al., 2024). In the preliminary design phase, a model is quickly constructed through intelligent matching of historical ship types, equipment parameters, and ship owner requirements in a virtual environment (Solheim et al., 2024). After completing this phase, the model is refined by comparing actual data discrepancies after simulation verification. Bucci et al. (2021) present an advanced methodological approach aimed at integrating digital twin technology into the initial stages of ship design, enhancing real-time data integration across CAD, CAE, and CAM tools. The research outlines the potential for DT to enable a seamless transition to virtual prototypes, ensuring real-time validation and modifications. Giering and Dyck (2021) discussed how digital twin can mature from an early-stage twin to a digital twin prototype in the detailed design phase. The digital twin enhances collaboration among designers, shipyards, clients, and classification agencies, and the digital twin system integrates extensive CAD tools for detailed modelling, like hull structures and electric systems, reducing redundancies and ensuring all components are virtually represented. The production design model, developed in the virtual space, describes the ship production requirements and undergoes virtual verification and optimization (Z. Cheng et al., 2022). Once these three design stages are complete, their virtual models are integrated to form a comprehensive digital twin model (DTM) of the ship design that satisfies the needs of shipowners. The high-fidelity simulation environment and computational tools of digital twins significantly enhance design verification, accelerate the design process, and improve accuracy.

The manufacturing stage of shipbuilding involves material processing and hull assembly, utilizing a digital twin model that includes both ships and shipbuilding workshops within a virtual space. Before actual production commences, the workshop model performs virtual manufacturing guided by the design digital twin model (DTM) developed in the design stage. This process allows for the identification and rectification of discrepancies between virtual manufacturing data and design expectations, optimizing ship design and avoiding potential losses during real-world manufacturing (Xiao et al., 2022). Moreover, the distributed, off-site construction of hull components often leads to information silos in the physical space. To combat this H. Zhou et al. (2024) proposed a three-dimensional (3D) visual monitoring method for marine intelligent mould beds using digital twin technology, which effectively monitors and controls subsection deformation during the shipbuilding process. This method not only enhances design by comparing real manufacturing data with design specifications but also ensures the production schedule alignment with delivery timelines. By dynamically updating and optimizing the production plan, coordinating material transportation, equipment maintenance, and material usage, the planning model ensures a streamlined manufacturing process (H. Zhang et al., 2024).

During the ship operation and maintenance phase, the ship digital twin model (DTM) is developed in the virtual space to reflect its actual operations. Integrating digital twins with advanced sensors, AI, and the Industrial Internet of Things (IIoT) enhances predictive maintenance by analysing ship sensor data to anticipate equipment failures, enabling proactive maintenance and reducing downtime (Bhagavathi et al., 2023). Utilizing historical maintenance data, maintenance knowledge, and intelligent perception data from the ship, a ship maintenance DTM is promptly generated in the virtual realm. Hasan et al. (2024) leveraged sensor array data combined with the ship's dynamic model, utilizing an advanced adaptive extended Kalman filter (AEKF) algorithm via digital twin technology to diagnose faults in autonomous ships. Furthermore, digital twin technology is instrumental in navigation decision support, enabling operational trend pre-dictions (G. Li et al., 2023).

Monitoring and analysing the ship structural stress is pivotal for developing the ship digital twin model. Liu and Ren (2023) presents an approach to rapidly obtain stress for structural strength assessment in ship digital twin models. It enables the quick acquisition of yield strength assessment stress at both monitoring and non-monitoring points. This method eliminates the need for load inversion work and directly identifies the actual loads experienced by the ship structural components through the characteristic features of stress states induced by specific loads. Neves et al. (2021) proposed a methodology for developing a DT ship using algorithms to modify traditional finite element method steps. This digital twin can be applied in ship operation, especially for monitoring and predicting the structural integrity of FPSO vessels. The methodology integrates information from the coupled system, updates the finite element model, performs simulations, and processes results to provide an overview of structural integrity. This approach helps optimize ship operations by evaluating various conditions and parameters, reducing unnecessary maintenance that could impact operations. Anyfantis (2021) proposed hull structural health monitoring system for the maritime industry, where deep learning is used to identify damage-sensitive attributes and optimize the placement of strain sensors. This allows determining structural states from specific strain measurements. The methodology improves real-time surveillance and predictive maintenance of hull structures, reduces the need for manual supervision, and lowers operational costs.

Regarding ship navigation, Lee et al. (2022) introduces a digital twin technology for ship operations in real-time sea conditions. This technology predicts risks and optimal routes by analysing ocean waves and ship performance in navigation and manoeuvring. A core component is predicting ocean waves using a sophisticated algorithm. This algorithm reconstructs waves from radar images and forecasts them in the ship path for 10 min. By integrating these elements, the system predicts risks and ship performance in various ocean settings. Fujikubo et al. (2024) presented a summary of an industry-academia joint R&D project on the digital twin for ship structures in Japan. The DT model aims to grasp the stress responses over the whole ship structure in waves by data assimilation that merges hull monitoring and numerical simulation. Three data assimilation methods, namely, the wave spectrum method, Kalman filter method, and inverse finite element method were used, and their effectiveness was examined through model and full-scale ship measurements. Methods for predicting short-term extreme responses and long-term cumulative fatigue damage were developed for navigation and maintenance support using statistical approaches. In comparison with conventional approaches, response predictions were significantly improved by DT using real response data in encountered waves. Utilization scenarios for DT in the maritime industry were presented from the viewpoints of navigation support, maintenance support, rule improvement, and product value improvement, together with future research needs for implementation in the maritime industry.

In addition to improving and assisting current navigation decisions, digital twin technology can collect and analyse ship response data to provide guidance for subsequent ship design optimization. Hinz et al. (2023) proves the application of DT technology in evaluating emission reduction by novel technologies, including optimizing ship operations, improving and optimizing design and equipment, and exploring low-carbon fuel alternative. The digital twin model can also be specifically applied to naval vessels by providing a virtual asset that gathers accurate digital models and simulators within a common environment (Gualeni et al., 2023), which includes a network of sensors deployed on board and a live connection between the real and virtual space. Bekker and De Koker (2023) applied DT technology to improve propulsion safety margins in ice. By combining measurements of shear strain and shaft speed from the operational propulsion shaft with an inverse model, the digital twin estimates the peak magnitude of the ice-induced propeller moment for comparison to a design limit. The value proposition of the digital twin is to assist navigators with situational awareness through calculation of the structural margin from measurements and algorithms which estimate the ice-induced propeller moment.

5.3 Design and Operation Tools Applied to Life Cycle Management

With the increasing integration of smart technologies and autonomous navigation in vessels through Information and Communications Technologies (ICT)—which combines the Internet of Things (IoT), artificial intelligence (AI), and Big Data—the shipbuilding industry is undergoing rapid transformation. To thrive in this evolving landscape, it is essential to acquire advanced technologies and enhance vessel quality. The study of Park and Huh (2022) seeks to explore methods for collecting and managing Big Data within the shipbuilding sector, catering to diverse data types and communication methods. Additionally, it will examine how to effectively utilize this Big Data to adapt to the swift changes in the industry. The proposed model in this research focuses specifically on the Asian shipbuilding industry.

With advancements in IoT technology, digital twins are increasingly being utilized across various fields. They play a crucial role in integrating physical information with the digital realm, serving as an effective tool for the intelligent enhancement of products. Digital twins offer a fresh perspective on managing the entire lifecycle of complex products. The lifecycle of a ship encompasses several stages: design, construction, operation, maintenance, and ultimately, scrapping and recycling. However, each of these stages operates with a degree of independence, often leading to the emergence of information silos. To overcome this problem, Xiao et al. (2022) introduces a vertical-horizontal design concept as explained in 5.2, which is a comprehensive model for the digital twin of the ship's entire lifecycle.

In January 2023, Cadmatic and CONTACT Software announced a collaboration aimed at enhancing PLM process integration tailored specifically for the shipbuilding sector (Grealou, 2023; Seppälä, 2023). The goal of this partnership is to encompass the complete lifecycle of shipbuilding projects, including design, engineering, prefabrication, production, maintenance, and operation. By introducing the first PLM solution specifically tailored for shipbuilding, the project aims to help customers significantly enhance quality and reduce time throughout the lifecycle of their projects.

Digital continuity, Industry 4.0, and Product Lifecycle Management (PLM) present significant challenges for industries managing the complete lifecycle of products, particularly in sectors like aerospace and shipbuilding. The research of Rubio et al. (2023) aims to integrate functional and industrial product structures along with process structure concepts into the shipbuilding sector. The objective of the paper is to propose a process-oriented approach for organizing naval manufacturing engineering. This approach involves the creation of a product structure for both design and manufacturing, including an engineering Bill of Materials (eBOM) and a manufacturing Bill of Materials (mBOM), as well as a process structure known as the Bill of Process (BOP). These elements are intended to facilitate digital continuity by leveraging advanced PLM tools.

Based on an analysis of approximately 1,400 articles, a systematic review (Jacquet et al., 2024) employing the snowball sampling method identified 32 relevant studies for bibliometric analysis. This analysis focused on key themes related to Life Cycle Assessment (LCA), emphasizing several critical elements: the definition of the functional unit, selection of boundaries, application of cutoff rules, methodologies concerning life cycle inventory, selection of impact categories, utilization of databases, frameworks for life cycle inventory modelling, available LCA tools, methods for characterization, as well as considerations of temporal and spatial factors. The analysis further addressed normalization, weighting, and both uncertainty and sensitivity analyses. Notably, the findings revealed that functional units were inadequately defined in 90% of the examined cases, and a lack of consistency about boundary definitions was prevalent. Additionally, a multi-criteria approach was clear in 85% of the studies analysed. In conclusion, while the establishment of a standardized LCA framework could be beneficial, it is imperative to conduct a thorough examination of the methodological trends associated with LCA in the maritime industry prior to such standardization efforts.

A paper (Nickpasand et al., 2024) examines the utilization of Machine Learning (ML) in Product Lifecycle Management (PLM) systems to reduce the duration of Engineering Change Orders (ECOs), thereby enhancing manufacturing responsiveness. The application of ML in PLM systems can significantly reduce the time required for Engineering Change Orders (ECOs). A classification method using neural networks and 3D design graphics assists in predicting design changes' impact on manufacturing. The evaluation of the model, developed using data from Siemens Gamesa Renewable Energy's offshore wind turbine hub, also discusses inherent challenges and limitations in other manufacturing contexts.

5.4 Dismantling

Ship dismantling, or ship breaking is an important process in the ship life cycle and plays a crucial role in ship recycling, especially for the material extracting and reclaiming, ecological footprint reduction and marine waste management. This consequently helps move the maritime industry to a circular economy model.

Ship recycling regulations include Basel Convention (1989), Hong Kong International Convention for the Safe and Environmentally Sound Recycling of Ships (the Hong Kong Convention) and European Union Ship-Recycling Regulation (EU, 2013). The Hong Kong Convention, adopted in May 2019, will enter into force on June 26, 2025. This was set in June 2023 when the entry-into-force conditions were met (IMO, 2023b).

Ship recycling research in general has seen an uptrend in recent years (Moussa et al., 2024). This systematic literature review and bibliometric analysis identified themes such as environment, technologies, health and safety, economics, legal studies and combinatory themes to guiding future research in this area and highlights the multidisciplinary nature of ship recycling research. Other review studies can be found in Tola et al. (2023) and Mannan et al. (2023).

To understand the gap between current regulations and the existing yards, Fariya et al. (2022) proposed a ship recycling framework for shipyards by discrete event simulation and demonstrated its efficiency in identifying problems, improving measures and optimizing resource. Jamaluddin et al. (2022) evaluated the Indonesian shipyard condition by Hazard Identification Risk Assessment Risk Control (HIRARC) method to assess the gap between the Hong Kong Convention and the existing conditions. The study showed that 55% of sub clauses out of the 56 does not exist or are not handled properly; and 17 out of 56 potential hazards or 28% are of high or extreme risk.

Önal (2023) compared different cutting methods for ships older than 30 years from a recycling standpoint and found the water jet cutting method can lower the environmental impacts than existing oxy-fuel cutting method. Meanwhile, new cutting methods need to be improved to further reduce environmental impact by using renewable energy sources, reducing block cutting time.

5.5 Offshore Structures

Liu, Zhang, et al. (2023) introduced a unified framework using digital twin (DT) for floating wind turbines (FWTs) across their life cycle. Employing a digital 3D model, the framework allows real-time synchronization and inversion of sensor data for global state simulation and analysis of FWTs. The framework is validated through a twin-barge float-over project case study, demonstrating its effectiveness in timely monitoring, construction plan visualization, early anomaly detection, and accurate posture and environment recognition. The results highlight the framework's potential for ensuring safety, optimizing project plans, and enhancing construction efficiency.

Xia and Zou (2023) reviewed latest research on using digital twin (DT) technology to improve OWF operations and maintenance (O&M), covering failure analysis, strategies, optimization models, and DT-based management. A DT-based O&M optimization framework is proposed to enhance O&M intelligence, along with a discussion on valuing DT models and suggestions for future research areas.

Mousavi et al. (2024) address the challenge of data acquisition in Structural Health Monitoring (SHM) by proposing a digital twin-based method for damage detection in structures. They utilize a digital twin generated through a physics-based model to simulate diverse damage scenarios in a Floating Wind Turbine (FWT) structure. A Deep Convolution Long Short-Term Memory Neural Network (DCLSTMNN) is trained with frequency data from the simulated FWT model under constant loads. The model is then evaluated under variable loading conditions, demonstrating its robustness. The proposed digital twin-based method accurately detects the location and severity of damage, outperforming other methods despite various uncertain-tie.

Brussa et al. (2023) conducted a case study for an offshore wind farm intended to be deployed in the Mediterranean Sea using the cradle-to-grave life cycle assessment (LCA) and pointed out that raw materials acquisition has the highest contributions to the overall impacts and is the main hotspot of the floating wind farm's life cycle. It also compared the modelled wind farm with the Italian national grid and found that wind energy is significantly better when environmental performance is considered. The paper concludes that floating offshore wind power can be considered a promising technology and is competitive.

LCA was also carried out for a United States based offshore wind farm by Moussavi et al. (2023) in their study to compare environmental sustainability between novel designs with concrete foundations and steel monopile foundations. The comparison shows that the novel design has a lower LCA profile and hence less overall life cycle environmental impact.

Kouloumpis and Azapagic (2022) developed a model for non-experts to estimate the life cycle impacts of offshore wind farm. The model relies on databases that follow the ISO 14040/44 LCA methodology. User can evaluate an installation's environment impact including global warming potential, depletion of resources, human toxicity and eco-toxicities, etc.

J. Zhou et al. (2024) designed a framework to help select life cycle design scheme of offshore wind turbine. In the framework, multiple criteria are included to encompass the life cycle of the wind turbine. Criteria correlation is established by fuzzy Technique for Order of Preference by Similarity to Ideal Solution (TOPSIS) method based on analytic network process (ANP). The proposed framework was used in a case study to illustrate its effectiveness along with sensitivity study and comparative analyses.

5.6 Conclusions

The integration of data streams in shipbuilding enhances stakeholder access to reliable information through advanced tools and standards like OCX. There is a significant need for updated standards for information exchange, particularly in the context of digital twin and cyber-physical system applications in shipbuilding. The ongoing Smart European Shipyard (SEUS) project aims to develop advanced computational tools to support lifecycle management and improve operational efficiency in shipbuilding.

Digital twin technology enables advanced ship health management through predictive maintenance and condition monitoring. The SWLC-DT (Ship Whole Life Cycle Digital Twin) framework integrates historical and real-time data, enhancing lifecycle management from design to maintenance. Digital twins optimize manufacturing processes by identifying discrepancies ahead of production, promoting efficiency and safety in shipbuilding.

The integration of IoT, AI, and Big Data is essential for advancing the shipbuilding industry and improving vessel quality. The proposed vertical-horizontal design model aims to create a comprehensive digital twin for the entire lifecycle of ships to prevent information silos. Collaborations in the sector focus on enhancing PLM processes, ensuring digital continuity across shipbuilding projects, and leveraging Machine Learning to improve responsiveness in manufacturing processes.

Ship dismantling is essential for material recovery and transitioning the maritime industry towards a circular economy. Water jet cutting is more environmentally friendly than oxy-fuel cutting for ships over 30 years old. There is a notable gap between existing regulations, like the Hong Kong Convention, and compliant practices in shipyards, with studies emphasizing the need for improved safety and efficiency measures.

Recent methodologies employing digital twin technology aim to optimize operations and maintenance in offshore wind farms through improved intelligence and predictive analysis during the monitoring and construction of floating wind turbines and providing advantages in anomaly detection and project efficiency. Life cycle assessments highlight the environmental benefits of floating offshore wind farms, showing lower life cycle impacts compared to traditional designs and supporting their sustainability as a competitive energy source.

6 Reliability and Risk Assessment

In this chapter, the critical developments in reliability and risk assessment for ships and offshore structures, including the integration of digital technologies to predict and mitigate structural failures, and the importance of incorporating lifecycle considerations into the design and maintenance of marine structures, are reviewed.

6.1 Structural Reliability Analysis (SRA)

6.1.1 Introduction to SRA

Over the past three years, the field of Structural Reliability Analysis (SRA) for ships and offshore structures has experienced significant advancements, evolving towards increasingly sophisticated, multidimensional and probabilistic methods, along with a focus on holistic and systemic approaches. These developments are crucial and the integral role of modern maritime operations in the global economy and the inherent complexities of marine structural systems, also considering significant evolution in fixed and floating wind turbine foundations. This shift is essential for addressing the unique challenges of maritime structures, enhancing their safety, efficiency, and sustainability in a rapidly evolving global context.

Central to this evolution is the shift from traditional bivariate statistical methods to more advanced reliability approaches tailored to multi-dimensional structural responses. Studies by J. Sun et al. (2024), Gaidai, Yakimov, et al. (2023), and X. Xu et al. (2023) highlight the need for methods that effectively manage the high-dimensionality and complex cross-correlations characteristic of maritime structures, especially under harsh environmental conditions. These advanced methods enable a more accurate prediction of system failure risks and are applicable to a wide range of marine systems beyond cargo ships.

Specific challenges in maritime operations, such as ice and corrosion, have also garnered attention. Deshpande et al. (2024) emphasize the need for better ice accumulation prediction models, acknowledging the complex interplay between environmental parameters and ship features. Similarly, studies by Woloszyk and Garbatov (2022) (2024) underscore the significance of understanding corrosion in ships, advocating for probabilistic frameworks and Bayesian inference methods to refine corrosion estimation, while Hlaing et al. (2022) studies similar methods for inspection and maintenance planning of offshore wind applications. Also, Yamamoto et al. (2024) proposed a new corrosion wastage model which can treat mechanical wear observed in some ship structural members.

Vibration-induced stresses, a major concern in ship durability, are addressed through reliability-based frameworks as discussed by Kleivane et al. (2024) and Sadeghian

(2022). They recognize the lack of universal solutions for various vibration problems onboard ships and propose stochastic models and probabilistic analyses to assess the consequences of vibrations, particularly from engine and propeller excitations.

With respect to computational analysis, the Finite Element Method (FEM) has been instrumental, especially for dealing with structural discontinuities in ship construction, as explored by Hoque and Islam (2023). Additionally, Cui et al. (2023) introduce groundbreaking reliability-based topology optimization frameworks that integrate failsafe models, enhancing the structural integrity of complex marine architecture, while Leimeister and Kolios (2021) and Al-Sanad et al. (2023) presented a reliability-based design optimization of a fixed and spar-type floating offshore wind turbine support structure.

A more holistic and systemic view of structural integrity is evident in recent research. Kurniawan et al. (2023) and Yang et al. (2023) advocate for lifecycle considerations in ship structures, emphasizing the need for optimal maintenance programs and robust design strategies to ensure long-term reliability and safety. Abed et al. (2024) further contribute to this paradigm shift by integrating differential geometry into structural optimization and reliability analysis, offering innovative tools to manage uncertainty in various structural forms, including ships.

Environmental considerations and inherent uncertainties in maritime operations have also been a focal point. X. Zhang et al. (2023) address the challenges posed by arctic environmental conditions, proposing robust control schemes for mooring systems. Additionally, the uncertainties surrounding wave-induced ship responses, as discussed by Parunov et al. (2022), highlight the need for improved modelling techniques in the context of the shipping industry.

6.1.2 Methodologies

The evolution of methodologies in Structural Reliability Analysis (SRA) of ships and offshore structures has marked a significant shift from traditional to advanced techniques, incorporating probabilistic and statistical methods to enhance accuracy and predictability, particularly Bayesian methods and Monte Carlo simulations. This transition is reflected in recent studies, emphasizing the importance of integrating diverse data sources, sophisticated simulations, and modern statistical approaches to address the complexities of marine structural systems. These methodologies, often integrating SHM (Structural Health Monitoring) data, Digital Twin technologies, offer a more comprehensive understanding of structural behaviour, accommodate uncertainties more effectively, and enhance the predictability and accuracy of reliability assessments in the complex and dynamic marine environment.

Traditional SRA methods have been the foundation for assessing ship structural reliability, focusing on the hull girder ultimate bending capacity (Li et al., 2022). However, these methods often involve uncertainties that are difficult to quantify due to a lack of real-scale data. Georgiadis et al. (2023) address this challenge through Bayesian statistical inference, demonstrating a novel approach to determine the probabilistic model parameters based on expert judgment and high-fidelity data. This method is particularly relevant for scenarios with specific structural characteristics, allowing for more tailored and accurate reliability assessments.

The integration of SHM systems has been a crucial advancement, providing real-time data to predict breakages and extend lifetimes of structures. Moynihan et al. (2023) showcase a comparison between deterministic and probabilistic (Bayesian) model updating in the context of offshore wind turbines, highlighting the effectiveness of Bayesian updating in estimating posterior distribution parameters with increasing certainty. The value of information extracted from SHM systems is quantified in Kamariotis et al. (2022), employing a Bayesian decision analysis framework. This comprehensive approach models the entire process from data generation to model updating and reliability calculation, applied to a deteriorating bridge system. Similarly, Zhu et al. (2022) and Kamariotis et al. (2023) apply Bayesian updating to integrate experimental data into reliability estimation, improving accuracy and addressing epistemic uncertainty in complex structural systems.

In high-dimensional model parameter spaces, Jerez et al. (2022) explore the application of reliability methods within Bayesian model updating. This approach, which uses subset simulation algorithms and parametric model reduction techniques, effectively addresses complex structural dynamic models. Bayesian methods have also been revisited in geotechnical engineering for spatially varying soil parameters estimation, as demonstrated by Jiang et al. (2022). Their comparative study of different Bayesian methods underscores the effectiveness of these approaches in both low- and high-dimensional problems. Kinne and Thöns (2023) present an innovative approach to updating the continuous stress range distribution of a welded connection in a wind turbine support structure using strain measurements. This method uses Bayesian probability theory to reconcile predicted and design stress ranges, contributing to a more accurate reliability calculation.

Digital Twin (DT) technology, as explored by M. Wang et al. (2022), represents a significant leap in offshore platform reliability analysis. Employing Markov Chain Monte Carlo (MCMC)-based Bayesian updating, this method demonstrates how natural frequencies and mode shapes obtained from structural dynamic responses can be utilized to create a highly precise virtual model, effectively reducing parameter uncertainty. Kleivane et al. (2024), Kitson et al. (2023) and Park and Kim (2022) further emphasize the importance of probabilistic methods in SRA, with Monte Carlo simulations playing a crucial role in estimating reliability indexes and addressing uncertainties in material properties, environmental conditions, and structural behavior.

6.1.3 Case Studies

Recent case studies in Structural Reliability Analysis (SRA) across applications for ships and offshore structures have provided critical insights and lessons, showcasing the importance of advanced methodologies in ensuring the safety and efficiency of these structures.

Studies on floating offshore wind turbines (FOWTs), like the one from Song et al. (2023), highlight the significance of fatigue reliability-based design. Their study adopts the C-vine copula method to model concurrent wind and wave conditions, emphasizing the importance of addressing uncertainties in fatigue estimation. The incorporation of a probability-based sampling method and bootstrap simulations to assess uncertainty quantitatively marks a notable advancement in the fatigue reliability evaluation of FOWTs.

Farhan et al. (2022) delve into maintenance optimization for deteriorating offshore structures. Their research outlines the optimal inspection and repair strategies for welded joints in offshore wind turbine structures, employing decision and value of information analyses to identify the most cost-effective maintenance schedule. Heo et al. (2022) explore the sustainable reuse of decommissioned offshore jacket platforms to support wind turbines. Their study employs a fatigue reliability-based framework to plan and optimize the repurposing of these structures, integrating life-cycle evaluation and revenue optimization. This approach underscores the potential for sustainable asset reuse in the offshore energy sector.

Othman and Mohd (2024) present a detailed study on the reliability of an offshore jacket platform, employing nonlinear finite element analysis (NLFEA) to evaluate the ultimate strength of the platform under varying water depths and environmental conditions. Their findings emphasize the critical role of response analysis in designing and operating offshore structures, particularly in identifying the structure's weakest parts under different load directions. Farhan et al. (2022) present a probabilistic approach to assess the reliability of offshore jacket structures. Utilizing the Monte Carlo Simulation method, their research demonstrates high reliability levels for these structures under various environmental conditions, reinforcing the importance of incorporating statistical methods in structural reliability assessments.

Gao et al. (2024) propose a life cycle structural integrity design method for offshore oil platforms and subsea production systems (SPS). Their approach combines operational reliability, risk controllability, and maintenance economy, extending the time dimension to the full life cycle of structures. This study exemplifies the need for comprehensive design methodologies that address the long service life and high-risk nature of offshore systems.

Olatunde et al. (2023) introduce a methodology for Multi Criteria Decision Making (MCDM) analysis in the context of deep-sea pipelines. By focusing on the most critical failure modes through expert judgment and existing literature, the study provides a streamlined process for predicting the longevity of offshore structures and informs key decisions for decommissioning and repurposing.

6.1.4 Technology Integration

The integration of advanced technologies such as artificial intelligence (AI), machine learning (ML), and computational tools in Structural Reliability Analysis (SRA) is revolutionizing the field, offering enhanced accuracy, efficiency, and innovation in design and analysis processes. These technologies are being increasingly applied to a range of structural applications, including bridges, ships, wind turbines, and offshore platforms.

Velasco-Parra et al. (2022) showcased the use of ANSYS software for structural analysis in evaluating the performance of environmentally sustainable composite structures in ship manufacturing. This approach illustrates the advances in computational tools for assessing new materials and designs. Gao (2022) and Hoque and Islam (2023) emphasized the significance of multifunctional physics coupling software, like COMSOL, and finite element method (FEM) programs in analysing the reliability and safety of ship components such as propellers and structural discontinuities. These tools provide detailed insights into structural behaviour under various operational conditions. Shi et al. (2022) developed a comprehensive 3D finite element model of a wind turbine using Abaqus and validated it through dynamic analyses. The integration of wind loads, wave loads, and servo-control commands in their framework exemplifies the use of advanced computational tools in simulating complex structural systems.

Q. Wang et al. (2022) presented a study on the deformation of blast-loaded metal plates, utilizing von Karman's large deformation theory and a computational fluid dynamics (CFD) solver. The integration of a machine learning regression using the sklearn package of Python to predict deformation outcomes highlights the potential of ML in enhancing prediction efficiency and reliability in structural analysis. Xu and Duan (2022) and Hong and Sheng (2022) focused on the seismic analysis of offshore wind turbines, employing software platforms that integrate aerodynamic, hydraulic, and seismic simulation capabilities. These studies underline the importance of considering multiple environmental factors in structural analysis to ensure reliability and safety.

Fu et al. (2022) developed an intelligent wireless monitoring system for bridges that employs ultralow-power event-triggered wireless sensors and an AI-based framework. This system provides rapid assessment of bridge conditions, including impact localization and force estimation, demonstrating the effectiveness of AI in real-time monitoring and damage assessment. Finally, Lu et al. (2023) and Veen et al. (2023) highlighted the integration of spectral representation and dimension-reduction methods for stochastic wind-wave field modelling and the enhancement of offshore operations simulation software for designing floating offshore wind turbines. These studies represent the cutting-edge integration of computational methodologies in structural reliability analysis.

6.2 Application of SRA

6.2.1 SRA in Offshore Structures

The integration of SRA in various industries signifies a major advancement in designing, monitoring, and maintaining structures, especially in offshore, maritime, and wind turbine sectors. By utilizing advanced computational models, sensor technologies, and innovative methodologies, these industries effectively manage uncertainties, optimize safety, and enhance the lifecycle of critical assets.

Jorgensen et al. (2022) developed a Gaussian Process surrogate model for predicting cumulative fatigue damage in bolts of offshore wind turbine structures. By considering uncertainties in bolt pre-load estimation, this model enables accurate predictions of annual probability of failure and reliability index under the fatigue limit state. This innovative approach informs practical inspection and maintenance strategies, thus integrating SRA into the lifecycle management of these structures.

In the offshore oil and gas sector, Wu et al. (2022) emphasized the significance of SRA. They reviewed theoretical models, numerical simulations, and experimental tests in hydrodynamic analysis of offshore platforms, structural mechanics of flexible pipes and cables, and monitoring technology under marine loads. These findings underscore the need for advanced SRA applications to address environmental load uncertainties in these industries. Gaidai, Xu, et al. (2023) and J. Sun et al. (2024) presented new techniques for risk and reliability analysis of multi-dimensional nonlinear dynamic systems, crucial for evaluating failure risks in complex structures. These methodologies signify the integration of advanced SRA methods in industry. Vergassola et al. (2023) introduced a low-fidelity model for lattice support structures in offshore wind applications to achieve computational efficiency without sacrificing prediction reliability. This approach is reflective of the industry's move towards advanced modelling techniques in wind energy asset management.

Agbakwuru et al. (2023) analysed SHM and maintenance technologies for Offshore Wind Turbine Systems (OWTS) using a multi-criteria evaluation method. The study prioritizes technologies based on various indicators, emphasizing the role of SHM in enhancing OWTS reliability and reducing maintenance costs. Eichner et al. (2023) presented a framework for optimizing the placement of vibration sensors on offshore jacket structures using a genetic algorithm. This method maximizes the value of information from the monitoring system, showcasing the role of SRA in guiding maintenance and inspection decisions.

Parunov et al. (2022) and Vanem et al. (2022) discussed modelling uncertainties in wave-induced loads and environmental conditions for maritime structures. These uncertainties impact ship structural reliability analysis and the development of ship digital twins, highlighting the integration of SRA in maritime industry practices. Neves et al. (2023) suggested a methodology for developing digital twins for the oil and gas industry that perform automated numerical analysis based on high-fidelity Finite Element Model Updating (FEMU). This method exemplifies the advanced integration of SRA in managing the lifecycle of offshore structures, particularly in counteracting aging effects. Wei et al. (2023) developed a framework to estimate extreme hydrodynamic loads induced by typhoons on bridges. This method, considering environmental correlations, is crucial for ensuring the safety of sea-crossing bridges under extreme weather conditions.

Yang et al. (2023) proposed a novel Reliability-based Design Optimization (RBDO) strategy for offshore engineering structures, aiming to enhance cost-efficiency and structural reliability. This strategy demonstrates the evolving application of SRA in optimizing marine engineering structure designs. Cui et al. (2023) introduced a reliability-based topology optimization framework that incorporates fail-safe models to enhance reliability in complex marine structures. This innovation reflects the industry trend towards sophisticated SRA techniques for ensuring structural integrity.

6.2.2 SRA in Industry

The integration of Structural Reliability Analysis (SRA) in rule development for maritime and offshore industries has seen substantial progress, impacted regulatory frameworks and fostered collaboration with classification societies. The integration of modern methods into regulatory frameworks and rule development processes is instrumental in enhancing safety, efficiency, and sustainability in these sectors. Motok et al. (2022) identified a gap in the sophistication of ship design methods compared to other industries like aeronautics or nuclear. They noted the implementation of reliability-based methods in modern rules and codes for the design of ocean-going vessels but highlighted a lack in such development for river-sea ships. Their research aimed to estimate the reliability index of rule-based formulas for still water bending moments, based on the analysis of over 400 load cases of 37 river-sea tankers. This study is a stepping stone for future investigations that could encompass wave-induced loads.

Gaidai, Storhaug, et al. (2022) discussed the need for more optimized structural design in the shipping industry, driven by sustainability and digitalization. Their work on hull stress monitoring systems for container ships revealed the challenges of providing reliable forecast predictions for hull girder loading. The study highlighted the importance of reliability assessment methods for on-board decision support, which could facilitate a balance between safety and structural design without compromising seamanship.

Adiputra et al. (2023) presented a reliability-based approach to assess the impact of structural and load uncertainties on ship hull girders. The study incorporated various uncertainties, including material properties, geometric properties, and environmental factors, to calculate the hull girder ultimate strength using Smith's method. The research highlighted sagging conditions as the dominant collapse mode, validating the approach's viability for assessing ship safety.

Hanif et al. (2023) conducted ultimate strength analyses of stiffened panels by varying several parameters, including initial geometric imperfection. Their research derived a new formula for predicting the ultimate strength of stiffened panels, emphasizing the considerable influence of varied parameter values on ultimate strength and collapse modes.

Katanyoowongchareon et al. (2023) applied quantitative risk assessment (QRA) and structural reliability analysis (SRA) in the context of offshore platforms in the Gulf of Thailand. Their approach to inspection strategy demonstrated how QRA and SRA can optimize operational costs while maintaining safety and sustainability standards.

Parunov et al. (2022), D'Antuono et al. (2023) and D. Liu et al. (2024) further elaborated on the integration of SRA in various contexts. Parunov et al. focused on uncertainties in wave-induced ship responses, D'Antuono et al. on the significance of signal analysis in Structural Health Monitoring (SHM) systems, and D. Liu et al. on a fatigue assessment method for offshore steel catenary risers under vortex-induced vibrations.

6.2.3 Challenges and Opportunities

Several studies underline the pressing challenges and emerging opportunities in SRA. Addressing the gap between research and practice requires innovative approaches, advanced analytical methods, and the integration of digital technologies to enhance the reliability and safety of structures in various industries.

Moseyko and Ol'Khovik (2023) discuss the complexities in predicting the service life of ship mechanical systems operating in the Arctic, considering cyclic loads and marine corrosive environments. Their work analyses the lifecycle of various ship systems to refine reliability calculations and develop maintenance and repair regulations. This reflects the increasing emphasis on the reliability of marine systems under challenging environmental conditions. Adumene and Ikue-John (2022) review the challenges in offshore oil and gas drilling operations in harsh arctic environments. They emphasize the importance of understanding environmental constraints and associated risks for the safe design, installation, and operation of offshore infrastructures. The paper suggests resilient innovation, IoT, and digitalization as potential solutions to address these challenges.

Gaidai, Storhaug, et al. (2022) highlight the shipping industry's focus on sustainability and the reduction of CO_2 emissions, which has led to an emphasis on optimizing structural design without compromising safety. The use of sensor technology, particularly in hull stress monitoring systems, is crucial for providing reliable predictions and enhancing safety in container ships. The study demonstrates the importance of balancing safety and structural design through advanced monitoring and analysis methods. Shi et al. (2023) focus on the development of floating photovoltaic (FPV) systems as a clean energy source. They summarize the latest progress in FPVs, emphasizing the critical structural design considerations and identifying significant challenges in survivability, long-term reliability, and environmental impact for marine FPV systems.

Salis et al. (2024) emphasize the growing number of offshore floating structures and the increasing frequency and severity of extreme waves due to climate change. Their research on modelling wave-structure interaction using the Smoothed Particle Hydrodynamics (SPH) technique and dynamic analysis of moorings and structures reflects the growing importance of numerical models in the design and optimization of marine structures. Gaidai, Fu, et al. (2022) describe a structural reliability method suitable for multi-dimensional structural responses, emphasizing the need for methods that can manage system high dimensionality and cross-correlation between different dimensions. The study focuses on container ships and highlights the challenges in modelling phenomena like whipping, a transient hull girder vibration phenomenon, underscoring the need for reliable methods to predict system failure probability.

Wu et al. (2024) introduce a dynamic analysis framework for offshore wind turbines in ice-covered sea areas, considering soil-structure interaction, stochastic ice loads, and aerodynamic loads. This study highlights the need for comprehensive approaches to ensure the structural safety and reliability of offshore wind turbines in complex marine environments. K. Xu et al. (2023) address the challenges in estimating offshore wind turbine loads, focusing on the effects of rotating blades on wind velocity and the interactions among wind, waves, and structures. They propose a multivariate coherence effect-based evaluation method to analyse the rotational effect on wind velocity, demonstrating the complexity and need for sophisticated analysis methods in offshore wind turbine load estimation.

Do (2023) highlights the challenges in assessing the structural integrity of Tension Leg Platforms (TLPs) after collisions. The development of computational models and empirical equations for predicting residual strength post-collision underscores the need for improved design methods to predict the ultimate strength of structures subjected to accidental impacts. Tayab et al. (2022) address the challenges in the oil and gas industry related to efficiency, safety, and asset integrity, especially in harsh environmental conditions. They review process safety events and emphasize the importance of integrating process safety and operational excellence frameworks to prevent incidents. Shenoy et al. (2022) discuss the challenges and strategies involved in extending the life of aging jetty berthing structures. The study reveals the difficulties in assessing the current condition of structures due to factors like limited existing documentation and thick marine growth. The successful restoration and integrity assurance for extended service life demonstrate the effectiveness of thorough inspection and structural analysis strategies.

6.3 Risk-Based Assessment for Ships and Offshore Structures

6.3.1 Risk Identification and Analysis

The field of risk identification and analysis for ships and offshore structures is expanding and adapting rapidly to address new challenges. The integration of sophisticated analytical tools, coupled with a focus on emerging risks such as autonomy, cybersecurity, and environmental changes, is driving the development of robust, predictive, and comprehensive risk assessment models.

The maritime industry has been navigating through change, largely driven by technological advancements and evolving environmental conditions. This period has seen the emergence of new risks, particularly with the advent of smart shipping and Maritime Autonomous Surface Ships (MASS). Researchers like Chaal et al. (2022) and Bolbot et al. (2022) have studied the complexities introduced by this shift towards automation, highlighting the challenges in ensuring navigation safety and managing the uncertainties associated with autonomous systems. The trend towards autonomous shipping is evident in the wealth of research focusing on the risks associated with MASS. This shift, as elaborated by Ventikos et al. (2023) and Y. Zou, Y. Zhang, et al. (2024), indicates a move towards more autonomous operations, paralleled by the integration of advanced technologies for predictive risk assessment.

Parallel to these technological risks, the industry is contending with natural hazards. Studies like those of Jiang, Fu, et al. (2023) point out the significant risk of fire accidents on passenger ships, characterized by their rapid spread and the difficulty in managing effective evacuations. The exploration of extreme environments, such as the Arctic, further compounds these risks, with studies by Adumene and Ikue-John (2022) and Xin et al. (2023) spotlighting the unique challenges posed by harsh weather conditions and sea ice.

Addressing these emerging risks, the field has seen innovative advancements in risk modelling and analysis techniques. The adoption of System Theoretic Process Analysis (STPA), as used in (Chaal et al., 2022) research, offers a solid foundation for hazard analysis in highly autonomous systems. Likewise, Bayesian Networks and machine learning tools are increasingly utilized for their robustness in assessing risks in complex and uncertain scenarios.

The modelling of collision risks has also evolved, using data from systems like AIS to predict risks in crowded waterways, considering variables like ship domain, manoeuvrability, and traffic density. Techniques like Event Trees and Failure Mode and Effects Analysis (FMEA), as discussed in the works of researchers like Z. Wang et al. (2023), L. Liu et al. (2022), and X. Li, P. Oh, et al. (2023), provide a quantitative approach to assessing and mitigating risks in diverse maritime scenarios, ranging from LNG bunkering to fire accidents.

Environmental changes, notably due to climate change, are also influencing risk assessment methods. The need for comprehensive risk management frameworks that incorporate various risk factors is highlighted in studies like those of Zhao et al. (2023) and W. Zhang, Y. Zhang, et al. (2022). These frameworks are being developed to encompass the entire lifecycle of ships and offshore structures, from design to decommissioning.

The offshore wind industry has also been instrumental in risk identification studies, as risk-based approaches, including risk-based inspection and maintenance attract significant attention. To this end, Lopez and Kolios (2022), López et al. (2023) and Lopez et al. (2024), have studied in depth risk identification and quantification of offshore wind turbine blades, while Kolios (2022) has looked into the risk identification of offshore cables through FMECA.

6.3.2 Decision-Making Frameworks

The maritime and offshore industries are increasingly utilizing sophisticated decision-making frameworks that intertwine risk assessment with cost-benefit analysis. These frameworks help optimize design, operation, and retrofitting decisions by balancing economic, environmental, and safety considerations. Advanced models, such as Bayesian networks and probabilistic methods, facilitate more informed, data-driven decision-making processes.

Recent studies in maritime and offshore engineering demonstrate an increasing emphasis on integrating risk assessments into the design and operational stages of maritime and offshore projects. Zhang and Zhou (2023) illustrated this through the Hangzhou Bay Cross-sea Railway Bridge project, where mathematical models facilitated optimized construction organization and investment control. Similarly, Christensen et al. (2022) evaluated the risks of Arctic shipping, providing insights into safer and more cost-effective maritime routes in response to global warming.

The importance of cost-benefit analysis, considering both economic and environmental factors, has become vital in decision-making processes. Studies like those by Garbatov, Georgiev, et al. (2023) on ship propulsion system retrofitting and Taghavifar and Perera (2023) on the potential of LNG as a greener fuel option for ships, used life cycle assessments and probabilistic models to underline benefits in terms of emission reductions and fuel efficiency.

With reference to structural design and retrofitting, innovative approaches that integrate risk assessments with cost considerations are notable. Garbatov, Palomba, et al. (2023) developed a risk-based formulation for ship hull structural design, considering environmental and operational risks alongside economic factors. Alzayedi et al. (2022) evaluated alternative propulsion technologies and fuels, highlighting their economic and environmental impacts.

The use of advanced models for comprehensive risk assessments has become more prevalent. Luoma et al. (2022) applied Bayesian networks for multi-criteria decision analysis in biofouling management, considering environmental and cost factors. Basnet et al. (2023) introduced a method combining Systems-Theoretic Process Analysis, Bayesian Network, and Influence diagram for risk-based decision-making in remote pilotage operations.

Emerging risks in maritime operations are addressed through robust decision-making frameworks. Acheampong and Kemp (2022) analysed health, safety, and environmental regulation in the offshore oil and gas industry, applying economic approaches for evaluating safety investments. Blindheim et al. (2023) proposed a method to transform qualitative risk analysis results into numeric optimal control problems for autonomous ship navigation, demonstrating the integration of risk management into autonomous control systems.

6.3.3 Environmental and Operational Considerations

The integration of environmental and operational considerations in maritime and offshore activities has grown increasingly complex, particularly in the context of environmental risk assessment for Arctic operations and operational risks in changing maritime routes. This complexity is driven by global environmental concerns, advances in maritime technology, and the evolving nature of maritime routes, particularly in sensitive regions like the Arctic. The integration of advanced analytical methods, comprehensive risk management strategies, and a focus on sustainable practices reflects the industry's commitment to navigating these complex challenges effectively and responsibly.

Environmental risk assessment for Arctic operations is a primary focus due to the unique challenges posed by this region's harsh climate, ice conditions, and ecological sensitivity. Research by Bergström et al. (2022) and Liu, Ma, et al. (2023) underscores the necessity of comprehensive risk management strategies that address both short-term operational risks and long-term extreme ice loads. The Polar Operational Limit Assessment Risk Indexing System (POLARIS) has been expanded to include the magnitude of consequences of potential adverse events, while Liu, Ma, et al. (2023) utilized fuzzy evidential reasoning for spatiotemporal dynamic assessment of navigable environmental risks. These approaches exemplify the shift towards more nuanced and data-driven risk management in Arctic operations.

The risks associated with changing maritime routes are further highlighted by Yao and Yang (2022) and Zhou et al. (2023) who emphasize the complexity of safety challenges in global shipping. The incorporation of spatial density analysis and machine learning techniques to investigate the patterns of maritime accidents reflects a growing need for innovative solutions to manage these risks effectively. The role of environmental risk factors, such as wave height, sea surface temperature, wind speed, and water depth, is pivotal in understanding and mitigating these risks. Several studies also highlight the need for effective risk control options (RCOs) for maritime accidents, particularly in ice-covered Arctic waters. Siyuan et al. (2023) conducted an identification of RCOs using text mining approaches, demonstrating the importance of comprehensive strategies that encompass environmental, technical, human, and organizational aspects in Arctic shipping.

The maritime industry's response to environmental pressures is critical, as noted by Chuah et al. (2023) in their assessment of greenhouse gas emission reduction initiatives for Malaysian-registered domestic ships. Similarly, Hill et al. (2022) explored the methods used to assess environmental risks associated with military shipwrecks, combining archival research with computational oil spill modelling and environmental sampling. These studies highlight the significance of considering both the immediate and long-term environmental impacts of maritime operations.

Operational risks in offshore platforms and shipyards are also crucial areas of focus. For instance, Wang, Zhou, et al. (2024) developed a dynamic risk assessment method for offshore platform systems using the Bayesian network, considering various subsystems like Human, Production and Storage, and Environment. This approach underscores the importance of a holistic view of risk management in offshore operations.

6.4 Uncertainty Modelling and Its Application to Reliability and Risk Assessment

6.4.1 Uncertainty Sources

In the domain of reliability and risk assessment for maritime and offshore structures, the modelling of uncertainties plays a pivotal role. The evolving research in uncertainty modelling underscores the dynamic and multifaceted nature of environmental and structural uncertainties in maritime and offshore engineering. The shift towards probabilistic models, sensitivity analysis, and the integration of climatic data reflects a deeper understanding of the inherent uncertainties and the need for advanced predictive tools. This approach not only enhances the accuracy of risk and reliability assessments but also informs better decision-making in the design and operation of maritime and offshore structures.

The unpredictability of sea conditions and loading sequences, as discussed by Tamimi et al. (2022), exemplifies the challenges in predicting fatigue crack growth in ship structures. Climate change, with its potential to alter sea state characteristics over the lifespan of ships, further complicates this issue. This necessitates the integration of probabilistic approaches and global climate models in predictive frameworks to account for such environmental uncertainties. Makris et al. (2023) address the stochastic nature of wave loading experienced by ship hulls, integrating the spectral approach into a probabilistic fatigue crack growth model. This model effectively represents both randomness in encountered sea states and the sequential nature of these states, providing a better understanding of fatigue accumulation in maritime structures. Hirdaris et al. (2023) reviewed about uncertainties in hull girder loads prediction focusing on the hydroelastic modelling and the simulation and model tests. It is concluded that challenges in realizing and modelling uncertainties are attributed to the limitations of numerical methods to suitably model nonlinearities, the ambiguity of model tests, and the systematic use of data emerging from computational, model- or full-scale methods. An approach is recommended to assess the uncertainty in the hydroelastic responses to wave loading and an example is provided to demonstrate the application of the procedure. The research by Wiley et al. (2023) on floating wind turbines introduces a unique perspective, focusing on the complex interplay of wind, waves, and current in generating loads. Their sensitivity analysis of multiple input parameters reveals key drivers in design, such as wind turbulence and current velocity, emphasizing the need for precise parameter estimation to reduce safety margins and cost implications.

Adiputra et al. (2023) take this a step further by investigating the effects of both structural and load uncertainties on ship hull girders. Their approach highlights the multifaceted nature of uncertainty, encompassing material properties, geometric properties, initial imperfections, and various environmental factors. This comprehensive view underscores the complexity of reliably estimating hull girder strength, necessitating sophisticated probabilistic methods like Monte Carlo simulations for better accuracy.

Tott-Buswell et al. (2023) and Han et al. (2024) both explore soil-structure interactions in their respective studies, albeit in different contexts. Tott-Buswell et al. focus on the nonlinear soil response to storm-induced loads on monopile-supported offshore wind turbines, using dynamic Winkler-type models. Han et al. take a numerical approach to model ice-structure interactions, considering the spatial heterogeneity of ice sheet materials. Both studies highlight the importance of accounting for complex environmental interactions and their impact on structural integrity.

Estrela et al. (2023)'s work on offshore well cement sheaths further illustrates the limitations of deterministic approaches in structural evaluation. By advocating a reliability-based methodology, they emphasize the necessity to consider statistical information about material properties and loads, providing insights into the design process that deterministic methods fail to capture.

6.4.2 Modelling Techniques

The review of uncertainty modelling in maritime and offshore engineering illustrates a field in transition, marked by the integration of advanced computational methods, probabilistic modelling, and AI-driven techniques. These studies collectively demonstrate the industry's move towards more advanced, data-informed approaches to tackle the inherent uncertainties in maritime environments. This shift is not only enhancing predictive capabilities but also informing safer, more efficient, and economically viable design and operational strategies.

In addition to Georgiadis et al. (2023) and Parunov et al. (2022), several other studies have made significant contributions to this field. For example, X. Li, Y. Zhang, et al. (2023) investigate the probabilistic modelling of ship hull girder strength, considering the inherent variability in material properties and fabrication processes. Their work emphasizes the importance of accounting for these variabilities in structural reliability assessments. L. Wang et al. (2022) focus on the uncertainty in hydrodynamic loading, which is critical for offshore platform design. Their study applies stochastic methods to model the complex interactions between ocean waves and structures, demonstrating the importance of robust probabilistic models in predicting structural responses to environmental loads.

Do (2023) research on submarine hulls post-collision is complemented by additional studies that utilize computational models for different purposes. Similarly, Braun et al. (2022) address the challenge of modelling ice loads on Arctic offshore structures. Their study utilizes a combination of empirical data and simulation techniques to better understand the impact of ice loads, which are notoriously difficult to predict due to their stochastic nature.

Beyond Silionis and Anyfantis (2023) there is growing interest in incorporating uncertainties into Structural Health Monitoring (SHM) systems. Zhang and Noshadravan (2023) present a novel SHM method for offshore platforms, employing vibration-based techniques to detect structural damage. Their method accounts for uncertainties in operational conditions and environmental factors, enhancing the reliability of damage detection algorithms.

Chen et al. (2023) proposed a new method called SWSE (Stochastic Wave Spectra Estimation) to estimate wave spectra based on ship response measurements, in which the frequency domain wave estimation method (FDWE) is extended into a probabilistic analytical framework to estimate the encountered sea states involving uncertainty in transfer functions of a ship. The Hermite polynomial chaos expansion (PCE) is used to represent the uncertainty in the transfer functions and the response surfaces.

Finally, Xiong et al. (2022)'s work on intelligent fire risk identification in cruise ships is augmented by similar studies using AI and machine learning for risk management.

6.4.3 Case Examples

Several case studies underscore a paradigm shift in maritime and offshore engineering toward embracing uncertainty as a central component of design, operation, and strategic planning. The increasing reliance on probabilistic models, data analytics, and machine learning reflects a growing acknowledgment of the complexities inherent in these environments. By integrating these advanced methodologies, the industry is enhancing its capability to navigate the uncertain waters of maritime and offshore operations, leading to safer, more efficient, and sustainable practices.

In maritime and offshore engineering, effectively managing uncertainty is a practical necessity. This is exemplified in studies like that of Maduka et al. (2022) who advanced seismic demand prediction for offshore platforms by integrating modal characteristics with uncertainties in material properties and geotechnical factors. Similarly, X. Li, Y. Zhang, et al. (2023) tackled the complex uncertainty in pitting corrosion of offshore pipelines using Bayesian Network and Hierarchical Bayesian Analysis, highlighting a trend toward probabilistic methodologies that capture temporal evolution and failure probabilities in structural integrity assessments.

In a few decades, probabilistic corrosion wastage models have been proposed to estimate the thickness reduction due to corrosion wear. However, thickness loss due to mechanical damage observed in some ship structural members has not been considered in the past corrosion wastage models. Yamamoto et al. (2024) proposed a new probabilistic model for the case where mechanical wear is superimposed to corrosion wear. By the proposed method, it is possible to evaluate the degree of mechanical wear superimposed into corrosion wear from the thickness measurement data.

Rosa et al. (2022) and Mao et al. (2022) addressed uncertainties in dynamic systems but from different perspectives. Rosa et al. research into reservoir modelling, balancing model fidelity with computational efficiency, which is crucial for both short-term operations and long-term strategic planning. On the other hand, Mao et al. demonstrated how deep neural networks could predict mooring line tensions in offshore platforms under failure conditions, showcasing the integration of advanced data analytics in operational safety.

In their work, Loake et al. (2022) and Aliyar et al. (2022) brought to the fore the importance of accurate environmental modelling. Loake et al. developed a Bayesian model to refine meteorological forecasts, thereby reducing uncertainties in metocean conditions that are vital for offshore operations. Aliyar et al. approached environmental modelling by simulating extreme wave interactions with structures, reinforcing the need for computational accuracy in assessing structural responses to natural forces.

Q. Sun, P. Olschewski, et al. (2024) and Díaz et al. (2022) illustrated the complexities involved in forecasting and decision-making under uncertainty. Sun et al. provided an in-depth analysis of typhoon behaviours using regional climate models, emphasizing the intricate interplay of various model settings and parameterization schemes. Díaz et al.

compared methodologies for wind farm location selection, demonstrating how different approaches cater to diverse project objectives, particularly in balancing computational complexity and decision-making agility.

Lastly, Kamidelivand et al. (2023), Lin et al. (2024), and Hwang et al. (2023) each contribute distinct perspectives on uncertainty management. Kamidelivand et al. evaluated operational and maintenance strategies for offshore wind projects, highlighting the role of planned maintenance in managing unexpected failures and costs. Lin et al. focused on the reliability-based design of monopiles for wind turbines, addressing the underestimated pile-soil stiffness in existing methods. Hwang et al.'s calibration of Cone Penetration Test methods for liquefaction potential assessment in Taiwan showcased regional model adaptations for global applications.

6.5 Conclusions

From the review on reliability and risk assessment, several key observations emerge, illustrating the depth and breadth of recent advancements in Structural Reliability Analysis (SRA) and Risk-based Assessment for ships and offshore structures. These observations reflect the evolving landscape of structural reliability and risk assessment in maritime and offshore engineering, underscoring a shift towards more comprehensive, probabilistic, and technologically integrated approaches.

- There's a significant shift from traditional bivariate statistical methods to more advanced, multidimensional reliability approaches, particularly for systems with high-dimensionality and complex cross-correlations.
- Attention is paid to specific challenges like ice accumulation and corrosion in ships. Probabilistic frameworks and Bayesian inference methods are being explored for more accurate predictions and estimations.
- The use of advanced computational analysis tools, such as the Finite Element Method and reliability-based topology optimization frameworks, is highlighted as crucial for managing structural discontinuities and enhancing the integrity of complex marine structures.
- A trend towards holistic approaches in structural integrity is observed, with a focus on lifecycle considerations and optimal maintenance programs.
- The importance of environmental considerations and the management of inherent uncertain-ties in maritime operations are emphasized, with specific references to arctic environmental conditions and wave-induced ship responses.
- Case studies demonstrate the practical application of advanced SRA methods in real-world scenarios, ranging from fatigue reliability in offshore wind turbines to the sustainable reuse of offshore structures.
- The integration of AI, ML, and computational tools into SRA processes marks a significant advancement, offering enhanced accuracy, efficiency, and innovation in design and analysis processes.
- The application of SRA in various industries, including its influence on regulatory frameworks and rule development processes, is significantly impacting safety, efficiency, and sustainability in these sectors.

- There's a notable emphasis on managing environmental and operational risks, especially in challenging conditions like Arctic operations and changing maritime routes, demanding innovative and data-driven approaches.

7 State-of-the-Art vs. State-of-the-Practice

7.1 Introduction

This chapter aims to identify and potentially narrow the gap between research and practice. One key aspect of the design process is coordination and communication between the various tools used for different purposes and disciplines. For example, the tools that perform calculations for structural or hydrodynamic aspects of the design need to be compatible and consistent with each other, as well as the tools that check compliance with the relevant rules and codes of the ship. This requires careful attention to the interface and integration between the distinctive design tools regarding data exchange and functionality.

The maritime industry is undergoing a rapid transformation driven by the development and application of emerging ICT (Information and Communication Technology), such as artificial intelligence, big data, cloud computing, blockchain and 5G. These technologies offer new opportunities and challenges for the design, operation, and maintenance of ships and offshore structures, subject to complex and dynamic environmental loads, safety regulations, and market demands. The current state-of-the-art and future trends of emerging ICT technologies in ships and offshore structures are reviewed, focusing on their potential benefits, limitations, and research gaps. The main issue is successfully implementing and integrating these technologies in the maritime domain. Some examples of emerging ICT technologies that are relevant for ship and offshore structures are.

- Artificial intelligence (AI) can be used to optimise the design of ships and offshore structures, improve the decision-making and situational awareness of operators and managers, and enable autonomous navigation and control of vessels and platforms.
- Big data can be used to collect, store, analyse and visualise substantial amounts of data from various sources, such as sensors, cameras, radars, satellites, and social networks. This can help to monitor the condition and performance of ships and offshore structures, detect anomalies and faults, predict failures and maintenance needs, and enhance safety and security.
- Cloud computing can provide scalable, flexible, and cost-effective computing resources and services for ships and offshore structures. This can facilitate the sharing and integration of data and information among stakeholders, such as owners, operators, regulators, insurers, and researchers.
- Blockchain can create secure, transparent and decentralised digital platforms for transactions and contracts in the maritime industry. This can improve the efficiency and trustworthiness of supply chains, logistics, trade finance, certification, and compliance.
- 5G can provide reliable high-speed, low-latency wireless communication for ships and offshore structures. This can enable real-time data transmission and remote control of vessels and platforms, supporting new applications such as augmented reality and virtual reality.

The following chapters will discuss the recent state-of-the-art and state-of-practice in design methodology and tools. They will then discuss the development of design optimisation and decision-making tools. Later, advancements in lifecycle management and integrated design approaches will be discussed.

7.2 Design Methods

Ship design has transformed from manual methods to computer-aided digital approaches, primarily since the 1970s, driven by advancements in computer hardware and software. Recent focus has been on implementing intelligent digital tools and systems under Industry 4.0 to improve the efficiency and quality of the ship design lifecycle (Papanikolaou et al., 2024). Key technologies such as parametric optimisation, Simulation-driven Ship Design (SDSD), product lifecycle management, digital twins, and artificial intelligence are frequently employed in the maritime industry for various stages of ship design, production, and operation. Simulation-driven ship design is emerging as a pivotal element within ship design procedures. An example of how these technologies can be implemented in a traditional design approach is illustrated in Fig. 8.

Bucci et al. (2021) discuss the complexities of modern ship design, emphasising the critical role of early decisions in determining ship functionality and design while considering various technical and regulatory requirements within budget constraints. Bottero and Gualeni (2022) discuss a problem-orientated approach using Systems Engineering (SE) methodology, seeking the most efficient and effective solution that meets the customers' needs. Dahlke and Schmelzer (2023) discuss past and present challenges in naval ship design through literature and industry surveys advocating for a holistic and coherent approach to ship design methodology and tool support, particularly during the concept and preliminary design phases (see Fig. 2 in Dahlke and Schmelzer (2023)).

Astrup et al. (2023) address the benefits of digitalising the class verification and approval process using 3D models, thus improving design quality and the review process for designers, shipyards, and classification societies. Sakagami et al. (2022) present the challenges of implementing 3D Model Based Approval (3DMBA) in ship design by using different software for tasks such as 3D modelling, rule calculation, direct strength calculation, and 2D drawing and propose integrating rule and strength calculations directly into 3D CAD systems and implementing viewer functions for efficient plan approval.

Silva-Campillo et al. (2021) present a two-stage design methodology to analyse various cut-out geometries in the transverse web of a torsion box structure in container ship design to identify the most optimised cut-out geometry. Yu et al. (2023) present a reliability-based design optimisation (RBDO) method that integrates three optimisation techniques, namely, topology, shape, and size optimisation, dealing with the uncertainties and demands of the marine environment. Kendibilir and Kefal (2023) present aerodynamics topology optimisation (PDTO) to design lighter and more crack-resistant ship structures with a particular application on an optimised bulkhead geometry, proving that PDTO can improve the strength-to-weight ratio of marine structures.

Wan et al. (2022) address the integrated design of submersible surface ships (SSSs) by employing simulation-based design (SBD) techniques to reduce hydrodynamic resistance using a self-fusion method to modify the hull cross-section shape and a fourth-order

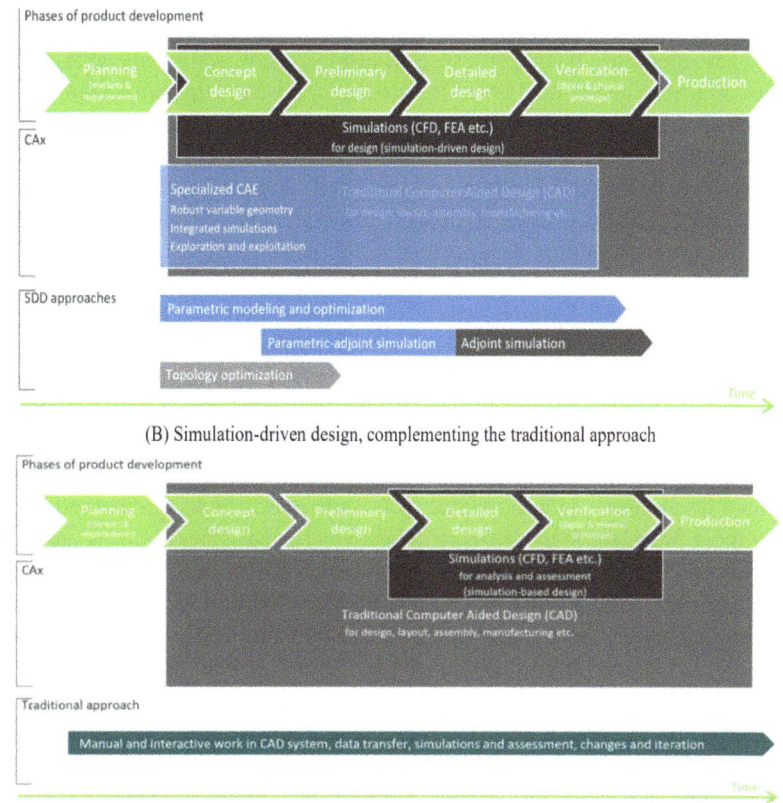

Fig. 8. SDSD use in ship design processes (Reused from Papanikolaou et al. (2024) under CC BY 4.0)

response surface model for hydrodynamic performance. Guan et al. (2022) present an automatic optimisation method for propeller design, combining Fluid-Structure Interaction (FSI), Design of Experiment (DoE), and the Non-dominated Sorting Genetic Algorithm II (NSGA-II). Du et al. (2024) discuss a hybrid simulation-based design (SBD) framework for dynamically adjusting the balance between efficiency and precision in optimising ship resistance by incorporating a particle swarm optimisation (PSO) optimiser, a multilevel Free-Form-Deformation modeller, a Star CCM + solver, and a fully connected neural network surrogate model. It can be concluded that simulation-driven ship design paves the way for parametric optimisation of ship structural design.

7.3 Design Tools

H. Li et al. (2023) proposed a Multi-Domain Feature Mapping (MDFM) tool to improve ship structure design, analysis, and virtual evaluation, enabling seamless collaboration among CAD, CAE, VR, and FEA data in virtual environments. Pérez-Arribas (2023) presents a parametric ship hull design method using Rhinoceros. This widely used CAD software allows easy shape changes based on geometric parameters to create ship hulls defined by key geometric parameters such as displacement, waterplane area, LCB and LCF. Perez-Martinez and Perez Fernandez (2023) propose the transition from 2D drawings to a 3D approach in shipbuilding to optimise resources and address environmental concerns, improving general arrangement, naval architecture calculations, and the primary design stage, allowing for better data reuse and early material and cost estimation. Chon et al. (2023) propose using multiview images from 3D CAD data to reduce scheduling disruptions in shipbuilding. Ryumin et al. (2023) discuss geometric modelling challenges of ship hull structures using SMTU CADS-Hull software to idealise stiffened shells and plate structural design based on local strength and buckling requirements.

Louvros et al. (2021) present a methodology for optimising ship design using a Holistic Design approach to enhance noise and evacuation flow functionalities by rearranging onboard spaces, which improves ship design. Reviewing the trend in the development of design tools, recent efforts have been to further enhance integrated and digitised systems by using emerging technologies such as digital twins, virtual reality, big data, and artificial intelligence. Incorporating 3D models into ship design methodologies is becoming more prevalent in the industry. System engineering perspectives are progressively advanced in the domain of tangible ship design. Computer-Aided Design (CAD) systems are pivotal in the domain of naval architecture, exhibiting a high degree of integration with Computer-Aided Engineering (CAE) and Product Lifecycle Management (PLM) systems. Contemporary developments within this field are primarily directed towards enhancing the efficiency of data exchange across diverse software platforms. The OCX standard, originating from the APPROVED initiative, has garnered substantial industry support with more than 30 maritime sector participants. This standard facilitates streamlining processes, such as vessel design and classification material approvals through 3D models (DAMEN, 2023; DELTAMARIN, 2022). The transfer quality of 3D models between platforms such as NAPA and CATIA has shown considerable potential in effectively managing structural data (Gušani et al., 2023). Furthermore, C. Li et al. (2023) introduce a novel Multi-Domain Feature Mapping (MDFM) tool designed to augment the design, analysis, and virtual evaluation of ship structures. This tool facilitates seamless collaboration between CAD, CAE, Virtual Reality (VR), and Finite Element Analysis (FEA) data within virtual environments.

The shift towards a 3D model-based approach in ship design has accelerated, enabling comprehensive product information models for manufacturing and operational purposes, including digital twins (Bitomsky et al., 2022). This model-based development is essential for tasks ranging from structural optimisation and rule checks to manufacturing processes (Amano et al., 2022). Notable implementations include the integration of 3D CAD with classification software to improve design productivity (Kim et al., 2022) and efficiency (Sakagami et al., 2022). The transition from 2D drawings to 3D approaches optimises resources and addresses environmental concerns, improving general arrangement and early material and cost estimation (Perez-Martinez & Perez Fernandez, 2023). Furthermore, using multiview images from 3D CAD data can reduce scheduling disruptions in shipbuilding (Chon et al., 2023). Other advancements include parametric ship hull design methods using Rhinoceros (Pérez-Arribas, 2023) and tackling geometric modelling challenges of ship hull structures using CADS-Hull software (Ryumin et al., 2023). Reviewing recent trends, emerging technologies such as digital twins, virtual reality, big data, and AI are increasingly incorporated into integrated and digitised systems, enhancing functionalities and advancing tangible ship design (Louvros et al., 2021).

Classification societies provide essential software tools for the shipbuilding industry, enhancing design accuracy and automating production processes. Recent updates highlight tools such as ABS's CSR Prescriptive Analysis and FE Analysis for compliance with multiple CSR versions, integrating with FEM tools like Nastran and Patran (Ivaldi et al., 2022). In May 2024, ABS launched Eagle Unified™, a single-platform tool that can support multiple vessel types, with containership and LNGC capabilities available at launch, and with future releases to incorporate other marine and offshore vessel types. BV offers MARS and Veristar Hull for geometric and strength calculations and 3D FE modelling. In 2023, CCS COMPASS-HCSR software comprehensively address CSR BC & OT requirements, facilitating hull girder strength assessments and direct strength analysis via MSC.Patran. DNV Nauticus Hull provide comprehensive CSR compliance and structural analysis integrated with Sesam GeniE. Other tools include LR's RulesCalc, KR's Sea Trust-RuleScant, and Class NK's PrimeShip-HULL for structural strength and FE analysis. An emerging trend is integrating optimisation tools with classification software, as demonstrated by MARS 2000 by BV.

Various calculation tools have emerged beyond the classification society tools for ship design and structural assessment. The HOLISHIP project uses the CAESES optimisation platform to explore parametric ship models, aiming for environmentally friendly and cost-effective designs. (Papanikolaou et al., 2022). CAESES helps to understand the impact of assorted designs on ship performance, which is crucial for meeting modern economic and environmental requirements. Additionally, MetOcean provides maritime engineers access to oceanographic and meteorological data, integrating historical and forecast data to improve ship design performance and safety evaluations (Dæhlen et al., 2023). These tools reflect a trend towards integrating advanced modelling and environmental data to optimise ship design and operation.

Recent developments in ship structural optimisation employ metaheuristic algorithms and machine learning to boost design efficiency and performance. Silva-Campillo et al. (2022) created an 'omega bracket' for a platform supply vessel, optimising weight without compromising buckling strength. Putra and Kitamura (2021) devised a three-stage optimisation for hatch covers, integrating GA and layout optimisation to reduce material costs. Jiang, Yang, et al. (2023) applied the D-ABC algorithm for multi-objective optimisation, significantly reducing riverboat structures' cross-sectional area and centre of gravity. Topology optimisation (TO) is gaining traction in ship design, with Chu et al. (2021) choosing a level-set method to optimise the weight, size, layout and topology of stiffened panels. Bakker et al. (2021) and Chen et al. (2021) also applied to the stiffener layout and the fatigue performance of the brackets, respectively. C. Zhang, K. Long, et al. (2022) proposed TO for offshore wind turbine jacket structures under aerodynamic and hydrodynamic loads. The integration of TO with additive manufacturing (AM) by Zou et al. (2021) and Amir and Amir (2021) allow the creation of complex structures. As Qiu et al. (2021) use a CNN-RNN model and C. Deng et al. (2022) with the SOLO algorithm, machine learning is advancing TO by accelerating nongradient optimisation processes.

Machine learning, particularly deep learning, has significantly advanced design and optimisation in recent years. Notable applications include structural optimisation for passenger ship hulls using model order reduction techniques (Tezzele et al., 2023) and graph neural networks for stress prediction in stiffened panels (Cai & Jelovica, 2024). AI has also been used to create parametric ship hull designs, achieving notable improvements in performance metrics (Bagazinski & Ahmed, 2023a). The use of machine learning in topology optimisation (TO) has streamlined processes and reduced computational costs, with methods such as finite element network analysis (FENA) and physics-informed neural networks (PINN) showing promising results (Jeong et al., 2023; Jokar & Semperlotti, 2021). Despite challenges such as reliance on large training datasets and scalability issues, these advances highlight the potential of integrating machine learning with traditional design and optimisation methods to achieve more efficient and effective results (Shin et al., 2023; Z. Zhang et al., 2021).

Data collection and storage in the maritime industry is rapidly evolving with the adoption of sensing, digitalisation, machine learning, and digital twins. This evolution presents challenges such as ensuring data quality, privacy, and security, as well as the need for efficient management of large datasets and integration of diverse data sources. Advances include development of scalable methods for realistic testbed scenarios using historical data (Bakdi et al., 2021). Furthermore, C. Wang et al. (2023) introduced an extensive data system for combat methods verification in surface ships.

Virtual reality (VR) and augmented reality (AR) technologies are increasingly integrated into the shipbuilding industry, improving design and operational processes. Gernez et al. (2023) explore VR's role in collaborative maritime design, noting its supplementary use alongside traditional communication methods like emails and 2D sketching. Choi and Park (2021) present AR solutions for efficient navigation and installation in offshore construction. Fernández-Caramés and Fraga-Lamas (2023) provide a comprehensive review of AR and mixed reality in shipbuilding, highlighting advancements in hardware and software and their potential to improve shipyard and ship operations.

Digital twins (DTs) are essential for modern ship design and production, providing a sophisticated method of integrating real-time data with virtual models. They enhance various stages of the life cycle of a ship by continuously learning and evolving through data integration, thus reducing the dependency on traditional physics-based models. Fonseca et al. (2022) emphasise the need for standardised digital twin data to maximise their effectiveness in maritime applications. Although DTs have shown significant promise in the manufacturing and operational phases, Mauro and Kana (2023) highlight a critical gap in their application during the design and decommissioning phases. Fujikubo et al. (2024) present a model that improves stress response predictions for ship structures through advanced data assimilation methods. Recent advances include frameworks for digital twin-based ship design and performance optimisation (Harries et al., 2022) and retrofit design (H. Zhang et al., 2023). However, Papanikolaou et al. (2024) note that there is still a lack of standardised methodologies for design and retrofit based on DT, underscoring the need for further research and development to exploit DT technologies in shipbuilding fully.

In offshore structures, recent advancements have focused on integrating digital twin (DT) technology to enhance monitoring and damage detection. Liu, Zhang, et al. (2023) present a unified DT framework for floating wind turbines (FWT) that enables real-time synchronisation of sensor data with digital models, facilitating global state simulation, visualisation of construction plans, and anomaly detection. This framework has been validated through practical case studies, demonstrating its effectiveness in improving safety and optimising construction processes. Moussavi et al. (2023) propose a DT-based structural health monitoring (SHM) approach that uses a physics-based model to simulate damage scenarios in FWTs. Their method employs a Deep Convolution Long Short-Term Memory Neural Network (DCLSTMNN) to detect and assess damage accurately, with superior performance in handling uncertainties compared to traditional methods. These developments highlight the growing importance of DTs in improving the reliability and efficiency of offshore structures.

7.4 Optimisation and Decision-Making Tools Developments

Y. Liu et al. (2024) introduce an online parameter identification approach called maximum likelihood multi-innovation recursive least squares (ML-MI-RLS) for nonlinear ship motion models. Shabakhty et al. (2024) discuss the structural optimisation of offshore jacket platforms by using a novel metaheuristic algorithm called Enhanced Colliding Bodies Optimisation (ECBO) to reduce computational costs compared to the Genetic Algorithm (GA). J. Liu et al. (2024) present a model-driven method for predicting welding quality in ship construction, addressing issues with existing methods that rely on hypothetical premises and subjective factors. Babadi and Ghassemi (2024) discuss the multi-objective optimisation of a patrol vessel's hull design to improve safety, economic efficiency, and technical performance by implementing strip theory for seakeeping analysis and a fuzzy model to generate the hull form. Y. Zhang et al. (2024) focus on improving ship hull design flexibility using T-spline technology by reducing control points while allowing for local and global modifications. Armanfar et al. (2024) explore integrating lattice structures through additive manufacturing (AM) in shipbuilding, highlighting their benefits, such as cost efficiency, design flexibility, and rapid prototyping.

Cui et al. (2024) propose the bidirectional adaptive conjugate gradient (BACG) algorithm enhanced with a concave vs. convex decision criterion (CCDC) to address inefficiencies and convergence issues in reliability assessment (RA) for optimisation. Kang et al. (2024) address the limitations of deterministic ship structural optimisation by introducing a Reliability-Based Design Optimisation (RBDO) approach. They develop a high-precision agent model for the ship hull structure's limit state using the BP neural network, the radial basis function neural network, and the support vector machine with the SMOTE oversampling algorithm. The Monte Carlo simulation method is used for reliability computation, focussing on a river-sea-going ship. The study integrates the definition of rules, the results of direct structural calculations, and lifecycle reliability requirements as boundary conditions. A simulated annealing algorithm optimises the lightweight structure within the RBDO framework. This system improves both efficiency and accuracy in ship structural optimisation, offering significant improvements for design processes.

Kim et al. (2024) propose a novel approach to optimise hull form design for small ships by parameterising it using 29 variables to create a range of characteristic curves and training an integrated genetic algorithm-Deep Neural Network (DNN) model to predict the hull performance. Feng et al. (2024) discuss the challenges in optimising ship resistance parameters by evaluating six distinct optimisation frameworks, focussing on integrating two surrogate models, namely, the Radial Basis Function (RBF) neural network and Kriging, with three evolutionary algorithms: Particle Swarm Optimisation (PSO), Differential Evolution (DE), and Artificial Bee Colony (ABC). Altunsaray et al. (2023) address the challenge of designing lightweight polymer-based composite high-speed marine vehicles by optimising material selection and incorporating advanced composite components using established supply chains and novel composite materials. H. Li et al. (2023) address the limitations of deterministic design optimisation in ship design by incorporating both aleatory and epistemic uncertainties to derive a unified analysis and propagation method for mixed uncertainty. Y. Zhang et al. (2023) studied a hull form optimisation method using Principal Component Analysis (PCA) for dimensionality reduction using the Multi-Island Genetic Algorithm (MIGA). Huang et al. (2023) propose a ship design optimisation framework that incorporates probabilistic compliance with decarbonisation regulations to enhance the profitability of shipping operations. Bagazinski and Ahmed (2023a) address the challenge of longship design cycles by developing a machine learning tool that uses a newly created dataset of 30,000 ship hulls, by which the hull drag is claimed to be reduced by 60% while preserving key design features.

Lin et al. (2022) study the design optimisation of offshore sandwich blast walls using honeycomb cores to resist oil and gas explosions using ANSYS/LS-DYNA for dynamic performance analysis. C. Sun et al. (2022) address issues in ship line design caused by using discrete data points from offset tables, which lead to fitting errors and difficulty in modification. The authors develop a B-spline curve fitting method using hunger predation optimisation (HPA), incorporating knot guidance, hungry predation optimisation, and adaptive algorithm parameter adjustments. HPA transforms discrete ship lines into continuous B-spline curves, enhancing accuracy and modifiability. The method achieves an optimal set of control points that meets error thresholds by iteratively adjusting the knot vector based on real-time feedback. The effectiveness of HPA is validated through comparisons with related research and engineering software. Y. Wang et al. (2022) address the challenge of designing and optimising hull shapes for superior hydrodynamic performance using deep learning methodologies, e.g., the Deep Neural Network (DNN) approach and Variational Autoencoder (VAE) to encode and synthesise hull designs based on hydrodynamic performance. Two data augmentation techniques, Perlin noise mapping and free-form deformation (FFD), are utilised to enhance the training data set derived from an initial parent hull. The trained VAE efficiently optimises hull designs to reduce drag coefficients. Validation experiments demonstrate that the Convolutional Neural Network (CNN) model accurately reconstructs input hulls and predicts drag coefficients.

F. Liu et al. (2022,) discuss optimising the shape and size of brackets within hull structures to simultaneously reduce mass while ensuring high-stress performance by developing a parametric finite element model and implementing the optimisation process on the Insight platform. The study highlights that equal weighting among dimensionless objectives enhances the effectiveness of the optimisation process, providing valuable information on structural optimisation within the hull design. Sun and Zhang (2022) addressed optimising the seakeeping performance of a container ship under regular waves using potential flow theory to design an energy-saving ship form for practical navigation. The optimisation aims to minimise wave-added resistance and the sum of heave and pitch amplitude, with the Sobol algorithm and improved non-dominated sorting genetic algorithm (NSGA-II) as objectives. This study validates the effectiveness and feasibility of the proposed optimisation approach to improve the characteristics of ship seakeeping under regular wave conditions.

Zhang and Chen (2022) apply a deep belief network (DBN) method for the multi-objective particle swarm optimisation (MOPSO) algorithm of ship design under non-linear wave conditions. This study underscores the efficacy of DBN-based methods in improving ship design optimisation under non-linear wave conditions.

F. Sun et al. (2022) address limitations in traditional marine CAD parametric drawing systems by introducing a configuration-based development approach. Cai and Jelovica (2022) perform a benchmark analysis of swarm and evolutionary algorithms to optimise the structural design of a 180-m-long chemical tanker to meet class society requirements. The algorithms evaluated include PSO, NSGA-II, MOEA/D, MVO, MOMVO, MOGWO, IGWO, and GSA, each tested with various constraint handling techniques (CHTs) such as static and dynamic penalty, adaptive threshold, and repair methods. Nazemian and Ghadimi (2022) study the modification of a trimaran ship hull form to improve resistance and wake field characteristics using a novel approach combining an adjoint solver and CAD-based optimisation platform. This platform facilitates multi-objective optimisation, focussing on shape optimisation through FFD and adjustments to side hull arrangement and length, encompassing eleven design variables.

Vettor et al. (2021) focus on probabilistically defining objectives and constraints crucial to route optimisation systems using ensemble forecast data, showcasing how uncertainties in weather forecasts can be integrated to enhance the reliability and efficiency of route optimisation decisions. Lang et al. (2021) study the impact of operational issues, such as voyage optimisations, on ship fatigue damage accumulations. Doğan and Özden (2022) address optimal welding parameters, such as currents and voltages for thin steel panels used in ship hull construction. Kondratenko et al. (2021) present a holistic multi-objective design approach for optimising Arctic Offshore Supply Vessels (OSVs) for cost- and eco-efficiency during the conceptual design phase, using an adaptation of the Artificial Bee Colony (ABC) algorithm for multi-objective optimisation. Case studies demonstrate the approach's feasibility and effectiveness, showing that icebreaker assistance significantly extends the design space, improving energy and cost efficiency.

Silionis and Anyfantis (2021) discuss steps towards developing an onboard Structural Health Monitoring System for ship hulls, using an optimisation-based approach to identify damage by processing static response data. Three optimisation algorithms, namely gradient-based, genetic algorithm-based, and statistics-based, are evaluated, with gradient and GA approaches proving more efficient. In contrast, the statistics-based method is less computationally demanding. Trinh et al. (2024) present a GAN-based 3D ship hull design optimisation method that bypasses traditional hydrodynamic simulations by training a separate model to predict ship performance indicators. Yu et al. (2024) present a reliability-based topology-topography optimisation (RBTTO) method for designing lightweight and reliable ship bulkheads, considering multiple failure modes. The method integrates material penalisation and node activation (MPNA) to optimise the combined topology and topography. A novel multifailure reliability assessment strategy uses probabilistic criteria and expert elicitation to account for uncertainties. Bertram et al. (2023) reviewed the evolution of practical ship hull optimisation over the past fifty years, examining key components such as geometric models, hydrodynamic models, optimisation algorithms, and objective functions with constraints.

Guan et al. (2021) present a numerical simulation-based optimisation approach for the design of the Twin Hull the design of the Twin Hull Small Waterplane Area (SWATH), integrating a parametric geometry model, numerical analysis of total resistance, response surface methodology (RSM), and evolutionary optimisation algorithm (EVOL). The parametric hull-form model uses characteristic curves and adjustable parameters for rapid automated geometry variation. Numerical simulations calculate resistance forces, which serve as an optimisation objective. A stepwise selection method constructs the RSM model to expedite the optimisation, and the EVOL algorithm resolves the optimisation problem.

7.5 Integrated Design Approach

Papanikolaou et al. (2023) discuss the results of the HOLISHIP project, aimed at reducing marine emissions through innovative design solutions. It focuses on a multi-objective optimisation approach for green shipping, showcasing its application through two case studies: a green RoPAX and a fully electric Double-Ended Ferry. Studies highlight the economic viability and environmental benefits of optimised designs. Amano et al. (2022) present a technology platform for rule-based structural optimisation in ship design, integrating the necessary modelling, prescriptive rule checks, FEM-based rule checks, and weight calculations into a single automated and streamlined process. This integration significantly reduces the iteration cycle time. Ebrahimi et al. (2021) review the effects of complexity on competitiveness in ship design, using archival data from 100 ship design projects by eight Norwegian designers and applying multivariate data analysis to demonstrate a correlation between complexity and competitiveness. The study reveals that the influence of complexity varies in magnitude and directionality among different factors, offering insights to improve complexity management in ship design. Park and Huh (2022) explore how the integration of ICT, including IoT, AI, and Big Data, can accelerate advances in the shipbuilding industry, focusing on methods for collecting and managing diverse data types and communication methods within the industry and on using this big data to adapt to rapid changes.

Ha et al. (2022) propose a method for optimising ship arrangement design using digital twin technology, enabling the consideration of various alternatives before production. This approach involves setting design variables, such as partition location and equipment, with objective functions that include installation cost, expert feasibility, and space availability. The optimised arrangement can be reviewed virtually as if it were manufactured, allowing for visual assessment by designers. Mombiela and Zadeh (2021) present an algorithm for designing ship power systems during the preliminary design phase, incorporating an embedded control at the Energy Management System (EMS) level to optimise cost, availability, safety, and emissions. The study examines alternatives such as full electric propulsion and fuel-based energy producers, evaluating different performance indicators for cost optimisation. The algorithm is applied to an offshore supply vessel (OSV), comparing the battery and generator set (GENSET) power contributions. Compares seven battery types for an all-electric ship, analysing maximum and instantaneous Depth of Discharge (DoD) for EMS control. This algorithm structure lays the foundations for future hybrid power plants.

7.6 Conclusions

Section 7 concludes by acknowledging that design tool advancements have significantly improved marine structures' safety and sustainability. In the current landscape, a significant trend is merging artificial intelligence (AI), machine learning (ML) and data analysis in designing marine structures. These advanced technological tools have the potential to drastically change how structural analysis is done, allowing for ongoing monitoring of a ship's strength and making it easier to create digital replicas using modelling methods. This shift can improve our understanding of how structures behave in real-world conditions, providing valuable insights that can help build better ships and even impact design rules.

Recent progress in deep learning (DL) has made it easier to use these methods, though they still require many data. Looking ahead, the focus will likely move toward examining ship structures with limited data because relevant information is often scarce. A promising approach is to merge data analysis and machine learning with physics-based methods, allowing for tweaking success criteria to match the fundamental physical properties.

Recent advances in Computer-Aided Design (CAD) and Computer-Aided Engineering (CAE) have led to significant improvements in the efficient design of ships and other marine structures. Using digital tools and 3D models has facilitated the design process, ensuring greater efficiency and precision. Although the viability of digital twin technologies (DX) is still in the experimental phase, they hold substantial promise for the future. Establishing a complete Product Lifecycle Management (PLM) is crucial to successfully implementing digital tools. It requires collaboration across many sectors, such as design offices, classification bodies, shipyards, and shipping and service companies. This collective effort helps to enhance overall operational efficiency. When these diverse fundamental technologies are synergised, there is immense potential to yield significant outcomes that individual isolated technologies could not deliver.

8 Benchmark Study of Topology Optimization

8.1 Introduction

Topology Optimization (TO) is known as an optimization technique, which can find the optimum distribution of material(s) within a given design domain to minimize or maximize some objective function(s), while satisfying a set of given conditions (constraints).

It is considered that TO is suitable to generate innovative but effective structures, which cannot be built using the conventional design procedure. For example, in recent years, additive manufacturing has become a common manufacturing technique, and TO has been used to obtain an effective geometry as reported in Sect. 4. The most popular approach in TO is the solid isotropic material with penalization (SIMP) due to its simplicity of implementation, which was initially introduced by Bendsøe (1989) and was applied to the design of compliant mechanisms by Sigmund (1997). The basic concept of TO can be found in the textbook by Bendsøe and Sigmund (2004). Though a lot of research of TO have been conducted for many years, there are not so many examples where TO has been applied to the actual design of ships and offshore structures. On the other hand, in recent years, the simple and easy programs are in public (for example, Sigmund (2001)) which can be used to develop an in-house program code of TO. Moreover, the computer program of TO has been implemented into some commercial FEM software, such as MSC. Nastran, OptiStruct, Ansys, etc. Now TO can become used as an accessible tool for every engineering designer.

Considering such a situation, it is worthwhile to investigate how TO can be used in the actual design of ship and offshore structures. In this committee report, a benchmark study of topology optimization is conducted for ship cross-section of a tanker under multiple loading conditions. 2-dimensional (2D) problem is considered for this benchmark study, to discuss the applicability of TO to practical problems by considering relatively simple 2D problem with which optimization results can be obtained without significant effort. It is noted that many loading conditions should be considered in usual design of ship structures, so that five loading conditions are considered for this benchmark study.

8.2 Formulation of Topology Optimization

8.2.1 Single Loading Condition

When considering single load case problems, typical simple TO problems can be formulated where the objective is to minimize the compliance under that single loading condition, subject to a volume constraint. The problem can be stated as follows:

$$\min_{x} c(x) = u^T K u \quad (1)$$

$$\text{subject to}: \frac{V(x)}{V_0} = f \quad (2)$$

$$K u = F \quad (3)$$

$$0 < x_{min} \leq x_i \leq 1, i = 1, 2, \ldots, n \quad (4)$$

where $c(x)$ is the compliance of the system associated with the load case; u is the global displacement vector; K is the global stiffness matrix; $V(x)$ and V_0 are the material volume and the volume of the design domain, respectively; f is a volume fraction, given to constrain the amount of material that can be used for the structure; F is the global force vector of the load case; $x = \{x_1 \ x_2 \ \cdots \ x_n\}^T$ is the design variables, which means the density of material for each element, n is the number of design variables (number of finite elements), and x_{min} is a positive small lower bound of the design variables to avoid the singularity issue. The stiffness matrix K is defined with respect to the design variables, which is $K = K(x)$.

8.2.2 Multiple Loading Conditions

To deal with multiple loading conditions, the traditional equally weighted-sum method (TEWS) is used in the benchmark study, which is usually implemented in the commercial TO software. In the TEWS method, objective function of the TO problem becomes as follows.

$$\min_{x} f(x) = \sum_{j=1}^{n_L} c_j(x), \quad (5)$$

where n_L is the number of given loading conditions; $c_j(x)$ is the compliance of the system associated with the j-th loading condition. The detail of the concept of TEWS is found in Díaz and Bendsøe (1992). Note that in the following benchmark results, for one in-house program (Program1), the results for multiple loading conditions are also obtained by BCM instead of TEWS, which is based on the min-max compliance concept (Pham-Truong et al., 2023)

8.3 Benchmark Problem (Ship Cross-Section Under Multiple Loading Conditions)

This problem aims to find the optimum design for a frame in the middle part of an oil tanker, considering different loading conditions according to the classification rules. The used regulations follow the Guidelines for Tanker Structures (ClassNK, 2001).

8.3.1 Target Structure

A crude oil tanker (Aframax class) is used as a target structure for this benchmark study. The target structure and its dimensions are completely imaginary ones which are deduced based on the principal particulars of the existing ship available on the internet (ClassNK, 2023). The principal particulars of this ship are given in Table 11. The cross-section design in the middle part is assumed to be similar to the design of the Suezmax DH oil tanker as shown in Fig. 9 (See ClassNK (2023) and Emi (2006)). Then, the structural arrangement of the midship section can be assumed as shown Fig. 10.

It is noted that the topology optimization (TO) problem for this structure can be defined either as a 2D problem or as a 3D (Shell or Solid) problem. However, to simplify the problem, it is assumed in this benchmark study that the TO problem is defined as a 2D problem, aiming to find the optimum distribution of material (steel) in the design domain of the cross section in the middle part of the oil tanker. Moreover, for simplicity, it is also assumed that the midship section of the tanker has the structure with right corners at the bilges and the bilge hopper plates are not arranged, to simplify the problem as shown in Fig. 10. The geometrical parameters were decided as shown in Table 12, which is decided according to the classification rules, as well as based on experience and knowledge about the real large ships.

Table 11. The target Aframax tanker

Gross tonnage	66,000
Deadweight tonnage (DWT)	120,000 tons
Overall length (L_{OA})	247 m
Length between perpendiculars (L_{pp})	235 m
Overall breadth (B)	44.4 m
Depth (D)	22.0 m
Draft (d)	15.4 m

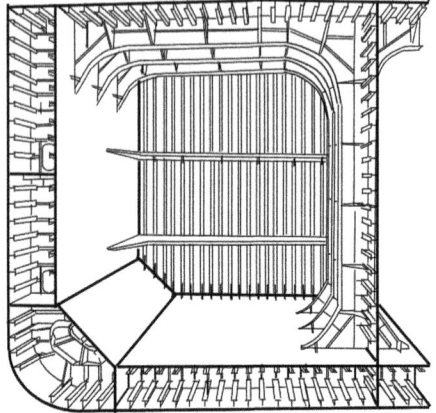

Fig. 9. Typical structure of Suezmax Tanker

The entire ship cross-sectional space is generally considered as the design domain (Case 1 in Table 3). However, as the typical design strategy, geometrical restrictions as like Case 2 and Case 3 shown in Table 13 are also considered in the benchmark study. In case 2, the double side and double bottom must be solid structures. Then, the design domain is the grey region, which is left off after setting the region of the double side and double bottom as always being solid structures (colour black). In case 3, there is a large opening in each of the cargo tanks of the ship. The design domain now is the grey domain, left off after setting the region of the openings as always being empty (colour white). It is noted that the longitudinal primary structural members arranged in longitudinal direction (bold lines in the drawing of the cross section in Fig. 13) contribute to support the loadings. To simulate the contribution of these members (the plating shells, bottom girders and keel, side stringers), bar elements or other structural members should be arranged along these plating.

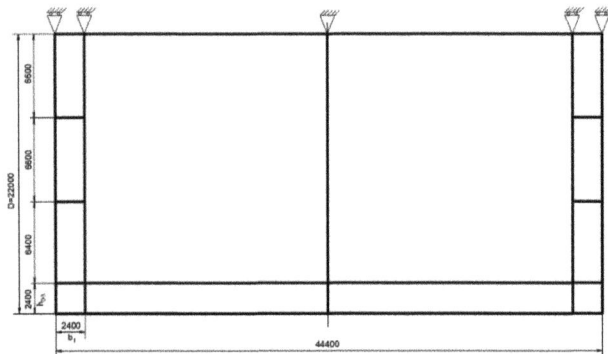

Fig. 10. Target area of TO (A simplified cross-section)

Table 12. Geometric parameters of the target structure

Parameter	Value [mm]	Notes
Double bottom height h_{DB}	2400	
Double side width b_{DS}	2400	
Spacing between side stringers l_1	6600	From the upper deck down
Plating thickness t	20	For all shell plating, including outer and inner side shells, longitudinal bulkheads, bottom keel and girders, outer bottom, inner bottom, upper deck, side stringers
Breadth of attached plates S	4000	This dimension is used to calculate the load magnitude from the pressure of water and oil on the structure
Thickness of the frame structure t_e	15	If the topology optimization problem is defined as the 2D problem, this is used for computation of the 2D finite element analysis.

Table 13. Case studies

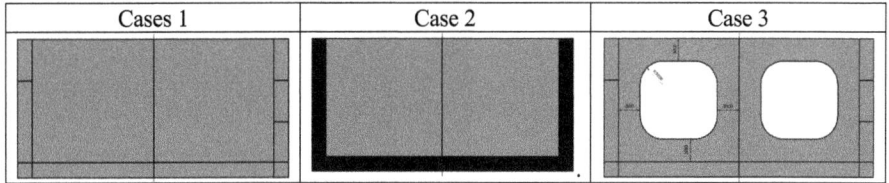

A volume constraint is considered for the topology optimization problem. The volume fraction of the structure is given as a constant,

$$\frac{V_{structure}}{V_0} = f, \qquad (6)$$

where, $V_{structure}$ is the material volume of the structure, V_0 is the volume of the whole domain, and it is assumed that $f = 0.4$. Based on the initial structure of the prototype.

8.3.2 Loading Conditions

To design ship structures, multiple loading conditions should be considered instead of single loading condition. In this study, according to the classification rule (ClassNK, 2001), 5 loading conditions in Table 14 are considered for the optimization. Note that LC2 is similar to the LC1, but the tank on the right-hand side is fulfilled with water in LC2 while that on the left-hand side is fulfilled in LC1.

To calculate the loads applied to the 2D finite element model from Table 14, the breadth of the attached plates ($S = 4000$ mm) is considered, which are attached perpendicularly to the cross-section plane (the design domain of the 2D problem), at the position of the platings (double sides, longitudinal bulkhead, upper deck plating, double bottoms, etc.).

Table 14: Loading conditions

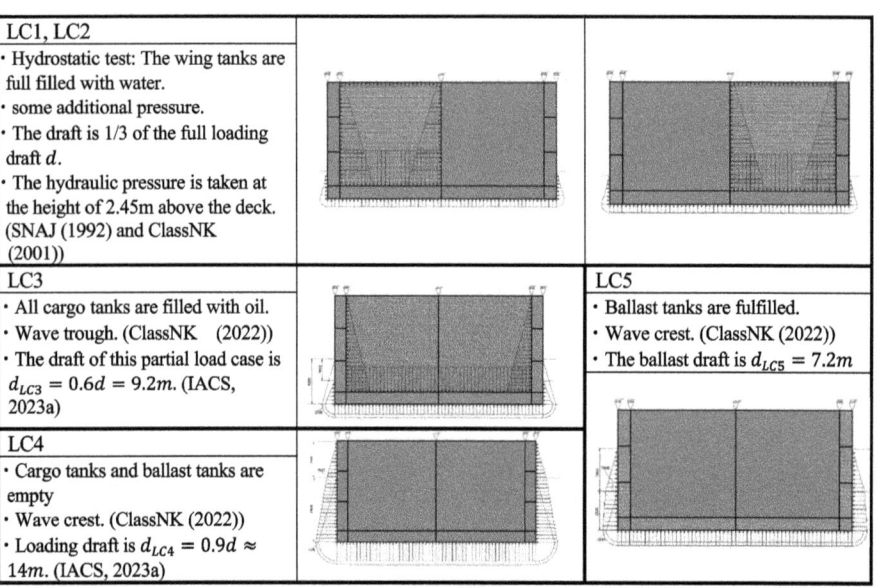

LC1, LC2	
• Hydrostatic test: The wing tanks are full filled with water. • some additional pressure. • The draft is 1/3 of the full loading draft d. • The hydraulic pressure is taken at the height of 2.45m above the deck. (SNAJ (1992) and ClassNK (2001))	
LC3	**LC5**
• All cargo tanks are filled with oil. • Wave trough. (ClassNK (2022)) • The draft of this partial load case is $d_{LC3} = 0.6d = 9.2m$. (IACS, 2023a)	• Ballast tanks are fulfilled. • Wave crest. (ClassNK (2022)) • The ballast draft is $d_{LC5} = 7.2m$
LC4	
• Cargo tanks and ballast tanks are empty • Wave crest. (ClassNK (2022)) • Loading draft is $d_{LC4} = 0.9d \approx 14m$. (IACS, 2023a)	

8.3.3 Finite Element Modelling

On finite element modelling in the benchmark study, it is assumed that that the problem can be modelled as a 2-D problem or as a 3-D shell model. However, the 2-D model of the cross section was defined as an example modelling, which can be regarded as the standard model of the benchmark problem. In this model, the design domain is discretized into 2D square elements. As described in 8.3.1, the shell plating (double sides, longitudinal bulkhead, bottom centre girder and bottom side girders, upper deck plating, double bottoms, side stringers) are modelled using bar elements in the 2D structure (see solid lines in Fig. 10). The cross-sectional area of the bar elements (A) is taken as the cross-sectional area of the attached plates to the structure, which is the product of the plate thickness (t) and the breadth of attached plates (S). The material used to build the ship is steel, the Young's modulus of which is equal to 206,000 N/mm^2, and the Poisson's ratio equals to 0.3, according to (IACS, 2023b).

8.4 Review of In-House Programs and Commercial Software Used in Benchmark Study

In this benchmark study, we solved the TO problem shown in the previous section using two in-house program codes and three commercial software. The features of these tools are as follows.

8.4.1 In-House Programs

- In-house program 1 (Program1(BCM))

  The first program used to solve the benchmark problem was developed by Pham-Truong et al. (2023). This TO program (Program1) was developed using MATLAB based on SIMP approach (Density method). Sequential Linear Programming (SLP) is employed to solve the optimization problem. In this program, firstly the conventional weighted-sum method (See Eq. (5)) is applied to obtain the initial optimum result. Then the structure is further optimized by minimizing the maximum compliance among all ones under each loading condition. The method called Bisection Constraint Method (BCM) was used to minimize the maximum compliance. Here in this report, the TO results obtained by BCM method are presented as "Program1". The volume fraction ($f = 0.4$) is considered as a constraint condition. For finite element mesh, 2D square (plane-stress) elements are used for the cross-sectional area, where the size of the square elements is 200 mm × 200 mm and the number of total elements is 222 × 110. The bar elements are arranged along with the longitudinal members as requested in 8.3.

- In-house program 2 (Program2)

  The second program is developed at UBC using MATLAB programming, based on the SIMP approach, but with an accelerated FEM analysis aimed at high-resolution, large-scale TO. The optimality criteria (OC) method (Hassani & Hinton, 1998) is used as the optimizer to minimize compliance as the objective function. For multiple loading conditions, the objective function is taken as the average compliance of all loading cases. Sensitivity information required includes the first-order derivatives of the objective and constraint (volume fraction) functions, obtainable directly through the chain rule. All parameter settings remain the same as Program1, except for the element size, thus affecting the number of elements. The square elements measure 20 mm × 20 mm, resulting in a total of 2220 × 1100 elements. To address the TO problem with over 20 million elements, the method of multigrid conjugate gradient (MGCG) (Amir et al., 2014) is employed to accelerate finite element analysis. Vectorization and sparse optimization within MATLAB also reduce the time required for global matrix and vector assembly. Additionally, parallel computing using MATLAB's parallel computing toolbox (MathWorks, 2024) allows for the simultaneous computation of compliance across all loading cases in multi-loading conditions.

8.4.2 Commercial Software

We used three TO software to solve the benchmark problem. They are (a) Patran/Nastran (MSC) (Hexagon, 2022) (b) OptiStruct (Altair) (Altair 2024b) and (c) Ansys (Ansys, 2024). Table 15 shows the functions for each software. The feature used to solve the benchmark problem is shown in bold and underline font in this Table. It is noted that Density Method (SIMP method) is used in all software. For the objective function of multiple loading conditions, Patran/Nastran can setup personalized objective. In this study, we consider Average Compliance as objective function. In other software, Sum of Compliance is considered as objective function of multiple loading conditions. Though different objective function is used in different software, the objective function of the optimization can be regarded as similar, because the average is the summation divided by the number of loading conditions.

In Patran/Nastran, Density Constraint (Checkerboard-free method) can be applied, and Minimum Member Size option can be used as Manufacturing Constraint. In the Minimum Member Size option, design variable (density) of the elements within a distance (set parameter, TDMIN) from an element with a design variable close to 1.0 are filtered to ensure they are not small. Value of TDMIN is problem-size dependent but not mesh size dependent. (See Hexagon (2022) for the information of these techniques). In the results of topology optimization by Patran/Nastran shown in the next section, both the Checkerboard-free method and Minimum Member Size Control are applied.

In the optimization by OptiStruct, Minimum Member Size (MINDIM) control option was used which is a manufacturing constraint (Zhou et al., 1999; Zhou et al., 2001). This technique can be used to obtain the simplified geometry. In this study, Minimum Member Size control was used with MINDIM = 600mm. Also, DISCRETE parameter was set as 5 to make the penalty power of SIMP approach $p = 6$, and DELTOP was set as 0.2 (Default value is 0.5) by which fractional move limit for design variables during the optimization is reduced, expecting avoidance of violation of constraints. In Case2, when the above parameter is used for TO, the performance of the final structures was extremely poor (high compliance), although the structures appear to be highly clear and realistic. For this reason, the results of Case2 by OptiStruct in this report was obtained by changing the DISCRETE parameter to 2, instead of 5.

In the optimization by Ansys, we also applied Minimum Member Size option (Ansys, 2023) as like OptiStruct. The value of the minimum member size was set as 1000 mm in this benchmark study. Also, the penalty number, p, was changed to 5 from the default value, 3.

We also tried to use another software, Inspire (Altair) (Altair, 2024a) to solve the benchmark problem. However, we found that it is difficult to use Inspire for this benchmark problem. This is because bar elements cannot be used in Inspire. Also, it is difficult to define complicated loading conditions by Inspire, because it cannot directly deal with the finite element mesh. On the other hand, the advantage of Inspire is that it does not require user to manage the FEM model directly, as it is automatically generated internally. This allows users to easily create geometric models and perform topology optimization. It can be concluded that Inspire is effective software for relatively simple TO problems, but not suitable for this benchmark study.

8.4.3 Summary of Topology Optimization Tools

Table 16 shows the summary of topology optimization tools used in this study. It is noted that different optimization results are obtained if different options are used when using commercial software. Thus, in the results of the topology optimization by the commercial

Committee IV.1: Design Methods, Principles and Criteria   815

software shown below, the only the results are shown selected by the members who participated in the benchmark considering the viewpoint of performance (compliance) or the practicality of the obtained geometry.

8.5 Results of Topology Optimization for Single Loading Conditions

In this section, the results of TO for each single loading condition are shown to compare the difference of the TO results between the TO tools.

(1) Case 1

Figure 11 shows the topology optimization (TO) results for Case 1. In Table 17, compliance for each loading condition for the topology obtained by different tools (programs and software) are shown. It is observed that the values of compliance are different with different tool because of different definition of compliance. One of the important observations is that the loading condition with highest compliance is LC4. This means that the most critical loading condition is LC4. On the other hand,

Table 15. Features for each Commercial Software

	(a) Patran/Nastran	(b) OptiStruct	(c) Ansys
(1) Objective Function	**Minimize Compliance** Maximize Frequency/Eigenvalue	**Minimize Compliance** Minimize Volume Maximize Frequency	**Minimize Compliance** Minimize Volume Maximize Frequency Minimize Stress (p-norm)
(2) Constraint	**Mass Fraction**, etc.	Volume, **Volume Fraction**, Mass, Displacement, von-Mises Stress, etc.	Mass/**Volume**, Compliance, Displacement, Stress(p-norm), etc.
(3) FE model (Element)	3D Solid Element, **2D Shell Element** (**Bar element** for longitudinal members)	**Shell** (**2D**,3D), Plane(2D), Solid (3D), **Bar**, Beam	**Shell**(**2D**,3D), Solid(3D) (for FEM: **Beam**, Spring, Bar, etc.)
(4) Possible Output	Element Density, Curve, surface or solid after fitting	Element Density, OSSmooth (IGES Surface, STL, Hyper3D) Responses (Compliance、 etc.), Volume Fraction	Element Density, STL, Responses (Compliance, Stress, etc.), Volume
(5) Method of TO	**Density Method** MSCADS or IPOPT as optimization algorithms	**Density Method** Level-set Method	**Density Method** Level-set Method

(*continued*)

Table 15. (*continued*)

	(a) Patran/Nastran	(b) OptiStruct	(c) Ansys
(6) Multiple Loading Conditions	**Average Compliance** (Possible to use personalized design objective)	**Sum of Compliance**	**Weighted Sum**
(7) Other Features	**Checkboard-Free Method** –Filtering algorithm –Density Constraint **Manufacturing Constraints** **–Minimum Member size** –Symmetric Constraints –Extrusion Constraints –Casting Constraints	**Manufacturing Constraints:** **–Minimum Member Size Control** –DISCRETE parameter (remove intermediate density) –Maximum Member Size Control –Die Cutting Direction Constraints –Extrusion Constraints · Pattern Repeat · Pattern Grouping	**Manufacturing Constraints:** **–Member Size** –Extrusion –Pull Direction –AM (Additive Manufacturing) Overhang Constraint

Table 16. Summary of Analysis Condition for Topology Optimization Programs and Software

	Program1	Program2	Patran/Nastran	OptiStruct	Ansys
Mesh Size	200mm × 200mm	20mm × 20mm	50mm × 50mm	200mm × 200mm	200mm × 200mm
Element type for longitudinal members	Bar element	2D element	Bar element	Bar element	Beam element
Main options	BCM method	Parallel computing, MGCG	Checkerboard-free method / Minimum member size (150mm)	Minimum member size (600mm)	Minimum member size(1000mm), penalty5

LC5 has smallest compliance, which may mean that LC5 is not important compared with others.

(2) Case 2, Case 3

Figure 12 shows the results for Case 2 in which TO is conducted assuming that the double sides and the double bottom are filled with solid materials. Figure 13 shows the results of Case 3, where large opening exists in each cargo tank. For these two cases, like the Case 1, it is observed that in-house programs (Program1 and Program2) give complicated geometries with many holes compared with the results obtained by the commercial software. Moreover, in Case2 (Fig. 12), much simpler

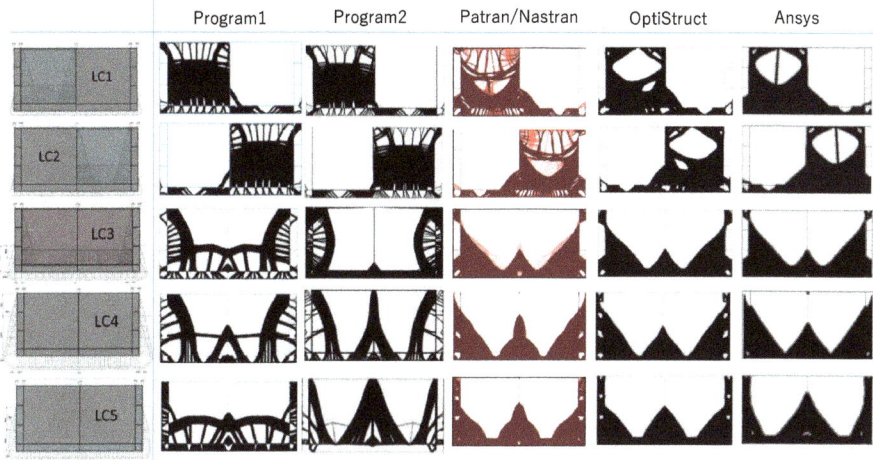

Fig. 11. Case 1: TO results for each single loading condition for different software

Table 17. Compliance for each loading condition for the topology obtained by different tools (programs and software): Case1

	Program1	Program2	Patran/Nastran	OptiStruct	Ansys
LC1	1.42E+08	8.72E+07	1.03E+08	4.36E+07	4.84E+07
LC2	1.42E+08	8.72E+07	1.04E+08	4.36E+07	4.85E+07
LC3	1.90E+08	4.20E+07	1.05E+08	5.05E+07	5.65E+07
LC4	**4.09E+08**	**2.69E+08**	**2.65E+08**	**1.39E+08**	**1.57E+08**
LC5	9.10E+07	4.91E+07	4.98E+07	2.51E+07	3.74E+07

geometry compared with Case 1 (Fig. 11) is obtained because of restricted design domain. For Case 3 (Fig. 13), the difference between different software seems to be reduced compared with Case 1 (Fig. 11) especially for LC3, LC4 and LC5. This might be because the flexibility of design is reduced by large openings.

It is noted that, when same TO tool (same program or same commercial software) is considered, it is also seen that the compliances for Case 2 and Case 3 is generally larger than those of Case 1, though the compliances for each result are not shown in this manuscript. This means that better performance from the viewpoint of compliance minimization is obtained for Case 1 compared with Case 2 and Case 3, because of larger flexibility of design in Case 1. Finally, it is important to conclude that the obtained geometry from a single loading condition cannot be suitable for other loading conditions. It is necessary to

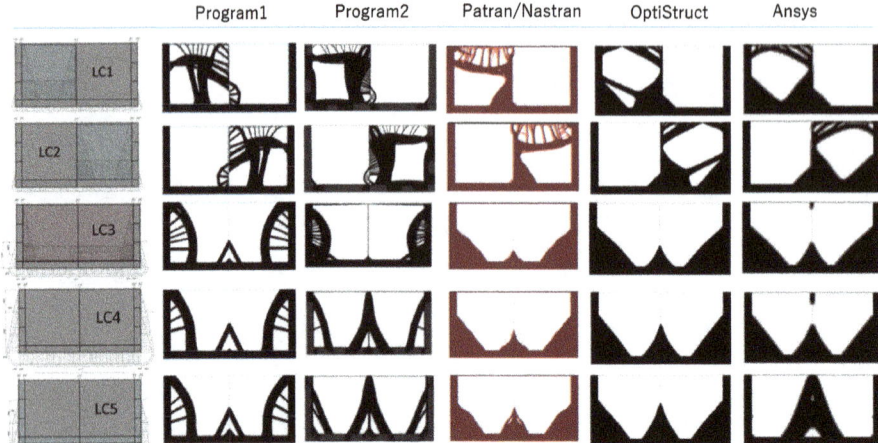

Fig. 12. Case 2: TO results for each single loading condition for different software

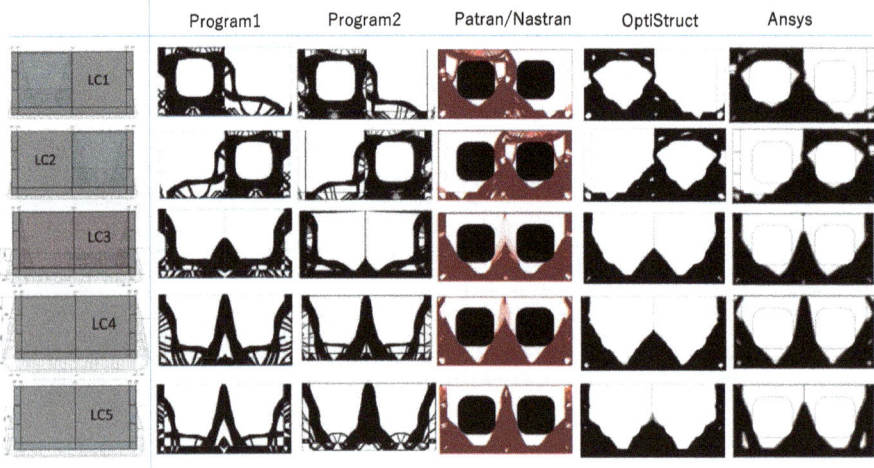

Fig. 13. Case 3: TO results for each single loading condition for different software

consider all necessary loading conditions to conduct topology optimization for practical design of ship structures.

8.6 Results of Topology Optimization for Multiple Loading Conditions

Topology optimization results by different tools for Case 1, Case 2 and Case 3 are shown in Fig. 14. It is noted that, different from the results for single loading conditions shown in 8.5, these geometries are obtained by considering all of loading conditions (LC1-LC5) in the definition of objective function. Figure 14(a) shows the TO results for Case 1, and Table 8 shows the compliances for each loading condition. It is observed in Case 1 that the different results are obtained by different TO tools. This might be caused from different techniques used in different tools. When the results of in-house programs (Program1 and Program2) are compared with the results by commercial software, the obtained geometries are rather complicated. If the result of Program2 is compared with that of Program1, the obtained result seems to be simpler in which finer mesh is used compared with Program1.

In the results of the commercial software (Nastran/Patran, OptiStruct, Ansys), minimum member size technique is used. By using the minimum member size control technique, rather simpler geometry can be obtained. For example, when using OptiStruct without minimum member size control, the geometries with intermediate density elements and checker board pattern are observed as shown in Fig. 15(a), whereas the OptiStruct results shown in Fig. 14(a) with the member size control give more practical results without intermediate density, checkerboard pattern and very long and narrow (truss-like) members. In addition, the compliance values when member size control is considered (See OptiStruct result in Fig. 14(a) and Table 18) appear to have increased compared with the default settings (Fig. 15(a)(b)). This means that, from the viewpoint of minimization of compliance, the results with default setting might be better compared with the results using minimum member size control. However, the compliance values increased only slightly for those with the minimum member size compared with the default setting. Therefore, it can be considered that adding these settings will provide the better result, which is closer to more practical structure, though long and narrow (truss-like) members and intermediate density elements still existed in the TO result.

When the results obtained by different commercial software in Fig. 14(a) are compared, it is observed that the geometries are not so different, but different structure is observed especially near the deck. It can be concluded that different results are obtained by different TO tools because different technics of topology optimization are used. It seems to be difficult to obtain the same result by different TO tools, and there might be a possibility that many local minimum solutions exist in the topology optimization.

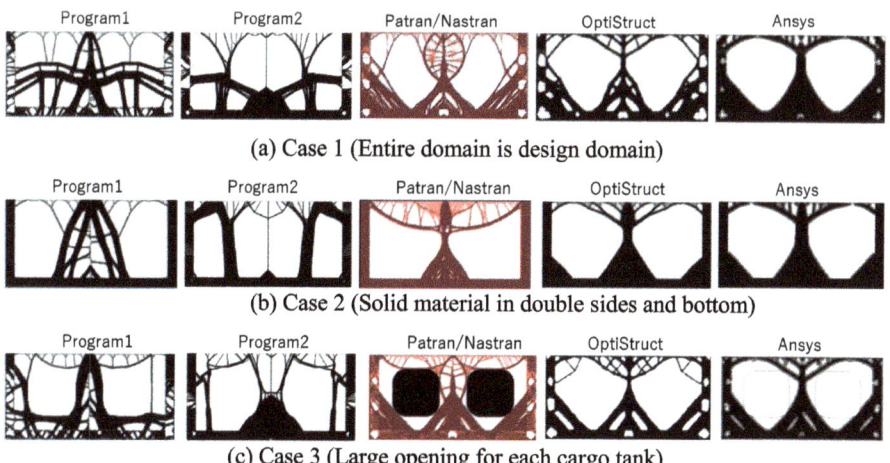

Fig. 14. TO results for multiple loading conditions for different programs and software

Table 18. Compliance for each loading condition for the multiple loading topology optimization results obtained by different tools (programs and software): Case1

	Program1	Program2	Patran/Nastran	OptiStruct	Ansys
LC1	**4.95E+08**	8.49E+07	1.82E+08	7.35E+07	7.51E+07
LC2	**4.95E+08**	8.49E+07	1.81E+08	7.35E+07	7.51E+07
LC3	3.13E+08	9.50E+07	1.21E+08	6.20E+07	6.37E+07
LC4	**4.95E+08**	**2.31E+08**	**3.05E+08**	**1.56E+08**	**1.55E+08**
LC5	2.19E+08	1.00E+08	5.88E+07	2.95E+07	3.24E+07

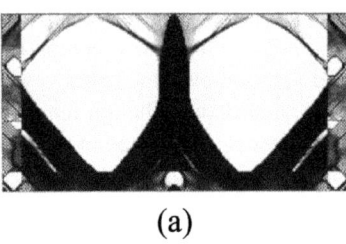

	OptiStruct
LC1	6.16E+07
LC2	6.16E+07
LC3	5.52E+07
LC4	**1.34E+08**
LC5	2.56E+07

(a) (b)

Fig. 15. OptiStruct results without member size control (default setting) for Case 1 ((a) Obtained geometry, (b) Compliances for each LC)

Table 19. Compliance for each loading condition for the multiple loading topology optimization results obtained by different tools (programs and software): Case2

	Program1	Program2	Patran/Nastran	OptiStruct	Ansys
LC1	**5.80E+08**	8.57E+07	2.90E+08	7.45E+07	7.51E+07
LC2	**5.80E+08**	8.57E+07	2.90E+08	7.45E+07	7.51E+07
LC3	3.20E+08	9.81E+07	2.35E+08	7.10E+07	6.39E+07
LC4	**5.80E+08**	**1.94E+08**	**4.79E+08**	**1.63E+08**	**1.53E+08**
LC5	1.08E+08	8.03E+07	8.83E+07	2.77E+07	2.14E+07

For the Program1, BCM (Pham-Truong et al. 2023) method is used by which the max compliance is minimized, resulting that the compliances of LC1, LC2 and LC4 are almost same as shown in Table 18 (Program1). For all other TO tools, the maximum compliance is observed in LC4, which is the most critical loading condition for optimization. On

Table 20. Compliance for each loading condition for the multiple loading topology optimization results obtained by different tools (programs and software): Case3

	Program1	Program2	Patran/Nastran	OptiStruct	Ansys
LC1	**5.06E+08**	8.45E+07	1.65E+08	6.90E+07	7.63E+07
LC2	**5.61E+08**	8.45E+07	1.65E+08	6.90E+07	7.67E+07
LC3	3.14E+08	1.43E+08	1.18E+08	6.30E+07	6.45E+07
LC4	**5.61E+08**	**1.64E+08**	**2.94E+08**	**1.56E+08**	**1.56E+08**
LC5	1.70E+08	5.88E+07	5.61E+07	2.80E+07	3.26E+07

the other hand, the compliance of LC5 is smallest so that LC5 can be regarded as not critical loading condition.

It is also noted that, before obtaining the results of TO, we expected that non-symmetric results can be obtained by TO. In the real structure as shown Fig. 9, the vertical transverse member, which attaches to the centre longitudinal bulkhead (CL-L-BHD), is arranged at one side of the CL-L-BHD, which means not symmetric regarding centre line. The reason for non-symmetric structure in real ship is that the CL-L-BHD can effectively support the bending of the vertical transverse member caused by the cargo pressure from the one side cargo tank. However, it is observed from Fig. 14 that the obtained TO results are almost symmetric for all cases and by all tools, though slightly non-symmetrical results are obtained by program 1 with BCM method. A probable reason for the symmetric results is that the loading conditions, LC1 and LC2, might be not so critical compared with LC4. Additionally, obtaining non-symmetric structure such like Fig. 12 might be difficult, because the obtained symmetric structure could be local minimum solution.

When the results of Case 2 (Fig. 14(b)) are observed, simpler geometries are obtained compared with Case 1. This is because the target domain of topology optimization is smaller by assuming the solid materials (shown in black) in the double side and double bottom. The material is arranged around the left-bottom corner of the left side cargo tank and around the right bottom corner of the right-side cargo tank in OptiStruct and Ansys, whereas no material is arranged around these corners in Program1, Program2 and Patran/Nastran. Like Case 1, different results are obtained by different tools. As for the results of Case 3 (Fig. 14(c)) in which large opening in the cargo tanks is assumed, the obtained structure become rather complicated.

Table 19 and Table 20 shows the compliances of the results of Case 2 and Case 3, respectively. The compliance showed a different trend if the TO tool is different. In Program1 and Program2, the compliance of Case 2 and Case 3 is larger than that of Case 1. This means that the performance of the Case 1 is better from the viewpoint of compliance, because design flexibility is larger in Case 1 compared with Case 2 and Case 3. On the other hand, when commercial software is used, the compliance of Case 1 is not always smaller compared with other Cases. The reason for this is currently unknown but may be related to the application of manufacturing constraints such as minimum member size control.

One of the disadvantages of optimization is that in the most of results, truss-like structure (long and narrow member) is obtained in the results. Though the obtained results have superior performance from the viewpoint of compliance, such truss-like member might not be used in practical ship structure, because the buckling might occur with such members. To overcome this problem, it is necessary to conduct strength analysis for the obtained structure by TO. Also it might be good idea for the topology optimization formulation to include the constraint related to buckling (Ferrari et al., 2021) though the buckling constraint is currently not implemented in the commercial software. Another straightforward way to overcome this problem is to apply fine-tuning. In past research, it is proposed that obtained structures by topology optimization should be modified to make them more realistic structures. These methods are called fine-tuning (Liu et al., 2018; Stankiewicz et al., 2021; Subedi et al., 2020) which modifies the structure to align it with the desired specifications.

Other disadvantage of this 2D topology optimization is that it is not possible to consider effective arrangement of faceplate. As shown in Fig. 9, in real structures, faceplate is attached along with the transverse members. Since this benchmark study dealt with a 2D problem, it is not possible to represent such members. It is necessary to formulate a topology optimization method that can take them into account to obtain more practical design of ship structures.

8.7 Conclusions

In this chapter, a benchmark study of topology optimization for the ship structure using the in-house programs and the commercial software is conducted. The following conclusions are obtained.

- By in-house programs and software for TO, it might be possible to obtain the idea of new and novel structures, which might be efficient from the viewpoint of compliance. It is noted that multiple loading conditions should be considered for practical design of ship structures.
- The obtained structures are usually not so practical due to the presence of truss-like members, checkerboard pattern, or intermediate density elements. By setting the minimum member size control which are functions implemented in the commercial software, these can be removed to some extent, and a more practical structure can be obtained. However, it is concluded that further improvement of the obtained results, such as fine-tuning, may be necessary to obtain more practical structures.
- One of the disadvantages of the 2D topology optimization in this study is that truss-like members, which may suffer from buckling, are arranged in the TO results. Also, it is not possible to consider face plates which are arranged along with the transverse frames in real ship structures. In the future, it is expected to study a TO method that can overcome such disadvantages and can obtain more practical designs.
- Also, as future study, it is expected to study about application of 3D topology optimization for designing new and innovative ship structures.

9 Conclusions

This Committee IV.1 report, discussed about recent progress and development in design methods, principles and criteria. The key findings and conclusions are as follows.

Section 2 highlights the increased pressure on the design process from new emissions regulations, including the need to lightweight vessels with new materials and approaches. It also highlights the potential benefits that digitalisation will have on design, especially the ability to perform Risk-Based Design with increased quantities of data and better storage. It highlights the difficulties in performing these types of approaches in autonomous vessels due to the lack of data, but that vessel design might not change substantially in the short term, as crews will be required on most vessels for the short-to-medium-term. Finally, there is increased interest in Digitalization and a range of techniques in the literature, with increasing applications in industry.

Section 3 highlights regulatory framework for ships and offshore structures. IMO has focused a need of improvement on design standard related to the alternative fuels. Thus, there is discussion about material compatibility and its strength criteria. There is interesting technical investigation in IACS for improvement of wave scatter diagram in REC. 34. It also includes the investigation not only of environmental conditions but also of ship's route and wave heading angles. In addition, the IACS annual review gives overview that there is important research on corrosion addition and buckling strength assessment procedure.

In Sect. 4, the review of classification societies' software tools highlights a trend towards integrating optimization tools to facilitate structural optimization. The OCX CAD exchange format stands out to streamline the exchange of models and related data between different software systems. Structural optimization tools have advanced to provide more reliable results under various conditions, with topology optimization gaining traction as a fundamental method for designing better-performing structures in the marine industry. Machine learning, particularly deep learning techniques, has seen numerous exciting advancements aimed at developing better surrogate models for design, enabling more effective analysis and optimization. While AR and VR technologies are increasingly being adopted in production, their potential in design remains largely untapped. Digital twins are a hot topic with many interpretations, making them a highly active area of research. Specific tools for optimizing offshore structures have also been proposed, offering faster design processes and better-integrated physics.

The review in Sect. 5 highlights the pressing need for updated standards concerning information exchange, especially in the realm of digital twin technology and cyber-physical systems within the shipbuilding industry. Digital twin technology helps sophisticated ship health management by integrating both historical and real-time data, thereby enhancing lifecycle management from the design phase through to maintenance. Collaborative efforts in the sector are focused on improving Product Lifecycle Management (PLM) processes, ensuring digital continuity throughout shipbuilding projects, and using Machine Learning to increase responsiveness in manufacturing operations. In the area of ship dismantling, there is a significant discrepancy between existing regulations, such as the Hong Kong Convention, and the actual practices seen in shipyards. Research shows a clear necessity for enhanced safety and efficiency measures. Additionally, lifecycle assessments reveal that floating offshore wind farms have a lower environmental impact compared to traditional designs, supporting their sustainability as a practical energy source.

Section 6 concludes with the recognition that advancements in structural reliability and risk assessment have significantly improved the safety and sustainability of marine structures. The shift towards multi-dimensional probabilistic methods, alongside the integration of technologies like SHM and Digital Twin, offers a more comprehensive understanding of structural behaviour under various environmental conditions. Case studies demonstrated the practical application of these advanced methods, particularly in offshore wind turbines and complex marine environments.

In Sect. 7, the gaps between research and practice in the maritime industry were reported, especially the growing trend of integrating AI, machine learning, and data analysis in the design of marine structures. Advanced tools enable continuous monitoring of a ship's strength and the creation of digital replicas, providing valuable insights for building superior vessels. Recent advancements in deep learning have made these techniques more accessible, and there is an increasing emphasis on analysing ship structures with limited data. Furthermore, recent advancements in Computer-Aided Design and Engineering have significantly enhanced the efficiency and precision of ship design.

In Sect. 8, a benchmark study of 2D topology optimization (TO) for the cross-section of the tanker is conducted using the in-house programs and the commercial software. Though the structure obtained by TO could not be directly used for practical design of the ship structure, it can be concluded that the idea of new and novel structures can be obtained from the result of TO.

Based on the above conclusions, the Committee IV.1 discussed critical issues that should be challenged in the field of naval architecture and ocean engineering in coming years. It is recommended that the following topics will be addressed.

Moving forwards, it will be important to balance the two key topics of emissions reductions and digitalization. Despite much progress in these areas, there seems to be substantially less focus on the application of these approaches to structures than in other areas. Additional re-search into the change in structural requirements due to new fuels and autonomy will be re-quired. As will the increase in sensor implementation to monitor the health of structures and techniques to assess the health, especially allowing new lightweight designs.

In regulatory framework point of view, the IACS's research for wave scatter diagram in REC.34, corrosion addition and buckling criteria may result revision of IACS unified requirement and CSR BC&OT rules. These changes may influence on design of ships. Also, as alternative fuels are expected to be selected commonly for cargo ships, the distinctive style of tank design or new material will be more interested in the cost efficiency and safety. Since digital means of equipment will accelerate digital twin and ships' autonomy, this movement will affect ship's design.

Methodologies and tools for designing ship and offshore structures should continue to advance with high-fidelity approaches, addressing increasingly complex scenarios under varied load conditions. The growing capabilities of advanced machine learning across various fields should be leveraged within the marine industry, with a focus on creating more efficient tools for the generative computational design of large-scale structures.

Digitalization and digital lifecycles encompass a broad range of specific technologies. The major challenge is to effectively organize, implement, and secure investment in these areas, particularly in relation to visualization, artificial intelligence, virtual vessels, and autonomy.

About reliability and risk assessment, we highlighted the need for continued research into computational tools and probabilistic models to manage uncertainties, which is crucial for the future of maritime and offshore engineering. This approach ensures that structures remain safe, reliable, and efficient throughout their lifecycle, even in harsh operating conditions.

To bridge the gap between research and practice in the maritime industry, it is recommended to continue to work for integrating AI, machine learning, and data analysis in the design of marine structures. Also, digital twin technologies have the potential for significant impact through cross-sector collaboration in the future.

About topology optimization, it is a promising idea for the ship designer to use TO tools to obtain new idea when designing new structures. Also in the future, it is expected to study about application of 3D topology optimization to design of ship structures, and to study about more advanced TO method which is applicable to more practical designs.

References

Abed, S.A., Ghassan, M., Qaes, S., Fiadh, M.S., Mohammed, Z.A.: Structural reliability and optimization using differential geometric approaches. Iraqi J. Comput. Sci. Math. **5**(1), 168–174 (2024). https://doi.org/10.52866/ijcsm.2024.05.01.012

ABS: Advisory on Structural Health Monitoring: The application of Sensor-Based approaches (2020a). https://ww2.eagle.org/content/dam/eagle/advisories-and-debriefs/abs-advisory-on-structural-health-monitoring-the-application-of-sensor-based-approaches.pdf. Accessed 14 Nov 2024

ABS: Ammonia as Marine Fuel (2020b)

ABS: Hydrogen as marine fuel (2021)

ABS: Rules for building and classing Marine Vessels July 2024 edition, Pt. 3 Hull Construction and Equipment (2024)

Acheampong, T., Kemp, A.G.: Health, safety and environmental (HSE) regulation and outcomes in the offshore oil and gas industry: Performance review of trends in the United Kingdom Continental Shelf. Saf. Sci. **148**, 105634 (2022). https://doi.org/10.1016/j.ssci.2021.105634

Adiputra, R., Yoshikawa, T., Erwandi, E.: Reliability-based assessment of ship hull girder ultimate strength. Curved Layer. Struct. **10**(1) (2023). https://doi.org/10.1515/cls-2022-0189. Article 20220189

Adumene, S., Ikue-John, H.: Offshore system safety and operational challenges in harsh Arctic operations. J. Saf. Sci. Resil. **3**(2), 153–168 (2022). https://doi.org/10.1016/j.jnlssr.2022.02.001

Agbakwuru, J.A., Nwaoha, T.C., Udosoh, N.E.: Application of CRITIC–EDAS-Based Approach in Structural Health Monitoring and Maintenance of Offshore Wind Turbine Systems. J. Mar. Sci. Appl. **22**(3), 545–555 (2023). https://doi.org/10.1007/s11804-023-00355-5

Akamatsu, D., Noguchi, Y., Matsushima, K., Sato, Y., Yanagimoto, J., Yamada, T.: Two-phase topology optimization for metamaterials with negative Poisson's ratio. Comp. Struct. **311** (2023). https://doi.org/10.1016/j.compstruct.2023.116800. Article 116800

Al-Sanad, S., Parol, J., Wang, L., Kolios, A.: Design optimisation of wind turbine towers with reliability-based calibration of partial safety factors. Energy Rep. **9**, 2548–2556 (2023). https://doi.org/10.1016/j.egyr.2023.01.090

Aliyar, S., Ducrozet, G., Bouscasse, B., Venkatachalam, S., Ferrant, P.: Efficiency and accuracy of the domain and functional decomposition strategies for the wave-structure interaction problem. Ocean Eng. **266** (2022). https://doi.org/10.1016/j.oceaneng.2022.112568. Article 112568

Altair: Altair® Inspire™ (Inspire, 2024) (2024a). https://altair.com/inspire/. Accessed 29 Feb 2024

Altair: Altair® OptiStruct® (2024b). https://altair.com/optistruct/. Accessed 29Feb 2024

Altunsaray, E., et al.: Ways of weight optimization for polymer-based composite high-speed marine vehicles. In: Progress in Marine Science and Technology. IOS Press (2023). https://doi.org/10.3233/pmst230025

Alzayedi, A.M.T., Sampath, S., Pilidis, P.: Techno–economic and risk evaluation of combined cycle propulsion systems in large container ships. Energies **15**(14) (2022). https://doi.org/10.3390/en15145178. Article 5178

Amano, M., Tanaka, Y., Masui, T., Hulkkonen, T.: Streamlining the ship structural optimization process by using an early 3d product model. In: International Conference on Computer Applications in Shipbuilding (ICCAS 2022), Yokohama, Japan, 13–15 September (2022)

Amir, E., Amir, O.: Concurrent high-resolution topology optimization of structures and their supports for additive manufacturing. Struct. Multidiscipl. Optim. **63**(6), 2589–2612 (2021). https://doi.org/10.1007/s00158-020-02835-6

Amir, O., Aage, N., Lazarov, B.S.: On multigrid-CG for efficient topology optimization. Struct. Multidiscip. Optim. **49**(5), 815–829 (2014). https://doi.org/10.1007/s00158-013-1015-5

Ansys: Manufacturing Constraint. Ansys 2023/R2 Mechanical User's Guide, p. 680 (2023)

Ansys: Ansys Toplogy Optimization (2024). https://www.ansys.com/applications/topology-optimization. Accessed 4 July 2024

Anyfantis, K.N.: An abstract approach toward the structural digital twin of ship hulls: a numerical study applied to a box girder geometry. Proc. Inst. Mech. Eng. Part M J. Eng. Marit. Environ. **235**(3), 718–736 (2021). https://doi.org/10.1177/1475090221989188

Ao, Y., Li, Y., Gong, J., Li, S.: An artificial intelligence-aided design (AIAD) of ship hull structures. J. Ocean Eng. Sci. **8**(1), 15–32 (2023). https://doi.org/10.1016/j.joes.2021.11.003

Armanfar, A., et al.: Embedding lattice structures into ship hulls for structural optimization and additive manufacturing. Ocean Eng. **301**(117601) (2024). https://doi.org/10.1016/j.oceaneng.2024.117601

Astrup, O., Aae, O., Gronlie, A., Uyanik, O., Gigernes, S.: Enhancing the 3D approval process using functional zones in ship design. In: 22nd Conference on Computer Applicationsand Information Technology in the Maritime Industries, COMPIT 2023, Drübeck, Germany, 23–25 May (2023)

Babadi, M.K., Ghassemi, H.: Optimization of ship hull forms by changing CM and CB coefficients to obtain optimal seakeeping performance. PLoS ONE **19**(5) (2024). https://doi.org/10.1371/journal.pone.0302054

Bagazinski, N.J., Ahmed, F.: Ship-D: ship hull dataset for design optimization using machine learning. In: Volume 3A: 49th Design Automation Conference (DAC) (2023a). https://doi.org/10.1115/detc2023-117003

Bagazinski, N.J., Ahmed, F.: ShipGen: a diffusion model for parametric ship hull generation with multiple objectives and constraints. J. Mar. Sci. Eng. **11**(12) (2023b). https://doi.org/10.3390/jmse11122215. Article 2215

Bakdi, A., Glad, I.K., Vanem, E.: Testbed scenario design exploiting traffic big data for autonomous ship trials under multiple conflicts with collision/grounding risks and spatio-temporal dependencies. IEEE Trans. Intell. Transp. Syst. **22**(12), 7914–7930 (2021). https://doi.org/10.1109/tits.2021.3095547

Bakker, C., Zhang, L., Higginson, K., Keulen, F.: Simultaneous optimization of topology and layout of modular stiffeners on shells and plates. Struct. Multidiscip. Optim. **64**(5), 3147–3161 (2021). https://doi.org/10.1007/s00158-021-03081-0

Balci, G., Phan, T.T.N., Surucu-Balci, E., Iris, Ç.: A roadmap to alternative fuels for decarbonising shipping: the case of green ammonia. Res. Transp. Bus. Management. **53** (2024). https://doi.org/10.1016/j.rtbm.2024.101100

Basnet, S., BahooToroody, A., Montewka, J., Chaal, M., Valdez Banda, O.A.: Selecting cost-effective risk control option for advanced maritime operations; Integration of STPA-BN-Influence diagram. Ocean Eng. **280** (2023). https://doi.org/10.1016/j.oceaneng.2023.114631. Article 114631

Bekker, A., De Koker, N.: Towards a digital twin to inform propulsion safety margins in ice. In: 22nd Conference on Computer Applications and Information Technology in the Maritime Industries, COMPIT 2023, Drübeck, Germany, 23–25 May (2023)

Bendsøe, M.P.: Optimal shape design as a material distribution problem. Struct. Optim. **1**(4), 193–202 (1989). https://doi.org/10.1007/BF01650949

Bendsøe, M.P., Sigmund, O.: Topology Optimization - Theory, Methods and Applications. Springer, Heidelberg (2004). https://doi.org/10.1007/978-3-662-05086-6

Bergström, M., et al.: A comprehensive approach to scenario-based risk management for Arctic waters. Ship Technol. Res. **69**(3), 129–157 (2022). https://doi.org/10.1080/09377255.2022.2049967

Bertram, V., Heimann, J., Hochkirch, K.: Ship hull optimization – past, present, prospects. Ship Technol. Res., 1–10 (2023). https://doi.org/10.1080/09377255.2023.2271710

Bhagavathi, R., Kufoalor, D.K.M., Hasan, A.: Digital twin-driven fault diagnosis for autonomous surface vehicles. IEEE Access **11**, 41096–41104 (2023). https://doi.org/10.1109/ACCESS.2023.3268711

Bitomsky, J., Grau, M., Lützenberger, J., Zerbst, C.: Digital enterprise platform - enabling efficient CAD, PLM, ERP integration on shipyards. In: International Conference on Computer Applications in Shipbuilding (ICCAS 2022), Yokohama, Japan, 13–15 September (2022)

Blindheim, S., Johansen, T.A., Utne, I.B.: Risk-based supervisory control for autonomous ship navigation. J. Mar. Sci. Technol. **28**(3), 624–648 (2023). https://doi.org/10.1007/s00773-023-00945-6

Bobeldijk, M., Dragt, S., Hoogeland, M., van Bergen, J.: Assessment of the technical safe limit speed of a non-ice-strengthened naval vessel with representative and alternative side shell designs in ice-infested waters. Ships Offshore Struct. **16**(S1), 275–289 (2021). https://doi.org/10.1080/17445302.2021.1912475

Bojinov, I.: Keep Your AI Projects on Track. Harvard Business Review. The Magazine, November–December 2023 (2023). https://hbr.org/

Bolbot, V., Gkerekos, C., Theotokatos, G., Boulougouris, E.: Automatic traffic scenarios generation for autonomous ships collision avoidance system testing. Ocean Eng. **254** (2022). https://doi.org/10.1016/j.oceaneng.2022.111309. Article 111309

Bottero, M., Gualeni, P.: Systems engineering and ship design: a synergy for getting the right design and the design right. In: Rizzuto, E., Ruggiero, V. (eds.) Technology and Science for the Ships of the Future, Proceedings of NAV 2022: 20th International Conference on Ship & Maritime Research (Vol. 6 - Progress in Marine Science and Technology, pp. 340–347) (2022). https://doi.org/10.3233/PMST220042

Braun, M., et al.: Development of combined load spectra for offshore structures subjected to wind, wave, and ice loading. Energies **15**(2), 559 (2022). https://www.mdpi.com/1996-1073/15/2/559

Braun, M., Ehlers, S.: Review of methods for the high-cycle fatigue strength assessment of steel structures subjected to sub-zero temperature. Mar. Struct. **82** (2022). https://doi.org/10.1016/j.marstruc.2021.103153. Article 103153

Brussa, G., Grosso, M., Rigamonti, L.: Life cycle assessment of a floating offshore wind farm in Italy. Sustain. Prod. Consump. **39**, 134–144 (2023). https://doi.org/10.1016/j.spc.2023.05.006

Bucci, V., Sulligoi, G., Chalfant, J., Chryssostomidis, C.: Evolution in design methodology for complex electric ships. J. Ship Prod. Des. **37**(04), 215–227 (2021). https://doi.org/10.5957/jspd.08190045

Burggräf, P., Adlon, T., Beyer, M., Schäfer, N., Fulterer, J.: Enabling individual part production planning in shipbuilding: machine-learning-assisted prediction of production times. In: 21st Conference on Computer Applications and Information Technology in the Maritime Industries, COMPIT 2022, Pontignano, 21–23 June (2022)

BV: Guidance note NI 641 DT R00 E. Guidelines for Autonomous Shipping, pp. 1–28 (2017)

BV: Methanol & ethanol fuelled ships (NR670, Issue August 2022) (2022a)

BV: Rules for the Classification of Steel Ships, Part B - Hull and Stability (2022b)

BV: Damen, NAPA and Bureau Veritas successfully deploy 3D Classification approvals for first ship design. Marine & Offshore,12 January (2023a). https://marine-offshore.bureauveritas.com/newsroom/damen-napa-and-bureau-veritas-successfully-deploy-3d-classification-approvals-first-ship. Accessed 12 Sept 2024

BV. (2023b). Hydrogen-fuelled ships

BV: Rules for the Classification of Steel Ships, Part B - Hull and Stability (2024)

Cai, Y., Jelovica, J.: Structural optimization of ships: Benchmark study of metaheuristic algorithms and constraint handling approaches. In: 41st International Conference on Offshore Mechanics and Arctic Engineering – OMAE 2022, Hamburg, Germany, 5–10 June (2022). https://doi.org/10.1115/omae2022-79199

Cai, Y., Jelovica, J.: Neural network-enabled discovery of mapping between variables and constraints for autonomous repair-based constraint handling in multi-objective structural optimization. Knowl. Based Syst. **280**(111032) (2023). https://doi.org/10.1016/j.knosys.2023.111032

Cai, Y., Jelovica, J.: Efficient graph representation in graph neural networks for stress predictions in stiffened panels. Thin-Walled Struct. **203**(112157) (2024). https://doi.org/10.1016/j.tws.2024.112157

Castaneda, M., Zapata, S., Cherni, J., Aristizabal, A.J., Dyner, I.: The long-term effects of cautious feed-in tariff reductions on photovoltaic generation in the UK residential sector. Renew. Energy **155**, 1432–1443 (2020). https://doi.org/10.1016/j.renene.2020.04.051

CCS: Rules for Classification of Sea-going Steel Ships, Part Two Hull (2024)

Chaal, M., Bahootoroody, A., Basnet, S., Valdez Banda, O.A., Goerlandt, F.: Towards system-theoretic risk assessment for future ships: a framework for selecting Risk Control Options. Ocean Eng. **259** (2022). https://doi.org/10.1016/j.oceaneng.2022.111797. Article 111797

Chen, X., Takami, T., Oka, M., Kawamura, Y., Okada, T.: Stochastic wave spectra estimation (SWSE) based on response surface methodology considering uncertainty in transfer functions of a ship. Mar. Struct. **90** (2023). https://doi.org/10.1016/j.marstruc.2023.103423. Article 103423

Chen, Y., et al.: Topology optimization design and experimental research of a 3D-printed metal aerospace bracket considering fatigue performance. Appl. Sci. **11**(15), 6671 (2021). https://doi.org/10.3390/app11156671

Cheng, J., Liu, Y., Cheng, M., Li, W., Li, T.: Optimum condition-based maintenance policy with dynamic inspections based on reinforcement learning. Ocean Eng. **261** (2022). https://doi.org/10.1016/j.oceaneng.2022.112058

Cheng, Z., Tan, E., Cai, M., Magee, A.R.: Concept design of a digital twin architecture for ship structural health management. J. Phys. Conf. Ser. **2311**(1) (2022). https://doi.org/10.1088/1742-6596/2311/1/012010

Choi, S., Park, J.-S.: Development of augmented reality system for productivity enhancement in offshore plant construction. J. Mar. Sci. Eng. **9**(2), 209 (2021). https://doi.org/10.3390/jmse9020209

Chon, H., Oh, D., Noh, J.: Classification of hull blocks of ships using CNN with multi-view image set from 3D CAD data. J. Mar. Sci. Eng. **11**(2), 333 (2023). https://doi.org/10.3390/jmse11020333

Chouche, O.: Smart digital shipyards with model-based manufacturing. In: International Conference on Computer Applications in Shipbuilding (ICCAS2022), Yokohama, Japan, 13–15 September (2022)

Christensen, M., Georgati, M., Arsanjani, J.J.: A risk-based approach for determining the future potential of commercial shipping in the Arctic. J. Mar. Eng. Technol. **21**(2), 82–99 (2022). https://doi.org/10.1080/20464177.2019.1672419

Chu, S., Featherston, C., Kim, H.A.: Design of stiffened panels for stress and buckling via topology optimization. Struct. Multidiscipl. Optim. **64**(5), 3123–3146 (2021). https://doi.org/10.1007/s00158-021-03062-3

Chuah, L.F., et al.: Implementation of the energy efficiency existing ship index and carbon intensity indicator on domestic ship for marine environmental protection. Environ. Res. **222** (2023). https://doi.org/10.1016/j.envres.2023.115348. Article 115348

ClassNK: Sec.5.3.2-1(2) Guidelines for Tanker Structures, p. 13 (2001)

ClassNK: Guidelines for Concept Design of Automated Operation/Autonomous Operation of ships (Provisional Version), pp. 1–20 (2018)

ClassNK: Rules for the Survey and Construction of Steel Ships, Part C Hull Construction and Equipment (2022)

ClassNK: Ship Registration, ClassNK (2023). https://www.classnk.or.jp/register/regships/regships.aspx. Accessed 31 July

Collette, M., et al.: Committee IV.2 - design methods. In: Guedes Soares, C., Garbatov, Y. (eds.) Proceedings of the 20th International Ship and Offshore Structures Congress (ISSC 2015). Taylor & Francis Group (2015)

Crawley, E., Cameronm, B., Selva, D.: System Architecture: Strategy and Product Development for Complex Systems. Pearson (2015)

Crupi, V., Epasto, G., Napolitano, F., Palomba, G., Papa, I., Russo, P.: Green composites for maritime engineering: a review. J. Mar. Sci. Eng. **11**(3) (2023). https://doi.org/10.3390/jmse11030599

Cui, Y., et al.: Novel methodology of fail-safe reliability-based topology optimization for large-scale marine structures. Struct. Multidisc. Optim. **66**(7) (2023). https://doi.org/10.1007/s00158-023-03614-9. Article 168

Cui, Y., et al.: A novel BACG inverse reliability algorithm for efficient and robust reliability-based topology optimization of marine structures. Ocean Eng. **298**(117165) (2024). https://doi.org/10.1016/j.oceaneng.2024.117165

D'Antuono, P., Weijtjens, W., Devriendt, C.: On the minimum required sampling frequency for reliable fatigue lifetime estimation in structural health monitoring. how much is enough? In: European Workshop on Structural Health Monitoring (EWSHM 2022). LNCE, vol. 253, pp. 133–142. Springer (2023)

da Cunha Jácome Vidal, P., Aguirre González, M.O., Cassimiro de Melo, D., de Oliveira Ferreira, P., Vasconcelos Sampaio, P.G., Lima, L.O.: Conceptual framework for the decommissioning process of offshore oil and gas platforms. Mar. Struct. **85**(103262) (2022). https://doi.org/10.1016/j.marstruc.2022.103262

Dæhlen, J., Heierli-Nesse, H., Bø, T.A.: Open-source web service for fast and reliable handling of weather data for use in maritime simulations. In: 22nd International Conference on Computer Applications and Information Technology in the Maritime Industries, COMPIT 2023, Drübeck, Germany, 23–25 May (2023)

Dahlke, P., Schmelzer, P.: Incentives to holistic methodology and computer aided tool support in naval vessels design. In: 22nd Conference on Computer and IT Applications in the Maritime Industries, COMPIT 2023, Druebeck, Germany, 23–25 May (2023)

DAMEN: Damen, NAPA and Bureau Veritas Successfully Deploy 3D Classification Approvals for First Ship Design (2023). https://marine-offshore.bureauveritas.com/newsroom/damen-napa-and-bureau-veritas-successfully-deploy-3d-classification-approvals-first-ship

DELTAMARIN: Deltamarin Receives 3D Model-based DNV Approval (2022). https://deltamarin.com/2022/11/deltamarin-receives-3d-model-based-dnv-approval/. Accessed 15 June 2024

Deng, C., Wang, Y., Qin, C., Fu, J., Lu, W.: Self-directed online machine learning for topology optimization. Nat. Commun. **13**(1), 388 (2022). https://doi.org/10.1038/s41467-021-27713-7

Deng, H., Vulimiri, P.S., To, A.C.: An efficient 146-line 3D sensitivity analysis code of stress-based topology optimization written in MATLAB. Optim. Eng. **23**(3), 1733–1757 (2022). https://doi.org/10.1007/s11081-021-09675-3

DeNucci, T., Brahan, D.: A digital twin for the formulation of ice accretion on vessels. In: 21th International Conference on Computer Applications and Information Technology in the Maritime Industries, COMPIT 2022, Pontignano, Italy, 21–23 June (2022)

Deshpande, S., Sæterdal, A., Sundsbø, P.A.: Experiments with sea spray icing: investigation of icing rates. J. Offshore Mech. Arct. Eng. **146**(1) (2024). https://doi.org/10.1115/1.4062255. Article 011601

Díaz, A.R., Bendsøe, M.P.: Shape optimization of structures for multiple loading conditions using a homogenization method. Struct. Optim. **4**(1), 17–22 (1992). https://doi.org/10.1007/BF01894077

Díaz, H., Loughney, S., Wang, J., Guedes Soares, C.: Comparison of multicriteria analysis techniques for decision making on floating offshore wind farms site selection. Ocean Eng. **248** (2022). https://doi.org/10.1016/j.oceaneng.2022.110751. Article 110751

DNV-GL: Class guideline DNVGL-CG-0264. Autonomous and remotely operated ships (2018)

DNV: Rules for Classification Ships Oct.2015 edition, Pt.3 Hull (2015)

DNV: Steel sandwich panel construction. DNV-CG-0154 (2022)

DNV: Alternative Fuels Insight (2024). https://services.veracity.com/. Accessed 23 Mar

Do, Q.T.: Residual ultimate strength formulations of submarine pressure hull subjected to collisions of attendant vessels or floating objects. In: AIP Conference Proceedings, vol. 2674, no. 1 (2023). https://doi.org/10.1063/5.0115016

Doğan, S.Ö., Özden, T.: Optimization of welding application parameters of thin sheet blocks used in the new-generation ship hull. Emerg. Mater. Res. **11**(1), 67–75 (2022). https://doi.org/10.1680/jemmr.20.00330

Doi, S., Sasaki, H., Igarashi, H.: Multi-objective topology optimization of rotating machines using deep learning. IEEE Trans. Magn. **55**(6) (2019). https://doi.org/10.1109/TMAG.2019.2899934. Article 8666164

Dolz, M., Martinez, X., Sá, D., Silva, J., Jurado, A.: Composite materials, technologies and manufacturing: current scenario of European Union shipyards. Ships Offshore Struct., 1–16 (2023). https://doi.org/10.1080/17445302.2023.2229160

DT4GS: The digital twin for green shipping. Horizon Europe project (2024). https://dt4gs.eu/

Du, L., Wu, Q., Shu, Y., Li, G.-N.: The effects of online-training artificial neural network mechanism and multi-stage parametric modeling method on simulation-based design system for ship optimization. Ocean Eng. **309**(118284) (2024). https://doi.org/10.1016/j.oceaneng.2024.118284

Dyer, A.S., et al.: Applied machine learning model comparison: predicting offshore platform integrity with gradient boosting algorithms and neural networks. Mar. Struct. **83**(103152) (2022). https://doi.org/10.1016/j.marstruc.2021.103152

Ebrahimi, A., Brett, P.O., Erikstad, S.O., Asbjørnslett, B.E.: Influence of ship design complexity on ship design competitiveness. J. Ship Prod. Des., 1–15 (2021). https://doi.org/10.5957/jspd.08200020

Eichner, L., Schneider, R., Baeßler, M.: Optimal vibration sensor placement for jacket support structures of offshore wind turbines based on value of information analysis. Ocean Eng. **288** (2023). https://doi.org/10.1016/j.oceaneng.2023.115407. Article 115407

Emi, H.: Illustrations of Hull Structures, 1st edn. (in English and Japanese). Tokyo: Seizando-Shoten Publishing Co., Ltd. (2006)

Erikstad, S.O., Lagemann, B.: Design methodology state-of-the-art report. In: SNAME, 14th International Marine Design Conference (2022). https://doi.org/10.5957/imdc-2022-301

Estrela, G.A., et al.: Ultimate strength reliability analysis applied to the design of cement sheaths in oil wells. In: Offshore Technology Conference Brasil, OTCB 2023, Rio de Janeiro, Brazil, 24–26 October (2023). https://doi.org/10.4043/32678-MS

EU: REGULATION (EU) No 1257/2013 on ship recycling and amending Regulation (EC) No 1013/2006 and Directive 2009/16/EC. OJ, L 330/1 (2013)

EU: Fit for 55: increasing the uptake of greener fuels in the aviation and maritime sectors (2023). https://www.consilium.europa.eu/en/infographics/fit-for-55-refueleu-and-fueleu/#0. Accessed 10 Mar 2024

Farhan, M., Schneider, R., Thöns, S.: Predictive information and maintenance optimization based on decision theory: a case study considering a welded joint in an offshore wind turbine support structure. Struct. Health Monit. **21**(1), 185–207 (2022). https://doi.org/10.1177/14759217209081833

Fariya, S., Gunbeyaz, S.A., Kurt, R.E., Turan, O.: Determining the effects of implementing IMO's Hong Kong Convention's requirements on the productivity of a ship recycling yard by using discrete event simulation. Ships Offshore Struct. **17**(11), 2508–2519 (2022). https://doi.org/10.1080/17445302.2021.2005355

Feng, J., Ye, S., Zhang, Y., Qi, L., Shen, Y.: A comparative study of combined surrogate models and evolutionary algorithms for ship resistance optimization. In: Fourth International Conference on Mechanical, Electronics, and Electrical and Automation Control (METMS 2024) (2024). https://doi.org/10.1117/12.3030217

Fernández-Caramés, T.M., Fraga-Lamas, P.: Augmented and Mixed Reality for Shipbuilding. Springer Handbooks, pp. 643–667 (2023). https://doi.org/10.1007/978-3-030-67822-7_26

Ferrari, F., Sigmund, O., Guest, J.K.: Topology optimization with linearized buckling criteria in 250 lines of Matlab. Struct. Multidiscip. Optim. **63**(6), 3045–3066 (2021). https://doi.org/10.1007/s00158-021-02854-x

Fonseca, Í.A., Gaspar, H.M.: Challenges when creating a cohesive digital twin ship: a data modelling perspective. Ship Technol. Res. **68**(2), 70–83 (2021). https://doi.org/10.1080/09377255.2020.1815140

Fonseca, Í.A., Gaspar, H.M., de Mello, P.C., Sasaki, H.A.U.: A standards-based digital twin of an experiment with a scale model ship. Comput. Aided Des. **145**(103191) (2022). https://doi.org/10.1016/j.cad.2021.103191

Forrester, A., Sobester, A., Keane, A.: Engineering Design Via Surrogate Modelling: A Practical Guide. Wiley (2008). https://doi.org/10.1002/9780470770801

Fu, Y., Zhu, Y., Hoang, T., Mechitov, K., Spencer, B.F.: xImpact: intelligent wireless system for cost-effective rapid condition assessment of bridges under impacts. Sensors **22**(15) (2022). https://doi.org/10.3390/s22155701. Article 5701

Fujikubo, M., et al.: A digital twin for ship structures—R&D project in Japan. Data-Centric Eng. **5**, e7 (2024). https://doi.org/10.1017/dce.2024.3

Gaidai, O., Fu, S., Xing, Y.: Novel reliability method for multidimensional nonlinear dynamic systems. Mar. Struct. **86**(103278), 103278 (2022). https://doi.org/10.1016/j.marstruc.2022.103278

Gaidai, O., et al.: On-board trend analysis for cargo vessel hull monitoring systems. In: Proceedings of the International Offshore and Polar Engineering Conference (ISOPE 2022), pp. 3628–3632 (2022). https://hdl.handle.net/11250/3039961

Gaidai, O., et al.: Novel methods for reliability study of multi-dimensional non-linear dynamic systems. Sci. Rep. **13**(1) (2023). https://doi.org/10.1038/s41598-023-30704-x. Article 3817

Gaidai, O., Yakimov, V., Wang, F., Hu, Q., Storhaug, G., Wang, K.: Lifetime assessment for container vessels. Appl. Ocean Res. **139** (2023). https://doi.org/10.1016/j.apor.2023.103708. Article 103708

Galparsoro, I., et al.: Reviewing the ecological impacts of offshore wind farms. npj Ocean Sustain **1**(1) (2022). https://doi.org/10.1038/s44183-022-00003-5

Gao, C., et al.: Life cycle structural integrity design approach for the components of subsea production system: SCSSV as a case study. IEEE/ASME Trans. Mechatron. **29**(4), 2768–2778 (2024). https://doi.org/10.1109/TMECH.2023.3329831

Gao, S.B.: Acoustic-structure interaction simulation of ship propellers based on COMSOL. In: Proc. SPIE 12261. Second International Conference on Mechanical Design and Simulation (MDS 2022), Wuhan, China, 20 September (2022)

Garbatov, Y., Georgiev, P., Yalamov, D.: Risk-based retrofitting analysis employing the carbon intensity indicator. Ocean Eng. **289** (2023). https://doi.org/10.1016/j.oceaneng.2023.116283. Article 116283

Garbatov, Y., Marchese, S.S., Palomba, G., Crupi, V.: Alternative hybrid lightweight ship hull structural design. Trends. Mar. Technol. Eng. **1**, 99–107 (2022)

Garbatov, Y., Palomba, G., Crupi, V.: Risk-based hybrid light-weight ship structural design accounting for carbon footprint. Appl. Sci. **13**(6), 3583 (2023). https://doi.org/10.3390/app13063583

Garbatov, Y., Sisci, F.: Sensitivity analysis of risk-based conceptual ship design. In: Progress in Maritime Technology and Engineering - Proceedings of the 4th International Conference on Maritime Technology and Engineering, MARTECH 2018, pp. 499–508 (2018). https://doi.org/10.1201/9780429505294-58

Garcia Agis, J.J., Brett, P., Ebrahimi, A., Kramel, D.: The potential of virtual reality (VR) tools and its application in conceptual ship design. In: 19th CInternational Conference on Computer and IT Applications in the Maritime Industries, Pontignano, Italy, 17–19 August, pp. 123–134 (2020)

Gartner: Gartner Says Nearly Half of CIOs Are Planning to Deploy Artificial Intelligence (2018). Accessed 20 Aug 2024. https://www.gartner.com/en/newsroom/press-releases/2018-02-13-gartner-says-nearly-half-of-cios-are-planning-to-deploy-artificial-intelligence

Gaspar, H., Seppälä, S., Koelman, H., Agis, J.J.G.: Can European shipyards be smarter? A proposal from the SEUS project. In: 22nd International Conference on Computer and IT Applications in the Maritime Industries, COMPIT 2023, Drübeck, Germany, 23–25 May 2023 (2023)

Georgiadis, D.G., Samuelides, E.S., Straub, D.: A Bayesian analysis for the quantification of strength model uncertainty factor of ship structures in ultimate limit state. Marine Struct. **92** (2023). https://doi.org/10.1016/j.marstruc.2023.103495. Article 103495

Gernez, E., Nordby, K., Archer Dreyer, S., Burås, T., Fauske, J.E.: How virtual reality is used in industrial maritime design processes: two case studies. Ocean Eng. **283**(115091) (2023). https://doi.org/10.1016/j.oceaneng.2023.115091

Giering, J.-E., Dyck, A.: Maritime Digital Twin architecture. A concept for holistic Digital Twin application for shipbuilding and shipping. at - Automatisierungstechnik **69**(12), 1081–1095 (2021). https://doi.org/10.1515/auto-2021-0082

Goodfellow, I., Bengio, J., Courville, A.: Deep Learning. The MIT Press (2016)

Grealou, L.: PLM and IoT integrate shipbuilding design and product engineering (2023). https://www.engineering.com/plm-and-iot-integrate-shipbuilding-design-and-product-engineering/. Accessed 27 Oct

Grieves, M.: Digital twin: manufacturing excellence through virtual factory replication.white Paper **1**, 1–7 (2014)

Gualeni, P., Petacco, N., Zini, A.: The role of the digital twin for a ship life cycle perspective: a focus on naval vessels. In: 22nd International Conference on Computer Applications and Information Technology in the Maritime Industries, COMPIT 2023, Drübeck, Germany, 23–25 May (2023)

Guan, G., Yang, Q., Wang, Y., Zhou, S., Zhuang, Z.: Parametric design and optimization of SWATH for reduced resistance based on evolutionary algorithm. J. Mar. Sci. Technol. **26**(1), 54–70 (2021). https://doi.org/10.1007/s00773-020-00721-w

Guan, G., Zhang, X., Wang, P., Yang, Q.: Multi-objective optimization design method of marine propeller based on fluid-structure interaction. Ocean Eng. **252**(111222) (2022). https://doi.org/10.1016/j.oceaneng.2022.111222

Gušani, S., Radonić, M., Puurula, J.: OCX standard and structural model reuse in the shipbuilding design. In: 22nd International Conference on Computer Applications and Information Technology in the Maritime Industries, COMPIT 2023, Drübeck, Germany, 23–35 May (2023)

Ha, J., Roh, M.-I., Kim, K.-S., Kong, M.-C.: Integrated method for the arrangement design of a ship for implementing digital twin in design. In: Practical Design of Ships and Other Floating Structures PRADS 2022, Dubrovnik, Croatia, 9–13 October (2022)

Hageman, R.B., van der Meulen, F.H., Rouhan, A., Kaminski, M.L.: Quantifying uncertainties for risk-based inspection planning using in-service hull structure monitoring of FPSO hulls. Marine Struct. **81** (2022). https://doi.org/10.1016/j.marstruc.2021.103100. Article 103100

Hammer, T.C., Willems, T., Hendrikse, H.: Dynamic ice loads for offshore wind support structure design. Marine Struct. **87** (2023). https://doi.org/10.1016/j.marstruc.2022.103335. Article 103335

Han, S., Yang, B., Yang, B., Zhang, G.: Numerical simulation of heterogeneous ice sheet-structure interaction based on cohesive element method. Appl. Ocean Res. **145** (2024). https://doi.org/10.1016/j.apor.2024.103942. Article 103942

Han, Y.-S., Lee, K., Lee, J., Lee, J., Nam, B.: A study on structural CAD data conversion between AVEVA MARINE ® and Intergraph Smart 3D ®. Ships Offshore Struct. **17**(11), 2396–2407 (2022). https://doi.org/10.1080/17445302.2021.1998852

Hanif, M.I., Adiputra, R., Prabowo, A.R., Yamada, Y., Firdaus, N.: Assessment of the ultimate strength of stiffened panels of ships considering uncertainties in geometrical aspects: finite element approach and simplified formula. Ocean Eng. **286** (2023). https://doi.org/10.1016/j.oceaneng.2023.115522. Article 115522

Harries, S., Thies, F., Hauschulz, S., Marzi, J., Gatchell, S.: Comparing hydrodynamics of original and remodeled hull forms and appendages for digital twins. In: 21th International Conference on Computer Applications and Information Technology in the Maritime Industries, COMPIT 2022, Pontignano, Italy, 21–23 June (2022)

Hasan, A., Asfihani, T., Osen, O., Bye, R.T.: Leveraging digital twins for fault diagnosis in autonomous ships. Ocean Eng. **292**, 116546 (2024). https://doi.org/10.1016/j.oceaneng.2023.116546

Hassani, B., Hinton, E.: A review of homogenization and topology optimization III - topology optimization using optimality criteria. Comput. Struct. **69**(6), 739 (1998). https://doi.org/10.1016/s0045-7949(98)00133-3

Heo, T., Liu, D.P., Manuel, L., Correia, J.A.F.O., Mendes, P.: Sustainable reuse of decommissioned jacket platforms for offshore wind energy accounting for accumulated fatigue damage. In: 41st International Conference on Offshore Mechanics and Arctic Engineering - OMAE2022, Hamburg, Germany, 5–10 June (2022). https://doi.org/10.1115/OMAE2022-79598

Hexagon: Design Sensitivity and Optimization User's Guide. MSC Nastran 2022.3 (2022)

Hill, P.G., Skelhorn, M., Leather, S.: Assessing the environmental risk posed by a legacy tanker wreck: a case study of the RFA War Mehtar. Env. Res. Commun. **4**(5) (2022). https://doi.org/10.1088/2515-7620/ac5bf0. Article 055005

Hinz, T., Mazerski, G., Krishnan, A., Molchanov, B., Elg, M., Wejberg, V.: Digital twin for evaluation of emission reduction by novel technologies. In: 22nd International Conference on Computer and IT Applications in the Maritime Industries, COMPIT 2023, Drübeck, Germany, 23–25 May, pp. 241–251 (2023)

Hirdaris, S., et al.: Review of the uncertainties associated to hull girder hydroelastic response and wave load predictions. Marine Struct. **89** (2023). https://doi.org/10.1016/j.marstruc.2023.103383. Article 103383

Hlaing, N., Morato, P.G., Nielsen, J.S., Amirafshari, P., Kolios, A., Rigo, P.: Inspection and maintenance planning for offshore wind structural components: integrating fatigue failure criteria with Bayesian networks and Markov decision processes. Struct. Infrastruct. Eng. **18**(7), 983–1001 (2022). https://doi.org/10.1080/15732479.2022.2037667

Hong, H.P., Sheng, C.: Reliability-based calibration of site-specific design typhoon wind and wave loads for wind turbine. Eng. Struct. **270** (2022). https://doi.org/10.1016/j.engstruct.2022.114885. Article 114885

Hoque, K. N., Islam, M.S.: Finite element approach to analyze structural discontinuities associated with ship hull. J. Naval Archit. Marine Eng. **20**(3) (2023). https://doi.org/10.3329/jname.v20i3.69756

Huang, J., Wei, Q., Liu, Y.: Ship design optimization considering probabilistic compliance of decarbonization regulations. In: Advances in the Analysis and Design of Marine Structures - Proceedings of the 9th International Conference on Marine Structures, MARSTRUCT 2023, Gothenburg, Sweden, 3–5 April, pp. 819–825 (2023). https://doi.org/10.1201/9781003399759

Hwang, J.H., Lu, C.C., Wang, J.S.: Characterized model uncertainties of CPT-based simplified procedures for assessing soil liquefaction and its application to Taiwan offshore wind farms. Appl. Ocean Res. **138** (2023). https://doi.org/10.1016/j.apor.2023.103645. Article 103645

IACS: IACS Technical Work, Annual Review, pp.14–27 (2019)

IACS: Annual Review (2020)

IACS: Annual Review (2021)

IACS: IACS Technical Work, Improved and enriched wave data, Annual Review, pp. 18–33 (2022a)

IACS: Rec 34. Standard Wave Data – Rev.2 Dec 2022 Complete Revision (2022b)

IACS: Annual Review (2023a)

IACS: Part 1 Chapter 3, Section 1, 2.1.1. In Common Structural Rules for Bulk Carriers and Oil Tankers, p. 83 (2023b)

IACS: Common Structural Rules for Bulk Carriers and Oil Tankers. Section 2- Rule Principles (2024)

IACS(HF&TB): History Files (HF) and Technical Background (TB) documents for Recommendations (2023)

IACS(RCN): Common Structural Rules for Bulk Carriers and Oil Tankers, Rule Change Notice 1 to 01 JAN 2022 version (2022)

IACS(RCN): Common Structural Rules for Bulk Carriers and Oil Tankers, Rule Change Notice 1 to 01 JAN 2023 version (2023)

IACS(UR): Implementation status 2023 (2023)

Ibrion, M., Paltrinieri, N., Nejad, A.R.: On risk of digital twin implementation in marine industry: learning from aviation industry. In: International Maritime and Port Technology and Development Conference and International Conference on Maritime Autonomous Surface Ships, Trondheim, Norway, 13–14 November, vol. 1357 (2019). https://iopscience.iop.org/article/10.1088/1742-6596/1357/1/012009

Ichinose, Y., Hayashi, M., Nomura, S., Moser, B., Hiekata, K.: Sustainable data centers in Southeast Asia: offshore, nearshore, and onshore systems for integrated data and power. Sustain. Cities Soc. **81** (2022). https://doi.org/10.1016/j.scs.2022.103867. Article 103867

IMO: Resolution A.963(23). IMO policies and practices related to the reduction of greenhouse gas emissions from ship (2004)
IMO: MSC.454(100), Revised guidelines for verification of conformity with goal-based ship construction standards for bulk carriers and oil tankers (2018)
IMO: MSC 104th session (2021)
IMO: IGC Code, Ch.6 Materials of construction and quality control (2022a)
IMO: MSC 105th session (2022b)
IMO: MSC 106th session (2022c)
IMO: Historic Background. Climate action and clean air in shipping (2023a). https://www.imo.org/en/OurWork/Environment/Pages/Historic-Background.aspx. Accessed 15 Apr 2024
IMO: Hong Kong ship recycling Convention set to enter into force (2023b). Accessed 28 Sept 2024. https://www.imo.org/en/MediaCentre/PressBriefings/pages/Hong-Kong-Convention-set-to-enter-into-force-.aspx
IMO: MSC 107th session (2023c)
IMO: Autonomous shipping (2024a). https://www.imo.org/en/MediaCentre/HotTopics/Pages/Autonomous-shipping.aspx. Accessed 07 Sept 2024
IMO: MSC 108th session (2024b)
IRS: Part 3, General hull requirements, Volume I and II. In Ships Rules and Regulations for the construction and classification of steel ships (2024)
ISO: ISO/IEC 81346-1. Industrial Systems, Installations and Equipment and Industrial Products - Structuring Principles and Reference Designations - Part 1: Basic Rules (2022)
ISO: ISO/TS 15926-11:2023(en). Industrial automation systems and integration — Integration of life-cycle data for process plants including oil and gas production facilities — Part 11: Simplified industrial usage of reference data based on RDFS methodology (2023). https://www.iso.org/obp/ui/en/#iso:std:79005:en
ISO: ISO 19848:2024. Ships and marine technology — Standard data for shipboard machinery and equipment (2024)
Ito, M.: Fundamentals and applications of risk-based Design. ClassNK Tech. J. **6**(II), 21–32 (2022)
Ivaldi, A., et al.: Committee IV.2: design methods. In: 21st International Ship and Offshore Structures Congress, (ISSC2022) Vancouver, Canada, September 2022 (2022)
Jacquet, L., le Duigou, A., Kerbrat, O.: A systematic literature review on holistic lifecycle assessments as a basis to create a standard in maritime industry. Int. J. Life Cycle Assess. **29**(4), 683–705 (2024). https://doi.org/10.1007/s11367-023-02269-4
Jamaluddin, Zaman, M.B., et al.: Safety analysis and ship recycling yard evaluation of Hong Kong international convention for the safe and environmentally sound recycling of ships. IOP Conf. Ser. Earth Env. Sci. **972**(1) (2022). https://doi.org/10.1088/1755-1315/972/1/012052
Jelovica, J., Cai, Y.:x Improved multi-objective structural optimization with adaptive repair-based constraint handling. Eng. Optim., 1–20 (2022). https://doi.org/10.1080/0305215x.2022.2147518
Jeong, H., Bai, J., Batuwatta-Gamage, C.P., Rathnayaka, C., Zhou, Y., Gu, Y.: A physics-informed neural network-based topology optimization (PINNTO) framework for structural optimization. Eng. Struct. **278**, 115484 (2023). https://doi.org/10.1016/j.engstruct.2022.115484
Jerez, D.J., Jensen, H.A., Beer, M.: An effective implementation of reliability methods for Bayesian model updating of structural dynamic models with multiple uncertain parameters. Reliab. Eng. Syst. Safety **225** (2022). https://doi.org/10.1016/j.ress.2022.108634. Article 108634
Jiang, C., Fu, S., Yu, Y.: Risk identification analysis of fire accidents for passenger ships by FMEA. In: 7th IEEE International Conference on Transportation Information and Safety, ICTIS 2023, Xi'an, China, 4–6 August, pp. 507–512 (2023). https://ieeexplore.ieee.org/document/10243781

Jiang, C., Yang, S., Nie, P., Xiang, X.: Multi-objective structural profile optimization of ships based on improved Artificial Bee Colony Algorithm and structural component library. Ocean Eng. **283** (2023). https://doi.org/10.1016/j.oceaneng.2023.115124. Article 115124

Jiang, S.H., Wang, L., Ouyang, S., Huang, J., Liu, Y.: A comparative study of Bayesian inverse analyses of spatially varying soil parameters for slope reliability updating. Georisk **16**(4), 746–765 (2022). https://doi.org/10.1080/17499518.2021.2010098

Jokar, M., Semperlotti, F.: Finite element network analysis: a machine learning based computational framework for the simulation of physical systems. Comput. Struct. **247** (2021). https://doi.org/10.1016/j.compstruc.2021.106484. Article 106484

Jones, D.R.: A taxonomy of global optimization methods based on response surfaces. J. Global Optim. **21**, 345–383 (2001)

Jorgensen, J., Hodkiewicz, M., Cripps, E., Hassan, G.M.: Probabilistic assessment of the effect of bolt pre-load loss over time in offshore wind turbine bolted ring-flanges using a gaussian process surrogate model. In: 41st International Conference on Offshore Mechanics and Arctic Engineering - OMAE2022, Hamburg, Germany, 5–10 June (2022). https://doi.org/10.1115/OMAE2022-79265

Kamariotis, A., Chatzi, E., Straub, D.: Value of information from vibration-based structural health monitoring extracted via Bayesian model updating. Mech. Syst. Signal Process. **166** (2022). https://doi.org/10.1016/j.ymssp.2021.108465. Article 108465

Kamariotis, A., Chatzi, E., Straub, D.: A framework for quantifying the value of vibration-based structural health monitoring. Mech. Syst. Signal Process. **184** (2023). https://doi.org/10.1016/j.ymssp.2022.109708. Article 109708

Kamidelivand, M., Deeney, P., McAuliffe, F.D., O'Connell, R., Mulcahy, I., Murphy, J.: Adapting optimal preventive maintenance strategies for floating offshore wind in atlantic areas by integrating O&M modelling and FMECA. In: OCEANS 2023, Limerick, Ireland, 5–8 June (2023)

Kang, Y., Pei, Z., Ao, L., Wu, W.: Reliability-based design optimization of river-sea-going ship based on agent model technology. Mar. Struct. **94**(103561) (2024). https://doi.org/10.1016/j.marstruc.2023.103561

Kapasakalis, K.A., Gkikakis, A.E., Sapountzakis, E.J., Chatzi, E.N., Kampitsis, A.E.: Multi-objective optimization of a negative stiffness vibration control system for offshore wind turbines. Ocean Eng. **303**(117631) (2024). https://doi.org/10.1016/j.oceaneng.2024.117631

Katanyoowongcharoen, S., et al.: Structural reliability analysis and quantitative risk assessment for optimizing cost of offshore structural integrity management and life extension programme without air diving operation. In: International Petroleum Technology Conference, IPTC 2023, Bangkok, Thailand, 1–3 March (2023). https://doi.org/10.2523/IPTC-22856-MS

Kendibilir, A., Kefal, A.: Enhanced ship cross-section design methodology using peridynamics topology optimization. Ocean Eng. **286**(115531) (2023). https://doi.org/10.1016/j.oceaneng.2023.115531

Khan, S., Goucher-Lambert, K., Kostas, K., Kaklis, P.: ShipHullGAN: a generic parametric modeller for ship hull design using deep convolutional generative model. Comput. Methods Appl. Mech. Eng. **411** (2023). https://doi.org/10.1016/j.cma.2023.116051. Article 116051

Kim, E., Amdahl, J., Wang, X.: Making sense of speed effects on ice crushing pressure-area relationships in IACS ice-strengthening rules for ships. Ocean Eng. **230** (2021). https://doi.org/10.1016/j.oceaneng.2021.109059. Article 109059

Kim, J.-H., Roh, M.-I., Yeo, I.-C.: Hull form optimization of fully parameterized small ships using characteristic curves and deep neural networks. Int. J. Nav. Archit. Ocean Eng. **16**(100596) (2024). https://doi.org/10.1016/j.ijnaoe.2024.100596

Kim, J.O., Lee, J.H., Park, H.G., Jang, T.G.: Interface development between the 3D CAD software and the structural strength assessment software for efficient classification approval. In: International Conference on Computer Applications in Shipbuilding (ICCAS 2022), Yokohama, Japan, 13–15 September (2022)

Kinne, M., Thöns, S.: Fatigue reliability based on predicted posterior stress ranges determined from strain measurements of wind turbine support structures. Energies **16**(5) (2023). https://doi.org/10.3390/en16052225. Article 2225

Kitson, M., Roberts, R., O'Brien, C., Sheppard, R., Moslehy, Y.: Wandoo a – a life extension case study implementing reliability methodologies. In: Annual Offshore Technology Conference, Houston, Texas, USA, 1–4 May (2023). https://doi.org/10.4043/32195-MS

Kleivane, S.K., Leira, B.J., Steen, S.: Reliability analysis of crack growth occurrence for a secondary hull component due to vibration excitation. J. Offshore Mech. Arct. Eng. **146**(5) (2024). https://doi.org/10.1115/1.4064499. Article 051702

Koelman, H.J., Veelo, B.N., Seppälä, L., Filius, P.: Closing the gap between early and detailed ship design models. In: Proceedings of 15th International Marine Design Conference (IMDC-2024), Amsterdam, the Netherlands, 2–6 June (2024). https://doi.org/10.59490/imdc.2024.837

Kolios, A.J.: Maintenance strategy development for subsea cables. In: Proceedings - Annual Reliability and Maintainability Symposium, Tucson, Arizona, USA, 24–27 January 2022 (2022)

Kondratenko, A.A., Bergström, M., Reutskii, A., Kujala, P.: A holistic multi-objective design optimization approach for Arctic Offshore Supply Vessels. Sustainability **13**(10), 5550 (2021). https://doi.org/10.3390/su13105550

Kondratenko, A.A., Kujala, P., Hirdaris, S.E.: Holistic and sustainable design optimization of Arctic ships. Ocean Eng. **275** (2023). https://doi.org/10.1016/j.oceaneng.2023.114095. Article 114095

Kouloumpis, V., Azapagic, A.: A model for estimating life cycle environmental impacts of offshore wind electricity considering specific characteristics of wind farms. Sustain. Prod. Consump. **29**, 495–506 (2022). https://doi.org/10.1016/j.spc.2021.10.024

KR: Rules for the Classification of Steel Ships, Pt.14 Structural Rules for Container Ship (2023a)

KR: Rules for the Classification of Steel Ships, Pt.15 Structural Rules for Membrane Type Liquefied Natural Gas carrier (2023b)

Kupwiwat, C.-T., Hayashi, K., Ohsaki, M.: Multi-objective optimization of truss structure using multi-agent reinforcement learning and graph representation. Eng. Appl. Artif. Intell. **129**(107594) (2024). https://doi.org/10.1016/j.engappai.2023.107594

Kurniawan, H., Pitana, T., Siswantoro, N.: Risk-Based Integrity Management System of Oil Tanker Hull Structure Due to Corrosion. Key Engineering Materials, pp. 121–129 (2023). https://www.scopus.com/inward/record.uri?eid=2-s2.0-85150510179&doi=10.4028%2fp-3z2t4s&partnerID=40&md5=c0d3a348c2b4bbcd509d6862c1a5e2bb

Lang, X., Wang, H., Mao, W., Osawa, N.: Impact of ship operations aided by voyage optimization on a ship's fatigue assessment. J. Mar. Sci. Technol. **26**(3), 750–771 (2021). https://doi.org/10.1007/s00773-020-00769-8

Larsen, M.L., Arora, V., Adhikari, S., Clausen, H.B.: Optimization of welded K-node in offshore jacket structure including the stochastic size effect. Mar. Struct. **82**(103128) (2022). https://doi.org/10.1016/j.marstruc.2021.103128

Lazakis, I., et al.: Committee IV.2 - Design Methods Proceedings of the 20th International Ship and Offshore Structures Congress (ISSC 2018), Netherlands, pp. 609–708 (2018)

Lee, J.-H., Nam, Y.-S., Kim, Y., Liu, Y., Lee, J., Yang, H.: Real-time digital twin for ship operation in waves. Ocean Eng. **266**, 112867 (2022). https://doi.org/10.1016/j.oceaneng.2022.112867

Leimeister, M., Kolios, A.: Reliability-based design optimization of a spar-type floating offshore wind turbine support structure. Reliab. Eng. Syst. Safety **213** (2021). https://doi.org/10.1016/j.ress.2021.107666. Article 107666

Lempitsky, V., Vedaldi, A., Ulyanov, D.: Deep image prior. In: 2018 IEEE/CVF Conference on Computer Vision and Pattern Recognition, Salt Lake City, UT, USA, 18–23 June, pp. 9446–9454 (2018). https://ieeexplore.ieee.org/document/8579082

Lemström, I., Polojärvi, A., Tuhkuri, J.: Model-scale tests on ice-structure interaction in shallow water, Part I: global ice loads and the ice loading process. Marine Struct. **81** (2022). https://doi.org/10.1016/j.marstruc.2021.103106. Article 103106

Li, C., Wei, P., Luo, X., Jiang, Z., Wang, D.: An unified CAD/CAE/VR tool for ship structure design and evaluation based on multi-domain feature mapping. Ocean Eng. **280**, 114888 (2023). https://doi.org/10.1016/j.oceaneng.2023.114888

Li, G., Holmeset, F.T., Zhang, H.: Toward remote control center for marine operation: a case study of R/V gunnerus. In: 11th International Conference on Control, Mechatronics and Automation (ICCMA), Agder, Norway, 1–3 November, pp. 183–188 (2023)

Li, H., Wei, X., Liu, Z., Feng, B., Zheng, Q.: Ship design optimization with mixed uncertainty based on evidence theory. Ocean Eng. **279**(114554) (2023). https://doi.org/10.1016/j.oceaneng.2023.114554

Li, S., Kim, D.K., Ringsberg, J.W., Liu, B., Benson, S.D.: Uncertainty of ship hull girder ultimate strength in global bending predicted by smith-type collapse analysis. Trans. Royal Inst. Naval Archit. Part A Int. J. Marit. Eng. **164**(2 A), A185-A205 (2022). https://doi.org/10.5750/ijme.v164iA2.1157

Li, X., Oh, P., Zhou, Y., Yuen, K.F.: Operational risk identification of maritime surface autonomous ship: a network analysis approach. Transp. Policy **130**, 1–14 (2023). https://doi.org/10.1016/j.tranpol.2022.10.012

Li, X., Zhang, Y., Zhang, L., Han, Z.: A probabilistic assessment methodology for pitting corrosion condition of offshore crude oil pipelines. Ocean Eng. **288** (2023). https://doi.org/10.1016/j.oceaneng.2023.116112. Article 116112

Li, Y., Zhang, Y., Li, W., Jiang, T.: Marine wireless big data: efficient transmission, related applications, and challenges. IEEE Wirel. Commun. **25**(1), 19–25 (2018). https://doi.org/10.1109/mwc.2018.1700192

Liang, K., Chen, Y., Zhang, Q.: A digital twin model construction method for ships. In: 2023 IEEE 11th International Conference on Computer Science and Network Technology (ICCSNT), Dalian, China, 21–22 October 2023

Lin, H., et al.: Numerical investigation on performance optimization of offshore sandwich blast walls with different honeycomb cores subjected to blast loading. J. Mar. Sci. Eng. **10**(11), 1743 (2022). https://doi.org/10.3390/jmse10111743

Lin, J., et al.: Characterization of model uncertainty of the P-Y method for reliability-based design of offshore wind turbines under the serviceability limit state. Ocean Eng. **293** (2024). https://doi.org/10.1016/j.oceaneng.2023.116662. Article 116662

Liu, D., Li, S., Ai, S., Sun, L., Soares, C.G.: Fatigue analysis of steel catenary risers under coupled cross-flow and in-line vortex-induced vibrations with oblique incoming flow. Marine Struct. **95** (2024). https://doi.org/10.1016/j.marstruc.2024.103578. Article 103578

Liu, F., Hu, Y.-M., Feng, G.-Q., Zhao, W.-D., Zhang, M.: A study on the multi-objective optimization method of brackets in ship structures. China Ocean Eng. **36**(2), 208–222 (2022). https://doi.org/10.1007/s13344-022-0018-7

Liu, J., Cheng, Y., Jing, X., Liu, X., Chen, Y.: Prediction and optimization method for welding quality of components in ship construction. Sci. Rep. **14**(1), 9353 (2024). https://doi.org/10.1038/s41598-024-59490-w

Liu, J., et al.: Current and future trends in topology optimization for additive manufacturing. Struct. Multidiscip. Optim. **57**(6), 2457–2483 (2018). https://doi.org/10.1007/s00158-018-1994-3

Liu, L., Zhang, M., Hu, Y., Zhu, W., Xu, S., Yu, Q.: A probabilistic analytics method to identify striking ship of ship-buoy contact at coastal waters. Ocean Eng. **266** (2022). https://doi.org/10.1016/j.oceaneng.2022.113102. Article 113102

Liu, Y., et al.: Maneuverability prediction of ship nonlinear motion models based on parameter identification and optimization. Measurement **236**(115033) (2024). https://doi.org/10.1016/j.measurement.2024.115033

Liu, Y., Ma, X., Qiao, W., Han, B.: On the determination and rank for the environmental risk aspects for ship navigating in the Arctic based on big Earth data. Risk Anal. **43**(11), 2186–2210 (2023). https://doi.org/10.1111/risa.13987

Liu, Y., Ren, H.: Rapid acquisition method for structural strength evaluation stresses of the ship digital twin model. Ocean Eng. **285**, 115323 (2023). https://doi.org/10.1016/j.oceaneng.2023.115323

Liu, Y., Zhang, J.-M., Min, Y.-T., Yu, Y., Lin, C., Hu, Z.-Z.: A digital twin-based framework for simulation and monitoring analysis of floating wind turbine structures. Ocean Eng. **283**, 115009 (2023). https://doi.org/10.1016/j.oceaneng.2023.115009

Liu, Y., Zhou, M., Wei, C., Lin, Z.: Topology optimization of self-supporting infill structures. Struct. Multidiscip. Optim. **63**(5), 2289–2304 (2021). https://doi.org/10.1007/s00158-020-02805-y

Loake, M.C.A., Astfalck, L.C., Cripps, E.J.: Modelling sea surface wind measurements on Australia's North-West Shelf. Ocean Eng. **244** (2022). https://doi.org/10.1016/j.oceaneng.2021.110308. Article 110308

Lopez, J.C., Kolios, A.: Risk-based maintenance strategy selection for wind turbine composite blades. Energy Rep. **8**, 5541–5561 (2022). https://doi.org/10.1016/j.egyr.2022.04.027

López, J.C., Kolios, A., Wang, L., Chiachio, M.: A wind turbine blade leading edge rain erosion computational framework. Renew. Energy **203**, 131–141 (2023). https://doi.org/10.1016/j.renene.2022.12.050

Lopez, J.C., Kolios, A., Wang, L., Chiachio, M., Dimitrov, N.: Reliability-based leading edge erosion maintenance strategy selection framework. Appl. Energy **358** (2024). https://doi.org/10.1016/j.apenergy.2023.122612. Article 122612

Louvros, P., Boulougouris, E., Coraddu, A., Vassalos, D., Theotokatos, G.: Multi-objective optimisation as an early design tool for smart ship internal arrangement. Ships Offshore Struct., 1–11 (2021). https://doi.org/10.1080/17445302.2021.1926142

LR: ShipRight Design and Construction, Structural Design Assessment, Primary Hull and Cargo Tank Structure of a Type A tank Liquefied Gas Carriers (2023)

Lu, K.X., Zhang, W.Y., He, B.N., Liu, Z.X.: Research on integrated dimension-reduction modeling of stochastic wind-wave and reliability of offshore wind turbines. Jisuan Lixue Xuebao/Chinese J. Comput. Mech. **40**(6), 879–884 (2023). https://doi.org/10.7511/jslx20220514001

Luoma, E., et al.: A multi-criteria decision analysis model for ship biofouling management in the Baltic Sea. Sci. Total Env. **852** (2022). https://doi.org/10.1016/j.scitotenv.2022.158316. Article 158316

Lv, Z., Lv, H., Fridenfalk, M.: Digital twins in the marine industry. Electronics **12**(9) (2023). https://doi.org/10.3390/electronics12092025. Article 2025

Maalouf, T., Norsetter, J.: ReedSmith: Big waves: global autonomous ships market on the rise (2024). https://www.shiplawlog.com/2024/03/08/big-waves-global-autonomous-ships-market-on-the-rise/. Accessed 07 Sept 2024

Maduka, M., Coughlan, K., Arwade, S., Schoefs, F., Bates, A., Thiagarajan, K.: Hydrodynamic effects of surface roughness on cylinders: literature review and research gaps. In: Proceedings of the ASME 2022 4th International Offshore Wind Technical Conference, IOWTC2022, Northeastern University, Boston, Massachusetts, 7–8 December 2022 (2022)

Madusanka, N.S., Fan, Y., Yang, S., Xiang, X.: Digital Twin in the maritime domain: a review and emerging trends. J. Mar. Sci. Eng. **11**(5), 1021 (2023). https://doi.org/10.3390/jmse11051021

Makris, P., Silionis, N.E., Anyfantis, K.N.: Spectral fatigue analysis of ship structures based on a stochastic crack growth state model. Int. J. Fatigue **176** (2023). https://doi.org/10.1016/j.ijfatigue.2023.107878. Article 107878

Mannan, B., Rizvi, M.J., Dai, Y.M.: Does end-of-life ships research trends change in last three decades? A review for the future roadmap. J. Int. Marit. Saf. Environ. Affairs Shipp. 7(1), 2187603 (2023). https://doi.org/10.1080/25725084.2023.2187603

Mao, Y., Wang, T., Duan, M.: A DNN-based approach to predict dynamic mooring tensions for semi-submersible platform under a mooring line failure condition. Ocean Eng. **266** (2022). https://doi.org/10.1016/j.oceaneng.2022.112767. Article 112767

Martin, J., Neumann, A., Ødegård, A.: Renewable hydrogen and synthetic fuels versus fossil fuels for trucking, shipping and aviation: a holistic cost model. Renew. Sustain. Energy Rev. **186** (2023). https://doi.org/10.1016/j.rser.2023.113637

MathWorks: Get Started with Parallel Computing Toolbox (2024). www.mathworks.com/help/parallel-computing/getting-started-with-parallel-computing-toolbox.html. Accessed 4 May 2024

Mauro, F., Kana, A.A.: Digital twin for ship life-cycle: a critical systematic review. Ocean Eng. **269**(113479) (2023). https://doi.org/10.1016/j.oceaneng.2022.113479

Meng, D., Chen, X., Wei, Z., Ji, S.: Prediction of flexural and uniaxial compressive strengths of sea ice with optimized recurrent neural network. Ocean Eng. **288** (2023). https://doi.org/10.1016/j.oceaneng.2023.115921. Article 115921

MEPC: RESOLUTION MEPC.377(80). 2023 IMO strategy on reduction of GHG emissions from ships (2023). Accessed 7 July 2023

Mokhtari, M., Kim, E., Amdahl, J.: Numerical simulation of an aluminium panel subject to ice impact load using a rate and pressure dependent elastoplastic material model for ice. In: 42nd International Conference on Offshore Mechanics and Arctic Engineering - OMAE2023, Melbourne, Australia, 11–16 June (2023). https://doi.org/10.1115/OMAE2023-104771

Molina Vargas, D.G., Vijayan, K.K., Mork, O.J.: Augmented reality for future research opportunities and challenges in the shipbuilding industry: a literature review. Procedia Manuf. **45**, 497–503 (2020). https://doi.org/10.1016/j.promfg.2020.04.063

Mombiela, D.C., Zadeh, M.: Integrated design and control approach for marine power systems based on operational data. In: Digital Twin to Design. 2021 IEEE Transportation Electrification Conference & Expo (ITEC), Chicago, Illinois, USA, 21–25 June, pp. 520–527 (2021)

Moseyko, E., Ol'Khovik, E.: Analysis of reliability of shipboard systems operating in the Arctic. In: IV International Scientific Conference on Advanced Technologies in Aerospace, Mechanical and Automation Engineering: (MIST: Aerospace-IV 2021), Krasnoyarsk, Russian Federation, 10–11 December (2023). https://doi.org/10.1063/5.0125271

Motok, M., Momčilović, N., Rudaković, S.: Reliability based structural design of river–sea tankers: still water loading effects. Marine Struct. **83** (2022). https://doi.org/10.1016/j.marstruc.2022.103202. Article 103202

Mousavi, Z., Varahram, S., Ettefagh, M.M., Sadeghi, M.H., Feng, W.-Q., Bayat, M.: A digital twin-based framework for damage detection of a floating wind turbine structure under various loading conditions based on deep learning approach. Ocean Eng. **292**, 116563 (2024). https://doi.org/10.1016/j.oceaneng.2023.116563

Moussa, A.A., Farag, Y.B.A., Gunbeyaz, S.A., Fahim, N.S., Kurt, R.E.: Development and research directions in ship recycling: a systematic literature review with bibliometric analysis. Mar. Pollut. Bull. **201**, 116247 (2024). https://doi.org/10.1016/j.marpolbul.2024.116247

Moussavi, S., Barutha, P., Dvorak, B.: Environmental life cycle assessment of a novel offshore wind energy design project: a United States based case study. Renew. Sustain. Energy Rev. **185**, 113643 (2023). https://doi.org/10.1016/j.rser.2023.113643

Mouzakitis, S., et al.: Enabling maritime digitalization by extreme-scale analytics, AI and digital twins: the vesselai architecture. Lecture Notes in Networks and Systems, vol. 544, pp. 246–256 (2023). https://doi.org/10.1007/978-3-031-16075-2_16

Moynihan, B., Mehrjoo, A., Moaveni, B., McAdam, R., Rüdinger, F., Hines, E.: System identification and finite element model updating of a 6 MW offshore wind turbine using vibrational response measurements. Renew. Energy **219** (2023). https://doi.org/10.1016/j.renene.2023. 119430. Article 119430

MSC: MSC 107/5/1. Development of a Goal-based instrument for Maritime Autonomous Surface Ships (MASS) (2023). https://www.imo.org/en/MediaCentre/HotTopics/Pages/Autonomous-shipping.aspx. Accessed 7 Sept 2024

NASA: NASA Systems Engineering Handbook: NASA/SP-2016-6105 Rev2 (2016). https://www.nasa.gov/wp-content/uploads/2018/09/nasa_systems_engineering_handbook_0.pdf. Accessed 20 July 2024

Nazemian, A., Ghadimi, P.: Multi-objective optimization of ship hull modification based on resistance and wake field improvement: combination of adjoint solver and CAD-CFD-based approach. J. Braz. Soc. Mech. Sci. Eng. **44**(1) (2022). https://doi.org/10.1007/s40430-021-03335-4

Nemmour, A., Inayat, A., Janajreh, I., Ghenai, C.: Green hydrogen-based E-fuels (E-methane, E-methanol, E-ammonia) to support clean energy transition: a literature review. Int. J. Hydrogen Energy **48**(75), 29011–29033 (2023). https://doi.org/10.1016/j.ijhydene.2023.03.240

Neves, K.L., et al.: A methodology for development of a digital twin ship. In: XLII Ibero-Latin American Congress on Computational Methods in Engineering, Rio de Janeiro, 9–12 November 2021 (2021)

Neves, K.L.S., Dotta, R., Malta, E.B., Gay Neto, A., Franzini, G.R., Bitencourt, L.A.G.: A digital twin to predict failure probability of an FPSO hull based on corrosion models. J. Mar. Sci. Technol. **28**(4), 862–875 (2023). https://doi.org/10.1007/s00773-023-00963-4

Nickpasand, M., Jaafari, H.E., Gaspar, H.M.: Machine learning in agile manufacturing; a usecase from offshore wind turbine product lifecycle management (PLM) system. In: 38th ECMS International Conference on Modelling and Simulation, pp. 26–33 (2024). https://doi.org/10.7148/2024-0026

Nikolopoulos, L., Boulougouris, E.: A novel method for the holistic, simulation driven ship design optimization under uncertainty in the big data era. Ocean Eng. **218** (2020). https://doi.org/10.1016/j.oceaneng.2020.107634. Article 107634

Oh, S., Jung, Y., Kim, S., Lee, I., Kang, N.: Deep generative design: integration of topology optimization and generative models. J. Mech. Des. N. Y. **141**(11), 1 (2019). https://doi.org/10.1115/1.4044229

Ojo, A., Collu, M., Coraddu, A.: Multidisciplinary design analysis and optimization of floating offshore wind turbine substructures: a review. Ocean Eng. **266**(112727) (2022). https://doi.org/10.1016/j.oceaneng.2022.112727

Olatunde, M., Sriramula, S., Siddiq, A.M., Akisanya, A.R., Brixton, A.D.: A generic methodology for predicting the longevity of offshore infrastructure. In: SPE Offshore Europe Conference & Exhibition, Aberdeen, Scotland, UK, 5–8 September (2023). https://doi.org/10.2118/215506-MS

Önal, M.: Ship recycling perspective on environmental impacts - a case study for the ships in service. Heliyon **9**(10), e21157 (2023). https://doi.org/10.1016/j.heliyon.2023e21157

Othman, N.A., Mohd, M.H.: Response of the offshore jacket platform at its ultimate strength under operating and extreme loads: a Malaysian waters case study. Ships Offshore Struct. (2024). https://doi.org/10.1080/17445302.2024.2312726

Papanikolaou, A.: Risk-Based Ship Design: Methods, Tools and Applications. Springer (2009). https://doi.org/10.1007/978-3-540-89042-3

Papanikolaou, A., Boulougouris, E., Erikstad, S.-O., Harries, S., Kana, A.A.: Ship design in the era of digital transition: a state-of-the-art report. In: 15th International Marine Design Conference (IMDC-2024), Amsterdam, the Netherlands, 2–6 June (2024). https://doi.org/10.59490/imdc.2024.784

Papanikolaou, A., et al.: A holistic approach to ship design: tools and applications. J. Ship Res. **66**(01), 25–53 (2022). https://doi.org/10.5957/josr.12190070

Papanikolaou, A., et al.: Holistic ship design for green shipping. Transp. Res. Procedia **72**, 1224–1231 (2023). https://doi.org/10.1016/j.trpro.2023.11.581

Park, M. J., Kim, Y.: MCMC based probabilistic wave spectrum estimation using onboard measurement data. In: 41st International Conference on Offshore Mechanics and Arctic Engineering - OMAE2022, Hamburg, Geermany, 5–10 June (2022). https://doi.org/10.1115/OMAE2022-80627

Park, S., Huh, J.-H.: Study on PLM and Big Data collection for the digital transformation of the shipbuilding industry. J. Mar. Sci. Eng. **10**(10), 1488 (2022). https://doi.org/10.3390/jmse10101488

Parunov, J., Guedes Soares, C., Hirdaris, S., Wang, X.: Uncertainties in modelling the low-frequency wave-induced global loads in ships. Mar. Struct. **86** (2022). https://doi.org/10.1016/j.marstruc.2022.103307. Article 103307

Pereira, T., Garbatov, Y.: Multi-attribute decision-making ship structural design. J. Mar. Sci. Eng. **10**(8) (2022). https://doi.org/10.3390/jmse10081046. Article 1046

Pérez-Arribas, F.: Parametric generation of small ship hulls with CAD software. J. Mar. Sci. Eng. **11**(5), 976 (2023). https://doi.org/10.3390/jmse11050976

Perez-Martinez, J., Perez Fernandez, R.: Material and production optimization of the ship design process by introducing CADs from early design stages. J. Mar. Sci. Eng. **11**(1), 233 (2023). https://doi.org/10.3390/jmse11010233

Pérez Fernández, R.: A new approach for using CAD and PLM integration. In: International Conference on Computer Applications in Shipbuilding, ICCAS 2022, Yokohama, Japan, 13–15 September (2022)

Pham-Truong, T., Kawamura, Y., Okada, T.: Bisection constraint method for multiple-loading conditions in structural topology optimization. Appl. Sci. **13**(24) (2023). https://doi.org/10.3390/app132413005. Article 13005

Purcell, E., Nejad, A. R., Bekker, A.: Detection of ice using ship propulsion and navigation measurements. Ocean Eng. **273** (2023). https://doi.org/10.1016/j.oceaneng.2023.113992. Article 113992

Putra, G.L., Kitamura, M.: Study on optimal design of hatch cover via a three-stage optimization method involving material selection, size, and plate layout arrangement. Ocean Eng. **219** (2021). https://doi.org/10.1016/j.oceaneng.2020.108284. Article 108284

Qiu, C., Du, S., Yang, J.: A deep learning approach for efficient topology optimization based on the element removal strategy. Mater. Des. **212**(110179) (2021). https://doi.org/10.1016/j.matdes.2021.110179

Raissi, M., Perdikaris, P., Karniadakis, G.E.: Physics-informed neural networks: a deep learning framework for solving forward and inverse problems involving nonlinear partial differential equations. J. Comput. Phys. **378**, 686–707 (2019). https://doi.org/10.1016/j.jcp.2018.10.045

Röltgen, D., Dumitrescu, R.: Classification of industrial Augmented Reality use cases. Procedia CIRP **91**, 93–100 (2020). https://doi.org/10.1016/j.procir.2020.01.137

Rosa, D.R., Schiozer, D.J., Davolio, A.: Impact of model and data resolutions in 4D seismic data assimilation applied to an offshore reservoir in Brazil. J. Pet. Sci. Eng. **216** (2022). https://doi.org/10.1016/j.petrol.2022.110830. Article 110830

Rubio, L.R., Mariscal, A.M., Alvarez, E.P., Mas, F.: A process-oriented approach for shipbuilding industrial design using advanced PLM tools. In: IFIP Advances in Information and Communication Technology, Grenoble, France, 10–13 July (2023)

Ryumin, S., Tryaskin, V., Plotnikov, K.: Algorithms for the recognition of the hull structures' elementary plate panels and the determination of their parameters in a ship CAD system. J. Mar. Sci. Eng. **11**(1), 189 (2023). https://doi.org/10.3390/jmse11010189

Sadeghian, M.: The reliability assessment of a ship structure under corrosion and fatigue, using structural health monitoring. Int. J. Eng. Trans. B **35**(9), 1765–1778 (2022). https://doi.org/10.5829/ije.2022.35.09c.13

Sakagami, M., Otaguro, T., Masui, T., Lin, P., Shimakawa, Y., Hisano, S.: Making 3D model-based approval a reality. In: International Conference on Computer Applications in Shipbuilding (ICCAS 2022), Yokohama, Japan, 13–15 September (2022)

Salis, N., Hu, X., Luo, M., Reali, A., Manenti, S.: 3D SPH analysis of focused waves interacting with a floating structure. Appl. Ocean Res. **144** (2024). https://doi.org/10.1016/j.apor.2024.103885. Article 103885

Sassanelli, C., Urbinati, A., Rosa, P., Chiaroni, D., Terzi, S.: Addressing circular economy through design for X approaches: a systematic literature review. Comput. Ind. **120** (2020). https://doi.org/10.1016/j.compind.2020.103245. Article 103245

Seppälä, L.: Using PLM to unlock shipbuilding's digital potential (2023). https://www.cadmatic.com/en/resources/articles/using-plm-to-unlock-shipbuildings-digital-potential/. Accessed 25 Oct

Shabakhty, N., Asgari Motlagh, A., Kaveh, A.: Optimal design of offshore jacket platform using enhanced colliding bodies optimization algorithm. Mar. Struct. **97**(103640) (2024). https://doi.org/10.1016/j.marstruc.2024.103640

Shenoy, M.A., Subramanian, T., Al Awadhi, I.: A structured approach to rejuvenate service life of marine structures. In: ADIPEC (Abu Dhabi International Petroleum Exhibition and Conference), Abu Dhabi, UAE, October 31–November 3 (2022). https://doi.org/10.2118/211 123-MS

Shi, S., Zhai, E., Xu, C., Iqbal, K., Sun, Y., Wang, S.: Influence of pile-soil interaction on dynamic properties and response of offshore wind turbine with monopile foundation in sand site. Appl. Ocean Res. **126** (2022). https://doi.org/10.1016/j.apor.2022.103279. Article 103279

Shi, W., et al.: Review on the development of marine floating photovoltaic systems. Ocean Eng. **286** (2023). https://doi.org/10.1016/j.oceaneng.2023.115560. Article 115560

Shi, Y., Yang, D., Li, Q., Qin, J.: Cabin noise analysis of polar transport vessels under ship–ice–water–air coupling continuous icebreaking based on the S-ALE algorithm. Marine Struct. **95** (2024). https://doi.org/10.1016/j.marstruc.2024.103601. Article 103601

Shin, S., Shin, D., Kang, N.: Topology optimization via machine learning and deep learning: a review. J. Computat. Des. Eng. **10**(4), 1736–1766 (2023). https://doi.org/10.1093/jcde/qwad072

Shu, D., et al.: 3D design using generative adversarial networks and physics-based validation. J. Mech. Des. N. Y. **142**(7), 1–51 (2020). https://doi.org/10.1115/1.4045419

Sigmund, O.: On the design of compliant mechanisms using topology optimization*. Mech. Struct. Mach. **25**(4), 493–524 (1997). https://doi.org/10.1080/08905459708945415

Sigmund, O.: A 99 line topology optimization code written in matlab. Struct. Multidiscip. Optim. **21**(2), 120–127 (2001). https://doi.org/10.1007/s001580050176

Silionis, N.E., Anyfantis, K.N.: An optimization based approach for damage identification of idealized ship structural assemblies. In: 40th International Conference on Ocean, Offshore and Arctic Engineering- OMAE2021, Virtual, Online, 21–30 June (2021). https://doi.org/10.1115/omae2021-60810

Silionis, N.E., Anyfantis, K.N.: Uncertainty quantification within strain-based SHM schemes used for detecting thickness loss in ship hulls. In: Advances in the Analysis and Design of Marine Structures - Proceedings of the 9th International Conference on Marine Structures, MARSTRUCT 2023, Gothenburg, Sweden, pp. 743–750 (2023). https://doi.org/10.1201/9781003399759

Silva-Campillo, A., Suárez-Bermejo, J.C., Herreros-Sierra, M.A., de Vicente, M.: Design methodology in transverse webs of the torsional box structure in an ultra large container ship. Int. J. Nav. Archit. Ocean Eng. **13**, 772–785 (2021). https://doi.org/10.1016/j.ijnaoe.2021.08.004

Silva-Campillo, A., Ulla-Campos, L., Suárez-Bermejo, J. C., Herreros-Sierra, M.A.: Effect of bow hull form on the buckling strength assessment of the corner bracket connection. Ocean Eng. **265**(112562) (2022). https://doi.org/10.1016/j.oceaneng.2022.112562

Siyuan, G., Shanshan, F., Zhang, Y.: A preliminary investigation of risk control options for maritime accidents in arctic waters. In: 42nd Proceedings of the International Conference on Offshore Mechanics and Arctic Engineering - OMAE2023, Melbourne, Australia, 11–16 June (2023). https://doi.org/10.1115/OMAE2023-105765

SOLAS: Regulation 3-10. Goal-based ship construction standards for bulk carriers and oil tankers. Added by Res.MSC.290(87) (2010)

Solheim, A.V., et al.: System-based ship design of a deep-sea mining vessel. Ship Technol. Res. (2024). https://doi.org/10.1080/09377255.2024.2396197

Song, Y., Sun, T., Zhang, Z.: Fatigue reliability analysis of floating offshore wind turbines considering the uncertainty due to finite sampling of load conditions. Renew. Energy **212**, 570–588 (2023). https://doi.org/10.1016/j.renene.2023.05.070

Stankiewicz, G., Dev, C., Steinmann, P.: Coupled topology and shape optimization using an embedding domain discretization method. Struct. Multidiscip. Optim. **64**(4), 2687–2707 (2021). https://doi.org/10.1007/s00158-021-03024-9

Subedi, S.C., Verma, C.S., Suresh, K.: A review of methods for the geometric post-processing of topology optimized models. J. Comput. Inf. Sci. Eng. **20**(6) (2020). https://doi.org/10.1115/1.4047429. Article 060801

Sullivan, B.P., Rossi, M., Terzi, S.: LINCOLN-lean innovative connected vessels. Holistic Approach Ship Des. **2**, 439–467 (2021). https://doi.org/10.1007/978-3-030-71091-0_14

Sun, C., Chen, Z., Yi, J., Li, D.: A data-driven approach to full-field stress reconstruction of ship hull structure using deep learning. Eng. Appl. Artif. Intell. **133** (2024). https://doi.org/10.1016/j.engappai.2024.108414

Sun, C., Liu, M., Ge, S.: B-spline curve fitting of hungry predation optimization on ship line design. Appl. Sci. **12**(19), 9465 (2022). https://doi.org/10.3390/app12199465

Sun, F., Zhou, J., Shi, H., Shen, S.: Implementation of marine CAD parametric drawing system based on configuration method. In: Second International Symposium on Computer Technology and Information Science (ISCTIS 2022) (2022). https://doi.org/10.1117/12.2653795

Sun, J., Gaidai, O., Wang, F., Yakimov, V.: Gaidai reliability method for fixed offshore structures. J. Braz. Soc. Mech. Sci. Eng. **46**(1) (2024). https://doi.org/10.1007/s40430-023-04607-x. Article 27

Sun, Q., Chen, J., Zhou, L., Ding, S., Han, S.: A study on ice resistance prediction based on deep learning data generation method. Ocean Eng. **301** (2024). https://doi.org/10.1016/j.oceaneng.2024.117467. Article 117467

Sun, Q., Olschewski, P., Wei, J., Tian, Z., Sun, L., Kunstmann, H., Laux, P.: Key ingredients in regional climate modelling for improving the representation of typhoon tracks and intensities. Hydrol. Earth Syst. Sci. **28**(4), 761–780 (2024). https://doi.org/10.5194/hess-28-761-2024

Sun, Y.-H., Zhang, B.-J.: Seakeeping design optimization of KRISO container ship (KCS) ship form considering wave action. Mar. Technol. Soc. J. **56**(2), 64–72 (2022). https://doi.org/10.4031/mtsj.56.2.6

Tadros, M., Ventura, M., Soares, C.G.: Review of current regulations, available technologies, and future trends in the green shipping industry. Ocean Eng. **280** (2023a). https://doi.org/10.1016/j.oceaneng.2023.114670

Tadros, M., Ventura, M., Soares, C.G.: Review of the decision support methods used in optimizing ship hulls towards improving energy efficiency. J. Marine Sci. Eng. **11**(4) (2023b). https://doi.org/10.3390/jmse11040835. Article 835

Taghavifar, H., Perera, L.P.: Life cycle emission and cost assessment for LNG-retrofitted vessels: the risk and sensitivity analyses under fuel property and load variations. Ocean Eng. **282** (2023). https://doi.org/10.1016/j.oceaneng.2023.114940. Article 114940

Tamimi, M.F., Khandel, O., Soliman, M.: A framework for quantifying fatigue deterioration of ship structures under changing climate conditions. Ships Offshore Struct. **17**(12), 2745–2760 (2022). https://doi.org/10.1080/17445302.2021.2018223

Tayab, M., Awan, Z., Shah, V., Saif, A., Al Hameli, F.: A novel approach to prevent process safety incidents by integrating & strengthening process safety barriers & operational excellence frameworks in upstream & downstream segments of oil & gas operations. In: Society of Petroleum Engineers - ADIPEC 2022, Abu Dhabi, UAE, October 31–November 3 (2022). https://doi.org/10.2118/210971-MS

Tezzele, M., Fabris, L., Sidari, M., Sicchiero, M., Rozza, G.: A multifidelity approach coupling parameter space reduction and nonintrusive POD with application to structural optimization of passenger ship hulls. Int. J. Numer. Methods Eng. **124**(5), 1193–1210 (2023). https://doi.org/10.1002/nme.7159

Tola, F., Mosconi, E. M., Marconi, M., Gianvincenzi, M.: Perspectives for the development of a circular economy model to promote ship recycling practices in the european context: a systemic literature review. Sustainability **15**(7) (2023)

Tott-Buswell, J., Hilton, J., Berberic, S., Jalbi, S., Prendergast, L.J.: Frequency analysis of monopiles with masing-type hysteresis damping under large-strain cyclic loading. In: Experimental Vibration Analysis for Civil Engineering Structures (EVACES 2023). LNCE, vol. 432, pp. 232–241 (2023). https://doi.org/10.1007/978-3-031-39109-5_24

Trinh, L.T., Hamagami, T., Okamoto, N.: 3D ship hull design direct optimization using generative adversarial network. J. Adv. Comput. Intell. Intell. Inform. **28**(3), 693–703 (2024). https://doi.org/10.20965/jaciii.2024.p0693

Trivyza, N.L., Rentizelas, A., Theotokatos, G.: A novel multi-objective decision support method for ship energy systems synthesis to enhance sustainability. Energy Convers. Manage. **168**, 128–149 (2018). https://doi.org/10.1016/j.enconman.2018.04.020

Tyflopoulos, E., Lien, M., Steinert, M.: Optimization of brake calipers using topology optimization for additive manufacturing. Appl. Sci. **11**(4), 1437 (2021). https://doi.org/10.3390/app11041437

Uddin, M.S., et al.: Ship deck segmentation in engineering document using generative adversarial networks. In: 2022 IEEE World AI IoT Congress (AIIoT) (2022). https://doi.org/10.1109/aiiot54504.2022.9817355

UN: Review of Maritime Transport 2023 (2023)

Vadyala, S.R., Betgeri, S.N., Matthews, J.C., Matthews, E.: A review of physics-based machine learning in civil engineering. Results Eng. **13** (2022). https://doi.org/10.1016/j.rineng.2021.100316

Vanem, E., Zhu, T., Babanin, A.: Statistical modelling of the ocean environment – a review of recent developments in theory and applications. Marine Struct. **86** (2022). https://doi.org/10.1016/j.marstruc.2022.103297. Article 103297

Veen, D., Pahos, S. J., Meng, S., Dillenburg, S.: A new method for the design and coupled analysis of floating offshore wind turbines. In: 42nd International Conference on Offshore Mechanics and Arctic Engineering - OMAE2023, Melbourne, Australia, 11–16 June (2023). https://doi.org/10.1115/OMAE2023-104576

Velasco-Parra, J.A., Ramon-Valencia, B.A., Lopez-Arraiza, A.: Effects of seawater immersion on a glass fibre reinforced bioepoxy mechanical properties and its application in the ship hull finite-element analysis. Proc. Inst. Mech. Eng. Part L. J. Mater. Des. Appl. **236**(1), 147–154 (2022). https://doi.org/10.1177/14644207211042987

Ventikos, N., Louzis, K., Sotiralis, P., Koimtzoglou, A., Annetis, M.: Integrating human factors in risk-based design: a critical review. In: Conference: ERGOSHIP 2021, 2–3 September (online confrence), Busan, South Korea (2021)

Ventikos, N.P., Chmurski, A., Louzis, K.: A systems-based application for autonomous vessels safety: hazard identification as a function of increasing autonomy levels. Safety Sci. **131**, 104919 (2020). https://doi.org/10.1016/j.ssci.2020.104919

Ventikos, N.P., Siokouros, P., Koimtzoglou, A.: Application of the EAST-BL method on a MASS system for Hazard Identification and Risk Assessment International Conference on Maritime Autonomous Surface Ships (ICMASS 2023) Rotherdam, the Netherlands, 8–9 November (2023). https://iopscience.iop.org/article/10.1088/1742-6596/2618/1/012014

Vergassola, M., Cabboi, A., van der Male, P., Colomés, O.: A low-fidelity model for the dynamic analysis of full-lattice wind support structures. Marine Struct. **92** (2023). https://doi.org/10.1016/j.marstruc.2023.103506. Article 103506

Vettor, R., Bergamini, G., Guedes Soares, C.: A comprehensive approach to account for weather uncertainties in ship route optimization. J. Mar. Sci. Eng. **9**(12), 1434 (2021). https://doi.org/10.3390/jmse9121434

Wan, Y., Hou, Y., Xiong, Y., Dong, Z., Zhang, Y., Gong, C.: Interval optimization design of a submersible surface ship form considering the uncertainty of surrogate model. Ocean Eng. **263**(112262) (2022). https://doi.org/10.1016/j.oceaneng.2022.112262

Wang, C., Zhang, G., Qiu, C., Zhang, X.: Combat methods verification system of surface ships based on big data technology. In: 9th International Conference on Big Data and Information Analytics (BigDIA), Haikou, China, 15–17 December 2023 (2023). https://doi.org/10.1109/bigdia60676.2023.10429396

Wang, G., et al.: Challenges and opportunities to floating wind: perspectives of naval architects. In: 15th International Symposium on Practical Design of Ships and Other Floating Structures, PRADS2022, Dubrovnik, Croatia, 09–13 October (2022)

Wang, L.: Research on large-scale ship data storage system. In: 2020 International Conference on Information Science, Parallel and Distributed Systems (ISPDS) (2020). https://doi.org/10.1109/ispds51347.2020.00028

Wang, L., et al.: Validation of CFD simulations of the moored DeepCwind offshore wind semisubmersible in irregular waves. Ocean Eng. **260** (2022). https://doi.org/10.1016/j.oceaneng.2022.112028. Article 112028

Wang, M., Leng, J., Feng, S., Li, Z., Incecik, A.: Precisely modeling offshore jacket structures considering model parameters uncertainty using Bayesian updating. Ocean Eng. **258** (2022). https://doi.org/10.1016/j.oceaneng.2022.111410. Article 111410

Wang, Q., Ren, H., Li, J.: Dynamic response of contact-blast-loaded free metal plate: theoretical model, experiments and numerical simulation. Thin-Walled Struct. **175** (2022). https://doi.org/10.1016/j.tws.2022.109228. Article 109228

Wang, Y., Joseph, J., Aniruddhan Unni, T.P., Yamakawa, S., Barati Farimani, A., Shimada, K.: Three-dimensional ship hull encoding and optimization via deep neural networks. J. Mech. Des. N. Y. **144**(10), 1–15 (2022). https://doi.org/10.1115/1.4054494

Wang, Y., Shimada, K., Barati Farimani, A.: Airfoil GAN: encoding and synthesizing airfoils for aerodynamic shape optimization. J. Comput. Des. Eng. **10**(4), 1350–1362 (2023). https://doi.org/10.1093/jcde/qwad046

Wang, Z., Mantey, S.K., Zhang, X.: A numerical tool for efficient analysis and optimization of offshore wind turbine jacket substructure considering realistic boundary and loading conditions. Mar. Struct. **95** (103605) (2024). https://doi.org/10.1016/j.marstruc.2024.103605

Wang, Z., Wu, Y., Chu, X., Liu, C., Zheng, M.: Risk identification method for ship navigation in the complex waterways via consideration of ship domain. J. Mar. Sci. Eng. **11**(12) (2023). https://doi.org/10.3390/jmse11122265. Article 2265

Wang, Z., Zhou, Y., Wang, T.: Dynamic risk assessment of oil spill accident on offshore platform based on the bayesian network. IEEE Trans. Eng. Manage. **71**, 9188–9201 (2024). https://doi.org/10.1109/TEM.2023.3327436

Wei, K., Shang, D., Zhong, X.: Integrated approach for estimating extreme hydrodynamic loads on elevated pile cap foundation using environmental contour of simulated typhoon wave, current, and surge conditions. J. Offshore Mech. Arct. Eng. **145**(2) (2023). https://doi.org/10.1115/1.4056037. Article 021702

Wei, P., Li, C., Jiang, Z., Wang, D.: Real-time digital twin of ship structure deformation field based on the inverse finite element method. J. Mar. Sci. Eng. **12**(2) (2024). https://doi.org/10.3390/jmse12020257. Article 257

Wei, Y., Zhao, W., Wan, D.: Parallel efficient global optimization algorithm for ship hull form optimization. In 15th International Symposium on Practical Design of Ships and Other Floating Structures (PRADS 2022), Dubrovnik, Croatia, 9–13 October (2022)

Wiley, W., Jonkman, J., Robertson, A., Shaler, K.: Sensitivity analysis of numerical modeling input parameters on floating offshore wind turbine loads. Wind Energy Sci. **8**(10), 1575–1595 (2023). https://doi.org/10.5194/wes-8-1575-2023

Woloszyk, K., Garbatov, Y.: Advances in modelling and analysis of strength of corroded ship structures. J. Mar. Sci. Eng. **10**(6) (2022). https://doi.org/10.3390/jmse10060807. Article 807

Woloszyk, K., Garbatov, Y.: A probabilistic-driven framework for enhanced corrosion estimation of ship structural components. Reliab. Eng. Syst. Safety **242** (2024). https://doi.org/10.1016/j.ress.2023.109721. Article 109721

Wu, T., Zhang, C., Guo, X.: Dynamic responses of monopile offshore wind turbines in cold sea regions: ice and aerodynamic loads with soil-structure interaction. Ocean Eng. **292** (2024). https://doi.org/10.1016/j.oceaneng.2023.116536. Article 116536

Wu, W.H., et al.: An overview of structural design, analysis and common monitoring technologies for floating platform and flexible cable and riser. China Ocean Eng. **36**(4), 511–531 (2022). https://doi.org/10.1007/s13344-022-0044-5

Xia, J., Zou, G.: Operation and maintenance optimization of offshore wind farms based on digital twin: a review. Ocean Eng. **268** (113322) (2023). https://doi.org/10.1016/j.oceaneng.2022.113322

Xiao, W., He, M., Wei, Z., Wang, N.: SWLC-DT: an architecture for ship whole life cycle digital twin based on vertical–horizontal design. Machines **10**(11) (2022). https://doi.org/10.3390/machines10110998. Article 998

Xiao, X., Cirak, F.: Infill topology and shape optimization of lattice-skin structures. Int. J. Numer. Meth. Eng. **123**(3), 664–682 (2022). https://doi.org/10.1002/nme.6866

Xin, X., Liu, K., Loughney, S., Wang, J., Yang, Z.: Maritime traffic clustering to capture high-risk multi-ship encounters in complex waters. Reliab. Eng. Syst. Safety **230** (2023). https://doi.org/10.1016/j.ress.2022.108936. Article 108936

Xiong, Z., Xiang, B., Chen, Y., Chen, B.: Research on the risk classification of cruise ship fires based on an attention-BP neural network. Pol. Marit. Res. **29**(3), 61–68 (2022). https://doi.org/10.2478/pomr-2022-0026

Xu, K., Liu, F., Liu, D.: A multifield loads evaluation method for offshore wind turbines considering multivariate coherence effect. Ocean Eng. **280** (2023). https://doi.org/10.1016/j.oceaneng.2023.114586. Article 114586

Xu, S., Liu, J., Ma, Y.: Residual stress constrained self-support topology optimization for metal additive manufacturing. Comput. Methods Appl. Mech. Eng. **389** (2022). https://doi.org/10.1016/j.cma.2021.114380. Article 114380

Xu, X., Gaidai, O., Yakimov, V., Xing, Y., Wang, F.: FPSO offloading operational safety study by a multi-dimensional reliability method. Ocean Eng. **281** (2023). https://doi.org/10.1016/j.oceaneng.2023.114652. Article 114652

Xu, Y., Duan, J.: Dynamic response analysis of offshore single pile wind turbine under near-field ground motion. Zhendong yu Chongji/J. Vibr. Shock **41**(23), 222–229+240 (2022). https://doi.org/10.13465/j.cnki.jvs.2022.23.026

Yamamoto, N., Bollero, A., Son, C.W., Koyama, H., Choi, J.I., Lv, Y.: A new model for predicting thickness loss due to both corrosion wear and mechanical wear. Mar. Struct. **97** (2024). https://doi.org/10.1016/j.marstruc.2024.103659. Article 103659

Yang, S., Meng, D., Guo, Y., Nie, P., Jesus, A.M.P.: A reliability-based design and optimization strategy using a novel MPP searching method for maritime engineering structures. Int. J. Struct. Integr. **14**(5), 809–826 (2023). https://doi.org/10.1108/IJSI-06-2023-0049

Yao, S., Yang, J.: Research on shipyard HSE management system based on PDCA cycle. In: Proceedings - 2022 3rd International Conference on Education, Knowledge and Information Management, ICEKIM 2022, Virtual, Online, 21–23 January, pp. 930–936 (2022). https://ieeexplore.ieee.org/document/10027294

Yu, Y., et al.: Reliability-based topology-topography optimization for ship bulkhead structures considering multi-failure modes. Ocean Eng. **293**(116681) (2024). https://doi.org/10.1016/j.oceaneng.2024.116681

Yu, Y., et al.: Reliability-based design method for marine structures combining topology, shape, and size optimization. Ocean Eng. **286**(115490) (2023). https://doi.org/10.1016/j.oceaneng.2023.115490

Yu, Z., Amdahl, J.: A Rayleigh-Ritz solution for high order natural frequencies and eigenmodes of monopile supported offshore wind turbines considering tapered towers and soil pile interactions. Mar. Struct. **92**(103482) (2023). https://doi.org/10.1016/j.marstruc.2023.103482

Yuen, S., Ezard, T.H.G., Sobey, A.J.: Epigenetic opportunities for evolutionary computation. R. Soc. Open Sci. **10**(5), 221256 (2023). https://doi.org/10.1098/rsos.221256

Yüksel, N., Börklü, H.R.: A generative deep learning approach for improving the mechanical performance of structural components. Appl. Sci. **14**(9), 3564 (2024). https://doi.org/10.3390/app14093564

Yüksel, N., Börklü, H.R., Sezer, H.K., Canyurt, O.E.: Review of artificial intelligence applications in engineering design perspective. Eng. Appl. Artif. Intell. **118**, 105697 (2023). https://doi.org/10.1016/j.engappai.2022.105697

Zaraphonitis, G., Kytariolou, A., Dafermos, G., Gatchell, S., Östman, A.: Hydrodynamic optimisation of a containership and a bulkcarrier for life-cycle operation. Holist. Approach Ship Des. **2**, 231–255 (2021). https://doi.org/10.1007/978-3-030-71091-0_8

Zehnder, J., Li, Y., Coros, S., Thomaszewski, B.: NTopo: mesh-free topology optimization using implicit neural representations. In: 35th International Conference on Neural Information Processing Systems, pp. 10368–10381 (2021). https://dl.acm.org/doi/10.5555/3540261.3541054

Zerbst, C.: OCX on the way from research to industry practice. In: 22nd International Conference on Computer and IT Applications in the Maritime Industries. COMPIT 2023, Drübeck, Germany, 23–25 May 2023 (2023)

Zerbst, C., Lutz, U., Danetzky, A.: Concept to reality: implementing a digital twin for ship production. In: 21th International Conference on Computer Applications and Information Technology in the Maritime Industries, COMPIT 2022, Pontignano, Italy, 21–23 June (2022)

Zhang, C., Long, K., Zhang, J., Lu, F., Bai, X., Jia, J.: A topology optimization methodology for the offshore wind turbine jacket structure in the concept phase. Ocean Eng. **266** (2022). https://doi.org/10.1016/j.oceaneng.2022.112974. Article 112974

Zhang, C., Zhang, D., Zhang, M., Zhang, J., Mao, W.: A three-dimensional ant colony algorithm for multi-objective ice routing of a ship in the Arctic area. Ocean Eng. **266** (2022). https://doi.org/10.1016/j.oceaneng.2022.113241. Article 113241

Zhang, H., et al.: A Digital twin of the research vessel gunnerus for lifecycle services: outlining key technologies. IEEE Robot. Autom. Mag. **30**(3), 6–19 (2023). https://doi.org/10.1109/MRA.2022.3217745

Zhang, H., Tian, S., Li, R., Xu, W., Hu, Y.: Knowledge-driven scheduling of digital twin-based flexible ship pipe manufacturing workshop. In: Advances in Remanufacturing. Proceedings of the VII International Workshop on Autonomous Remanufacturing, pp. 293–306 (2024). https://doi.org/10.1007/978-3-031-52649-7_23

Zhang, J., Zhou, C.: Research on cost analysis and construction optimization of cross-sea railway bridge based on FCM-CSIs. J. Railway Sci. Eng. **20**(1), 347–355 (2023). https://doi.org/10.19713/j.cnki.43-1423/u.T20220199

Zhang, S., Chen, Y.: Research on deep learning method in ship hull form optimization. In: Third International Conference on Computer Science and Communication Technology (ICCSCT 2022) (2022). https://doi.org/10.1117/12.2662508

Zhang, S., Tezdogan, T., Zhang, B., Lin, L.: Research on the hull form optimization using the surrogate models. Eng. Appl. Comput. Fluid Mech. **15**(1), 747–761 (2021). https://doi.org/10.1080/19942060.2021.1915875

Zhang, W., Li, J., Li, L., Yang, Q.: A systematic literature survey of the yield or failure criteria used for ice material. Ocean Eng. **254** (2022). https://doi.org/10.1016/j.oceaneng.2022.111360. Article 111360

Zhang, W., Zhang, Y., Qiao, W.: Risk scenario evaluation for intelligent ships by mapping hierarchical holographic modeling into risk filtering, ranking and management. Sustainability **14**(4) (2022). https://doi.org/10.3390/su14042103. Article 2103

Zhang, X., Noshadravan, A.: Efficient lifecycle reliability assessment of offshore wind turbines using digital twin structural health monitoring 2023: designing SHM for sustainability, maintainability, and reliability. In: Proceedings of the 14th International Workshop on Structural Health Monitoring, Stanford, California, US, 12–14 September, pp. 675–682 (2023). https://dpi-proceedings.com/index.php/shm2023/article/view/36801

Zhang, X., Wang, Y., Chemori, A.: Structural reliability based energy-efficient arctic position mooring control of moored offshore structures under ice loads. Ocean Eng. **268** (2023). https://doi.org/10.1016/j.oceaneng.2022.113435. Article 113435

Zhang, Y., Ma, N., Gu, X., Shi, Q.: A dimensionality reduction method based on principal component analysis for ship hull form optimization. In: The 33rd International Ocean and Polar Engineering Conference (ISOPE 2023), Ottawa, Canada, 19–23 June (2023)

Zhang, Y., Ma, N., Gu, X., Shi, Q.: Geometric space construction method combined of a spline-skinning based geometric variation method and PCA dimensionality reduction for ship hull form optimization. Ocean Eng. **302**(117604) (2024). https://doi.org/10.1016/j.oceaneng.2024.117604

Zhang, Z., Li, Y., Zhou, W., Chen, X., Yao, W., Zhao, Y.: TONR: an exploration for a novel way combining neural network with topology optimization. Comput. Meth. Appl. Mech. Eng. **386** (2021). https://doi.org/10.1016/j.cma.2021.114083. Article 114083

Zhao, C., Cao, X., Ren, Y.: Risk analysis of bridge ship collision based on AIS data model and nonlinear finite element. Nonl. Eng. **12**(1) (2023). https://doi.org/10.1515/nleng-2022-0324. Article 20220324

Zheng, S., Li, C., Xiao, Y.: Efficient optimization design method of jacket structures for offshore wind turbines. Mar. Struct., **89**(103372) (2023). https://doi.org/10.1016/j.marstruc.2023.103372

Zheng, Y., Wang, Y., Lu, X., Liao, Z., Qu, J.: Evolutionary topology optimization for mechanical metamaterials with auxetic property. Int. J. Mech. Sci. **179** (2020). https://doi.org/10.1016/j.ijmecsci.2020.105638. Article 105638

Zhou, H., et al.: Research on the three-dimensional visual management and control for marine intelligent mould bed based on digital twin. Ships Offshore Struc. **19**(6), 756–768 (2024). https://doi.org/10.1080/17445302.2023.2208495

Zhou, J., et al.: A life cycle decision framework of China offshore wind turbines with ANP-Intuitionistic fuzzy TOPSIS method. Green Manuf. Open **2**(1), 3 (2024)

Zhou, M., Shyy, Y.K., Thomas, H.L.: Checkerboard and minimum member size control in topology optimization. In: Proceedings of the 3rd WCSMO, Buffalo, NY (1999)

Zhou, M., Shyy, Y.K., Thomas, H.L.: Checkerboard and minimum member size control in topology optimization. Struct. Multidiscip. Optim. **21**(2), 152–158 (2001). https://doi.org/10.1007/s001580050179

Zhou, X., Ruan, X., Wang, H., Zhou, G.: Exploring spatial patterns and environmental risk factors for global maritime accidents: a 20-year analysis. Ocean Eng., **286** (2023). https://doi.org/10.1016/j.oceaneng.2023.115628. Article 115628

Zhu, L., Huang, X., Yuan, C., Du, Z.: Structural reliability updating using experimental data. J. Mech. Sci. Technol. **36**(1), 135–143 (2022). https://doi.org/10.1007/s12206-021-1212-x

Zou, J., Zhang, Y., Feng, Z.: Topology optimization for additive manufacturing with self-supporting constraint. Struct. Multidiscip. Optim. **63**(5), 2341–2353 (2021). https://doi.org/10.1007/s00158-020-02815-w

Zou, P., Bricker, J.D., Fujisaki-Manome, A., Garcia, F.E.: Characteristics of ice-structure-soil interaction of an offshore wind turbine. Ocean Eng. **295** (2024). https://doi.org/10.1016/j.oceaneng.2024.116975. Article 116975

Zou, Y., et al.: An advanced machine vision-based method for abnormal detection of transverse vibrations in ship propulsion shafting. Ocean Eng. **314** (2024). https://doi.org/10.1016/j.oceaneng.2024.119724

Zou, Y., Zhang, Y., Wang, S., Jiang, Z., Wang, X.: Ship regulatory method for maritime mixed traffic scenarios based on key risk ship identification. Ocean Eng. **298** (2024). https://doi.org/10.1016/j.oceaneng.2024.117105. Article 117105

Open Access This chapter is licensed under the terms of the Creative Commons Attribution-NonCommercial-NoDerivatives 4.0 International License (http://creativecommons.org/licenses/by-nc-nd/4.0/), which permits any noncommercial use, sharing, distribution and reproduction in any medium or format, as long as you give appropriate credit to the original author(s) and the source, provide a link to the Creative Commons license and indicate if you modified the licensed material. You do not have permission under this license to share adapted material derived from this chapter or parts of it.

The images or other third party material in this chapter are included in the chapter's Creative Commons license, unless indicated otherwise in a credit line to the material. If material is not included in the chapter's Creative Commons license and your intended use is not permitted by statutory regulation or exceeds the permitted use, you will need to obtain permission directly from the copyright holder.

Committee IV.2: Material and Fabrication Technology

Agnes Marie Horn[1(✉)], Jean-David Caprace[2], Matthias Krause[3], Dora Tsiourva[4], Alessandro Caleo[5], Kunihiro Hamada[6], Mojtaba Mokhtari[9], Iraklis Lazakis[7], Stéphane Paboeuf[8], and Bernt Leira[1]

[1] DNV, Oslo, Norway
Agnes.Marie.Horn@dnvgl.com
[2] Federal University of Rio de Janeiro, Rio de Janeiro, Brazil
[3] Hamburg, Germany
[4] Athens, Greece
[5] Genoa, Italy
[6] Hiroshima University, Hiroshima, Japan
[7] University of Strathclyde, Glasgow, United Kingdom
[8] Bureau Veritas, Paris, France
[9] Trondheim, Norway

Committee Mandate. Concern for the current developments in materials and fabrication technologies, knowledge gaps, recommendations on how to improve design, qualification and approval processes in order to keep pace with upcoming innovations.

Further, the Committee shall raise the awareness on challenges from new materials, advanced fabrication processes and increased digitization. Special attentions to be given to advanced material and fabrication models to support structural analysis during design and operation.

Attentions are also to be given to material aging and assessment methods during structural life cycle.

Keywords: Material for the energy transition · lightweight · composites · adhesive · welding · residual stress · qualification · approval · fabrication

1 Introduction

The report of the Committee IV.2 Material and Fabrication technology provides an in-depth overview of recent advancements in materials and fabrication technologies, particularly in the ship and energy sector. This is the first period in which this committee is one of the eight technical committees and not a specialist committee. The scope of the mandate has increased, and the report shall also cover advanced material and fabrication models to support structural analysis during design and operation. In addition to material aging and assessment methods during structural life cycle.

There is a shortage of material supply in the world today. Steel production has doubled in the past two decades, mainly due to demand in China and India. Steel production is very energy demanding and estimated to contribute to 8% of global carbon emissions and about 30% of emissions from industry. There is a push by the steel industry to reduce their CO_2 footprint by change from coal to the more efficient electric arc furnace (EAF) methods. Decarbonizing steel production by use of hydrogen, will be an important game changer for this industry, and several steel makers are looking into this technology to produce "green steel". The same trend is seen for cement, with 8% of manufacturing energy demand in 2023. Currently, cement production is predominantly taking place in China which accounts for half of the production worldwide. Hence, technologies on reducing the carbon footprint of material production sectors and recycling will be key enablers for a more sustainable growth within the Maritime and Energy sector the years to come.

Today, the common approach for structural engineers in the Maritime and Energy sectors is to use simple material models with tabulated yield and tensile properties. The selection of materials tends to be governed by prescriptive rules and standards that follow a traditionally conservative approach that does not reflect either the development of materials or the given structure's full utilization or response in case of accidental events. Yet, from a safety and sustainability perspective, it has never been more important to investigate the potential of new and improved material and fabrication processes to obtain an even better design solution. A better fundamental understanding of materials, combined with digital solutions and new manufacturing processes can revolutionize our approach towards design. Hence, focus on different materials from metallic to non-metallic are presented in Sect. 2, while Sect. 3 addresses joining and fabrication processes.

New technology and new materials are often hindered by strict regulations, and having a framework that systematically enables a streamlined qualification process is of outermost importance for fast adoptions of the latest technologies. The fast transformation to a more emission reduction industry requires a fast and transparent qualification and approval process, this has been discussed in Sect. 4.

The energy transition will play a key role in the years to come, challenging the need for more use of steel material to build wind towers, and e.g. pipelines to transport H_2 and CO_2. Hence, in Sect. 5 the durability of materials is discussed with focus on the environmental impact of materials and the new energy trends that pose enhanced challenges like high and low temperatures, high strain rates and exposure to fatigue.

Understanding materials footprint from a life-cycle perspective involves evaluating their environmental impact from extraction to disposal. Section 6 discusses the material in the content of a life cycles perspective, where both environmentally friendly materials, implementation of new assessment methods and scientific tools are possible enablers to reduce environmental footprints for the Maritime and Energy sectors. The chapter also evaluates different recycling processes.

Digital twin technology has revolutionized the shipbuilding and offshore platform manufacturing industries by creating virtual replicas of physical assets, processes, and systems. As described in Sect. 7, this innovation enhances design, production, simulation, monitoring, and predictive maintenance, driving significant digital transformation in the maritime industry. The need for increased efficiency, safety, and sustainability has propelled this shift.

In recent years, dedicated meetings between committee members and external stakeholders were successfully held to enhance discussion on subjects of specific interest between experts. Several chapters of the report prepared by the committee IV.2 address numerous subjects related to the topic of more environmentally friendly materials and processes in the maritime industry. A workshop between the ISSC committee members and industry experts was held in Hamburg in September 2024 with a focus on novel production, repair and recycling processes with a potential to reduce emissions in non-operational processes of ship's lives, a detailed summary of the meeting is found in Appendix A.

1.1 General Trends in Material and Fabrication Research

The majority of recent EU funded maritime R&D projects which addressed materials and fabrication were focusing on digitalization aspects. However, there are also examples of projects dedicated to R&D on specific materials or fabrication technology taking place within EU like Fibre4Yards (FIBRE composite manufacturing technologies FOR the automation and modular construction in shipYARDS) and CirclesOfLife and EcoShipYard which addresses Concepts for Shipyard Environmental Performance Index (SEPI) and Cradle to cradle (C2C) digital ship passport, these projects are addressed in Sect. 6 of the report.

In Germany, the Ministry for Economic Affairs and Climate Action (BMWK) is continuously funding application-oriented R&D projects on maritime subjects, including fabrication, e.g. smartBOND on adhesive bonding and FOLAMI on Laser welding. In addition, BMWK launched a trans sectoral funding program dedicated to lightweight construction, which also enabled projects on maritime applications, e.g. LESSMAT and an innovation cluster which supports the development of the maritime lightweight network MariLight.

In Norway, most of the material and fabrication research is related to the Energy sector within the field of new innovative materials solutions in cost-efficient, low-carbon production chains by enabling more use of recycled metal, e.g. by increasing the understanding of how alloying elements accumulated by recycling affect microstructure and properties addressed in SFI PhysMet. While SFI BLUES (Floating Structures for the Next Generation of Ocean Industries), addresses research on lightweight materials like composites and aluminum, including multi-material and fabrication technologies.

In Japan, research on the automation and digitization of the painting process was carried out with the support of the Nippon Foundation. In this research, in order to reduce man-hours and improve painting quality, quantitative measurement equipment for pre-painting treatment conditions and equipment for measuring the thickness of the paint film were developed, as well as equipment for automating the striped coating process. In addition, as part of the NEDO (New Energy and Industrial Technology Development Organization) project, research of work support and automation system for plate bending by AI line heating research was carried out. In this research, a small automatic linear heating robot and a plate bending support system using AR (augmented reality) technology were developed. Furthermore, by combining simulation technology and AI technology, an AI heating plan generation system was developed to automate the generation of heating plans, which had previously relied on the experience and know-how of skilled worker.

2 Materials

The world of shipbuilding and offshore construction is constantly evolving, driven by the relentless pursuit of weight savings, increased structural strength, durability, and reduced environmental impact. This push research and experimentation with new materials that can offer the required characteristics.

In addition to well-known metallic materials (mainly steel and aluminum), there has been a strong trend toward using non-metallic materials, such as composites, polymers, and fiber-reinforced polymers. These new materials offer excellent mechanical properties, corrosion resistance, and a low environmental impact. Solutions that combine different materials to optimally leverage their properties are also being studied.

In terms of composite materials and polymers, research is focused on new sustainable fibers or materials derived from biotechnology (e.g., natural fabrics and fibers made entirely from recycled plastics). The attention to new materials is also demonstrated by research projects like RAMSSES (Realisation and Demonstration of Advanced Material Solutions for Sustainable and Efficient Ships - https://www.ramsses-project.eu/).

2.1 Non Metallic Materials

2.1.1 Composites in Ship Building

In various European countries, stakeholders are making efforts to bring the use of fibre composite materials into maritime applications. A study in Swedish comprehensively addressed the topic of the fire properties of various fibre and matrix materials as well as additives and proposed a material structure with an intumescent coating, which, according to the report, proved itself in mechanical and fire tests (Sandinge et al., 2022). In Germany, the MariLight network has been established as a network for stakeholders which are interested in exchanging views on lightweight construction in the ship and boat building sector, and furthering its application ("MariLight.Net - Das maritime Leichtbaunetzwerk," 2024). With a specific focus on composite materials and their possibilities and challenges, but not limited to that group of materials, the network members have developed a strategic innovation roadmap on maritime lightweight construction (Krause and Steinlein, 2024).

Regarding the use of Composites, in 2024 MariLight members have been contributing to the ongoing discussion on a possible update of relevant IMO regulations. A Correspondence Group (CG) focused on revising interim guidelines for using Fibre-Reinforced Plastic (FRP) in ship structures (International Maritime Organization (IMO), 2017). Discussion topics included fire safety, load-bearing capability, and recycling. Members agreed on the need to develop guidelines for fire testing, addressing issues like fire growth, smoke, and toxicity. There was also a consensus on including provisions for load-bearing elements in the revised guidelines. However, the group was divided on whether FRP load-bearing fire divisions should be included, with some members expressing that this might be outside the guidelines' original scope. Additionally, while there was support for including recycling considerations, the group chose not to prioritize a full Life Cycle Assessment. The CG's report was presented and discussed at IMO Ship design and construction sub committee's annual session (SDC11) in January 2025. As a result, another CG was started and given the task to prepare a report for SDC12. Among others, the new CG's Terms or Reference include tasks 'to address also load bearing divisions and elements, in addition to other aspects, as part of its revision work, within the scope of SOLAS chapter II-2 from fire safety perspective'. The SG's mandate was also to invite the IMO Maritime Safety Committee (MSC) 'to confirm whether or not load-bearing divisions and elements contributing to global strength are considered as part of the scope of the existing output.' Furthermore, the CG was asked to consider fire and recycling matters as a priority and to provide discussion points which fall within the remits of bodies other than SDC.

2.1.2 Composites in Energy Sector

There is a growing need for further developing technologies where polymer-based composites, particularly fiber reinforced plastics (FRPs), will play a crucial role in the energy transition. These composites are especially important in advancing offshore wind, hydrogen production, carbon capture and storage (CCS), and maritime applications, (Slater, 2022). One of the most recent challenges introduced by the energy transition is the need for a reliable and cost-effective infrastructure for large-scale transportation of gaseous hydrogen. Due to hydrogen embrittlement issue with steel (see Sect. 5.2.4), flexible composite pipes, including thermoplastic composite pipes (TCPs) and reinforced thermoplastic pipes (RTPs), have emerged as promising alternatives. TCPs and RTPs (Fig. 1) have long been used in the oil and gas industry, primarily due to their significantly higher strength-to-weight ratio compared to steel, greater flexibility, better fatigue properties, and resistance to corrosion. These characteristics are particularly valuable for offshore applications considering the highly corrosive and dynamic environment. They even facilitate the offshore transportation and installation, given their lighter weight and higher flexibility that enables spooling. Over the past decade, flexible composite pipes have been used in thousands of kilometers of pipelines for transporting hydrocarbons and water.

Despite their advantages, RTPs introduce several challenges that have limited their widespread adoption compared to steel pipelines. A significant concern is the uncertainty surrounding the long-term performance of RTPs, which arises from their complex degradation mechanisms, further compounded by their relatively more intricate nonlinear mechanical behavior. They are prone to hygrothermal effects, where moisture absorption, accelerated by rising temperatures, leads to swelling of epoxy and polyester resins. This degradation reduces their mechanical properties, increases viscoelastic behavior (creep), and ultimately shortens their fatigue life. Additionally, their lower stiffness compared to metallic pipes makes them more susceptible to buckling, especially in subsea applications with high external pressures. They are also vulnerable to punctures from sharp objects like rocks, have recycling challenges, and generally exhibit poorer fire retardancy than steel pipelines. Another significant concern is ultraviolet (UV) degradation of plastic materials. Although manufacturers typically incorporate UV stabilizers into the pipe's outer layer, the effectiveness of these additives under prolonged sunlight exposure, especially in offshore environments where other degradation mechanisms may interact with UV-induced damage, remains an area requiring further study.

Cost has been another main barrier for flexible composite pipes, as they have traditionally been more expensive than steel pipelines. However, as flexible composite pipes production costs decrease with technological advancements and consider the high maintenance costs associated with steel pipelines due to corrosion, the life-cycle cost of flexible composite pipelines are becoming more competitive with steel pipelines. DNV is currently conducting a cost-benefit analysis comparing composite and steel pipes, alongside risk assessments, which could address gaps in existing design and qualification standards for composite pipes (i.e., DNV-ST-F119 and API 15S).

A particular challenge in hydrogen transportation is its low energy density compared to hydrocarbons, necessitating transportation under higher operating pressure. One potential solution is the use of hybrid steel-composite pipes, where the internal surface of a steel pipeline is lined with polymer-based composites. This approach is applicable not only for new pipelines but also for repurposing existing hydrocarbon pipelines for hydrogen transport. Polymeric, composite, and metallic liners have been traditionally applied for corrosion protection and are not new. However, for hydrogen transportation challenges regarding potential hydrogen embrittlement of metallic liners and failure of the polymeric/composite ones at high pressures remain. Also, hydrogen permeation through polymers used in composite pipes and liners can still lead to hydrogen embrittlement of the steel layer, although at a reduced rate. Therefore, current research in this area focuses on optimizing liner options (polymeric, composite, and metallic) in terms of strength, fatigue resistance, permeation performance, and cost.

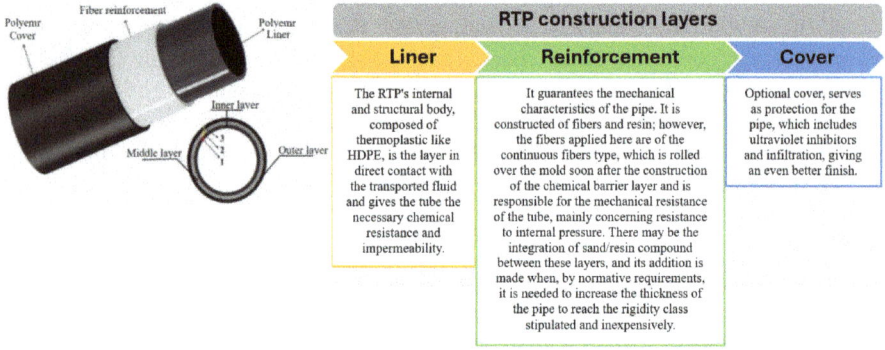

Fig. 1. RTP structure and construction layers (Karim et al., 2023).

2.1.3 Smart Materials

Smart materials refer to materials that can respond dynamically to external stimuli such as mechanical stress, magnetic fields, temperature, or electrical inputs, resulting in changes to their properties like stiffness, shape, or damping. These adaptive characteristics make smart materials a critical component in several industries, including marine and offshore where harsh environmental conditions and dynamic loads demand responsive and resilient structures.

In the offshore industry, smart materials are increasingly being integrated into structures to improve reliability, durability, and performance. For instance, piezoelectric materials are used in sensor systems to monitor the health of marine structures, detecting damage such as cracks (e.g., by using inverse Finite Element Method based on the strain field measurement changes) before they become critical. Shape Memory Alloys (SMAs), which return to their original shape when exposed to specific temperatures or stress levels, are implemented in actuators and coupling mechanisms to provide flexibility and adaptive responses under load (Huang et al., 2020).

Among smart materials, Magnetorheological Elastomers (MREs) are emerging as promising solutions for load mitigation in offshore structures. MREs are composite materials composed of a flexible elastomer matrix embedded with micron-sized magnetic particles. These materials exhibit controllable stiffness when subjected to magnetic fields, making them ideal for adaptive systems in which mechanical and consequently dynamic properties need to adjust based on environmental or operational demands. MREs can be used in vibration absorbers for floating platforms, helping to reduce the impact of wave-induced vibrations on structural components (Huang et al., 2020), due to their potential in vibration isolation and adaptive load-bearing systems.

Additionally, MREs are being explored for tuneable stiffness joints in modular marine structures, where their ability to change stiffness under a magnetic field can optimize load distribution across floating platforms in varying sea conditions (Danemyr, 2024).

In the context of marine renewable energy systems, the flexibility and adaptability of MREs have notable implications for structural resilience and efficiency. As highlighted by Danemyr (2024), MREs can be particularly effective in controllable connectors for modular marine renewable energy platforms. In these systems, connectors between different modules must adapt to shifting loads and dynamic sea states, and MREs offer the capability to modify their stiffness in real-time based on the intensity of external forces. This adaptability minimizes stress concentrations at the connectors, enhancing the longevity and durability of marine structures exposed to waves and wind.

However, challenges remain in assessing the detailed design and cost-effectiveness of such stiffness-adjustable joints, particularly when considering the added complexity and potential degradation of MREs in harsh marine environments. Long-term performance and resistance to environmental factors, such as saltwater and mechanical wear, are key areas for further research to ensure the sustainable and economical use of MREs in offshore applications. Moreover, quantitative studies on the ageing of magnetic forces during long-term use and the durability of MREs, especially when applied to the seawater splash zone, would be desirable for their application to offshore structures.

2.2 Metallic Materials

2.2.1 Aluminum

Aluminum, known for its lightweight and corrosion-resistant characteristics and excellent formability, make it a material of choice for reducing weight while maintaining structural integrity. Although its good corrosion resistance, aluminum is not used to a large extent in the energy sector. Typical structures are helideck (6082 alloy) and living quarters (5083) where galvanic corrosion in contact with steel structure and bolts have been seen (Saai et al., 2023).

The project MARINAL (Knudsen, O. Ø and Bertram, J 2022) investigated use of aluminum for extended use in marine constructions. One challenge pointed out in the project that limits the use of aluminum, is lack of rules and regulations, as standardizing bodies have a focus on steel. MARINAL focused on closing the gap of corrosion properties of standard aluminum alloys, including recycled alloys.

LNG tanks are often made of aluminum due to excellent low temperature properties, however they are prone to fatigue due to sloshing and thermal loads. Fueki (2024) investigated the fatigue performance of needle peened aluminum JIS A5083P-O MIG welded butt welds and concluded that the fatigue improvement method only marginally increased the fatigue strength due to relatively high local stresses at the striking groove. However, the specimens with a 1 mm-deep semicircular slit showed 100% improvement in stress amplitude at 2 mill cycles compared to untreated specimen. See Sect. 3.1.2 for joining aluminum and steel by use of friction stir welding.

2.2.2 High Strength Steel

HSLA (High Strength Low Alloy) steels are crucial in modern shipbuilding and in the energy sector due to their excellent mechanical properties and corrosion resistance. HSLA is typically used for heavily utilized structures to reduce wall thickness and hence weight reduction where this is critical. HSLA steels have a high yield strength which allows for using less material, making them suitable for large structural components that endure extreme conditions, such as those found in the marine. HSLA steels are more resistant to the harsh marine environment than conventional steels, lowering maintenance costs and reducing the risk of structural failure. HSLA steels require advanced technology, leading to higher upfront costs compared to standard steels. Welding HSLA steels can be more complex, necessitating specialized skills and advanced welding techniques to prevent defects.

The European research project MOSAIC (Materials Onboard: Steel Advancements and Integrated Composites) conducted an in-depth analysis of HSLA (High Strength Low Alloy) steels to assess their applicability in naval structures. The primary goal was to enhance mechanical performance and corrosion resistance in ships while reducing structural weight. The project (cordis.europa.eu) looked into three types of HSLA steels: S460 and S690 steel which are already used in civil constructions, bridges etc. in comparison with the already used in shipping industry AH36 steel. All of them were tested to determine their mechanical properties and corrosion resistance. The project developed also welding methods for these steels, employing different welding combinations, similar and dissimilar ones, with innovative welding techniques, Laser Hybrid Welding (LHW) and Friction Stir Welding (FSW), as well as conventional Arc Welding. The material fatigue and fracture toughness properties were established. The MOSAIC project showcased that integrating HSLA steels and innovative welding techniques into naval constructions could provide significant advantages in terms of performance, efficiency, and sustainability due to less material needed compared to lower strength steels, see section for Sect. 3.1.2 for friction stir welding of HSLA steels.

2.3 Coatings

The protective coating industry for naval and offshore sectors is undergoing constant transformation, driven by the need to address corrosion, biofouling, and environmental sustainability. These advancements aim to enhance durability, energy efficiency, and reduce operational costs. A combination of evolving regulations, like the IMO's Energy Efficiency Existing Ship Index (EEXI) and the Carbon Intensity Indicator (CII), push towards decarbonization encouraging shipowners to adopt high-performance coating technologies. Among the most significant innovations are biocide-free silicone-based coatings and anti-fouling systems designed to minimize environmental impact while improving vessel performance.

Anti-Fouling Coatings: These are crucial in preventing marine organisms from attaching to the hull, which reduces drag and optimizes fuel consumption. Innovations like hybrid silicone and hard foul-release coatings have been developed to reduce friction and cut fuel consumption by up to 10%.

Nano-Coatings: The incorporation of nanoparticles such as graphene and nanosilver into marine coatings strengthens corrosion resistance while offering superior protection against UV radiation and chemical exposure. These nano-coatings also help reduce drag, improving overall fuel efficiency and providing long-term environmental benefits.

Self-Polishing Coatings (SPCs) are known for their ability to control the rate of biocide release, thereby maintaining a smooth and polished surface over time. This continuous renewal process not only reduces maintenance costs but also enhances vessel speed by maintaining a low drag surface. The energy savings and fuel efficiency gains have been significant, with reductions of up to 40% in fuel consumption noted in long-term studies. Despite their effectiveness, the reliance on biocides poses significant environmental challenges. Heavy metals and toxic organic compounds released into the marine environment can harm non-target species and disrupt marine ecosystems.

Biomimetic Coatings: Biomimetic coatings, inspired by natural surfaces like shark skin or lotus leaves, exhibit an environmentally friendly profile by avoiding toxic chemicals. These coatings reduce drag, improve energy efficiency, and minimize fuel consumption, potentially offering up to a 60% reduction in drag friction, as cited in several studies. They are particularly effective at reducing the adhesion of marine organisms while in motion, which makes them suitable for dynamic applications like ships in frequent operation. However, biomimetic coatings are less effective for stationary vessels or equipment, where biofouling can still accumulate. Mechanical durability remains a concern, limiting their application for longer-term uses, as noted by Schultz et al. who reported a significant performance decrease in static conditions.

Proactive Hull Cleaning: Technologies like EverClean, a robotic cleaning solution, help prevent the accumulation of biofouling on hulls by performing regular cleaning operations without interrupting the ship's function. This approach minimizes the need for harmful biocidal coatings and helps maintain optimal fuel efficiency. Current research in robotic cleaning systems attempts to incorporate artificial intelligence technologies to recognize and guide the cleaning process.

The shift towards more environmentally friendly coatings is a focus area for development to meet stringent environmental regulations. One example is the non-biocidal anti-fouling coatings, which rely on smooth surfaces to prevent the attachment of marine organisms.

Self-Healing Coatings: These advanced coatings autonomously repair minor damages, thus preventing corrosion from spreading. They are particularly valuable in harsh marine environments where routine maintenance can be challenging. This technology ensures prolonged protection and reduces downtime.

SLIPS (Slippery Liquid-Infused Porous Surfaces) use a thin lubricating layer that creates a "slippery" surface, minimizing friction and preventing organisms from adhering. This technology has demonstrated high effectiveness in resisting biofilm formation and macrofouling. One notable advantage is the minimal need for frequent reapplication, which reduces maintenance efforts and costs. According to recent experimental data, SLIPS coatings demonstrated a reduction in biological adhesion by up to 80% in dynamic water conditions.

However, their efficiency drops considerably for vessels or equipment that remain stationary for extended periods. Maintaining the liquid-infused layer requires energy input or active renewal, which can complicate implementation and increase costs in remote areas or for static applications. Furthermore, the need for specialized materials and application techniques increases initial costs compared to more conventional coatings.

Predictive Maintenance: The integration of IoT and predictive analytics allows real-time monitoring of coating degradation. This helps identify early signs of wear, allowing for timely maintenance, which minimizes repair costs and extends the lifespan of vessels.

Real time corrosion detection system are systems that use sensors to detect changes in environmental conditions, such as moisture, temperature, and chemical exposure, which can accelerate corrosion. The data collected allows operators to take proactive maintenance measures, reducing unexpected failures and extending the lifespan of infrastructure. This technology is particularly valuable in harsh environments, like offshore facilities, where corrosion can rapidly degrade materials.

2.4 Concrete

Concrete is one of the most widely used construction materials in the world, recognized for its durability and strength. Concrete has been used for large offshore installations, especially in harsh environment and has lately been used for the Spar-type floating foundation for the 88MW Hywind Tampen offshore wind installation. Cement is also used for grouted connections which transfer load from the turbine system to the foundation in monopile wind towers, and to reduce cost for installation, the curing time is essential to reduce installation time, K.-Y. Song et al., (2022). The production of cement, a key ingredient in concrete, poses significant environmental challenges, particularly concerning CO_2 emissions and contributes to approximately 8% of global emissions.

Self-healing concrete incorporates materials that can repair cracks autonomously, Dong et al., (2023). This innovative technology enhances the longevity of concrete structures and reduces maintenance costs. Shivanshi et al., (2023) investigated the various self-healing mechanisms, such as the use of encapsulated bacteria and they concluded that the concrete strength was not adversely affected by the added bacteria Microbially Induced Calcite Precipitation (MICP).

Geopolymer concrete is an environmentally friendly alternative to traditional Portland cement concrete which is used e.g. offshore. It is produced using industrial byproducts like fly ash and slag, which significantly reduce CO_2 emissions. Research in geopolymer concrete aims to understand its long-term performance and potential applications, with focus of the microstructure to improve durability characteristics, (Amran et al., 2021). Coffetti et al., (2022) explore various strategies to achieve sustainability in the concrete industry. These approaches include the use of alternative materials such as supplementary cementitious materials (SCMs), the development of carbon capture and storage (CCS) technologies, including promotion of recycling and reusing concrete waste. The research emphasizes the importance of innovation and collaboration among industry stakeholders to implement these sustainable practices effectively.

2.5 Summary and Recommendations for Future Work

The energy transition is driving demand for polymer-based composites such as FRPs in hydrogen transportation. RTPs offer advantages over steel pipes, such as higher strength-to-weight ratios, flexibility, and corrosion resistance. However, challenges such as complex long-term degradation, hydrogen permeation, and high operating pressures needed for hydrogen transport still hinder full adoption of RTPs. Some areas requiring further research include:

- Long-term performance and degradation of RTPs for hydrogen transportation especially in aggressive marine environments.
- Polymer liners with reduced hydrogen permeation for steel pipes.
- Cost-effective manufacturing processes for RTPs.
- RTP fire safety.
- UV-induced degradation of RTPs in offshore environments.
- Standards for hydrogen transport with RTPs and polymer-lined steel pipelines.

Long-term performance studies on smart materials are necessary to assess durability under harsh marine conditions like saltwater exposure and cyclic loads. Additionally, while smart materials can have higher initial costs, a comprehensive cost-benefit analysis is needed to weigh their potential life-cycle savings from reduced maintenance and increased efficiency.

Integrating smart materials with digital systems, such as digital twins, would enhance their application, allowing for real-time monitoring and predictive maintenance. Standardizing testing protocols is also crucial to ensure consistent quality and foster broader industry acceptance. Research to improve their mechanical properties, especially under extreme loads, will expand their use in critical offshore components.

There is a focus on research towards more environmentally friendly coatings for development to meet stringent environmental regulations. One example is the non-biocidal anti-fouling coatings, which rely on smooth surfaces to prevent the attachment of marine organisms.

While HSLA steels offer significant advantages in weight reduction and structural performance, challenges remain in terms of cost and the specialized skills required for their processing.

Significant research areas include high-performance concrete (HPC), self-healing concrete, and geopolymer concrete. HPC is designed to offer superior strength, durability, and workability compared to conventional concrete. Research in HPC focuses on optimizing the mix design to achieve these properties while also improving the material's environmental footprint.

3 Joining and Fabrication

The Committee observed various progresses particularly for multi-material combinations and for non-conventional steel types. Further relevant developments were identified in the area of additive manufacturing and production management. The results can contribute to enhancing the capability to produce sustainable, lightweight ships and to ensure efficient, high quality production processes. Most of the published work the Committee is aware of deals with shipbuilding processes, while little of the material is related to offshore. Therefore, the latter area of application is represented to a small extent only in this chapter.

3.1 Advances in Joining Technology

Production efficiency and weld joint quality is crucial, two methods of interest where research has taken place in recent years make use of laser welding for joining hybrid joints and friction stir welding, which are discussed more in detail.

3.1.1 Joining Hybrid Joints with Laser Welding

Due to low heat input and therefore small deformation potential, laser beam welding processes are widely used by shipyards to join large ship components consisting of similar materials. For hybrid joints made of steel and aluminum (e.g. connectors between ship hulls and superstructures), the currently used technology is explosion welding. This technology is not only complex, time-consuming and cost-intensive, but also restricts the freedom of design and the joints are limited. The German project FOLAMI is developing a new welding method involving two intersecting laser beams to produce joints that fulfil the requirements more efficiently (Lahdo et al., 2024). The aim is to produce an overlap-welded joint with an undercut that achieves high strength. The approach is expected to increase the formability of the produced connectors. FOLAMI will end in December 2024. More detailed results are likely to be published at the time of the 2025 ISSC Congress.

3.1.2 Friction Stir Welding

Another technology which offers advantages related to low heat input and which also is suitable for combining dissimilar materials is Friction Stir Welding (FSW). Originally developed and well established as a means for joining lightweight metals like aluminum, previous reports of the ISSC Committee for materials and fabrication addressed progress in the development of solutions for other material combinations. While encouraging results in terms of process performance and mechanical properties could already been demonstrated, a commercially successful introduction was hindered by high costs for tools accompanied with a short tool lifetime. Focusing on applications in the automotive industry, Werz investigated an approach for hybrid aluminum/steel joints. The setup foresees that the tool is in direct touch only with the aluminum part, resulting in reduced wear and tear of the tool (Werz, 2020). The technology is said to bear potential for upscaling to dimensions typical for the maritime industry. The EU funded project RESURGAM (finished in 2024) addressed technologies to join steel with steel in typical shipbuilding application areas, related to newbuilding and underwater repair. The consortium aimed to overcome the issues of the past with novel diamond-based tools. Reportedly, the solution works well for the newbuild application. Unfortunately, no reports which supported this finding were available at the time of writing the Committee's report ("RESURGAM Project website," 2024).

Apart from the applications described above, FSW is particularly beneficial for joining high-strength low-alloy (HSLA) steels, which are commonly used in shipbuilding due to their excellent strength-to-weight ratio. Research into the effects of tool rotational speed during FSW on DMR249A HSLA steel has shown that optimizing this parameter can significantly improve the microstructure and mechanical properties of the welded joints. For instance, FSW conducted at an optimal speed of 600 rpm resulted in joints with superior tensile strength and impact toughness compared to those welded at other speeds (Ragu Nathan et al., 2023). This enhancement is attributed to the formation of a refined microstructure within the weld zone, which contributes to the overall durability and performance of the ship's structure. The findings underscore the importance of fine-tuning welding parameters to achieve the desired structural outcomes, ensuring that the joints can withstand the demanding conditions encountered in maritime environments.

3.1.3 Adhesive Bonding

In shipbuilding, the predominant choice for joining dissimilar materials is still adhesive bonding. Traditionally dominated by welding, the shipbuilding industry used to be cautious in adopting adhesive bonding, primarily due to a lack of trust in this alternative joining method. However, the technology has become increasingly vital in European shipbuilding due to growing demands for lightweight construction, enhanced aesthetics, and shortened planning and production cycles. Progress in R&D has widened the fields of application and furthered the acceptance. The most recent key projects reported by Fröck and Krause (2023) include the following ones:

- **Underwater Bonding** - (Lemmerich and Vaccari, 2023) developed mounting systems capable of being bonded underwater, and established parameters for automated underwater bonding, marking a significant step forward in maintenance and repair capabilities in the maritime industry. Results from underwater trials indicated that specially formulated adhesives could retain structural integrity under submersion, opening possibilities for efficient underwater repairs without the need for dry-docking.

- **Standardized Aging Cycles for Marine Adhesive Bonds** – The project "Development and Validation of Laboratory Aging Cycles for the Approval of Marine Adhesive Joints" (IGF No. 21189 BG) addresses a critical need for standardized aging assessments. The project investigates aging effects on adhesive joints by exposing samples to marine environments for extended periods and subsequently developing standardized aging protocols. These cycles are expected to support certification processes and foster confidence in the durability and reliability of adhesive bonds in marine applications (Kayer and Klapp, 2024).
- **Hybrid Lightweight Structures for Maritime Applications** – The European Interreg-funded international project QUALIFY focused on creating hybrid lightweight structures for marine use, emphasizing combinations of composites with traditional materials (Stefanov, 2024). The primary aim was to establish guidelines for design, manufacturing, testing, and quality control for large bonded structures, such as composite deckhouses on steel hulls and offshore wind turbine connections. The guidelines led to updated certification standards, with Bureau Veritas working on the new rules note, NR613, specifically addressing adhesive joints in maritime contexts. Publication of NR613 is planned for end of 2024.
- **Advanced process automation** – The ongoing project smartBOND addresses the need for advanced, versatile bonding solutions in shipbuilding, especially where compact working conditions challenge traditional bonding methods. A central focus is on developing adaptable and automated bonding systems that support high-quality production with increased worker safety and productivity. The initiative also emphasizes digital integration, employing real-time process monitoring and data management systems aligned with Industry 4.0 principles. Such systems ensure quality control and regulatory compliance, helping to eliminate errors from manual adhesive applications (Fröck, 2024).
- **Process certification** – Progress has been observed in terms of development of standards which encourage wider use of the technology. By establishing clear guidelines and quality management requirements, the draft standard DIN 2304-2:2024 supports the wider and more straightforward adoption of adhesive bonding technology in shipbuilding. Hence, regulatory compliance can be simplified, and technical barriers reduced (Wang, 2023).

3.2 Fabrication Processes

Recent advancements in shipbuilding have focused on enhancing structural integrity, precision, and fatigue resistance to meet the evolving demands of modern maritime operations. The industry's efforts are particularly concentrated on refining welding techniques, understanding the impact of environmental conditions on materials, and improving quality control processes in ship fabrication.

One of the primary challenges in shipbuilding is maintaining dimensional precision during the fabrication of ultra-large container ships (ULCS), especially for those with capacities larger than 20,000 TEU. These ships require precise structural components such as watertight transverse bulkheads and torsion boxes, where out-of-plane welding distortions can compromise overall structural integrity. A study addressing this challenge used elastic finite element (FE) analysis in conjunction with welding inherent deformation theory to predict these distortions. The results demonstrated that the predicted distortions closely align with actual measurements, validating the accuracy of the model (Wang et al., 2020a). Furthermore, the presented research study highlighted the importance of optimizing welding sequences and groove designs to significantly reduce distortions, thus ensuring the dimensional precision crucial for ULCS construction. This precision is essential not only for maintaining structural integrity but also for ensuring the efficient loading and unloading of containers, which is critical for the operational effectiveness of these massive vessels.

Quality control in ship fabrication, particularly in the production of curved plates used in bow sections, is another area where significant advancements have been made. The bow sections of ships and offshore structures often incorporate complex curved surfaces designed to minimize propulsive resistance and enhance fuel efficiency. Traditionally, fabricating these curved surfaces has been challenging, with conventional methods often leading to inaccuracies. However, a novel methodology utilizing lightweight models and point cloud data from 3D scanners has been developed to assess the dimensional accuracy of these plates. By comparing the measured surfaces with design data, this approach allows for the identification of dimensional discrepancies and ensures that the fabricated components meet stringent quality standards (Kim et al., 2024). The successful application of this methodology in shipyards has proven its effectiveness in improving the quality of curved plates, which are crucial for optimizing the performance and fuel efficiency of ships.

The structural integrity of ships is also a critical consideration when accounting for potential accidental damage and in-service corrosion. In this context, a probabilistic framework has been proposed to estimate the probability of hull girder failure by incorporating advanced corrosion degradation models and accidental impact scenarios. The suggested study utilized Monte Carlo simulations and novel corrosion modelling techniques to account for the uncertainties associated with these factors (Woloszyk et al., 2024). The results indicate that corrosion significantly weakens the structure over time, particularly when combined with accidental damage, such as collisions or groundings. This framework emphasizes the importance of considering corrosion effects during the design phase, as it can greatly influence the long-term safety and durability of ships. By incorporating these considerations, ship designers can develop more robust structures that are better equipped to withstand the harsh conditions of maritime operations.

In a related study, the structural response of U-type corrugated core sandwich panels, which are commonly used in ship structures, was analyzed under lateral quasi-static compression loads. This research work provided both experimental and analytical insights into the deformation characteristics of these panels. The study found that the deformation process of these panels could be divided into distinct stages, with each stage contributing to the overall structural resistance (K. Liu et al., 2022a). The development of a theoretical deformation model that aligns well with experimental and numerical results further supports the effective use of these panels in ship construction. This model is crucial for predicting the behavior of these panels under load, ensuring that they perform as expected in real-world applications.

3.2.1 Additive Manufacturing

Additive Manufacturing (AM), commonly known as 3D printing, has transformed the production landscape, providing innovative solutions across various industries, but it is still not used to its full extent for offshore and Maritime industry. Class Societies have developed standards and guidelines encompassing every stage of the AM lifecycle, from material selection to final product inspection. As an example, BV provides guidelines describing the certification procedures for products to be used on ships or offshore and manufactured with metallic materials by wire arc additive manufacturing (WAAM) layer by layer process (3D printing) based on a 3D model data (BV, 2019). WAAM optimization algorithms are studied by (Mohd Mansor et al., 2024) who investigated critical process parameters like travel speed, heat input, shielding gas, wire selection and the optimized deposition strategy (oscillation, parallel, and weaving) and their impact on the mechanical properties like yield and tensile strength. In the ongoing research project IAMfat run by Delft university, the aim is to improve the fatigue strength by optimizing the weld geometry of K and T-joint made by WAAM compared to e.g. Submerged Arc Welding (SAW) most commonly fabrication method used.

Moreover, the MariLight project focuses on advancing the shipping industry by reducing the weight of ship structures through topology optimization and Large-Scale Additive Manufacturing (LSAM) (MariLight, 2024). The project aims to lower CO_2 emissions, reduce material usage, and enhance manufacturing efficiency. Some key areas of application incorporate developments in the bulbous bow are of the ship as well as smaller foundation items in the forecastle deck area. Key achievements include a 25% reduction in structural weight, a 60% environmental impact decrease, and a 90% reduction in lead time. MariLight has also instigated the effort for the development and introduction of the appropriate regulatory framework for certification and digital traceability, setting the stage for sustainable and efficient marine fabrication, with deployment targeted by 2027.

3.2.2 Fabrication Using IT

ISSC2022 COMMITTEE V.3 pointed out that the smart shipyards of the future will use digital technologies to transform the processes to obtain the various merits (Josefson et al., 2022). In recent years, the trend towards digitalization has been greatly accelerated.

One of such movements is the study of the system frameworks to the smart shipyards. Dassault Systèmes pays attention to "Collaborative platform", "Data Driven", "Model, based manufacturing" and "Integrated Tools" which are keywords to realize smart shipyards and is developing a system environment to support them (Chouche, 2022).

Muñoz and Perez-Fernandez (2021) pointed out the importance of "Digital Platform", "VR (Virtual Reality) and AR (Augmented Reality)" and "AI (Artificial Intelligence)" for the introduction of Industry 4.0 technologies in shipyards. As indicated in these papers, the importance of model-based planning and manufacturing using VR, AR and AI on an integrated digital platform become a common understanding worldwide to realize the smart shipyards.

Woo et al., (2021) developed a new diagnostic framework for smart shipyard maturity level assessment based on the keywords of connectivity, automation, and intelligence. The framework was applied to eight shipyards in South Korea to diagnose their smartness maturity level, and the results with the smart level as an input represents the actual efficiency of shipyards.

Some specific efforts to support the application of smart shipyards using IT technologies are also reported in the literature. Munín-Doce et al., (2020) analyzed the possibilities offered by an IoT platform intended for the pre-outfitting workshop. RFID was used to collect information about the plant, and the simulation was conducted using this information. it is shown that maintenance cost of equipment can be reduced by 20% by analyzing information from RFIDs and various sensors attached to the equipment.

Research on application of VR/AR/MR technologies to shipyards are also executed. Mimori and Hiraki (2022) developed a system that visualizes work instructions using MR (Mixed Reality). By applying the system to the installation of outfitting work, they confirmed that the system clarified instructions to workers, improved workability, and reduced the number of errors. Kunkera et al., (2022) use VR technology and 3D models to simulate the installation of outfitting equipment. By monitoring the installation work and performing assembly simulation based on the monitored data, they achieved a 10% reduction in work time, 90% prevention of errors during assembly, and 70% pre-outfitting. Li et al. (2019) focused on crane operations for moving and loading blocks, and developed a simulator using dynamic analysis and VR technology that can simulate the effects of crane operations on the behavior of blocks and the collaborative operations of multiple cranes.

Discrete Event Simulation (DES) is essential for model-based manufacturing and has been continuously studied. Lee et al. (2020) proposed a simulation-based master planning method by linking a production planning system and a DES system and applied the method to an actual shipyard. Okubo and Mitsuyuki (2022) proposed a method to represent the shipbuilding process by four system models i.e. Product, Workflow, Workplace and Team, and generated the realistic production plan automatically using the developed process simulation. Taniguchi and Matsuo (2022) represent each worker as an agent and propose a detailed simulation model and model generation method that takes realistic constraints and operations into account. A specialized DES library, the Simulation Toolkit for Shipbuilding (STS) from Simplan in Germany, was discussed in previous reports of the Committee Materials and Fabrication Technology. Steinhauer et al. recently reported on the latest developments and applications of the STS Library (Steinhauer et al., 2024). Key applications include backward scheduling with STS_Scheduler (for identifying optimal timelines and reducing risks in early planning phases), centralized downtime management (for precise control of scheduled maintenance and other downtimes), offshore production simulation (considering this technology field's specific restrictions and operational complexities in simulation modelling), enhanced AGV and 3D Management as well as site and space management. Looking ahead, the developers of STS expect that DES applications will be expanded to include sustainability and circular economy considerations, with new tools under development to assess environmental impacts throughout a ship's life cycle. Projects like CirclesOfLife (Krause and Hübler, 2024) are creating environmental performance indices for shipyards, and future

DES modules will likely support the dismantling, recycling, and repurposing of ships, enhancing circular economy practices in shipbuilding.

The shipbuilding industry has been actively introducing CAD/CAM/CAE systems and ERP (Enterprise Resource Planning). In recent years, the importance of information sharing in a wider range of areas has been recognized, and PLM (Product Life Cycle Management) system has been paid attention. PLM integrally manages various technical information generated in each process, including planning, design, procurement, production, sales, and maintenance. PLM is said to bring many benefits such as improved operational efficiency, strengthened manufacturing systems, and cost reduction.

Bitomsky et al. (2022) pointed out that PLM can be used in a variety of ways, including integrated management of design and production information, information transfer between systems, information management between different CAD systems, and migration of legacy systems to modern systems. Integrating existing or additional tools could become a business enabler for shipyards because these integrated digital representations of the ship allow for a new level of product lifecycle support including manufacturing methods, field data collection, and so on.

Siemens insisted that there are various levels of integration of CAD and PLM data and that PLM data is currently updated in synchronization with design changes, improving the reliability of PLM data (Perez Fernandez, 2022b).

Taniguchi et al. examined PLM systems that can be effectively used in production planning and production control and proposed the representation method of M-BOM (Bill of Materials) and BOP (Bill of Process) and how to generate such information (Taniguchi et al., 2022). By using these, the data flow from the PLM system to the production simulator is introduced so that the production simulator can be operated as soon as the data is prepared in the PLM system.

3.2.3 Monitoring System for Shipyards

At ISSC 2022, the importance of monitoring system for shipyards was pointed out and the latest status of monitoring in shipyards was reported, with a particular focus on the welding process and weld quality monitoring. Since then, various monitoring technologies for shipyards have been studied and developed.

Aoyama et al. (2021) developed the Monitoring platform for cutting and subassembly processes. This system uses monitoring data from fixed-point cameras and workers' smart phones to identify the work type, worker, work position, target workpiece, and work time. By inputting this information into DES and performing simulations, it is possible to analyze the deviations from the production plan and the factors that may cause such deviations.

Research on monitoring systems using wearable sensors is also executed in a wide area. Pribadi and Shinoda, (2022) showed that the analysis of wrist movements obtained from wearable sensors can be used to identify differences of skill levels between skilled and novice workers and it is effective to educate workers.

Mitsuyuki et al. developed a system that supports plate bending operations using 3D models and a laser scanner and installed the system on an actual shipyard line (Mitsuyuki et al., 2020). The point cloud data obtained from the laser scanner is compared with the

3D surface on the CAD model. The system assists the operator by visualizing the errors in the curved surface shape and the necessary heating lines.

Kanno et al. (2021) demonstrated that by using machine learning models using information from wearable sensors and smartphones, it is possible to identify the operation type and operation time automatically and immediately.

Park et al. (2021) developed a system to identify the location of workers in shipyards to improve worker safety. There are so many large steel structures in a shipyard, and it is difficult to identify the location of workers by conventional methods. This problem has been solved by developing a location identifying system using ultrasound signals. Watanabe et al., (2022) developed a system to estimate leg length and undercut using Deep Learning. For actual use in shipyards, the input data is not image data but welder log data (current, voltage, and welding wire feed rate) and fixed welding conditions (torch angle, etc.) and sufficient accuracy for practical use is obtained.

3.3 Summary and Recommendations for Future Work

Joining techniques for dissimilar materials should be empowered to unlock lightweighting potential; therefore, upscaling of technologies (adhesive bonding, FSW and novel laser welding processes) and demonstration of suitability for large maritime structures is required; particularly regarding FSW, demonstration of cost-efficient processes/tools should be addressed.

Although substantial progress has been made on adhesive bonding, further development is essential to match the technology's reliability, versatility, and efficiency with that of welding. Innovations such as automated adhesive application systems, real-time process monitoring, and the integration of Industry 4.0 standards in data management are anticipated to streamline adhesive processes and enhance product quality. As new standards emerge, supported by comprehensive testing and international collaborations, adhesive technology is positioned to become a cornerstone in sustainable, lightweight shipbuilding.

Other joining techniques, such as high energy beam welding, resistance spot welding (RSW) and ultrasonic welding are potentially usable in the shipbuilding and offshore industry. It is important to point out that to maximize the reliability of the joint and the production efficiency it is necessary to carry out an integrated study between the production process of the shipbuilding and offshore industry considering the joining technique to be used (e.g. for the well-known bolted connection, it may be necessary to guarantee a watertight connection; tighter joint tolerances than those typically used in maritime carpentry, etc.

In the past three years, various production support methods using IT technology have been proposed for the realization of Smart Shipyards. The importance of model-based planning and manufacturing using an integrated digital platform becomes a common understanding worldwide to realize the Smart Shipyards. However, these studies have focused on areas where information technology is relatively easy to apply and have mainly aimed at improving efficiency. In such an environment, it is necessary to provide various types of support for information generation and utilization. For this reason, various support systems using VR, AR, and AI have been developed and used aggressively.

Recent studies emphasize the need for precision, reliability, and adaptability in the face of evolving maritime demands. Especially in the case of building ultra-large container ships, it is important to optimize welding sequences and groove design to reduce distortions. The above is critical for maintaining both structural integrity and ensuring the efficient loading and unloading of containers, which is a critical aspect for these ship types. Additionally, by integrating advanced welding techniques, rigorous fatigue assessments, and innovative quality control methods, the shipbuilding industry is better equipped to produce vessels that are not only efficient and durable but also capable of withstanding the harsh conditions of global maritime operations. The continuous improvement of these technologies and methodologies will ensure that future ships are safer, more reliable, and better suited to the demands of international trade and exploration.

The importance of information integration in a wider range has been recognized in recent years, and the requirements and effectiveness of PLM systems in shipbuilding have begun to be discussed. On the other hand, considering the digitalization of the shipbuilding industry in the future, it is necessary to consider information management and information sharing over a wider area, including monitoring data during construction and operation. Another important issue is how to manage environmental impact information. The ISSC committee hope that research and development on these issues will be accelerated in the near future.

The development of monitoring systems in shipbuilding is progressing and is becoming possible to obtain various data on product accuracy and work information. For effective use of monitoring data, it is necessary to analyze the data and convert it into useful information. However, at present, the accuracy of monitoring and/or processed data is not sufficient in some areas due to the short monitoring period. In addition to the continuous acquisition of monitoring data, it is necessary to deepen discussion on what kind of information can be obtained through data analysis and how to improve the competitiveness of shipyards in the future.

More emphasis should be placed on sustainability in shipbuilding. While efforts in the past were related to reducing emissions in ship operation, future work should also address environmentally friendly shipyard processes, and shipbuilding technologies for enhanced circularity.

4 Qualification and Approval

This chapter concerns the qualification and approval of materials and processes and deals with:

- Qualification and approval processes by the class societies
- Technology qualification
- New processes
- Importance of domain knowledge in data generation and interpretation of historical data (data driven models))
- Concepts for the approval procedure for novel (FRP) materials
- Use of advanced material models.

4.1 Qualification and Approval Processes by the Class Societies

The qualification and approval processes by the Class Societies are crucial in ensuring the safety and reliability of materials, components and systems used in the marine and offshore industries. Such processes are more or less similar for all IACS members (International Association of Classification Societies) and are based on following stages:

1) Type Approval Certification

Type approval certification is required by Class Societies and depends on the nature of the material, component or system. Performed tests can be various to evaluate factors such as mechanical strength, fire resistance, corrosion resistance, and environmental performance.

2) Design Review

The design is reviewed by the Class Societies to meet requirements in accordance with class rules.

3) Construction Survey

The Class Society inspects and surveys the production or the construction of materials, components and systems. The survey is based on manufacturing processes and quality control.

4) Onboard Inspection

For materials, components or systems installed onboard, Class Societies verify their installation, functionality and compliance with regulations.

5) Class Certificates

Upon successful completion of the process, a class certificate is issued.

Processes are always under improvement and Classification Societies are developing 3D classification platforms to eliminate 2D drawings at the Design Review stage. A unique 3D model of a ship or asset will be used to perform calculations, exchange information and address classification comments. Ship owners and managers further maintain access to the 3D model developed at the design stage throughout the asset's lifetime, building a complete and evolving model of their vessel and its history (Bureau Veritas, 2020).

4.2 Technology Qualification

The technology qualification basis describes the technology, defines how the technology will be used, the environment in which it is intended to be used and specify its required functions, acceptance criteria and performance expectations. This includes the performance requirements throughout the life cycle of the technology. Technology categorization below is based on DNV-RP-A203 (DNV, 2019) and tabulated in Table 1.

Table 1. Technology categories according to DNV-RP-A203

Application area	Technology status		
	Proven	Limited field history	New or unproven
Known	1	2	3
Limited knowledge	2	3	4
New	3	4	4

This categorization implies the following:

1. No new technical uncertainties
2. New technical uncertainties
3. New technical challenges
4. Demanding new technical challenges

To accelerate the digital transformation of shipyards, ABS published in 2023 a new Guide for Smart Technologies for Shipyards (ABS, 2023a). ABS gives the following examples of implementation of Smart Technologies in shipyards:

- Location intelligence (used for tracking location of personnel within the shipyard environment)
- Artificial Intelligence and data analytics (e.g., robotics or AI-driven applications in production processes, using real-time production and inventory data to optimize inventory management, using image recognition algorithms for welding quality inspection, and using real-time equipment sensor data to drive predictive and cognitive maintenance analytics)
- Worker on-the-job health monitoring (e.g., using real-time sensor data to monitor the workforce safety and health by the safety department)
- Augmented Reality – to connect assets and facilities, making sense of data and digitize business operations (e.g., to assist maintenance personnel in maintenance and repair of equipment or assets)

This guide:

i) Describes the approach for recognition of any shipyard that incorporates Smart Technologies into their operational process(es).
ii) Introduces a systematic process to verify and validate the Smart Technologies which have been adopted into the yard process(es).
iii) Documents the process for utilizing the technologies to meet Class requirements, when applicable.

In order to support new manufacturing processes and especially for 3D printing, a comprehensive review on the status and pathway of the qualification and certification has been carried out by. Chen et al., (2022). An example of the certification scheme for Additive Manufacturing process proposed by Bureau Veritas is presented in Fig. 2.

A similar certification scheme can also be found in DNV-CG-0197 "Additive manufacturing - qualification and certification process for materials and components" (DNV, 2021), whereas guidance on approval of Additive manufacturing is specified in DNV-CP-0267 (DNV, 2022).

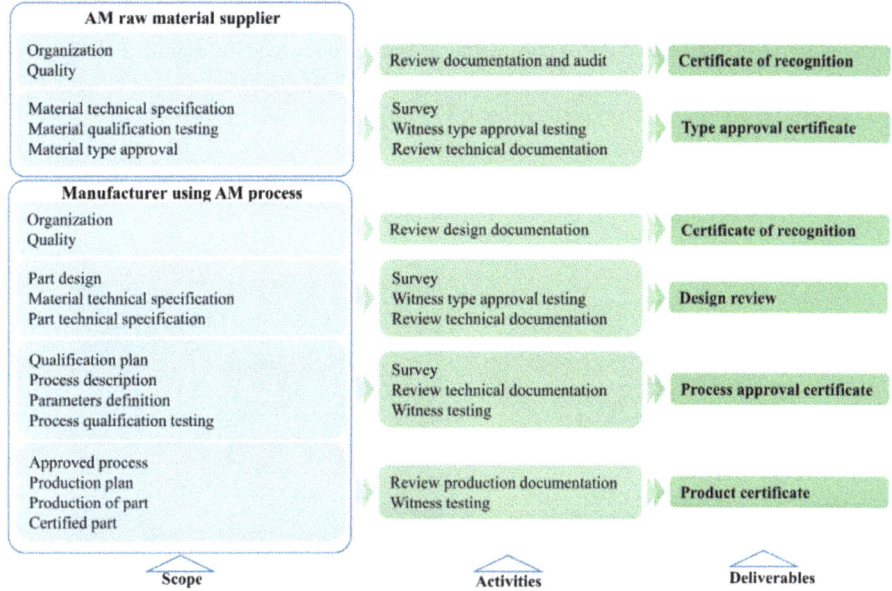

Fig. 2. Certification scheme for products made of metallic materials by Wire Arc Additive Manufacturing

Lloyd's Register also published the Guidance Note for Polymer Additive Manufacturing Certification in April 2021 (Lloyd's Register, 2021).

4.3 New Processes

A new joining method Local vacuum Electrode Beam welding (LVEB) is seeking to replace ordinary Submerged Arc Welding (SAW), with a more efficient method for longitudinal seam welding of wind turbine foundations. The local vacuum electron beam welding system was developed to exploit the advantages of high-power EB welding on large structures without the need for a large vacuum enclosure, see Fig. 3. The reduced pressure electron gun which allows the beam to perform high quality welding in a relatively coarse vacuum (1mbar) or 'reduced pressure' atmosphere in contrast to conventional Electron Beam (EB) welding which has to be carried out at high vacuum (10–4 mbar range), necessitating use of a vacuum chamber. LVEB can join thick sections in a single weld pass without requirement for pre-heat, which can increase weld throughput up to 5 times compared to submerged arc welded can-sections, see Fig. 4 (TWI, n.d). This allowed manufacture and installation of two welds in a monopile transition piece cans in the 3.6GW Dogger Bank Wind Farm.

Standards for qualification of electron beam welding, such as ASME IX, ISO 15614-11 and ISO 15609-3, define essential variables, which shall be controlled during welding. LVEB is currently not covered in fabrication codes today and the current qualification was based on requirements in DNV-OS-C401, ASME IX, ISO 15614-11.

Ongoing research took place in UK, and the Rapidweld UK research project won the Energy Institute's International Energy Technology award in 2024 (TWI, n.d) The welding process was documented feasible by use of the Technical Quality (TQ) process in DNV-SE-0160 and DNV-RP-A203, with the goal to document adequate weld quality according to requirements in DNV-OS-C401 and DNV-ST-0126.

The LVEB welding technique is more efficient than traditional submerged arc welding and provides great reductions in energy use, which helps further advance the environmental benefits of wind energy. It was proven that the process used 90% less energy, costs 88% less, and produces 97% less CO_2 emissions, (TWI, n.d)

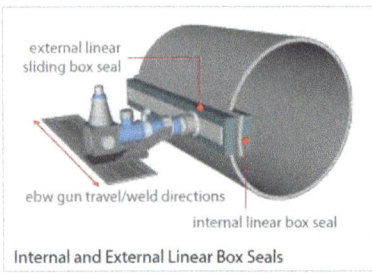
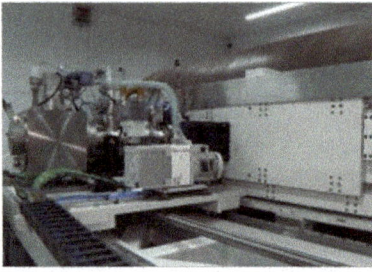
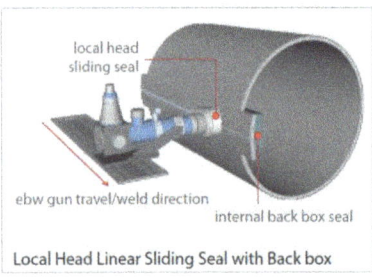

Fig. 3. A schematic of the LVEB welding for longitudinal seam welds ("Camvaceng-Ebflow," 2023)

On the same topic, TWI are currently running the Joint Industry Project (JIP) "Validation of Out-of-Chamber Electron Beam Welding for the Fabrication of Offshore wind Turbine Support Structures" to gain more knowledge of the process with focus on the fatigue performance.

New processes for the manufacturing of vessels in composite materials have been investigated within the European research project Fiber4yards (https://www.fibre4yards.eu/). During the project, the following "new" processes have been investigated, "new" meaning innovative process or existing process adapted for marine construction (Martínez et al., 2022). A demonstrator was built at the end of the project using all investigated processes, see Fig. 5.

- **Curved Pultruded Profiles:** Curing the pultruded profiles at exit of the die with UV radiation, enables the reduction of pulling forces, simplifies the dies required by the pultrusion process and, more importantly, allows the manufacturing of profiles with a curved longitudinal axis, using a robotic pulling system. These profiles can be an

Committee IV.2: Material and Fabrication Technology 875

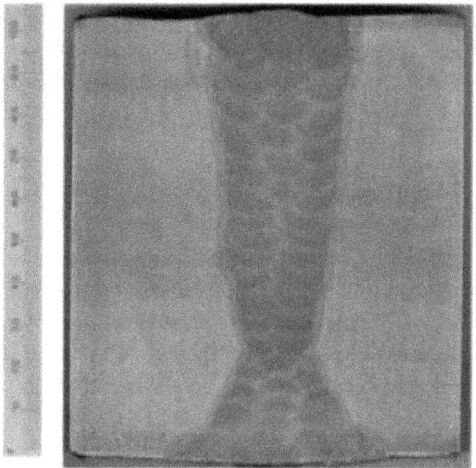

Fig. 4. Single pass electron beam weld (Right) compared to multi-pass submerged arc weld (left) ("Camvaceng-Ebflow," 2023)

excellent solution for longitudinal and transversal stiffeners for the ship hull, as their geometry can be adapted perfectly to any ship configuration.

- **Automated Fibre Placement (AFP):** AFP places multiple individual tows onto a highly curved mandrel at high speed, using a robotic numerically controlled placement head, one ply (layer) at a time. AFP allows fabrication of highly customized parts (with thermoset, thermoplastic or dry fibre tows) where each ply can be placed at different angles to best carry the required loads
- **Automated Tape Laying (ATL):** ATL is used for flat or mildly contoured parts. ATL usually lays down 3, 6, or 12 inches wide unidirectional tape, depending on whether the application is for flat structure or mildly contoured structure. ATL uses gantry style machines. Tape material can be both thermoplastic and thermoset.
- Similarly to AFP, ATL allows the manufacturing of high-quality composites. It is a common procedure in other industries such as aeronautical or for manufacturing windmill blades.
- **Adaptative Curved Moulds:** These are moulds that can be adjusted numerically to any given geometry. These are used for manufacturing composites by infusion.
  One of the larger costs of composite shipbuilding, in terms of time and economy, is the fabrication of the moulds. Quite often these are used only once because of geometry variations from ship to ship. Having adaptive, reusable curved moulds will minimize enormously this cost, as the mould can be used for all ship geometries with a minimal time to adapt its geometry to the new configuration.
- **Thermoplastic Hot Stamping:** In this process, 2D blank thermoplastic composite sheets, manufactured by a full automated line (e.g. ATL) can be hot stamped in a mould to give them any specific shape. This is a procedure with short production cycles that allows giving complex geometries to the composite with a high-quality finish.

Thermoplastic hot stamping can be used to manufacture some ship components requiring high geometric accuracy and will be especially useful for parts that require a large number of units. A possible example is the use of this technology to manufacture connection blocks where the different components are joined.

- **Additive Manufacturing:** In Additive Manufacturing (3D printing) a 3D object is built from a CAD model by successively adding material layer by layer. This covers a variety of processes in which material is joined or solidified, such as Fused Deposition Modelling (FDM), stereolithography (SLA) and selective laser sintering (SLS). Actually, the composites made with additive manufacturing can contain short fibres or long fibres, which are deposited in the final part at the same time as the resin. 3D printing can be used to manufacture nearly any geometry without the need of a mould, which makes the procedure very versatile for the construction of ship parts or components. Currently the strength of the resulting material does not allow using this procedure for structural parts with high strength requirements. However, the finished structure can be reinforced with an ATL method to achieve the required strength and stiffness.

Fig. 5. Demonstrator built by Fibre4yards consortium (https://www.fibre4yards.eu/).

4.4 Data Driven Models: The Importance of Domain Knowledge in Data Generation and Interpretation of Historical Data

In recent years, data-driven models (DDMs) grounded in historical data have gained significant attention across many disciplines, including the qualification and approval of materials and processes. Research studies on artificial intelligence (AI) methods, particularly machine learning (ML), are proliferating across nearly all engineering fields. This surge is driven by advancements in computational power, data accessibility, and algorithms, enabling ML to uncover complex relationships that classical methods may not fully capture. However, some issues observed in recent studies have raised serious concerns about the credibility of these approaches and even their necessity for certain applications.

The development of ML models hinges critically on the availability, size, and quality of data. The accuracy and reliability of these models are directly correlated with the accuracy and relevance of the training data, making it imperative that model developers possess a robust understanding of the specific domain to which the ML model is applied. Some claims, such as those in (Cai et al., 2022), suggest that data-driven methods require less domain knowledge compared to other approaches like finite element methods (FEM) and semi-empirical methods in classification rules. Thus, some fundamental issues in (Cai et al., 2022) and two other studies, (González-Arévalo et al., 2021; Velázquez et al., 2022a), are used herein as examples to showcase the significance of domain knowledge. These issues are listed below:

1) Methodological errors in generating data.
2) Misinterpretation of data used to train ML models.
3) Limited generalization of models to real-world scenarios due to using data from oversimplified tests/simulations or data scarcity.
4) Unnecessary resource expenditure on developing data-driven models where fast, accurate, and reliable solutions already exists.
5) Trustworthiness of the model in high-risk applications.

Examples of issue #1 are references (González-Arévalo et al., 2021; Velázquez et al., 2022b) where stress-strain curves obtained from uniaxial tensile testing of API 5L steel specimens exhibit significant flaws (Fig. 6). The strain values in these studies appear to have been derived from the crosshead displacements of the test machine causing the J shape elastic response instead of linear response leading to a Young's modules of 10 GPa instead of the typical 210 GPa for steel and an oddly large yield strain. The J shape elastic response is because crosshead displacements include those in the gripping system, the machine itself, and the initial slippage of the specimen when "sitting" itself in the grips. As a result, these factors may introduce significant artificial displacements, and consequently, strains, which the specimen does not actually experience. These erroneous data were then used by Velázquez et al., (2022) to develop finite element models of pitting-corroded steel pipes for burst pressure estimation. The simulations, marred by flawed data, were subsequently employed in a benchmark study comparing the performance of semi-empirical burst capacity models for corroded pipelines. The resulting conclusions by Velázquez et al., 2022b) suggested that burst capacity models developed by classification societies such as DNV are significantly unsafe. This odd assertion stems from the flawed stress-strain curves used as input to the models and other modelling errors, such as unrealistic corrosion defect profiles and incorrect boundary conditions. Proper domain expertise is crucial for identifying and excluding such erroneous data during the data collection phase for data-driven models.

Issues #2 through #5 are evident in (Cai et al., 2022), as critiqued in detail by (Mokhtari and Melchers, 2024). In (Cai et al., 2022) similar problematic conclusions

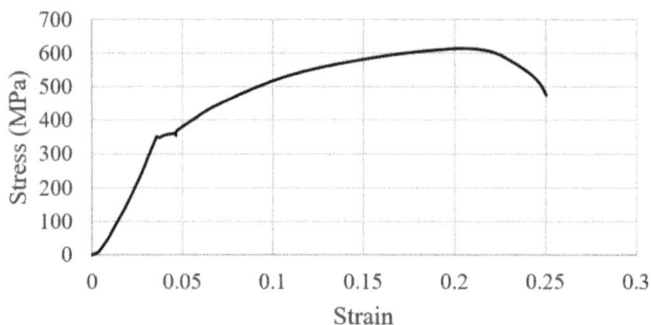

Fig. 6. A significantly flawed stress-strain curve of API 5L steel due to methodological errors in material testing in (González-Arévalo et al., 2021; Velázquez et al., 2022b).

to those by Velázquez et al., 2022b) were drawn regarding the safety of burst capacity models developed by classification societies such as DNV and ASME. Cai et al., (2022) mistakenly used the "true" ultimate tensile strength (UTS) instead of the "engineering" UTS in the burst capacity models recommended by DNV-RP-F101 and ASME B31G. Additionally, they misinterpreted the sizes of corrosion defects reported in the literature, such as those in Mok et al. (1991). For instance, in two cases, the defect lengths were incorrectly entered as 10 mm, whereas the actual lengths in (Mok et al., 1991) were 102 mm and 203 mm. These significant errors led to an overestimation of burst pressure by as much as 70% in the DNV-RP-F101 model. Consequently, the use of these erroneous data for training their ML models severely compromises the validity of the models presented in (Cai et al., 2022).

The most concerning aspect of the study by Cai et al. (2022) is that, despite the incorporation of erroneous data, the ML models demonstrated high accuracy during the validation with test datasets. It is well-established that the predictive performance of ML models is intrinsically linked to the quality and accuracy of their training data. Therefore, models trained on flawed data should yield inaccurate predictions, at least in certain instances. If these instances are omitted from the validation process, and the model is consequently deemed reliable, this misplaced trust can lead to catastrophic outcomes in practical applications. This situation highlights the paramount importance of ensuring both the quality and diversity of test data used in the validation of data-driven models.

Regarding issue #3, the ML models in (Cai et al., 2022) were trained on data from pipes subjected to a single, external rectangular corrosion defect (not common in practice) under internal pressure. Even for such a simple case, the authors faced a shortage of relevant data, resulting in a small training dataset. This is followed by issue #4, considering recently developed and powerful automated finite element analysis (FEA) tools, employed in finite element based digital twins. These tools can quickly, accurately and automatically develop a digital twin of the corroded pipe and calculate the burst pressure using cloud computing. They do not have the ML models' generalization issue as they are physics-based, validated, and can automatically develop multiple internal and/or external defects with different shapes (not only rectangular) and can also account for various operational and environmental loads, effects of temperature on material properties, different boundary conditions, etc. Therefore, it seems unreasonable to undergo the immensely labour-intensive, time-consuming and expensive process of data curation to develop an ML model free of the aforementioned errors capable of considering various corrosion defect types/shapes/numbers and operational and boundary conditions when, at best, such an effort can only reach the same capability as that already available in rapidly improving FEA tools. A good example where an ML model could be useful is when an accurate and quick solution is not yet available and/or there are significant uncertainties in existing solutions. In the context of (Cai et al., 2022), an ML model for time-dependent burst capacity prediction of corroded pipelines could be valuable, given the uncertainties in predicting long-term pitting corrosion development.

Finally, issue #5 addresses the risks associated with using ML models trained on erroneous data, as seen in (Cai et al., 2022). In high-risk applications, such as predicting the burst pressure of steel pipelines that transport hazardous and flammable fluids under high pressure, even a single overprediction could have catastrophic consequences.

4.5 Concepts for the Approval Procedure for Novel (FRP) Materials

The approval procedure for novel (FRP) materials has been investigated in the European research project Fibre4Yards (https://www.fibre4yards.eu/). Indeed, the use of raw materials that are not known by the Class Society, will imply to carry out a testing campaign in order to qualify the materials regarding their mechanical and physical properties and to evaluate the influence of environmental conditions such as temperature, water immersion, etc. Figure 7 shows the flowchart for the homologation of raw materials following the classic or the equivalent homologation procedure.

Depending on the raw materials to be tested, the following tests as per Table 2 are to be performed. On a case-by-case basis, in addition to these mechanical tests, following tests might also be performed:

- Density
- Glass transition temperature
- Heat deflection temperature
- Water absorption
- Coefficient of thermal expansion
- Hardness
- Resistance to UV
- Resistance to chemicals

4.6 Summary and Recommendations for Future Work

Qualification and approval processes are also evolving to consider the development of new technologies such as the digitalization, new processes or the use of new materials. The "classical" approach remains the same but to be adapted or completed to support innovation and development of the maritime industry. Classification Societies must work closely with shipyards and shipowners to ensure a seamless alignment between rules, guidelines, and innovative technologies being implemented. This collaborative effort is

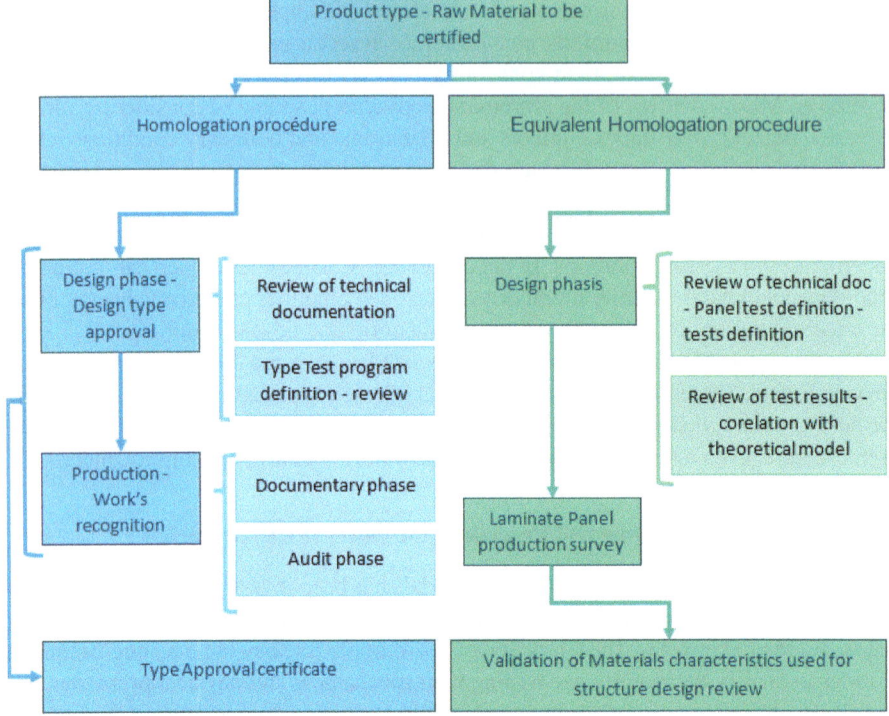

Fig. 7. Flowchart for raw material homologation, Fibre4Yards project.

Table 2. Raw materials mechanical tests

Properties	Standard / test methods	Comments
Tensile	ISO 527 or equivalent	Used for obtaining tensile modulus, elongation at break, tensile stress at break
Compression	ISO 7743, ISO 604 or equivalent	Used for obtaining compressive modulus, elongation at break, compressive stress at break
Shear	ISO1827, ISO11003 or equivalent	Used for obtaining shear modulus of material, shear elongation at break, shear stress at break

crucial for maintaining safety standards while fostering progress and embracing cutting-edge solutions within the maritime sector.

While some recent studies suggest that, compared to the conventional methods, data-driven methods require less domain knowledge, it remains a critical factor, particularly in developing machine learning models. A lack of domain expertise has led to significant issues in literature, including methodological flaws in data generation, misinterpretation of historical data and established models, limited applicability to real-world scenarios, and unnecessary resource expenditure when existing solutions are sufficient. Addressing these shortcomings requires sector-specific regulations, such as those provided by classification societies, to establish risk-based guidelines for the development and use of data-driven models. Such guidelines would ensure the safe and efficient deployment of these models, while avoiding unnecessary resource expenditure when efficient, safe, and proven solutions are already available.

Class rules are mainly based on "classic" materials limiting the use on non-conventional solutions. Recent development proposed the certification scheme leading to the characterization of material in the scope of the design assessment. By expanding the scope of class rules to encompass a wider range of materials, the maritime industry can leverage the benefits of advanced material technologies, improving efficiency, performance, and sustainability in maritime sector.

The next challenge for qualification and approval of materials and fabrication processes will be the consideration of environmental impact. Factors like carbon footprint, recycling, and environmental contamination will need to be evaluated and could become important new criteria in the qualification and approval procedures going forward.

This aligns with the increasing focus on sustainability and reducing the environmental impact of industrial processes and products across many sectors. Incorporating environmental considerations into materials and fabrication approvals is a logical next step as companies and industries work to minimize their ecological footprint.

5 Durability of Materials

5.1 Environmental Influence on Material Performance

Environmental factors significantly impact the performance and longevity of materials. These influences can lead to various forms of degradation from loss of strength and toughness, affecting the structural integrity and functionality of a component or structure.

5.1.1 Environmental Assisted Cracking

Thermo-Mechanically Controlled Processed (TMCP) steel pipe has been used for sour service for decades. Despite the Kashagan failure (Schwab, 2024) raising doubts about TMCP pipe, many projects have successfully used X60 or X65 material with ferritic-pearlitic or bainitic microstructures in Region 3 of the NACE diagram (Fig. 8) for sour gas. The root cause of the failure and leaks detected only eight days after start-up in September 2013 of the Kashagan pipelines was attributed to the plate rolling and manufacturing process, which introduced local hard zones (LHZ) in the TMCP plates used for the line pipe. The leaks occurred due to longitudinal sulfide stress corrosion cracking initiated in these local hard zones (Yue et al., 2020). Following the failure, a question appeared about the suitability limits of carbon steel pipes produced by both TMCP and quenched and tempered (QT) processes in the upper range of Region 3 (Fig. 8) in the NACE diagram, extensive research including that by (Schruff et al., 2017) has been conducted during recent years to try to define a safe use. The research has focused on the effects of stress, hardness, and sulfur content on corrosion and stress corrosion cracking (SCC).

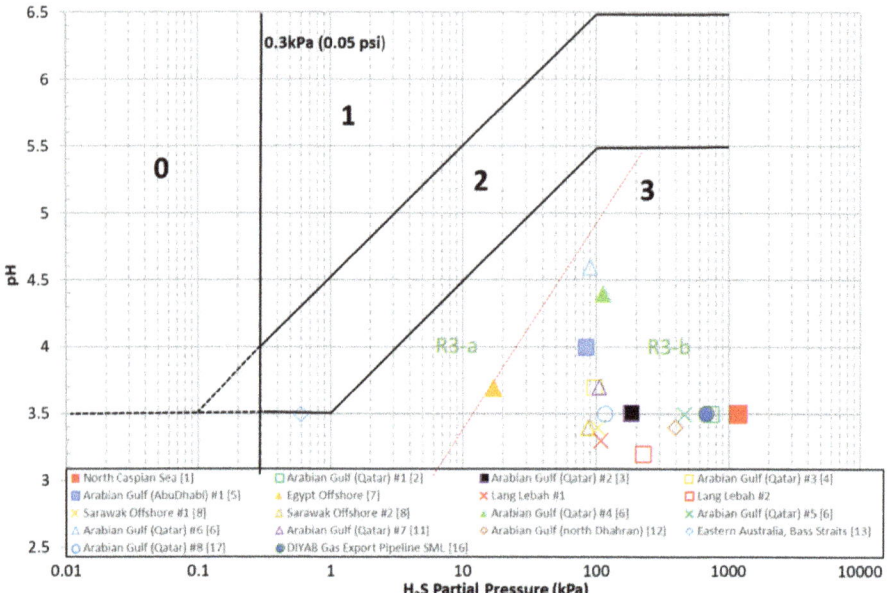

Fig. 8. NACE MR0175/ISO15156-2 pH-H2S partial pressure diagram with location of sour service project environments. Subdivision into Region 3-a and 3-b as proposed by (Yue et al., 2020) has been indicated.

Key findings and contributions from various studies are summarized below:

- Stress and Hardness: Research by Shimamura (2022) highlights how stress and hardness influence the formation of corrosion grooves in sour linepipe steel, especially in environments with low H2S content.
- Corrosion Groove Formation and SSC Initiation: Izumi et al., (2020) studied the impact of H2S presence, surface hardness, and applied stress on corrosion groove formation and SSC initiation in low alloy steel.
- Loading Profile and Cracking Behavior: Huggins-Gonzalez et al., (2023) investigated how different loading profiles and sour environments affect the cracking behavior of C110 steel.
- Unified Corrosion Mechanism: Kahyarian et al., (2020) reviewed recent developments in understanding mild steel corrosion in aqueous weak acid solutions, including the presence of carboxylic acids, CO2, and H2S.
- Newbury et al., (2019) focused on qualifying TMCP pipe for severe sour service and mitigating local hard zones to enhance material performance.

Currently the on-going DNV JIP "Sour" is looking into providing requirements and acceptance criteria for use of carbon steel pipes for sour service in Region 3 (NACE criteria). For sour cases, where carbon steel pipes are not an option, clad and lined pipes are a viable option. Due to low weld metal and HAZ toughness properties in sour service for carbon steel pipes, a deterministic engineering critical assessment (ECA) approach would lead to very small weld acceptance criteria, which would be challenging to detect and most likely many repairs. Tronskar et al., (2024) developed a probabilistic framework for ECA of weld defects of offshore pipelines subjected to sour service Region 1 that documented acceptable annual target probability of 10^{-5} pr ("DNV-ST-F101 Submarine Pipeline Systems," 2021) for detectable weld defect sizes.

5.1.2 Long-Term Pitting Corrosion Morphology

General corrosion is typically addressed in design by incorporating a corrosion allowance, with a minimum of 3 mm commonly recommended for static utilization checks of C-Mn steel structures. For fatigue analysis, it is standard to assume half of this allowance (DNV-ST-F101 Submarine Pipeline Systems). Standards such as API 579-1/ASME FFS provide guidelines for assessing corrosion, either through wall thickness reduction (Level 4) or pitting corrosion (Level 5), though these assessments generally do not account for severe fatigue loading. For evaluating the remaining strength/life of pitting-corroded assets, where localized corrosion defect data is available from inspections, semi-empirical models such as those in DNV-RP-F101 offer more detailed assessments, incorporating basic geometrical properties of defects. For critical defects, higher-fidelity methods, such as FEM, are used, particularly in research studies. Regardless of the approach, it is common practice in analytical, experimental, and numerical studies to model the corrosion defects using simplified geometries (e.g., hemispherical, cylindrical, conical, and cuboidal geometries or circular holes in 2D models).

Recent studies have found that such oversimplified shape idealization of pitting corrosion defects may lead to significantly inaccurate structural integrity assessments. Failures driven by plastic fracture are less sensitive to the details of pit morphology (Mokhtari et al., 2023d; Mokhtari and Melchers, 2018, 2019). For example, defect shape idealization for pitting corroded pipelines often causes conservative remaining strength estimations given that such simple geometries (e.g., a cuboid) usually represent larger metal loss compared to the actual irregular pitting corrosion defects with the same dimensions (Mokhtari and Melchers, 2019, 2018). However, when fatigue or brittle fracture is the dominant failure mode, the results are quite sensitive to the geometrical details of the defect morphology as these details significantly affect the intensity and location of stress/strain concentration. Thus, defect shape simplifications using smooth profiles could cause significant overestimation of remaining strength or fatigue life (Mokhtari et al., 2023e; Mokhtari and Melchers, 2020).

In a numerical example applying the Brown-Miller strain criterion with Morrow mean stress correction (Mokhtari et al., 2023e), it was demonstrated that the crack initiation life from an ellipsoidal pit was 17.2 times larger than that from an irregular pit with the same dimensions (Fig. 9). The sharper features within the irregular pit induced 279% higher equivalent plastic strain in the first load cycle compared to that in the ellipsoidal pit, with this difference growing in subsequent cycles. The residual stress in the irregular pit was significantly higher and more localized, with von Mises stress being 338% greater than that in the ellipsoidal pit. In the elastic range, before plastic deformation occurs, the maximum von Mises stress in the irregular pit was up to 60%

higher than in the ellipsoidal pit. Despite these substantial differences, maximum local stress in both the irregular and ellipsoidal pits reached nearly the same value once the yield stress was exceeded in the first cycle. Consequently, when plastic deformation develops within pits, a stress-based monotonic load analysis fails to show the effects of pit shape idealization and may lead to misleading conclusions that an irregular pit can be effectively represented by an overly simplified defect shape like an ellipsoid.

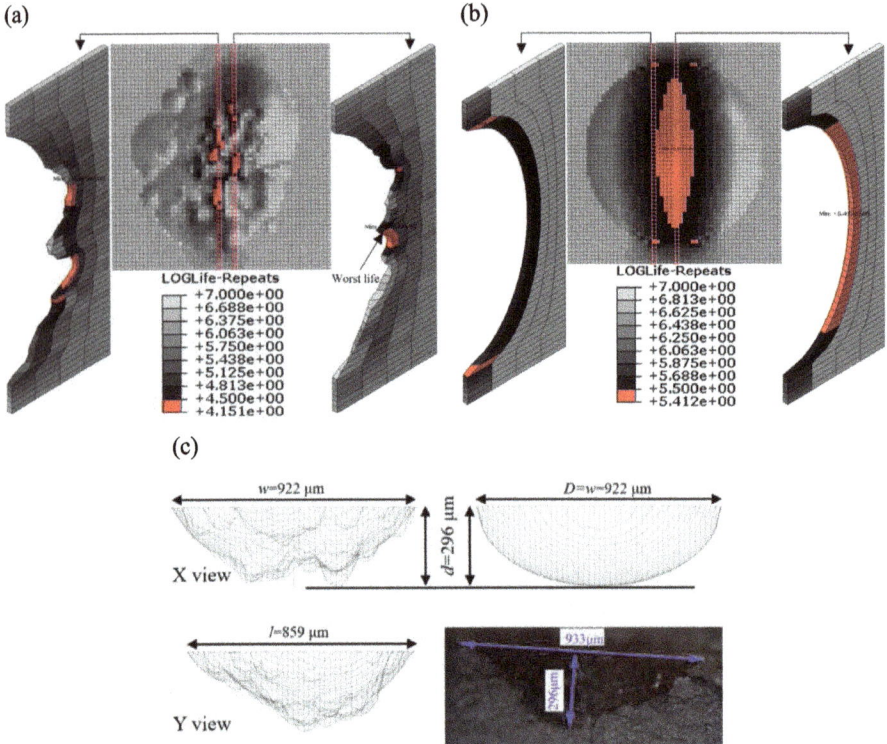

Fig. 9. Logarithmic fatigue life distribution in an irregular pit (a) and its ellipsoidal counterpart (b). The results are from the finite element analysis of a fatigue test on a standard Q235 steel specimen (Mokhtari et al., 2023e). The irregular pit was generated using a random pit generation algorithm, which appears to have closely modelled the typical irregular morphology of pitting corrosion defects (c) with some simplifications necessary for engineering applications (the physical pit image is from (Xu and Wang, 2015)).

It is clear that accurate modelling of pitting corrosion morphology is crucial in certain applications. However, this comes with several challenges. The morphology of these defects can be highly complex, and due to the stochastic nature of pitting corrosion, both the size and shape of the defects can vary randomly from one case to another. Furthermore, the level of detail achievable in the model is constrained by the size of the finite elements that can practically be used in the analysis. This challenge is further compounded when only shell elements can be employed, which is often the case in many marine engineering applications, as opposed to solid elements.

To improve the geometrical modelling of irregular pitting corrosion defects, it is essential to understand both the development mechanism and the morphology of naturally occurring complex-shaped pits. Most irregular pitting corrosion defects seem to result from the coalescence of smaller, rounded pits (Fig. 10). Melchers has explained the development mechanism of short- and long-term pitting corrosion in multiple studies (e.g., (Melchers, 2021, 2020, 2019; Melchers and Ahammed, 2021)). A summary of this mechanism is provided below from (Melchers, 2020).

Pitting corrosion depth has been shown to develop in an incremental, step-wise pattern, particularly in seawater environments (Fig. 11). In the first increment, initial pits nucleate and grow on a near-perfect surface. The dissolution leading to pit initiation is driven by small differences in local electrochemical potential induced by material imperfections, inclusions and/or local alloy constituents. Only some of the initial pits become stable and propagate with time. These stable pits keep growing in depth until the driving potential becomes exhausted and the growth rate of pit depth slows down. However, the pits continue to expand laterally as the pit walls remain within the potential range. Eventually, this lateral growth can lead to the coalescence of neighboring pits, forming a plateau, referred to as a "broad pit" in some references (Phull et al., 1997). This cycle of pit initiation, growth, and coalescence repeats in subsequent increments, but on an increasingly corroded surface. As a result, new pits can initiate and grow on the floor of existing pits, consistent with field observations (Fig. 10 and Fig. 11). With longer exposure times, these cycles lead to the formation of deeper broad pits with an incremental growth pattern. It should be noted that pit development (both initiation and depth development) varies from point to point on the metal surface. This is because the potential difference driving pit development is not exactly the same at all points on a metal surface where anodic dissolution occurs. Therefore, asymptotically, pit depths on a corroding surface can be considered statistically independent. The continuous repetition of the pit development and coalescence cycle, whether on the uncorroded metal surface or within existing pits, ultimately results in an irregular morphology, often characterized by a dimpled pattern at the base of the corrosion defect. For more information, particularly regarding the different phases involved in corrosion development including oxic and anoxic conditions due to the development of corrosion product layer, refer to (Melchers, 2020).

Several studies have simulated the irregular shapes of pitting corrosion defects by randomly distributing spherical cap defects in multiple hierarchical levels (e.g., Fig. 12 for long-term and Fig. 9c for shorter-term pitting corrosion). At the first level, spherical caps with a random spatial distribution and size, within a specified range, are removed from an intact surface to simply represent amalgamation of the oldest and largest pits including broad pits (also termed "historical" pits). Subsequent levels introduce smaller spherical caps on the existing corroded surface. This process is continued until the desired level of detail is achieved. However, several open questions remain, particularly concerning the probability density functions governing the size and spatial distribution of these caps at different levels, which could vary depending on exposure environments and duration as well as the specific alloys involved. The application of image processing and machine learning techniques could play a significant role in determining these

Fig. 10. Small, rounded pits can grow and coalesce to form a larger irregular pit both on an intact and on an already corroded surface (Jeffrey and Melchers, 2007; Melchers, 2020).

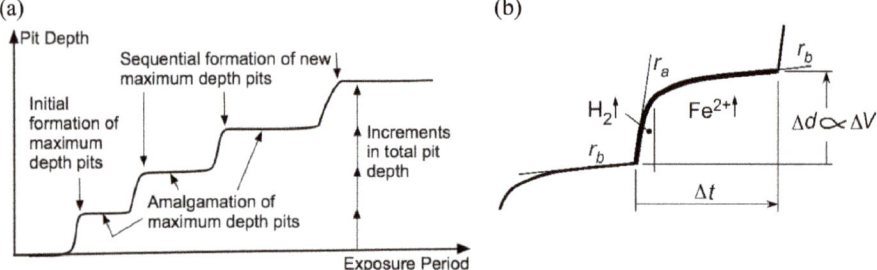

Fig. 11. (a) Schematic plot of deepest pit depths development with an incremental step-wise pattern. As the pitting corrosion develops in time, the duration at which pits with maximum depth coalesce increases due to increased rust build-up (Melchers and Ahammed, 2021). (b) Proposed model for each increment with the instantaneous corrosion rate changing from r_b to r_a at the transition point. Δd is the pit depth increment for time increment Δt, and ΔV is the driving electrochemical potential at this increment (Melchers, 2020).

distributions. More importantly, to accurately simulate the combined effects of loading and pitting corrosion growth – which can alter stress history and introduce residual stresses – it is essential to investigate the time-varying distribution of the parameters noted above. Image processing and machine learning combined with other techniques in computer-aided geometric design could also be used to develop other approaches for the 3D simulation of pitting corrosion morphology.

5.1.3 Nickel-Based Alloys

Most nickel-based alloys were considered an excellent and reliable material selection for subsea components exposed to seawater and cathodic protection. However, the offshore industry has seen several failures in recent decades in nickel-based components (Alloy 716, Alloy 625 grade 1, Alloy 725 and Alloy 718). For most of the components that failed, there were areas with local stress concentrations and cathodic protection was concluded

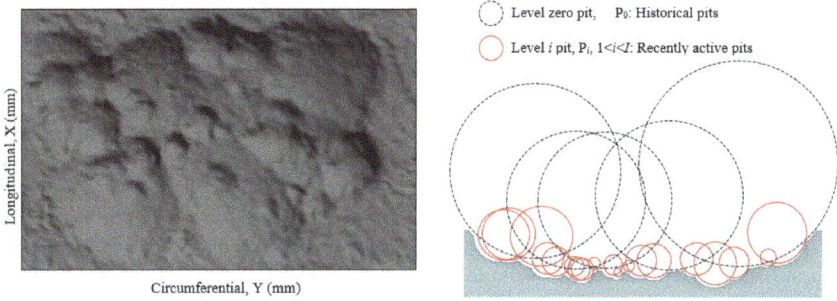

(a) Plan view of an isolated irregular corrosion defect

(b) The spheres approximating the 'river-bottom' profile of the defect

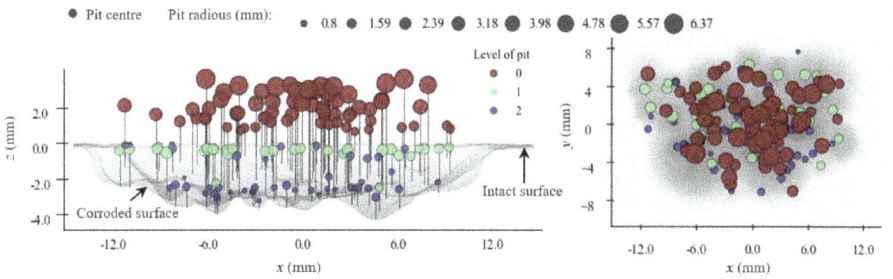

(c) Side and plan views of the irregular defect with centers, radii, levels and coordinates of the approximating spheres.

Fig. 12. An example of how the irregular pitting corrosion morphology (long-term in this example) can be approximated by merging many spherical caps in multiple hierarchical levels.

to be the likely hydrogen source. Hydrogen-induced stress cracking (HISC) is a potential failure mechanism for nickel-based alloys and the materials fracture resistance may be significantly reduced when exposed to tensile stress and a source of hydrogen. Some Alloy 625 grade 1 failures are reported to have been initiated from areas with geometrical stress concentrations. Some other failures of Alloy 625 grade 1 were initiated in areas with residual stress.

Haag et al., (2024) investigated the hydrogen embrittlement in UNS N07718 for two different heat-treating schedules with yield strength of 120 ksi (827 MPa) and 150 ksi (1034 MPa) respectively. The material testing (slow strain rate and toughness testing) revealed that the 150 ksi yield strength was more susceptible to hydrogen embrittlement (HE) than the 120 ksi material, but the latter is also substantially affected by HE. Liu et al. (2024) investigated the ductility loss of Alloy 625 and found that the ductility loss was highest at conventional strain rate with a reduction of 20.1% while for slower strain testing the reduction was 11.9% compared to air test properties. Yu et al, (2024) studied the impact of $Cr_{23}C_6$ carbides on the susceptibility of nickel-based alloys to HE and concluded that they had an adverse effect. Similar effect is also found for the $M_{23}C_6$ carbides in nickel-based alloys.

API 6ACRA is a standard that has become broadly used and has been a contributor to increasing the quality of nickel-based alloys, however, for nickel alloys used in a hydrogen charging environment, there are no established HISC design criteria specified.

5.2 Material Challenges in the Energy Transition

The unprecedented energy transition presents significant challenges, with material issues emerging as a key factor shaping the path forward. The microstructure, strength, and corrosion resistance of materials will play a critical role in determining the feasibility of new technologies. In carbon capture and storage (CCS) systems, the selection of materials for the process and CO2 storage will be pivotal to the system's success. Similarly, in the realm of alternative fuels, the ability of materials to support the technologies adopted, i.e. resist hydrogen embrittlement in hydrogen storage applications, will dictate whether these technologies can be effectively implemented. Current research is focused on developing advanced steels and other innovative materials to support energy transition.

5.2.1 CCS (Carbon Capture and Storage)

CO_2 storage technologies are already integrated in various industries, e.g., oil and gas, and hydrogen production, however these are all land-based operations. On the other hand, in marine vessels, although carbon storage would be particularly pertinent due to their high fuel emissions, this technology is still under research. That is because on-board carbon capture and storage (OCCS) comes together with challenges like space – weight limitations, design and safety issues, material selection and the nature of CO_2 ("A case study of the largest shipping segments, main carbon-based fuels, and full and partial application as part of a newbuild or retrofit," 2022).

CO_2 only exists as either solid or gas at atmospheric pressure and therefore should be pressurized to reach a liquid state. According to (Orchard et al., 2021) CO_2 shipping is expected to be most cost-effective in either low pressure and low temperature conditions: -55 to -40 °C, 5–10 bar, or medium temperature and medium pressure conditions: -30 to -20 °C, 15–20 bar. Nowadays, medium P condition is applied during small-scale CO2 transportation and has been adopted as the transport pressure for early CCS projects. However, the low P condition is considered as the most cost-effective and viable option for ship sizes above 10,000 t CO_2 (Orchard et al., 2021). Semi-refrigerated Type C tanks are commonly chosen for the transportation and storage of liquefied gases due to their ability to balance pressure and temperature requirements effectively (Al Baroudi et al., 2021).

Liquefied CO_2 presents a density of 1.17 t/m3 at -52 °C, (twice the density of liquefied petroleum gases, LPG). Pure CO_2 has a "triple-point" at 5.12 bar and -56.6 °C, hence cannot be transported in liquid form at pressure below the triple point (dry-ice (Fig. 13)). This high density enhances transport efficiency by maximizing the amount of CO_2 carried per voyage, reducing transportation costs. Several studies have examined optimal conditions for the liquefaction process. These studies suggested operating pressure and temperature close to the triple point, thus ranging temperature from -30 °C to -52 °C and the pressure ranging from 5.2 bar to 15 bar (Fig. 13). In particular, (Seo et al., 2016) performed analysis of seven CO_2 ship-based CCS chains with liquefaction pressures between the triple and critical point and concluded that 15 bar (-27.7 °C) is the optimal pressure. While 6 bar (-52.3 °C) had lower costs for storage tanks and carriers due to lower pressure, it required higher energy consumption, raising the liquefaction system's cost.

The low-pressure ship transport of CO_2 offers a promising solution to scale up CCS deployment, with significant technical, economic, and environmental benefits. However, it also presents challenges in materials selection, containment systems, operational safety, and supply chain integration, requiring careful consideration and technological innovation.

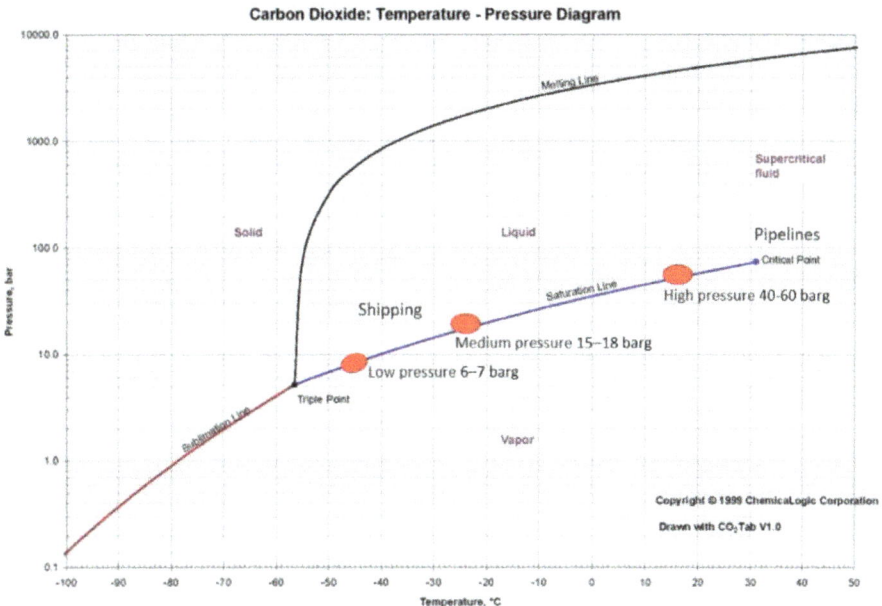

Fig. 13. Carbon dioxide, CO2, temperature – pressure diagram (Notaro et al., 2022).

The transport of CO_2 in shipping systems involves complex thermodynamic and material considerations to ensure safe, efficient, and cost-effective operations. Thermophysical properties like density, solubility, phase equilibria, and impurity effects are critical parameters for designing effective shipping chains. Density affects significantly cargo capacity, voyage stability, and overall costs. High-density states near the triple point maximize cargo efficiency, but impurities reduce CO2 density, lowering storage capacity and increasing injection pressure. Tank size is further constrained by pressure-related wall thickness limitations, emphasizing the need for precise impurity control. Water solubility in CO_2 is influenced by temperature, pressure, and impurities. At cryogenic temperatures, solubility decreases significantly, minimizing corrosion and hydrate risks. However, impurities like NO_2 and SO_2 reduce water solubility, while H_2S increases it. Impurities like N_2 and H_2 elevate vapor pressure and widen the two-phase envelope, complicating transport. Impurities impact phase equilibria, material integrity, and operational safety and especially at conditions typical of shipping H_2S presence could enhance embrittlement in the steel, while SO_x, NO_X and O_2 favor corrosion phenomena (Al Baroudi et al., 2021).

Corrosion presents a significant challenge in CO_2 transportation, especially when water is present. Experience with CO_2 pipelines has demonstrated that many metals are susceptible to corrosion under these conditions, as the interaction of CO_2 with water can lead to the formation of carbonic acid, which is highly corrosive to common construction materials. The most problematic impurity is free water (H_2O), which reacts with CO_2 and H_2S and forms the highly corrosive carbonic and sulfuric acids; the latter is formed together with oxygen (Ma et al., 2021). The presence of impurities such as oxygen (O_2), hydrogen sulfide (H_2S), or sulfur oxides (SO_x) can further exacerbate the corrosive environment, potentially leading to the formation of other aggressive acids like sulfuric acid (Fig. 14).

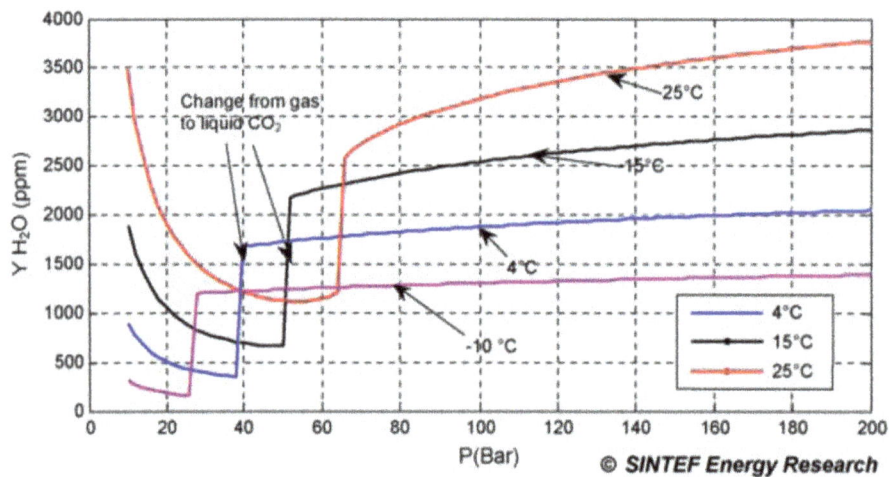

Fig. 14. Solubility of water in pure carbon dioxide (Austegard, A., Barrio, M, 2006).

The usual design of a CO_2 containment system is based on a Cylindrical Type-C tank in volumes up to 6000 m^3. Intermediate CO_2 storage cylindrical vertical tanks, with capacity of 3,000 t of CO_2 are currently employed in commercial-grade projects by Yara Praxair (Al Baroudi et al., 2021).

Semi-pressurized tanks for CO_2 storage and transportation are available in cylindrical, bi-lobate, and spherical designs. Cylindrical tanks are favored for their straightforward construction and scalability, with optimal designs, using 9% Ni steel with a 10 mm wall thickness and 4,500 m^3 capacity. Spherical tanks are structurally efficient, and present better pressure distribution and rather lower installation costs despite higher fabrication challenges. Large capacities, of 20,000 m^3 carbon steel tanks, are proposed in the literature, for high-volume applications. Material choices include carbon steel,

304L/316L stainless steel, and aluminum 1050, selected based on corrosion resistance and cost considerations. Material and design must address CO_2-induced corrosion risks, especially in the presence of impurities and free water (Al Baroudi et al., 2021).

Materials considered in the literature include Fine-Grained Carbon-Manganese Steels (e.g., NV 4-4L) which are cost-effective for temperatures down to −55 °C, Nickel Steels (e.g., NV 5% Ni) which provide excellent performance in low-temperature environments and Austenitic Manganese Alloy Steels (e.g., NV Mn 400), which offer superior toughness and thus can perform in extreme conditions. In general, materials with enhanced strength, which will enable the construction of larger tanks while ensuring safety and structural integrity need to be studied.

Typical storage tanks on-land are made of conventional low carbon steels, aluminum or austenitic stainless steels. These materials can retain toughness at cryogenic temperatures, but they all have poor strength-to-weight ratios. Several modern SS can surpass these values, without damaging corrosion resistance or fracture toughness.

To ensure the quality of materials, specific tests are proposed in the literature. These include Impact Test, which evaluates material performance at −75 °C in parent materials, welded conditions, and after post-weld heat treatment (PWHT); and Crack Tip Opening Displacement (CTOD) Testing, which assesses the fracture resistance at −55 °C to comply with IMO MSC.1/Circ.1622 standards and fatigue assessment, which will ensure durability against cyclic loading in large containment systems.

Large-scale CO_2 shipping can greatly benefit from the well-established LNG and LPG sectors, particularly during initial implementations. While CO_2 differs in pressure and temperature conditions, liquid CO_2 near its triple point shares a similar liquid-to-gas density ratio with LNG, allowing for more meaningful comparisons. However, there are several drawbacks like the fact that LPG carriers face limitations for CO_2 transport, due to CO_2's higher density (1,050–1,200 kg/m^3 vs. 550–700 kg/m^3 for LPG) and low maximum design pressures (≤0.8 MPa). Smaller carriers, operating at 1.1–1.9 MPa, could transport 2,000–3,000 tons of CO_2 at medium pressures (2020-10, IEA).

Overall, the unique demands of CO_2 transportation, especially at cryogenic conditions, require careful material selection and qualification. Balancing cost, operational efficiency, and durability remains a significant challenge, underscoring the need for further research and testing, particularly for materials exposed to impurities and harsh storage conditions. These efforts will support the development of safe, efficient, and economically viable CO_2 shipping systems.

5.2.2 Classification Societies on Carbon Capture Systems

A number of Classification Societies have also delved into the development of guidelines and recommendations for Onboard ships Carbon Capture and Storage (OCCS) systems. Bureau Veritas (BV) (2023) has specified the processes for CO_2 capture, compression, liquefaction, and storage. These regulations outline required documentation, safety aspects, system design, construction, installation, and certification processes. Similarly, Lloyd's Register (LR) incorporated rules under "Emissions Abatement Plant for Combustion Machinery", detailing the design, construction, and installation surveys of Emissions Abatement Carbon Capture and Storage (EACSS) systems, including ship preparations for EACSS installation (LR, 2023). The rules also cover ship preparations for EACSS installation. The American Bureau of Shipping (ABS) has mandated the EGC-OCCS notation for ships with OCCS systems and introduced the optional notation EGC-OCCS Ready, indicating ships suitable for future OCCS installation (ABS, 2022). ABS's requirements encompass aspects such as system arrangement, CO_2 compression, liquefied CO_2 storage, and monitoring/control systems.

DNV has also published guidelines addressing various aspects of carbon capture processes including exhaust pre-treatment, absorption, liquefaction, storage, transfer systems, solidification, and separation (DNV 2021). The guidelines focus on absorption, desorption, and solidification methods using amines/chemicals, allowing flexibility for specific system designs based on separate evaluations. With regards to ClassNK they have developed a number of guidelines regarding new technologies, and more specifically the ones for shipboard CO_2 capture and storage for approval in principle and general design approval, ones for technology qualification and finally guidelines for shipboard CO_2 capture and storage systems (ClassNK, 2023). Regarding the latter, the guidelines include the requirements for the CO_2 capture systems and associated equipment and the CO_2 storage systems and associated equipment. Overall, it can be said that, in an effort to provide clear guidance and way forward of the onboard CO_2 systems, Classification Societies have provided a thorough path for assessment of new technologies and their components including the ways these will be integrated especially within a marine environment and onboard a ship, either at existing or new-built stage.

5.2.3 Onboard CO_2 Capture Industry and Research Projects

The project Realising Maritime Carbon Capture to demonstrate the Ability to Lower Emissions (REMARCCABLE) has been granted Approval in Principle by ABS to test an onboard carbon capture system on a medium range tanker of Stena Bulk (McLellan, 2023). It should be mentioned that the first phase of the project was initiated in 2022. During the sea trials, 500 h of operation of the CCS system was expected to be performed, from burning high-sulfur fuel oil or very-low sulfur fuel oil to the offloading of the captured CO2 and liquefaction at ports along the route of a 10-day deep-sea voyage. The Stena Bulk was expected to retain the carbon capture system onboard the Stena Bulk and extend its use beyond this pilot project too.

Another industry application of onboard CO_2 capture is the Carbon Capture on the Ocean (CC-Ocean) project initiated by Mitsubishi Shipbuilding Co. Ltd in cooperation with Kawasaki Kisen Kaisha Ltd and the support of the Japan's Ministry of Land, Infrastructure, Transport and Tourism (ClassNK, 2022). The 88,000 tons type bulk carrier (coal carrier) Corona Utility, registered in ClassNK, was chosen as the ship to carry out the world's first demonstration of a CO_2 capture system within a commercial setting. The demonstration lasted for two years, after a HAZID approach had been carried out of the demo plant. Following 6 months of operation, the amount, ratio, and purity of the captured CO_2 were all in line with the original plan, validating the feasibility of capturing CO_2 from the flue gas of marine engines onboard ships.

The EverLoNG project is another industry effort to assess the uptake of onboard carbon capture systems, demonstrating their use on board LNG fueled ships and moving this technology closer to market readiness. In this case, an LNG-powered LNG carrier was equipped with an onboard carbon capture prototype, under the EverLoNG project

(Ovcina Mandra, 2023). Following the installation of the prototype, the aim is to have the system operating during a 3000-h test campaign. After the trial on the LNG carrier, the prototype equipment will be removed and installed on an LNG-powered crane ship and will conduct 500 h of CO_2 capture operations. As a result, the comparison of the system performance on both ships will be able to provide a holistic picture of its capabilities. The goal is to achieve a 70% reduction in CO_2 emissions onboard.

Another interesting application of onboard CO_2 systems is the delivery of an AiP by BV on a CO_2 capture system installed onboard two bulk carrier ships (BV, 2023). Initially, the technical feasibility of the onboard CCS system was validated compared to the IMO's Carbon Intensity regulation. Two bulk carriers were chosen for the initial application; Tianjin Venture of 53,000 dwt and CSSC Wan Mei that reaches 176,000 dwt. The system consists of an organic amine solution which extracts CO_2 from exhaust flue gas prior to being liquefied via cooling and stored in a low temperature storage tank. Laboratory tests demonstrated a rate of CO2 capture over 85%.

Furthermore, Lomarlabs (2024) presented sea trials of a carbon capture system on board a Lomar container vessel testing the Seabound technology capturing roughly one tonne of CO_2 per day. The prototype system was retrofitted on the aft deck of the ship next to the exhaust funnel, transforming CO_2 emissions into solid calcium carbonate pebbles. Sea trials of the new system lasted for two months reaching a level of 78% CO_2 capture working in conjunction of removing sulfur emissions acting as a dual-purpose scrubber. Overall, the test successfully captured roughly one ton of CO_2 per day in the prototype system, demonstrating the feasibility of this technology.

At a further research level, the Green Marine (2024) project was initiated in February 2023 to examine the feasibility, test and assess the installation of technologies related to carbon capture onboard ships. Two technologies are suggested for incorporation onboard existing vessels; the first one is based on a membrane- CO_2 capture system; while the second one is a carbon capture solution based on an alkaline solution with Ca and Mg from seawater. This partially funded EU/Innovate UK project aims to initially implement the above-mentioned technologies on a land-based setting simulating the marine environment. The second step of implementation will see the technologies installed and tested onboard a car/passenger ferry operating in Scotland.

Value Maritime (2024) provides another example of applied research related to carbon capture and storage. In this respect, Value Maritime (VM) and Japanese shipping company Mitsui O.S.K. Lines, Ltd. (MOL) will spearhead the implementation of an Exhaust Gas Cleaning System (EGCS) and Carbon Capture feature for an LR1 Product Tanker. The suggested application will involve a 15MW EGCS Filtree system with the addition of a Carbon Capture unit with the aim of capturing 10–30% of CO_2 emissions produced onboard.

5.2.4 Hydrogen–Pipeline Transport

There is currently considerable interest in using hydrogen as a fuel to facilitate the clean energy transition. It is expected that hydrogen from both green and blue sources will play a role in this transition, and pipelines will most likely be the most viable solution for transport to end-users. There is a great interest by the energy sector to look at both new and re-purposed steel pipelines. Use of pipelines for storage purposes are a viable option among options like pressure vessels and geological storage of hydrogen (Ma et al., 2024). From a transport perspective it is desirable with high transport pressures as hydrogen is less energy dense than natural gas (1/3 the energy). However, it is challenging to document adequate steel material toughness and fatigue properties for relatively high H2 pressures, and using large diameter seam welded pipelines for storage will introduce high fatigue cycling from filling and emptying which may be detrimental from a fatigue point of view. It is well recognized that hydrogen may promote hydrogen embrittlement in steel (Østby et al., 2023; Ronevich et al., 2021), which can have an adverse effect on the pipeline integrity. In H2 environment the crack growth in steel is severely accelerated compared to that of air environment, and hence the fatigue capacity reduces. Some general observations are shown in Fig. 15 (Østby et al., 2024).

- For low ΔK, , the fatigue crack growth (FCGR) in air and H2 appears to be similar.
- At a certain ΔK the FCGR in H2 starts to accelerate quickly. This is typically observed in the ΔK range of 5 to 10 MPa$\sqrt{}$m.
- The rapid increase in FCGR continues for a limited increase in ΔK, typically up to 10–20 MPa$\sqrt{}$m, after which the slope of the curves changes again.
- After this the continued crack growth takes place with a slope of the Paris curve, which is similar (in the log-log plot) to the in-air curve, however, with up to 30–40 times the crack growth rate per cycle.

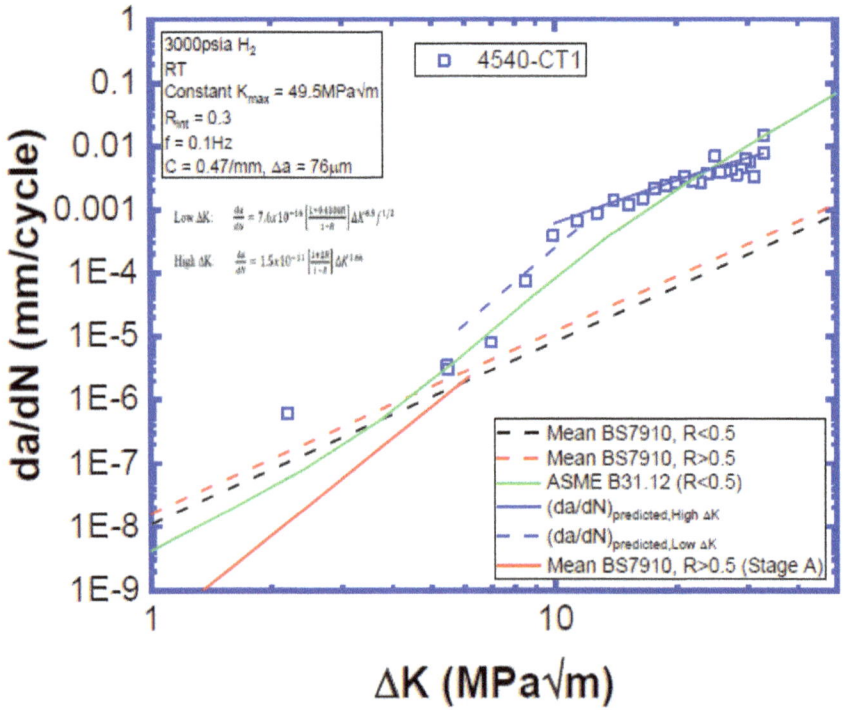

Fig. 15. FCGR curve obtained for an X65 base material at 210 bar H_2.

The onset of rapid crack growth as seen in Fig. 15 is dependent on the magnitude of the H2 pressure, fugacity and the R-ratio (R = σmin/σmax) and this has led to a proposal of general curves for FCGR in H2, e.g. the curves proposed by (San Marchi and Ronevich, 2022), also known as the Sandia crack growth curve:

$$\text{for high } \Delta k : \frac{da}{dN} = 1.5 \times 10^{-11} \left[\frac{1+2R}{1-R} \right] \Delta^{3.66} \quad (1)$$

$$\text{for low } \Delta k : \frac{da}{dN} = 7.6 \times 10^{-16} \left[\frac{1+0.4286R}{1-R} \right] \Delta^{6.5} * f^{0.5} \quad (2)$$

The hydrogen environment does not significantly affect the shape of the stress-strain curve up to the UTS as seen in Fig. 16a (Østby et al., 2023). Both specimens tested at 20 and 200 bar H2 exhibit a strongly reduced area (RA) compared to the inert N2 references, and the RAR (relative area reduction, i.e. the ratio between the RA for the H2 and N2 case) comparison is shown in Fig. 16b. The post-test fracture surfaces support this finding shown in Fig. 16c, where much larger fracture surface, i.e. less necking before failure for the specimens tested in 200 bar compared to the 2 bar hydrogen and N2 tests.

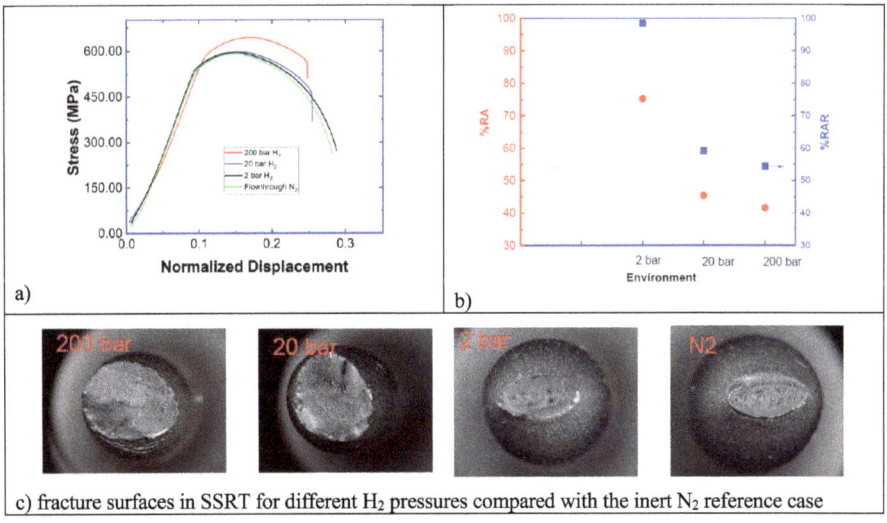

c) fracture surfaces in SSRT for different H$_2$ pressures compared with the inert N$_2$ reference case

Fig. 16. The effect of H$_2$ pressure on the a) shape of stress-strain curve and b) reduction of area, for samples tested at 10^{-6} s^{-1} (Østby et al., 2023).

The fracture toughness is significantly reduced in hydrogen environments. There are currently uncertainties regarding the definition and measurements of the fracture toughness in hydrogen environment and dedicated fracture toughness testing under representative conditions should be carried out, and both rising displacement testing, and constant loading testing may be needed to document the material toughness and to verify that static crack growth is not taking place. The rate at which a load is applied to a material influences how hydrogen affects its performance. With slower load increases, hydrogen has more time to weaken the crack tip, reducing fracture toughness and potentially causing significant crack growth.

In the ongoing DNV JIP H2 pipe guidance on how to derive a toughness value that can be input to an engineering critical assessment to define weld flaw acceptance is provided in the DNV -RP 123 «Hydrogen pipeline systems – supplementary recommendations» which is currently under review by the JIP sponsors.

Li et al., (2024) investigated the influence of specimens' size of Single Edge Notch Tensile (SENT) specimens tested in air and electrochemical charging to simulate hydrogen environment of ferritic welded steel. They found that the fracture toughness decreased for the thicker specimens, see Fig. 17 for the Crack Moth Opening Displacement curves (CMOD). The obtained CTOD in air for the three specimen geometries (B/W = 0.5, B/W = 1 and B/W = 2) in air was 0.85 mm, 0.81 mm, 0.76 mm respectively while in hydrogen charged environment the values dropped to 0.12 mm, 0.07 mm and 0.045 mm respectively, hence a significant drop in toughness.

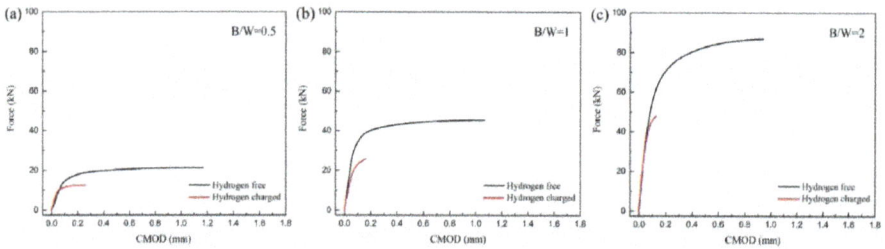

Fig. 17. Load–CMOD curves of weld metal SENT specimens during fracture toughness tests obtained from lower clip gauge.

General wall thickness design according to e.g. ("DNV-ST-F101 Submarine Pipeline Systems," 2021) is provided to prevent burst without accounting for weld defects. However, due to reduced ductility and increased crack growth in H2 environment, allowable pressure may be governed by the weld defects in the pipelines. In a fracture mechanics assessment, several parameters are needed as input, one of them being the weld residual stresses. BS7910 recommend using 100% of the yield stress of the material, and this has generally been found feasible for natural gas pipelines. However, using the same approach for evaluating weld flaws in H2 pipelines will lead to very low utilization (reduced pressure) due to the combination of low toughness and residual stresses, this is especially a showstopper for re-purposing of existing pipelines. Hence, it is recommended to study the influence of residual stresses on the fracture resistance of pipelines transporting H_2.

There are different initiatives worldwide related to research on the use of hydrogen gas. Some of them are provided below:

DNV H2Pipe JIP is managed by DNV. The JIP performs both small-scale testing in hydrogen environment and looks into applicable subsea pipeline integrity assessment procedures (Østby et al., 2023). The research project is aiming to develop a new code for the design, re-qualification, construction and operation of offshore pipelines to transport hydrogen – either pure or blended with natural gas. The DNV-RP-123 on "Hydrogen pipeline system-supplementary recommendations" is planned to be sent to hearing early 2025. This document will provide supplementary recommendations to DNV-ST-F101 Submarine Pipeline Systems.

The HyLINE project is managed by SINTEF and is an ongoing research project that started in 2019. The work carried out in HyLINE has mainly been based on testing in simulated cathodic protection (CP) environments and testing in actual hydrogen environments has now been started, ("HyLINE - Safe Pipelines for Hydrogen Transport," 2019).

The European Pipeline Research Group (EPRG) is currently performing several activities looking into pipelines for hydrogen transport. EPRG have recently released a white paper with proposal for frameworks for assessment associated to possible repurposing of pipelines for hydrogen transport, including a quite extensive list of gaps ("EPRG Hydrogen Pipelines Integrity Management and Repurposing Guideline White Paper," 2023). There are also both small- and large-scale testing being performed by EPRG. The large-scale testing is considering pressure cycling of a pipe with pre-exiting flaws.

DVGW (Deutscher Vereins des Gas- und Wasserfaches) has performed a series of fatigue and fracture toughness test in hydrogen environments for different pipeline materials applied onshore in Germany.

There are also other major initiatives. In the US both Sandia and NIST have dedicated activities looking into effect of hydrogen environment on C-Mn steel.

ASME B31.12 - Hydrogen Piping and Pipelines was updated in 2023. The update was mainly editorial. PRCI in US is currently involved in a research project with the goal to re-write ASME B31.12, moving the pipeline part to ASME B31.8.

5.2.5 Hydrogen – Shipping

Long-term solutions to minimize shipping emissions include hydrogen as a promising option among the alternative fuels. However, storing large quantities of hydrogen onboard ships in harsh marine environments poses several risks. While research exists on hydrogen storage and fuel cells, there is limited understanding of material selection and failure mechanisms for storage tanks and insulation methods for marine applications.

Shipping hydrogen in its gaseous state is considered the simplest method due to low energy consumption for compression, which requires about 1.1 kWh/kg to compress hydrogen from 2 to 25 MPa. This makes it the most energy-efficient option compared to other methods. The infrastructure needed is minimal, since compressed hydrogen can be directly injected into ships, simplifying logistics. However, a significant challenge is its low volumetric storage density, meaning that compressed gaseous hydrogen (CGH2) occupies more space compared to other forms of hydrogen storage, limiting the amount transported per shipment. Despite challenges like low storage density, compressed hydrogen shipping could play a vital role in regional hydrogen supply chains, particularly within shorter shipping distances.

Global Energy Ventures Ltd (GEV) is developing the world's first compressed hydrogen cargo ship, the C-H2, equipped with two cylindrical tanks, each 12 m in diameter, operating at 25 MPa, capable of transporting 430 tons of hydrogen. A demonstration project is planned for 2023, with commercial use aimed for 2025. GEV also plans to develop a larger ship with two 20-m diameter tanks to transport 2,000 tons of hydrogen. GEV has identified that CGH2 shipping is competitive for distances up to 2,000 nautical miles (3,700 km), such as between Australia and Singapore, and remains competitive for routes to Japan, South Korea, and China (d'Amore-Domenech et al., 2023).

Liquid hydrogen (LH2) shipping has gained traction due to its high volumetric density and the ability to leverage existing cryogenic technologies. The technology for LH2 shipping draws from liquefied natural gas (LNG) shipping methods, which already manage cryogenic transport at -162 °C. With many large-scale LNG terminals in Europe—37 operational and 27 more planned by the end of 2021—existing infrastructure can be upgraded for LH2 import and export due to the compatibility of facilities.

LH2 shipping has advantages over other hydrogen carriers like ammonia and liquid organic hydrogen carriers (LOHCs). Liquefaction plants, located on the exporter's side where cheap renewable energy is available, minimize costs. Additionally, minimal energy is needed on the importer's side for storage and regasification, providing high-purity hydrogen directly without further purification. However, LH2 shipping also faces challenges, notably the issue of boil-off loss due to the cryogenic conditions required. Since re-liquefying evaporated hydrogen on the ship is not currently feasible due to the space and equipment needed, excess vapor must be vented into the atmosphere once it exceeds tank pressure limits. To address this, current methods involve using the evaporated hydrogen as a fuel source, either by converting it to electric power using fuel cells or burning it directly in engines adapted from LNG ships. These methods help manage boil-off losses but limit the maximum sailing time and distance of LH2 ships. The Joint Research Centre (JRC) of the European Commission has determined that LH2 shipping is ideal for distances between 2,500 and 16,000 km, such as from Saudi Arabia to Rotterdam.

The HySTRA project in Japan has a framework in place where it's been reported that the Suiso Frontier, with a weight of 8000 tonnes, is able to transport large quantities of LH2 over extended distances via the sea. The LH2, which is reduced to a volume of just 1/800th of its original gaseous state, is cooled to a frigid -253 °C. The HyShip project aims to develop and validate approaches to build (3 MW) and scale (up to 20 MW) fuel cell systems for ships using liquid hydrogen (LH2) as fuel, with the goal of lowering operational and design costs. The project will integrate technical solutions into a larger socio-technical system to create the first European maritime supply chain for LH2. This will involve building two sister ships that will connect a new hydrogen production facility with LH2 demand in a series of vessels. The project will use state-of-the-art ship design and intelligent energy management systems to lower costs and generate value for Europe by creating a scalable distribution system for Certified Green H2.

LH2CRAFT project aims to develop the technology for long-term storage and long-distance transportation of LH_2 on ships by developing new design solutions for storage at 20 K and demonstrating it on a 180 m^3 containment system. The project entails the design, manufacturing, and prototype demonstration of approximately 10 tons of LH2 stored within a membrane-type tank, alongside the development of a large-capacity vessel, exemplified by a 200,000 m^3 LH2 storage solution.

NASA and four industry partners (Boeing, ATK, Lockheed Martin and Northrup Grumman) have been advancing composite materials and processes to reduce the overall weight and cost of LH2 tanks via the Composite Cryotank Technology Demonstration (CCTD) Project. Composite LH2 tank technology continues to be an active research area to face the challenges of metallic tanks (Olsson et al., 2024),

Shipping hydrogen as ammonia is another option due to less stringent pressure and temperature requirements. Ammonia transport at atmospheric pressure and −33 °C or at 25 °C and 10 bar is well-established, with about 17.5 million tons shipped annually using 170 ships and extensive infrastructure. However, converting hydrogen to ammonia is energy-intensive, and cracking ammonia back to hydrogen requires additional energy (~20% of hydrogen's lower heating value) and further separation of nitrogen and hydrogen. Due to these energy costs, ammonia-based hydrogen shipping is currently less attractive, with development plans still largely theoretical.

Shipping hydrogen using liquid organic hydrogen carriers (LOHC) is an alternative to the former high-pressure or low-temperature transport methods. LOHCs, like H18-DBT, can carry hydrogen through hydrogenation and release it through dehydrogenation, offering easier handling and higher hydrogen density than compressed gas (CGH2). With 56 kg/m^3, LOHC can transport up to 78% of the amount that liquid hydrogen (LH2) can. LOHCs have properties similar to diesel, enabling use of existing transport infrastructure. However, dehydrogenation requires significant energy (9 kWh/kgH2), increasing costs on the import side. The absence of boil-off and safer storage benefits LOHC shipping, but drawbacks include additional fuel systems, back-transport of depleted LOHC, and potential health hazards. While LOHC could be economically favorable for long-distance transport, no large-scale LOHC shipping systems currently exist, requiring further technological advancements (Yang et al., 2023).

Comparing the different hydrogen transport technologies, it can be easily derived that LH2 shipping allows the transport of large energy amounts, with liquefaction done where energy is cheaper, such as the exporter's side. However, it faces challenges like cryogenic storage, boil-off losses, and high initial costs for infrastructure. CGH2 shipping requires no conversion, but its low volumetric density makes it less favorable for long distances. Ammonia and LOHC offer more convenient storage and handling conditions, reducing initial shipping costs. However, both require significant energy for dehydrogenation. Ammonia is more versatile than LOHC because it can release hydrogen or be directly used as fuel or feedstock in the chemical industry.

Handling and storing hydrogen at cryogenic temperatures on ships presents several hazards, primarily related to fire safety due to hydrogen's wide range of ignition limits in air (4.1–74.2 vol%). Fire or explosions may result from leaks or spills, mechanical or thermal ignition, and material failures, particularly if compressed gaseous hydrogen (GH2) is stored. Key challenges include hydrogen permeation, embrittlement, and heat transfer issues. Increased temperatures can raise the permeability of storage tank materials, leading to leaks, potential fires, and micro-cracks. Hydrogen embrittlement reduces the fracture toughness of metals, causing premature failures. Heat transfer is especially critical for liquid hydrogen (LH2) storage due to the extreme temperature difference between LH2 (20 K) and the ambient environment (300 K), leading to boil-off gas that must be safely managed. Excessive boil-off can cause fires and reduce ship range. Spherical tanks minimize heat transfer by reducing surface area, but heat transfer may also cause thermal expansion, potentially separating composite cryogenic tanks from metallic liners. LH2 and cryo-compressed hydrogen (CcH2) storage methods have 6 to 15 times lower adiabatic expansion energy than compressed GH2, offering potentially higher safety in the event of sudden pressure vessel failure (Depken et al., 2022).

5.2.6 Materials and Insulation for Storing Cryo-Compressed and Liquid Hydrogen

Due to the extremely low temperature of liquid hydrogen (-253 °C), special materials are required for storage and transportation containers. The main requirements are the adaptability of materials in a liquid hydrogen environment, resistance to hydrogen embrittlement, mechanical properties, and thermophysical properties at cryogenic temperatures.

Stainless steel is the most widely used material for liquid hydrogen vessels in cryogenic applications. Austenitic stainless steels are generally preferred for liquid hydrogen transportation due to their behavior at cryogenic conditions. The face-centered cubic crystal structure of stainless steel presents excellent plastic deformation ability. As temperature decreases, these steels exhibit increased strength, while maintaining adequate plasticity and impact resistance. Adding higher levels of nickel (Ni) and chromium (Cr) enhances their stability at low temperatures. The 300-series Cr-Ni austenitic stainless steels are extensively used for storing cryogenic liquefied gases due to their superior overall properties.

The alloy elements of stainless-steel alloys directly guides their applications, for instance, 316L stainless steel, with high molybdenum (Mo) content, presents excellent resistance to chloride-induced corrosion, making it suitable for marine environments. Meanwhile, 321 stainless steels, containing titanium (Ti), ans provides resistance to intergranular corrosion and higher strength at elevated temperatures, making it ideal for high-corrosion and high-heat conditions. Table 3 summarizes metallic materials suitable for cryogenic and hydrogen applications.

Metals exposed to cryogenic temperatures typically present increased elastic modulus, tensile strength, yield strength, and fatigue performance. However, their plasticity is significantly influenced by ductile-brittle transition behaviour at low temperatures. Materials with Body-Centered Cubic (BCC) structures often present a sharp reduction in plasticity below their ductile-brittle transition temperature, making them prone to brittle fractures and unsuitable for cryogenic conditions. In contrast, materials that do not undergo a brittle transition generally show improved elongation as temperatures decrease.

To evaluate material suitability for cryogenic environments, common tests include low-temperature impact toughness tests, drop weight tests, full-thickness tests, and fracture mechanics tests. These assessments are performed on base metals, welds, and heat-affected zones (HAZs), as weldments often exhibit weaker performance than base metals.

In particular, the tensile and yield strength of stainless steel increase as temperature decreases, while its plasticity slightly reduces, as shown in Fig. 18 and Fig. 19. This reduction in plasticity is not severe, allowing stainless steel to retain sufficient ductility at cryogenic temperatures. Materials that exhibit significant brittle fracture at low temperatures are unsuitable for cryogenic applications due to the increased risk of crack

Table 3. Materials suitable for liquid hydrogen service (Ustolin et al., 2022)

Material	Alloy Grade	Remarks
Aluminium alloys	2029	Suitable for aerospace applications
	2219	Suitable for aerospace applications
Titanium alloys	-	Elongation, toughness, and fracture toughness decrease at cryogenic temperature
Copper Alloys	-	Highly ductile and toughness at cryogenic temperature
Austenitic stainless steels	304	Susceptible to hydrogen embrittlement
	310	Not susceptible to hydrogen embrittlement
	316	Not susceptible to hydrogen embrittlement
	316L	Suitable for the marine environment
	321	Highly resistant to corrosion

initiation and fracture. Materials with face-centered cubic (FCC) structures, such as austenitic stainless steel and aluminum alloys, do not experience brittle fractures at low temperatures, i.e. 316 stainless steel exhibits increased fracture toughness down to 77 K (-196 °C), with a slight decrease at 20 K (-253°K), yet remaining higher than at room temperature. In contrast, body-centered cubic (BCC) materials like martensitic steel exhibit decreased plasticity and brittle fractures as temperature declines. Materials with hexagonal close-packed (HCP) structures, such as titanium alloys, fall between these extremes but the low-temperature plasticity can be improved employing compositional and microstructural adjustments. Obtaining the tensile properties of various materials at 20 K (-253°K) is crucial for the design of storage and transportation containers. Therefore, conducting systematic and comprehensive testing of tensile properties at ultra-low temperatures (≤ -196 °C) is essential.

For cryo-compressed (CcH2) and liquid hydrogen (LH2) applications, four hydrogen storage tank types are approved: Type I, II, III, and IV. Type I tanks, made entirely of metal, are the most technically mature method for CcH2 storage. Type II tanks consist of a metallic liner hoop-wrapped with lightweight fiber-reinforced composite material using the filament winding method. Type III tanks, similar to Type II, are fully wrapped with fiber-reinforced composites and are suitable for storage pressures under 350 bar, with development ongoing for 700 bar. Type IV tanks are constructed with a polymeric liner fully wrapped in fiber-reinforced composite; common liner materials include polyethylene and polyamide and Type V tanks are linerless composite tanks, (Cheng et al., 2024) (Olsson et al., 2024).

Maintaining liquid hydrogen (LH2) at cryogenic temperatures (-253 °C) presents challenges like heat leakage, which causes partial evaporation of LH2, leading to boil-off gas (BOG) and complex thermal behaviors in storage tanks. Research efforts are directed toward developing advanced insulation materials to mitigate these issues (Yin et al., 2024).

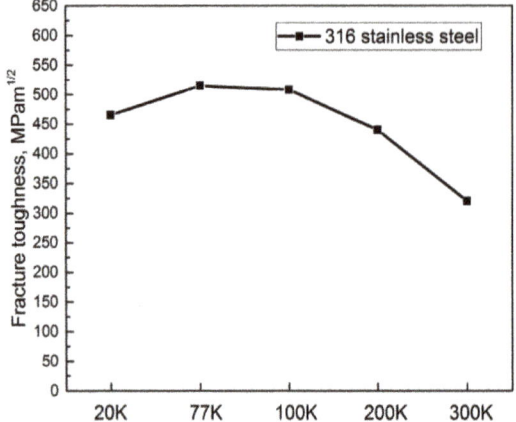

Fig. 18. Cryogenic fracture toughness of 316 stainless steel (Qiu et al., 2021)

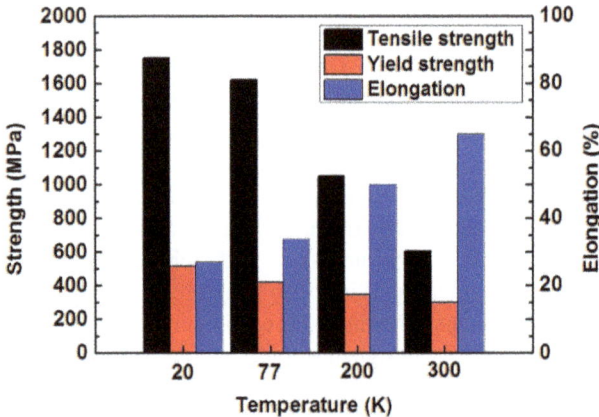

Fig. 19. Cryogenic properties of 18Cr-8Ni stainless steel (Qiu et al., 2021)

In particular, insulating cryo-compressed (CcH2) and liquid hydrogen (LH2) storage tanks is essential for safe and economical operations, to maintain cryogenic temperatures, minimize hydrogen release, and prevent condensation or solidification of atmospheric gases that could cause corrosion. Most existing experience comes from land-based applications, but these methods are also applicable to marine environments. Three main heat transfer mechanisms are identified in cryogenic hydrogen storage and transfer: radiation heat transfer, contact heat transfer, and convection. Radiation heat transfer is proportional to the emissivity of the insulation shields and inversely proportional to the number of layers, making multilayer insulation (MLI) a common choice for CcH2 and LH2 tanks. MLI typically consists of 30–80 layers of materials like aluminum foil, Mylar, and low-conductivity spacers, offering thermal conductivity up to three orders of magnitude lower than fiberglass insulation.

A vacuum between the tank's outer and inner shells further reduces conduction and convection. However, maintaining a high vacuum level is challenging due to deformation of the jacket vacuum system; one solution involves using low-density polymer foams. Variable-density multilayer insulation (VDMLI) combines spray-on foam insulation (SOFI) with variable-density layers to improve thermal performance, where lower-density layers are inside and higher-density layers outside. VDMLI with a vapor-cooled shield (VDMLI + VCS), which uses self-evaporating gases from LH2 containment to absorb heat, significantly reduces heat flux. The VDMLI + VCS method achieves the lowest heat flux (0.14 W/m^2) when the VCS is at 30% of the distance between the tank and ambient environment. It also reduces hydrogen boil-off losses by over 58% compared to MLVI in a 45-day test and is 74% lighter than MLVI.

Other insulation methods, like microsphere insulation (MI), MLI with vapor-cooled shields (MLI + VCS), and VDMLI with VCS, offer varying levels of performance. The MLI + VCS method, for example, reduces heat flux by 16% compared to VDMLI and over 50% daily hydrogen boil-off losses compared to MLVI. Despite the higher initial cost, the low heat flux and reduced hydrogen losses of VDMLI + VCS could offset costs over long-term operations (Yin et al., 2024).

A numerical model was developed by D.-H. Kang et al. (2024) that represents the physical shape of a spherical-shaped hydrogen storage tank to reduce insulation thickness. The optimal placement of VCS insulation was identified, highlighting its integration directly onto the solid insulation. This integration resulted in a significant reduction in the number of MLI layers required—64.2% at a vacuum pressure of 5 Pa and 53.6% at 1 Pa. Additionally, when MLI and VCS replaced solid insulation, the total insulation thickness was reduced by 51.4% at a vacuum pressure of 1 Pa (Fig. 20).

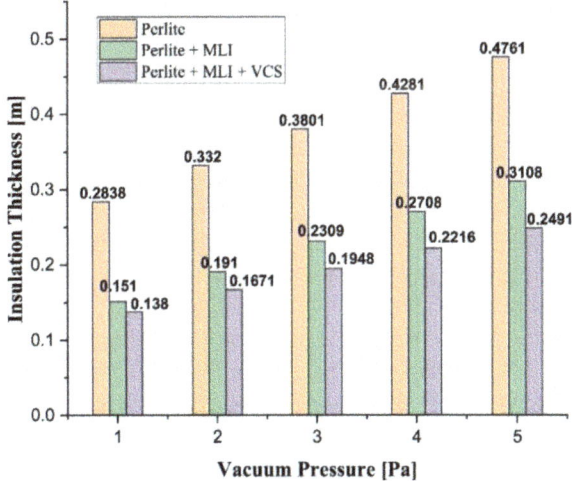

Fig. 20. VCS can effectively decrease the insulation thickness as the vacuum pressure increases (Kang et al., 2024).

While most hydrogen insulation techniques are developed for aviation and land vehicles, they have potential for marine applications. Materials like polyurethane foam, used in LNG carriers and spacecraft, and phenolic resin foam panels, could also insulate marine hydrogen tanks. Existing methods can reduce heat fluxes significantly, but advanced insulation like VDMLI + VCS remains superior, offering very low heat flux (0.1–10 W/m^2) (Liu et al., 2023).

In conclusion, it is necessary to carry out systematic and comprehensive testing of ultra-low temperature tensile properties to ensure the safe, efficient, and practical use of materials in cryogenic applications. These tests provide critical insights into how materials perform under extreme conditions, such as those encountered in liquid hydrogen (LH2) storage and transportation systems, which involve challenges like weight constraints, safety concerns, and the need for innovative material solutions.

5.2.7 Cryogenic Heat Transfer for Brittle Failure Analysis of Marine/offshore Structures Subject to Accidental Cryogenic Spill

Material embrittlement at low temperatures, particularly in steel, has been well-documented for decades (Fig. 21). Beyond operations in low-temperature environments, such as Arctic waters, accidental spills of cryogenic liquids like LNG can cause a rapid and significant drop in temperature, necessitating thermomechanical analysis for integrity assessments. Accurate modeling of the heat flux curve is critical in such analyses to predict temperature distribution and thermal stresses in marine structural steel. Mokhtari et al. (2022b) developed a heat flux curve for EH36 marine steel under liquid nitrogen (LN$_2$) pool boiling conditions using finite element thermal analysis (FETA), based on the experiments conducted in (Nam et al., 2021). LN$_2$ was used instead of LNG due to its similar cryogenic properties, ease of handling in experimental setups, and safety considerations, while still providing valuable insights into the thermal behavior relevant to cryogenic spills.

The experimental study in (Nam et al., 2021) established a basis for understanding the heat transfer characteristics between LN$_2$ and EH36 steel. During the pool boiling tests, several thermocouples were placed on the steel specimens to record the temperature histories (Fig. 22). The tests captured key boiling regimes, including film boiling, transition boiling, and nucleate boiling, and helped estimate critical points like the Leidenfrost point and the Critical Heat Flux (CHF). The recorded data from these tests provided the groundwork for further computational analysis aimed at refining the heat flux curve in (Mokhtari et al., 2022b).

The FETA results in (Mokhtari et al., 2022b) led to the development of the new heat flux curve for LN$_2$-EH36 steel interactions (Fig. 23). This curve differed from conventional boiling curves reported in the literature (e.g., (Barron and Nellis, 2017)) by predicting an earlier transition from film boiling to nucleate boiling, which resulted in a higher Leidenfrost point. This shift was attributed to the specific characteristics of the EH36 steel surface, including its roughness and crack features, as well as the influence of external conditions like ambient air convection.

The development of this more accurate heat flux curve has significant implications for the Accidental Limit State (ALS) design of marine structures exposed to cryogenic spills. The refined curve provides a more reliable foundation for thermal and structural analysis,

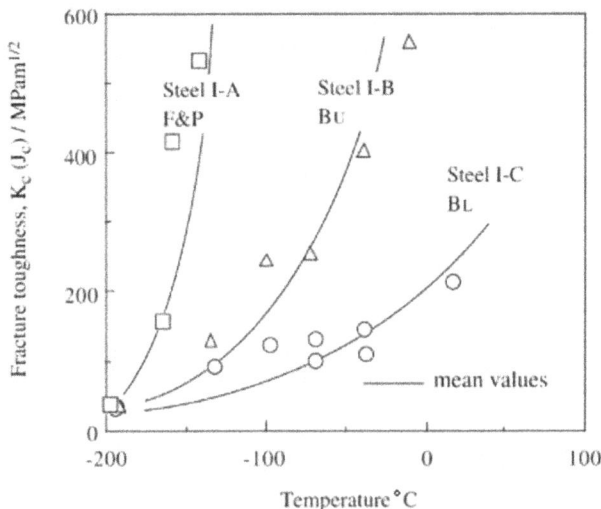

Fig. 21. Significant fracture toughness reduction of low carbon steels at low temperatures (Miyata and Tagawa, 2002).

helping engineers predict the local thermal stresses that can lead to brittle fracture (Nam et al., 2023). This is especially critical in scenarios where marine structures are exposed to LNG spills.

Future research should explore the impact of accidental spillage of liquefied hydrogen (LH_2) on the brittle fracture of marine steels, particularly as LH_2 has a significantly lower boiling point than LNG and LN_2, which may introduce additional complexities in thermal behavior and fracture risks. Additionally, with the growing demand of liquid CO2 (LCO2) transport for CCS, similar studies on its accidental spillage on marine steel are required. While the structural integrity risks may be less severe due to the much higher boiling point of LCO2, understanding its potential impact remains crucial for comprehensive risk assessments.

5.3 Temperature Effect on Materials

5.4 Mechanical Behavior of Materials Under Elevated Strain Rates

Both man-made and natural materials exhibit altered mechanical behavior under elevated strain rates compared to quasi-static loading conditions. In marine engineering, especially in design and operation, high loading rates for widely used man-made materials are typically associated with accidental scenarios such as collisions and explosions. However, for natural materials like ice, rate-dependent mechanical behavior can occur even under normal loading conditions due to their viscoelastic nature. The following sections discuss these phenomena within the context of typical design scenarios for marine structures.

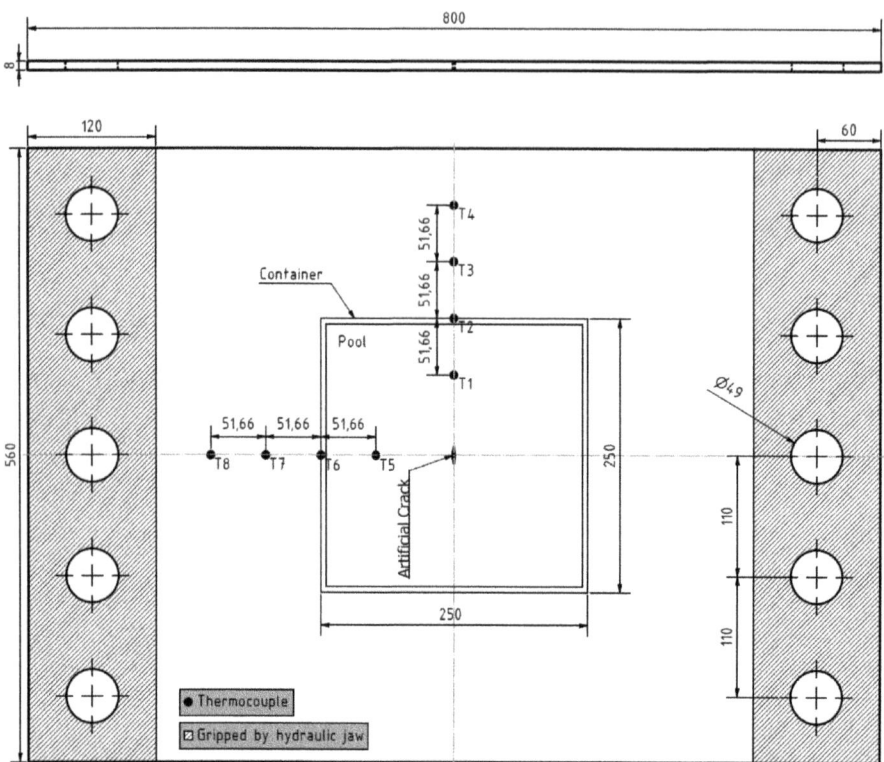

Fig. 22. Dimensions of the specimen and the container as well as the location of thermocouples.

5.4.1 Steel and Aluminum Alloys

Steel and aluminum alloys are the most used materials in marine applications that could undergo high strain rates in accidental scenarios. They are also heavily used in other industries such as automotive, aerospace, and defense, and thus significant research has been conducted on their constitutive modelling. The Johnson-Cook (J-C) and Cowper-Symonds (C-S) models, both plasticity-based, are widely recognized for simulating the strain-rate dependent mechanical behaviors of these and other alloys.

The J-C model (Johnson and Cook, 1983) accounts for strain hardening, strain rate hardening, and thermal softening, and defines the von Mises flow (yield) stress, s_y, by.

$$s_y = \left[a + b(\bar{\varepsilon}^p)^n\right]\left[1 + c\ln\left(\frac{\dot{\bar{\varepsilon}}^p}{\dot{\varepsilon}_0}\right)\right]\left[1 - \left(\frac{T - T_0}{T_m - T_0}\right)^m\right] \quad (3)$$

a Initial yield stress
  b Work (strain) hardening modulus of the J-C model
  c Strain rate hardening coefficient of the J-C model
  m Thermal softening exponent
  n Work hardening exponent
  T Temperature

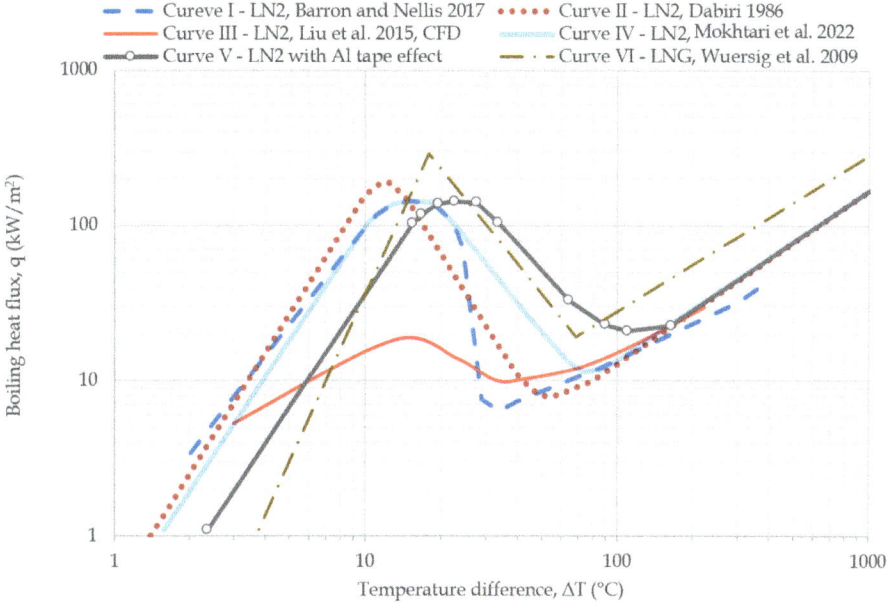

Fig. 23. Comparison of LN$_2$ and LNG heat flux curves (Mokhtari et al., 2022b).

T_0 Reference temperature
T_m Melting temperature
$\bar{\varepsilon}^p$ Equivalent plastic strain
$\dot{\varepsilon}_0$ Reference strain rate
$\dot{\bar{\varepsilon}}^p$ Equivalent plastic strain rate

The parameters for the J-C model are typically obtained from stress-strain curves generated through split Hopkinson bar tests, which are designed to measure material behavior at high strain rates. In contrast, the C-S model is more straightforward, with fewer parameters to determine. It is expressed as.

$$s_y = s_{y-\text{static}} \left[1 + \left(\frac{\dot{\bar{\varepsilon}}^p}{d} \right)^{\frac{1}{q}} \right] \qquad (4)$$

$s_{y-\text{static}}$ Yield stress under quasi-static condition
d and q Model parameters

By idealizing the plastic phase of the quasi-static stress-strain curve in a manner similar to the J-C model, the C-S model can be reformulated to resemble the structure of the J-C model more closely such that.

$$s_y = \left[a + b(\bar{\varepsilon}^p)^n \right] \left[1 + \left(\frac{\dot{\bar{\varepsilon}}^p}{d} \right)^{\frac{1}{q}} \right] \qquad (5)$$

Under quasi-static conditions, excluding temperature softening in the J-C model, both J-C and C-S models converge towards the material's quasi-static plastic stress-strain curve, producing increasingly more similar results as strain rates decrease.

Equation (3), widely implemented in general-purpose finite element codes, assumes a separable (multiplicative) form in which strain hardening, strain rate sensitivity, and thermal softening act independently. However, real material behavior often exhibits coupled effects, most notably, strain rate influences heat generation, which in turn affects the yield stress. In strongly coupled thermomechanical processes, such as adiabatic heating at high strain rates, this separability assumption becomes limiting. As a result, Eq. (3) alone without thermomechanical coupling may not accurately capture the material's true mechanical response.

Furthermore, many experimental studies do not account for the fact that the strain rate at a given material point evolves continuously during deformation. It is often approximated from the crosshead speed and gauge length, assuming it remains constant throughout the test. Likewise, validation of the J-C model using split Hopkinson pressure bar tests typically relies on simplified assumptions about uniform strain rate, without directly measuring its temporal variation within the specimen.

The definition of "quasi-static conditions", particularly the yield stress used in Eq. (4), can also be somewhat ambiguous. This parameter is usually obtained from standard uniaxial tensile tests (e.g., ASTM E8) conducted at low strain rates on the order of 10^{-3} to 10^{-4} s^{-1}.

In general, the J-C model is considered more accurate, particularly at higher strain rates or when temperature effects are involved, such as in ballistic impacts or machining. A recent comparative study (W. Liu et al., 2022a) demonstrated that in the electromagnetic forming process of an aluminum alloy sheet, the J-C model outperformed the C-S model at strain rates between 500 s^{-1} and 3500 s^{-1}. The study did not involve any temperature softening, indicating that the J-C model more effectively captures strain rate hardening effects compared to the C-S model, even when there are no considerable changes in material temperature.

The J-C model has also shown accuracy at lower strain rates, as evidenced by its application in simulating an ice drop test on an aluminum stiffened panel with low impact energy (250 J) (Mokhtari et al., 2023a). The strain rate conditions experienced by ships and offshore structures during collisions or natural events such as wave slamming and iceberg impacts are typically moderate, and not the extreme conditions of ballistic impact. Therefore, it is crucial to model material behavior under strain rates that reflect realistic service conditions. For this purpose, Gotoh (2015) proposed applying a rate-temperature parameter $R = T \ln(10^8/\dot{\varepsilon}))$ to obtain the yield stress in the form $s_y = A\exp(B/R)$, where A and B are material constants.

In ice-structure collision simulations, the accuracy of the simulation relies heavily on the material models used for both the structures and, more critically, the ice, since most of the energy is expected to be dissipated through the crushing and other failure mechanisms of ice. However, accurately modelling the behavior of ice remains a major challenge in marine engineering due to its complex and variable nature.

5.4.2 Ice

In the ISSC2018 report, a special committee on Arctic Technology focused extensively on ice-related topics. This committee has now included a summary of the latest research advancements in the material modelling of ice.

Material modelling of ice – specifically sea, glacial, and lake ice – is a challenge somewhat unique to marine/offshore engineering, unlike the modelling of steel and aluminum, which applies across many engineering disciplines. Consequently, ice material modelling is not regulated and has received significantly less research compared to metals and even other material categories. This, combined with the highly complex mechanical behavior and failure mechanisms of ice, as well as its greatly varied microstructure, sizes, and shapes, has made ice modelling particularly challenging.

The main microstructural factors affecting ice mechanical behavior include grain size and type, porosity, and brine volume. Ice is sensitive not only to strain rate but also to temperature, hydrostatic pressure, and load direction (for non-isotropic ice), which are among the primary loading and environmental factors. Under compression and crushing, ice undergoes microcracking, dynamic recrystallisation, and phase transitions, modelling of which requires a deep understanding of both continuum and damage mechanics. Flexural failure of level ice under bending introduces further complexities, often necessitating fracture mechanics approaches, which challenge continuum mechanics based numerical methods like finite element methods. Unlike metals, where strain-rate sensitivity is mainly a concern at high strain rates, ice demands consideration of strain-rate effects even at low strain rates due to its viscoelastic behavior (Fig. 24). Nevertheless, plasticity-based constitutive models are often used to simulate ice crushing (mostly in collision scenarios), typically treating ice as an elastoplastic material. These elastoplastic models are applied in "design-ice" scenarios, where the objective is to approximate ice loads on marine structures during collisions or high-rate loading with computational efficiency.

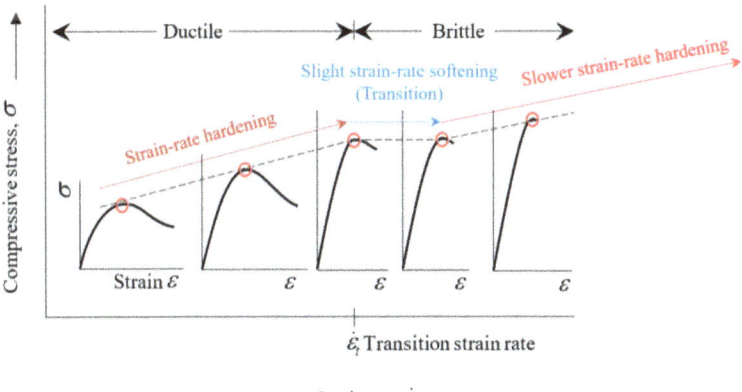

Fig. 24. Schematic illustration of ice stress-strain curve dependency on the strain rate under uniaxial compressive loading condition. Strain rate hardening occurs in the ductile region ($\dot{\varepsilon} < \dot{\varepsilon}_t$), , followed by slight strain rate softening in the transition range, which is then succeeded by another strain rate hardening region but with a lower increase rate (Mokhtari and Leira, 2024).

Given this background, the literature lacks a consensus on the appropriate approach or approaches for ice material modelling in ice-structure interaction (ISI) simulations, and there is no standard/guidance on model selection/development. To address this gap, a critical review was recently conducted (Mokhtari and Leira, 2024), focusing on models used in ice crushing simulations, which significantly affect the results in both pure ice crushing and concurrent ice crushing with bending failure simulations. The review, grounded in established principles of ice mechanics and empirical evidence, provides insights into the advantages and limitations of various models. The key conclusions from this review are listed in Table 4 and summarized in the following paragraphs.

Table 4. Overview of major limitations of the main constitutive models reviewed. "Y" marks where a limitation exists while empty green cells indicate no such limitation for the model (Mokhtari and Leira, 2024).

Major limitations	Elastoplasticity					Viscoelasticity		
	Mohr-Coulomb	Drucker-Prager	Crushable Foam	Liu et al. (2011)	Mokhtari et al. (2023b)	DMa by Xiao (1997)	DMb by Turner (2018)	iDM by Mokhtari et al. (2023c)
Limited to brittle range above transition strain rate	Y	Y	Y	Y	Y			
Strain-rate independent elastic stiffness	Y	Y	Y	Y	Y			
Does not account for the effects of pulverized ice viscous flow	Y	Y	Y	Y	Y			
Strain-rate independent strength	Y	Y	Y	Y				
Does not include pressure softening of ice	Y	Y						
Cannot capture confining/hydrostatic pressure in ice				Y				
Limited to ductile range below transition strain rate						Y		
Inaccurate in simulating progressive crushing of ice						Y	Y	
Complex, parameter-intensive, and computationally expensive						Y	Y	Y

Yield and failure criteria borrowed from rock, soil, and concrete mechanics (a common trend in ISI simulations) including Mohr-Coulomb and Drucker-Prager are not appropriacies in their original forms for capturing pressure softening of ice as shown in Fig. 25. Therefore, when using constitutive models based on these or similar criteria that omit pressure softening, it is crucial to proceed with caution. Their use should be restricted to situations where the hydrostatic pressure in ice remains well below the transition pressure from hardening to softening (i.e., the p range in Fig. 25b, where Mohr-Coulomb and Drucker-Prager surfaces are close to the elliptical surface).

In plasticity-based models, the elliptical yield envelope (Fig. 25) – a simplified version of the Tsai-Wu criterion – has gained popularity for its alignment with physical compression test data, accounting for both pressure hardening and pressure softening in ice. As a result, the Crushable Foam plasticity model, which incorporates this yield envelope, is frequently used to model ice crushing. This model's widespread use is also

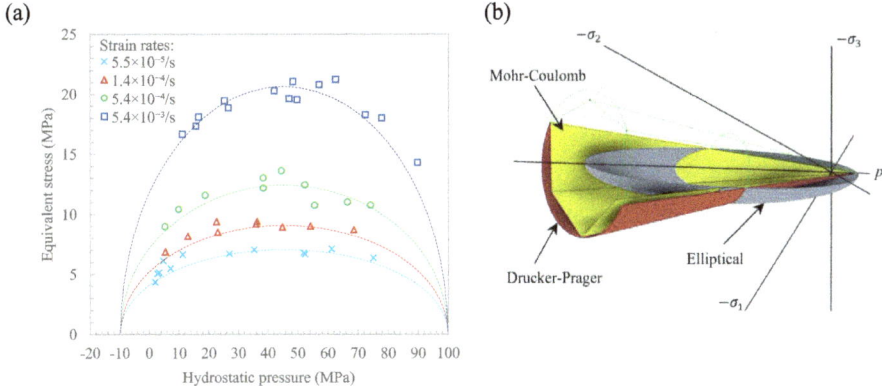

Fig. 25. (a) Ice strength vs. confining pressure data from Jones (1982) for freshwater isotropic ice and the elliptical failure envelope fitting proposed by Derradji-Aouat (2000). Ice strength here is the highest equivalent stress recorded during compressive mechanical testing of polycrystalline cylindrical specimens at $-10\ °C$ (see (Jones, 1982)). (b) Mohr–Coulomb, Drucker–Prager, and the elliptical strength envelopes in the principial stress space. Note: this is a schematic presentation only, to compare the shapes. The location, aspect ratio, and size of each envelope should be determined by calibrating the model parameters. Refer to (Mokhtari and Leira, 2024) for further details.

due to its availability in commercial finite element analysis (FEA) software. However, a recent study (Mokhtari et al., 2022a) has highlighted its inability to accurately replicate the effects of confining pressure in ice due to its flow rule, which assumes a zero plastic Poisson's ratio. As a workaround to this shortcoming, researchers often manually partition the ice domain into zones with high and low strength properties during pre-processing to mitigate the limitations of the Crushable Foam model, which fails to correctly model the variations in ice strength caused by changes in hydrostatic pressure during the ice crushing process. This manual partitioning method, however, is problematic due to the lack of precise information on the size and shape of these zones across different ice-structure interaction scenarios, which involve varying ice types, geometries, loading rates, and environmental conditions. Given these uncertainties, manual partitioning is not a practical or reliable approach.

An alternative plasticity model, based on the elliptical yield criterion, originally developed by Liu et al. (2011), has been benchmarked by Mokhtari et al. (2022a) (the subroutine is available on GitHub (Mokhtari, 2022)). This model features an associated flow rule that effectively captures confining pressure effects, surpassing the Crushable Foam model in ice crushing simulations. The model has been further developed by Mokhtari et al. (2023a, 2023b) to incorporate strain rate effects as the original model did not account for strain rate effects.

The idealization of ice mechanical behavior using plastic rheology is applicable only when strain rates are above the ductile-to-brittle transition point (i.e., the brittle regime in Fig. 24). Below this threshold, viscoelastic effects become too significant to ignore. A viscoelastic model (e.g., (Mokhtari et al., 2024, 2023c)) is applicable across all strain rates (i.e., in both ductile and brittle regimes), and can be used universally in ISI simulations. These models also capture phase transitions, (i.e., from the solid intact state to the fluid-like pulverized ice), including the viscous flow of pulverized ice extruding from ice-structure interface.

A viscoelastic model combined with the cohesive zone method for simulating ice bending failure (Mokhtari et al., 2024) has been shown to effectively capture the fracture patterns of level ice, as well as the magnitudes and frequencies of ice loads during interactions with fixed offshore structures, all within a single simulation (Fig. 26). Despite all these advantages, high-fidelity viscoelastic models are complex, require extensive calibration, and demand significant computational resources, making them less practical for large-scale marine engineering projects.

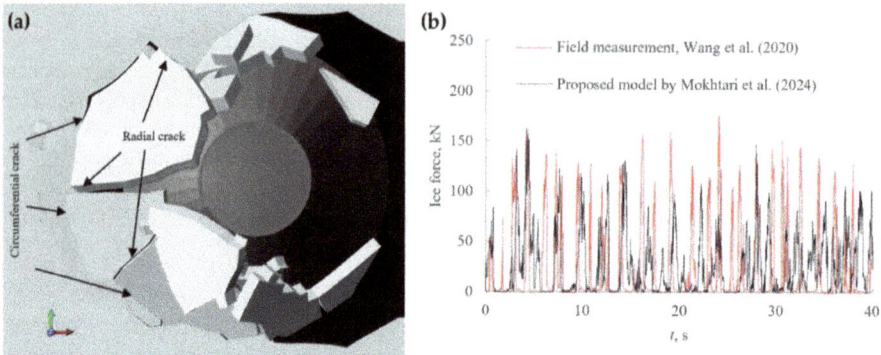

Fig. 26. (a) Ice-breaking pattern in the numerical simulation (Mokhtari et al., 2024) and (b) the ice forces from field measured data (Y. Wang et al., 2020) and the numerical simulation (Mokhtari et al., 2024).

5.5 Fatigue of Metallic Materials

Fatigue and fracture are addressed in ISSC Committee III.2 Fatigue and Fracture; however, as fatigue is a material degradation mechanism, the topic is briefly addressed below at a high level in the context of environmental impact on fatigue properties.

For metallic materials, fatigue resistance curves are provided in several class guidelines, including DNV, BS, API, IIW, and Eurocode. Recommendations for slope transitions in steel structures are generally based on two slope curves. In DNV-RP-C203 (2024) and BS7608 (2014)BS 7608, (2014), the slope changes from m to m + 2 for most of the fatigue Classes for welded joints, which translates to a change from a slope of m = 3 to m = 5. What is common for all the S-N curves is that they are mostly based on constant amplitude testing, while the environmental loading is variable in magnitude. The fatigue design S-N curves for welded joints include a change in slope to enable fatigue life under variable amplitude loading to be calculated. The slope transition point (STP) is located at 1×10^7 cycles in DNV-RP-C203 but at 5×10^7 cycles in BS7608 for a D to W3 class joints. Recent fatigue test data indicate that the fatigue limit is higher and for details showing a high fatigue strength, this has been incorporated in the 2024 revision of DNV-RP-C203 and Eurocode 3 part 1–9 (EN-1993-1- 9 where for Classes B1 to C2 the slope change is moved to 5×10^6 cycles from 1×10^7 cycles. The different STP will result in different calculated fatigue lives and the difference in calculated fatigue life between the two fatigue guidelines can be significant, and for fatigue critical risers a factor of approximately 2 on life was estimated. In order to provide better guidance to industry on the approach to use for fatigue analysis of girth welds in pipelines and risers specifically, TWI and DNV carried out a project where the ultimate aim was to provide guidance on a common S-N curve suitable for calculating the fatigue life of girth welds with a focus on the slope change area of the S-N curves (Horn and Johnston, 2024; Johnston and Horn, 2024). However, the conclusion of the work was that to unify the position of the STP in the BS 7608 and DNV-RP-C203 documents, through a better understanding of the actual effect of variable amplitude loading (VAL) on the fatigue life of welds in the high endurance regime, more investigation of the effect of VAL is needed.

Strain-based fatigue reliability assessment is an important aspect of shipbuilding research, particularly for welded joints in ship structures. This method involves analyzing the fatigue life of joints by considering low and high-cycle fatigue regimes and incorporating factors such as secondary notches, misalignment, and residual stresses into the assessment. A study applying this approach found that these factors significantly influence the fatigue crack initiation life, which is the point at which a detectable crack forms (Dong et al., 2021). By accounting for these variables, the strain-based fatigue reliability assessment provides a more accurate prediction of fatigue life, thereby enhancing the safety and reliability of ship structures over their operational lifespan. This method is especially valuable for identifying potential weak points in welded joints that could lead to failures, allowing for more targeted maintenance and inspection strategies.

The evaluation of fatigue behavior extends beyond just environmental considerations, encompassing the performance of various welding techniques under high-stress conditions. For example, flux-cored arc welding (FCAW) is widely used in the fabrication of naval-grade DMR249A steel joints, which are critical for ship hull structures. A comprehensive study by Hariprasath et al. (2023) assessed the fatigue properties of FCAW joints and found that these joints retain a high percentage of the fatigue strength and tensile strength of the parent metal. Specifically, the FCAW joints achieved a fatigue strength that is 92.8% that of the parent metal, demonstrating their robustness and suitability for high-stress maritime applications. These results are significant for shipbuilders as they confirm the reliability of FCAW for producing durable and fatigue-resistant joints, which are essential for maintaining the structural integrity of ship hulls over their operational life.

5.6 Summary and Recommendations for Future Work

Studies have highlighted the effects of stress, hardness, and sulfur content on corrosion and SSC, while exploring the role of loading profiles and unified corrosion mechanisms. Mitigation strategies, such as reducing LHZ and using clad or lined pipes for severe sour conditions, are also being investigated. A probabilistic engineering critical assessment (ECA) framework has been developed to address weld defect acceptance in sour environments, offering improved reliability over deterministic approaches. The ongoing

DNV JIP "Sour" aims to establish clear criteria for carbon steel pipe applications in challenging sour service conditions.

The morphology of localized defects caused by pitting corrosion could largely influence the remaining strength assessment of corroded structures. The morphology of pitting corrosion defects has been shown to evolve through incremental stages of pit deepening and lateral expansion. Over time, pits coalesce into larger depressions, creating variability in defect shapes and sizes, which intensifies stress concentration and affects crack initiation. Simplified defect depth profiles (e.g., rectangular, parabolic, and elliptical) fail to capture these irregularities, leading to inaccuracies in structural integrity assessments, particularly in fatigue-sensitive applications. Future research should focus on precise modeling of pitting morphology using stochastic methods, image processing, and machine learning, while also studying environmental factors and statistical distributions of pit sizes and locations. If the defect geometry used for assessing pitting corrosion-induced failures can be standardized, it is expected that pit/defect geometry databases will be developed within broader frameworks such as IACS. This standardization would enable more quantitative and consistent assessments of the damage caused by pitting corrosion.

Nickel-based alloys, long considered reliable for subsea components exposed to seawater and cathodic protection, have experienced several failures in recent decades, including Alloy 716, Alloy 625 grade 1, Alloy 725, and Alloy 718. These failures often involved local stress concentrations and hydrogen-induced stress cracking (HISC), with cathodic protection identified as a hydrogen source. Research highlights the susceptibility of high-yield-strength alloys to hydrogen embrittlement (HE). For instance, UNS N07718 with a yield strength of 150 ksi is more prone to HE than lower-strength variants. Similarly, strain rate and the presence of carbides like $Cr_{23}C_6$ and $M_{23}C_6$ exacerbate HE and reduce ductility. While API 6ACRA has improved nickel alloy quality, no specific HISC design criteria exist for hydrogen-charged environments, leaving a critical gap in material reliability standards.

Related to Carbon Capture technologies and industry guidelines and recommendations, recent developments include several initial efforts across different technologies and their demonstration onboard ships. Chemical absorption and membranes technology seem to be some of the most popular ones at the moment. Challenges still apply, especially considering capital and operational expenditure, installation issues and retrofitting onboard ships. Classification Societies are also moving fast to develop guidelines and recommendations for the introduction of such new technologies' onboard ships. It would be expected that additional and more specific measures and regulations will need to come into force considering carbon capture and storage onboard ships.

Hydrogen is seen as one enabler for the energy transition and there is a strong focus on hydrogen research. Transport of hydrogen will most likely be through pipelines, newbuilt or repurposed, however defects in C-Mn pipelines transporting hydrogen are more critical, potentially affecting pipeline integrity and leading to stricter utilization and defect acceptance criteria. Additionally, there may be a need for special considerations in integrity assessments and material toughness testing. Unlike regular pipeline assessments where loading rate is a secondary factor, hydrogen environments may require more stringent guidelines and requirements.

Accidental cryogenic liquid spills, such as LNG, in marine and offshore structures pose significant risks due to rapid cooling, which induces thermal stresses and potential brittle failure in steels not designed for such conditions. Accurate modeling of heat transfer during spills, including boiling regimes, is critical for predicting temperature distribution and thermal stress effects. Recent advancements have refined heat flux curves for specific materials, improving failure risk assessments. Future research should focus on expanding material coverage, including those needed for emerging fuels like liquid hydrogen (LH2) with lower boiling points, and enhancing computational models to better represent cryogenic heat transfer and mechanical properties at extremely low temperatures.

Understanding the behavior of materials under elevated strain rates is crucial for the design and safety of marine structures, as both man-made and natural materials respond differently compared to slower loading scenarios. Commonly, the Johnson-Cook and Cowper-Symonds models are used to simulate the mechanical behavior of metallic materials like steel and aluminum alloys under high strain rates, while natural materials such as ice, with their viscoelastic properties, require more sophisticated models that account for both high and low strain rate effects. The key challenge lies in developing accurate models that capture the viscoelastic and viscoplastic behavior of materials under diverse environmental conditions and a wide range of strain rates, enabling more reliable simulations of real-world scenarios.

Metallic materials rely on fatigue resistance curves provided by guidelines such as DNV-RP-C203, BS7608, API, IIW, and Eurocode, though these are primarily derived from constant amplitude testing, whereas environmental loads vary in magnitude. Slope transition points (STPs) in S-N curves differ among guidelines, with recent revisions reflecting higher fatigue strength for certain steel classes. Variations in STPs significantly impact calculated fatigue life, as evidenced by differences in pipeline riser fatigue life predictions between guidelines. Collaborative efforts by TWI and DNV to unify S-N curve approaches highlighted the need for further research into variable amplitude loading effects. Strain-based fatigue reliability assessments, incorporating factors like secondary notches and residual stresses, provide more accurate predictions for welded joints, enhancing ship structure safety. Studies also confirm the robustness of flux-cored arc welding (FCAW) for naval-grade steel, retaining 92.8% of the parent metal's fatigue strength, demonstrating its suitability for high-stress maritime applications.

6 Materials in a Life-Cycle perspective

In her opening speech to the ISSC congress 2022 in Vancouver, Christina Wang underlined that the '2Ds' (Digitalisation and Decarbonisation) are currently the mega trends which dominate the work on ship and offshore structures. Ms Wang suggested that each of the 2025 committees should link its activities to these two topics. For Committee IV.2, this raises the question how to assess RD&I on materials and fabrication with respect to their contribution towards a maritime industry which produces less emissions and is more sustainable. It was decided to devote a separate chapter to the lifecycle perspective of materials. The following three sub sections will address progress in fabrication and design of maritime assets, their operation and recycling.

When working on the subject, it quickly became obvious that isolated consideration of technologies and their immediate environmental impact falls short and needs to be replaced by holistic and lifecycle-oriented approaches and should be discussed with researchers from various disciplines. Therefore, the Committee decided to team up with external experts, and organized a workshop to discuss topics like.

- particular challenges (acceptance of new materials and knowledge regarding their potential, size, complexity and long expected lifetime of products, lacking knowledge about actually applied materials in a given ship, business cases/incentives for circular economy…)
- existing progress and further needs (recycling technologies, use of biobased/recycled materials and structures, IMO activities, growing customer demand)

The full documentation of this 'Hamburg meeting' is included in appendix A.

6.1 Fabrication and Design of Materials in a Life Cycle Perspective

In the following materials and fabrications are addressed in a holistic manner addressing environmentally friendly materials, use of material modeling and technologies which are enablers for a more environmental friendly shipyard process.

6.1.1 Environmentally Friendly Materials with Reduced Carbon Footprint

When it comes to composites, raw materials from renewable sources can be an environmentally friendly solution. This applies both for the fibres and the matrices. With respect to fibres, companies like Greenboats (see report from the Hamburg meeting) are already on the market, offering solutions to the boat industry and other sectors. However, there is also research on alternative matrix systems from natural sources. For instance, polybenzoxazines are suitable, as they can be produced from corn or sesame seed. Furthermore, they offer high temperature resistance and low heat release rates in case of fire. Smoke gas density and toxicity is also reportedly low. In the project Greenlight (Ohle, 2021), final report available at the ISSC congress), composites with a polybenzoxazines matrix have been researched with respect to their suitability to meet requirements for maritime lightweight applications on SOLAS vessels.

Oyster shell is a waste product and Ambika Harikumar et al., (2023) study the use of oyster shell powder as a reinforcement material in nylon-6 (PA6) composites. The study revealed that the tensile strength increased from 76 MPa up to 83 MPa, while the strain at break decreased significantly, the wear rate increased with approximately 35% and higher. The author suggested further study the impact of the particle size of the oyster shell powder. The committee believes that eco-initiative will see more and more focus in order to develop decarbonizing materials for the future.

6.1.2 Material Models - Structural Design Practices

In the design of ships and offshore structures, material models play a critical role in predicting structural behavior under diverse loading conditions and environments. Material modeling can be approached through two different methods: deterministic and probabilistic models. While deterministic approaches assume constant and well-known material properties, probabilistic approaches account for the inherent variability in materials.

Deterministic models use fixed material properties such as yield strength, tensile strength, and Young's modulus, which are typically based on standardized tests. These models apply higher safety factors to cover uncertainties in material behavior, potentially leading to over-conservative designs.

Probabilistic models recognize that material properties can vary due to factors such as manufacturing processes, environmental exposure, aging, and operational conditions. These models use probability distributions to represent these uncertainties, leading to more accurate predictions of structural behavior over time. Probabilistic approaches are especially useful in fatigue analysis and reliability-based design, where Monte Carlo simulations, among other reliability methods, can be employed to assess the probability of failure under varying conditions. This probabilistic approach, which accounts for variations in material properties from one sample to another, has been widely explored.

However, a more recent and less explored area is probabilistic material modeling for materials with inherent inhomogeneity. One example of such materials is metal foams, which exhibit spatial variability in mechanical properties within a single specimen due to differences in pore size and distribution. In applications such as ballistic shielding, where the projectile size could be comparable to the pore size, the material's ballistic performance can vary depending on the exact impact location. For instance, the ballistic limit, which is the velocity at which a projectile can penetrate the shield, may differ based on whether the projectile strikes a dense region or a more porous one. This inhomogeneous structure introduces additional complexity into the material's performance, making probabilistic material models even more essential for accurately predicting behavior under impact or stress.

Natural materials, such as ice, exhibit significantly higher variability in mechanical properties than many man-made materials. This is due to the complex and uncontrolled conditions under which natural materials form. For instance, the strength and stiffness of ice can vary widely depending on factors such as temperature, salinity, grain size, and the presence of impurities or air pockets, as well as geometrical factors such as random changes in ice sheet thickness. In contrast, man-made materials like metals and composites are manufactured under controlled conditions, allowing for more uniform properties. This fundamental difference means that when designing structures that interact with natural materials—such as ships navigating icy waters—the special variability of material properties must be accounted for in a probabilistic manner in each simulation model.

The review by (Feng et al., 2023) explores advancements and challenges in non-deterministic fracture mechanics, emphasizing the significance of addressing uncertainties in material behavior and fracture mechanisms. The study categorizes computational fracture models into embedded, smeared, and regularized approaches, highlighting methods like the extended finite element method (XFEM), crack band model, and phase field method (PFM). Each model is evaluated for its ability to address variability in material properties and external loading. Importantly, the review study discusses integrating machine learning algorithms, such as neural networks and support vector regression, to improve efficiency and accuracy in predicting fracture responses.

One of the interesting studies reviewed in (Feng et al., 2023) is the work by (Huang et al., 2021), which focused on probabilistic modeling of concrete material heterogeneity using high-fidelity random fields derived from micro-CT imaging. Continuous random fields, representing spatial variability in material properties like tensile strength, were generated by statistical reconstruction algorithms and the Karhunen-Loève expansion technique. The method integrated these random fields into finite element models for phase-field-based fracture analysis, providing a framework for simulating crack initiation and propagation in heterogeneous materials. By coupling these detailed models with Monte Carlo simulations, the study captured the variability in mesoscale fracture processes, crack paths, and load-displacement curves, achieving good correlations with experimental data. Various crack growth paths from their models under different random field conditions are illustrated in Fig. 27. Figure 28 shows the comparison between numerically simulated and experimentally observed crack growth paths and load–displacement curves for 100 fracture simulations with random fields in the L-shaped concrete specimen.

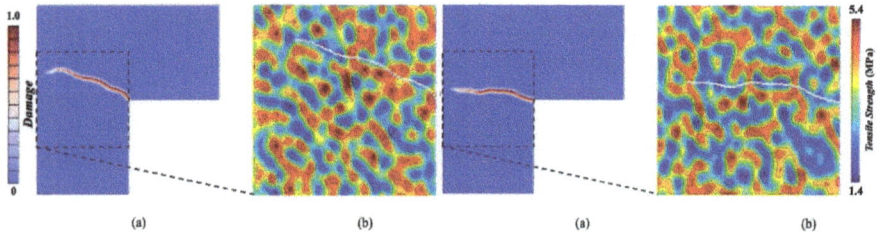

Fig. 27. (a) various damage patterns in L-shaped concrete and (b) detailed crack growth paths influenced by different random field conditions (Huang et al., 2021).

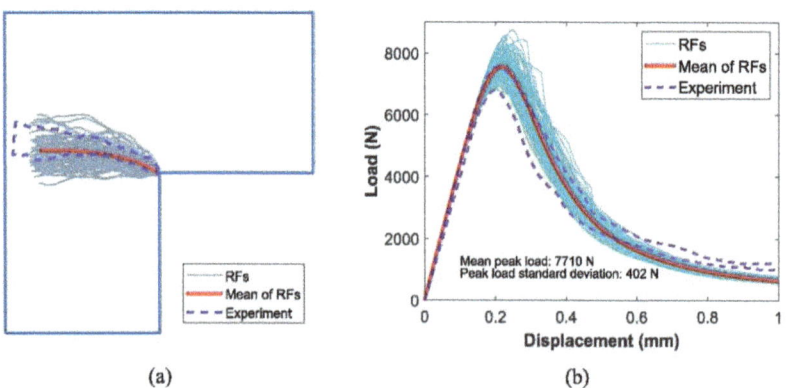

Fig. 28. Comparison between numerical simulations and experimental results: (a) crack growth paths and (b) load–displacement curves for L-shaped concrete subjected to 100 random field conditions (Huang et al., 2021).

It is important to emphasize that the accuracy of probabilistic models is fundamentally dependent on the fidelity of the underlying constitutive material model. Therefore, before attempting to improve simulation accuracy through probabilistic approaches, it is essential to ensure that the constitutive model accurately captures the relevant physical behavior, especially under dynamic loading. If observed that inaccuracies are due to limitations in the material model, efforts should focus on improving it, for example, by incorporating relevant physical phenomena.

6.1.3 Green Shipyard Processes

Two European projects, EcoShipYard and CirclesOfLife, are currently addressing in detail methodologies to assess in a structured, comprehensive and transparent way to which extent shipyards are applying measures and technologies to reduce emissions and energy consumptions. For more information about these projects, see the report of the Hamburg meeting in Appendix A.

The study by (Zhai, 2023) elaborates on the background of green manufacturing in China's shipbuilding industry and analyses the types of energy consumption that need to be invested in the implementation of green manufacturing in the shipbuilding industry based on the environmental value chain. Then, based on the panel data of shipbuilding industry energy consumption of provinces, municipalities and autonomous regions, the relationship between energy consumption and pollutant emission and shipbuilding cost changes in shipbuilding process was studied by using Ridge Regression model. It is found that changes in the amount of energy consumption and the amount of pollutant emissions in the shipbuilding industry will lead to changes in shipbuilding costs to varying degrees. The conclusion points out that in order to develop green economy, shipbuilding enterprises must actively reduce energy consumption in shipbuilding process and gradually promote green energy instead of traditional energy to achieve the purpose of energy saving and cost reduction.

Md Daud et al. (2024) investigate the advancement of modern shipyards considering the current regulatory regime through environmental-friendly approaches. They suggest that green shipyards, incorporating best practices and technology, can reduce the pollution generated during vessel construction. Their paper investigates the current trends of green shipbuilding and proposes potential practices of a green shipyard. A systematic review is conducted concluding Industry 4.0 (I4.0) technologies are essential to be implemented in green shipyards, advising 16 potential practices and technology enhancements. The study also identifies that automation is the most discussed topic in I4.0 technologies followed by alternative fuel (LNG/LPG fueled engine) and 3D printing and design.

Baihaqi et al. (2023) suggest the application of a novel integrated Value Engineering and Risk Assessment (VENRA) framework for measuring shipyard performance based on the combination of fuzzy Decision-Making Trial and Evaluation Laboratory (DEMATEL) and Analytic Hierarchy Process (AHP) tools to improve shipyard productivity and reduce the shipyards environmental footprint. The tools discussed in their paper are used to assess the criteria cause-effect and weight ranking analysis, considering the causal and affected groups while prioritizing the criteria and sub-criteria ranking. A shipyard case study was used to apply the proposed framework, showing that the shipyard with a high personnel's safety group is majorly more important than the environmental impact. The combination of hybrid MCDM tools has enhanced the process of determining the criteria analysis. Waste management has become the most impacting attribute amongst the criteria group, while the HSE department is the most critical criterion. However, the green energy used is still a minor factor as it is still not fully exploited within the existing shipyard and has not been fully supported by existing regulations yet.

Vakili et al. (2023) investigate the issues related to the decarbonization of the operational phase of a ship's life cycle, especially as these are affected by the shipbuilding stage. They further suggest that holistic and transdisciplinary studies of the shipbuilding energy sector are not currently present, and a holistic approach is needed to discuss the potential of measures and tools to improve the shipbuilding industry to include zero emissions. Their approach aims to provide trends, recommendations and policies for decarbonization of the shipping industry from a life cycle perspective. Taking into account a holistic and transdisciplinary approach, the energy sector in shipbuilding is categorized into an energy supply system, an energy economic system and an energy ecosystem, and the main disciplines for improving energy efficiency and promoting "zero emissions" for shipyards are identified, measures and tools within each discipline are proposed, and their mitigation potential and key issues for improving energy efficiency and reducing air emissions from shipyard activities are discussed. The suggested case study highlights the economic, environmental and sustainability benefits of implementing the proposed modern energy system within an Italian shipyard. The authors conclude by suggesting that the implementation of the energy management framework can accelerate the transition to a zero-emission shipbuilding industry and assist in eliminating air emissions in the which are currently provided due to the complexity, the different reduction potentials, the costs and the relationship and interaction between measures and tools.

In a follow up paper, (Baihaqi et al., 2024b) continue their work on exploring and implementing the VENRA framework on shipyard productivity as a means to strategically evaluate shipyard performance. The framework five key elements (Technical, Business, External, Personnel Safety and Environment) are examined and a case study is provided demonstrating the business element in more detail. Integrated fuzzy Decision Making Trial and Evaluation Laboratory (DEMATEL) and Weighted Evaluation Technique (WET) are used to assess the cause-effect and weight analysis criteria, considering the causal and effected criteria while prioritizing weight ranking. An objective grading system is developed to determine the shipyard score based on multi-resource qualitative and quantitative data. The case study demonstrates that the shipyard's 'delivery time' remains the most critical and influential aspect of the business elements' performance. In addition, the top three most important factors, 'delivery time', 'financial report condition,' and 'ship manufacturing cost,' must be considered, as they directly impact shipyard performance. Despite being a minor element, 'innovation and human resources' is the second most influential factor after 'delivery time'. The case study demonstrates that the framework can identify cause-and-effect criteria and prioritize via methodologies.

6.2 Materials in Operation in a Life Cycles Perspective

6.2.1 Effects of Ship Structure and Operation Activities on the Materials Behavior

The push towards decarbonizing the maritime industry, guided by the International Maritime Organization's (IMO) 2050 target, has accelerated the adoption of alternative propulsion systems in ships. This transformation primarily involves integrating zero-emission technologies like batteries and fuel cells into conventional ship structures. The shift has profound implications on material behaviour, structural integrity, and operational efficiency, requiring a reassessment of traditional design and fabrication processes.

Battery-powered propulsion systems offer a pathway to zero-emission shipping, but their practicality is often limited by their weight and volume. For example, a study highlights that a battery-powered system, even with speed reduction and the addition of charging stations, occupies a significant portion of the ship's deadweight, reducing cargo capacity by a considerable margin. This challenge has led to the exploration of hybrid systems where hydrogen fuel cells are integrated with batteries to offset these limitations. By reducing the overall footprint of the propulsion system to approximately 20% of the deadweight, this hybrid approach becomes more feasible for retrofitting existing vessels (Arabnejad et al., 2024).

Comparative studies of hybrid vessels, such as research vessels, reveal that hydrogen fuel cells can outperform battery systems in several key areas. For instance, hydrogen hybrid variants offer significantly longer zero-emission runtime and can meet a higher percentage of operational requirements using hydrogen power alone. Despite the higher initial capital costs, the long-term environmental and operational benefits of hydrogen fuel cells make them a compelling alternative to conventional diesel-electric powertrains (Klebanoff et al., 2021). The Scripps Institution of Oceanography study comparing a traditional diesel-electric research vessel with battery and hydrogen hybrid alternatives demonstrates that hydrogen hybrid systems offer superior zero-emission runtime and efficiency, making them more suitable for research missions (Klebanoff et al., 2021).

Further research into hybrid power systems combining different energy sources underscores the complexities of integrating these technologies into ship structures. A study on gas-electric hybrid systems, for instance, emphasizes the importance of optimizing energy density and load response. By balancing power distribution and system mass, researchers developed methods that enhance energy efficiency, reduce fuel consumption and emissions by up to 17.1%, and improve system reliability during transient conditions—a crucial factor in maintaining operational integrity over time (S. Wu et al., 2024a).. Similarly, research on the optimal sizing and utilization of fuel cells in ferries illustrates how careful component selection and dynamic programming can lead to substantial operational efficiency improvements and cost reductions (Cha et al., 2024).

The potential of proton exchange membrane fuel cells (PEMFCs) as a primary energy source for ship propulsion is increasingly being recognized. For example, the integration of PEMFCs with auxiliary batteries in unmanned surface vehicles (USVs) shows promise in extending operational range and enhancing power management. A study demonstrated that replacing traditional lithium-ion batteries with PEMFCs could significantly increase the endurance mileage of USVs, indicating that PEMFCs could be a viable solution for various maritime vessels (J. Wu et al., 2024b).

In addition to these operational advancements, hydrogen fuel cells offer strategic benefits in terms of compliance with emerging technical standards and regulations. A study exploring the status and prospects of hydrogen-powered ships highlights the critical role of developing robust regulations and technical standards to facilitate the broader adoption of hydrogen as a maritime fuel. This paper emphasizes the need for international classification societies to lead the development of a ship-based hydrogen energy standard system, which is crucial for ensuring safety and reliability in hydrogen fuel cell applications (Z. Wang et al., 2024a). This regulatory framework is essential to advancing the maritime sector's zero-carbon transformation and ensuring the safe and effective deployment of hydrogen fuel cell technology.

Moreover, studies have compared different fuel storage systems for fuel cells, such as Liquid Organic Hydrogen Carriers (LOHCs) and ammonia, to determine their feasibility for use in maritime propulsion. One such study concluded that while LOHCs offer some benefits, ammonia-based systems are generally more effective due to their lower mass and volume, making them a more competitive option for hydrogen storage in the maritime sector (Gambini et al., 2024). The research highlights the importance of optimizing fuel storage solutions to maximize the efficiency and viability of fuel cell-powered ships. The research highlights the importance of optimizing fuel storage solutions to maximize the efficiency and viability of fuel cell-powered ships.

Finally, practical applications of these technologies are illustrated by projects like Project Nautilus, which successfully integrated hydrogen fuel cell technology into a diesel-battery hybrid passenger ferry. The project addressed regulatory and safety concerns, providing a blueprint for the broader adoption of hydrogen as a clean energy source in the maritime sector. The insights gained from this project are invaluable for understanding the practical challenges and solutions involved in retrofitting ships with advanced fuel cell technology (Pal et al., 2024).

These studies collectively underscore the potential of hydrogen fuel cells as a promising alternative to batteries in maritime applications, particularly for long-haul and high-power scenarios. However, they also highlight the challenges in integrating these systems into existing ship structures, necessitating advancements in material science, fabrication technologies, and regulatory frameworks to fully realize their benefits. The limitations of battery-electric systems, as demonstrated in studies assessing their application to container ships, also underline the need for continued innovation and optimization in this field. Although battery propulsion systems can be technically feasible for smaller vessels and shorter passages, their economic and environmental performance compared to conventional diesel engines remains limited (Kistner et al., 2024). Although battery propulsion systems can be technically feasible for smaller vessels and shorter passages, their economic and environmental performance compared to conventional diesel engines remains limited.

In summary, integrating batteries and hydrogen fuel cells into maritime vessels is key to achieving IMO's decarbonization goals. While batteries offer zero emissions, their weight and volume limitations make them less viable for widespread use. Hydrogen fuel cells, especially in hybrid systems, provide better range, efficiency, and environmental benefits. However, to fully harness these technologies, advancements in materials, fabrication, and regulations are necessary. Continued innovation is essential for overcoming these challenges and driving the sustainable transformation of the shipping industry.

6.2.2 Implementation of New Scientific Tools in the Ship and Offshore Structure Life Cycle

Offshore asset maintenance is vital for maintaining the structural integrity, operational safety, and longevity of critical infrastructures, such as vessels and platforms, that serve key roles in maritime sectors, including oil and gas, marine transportation, and renewable energy (Abbas and Shafiee, 2020). These assets are exposed to the harsh and corrosive marine environment, necessitating strict maintenance protocols to prevent structural deterioration and ensure consistent productivity, safety, and environmental protection. Given the significant investments involved, proactive maintenance strategies are essential to prevent issues such as corrosion, cracks, wear, and structural anomalies that can lead to costly failures or accidents.

Traditional offshore inspection methods, such as visual and manual surveys, present considerable challenges that limit their effectiveness (frequency). These methods are labor-intensive, time-consuming, and susceptible to human error, relying heavily on the judgment of the inspector, which can lead to inconsistencies in detection and assessment. Environmental conditions, such as adverse weather, further complicate inspections and can obscure corrosion indicators, especially those beneath the surface. The limitation of traditional techniques to detect only surface-level corrosion and the risks associated with scaffolding or rope access to tall structures also underscore their inadequacy, (Bonnin-Pascual and Ortiz, 2019) and (Alum and Eze, 2020). The reactive nature of these inspections often results in the addressing of corrosion only after significant damage has occurred, necessitating the exploration of more effective and sustainable alternatives, (Poggi et al., 2022).

Unmanned Aerial Vehicles (UAVs) have emerged as a promising solution for offshore asset monitoring, transforming data collection and inspection practices, (Wanasinghe et al., 2020). UAVs are capable of accessing confined spaces, high altitudes, and other challenging areas that are typically hazardous for human inspectors, thereby reducing the risk to personnel. Equipped with advanced sensors, such as high-resolution cameras and LiDAR systems, UAVs offer a detailed assessment of offshore installations, allowing for rapid and efficient inspection, (Feroz and Abu Dabous, 2021). This capability supports a preventative maintenance approach by facilitating early detection of structural issues, including corrosion, cracks, and deformations, and enabling real-time monitoring. UAVs also contribute to the reduction of inspection costs while enhancing the safety and timeliness of inspections, (Sreenath et al., 2020) and (Carrillo et al., 2021).

The integration of UAVs with computer vision and machine learning algorithms significantly enhances their ability to detect and assess structural anomalies. Computer vision, combined with machine learning, is increasingly applied to automated corrosion detection, offering a sophisticated approach to identifying and quantifying corrosion in offshore structures. Convolutional Neural Networks (CNNs) have proven effective in learning and identifying patterns in image data, allowing for accurate differentiation between corroded and non-corroded surfaces under varying environmental conditions,

(Yao et al., 2019). The combination of UAVs and machine learning creates a powerful tool for large-scale, automated inspections, which enhances the consistency and accuracy of corrosion detection while reducing the reliance on manual inspections, (Forkan et al., 2022).

Advances in automated corrosion detection have also seen the use of machine learning models to predict corrosion rates and evaluate influencing factors, achieving high prediction accuracy (W. Liu et al., 2022a), (Nguyen et al., 2022), (Bobadilla et al., 2024), (Wang et al., 2023), (Yu and Han, 2024) and (Pereira et al., 2024). CNNs, along with fusion techniques that combine vision and infrared imagery, have been shown to improve the reliability of automated corrosion detection, (Jin Lim et al., 2021). These advancements help in identifying potential issues before they become significant, which supports predictive maintenance strategies essential for offshore assets, (Jiang et al., 2024) and (Wen et al., 2020).

However, there are several challenges to implementing machine learning for corrosion detection in offshore settings. A primary challenge lies in the need for large and diverse datasets to effectively train these models. High-quality data is essential, but the variability in corrosion patterns and environmental conditions makes data collection challenging. Furthermore, computer vision systems are susceptible to environmental factors, such as lighting, shadows, and reflections, which can affect corrosion detection accuracy. The computational demands of machine learning algorithms, particularly for real-time processing of high-resolution images, require significant hardware and software resources. Additionally, the interpretability of machine learning models is a major limitation in safety-critical applications, as understanding the rationale behind their predictions is crucial.

The complexity of corrosion processes, influenced by numerous environmental variables, presents further difficulties for machine learning models. Predicting localized corrosion in certain alloys, for instance, remains a significant challenge, (Jiang et al., 2020) and (Thanush et al., 2022). Despite these challenges, the integration of advanced technologies, such as UAVs, computer vision, and machine learning, represents a promising direction for improving offshore asset monitoring, offering more precise, efficient, and safer inspection solutions compared to traditional methods.

6.2.3 Assessment Methods During Structural Life Cycle

Recent research into ship structural lifecycle assessment methodologies underscores the growing importance of addressing environmental, technological, and operational factors to enhance sustainability and efficiency throughout a vessel's life. The following review synthesizes findings from key studies, highlighting advancements in lifecycle analysis, digital twin technology, end-of-life utilization, fatigue crack management, and energy solutions.

One of the core areas of research focuses on the integration of lifecycle assessment (LCA) with energy solutions to achieve net-zero emissions in the maritime sector. A comprehensive cradle-to-grave assessment reveals that operational phases of ships significantly contribute to total greenhouse gas emissions, particularly when using conventional marine fuels (Bellot et al., 2024). The study emphasizes that synthetic methanol, if produced from renewable sources, can potentially reduce carbon emissions substantially.

However, practical challenges such as the impact on energy infrastructure and the need for substantial carbon offset credits must be addressed. Similarly, the lifecycle analysis of zero-carbon fuels—ammonia, hydrogen, and electricity—demonstrates varying impacts on emissions and feasibility. Hydrogen and battery-electric systems offer notable reductions in greenhouse gases but face practical limitations in terms of energy storage and space onboard. Ammonia and methanol present more immediate practical alternatives, though they have higher environmental impacts and cost fluctuations (Park et al., 2022). These findings highlight the need for a nuanced approach to fuel selection, integrating LCA with energy modelling to evaluate different fuel options comprehensively.

Digital twin technology represents a significant advancement in managing ship lifecycles. This technology involves creating a dynamic virtual model of a ship that mirrors its physical counterpart, facilitating real-time monitoring and optimization (Mauro and Kana, 2023). Despite its potential, the current application of digital twins in the maritime industry often lacks the necessary data exchange between physical and virtual environments. This discrepancy limits the effectiveness of digital twins, which are frequently confused with simpler virtual models. A critical review of digital twin applications underscores the need for better integration, particularly in the design and decommissioning phases. Addressing these gaps can enhance the accuracy and utility of digital twins in lifecycle management, offering more robust predictive maintenance and operational optimization.

Innovative approaches to end-of-life ship utilization include repurposing retired vessels as floating seawalls. This method aims to extend the lifecycle of decommissioned ships while mitigating environmental and human health risks associated with traditional shipbreaking practices (G. Wang et al., 2024b). The research demonstrates that floating seawalls made from retired vessels can effectively protect coastal areas by leveraging the concept of gap resonance for wave attenuation. This approach not only provides a new function for end-of-life ships but also contributes to sustainability by reducing the environmental footprint of shipbreaking. The study highlights the potential of such innovative uses for retired ships, suggesting that they can serve valuable roles beyond their initial operational life.

Effective management of fatigue cracks in ship structures is crucial for maintaining structural integrity and safety throughout a vessel's life. Recent research has introduced an entropy-based framework for optimizing the inspection and repair scheduling of multiple fatigue cracks (Chulahwat and Mahmoud, 2023). This probabilistic framework improves upon traditional methods by considering the propagation rates of cracks as random variables with predefined statistical distributions. The approach offers a more cost-effective and efficient scheduling strategy for maintenance and repair, reducing the financial burden while enhancing safety. This framework represents a significant advancement in lifecycle cost management, ensuring that structural integrity is maintained without incurring excessive costs.

The techno-economic feasibility and environmental performance of various fossil-free fuels are critical considerations for future maritime operations. A comparative life cycle study evaluates the viability of different energy carriers—battery-electric systems, hydrogen, methanol, and ammonia—across different ship types, including RoPax ferries, tankers, and service vessels (Kanchiralla et al., 2023). The study reveals that while battery-electric and hydrogen options can reduce greenhouse gas emissions significantly, they face practical constraints related to energy storage and cost. Ammonia and methanol offer more immediate practicality but come with higher environmental impacts and economic variability. This research underscores the complexity of selecting optimal energy solutions, highlighting the need for a balanced approach that considers both environmental benefits and economic feasibility.

Overall, advancements in ship lifecycle assessment methods provide a comprehensive understanding of the factors influencing maritime sustainability and efficiency. Integrating lifecycle analysis with innovative energy solutions, advancing digital twin technology, repurposing end-of-life vessels, and optimizing fatigue crack management all contribute to a more sustainable and resilient maritime industry. These approaches collectively support the transition towards cleaner, more efficient shipping practices, ensuring that future maritime operations are environmentally friendly and economically viable.

6.2.4 Statistics of Failure Modes

To learn from failures is crucial in order to design safe and optimized structures and vessels, The general failure modes of steel structures are loss of structural capacity, and the failure mechanics consist of material degradation including embrittlement, material loss, fatigue, and excessive loading. The most common influencing factors causing these failure mechanisms are shown in Fig. 29.

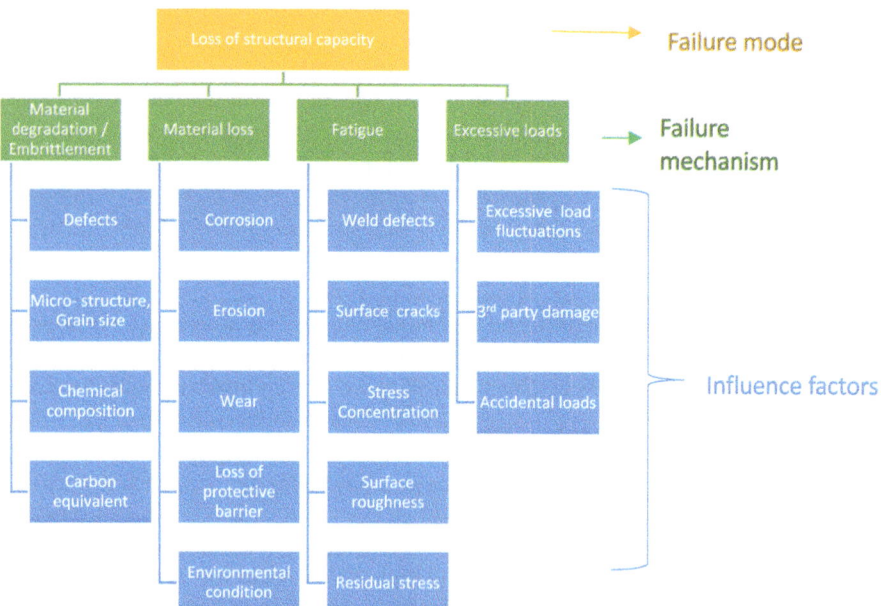

Fig. 29. Illustration of failure mechanisms and influencing factors of steel structures (Horn et al., 2023)

In the following the different failure incidences and severity of these based on the CODAM dataset is plotted in Fig. 30.

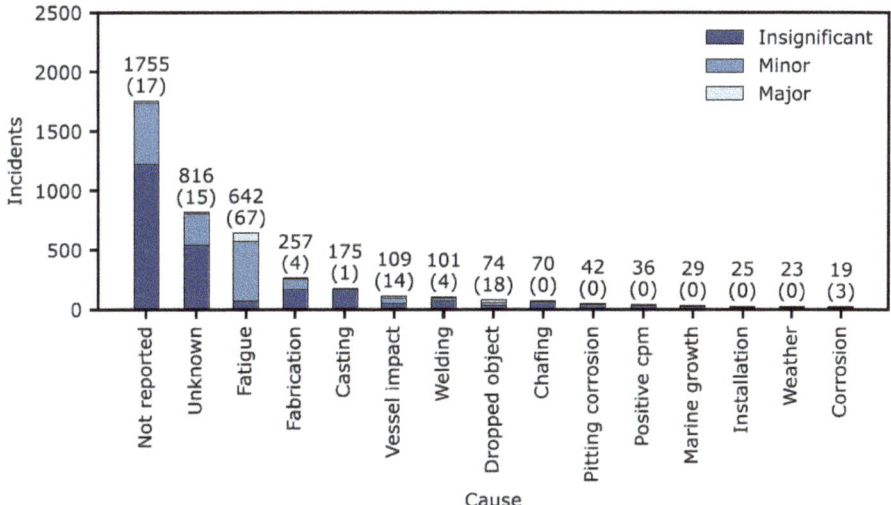

Fig. 30. Incident causes grouped by severity from the CODAM dataset in the period of 1974–2021. The numbers in parentheses indicate major accidents (Svendsen et al., 2022)

A high-level summary of the most common offshore regulators and their reporting of major structural accidents are summarized in Table 5 or the past few years. As can be seen, only one major accident has been reported by EU which was the INA's missing platform.

Limited research papers have been found addressing failures and failure statistics and it is recommended to have a more open communications and sharing of knowledge to prevent major accidents within the Maritime and Energy sector.

6.3 Recycling of Composite Material

For ships with structures made of conventional structural materials, there is an established recycling industry, whereas recycling of ships made of composite materials is rarely observed. The related discussion in the Hamburg meeting (summary in A.2.2 Material in a life-cycle perspective) came to the conclusion that the absence of a composite ship recycling industry is rather due to a current lack of viable business cases than a lack of suitable technologies. Several recycling techniques exist for composites materials, and a summary is provided below.

Pyrolysis thermal treatment aims to recover the reinforcing fibres by thermally decomposing and degrading the polymer chains of the organic matrix in an anoxic chamber at temperatures of around 300 to 800 °C for a period of several hours. At the end of the treatment, the matrix is generally recovered in the form of oil, gas or solid coal products. These elements can then be reused as inputs for other chemical processes or simply used as an energy source, while the recycled fibers can be re-used.

Table 5. Reported incidents defined as major the past five years (Horn et al., 2023).

Location	Reported incidents
EU: Safety of offshore oil and gas operations (europa.eu)	Report instances with loss of structural integrity. Data is 2016 (0), 2017 (0), 2018 (0), 2019 (2 – not possible to identify what these were), 2020 (1 – total loss of platform in Croatia - INA's missing gas platform found on seabed l Croatia Week), 20.21 (0)
UK: Mandatory with RIDDOR to report dangerous occurrences i	Operators shall report accidents related to: Collapses: "Any unintentional collapse or partial collapse of any offshore installation or of any plant on an offshore installation which jeopardizes the overall structural integrity of the installation" Subsidence or collapse of seabed: "Any subsidence or collapse of the seabed likely to affect the foundations or the overall structural integrity of an offshore installation" In informal discussion with regulator, no records of any of the major failures above. Some failures have occurred, but not to the level to be reported, though very few known of
IOGP	None reported accidents, informal feedback that there are no instances of structural failure
US: Offshore Incident Statistics l Bureau of Safety and Environmental Enforcement (bsee.gov), Regulator collects offshore incident data	Search for major incidents in 2018, 2019, 2020 and 2021, only reported various crane incidents, and some boat impacts with jacket leg damage are noted
Australia: NOPSEMA	Zero regulatory actions relating to structural (and well) integrity in 2019–2020

Chemical recycling, also called solvolysis, is mainly used in the context of thermosetting composites where the principle consists of decomposing the organic portion of the material by depolymerization and thus freeing the fibres from their polymeric matrix. Water appears to be the most widely used solvent in addition to ethanol, methanol, propanol or acetone. Concerning carbon fiber, chemical recycling makes it possible to recover fibres of variable length with a surface topology almost identical to virgin fibres.

Mechanical recycling is potentially the most economical method of composite processing because it allows the recycling of the entire processed material. The recycling process takes place in a high-speed crusher in which the material is cut into finer elements typically ranging from 10 mm to micronized particles of less than 50 μm. This phase is most often followed by two complementary steps: the first to recover coarse-sized pieces and the second to refine the particle size according to the intended use. A final screening operation finally makes it possible to split the obtained product into batches of different sizes.

Storage or burning is undoubtedly the most unfavorable solution in the waste treatment cycle.

Composite waste can be used as substitute fuels in conventional incineration units for energy recovery as found in the management of non-recoverable household waste.

6.4 Summary and Recommendations for Future Work

Enhanced material modelling capabilities will be an enabler for safer and more sustainable structures. As presented in this chapter and also discussions at the Hamburg meeting, there are two pathways to improve the sustainability of marine structures: Aiming for a long operational life of each asset and reducing its overall ecological footprint. Hence, the recommendation is twofold:

1. Ensuring long operational life of maritime assets

In marine and offshore design, material models are essential for predicting mechanical behavior under various conditions. Deterministic models use fixed values, often requiring conservative safety factors, while probabilistic models account for variability, offering more accurate predictions by using statistical distributions. This probabilistic approach is especially valuable for assessing fatigue and reliability, allowing for better risk management in design.

Enhancing prediction accuracy requires developing probabilistic material models that incorporate the spatial distribution of mechanical properties within individual samples. This is especially important for local mechanical behaviors that are affected by factors such as the nonuniform distribution of manufacturing defects, variations in pore size in metallic foams, and the inherent inhomogeneity found in concrete and natural materials like ice.

2. Contribute to enhanced low-emission shipyard processes and circular economy

Advancement of technologies and methods – To improve the capability of measuring and applying sustainability principles across the ship's lifecycle, there is a need for tools which encompass concepts such as the Shipyard Environmental Performance Index (SEPI) and Digital Product Passports (DPP). To develop sustainable products, effective design tools are required, capable of creating lifecycle-optimized solutions. Optimization criteria comprise product performance and overall emissions, repairability, readiness for retrofit and recycling or reuse, and the solution space should include the possibilities of new materials (such as composites) and manufacturing techniques (such as additive manufacturing).

Improved Frameworks and Collaboration – Creating favorable conditions through regulatory adaptation, market incentives, and knowledge-sharing platforms is critical. This includes green certifications (based on SEPI), collaboration between stakeholders in projects like CirclesOfLife, and targeted training programs to bridge knowledge gaps and foster the adoption of sustainable practices across the industry.

It is suggested that future committees compare different choices of nature-based composite materials, not only in terms of their technical performance, but also regarding their availability and sustainability (e.g. is the raw material part of the food chain).

Limited research papers have been found addressing failures and failure statistics and it is recommended to have a more open communications and sharing of knowledge to prevent major accidents within the Maritime and Energy sector.

7 Digitalization Related to Material and Fabrication

Digital twin (DT) technology has emerged as a transformative force in the shipbuilding and offshore platform manufacturing industries over the past years. By creating virtual replicas of physical assets, processes, and systems, digital twins enable enhanced design, production, simulation, monitoring, and predictive maintenance. The maritime industry is undergoing a significant digital transformation, driven by the need for increased efficiency, safety, and sustainability (Jones et al., 2020).

The application of digital twin technology extends beyond individual shipbuilding processes to encompass entire shipyards. By creating comprehensive virtual models of shipyard facilities, operations, and workflows, digital twins provide an integrated approach to managing shipyard activities. This holistic perspective enables enhanced planning, optimization of resource allocation, and a more cohesive management strategy that aligns with the principles of Industry 4.0. Digital twins of shipyards facilitate improved decision-making, predictive maintenance, and overall efficiency, driving progress towards a more intelligent and responsive shipyard ecosystem. Over the last years, advancements in computational power, data analytics, and the Internet of Things (IoT) have propelled the development of sophisticated digital twin models. These models now integrate complex data streams from sensors, enabling high-fidelity simulations and real-time analytics. The convergence of digital twins with Industry 4.0 principles has further enhanced their applicability in manufacturing and operational contexts.

7.1 Digital Twins of Shipbuilding Processes

Digital twins have been implemented across various stages of the shipbuilding process, enhancing specific tasks by integrating real-time monitoring, modeling, and simulation tools. Below are some notable examples of digital twins employed in shipbuilding processes.

The plate bending support system developed by Mitsuyuki et al., (2020) is specialized DT for the plate bending process. This system enhances quality and efficiency by comparing plate shapes acquired from monitoring data with CAD data in real-time and determining additional heating positions using a physical model. Kunkera et al. (2022) employed VR technology and 3D models to simulate outfitting equipment installation. By monitoring the installation work and performing assembly simulations based on the monitored data, they achieved a 10% reduction in work time, a 90% reduction in errors during assembly, and a 70% increase in pre-outfitting efficiency.

Aoyama et al. (2021) developed a DT system for the cutting and sub-assembly process, which uses monitoring data from fixed-point cameras and workers' smartphones to identify the work type, worker, work position, target workpiece, and work time. By integrating this information into Discrete Event Simulation (DES) models, it became possible to analyze deviations from the production plan and identify factors that may cause such deviations i.e. information on unreasonable (Muri), waste (Muda), and unevenness (Mura) in the actual production process.

While many examples of DT have been introduced within a limited scope of shipbuilding processes, challenges remain in realizing a fully integrated digital twin for the entire shipbuilding process (Pang et al., 2021). Although frameworks for system design and monitoring have been proposed, a comprehensive integration of monitoring results with simulation models is still required to fully harness DT's potential.

Moreover, implementing Model-Based Systems Engineering (MBSE) is crucial to effectively analyze and improve the complex shipbuilding processes. MBSE is a methodology that uses models to support various decision-making processes, such as requirements, design, analysis, and verification, throughout the system lifecycle. Given the large-scale and intricate nature of shipbuilding, strong support is needed for the generation and comparative analysis of simulation models, which is essential for realizing MBSE in shipbuilding. The development of various monitoring methods, as well as the generation of effective simulation models, remains an important area for further research and development.

7.2 Digitalization in Ship Manufacturing

Mauro and Kana (2023) explore digitalization in ship manufacturing, introducing digital twins to support the entire ship-life cycle, from design to decommissioning. They emphasize that the term "digital twin" is often misused in the shipping industry to describe any virtual model, while a true digital twin requires continuous data exchange between the physical and virtual environments, a feature absent in most current implementations. The authors note a significant gap in research on digital twins specifically for ship design and decommissioning phases, urging further studies to address these overlooked areas.

Jagusch et al. (2021) present an approach that uses digital threads to map and optimize shipbuilding processes, from design to production. The iterative nature of shipbuilding requires a consistent data model, which the digital thread provides, a foundational component for an effective digital twin. This consistent data model helps manage the continuous changes and adjustments in shipbuilding, ensuring that every iteration is efficiently tracked and integrated into the overall process.

Yi et al. (2023) examine technological innovation in Chinese shipyards, specifically addressing health and safety issues for frontline workers caused by dangerous conditions, such as noise and paint mist pollution. These hazardous environments have led to low production efficiency and personnel allocation difficulties. To mitigate these risks, the authors proposed the introduction of a smart shipyard model. The proposed model aims to create safer working conditions, increase efficiency, and reduce costs through the adoption of advanced technologies. Case studies demonstrate significant improvements: 100% digital modeling of ships, a 70% increase in production efficiency, and considerable cost reductions, including savings in labor and fuel.

Furthermore, Dolz et al., (2024) analyze processes and technologies employed in EU shipyards, focusing on composite materials and advanced construction techniques like Adaptive Moulds, Automated Fibre Placement, and Additive Manufacturing. Their study introduces new indexes to evaluate shipyard technological levels, revealing a strong interest in digitalization among shipyards, despite financial constraints and concerns about the impact on operational efficiency. The study finds that almost 95% of shipyards are either using or planning to use composite materials, particularly fiberglass and polyester resin, primarily manufactured through manual lamination and vacuum infusion techniques. The authors analyze processes and technologies employed in EU shipyards, focusing on composite materials and advanced construction techniques like Adaptive Moulds, Automated.

Moreover, Fonseca and Gaspar (2021) discuss the use of digital twins for specialized ship systems, identifying challenges related to data modeling and proposing an open standard for digital twin data in the maritime context. The proposed standardization is aimed at overcoming existing challenges in data integration, thereby facilitating a broader and more consistent implementation of digital twins across various shipyard functions and improving interoperability between systems.

7.3 Advancements Toward Intelligent Shipyards

To maintain competitiveness, shipyards need ongoing technological improvements, particularly in automation and intelligent planning systems. In this respect, Lee et al. (2020) present a simulation-based shipbuilding planning system that integrates planning and scheduling through discrete event simulation. This approach allows for a process-centric view, enhancing the quality of production planning and improving overall shipbuilding productivity by evaluating different scenarios and optimizing resource allocation.

On the other hand, Giskeødegård et al. (2023) focus on digital assistance systems in Norwegian shipyards, highlighting the necessity of aligning digital tools with workers' needs to maximize their value. Their study reveals a positive attitude among production workers towards digital solutions, especially when these tools enhance their daily tasks. The authors emphasize the importance of user-centered design to ensure that digital solutions are not only effective but also accepted and efficiently utilized by workers, thus driving productivity gains.

Additionally, Iwańkowicz and Rutkowski (2023) review the application of digital twins across various industries, emphasizing that shipbuilding's traditionally low level of digitalization presents significant challenges for implementation. They propose a comprehensive digital twin concept for ship design and production, which includes a dimensional quality management metasystem (DQMM) to enhance control over shipbuilding processes. The DQMM framework aims to improve dimensional accuracy and quality assurance, facilitating better monitoring and precision in production activities.

Finally, Giering and Dyck (2021) discuss the crucial role of digital twins in addressing global challenges in the maritime sector, such as rising costs and pressures for decarbonization. They propose a Maritime Digital Twin Architecture (MDTA) to guide the development and deployment of digital twin solutions across the ship lifecycle. The MDTA framework is intended to standardize the integration of digital twin technologies, helping shipbuilders and operators manage complexities and meet sustainability goals.

Committee IV.2: Material and Fabrication Technology 933

7.4 Summary and Recommendations for Future Work

Digital twin technology has emerged as a transformative force in the shipbuilding and offshore platform manufacturing industries over the past years. By creating virtual replicas of physical assets, processes, and systems, digital twins enable enhanced design, production, simulation, monitoring, and predictive maintenance.

The maritime industry is undergoing a significant digital transformation, driven by the need for increased efficiency, safety, and sustainability. Digital twin technology has gained prominence as a tool to achieve these objectives by facilitating real-time monitoring, predictive maintenance, and optimized design processes, ultimately offering substantial economic and operational benefits. The convergence of digital twins with Industry 4.0 principles has further enhanced their applicability in manufacturing and operational contexts.

Physical assets or systems in shipbuilding include ships, factories, processes, workflows, and operators. To realize digital twins in shipbuilding, modeling and monitoring of a wide range of information and phenomena is necessary. Recent advancements in computational power, data analytics, and the Internet of Things (IoT) have propelled the development of sophisticated digital twin models that integrate complex data streams from sensors, enabling high-fidelity simulations and real-time analytics.

Despite these advancements, challenges remain in fully integrating digital twin technology across all stages of shipbuilding and shipyard operations. The complexity of physical and virtual environment integration, financial barriers, and limited standardization continue to hinder widespread adoption. Addressing these challenges will require further research in refining digital twin concepts, particularly in ship design, decommissioning, and smart shipyard development aligned with Industry 4.0 standards.

The potential benefits of digital twins for shipyards are clear, including among others enhanced resource efficiency, safety, and productivity. Moving forward, a focus on comprehensive integration, robust data exchange mechanisms, and the implementation of Model-Based Systems Engineering (MBSE) will be essential to fully harness the transformative capabilities of digital twins. By overcoming current limitations and driving innovation, the maritime industry can better adapt to technological advancements and maintain competitiveness in a rapidly evolving environment.

8 Conclusions and Recommendations

This report addresses the developments in materials and fabrication technologies, knowledge gaps, recommendations on how to improve design, qualification and approval processes to keep pace with upcoming innovations in the Maritime and Energy sectors.

This is the first period in which this committee is one of the eight technical committees in ISSC, and among the new topics that have been included in the mandate from previous years are the focus on advanced material and fabrication models to support structural analysis during design and operation and material aging and assessment methods during structural life cycle. The report has specifically addressed research linked to lightweight materials and materials for the energy transition, including advances in fabrication processes related to shipyards.

The Maritime and Energy sectors will see significant increases in material demand in the years to come, especially related to the renewable sector, demand for newbuilt pipelines for CO_2 and H_2 transport, and lightweight vessels among others. There is a strong political support for renewable energy and decarbonization, However, the wind energy market requires large amounts of steel and cement, and it is a pressure on the production sectors to reduce carbon footprint. The industry is pushing the use of material modelling capabilities with the aim to shorten the time to market for new technologies. Use of more analytical modeling capabilities may also enable the use of more recycled materials with the aim of improving design and operation of assets. In a life cycle perspective of assets, material selection becomes crucial and material aging and assessment methods during structural life cycle are of utmost importance, to document safe and sustainable operation. Some key trends seen within material and fabrication technologies in the last 3 years and recommendations are highlighted below:

- The energy transition is driving demand for polymer-based composites such as FRPs in hydrogen transportation. However, challenges such as complex long-term degradation, hydrogen permeation, and high operating pressures needed for hydrogen transport still hinder full adoption of RTPs and further research is needed.
- It is suggested that future committees compare different choices of nature-based composite materials, not only in terms of their technical performance, but also regarding their availability and sustainability (e.g. is the raw material part of the food chain).
- The cement industry is very energy demanding, significant research areas include high-performance concrete (HPC), self-healing concrete, and geopolymer concrete. Research focuses on optimizing the mix design of concrete to achieve adequate mechanical properties while also improving the material's environmental footprint.
- To unlock the lightweighting potential in the shipbuilding and energy sectors, it is crucial to upscale joining technologies such as adhesive bonding, friction stir welding (FSW), and novel laser welding and electron beam welding processes. Further development is needed to match the reliability, versatility, and efficiency of adhesive bonding with welding. Innovations like automated adhesive application systems, real-time process monitoring, and Industry 4.0 standards integration are expected to streamline adhesive processes and improve product quality.
- Recent advancements in IT technology have led to the development of Smart Shipyards, emphasizing model-based planning and manufacturing using integrated digital platforms. Precision, reliability, and adaptability are critical for building ultra-large container ships and optimizing welding sequences and groove design is essential to reduce distortions and maintain structural integrity. Support systems utilizing VR, AR, and AI have been aggressively developed to enhance information generation and utilization.
- Future digitalization efforts should focus on information management and sharing, including monitoring data during construction and operation. The development of monitoring systems is progressing, but the accuracy of monitoring data needs improvement. Continuous data acquisition and analysis are necessary to enhance shipyard competitiveness.
- Recent literature has suggested that data-driven methods require less domain knowledge than conventional approaches. However, the absence of domain expertise in such

studies has resulted in critical issues, including flawed data generation, misinterpretation of historical data and established models, limited applicability to real-world scenarios, and unnecessary resource expenditure when proven solutions are already available. To mitigate these issues, sector-specific, risk-based regulations are essential to ensure the safe, efficient, and effective development and use of data-driven models.
- Sustainability in shipbuilding should be emphasized, with future efforts addressing environmentally friendly shipyard processes and technologies for enhanced circularity.
- Nickel-based alloys, used in subsea components, have experienced failures due to hydrogen-induced stress cracking (HISC) and hydrogen embrittlement (HE). High-yield-strength alloys are particularly susceptible. Despite improvements in nickel alloy quality, specific HISC design criteria for hydrogen environments are lacking.
- In carbon capture technologies, chemical absorption and membranes are popular, but challenges remain in terms of costs, installation, and retrofitting on ships. Classification societies are developing guidelines for these technologies.
- Hydrogen transport through pipelines faces challenges due to the critical nature of defects in C-Mn pipelines, requiring stricter utilization and defect acceptance criteria. Special considerations in integrity assessments and material toughness testing are needed.
- The morphology of pitting corrosion defects significantly influences the remaining strength of corroded structures. Pitting evolves through stages, leading to variability in defect shapes and sizes, which affects stress concentration and crack initiation. Future research should focus on precise modeling of pitting morphology using advanced methods.
- Accidental cryogenic liquid spills, such as LNG, pose risks due to rapid cooling and thermal stress. Accurate modeling of heat transfer during spills is crucial. Future research should expand material coverage and improve computational models for cryogenic conditions.
- Understanding material behavior under elevated strain rates is essential for marine structures. Accurate models for both manmade and natural materials are needed to simulate real-world scenarios.
- Enhanced material modelling capabilities will shorten the time to market for new technologies and enable the use of more recycled materials in construction.
- In marine and offshore design, material models are essential for predicting mechanical behavior under various conditions. Deterministic models use fixed values, often requiring conservative safety factors, while probabilistic models account for variability, offering more accurate predictions by using statistical distributions. This probabilistic approach is especially valuable for assessing fatigue and reliability, allowing for better risk management in design.
- Limited research papers are found addressing failures and failure statistics and it is recommended to have more open communication and sharing of knowledge to prevent major accidents within the Maritime and Energy sector.
- Class rules are mainly based on "classic" materials limiting the use of non-conventional solutions. Recent development proposed the certification scheme leading to the characterization of material in the scope of the design assessment. By

expanding the scope of class rules to encompass a wider range of materials, the maritime industry can leverage the benefits of advanced material technologies, improving efficiency, performance, and sustainability in maritime sector.
- Digital twin technology has revolutionized shipbuilding and offshore platform manufacturing by creating virtual replicas of physical assets, processes, and systems. This technology enhances design, production, simulation, monitoring, and predictive maintenance, driving efficiency, safety, and sustainability in the maritime industry. The integration of digital twins with Industry 4.0 principles has further improved their application in manufacturing and operations. Future efforts should focus on comprehensive integration, robust data exchange, and Model-Based Systems Engineering (MBSE) to fully leverage the potential of digital twins and maintain competitiveness in the evolving maritime industry.
- Despite advancements, challenges such as integrating physical and virtual environments, financial barriers, and limited standardization hinder widespread adoption of digital-twin-technologies. Addressing these issues requires further research and development, particularly in design, decommissioning, and smart shipyard development.
- Improved Frameworks and Collaboration – Creating favorable conditions through regulatory adaptation, market incentives, and knowledge-sharing platforms is critical. This includes green certifications (based on SEPI), collaboration between stakeholders in projects like CirclesOfLife, and targeted training programs to bridge knowledge gaps and foster the adoption of sustainable practices across the industry.
- In recent years, dedicated meetings between committee members and external stakeholders were successfully held to enhance discussion on subjects of specific interest between experts. A workshop between the ISSC committee members and industry expert was held in Hamburg September 2024 with focus on novel production, repair and recycling processes with a potential to reduce emissions in non-operational processes of ship's lives. The committee also encourages the next committee to take the initiative for industrial cooperation with specialists in material technology and fabrication

These highlights underscore the importance of material and fabrication technologies in the energy transition, emphasizing the need for sustainable practices, technological advancements, and strong policy support to meet future energy demands.

Appendix A: Hamburg Meeting 2024

Documentation of an expert workshop
Topics: Low emission shipyard processes and circularity ready products – opportunities and challenges on the way towards a sustainable maritime industry
Format: Face-to-face meeting.
Date: September 17th, 2024.
Place and Venue: DNV | Brooktorkai 18 | 20457 Hamburg, Germany.
A.1.1. Focus Topic 'Materials and Processes in a Lifecycle Perspective'
In recent years, dedicated meetings between committee members and external stakeholders were successfully held to enhance discussion on subjects of specific interest

between experts. Several chapters of the report currently being prepared by the committee IV.2 will address numerous subjects related to the topic of more environmentally friendly materials and processes in the maritime industry, including:

1. Methodologies to assess and compare transparently the environmental friendliness of life cycle processes other than ships' operation (newbuilding, repair/conversion, end-of-life)
2. Methodologies to assess the circularity readiness of materials and components used in ships
3. Novel and eco-friendly manufacturing, joining and repair technologies,
4. Development of new materials, especially fibre reinforced plastics for lightweight construction,
5. Approval processes for new materials and technologies.

There are currently various initiatives on international and national level that cover the abovementioned aspects, for instance:

- **R&D&I:** In the EU funded project CirclesOfLife (start in January 2024), industry, research organisations and NGOs will develop concepts for a *Shipyard Environmental Performance Index* (SEPI) and a digital *Cradle to Cradle (C2C) Ship Circularity Passport*. The goals include overcoming the fact that the non-operational footprint of the ship, i.e. shipyard processes as well as materials and components integrated into the ship, remain a black box, and paving the way for a greener and more circular shipbuilding and shipping industry, Krause and Hübler, (2024).
- **Regulation:** The use of materials other than steel in ship structures is governed by IMO regulations. For the use of composites in SOLAS vessels, a dedicated Interim Guideline (IG) is in place (International Maritime Organization (IMO), 2017). Recently, the IMO Sub Committee Ship Design and Construction (SDC) created a Correspondence Group to formalise the dialogue between IMO and industry stakeholders about a possible review of the IG, up to the potential expansion of its scope to load bearing structures made of composite materials.
- **Networks of stakeholders:** Industry associations and policy makers have the greening of transport and the enhancement of material circularity on their agenda and pursue these goals actively in their responsible bodies. Actors in Germany include MariLight (the network for lightweight design in the German maritime industry) and the Federal Ministry for Economic Affairs and Climate Action (*Bundesministerium für Wirtschaft und Klimaschutz* – BMWK) ("MariLight.Net - Das maritime Leichtbaunetzwerk," 2024).

Since the activities are interrelated in many ways and demand to be considered in their entirety, an expert meeting with representatives from the abovementioned groups was suggested.

A.2 Participants

A.2.1 ISSC – Committee IV.2 Material and Fabrication Technology

Name	Country, Organisation, comments
Agnes Marie Horn	Norway – DNV AS; Committee chair
Matthias Krause	Germany – Center of Maritime Technologies gGmbH; Link to CirclesOfLife
Iraklis Lazakis	UK – Department of Naval Architecture, Ocean & Marine Engineering	University of Strathclyde
Stéphane Paboeuf	France – Bureau Veritas; Link to IACS and to the CirclesOfLife and RAMSSES projects

A.2.2 Consortium members of the EU project CirclesOfLife

The workshop was attended by various representatives of the project consortium:

Name	Organisation, role in the project
Jorinus Kalis	Damen Research Development & Innovation BV; project coordinator
Conrad Plange	Flensburger Schiffbau-Gesellschaft mbH; leading the concept and methodology definition
Stephan Wurst	BALance Technology Consulting GmbH; leading the implementation of means for assessment
Michael Hübler	Center of Maritime Technologies gGmbH; leading the validation and assessment of actual emissions, and best practice guide to reduce them
Rosanne van Houwelingen	Leading the dissemination, communication, exploitation and training activities
Romain Benoit	Surfrider Foundation Europe; Representing the Green Marine Europe initiative for environmental self-assessment of shipyards
Benedetta Mantoan	NGO Shipbreaking Platform; Representation of/liaison with the stakeholder group of the ship recycling industry
Theresa Wilson	Flensburger Schiffbau-Gesellschaft mbH; contributing to the concept and methodology definition and to a case study 'new build'

A.2.3 Further Experts

Further experts who participated in the workshop are representing the following organisations and initiatives:

Name	Organisation, relevance
Jon Steinlein	Center of Maritime Technologies gGmbH, Germany; Coordinator of the network for maritime lightweight construction MariLight.Net
Bastian Brenken	Composites United; Managing director of one of the world's largest network for fiber-based multi-material lightweight design

(continued)

Committee IV.2: Material and Fabrication Technology 939

(*continued*)

Name	Organisation, relevance
Paul Riesen	Greenboats GmbH; Supplier of components and products made of lightweight and sustainable materials
Katharina Koschek	Fraunhofer Institute for Manufacturing Technology and Advanced Materials (IFAM), Department Adhesive Bonding and Polymeric Materials
Frank Roland	German Aerospace Center (DLR), Institute of Maritime Energy Systems; partner in the EU project EcoShipYard, former ISSC Committee member
Jens Bottke	Abeking & Rasmussen Schiffs- und Yachtwerft SE; Vice Chair of the German Shipbuilding and Ocean industries association (VSM e. V.) ESG committee
Philippe Noury	DNV AS; classification society with a considerable track record on R&D in shipbuilding and in composites materials application
Abhiram Kakkollilrajan	German Aerospace Center (DLR), Institute of Maritime Energy Systems; partner in the EU project EcoShipYard
Agnieszka Dzielendziek	DLR, Institute of Maritime Energy Systems; PhD in sustainable composites

A.3 Workshop Report

A.3.1 Aession I Greening of Shipyard Processes

Focus on novel production, repair and recycling processes with a potential to reduce emissions in non-operational processes of ship's lives. The processing of any kind of materials can be covered.

The session commenced with a pitch presentation, 'SEPI – Concept to measure, determine and benchmark the footprint of shipyard processes', about the mission and approach of the ongoing European R&D project CirclesOfLife. The project aims to contribute to enhanced sustainability of the maritime sector by addressing lifecycle processes other than the ship's operation. The scope covers two major fields: (a) shipbuilding processes in newbuild, repair/conversion and end-of-life. The aim is to provide a blueprint for a Shipyard Environmental Performance Index (SEPI), a scheme which allows to rate the environmental performance of shipyards and to assess the potential to achieve an improved SEPI score through technical innovation. (b) circularity in shipbuilding. It is envisaged to deliver a concept for a cradle to cradle digital ship passport (C2C DPP) which informs about the materials and components a ship actually consists of, and whether and how they can be recycled, reused, repurposed etc. Both the SEPI and the C2C DPP concept will be assessed and demonstrated by means of dedicated case studies. The project's storyline is represented in Fig. 31.

A second EU funded project named EcoShipYard, dedicated to exactly the same subject, is running in parallel with and independently from CirclesOfLife ("EcoShipYard,"

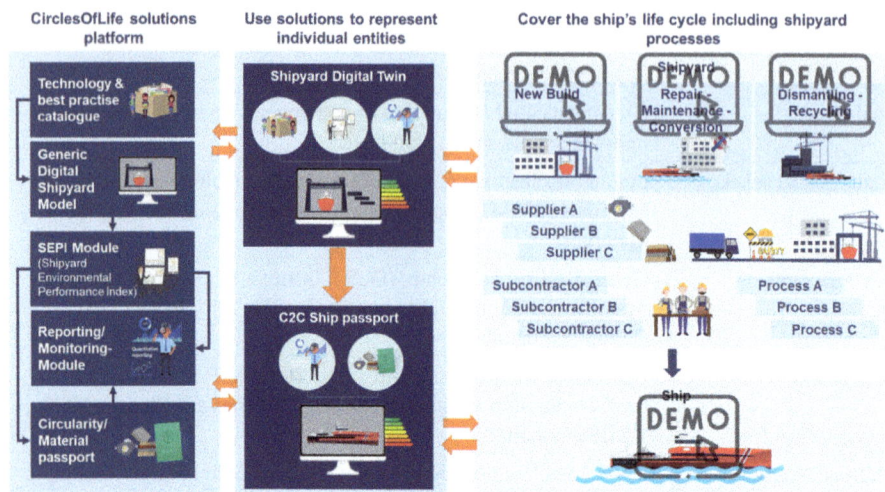

Fig. 31. CirclesOfLife Deliverables and Implementation

2024). The two projects are collaborating on several fields, including the organisation of joint public conferences.

The second pitch presentation came from Eriks BV in the Netherlands, a supplier of shipyard equipment and member of the CirclesOfLife consortium. Being aware of the public demand for climate action, the company is currently reorganising its business strategy with a focus on emissions. A quite comprehensive as-is assessment is ongoing, including a detailed LCA for all products (the portfolio comprises some five thousand different products). While it is challenging to exactly quantify the emissions, it has become obvious that the so-called scope 3 emissions stemming from upstream processes (particularly obtaining raw materials, semi-finished products, and transport) are representing the most critical hot spots. Therefore, Eriks is focusing on material-related approaches (alternative sources of supply or substitution of materials). Regarding alternative materials, each option has to be checked carefully for potential negative side effects, be it deterioration of product properties or additional emissions elsewhere in the process chain. Alternative manufacturing technologies within the manufacturing process can be considered as well, but again the impact on product properties needs to be considered.

In the following discussion, it was widely agreed that an isolated assessment of production processes' footprints without considering materials' and products' life cycle properties makes limited sense. However, sufficiently accurate knowledge about the emissions and the consumed resources along the process chain is considered necessary as well as insight into the potential levers for improvement.

Processes and Technologies for Improved Sustainability

It is agreed that various low-emission technologies for shipbuilding are around. To narrow down where the greatest effect can be achieved, it was suggested to look at what contributes most to the overall lifecycle footprint of the ship first and then to derive the subjects to concentrate on. Two fields of action were identified, and possible solutions were suggested:

(a) A long lifetime of a ship is a highly desirable and sustainable choice, as it means a longer time till the ship is going to be scrapped and needs to be replaced. Suitable processes, technologies and concepts contributing to a long life of a ship include additive manufacturing of spare parts, structural health monitoring, design for easy repair, conversion and disassembly, modularity of the design, easy access to maintenance and repair (e.g. on-board kits), reusability of materials, structures and components, and dedicated crew training.

(b) Alternative fuels and propulsion techniques are the main keys to reduce the footprint of ships in operation. Following this line of thinking, the focus should be on materials and processes which support efficient and safe application of such concepts (e.g. lightweight construction to maintain stability when using towering wind assisted propulsion facilities or to increase the range of battery powered ships). One concern regarding this suggestion was the question whether it is desirable to create a long-living ship while at the same time sacrificing opportunities for design optimisation. In consequence, a newbuilt ship would consume some percentages of fuel more than technically possible. Over time, witnessing more innovation cycles than a short-living ship, the gap between possible and actual consumption would become even worse. This illustrates the scope of the issue, which is clearly an optimisation problem. Furthermore, it should be noted that design for easy conversion is probably the most important aspect among the suggested measures under (a). Refurbishing or changing the engine during the ship's life can help, and the probable need for such modifications should be anticipated by retrofit-ready, modular design.

Further discussion dealt with specific technologies, material groups and industry sectors. An increasing interest in lifetime extension of composite ships is being observed, and the required technologies are available. As an example, typically only the first layer of composite parts suffer from the typical wear and tear in maritime environments. Under the first layer, the material is usually still in excellent condition. By simply grinding the affected zone and then remaking the uppermost layer, parts can virtually be restored to new condition. As well, oil & gas companies want to extend the lifetime of their steel platforms. Here, composite patch repair is a solution which is gaining attention and use of composite wrapping of e.g. corroded process pipes.

Means to Measure, Document and Improve the Performance

While the CirclesOfLife project which was briefly introduced in the pitch presentation is aiming to elaborate consistent and viable SEPI and C2C DPP concepts, various maritime stakeholders have already been active systematically assessing environmental footprints. Some examples for current practices:

- Life Cycle Assessment (LCA) is gaining importance in the ship design process, for instance at GREENBOATS. However, the decisive factor is the purpose of the vessel and its performance requirements; the optimisation of LCA properties is subordinate, and even more so the selection of low-emission technologies. Transparency is an issue; industry stakeholders are usually only willing to issue the results of their assessment, but not the details they are based on.

- Certain shipyards have implemented strategies to reduce their footprint and to – at least qualitatively – document the progress. For instance, LISNAVE in Portugal adopted the assessment scheme of Green Marine Europe (see session IV –(A.3.4)). Local factors such as neighbourhood to a nature reserve were part of the motivation here.
- In the course of investment planning, many European shipyards are obliged to assess the ecological impact. There is a lack of suitable tools which take into consideration both the economic and ecological aspects when undergoing an investment planning process.

The list above includes some of the observed shortcomings and challenges. It was agreed that assessing entire shipyards in terms of their footprint is even harder. It begins with the general question to determine what an eco-friendly shipyard actually is, due to the complexity and heterogeneity of the matter (very different shipyard types, sizes, maturity levels, very different types of ships, complexity of the structures and components). The complexity and heterogeneity also make it challenging to compare shipyards against each other. Even more difficult is the idea to compare shipyards which are specialising into different lifecycle processes, e.g. newbuild vs. repair. Several studies and approaches were discussed:

- A recent PhD Thesis at Strathclyde University proposes a shipyard assessment framework, looking at the performance not only in terms of efficiency, but also environmental and human aspects (Baihaqi et al., 2024a). The concept is based on a value engineering approach.
- A study in South Korea (Hun Woo et al., 2021) analysed and compared the development status of shipyards there in terms of 'smartness' (e.g. maturity of digitalisation and application of Industry 4.0); the level of a shipyard's smartness can be considered as an important indicator for the ability to assess its ecological footprint and to identify the potential for improvement.
- An existing ABS guideline (ABS, 2023b) addresses smartness as well; however it is questioned whether it is applicable to all shipyards. It was suggested to look at the different approaches for comparing shipyards. However, it should be avoided to compare yards that are completely unlike each other.
- The yachting industry did some studies ("Yacht Environmental Transparency Index," 2024), facing many issues with stakeholders unwilling to share information. However, there is a need to create transparency, which again requires collaboration between the stakeholders along the supply chain.
- The question was raised whether the classification societies should come up with rules that enforce such a collaboration. BV approves designs regarding compliance with IMO and own rules. DNV operates a platform (Veracity) where related properties can be shared without disclosing the detailed data.

Finally, the discussion touched some of the business-inherent aspects which currently prevent a further uptake of sustainability and circularity in maritime, and which were to be discussed in session IV:

- Limited control – Even though a shipyard might be willing to take comprehensive measures towards low-footprint processes, it is usually not entirely responsible for

the situation, since the customer's specification often prohibits eco-friendly designs, material choices or processes and technologies.
• The classical (linear economy based) mindset of shipyards is not too much interested in long-living products, but to offer a replacement once a used ship is outdated.
• A clear need is seen to develop viable, circularity-based business concepts. Future business models could include leasing models, or the idea of shipyards purchasing back used products.

Survey

After session I, the attendees were asked to answer a survey:

• Question: 'What is the biggest lever for reducing emissions in shipyard processes?'
• Mode: Multiple choice; each participant was given three votes which could be accumulated to a single option or distribute over several of those.
• Possible answers: Eleven possible answers were suggested which had been created spontaneously by the moderator of the session, summarising the contributions to the discussion. It appears that some of the answers received few votes because of unclear formulation. Example: 'Assignment of measurements to meaningful references' should read 'Choosing suitable reference units to make measurements comparable'.
• Outcome: However, the answers allow some conclusions, especially if categorised as shown in Fig. 32 (the categorisation took place after the voting). The workshop participants were most likely to agree that improved knowledge and transparency (cluster 2) regarding actual emissions can contribute to their reduction. Creating an acknowledged assessment framework (cluster 3) and extending the ships' lifetime (cluster 4) are also considered relevant levers. On the other hand, there was only little support for cluster 1 – shipyard equipment and technology.

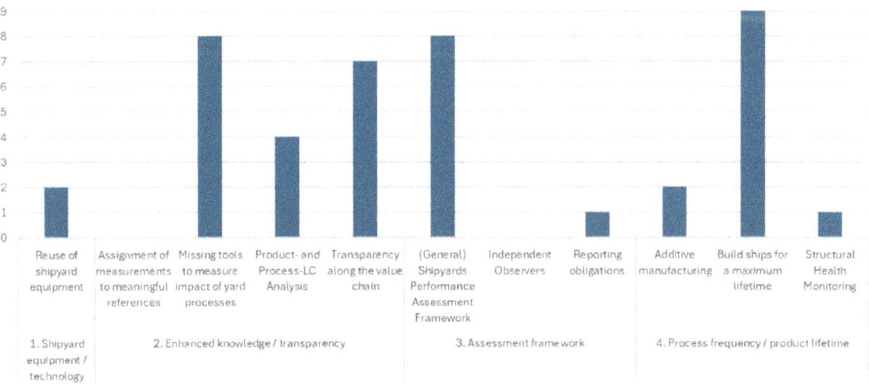

Fig. 32. Survey session 1 – clustered results

A.3.2 Session II: Material in a Life-Cycle Perspective
Focus on 'New' Materials

Two pitch presentations were given in the beginning of this session. The first was titled 'Recycling technologies for FRP' and was delivered by Composites United. It began with a quick look into legislation in Germany (Circular Economy Act – Kreislaufwirtschaftsgesetz KrWG). Its underlying principle is the so-called waste hierarchy (Fig. 33), which suggests decreasing levels of desirability for different end-of-life approaches to products, ranging from *reduce* over *reuse*, *recycling* and *recovery* to *disposal* ("Reduce, Reuse, Recycle and Recover," 2014). After that, several technologies for fibre composites were addressed, including economic aspects. In the second pitch, 'Material and business model innovation in the boat building industry', Greenboats gave an understanding of the company's philosophy and multiannual strategy. Sustainability is considered as one of its 'horizons' for success. The ambition is to offer products which come with a lower ecological footprint than conventional ones, but which are also on a par in terms of operational performance and price (Fig. 34). Current products include different boat hatches made of nature based materials, and circular structures which can be found in recreation vehicles.

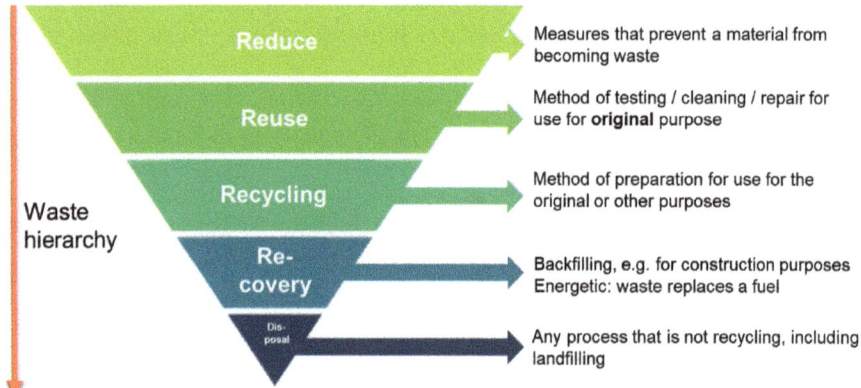

Fig. 33. The waste hierarchy

Fig. 34. Greenboat's roadmap

Technologies for Used Materials

The following discussion of end-of-life technologies for composites focused on GFRP, as this type of material is considered more likely to be used in shipbuilding than others. It was generally agreed that suitable technologies are already in place, while the main challenges were seen on the economical area. Figure 35 gives an overview of available technologies, classifying different types of processes which lead to different results (glass fiber reinforced thermoplastic parts, new glass fibers, cement, energy), which correspond with different levels on the waste hierarchy pyramid (Kühne, Dr. Christian et al., 2022). Currently, the most widespread end-of-life principles in the transport industry are recycling and recovery, whereas reuse is rarely observed. Particularly the application of mechanical processes is gaining momentum in the wind energy industry. New solutions include shredding and recycling of GFRP from wind turbine blades (EURECUM GmbH) ("Recyceln von Windenergieanlagen - EURECUM," 2024), recyclable decking boards with GFRP remnants (Novo-Tech), and recyclate compounds from GFRP residues (iwas-concepts AG) ("iwas-concepts AG - innovative recycling solutions," 2024).

In principle, thermochemical processes are suitable for utilising both the fibers and the resins of used FRP. However, technical development and existing applications are focusing on recycling the fibers, as they are more valuable. From used resins, oil and gas could be extracted and new polymers could be generated from those, but the costs would not be competitive. Thermochemical processes are currently applied on CFRP materials, but not on GFRP, since the price for GFRP recycled by means of e.g. pyrolysis technology cannot compete with juvenile GFRP.

Several R&D projects have looked into the subject with a view on technology uptake in the wind energy sector. Basically, two paths have been addressed:

- Recycling used materials from existing assets – The Danish funded project CETEC (Windkraft-Journal, 2021) looked into the recycling of rotor blades based on conventional thermoset resin systems. According to reports by project partner Vestas, a technology has been invented that allows extracting new raw materials from used epoxy-based blades, thus avoiding landfill storage which has long been considered unavoidable (Vestas Wind Systems A/S, 2024).
- Provide new materials and products which are ready for reuse, recycling etc. – In the ZEBRA project, new recyclable resin systems with intrinsic recyclability have been developed. Process development and the production of 77 m rotor blades has been done which are currently being tested (Arkema, 2024).

Pathways and Obstacles Towards Practical Circularity

Business cases – For any business that involves recycling, economy of scale is a key criterion for viability of business models. It takes a reliable minimum material volume both on the supply and the demand side – if either of both occurs only sporadically and scattered over the world, it will be hard to create a business. In this respect it is assumed that markets like China will benefit from more favourable framework conditions than e.g. Europe. Looking at industry branches, recycling is gaining momentum in the wind energy sector, as there is a rising volume of GFRP waste. This is not only demonstrated by the numerous R&D activities mentioned above. A cement producer near Hannover is ready to accept 40.000 tonnes of GFRP per year, which corresponds with the annual waste of German wind turbines, representing a reasonable material demand and supply.

There are quite some popular examples for repurposing of used GFRP, including in the general press, e.g. benches or bicycle shelters made of used wind rotor blades.

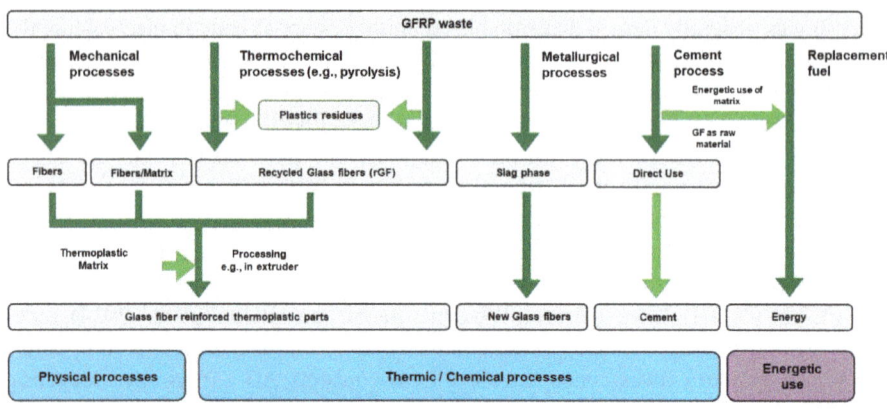

Fig. 35. Overview of GRP recycling / recovery

Projects like this are considered reasonable pilot or niche applications, but they cannot be the ultimate solution to the problem, for instance because of the inability of the garden furniture industry to absorb large amounts of GFRP waste from the wind sector in the long run. Also, new windmills would still require juvenile GFRP. Therefore, and in line with the waste hierarchy idea discussed above, reuse should be preferred to recycling.

Long life cycles – Once more, the typically high life expectancy of ships and the complex interrelationships within the maritime industry were highlighted, now with their implications and challenges on creating a circular economy.

The ownership and liability issue – Any novel business model needs to be attractive for the owner of the asset. Nowadays, a ship's owner changes several times over its lifetime, and quite often there is not one single owner at a time but a rather fragmented ownership structure, which adds to the complexity of the endeavour to create attractive incentives for circularity and sustainable recycling. In view of this problem, new business models could become interesting which turn the liability of the manufacturer into an asset. The shipyard and/or the system suppliers remain the owner of the vessel and merely hand it over to the ship operator, offering lifecycle services and finally take care of its scrapping or recycling. In times of increasingly scarce resources, there is a chance that the materials a ship is made of will be seen as an investment with rising value over a vessel's lifetime, and stakeholders in the manufacturing industry could see their role more as a material bank rather than a vendor.

Lacking availability of knowledge and data – There is a widespread lack of knowledge among ship owners on where and how ship recycling could be done, and there are no sufficient means to make such knowledge available on international level. Standardised and agreed methods to characterise materials in terms of their recyclability could help. For the designers of new ships, it would both be important to have access to such material data, but also to have knowledge and methods at hand for creating circularity-ready ships, e.g. standardisation and modularisation in design or easy to disassemble structures.

Regulatory aspects – It is considered that regulations can generally be a useful instrument to enhance recycling. For instance, by making it mandatory to collect waste, the volume of material which is available for recycling can be increased. On the other hand, regulations can also act as an obstacle, e.g. when penalties discourage the introduction of innovative materials.

More specifically with respect to the use of FRP in SOLAS ships, current developments in the International Maritime Organisation (IMO) were discussed, e.g. the International Maritime Organization's (IMO) Interim Guidelines which govern the use of FRP elements within ship structures (International Maritime Organization (IMO), 2017). A Correspondence Group at IMO's Sub Committee on Ship Design and Construction (SDC) is currently discussing a possible revision of the Interim Guideline. On one side there are considerable concerns against opening the regulations to a wider use of composite materials, arguing that such kind of material should not be used in ships at all because the relatively high difficulty of recycling, let aside fire safety aspects. On the other hand, proponents are pointing out that the issues of FRP regarding recycling are clearly neglectable given their clear and proven advantages such as lightweight or better performance in corrosive environments. The impression is that discussions are often driven by different interests rather than technical facts, which does not really help to speed up the already time-consuming rule making process.

Situations like this have not been observed in the maritime sector only. In the automotive industry for instance, it took a long time until the insight that 98 percent of the lifetime CO_2 equivalent of a car is attributed to its operational phase ensured that a certain break-even point in the discussion was achieved, ensuring a friendlier attitude towards FRP.

Note: Incentives and regulatory instruments as a possible means to enhance circular economy were raised, see documentation to session IV (A.3.4).

A.3.3 Session III: 'Composite Structures Nowadays and in the Years to Come'

Focus on Applications of Composites in the Maritime Industry

A presentation by DNV titled 'Advancing Composite Structures: DNV's Current and Future Role in Technology Qualification and R&D' served as introduction to this session. Featuring nine examples from the company's involvement in various research fields, the range of activities and application areas was outlined, along with an assessment of their advantages and challenges as well as the contributions from DNV and other classification societies and regulatory bodies, see Table 6.

The examples reminded that much has been done in the past years, and a variety of application areas for composite materials in the maritime industry were created. Besides the obvious benefits of reduced weight (less fuel consumption and/or better payload/deadweight ratio), further advantages are offered by innovative composite materials and related processes. Many of the innovative solutions listed in the presentation have already been implemented. The composite tween deck had to be stopped during the Covid-19 pandemic in the middle of the construction of its first commercial application. It was added that a similar solution for hoistable decks is also conceivable, and Uljanik shipyard in Croatia built a car carrier with removable composite panels on the upper decks, allowing for adding one more deck thanks to reduced weight and a lowered center of gravity (Composites World, 2018).

The question was raised whether it can be showcased that those applications which are stemming from naval shipbuilding or the yacht industry could be transferred to the commercial shipbuilding sector. Here, it was pointed out that innovative solutions are

Table 6. Summary of Technology areas presented by DNV

Subject	Main benefits ('+') and challenges ('!')	Rules, R&D
1. Composite Superstructure for Large Naval Vessels	+ payload, speed, stability, signature ! blast/ballistic resistance, streamlining the engineering process	DNV Naval rules EU project MaJoR
2. Composite Pressure Vessels for H_2 and CO2 Maritime Transport – for fuel & cargo tanks	+ enhancing decarbonisation (alternative fuels) ! exotic concepts not fitting in existing standards	New upcoming DNV rules; DNV-CG-0672 Class Guidelines (Rules) for Composite tanks for compressed hydrogen DNV-CP-0673 Class Programme (TAC) for Composite tanks for compressed hydrogen
3. Composite Patch Repair – for Pipes, structures (FRP and steel) in FPSOs etc	+ cold repairs ! general technical misunderstandings wrt bonded connections	More efficient qualification framework needed, starting with less critical cases
4. Composite Propeller – for Research and naval vessels	+ efficiency, cavitation, noise & vibration, avoid scarce materials (Cu)	rules by BV, Class NK no DNV rules
5. Composite Hatch Cover and Tween Decks – for newbuild and refit of used ships; also hoistable decks	+ simple and scalable design, cheap, non-corrosive, easy repair, stability	DNV rules flag state approval
6. Fire Safety of Composite Structures for Large Ships – for Commercial and naval ships	+ long R&D track at DNV and BV ! Wartime of Naval ships	IMO regulations (SOLAS) DNV drafting MSC.1/Circ.1574
7. Composite Wind Propulsion Systems – for Bulk carrier, Tanker, Roro, Car carrier, MPV, PAX	+ enhancing decarbonisation (alternative propulsion) ! deck space required	DNV Rules [DNV-RU-Ship Pt. 6. Ch 2 July 2024 and Standard [DNV-ST-0511 Wind assistance population system 2023]

(*continued*)

Table 6. (continued)

Subject	Main benefits ('+') and challenges ('!')	Rules, R&D
8. 3D Printing of Composite Structures – for propellers, impellers, composite repairs, gangway, passerelle	+ New geometric design options, lifetime extension ! trust between stakeholders to be built	DNV-ST-B203 rules for AM of metals and additional RPs and CP; DNV JIP to extend scope to polymers and composites
9. Thermoplastic Composites – for subsea transport of water, gas, HC, H_2 and CO_2	+ Recyclability and reversibility (melting, reshaping retaining material properties), easy, fast and low-cost processing (injection moulding, extrusion, blow moulding, winding), impact resistance	DNV-ST-F119 Future R&D&I on pressure vessels, piping systems for ballast, scrubber systems, wind turbine blades

easier to be realised in naval applications thanks to the absence of prescriptive rules in this field. Also, the market pull for composites is often missing in the civil sector. Ship owners and also some shipyards shy away from the material for several reasons, including:

- unawareness of shipowners regarding repair methods and other life cycle processes,
- lacking knowledge about R&D results and application possibilities for composites (this can be observed at all stakeholders in the maritime industry),
- lack of ability to demonstrate overall lifecycle benefits (both economically and ecologically),
- alternative design approval process involving extensive risk assessment, instead of accepting risk based design,
- uncertainty whether unconventional solutions would eventually be approved,
- physical limitations of composites which make their application in very large ships difficult, and require multimaterial designs with suitable joining technologies,
- issues ensuring the prefinancing of newbuild contracts when technical details are unknown at the time of signing the contract.

Notwithstanding these obstacles, there was a great confidence that technology transfer is possible. Again, it is key to identify the suitable business cases. Those should come with low cost, low risk and high potential for scaling. The discussion was concluded with the view that despite some stagnation or backlashes, there should be no reason for being discouraged about a wider use of composites in the maritime field – technologies will find their way, provided that there are people who see viable business cases for them. Once those are clearly demonstrated, even the creation of more composite-friendly IMO regulations would become possible.

A.3.4 Session IV – Creating Incentives, Overcoming/removing Barriers

Focus on Activities and Needs Regarding Qualification, Approval Processes/Regulations, and Acceptance

The first pitch presentation to this session, 'Thoughts on Digital Product Passports' by Composites United, addressed the issue of lacking data on materials and their recyclability of actual ships. One approach to overcome this is the development of Digital Product Passports (DPPs). In the European funded Project DigiPass ("DigiPass - Project Website," 2024), DPP concepts are currently being developed, with one of the foci being Composite Materials and their particular properties.

The second pitch presentation focused on the process chain in the maritime industry. 'Environmental certification of shipyards' by Green Marine Europe (GME) introduced an initiative which stems from Canada, in the first place creating a sector-specific label for ship operators and ports, allowing to get certified for their efforts for ecological transition. In a joint effort with the CirclesOfLife project, the initiative is currently rolled out to Europe, at the same time also involving the stakeholders of the ship production, repair/conversion and end-of-life processes. Interested stakeholders need to undergo a certification process (Fig. 36) in which several performance indicators are assessed (Fig. 37).

Fig. 36. Certification process by GME

As already mentioned in the sessions before, the transition from linear to circular economy requires that those who dismantle a ship at the end of its life need to have all relevant information about the materials and the components used, and it should not be their task to find those out. All stakeholders need to be involved in the creation and maintenance of such a data base, including the production supply chain and the lifecycle phases during the ship's service life as well. Sharing data is an issue which is difficult to overcome. Identified requirements include IT (create a neutral unit which can centrally save and provide the relevant data), IPR (sharing material information without disclosing trade secrets), and again viable business models.

With respect to both pitch presentations, it was welcomed that two European Projects (CirclesOfLife and EcoShipYard) are currently addressing the problems, at the same time raising the issue that the projects could come up with different, in the worst case contradicting conclusions and competing solutions. Such a scenario should be avoided by collaboration and exchange in an early phase. Moreover, with respect to sustainability reporting, it would be of utmost importance to create solutions which help stakeholders fulfil their obligations rather than creating new burdens.

GME replied that the idea is to enable stakeholders to demonstrate that they are doing more than what is asked for by existing regulations, and that the certification scheme is not intended to compete with classification societies (GME collaborates with BV, DNV and others). Damen added that they even nowadays get cheaper loans when the fulfilment of green standards is demonstrated. Hence, a business case is already given.

Fig. 37. Performance indicators by GME and their applicability to different sectors in the maritime industry

Survey

After session IV, the attendees were asked to answer a survey:

- Question: 'Which of the discussed measures could create the biggest market pull?'
- Mode: Same as survey after session 1
- Possible answers: Eleven possible answers were suggested which had been created spontaneously by the moderator of the session, summarising the contributions to the discussion.
- Outcome: As Fig. 38 shows, the answer 'Digital Product Passport' clearly won the most votes. Hence, again the cluster 'enhanced knowledge/transparency' was considered most relevant, together with 'assessment framework' (Cluster 3), while the cluster 'business models' scored slightly worse.

A.3.5 Conclusions

Based on the discussions held during the workshop on sustainability and circular economy in shipbuilding, the following recommendations have been proposed to guide future development in this field:

1. **Enhance Lifecycle Assessment (LCA) Framework** – There is a clear need for robust methodologies to assess and compare the environmental impacts of shipbuilding processes, from design and production to end-of-life stages. Tools like the Shipyard

Environmental Performance Index (SEPI) and Digital Product Passports (DPP) should be refined and implemented widely to provide transparent metrics for stakeholders.

2. **Promote Circularity in Material and Design Choices** – Developing materials and components with high recyclability and minimal environmental impact is essential. New initiatives should focus on modular and disassemble designs that facilitate reuse, repair, and recycling. The integration of cradle-to-cradle concepts into ship design and production should be prioritized.

3. **Support the Adoption of Alternative Technologies** – Investment in novel manufacturing processes, such as composite patch repair and additive manufacturing, can extend ship lifespans and reduce emissions. Further research and development in eco-friendly alternatives, such as lightweight materials for fuel-efficient designs, should be incentivized.

4. **Regulatory and Market Incentives** – Governments and industry bodies must create incentives for adopting sustainable practices, such as green certifications for shipyards and financial benefits for low-carbon technologies. Collaborating with IMO and classification societies to adapt and expand regulations for sustainable materials and processes is critical.

5. **Strengthen Collaboration and Knowledge Sharing** – Platforms for exchanging data and best practices among shipyards, suppliers, and regulatory bodies are vital. Collaborative projects, like CirclesOfLife and EcoShipYard, should continue to align objectives to avoid duplication and foster innovation.

6. **Foster Education and Training** – To overcome knowledge gaps, comprehensive training programs should be developed for ship designers, builders, and operators, focusing on sustainability, lifecycle thinking, and innovative materials.

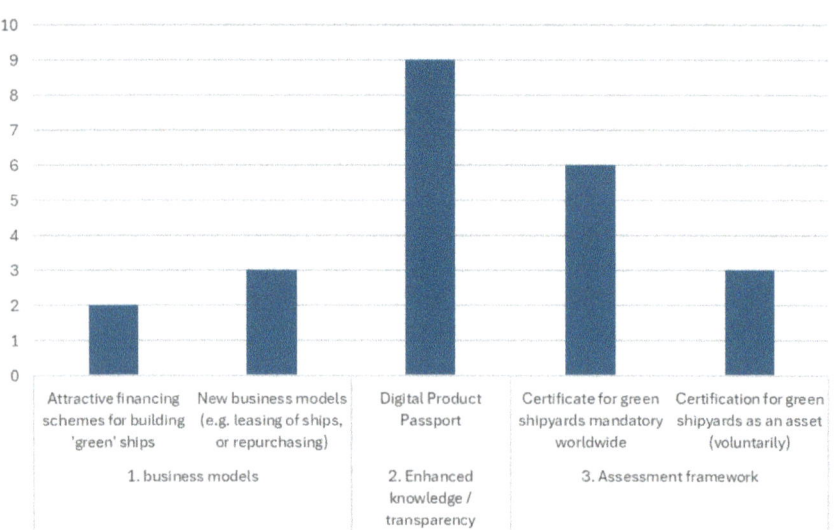

Fig. 38. Survey session IV – clustered results

References

Abbas, M., Shafiee, M.: An overview of maintenance management strategies for corroded steel structures in extreme marine environments. Mar. Struct. **71**, 102718 (2020). https://doi.org/10.1016/j.marstruc.2020.102718

ABS: Guide for Smart Technologies for Shipyards (2023)

Al Baroudi, H., Awoyomi, A., Patchigolla, K., Jonnalagadda, K., Anthony, E.J.: A review of large-scale CO_2 shipping and marine emissions management for carbon capture, utilisation and storage. Appl. Energy **287**, 116510 (2021). https://doi.org/10.1016/j.apenergy.2021.116510

Alum, M., Eze, T.: The new faces of corrosion and damage detection in oil and gas facilities: a brief of what has worked so far and how it can work for you, in: Day 1 Tue, August 11, 2020. Presented at the SPE Nigeria Annual International Conference and Exhibition, SPE, Virtual, p. D013S004R022 (2020). https://doi.org/10.2118/203745-MS

Ambika Harikumar, A.K., Heitzmann, M.T., Basnayake, A., Khakbaz, H., Martin, D.: Tribological performance of Nylon-6/Oyster shell composites. In: International Conference on Offshore Mechanics and Arctic Engineering. American Society of Mechanical Engineers, p. V003T03A002 (2023)

Amran, M., et al.: Long-term durability properties of geopolymer concrete: an in-depth review. Case Stud. Constr. Mater. **15**, e00661 (2021). https://doi.org/10.1016/j.cscm.2021.e00661

Aoyama, K., Yotsuzuka, T., Tanaka, Y., Tanabe, Y.: "Monitoring Platform" of monitoring and visualizing system for shipyard: application to cutting and subassembly processes. In: Okada, T., Suzuki, K., Kawamura, Y. (eds.) Practical Design of Ships and Other Floating Structures. LNCE, pp. 321–337. Springer, Singapore (2021). https://doi.org/10.1007/978-981-15-4680-8_23

Arabnejad, M.H., Thies, F., Yao, H.-D., Ringsberg, J.W.: Zero-emission propulsion system featuring, Flettner rotors, batteries and fuel cells, for a merchant ship. Ocean Eng. **310**, 118618 (2024). https://doi.org/10.1016/j.oceaneng.2024.118618

Arkema: ZEBRA consortium unveils second recyclable wind turbine blade, advances sustainability in wind energy [WWW Document] (2024). https://www.arkema.com/global/en/media/newslist/news/global/corporate/2023/20231220-zebra-unveils-second-recyclable-wind-turbine-blade/. Accessed 17 Oct 24

Austegard, A., Barrio, M.: Project Internal Memo DYNAMIS: Inert components, Solubility of Water in CO_2 and Mixtures of CO_2 and CO_2 Hydrates, p. 2005. Trondheim, Norway (2006)

Baihaqi, I., Lazakis, I., Kurt, R.E.: Development of a novel integrated value engineering and risk assessment (VENRA) framework for shipyard performance measurement: a case study for an Indonesian shipyard. Ships Offshore Struct. **19**, 1118–1133 (2024). https://doi.org/10.1080/17445302.2023.2228115

Baihaqi, I., Lazakis, I., Supomo, H.: A novel shipyard performance measurement approach through an integrated Value Engineering and Risk Assessment (VENRA) framework using a hybrid MCDM tool. Proc. Inst. Mech. Eng. Part M J. Eng. Marit. Environ. **238**, 911–932 (2024). https://doi.org/10.1177/14750902231219533

Barron, R.F., Nellis, G.F.: Cryogenic Heat Transfer. CRC Press (2017)

Bellot, A., Baumler, R., Bouallou, C., Nemer, M., Ölcer, A.: A cradle-to-grave assessment of carbon and energy footprints of ships using synthetic fuel for net-zeros operations. Ocean Eng. **310**, 118392 (2024). https://doi.org/10.1016/j.oceaneng.2024.118392

Bobadilla, R.B., Miranda Barbosa, N., Vera Muñoz, S.O.A., Lourenço De Souza, M.I., Caprace, J.-D.: Deep learning's role in transforming offshore asset corrosion monitoring. In: Volume 3: Materials Technology; Subsea Technology. Presented at the ASME 2024 43rd International Conference on Ocean, Offshore and Arctic Engineering, American Society of Mechanical Engineers, Singapore, Singapore, p. V003T03A011 (2024). https://doi.org/10.1115/OMAE2024-127615

Bonnin-Pascual, F., Ortiz, A.: On the use of robots and vision technologies for the inspection of vessels: a survey on recent advances. Ocean Eng. **190**, 106420 (2019). https://doi.org/10.1016/j.oceaneng.2019.106420

BS 7608, B.S.: Guide to fatigue design and assessment of steel products. British Standard Institute (2014)

Veritas, B.: Digital Classification - Industry 4.0 & The Future of Classification (2020)

Cai, J., Jiang, X., Yang, Y., Lodewijks, G., Wang, M.: Data-driven methods to predict the burst strength of corroded line pipelines subjected to internal pressure. J. Mar. Sci. Appl. **21**, 115–132 (2022)

Camvaceng-Ebflow (2023)

Carrillo, D., Mikhaylov, K., Nardelli, P.J., Andreev, S., Da Costa, D.B.: Understanding UAV-based WPCN-aided capabilities for offshore monitoring applications. IEEE Wireless Commun. **28**, 114–120 (2021). https://doi.org/10.1109/MWC.001.2000218

Cha, M., Enshaei, H., Nguyen, H., Jayasinghe, S.G.: Optimal sizing and evaluation of efficient fuel cell utilization for fuel cell battery hybrid electric ferry. Energy Convers. Manage. **315**, 118723 (2024). https://doi.org/10.1016/j.enconman.2024.118723

Chen, Z., Han, C., Gao, M., Kandukuri, S.Y., Zhou, K.: A review on qualification and certification for metal additive manufacturing. Virtual Phys. Prototyping **17**, 382–405 (2022). https://doi.org/10.1080/17452759.2021.2018938

Chouche, O.: Smart digital shipyards with model-based manufacturing. In: ICCAS 2022 (2022). https://doi.org/10.3940/rina.iccas.2022.05

Chulahwat, A., Mahmoud, H.: Entropy-based framework for optimal life-cycle management of multiple fatigue cracks in ship structures. Struct. Saf. **105**, 102379 (2023). https://doi.org/10.1016/j.strusafe.2023.102379

ClassNK: "CC-Ocean" marine-based CO_2 capture system demonstration project receives "Marine Engineering of the Year 2021" Award from the Japan Institute of Marine Engineering (Press release) (2022)

Coffetti, D., Crotti, E., Gazzaniga, G., Carrara, M., Pastore, T., Coppola, L.: Pathways towards sustainable concrete. Cem. Concr. Res. **154**, 106718 (2022)

Composites World: Low weight on the high seas I A new, award-winning composite shipbuilding material saves fuel, increases car shipping capacity. Composites World (2018)

Danemyr, L.: Material Characterization of Magnetorheological Elastomers for Controllable Connectors in Intelligent Multi-Modular Structures (Master thesis). NTNU (2024)

Derradji-Aouat, A.: A unified failure envelope for isotropic fresh water ice and iceberg ice. In: Proceedings of ETCE/OMAE Joint Conference Energy for the New Millenium (2000)

DigiPass - Project Website [WWW Document] (2024). https://www.digi-pass.eu/ps://ms.hereon.de/digipass/. Accessed 11 Apr 2024

DNV: DNV-CP-0267 Class programme - Approval of manufacturers (2022)

DNV: DNV-GG-0197 Additive manufacturing - Qualification and certification process for materials and components (2021)

DNV: DNV-RP-A203 Technology qualification - Recommended practice (2019)

DNV-ST-F101 Submarine Pipeline Systems (2021)

Dolz, M., Martinez, X., Sá, D., Silva, J., Jurado, A.: Composite materials, technologies and manufacturing: current scenario of European Union shipyards. Ships Offshore Struct. **19**, 1157–1172 (2024). https://doi.org/10.1080/17445302.2023.2229160

Dong, S., Zhang, W., Wang, X., Han, B.: New-generation pavement empowered by smart and multifunctional concretes: a review. Constr. Build. Mater. **402**, 132980 (2023). https://doi.org/10.1016/j.conbuildmat.2023.132980

Dong, Y., Garbatov, Y., Guedes Soares, C.: Strain-based fatigue reliability assessment of welded joints in ship structures. Mar. Struct. **75**, 102878 (2021). https://doi.org/10.1016/j.marstruc.2020.102878

EcoShipYard: Empowering EU shipyards/shipowners with tools for assessing ships' environmental impact & shipyard processes [WWW Document] (2024). EcoShipYard. https://ecoshipyard.eu/. Accessed 17 Oct 2024

EPRG Hydrogen Pipelines Integrity Management and Repurposing Guideline White Paper (2023)

Feng, Y., Wu, D., Stewart, M.G., Gao, W.: Past, current and future trends and challenges in non-deterministic fracture mechanics: a review. Comput. Methods Appl. Mech. Eng. **412**, 116102 (2023). https://doi.org/10.1016/j.cma.2023.116102

Feroz, S., Abu Dabous, S.: UAV-based remote sensing applications for bridge condition assessment. Remote Sens. **13**, 1809 (2021). https://doi.org/10.3390/rs13091809

Fonseca, Í.A., Gaspar, H.M.: Challenges when creating a cohesive digital twin ship: a data modelling perspective. Ship Technol. Res. **68**, 70–83 (2021). https://doi.org/10.1080/09377255.2020.1815140

Forkan, A.R.M., et al.: CorrDetector: a framework for structural corrosion detection from drone images using ensemble deep learning. Expert Syst. Appl. **193**, 116461 (2022). https://doi.org/10.1016/j.eswa.2021.116461

Fröck, L.: Fügen von innovativen Materialien in der schiffbaulichen Fertigung mittels automatisiertem Klebeprozess [WWW Document]. Fraunhofer-Institut für Großstrukturen in der Produktionstechnik IGP (2024). https://www.igp.fraunhofer.de/de/kompetenzfelder/klebtechnik/smartbond.html. Accessed 26 Aug 2024

Fröck, L., Krause, M.: Strategien zur Etablierung der Klebtechnik im Schiffbau. adhäsion Kleben+Dichten 18–22 (2023)

Fueki, R.: Effects of needle peening on fatigue strength of aluminum alloy welded joint containing a surface defect. In: Volume 3: Materials Technology; Subsea Technology. Presented at the ASME 2024 43rd International Conference on Ocean, Offshore and Arctic Engineering, American Society of Mechanical Engineers, Singapore, Singapore, p. V003T03A003 (2024). https://doi.org/10.1115/OMAE2024-125415

Gambini, M., Guarnaccia, F., Manno, M., Vellini, M.: Feasibility study of LOHC-SOFC systems under dynamic behavior for cargo ships compared to ammonia alternatives. Int. J. Hydrogen Energy **81**, 81–92 (2024). https://doi.org/10.1016/j.ijhydene.2024.07.224

Giering, J.-E., Dyck, A.: Maritime digital twin architecture: a concept for holistic Digital Twin application for shipbuilding and shipping. Automatisierungstechnik **69**, 1081–1095 (2021). https://doi.org/10.1515/auto-2021-0082

Giskeødegård, M., Kjersem, K., Jahn, N., Rost, R.: Tool or hassle?- Production workers evaluation of the potential of digital assistance systems on the shopfloor in shipbuilding projects. Cogent Eng. **10**, 2161763 (2023). https://doi.org/10.1080/23311916.2022.2161763

González-Arévalo, N.E., et al.: Influence of aging steel on pipeline burst pressure prediction and its impact on failure probability estimation. Eng. Fail. Anal. **120**, 104950 (2021)

Gotoh, K.: Practical evaluation of the strain rate and temperature effects on fracture toughness of steels. In: International Conference on Offshore Mechanics and Arctic Engineering. American Society of Mechanical Engineers, p. V004T03A015 (2015)

Green Marine: Retrofitting towards climate neutrality. The Green Marine project, funded by EU Horizon Europe and Innovate UK (2024)

Haag, J., Scheid, A., Correia Freire Ferreira, D., Eduardo Fortis Kwietniewski, C.: Hydrogen embrittlement of the nickel alloy UNS N07718 for two different heat-treating schedules. Eng. Fract. Mech. **304**, 110171 (2024). https://doi.org/10.1016/j.engfracmech.2024.110171

Hariprasath, P., Sivaraj, P., Balasubramanian, V., Pilli, S., Sridhar, K.: Evaluation of high cycle fatigue behavior of flux cored arc welded naval grade DMR249 a grade steel joints for ship hull structures. Forces Mech. **11**, 100189 (2023). https://doi.org/10.1016/j.finmec.2023.100189

Horn, A.M., Johnston, C.: Fatigue performance of riser quality girth welds: fracture mechanics study of the slope transition point using DNV's girth weld database. In: International Conference on Offshore Mechanics and Arctic Engineering. American Society of Mechanical Engineers, p. V003T03A004 (2024)

Horn, A.M., Storhaug, G., Håland, J., Jagite, G.: Structural health monitoring by use of sensor data (2023)

Huang, R., Zheng, S., Liu, Z., Ng, T.Y.: Recent advances of the constitutive models of smart materials — hydrogels and shape memory polymers. Int. J. Appl. Mech. **12**, 2050014 (2020). https://doi.org/10.1142/S1758825120500143

Huang, Y., Zhang, H., Li, B., Yang, Z., Wu, J., Withers, P.J.: Generation of high-fidelity random fields from micro CT images and phase field-based mesoscale fracture modelling of concrete. Eng. Fract. Mech. **249**, 107762 (2021). https://doi.org/10.1016/j.engfracmech.2021.107762

Huggins-Gonzalez, A.S., Thodla, R., Gui, F., Bezensek, B., Chambers, B.: Effect of loading profile and sour environment on cracking behavior of C110. In: AMPP CORROSION. AMPP, p. AMPP-2023-19401 (2023)

Hun Woo, J., Zhu, H., Kun Lee, D., Chung, H., Yongkuk, J.: Assessment framework of smart shipyard maturity level via data envelopment analysis. Sustainability **2021**, 1 (2021)

HyLINE - Safe Pipelines for Hydrogen Transport (2019)

International Maritime Organization (IMO): MSC.1/Circ.1574 Interim Guidelines for Use of (FRP) Elements within SHIP Structures: Fire Safety Issues (2017)

Iwańkowicz, R., Rutkowski, R.: Digital twin of shipbuilding process in shipyard 4.0. Sustainability **15**, 9733 (2023). https://doi.org/10.3390/su15129733

iwas-concepts AG - innovative recycling solutions [WWW Document] 2024. . iwas-concepts AG (2024). https://www.iwas-concepts.ch/. Accessed 8 Oct 24

Izumi, D., Samusawa, I., Shimamura, J., Igi, S., Ishikawa, N., Kondo, J.: Effect of stress and hardness on corrosion groove of sour linepipe steel in low H2S content sour environment. In: ISOPE International Ocean and Polar Engineering Conference. ISOPE, p. ISOPE-I-20-4114 (2020)

Jagusch, K., Sender, J., Jericho, D., Flügge, W.: Digital thread in shipbuilding as a prerequisite for the digital twin. Procedia CIRP **104**, 318–323 (2021). https://doi.org/10.1016/j.procir.2021.11.054

Jeffrey, R., Melchers, R.E.: The changing topography of corroding mild steel surfaces in seawater. Corros. Sci. **49**, 2270–2288 (2007)

Jiang, W., Liu, T., Chen, Y., Chen, Q., Zhang, X.-X., Gui, G.: Lightweight convolutional neural network-based method for crane safety inspection. In: 2020 7th International Conference on Dependable Systems and Their Applications (DSA). Presented at the 2020 7th International Conference on Dependable Systems and Their Applications (DSA), Xi'an, China, pp. 324–329. IEEE (2020). https://doi.org/10.1109/DSA51864.2020.00058

Jiang, Y., Liu, L., Yan, J., Wu, Z.: Room-to-low temperature thermo-mechanical behavior and corresponding constitutive model of liquid oxygen compatible epoxy composites. Compos. Sci. Technol. **245**, 110357 (2024). https://doi.org/10.1016/j.compscitech.2023.110357

Jin Lim, H., Hwang, S., Kim, H., Sohn, H.: Steel bridge corrosion inspection with combined vision and thermographic images. Struct. Health Monit. **20**, 3424–3435 (2021). https://doi.org/10.1177/1475921721989407

Johnson, G.R., Cook, W.H.: A constitutive model and data for metals subjected to large strains, high strain rates and high temperatures (1983)

Johnston, C., Horn, A.M.: Fatigue performance of riser quality girth welds: analysis of TWI and DNV's databases. Fatigue Fract. Eng. Mater. Struct. **47**, 1482–1492 (2024)

Jones, D., Snider, C., Nassehi, A., Yon, J., Hicks, B.: Characterising the digital twin: a systematic literature review. CIRP J. Manuf. Sci. Technol. **29**, 36–52 (2020). https://doi.org/10.1016/j.cirpj.2020.02.002

Jones, S.J.: The confined compressive strength of polycrystalline ice. J. Glaciol. **28**, 171–178 (1982)

Josefson, L., et al.: Committee V.3: materials and fabrication technology. In: Day 1 Sun, September 11, 2022. Presented at the 21st International Ship and Offshore Structures Congress, Volume 2, SNAME, Vancouver, Canada, p. D011S001R004 (2022). https://doi.org/10.5957/ISSC-2022-COMMITTEE-V-3

Kahyarian, A., Brown, B., Nešić, S.: The unified mechanism of corrosion in aqueous weak acids solutions: a review of the recent developments in mechanistic understandings of mild steel corrosion in the presence of carboxylic acids, carbon dioxide, and hydrogen sulfide. Corrosion **76**, 268–278 (2020)

Kanchiralla, F.M., Brynolf, S., Olsson, T., Ellis, J., Hansson, J., Grahn, M.: How do variations in ship operation impact the techno-economic feasibility and environmental performance of fossil-free fuels? A life cycle study. Appl. Energy **350**, 121773 (2023). https://doi.org/10.1016/j.apenergy.2023.121773

Kang, D.-H., An, J.-H., Lee, C.-J.: Numerical modeling and optimization of thermal insulation for liquid hydrogen storage tanks. Energy **291**, 130143 (2024). https://doi.org/10.1016/j.energy.2023.130143

Kanno, M., Hirata, N., Hamada, K.: Study on shipbuilding worker's operation classification using machine learning with minimal sensor modules. In: Proceedings of TEAM2021 62 (2021)

Karim, M.A., Abdullah, M.Z., Deifalla, A.F., Azab, M., Waqar, A.: An assessment of the processing parameters and application of fibre-reinforced polymers (FRPs) in the petroleum and natural gas industries: a review. Results Eng. **18**, 101091 (2023). https://doi.org/10.1016/j.rineng.2023.101091

Kayer, N., Klapp, O.: Entwicklung und Validierung von Laboralterungszyklen für die Zulassung von schiffbaulichen Klebverbindungen. Schlussbericht zu dem IGF-Vorhaben Nr. 21189 BG. Rostock, Bremen (2024)

Kim, H., Ahn, C., Jeoung, Y., Mun, S., Hong, K., Kwon, K.: Assessment of fabrication completeness for curved plates in ships and offshore plants using lightweight models and point cloud data. Ocean Eng. **292**, 116438 (2024). https://doi.org/10.1016/j.oceaneng.2023.116438

Kistner, L., Bensmann, A., Hanke-Rauschenbach, R.: Potentials and limitations of battery-electric container ship propulsion systems. Energy Convers. Manag. X **21**, 100507 (2024). https://doi.org/10.1016/j.ecmx.2023.100507

Klebanoff, L.E., Caughlan, S.A.M., Madsen, R.T., Conard, C.J., Leach, T.S., Appelgate, T.B.: Comparative study of a hybrid research vessel utilizing batteries or hydrogen fuel cells. Int. J. Hydrogen Energy **46**, 38051–38072 (2021). https://doi.org/10.1016/j.ijhydene.2021.09.047

Krause, M., Hübler, M.: CirclesOfLife - Enhancing material CIRCularity and Lower Emissions of Ship building processes in all phases of the LIFE cycle. CMT (2024). https://www.cmt-net.org/circlesoflife/?lang=en. Accessed 15 Oct 24

Krause, M., Steinlein, J.: Strategische Roadmap zum maritimen Leichtbau - Bestandsaufnahme, Potenziale und Ziele für den Einsatz von Leichtbaulösungen im maritimen Umfeld in Deutschland und Europa sowie Vorschläge zur Umsetzung (2024)

Kühne, C., et al.: Entwicklung von Rückbau- und Recyclingstandards für Rotorblätter (2022)

Kunkera, Z., Opetuk, T., Hadžić, N., Tošanović, N.: Using digital twin in a shipbuilding project. Appl. Sci. **12**, 12721 (2022). https://doi.org/10.3390/app122412721

Lahdo, R., et al.: Folami - Formschlüssiges Laserstrahlschweissen der Mischverbindung aus Stahl und Aluminium für betriebsfeste Halbzeuge im Schiffbau. In: Statustagung Maritime Technologien 2024. Presented at the Statustagung Maritime Technologien 2024, Tagungsband der Statustagung 2024, pp. 255–269 (2024)

Lee, Y.G., Ju, S., Woo, J.H.: Simulation-based planning system for shipbuilding. Int. J. Comput. Integr. Manuf. **33**, 626–641 (2020). https://doi.org/10.1080/0951192X.2020.1775304

Lemmerich, L., Vaccari, L.: Untersuchung und Optimierung der Prozessparameter und Werkzeuge zum Unterwasserkleben von Halterungssystemen. Schlussbericht zu dem IGF-Vorhaben Nr. 21.002 BG (2023)

Li, X., Nie, J., Wang, X., Zhang, H.: Effects of hydrogen and specimen thickness on fracture toughness of ferritic steel welded joint. Int. J. Hydrogen Energy **82**, 11–24 (2024). https://doi.org/10.1016/j.ijhydene.2024.07.357

Li, X., Roh, M.-I., Ham, S.-H.: A collaborative simulation in shipbuilding and the offshore installation based on the integration of the dynamic analysis, virtual reality, and control devices. Int. J. Naval Archit. Ocean Eng. **11**, 699–722 (2019). https://doi.org/10.1016/j.ijnaoe.2019.02.010

Liu, J., Qin, L., Rong, L., Zhao, M.: Abnormal difference of hydrogen-induced ductility loss in nickel-based alloy 625 at conventional and slow strain rate. Int. J. Hydrogen Energy **83**, 23–28 (2024). https://doi.org/10.1016/j.ijhydene.2024.08.030

Liu, K., Zong, S., Li, Y., Wang, Z., Hu, Z., Wang, Z.: Structural response of the U-type corrugated core sandwich panel used in ship structures under the lateral quasi-static compression load. Mar. Struct. **84**, 103198 (2022). https://doi.org/10.1016/j.marstruc.2022.103198

Liu, W., Zhou, H., Li, J., Meng, Z., Xu, Z., Huang, S.: Comparison of Johnson-Cook and Cowper-Symonds models for aluminum alloy sheet by inverse identification based on electromagnetic bulge. Int. J. Mater. Form. **15**, 10 (2022)

Liu, Z., Amdahl, J., Løset, S.: Plasticity based material modelling of ice and its application to ship–iceberg impacts. Cold Reg. Sci. Technol. **65**, 326–334 (2011)

Lloyd's Register: Guidance Note for Polymer Additive Manufacturing Certification (2021)

Lomarlabs: Lomarlabs' portfolio company SEABOUND demonstrates potential of carbon-capture device following sea trials on a Lomar vessel (2024)

Ma, N., Zhao, W., Wang, W., Li, X., Zhou, H.: Large scale of green hydrogen storage: Opportunities and challenges. Int. J. Hydrogen Energy **50**, 379–396 (2024)

Ma, Y., Xing, Y., Ong, M.C., Hemmingsen, T.H.: Baseline design of a subsea shuttle tanker system for liquid carbon dioxide transportation. Ocean Eng. **240**, 109891 (2021). https://doi.org/10.1016/j.oceaneng.2021.109891

MariLight.Net - Das maritime Leichtbaunetzwerk [WWW Document], 2024. . Marilight.Net | Netzwerk zur Förderung von maritimen Leichtbauanwendungen. https://marilight.net/. Accessed 15 Oct 2024

Martínez, X., Sá, D., Silva, J., Alvarez-Buylla, S.: FIBRE4YARDS: fibre composite manufacturing technologies for the automation and modular construction in shipyards. In: MATCOMP, vol. 6 (2022). https://doi.org/10.23967/r.matcomp.2022.11.06

Mauro, F., Kana, A.A.: Digital twin for ship life-cycle: a critical systematic review. Ocean Eng. **269**, 113479 (2023). https://doi.org/10.1016/j.oceaneng.2022.113479

McLellan, A.: Navigating the shipping route to net zero with REMARCCABLE shipboard carbon capture technology. ImarEST online Mag. (2023)

Md Daud, A.H., Harun, M., Jeevan, J., Mohd Salleh, N.H.: A systematic review of the current trends and potential practices of green shipbuilding. Aust. J. Marit. Ocean Aff. **16**, 105–126 (2024). https://doi.org/10.1080/18366503.2023.2194120

Melchers, R.E.: New insights from probabilistic modelling of corrosion in structural reliability analysis. Struct. Saf. **88**, 102034 (2021)

Melchers, R.E.: A review of trends for corrosion loss and pit depth in longer-term exposures. Corros. Mater. Degrad. **1**, 42–58 (2020)

Melchers, R.E.: Predicting long-term corrosion of metal alloys in physical infrastructure. npj Mater. Degrad. **3**, 4 (2019)

Melchers, R.E., Ahammed, M.: Estimating the long-term reliability of steel and cast iron pipelines subject to pitting corrosion. Sustainability **13**, 13235 (2021)

Mimori, Y., Hiraki, T.: Application of mixed reality technology to ship manufacturing process. In: ICCAS 2022 (2022). https://doi.org/10.3940/rina.iccas.2022.24

Mitsuyuki, T., Hiekata, K., Kasahara, T.: Development of manufacturing support system for ship curved shell plate using laser scanner. Results Eng. **7**, 100157 (2020). https://doi.org/10.1016/j.rineng.2020.100157

Miyata, T., Tagawa, T.: Mezzo-scopic analysis of fracture toughness in steels. Mater. Res. **5**, 85–93 (2002)

Mohd Mansor, M.S., et al.: Integrated approach to Wire Arc Additive Manufacturing (WAAM) optimization: harnessing the synergy of process parameters and deposition strategies. J. Market. Res. **30**, 2478–2499 (2024). https://doi.org/10.1016/j.jmrt.2024.03.170

Mok, D.H.B., Pick, R.J., Glover, A.G., Hoff, R.: Bursting of line pipe with long external corrosion. Int. J. Press. Vessels Pip. **46**, 195–216 (1991). https://doi.org/10.1016/0308-0161(91)90015-T

Mokhtari, M.: Ice-Elastoplastic-Material-Model (2022)

Mokhtari, M., Kim, E., Amdahl, J.: Numerical simulation of concurrent flexural and crushing failure of level ice. In: International Conference on Offshore Mechanics and Arctic Engineering. American Society of Mechanical Engineers, p. V006T07A023 (2024)

Mokhtari, M., Kim, E., Amdahl, J.: Numerical simulation of an aluminium panel subject to ice impact load using a rate and pressure dependent elastoplastic material model for ice. In: International Conference on Offshore Mechanics and Arctic Engineering. American Society of Mechanical Engineers, p. V006T07A016 (2023a)

Mokhtari, M., Kim, E., Amdahl, J.: A rate and pressure dependent elastoplastic material model for glacial ice colliding with marine structures. In: Advances in the Analysis and Design of Marine Structures, pp. 597–603. CRC Press (2023b)

Mokhtari, M., Kim, E., Amdahl, J.: A non-linear viscoelastic material model with progressive damage based on microstructural evolution and phase transition in polycrystalline ice for design against ice impact. Int. J. Impact Eng **176**, 104563 (2023)

Mokhtari, M., Kim, E., Amdahl, J.: Pressure-dependent plasticity models with convex yield loci for explicit ice crushing simulations. Mar. Struct. **84**, 103233 (2022)

Mokhtari, M., Leira, B.J.: A critical review of constitutive models applied to ice-crushing simulations. J. Mar. Sci. Eng. **12**, 1021 (2024)

Mokhtari, M., Melchers, R.E.: Discussion on "data-driven methods to predict the burst strength of corroded line pipelines subjected to internal pressure https://doi.org/10.1007/s11804-022-00263-0." J. Marine. Sci. Appl. (2024). https://doi.org/10.1007/s11804-024-00565-5

Mokhtari, M., Melchers, R.E.: Reliability of the conventional approach for stress/fatigue analysis of pitting corroded pipelines – development of a safer approach. Struct. Saf. **85**, 101943 (2020). https://doi.org/10.1016/j.strusafe.2020.101943

Mokhtari, M., Melchers, R.E.: Next-generation fracture prediction models for pipes with localized corrosion defects. Eng. Fail. Anal. **105**, 610–626 (2019). https://doi.org/10.1016/j.engfailanal.2019.06.094

Mokhtari, M., Melchers, R.E.: A new approach to assess the remaining strength of corroded steel pipes. Eng. Fail. Anal. **93**, 144–156 (2018). https://doi.org/10.1016/j.engfailanal.2018.07.011

Mokhtari, M., Nam, W., Amdahl, J.: Thermal analysis of marine structural steel EH36 subject to non-spreading cryogenic spills. Part II: finite element analysis. Ships Offshore Struct. **17**, 2176–2185 (2022). https://doi.org/10.1080/17445302.2021.1979920

Mokhtari, M., Wang, X., Amdahl, J.: Buckling and post-buckling behaviour of extruded aluminium panels subject to the combined effects of welding and pitting corrosion. In: OMAE2023. Volume 2: Structures, Safety, and Reliability (2023d). https://doi.org/10.1115/OMAE2023-105048

Mokhtari, M., Wang, X., Amdahl, J.: Numerical analysis of pit-to-crack transition under corrosion fatigue using a stochastic pit generation algorithm. In: Advances in the Analysis and Design of Marine Structures (2023e)

Munín-Doce, A., Díaz-Casás, V., Trueba, P., Ferreno-González, S., Vilar-Montesinos, M.: Industrial Internet of Things in the production environment of a Shipyard 4.0. Int. J. Adv. Manuf. Technol. **108**, 47–59 (2020). https://doi.org/10.1007/s00170-020-05229-6

Muñoz, J.A., Perez-Fernandez, R.: Adopting industry 4.0 technologies in shipbuilding through CAD systems. IJME **163**, 41–49 (2021). https://doi.org/10.5750/ijme.v163iA1.4

Nam, W., Hopperstad, O.S., Amdahl, J.: Thermal analysis of marine structural steel EH36 subject to non-spreading cryogenic spills: Part III: structural response assessment. Ships Offshore Struct. **18**, 1505–1518 (2023). https://doi.org/10.1080/17445302.2022.2126125

Nam, W., Mokhtari, M., Amdahl, J.: Thermal analysis of marine structural steel EH36 subject to non-spreading cryogenic spills. Part I: Exp. Stud., 1–9 (2021)

Newbury, B.D., Fairchild, D.P., Prescott, C.A., Anderson, T.D., Wasson, A.J.: Qualification of TMCP pipe for severe sour service: mitigation of local hard zones. In: International Conference on Offshore Mechanics and Arctic Engineering. American Society of Mechanical Engineers, p. V004T03A016 (2019)

Nguyen, T.C., So, Y.-S., Yoo, J.-S., Kim, J.-G.: Machine learning modeling of predictive external corrosion rates of spent nuclear fuel canister in soil (2022). https://doi.org/10.21203/rs.3.rs-1928202/v1

Notaro, G., Belgaroui, J., Maråk, K., Tverrå, R., Burthom, S., Sørhaug, E.M.: CETO: technology qualification of low-pressure Co2 ship transport. SSRN J. (2022). https://doi.org/10.2139/ssrn.4272083

Ohle, M.: Greenlight project [WWW Document]. Fire-resistant and bio-based fiber-reinforced composites for structural lightweight design in ships (2021). https://www.ifam.fraunhofer.de/en/Press_Releases/frp-lightweight-design-ships.html. Accessed 26 Nov 2024

Okubo, Y., Mitsuyuki, T.: Ship production planning using shipbuilding system modeling and discrete time process simulation. JMSE **10**, 176 (2022). https://doi.org/10.3390/jmse10020176

Olsson, R., Cameron, C., Moreau, F., Marklund, E., Merzkirch, M., Pettersson, J.: Design, manufacture, and cryogenic testing of a linerless composite tank for liquid hydrogen. Appl. Compos. Mater. **31**, 1131–1154 (2024). https://doi.org/10.1007/s10443-024-10219-y

Orchard, K., et al.: The status and challenges of CO_2 shipping infrastructures. SSRN J. (2021). https://doi.org/10.2139/ssrn.3820877

Østby, E., Thodla, R., Helgaker, J.F., Collberg, L., Horn, A.M.: H2Pipe JIP: C-Mn pipelines for hydrogen transport–What is governing the fracture toughness? In: ISOPE International Ocean and Polar Engineering Conference. ISOPE, p. ISOPE-I-24-452 (2024)

Østby, E., Thodla, R., Helgaker, J.F., Collberg, L., Horn, A.M.: H2Pipe JIP–Development of guidelines for design of offshore hydrogen pipelines. In: ISOPE International Ocean and Polar Engineering Conference. ISOPE, p. ISOPE-I-23-451 (2023)

Ovcina Mandra, J.: Total Energies equips LNG carrier with EverLoNG's carbon capture system. EverLoNG project (2023)

Pal, N., et al.: Project nautilus: introducing a hydrogen fuel cell system as a retrofit for a hybrid electric vessel. Int. J. Hydrogen Energy **53**, 1457–1476 (2024). https://doi.org/10.1016/j.ijhydene.2023.11.309

Pang, T.Y., Pelaez Restrepo, J.D., Cheng, C.-T., Yasin, A., Lim, H., Miletic, M.: Developing a digital twin and digital thread framework for an 'industry 4.0' shipyard. Appl. Sci. **11**, 1097 (2021). https://doi.org/10.3390/app11031097

Park, C., Jeong, B., Zhou, P.: Lifecycle energy solution of the electric propulsion ship with Live-Life cycle assessment for clean maritime economy. Appl. Energy **328**, 120174 (2022). https://doi.org/10.1016/j.apenergy.2022.120174

Park, J., Kim, H., Yoon, J., Kim, H., Park, C., Hong, D.: Development of an ultrasound technology-based indoor-location monitoring service system for worker safety in shipbuilding and offshore industry. Processes **9**, 304 (2021). https://doi.org/10.3390/pr9020304

Pereira, A.A., Neves, A.C., Ladeira, D., Caprace, J.-D.: Corrosion prediction of FPSOs hull using machine learning. Mar. Struct. **97**, 103652 (2024). https://doi.org/10.1016/j.marstruc.2024.103652

Perez Fernandez, R.: A new approach for using CAD and PLM integration. In: ICCAS 2022 (2022a). https://doi.org/10.3940/rina.iccas.2022.30

Phull, B.S., Pikul, S.J., Kain, R.M.: Seawater corrosivity around the world: results from five years of testing. ASTM Spec. Tech. Publ. **1300**, 34–73 (1997)

Poggi, L., Gaggero, T., Gaiotti, M., Ravina, E., Rizzo, C.M.: Robotic inspection of ships: inherent challenges and assessment of their effectiveness. Ships Offshore Struct. **17**, 742–756 (2022). https://doi.org/10.1080/17445302.2020.1866378

Qiu, Y., Yang, H., Tong, L., Wang, L.: Research progress of cryogenic materials for storage and transportation of liquid hydrogen. Metals **11**, 1101 (2021). https://doi.org/10.3390/met11071101

Ragu Nathan, S., Balasubramanian, V., Rao, A.G., Sonar, T., Ivanov, M., Suganeswaran, K.: Effect of tool rotational speed on microstructure and mechanical properties of friction stir welded DMR249A high strength low alloy steel butt joints for fabrication of light weight ship building structures. Int. J. Lightweight Mater. Manuf. **6**, 469–482 (2023). https://doi.org/10.1016/j.ijlmm.2023.05.004

Recyceln von Windenergieanlagen – EURECUM: EURECUM - spezialisiert auf maßgeschneiderte Versorgungs- und Entsorgungskonzepte (2024). https://www.eurecum-gmbh.de/windenergieanlagen-recycling/. Accessed 17 Oct 2024

Reduce, Reuse, Recycle and Recover (2014). https://csrno.ca/en/solid-waste/the-4rs/. Accessed 8 Oct 2024

RESURGAM | Robotic Survey, Repair and Agile Manufacture | Resurgam will introduce high productivity Friction Stir Welding of steel to European shipyards. [WWW Document] (2024). https://www.resurgamproject.eu/index.html. Accessed 10 Nov 2024

Ronevich, J.A., Song, E.J., Somerday, B.P., San Marchi, C.W.: Hydrogen-assisted fracture resistance of pipeline welds in gaseous hydrogen. Int. J. Hydrogen Energy **46**, 7601–7614 (2021)

Saai, A., Delhaye, V., Lange, T., Furu, T., Aamot, K., Lein, J.: Comparative study of the effects of galvanic corrosion on the strength and the failure of aluminium and stainless steel bolted joints. J. Adv. Joining Process. **8**, 100163 (2023). https://doi.org/10.1016/j.jajp.2023.100163

San Marchi, C., Ronevich, J.A.: Fatigue and fracture of pipeline steels in high-pressure hydrogen gas. In: Pressure Vessels and Piping Conference. American Society of Mechanical Engineers, p. V04BT06A034 (2022)

Sandinge, A., Ukaj, K., Sjögren, A.: SÄKERHET OCH TRANSPORT BRANDFORSKNING (2022)

Schruff, C., Kalwa, C., Hillenbrand, H.G.: Application of TMCP material for large diameter pipelines under sour service conditions. In: Day 3 Thu, October 26, 2017. Presented at the OTC Brasil, OTC, Rio de Janeiro, Brazil, p. D031S023R001 (2017). https://doi.org/10.4043/28082-MS

Schwab, J.: Oil as the villain? How the Kazakhstani media unsuccessfully framed a pipeline leak at the giant Kashagan oil field. Energy Res. Soc. Sci. **118**, 103748 (2024). https://doi.org/10.1016/j.erss.2024.103748

Seo, Y., Huh, C., Lee, S., Chang, D.: Comparison of CO_2 liquefaction pressures for ship-based carbon capture and storage (CCS) chain. Int. J. Greenhouse Gas Control **52**, 1–12 (2016). https://doi.org/10.1016/j.ijggc.2016.06.011

Shimamura, J., Morikawa, T., Yamasaki, S., Tanaka, M.: Sulfide Stress Cracking (SSC) of low alloy linepipe steels in low H_2S content sour environment. ISIJ Int. **62**, 2095–2106 (2022). https://doi.org/10.2355/isijinternational.ISIJINT-2022-236

Shivanshi, S., Chakraborti, G., Sandesh Upadhyaya, K., Kannan, N.: A study on bacterial self-healing concrete encapsulated in lightweight expanded clay aggregates. Mater. Today Proc., S2214785323016073 (2023). https://doi.org/10.1016/j.matpr.2023.03.541

Slater, N.J.: DNV issues new release of composite components standard that aims to ensure safety, cost savings and smaller environmental footprint in the energy and maritime sectors [WWW Document]. DNV (2022). https://www.dnv.com/news/dnv-issues-new-release-of-composite-components-standard-that-aims-to-ensure-safety-cost-savings-and-smaller-environmental-footprint-in-the-energy-and-maritime-sectors-226799/. Accessed 1 Sept 2024

Song, K.-Y., You, Y.-S., Sun, M.-Y.: Ultimate load transfer characteristics of a typical grouted connection used in the monopile foundations of offshore wind turbines. Ocean Eng. **260**, 111988 (2022). https://doi.org/10.1016/j.oceaneng.2022.111988

Sreenath, S., Malik, H., Husnu, N., Kalaichelavan, K.: Assessment and use of unmanned aerial vehicle for civil structural health monitoring. Procedia Comput. Sci. **170**, 656–663 (2020). https://doi.org/10.1016/j.procs.2020.03.174

Stefanov, T.: Enabling qualification of hybrid structures for lightweight and safe maritime transport [WWW Document] (2024). https://www.interreg2seas.eu/en/qualify. Accessed 22 Nov 2024

Svendsen, et al.: The use of digital solutions and structural health monitoring for integrity management of offshore structures (2022)

Taniguchi, T., Matsuo, K.: A study of process simulation based on a multi-agent system for shipbuilding. In: ICCAS 2022 (2022). https://doi.org/10.3940/rina.iccas.2022.42

Taniguchi, T., Morishita, M., Matsuo, K.: Development of PLM system for production planning and production control in shipbuilding. In: ICCAS 2022 (2022). https://doi.org/10.3940/rina.iccas.2022.21

Thanush, A.A., Chitra, P., Kasinath, J., Surya Prakash, R.: Atmospheric corrosion rate prediction of low-alloy steel using machine learning models. IOP Conf. Ser. Mater. Sci. Eng. **1248**, 012050 (2022). https://doi.org/10.1088/1757-899X/1248/1/012050

Tronskar, J., Kit Wui, L., Sigurdsson, G.: Deterministic and probabilistic pipeline girth weld ECA considering internal surface flaw in sour service region 1. In: International Conference on Offshore Mechanics and Arctic Engineering. American Society of Mechanical Engineers, p. V003T03A009 (2024)

Turner, J.: Constitutive behaviour of ice under compressive states of stress and its application to ice-structure interactions. Memorial University of Newfoundland (2018)

TWI: Rapidweld UK research (n.d.)

Ustolin, F., Campari, A., Taccani, R.: An extensive review of liquid hydrogen in transportation with focus on the maritime sector. J. Mar. Sci. Eng. **10**, 1222 (2022). https://doi.org/10.3390/jmse10091222

Vakili, S., Schönborn, A., Ölçer, A.I.: The road to zero emission shipbuilding Industry: a systematic and transdisciplinary approach to modern multi-energy shipyards. Energy Convers. Manag. X **18**, 100365 (2023). https://doi.org/10.1016/j.ecmx.2023.100365

Velázquez, J.C., González-Arévalo, N.E., Díaz-Cruz, M., Cervantes-Tobón, A., Herrera-Hernández, H., Hernández-Sánchez, E.: Failure pressure estimation for an aged and corroded oil and gas pipeline: a finite element study. J. Nat. Gas Sci. Eng. **101**, 104532 (2022)

Vestas Wind Systems A/S: Vestas unveils circularity solution to end landfill for turbine blades [WWW Document] (2024). https://www.vestas.com/en/media/company-news/2023/vestas-unveils-circularity-solution-to-end-landfill-for-c3710818. Accessed 17 Oct 2024

Wanasinghe, T.R., Gosine, R.G., De Silva, O., Mann, G.K.I., James, L.A., Warrian, P.: Unmanned aerial systems for the oil and gas industry: overview, applications, and challenges. IEEE Access **8**, 166980–166997 (2020). https://doi.org/10.1109/ACCESS.2020.3020593

Wang, G., Bar, D., Schreier, S.: The potential of end-of-life ships as a floating seawall and the methodical use of gap resonance for wave attenuation. Ocean Eng. **298**, 117246 (2024). https://doi.org/10.1016/j.oceaneng.2024.117246

Wang, J., Shi, X., Zhou, H., Yang, Z., Liu, J.: Dimensional precision controlling on out-of-plane welding distortion of major structures in fabrication of ultra large container ship with 20000TEU. Ocean Eng. **199**, 106993 (2020). https://doi.org/10.1016/j.oceaneng.2020.106993

Wang, P.: Normentwurf DIN 2304-2 zum Kleben im Schiffbau. adhäsion Kleben+Dichten 12–17 (2023)

Wang, X., Zhang, W., Zhang, W., Ai, Y.: A machine learning method for predicting corrosion weight gain of uranium and uranium alloys. Materials **16**, 631 (2023). https://doi.org/10.3390/ma16020631

Wang, Y., Yao, X., Teo, F.C., Zhang, J.: Cohesive element method to level ice-sloping structure interactions. Int. J. Offshore Polar Eng. **30**, 385–394 (2020)

Wang, Z., Li, M., Zhao, F., Ji, Y., Han, F.: Status and prospects in technical standards of hydrogen-powered ships for advancing maritime zero-carbon transformation. Int. J. Hydrogen Energy **62**, 925–946 (2024). https://doi.org/10.1016/j.ijhydene.2024.03.083

Watanabe, N., Yamasaki, K., Gotoh, K.: A study for automatic inspection of leg length and undercut in the T-shaped joint using deep learning. In: Proceedings of International Conference on Welding and Joining, vol. 185 (2022)

Wen, F., Gu, H., Wang, B., Vitali, T., Rasberry, H.: The development of a machine learning-based image recognition tool for coating inspections. In: Day 3 Wed, May 06, 2020. Presented at the Offshore Technology Conference, OTC, Houston, Texas, USA, p. D031S032R003 (2020). https://doi.org/10.4043/30504-MS

Werz, D.-I.M.J.: Experimentelle und numerische Untersuchungen des Rührreibschweißens von Aluminium- und Aluminium-Stahl-Verbindungen zur Verbesserung der mechanischen Eigenschaften. Universität Stuttgart (2020)

Windkraft-Journal: Vestas und Partner arbeiten an der vollständigen Recyclingfähigkeit von Windturbinenblättern. Windkraft-Journal (2021). https://www.windkraft-journal.de/2021/05/17/vestas-und-partner-arbeiten-an-der-vollstaendige-recyclingfaehigkeit-von-windturbinen blaettern/162134. Accessed 17 Oct 2024

Woloszyk, K., Goerlandt, F., Montewka, J.: A framework to analyse the probability of accidental hull girder failure considering advanced corrosion degradation for risk-based ship design. Reliab. Eng. Syst. Saf. **251**, 110336 (2024). https://doi.org/10.1016/j.ress.2024.110336

Woo, J.H., Zhu, H., Lee, D.K., Chung, H., Jeong, Y.: Assessment framework of smart shipyard maturity level via data envelopment analysis. Sustainability **13**, 1964 (2021). https://doi.org/10.3390/su13041964

Wu, J., Cai, S., Guan, Y., Li, S., Tu, Z.: Design and performance evaluation of power system for unmanned ship based on proton exchange membrane fuel cell. Int. J. Hydrogen Energy **59**, 730–741 (2024). https://doi.org/10.1016/j.ijhydene.2024.02.063

Wu, S., Li, T., Xu, F., Chen, R., Zhou, X., Wang, B.: Configuration size optimization of gas-electric hybrid power systems on ships considering energy density and engine load response. Energy Convers. Manage. **301**, 118069 (2024). https://doi.org/10.1016/j.enconman.2024.118069

Xiao, J.: Damage and fracture of brittle viscoelastic solids with application to ice load models. Memorial University of Newfoundland (1997)

Xu, S., Wang, Y.: Estimating the effects of corrosion pits on the fatigue life of steel plate based on the 3D profile. Int. J. Fatigue **72**, 27–41 (2015)

Yacht Environmental Transparency Index: Water Revolution Foundation (2024). https://waterrevolutionfoundation.org/programmes/yacht-environmental-transparency-index/. Accessed 17 Oct 2024

Yao, Y., Yang, Y., Wang, Y., Zhao, X.: Artificial intelligence-based hull structural plate corrosion damage detection and recognition using convolutional neural network. Appl. Ocean Res. **90**, 101823 (2019). https://doi.org/10.1016/j.apor.2019.05.008

Yi, Z., et al.: Intelligent initial model and case design analysis of smart factory for shipyard in China. Eng. Appl. Artif. Intell. **123**, 106426 (2023). https://doi.org/10.1016/j.engappai.2023.106426

Yu, C., Kawabata, T., Okita, T., Uranaka, S.: First-principles study of hydrogen solubility and embrittlement of Cr23C6 in nickel-based alloys. Comput. Mater. Sci. **245**, 113304 (2024). https://doi.org/10.1016/j.commatsci.2024.113304

Yu, Q., Han, Y.: Recognition of surface corrosion morphology on coastal engineering structures using machine vision technology. In: Volume 3: Materials Technology; Subsea Technology. Presented at the ASME 2024 43rd International Conference on Ocean, Offshore and Arctic Engineering, American Society of Mechanical Engineers, Singapore, Singapore, p. V003T03A012 (2024). https://doi.org/10.1115/OMAE2024-127951

Yue, X., et al.: Sulfide stress cracking test of TMCP pipeline steels in NACE MR0175 region 3 conditions. In: NACE CORROSION. NACE, p. NACE-2020-14446 (2020)

Zhai, Y.: Study on the relationship between energy consumption of shipbuilding and shipbuilding costs. In: Baeyens, J., Dewil, R., Rossi, B., Deng, Y. (eds.) Proceedings of 2022 4th International Conference on Environment Sciences and Renewable Energy, Environmental Science and Engineering, pp. 175–188. Springer Nature Singapore (2023). https://doi.org/10.1007/978-981-19-9440-1_14

Open Access This chapter is licensed under the terms of the Creative Commons Attribution-NonCommercial-NoDerivatives 4.0 International License (http://creativecommons.org/licenses/by-nc-nd/4.0/), which permits any noncommercial use, sharing, distribution and reproduction in any medium or format, as long as you give appropriate credit to the original author(s) and the source, provide a link to the Creative Commons license and indicate if you modified the licensed material. You do not have permission under this license to share adapted material derived from this chapter or parts of it.

The images or other third party material in this chapter are included in the chapter's Creative Commons license, unless indicated otherwise in a credit line to the material. If material is not included in the chapter's Creative Commons license and your intended use is not permitted by statutory regulation or exceeds the permitted use, you will need to obtain permission directly from the copyright holder.

Author Index

A
Abbasnia, Arash 133
Alley, Erick 253
Amlashi, H. 729
An, G. 578
Andric, Jerolim 253

B
Bernardino, Mariana 1
Brubak, L. 452

C
Caire, Marcelo 253
Caleo, Alessandro 852
Campos, Ricardo Martins 1
Caprace, Jean-David 852
Chen, B.-Q. 452
Chen, X. 452
Chun, M. 452

D
Darie, I. 452
De Diego, V. 337
de Hauteclocque, Guillaume 1
de Souza, M. I. L. 452
del Amo, M. Vicente 578
Dessi, D. 337
Deul, M. 578
Dhavalikar, S. 337
Diebold, Louis 133
Dong, P. 578
Dong, Sheng 1

F
Fujarra, Andre 133

G
Gaggero, Tomaso 133
Gaiotti, M. 452
Gao, Zhen 253
Georgiadis, D. 452
Georgiev, P. 729

H
Hamada, Kunihiro 852
Haselbach, P. 578
Heggelund, S. 578
Hirdaris, Spyros 133
Holtmann, M. 337
Horn, Agnes Marie 852

I
Ilman, Eko Charnius 253

J
Jelovica, J. 729
Jiang, Zhiyu 253
Johannessen, Thomas Berge 1
Jurisic, P. 578

K
Kahl, A. 578
Kawabata, T. 578
Kawamura, Y. 729
Kaydihan, L. 337
Kim, S. J. 337
Kim, Sangyeob 133
Kim, Yooil 253
Kolios, A. 729
Konispoliatis, Dimitrio S 133
Kõrgesaar, M. 452
Krause, Matthias 852
Kwon, J. M. 729

L
Lataire, Evert 1
Lazakis, Iraklis 852
Leira, Bernt 852
Liu, J. 578
Luo, Hanbing 253

M
Magoga, T. 452
Maljaars, J. 578
Mokhtari, Mojtaba 852

Moon, Joong Soo 1
Moro, L. 337

N
Nahshon, K. 452
Nakayama, S. 578
Niraula, A. 578

O
Oka, Masayoshi 133
Okafuji, T. 452
Ozdemir, M. 578

P
Paboeuf, Stéphane 852
Pais, T. 337
Paiva, A. 337
Paredes, M. 452
Polojärvi, Arttu 1
Prasetyo, F. 578
Prpic-Orsic, Jasna 1

R
Remes, H. 578
Ringsberg, J. W. 452
Rodenburg, J. 578
Rodrigues, Jose Miguel 133
Roh, J. 578
Romanoff, J. 452

S
Sacchet, Alessandro 253
Sales, J. 729
Sazidy, Mahmud 133
Schipperen, I. 452
Sicchiero, M. 578
Sidari, M. 729
Sirkar, J. 729
Sobey, A. 729

Soliman, M. 578
Song, X. 578
Sprenger, Florian 133
Stanley, Aaron 253
Storhaug, G. 337

T
Takahashi, H. 337
Tang, W. 729
Tavakoli, S. 337
Tsiourva, Dora 852

V
van Essen, Sanne 1
Vanem, Erik 1
Vhanmane, S. 729
von Bock und Polach, Franz 1

W
Wang, Jungyong 133
Wang, S. 337
Wang, Y. 452
Wellens, Peter 133
Wołoszyk, Krzysztof 253
Wu, S. 578

Y
Yan, Y. 729
Yanagihara, Daisuke 253
Yanagimoto, F. 578
Yeter, B. 578

Z
Zamarin, A. 452
Zao, B. 337
Zhan, Z. 452
Zhang, Guiyong 133
Zhu, Tingyao 1
Zubar, A. M. Mohammad 452

GPSR Compliance
The European Union's (EU) General Product Safety Regulation (GPSR) is a set
of rules that requires consumer products to be safe and our obligations to
ensure this.

If you have any concerns about our products, you can contact us on

ProductSafety@springernature.com

In case Publisher is established outside the EU, the EU authorized
representative is:

Springer Nature Customer Service Center GmbH
Europaplatz 3
69115 Heidelberg, Germany

www.ingramcontent.com/pod-product-compliance
Ingram Content Group UK Ltd.
Pitfield, Milton Keynes, MK11 3LW, UK
UKHW022202230426
470311UK00001BA/1